Mountains, Climate and Biodiversity

Mountains, Climate and Biodiversity

Edited by

Carina Hoorn
University of Amsterdam
Amsterdam, The Netherlands

Allison Perrigo
Gothenburg Global Biodiversity Centre
Sweden

Department of Biological and Environmental Sciences,
University of Gothenburg, Sweden

Forest Cat Editing
Uppsala, Sweden

Alexandre Antonelli
Gothenburg Global Biodiversity Centre
Sweden

Department of Biological and Environmental Sciences,
University of Gothenburg, Sweden

Gothenburg Botanical Garden
Sweden

WILEY Blackwell

This edition first published 2018
© 2018 John Wiley & Sons Ltd

Registered Office(s)
John Wiley & Sons, Inc., 111 River Street, Hoboken, NJ 07030, USA
John Wiley & Sons Ltd, The Atrium, Southern Gate, Chichester, West Sussex, PO19 8SQ, UK

Editorial Office
9600 Garsington Road, Oxford, OX4 2DQ, UK

For details of our global editorial offices, customer services, and more information about Wiley products visit us at www.wiley.com.

Wiley also publishes its books in a variety of electronic formats and by print-on-demand. Some content that appears in standard print versions of this book may not be available in other formats.

Library of Congress Cataloging-in-Publication Data

Names: Hoorn, C. (Carina), editor. | Perrigo, Allison, editor. | Antonelli,
 Alexandre, 1978– editor.
Title: Mountains, climate and biodiversity / edited by Dr. Carina Hoorn, Dr. Allison Perrigo,
 Prof. Alexandre Antonelli.
Description: Hoboken, NJ : John Wiley & Sons, 2018. | Includes bibliographical references and index. |
Identifiers: LCCN 2017039551 (print) | LCCN 2017047780 (ebook) | ISBN 9781119159896 (pdf) |
 ISBN 9781119159889 (epub) | ISBN 9781119159872 (cloth)
Subjects: LCSH: Orogeny. | Mountain biodiversity. | Climatic changes.
Classification: LCC QE621 (ebook) | LCC QE621 .M684 2018 (print) | DDC 577.5/3–dc23
LC record available at https://lccn.loc.gov/2017039551

Cover Design: Wiley
Cover Image: Courtesy of Maurico Diazgranados

The superpáramo of the Sierra Nevada del Cocuy National Park, in Colombia, with rocky slopes of Cretaceous (Albian-Aptian) sedimentary quartzitic sandstone bedrock and shales with occasional limestone inclusions. Espeletia lopezii Cuatrec. grows here in a pit at the base of the vertical slopes of the Ritacuba Blanco peak.

Set in 10/12pt Warnock by SPi Global, Pondicherry, India
Printed in Singapore by Markono Print Media Pte Ltd

10 9 8 7 6 5 4 3 2

In memory of Alexander von Humboldt

Let his passion for the natural world in general, and the links between mountains, climate and biodiversity in particular, continue to inspire many generations of students to come.

Contents

List of Contributors *xi*
Acknowledgments *xvii*
Foreword by Peter Raven *xix*
Biography of Editors *xxiii*
Glossary *xxv*
About the Companion Website *xxxv*

1 Mountains, Climate and Biodiversity: An Introduction *1*
 Carina Hoorn, Allison Perrigo and Alexandre Antonelli

 Part I Mountains, Relief and Climate *15*

2 Simple Concepts Underlying the Structure, Support and Growth of Mountain Ranges, High Plateaus and Other High Terrain *17*
 Peter Molnar

3 An Overview of Dynamic Topography: The Influence of Mantle Circulation on Surface Topography and Landscape *37*
 Caroline M. Eakin and Carolina Lithgow-Bertelloni

4 Mountain Relief, Climate and Surface Processes *51*
 Peter van der Beek

5 Dating Mountain Building: Exhumation and Surface Uplift *69*
 Matthias Bernet, Verónica Torres Acosta and Mauricio A. Bermúdez

6 Stable Isotope Paleoaltimetry: Paleotopography as a Key Element in the Evolution of Landscapes and Life *81*
 Andreas Mulch and C. Page Chamberlain

7 Phytopaleoaltimetry: Using Plant Fossils to Measure Past Land Surface Elevation *95*
 Robert A. Spicer

8 Cenozoic Mountain Building and Climate Evolution *111*
 Phoebe G. Aron and Christopher J. Poulsen

9 Paleoclimate *123*
 Hemmo A. Abels and Martin Ziegler

Part II When Biology Meets Mountain Building *135*

10 Mountain Geodiversity: Characteristics, Values and Climate Change *137*
John E. Gordon

11 Geodiversity Mapping in Alpine Areas *155*
Arie C. Seijmonsbergen, Matheus G.G. De Jong, Babs Hagendoorn, Johannes G.B. Oostermeijer and Kenneth F. Rijsdijk

12 Historical Connectivity and Mountain Biodiversity *171*
Suzette G.A. Flantua and Henry Hooghiemstra

13 The Environmental Heterogeneity of Mountains at a Fine Scale in a Changing World *187*
Andrés J. Cortés and Julia A. Wheeler

14 Mountains, Climate and Mammals *201*
Catherine Badgley, Tara M. Smiley and Rachel Cable

15 Inferring Macroevolutionary Dynamics in Mountain Systems from Fossils *217*
Daniele Silvestro and Jan Schnitzler

16 The Interplay between Geological History and Ecology in Mountains *231*
Catherine H. Graham, Mauricio Parra, Andrés Mora and Camilo Higuera

17 Mountains and the Diversity of Birds *245*
Jon Fjeldså

18 Teasing Apart Mountain Uplift, Climate Change and Biotic Drivers of Species Diversification *257*
Fabien L. Condamine, Alexandre Antonelli, Laura P. Lagomarsino, Carina Hoorn and Lee Hsiang Liow

19 Upland and Lowland Fishes: A Test of the River Capture Hypothesis *273*
James S. Albert, Jack M. Craig, Victor A. Tagliacollo and Paulo Petry

20 Different Ways of Defining Diversity, and How to Apply Them in Montane Systems *295*
Hanna Tuomisto

21 A Modeling Framework to Estimate and Project Species Distributions in Space and Time *309*
Niels Raes and Jesús Aguirre-Gutiérrez

Part III Mountains and Biota of the World *321*

22 Evolution of the Isthmus of Panama: Biological, Paleoceanographic and Paleoclimatological Implications *323*
Carlos Jaramillo

23 The Tepuis of the Guiana Highlands *339*
Otto Huber, Ghillean T. Prance, Salomon B. Kroonenberg and Alexandre Antonelli

24 Ice-Bound Antarctica: Biotic Consequences of the Shift from a Temperate to a Polar Climate *355*
Peter Convey, Vanessa C. Bowman, Steven L. Chown, Jane E. Francis, Ceridwen Fraser, John L. Smellie, Bryan Storey and Aleks Terauds

25 The Biogeography, Origin and Characteristics of the Vascular Plant Flora and Vegetation
of the New Zealand Mountains   *375*
Matt S. McGlone, Peter Heenan, Timothy Millar and Ellen Cieraad

26 The East African Rift System: Tectonics, Climate and Biodiversity   *391*
Uwe Ring, Christian Albrecht and Friedemann Schrenk

27 The Alps: A Geological, Climatic and Human Perspective on Vegetation
History and Modern Plant Diversity   *413*
*Séverine Fauquette, Jean-Pierre Suc, Frédéric Médail, Serge D. Muller, Gonzalo Jiménez-Moreno,
Adele Bertini, Edoardo Martinetto, Speranta-Maria Popescu, Zhuo Zheng and
Jacques-Louis de Beaulieu*

28 Cenozoic Evolution of Geobiodiversity in the Tibeto-Himalayan Region   *429*
Volker Mosbrugger, Adrien Favre, Alexandra N. Muellner-Riehl, Martin Päckert and Andreas Mulch

29 Neogene Paleoenvironmental Changes and their Role in Plant Diversity in Yunnan,
South-Western China   *449*
Zhe-Kun Zhou, Tao Su and Yong-Jiang Huang

30 Influence of Mountain Formation on Floral Diversification in Japan, Based
on Macrofossil Evidence   *459*
Arata Momohara

31 The Complex History of Mountain Building and the Establishment of Mountain Biota
in Southeast Asia and Eastern Indonesia   *475*
Robert J. Morley

Index   *495*

List of Contributors

Hemmo A. Abels
Department of Geosciences and Engineering
Delft University of Technology
Delft, Netherlands

Jesús Aguirre-Gutiérrez
Naturalis Biodiversity Center
Leiden, Netherlands
and
Institute for Biodiversity and Ecosystem Dynamics,
University of Amsterdam, Amsterdam, Netherlands
and
Environmental Change Institute, School of Geography
and the Environment, University of Oxford, Oxford, UK

James S. Albert
Department of Biology
University of Louisiana at Lafayette
Lafayette, LA, USA

Christian Albrecht
Department of Animal Ecology and Systematics
Justus Liebig University
Giessen, Germany
and
Department of Biology
Mbarara University of Science and Technology
Mbarara, Uganda

Alexandre Antonelli
Gothenburg Global Biodiversity Centre
Gothenburg, Sweden
and
Department of Biological and Environmental
Sciences
University of Gothenburg, Gothenburg, Sweden
and
Gothenburg Botanical Garden
Gothenburg, Sweden

Phoebe G. Aron
Department of Earth and Environmental Sciences
University of Michigan
Ann Arbor, MI, USA

Catherine Badgley
Department of Ecology and Evolutionary Biology
University of Michigan
Ann Arbor, MI, USA

Mauricio A. Bermúdez
Faculty of Natural Sciences and Mathematics
University of Ibagué
Ibagué, Colombia

Matthias Bernet
Institut des Sciences de la Terre (ISTerre)
Grenoble Alps University
Grenoble, France

Adele Bertini
Department of Earth Sciences
University of Florence
Florence, Italy

Vanessa C. Bowman
British Antarctic Survey
Cambridge, UK

Rachel Cable
Department of Ecology and Evolutionary Biology
University of Michigan
Ann Arbor, MI, USA

C. Page Chamberlain
Earth System Sciences
Stanford University
Stanford, CA, USA

Steven L. Chown
School of Biological Sciences
Monash University
Melbourne, Australia

Ellen Cieraad
Institute of Environmental Sciences (CML)
Leiden University
Leiden, Netherlands

Fabien L. Condamine
CNRS, Institut des Sciences de l'Evolution
de Montpellier
University of Montpellier
Montpellier, France
and
Gothenburg Global Biodiversity Centre
Gothenburg, Sweden

Peter Convey
British Antarctic Survey
Cambridge, UK

Andrés J. Cortés
Department of Ecology and Genetics
Uppsala University
and
Department of Plant Biology
Swedish Agricultural University
Uppsala, Sweden
and
Department of Biological and Environmental Sciences
University of Gothenburg
Gothenburg, Sweden

Jack M. Craig
Department of Biology
University of Louisiana at Lafayette
Lafayette, LA, USA

Jacques-Louis de Beaulieu
Institut Méditerranéen de Biodiversité et d'Ecologie
marine et continentale (IMBE)
Aix-Marseille University/CNRS/IRD/AU
Aix-en-Provence, France

Matheus G.G. De Jong
Research Foundation for Alpine and Subalpine
Environments
Amsterdam, Netherlands

Caroline M. Eakin
Research School of Earth Sciences
Australian National University
Canberra, Australia
and
University of Southampton
National Oceanography Centre Southampton
Southampton, UK

Séverine Fauquette
Institut des Sciences de l'Evolution
CNRS
University of Montpellier
Montpellier, France

Adrien Favre
Department of Molecular Evolution and Plant
Systematics & Herbarium (LZ)
Institute of Biology
Leipzig University
Leipzig, Germany
and
Institute for Ecology, Evolution and Diversity
Goethe University
Frankfurt, Germany

Jon Fjeldså
Center for Macroecology, Evolution and Climate
at the Natural History Museum of Denmark
University of Copenhagen
Copenhagen, Denmark

Suzette G.A. Flantua
Institute for Biodiversity and Ecosystem Dynamics
University of Amsterdam
Amsterdam, Netherlands

Jane E. Francis
British Antarctic Survey
Cambridge, UK

Ceridwen Fraser
Fenner School of Environment and Society
Australian National University
Canberra, Australia

John E. Gordon
School of Geography and Sustainable Development
University of St. Andrews
St. Andrews, UK

Catherine H. Graham
Department of Ecology and Evolution
Stony Brook University
Stony Brook, NY, USA
and
Swiss Federal Research Institute WSL
Unit of Biodiversity and Conservation Biology
Birmensdorf, Switzerland

Babs Hagendoorn
Institute for Biodiversity and Ecosystem
Dynamics
University of Amsterdam
Amsterdam, Netherlands

Peter Heenan
Landcare Research
Lincoln, New Zealand

Camilo Higuera
Ecopetrol
Exploration Division
Bogota, Colombia

Henry Hooghiemstra
Institute for Biodiversity and Ecosystem Dynamics
University of Amsterdam
Amsterdam, Netherlands

Carina Hoorn
Institute for Biodiversity and Ecosystem Dynamics
University of Amsterdam
Amsterdam, Netherlands

Yong-Jiang Huang
Key Laboratory for Plant Diversity and Biogeography
of East Asia
Kunming Institute of Botany
Chinese Academy of Sciences
Kunming, China

Otto Huber
Instituto Experimental Jardín Botánico de Caracas (UCV)
Caracas, Venezuela

Carlos Jaramillo
Smithsonian Tropical Research Institute
Panama, Republic of Panama

Gonzalo Jiménez-Moreno
Departamento de Estratigrafía y Paleontología
University of Granada
Fuenta Nueva S/N
Granada, Spain

Salomon B. Kroonenberg
Delft University of Technology
Department of Geosciences
Delft, Netherlands
and
Anton de Kom University of Suriname
Tammenga, Suriname

Laura P. Lagomarsino
Gothenburg Global Biodiversity Centre
and
Department of Biological and Environmental Sciences
University of Gothenburg
Gothenburg, Sweden
and
Department of Biology
University of Missouri St. Louis
St. Louis, MO, USA

Lee Hsiang Liow
Natural History Museum
and
Centre for Ecological and Evolutionary Synthesis (CEES)
Department of Biosciences
University of Oslo
Oslo, Norway

Carolina Lithgow-Bertelloni
University College London
London, UK

Edoardo Martinetto
Department of Earth Sciences
University of Torino
Torino, Italy

Matt S. McGlone
Landcare Research
Lincoln, New Zealand

Frédéric Médail
Institut Méditerranéen de Biodiversité et d'Ecologie
marine et continentale (IMBE)
Aix-Marseille University/CNRS/IRD/AU
Aix-en-Provence, France

Timothy Millar
Plant & Food Research
Christchurch, New Zealand

Peter Molnar
Department of Geological Sciences and Cooperative
Institute for Research in Environmental
Sciences (CIRES)
University of Colorado at Boulder
Boulder, CO, USA

Arata Momohara
Graduate School of Horticulture
Chiba University
Matsudo, Chiba, Japan

Andrés Mora
Ecopetrol
Exploration Division
Bogota, Colombia

Robert J. Morley
Palynova UK and Royal Holloway
University of London
London, UK

Volker Mosbrugger
Senckenberg Research Institute and Natural
History Museum
Frankfurt, Germany

Alexandra N. Muellner-Riehl
Department of Molecular Evolution and Plant
Systematics & Herbarium (LZ)
Institute of Biology
Leipzig University
Leipzig, Germany
and
German Centre for Integrative Biodiversity
Research (iDiv)
Halle-Jena-Leipzig
Leipzig, Germany

Andreas Mulch
Senckenberg Biodiversity and Climate Research
Centre (BiK-F)
Frankfurt, Germany
and
Institute of Geosciences
Goethe University
Frankfurt, Germany

Serge D. Muller
Institut des Sciences de l'Evolution
CNRS
University of Montpellier
Montpellier, France

Johannes G.B. Oostermeijer
Institute for Biodiversity and Ecosystem Dynamics
University of Amsterdam
Amsterdam, Netherlands

Martin Päckert
Senckenberg Natural History Collections
Museum für Tierkunde
Dresden, Germany

Mauricio Parra
Institute of Energy and Environment
University of Sao Paulo
Sao Paulo, Brazil

Allison Perrigo
Forest Cat Editing
Uppsala, Sweden
and
Gothenburg Global Biodiversity Centre
Gothenburg, Sweden

and
Department of Biological and Environmental Sciences
University of Gothenburg
Gothenburg, Sweden

Paulo Petry
The Nature Conservancy
Arlington, VA, USA
and
Museum of Comparative Zoology
Harvard University
Cambridge, MA, USA

Speranta-Maria Popescu
GeoBioStratData.Consulting
Rillieux la Pape, France

Christopher J. Poulsen
Department of Earth and Environmental Sciences
University of Michigan
Ann Arbor, MI, USA

Ghillean T. Prance
Herbarium
Royal Botanic Gardens
Kew, Richmond, UK

Niels Raes
Naturalis Biodiversity Center
Leiden, Netherlands

Kenneth F. Rijsdijk
Institute for Biodiversity and Ecosystem Dynamics
University of Amsterdam
Amsterdam, Netherlands

Uwe Ring
Institutionen för geologiska vetenskaper
Stockholm University
Stockholm, Sweden

Jan Schnitzler
Institute of Biology
Leipzig University
Leipzig, Germany

Friedemann Schrenk
Senckenberg Research Institute and Institute for
Ecology, Evolution and Diversity
Goethe University
Frankfurt, Germany

Arie C. Seijmonsbergen
Institute for Biodiversity and Ecosystem Dynamics
University of Amsterdam
Amsterdam, Netherlands

Daniele Silvestro
Department of Biological and Environmental Sciences
University of Gothenburg
and
Gothenburg Global Biodiversity Centre
Gothenburg, Sweden

John L. Smellie
Department of Geology
University of Leicester
Leicester, UK

Tara M. Smiley
Department of Earth and Environmental Sciences
University of Michigan
Ann Arbor, MI, USA
and
Integrative Biology
Oregon State University
Corvallis, OR, USA

Robert A. Spicer
School of Environment, Earth and Ecosystem Sciences
The Open University
Milton Keynes, UK

Bryan Storey
Gateway Antarctica
University of Canterbury
Christchurch, New Zealand

Tao Su
Key Laboratory of Tropical Forest Ecology
Xishuangbanna Tropical Botanical Garden
Chinese Academy of Sciences
Mengla, China

Jean-Pierre Suc
Sorbonne University
CNRS
Institut des Sciences de la Terre Paris (iSTeP)
Paris, France

Victor A. Tagliacollo
Universidade Federal do Tocantins
Palmas, Tocantins, Brazil

Aleks Terauds
Australian Antarctic Division
Kingston, Tasmania, Australia

Verónica Torres Acosta
Institut für Erd- und Umweltwissenschaften
University of Potsdam
Potsdam, Germany

Hanna Tuomisto
Department of Biology
University of Turku
Turku, Finland

Peter van der Beek
Institut des Sciences de la Terre (ISTerre)
Grenoble Alps University
Grenoble, France

Julia A. Wheeler
WSL Institute for Snow and Avalanche Research SLF
Davos, Switzerland
and
Institute of Botany
University of Basel
Basel, Switzerland
and
Department of Environmental Conservation
University of Massachusetts
Amherst, MA, USA

Zhuo Zheng
Department of Earth Sciences
Sun Yat-sen University
Guangzhou, China

Zhe-Kun Zhou
Key Laboratory of Tropical Forest Ecology
Xishuangbanna Tropical Botanical Garden
Chinese Academy of Sciences
Mengla, China
and
Key Laboratory for Plant Diversity and Biogeography
of East Asia
Kunming Institute of Botany
Chinese Academy of Sciences
Kunming, China

Martin Ziegler
Department of Earth Sciences
Utrecht University
Utrecht, Netherlands

Acknowledgments

We warmly acknowledge all those who shared their time with us by reviewing the book chapters and by providing advice on how to improve the quality of this volume. In alphabetical order:

Hemmo Abels, Josué Anderson, Bob Anderson, Christine Bacon, Catherine Badgley, Peter Barrett, Peter van der Beek, Mauricio Bermúdez, Stephen Blackmore, Bert van Bocxlaer, David Cantrill, Jose S. Carrion, Page Chamberlain, Richard Corlett, Katharine Dickinson, Guillaume Dupont-Nivet, Søren Faurby, Alexandre Diniz-Filho, Lei Fumin, John Gordon, Murray Gray, Stephan Halloy, Douwe van Hinsbergen, Henry Hooghiemstra, Colin Hughes, Jose Joordens, Sander van der Kaars, Yoshiaki Kameyama, Michael Kessler, Christian Körner, Holger Kreft, Salomon Kroonenberg, Laura Lagomarsino, Holly Lutz, Paul Mann, Colin Meurk, Herbert Meyer, Peter H. Molnar, Andres Mora, Lewis Owen, Jay Quade, Willem Renema, Susanne Renner, Robert Ricklefs, Brett Riddle, William F. Ruddiman, Jan Schnitzler, Daniele Silvestro, Petr Sklenář, Ed Sobel, Bob Spicer, Richard A. Spikings, Hans ter Steege, Masaki Takahashi, Torsten Utescher, Luis Valente, Henry Wichura, Sean Willet, Jack Williams, Rafael O. Wüest and Zhe Khun Zhou.

We also thank the reviewers who chose to remain anonymous, and many other colleagues and students for numerous discussions and support.

Suzette Flantua is thanked for helping with many figures in the book, as well as Klaudia Kuiper, for producing the geological time scales.

James Albert is thanked for providing many of the definitions found in the Glossary.

Funding: AA is supported by grants from the Swedish Research Council (B0569601), which covered a substantial part of ALP's editorial costs for this book, the European Research Council under the European Union's Seventh Framework Programme (FP/2007-2013, ERC Grant Agreement n. 331024) and a Wallenberg Academy Fellowship. We also thank the sFossil workshop at the Synthesis Centre for Biodiversity Sciences sDiv (DFG grant FZT 118).

Foreword

Peter Raven

Mountains are stony ripples in Earth's crust, cast up to heights by the processes of geology and then worn down again over millions of years. Until about 10 000 years ago, when agriculture made it possible for us to accumulate food and build towns and cities as centers of civilization, mountains remained the highest and most obvious physical features that we knew. It was in mountains that we found inspiration and met with the gods; they soon became both our goals and our boundaries. We have always been filled with the desire to go beyond the ranges to see what we might find, but at the same time we feel comfortable when mountains enclose us within limits we can see and understand.

As soon as people began to think about mountains, they were mystified by them. Volcanoes came boiling out of Earth's crust, while other mountains seemed to arise for no apparent reason. The relationship between masses of heavy rocks and uplift provided part of the answer, but then plate tectonics provided a much more complete understanding. We have learned the lithospheric plates consist of an upper, lighter (~granitic) crust and lower, denser (~basaltic) rocks, which together move on top of the asthenosphere as a result of convection currents in the mantle. When two plates collide, the lighter rocks are often thrown upward, as in the origin of the Himalaya following the collision of India with the Asian plate. With that understanding, we can do a better job of dating the ranges that loom above us and explore when and how they acquired their characteristic biota.

Alexander von Humboldt (1769–1859), the great Prussian nature philosopher, was born with an incredible desire to see and explore as much of the world as he could. His wanderlust led him to set out for Latin America in 1799 and to remain there, traveling widely, for 5 years. He is justifiably credited with the invention of a comprehensive concept of nature that encompassed every aspect of our planet. Among his many findings was the key insight about the global relationship between biota and elevation, as well as latitude, which we have taken as a cornerstone of the distribution of biomes and their constituent plants and animals ever since.

The organization of life on land: biomes

Over the 4.54 billion-year history of our planet, life first appeared, and then evolved and differentiated. All organisms were aquatic until about 485–475 million years ago, when the ancestors of plants, fungi, vertebrates and arthropods emerged onto land and began to differentiate. In the diverse and often sharply defined habitats that existed on land, partly because of the presence and distribution of mountains, organisms evolved to become incredibly diverse within these new boundaries. Vascular plants, which originated some 430 million years ago, grew into trees and formed forests that were homes for a variety of organisms, and eventually for dinosaurs, mammals, birds and modern groups of arthropods. Flowering plants, which appear in the early Cretaceous period, 130 million years ago, diversified to dominate most of the world's forests by the end of that period, although vast stretches dominated by conifers still ranged through the northern lands. Animals, fungi and protists differentiated with them and formed the myriad relationships by means of which the biomes function. Mountains divided the biomes and provided sharp habitat differences locally, into which organisms migrated, differentiated and evolved.

During the Mesozoic era, and until approximately the start of the Eocene period, about 55 million years ago, the world's biomes were less clearly differentiated than they have become subsequently. Warm temperate to subtropical forests occupied the world's lower latitudes, differentiating over the course of the Tertiary era into temperate forests, shrublands, grasslands and deserts. The steepening of the equator-to-pole climate gradient produced modern tropical rainforests and polar habitats as a result of the overall development of regional climates. As habitats differentiated, mountains became increasingly significant in determining patterns of distribution for organisms. In both the north and the south, biomes and the organisms that lived in them extended equatorward at ever higher elevations, with other biomes flanking them below.

The opening of the Drake Passage between southern South America and Antarctica, which began some 50 million years ago in the Eocene, had a profound effect on world climate. Air and water circulation around the southern hemisphere rapidly cooled the region, resulting in the gradual differentiation of the "modern" biomes and the formation of continental glaciers that spread over Antarctica at least some 34 million years ago. Closed as a path for migration between South America and Australia, Antarctica became the world's ice box – although remnant forests and other vegetation persisted there until a few million years ago. Eventually, continental glaciers were formed in the north, too, and polar and boreal biomes originated in the north and extended down along the mountains, with alternating glacial expansion and retreat starting around 2.6 million years ago and extending to the last period of retreat of the ice about 10 800 years ago.

As the patterns of biome distribution in the north fluctuated as the ice periodically expanded and retreated, the outlines of the natural world that we know formed gradually. Before the middle Miocene period, about 15 million years ago, Eurasia and North America were clothed with warm-temperate to semitropical biomes, in which plants and animals that have survived to the present occurred together with others now extinct in fewer, broader and biologically simpler biomes than today.

The ways in which the locations and forms of mountains and mountain ranges have played and continue to play a major role in determining present-day patterns of distribution are many. The relationship is illustrated by the numbers of species of vascular plants in three north temperate political units of approximately equal size: Europe, about 11 500 species; the USA, about 19 000 species; and China, about 32 500 species. The patterns of differentiation and richness of many other groups, such as amphibians, follow similar patterns, which have resulted from their shared history.

The movement of tectonic plates that plowed the Indian subcontinent into southern Eurasia elevated a series of mountain ranges in China that extends southward into Southeast Asia. When continental ice sheets reached their maximum extent, biotic communities could migrate southward relatively easily, along with the climates to which they were adapted. As the climate warmed, those communities returned northward, often separating into disjunct areas on the many distinct mountain ranges of China and further evolving, thus adding to the overall diversity and richness of the Chinese biota. Fully 40% of China lies above 2000 m elevation! Moreover, fully tropical vegetation reaches southern China, but not Europe or the USA.

The broad outlines of vegetation were similar across Eurasia and North America, but biomes and individual species fared differently depending on the options they had when faced with changing climates. In western Eurasia (Europe), migration southward during the glacial advances was blocked by the east–west-running Alps and then by the Mediterranean, so that many species simply perished as the ice reached its maximum extent. Patterns of survival during glacial times along the Mediterranean Sea and elsewhere provide records of the past that are of great interest to biogeographers. It is in the Caucasus that the original biota of Europe survived best, in a region where the mountains extend farther south than they do to the west.

In North America, areas of survival for the original mixed forest biota, determined by studies of fossil pollen, were located along the coast of North and South Carolina and in the Edwards Plateau region of Texas. As the climate warmed, these forests migrated back into the coves and slopes of the Appalachians, Ozarks and other areas of eastern North America, forming the rich relict forests characteristic of these regions. Some of the original communities ranged further south along the Sierra Madre Occidental of Mexico and even into South America. Greater opportunities for survival left temperate North America with a much richer biota than Europe; in contrast, fewer opportunities for survival left the region biologically not nearly as rich as China. Thus, while *Metasequoia*, *Sequoia*, *Ginkgo* and *Cercidiphyllum* once occurred in rich mixed forests right across North America and Eurasia, they are now confined to much more limited regions defined by the patterns just discussed.

In the southern hemisphere, Antarctica originally connected South America with Australia. The combined land mass was clothed with a continuous warm-temperate forest that has survived in disjunct areas of South America and Australasia. The separation of this supercontinent into its constituent parts also broke up this forest and ended opportunities for migration across the south, where the geography is very different from that in the north. Of the three southern continents aside from Antarctica, South America extends farthest south, at about 55° S latitude – roughly the same distance from the equator as London – whereas the southernmost point of Australia, a continent that is continuing to move north, lies at about the same latitude as Oregon or southern France. Southernmost South Africa is located at about the same latitude as Los Angeles, and a lower one than southernmost Europe. Southern climates are milder than those in the north, with relict austral forests surviving in Australia, New Zealand, New Caledonia and southern South America. These forests have a composition reflecting the widespread forest that covered these lands when they were connected 80–50 million years ago.

Thus, continental movements and orogeny have shaped the patterns of our modern biomes, but human activities have changed and continue to change them profoundly. We are driving to extinction a huge proportion of the species of organisms that make our life on Earth possible. By providing refuges for biomes that require cool habitats, mountains will play an important part in determining which organisms will survive the effects of our devastating and increasingly dominant activities.

Mountains as cradles of biodiversity

Mountains provide a great deal of the stimulus for the origin of the incredible diversity of terrestrial life. Ascending them, one encounters habitats not found on the plains below; crossing them, depending on their orientation and latitude, one finds drier habitats and deserts that are associated with the formation of rain shadows. In the zones of the prevailing westerlies, roughly 30 to 60°N and S latitude, north–south-running mountains such as the Andes and the Sierra Nevada of California have forests on their western sides and dry biomes to the east. These patterns form a framework for the evolution and migration of organisms that is greatly altered by regional climate change, as this book shows amply. As the biomes start to move, they are enriched or become depauperate, depending on which source areas, harboring different biomes or constellations of species, are near enough to provide source populations.

All the while, mountains are, in effect, engines of biodiversity. Organisms that reach them gain access thereby to new kinds of habitats into which they may evolve, forming new species readily. The borders between biomes have long been recognized as active areas of evolution, and when mountains are present, the opportunities are multiplied many times. In a sense, the panorama of life on Earth may be likened to the patterns that form in a kaleidoscope, changing as the tube rotates. On land, the mountains and other barriers (like oceans, lakes and rivers) interrupt the patterns of differentiation and distribution unevenly and often with jagged borders, and the opportunities for evolution are enhanced. In this sense, mountains may be said to be cradles of biodiversity on all of the continents.

Our influence on the future

As the ice from the last expansion of glaciers on a continental scale withdrew, our ancestors, earlier living as hunter-gatherers, developed agriculture, starting about 12 000 years ago. Our species, *Homo sapiens*, which is first known from African fossils about 300 000 years old, migrated out of Africa into Eurasia some 70 000 years ago. It interbred with some of the earlier hominins, like Neanderthals, that it encountered in the north and east. However, after about 30 000 years, modern humans were the only human inhabitants of the vast continent of Eurasia. At that time, our whole population is estimated to have consisted of about 1 million people; with the development of agriculture and the organization of villages, towns and cities, our numbers and our influence on all parts of the world began to grow rapidly. When humans crossed to the New World some 15 000 years ago, they brought dogs with them, and invented agriculture independently as regional climates improved. Wherever people went, they appear to have killed off most of the remaining large mammals and flightless birds, but it is the conversion of about a third of the world's surface to crops and grazing that has had the greatest effect on the survival of biodiversity. As this process occurred, our numbers grew rapidly: there are now about 7.4 billion people on Earth, increasing by 200 000 additional people every day. The population is projected to grow to 9.9 billion within 33 years, by mid-century (www.prb.org). We are estimated to be consuming 164% of the world's capacity for sustainable productivity (www.footprintnetwork.org), with some 800 million of us malnourished. Given these relationships, it should not be surprising that we are destroying species rapidly along with their habitats.

Global climate change is also having a major effect on the extinction of species, which is almost certain to increase in the future. The development of our agriculture and the consequent flowering of our civilization have taken place in a time of comparatively constant climates since the last withdrawal of continental ice sheets. But now, temperatures are increasing more rapidly, to some of the highest levels that the planet has experienced for hundreds of thousands, and probably millions, of years, with human activities being the major cause. As a result, we are pushing ourselves out of the temperature ranges in which we have so rapidly expanded our population and increased our pressures on Earth during the past 10 000 years. The survival of huge numbers of species depends on the outcome of international negotiations and the degree to which nations are successful in meeting the goals they set. At any rate, the current IPCC report estimates that between 20 and 30% of all species may be lost as a result of climate change alone.

Mountains and their interplay with climate and life constitute the major thrust of this book, which examines these interactions in many meaningful ways. Just as mountains have played a key role in the evolution and diversification of life, they will play a key role in determining which species survive and which are lost over the coming decades. Because of the sharply contrasting patterns of continental distribution in the northern and southern hemispheres, the loss of species from climate change is very likely to be more extensive in the south. For cool-adapted species occurring near the poleward ends of the southern continents, there will literally be no place to go. For example, the hundreds of endemic species of animals and plants that occur on Table Mountain near Cape Town, South Africa and those that occur along the southern edge of Australia are likely to perish as the global climate warms. In the northern hemisphere, species can, in principle, move northward or to higher elevations in the mountains: a phenomenon that is already being observed.

Biological extinction is now estimated to be proceeding at about 1000 times the background rate for the past 66 million years, and is continuing to climb along with our numbers and our zeal for ever-greater levels of consumption. Consequently, we could lose as much as half of all species on Earth by the end of the century: a massive change that will have immense consequences for our descendants and for the functioning of our planet. Just as they have played a major role in the evolution of species in the past, mountains will play a key role in their survival now and in the future, making the interrelationships that are well explored in this book of major importance as we consider how to attempt to regain global sustainability.

This book should provide encouragement to those who wish to understand mountains, climate and biological communities as parts of a great interlocking puzzle, much as Alexander von Humboldt did so magnificently with far less information at his disposal than we possess now. From a spiritual point of view, mountains contain some of the most appreciated, inspiring and delightful places on Earth, ones that will retain a special interest for as long as our civilization exists. For me, the special pleasure that we feel in mountains is well expressed by these lines by the American writer Wallace Stegner, from his remarkable 1971 novel *Angle of Repose*:

The mountains of the Great Divide are not, as everyone knows, born treeless, though we always think of them as above timberline with the eternal snows on their heads. They wade up through ancient forests and plunge into canyons tangled up with watercourses and pause in little gem-like valleys and march attended by loud winds across high plateaus, but all such incidents of the lower world they leave behind them when they begin to strip for the skies: like the Holy Ones of old, they go up alone and barren of all circumstance to meet their transfiguration.

Biography

Prof. Peter H. Raven, ForMemRS, President Emeritus of the Missouri Botanical Garden, St. Louis, and George Engelmann Professor of Botany Emeritus, Washington University in St. Louis has devoted his scientific career to developing the understanding of the evolution and diversity of plants, their interactions with other organisms and their conservation. A recognized authority on biogeography, he was one of the first to apply the findings of plate tectonics to the understanding of the field. He is a Trustee of the National Geographic Society and Chairman of the Board of Trustees of the Center for Plant Conservation, co-editor of the 49-volume *Flora of China* and co-author of *The Biology of Plants*, a best-selling textbook on botany

for nearly 50 years. In 2001, Dr. Raven received the National Medal of Science, the highest award for scientific accomplishment in the USA. He has been president of the American Association for the Advancement of Science, Sigma Xi, the American Institute of Biological Sciences and a number of other organizations. He served for 12 years as Home Secretary of the National Academy of Sciences, to which he was elected in 1977. He is also a member of the American Academy of Arts and Sciences and of the American Philosophical Society, as well as of the academies of science in a number of other countries, including China, India, Russia, Brazil, Sweden, Denmark, Argentina, Mexico and Australia.

Biography of Editors

Carina Hoorn is geologist/palaeoecologist and Research Associate at the University of Amsterdam in The Netherlands. She holds a PhD from this university, and an MSc in Science Communication from Imperial College in London. Her work focuses on the interaction between geological and biological processes through time and the significance of this relation for biodiversity. She is particularly interested in the Andes-Amazonian system and the Tibeto-Himalayan region. Previously she edited *Amazonia, Landscape and Species Evolution* (C. Hoorn & F. Wesselingh, eds, Wiley).

Allison L. Perrigo is an academic editor, and is the owner and operator of Forest Cat Editing. She is currently the coordinator of the Gothenburg Global Biodiversity Centre in Gothenburg, Sweden. She holds a PhD in systematic biology from Uppsala University in Sweden where she studied the diversity and distribution of protists and, later, plants. Her research interests include biogeography, molecular phylogenetics and the biodiversity of poorly explored regions. She is broadly involved in scientific communication and presenting science to a wider audience.

Alexandre Antonelli is Professor in Biodiversity and Systematics at the University of Gothenburg in Sweden, Scientific Curator at Gothenburg Botanical Garden and Director of the Gothenburg Global Biodiversity Centre. His research focuses on understanding the origins and evolution of biodiversity, with a focus on cross-taxonomic and methodological studies in the American tropics. His team combines molecular, palaeontological and distribution data for inferring biodiversity patterns and the dynamics of speciation, extinction and migration across time and space. For more information please visit http://antonelli-lab.net.

Glossary

This glossary is intended to clarify terminology specific to the biological and geological sciences that is used within this book. While the definitions aim to clarify the text herein, many of these words have additional meanings, within and outside of the biological and geological sciences, that are not included within the scope of this text, and are therefore omitted. For further discussion on any of the terms listed in this section, we refer the reader to the Oxford University Press's *Dictionary of Biology*, 7th edn. and *A Dictionary of Geology & Earth Sciences*, 3rd edn.

Abiotic Relating to nonliving (physical) factors.
Ablation The erosive removal of ice through melting, evaporation, sublimation or wind erosion.
Accretion The addition of material to a land mass or tectonic plate by tectonic processes.
Adaptive radiation Rapid diversification of an ancestral species into several different daughter species or subspecies, which are typically adapted to different ecological niches (e.g., Darwin's finches). This occurs when the evolution of a new trait (or set of traits) or the emergence of a new habitat promotes diversification.
Adiabatic A process during which no heat is gained or lost. Adiabatic cooling of Earth's atmosphere occurs when air ascends and pressure decreases. Adiabatic heating is the inverse of this process. See also *lapse rate*.
Advection The horizontal movement of an ocean current, air mass or solid mass (i.e., rock advection in tectonics), transporting heat/cold, humidity or salinity.
Aeolian Relating to the wind.
African Superswell A topographic anomaly in the southern half of Africa that is caused by upwelling in the mantle and/or abnormal heating of the lithosphere.
Albedo A reflection coefficient indicating the fraction of solar radiation (energy) that is reflected from Earth back into space. Reflective surfaces – such as snow and ice – have a high albedo, whereas absorptive surfaces – such as oceans and forests – have a low albedo.

Allele The two or more variations of a gene that can be found in the same location on a specific chromosome. The possession of different alleles can result in different phenotypic expressions. A diploid organism may be homozygotic for an allele (have two similar copies of the allele) or heterozygotic (have two different copies).
Allopatric Literally, "other country." Refers to distribution areas of different taxa whose ranges do not overlap. See also *sympatry*.
Allopatric speciation The formation of a new species following the physical isolation of populations by an extrinsic spatial barrier (i.e., geographic speciation). The frequency of allopatric speciation (vs. sympatric or parapatric speciation) is debated, but all evolutionary biologists agree that allopatry is a common mode by which new species arise.
Alluvium The residual mineral deposit (clay, silt, sand and gravel) in a river bed, floodplain, valley or delta left by flowing water.
Antarctic Circumpolar Current (ACC) The clockwise (west to east) ocean current that encircles Antarctica.
Aspect (geology) The compass direction of a slope face, indicating orographic orientation. Aspect strongly influences surface temperatures, as it directly relates to the angle at which the sun hits the slope.
Asthenosphere The weak layer of Earth's upper mantle, directly below the *lithosphere*.
Atlantic Meridional Overturning Circulation (AMOC) The circulating northward flow of warm, highly saline, shallow Atlantic Ocean water and the counteracting southward flow of cooler, less saline waters from the deep Atlantic. The AMOC is an important component of *thermohaline circulation* and relies heavily on the *North Atlantic Deep Water*.
Autecological traits Intrinsic characteristics relating to the way a species interacts with its environment.
Authigenic Minerals or rocks that were formed in their current location.

Autochthonous (biology) Indigenous or native. Applied to species, food or nutrient input, or to sediment that was both produced and deposited within the area of reference.

Autochthonous (geology) Native in the sense of having originated (evolved) in the place in question.

Avulsion (river) The geologically sudden rerouting of a river from its original channel to a different direction, where a new river channel forms.

Basement (rock) Metamorphic or igneous rocks on which a sedimentary platform or cover is found.

Batholith A large ($100\,km^2$ or more) igneous rock body produced by the cooling of magma deep in Earth's crust.

Bedrock The solid layer of rock at Earth's surface, covered by the *regolith*.

Biodiversity The variation in life at all levels, from genes to species, as well as at higher taxonomic levels and throughout larger ecological spaces, such as biospheres. It can be measured with *species richness* or *species diversity*, among other metrics. Abbreviated form of "biological diversity."

Biogeography The documenting and understanding of spatial patterns of biological diversity.

Biological pump The part of the carbon cycle whereby carbon is moved from the atmosphere into the deep sea through biological processes. Not to be confused with a *species pump*.

Biome A broadly defined biological assemblage occurring in habitats with similar climates over a large area. It is equivalent to an *ecosystem*.

Biotic Relating to living (especially ecological) factors.

Biotope An area with a combination of environmental conditions that hosts a specific biological community.

Broadleaved forest A forest type dominated by wide-leaf *deciduous* trees found in humid regions in both the northern and southern hemispheres.

Central American Seaway (CAS) The ocean gap that occurred along the tectonic boundary between the South American plate and the Panama microplate, connecting the Pacific Ocean and the Caribbean. Other sea passages besides the CAS may also have connected these bodies.

Cirque A steep-walled semi-circular mountain hollow or valley head with a lower-gradient floor, formed by mountain glacier erosion.

Clade In evolutionary biology, a branch on a cladogram or phylogenetic tree. A monophyletic group.

Cladogenesis The origin of a new clade through the splitting of a single parental lineage into two distinct daughter lineages.

Climate The statistics of weather averaged over periods of years or longer.

Coeval Of the same age.

Commensalistic A relationship between two organisms whereby one gains a benefit, while the other is neither benefited nor harmed.

Condensation level Elevation above Earth's surface at which the relative humidity (RH) of an air parcel reaches saturation when it is cooled by dry adiabatic lifting. Precipitation will form once a rising air parcel reaches the condensation level. Also called "cloud condensation level" or "lifting condensation level."

Conifer A group of gymnosperms belonging to the division Pinophyta that produce seeds but, unlike Angiosperms, do not produce flowers. Well-known conifers include pines, yews and cypress.

Contact (geology) The point at which two geological entities of different types (usually rocks) touch one another.

Convection See *moist convection*.

Cordillera An expansive series of mountain ranges (chains), often marked by the inclusion of other features – such as valleys, basins, rivers, plains and plateaus – between parallel chains.

Coterminous Having the same border, area or range.

Craton A portion of Earth's continental crust that has been geologically stable since Precambrian times.

Crust Earth's outermost layer, composed of low-density silicate rocks. Earth's crustal thickness ranges from $5–10\,km$ beneath the oceans up to $\sim75\,km$ beneath some parts of the Tibetan Plateau.

Cryosphere The parts of Earth where water is frozen, including glaciers, ice sheets, permafrost, sea ice, lake ice and river ice.

Deciduous (tree) A tree that sheds and regrows its leaves annually, usually in response to winter cooling and dry seasons.

$\delta^{18}O$ The ratio of oxygen-18 and oxygen-16, relative to a standard, in a sample used as a proxy for past climate conditions. The ratio can be determined from *biotic* (e.g., coral, foraminifera) or *abiotic* (e.g., ice core) remains.

Denudation The removal of soil or rock by the combined effect of physical and chemical surface processes, especially erosion.

Diabatic A thermodynamic change in a system marked by the exchange of heat (energy) with its surroundings. Opposite of *adiabatic*.

Diagenesis The physical and chemical changes that convert a sediment into a sedimentary rock.

Diapause An extended state of dormancy initiated and terminated by specific environmental conditions.

Dioecious Organisms, especially plants, in which the male and female reproductive organs occur in separate individuals. See also *monoecious*.

Disjunct distribution A *species* or higher taxonomic unit (genus, family) that is widely separated geographically, such as *Rhipsalis baccifera* (a cactus), which is native in both South America and Africa.

Dispersal Range expansion, by the passive or active transport of organisms or propagules (e.g., seeds), beyond the limits of a species' distributional area.

Dispersive A plant or animal that moves from the location of its birth to a breeding area to reproduce.

Diversification An increase in the species richness and/or phenotypic disparity of a clade. Diversification may be due to natural selection or to macroevolutionary processes that result in a net excess of speciation over extinction.

Diversification rate (net) The rate of speciation minus the rate of extinction.

Dynamic topography Topography due to flow in the *mantle*.

Ecosystem A community of organisms living in conjunction with the *abiotic* components of their environment, interacting as a system. See also *biome*.

Ecotone The relatively narrow and sharply defined transition zone between two or more communities.

Edaphic Relating to the soil.

Eddy diffusion Atmospheric mixing caused by turbulent flow around and behind obstacles to atmospheric circulation.

Endemic A taxon geographically restricted to a particular area or region. The term "endemic" should always be used in direct connection with a region, such as "endemic to Mount Kinabalu" or "endemic to Australia" (ultimately, all species are endemic to Earth). Species with a narrow geographic range often have a small population size and are vulnerable to extinction.

Endogenic Formed or occurring beneath Earth's surface (internally). See also *exogenic*.

Epigenetic (biology) Nongenetic factors that influence phenotype by turning genes "on" and "off."

Epistatic interactions An interaction between two mutations, wherein one directly affects the expression of the other.

Erosion The mechanical or chemical removal of rock at Earth's surface, often due to continuous contact of the surface with wind, water or ice.

Eurybiomic Tolerating variable habitats.

Exhumation The movement of rocks towards Earth's surface.

Exogenic Formed or occurring on Earth's surface (externally). See also *endogenic*.

Extant Currently in existence/living, as in "extant species." Opposite of *extinct*.

Extinct No longer in existence/surviving, as in "extinct families." Opposite of *extant*.

Extinction The global disappearance of a species or a population. When all the members of a lineage or taxon die, the group is said to be *extinct*. See also *extirpation*.

Extirpation The total disappearance of a species in a single area, although the species still persists elsewhere. See also *extinction*.

Fall line The boundary between an upland region (usually made of hard *basement* rock) and a lowland region (plain, usually *sedimentary* rock). The usage of "fall" refers to the rapids or waterfalls that generally mark a river's crossing of the fall line.

Fault A fracture in Earth's crust with relative motion between the two sides.

Fauna Animal life, often used to distinguish from plant life (*flora*).

Fellfield An alpine or tundra slope that is exposed to frost dynamics and winds, resulting in a characteristic appearance and *flora*.

Fitness The relative measure of reproductive success of an organism based on the frequency with which it passes on its genetic material to the successive generation.

Flora Plant life, often used to distinguish from animal life (*fauna*). Also a list of all plant taxa that occur in a chosen area.

Fluvial Of or referring to rivers or river valley ecosystems.

Fold The result of a geological *stratum* deforming and bending. Folds can be used to infer past geological movements in a region.

Forb A non-woody flowering plant that is not a grass, sedge or rush.

Founder (geology) To sink or become submerged.

Founder effect (biology) The process whereby a new *population* establishes from a few individuals, resulting in a loss of genetic variation.

Fumarole A crustal opening that exudes hot gasses. Often found in association with volcanoes.

Gene flow The movement of genes within or among populations by interbreeding.

Genetic drift Changes in the frequencies of alleles in a population that occur by chance, rather than as a result of natural selection.

Genetic isolation The limiting or interruption of *gene flow*.

Genome The entirety of the hereditary information of an organism that is encoded in the DNA, including both the coding and noncoding regions. Genomic data are increasingly used to assess the relationships among living organisms.

Genotype The set of two genes possessed by an individual at a given locus. More generally, the genetic profile of an individual. See also *phenotype*.

Geodiversity The variation in Earth's physical makeup. This includes the local variation in minerals, rocks, sediments, fossils, soils and water and the geomorphological processes that shape Earth's surface.

Geographic Information System (GIS) Computer-based system for the capture, storage, retrieval, analysis and display of spatial (i.e., locationally defined) data.

Geoid An equipotential reference surface that is always normal to the local gravity vector. It is the mean sea level over the oceans.

Geothermal gradient The gradient with which the temperature increases with depth beneath Earth's surface. A commonly cited rate of change is 30 °C/km.

Glacial (period) A period dominated by relatively cool temperatures and extensive glaciation. Global temperatures vary during a glacial period, leading to *stadial* (cooler) and *interstadial* (warmer) periods throughout the glacial.

Glacial lake outburst flood (GLOF) The catastrophic flooding that results when a morainic dam that contains a glacial lake breaks.

Graben A depressed ("dropped") block of land that is bordered by normal faults on either side, moving it downward relative to the surrounding horsts (i.e., adjacent higher land), forming a *rift*.

Graphical user interface (GUI) A visual system that allows a user to intuitively operate a computer program using windows, menus and buttons, as opposed to a text-based (command-line) interface.

Graywacke A hard, coarse *sedimentary* rock (*sandstone*) with over 15% clay.

Great American Biotic Interchange (GABI) The event marking the migration of terrestrial species between the Americas, across the Isthmus of Panama, following the connection of South America and the Panama Bloc after a long period of reciprocal isolation.

Hadley cell A major atmospheric convection cell rising at the equator and sinking at about 30° of latitude to the north or south.

Heritability An estimate of how much variation in a phenotypic trait is due to genetics. Heritability is observed as a proportion ranging from 0 (genetics play no role in the observed phenotype) to 1 (the phenotype is based entirely on genetics).

Heterozygous Describes a diploid individual possessing two different *alleles* of a gene at a given locus. Opposite of *homozygous*.

Holarctic A major biogeographic zone that covers most of the northern hemisphere, including North America as far south as the Mexican Chihuahuan Desert, nontropical Europe and Asia, and Africa north of the Sahara Desert. The Holarctic is often divided into the *Palearctic* (*Old World* part) and *Nearctic* (*New World* part).

Hominin A member of the tribe Hominini, which includes the *extant* genera *Homo* (humans) and *Pan* (chimpanzees and bonobos), as well as *extinct* genera such as *Australopithecus* and *Kenyanthropus*, among others.

Homozygous Describes a diploid individual possessing two of the same *allele* of a gene at a given locus. Opposite of *heterozygous*.

Horizontal isotopic moisture gradients Changes in $\delta^{18}O$ (change in oxygen isotopes) or δD (change in hydrogen isotopes) in precipitation across continental interiors along a single air-mass trajectory.

Hot spot (geology) An unusually active volcanic region related to a mantle plume. Hot spots commonly occur within lithospheric plates (e.g., Hawaiian Islands), but some, like Iceland, lie along mid-ocean ridges.

Hotspot (biology) An area that is both biologically mega-diverse and immediately threatened with anthropogenic destruction.

Hybridization Interbreeding between two different *species*. The establishment of hybrid populations is more common in plants than in animals.

Hypsometry The elevation of the land relative to mean sea level.

Igneous (rock) Rock formed from lava or *magma*. One of three main classes of rocks, along with *sedimentary* and *metamorphic*. Examples of igneous rock types are basalt, granite and pumice.

Incision (river) The process by which a river erodes bedrock, cutting down through the river bed.

Indigenous Native to a given region or ecosystem, as a result of natural processes with no human intervention. Indigenous taxa are not necessarily *endemic*, and may be native to other regions as well.

Indochina The biogeographic region bounded by the Indian subcontinent to the west and China to the north, encompassing mainland Southeast Asia: Myanmar, Thailand, Laos, Cambodia, Vietnam and peninsular Malaysia.

Interstadial A warmer interval during a *glacial* period, marked by temporary warming and ice retreat. See also *stadial*.

Intertropical Convergence Zone (ITCZ) The area encircling Earth near the equator, where winds originating in the northern and southern hemispheres come together. Annual movements of the ITCZ result in the wet and dry seasons of northern South America. The ITCZ is located over the oceans; the equivalent over land is called the monsoon front.

Isopach The lines on a map indicating layers or strata of equal thickness, measured perpendicular to the layer boundary.

Isopleth The lines on a map indicating places where a specific condition (e.g., surface elevation, amounts of precipitation or atmospheric pressure) is equivalent.

Isostasy The idealized state in which the mass of rock in any column above some depth (the depth of compensation) is the same everywhere. This is Archimedes' Principle applied to the crust and upper mantle of Earth.

ka (kilo-annum) Thousand years ago, indicative of an absolute date. See also *ky* and *Ma*.

Kinematics Movement of material without specification of forces; hence, different from dynamics.

ky Thousand years, indicative of a duration. See also *ka* and *My*.

Lacustrine Relating to lakes.

Lahar A volcanic debris or mudflow. Lahars are characterized by their extremely destructive nature.

Lapse rate The vertical temperature gradient in the atmosphere, determined by the rate at which the temperature of an air parcel decreases with height, primarily controlled by adiabatic heating or cooling. For an ascending or descending unsaturated (dry) particle, the average dry adiabatic lapse rate is 9.8 °C/ km. See also *adiabatic*.

Last Glacial Maximum (LGM) The period marking the lowest temperature during the most recent glacial period in Earth's history, spanning from ~26.5 to 19.0 ka.

Leeward The downwind side (of a mountain, island, etc.). The leeward side of a feature is protected from prevailing winds and is typically drier than the *windward* side.

Linnean shortfall The discrepancy between the number of formally described species and the actual number of species. Named after the Swedish naturalist Carl von Linné (Linneaus), who formalized the now-standard system of naming species (binomial nomenclature).

Lithosphere The rigid outer layer of Earth, which is broken up into various tectonic plates. It includes the *crust* and the upper part of the *mantle*. *Plate tectonics* describes the relative movement of these essentially rigid plates.

Local extinction See *extirpation*.

Lowstand A time when sea levels are at a relative minimum.

Ma (mega-annum) Million years ago, indicative of an absolute date. See also *My* and *ka*.

Macroevolution The evolutionary process taking place at or above the species level, usually involving time scales of millions of years. See also *microevolution*.

Magma The high-temperature mixture of molten or semi-molten rock, volatiles and solids occurring beneath Earth's surface. Once magma reaches Earth's surface through a volcanic vent, it is known as lava.

Malesia The biogeographic zone spanning the equator between Asia and Australia, encompassing Sundaland (the Malay Peninsula and the islands of Sumatra, Java, Bali and Borneo), *Wallacea* (all of the Indonesian islands east of the *Wallace Line*, including Sulawesi, Lombok, Flores and Timor, as well as East Timor) and Papua New Guinea.

Mantle The layer of Earth between the crust and the outer core, making up over 80% of Earth's volume. The mantle is predominantly solid, but acts like a viscous fluid over geological time. On average, the mantle is nearly 3000 km thick. It can be further subdivided into the upper mantle (made up of the *asthenosphere* and the *lithosphere*) and the lower mantle.

Marl A *sedimentary* rock type with varying amounts of calcium carbonate ($CaCO_3$) and clay.

Mega-mesothermal plants Subtropical plants growing in areas with a mean annual temperature (MAT) of 20–24 °C.

Megathermal plants Plant taxa that are wholly frost-intolerant, and hence at the present time restricted to the frost-free tropics.

Meromictic Describes a lake with layers of water that do not intermix.

Meso-microthermal plants Cool-temperate plants growing in areas with a mean annual temperature (MAT) of 12–14 °C.

Mesothermal plants Warm-temperate plants growing in areas with a mean annual temperature (MAT) of 14–20 °C.

Metamorphic (rock) Rock formed through transformation due primarily to either heat or pressure. One of three main classes of rocks, along with *igneous* and *sedimentary*. Examples of metamorphic rock types are marble, schist and slate.

Metamorphism The process by which a rock is compositionally or structurally altered, usually by heat or pressure.

Microevolution The evolutionary process acting below the species level (e.g., within or among populations of the same species), usually involving time scales shorter than millions of years and expressed in number of generations. See also *macroevolution*.

Microsatellite A highly variable, short (two to five nucleotides long) repeating segment in a DNA sequence that can be used to assess population-level patterns within a species.

Microthermal plants Plants that are tolerant of frosts, and hence are widespread in temperate regions and in tropicalpine vegetation.

Miocene Climatic Optimum (MCO) A period between ca. 17 and 15 Ma, marked by a significantly warmer climate than today (~4–5 °C). The MCO deviates significantly from the overall cooling trend of the last 50 My. Also called the middle Miocene Climatic Optimum or mid-Miocene Climatic Optimum.

Moist convection Rising of moisture-laden air, leading to condensation of clouds and precipitation.

Moist enthalpy A measure of the energy in a parcel of air in terms of sensible and latent heat and water vapor.

Molecular phylogeny A tree-like diagram constructed using DNA sequence similarities among taxa to infer evolutionary relationships.

Monecious Plants in which the individuals produce both male and female flowers, although the organs are separate. See also *dioecious*.

Mountain A steep, protruding landmass that is significantly higher than the surrounding geological features, formed by tectonic forces or volcanism. A minimum height of 600 m has been proposed by Whittow's *Dictionary of Physical Geography*, but this can be considered a general guideline, as lower protrusions have been labelled as mountains, and steepness must also be taken into account (e.g., so that a high-altitude plateau is not labelled a mountain). Mountains often have a number of *biomes* that occur along their slopes, making them especially interesting to those studying *biodiversity*.

My Million years, indicative of a duration. See also *Ma* and *ky*.

Nappe A rock mass that is thrust over another mass as a result of folding or *overthrusting* on a regional scale.

Nearctic The biogeographic region comprising North America from Greenland to the middle of Mexico. It shares a similar fauna with the *Palearctic*, and together the two regions make up the *Holarctic*. The *Nearctic* and *Neotropics* make up the *New World*.

Neontology The systematic study of living taxa.

Neotropics The biogeographic region comprising South America, Central America and North America up through the middle of Mexico. Together, the *Neotropics* and *Nearctic* make up the *New World*.

New World The collective name for North and South America. See also *Old World*.

Niche The functional position of an organism in a community, including its interaction with all physical, chemical and biological parameters of the environment.

North Atlantic Deep Water (NADW) A highly saline, high-oxygen content, and nutrient-poor water mass that forms in the North Atlantic and subsequently sinks, playing a role in *thermohaline circulation*. It is a key component of the *Atlantic Meridional Overturning Circulation*.

Northern hemisphere glaciation (NHG) The development of glaciers in the northern hemisphere, specifically from ca. 3 Ma.

Nunatak An isolated mountaintop projecting above an extensive area of surrounding snow or ice, possibly serving as a *refugium* for cold-adapted species during cool or glacial periods.

Obliquity The changes of the axial tilt of Earth's spin (between 22.1 and 24.5°) with respect to the plane of the orbit, resulting in cycles of 41 ky.

Old World The collective name for Asia, Africa and Europe. See also *New World*.

Orocline The curvature of an *orogen* that occurred after its formation.

Orogen A crustal belt involved in mountain formation.

Orogenesis The process of mountain formation.

Orogeny The process of mountain formation, especially by folding and faulting of Earth's crust and by plastic folding, metamorphism and the intrusion of magma into the lower parts of the crust.

Orographic precipitation Precipitation that has been enhanced or modified by the flow of air over topography, typically through the forcing of vertical atmospheric motions.

Outgassing The release of gasses from Earth's surface into the atmosphere.

Overthrusting The movement of one body or rock over another. May result in the formation of a *nappe*.

Palearctic The biogeographic region comprising Europe, North Africa and the part of Asia to the north of the Himalaya. It has a similar faunal composition to the *Nearctic*, and together the two regions make up the *Holarctic*.

Paleoendemic An *endemic* taxon that has been in the same region for a relatively long time, usually on the order of many tens of millions of years.

Paleosol A *stratum* or soil horizon formed during a past geological period that is presently buried under more recent layers.

Palynoflora The floral assemblage deduced from the study of pollen and spores.

Palynology The study of pollen and spores from angiosperms, gymnosperms, pteridophytes, bryophytes and fungi. Literally, the study of "dust."

Parapatry The condition whereby two taxa have adjacent but generally non-overlapping ranges.

Passerine A bird of the order Passeriformes, which accounts for over half of all bird species.

pCO$_2$ The partial pressure of carbon dioxide (CO$_2$), generally measured from water or ice. CO$_2$ in the atmosphere is expected to be in equilibrium with dissolved CO$_2$ in Earth's surface waters (such as the oceans), and thus ice-bound pCO$_2$ will indicate past CO$_2$ levels in the atmosphere.

Pedogenic Describes processes occurring in the soil, or the formation of soil.

Periglacial Describes an environment that is directly affected by cold, nonglacial processes, including repeated freezing and thawing, either directly adjacent to a glacier or ice sheet or in an area that experiences similar conditions.

Phenology The study of the short-term effects of climatic (often seasonal or annual) cycles on the timing of biological phenomena such as flowering, leaf shedding and bird migrations.

Phenotype The physical or structural characteristics of an organism, produced by the interaction of genotype and environment during growth and development. See also *genotype*.

Phenotypic plasticity Nongenetic variation in organisms in response to environmental factors.

Phylogeny A branching tree-like diagram that represent the evolutionary relationships among species. See also *molecular phylogeny*.

Plate tectonics The theory that the surface of Earth is made of a number of plates, which have moved throughout geological time, resulting in the present-day positions of the continents.

Plateau A flat, high-elevation expanse.

Pleiotropy When a single gene influences two or more apparently unrelated phenotypic traits.

Population A local group of organisms that interbreed and share a gene pool. A *species* is made up of one or more populations.

Posterior probability The statistical probability that a hypothesis is true, given the data.

Precession Changes in the direction of Earth's axis of rotation relative to the fixed stars. The climatic precession cycle depicts the net impact of solar insolation on Earth, and currently has primary periods of 18.9, 22.3 and 23.6 ky.

Prograde metamorphism The physical change that occurs when a rock is buried and heated. See also *metamorphism*.

Pyroclastic Relating to the rock fragments expelled by a volcano. See also *magma*.

Quiescent In an inactive or dormant state.

Radiation An event of rapid *diversification* in which many new *species* arise in a relatively brief period of time. Radiations can be triggered by a novel adaptation (*adaptive radiation*) or by the opening of novel or previously inaccessible ecospace.

Rain shadow The relatively dry area on the *leeward* side of a mountain.

Refugium An area or environment in which a displaced *species* or *population* survives through an unfavourable climatic change, such as a period of glaciation or aridification.

Regolith The unconsolidated, heterogeneous rocky layer covering *bedrock*.

Relict A *species* or *clade* that has persisted relatively unchanged since a much earlier geological age.

Relief The difference in elevation between two points.

Rift An area where the *lithosphere* has pulled apart.

Ring of Fire Belts of volcanoes surrounding the Pacific Ocean where one plate of *lithosphere* plunges beneath another. This is where the majority of Earth's earthquakes and active volcanos occur. Also known as the circum-Pacific belt.

Riparian Relating to the area adjacent to a river or stream.

Sandstone A *sedimentary* rock type composed primarily of sand-sized mineral or rock particulate.

Savannah A tropical and subtropical ecosystem type dominated by grassy plains, with few trees.

Saxicolous A term describing an organism that inhabits the area on or among rocky outcroppings.

Sclerophyllous Scrubby vegetation with leaves hardened by woody tissue that have small distances between them.

Scree The collected small, loose rocks that fall from a steep slope or cliff, forming a debris pile or shingle-covered slope at the base of the feature. Also called talus.

Sea level The average level of the oceans across Earth, used as a standard to determine elevations and altitudes. Over short periods (hours, days or months), local sea level may vary due to tides, wind, gravitational forces or other *abiotic* factors. In most cases, sea level is used interchangeably with mean sea level (MSL).

Seasonality An annually occurring predictable change, often used in relation to variations in temperature and precipitation throughout the year.

Sedimentary (rock) Rock formed from mineral and organic matter (sediment) that is deposited and then cemented. One of three main classes of rocks, along with *metamorphic* and *igneous*. Examples of sedimentary rock types are limestone, *sandstone* and shale.

Shield An ancient and tectonically stable portion of continental crust that has survived the merging and splitting of continents and supercontinents for at least 500 million years and is distinguished from regions of more recent geological origin that are subject to subsidence or down-warping.

Sill A sheet-like intrusion between older layers of rock.

Single-nucleotide polymorphism (SNP) A variation in a single base pair in an organism's DNA.

Sky island A mountaintop isolated from surrounding mountains or areas with a similar ecology. The ecologically dissimilar area surrounding the mountaintop serves as a dispersal barrier.

Slip The displacement vector describing the relative movement of features on either side of a *fault*.

Speciation The evolutionary formation of new species, usually by the division of a single ancestral species into two genetically distinct daughter species. Other processes can also lead to speciation, such as *hybridization* and anagenetic evolutionary change.

Species A fundamental unit of biodiversity, biogeography, evolution and ecology. Species can be variously defined by the biological species concept (BSC), the cladistic species concept, the ecological species concept, the phylogenetic species concept or the recognition species concept, among others. The BSC, according to which a species is a set of interbreeding organisms, is one of the most widely used definitions, at least by biologists who study vertebrates. A species is referred to by a Linnean binomial, such as *Homo sapiens* for human beings.

Species distribution model (SDM) A model that characterizes the multivariate ecological space delimiting species' distributions, and projects this subset of ecological space back onto geography, resulting in habitat suitability maps. Also known as an ecological niche model.

Species diversity A metric combining *species richness* and species abundance distribution.

Species pump A situation in which species diversity increases over time as a result of the cyclical connectivity and isolation of sky islands during climate change.

Species richness The number of species present in a location. See also *species diversity*.

Species sink A low-quality habitat that has higher than expected diversity due to the dispersal of organisms from a *species source*.

Species source A high-quality habitat where species can form large and diverse populations, and from which they may disperse to a *species sink*.

Stadial A cooler interval during a *glacial* period, marked by cooling and the advance of ice. See also *interstadial*.

Stratigraphy The study of rock layers and layering (stratification), and the temporal correlation of rocks from different localities.

Stratum A rock layer.

Strike The orientation or azimuth of the intersection of a planar geological feature, such as a *fault*, with Earth's surface.

Subduction zone The site where one plate of oceanic *lithosphere* bends down and plunges into the *mantle*.

Subsidence The downward motion of Earth's surface relative to the *geoid*. See also *uplift*.

Surface uplift The upward movement of Earth's surface relative to the *geoid* or mean sea level.

Sympatry Living in the same geographic region.

Systematics The study of the evolutionary relationships among taxa, including the delimitation of species. See also *taxonomy*.

Talus See *scree*.

Taphonomy A field within paleontology that studies biases in the fossil record arising from the processes of fossilization and preservation.

Taxon A species or a group of *species* (e.g., a genus, family or order) recognized as a unit of classification. A taxon should, according to most biologist, be monophyletic (i.e., share a common ancestor.) See also *clade*.

Taxonomy The science of classifying and naming organisms. See also *systematics*.

Teleconnections On geological or evolutionary time scales, teleconnections link climate changes occurring in widely separated regions of our planet. This can occur, for example, through heat transport in the oceans.

Tepui A table-top *sandstone* mountain, generally isolated from other, similar features. Tepuis are characteristic of the Guiana Highlands of north-eastern South America. "Tepui" has been traced to a term from the indigenous Pemón inhabitants in eastern Guiana, and may indicate the structures as the "houses of the Gods."

Terrane A fragment of crustal material formed on, or broken off from, one tectonic plate and accreted (sutured) to crust lying on another. The crustal block or fragment preserves its own distinctive geologic history, which is different from that of the surrounding areas.

Thalweg A line drawn along the course of a valley or river, indicating its lowest elevation along its length.

Thermohaline circulation (THC) See *Atlantic Meridional Overturning Circulation*.

Till The unsorted sediment deposited by a glacier during its movement.

Time-calibrated (dated) phylogeny A phylogeny in which branch lengths correspond to absolute time, thus providing the timing of divergence events based on a number of calibration points, such as fossil occurrences of species or geological events.

Tomography The rendering of a volume of Earth's interior based on the behaviour of waves produced by earthquakes or explosions. Also called "seismic tomography" and analogous to a CT scan of the human brain. Inferences of lateral variations in the Earth structure deduced from seismic waves passing through that structure from different directions.

Topography The three-dimensional (3D) arrangements of Earth's features on a local or global scale.

Torpor An inactive state.

Transform fault A *fault* where two lithospheric plates slide horizontally past one another. They are found most frequently as part of mid-oceanic ridges.

Tropical A climate type found in the *tropics*, consisting of two seasons (wet and dry), and where all months have a mean monthly temperature of >18 °C.

Tropics The portion of Earth between 23°30′ S and 23°30′ N latitude. The tropics include all the areas on Earth where the sun reaches a point directly overhead at least once during the solar year.

Troposphere The lowest layer of Earth's atmosphere. This is where almost all weather patterns occur. It is thickest in the tropics (up to 20 km) and thins towards the poles (7 km).

Tuya A flat-topped volcano that is formed when an eruption occurs directly below a glacier, causing the lava to spread horizontally.

Uplift The upward motion of a portion of Earth's surface. See also *subsidence*.

Vascular plant Plants that have a vascular system consisting of xylem and phloem. They include angiosperms (flowering plants), ferns and conifers, among others. In contrast, nonvascular plants do not have xylem or phloem; they include mosses and liverworts. Sometimes also called "higher plants."

Vicariance Speciation that occurs as a result of the physical separation and subsequent isolation of portions of an original population. This can be due to tectonic movement, or to smaller events such as the change of a river's course.

Walker Circulation A model explaining air circulation in the tropics based on longitudinal differences in surface pressure and temperature in the Pacific Ocean, which strongly influences global weather patterns. It's observed as the east–west circulation of the tropical atmosphere, consisting of ascent over the western Pacific and Indonesia, divergence aloft (to the east over the Pacific and to the west over the Indian Ocean) and convergence near the surface.

Wallace Line A faunal boundary between *Wallacea* and Sundaland (Asia), noted by naturalist Alfred Russell Wallace in his 1876 book *The Geographic Distribution of Animals*.

Wallacea A complex plate boundary zone between the Indo-Australian and Eurasian plates, bounded by the Wallace and Lydekker Lines. It represents a transition zone of *faunas* and *floras* from both Asia and Australia.

Wallacean shortfall The knowledge gap regarding species distributions.

Weathering The gradual breakdown of a rocky surface due to physical, chemical or biological processes.

Westerlies Prevailing winds blowing from west to east, primarily found between 30 and 60 degrees of latitude.

Windward The upwind side (of a mountain, island, etc.). The windward side of a feature is exposed to the prevailing winds and is typically wetter than the *leeward* side.

About the Companion Website

This book is accompanied by a companion website:

www.wiley.com\go\hoorn\mountains,climateandbiodiversity

The website includes:

- Powerpoints of all figures from the book for downloading
- Videos

1

Mountains, Climate and Biodiversity: An Introduction

Carina Hoorn[1], Allison Perrigo[2,3,4] and Alexandre Antonelli[3,4,5]

[1] *Institute for Biodiversity and Ecosystem Dynamics, University of Amsterdam, Netherlands*
[2] *Forest Cat Editing, Uppsala, Sweden*
[3] *Gothenburg Global Biodiversity Centre, Gothenburg, Sweden*
[4] *Department of Biological and Environmental Sciences, University of Gothenburg, Gothenburg, Sweden*
[5] *Gothenburg Botanical Garden, Gothenburg, Sweden*

Abstract

Mountains harbor about one-quarter of all terrestrial species in about a tenth of the world's continental surface outside Antarctica. This disproportionate diversity makes mountains a focal point for research on the generation and maintenance of biodiversity. Some of the key features that make mountains so biologically diverse are the elevational gradient, physiographic and climatic diversity, and prolonged isolation of their peaks and valleys. These features reflect the complex interactions between plate tectonics and mountain building, climate change and erosion over time scales extending to millions of years. It is now widely accepted that these large-scale processes play a fundamental role in biotic evolution across space and time. Together with ecological interactions among organisms, they form the basis for modern biogeography. But why, when and how the interactions between the geosphere, biosphere and atmosphere resulted in such high biodiversity in mountains is insufficiently understood. In this book, a multidisciplinary team of authors discusses the state of research at the interface between the geo- and biospheres and addresses these and other questions, while presenting examples from mountain systems around the world.

Keywords: *mountain building, biodiversity, plate tectonics, geo-biodiversity*

1.1 Introduction

Can you imagine a world without mountains? It would undoubtedly be a much less diverse place in terms of biomes, habitats and species. Mountains are the cradles of all major river systems, they are the central determinants of regional- and continental-scale climate and they comprise many unique biomes (Figure 1.1). They generate massive influxes of sediment that are divulged into adjacent territories (e.g., from the Andes across the Amazon basin, and from the Rockies into the Great Plains). For these reasons, the effects of mountains reach well beyond their immediate slopes (Gentry 1982; Finarelli & Badgley 2010; Hoorn et al. 2010).

Mountains also have a dual role in that they both generate and receive biodiversity (Hoorn et al. 2013). On one hand, they can generate diversity through in situ adaptations and diversification, subsequently providing neighboring regions with new lineages (e.g. Antonelli et al. 2009; Santos et al. 2009). On the other, they are able to support pre-adapted lineages from other similar montane regions that arrive via long-distance dispersal (Merckx et al. 2015). Nevertheless, teasing apart the relative contributions of in situ diversification versus dispersal (Antonelli 2015), and assessing how and when different climatic and geological conditions influenced different regions, is a matter of intense research. Likewise, we are just beginning to understand how and when climate and tectonism interact, and how they together affect biodiversity.

The effect mountain building has on climate, and how these processes together influence the speciation, extinction and migration of different taxa, is hotly debated.

Mountains, Climate and Biodiversity, First Edition. Edited by Carina Hoorn, Allison Perrigo and Alexandre Antonelli.
© 2018 John Wiley & Sons Ltd. Published 2018 by John Wiley & Sons Ltd.
Companion website: www.wiley.com\go\hoorn\mountains,climateandbiodiversity

Figure 1.1 The center section of Humboldt's classical tableau, illustrating a cross-section of the Chimborazo volcano in Ecuador, the highest mountain peak as measured from the center of the globe. This detailed drawing depicts one of the earliest studies of how the mountain biota is structured along an elevation gradient. Humboldt recognized the existence of distinct vegetation zones at different elevations with largely unique sets of species, constrained by climatic and physiological adaptations. This pioneering work is often considered a landmark in biogeography. *Source:* Humboldt & Bonpland (1807). See also Plate 1 in color plate section.

A number of studies have touched on some or many aspects of this set of interactions (Hughes & Atchison 2015; Hughes 2016; Lagomarsino et al. 2016), yet none has fully addressed the complexity of the field in a single work. We have therefore commissioned 31 peer-reviewed chapters that, when taken as a whole, address this need.

One of the fundamental questions to address is: *How can we untangle mountain building and climate change, and what influence did each of these processes have on biological diversification*? It has been known for some time that Plio–Pleistocene climate change is responsible for pronounced changes in relief and a vast increase in global erosion rates (Molnar & England 1990). However, in recent years, mountain uplift, rates of erosion and paleoaltitude have begun to be measured more accu-

rately, thanks to advances in analytical methods in geosciences (e.g., in the fields of isotope and fission-track analysis) (Gosse & Stone 2001; McElwain 2004; Reiners & Brandon 2006; Forest 2007; Polissar et al. 2009; Lomax et al. 2012; Mulch 2016). These developments have enabled a global assessment of the timing and geographic extent of mountain building, erosion and relief (Herman et al. 2013; Herman & Champagnac 2015). Together, the data thus obtained provide a geohistorical guideline that helps to improve models of biotic evolution. Accurate mountain uplift ages have already been successfully applied in the context of molecular phylogenetic studies that test for the influence of surface uplift on species diversification (Lagomarsino et al. 2016).

Other major questions include: *When did taxa evolve, how did they respond to the ecological opportunities that followed from mountain building and what were their geographic distributions through time?* The generation of novel, carefully sampled biological data from extant species, together with improved databases on the fossil record – including enhanced geochronology from the Neotoma Paleoecology database, the Paleobiology database and Neclime, to name but a few –offers new perspectives on biotic evolution in mountain regions (e.g., Favre et al. 2015; Flantua et al. 2015). This, combined with new methods for predicting diversification and range evolution based on fossil records (e.g. Silvestro et al. 2016) and molecular sequences (e.g. Antonelli et al. 2016; Morlon et al. 2016) and methods for cleaning and processing vast amounts of extant species-occurrence data (Töpel et al. 2016), provides researchers with valuable tools and data for testing specific hypotheses on the evolution of mountain biodiversity.

Finally, determining the relative roles of abiotic and biotic processes in the assembly, generation and maintenance of biodiversity is a central task in understanding biological distributions, and it forms the core of this book. Here, specialists from different disciplines have joined forces to synthesize the current knowledge on mountain building, climate and biodiversity. To help the reader through this cross-disciplinary volume, the text is accompanied by a glossary of terms and a geological time scale (see back-cover inset).

1.2 What are Mountains?

Mountains are defined as "landforms that rise prominently above their surroundings, generally exhibiting steep slopes, a relatively confined summit area, and considerable local relief" (Molnar 2015). They cover over a tenth of the continental surface of the Earth (Figure 1.2), based on a recent evaluation by Körner et al. (2016), who used the ruggedness of Earth's terrestrial surface, excluding Antarctica, as the constraining feature for identifying "mountains" (Figure 1.3). Based on this estimate, they calculated that mountains cover

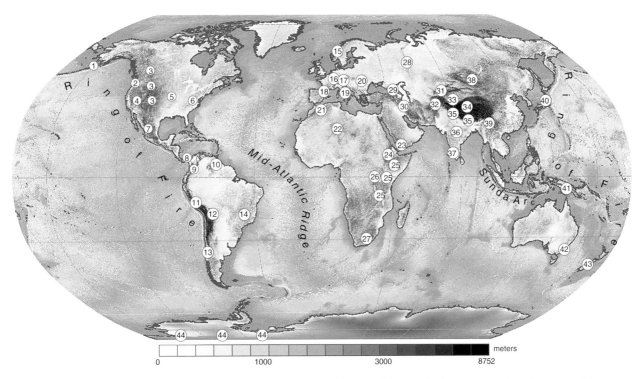

Figure 1.2 A selection of the most prominent mountain systems on Earth, as well as several other major geologic features and systems discussed throughout this book (image courtesy Suzette Flantua). Americas: 1, Aleutian Arc; 2, Cascades; 3, Rocky Mountains (Rockies); 4, Basin and Range Province; 5, Great Plains; 6, Appalachians; 7, Sierra Madre; 8, Panama Isthmus; 9, Northern Andes; 10, Guiana Highlands; 11, Central Andes; 12, Bolivian Altiplano; 13, Southern Andes; 14, Brazilian Highlands. Europe: 15, Scandinavian Mountains; 16, Jura Mountains; 17, Alps; 18, Pyrenees; 19, Apennines; 20, Carpathians. Africa-Arabia: 21, Atlas Mountains; 22, Ahaggar (Hoggar) Mountains; 23, Yemen Highlands; 24, Ethiopian Highlands; 25, East African Rift System (EARS); 26, Rwenzori Mountains; 27, Drakensberg. Asia: 28, Ural Mountains; 29, Caucasus Mountains; 30, Zagros Mountains; 31, Tien Shan; 32, Hindu Kush; 33, Kunlun Shan; 34, Tibetan Plateau; 35, Himalaya; 36, Deccan Plateau; 37, Western and Eastern Ghats; 38, Altai Mountains; 39, Hengduan Mountains; 40, Japanese Alps. Oceania: 41, New Guinea Highlands; 42, Eastern Highlands (Australia); 43, Southern Alps. Antarctica: 44, Transantarctic Mountains. See also Plate 2 in color plate section.

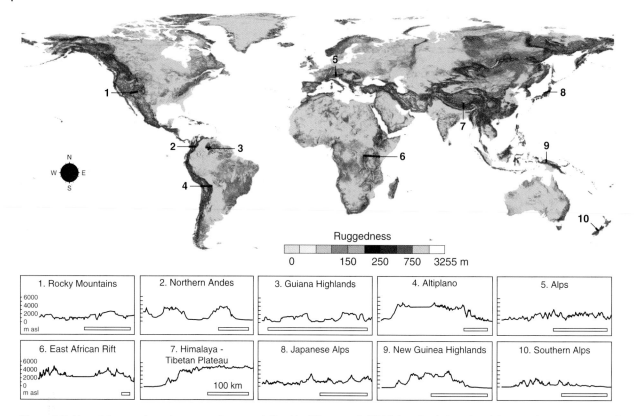

Figure 1.3 Mountain areas based on ruggedness, as defined by Körner et al. (2011) (maximal elevational distance between nine grid points of 30″ in 2.5′ pixel; for a 2.5′ pixel to be defined as "rugged" (i.e., mountainous), the difference between the lowest and highest of the nine points must exceed 200 m). Numbers indicate topographic profiles of selected mountain ranges around the world. The characteristic topography of a mountain directly relates to the potential impact and frequency of connectivity breaks caused by Pleistocene glacial cycles, and thus the expression of the flickering connectivity system. Bars below profiles indicate a 100 km distance proportional to the profile shown. *Source:* Adapted from Körner et al. (2011). Figure from Chapter 12. See also Plate 3 in color plate section.

13.8 million km^2 of Earth's land surface (12.5%), of which 3.3 million km^2 comprises alpine and nival belts.

Mountains are topographically complex, and are often rich in biodiversity. Barthlott et al. (1996, 2005) related this topographic complexity – or geodiversity (see also Gray 2004) – directly to biodiversity. Based on 100 × 100 km species-richness data, Barthlott et al. (1996, 2005) identified five global centres of vascular plant diversity, all of which are situated in or adjacent to mountainous regions (Figure 1.4a). Körner & Ohsawa (2006) and Körner et al. (2016) further estimated that half of all biodiversity hotspots (as defined by Myers et al. 2000) and a quarter of all terrestrial biodiversity is situated in mountains. Moreover, mountains host some of the most diverse ("hottest") biodiversity hotspots on the planet, including the tropical Andes and the Hengduan Mountains (Barthlott et al. 1996, 2005; Hughes & Eastwood 2006; Spehn et al. 2010, 2011; Madriñán et al. 2013; Hughes & Atchinson 2015). Recently, Badgley et al. (2017) presented a four-part framework under which

the processes that lead to this relationship between topographic complexity and high levels of biodiversity can be robustly assessed.

The connection between mountains and biodiversity is also seen in several taxa with well-understood species-richness patterns. These include mammals, birds and amphibians (extensive data on species ranges are accessible from www.iucnredlist.org and www.birdlife.org). However, as with hotspots, in making such connections it is important to consider that humans have influenced the distribution of many species (Faurby & Svenning 2015). The European brown bear *Ursus arctos*, for instance, once had a much wider distribution in Europe, but is presently mainly confined to mountainous regions due to anthropogenic pressure. However, when the anthropogenic modifications to species ranges are excluded, the correlation between mountainous regions and mammal diversity is evident (Figure 1.4b). While humans may have exaggerated the pattern of montane diversity, we clearly did not create it.

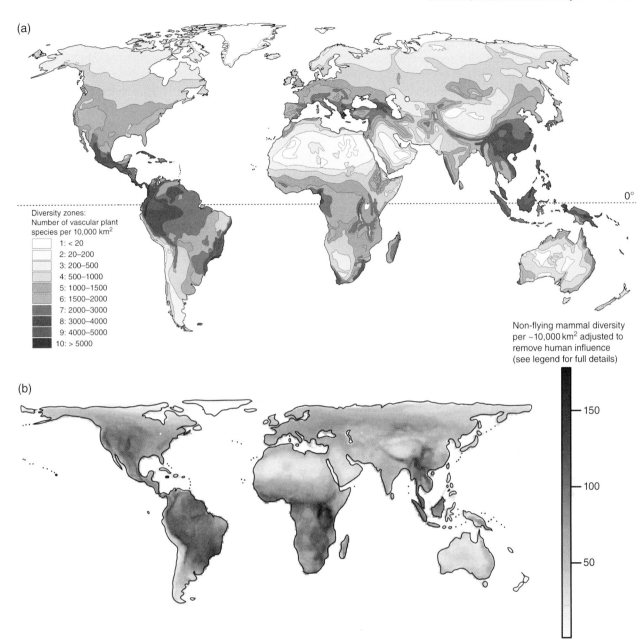

Figure 1.4 (a) Global plant diversity measured by estimated number of vascular plant species per 10,000 km². *Source:* Barthlott et al. (1996, 2007), with permission from Wilhelm Barthlott and Jens Mutke. (b) Estimated patterns of species diversity for terrestrial mammals. This map was constructed using estimated natural species ranges without human influences, and plotted using an equal-area Behrmann projection with colors proportional to the number of expected species. Several mountain regions have noticeably higher diversity than surrounding areas (e.g., the Rocky Mountains and Sierra Nevada in North America, the Andes in South America and, in particular, the East African Rift System). *Source:* Modified from Faurby & Svenning (2015), with permission from Søren Faurby. See also Plate 4 in color plate section.

1.3 The Physiography of Mountains and Patterns of Biodiversity

Mountains are natural barriers, characterized by their elevation gradients, their physiographic complexity and the changes in climate and environment that they cause through time. These factors influence the distribution and diversification of species, and also make mountains suitable for maintaining biodiversity over time. This is particularly pronounced in the tropics, where a species only needs to move relatively small distances to remain within its preferred niche space during periods of rapid climate change, as compared to species in the lowlands (Sandel et al. 2011; Condamine et al. 2016).

By the 19th century, Alexander von Humboldt had already recognized the important relationship between plant diversity, complex topography and climate (Figure 1.1) (Mutke 2010). Over the course of the 20th century, however, the role of geological processes was gradually segregated from biodiversity research, with paleontology, geology and biogeography seldom considered in ecological theory (Ricklefs 1987, 2004). In spite of this general trend, certain themes were recognized and expanded on during the 20th century, which can now be integrated in order to provide a better understanding of the complex biodiversity patterns in mountains.

It has been known since the 19th century that physiography relates to species richness, but the relevance of this notion has become eclipsed over time. Inspired by the theory of island biogeography proposed by MacArthur and Wilson (1963, 1967), Ricklefs (1977) coined the term "environmental heterogeneity" (EH), reviving the idea that a species can be excluded from a region due to environmental or evolutionary change. Ricklefs' (re)introduction of EH was a revolutionary step in ecology, as it renewed the debate on the relevance of geological processes in time and space as drivers for species richness and distribution. Later, Ricklefs (1987, 2004, 2005) concluded that despite long-term resistance to the idea from ecologists, large-scale regional and historical processes influence regional biodiversity, and that regional and local diversity are directly connected. Previously, large-scale processes had been considered too weak to influence the equilibrium that was achieved by local processes. Nevertheless, Ricklefs explained that patterns of biodiversity can only be properly interpreted within the broad context of regional and historical influences.

Many studies now recognize the importance of EH (see Tews et al. 2004 and Stein & Kreft 2015 for overviews) – including variation along the elevational gradient (EG) – in mountain biodiversity. The EG consists of gradual changes in the environment with increasing elevation along a mountain slope. Along an EG, biodiversity can be quantified in relation to hydrological, geochemical, climatological, topographical and edaphological factors, among others (Hughes & Eastwood 2006; Kreft & Jetz 2007; Kupfer 2010; Mutke 2010; Mutke et al. 2011; Moeslund et al. 2013; Flantua et al. 2014; Stein & Kreft 2015). However, the taxonomic coverage of a study is important to its interpretation. Recently, Peters et al. (2016) showed that at community level, temperature came out as the strongest driver for species diversity along the EG on Mount Kilimanjaro. Although spatial EH and EG are now well studied, their changes through time and parallel impact on mountain biodiversity remain poorly understood.

Early naturalists, such as Humboldt, Linnaeus and also Tournefort, recognized the importance of the EG as an abiotic determinant of biodiversity in mountain regions. This gradient is most pronounced in the tropics and loses its dramatic effect at higher latitudes. As Janzen (1967) stated, "In respect to temperature, valleys may be figuratively deeper to an organism living on the ridge top in the tropics than in a temperate area". The EG is characterized by a consistent decrease in temperature and air pressure as one moves upslope, whereas other climatic variables – such as solar radiation and precipitation – have a less consistent pattern (Körner 2003, 2007; McCain & Grytnes 2010). This differentiation in physical parameters along the EG is reflected in vegetation zones or belts.

Mountains often are conical, decreasing in area upslope (but see Elsen & Tingley 2015). The original concept derived from the south-western USA, where forested mountain peaks – or "sky islands" – were separated from a "sea" of desert; it is now expanded to include other regions with similar contrasting settings (McCormack et al. 2009). Although the classical species/area relationship would demand a decline in species towards the top of a mountain (see Lomolino 2000), in reality this does not occur. Meta-analyses of large floral and faunal data sets show that repeated isolation and connectivity between biomes favors endemism and colonization in sky islands (e.g., Steinbauer et al. 2016). In spite of this, the highest biodiversity is found in mountains at mid-elevation, where species of lower and higher altitudes meet (see Lomolino et al. 2010).

Mountains directly affect regional climate, because of the orographic barrier effect (Houze 2012). The orographic barrier redirects atmospheric masses, often by directing clouds upslope. At higher altitudes, these clouds cool, leading to precipitation. This process is particularly well studied in the Andes, where the South Monsoon and the South American Low-Level Jet are associated with the high Andes (e.g., Garreaud et al. 2008; Poulsen et al. 2010; Rohrmann et al. 2016). The effects of orographic rains extend into the tropical lowlands, leading to drainage modifications and high biodiversity (Hoorn et al. 2010). Mid-elevations in mountains are especially exposed to orographic rainfall, which makes them a suitable home for cloud forests rich in ferns (Kessler et al. 2016). For further details on mountain climatology in relation to modern plant assemblages and distribution, we refer the reader to Körner (2003, 2012) and Scherrer & Körner (2010).

Biota are not just passive recipients of physical and chemical changes in the environment and can also substantially influence the local and regional climate. These feedback effects occur through increases or decreases in albedo and evapotranspiration, and by fixing carbon

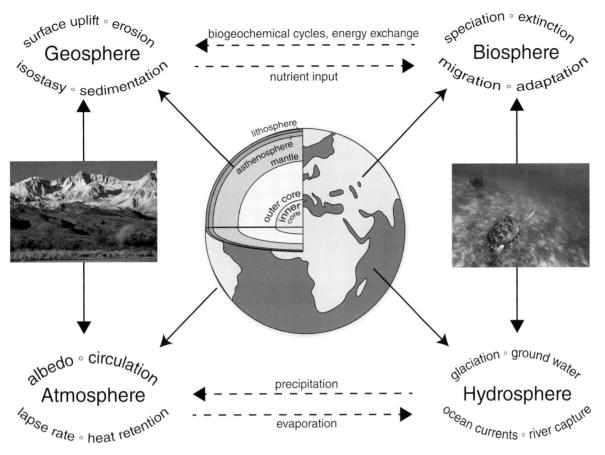

Figure 1.5 Schematic representation of the complex interaction between four of Earth's "spheres": the geosphere, biosphere, atmosphere and hydrosphere. Each sphere is surrounded by a selection of words indicating the terminology and processes associated with it. The Earth's layers are indicated in the cut-out, but are not drawn to scale. The lithosphere comprises the crust and the uppermost solid mantle. *Source:* Adapted from Senckenberg Magazine (2015). See also Plate 5 in color plate section.

(e.g., Malhi et al. 2002). Moreover, species also have a profound effect on the development of soils, on the biogeochemical cycles of nitrogen and other nutrients and on erosion and sedimentation processes (e.g., Durán-Zuazo & Rodríguez-Pleguezuelo 2007). These feedback mechanisms between the atmosphere, hydrosphere, geosphere and biosphere are illustrated in Figure 1.5.

1.4 Plate Tectonics, Mountain Building and the Biological (R)evolution

Plate tectonics are crucial for understanding biogeography and answering the question: *Why aren't all species found everywhere with a suitable habitat?* Although the breakup of the supercontinent Pangea and the later split of Gondwana and Laurasia seem something from the distant past, the biota of the modern world still carry their signatures. For example, they provide a plausible

explanation for why marsupials are found in both South America and Australia, and why penguins are not found on Greenland. Incidentally, both are derived from taxa that originated on the Gondwanan landmass and never dispersed with successful establishment on Laurasian landmasses. This is most evident in the recognition of biotic realms: continent-level regions containing similar sets of organisms at higher taxonomic levels. Their number, placement and limits were mapped by Wallace (1876), and later revised and refined by modern methods and large amounts of data (e.g., Holt et al. 2013).

Alfred Wegener was the first to recognize the relevance of plate tectonics in biological and climatic evolution (Wegener 1912; Köppen & Wegener 1924). Much later, Raven & Axelrod (1974) realized the importance of this global geological mechanism for biological distribution patterns. They presented a comprehensive review in which they evaluated the effects of continental drift on the current distribution of biodiversity, which constitutes a major contribution to the field now known as historical biogeography. More recently, the effect of plate tectonics on determining deep biological differences among continents was confirmed by spatial regression analyses (Ficetola et al. 2017).

Areas of active tectonics are tightly associated with vicariance and dispersal, two processes by which a taxon comes to exist in its current range. Taxa that passively "rode" the diverging landmasses during continental breakups are considered vicariant, as are those whose distributions were physically disconnected by smaller-scale processes such as river re-routing (e.g., some Amazonian fishes: Albert et al. 2006). Vicariance occurs on both large (e.g., continental) and small (e.g., intra-mountain) scales, and is simply the result of a taxon breaking into several populations due to an emerging physical or ecological barrier, leading often to independent evolution and speciation. In contrast, species that actively or passively colonized novel environments – such as oceanic islands or sky islands, or via migration over land bridges – are said to have dispersed. In addition to these processes, range expansion and contraction, sympatric speciation and local extinction further influence local species richness and diversity.

Mountain building is an important driver of speciation through time and space (Van der Hammen et al. 1973; Raven & Axelrod 1974; Van der Hammen & Cleef 1986). In the past two decades, there has been an increased focus on the role of geological processes in montane biotic evolution (Winkworth et al. 2005; Linder 2008; Antonelli et al. 2009; Potter & Szatmari 2009; Djamali et al. 2012; Särkinen et al., 2012; Heenan & McGlone 2013; Hoorn et al. 2013; Madriñán et al. 2013; Luebert & Weigend 2014; Favre et al. 2015; Hughes & Atchison 2015; Merckx et al. 2015; Craw et al. 2016; Hughes 2016; Lagomarsino et al. 2016; Renner 2016), the effects of which extend well into adjacent continental basins (Gentry 1982; Kohn & Fremd 2007, 2008; Badgley 2010; Finarelli & Badgley 2010; Hoorn et al. 2010; Gates et al. 2012; Jacques et al. 2014) and marine environments (Clague 1996; Renema et al. 2008; Wright & Stigall 2013; Yasuhara et al. 2016).

The mechanisms of mountain building and their effect on geomorphology and on peripheral sedimentary basins are further summarized in Owens & Slaymaker (2004) and Johnson & Harley (2012).

1.5 Mountains, Climate and Biodiversity: A Short Overview

This book is divided into three parts. The first two deal with processes and theory, while the third presents location-specific examples from around the world. Part I primarily addresses geological and hydrological processes, and biological mechanisms are incorporated in Part II. Part III is made up of chapters that apply concepts from the first two parts to exemplify how – by combining mountains, climate and biodiversity – we can arrive at a more holistic view of how montane biodiversity evolved.

Part I begins with a review of mountain and volcanic arc formation and isostasy – the equilibrium between the Earth's crust and mantle – and how these processes control continental elevation (Chapters 2 and 3). Climate, weathering and erosion are important factors that continuously reshape relief and promote uplift, particularly in the past 3 million years (Chapter 4). The methods used to quantify exhumation, surface uplift and mass removal through novel geological techniques and methodologies, such as fission-track and isotope analysis, are introduced (Chapter 5), along with an explanation of stable isotope paleoaltimetry, used to infer past altitudes (Chapter 6). Information gleaned from these techniques, when included in biological models, can lead to novel insights into the coupling (and timing) of abiotic and biotic processes. In addition to geochemical methods, fossil leaves also make excellent indicators of past precipitation and temperature (Chapter 7), which give insights into past topographic complexity, altitudes and climatic gradients. Mountains alter the regional climate and have distinct climatic gradients along their slopes that are directly connected to biotic evolution (Chapter 8). On an even larger scale, global drivers of climatic change through time include the Earth's surface reflectiveness (albedo), incoming solar radiation and atmospheric greenhouse gas contents, as they change over decades to million-year time scales (Chapter 9). The implications of this for the modern landscape and for biotic evolution are not yet well understood, but must be enormous.

In Part II, the key connections between topographic complexity, EG and biodiversity are explored. The concept of geodiversity underpins this part, and it is broadly defined and explored (Chapter 10). This is followed by a methodological introduction to geodiversity analysis and examples of how it can be used to predict biodiversity (Chapter 11). Over time, the highly diverse nature of mountains can be explained in a more process-oriented manner, focusing on the importance of shifting connectivity between mountain peaks during climatic change in the Pleistocene (Chapter 12). Different facets of EH and how isolation drives diversification patterns across mountains over shorter time scales play a major role (Chapter 13). High geodiversity is linked to high levels of biodiversity. The effects of the landscape and climate on past and future rates of diversification are explored in North American mammals (Chapter 14), and the components of this process are further decoupled by the separate estimations of speciation and extinction rates, as these do not necessarily respond similarly to environmental changes (Chapter 15). However, the time scales on which geological and evolutionary processes act may differ by several orders of magnitude, and broad correlations of regions with similar geodiversity over time do not always have presently analogous patterns of biodiversity (Chapter 16). With this in mind, the focus moves to the relationship between biota and physiography. When looking at global-scale patterns, climatic stability directly relates to those areas with the highest overall endemism in bird taxa (Chapter 17), whereas for hummingbirds specifically, it is instead temperature that correlates best with speciation rates over time (Chapter 18). As topography changes, catastrophic events leave their mark on major features, with equally major effects on the local biodiversity. River capture is a rare but significant result of physiographic change, and can lead to increased diversity in lowland river basins, which are already among the most species-dense ecosystems on Earth (Chapter 19). Finally, the key concepts and pitfalls in measuring biodiversity across regions are explained (Chapter 20), leading in to the theory behind how species distributions are estimated and projected across space and time (Chapter 21), which can be used to forecast the effects of climate change and the potential of mountains as climate refugia.

Part III is organized following the historical landmasses – Gondwana and Laurasia – and presents case studies from mountain systems around the world. The large-scale relationships between mountains, climate and biodiversity become apparent. The selected mountain systems are mostly young, formed in the Cenozoic. However, examples of older systems, such as the tepuis in South America and the Transantarctic Mountains, are also included. The geologically recent uplift of the Isthmus of Panama led to the Great American Biotic Interchange, and united former Laurasian and Gondwanan elements: North and South America, respectively. However, the timing of the biotic interchange does not directly coincide with the uplift, as there was significant habitat disparity between the newly uplifted region and the areas it connected (Chapter 22). Likewise, the biodiversity found on top of the table-like tepui of northern South America does not entirely match with the expected "Lost World" scenario once envisioned: despite significant endemicity in the region, there is also evidence of ongoing biotic exchange with surrounding ones (Chapter 23). Antarctica became an isolated continent starting about 50 million years ago. Once disconnected from South America and Australia, the temperate (almost tropical) setting in the Paleogene changed into the icy, biologically depleted environment of the present. The montane and alpine biota in the Transantarctic Mountains are a relic from a biodiverse past (Chapter 24). Another Gondwanan fragment, New Zealand, was mostly submerged after the breakup of the supercontinent. Significant parts of its landmass did not re-emerge until the late Neogene, and it was around that time that the Southern Alps and their biota formed, as shown by the high rates of speciation in certain taxa (Chapter 25). Africa is the largest continent that derived from Gondwana, and is primarily formed by old continental crust. This stable cratonic crust is, however, breaking up along one of the most striking geological features in the world: the East African Rift System. The geological evolution of this system is intimately connected to the paleoenvironment, climate, biodiversity and human evolution on the continent (Chapter 26). Meanwhile, in the Northern Hemisphere, a close biological connection persisted across Eurasia throughout the Meso-Cenozoic. EGs in the Alps and the mountains of Eastern Asia are well recorded in floral records, and the regions maintained similar biotas until the onset of recent global climate variations. In the past ca. 3 My, climate variations drove many of the European taxa to extinction (Chapter 27). In the wake of the reduction in European taxa, the Tibeto-Himalayan region witnessed a period of diversification (Chapter 28). Likewise, Eastern China and Japan were united by land until relatively recently. Around 10 Ma, rifting separated these regions, but the common history of the mountain biota of Japan and Eastern China is still preserved in the resemblance between their modern floras (Chapters 29 and 30). The book concludes with a review of vegetation dynamics in Southeast Asia, a region that – like New Zealand and Japan – is dramatically deforming at the intersection of two tectonic plates. This region constitutes a crossroads of migration, but also includes long-recognized, distinct biogeographic barriers. Detailed reconstructions of the natural history of this region are now possible through the integration of molecular and geological data (Chapter 31).

1.6 Outlook

Biodiversity loss through climate change and the destruction of habitats is seen as a major societal challenge (Rockström et al. 2009). On humanity's time scale, spanning hundreds to thousands of years, our actions are a threat to biodiversity through both landscape and climate changes. A large part of Earth's biodiversity may be destroyed before we have even begun to understand its origins and evolution. Even though we are not the only players shaping the world around us, we must strive to understand our impact on global processes before it is too late.

The effect of current climatic and landscape changes on biodiversity is a hot topic in international scientific research and political agendas. Large amounts of funding have been devoted to reconstructing the relationship between climate and biological diversity in order to make accurate projections of future scenarios. However, while climate is a dominant process that is visibly changing during our lifetimes, it is not the only mega-scale process affecting biodiversity. Over a longer time scale, spanning millions of years, processes such as plate tectonics, global climate and environmental changes have formed the world that we know, and these must be considered in order to properly understand present diversity and ongoing changes.

Biodiversity is transient and evolves continuously over different temporal and spatial scales (Benton 2009). Abiotic, mega-scale geodynamic processes, including mountain building, operate on the scales of entire continents and millions of years, heavily influencing the three major processes responsible for determining biodiversity: migration, speciation and extinction.

This book presents a compilation of contributed chapters from researchers in a variety of fields, studying different organisms and systems, who are all working towards the common goal of promoting a broader, combined view of geologic, climatic and biological processes and how these shape diversity. The advances showcased in this book permit us to develop a more informed and comprehensive view on the topic, and to expand on the early ideas proposed in the 19th and 20th centuries by the pioneers of biogeography. Taken as a whole, *Mountains, Climate and Biodiversity* is intended as a handbook for current and future students, a reference and an inspiration for how the field can grow and develop.

Acknowledgments

We thank Colin Hughes, Christian Körner, Salomon Kroonenberg and Robert Ricklefs for their constructive reviews, and Christine Bacon, Søren Faurby, Suzette Flantua, Henry Hooghiemstra and Daniele Silvestro for their comments on an earlier version of this chapter. Jens Mutke and Søren Faurby are also acknowledged for kindly producing Figure 1.4a and b. AA is supported by grants from the Swedish Research Council (B0569601), the European Research Council under the European Union's Seventh Framework Programme (FP/2007-2013, ERC Grant Agreement n. 331024), the Swedish Foundation for Strategic Research and a Wallenberg Academy Fellowship.

References

Albert, J., Lovejoy, N.R. & Crampton, W.G.R. (2006) Miocene tectonism and the separation of cis- and trans-Andean river basins: evidence from Neotropical fishes. *Journal of South American Earth Sciences* **21**, 14–27.

Antonelli, A. (2015) Multiple origins of mountain life. *Nature* **524**, 300–301.

Antonelli, A., Nylander, J.A.A., Persson, C. & Sanmartín, I. (2009) Tracing the impact of the Andean uplift on Neotropical plant evolution. *Proceedings of the National Academy of Sciences* **106**, 9749–9754.

Antonelli, A., Hettling, H., Condamine, F. et al. (2016) Towards a Self-Updating Platform for Estimating Rates of Speciation and Migration, Ages, and Relationships of Taxa (SUPERSMART). *Systematic Biology* syw066.

Badgley, C. (2010) Tectonics, topography, and mammalian diversity. *Ecography* **33**, 220–231.

Badgley, C., Smiley, T.M., Terry, R. et al. (2017). Biodiversity and topographic complexity: modern and geohistorical perspectives. *Trends in Ecology & Evolution* **32**, 211–226.

Barthlott, W., Lauer, W. & Placke, A. (1996) Global distribution of species diversity in vascular plants: towards a world map of phytodiversity. *Erdkunde* **50**, 317–327.

Barthlott, W., Mutke, J., Rafiqpoor M.D. et al. (2005) Global centers of vascular plant diversity. *Nova Acta Leopoldina NF* **92**, 61–83.

Barthlott, W., Hostert, A., Kier, G. et al. (2007) Geographic patterns of vascular plant diversity at continental to global scales. *Erdkunde* **61**, 305–315.

Benton, M.J. (2009) The Red Queen and the Court Jester: species diversity and the role of biotic and abiotic factors through time. *Science* **323**, 728–732.

Clague, D.A. (1996) The growth and subsidence of the Hawaiian-Emperor volcanic chain. In: Keast, A. & Miller, S.E. (eds.) *The Origin and Evolution of Pacific Island Biotas, New Guinea to Eastern Polynesia: Patterns and Processes*. Amsterdam: SPB Academic Publishing, pp. 35–50.

Condamine, F., Leslie, A.B. & Antonelli, A. (2016) Ancient islands acted as refugia and pumps for conifer diversity. *Cladistics* **33**, 69–92.

Craw, D., Upton, P., Burridge, C.P., Wallis, G.P. & Waters, J.M. (2016) Rapid biological speciation driven by tectonic evolution in New Zealand. *Nature Geoscience* **9**, 140–144.

Djamali, M., Baumel, A., Brewer, S. et al. (2012) Ecological implications of *Cousinia* Cass. (Asteraceae) persistence through the last two glacial–interglacial cycles in the continental Middle East for the Irano-Turanian flora. *Review of Palaeobotany and Palynology* **172**, 10–20.

Durán-Zuazo, V.H. & Rodríguez-Pleguezuelo, C.R. (2007) Soil-erosion and runoff prevention by plant covers. *A review. Agronomy for Sustainable Development* **28**, 65–86.

Elsen, P.R. & Tingley, M.W. (2015) Global mountain topography and the fate of montane species under climate change. *Nature Climate Change* **5**, 772–776.

Faurby, S. & Svenning, J.-C. (2015) Historic and prehistoric human-driven extinctions have reshaped global mammal diversity patterns. *Biodiversity Research* **21**, 1155–1166.

Favre, A., Päckert, M., Pauls, S.U. et al. (2015) The role of the uplift of the Qinghai-Tibetan Plateau for the evolution of Tibetan biotas. *Biological Reviews* **90**, 236–253.

Ficetola, G. F., Mazel, F. & Thuiller, W. (2017) Global determinants of zoogeographical boundaries. *Nature Ecology and Evolution* **1**, 0089.

Finarelli, J.A. & Badgley, C. (2010) Diversity dynamics of Miocene mammals in relation to the history of tectonism and climate. *Proceedings of the Royal Society B* **277**, 2721–2726.

Flantua, S.G.A., Hooghiemstra, H., Van Boxel, J.H., Cabrera, M., González-Carranza, Z. & González-Arango, C. (2014) Connectivity dynamics since the last glacial maximum in the Northern Andes. In: *Paleobotany and Biogeography: A Festschrift for Alan Graham in his 80th Year*. Monographs in Systematic Botany from the Missouri Botanical Garden 128, Missouri Botanical Garden, St. Louis, pp. 98–123.

Flantua, S., Hooghiemstra, H., Grimm, E.C. et al. (2015) Updated site compilation of the Latin American Pollen Database; challenging new research. *Review of Palaeobotany and Palynology* **223**, 104–115.

Forest, C.E. (2007) Paleoaltimetry: a review of thermodynamic methods. *Reviews in Mineralogy & Geochemistry* **66**, 173–193.

Garreaud, R.D., Vuille, M., Compagnucci, R. & Marengo, J. (2008) Present-day South American climate. *Palaeogeography, Palaeoclimatology, Palaeoecology* **281**, 180–195.

Gates, T.A., Prieto-Marquez, A. & Zanno, L.E. (2012) Mountain building triggered Late Cretaceous North American megaherbivore dinosaur radiation. *PLoS ONE* **7**, e42135.

Gentry, A. (1982) Neotropical floristic diversity: phytogeographical connections between Central and South America, Pleistocene climatic fluctuations, or an accident of the Andean orogeny? *Annals of the Missouri Botanical Garden* **69**, 557–593.

Gosse, J.C. & Stone, J.O. (2001) Terrestrial cosmogenic nuclide methods passing milestones toward paleo-altimetry. *Earth and Space Science News* **82**, 82–89.

Gray, M. (2004) Defining Geodiversity. In: Grey, M. (ed.) *Geodiversity*. Chichester: John Wiley & Sons Ltd., pp. 1–9.

Heenan, P.B. & McGlone, M.S. (2013) Evolution of New Zealand alpine and open-habitat plant species during the late Cenozoic. *New Zealand Journal of Ecology* **37**, 105–113.

Herman, F. & Champagnac, J.-D. (2015) Plio-Pleistocene increase of erosion rates in mountain belts in response to climate change. *Terra Nova* **28**, 2–10.

Herman, F., Seward, D., Valla, P.G. et al. (2013) Worldwide acceleration of mountain erosion under a cooling climate. *Nature* **504**, 423–426.

Holt, B.G., Lessard, J.-P., Borregaard, M.K. et al. (2013) An update of Wallace's zoogeographic regions of the world. *Science* **339**, 74–78.

Hoorn, C., Wesselingh, F.P., ter Steege, H. et al. (2010) Amazonia through time: Andean uplift, climate change, landscape evolution and biodiversity. *Science* **330**, 927–931.

Hoorn, C., Mosbrugger, V., Mulch, A. & Antonelli, A. (2013) Mountain building and biodiversity. *Nature Geoscience* **6**, 154.

Houze, R.A. (2012) Orographic effects on precipitating clouds. *Reviews of Geophysics* **50**, RG1001.

Hughes, C.E. (2016) The tropical Andean plant diversity powerhouse. Commentary on Lagomarsino et al. *New Phytologist* **210**, 1152–1154.

Hughes, C.E. & Atchison, G.W. (2015) The ubiquity of alpine plant radiations: from the Andes to the Hengduan Mountains. *New Phytologist* **207**, 275–282.

Hughes C.E. & Eastwood R. (2006) Island radiation on a continental scale: exceptional rates of plant diversification after uplift of the Andes. *Proceedings of the National Academy of Sciences USA* **103**, 10 334–10 339.

Humboldt, A. & Bonpland, A. (1807) *Geographie der Pflanzen in den Tropenländern*. Paris: Tübingen.

Jacques, F.M.B., Su, T., Spicer, R.A. et al. (2014) Late Miocene southwestern Chinese floristic diversity shaped by the southeastern uplift of the Tibetan Plateau. *Palaeogeography, Palaeoclimatology, Palaeoecology* **411**, 208–215.

Janzen, D.H. (1967) Why mountain passes are higher in the tropics. *The American Naturalist* **101**, 233–249.

Johnson M.R.W. & Harley, S. (2012) *Orogenesis: The Making of Mountains*. Cambridge: Cambridge University Press.

Kessler, M., Karger, D.N. & Kluge, J. (2016) Elevational diversity patterns as an example for evolutionary and ecological dynamics in ferns and lycophytes. *Journal of Systematics and Evolution* **54**, 617–625.

Kohn, M.J. & Fremd, T.J. (2007) Tectonic controls on isotope compositions and species diverification, John Day Basin, central Oregon. *PaleoBios* **27**, 48–61.

Kohn, M.J. & Fremd, T.J. (2008) Miocene tectonics and climate forcing of biodiversity, western United States. *Geology* **36**, 783–786.

Köppen, W. & Wegener, A. (1924) *Die Klimate der geologischen Vorzeit*. Berlin: Verlag von Gebrüder Bortraeger.

Körner, C. (2003) *Alpine Plant Life: Functional Plant Ecology of High Mountain Ecosystems*. Berlin: Springer Science & Business Media.

Körner, C. (2007) The use of "altitude" in ecological research. *TRENDS in Ecology and Evolution* **22**, 569–574.

Körner, C. (2012) *Alpine Treelines: Functional Ecology of the Global High Elevation Tree Limits*. Berlin: Spriger-Verlag.

Körner, C. & Ohsawa, M. (2006) Mountain systems. In: Hassan, R., Scholes, R. & Ash, N. (eds.) *Ecosystems and Human Well-Being: Current State and Trends, 1*. Washington, DC: Island Press, pp. 681–716.

Körner, C., Paulsen, J. & Spehn, E.M. (2011) A definition of mountains and their bioclimatic belts for global comparisons of biodiversity data. *Alpine Botany* **121**, 73–78.

Körner, C., Jetz, W., Paulsen, J., Payne, D., Rudmann-Maurer, K. & Spehn, E.M. (2016) A global inventory of mountains for bio-geographical applications. *Alpine Botany* **127**, 1–15.

Kreft, H. & Jetz, W. (2007) Global patterns and determinants of vascular plant diversity. *Proceedings of the National Academy of Sciences* **104**, 5925–5930.

Kupfer, J.A. (2010) Theory in landscape ecology and its relevance to biogeography. In Millington, A., Blumler, M. & Schickhoff, U. (eds.) *The SAGE Handbook of Biogeography*. Thousand Oaks, CA: Sage, pp. 57–74.

Lagomarsino, L.P., Condamine, F.L., Antonelli, A. et al. (2016) The abiotic and biotic drivers of rapid diversification in Andean bellflowers (Campanulaceae). *New Phytologist* **210**, 1430–1442.

Linder, H.P. (2008) Plant species radiations: where, when, why? *Philosophical Transactions of the Royal Society B* **363**, 3097–3105.

Lomax, B.H., Fraser, W.T., Harrington, G. et al. (2012) A novel palaeoaltimetry proxy based on spore and pollen wall chemistry. *Earth and Planetary Science Letters* **353**, 22–28.

Lomolino, M.V. (2000) Ecology's most general, yet protean pattern: the species-area relationship. *Journal of Biogeography* **27**, 17–26.

Lomolino, M.V., Riddle, B.R., Whittaker, R.J. & Brown, J.H. (2010) *Biogeography*, 4th edn. Sunderland, MA: Sinauer Associates.

Luebert, F. & Weigend, M. (2014) Phylogenetic insights into Andean plant diversification. *Frontiers in Ecology and Evolution* **2**, 27.

MacArthur, R.H. & Wilson, E.O. (1963) An equilibrium theory of insular zoogeography. *Evolution* **17**, 373–387.

MacArthur, R.H. & Wilson, E.O. (1967) *The Theory of Island Biogeography*. Princeton, NJ: Princeton University Press.

Madriñán, S., Cortés, A.J. & Richardson, J.E. (2013) Páramo is the world's fastest evolving and coolest biodiversity hotspot. *Frontiers in Genetics* **4**, 192.

Malhi, Y., Meir, P. & Brown, S. (2002) Forests, carbon and global climate. *Philosophical Transactions of the Royal Society of London A* **360**, 1567–1591.

McCain, C.M & Grytnes, J.-A. (2010) Elevational gradients in species richness. In: *Encyclopedia of Life Sciences*. Chichester: John Wiley & Sons Ltd.

McCormack, J.E., Huang, H. & Knowles, L.L. (2009) Sky islands. In: Gillespie, R. & Clague, D. (eds.) *Encyclopedia of Islands*. Berkeley, CA: University of California Press, pp. 841–843.

McElwain, J.C. (2004) Climate-independent paleoaltimetry using stomatal density in fossil leaves as a proxy for CO2 partial pressure. *Geology* **32**, 1017–1020.

Merckx, V.S.F.T., Hendriks, K.P., Beentjes, K.K. et al. (2015) Evolution of endemism on a young tropical mountain. *Nature* **524**, 347–350.

Moeslund, J.E., Arge, L., Bøcher, P.K. et al. (2013) Topography as a driver of local terrestrial vascular plant diversity patterns. *Nordic Journal of Botany* **31**, 129–144.

Molnar, P. (2015) Mountain (landform). Available from: https://www.britannica.com/science/mountain-landform (last accessed September 1, 2017).

Molnar, P. & England, P. (1990) Late Cenozoic uplift of mountain ranges and global climate change: chicken or egg? *Nature* **346**, 29–34.

Morlon, H., Lewitus, E., Condamine, F.L. et al. (2016) RPANDA: an R package for macroevolutionary analyses on phylogenetic trees. *Methods in Ecology and Evolution* **7**, 589–597.

Mulch, A. (2016) Stable isotope paleoaltimetry and the evolution of landscapes and life. *Earth and Planetary Science Letters* **433**, 180–191.

Mutke, J. (2010) Biodiversity gradients. In: Millington, A., Blumler, M. & Schickhoff, U. (eds.) *The SAGE Handbook of Biogeography*. Thousand Oaks, CA: Sage, pp. 170–190.

Mutke, J., Sommer, J. H., Kreft, H. et al. (2011) Vascular plant diversity in a changing world: global centres and biome-specific patterns. In: Zachos, F.E. & Habel, J.C. (eds.) *Biodiversity Hotspots – Evolution and Conservation*. Berlin: Springer-Verlag, pp. 83–96.

Myers, N., Russell A. Mittermeier, R.A. et al. (2000) Biodiversity hotspots for conservation priorities. *Nature* **403**, 853–858.

Owens, P.N. & Slaymaker, O. (2004) *Mountain Geomorphology*. Oxford: Oxford University Press.

Peters, M.K., Hemp, A., Appelhans, T. et al. (2016) Predictors of elevational biodiversity gradients change from single taxa to the multi-taxa community level. *Nature Communications* **7**, 13736.

Polissar, P.J., Freeman, K.H., Rowley, D.B. et al. (2009) Paleoaltimetry of the Tibetan Plateau from D/H ratios of lipid biomarkers. *Earth and Planetary Science Letters* **287**, 64–76.

Potter, P.E. & Szatmari, P. (2009) Global Miocene tectonics and the modern world. *Earth-Science Reviews* **96**, 279–295.

Poulsen, C.J., Ehlers, T.A. & Insel, N. (2010) Onset of convective rainfall during gradual Late Miocene rise of the Central Andes. *Science* **328**, 490–493.

Raven, P.H. & Axelrod, D.I. (1974) Angiosperm biogeography and past continental movements. *Annals of the Missouri Botanical Gardens* **61**, 539–673.

Reiners, P.W. & Brandon, M. (2006) Using thermochronology to understand orogenic erosion. *Annual Review of Earth and Planetary Sciences* **34**, 419–466.

Renema, W., Bellwood, D., Braga, J.-C. et al. (2008) Hopping hotspots: global shifts in marine biodiversity. *Science* **321**, 654–657.

Renner, S. (2016) Available data point to a 4-km-high Tibetan Plateau by 40 Ma, but 100 molecular-clock papers have linked supposed recent uplift to young node ages. *Journal of Biogeography* **43**, 1479–1487.

Ricklefs, R.E. (1977) Environmental heterogeneity and plant diversity: a hypothesis. *The American Naturalist* **111**, 376–381.

Ricklefs, R.E. (1987) Community diversity: relative roles of local and regional processes. *Science* **235**, 167–171.

Ricklefs, R.E. (2004) A comprehensive framework for global patterns in biodiversity. *Ecology Letters* **7**, 1–15.

Ricklefs, R.E. (2005) Historical and ecological dimensions of global patterns in plant diversity. *Biologiske Skrifter* **55**, 583–603.

Rockström, J., Steffen, W., Noone, K. et al. (2009) A safe operating space for humanity. *Nature* **461**, 472–475.

Rohrmann, A., Sachse, D., Mulch, A. et al. (2016) Miocene orographic uplift forces rapid hydrological change in the southern central Andes. *Scientific Reports* **6**, 35678.

Sandel, B. Arge, L. Dalsgaard, B. et al. (2011) The influence of Late Quaternary climate-change velocity on species endemism. *Science* **334**, 660–664.

Santos, J. C., Coloma, L.A., Summers, K. et al. (2009) Amazonian amphibian diversity is primarily derived from late Miocene Andean lineages. *PLoS Biology* **7**, e56.

Särkinen, T., Pennington, R.T., Lavin, M. et al. (2012) Evolutionary islands in the Andes: persistence and isolation explain high endemism in Andean dry tropical forests. *Journal of Biogeography* **39**, 884–900.

Scherrer, D. & Körner, C. (2010) Topographically controlled thermal-habitat differentiation buffers alpine plant diversity against climate warming. *Journal of Biogeography* **38**, 406–416.

Senckenberg Magazine (2015) *Senckenberg – World of biodiversity/Blick in die zukunft – Senckenberg geobiodiversitätsforschung.*

Silvestro, D., Zizka, A., Bacon, C.D. et al. (2016) Fossil biogeography: a new model to infer dispersal, extinction and sampling from palaeontological data. *Proceedings of the Royal Society B* **371**, 20150225.

Spehn, E.M., Rudmann-Maurer, K., Körner C. & Maselli, D. (2010) *Mountain Biodiversity and Global Change.* Basel: GMBA-DIVERSITAS.

Spehn, E.M., Rudmann-Maurer, K. & Körner, C. (2011) Mountain biodiversity. *Plant Ecology & Diversity* **4**, 301–302.

Stein, A. & Kreft, H. (2015) Terminology and quantification of environmental heterogeneity in species-richness research. *Biological Reviews* **90**, 815–836.

Steinbauer, M.J., Field, R., Grytnes, J.-A. et al. (2016) Topography-driven isolation, speciation and a global increase of endemism with elevation. *Global Ecology and Biogeography* **25**, 1097–1107.

Tews, J., Brose, U., Grimm, V. et al. (2004) Animal species diversity driven by habitat heterogeneity/diversity: the importance of keystone structures. *Journal of Biogeography* **31**, 79–92.

Töpel, M., Zizka, A., Calió, M.F. et al. (2016) SpeciesGeoCoder: fast categorisation of species occurrences for analyses of biodiversity, biogeography, ecology and evolution. *Systematic Biology* syw064.

Van der Hammen, T. & Cleef, A. M. (1986) Development of the High Andean Paramo flora and vegetation. In: Vuilleumier, F. & Monasterio, M. (eds.) *Tropical Biogeography.* New York: Oxford University Press, pp. 153–201.

Van der Hammen, T., Werner, J.H. & Van Dommelen, H. (1973) Palynological record of the upheaval of the Northern Andes: a study of the Pliocene and Lower Quaternary of the Colombian Eastern Cordillera and the early evolution of its high-Andean biota. *Review of Palaeobotany and Palynology* **16**, 1–122.

Wallace, A.R. (1876) *The Geographical Distribution of Animals.* London: Macmillan.

Wegener, A. (1912) Die Entstehung der Kontinente. *Geologische Rundschau* **3**, 276–292.

Winkworth R.C., Wagstaff, S.J., Glenny, D. & Lockhart, P.J. (2005) Evolution of the New Zealand mountain flora: origins, diversification and dispersal. *Organisms, Diversity & Evolution* **5**, 237–247.

Wright, D.F. & Stigall, A.L. (2013) Geologic drivers of Late Ordovician faunal change in Laurentia: investigating links between tectonics, speciation, and biotic invasions. *PLoS ONE* **8**, e68353.

Yasuhara, M., Iwatani, H., Hunt, G. et al. (2016) Cenozoic dynamics of shallow-marine biodiversity in the Western Pacific. *Journal of Biogeography* **44**, 567–578.

Part I

Mountains, Relief and Climate

2

Simple Concepts Underlying the Structure, Support and Growth of Mountain Ranges, High Plateaus and Other High Terrain

Peter Molnar

Department of Geological Sciences and Cooperative Institute for Research in Environmental Sciences (CIRES), University of Colorado at Boulder, Boulder, CO, USA

Abstract

Two geodynamic processes build most high terrain. First, convergence between two expanses of Earth's crust thickens that crust, so that, in a state of isostatic equilibrium (for which columns of mass in the upper 100–200 km of earth are equal), the mean surface elevation is proportional to the degree that the crust has been compressed horizontally and thickened. Regions 5 km high (like Tibet and the Central Andes) are underlain by especially thick crust: ~70 km, compared with the more normal 35–40 km. Most mountain belts and plateaus that have formed by crustal thickening reach limiting elevations and then grow wider, not higher. Second, because hot material occupies more space (is less dense) than cold material, removal of the cold uppermost part of the mantle (mantle lithosphere) beneath a region and its replacement by hot mantle material (asthenosphere) can create high terrain, but rarely with as much as 2 km of excess height. Large expanses, such as that surrounding the East African Rift System (EARS) or much of the western USA, are high primarily because unusually hot material underlies them. Except on the flanks of outwardly growing mountain ranges, million-year average rates of surface uplift rarely if ever reach 1 mm/y (=1 km/My); accordingly, many millions of years to a few tens of millions of years have elapsed since most present-day mountain ranges began to form. Three processes – erosion, horizontal extension and thinning of crust, and cooling of a hot uppermost mantle – lower regional elevations, all in a state of isostatic equilibrium. Because of isostasy, erosion can carve deep canyons and remove sufficient mass to allow adjacent peaks to rise, despite average regional elevations decreasing. Crustal extension also creates relief, and in some regions, again, high peaks or ridges can grow, while average elevations decrease. In general, rates of erosion, crustal extension and especially cooling of the uppermost mantle are so slow that most high terrain has a long lifetime – measured in tens, if not hundreds, of millions of years.

Keywords: *geodynamics, mountain building, crust, lithosphere, plateau, tectonics*

2.1   Introduction

Even geologists know that plants and animals, not to mention bacteria and archaea, occupy different niches that vary with altitude, and that with changes in climate those habitats can vary in elevation. It follows that changing elevations on geologic time scales should affect how life indigenous to mountainous regions evolves, and that biologists might find themselves confronted with a plethora of possible elevation changes or past elevation distributions that can match their observations. Basic concepts in geology, however, limit plausible distributions of elevations, the rates at which they change (both upward and downward) and the amounts of change in different settings. My goal here is to present such basic concepts and to show how they place limits on possible elevation distributions and rates of change.

If the processes building high terrain were shut off, erosion would destroy high terrain. Rivers, glaciers and winds would transport eroded material to the seas, and the Earth would become flat. It is difficult to estimate how many millions of years must elapse before the land would flatten. Yet, only if a remarkably low erosion rate of ~1 μm/y prevailed widely might terrain stand as high as a few meters above sea level; such low rates have been measured from only a few special localities, but they do not characterize most regions. It follows that geodynamic processes continually build high terrain.

Few doubt that the geodynamic engine lies in the mantle (see Box 2.1 for a brief discussion of deep earth structure, as well as Figure 2.1); gravity acting on lateral variations in density forces flow, which is resisted by stresses associated with viscous flow (e.g., Schubert et al. 2001). The most obvious manifestation of such flow is plate tectonics: the movement of large "plates" of lithosphere – the strong, outer layer of the Earth, roughly 100 km thick (Figure 2.1), but somewhat thicker

in stable continental interiors (continental shields or cratons) and negligibly thin at mid-ocean ridges, where two plates diverge from each other (Figure 2.2). Reduced to its essence, plate tectonics includes not only the divergence of two plates at mid-ocean ridges, but also the sliding of one plate horizontally past another at transform faults and the subduction of one plate beneath another at subduction zones or "island arcs" (Figure 2.2). In its simplest form, plate tectonics creates relatively high terrain in two ways. First, volcanoes erupt along mid-ocean ridges and at subduction zones – for the islands of island arcs are virtually all volcanoes. Second, the upwelling of hot material in the mantle beneath mid-ocean ridges, and its subsequent cooling, makes mid-ocean ridges higher than the surrounding ocean floor (or, more precisely, such ridges lie beneath shallower water).

Although most high terrain results from processes whose engine operates in the mantle, such processes are not part of the basic plate-tectonics canon. Most mountain ranges and high plateaus arise because crust has been compressed horizontally and thickened – in some ways like snow plowed up into thick piles. This chapter focuses primarily on how crust is thickened to build high terrain. Some high terrain, however, has formed where the uppermost mantle is especially hot, including in some cases the outpouring of vast amounts of lava: "large igneous provinces" that are associated with relatively localized zones of upwelling in the underlying mantle. The high Ethiopian Plateau (Figure 2.3), with its cover of volcanic rock that reaches as far as Yemen on the Arabian Peninsula, offers an example.

The most fundamental concept underlying our understanding of high terrain is isostasy, which is Archimedes' Principle applied to the Earth; it requires that the mass of rock above sea level be balanced by a deficit of mass below it, so that the mass per unit area in a column from the surface to great depth is the same everywhere. The common analogy is with icebergs and ice cubes: for both, approximately 10% of the ice lies above the water level and 90% is submerged. In simple terms, isostasy comes in two forms depending on the thickness of the Earth's low-density crust and on how hot the uppermost mantle is. Isostasy prevails throughout the Earth and manifests itself not just when mountains are built, but also when they erode away. Just as mountain ranges and high plateaus can be built by either thickening the crust or warming the uppermost mantle, so regional subsidence can result from horizontal extension and thinning of crust, erosion and removal of the top layers of crust and cooling of hot regions of the mantle.

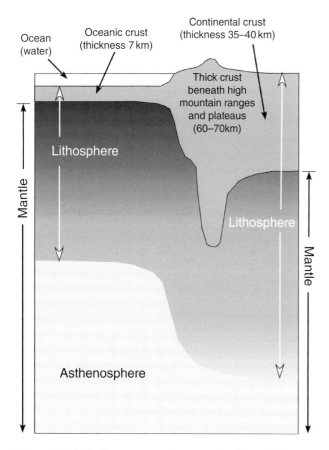

Figure 2.1 Vertically exaggerated cross-section through the upper mantle, showing crust and mantle, which differ from each other in chemical and mineralogical composition. The asthenosphere is a weak layer that underlies the stronger mantle lithosphere. The lithosphere includes the coldest uppermost part of the mantle and the crust. Note that the thicknesses of both crust and mantle lithospheres vary from place to place, but in general, where the crust is thick, the surface stands high to form a mountain belt or high plateau. The gradation in tone through the lithosphere is meant to show how strength decreases downward through it to the asthenosphere.

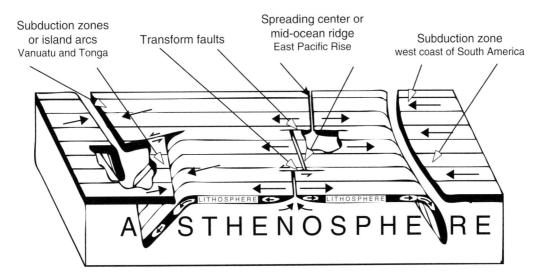

Subduction zones
or island arcs
Vanuatu and Tonga

Transform faults

Spreading center or
mid-ocean ridge
East Pacific Rise

Subduction zone
west coast of South America

Figure 2.2 Block diagram illustrating the essentials of plate tectonics – specifically, the relative motion of plates of lithosphere ~100 km thick with respect to one another. Separate plates of lithosphere form and move apart at spreading centers or mid-ocean ridges, and slide past one another at transform faults. At subduction zones, one plate plunges beneath another. This diagram was constructed to show a simplified view of the South Pacific and its margins. Two plates form at the East Pacific Rise: one is subducted beneath South America, and the other beneath the Tongan Islands. In this simplification, a plate on which the Tongan Islands lie plunges beneath Vanuatu (previously the New Hebrides Islands), which is shown here to lie on the Pacific Plate; actually, Vanuatu lies on a separate, small plate, which we ignore here. The Tonga and Vanuatu subduction zones are connected by a transform fault. *Source:* Isacks et al. (1968). Reproduced with permission of the American Geophysical Union.

Box 2.1 Basic Earth Structure

The outer few hundred kilometers of the Earth consist of layers that can be distinguished in one of two ways, depending either (i) on chemical and mineralogical composition or (ii) on temperature and strength, because temperature and strength are intimately linked (Figure 2.1).

The crust, which is rich in light elements like aluminum, calcium, silicon and sodium, and which has a mean density of ~2.8×10^3 kg/m^3, overlies the denser (~3.3×10^3 kg/m^3) mantle, which is rich in the heavier iron and magnesium (Figure 2.1). To a first (crude) approximation, the crust "floats" on the mantle. Thus, as discussed in Box 2.2, where the crust is thick, the surface stands high, and where it is thin, the surface commonly is low (Figure 2.4a). Beneath continents, the normal crustal thickness is 35–40 km where the surface is near sea level, but it can reach ~70 km where the surface stands 5 km high. A good rule of thumb is that 1 km of surface elevation is worth ~7 km of additional crustal thickness. Beneath oceans – except near their edges – the crust is thin (7 km), and the surface lies a few kilometers below sea level (Figure 2.1).

Like butter or ice cream, the strength of rock-forming minerals – and hence of rock itself – decreases as temperature increases. Throughout most of the mantle, slow flow at speeds of 1–10 mm/y carries heat upward, and maintains temperatures sufficiently high that the rock is weak. Because of the strong dependence of strength on temperature, the uppermost, cold part of the mantle is especially strong. Together, this cold mantle and its overlying crust form a strong "lithosphere" that overlies the weaker "asthenosphere." Of course, the boundary between them is gradational (Figure 2.1), but a separation into two layers – lithosphere and asthenosphere – allows simple descriptions of many processes, including plate tectonics (e.g., Molnar 2015). The thickness of the lithosphere, however, varies by large amounts, from arguably negligible thickness where the uppermost mantle is especially hot (such as beneath the mid-ocean ridges or beneath East Africa) to hundreds of kilometers of thickness where the upper mantle is cold (such as beneath Canada and most of Siberia) and where elevations are low. Like most material, rock occupies more space (and is less dense) when it is hot than when it is cold. Thus, regions with a thin lithosphere, and hence an atypically hot uppermost mantle, stand high (Figure 2.4b).

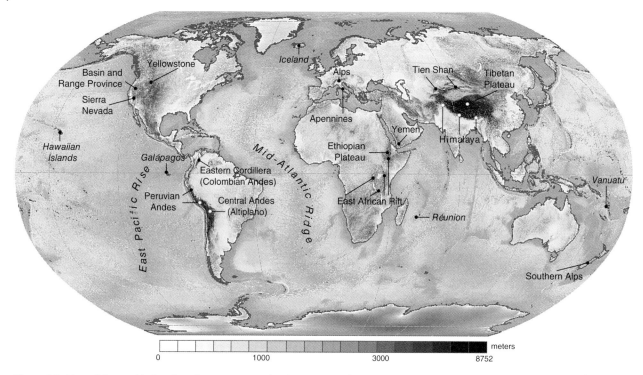

Figure 2.3 Map of the world, showing places mentioned in the text. See also Plate 2 in color plate section.

2.2 Support of High Terrain: Isostasy

After Pratt (1855) had reanalyzed the data from Everest's survey of India and shown that mass must be missing beneath the Himalaya and southern Tibet (Figure 2.3), Airy (1855) offered an explanation, which we now call "Airy isostasy" (Figure 2.4). Airy recognized that the strength of rock was insufficient to support the mass within the Himalaya and southern Tibet. He proposed that the crust must be atypically thick beneath high terrain. As already mentioned, a simple rule of thumb is that for 1 km of surface elevation, the crust should be approximately 7 km thicker than normal.

Allegedly miffed that Airy (1855) had converted his thorough analysis of Everest's data into a simple explanation, Pratt (1859) offered a different view: that the mean density of the rock beneath high terrain was lower than that beneath lower regions, much as the tops of wooden blocks of different densities but of the same thickness stand at different heights when immersed in water. When Airy and Pratt were working, the basic structure of the Earth – whether described in terms of crust and mantle or of lithosphere and asthenosphere (see Box 2.1) – had not yet been recognized, but we now know that lateral differences in temperature provide the most sensible explanation of the lateral variations in density that Pratt imagined, in what we now call "Pratt isostasy" (Figure 2.4).

Both forms of isostasy lend themselves to simple algebraic formulae that relate mean elevations to deep structure. See Box 2.2 for a discussion of such formulae.

Whereas crustal thickening – such as the doubling of crust initially 35 km thick to 70 km thickness – can create regions as high as 5 km, variations in the thermal structure of the upper mantle are not likely to account for elevation differences of more than 1–2 km. Indeed, the surface of the Tibetan Plateau at a mean elevation of nearly 5 km is underlain by crust ~35 km thicker than normal (although, see later for further discussion of the details). By contrast, the regional elevations of ~2000 m for much of Ethiopia (Figure 2.3) are almost surely compensated by a hot upper ~200 km of the mantle.

Of course, this discussion in terms of either thickened crust or a hot upper mantle is an oversimplification.

First, both phenomena can exist together, with thickened crust and a hot upper mantle. For example, the Tibetan Plateau stands ~5 km high over a huge region (Figure 2.3), but the crust in northern Tibet is 10–15 km thinner than that in southern Tibet (e.g., Brandon & Romanowicz 1986; Owens & Zandt 1997; Tseng et al. 2009). Were Airy isostasy to apply strictly, northern Tibet would stand 1.5–2.0 km lower than southern Tibet, but essentially no elevation difference exists. A wealth of seismological data, however, show that seismic wave speeds beneath northern Tibet are atypically low (e.g., Molnar & Chen 1984; Lyon-Caen

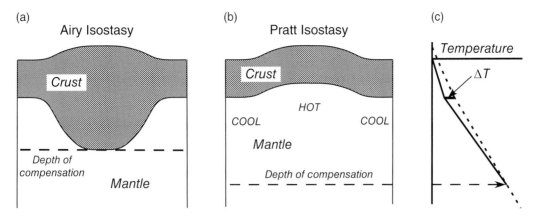

Figure 2.4 Simple schematic representations of isostasy. (a) In Airy isostasy, crust has thickened, and the mass above sea level is compensated by light crust that reaches into the mantle. (b) For Pratt isostasy, density varies laterally, and as shown here, lateral variation occurs largely in the mantle, because under high terrain, the mantle is hot. (c) Plots of temperature versus depth for two parts of (b): the dashed line shows the higher temperatures beneath the hot part of (b) and the solid line shows the cold regions flanking it. At any depth below the depth of compensation, the mass per unit area – and hence pressure – is treated as constant.

1986; Molnar 1990; Beghoul et al. 1993; Zhao & Xie 1993; Woodward & Molnar 1995; McNamara et al. 1997; Debayle et al. 2001; Dricker & Roecker 2002; Shapiro & Ritzwoller 2002; Liang & Song 2006), and that attenuation of seismic waves there is atypically high (Barazangi & Ni 1982; Ni & Barazangi 1983). Wave speeds decrease and attenuation increases with increasing temperature, and hence these data, plus scattered young volcanic rock (e.g., Turner et al. 1993, 1996; Ding et al. 2003), suggest that an unusually hot uppermost mantle underlies northern Tibet (e.g., Molnar 1988; Hatzfeld & Molnar 2010).

Second, the lithosphere has finite strength, and as a result, it behaves like a diving board with a diver on the end. High topography overlying the lithosphere can add weight (like the diver), which flexes it down (like the diving board). The lithosphere behaves as an effectively elastic plate, and distributes the load of the high topography across its surface (e.g., Watts 2001). For example, if the unusually stiff Indian Plate had not been underthrust beneath the Himalaya, and the Himalaya were isostatically compensated by thicker underlying crust according to Airy isostasy, the high Himalayan peaks might stand ~1000 m lower than they do (e.g., Lyon-Caen & Molnar 1983, 1985).

Finally, not only does flow in the asthenosphere move heat so as to maintain a nearly constant temperature at depth (or, more precisely, a temperature that increases with depth at the relatively small adiabatic gradient of ~0.5 °C/km), but flow-induced stresses apply tractions to the base of the lithosphere and deflect it upward or downward. However, such topography, dubbed "dynamic topography," is small – <300 m, except in narrow belts like deep-sea trenches (e.g.,

Braun 2010; Molnar et al. 2015) – and not likely to matter to biologists except at low elevations, where it could affect sea level.

2.3 Plate Tectonics and High Terrain

Mid-ocean ridges offer quintessential examples of Pratt isostasy. Hot material intrudes between diverging plates of lithosphere at mid-ocean ridges (Figure 2.2) and sticks to one of the plates. As this material moves away from the ridge, it slowly cools by losing heat to the overlying ocean, and the cooling material contracts. This contraction manifests itself as subsidence, so that the shallowest sea floor, ~2.5 km below sea level, marks the locus of divergence. It follows that where divergence is rapid, the "ridge" is wide, and where slow, it is relatively narrow. The highest rates of divergence are in the eastern Pacific, and before plate tectonics was understood, that region was named the "East Pacific Rise" (Figure 2.3), for the relatively shallow sea floor spanned a much broader width than that of the Mid-Atlantic Ridge.

The temperatures at which most rock-forming minerals melt increase as pressure increases. Thus, hot rock, as it rises and decompresses, can pass from being too cold to being sufficiently warm to melt despite the rock receiving no heat at all (and, in fact, cooling slightly because of adiabatic decompression). Accordingly, as hot rock rises at a mid-ocean ridge, some of it melts, and volcanoes erupt on the sea floor. Because the mid-ocean ridges lie ~2.5 km below sea level, however, only a triflingly few such volcanoes emerge as islands.

Surely, the islands of island arcs are much more important for biologists than are the few volcanoes above sea

Box 2.2 Isostasy

For Airy isostasy (Figure 2.4), where rock with density ρ_c stands at an excess height of h, the crust must extend deeper into the mantle than normal by ΔH. Let the density of the uppermost mantle be ρ_m. Isostasy states that mass per unit area of rock above sea level, which is given by the product of ρ_c and h, must equal a compensating mass deficit that is given by the product of the thickness of the crustal root, ΔH, and the difference in density between mantle and crust, $\rho_m - \rho_c$. Thus, $\rho_c\, h = (\rho_m - \rho_c)\,\Delta H$, and the thickness of the crustal root is given by:

$$\Delta H = \frac{\rho_c}{\rho_m - \rho_c} h \qquad (2.2.1)$$

Note that the crust is thicker than normal by $\Delta H + h$, and therefore the thickness of crust beneath the high terrain is greater than that beneath lowlands by:

$$\Delta H + h = \frac{\rho_m}{\rho_m - \rho_c} h \qquad (2.2.2)$$

Putting numbers into Equation 2.2.2, $\rho_c = 2.8 \times 10^3$ kg/m^3 and $\rho_m = 3.3 \times 10^3$ kg/m^3, gives the rule of thumb that an average surface height of ($h =$) 1 km implies that the crustal root reaches $\Delta H = 5.6$ km into the mantle, and that crust is thicker than normal by $\Delta H + h = 6.6$ km ≈ 7 km.

The simplest case for Pratt isostasy arises when the temperature gradient through the uppermost mantle is constant, which is a good approximation for the mantle portion of the lithosphere. Then, suppose that the gradients beneath a hot region and a cold region are different (Figure 2.4). At the base of the lithosphere, the temperature is essentially the same everywhere, because convection within the asthenosphere moves material fast enough to minimize lateral variations in temperature. Thus, the major difference between the two regions is the temperature at the base of the crust. Of course, if the temperature at the base of the crust differs in two different regions, so must the temperature within the crust, but let's ignore this, because the crust is generally thin compared with the thickness – or typical depth range – of mantle lithosphere.

First, imagine a layer with a thickness L and temperature T_1. If we warm that layer to a temperature of T_2, the layer will expand by an amount $\Delta L = \alpha\,(T_2 - T_1)\,L$, where α ($= 3 \times 10^{-5}\,°C^{-1}$) is the coefficient of thermal expansion. By analogy, suppose that the mantle portion of the lithosphere were heated so that its top, just below the crust, became warmer by ΔT. Because the temperature at the bottom of the mantle lithosphere is that of the asthenosphere and does not change, suppose the gradient changed to a new constant value through the layer. The average temperature through the layer would increase by only $\Delta T/2$. Thus, we would expect our layer of thickness L to become thicker by:

$$\Delta L = \alpha \frac{\Delta T}{2} L \qquad (2.2.3)$$

Although making the layer hotter in one place than in another makes it thicker, no mass is added to the layer. Thus, the weight of the layer at its base does not change, and all of the additional thickness, ΔL, manifests as a higher surface elevation.

In reality, the mantle lithosphere is not heated, but replaced by hot asthenosphere. So, for example, in the case where the entire mantle lithosphere below the crust was removed, the temperature at the top of the mantle lithosphere would become higher, but that at the bottom would remain the same, and Equation 2.2.3 would apply.

Consider an example. The temperature of the asthenosphere is ~1300 °C, and a representative temperature at the boundary between the crust and mantle is 600 °C. If all of the mantle lithosphere were removed and replaced by asthenosphere, so that the temperature at the base of the crust became 1300 °C, then $\Delta T = 700$ °C. For a thickness of the mantle lithosphere – before it was removed – of $L = 100$ km, Equation 2.2.3 gives a change in thickness and hence in elevation: $\Delta L = 3 \times 10^{-5}\,°C^{-1} \times 0.5 \times 700\,°C \times 100$ km ≈ 1 km. If the difference in thermal structure extended to a depth of 200 km, removal of mantle lithosphere beneath such a region might account for an increase in elevation of 2 km. Smaller temperature changes obviously will lead to smaller elevation changes, and 2 km is a maximum plausible change due to removal of mantle lithosphere.

level on the mid-ocean ridges. Two processes conspire to make volcanoes form where one plate of lithosphere underthrusts another (Figure 2.2).

First, as a slab of oceanic lithosphere plunges into the asthenosphere, it drags some asthenosphere down with it, and this flow requires counterflow above it to replace the material dragged down by the subducting lithosphere. This "corner flow" is understood well in fluid mechanics (Batchelor 1967; McKenzie 1969). Faster subduction causes stronger corner flow, which brings hotter asthenosphere to shallower depths in the corner between the downgoing and overriding lithosphere than does slower subduction (England & Wilkins 2004; England & Katz 2010). The hotter the rock, the more that can melt.

Second, the melting temperatures of most solids depend on the degree to which impurities have been added; recall that salt spread on ice lowers its melting temperature and can convert it into water even where its temperature is several degrees Celsius below zero. Hydrogen, whether as pure atoms, as the ion OH^- or as water (H_2O), lowers the melting temperature of many rock-forming minerals. The oceanic crust that forms as two plates diverge has been pervasively fractured, and water has filled those cracks. Moreover, water is trapped in the interstices between sediment grains deposited on the ocean floor. Thus, when sea floor and its sedimentary cover are subducted at an island arc, water is transported to depths of ~100 km. When put in contact with the hot asthenosphere, the water then lowers the melting temperature of the asthenospheric material (Gill 1981; Kelley et al. 2006; Plank 2018). Because of its low viscosity, the molten rock with low density can then rise buoyantly through the asthenosphere, despite the corner flow that otherwise might drag it down. Famous volcanoes such as Mount Fuji in Japan and Mount St. Helens in the US state of Washington have formed in such settings.

Although plate tectonics is arguably the most important large-scale geodynamic process occurring within the Earth, its role in the creation of mountains is somewhat abstract. Most high topography develops within continents, but the essential element of plate tectonics – rigid, undeforming plates of lithosphere (Figure 2.2) – works best beneath the oceans. Moreover, where the interiors of continents behave as parts of rigid plates, neither deformation of the crust nor creation of mountain ranges occurs.

2.4 The Growth of Mountain Ranges and High Plateaus

From the discussion of isostasy in the preceding sections and in Box 2.2, it should be obvious that two processes create high terrain: thickening of crust and increasing temperature of the uppermost mantle. These processes are so different that it makes sense to treat them separately.

2.4.1 Crustal Thickening

In most mountain belts, crust has thickened because two plates of lithosphere, at least one of which is capped by crust of thickness 35–40 km, have moved towards one another, and crust on one has been thrust atop the crust of the other. Because rock is essentially incompressible, if a layer of crust is compressed horizontally, it must thicken (Figure 2.5).

Such thrusting of pieces of crust atop one another can occur where a continent follows oceanic lithosphere into a subduction zone. A good example is the Himalaya (Figure 2.6a). India was once isolated in the middle of the Indian Ocean, and moved northward as part of a large

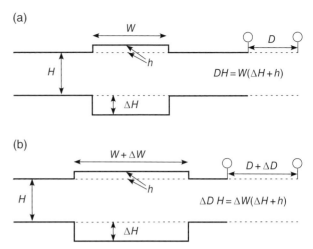

Figure 2.5 Simple schematics illustrating crustal thickening in a state of Airy isostasy. (a) Crust is compressed horizontally by an amount D, and thickens over a region of width W. The relationships of the height of the range, h, the thickness of excess crust in the crustal root, ΔH, and the amount of thickening, $\Delta H + h$, are given by Equations 2.2.1 and 2.2.2 (see Box 2.2). The crustal budget requires that $DH = W(\Delta H + h)$. (b) Once a mountain belt has been built, in order to accommodate continued convergence by an amount shown here as ΔD, the belt grows wider. The crustal budget requires that $\Delta DH = \Delta W(\Delta H + h)$.

plate of mostly oceanic lithosphere towards the rest of Eurasia. The oceanic lithosphere was subducted beneath the southern edge of Asia (what is now southern Tibet). When the northern edge of India reached the subduction zone, slices of sedimentary rock that had been deposited on India's northern continental margin were scraped off the margin and piled on top of the intact part of the subcontinent as it continued to move northward and plunge beneath southern Asia. With continued northward movement of India, additional slices were removed, and the stack of slices grew thicker as new ones were added to the bottom. Today, the rock of the Himalaya consists of such a stack of slices, with all Himalayan rock having been derived from India, not the rest of Eurasia, beneath which oceanic lithosphere had been subducted (Figure 2.6a).

Underthrusting of crust can also occur deep within continents, as in the Tien Shan of Asia (Figure 2.3), and at subduction zones on the landward side of volcanoes, as in the Andes (Figure 2.6b). While oceanic lithosphere beneath the eastern Pacific (the Nazca Plate) underthrusts beneath the Andes at the relatively high rate of 60–70 mm/y, slower (~10 mm/y) westward underthrusting of the Brazilian Shield beneath the Andes thickens the crust there to build the high, wide mountain belt. As in the Himalaya, the underthrusting of the Brazilian Shield beneath the Andes stacks slices of crust to build a high, widening mountain belt. In addition, during the subduction process, volcanoes have grown above molten rock that has solidified at depth to form granitic

(a)

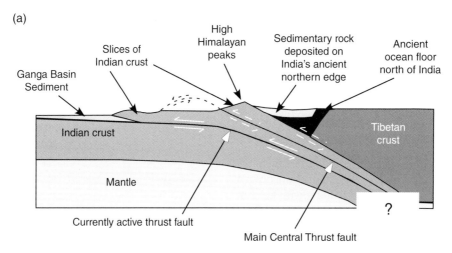

(b)

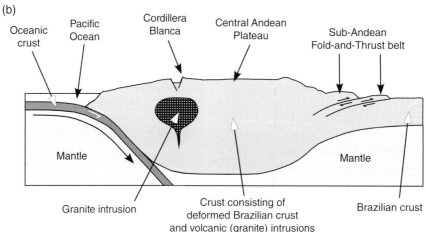

Figure 2.6 Simple cross-sections across the Himalaya and the Peruvian Andes. (a) The Himalaya have been built by slices of India's crust being scraped off its northern edge as the subcontinent moves northward and thrusts beneath southern Tibet. The question mark covers a region where ignorance would risk making any image misleading. (b) The Peruvian Andes have been built by the westward thrusting of the Brazilian Shield beneath the Andes and by the intrusion of granite formed both by melting of mantle beneath the regions as oceanic lithosphere has been thrust beneath the western edge of the belt and by melting Brazilian crust thrust beneath the belt.

batholiths. This granite includes both molten rock associated with that subduction and melted South American crust that has been thrust westward beneath the range.

Because of isostasy, the thrusting of slices of crust atop one another and the thickening of the crust make the surface of thickened crust stand higher than that above thin crust. Suppose crust of thickness H converges an amount D, so that a mountain range of width W forms (Figure 2.5a). The crust beneath the range should have thickened by $\Delta H + h$, and using Equation 2.2.2 (see Box 2.2), the budget is given by:

$$DH = W\left(\Delta H + h\right) = W \frac{\rho_m}{\rho_m - \rho_c} h \qquad (2.1)$$

We may ask: How much convergence, D, of crust with thickness $H = 35$ km must occur to build a range of width $W = 200$ km and height $h = 2$ km? From Equation 2.1, $D \approx 90$ km.

Present-day convergence of expanses of crust towards one another occurs at rates that range from only a few mm/y, or km/My (e.g., ~3 mm/y across the Apennines (D'Agostino et al. 2014), ~4 mm/y across the Eastern Cordillera of Colombia (Figure 2.3) (Mora-Páez et al. 2016)), to as fast as ~20 mm/y across the Tien Shan (Abdrakhmatov et al. 1996; Reigber et al. 2001; Zubovich et al. 2010). As discussed later, the Eastern Cordillera must have begun to form tens of millions of years ago. Although erosion of the Tien Shan was well underway at 25 Ma, which suggests that high topography was present at that time (e.g., Sobel & Dumitru 1997), most of the high terrain over a width of 200 km must have been built in a relatively short time, since ~10 Ma.

Where erosion removes material rapidly from a range, obviously more time is needed to build high mean elevations compared to where erosion is slow. For example, rock in the Southern Alps of New Zealand (Figure 2.3) seems to be eroding as rapidly as convergence across the range can supply new material (e.g., Adams 1980a,b; Koons 1990). Rock rises relative to sea level as fast as 10 mm/y (=10 km/My), but erosion removes it at the same rate.

Although some high mountain ranges that have been built by horizontal compression and thickening of crust might have grown in fewer than 10 My, most require a few tens of millions of years to become major mountain belts.

The preceding discussion obscures an aspect of mountain building that might be particularly important to biologists. In general, where sustained convergence of crustal terrains builds mountain belts, the high terrain does not rise en masse at a rate given by the budgets in Equation 2.1. Rather, most such belts reach limiting elevations and then grow outward (Figure 2.5b). Thus, the mean elevation of the high region need not increase with continued convergence; instead, the belt grows wider as continued convergence of surrounding lowland terrain thickens the crust on the flanks of the widening range.

This process, by which mountain belts grow wider, not higher, can be understood in terms of the simple energy budget of mountain building (e.g., Molnar & Lyon-Caen 1988). The forces that push two expanses of crust together must expend energy in two ways: (i) deformation of rock by folding and faulting dissipates energy by frictional or viscous processes, and ultimately by the production of heat; (ii) crustal thickening and the raising of the Earth's surface create gravitational potential energy that is stored in the high terrain and its crustal root. Moreover – and most importantly here – the rate of increase in potential energy is greater when the surface is already high than when the surface is low. It follows that less work must be done to lift low terrain than to lift terrain that is already high. Thus, outward growth of high terrain is energetically favored over the sustained rising of high terrain yet higher (e.g., England & Houseman 1986; Houseman & England 1986; Molnar & Lyon-Caen 1988).

The growth of the Tibetan Plateau serves as an example of such outward growth. Although the northern margin of the Tibetan Plateau became active shortly after India collided with southern Asia (e.g., Fang et al. 2003; Yin et al. 2008; Clark et al. 2010; Duvall et al. 2011), the locus of intense deformation seems to have migrated northward as it penetrated deeper into the the continent (e.g., England & Houseman 1986; Métivier et al. 1998; Tapponnier et al. 2001). The crust on the north-eastern edge of Tibet began to be compressed horizontally and its surface began to rise only since ~1–2 Ma (Zheng et al. 2013), tens of millions of years after India and Eurasia collided.

The Eastern Cordillera of Colombia, ~200 km wide, serves another good example. Thorough work by Mora et al. (2006, 2008, 2010) shows that the eastern flank of the belt has grown since 3 Ma. Quantitative bounds on rates of active faulting on the margin (Veloza et al. 2015) are consistent with such recent growth. Yet, the thick

crust beneath the Eastern Cordillera (Poveda et al. 2015) and the current rate of horizontal compression across the entire belt of only ~4 mm/y require tens of millions of years to have elapsed for the entire mountain belt to have been built (e.g., Mora-Páez et al. 2016). Geologic evidence from much of the interior of the belt indeed indicates that the mountain range is not young (i.e., 3–6 My old), but started to form tens of millions of years ago (e.g., Cediel et al. 2003; Gómez et al. 2003; Parra et al. 2009, 2012; Saylor et al. 2011; Ochoa et al. 2012; Sánchez et al. 2012)

2.4.2 Changes in the Thermal Structure of the Uppermost Mantle

Two processes can alter the thermal structure of the uppermost mantle: conduction of heat into the lithosphere and replacement of lithosphere by asthenosphere. As discussed later, when subsidence of high terrain is considered, conduction of heat is slow; many tens, if not hundreds of millions of years would be required to build high terrain. We may ignore this process. Replacement of lithosphere by asthenosphere, and therefore the advection of hot asthenosphere into space previously occupied by colder lithosphere, can be fast, however, so that increases in elevation of 1–2 km can occur in periods as short as a few million years.

It helps to separate the convective processes that can remove mantle lithosphere into two. First, especially hot material from deep in the Earth can rise to the base of the lithosphere, erode some of the lithosphere, and replace it with the hotter material. Second, thickened mantle lithosphere can become gravitationally unstable and sink into the asthenosphere to be replaced by it.

2.4.2.1 Hot Spots, Plumes from the Deep Mantle and Large Igneous Provinces

Morgan (1971, 1973) proposed that plumes of hot mantle rise from the core–mantle boundary deep in the Earth, and that when the hot material reaches the base of the lithosphere, it manifests itself as a "hot spot"; the islands of Hawaii and Iceland (Figure 2.3) provide the prototypes of surface manifestations of such hot spots, but for biologists, the Galápagos Islands probably are the most important example. A plume of smoke rising from a smokestack through relatively still air offers a homely image of the flow at depth in the mantle. The hot rising material spreads out beneath the lithosphere and erodes its base somewhat, so that the combination of especially hot material from below, the removal of the relatively cool base of the lithosphere, and basal tractions applied to it by the rising material can elevate the surface as much as 1000 m or so, over a broad area (Figure 2.7) (e.g., McKenzie 1994; Ribe & Christensen 1994; Crosby & McKenzie 2009). The expanses of high terrain over

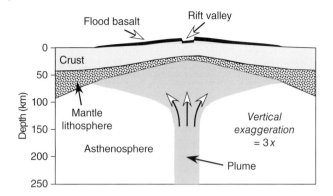

Figure 2.7 Simple, vertically exaggerated cross-section through a mantle plume of rising hot material that spreads out at the base of the crust; high topography, extensive basaltic cover or "flood basalt" overlies the "plume head," and in some cases – as shown here – a rift valley forms. The Ethiopian Plateau, crossed by the northern branch of the eastern rift of the East African Rift System (Figure 2.3), serves as an example.

Yellowstone in the western USA (Figure 2.3) and over Ethiopia and East Africa offer good examples of this process; the Hawaiian Islands form a linear volcanic chain over the axis of a wide region, which extends out to ~1000 km from the islands, where the surrounding sea floor is ~1000 m shallower than normal sea floor. Abundant seismological data show low seismic wave speeds in the mantle beneath these regions, and therefore suggest that hot material has replaced lower lithosphere (e.g., Bastow et al. 2008; Wolfe et al. 2009; Levandowski et al. 2014).

The rising hot material commonly leads to volcanism, which can be more localized than the wide expanse of hot material at the base of the lithosphere. The Hawaiian Islands formed as the Pacific Plate moved rapidly over such a plume in the underlying mantle, and the dates when volcanic rock erupted to form islands increase westward. Volcanism in such settings – if not in Hawaii – can be volumetrically large and can create "large igneous provinces." The Deccan Traps – widespread basaltic volcanic rock over much of central India – erupted at ~65 Ma, when India overlay a hot spot now beneath the island of La Réunion (Figure 2.3).

Visitors to volcanoes cannot fail to see the large difference between the classic cones of volcanoes like Fuji, Rainier or Etna and the much more gently sloping Hawaiian volcanoes of Mauna Loa and Mauna Kea, called "shield volcanoes." The difference stems from the viscosity of the lava. Because of the high temperature associated with the hot spot beneath Hawaii, the viscosity of the basalt that erupts there is much lower than that of the lavas that commonly form in island arc settings. Just as olive oil spilled on a table spreads faster than does honey, so the hot basalt at Hawaii reaches farther from

the eruptive center before it freezes solid than does the volcanic rock erupted at island arcs.

2.4.2.2 Removal of Dense Mantle Lithosphere

Insofar as mantle lithosphere differs little either chemically or mineralogically from asthenosphere, the lower temperature of mantle lithosphere should make it denser, and hence gravitationally unstable (prone to sink into the underlying less dense asthenosphere). Viscosity increases strongly with decreasing temperature, but by amounts that are difficult to quantify accurately. Both the extent to which mantle lithosphere is gravitationally unstable and the extent to which instability grows remain topics of study. Moreover, there definitely exist regions where mantle lithosphere is chemically different from asthenosphere. Because the mantle of such regions is depleted in heavy elements like iron, it is buoyant and gravitationally stable (e.g., Jordan 1975). Consequently, controversy surrounds the ideas discussed in this section.

In some subduction zones, deeper parts of the downgoing slab of lithosphere seem to have broken off and sunk rapidly into the asthenosphere (e.g., Isacks & Molnar 1971). The removal of the excess weight of lithosphere might then allow the overlying surface to rise, as seems to be occurring today in the Vanuatu Islands (e.g., Chatelain et al. 1992). Some have suggested that slab breakoff has contributed to the building of narrow mountain ranges, but because most such inferences impress me as deus ex machina, and following the admonition in the previous paragraph, I discuss this no further here.

Bird (1978; Bird & Baumgardner 1981) proposed that mantle lithosphere might peel away from its overlying crust, and that this "delamination" of crust and mantle lithosphere would create a gap into which hotter, less dense asthenosphere could intrude. Bird (1979) further suggested that the Colorado Plateau gained its present-day mean elevation by such delamination of lithosphere, and the insertion of hot asthenosphere into space once occupied by cold lithosphere.

Houseman et al. (1981) offered another view of how mantle lithosphere might be removed: when crust thickens because of the convergence of two regions, as in Figure 2.5, the mantle lithosphere beneath it must also thicken (Figure 2.8). Thus, the intrinsic instability associated with mantle lithosphere being colder and denser than asthenosphere should be enhanced. Gravity acting on the negative buoyancy of the thickened lithosphere will pull it down, so that the high terrain becomes lower than it would be in strict Airy isostasy. Then, if some of that dense mantle lithosphere sinks into the asthenosphere, it will be replaced by hot asthenosphere. The painting of a ceiling offers an analogy: if the viscosity of the paint is low, the paint will coalesce in drips,

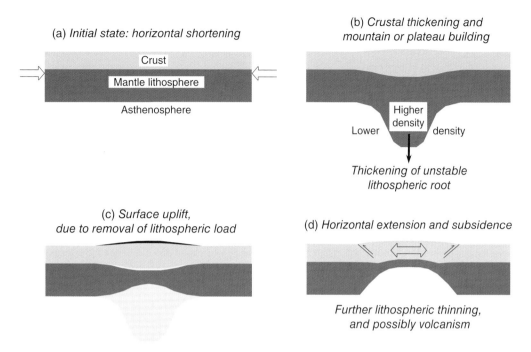

Figure 2.8 Sequence of cartoons showing the growth and collapse of a mountain range built by horizontal compression and thickening of lithosphere. (a) A region of lithosphere, including both the crust and its mantle part, is put under horizontal compression, where two adjacent plates of lithosphere are pushed towards one another. (b) The crust has thickened in response to the horizontal compression, and the surface above the thickened crust has risen. To a first approximation, this compression and thickening of crust accounts for the mean elevations of high terrain in both narrow mountain belts and some high plateaus. The mantle lithosphere has thickened, for the horizontal compression of the crust requires a comparable amount of horizontal compression of the mantle lithosphere. The thickened mantle lithosphere acts as a heavy load on the crust above it, and in principle the surface above it should not stand as high as it would in the absence of lateral variations in density in the mantle, much as a ship floats lower in the water when filled with cargo. Airy isostasy is therefore imperfect in the sense that the surface is lower than that expected simply from the thickening of the crust. (c) Following removal of a heavy load of thick, cold, dense mantle lithosphere – indicated by the light shading where thickened mantle lithosphere is shown in (b) – the surface should have risen and should stand higher than it would for strict Airy isostasy. The thin layer of dark shading at the top of the thick crust shows the modest gain in elevation, and the thin layer of light shading at the bottom shows, too, that the base of the crust should have risen by the same amount. Thus, both Pratt and Airy isostasy contribute to the elevation of the high terrain. (d) As a consequence of this excess elevation, the high terrain spreads apart, somewhat like ripe Camembert cheese removed from its box. The horizontal forces that thickened the crust no longer suffice to support the crust that has been elevated higher than it would be in strict Airy isostatic equilibrium. If the lateral forces imposed by converging plates are insufficient, the high terrain spreads apart, the mantle lithosphere thins yet more and volcanism becomes likely.

drawing adjacent paint into the drips, and ultimately thinning the layer of paint. In the case of the lithosphere, a result of its removal will be that the surface will rise; if most or all of the mantle lithosphere is removed, the surface will become higher than that for strict Airy isostasy, because both Airy and Pratt isostasy contribute to the support of the elevated crust (Figure 2.8).

Numerical experiments carried out on layers analogous to lithosphere and doubled in thickness suggest that much, if not all, of the mantle lithosphere might be removed in intervals as short as a few million years (e.g., Houseman & Molnar 1997; Conrad & Molnar 1999; Conrad 2000). The pattern of flow of mantle lithosphere as it sinks into the asthenosphere (as sheets of sinking material, localized drips, etc.) depends on the physical properties of the crust and mantle lithosphere and on the rate of convergence (e.g., Houseman et al. 2000; Molnar & Houseman 2004). Ignorance, especially of the viscosity of lithosphere, prohibits both precise estimates of how rapidly mantle lithosphere is removed and an understanding of how this process occurs. For example, for some viscosity and density structures, calculations show sheets of dense, unstable mantle lithosphere sinking into the asthenosphere beneath the flanks of a belt whose crust has undergone horizontal compression (Figure 2.9) (Molnar & Houseman 2004). Consistent with these calculations, in a seismological study of the Tien Shan (Figure 2.3), Li et al. (2009) reported high P-wave speeds (suggestive of cold material) in vertical, tabular zones reaching to depths of 400 km beneath the flanks of the range and low speeds (presumably indicating hot mantle material) beneath its axis of the belt.

Region of crustal shortening
and thickened crust

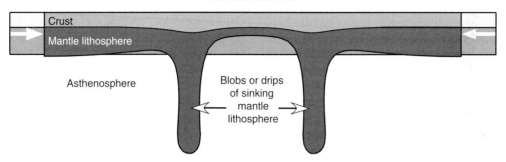

Figure 2.9 Simple cartoon illustrating an example of removal of mantle lithosphere. Horizontal compression of the lithosphere, indicated by arrows on the flanks of the diagram, induces thickening of the crust and mantle lithosphere. The mantle lithosphere, denser than the underlying asthenosphere, is gravitationally unstable, and blobs (or drips or sheets) of thickened mantle lithosphere sink into the asthenosphere. For the parameters used for this calculation, the sinking blobs developed not beneath the locus of maximum shortening, but on the flanks. *Source:* Adapted from Molnar & Houseman (2004).

The most widely cited example of surface uplift in response to removal of mantle lithosphere – although it is by no means accepted by everyone – comes from the Altiplano of the Central Andes (Figure 2.3). Garzione et al. (2006, 2008), using $\delta^{18}O$ in pedogenic carbonate sediment, and Ghosh et al. (2006), using the clumping of isotopes ^{18}O and ^{13}C in $CaCO_3$ to estimate past surface temperatures, inferred that 2000–3000 m of surface uplift occurred in only 3–5 My, between 11 and 6 Ma (see Chapter 6 for a discussion of these methods.) Garzione et al. (2006) argued that the surface uplift was too rapid to be the result of horizontal compression and thickening of crust, and therefore that mantle lithosphere must have been removed. Although this inference has been challenged with a variety of arguments, other observations concur with rapid surface uplift since ~10 Ma; in particular, numerous studies show evidence of abrupt increases in river incision into the flanks of the Altiplano at that time (e.g., Gubbels et al. 1993; Kennan et al. 1997; Victor et al. 2004; von Rotz et al. 2005; Kober et al. 2006; Hoke et al. 2007; Schildgen et al. 2007; Thouret et al. 2007; Farías et al. 2008).

In regions such as the Central Andes, where dense volcanic rock may have intruded the lower crust while volcanoes erupted lavas at the surface, the unusually dense lower crust can enhance both the rate at which dense material sinks into the asthenosphere (e.g., Kay & Kay 1993; Jull & Kelemen 2001) and the amount that the surface rises in response to its removal. Accordingly, the composition of volcanic rock exposed at the surface has been used as an argument for removal of mantle lithosphere (e.g., Kay & Kay 1993; Kay et al. 1994).

When (and if) mantle lithosphere is removed, the surface should rise higher than it would be for strict Airy isostasy (for which all mantle has the same density).

When this happens, there is a tendency for the crust to spread apart (Figure 2.8d) and extend: a phenomenon observed today in Tibet and the Peruvian Andes. Again, ripe Camembert cheese lifted from its box provides a homely analogy.

I should note, however, that there is some doubt that such removal of lithosphere occurs (e.g., McKenzie & Priestley 2008).

2.5 Destruction of Mountain Ranges and Other High Terrain

Three processes can destroy high terrain: erosion, horizontal extension and thinning of crust, and cooling of hot lithosphere – all in a state of isostatic equilibrium. Although all lead to lower mean elevations, the different processes affect the landscape differently.

2.5.1 Erosion

Geodynamic processes, like crustal thickening or removal of mantle lithosphere, affect mean elevations over large areas, but erosion enhances relief, as glaciers and rivers incise valleys – if not canyons – into surrounding high terrain. By analogy with statistics, geodynamic processes affect the mean elevation, but erosion determines its variance or standard deviation. As recognized long ago (e.g., Wager 1933, 1937; Holmes 1944, 1965), because of isostasy, rapid removal of rock from valleys and canyons results in peaks rising higher than they would in the absence of valley incision. Bird (1978) showed that the mass of rock missing from deep valleys along the Himalaya matches that present in the high peaks above the mean ~5000 m elevation of the region.

Stern et al. (2005) showed the same for the Transarctic Mountains, which have been deeply incised by glaciers. It follows that even in the absence of geodynamic processes, accelerated erosion and incision of valleys can, in a state of isostatic equilibrium, raise adjacent peaks. Champagnac et al. (2007) exploited this logic to account for the present-day rate at which geodetic benchmarks within the Swiss Alps rise.

Rates of erosion correlate well with relief (e.g., Ahnert 1970), as might be expected given that rivers and glaciers starting at high elevations have more potential energy (which can be used to erode and transport debris) than do those starting at low elevations. Such correlations have been used, in many cases with good reason, to justify assertions that deep incision of terrain reflects recent, tectonically induced uplift. Erosion, however, depends on the prevailing climate, and climate change can also alter erosion rates. Consequently, in some regions, recent deep incision can give the false impression that recent tectonic activity has created high terrain (Molnar & England 1990). Deep incision within high terrain, thermochronological constraints on erosion rates within high terrain, and abundant sediment in surrounding lowlands dating from 2–4 Ma characterize elevated areas throughout the Earth, regardless of the age of the most recent tectonic activity (e.g., Hay et al. 1988, 1989; Zhang et al. 2001; Molnar 2004; Champagnac et al. 2009; Herman et al. 2013; Herman & Champagnac 2016; Olson et al. 2016). Global cooling since ~4 Ma led to increased alpine glaciation, which is one possible cause of accelerated erosion.

Rates of erosion span several orders of magnitude, from as fast as 5–10 mm/y in parts of the Southern Alps of New Zealand (Griffiths & McSaveney 1983; Hicks et al. 1990; Herman et al. 2010), to as little as 10 μm/y (Edmond et al. 1995), or perhaps only 1 μm/y (Brown et al. 1992), in the steep-sided, 3000 m tepui of the Venezuelan Shield, and even more slowly in the dry valleys of Antarctica (Brown et al. 1991; Brook et al. 1995). At the low rates of 1–10 μm/y, or 1–10 km/Ga, high terrain could last nearly forever – at least on the time span over which nonmicrobial life has existed on Earth. Using erosion rates to estimate the lifetime of high terrain – even ignoring the likelihood of changes over time – requires consideration of isostasy. As for the thickening of crust in Equation 2.2.1 (see Box 2.2), if erosion removes a mass of rock equivalent to a layer of crust of thickness δH, so that the crust thins by δH, the weight per unit area will decrease by $\rho_c \, \delta H$. This reduction in mass must then be compensated by a rise of the underlying crust and mantle. The mean elevation of Earth's surface should decrease by only $\delta h < \delta H$, so that mass in the column remains balanced: $(\rho_m - \rho_c) \, \delta h = \rho_c \, \delta H$. Thus, erosion of an amount

δH leads to a decrease in the mean elevation of the surface of:

$$\delta h = \frac{\rho_m - \rho_c}{\rho_c} \delta H \qquad (2.2)$$

With the numbers used above, if an average of 1000 m of rock is eroded away, the mean elevation of the surface will drop only ~180 m. Again, however, because erosion surely will not be uniform in the real world, some portions of a region may rise in response, even though the mean elevation decreases, as Wager (1933, 1937) pointed out for high peaks in the Himalaya adjacent to deep canyons.

2.5.2 Crustal Extension and Thinning

Just as crustal thickening, via isostasy, makes the surface rise, so crustal thinning requires that it drop. Crust will thin when it extends horizontally and its areal extent increases. The same logic used for horizontal compression and thickening of crust expressed in Equation 2.2.1 (see Box 2.2) applies.

Crustal thinning is ongoing in many high regions, and specifically in two regions undergoing widespread crustal extension: the Tibetan Plateau and the Basin and Range Province of the western USA (Figure 2.3). In both regions, horizontal compression of crust has thickened crust and created high terrain, but now both regions are undergoing east–west crustal extension and thinning.

The mean elevation of Tibet approaches 5000 m, and the crustal thickness exceeds 60 km in the north and reaches 75 km in the south (e.g., Owens & Zandt 1997; Tseng et al. 2009). Yet, today, crustal thinning is widespread (e.g., Molnar & Tapponnier 1978; Ni & York 1978; Armijo et al. 1986; Elliott et al. 2010; Ge et al. 2015). Some change in the development of Tibet must have occurred for the region to have undergone a shift from horizontal compression and thickening of crust to horizontal extension and thinning of it. England & Houseman (1989) showed that the simplest mechanism likely to have operated is that thickened mantle lithosphere beneath Tibet was removed, and with the removal of the dense load beneath it, the surface rose (as happens when ballast is tossed overboard from a ship). The additional gravitational potential energy given to the plateau can then power crustal extension (Figure 2.8). We often liken Tibet to a humongous piece of ripe Camembert cheese that thins as it spreads out over a plate – over the India Plate, in this case. The rate at which the elevation decreases is too low to measure even with modern geodetic techniques, but with the sensible assumption that rock is incompressible, GPS velocities of control points require that the crust of Tibet be thinning at a rate of nearly 1% per million years (Ge et al. 2015). Dating of the normal faults – along which blocks of crust move apart – shows that the crustal thinning began between 15 and 10 Ma (see Ge et al. 2015 for a summary). Thus, because of isostasy, the surface of Tibet at 15–10 Ma may have been ~1000 m higher than it is today, and therefore near 6000 m.

For the Basin and Range Province, currently with a mean elevation of ~1500 m, paleoaltimetry shows early Cenozoic elevations of 3000–4000 m. Estimates of past elevations based on leaf physiognomy (Wolfe et al. 1997, 1998) (see Chapter 7 for a discussion of the method), on stable isotopes of $\delta^{18}O$, and on clumping of ^{18}O and ^{13}C from carbonate sediment (e.g., Davis et al. 2009; Mix et al. 2011; Lechler et al. 2013; Snell et al. 2014) (see Chapter 6 for a discussion of the method) suggest such past elevations. Inferred past drainage patterns of rivers crossing the region are consistent with elevations 1000–2000 m higher than today (e.g., Henry 2008; Henry et al. 2012; Cassel et al. 2014). The Basin and Range Province appears to be an analogue for what the Central Andes will become 10–20 My in the future (Molnar & Chen 1983; Molnar & Lyon-Caen 1988).

The lowering of the surface by crustal thinning is not uniform across regions expanding in area, like Tibet, the Basin and Range Province and segments of the Andes. Crustal thinning and extension occur by slip on faults – "normal faults" – that dip steeply through the upper crust; one block of crust drops relative to the other and creates a steep mountain front, like the eastern edge of the Sierra Nevada in California. Because of isostasy, the block of crust that lies beneath such a fault – the foot-wall – rises a small amount in response to the removal of the weight of the upper block – the hanging wall – that slides off of it. The Cordillera Blanca in Peru (Figure 2.6b) (Dalmayrac & Molnar 1981; Giovanni et al. 2010) and the 7694 m peak Gurla Mandata in southern Tibet (Murphy et al. 2002) offer spectacular examples of such "foot-wall" uplift.

2.5.3 Cooling of Hot Lithosphere

As already noted, warming of lithosphere by conduction of heat from a nearby source occurs slowly, and replacement of cold lithosphere with hot asthenosphere can occur at a much faster rate. Lithosphere that has become hot, however, can cool only by conduction of heat from its hot interior to the colder surface of the Earth. Cold material, which is denser than surrounding warmer material, will tend to sink into the latter, but cold, dense material cannot spontaneously rise to replace hot, low-density material at shallow depths.

Cooling of a slab of hot material by conduction of heat from it occurs exponentially in time. Suppose that a slab has been heated so that its temperature gradient increased, and the temperature at the base of the crust increased by an amount ΔT. Suppose also that the temperatures on the boundaries are fixed. Then, after a short time, when localized steep temperature gradients have been erased, this perturbation to temperature will decay according to:

$$\Delta T(t) = \Delta T \exp\left(-\frac{t}{\tau}\right) \tag{2.3}$$

(e.g., Carslaw & Jaeger 1959; Turcotte & Schubert 2002). The time constant, τ, is proportional to the square of the thickness of the layer, L, and inversely proportional to the coefficient of thermal diffusivity, $\kappa = 10^{-6} \, m^2/s$.

$$\tau = \frac{L^2}{\kappa \pi^2} \tag{2.4}$$

For $L = 100$ km, $\tau = 10^{15}$ s $= 31$ My, and for $L = 200$ km, $\tau = 124$ My.

Suppose a region stands higher than its surroundings by ΔL, given by Equation 2.2.3 (see Box 2.2), because the uppermost 100–200 km of mantle beneath it is hotter than its surroundings. Consider the simple case where the temperature at the top of the hotter lithosphere is greater by ΔT than that beneath the region with a normal temperature profile. Then, as the uppermost mantle cools so that the temperature profile approaches that beneath normal regions in the surrounding region, the surface will subside, reaching 36.8% ($= e^{-1}$) of its initial elevation when a time τ has elapsed. The thickness of old continental lithosphere is at least 200 km (e.g., Jaupart et al. 1998), and may exceed 300 km in some regions (e.g., Jordan 1975). Even with $L = 200$ km, for example, a region elevated 1000 m because of warmed lithosphere will still stand 670 m high after 50 My have elapsed. High terrain underlain by hot uppermost mantle can remain high for a duration that is much longer than the time needed to elevate it.

2.6 Conclusion

The fundamental concept that ties all surface topography together is isostasy (see Figures 2.4 and 2.5, and Box 2.2). High terrain must be balanced by a deficit of mass below it, so that the weight of a column of rock is essentially the same everywhere. The deficit can arise because crust – which is less dense than mantle (Box 2.1) – has thickened where the surface stands high, or because cold mantle lithosphere has been removed and replaced by hotter – and hence, less dense – asthenosphere.

Many of the concepts discussed in this chapter can be combined into a simple pattern of growth and decay of mountain belts that seems to apply to many regions in different stages of development (Figure 2.8). Horizontal compression thickens crust, and because of isostasy, the surface will stand high. Thickened mantle lithosphere might keep the surface lower than it would be if mantle lithosphere were not denser than the underlying asthenosphere. The thickening of mantle lithosphere, however, also enhances its gravitational instability, and if some or all of the mantle lithosphere sinks – by dripping or foundering into the asthenosphere, or by peeling away from the crust – the removal of its weight from the base of the high terrain will enable the surface to rise.

In fact, the surface can rise higher than it would if the densities of mantle lithosphere and asthenosphere were the same. If the surface rises high enough, the crust will extend horizontally and thin, and the surface will subside. The Tibetan Plateau and the Basin and Range Province of the western USA seem to have reached this stage, where the crust is extending horizontally and thinning.

Acknowledgments

I thank C. Hoorn, A. Perrigo and D.J.J. van Hinsbergen for constructive reviews.

References

Abdrakhmatov, K.Ye., Aldazhanov, S.A., Hager, B.H. et al. (1996) Relatively recent construction of the Tien Shan inferred from GPS measurements of present-day crustal deformation rates. *Nature* **384**, 450–453.

Adams, J. (1980a) Contemporary uplift and erosion of the Southern Alps, New Zealand: summary. *Geological Society of America Bulletin Part I* **91**, 2–4.

Adams, J. (1980b) Contemporary uplift and erosion of the Southern Alps, New Zealand. *Geological Society of America Bulletin Part II* **91**, 1–114.

Ahnert, F. (1970) Functional relationships between denudation, relief, and uplift in large mid-latitude drainage basins. *American Journal of Science* **268**, 243–268.

Airy, G.B. (1855) On the computation of the effect of the attraction of mountain-masses, as disturbing the apparent astronomical latitude of stations of geodetic surveys. *Philosophical Transactions of the Royal Society of London* **145**, 101–104.

Armijo, R., Tapponnier, P., Mercier, J.L. & Han Tong-Lin (1986) Quaternary extension in southern Tibet: field observations and tectonic implications. *Journal of Geophysical Research* **91**, 13 803–13 872.

Barazangi, M. & Ni, J. (1982) Velocities and propagation characteristics of Pn and Sn beneath the Himalayan arc and Tibetan Plateau: possible evidence for underthrusting of the Indian continental lithosphere beneath Tibet. *Geology* **10**, 179–185.

Bastow, I.D., Nyblade, A.A., Stuart, G.W. et al. (2008) Upper mantle seismic structure beneath the Ethiopian hot spot: Rifting at the edge of the African low-velocity anomaly. *Geochemistry Geophysics Geosystems* **9**, Q12022.

Batchelor, G.K. (1967) *An Introduction to Fluid Mechanics.* Cambridge: Cambridge University Press.

Beghoul, N., Barazangi, M. & Isacks, B.L. (1993) Lithospheric structure of Tibet and western North America: mechanisms of uplift and a comparative study. *Journal of Geophysical Research* **98**, 1997–2016.

Bird, P. (1978) Initiation of intracontinental subduction in the Himalaya. *Journal of Geophysical Research* **83**, 4975–4987.

Bird, P. (1979) Continental delamination and the Colorado Plateau. *Journal of Geophysical Research* **84**, 7561–7571.

Bird, P. & Baumgardner, J. (1981) Steady propagation of delamination events. *Journal of Geophysical Research* **86**, 4891–4903.

Brandon, C. & Romanowicz, B. (1986) A "no-lid" zone in the central Chang-Thang platform of Tibet: evidence from pure path phase velocity measurements of long period Rayleigh waves. *Journal of Geophysical Research* **91**, 6547–6564.

Braun, J. (2010) The many surface expressions of mantle dynamics. *Nature Geosciences* **3**, 825–833.

Brook, E.J., Brown, E.T., Kurz, M.D. et al. (1995) Constraints on age, erosion and uplift rates of Neogene glacial deposits in the Transantarctic Mountains determined from *in situ* cosmogenic 10Be and 26Al. *Geology* **23**, 1063–1066.

Brown, E.T., Edmond, J.M., Raisbeck, G.M. et al. (1991) Examination of surface exposure ages of Antarctic moraines using *in situ* produced ^{10}Be and 26Al. *Geochimica et Cosmochimica Acta* **55**, 2269–2283.

Brown, E.T., Stallard, R.F., Raisbeck, G.M. & Yiou, F. (1992) Determination of the denudation rate of Mount Roraima, Venezuela using cosmogenic ^{10}Be and ^{26}Al. *Eos: Transactions of the American Geophysical Union* **73**, 170.

Carslaw, H.S. & Jaeger, J.C. (1959) *Conduction of Heat in Solids.* Oxford: Oxford University Press.

Cassel, E.J., Breecker, D.O., Henry, C.D. et al. (2014) Profile of a paleo-orogen: high topography across the present-day Basin and Range from 40 to 23 Ma. *Geology* **42**, 1007–1010.

Cediel, F., Shaw, R.P. & Cáceres, C. (2003) Tectonic assembly of the Northern Andean Block. In: Bartolini, C., Buffler, R.T. & Blickwede, J. (eds.) *The Circum-Gulf of Mexico and the Caribbean: Hydrocarbon Habitats, Basin Formation, and Plate Tectonics*. American Association of Petroleum Geologists Memoirs **79**, pp. 815–848.

Champagnac, J.D., Molnar, P., Anderson, R.S. et al. (2007) Quaternary erosion-induced isostatic rebound in the Western Alps. *Geology* **35**, 195–198.

Champagnac, J.D., Schlunegger, F., Norton, K. et al. (2009) Erosion-driven uplift of the modern Central Alps. *Tectonophysics* **474**, 236–249.

Chatelain, J.L., Molnar, P., Prévot, R. & Isacks, B.L. (1992) Detachment of part of the downgoing slab and uplift of the New Hebrides (Vanuatu) Islands. *Geophysical Research Letters* **19**, 1507–1510.

Clark, M.K., Farley, K.A., Zheng, D.-W. et al. (2010) Early Cenozoic faulting of the northern Tibetan Plateau margin from apatite (U-Th)/He ages. *Earth and Planetary Science Letters* **296**, 78–88.

Conrad, C.P. (2000) Convective instability of thickening mantle lithosphere. *Geophysical Journal International* **143**, 52–70.

Conrad, C.P. & Molnar, P. (1999) Convective instability of a boundary layer with temperature- and strain-rate-dependent viscosity in terms of "available buoyancy." *Geophysical Journal International* **139**, 51–68.

Crosby, A.G. & McKenzie, D. (2009) An analysis of young ocean depth, gravity and global residual topography. *Geophysical Journal International* **178**, 1198–1219.

D'Agostino, N., England, P., Hunstad, I. & Selvaggi, G. (2014) Gravitational potential energy and active deformation in the Apennines. *Earth and Planetary Science Letters* **397**, 121–132.

Dalmayrac, B. & Molnar, P. (1981) Parallel thrust and normal faulting in Peru and constraints on the state of stress. *Earth and Planetary Research Letters* **55**, 473–481.

Davis, S.J., Mulch, A., Carroll, A.R. et al. (2009) Paleogene landscape evolution of the central North American Cordillera: Developing topography and hydrology in the Laramide foreland. *Geological Society of America Bulletin* **121**, 100–116.

Debayle, E., Leveque, J.-J. & Cara, M. (2001) Seismic evidence for a deeply rooted low-velocity anomaly in the upper mantle beneath the northeastern Afro-Arabian continent. *Earth and Planetary Research Letters* **193**, 423–436.

Ding, L., Kapp, P., Zhong, D. & Deng, W. (2003) Cenozoic volcanism in Tibet: evidence for a transition from oceanic to continental subduction. *Journal of Petrology* **44**, 1833–1865.

Dricker, I.G. & Roecker, S.W. (2002) Lateral heterogeneity in the upper mantle beneath the Tibetan plateau and its surroundings from SS-S travel time residuals. *Journal of Geophysical Research* **107**, 2305.

Duvall, A.R., Clark, M.K., van der Pluijm, B.A. & Li, C.-Y. (2011) Direct dating of Eocene reverse faulting in northeastern Tibet using Ar-dating of fault clays and low-temperature thermochronometry. *Earth and Planetary Science Letters* **304**, 520–526.

Edmond, J.M., Palmer, M.R., Measures, C.I. et al. (1995) The fluvial geochemistry and denudation rate of the Guyana Shield in Venezuela, Colombia, and Brazil. *Geochimica et Cosmochimica Acta* **59**, 3301–3325.

Elliott, J.R., Walters, R.J., England, P.C. et al. (2010) Extension on the Tibetan Plateau: recent normal faulting measured by InSAR and body wave seismology. *Geophysical Journal International* **183**, 503–535.

England, P. & Houseman, G. (1986) Finite strain calculations of continental deformation. 2. Comparison with the India-Asia collision zone. *Journal of Geophysical Research* **91**, 3664–3676.

England, P.C. & Houseman, G.A. (1989) Extension during continental convergence, with application to the Tibetan Plateau. *Journal of Geophysical Research* **94**, 17561–17579.

England, P.C. & Katz, R.F. (2010) Melting above the anhydrous solidus controls the location of volcanic arcs. *Nature* **467**, 700–703.

England, P. & Wilkins, C. (2004) A simple analytical approximation to the temperature structure in subduction zones. *Geophysical Journal International* **159**, 1138–1154.

Fang, X.-M., Garzione, C., Van der Voo, R. et al. (2003) Flexural subsidence by 29 Ma on the NE edge of Tibet from the magnetostratigraphy of Linxia Basin, China. *Earth and Planetary Science Letters* **210**, 545–560.

Farías, M., Charrier, R., Carretier, S. et al. (2008) Late Miocene high and rapid surface uplift and its erosional response in the Andes of central Chile (33°–35°S). *Tectonics* **27**, TC1005.

Garzione, C.N., Molnar, P., Libarkin, J.C. & MacFadden, B.J. (2006) Rapid Late Miocene rise of the Bolivian Altiplano: evidence for removal of mantle lithosphere. *Earth and Planetary Research Letters* **241**, 543–556.

Garzione, C.N., Hoke, G.D., Libarkin, J.C. et al. (2008) Rise of the Andes. *Science* **320**, 1304–1307.

Ge, W.-P., Molnar, P., Shen, Z.-K. & Li, Q. (2015) Present-day crustal thinning in the southern and northern Tibetan Plateau revealed by GPS measurements. *Geophysical Research Letters* **42**, 5227–5235.

Ghosh, P., Garzione, C.N. & Eiler, J.M. (2006) Rapid uplift of the Altiplano revealed through ^{13}C-^{18}O bonds in paleosol carbonates. *Science* **311**, 511–515.

Gill, J.B. (1981) *Orogenic Andesites and Plate Tectonics.* New York: Springer.

Giovanni, M.K., Horton, B.K., Garzione, C.N. et al. (2010) Extensional basin evolution in the Cordillera Blanca, Peru: stratigraphic and isotopic records of detachment faulting and orogenic collapse in the Andean hinterland. *Tectonics* **29**, TC6007.

Gómez, E., Jordan, T.E., Allmendinger, R.W. et al. (2003) Controls on architecture of the Late Cretaceous to Cenozoic southern Middle Magdalena Valley Basin, Colombia. *Geological Society of America Bulletin* **115**, 131–147.

Griffiths, G.A. & McSaveney, M.J. (1983) Hydrology of a basin with extreme rainfalls – Cropp River, New Zealand. *New Zealand Journal of Science* **26**, 293–306.

Gubbels, T., Isacks, B. & Farrar, E. (1993) High level surfaces, plateau uplift, and foreland basin development, Bolivian central Andes. *Geology* **21**, 695–698.

Hatzfeld, D. & Molnar, P. (2010) Comparisons of the kinematics and deep structures of the Zagros and Himalaya and of the Iranian and Tibetan plateaus and geodynamic implications. *Reviews of Geophysics* **48**, RG2005.

Hay, W.W., Sloan, J.L. & Wold, C.N. (1988) Mass/age distribution and composition of sediments on the ocean floor and the global rate of sediment subduction. *Journal of Geophysical Research* **93**, 14 933–14 940.

Hay, W.W., Shaw, C.A. & Wold, C.N. (1989) Mass-balanced paleogeographic reconstructions. *Geologische Rundschau* **78**, 207–242.

Henry, C.D. (2008) Ash-flow tuffs and paleovalleys in northeastern Nevada: implications for Eocene paleogeography and extension in the Sevier hinterland, northern Great Basin. *Geosphere* **4**, 1–35.

Henry, C.D., Hinz, N.H., Faulds, J.E. et al. (2012) Eocene-Early Miocene paleotopography of the Sierra Nevada-Great Basin-Nevadaplano based on widespread ash-flow tuffs and paleovalleys. *Geosphere* **8**, 1–27.

Herman, F. & Champagnac, J.-D. (2016) Plio-Pleistocene increase of erosion rates in mountain belts in response to climate change. *Terra Nova* **28**, 2–10.

Herman, F., Rhodes, E.J., Braun, J. & Heiniger, L. (2010) Uniform erosion rates and relief amplitude during glacial cycles in the Southern Alps of New Zealand, as revealed from OSL-thermochronology. *Earth and Planetary Science Letters* **297**, 183–189.

Herman, F., Seward, D., Valla, P.G. et al. (2013) Worldwide acceleration of mountain erosion under a cooling climate. *Nature* **504**, 423–426.

Hicks, D.M., McSaveney, M.J. & Chinn, T.J.H. (1990) Sedimentation in proglacial Ivory Lake, Southern Alps, New Zealand. *Arctic and Alpine Research* **22**, 26–42.

Hoke, G.D., Isacks, B.L., Jordan, T.E. et al. (2007) Geomorphic evidence for post-10 Ma uplift of the western flank of the central Andes 18°30′–22°S. *Tectonics* **26**, TC5021.

Holmes, A. (1944) *Principles of Physical Geology.* London: Thomas Nelson and Sons.

Holmes, A. (1965) *Principles of Physical Geology*, 2nd edn. New York: Ronald Press.

Houseman, G.A. & England, P.C. (1986) Finite strain calculations of continental deformation: 1. *Methods and general results for convergent zones. Journal of Geophysical Research* **91**, 3651–3663.

Houseman, G.A. & Molnar, P. (1997) Gravitational (Rayleigh-Taylor) instability of a layer with non-linear viscosity and convective thinning of continental lithosphere. *Geophysical Journal International* **128**, 125–150.

Houseman, G.A., McKenzie, D.P. & Molnar, P. (1981) Convective instability of a thickened boundary layer and its relevance for the thermal evolution of continental convergent belts. *Journal of Geophysical Research* **86**, 6115–6132.

Houseman, G., Neil, E.A. & Kohler, M.D. (2000) Lithospheric instability beneath the Transverse Ranges of California. *Journal of Geophysical Research* **105**, 16 237–16 250.

Isacks, B. & Molnar, P. (1971) Distribution of stresses in the descending lithosphere from a global survey of focal-mechanism solutions of mantle earthquakes. *Review of Geophysics and Space Physics* **9**, 103–174.

Isacks, B., Oliver, J. & Sykes, L.R. (1968) Seismology and the new global tectonics. *Journal of Geophysical Research* **73**, 5855–5899.

Jaupart, C., Mareschal, J.C., Guillou-Frottier, L. & Davaille, A. (1998) Heat flow and thickness of the lithosphere in the Canadian Shield. *Journal of Geophysical Research* **103**, 15 269–15 286.

Jordan, T.H. (1975) The continental tectosphere. *Reviews of Geophysics and Space Physics* **13**, 1–12.

Jull, M. & Kelemen, P.B. (2001) On the conditions for lower crustal convective instability. *Journal of Geophysical Research* **106**, 6423–6446.

Kay, R.W. & Kay, S.M. (1993) Delamination and delamination magmatism. *Tectonophysics* **219**, 177–189.

Kay, S.M., Coira, B. & Viramonte, J. (1994) Young mafic back arc volcanic rocks as indicators of continental lithospheric delamination beneath the Argentine Puna plateau, Central Andes. *Journal of Geophysical Research* **99**, 24 323–24 339.

Kelley, K.A., Plank, T., Grove, T.L. et al. (2006) Mantle melting as a function of water content beneath back-arc basins. *Journal of Geophysical Research* **111**, B09208.

Kennan, L., Lamb, S. & Hoke, L. (1997) High altitude paleosurfaces in the Bolivian Andes: evidence for Late Cenozoic surface uplift. In: Widdowson, M. (ed.)

Paleosurfaces: Recognition, Reconstruction and Paleoenvironmental Interpretation. Geological Society of London Special Publications 120, pp. 307–324.

Kober, F., Schlunegger, F., Zeilinger, G. & Schneider, H. (2006) Surface uplift and climate change: the geomorphic evolution of the western escarpment of the Andes of northern Chile between the Miocene and present. In: Willett, S.D., Hovius, N., Brandon, M.T. & Fisher, D.M. (eds.) *Tectonics, Climate, and Landscape Evolution.* Geological Society of America Special Paper **398**. Boulder, CO: Geological Society of America, pp. 75–86.

Koons, P.O. (1990) Two-sided orogen: collision and erosion from the sandbox to Southern Alps, New Zealand. *Geology* **18**, 679–682.

Lechler, A.R., Niemi, N.A., Hren, M.T. & Lohmann, K.C. (2013) Paleoelevation estimates for the northern and central proto-Basin and Range from carbonate clumped isotope thermometry. *Tectonics* **32**, 295–316.

Levandowski, W., Jones, C.H., Shen, W. et al. (2014) Origins of topography in the western US: mapping crustal and upper mantle density variations using a uniform seismic velocity model. *Journal of Geophysical Research: Solid Earth* **119**, 2375–2396.

Li, Z., Roecker, S., Li, Z. et al. (2009) Tomographic image of the crust and upper mantle beneath the western Tien Shan from the MANAS broadband deployment: Possible evidence for lithospheric delamination. *Tectonophysics* **477**, 49–57.

Liang, C.-T. & Song, X.-D. (2006) A low velocity belt beneath northern and eastern Tibetan Plateau from Pn tomography. *Geophysical Research Letters* **33**, L22306.

Lyon-Caen, H. (1986) Comparison of the upper mantle shear wave velocity structure of the Indian Shield and the Tibetan Plateau and tectonic implications. *Geophysical Journal of the Royal Astronomical Society* **86**, 727–749.

Lyon-Caen, H. & Molnar, P. (1983) Constraints on the structure of the Himalaya from an analysis of gravity anomalies and a flexural model of the lithosphere. *Journal of Geophysical Research* **88**, 8171–8191.

Lyon-Caen, H. & Molnar, P. (1985) Gravity anomalies, flexure of the Indian plate, and the structure, support and evolution of the Himalaya and Ganga Basin. *Tectonics* **4**, 513–538.

McKenzie, D.P. (1969) Speculations on the consequences and causes of plate motions. *Geophysical Journal of the Royal Astronomical Society* **18**, 1–32.

McKenzie, D. (1994) The relationship between topography and gravity on Earth and Venus. *Icarus* **112**, 55–88.

McKenzie, D. & Priestley, K. (2008) The influence of lithospheric thickness variations on continental evolution. *Lithos* **102**, 1–11.

McNamara, D.E., Walter, W.R., Owens, T.J. & Ammon, C.J. (1997) Upper mantle velocity structure beneath the Tibetan Plateau from Pn travel time tomography. *Journal of Geophysical Research* **102**, 493–505.

Métivier, F., Gaudemer, Y., Tapponnier, P. & Meyer, B. (1998) Northeastward growth of the Tibet plateau deduced from balanced reconstruction of the two areas: the Qaidam and Hexi corridor basins, China. *Tectonics* **17**, 823–842.

Mix, H.T., Mulch, A., Kent-Corson, M.L. & Chamberlain, C.P. (2011) Cenozoic migration of topography in the North American Cordillera. *Geology* **39**, 87–90.

Molnar, P. (1988) A review of geophysical constraints on the deep structure of the Tibetan Plateau, the Himalaya, and the Karakorum and their tectonic implications. *Philosophical Transactions of the Royal Society of London A: Mathematical, Physical and Engineering Sciences* **326**, 33–88.

Molnar, P. (1990) S-wave residuals from earthquakes in the Tibetan region and lateral variations in the upper mantle. *Earth and Planetary Science Letters* **101**, 68–77.

Molnar, P. (2004) Late Cenozoic increase in accumulation rates of terrestrial sediment: how might climate change have affected erosion rates? *Annual Review of Earth and Planetary Sciences* **32**, 67–89.

Molnar, P. (2015) *Plate Tectonics: A Very Short Introduction.* Oxford: Oxford University Press.

Molnar, P. & Chen, W.-P. (1983) Focal depths and fault plane solutions of earthquakes under the Tibetan plateau. *Journal of Geophysical Research* **88**, 1180–1196.

Molnar, P. & Chen, W.-P. (1984) S-P wave travel time residuals and lateral inhomogeneity in the mantle beneath Tibet and the Himalaya. *Journal of Geophysical Research* **89**, 6911–6917.

Molnar, P. & England, P. (1990) Late Cenozoic uplift of mountain ranges and global climate change: chicken or egg? *Nature* **346**, 29–34.

Molnar, P. & Houseman, G.A. (2004) Effects of buoyant crust on the gravitational instability of thickened mantle lithosphere at zones of intracontinental convergence. *Geophysical Journal International* **158**, 1134–1150.

Molnar, P. & Lyon-Caen, H. (1988) Some simple physical aspects of the support, structure, and evolution of mountain belts. In: Clark, S.P., Burchfiel, B.C. & Suppe, J. (eds.) *Processes in Continental Lithospheric Deformation.* Geological Society of America Special Paper **218**, pp. 179–207.

Molnar, P. & Tapponnier, P. (1978) Active tectonics of Tibet. *Journal of Geophysical Research* **83**, 5361–5375.

Molnar, P., England, P.C. & Jones, C.H. (2015) Mantle dynamics, isostasy, and the support of high terrain. *Journal of Geophysical Research: Solid Earth* **120**, 1932–1957.

Mora, A., Parra, M., Strecker, M.R. et al. (2006) Cenozoic contractional reactivation of Mesozoic extensional structures in the Eastern Cordillera of Colombia. *Tectonics* **25**, TC2010.

Mora, A., Parra, M., Strecker, M.R. et al. (2008) Climatic forcing of asymmetric orogenic evolution in the Eastern Cordillera of Colombia. *Geological Society of America Bulletin* **120**, 930–949.

Mora, A., Parra, M., Strecker, M.R. et al. (2010) The eastern foothills of the Eastern Cordillera of Colombia: an example of multiple factors controlling structural styles and active tectonics. *Geological Society of America Bulletin* **122**, 1846–1864.

Mora-Páez, H., Mencin, D.J., Molnar, P. et al. (2016) GPS velocities and the construction of the Eastern Cordillera of the Colombian Andes. *Geophysical Research Letters* **43**, 8407–8416.

Morgan, W.J. (1971) Convection plumes in the lower mantle. *Nature* **230**, 42–43.

Morgan, W.J. (1973) Plate motions and deep mantle convection. *Geological Society of America Memoirs* **132**, 7–22.

Murphy, M.A., Yin, A., Kapp, P. et al. (2002) Structural evolution of the Gurla Mandhata detachment system, southwest Tibet: implications for the eastward extent of the Karakoram fault system. *Geological Society of America Bulletin* **114**, 428–447.

Ni, J. & Barazangi, M. (1983) High-frequency seismic wave propagation beneath the Indian Shield, Himalayan arc, Tibetan Plateau, and surrounding regions: high uppermost mantle velocities and efficient Sn propagation beneath Tibet. *Geophysical Journal of the Royal Astronomical Society* **72**, 665–689.

Ni, J. & York, J.E. (1978) Late Cenozoic extensional tectonics of the Tibetan Plateau. *Journal of Geophysical Research* **83**, 5377–5387.

Ochoa, D., Hoorn, C., Jaramillo, C. et al. (2012) The final phase of tropical lowland deposition in the axial zone of the Eastern Cordillera: evidence from three palynological records. *Journal of South American Earth Sciences* **39**, 157–169.

Olson, P., Reynolds, E., Hinnov, L. & Goswami, A. (2016) Variation of ocean sediment thickness with crustal age. *Geochemistry Geophysics Geosystems* **17**, 1349–1369.

Owens, T.J. & Zandt, G. (1997) Implications of crustal property variations for models of Tibetan plateau evolution. *Nature* **387**, 37–43.

Parra, M., Mora, A., Jaramillo, C. et al. (2009) Orogenic wedge advance in the northern Andes: evidence from the Oligocene-Miocene sedimentary record of the Medina Basin, Eastern Cordillera, Colombia. *Geological Society of America Bulletin* **121**, 780–800.

Parra, M., Mora, A., Lopez, C. et al. (2012) Detecting earliest shortening and deformation advance in thrust belt hinterlands: example from the Colombian Andes. *Geology* **40**, 175–178.

Plank, T. (2018) The geochemistry of subduction zones, In White, W.M. (ed.) *Encyclopedia of Geochemistry*. Cham: Springer International Publishing. DOI 10.1007/978-3-319-39193-9_268-1.

Poveda, E., Monsalve, G. & Vargas, C.A. (2015) Receiver functions and crustal structure of the northwestern Andean region, Colombia. *Journal of Geophysical Research: Solid Earth* **120**, 2408–2425.

Pratt, J.H. (1855) On the attraction of the Himalaya mountains, and of elevated regions beyond them, upon them, upon the plumb-line in India. *Philosophical Transactions of the Royal Society of London* **145**, 53–100.

Pratt, J.H. (1859) On the deflection of the plumb-line in India, caused by the attraction of the Himmalaya mountains and of elevated regions beyond; and its modification by the compensating effect of a deficiency of matter below the mountain mass. *Philosophical Transactions of the Royal Society of London* **149**, 745–778.

Reigber, Ch., Michel, G.W., Galas, R. et al. (2001) New space geodetic constraints on the distribution of deformation in Central Asia. *Earth and Planetary Science Letters* **191**, 157–165.

Ribe, N.M. & Christensen, U.R. (1994) Three-dimensional modeling of plume-lithosphere interaction. *Journal of Geophysical Research* **99**, 669–682.

Sánchez, J., Horton, B.K., Tesón, E. et al. (2012) Kinematic evolution of Andean fold-thrust structures along the boundary between the Eastern Cordillera and Middle Magdalena Valley basin, Colombia. *Tectonics* **31**, TC3008.

Saylor, J.E., Horton, B.K., Nie, J.-S. et al. (2011) Evaluating foreland basin partitioning in the northern Andes using Cenozoic fill of the Floresta basin, Eastern Cordillera, Colombia. *Basin Research* **23**, 377–402.

Schildgen, T.F., Hodges, K.V., Whipple, K.X. et al. (2007) Uplift of the western margin of the Andean plateau revealed from canyon incision history, southern Peru. *Geology* **35**, 523–526.

Schubert, G., Turcotte, D.L. & Olson, P. (2001) *Mantle Convection in the Earth and Planets*. Cambridge: Cambridge University Press.

Shapiro, N.M. & Ritzwoller, M.H. (2002) Monte-Carlo inversion for a global shear-velocity model of the crust and upper mantle. *Geophysical Journal International* **151**, 88–105.

Sobel, E.R. & Dumitru, T.A. (1997) Thrusting and exhumation around the margins of the western Tarim Basin during the India-Asia collision. *Journal of Geophysical Research* **102**, 5043–5064.

Snell, K.E., Koch, P.L., Druschke, P. et al. (2014) High elevation of the "Nevadaplano" during the Late Cretaceous. *Earth and Planetary Science Letters* **386**, 52–63.

Stern, T.A., Baxter, A.K. & Barrett, P.J. (2005) Isostatic rebound due to glacial erosion within the Transantarctic Mountains. *Geology* **33**, 221–224.

Tapponnier, P., Xu, Z., Roger, F. et al. (2001) Oblique stepwise rise and growth of the Tibet Plateau. *Science* **294**, 1671–1677.

Thouret, J.-C., Wörner, G., Gunnell, Y. et al. (2007) Geochronologic and stratigraphic constraints on canyon incision and Miocene uplift of the Central Andes in Peru. *Earth and Planetary Science Letters* **263**, 151–166.

Tseng, T.-L., Chen, W.-P. & Nowack, R.L. (2009) Northward thinning of Tibetan crust revealed by virtual seismic profiles. *Geophysical Research Letters* **36**, L24304.

Turcotte, D.L. & Schubert, G. (2002) *Geodynamics: Applications of Continuum Mechanics to Geological Problems*, 2nd edn. Cambridge: Cambridge University Press.

Turner, S., Hawkesworth, C., Liu, J. et al. (1993) Timing of Tibetan uplift constrained by analysis of volcanic rocks. *Nature* **364**, 50–54.

Turner, S., Arnaud, N., Liu, J. et al. (1996) Post-collision, shoshonitic volcanism on the Tibetan Plateau: implications for convective thinning of the lithosphere and the source of ocean island basalts. *Journal of Petrology* **37**, 45–71.

Veloza, G., Taylor, M., Mora, A. & Gosse, J. (2015) Active mountain building along the eastern Colombian Subandes: A folding history from deformed terraces across the Tame anticline, Llanos Basin. *Geological Society of America Bulletin* **127**, 1155–1173.

Victor, P., Oncken, O. & Glodny, J. (2004) Uplift of the western Altiplano plateau: evidence from the Precordillera between 20° and 21°S (northern Chile). *Tectonics* **23**, TC4004.

von Rotz, R., Schlunegger, F., Heller, F. & Villa, I. (2005) Assessing the age of relief growth in the Andes of northern Chile: magneto-polarity chronologies from Neogene continental sections. *Terra Nova* **17**, 462–471.

Wager, L.R. (1933) The rise of the Himalaya. *Nature* **132**, 28.

Wager, L.R. (1937) The Arun River drainage pattern and the rise of the Himalaya. *Geographical Journal* **89**, 239–250.

Watts, A.B. (2001) *Isostasy and Flexure of the Lithosphere*. Cambridge: Cambridge University Press.

Wolfe, J.A., Schorn, H.E., Forest, C.E. & Molnar, P. (1997) Paleobotanical evidence for high altitudes in Nevada during the Miocene. *Science* **276**, 1672–1675.

Wolfe, J.A., Forest, C.E. & Molnar, P. (1998) Paleobotanical evidence of Eocene and Oligocene paleoaltitudes in midlatitude western North America. *Geological Society of America Bulletin* **110**, 664–678.

Wolfe, C.J., Solomon, S.C., Laske, G. et al. (2009) Supporting Online Material: Mantle Shear-Wave Velocity Structure beneath the Hawaiian Hot Spot. Available from: http://www.sciencemag.org/cgi/content/full/326/5958/1388/DC1 (last accessed September 1, 2017).

Woodward, R.L. & Molnar, P. (1995) Lateral heterogeneity in the upper mantle and *SS-S* traveltime intervals for *SS* rays reflected from the Tibetan Plateau and its surroundings. *Earth and Planetary Science Letters* **135**, 139–148.

Yin, A., Dang, Y.-Q., Wang, L.-C. et al. (2008) Cenozoic tectonic evolution of Qaidam basin and its surrounding regions (Part 1): the southern Qilian Shan-Nan Shan thrust belt and northern Qaidam basin. *Geological Society of America Bulletin* **120**, 813–846.

Zhang Peizhen, Molnar, P. & Downs, W.R. (2001) Increased sedimentation rates and grain sizes 2–4 Myr ago due to the influence of climate change on erosion rates. *Nature* **410**, 891–897.

Zhao, L.-S. & Xie, J. (1993) Lateral variations in compressional velocities beneath the Tibetan Plateau from *Pn* traveltime tomography. *Geophysical Journal International* **115**, 1070–1084.

Zheng, W.-J., Zhang, H.-P., Zhang, P.-Z. et al. (2013) Late Quaternary slip rates of the thrust faults in western Hexi Corridor (Northern Qilian Shan, China) and their implications for northeastward growth of the Tibetan Plateau. *Geosphere* **9**, 342–354.

Zubovich, A.V., Wang, X.-Q., Scherba, Y.G. et al. (2010) GPS velocity field for the Tien Shan and surrounding regions. *Tectonics* **29**, TC6014.

3

An Overview of Dynamic Topography: The Influence of Mantle Circulation on Surface Topography and Landscape

Caroline M. Eakin[1,2] and Carolina Lithgow-Bertelloni[3]

[1] *Research School of Earth Sciences, Australian National University, Canberra, Australia*
[2] *University of Southampton, National Oceanography Centre Southampton, Southampton, UK*
[3] *University College London, London, UK*

Abstract

The topography of Earth's surface is a direct reflection of our dynamic planet. It represents the balance between vigorous internal and external processes that act to either build topography – such as orogenic uplift – or to destroy it – such as erosion and sedimentation. The long-term cooling of the planet results in convection of the viscous mantle interior and plate tectonics at the surface. In this chapter, we focus on this internal mantle convection and the lesser known concept of dynamic topography, whereby mantle flow can cause vertical deflections of Earth's surface. We explore the methods commonly used to measure and/or model such dynamic topography and discuss the case study of Amazonia, where the effect of dynamic topography and its consequences can be found in the geological record. Overall, dynamic topography is found to be transient and cyclic, changing with the rate of mantle flow (mm/y) and resulting in fluctuations of the surface over long wavelengths (up to thousands of kilometers). The potential therefore exists to alter topography on a continental scale and to drive landscape changes over evolutionary timescales (on the order of ~1 My). Considering all sources of topography, both tectonic driven deformation and dynamic influences therefore help us better understand not only the shape of our planet, but also the Earth system as a whole, with interactions among landscape, life and the dynamics of the deep interior.

Keywords: *surface–interior interactions, mantle convection, plate tectonics, subduction, Amazonia*

3.1   Introduction

Topography of Earth's surface is a defining characteristic of our planet; from mountains and valleys, to continents and oceans, the lumps and bumps of the land are immediately recognizable to all of us. What is less widely appreciated, however, is why that topography is there, and how it directly reflects activity and dynamics within Earth's deep interior. Without this deep activity, we'd have a geologically dead planet: flat and barren except for the scars of impact cratering. Instead, long-term cooling of Earth results in viscous convection of the rocky mantle, the region between Earth's hot core and the cold outer surface.

The surface expression of mantle convection gives rise to plate tectonics: the process whereby Earth's oceanic crust is continually recycled (see Chapter 2). As Earth cools, it forms a rigid outermost layer, known as a thermal boundary layer, or the lithosphere. The lithosphere is broken up into around a dozen or so major tectonic plates, which move relative to one another, riding on top of the underlying weak mantle (i.e., asthenosphere). A variety of different plate boundary types occur as a result. For example, at mid-ocean ridges, the plates diverge and the mantle is drawn upwards to fill the gap, and new crust is generated. Similarly, at subduction zones, where plates converge, one plate (always an oceanic plate) descends beneath the other (either a continental or an oceanic plate) and is consumed back into the mantle. The dense descending plate is referred to as the subducting slab. If two buoyant continental plates converge, then a collision zone is formed, as neither plate can subduct.

Surface topography is typically defined as the deviation from the geoid: an equipotential reference surface that is

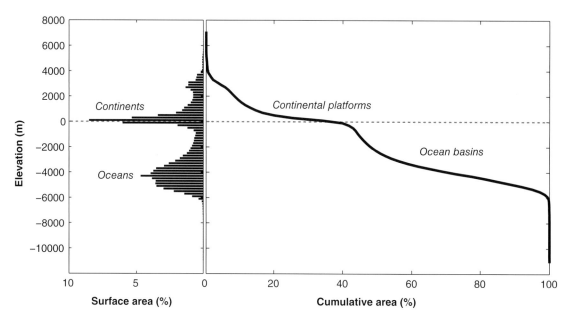

Figure 3.1 Variation of Earth's topography. Histograms of surface elevations using data from ETOPO1 (Amante & Eakins 2009). The distribution is dominated by the ocean–continent dichotomy. High topography (>1 km) – the focus of this book – covers less than 15% of Earth's surface.

always normal to the local gravity vector. The mean sea level (m.s.l.) is equivalent to the geoid, as water will continually flow until a constant surface potential is reached. Overall, surface elevations span approximately ±10 km from m.s.l., which is less than 0.2% of the total Earth radius. The deepest point lies at the bottom of the Mariana Trench, at 10.9 km below sea level, while the highest point, at the top of Mount Everest, is 8.8 km above it. These are very much the extremes, with most elevations either close to sea level or around 4 km below it, which reflects the primary continent–ocean division (Figure 3.1).

Most of Earth's surface is in equilibrium, as there exists a continual competition between processes that build topography – such as plate convergence, crustal thickening and tectonic uplift – and those that act to destroy it – such as erosion, transportation and sedimentation. At its most basic level, this is a balance between Earth's internal heat engine driving mantle convection and the external solar heating that drives the hydrological cycle and associated erosion and sediment transport. The balance between uplift and erosion results in a given loading (or unloading) of Earth's surface. The height at which this load sits is primarily governed by the principle of isostasy, and the span over which the load has been distributed due to lithospheric flexure (i.e. the ability of the plate to bend in response to the load). These topics are covered in detail by Molnar in Chapter 2. In this chapter, we instead focus on the field of "dynamic topography": a second-order effect to surface topography

that is directly related to the motion of the underlying mantle. Unlike variations in surface elevations due to isostasy, dynamic topography is not associated with any changes in crustal or lithospheric thickness. In the following sections, we provide an overview of dynamic topography and its importance, describe current methods of constraining or modeling dynamic topography, and discuss examples where dynamic topography is thought to have played a role in landscape evolution.

3.2 What is Dynamic Topography?

The simplest definition of dynamic topography is that it is the deflection of a boundary between two different layers due to tractions or stresses normal to the plane of the boundary that are imposed by flow acting in either layer. For Earth, dynamic topography manifests at the surface, where flow or motion of the underlying mantle generates normal stresses at the base of the lithospheric lid, inducing vertical motion (Figure 3.2). This motion derives from mantle convection, driven by the temperature difference between the hot planetary interior and the cold outer surface. Such mantle convection involves buoyant plumes of rising mantle material that will push up on the surface, producing dynamic uplift. Likewise, dense subducting slabs sink, dragging the mantle and the surface downwards, producing dynamic subsidence.

Given that Earth's surface is constantly evolving, and therefore that all topography can be described as

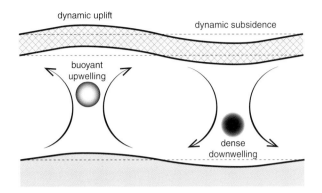

dynamic uplift

dynamic subsidence

buoyant
upwelling

dense
downwelling

Figure 3.2 Illustration of dynamic topography. Mass anomalies induce convective flow in the mantle (arrows) due to density differences. This flow exerts forces on the base of tectonic plates, causing deflection of the surface topography.

"dynamic" in some way, there is ongoing contention over the precise definition of "dynamic topography." A lot of this seems to stem from a difference between what it is we are measuring when we try to observe dynamic topography (Section 3.3) and how we treat dynamic topography in mantle-convection simulations (Section 3.4). Readers should therefore exercise caution when interpreting dynamic topography assertions between different studies. In this chapter, when we use the term "dynamic topography," we are referring to the definition provided in the previous paragraph, unless otherwise stated.

In theory, dynamic topography, as we have described it, can operate anywhere and everywhere there is mantle flow. It can also occur across a wide range of spatial scales, from 10s to 1000s of kilometers, depending on the wavelength of the underlying mantle convection (Flament et al. 2013; Gérault et al. 2015; Hoggard et al. 2016). In practice, however, it is difficult to isolate the dynamic component from surface topography. This is because dynamic deflections are relatively small compared to total topographic variations (Braun 2010). To counteract this, studies of dynamic topography tend to focus on longer wavelength patterns (~1000 km) that are difficult to explain based purely on tectonics or variations in crustal/lithospheric properties. For example, it has been shown that the long wavelength pattern of the geoid requires dynamic deflection of both Earth's internal boundaries and its surface (Hager et al. 1985).

While everyone tends to agree that dynamic topography is relatively small compared to the total topography, there is ongoing debate and controversy over what its maximum absolute amplitude can be, and at what wavelengths. This is partly attributable to the different definitions of "dynamic topography," as mentioned earlier, and partly to uncertainties associated with its

estimation. Broadly speaking, most models typically predict peak amplitudes on the order of 1–2 km at the longest wavelengths (although the locations of peak dynamic topography may differ somewhat between models), as reported by several recent review papers on the subject (Braun 2010; Flament et al. 2013; Liu 2015). At the most conservative end of the scale, Molnar et al. (2015) argue that dynamic topography is limited to at most 300 m, based on free-air and isostatic gravity anomalies. Additionally, a recent detailed study of global residual topography (see Section 3.3) argues for lower amplitudes at the longest wavelengths (~10 000 km) and higher amplitudes at shorter ones (~1000 km or less) compared to standard convection-model estimates (Hoggard et al. 2016).

Large deviations in surface topography are usually restricted to tectonically active regions, such as plate boundaries. The search for dynamic topography therefore tends to focus on tectonically quiet and peaceful regions – such as stable continental interiors – that remain sheltered from the intense modifications of plate tectonics (Mitrovica et al. 1989; Lithgow-Bertelloni & Gurnis 1997; Lithgow-Bertelloni & Silver 1998; Heine et al. 2008; Liu et al. 2008; DiCaprio et al. 2009; Shephard et al. 2010; Dávila & Lithgow-Bertelloni 2013). It is here that variations in topography due to mantle dynamics are more likely to be visible, where we do not expect abrupt changes in the crustal and lithospheric thicknesses, which would result in strong isostatic variations. However, this does not mean that dynamic topography only occurs in such settings. On the contrary, dynamic topography is likely ubiquitous, and nowhere on Earth – including continental platforms – can be assumed to be of fixed or stable elevation over geological time (Moucha et al. 2008).

A characteristic feature of dynamic topography is its transient and cyclic nature. The sequences of dynamic uplift and subsidence change with the mantle flow. The term "dynamic" in itself indicates something that is constantly in motion. Dynamic topography therefore offers an important pathway for understanding the temporal evolution of mantle convection. Typically, we rely on geophysical observations to probe the deep Earth, but these tools (e.g., seismic tomography or gravity surveys) provide just a single snapshot of what the mantle looks like today. The variable of time inherent within dynamic topography therefore has great potential to unravel not just the present-day state of the mantle, but also the past history of our dynamic Earth. Constraining and interpreting dynamic topography, however, can be a tricky business. In the following sections, we discuss the two main research avenues for investigating dynamic topography.

3.3 Residual Topography

It is thought that the isostatic component (Section 3.2.1) dominates Earth's topographic variations by as much as 90% (Molnar et al. 2015). It is relatively easy to make an estimate of this component using the principle of local isostasy and subtracting it from the actual topography (Figure 3.3). If the isostatic estimate is accurate, then what is left is categorized as the residual topography, and in theory should represent flexure (at short wavelengths) and dynamic topography (at all wavelengths, including long wavelengths). Residual topography is therefore our most direct method for estimating the present-day extent of dynamic topography.

While the isostatic calculation is straightforward, in reality the extent to which residual topography reflects true dynamic topography (i.e., flow at the base of the plate) is questionable, and overall the quantity is poorly constrained due to large uncertainties in the input parameters. Nevertheless, many different estimates and global maps of residual topography have been attempted by various groups (e.g. Panasyuk & Hager 2000; Kaban et al. 2003; Steinberger 2007; Flament et al. 2013; Hoggard et al. 2016). These tend to agree on the main long-wavelength (~1000 km) features present (Flament et al. 2013), and on relatively low amplitudes (on the order of 1 km) (Braun 2010). The locations of anomalies and the correlation between shorter-wavelength features can vary considerably, however. Differences mainly occur due to the large uncertainties inherent in the isostatic calculation. The effect of sediments, ice, the crust and the lithosphere must be considered, but small changes in density or layer thickness can propagate into large errors in the predicted isostatic topography.

Unfortunately, variations in the physical properties of Earth's layers, especially the lithospheric structure and density, are not well known. For the oceans, the density and thickness of oceanic plates are both a function of temperature and sea-floor age (see Section 2.5.3), and are typically calculated using either half-space cooling (Carslaw & Jaeger 1959) or plate models (e.g., Parsons & Sclater 1977). Even though lithosphere formation can be understood relatively simply in the oceans compared to

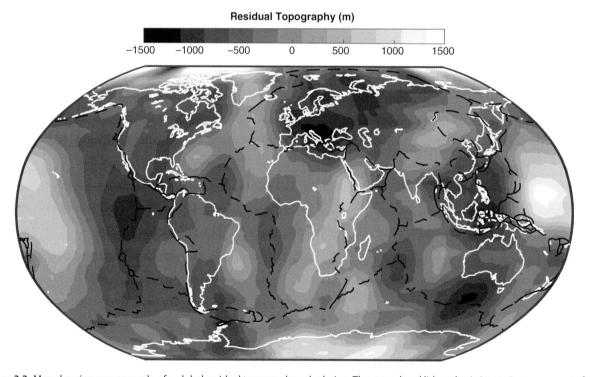

Figure 3.3 Map showing one example of a global residual topography calculation. The crustal and lithospheric isostatic component of topography has been subtracted from the present-day observed topography. This calculation was based on a model of lithospheric structure from Naliboff et al. (2012) and has been smoothed (averaged) over 10×10° bins to focus on longer-wavelength features. Prominent residual highs (light grays) are seen over the Western Pacific, the East African Rift, Antarctica and the North Atlantic. Meanwhile, residual lows (dark grays) are visible over Europe, the East Pacific Rise and the Australian–Antarctic Discordance (south of Australia). Thin white lines on the map delineate land surfaces, and black dashed lines show plate boundaries. See also Plate 7 in color plate section.

the continents, we are still unsure as to the nature of the base of the lithosphere, whether it is a purely thermal or a compositional boundary (Kawakatsu et al. 2009; Fischer et al. 2010; Schmerr 2012; Naif et al. 2013) or what is its thickness at old ages >70 Ma (Korenaga & Korenaga 2008; Adam & Vidal 2010; Hillier 2010; Goes et al. 2013). The crustal structure can be constrained better using seismic refraction and reflection surveys, but even for the most comprehensive catalogs, global coverage remains sparse, particularly in the middles of ocean basins (Hoggard et al. 2016). Residual topography estimates may therefore reflect unmapped variations in crustal and/or lithospheric properties, rather than be directly related to flow. For example, heating of the plate from below would not only induce dynamic uplift due to the buoyant convective motion, but would also alter the thermal structure of the plate, which would have an isostatic response. Part of the controversy surrounding the definition of dynamic topography (see Section 3.2) is centered upon this argument. In the future, however, there is hope that improvements in the imaging of global crustal and lithospheric structure may lead to better estimates of residual topography, and to the ability to separate out true dynamic topography associated only with mantle flow.

3.4   Modeling of Mantle Flow

A second – and popular – approach to studying dynamic topography is to model convection within the mantle system. The normal stresses that impinge on the base of the lithosphere as a result of this modeled mantle flow can be easily calculated, and from these the corresponding deflection of the surface can be estimated. A brief overview of such modeling is given in this section; for a comprehensive review, we refer the reader to Flament et al. (2013).

When modeling mantle convection, there are two main inputs to consider: the density structure and the viscosity structure of the mantle. Lateral variations in density or buoyancy are what drive the flow, while the rheology or viscosity of the mantle acts to resist flow and controls the rate of movement. The mantle density structure is typically prescribed either by the known location of subducted slabs or by inferences of seismic wavespeeds. Each of these approaches comes with its own strengths and weaknesses. For example, for subducted slabs, their location can be recreated based on our knowledge of subduction zones worldwide and the history of plate motions. The age of the subducting sea floor can also be easily converted into a thermal thickness, and therefore a density anomaly. As the negative buoyancy of cold, dense, subducting slabs is the primary driving force for

the circulation of Earth's interior, this approach allows for focused investigation of the strongest dynamic influence and the details that drive the flow (Lithgow-Bertelloni & Richards 1998). It is lacking, however, in active dynamic upwelling (e.g., mantle plumes) or any associated positive (upward) deflections.

Alternatively, to incorporate both upwelling and downwelling, seismic tomography is frequently deployed, with global shear wave models particularly popular (e.g., Grand 2002; Ritsema et al. 2011). This provides a seismic velocity structure of the whole mantle based on the travel times of various seismic waves that traverse it. In general, seismic waves will tend to travel faster through cold, dense material and slower through warm, buoyant material. However, the conversion between velocity and density is not straightforward, and depends on whether a velocity anomaly is considered mostly thermal or mostly compositional in origin (Simmons et al. 2009). This is particularly important at deep lower-mantle depths, where the same velocity anomaly can be interpreted as either a positive or a negative density anomaly, depending upon the thermochemical properties inferred (Karato & Karki 2001; Stixrude & Lithgow-Bertelloni 2007).

Aside from density anomalies, the other important factor to consider is the mantle viscosity. Uncertainties in present-day viscosity estimates are large, but thankfully only the relative viscosities of the various mantle layers are needed to calculate dynamic topography. The simple one-dimensional (1D) mantle structure is widely known and is characterized by a high-viscosity lithospheric lid over the low-viscosity asthenosphere, as well as a higher viscosity in the lower mantle compared to the upper mantle (e.g., Ricard et al. 1993). This simple layering – as well as more detailed variants – has been constrained over the years from a range of geophysical measurements, such as post-glacial rebound (Haskell 1935), modeling of the geoid (Hager 1984; Hager et al. 1985; Hager & Richards 1989; Ricard et al. 1993) and modeling of plate motions (Ricard et al. 1989; Forte et al. 1991; Ricard & Wuming 1991).

It is also known from laboratory experiments that the viscosity of upper-mantle materials, such as olivine, is strongly temperature-dependent, as well as grain size- and strain rate-dependent (King 1995; Hirth 2002). This can result in lateral viscosity variations (LVV), which affect dynamic topography predictions. The strongest LVV are likely to occur in subduction zones, as the cold and dense subducting slab should have a higher viscosity compared to the surrounding mantle (King & Hager 1994; Moresi & Gurnis 1996; Zhong & Davies 1999; Čadek & Fleitout 2003). Comparatively, melt and fluids released into the mantle wedge above subducting slabs will drastically lower the wedge viscosity (Billen & Gurnis 2001). Given such competing processes, the overall impact of short-wavelength LVV on dynamic topography – even where it is strongest – has yet to be fully understood (Billen et al. 2003).

3.5 Interaction of Dynamic Topography with the Landscape

3.5.1 Dynamic Topography and Mountainous Regions

As previously established, most of Earth's surface, including areas of high topography, is in isostatic equilibrium, meaning that the excess mass of a mountain above the surface is compensated by a low-density crustal root at depth. Dynamic topography therefore clearly does not build mountains, but we can still expect that it may influence how they evolve over time. It is difficult to directly quantify the amount that dynamic topography contributes to the development of mountainous regions, as it is typically dwarfed by complex tectonic processes, such as crustal shortening and faulting. Nonetheless, examples can be found in the literature that try to consider the role mantle convection and dynamic topography might play in mountainous regions.

At the most general level, we expect dynamic topography to be greatest over subduction zones where the largest density anomalies (cold subducting slabs) are present in the upper mantle. Subduction zones also build mountain ranges, such as the Andes (Section 3.5.3), and mountainous volcanic landmasses, such as Japan (Chapter 30) and Indonesia (Chapter 31). The sinking slabs beneath these regions should exert a downwards force or "pull" on the surface above. Such mountain ranges could therefore, in theory, be several hundreds of meters higher if such a force were removed (Flament 2014). An example of the influence of mantle dynamics on mountain belts can be found in the uplift history of the Himalaya (Chapter 28). During the Miocene, the Himalaya rose rapidly, and this is often associated with delamination (i.e., removal) of the mantle lithosphere beneath Tibet at this time (Molnar et al. 1993). Alternatively, it has been proposed that the northward drift of India over past subducted slabs in the mantle could explain fluctuations in the elevation of the region, including dynamic support for uplift, and subsidence of the adjacent foreland (Husson et al. 2014).

Similarly, the closure of the ancient Tethys Ocean as India and Africa moved north has left behind a complex geometry of various subducting slab segments and a broad zone of deformation in the Mediterranean region and the Alpine orogenic belt (Chapter 27). These sinking slab fragments are thought to induce vigorous small-scale convection in the mantle below this region, contributing to an alternating pattern of vertical deflections and complex micro-plate motions (Boschi et al. 2010; Faccenna & Becker 2010).

Prominent upwellings of the mantle, such as those associated with mantle plumes, should also generate dynamic uplift and build topography from volcanic outpourings and rifting. One good example of this is the East African Rift System (EARS) (Chapter 26), which is thought to be underlain by a large-scale mantle plume. Moucha & Forte (2011) proposed that the current topographic swell and southern propagation of the rift could be attributed to the northward motion of Africa over the plume, causing a migration of dynamic uplift.

3.5.2 Preservation in the Geological Record

Given the large uncertainties involved with modeling mantle flow, dynamic-topography model predictions can vary widely even for the same study region, including in the examples highlighted in the previous section (Braun 2010; Liu 2015). It is therefore imperative to constrain such models and predictions using available data. Given the temporal evolution of dynamic topography, a data set that spans a long time frame, such as the rock record, is arguably the most useful for comparison.

While most of the geological record is dominated by short-wavelength (100s of kilometers) tectonic events, long-wavelength (1000s of kilometers), low-amplitude (100s of meters) dynamic fluctuations can have a significant impact on surface processes, which affect the erosion and deposition of sediment. The ways in which dynamic topography can affect surface processes are numerous, from continental tilting and flooding (Mitrovica et al. 1989; Gurnis 1990; Pysklywec & Mitrovica 1998) to sedimentary basin evolution (Burgess et al. 1997; Heine et al. 2008) and drainage patterns (Shephard et al. 2010; Eakin et al. 2014; Liu 2015).

In general, dynamic subsidence is easier to constrain than dynamic uplift. This is because subsidence tends to induce basin formation and sediment deposition, which have a much better preservation potential than erosion, the most likely outcome of uplift. The best preservation chances of all are found in stable continental interiors such as North America and Australia, which are unaffected by the large signals from tectonically driven geomorphology.

3.5.3 Case Study: Peruvian Flat-Slab Subduction and the Evolution of Western Amazonia

One exemplary region where it is possible to explore the relationship between mountainous landscapes, dynamic topography and the sedimentary record is in South

Figure 3.4 Tectonic setting of Peruvian flat-slab subduction and western Amazonia. The total area thought to be covered by the Solimões Formation is outlined by the thick dashed white line, based on the maps of Hoorn (1994). Isopachs of the Solimões Formation thickness in the Acre and Solimões Basins are available; these are shown by thin white solid lines, after Latrubesse et al. (2010). Volcano locations (white triangles) are from Siebert & Simkin (2002). The gray arrow represents the absolute rate of motion of the Nazca Plate, from HS3-NUVEL1A (Gripp & Gordon 2002). The dotted line, a–a′, represents the location of the cross-section in Figure 3.5. Gray shading in the background represents topography/bathymetry, and the thin black lines are rivers. *Source:* Eakin et al. (2014). Reproduced with permission of Elsevier.

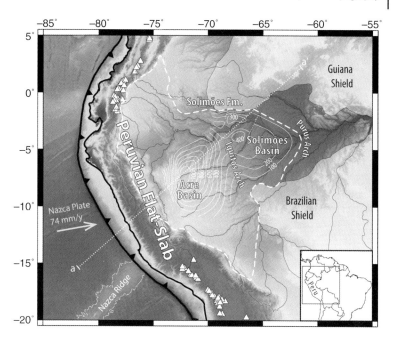

America. In this case study, largely based on the work of Eakin et al. (2014), we investigate the interaction between the Andes and the Amazonian Basin.

The largest flat slab in the world today exists beneath much of Peru, from 3 to 15°S (Figure 3.4). Here, the incoming Nazca Plate subducts at the offshore trench to a depth of around 100 km and then proceeds to slide horizontally beneath the overlying South American continent for several hundred kilometers, before resuming a steep descent into the mantle (Barazangi & Isacks 1976; Cahill & Isacks 1992). While subduction has been operating off the west coast of South America since the beginning of the Jurassic (~200 Ma) (Coira et al. 1982), the Peruvian flat slab is only a relatively recent feature, developing in the mid to late Miocene (~15–10 Ma) (Bissig et al. 2008; Ramos & Folguera 2009).

At the same time that this flat slab was developing, the surface above was going through its own transformation, which would eventually give way to the largest drainage basin in the world – the Amazon River system – and the most diverse habitat on the planet – the Amazon Rainforest. Prior to the formation of the Amazon River, drainage of the region is thought to have escaped to the north, into the Caribbean Sea (Hoorn et al. 1995), or to the west, into the Pacific Ocean (Potter 1997). During the late Miocene, however, the Andes grew rapidly, shutting off the drainage pathways to the north and west (Jordan & Gardeweg 1989; Kroonenberg et al. 1990). Meanwhile, a wide but shallow subsiding system developed in the western Amazon, stretching from eastern Peru to the Purus Arch (Figure 3.4). This system, known as the Pebas Mega-Wetland, was deposited in the sedimentary record as the Solimões Formation (Hoorn 1993, 1994, 2010a). Eventually, large-scale deposition across western Amazonia basins ceased, and instead sediment was transported across the continent from the Andes to the Atlantic by the newly formed Amazon River (Campbell et al. 2006; Figueiredo et al. 2009; Latrubesse et al. 2010).

For the Solimões Formation, isopachs are available that map the spatial distribution of sediment accumulation (i.e., layer thickness) in several Amazonia basins (Latrubesse et al. 2010), and thus provide a record of the landscape change over time with which to compare predictions of dynamic topography. Modeling the dynamic component of topography that results from the transition from "normal" subduction, with an average slab dip of 30°, to the present-day flat-slab morphology predicts a long-wavelength (~1500 km) dynamic subsidence of around 1–2 km (Figures 3.5 and 3.6). The locus of this dynamic subsidence migrates with the leading edge of subduction, and agrees well with the geometry of Solimões Formation deposits. The predicted amplitude of the dynamic subsidence from the flat-slab model is larger than what is observed (i.e., the depth of the basin), as mantle-flow models have a well-known tendency to overpredict amplitude due to uncertainties in some physical properties of Earth, such as the internal viscosity structure (Bertelloni & Gurnis 1997). The lateral position of the peak dynamic low, however, is better constrained – and this is well matched by the spatial distribution of sediment accumulation, which importantly extends well beyond the 300 km range of flexural subsidence due to the weight of the Andes (Figure 3.5).

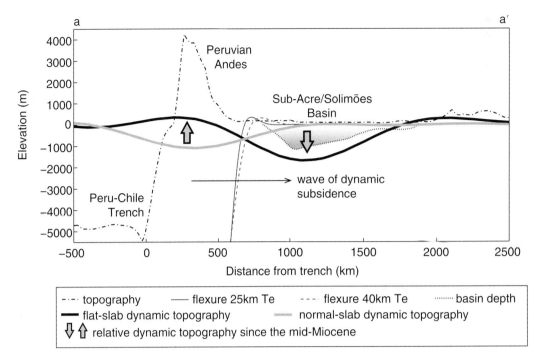

Figure 3.5 Profiles illustrating the different contributions from dynamic topography and flexure to the total topography across the Amazonian foreland. The thick profiles depict the impact of relative dynamic topography, given a transition from "normal" subduction (thick gray line) to flat-slab subduction (thick black line) (see the location of profile a–a' in Figure 3.4). This results in relative dynamic uplift beneath the Peruvian Andes but relative dynamic subsidence beneath the western Amazon Basin (distal foreland). The Sub-Acre/Solimões Basin (gray shaded region) is outside the influence of flexure driven by the mountain loading (thin gray lines), but does coincide with relative dynamic subsidence from the arrival of the flat slab. Two flexure profiles are shown for lithospheres of two different strengths or effective elastic thicknesses (Te). The depth of the basin (dotted black line) was determined from an isopach map (Latrubesse et al. 2010). Present-day topography (dot-dash line) was taken from ETOPO1 (Amante & Eakins 2009). *Source:* Eakin et al. (2014). Reproduced with permission of Elsevier.

Based on the preceding, the following model is proposed, in which both tectonic (e.g., mountain building, flexure) and dynamic processes played a role in the evolution of the present-day western Amazonian landscape and ecosystem. To elaborate, when the Pebas wetland began to form, subduction was likely steeper beneath Peru, the Andes had just started to grow and the first and oldest deposits of the Solimões Formation had been slowly laid down in the flexural and dynamic depressions adjacent to the orogen (Figure 3.6a,b). As flat-slab subduction took hold, plate coupling and tectonic deformation intensified, corresponding with a period of rapid uplift for the northern Andes (Jordan & Gardeweg 1989; Kroonenberg et al. 1990; Hoorn et al. 2010b). An increase in orogenic loads, erosional rates and sediment run-off resulted, as well as a further deepening and broadening of the flexural basins (Figure 3.6c,d). In addition, the longer-wavelength dynamic subsidence propagated further inland as the flat slab developed.

As uplift progressed and the Andes formed a continuous mountain belt, the drainage pathways to the Pacific and/or Caribbean were shut off, driving an eastward sediment flux (Hoorn et al. 1995, 2010b; Hungerbühler et al. 2002; Campbell et al. 2006). Easterly drainage would, however, have been partially blocked by structural highs such as the Purus Arch (Figures 3.4 and 3.6) (Figueiredo et al. 2009), trapping sediment in Amazonian lowland basins formed by the flexure and dynamic subsidence (Figure 3.6c,d). As the basins filled up, the paleoenvironmental conditions would have changed from a low-energy (mega-)wetland-type system to more energetic fluvial domains; this is corroborated by the depositional sequence of the Solimões Formation, which transitions from muddy to sandy deposits (Hoorn et al. 2010a; Figueiredo 2012).

Once the basins became full, the eastern rim (i.e., Purus Arch) may have been breached (Campbell et al. 2006), thus establishing the present-day connection between the Andes and the Atlantic Ocean of the Amazon River system (Figure 3.6e). Around the same time (~5 Ma) that deposition of the Solimões Formation in western Amazonia slowed (Latrubesse et al. 2010), the accumulation rate in the Amazon offshore fan took off (Dobson et al. 2001; Gorini et al. 2014), indicating a switch in the primary deposition of Andean-derived material from the Amazonian lowlands to the Atlantic Ocean, and a shift to the present-day configuration of Amazonia.

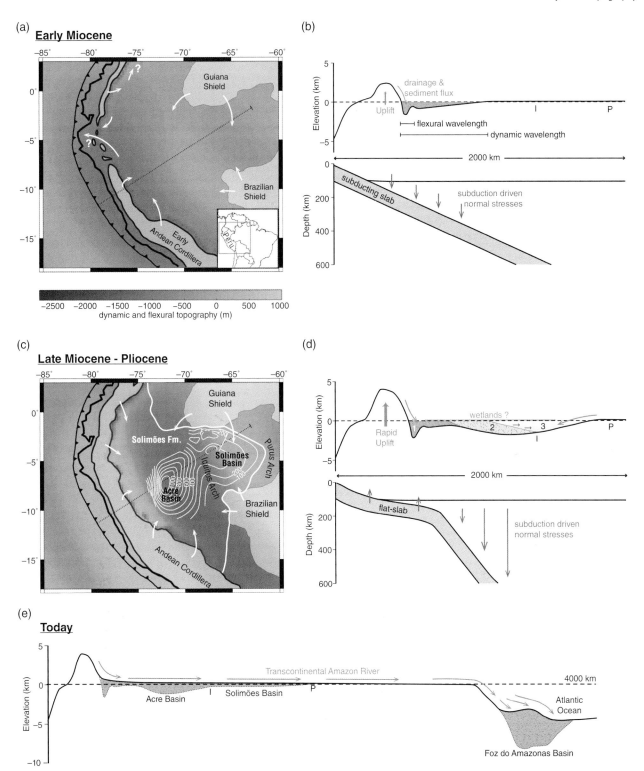

Figure 3.6 Schematic illustrations demonstrating how a change in subduction style could have driven the topographic and sedimentary evolution of western Amazonia since the Miocene. (a,c) Maps of the predicted topography from flexural and dynamic calculations both before (a) and on the arrival (c) of the flat slab. The scale bar for both maps is plotted below (a). White arrows represent drainage directions. The thin dotted lines in (a) and (c) represent the approximate locations of the cartoon cross-sections shown in (b) and (d), respectively. In (b) and (d), gray dotted regions indicate sedimentary infill, with region 1 indicating the oldest deposits and region 3 the youngest. The letters I and P mark the estimated locations of the Iquitos and Purus Arches, respectively. (e) Similar cross-section, extending from the trench to the Atlantic Ocean, showing the present-day configuration of drainage and the locations of the sedimentary deposits that preserve the record of landscape evolution hypothesized in (b) and (d). *Source:* Eakin et al. (2014). Reproduced with permission of Elsevier.

3.6 Conclusion

As exemplified by the case of Amazonia, considering dynamic topography alongside tectonic deformation is important for understanding landscape change over time – particularly long wavelength patterns and basin evolution in continental interiors. The temporal and cyclic nature of dynamic topography can have far-reaching consequences on everything from climate (Steinberger et al. 2015) to sea level (Conrad & Husson 2009; Spasojevic & Gurnis 2012) to drainage (Wegmann et al. 2007; Shephard et al. 2010), and thus has the power to affect ecosystems and environments through time. It is therefore possible that the conditions for life, evolution and the biosphere are connected to the mysterious dynamics of the deep mantle.

Acknowledgments

We thank Peter Molnar, Sean Willett and an anonymous reviewer for their comments and feedback on this chapter. We also thank the editors – Carina Hoorn, Alexandre Antonelli and Allison Perrigo – for their patience and support throughout this process.

References

Adam, C. & Vidal, V. (2010) Mantle flow drives the subsidence of oceanic plates. *Science* **328**, 83–5.

Amante, C. & Eakins, B.W. (2009) ETOPO1 1 Arc-Minute Global Relief Model: Procedures, Data Sources and Analysis. NOAA Technical Memorandum NESDIS NGDC-24, 19.

Barazangi, M. & Isacks, B.L. (1976) Spatial distribution of earthquakes and subduction of the Nazca plate beneath South America. *Geology* **4**, 686–692.

Bertelloni, C. & Gurnis, M. (1997) Cenozoic subsidence and uplift of continents from time-varying dynamic topography. *Geology* **25**, 735–738.

Billen, M.I. & Gurnis, M. (2001) A low viscosity wedge in subduction zones. *Earth and Planetary Science Letters* **193**, 227–236.

Billen, M.I., Gurnis, M. & Simons, M. (2003) Multiscale dynamics of the Tonga–Kermadec subduction zone. *Geophysical Journal International* **153**, 359–388.

Bissig, T., Ullrich, T.D., Tosdal, R.M. et al. (2008) The time-space distribution of Eocene to Miocene magmatism in the central Peruvian polymetallic province and its metallogenetic implications. *Journal of South American Earth Sciences* **26**, 16–35.

Boschi, L., Faccenna, C. & Becker, T.W. (2010) Mantle structure and dynamic topography in the Mediterranean Basin. *Geophysical Research Letters* **37**.

Braun, J. (2010) The many surface expressions of mantle dynamics. *Nature Geoscience* **3**, 825–833.

Burgess, P.M., Gurnis, M. & Moresi, L. (1997) Formation of sequences in the cratonic interior of North America by interaction between mantle, eustatic, and stratigraphic processes. *Geological Society of America Bulletin* **109**, 1515–1535.

Čadek, O. & Fleitout, L. (2003) Effect of lateral viscosity variations in the top 300 km on the geoid and dynamic topography. *Geophysical Journal International* **152**, 566–580.

Cahill, T. & Isacks, B.L. (1992) Seismicity and shape of the subducted Nazca Plate. *Journal of Geophysical Research* **97**, 17 503–17 529.

Campbell, K.E., Frailey, C.D. & Romero-Pittman, L. (2006) The Pan-Amazonian Ucayali Peneplain, late Neogene sedimentation in Amazonia, and the birth of the modern Amazon River system. *Palaeogeography, Palaeoclimatology, Palaeoecology* **239**, 166–219.

Carslaw, H.S. & Jaeger, J.C. (1959) *Conduction of Heat in Solids*, 2nd edn. Oxford: Clarendon Press.

Coira, B., Davidson, J., Mpodozis, C. & Ramos, V. (1982) Tectonic and magmatic evolution of the Andes of northern Argentina and Chile. *Earth-Science Reviews* **18**, 303–332.

Conrad, C.P. & Husson, L. (2009) Influence of dynamic topography on sea level and its rate of change. *Lithosphere* **1**, 110–120.

Dávila, F. & Lithgow-Bertelloni, C. (2013) Dynamic topography in South America. *Journal of South American Earth Sciences* **43**, 127–144.

DiCaprio, L., Gurnis, M. & Müller, R.D. (2009) Long-wavelength tilting of the Australian continent since the Late Cretaceous. *Earth and Planetary Science Letters* **278**, 175–185.

Dobson, D.M., Dickens, G.R. & Rea, D.K. (2001) Terrigenous sediment on Ceara Rise: a Cenozoic record of South American orogeny and erosion. *Palaeogeography, Palaeoclimatology, Palaeoecology* **165**, 215–229.

Eakin, C.M., Lithgow-Bertelloni, C. & Dávila, F.M. (2014) Influence of Peruvian flat-subduction dynamics on the evolution of western Amazonia. *Earth and Planetary Science Letters* **404**, 250–260.

Faccenna, C. & Becker, T.W. (2010) Shaping mobile belts by small-scale convection. *Nature* **465**, 602–605.

Figueiredo, J.J.P. (2012) Comment by J.P. Figueiredo & C. Hoorn on "Late Miocene sedimentary environments in south-western Amazonia (Solimões Formation; Brazil)" by Martin Gross, Werner E. Piller, Maria Ines Ramos, Jackson Douglas da Silva Paz. *Journal of South American Earth Sciences* **35**, 74–75.

Figueiredo, J., Hoorn, C., van der Ven, P. & Soares, E. (2009) Late Miocene onset of the Amazon River and the Amazon deep-sea fan: evidence from the Foz do Amazonas Basin. *Geology* **37**, 619–622.

Fischer, K.M., Ford, H.A., Abt, D.L. & Rychert, C.A. (2010) The Lithosphere–Asthenosphere boundary. *Annual Review of Earth and Planetary Sciences* **38**, 551–575.

Flament, N. (2014) Linking plate tectonics and mantle flow to Earth's topography. *Geology* **42**, 927–928.

Flament, N., Gurnis, M. & Muller, R.D. (2013) A review of observations and models of dynamic topography. *Lithosphere* **5**, 189–210.

Forte, A.M., Peltier, W.R. & Dziewonski, A.M. (1991) Inferences of mantle viscosity from tectonic plate velocities. *Geophysical Research Letters* **18**, 1747–1750.

Gérault, M., Husson, L., Miller, M.S. & Humphreys, E.D. (2015) Flat-slab subduction, topography, and mantle dynamics in southwestern Mexico. *Tectonics* **34**, 1892–1909.

Goes, S., Eakin, C.M. & Ritsema, J. (2013) Lithospheric cooling trends and deviations in oceanic PP–P and SS–S differential traveltimes. *Journal of Geophysical Research: Solid Earth* **118**, 996–1007.

Gorini, C., Haq, B.U., dos Reis, A.T. et al. (2014) Late Neogene sequence stratigraphic evolution of the Foz do Amazonas Basin, Brazil. *Terra Nova* **26**, 179–185.

Grand, S.P. (2002) Mantle shear-wave tomography and the fate of subducted slabs. *Philosophical Transactions. Series A, Mathematical, Physical, and Engineering Sciences* **360**, 2475–2491.

Gripp, A.E. & Gordon, R.G. (2002) Young tracks of hotspots and current plate velocities. *Geophysical Journal International* **150**, 321–361.

Gurnis, M. (1990) Bounds on global dynamic topography from Phanerozoic flooding of continental platforms. *Nature* **344**, 754–756.

Hager, B. (1984) Subducted slabs and the geoid: constraints on mantle rheology and flow. *Journal of Geophysical Research: Solid Earth* **89**, 6003–6015.

Hager, B.H. & Richards, M.A. (1989) Long-wavelength variations in Earth's geoid: physical models and dynamical implications. *Philosophical Transactions. Series A, Mathematical, Physical, and Engineering Sciences* **328**, 309–327.

Hager, B., Clayton, R., Richards, M. et al. (1985) Lower mantle heterogeneity, dynamic topography and the geoid. *Nature* **313**, 541–545.

Haskell, N.A. (1935) The motion of a viscous fluid under a surface load. *Journal of Applied Physics* **6**, 265–269.

Heine, C., Dietmar Müller, R., Steinberger, B. & Torsvik, T.H. (2008) Subsidence in intracontinental basins due to dynamic topography. *Physics of the Earth and Planetary Interiors* **171**, 252–264.

Hillier, J.K. (2010) Subsidence of "normal" seafloor: observations do indicate "flattening." *Journal of Geophysical Research* **115**, B03102.

Hirth, G. (2002) Laboratory constraints on the rheology of the upper mantle. *Reviews in Mineralogy and Geochemistry* **51**, 97–120.

Hoggard, M.J., White, N. & Al-Attar, D. (2016) Global dynamic topography observations reveal limited influence of large-scale mantle flow. *Nature Geoscience* **9**, 456–463.

Hoorn, C. (1993) Marine incursions and the influence of Andean tectonics on the Miocene depositional history of northwestern Amazonia: results of a palynostratigraphic study. *Palaeogeography, Palaeoclimatology, Palaeoecology* **105**, 267–309.

Hoorn, C. (1994) An environmental reconstruction of the palaeo-Amazon River system (Middle–Late Miocene, NW Amazonia). *Palaeogeography, Palaeoclimatology, Palaeoecology* **112**, 187–238.

Hoorn, C., Guerrero, J., Sarmiento, G. & Lorente, M. (1995) Andean tectonics as a cause for changing drainage patterns in Miocene northern South America. *Geology* **23**, 237–240.

Hoorn, C., Wesselingh, F.P., Hovikoski, J. & Guerrero, J. (2010a) The development of the Amazonian mega-wetland (Miocene; Brazil, Colombia, Peru, Bolivia). In: Hoorn, C. & Wesselingh, F.P. (eds.) *Amazonia, Landscape and Species Evolution: A Look into the Past.* Chichester: Wiley-Blackwell, pp. 123–142.

Hoorn, C., Wesselingh, F.P., Ter Steege, H. et al. (2010b) Amazonia through time: Andean uplift, climate change, landscape evolution, and biodiversity. *Science* **330**, 927–931.

Hungerbühler, D., Steinmann, M., Winkler, W. et al. (2002) Neogene stratigraphy and Andean geodynamics of southern Ecuador. *Earth-Science Reviews* **57**, 75–124.

Husson, L., Bernet, M., Guillot, S. et al. (2014) Dynamic ups and downs of the Himalaya. *Geology* **42**, 839–842.

Jordan, T.E. & Gardeweg, P.M. (1989) Tectonic evolution of the late Cenozoic central Andes (20°–33°S). In: Ben-Avraham, Z. (ed.) *The Evolution of the Pacific Ocean Margins.* New York: Oxford University Press, pp. 193–207.

Kaban, M.K., Schwintzer, P., Artemieva, I.M. & Mooney, W.D. (2003) Density of the continental roots: compositional and thermal contributions. *Earth and Planetary Science Letters* **209**, 53–69.

Karato, S. & Karki, B. (2001) Origin of lateral variation of seismic wave velocities and density in the deep mantle. *Journal of Geophysical Research* **106**, 21771–21783.

Kawakatsu, H., Kumar, P., Takei, Y. et al. (2009) Seismic evidence for sharp lithosphere-asthenosphere boundaries of oceanic plates. *Science* **324**, 499–502.

King, S. (1995) Models of mantle viscosity. In Ahrens, T.J. (ed.) *Mineral Physics & Crystallography: A Handbook of Physical Constants.* Washington, DC: American Geophysical Union, pp. 227–236.

King, S. & Hager, B. (1994) Subducted slabs and the geoid. 1. Numerical experiments with temperature-dependent viscosity. *Journal of Geophysical Research: Solid Earth* **99**, 19843–19852.

Korenaga, T. & Korenaga, J. (2008) Subsidence of normal oceanic lithosphere, apparent thermal expansivity, and seafloor flattening. *Earth and Planetary Science Letters* **268**, 41–51.

Kroonenberg, S.B., Bakker, J.G.M. & van der Wiel, A.M. (1990) Late Cenozoic uplift and paleogeography of the Colombian Andes: constraints on the development of high-andean biota. *Geologie en Mijnbouw* **69**, 279–290.

Latrubesse, E.M., Cozzuol, M., da Silva-Caminha, S.A.F. et al. (2010) The Late Miocene paleogeography of the Amazon Basin and the evolution of the Amazon River system. *Earth-Science Reviews* **99**, 99–124.

Lithgow-Bertelloni, C. & Gurnis, M. (1997) Cenozoic subsidence and uplift of continents from time-varying dynamic topography. *Geology* **25**, 735–738.

Lithgow-Bertelloni, C. & Richards, M. (1998) The dynamics of Cenozoic and Mesozoic plate motions. *Reviews of Geophysics* **36**, 27–78.

Lithgow-Bertelloni, C. & Silver, P. (1998) Dynamic topography, plate driving forces and the African superswell. *Nature* **395**, 269–272.

Liu, L. (2015) The ups and downs of North America: evaluating the role of mantle dynamic topography since the Mesozoic. *Reviews of Geophysics* **53**, 1022–1049.

Liu, L., Spasojević, S. & Gurnis, M. (2008) Reconstructing Farallon Plate subduction beneath North America back to the late Cretaceous. *Science* **322**, 934–938.

Mitrovica, J.X., Beaumont, C. & Jarvis, G.T. (1989) Tilting of continental interiors by the dynamical effects of subduction. *Tectonics* **8**, 1079–1094.

Molnar, P., England, P. & Martinod, J. (1993) Mantle dynamics, uplift of the Tibetan Plateau, and the Indian Monsoon. *Reviews of Geophysics* **31**, 357.

Molnar, P., England, P. & Jones, C. (2015) Mantle dynamics, isostasy, and the support of high terrain. *Journal of Geophysical Research: Solid Earth* **120**, 1932–1957.

Moresi, L. & Gurnis, M. (1996) Constraints on the lateral strength of slabs from three-dimensional dynamic flow models. *Earth and Planetary Science Letters* **138**, 15–28.

Moucha, R. & Forte, A.M. (2011) Changes in African topography driven by mantle convection. *Nature Geoscience* **4**, 707–712.

Moucha, R., Forte, A.M., Mitrovica, J.X. et al. (2008) Dynamic topography and long-term sea-level variations: there is no such thing as a stable continental platform. *Earth and Planetary Science Letters* **271**, 101–108.

Naif, S., Key, K., Constable, S. & Evans, R.L. (2013) Melt-rich channel observed at the lithosphere-asthenosphere boundary. *Nature* **495**, 356–359.

Naliboff, J.B., Lithgow-Bertelloni, C., Ruff, L.J. & de Koker, N. (2012) The effects of lithospheric thickness and density structure on Earth's stress field. *Geophysical Journal International* **188**, 1–17.

Panasyuk, S. & Hager, B. (2000) Models of isostatic and dynamic topography, geoid anomalies, and their uncertainties. *Journal of Geophysical Research: Solid Earth* **105**, 28199–28209.

Parsons, B. & Sclater, J.G. (1977) An analysis of the variation of ocean floor bathymetry and heat flow with age. *Journal of Geophysical Research* **82**, 803–827.

Potter, P.E. (1997) The Mesozoic and Cenozoic paleodrainage of South America: a natural history. *Journal of South American Earth Sciences* **10**, 331–344.

Pysklywec, R. & Mitrovica, J. (1998) Mantle flow mechanisms for the large-scale subsidence of continental interiors. *Geology* **26**, 687–690.

Ramos, V.A. & Folguera, A. (2009) *Andean Flat-Slab Subduction through Time. The Geological Society of London, Special Publications* **327**, pp. 31–54.

Ricard, Y. & Wuming, B. (1991) Inferring the viscosity and the 3-D density structure of the mantle from geoid, topography and plate velocities. *Geophysical Journal International* **105**, 561–571.

Ricard, Y., Vigny, C. & Froidevaux, C. (1989) Mantle heterogeneities, geoid, and plate motion: a Monte Carlo inversion. *Journal of Geophysical Research* **94**, 13739–13754.

Ricard, Y., Richards, M., Lithgow-Bertelloni, C. & Le Stunff, Y. (1993) A geodynamic model of mantle density heterogeneity. *Journal of Geophysical Research* **98**, 21895–21909.

Ritsema, J., Deuss, A., van Heijst, H.J. & Woodhouse, J.H. (2011) S40RTS: a degree-40 shear-velocity model for the mantle from new Rayleigh wave dispersion, teleseismic traveltime and normal-mode splitting function measurements. *Geophysical Journal International* **184**, 1223–1236.

Schmerr, N. (2012) The Gutenberg discontinuity: melt at the lithosphere-asthenosphere boundary. *Science* **335**, 1480–1483.

Shephard, G., Müller, R., Liu, L. & Gurnis, M. (2010) Miocene drainage reversal of the Amazon River driven by plate-mantle interaction. *Nature Geoscience* **3**, 870–875.

Siebert, L. & Simkin, T. (2002) *Volcanoes of the World: An Illustrated Catalog of Holocene Volcanoes and Their Eruptions.* Global Volcanism Program Digital Information Series, GVP-3. Washington, DC: Smithsonian Institution.

Simmons, N.A., Forte, A.M. & Grand, S.P. (2009) Joint seismic, geodynamic and mineral physical constraints on three-dimensional mantle heterogeneity: implications for the relative importance of thermal versus compositional heterogeneity. *Geophysical Journal International* **177**, 1284–1304.

Spasojevic, S. & Gurnis, M. (2012) Sea level and vertical motion of continents from dynamic earth models since the Late Cretaceous. *AAPG Bulletin* **96**, 2037–2064.

Steinberger, B. (2007) Effects of latent heat release at phase boundaries on flow in the Earth's mantle, phase boundary topography and dynamic topography at the Earth's surface. *Physics of the Earth and Planetary Interiors* **164**, 2–20.

Steinberger, B., Spakman, W., Japsen, P. & Torsvik, T.H. (2015) The key role of global solid-Earth processes in preconditioning Greenland's glaciation since the Pliocene. *Terra Nova* **27**, 1–8.

Stixrude, L. & Lithgow-Bertelloni, C. (2007) Influence of phase transformations on lateral heterogeneity and dynamics in Earth's mantle. *Earth and Planetary Science Letters* **263**, 45–55.

Wegmann, K.W., Zurek, B.D., Regalla, C.A. et al. (2007) Position of the Snake River watershed divide as an indicator of geodynamic processes in the greater Yellowstone region, western North America. *Geosphere* **3**, 272.

Zhong, S. & Davies, G. (1999) Effects of plate and slab viscosities on the geoid. *Earth and Planetary Science Letters* **170**, 487–496.

4

Mountain Relief, Climate and Surface Processes

Peter van der Beek

Institut des Sciences de la Terre (ISTerre), Grenoble Alps University, Grenoble, France

Abstract

The influence of climate on mountain-belt development, in terms of relief, rates and distribution of erosion, as well as the potential feedbacks on mountain building itself, has been heavily discussed within the geoscience community. Our theoretical understanding of the response of mountain belts to changes in climatic and tectonic forcing suggests that diagnostic signals of one or the other should exist, but we have not as yet been able to unambiguously extract such signals from the geological record. It has been suggested that the apparent global increase in sediment flux from mountain belts since the Plio–Pleistocene is a response to climatic cooling and increased climatic instability. Although recent work suggests that at least part of the signal may be intrinsic to the nature of the sedimentary record, independent compilations of thermochronology data, which record long-term exhumation of mountain belts, appear to show a similar increase in cooling and exhumation rates in the Pleistocene. Nonetheless, both the existence and the exact timing of such a general increase in mountain-belt erosion remain conjectural and require further, more detailed analysis. The potential link between tectonics, climate and the relief of mountain belts has attracted less attention. However, recent work has shown that climate – in particular, the efficiency of glacial erosion processes – can profoundly impact the relief of mountain belts. Several studies have reported significant increases in mountain-belt relief in the Quaternary, often explicitly related to glacial erosion and the carving of glacial valleys. Such relief development should be linked to the inferred increase in erosion rates in – and sediment flux from – mountain belts, and may be the primary explanation of this increase. This chapter reviews these ideas and the main contributions over the last several decades.

Keywords: *erosion, sediment flux, tectonic-climate interactions, glaciation, Pliocene–Quaternary, climate change, thermochronology, cosmogenic nuclides*

4.1 Introduction

The evolution of Earth's climate on geological time scales is intimately linked to solid-Earth processes, as the ultimate sources and sinks for atmospheric CO_2 are volcanic outgassing and sequestration in carbonaceous rocks, respectively (Berner et al. 1983). The temperature-dependence of the kinetics of silicate weathering, the main geological process that removes the greenhouse gas CO_2 from the atmosphere, provides a powerful negative feedback that stabilizes Earth's surface temperature on geological time scales (Kump et al. 2000). Long-term climatic cooling during the Cenozoic, culminating in the Quaternary glaciations (Zachos et al. 2001), has been linked to mountain uplift, particularly of the Himalaya, which would have led to increased erosion and weathering rates, and concomitant transfer of carbon from the atmosphere to the lithosphere (Figure 4.1) (Raymo & Ruddiman 1992).

At about the same time that Raymo & Ruddiman (1992) suggested the existence of a link between Cenozoic climate change and mountain uplift, the evidence for widespread late-Cenozoic uplift of mountain belts was questioned (Molnar & England 1990). Molnar & England (1990) pointed out that most of the "classical" geomorphological inferences of surface uplift in fact only implied enhanced erosion and incision of mountain belts. They argued that both of these could be climatically controlled,

Mountains, Climate and Biodiversity, First Edition. Edited by Carina Hoorn, Allison Perrigo and Alexandre Antonelli.
© 2018 John Wiley & Sons Ltd. Published 2018 by John Wiley & Sons Ltd.
Companion website: www.wiley.com/go/hoorn\mountains,climateandbiodiversity

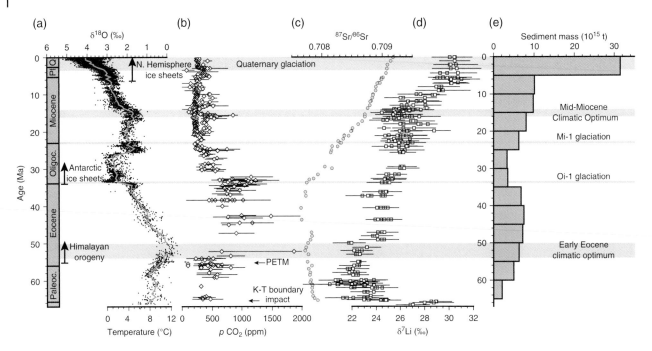

Figure 4.1 Cenozoic evolution of climate and erosional/weathering fluxes to the global ocean. (a) Deep-sea oxygen-isotope record. Dots are individual data, gray line is running average, bottom scale shows average seawater temperature calculated for an ice-free ocean. *Source:* Data from Zachos et al. (2001). (b) Estimates of atmospheric CO_2 partial pressure (open diamonds with error bars; scale on bottom) reconstructed from different marine and terrestrial proxies, with associated uncertainties. *Source:* Data from Beerling & Royer (2011). (c,d) Seawater Sr- (shaded circles) and Li-isotope (open squares) records, tracking the continental weathering flux to the global ocean. Li-isotope data are reported as δ^7Li (scale on bottom) with respect to NIST standard and include 2σ error bars. *Source:* Data from Misra & Froelich (2012). (e) Histogram of sediment mass deposited in the global ocean over 5 My intervals. Major tectonic and climatic events during the Cenozoic are indicated. K-T, Cretaceous–Tertiary; PETM, Paleocene–Eocene Thermal Maximum. *Source:* Modified from Zhang et al. (2001); data from Hay et al. (1988).

and moreover suggested a feedback mechanism by which uplift of mountain peaks could be the isostatic response to climatically enhanced erosion of mountain belts (see Chapter 2).

Molnar & England's (1990) model posed two hypotheses that have sparked considerable controversy and debate over the last 25 years. First, they suggested a climatic control on erosion rates; that is, the change to an (on average) colder and temporally more variable climate during the late Cenozoic would have enhanced continental erosion rates (Figure 4.1) (Zhang et al. 2001). The observational basis for this inference has been strongly contested in recent years (e.g., Schumer & Jerolmack 2009; Willenbring & von Blanckenburg 2010). Second, their mechanism for uplift of mountain peaks requires that the relief[1] of mountain belts has increased during the late Cenozoic. As the amount of isostatic uplift is by definition smaller than the average surface lowering by

erosion, it can only outstrip erosion on mountain peaks (leading to peak uplift) if the erosion of mountain belts is concentrated in the valleys, thereby increasing relief (Figure 4.2; see Chapter 2).

Going beyond the simple isostatic response to potentially enhanced erosion during the late Cenozoic proposed by Molnar & England (1990), Dahlen & Suppe (1988) and Beaumont et al. (1992) were the first to propose that erosion can also modify the tectonic structure and evolution of a mountain belt. The basis for this revolutionary idea lies in the description of orogenic belts as "critically tapered wedges"; that is, wedges of crustal material that are at the verge of brittle failure everywhere (Dahlen 1990). Because the state of stress within such a wedge depends on its thickness, which can be modified by surface processes, it follows that the wedge geometry and internal deformation should respond to variations in erosional efficiency, and therefore, potentially, climate: increased erosion will lead to a narrower wedge, while decreased erosion will lead to a wider wedge (Dahlen & Suppe 1988; Beaumont et al. 1992). Theoretically, therefore, active mountain belts should respond to a climatically modulated change in erosional efficiency by changing their geometry and

---

1 The notion of "relief" has many different definitions, and even very different meanings in different languages. Champagnac et al. (2014) provide a detailed review of the different definitions. Here, I will use "relief" to indicate any measure of topographic (altitudinal) variability within a defined area.

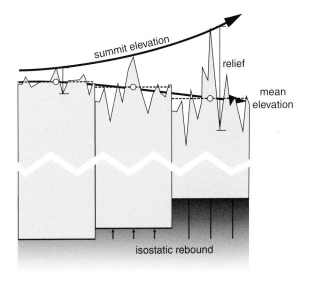

Figure 4.2 Uplift of mountain summits by the isostatic response to erosion, following the model of Molnar & England (1990). Note that summit uplift can only take place if relief increases, because erosion is concentrated in the valleys and is associated with a decrease in the average elevation of the region. *Source:* Adapted from Burbank & Anderson (2011).

kinematics in a predictable manner, as shown by analytical (e.g., Whipple & Meade 2006), numerical (e.g., Stolar et al. 2006) and experimental (e.g., Graveleau et al. 2012) models (Figure 4.3).

Another potential feedback between surface erosion and the tectonic evolution of a mountain belt occurs through the effect of erosion and exhumation on the thermal structure of the crust: rapid exhumation brings hot rocks to the surface, heating the crust through advection (England & Thompson 1984; Koons 1987). Since the ductile strength of rock is strongly dependent on temperature, this heat advection leads to local weakening of the orogenic crust, thereby focusing deformation (Koons 1987; Batt & Braun 1997). In this way, an efficient feedback loop may develop, in which rapid surface erosion leads to heating, weakening and rapid inflow of crustal material, in what has been termed a "tectonic aneurism" (Zeitler et al. 2001; Koons et al. 2002). These potentially important couplings between climate and tectonics, through topography, exhumation and surface processes, appear conceptually sound. However, field evidence for such coupling has remained both ambiguous and controversial (Whipple 2009).

Such issues are central to our understanding of Earth's coupled tectonic–climate system (Figure 4.4). In this chapter, I will review our current understanding of the potential couplings between climate, erosion and the relief of mountain belts, from both a theoretical and an observational point of view.

4.2 Relationships Between Climate, Erosion and Relief: Models and Concepts

The erosion of mountain belts involves a variety of processes. First, bedrock has to be transformed into transportable mobile material through physical, chemical and biological processes collectively known as weathering. This mobile material is subsequently transported by hillslope, fluvial or glacial processes that are driven by gravitational energy. Over the last 20 years, the geomorphic community has aimed to develop appropriate rules to describe these processes and their dependence on topography (relief), lithology, vegetation and climate (see, for instance, Dietrich et al. 2003; van der Beek 2013).

It has been shown for nearly half a century that a strong correlation exists between mountain-belt relief and erosion rates (Ahnert 1970; Pinet & Souriau 1988; Summerfield & Hulton 1994). The physical basis for this correlation lies in the dependence of the efficiency of erosional transport processes, both on hillslopes and in fluvial channels, on some metric of relief, such as slope or curvature (Dietrich et al. 2003). This correlation breaks down, however, in active high-relief landscapes with high erosion rates (Burbank et al. 1996; Montgomery & Brandon 2002), where bedrock landsliding becomes the major process of slope erosion, leading to threshold hillslopes that are maintained at their limit of stability. Thus, the relationship between relief and erosion rate is nonlinear (Figure 4.5); this relationship is understood theoretically (see Dietrich et al. 2003) and has been observed in numerous field studies (e.g., Wittmann et al. 2007; Ouimet et al. 2009; DiBiase et al. 2010).

The effect of climate on erosion rates is much less clear. First of all, the specific meaning of the generic term "climate" needs to be clarified. In many instances, the parameter that has been studied is precipitation rate. The simple and widely used "stream-power" law for river incision (e.g., Whipple & Tucker 1999; Whipple 2004) states that incision, i, is a nonlinear function of contributing drainage area, A (a proxy for discharge), and local river gradient, S:

$$i = KA^m S^n \tag{4.1}$$

where K is a lumped parameter that includes precipitation rate and bedrock erodibility, among other things, and m and n are positive exponents, typically taken to be ≤1. This law would suggest a simple linear relationship between precipitation (embedded in K) and incision rates, whereas natural data generally show no significant correlation between mean annual precipitation and erosion rates in global analyses of drainage basins

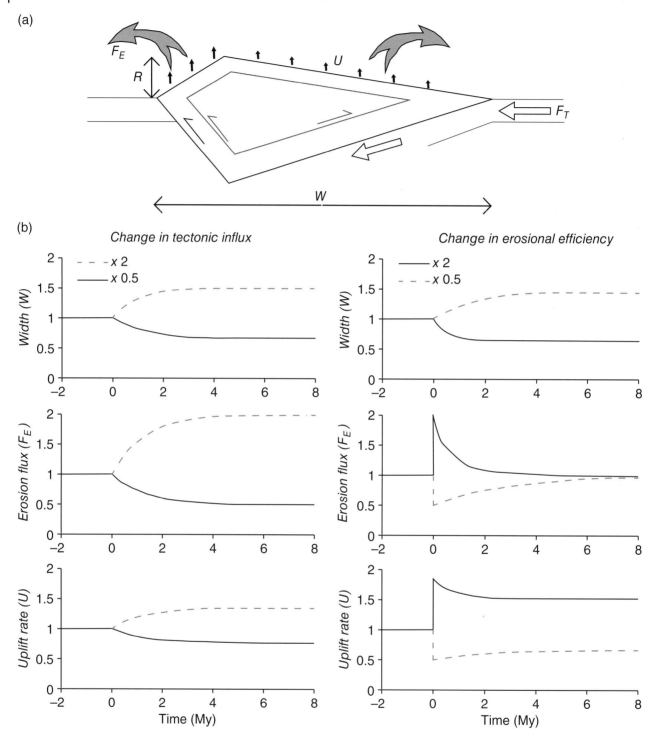

Figure 4.3 Evolution of a doubly vergent critically tapered wedge in response to changes in tectonics or climate. (a) Conceptual model of an orogenic wedge developing in response to tectonic accretion with a tectonic influx, F_T. The wedge is characterized by its width, W, height (relief), R, and rock-uplift rate, U. Erosion of the wedge feeds an erosional outflux, F_E. (b) Responses of the wedge width, W, erosional flux, F_E, and uplift rate, U, to step changes in tectonic influx (convergence rate) or climatically modulated erosional efficiency (normalized to a value of 1 before the change): left panels show response to a twofold increase (dashed gray lines) or decrease (continuous black lines) in tectonic influx; right panels to a twofold increase (continuous black lines) or decrease (dashed gray lines) in erosional efficiency. Results are predicted by an analytical model of a doubly vergent critical wedge coupled to a stream-power erosion law. *Source:* Adapted from Whipple & Meade (2006). Reproduced with permission of Elsevier.

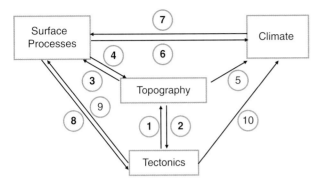

Figure 4.4 Schematic showing the different couplings and feedback loops inferred to control Earth's tectonic–erosion–climate system. The arrows and numbered circles indicate the different proposed couplings; bold numbers are couplings discussed in this chapter. (1) Tectonics controls topography, because crustal thickening due to tectonic accretion leads to surface uplift and an increase in relief. (2) Topography feeds back into tectonic deformation, through its influence on the crustal stress field. (3) Erosion rates are strongly dependent on relief. (4) Surface processes also modify topographic relief. (5) Topography directly controls climate, through its effect on atmospheric and oceanic circulation patterns. (6) Erosionally controlled continental silicate weathering and organic carbon burial act as sinks for atmospheric CO_2, thereby modulating climate. (7) Climate may influence erosion rates through the amount and temporal distribution of precipitation, and through the influence of glaciations. (8) A direct link between erosion and tectonics occurs through the thermal effect of exhumation and associated rock advection. (9) Tectonics may also directly affect erosion through rock fracturing. (10) Deep-Earth processes may affect climate directly through volcanic and metamorphic outgassing of CO_2. Closed circuits of arrows constitute potential feedback loops.

(Summerfield & Hulton 1994; Portenga & Bierman 2011). There are at least two reasons for this. First, the stream-power law represents an overly simplified view of the processes of bedrock river incision; it takes into account neither the role of sediment flux in rivers, nor that of thresholds for incision (Whipple 2004; Lague 2014). Including the threshold term in the river-incision law leads to the prediction that it is not so much the mean annual precipitation rate as the temporal variability of precipitation and discharge (i.e., the "storminess" of the climate) that should dictate river incision rates (Lague 2014); there is some field evidence to support this prediction (Dadson et al. 2003). Second, there is no theoretical model for the control of precipitation rate on hillslope erosion, which would mostly express itself through the effects of soil pore pressure on erodibility. Recent empirical evidence has shown that increasing vegetation cover can limit or even invert the potential control of precipitation on erosion rates (Torres Acosta et al. 2015).

A potentially more pronounced effect of climate on erosion is the transition to periodic glaciations, which

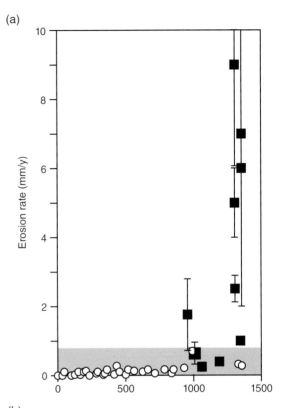

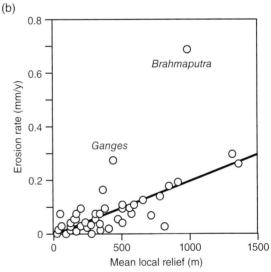

Figure 4.5 (a) Plot of erosion rate as a function of mean local relief (calculated in a 10 km-radius window) from sediment-flux data from rivers draining mostly tectonically inactive areas (open circles) and from diverse data (mostly thermochronology, but also cosmogenic data and landslide mapping) from tectonically active areas, such as the South Island of New Zealand, Taiwan and various parts of the Himalaya, including the very rapidly eroding syntaxes (black squares with error bars). The combined data clearly show the nonlinear relationship between relief and erosion rate, with mean local relief saturating at ~1500 m. (b) Magnified view of river sediment-flux data, showing the linear relationship for erosion rates <1 mm/y (corresponding to gray shaded box in (a)). *Source:* Adapted Montgomery & Brandon (2002); sediment-flux data from Summerfield and Hulton (1994).

have affected numerous high- and mid-latitude mountain belts since the Plio–Pleistocene. The relative effectiveness of glacial versus fluvial erosion has been the subject of several studies, with conflicting findings. Theoretical models predict that both main mechanisms of glacial erosion – abrasion and quarrying – are controlled by the basal sliding rate of a glacier (Hallet 1979, 1996), which is directly linked to ice flux and the presence of subglacial meltwater (Herman et al. 2011, 2015; Egholm et al. 2012; Iverson 2012). Present-day sediment fluxes from glaciated catchments appear to be significantly higher than those from nonglaciated catchments with otherwise similar characteristics (Hallet et al. 1996). However, as practically all glaciers worldwide are currently retreating, the modern erosion rates estimated from these fluxes could be severely affected by short-term transient effects related to such retreat (Koppes & Hallet 2002). Over longer time scales, glacial and fluvial erosion rates appear rather similar (Koppes & Montgomery 2009). However, as the landscape adapts to the change in erosional processes affecting it during glaciation and deglaciation, transient erosion rates may become significantly higher (Koppes & Montgomery 2009). Finally, the strong dependence of glacial erosion on basal sliding rate and the presence of subglacial meltwater means that cold-based glaciers, which are frozen to their beds and flow entirely by internal deformation, are essentially non-erosive. This protective property is confirmed by data on both short- and long-term erosion rates from cold-based glaciers (Thomson et al. 2010; Godon et al. 2013; Koppes et al. 2015). Thus, rather than being simply more or less effective than fluvial erosion, glacial erosion appears strongly selective: capable of rapidly carving out deep valleys where the ice flux is large, but not affecting the landscape strongly where ice is thin and/or frozen to the bedrock (e.g., Kessler et al. 2008; Valla et al. 2011a; Hall et al. 2013). This selective erosion leaves a strong imprint on the landscape, significantly steepening it locally and potentially leading to higher postglacial erosion rates (e.g., Norton et al. 2010).

Climate change may thus potentially affect erosion rates through the different processes described in the previous paragraphs; that is, erosion rates may be expected to increase as the regional climate becomes stormier, colder or generally more variable. In actively uplifting mountain belts, the erosional response to such changes should be transient, as the relief (and, in the case of a critically tapered orogenic wedge, the wedge width) will adjust to the new erosional environment and reach a new steady state in which the tectonic uplift rates set the erosion rates, typically with a response time of the order of 1 My (Whipple 2001). Such behavior is clearly demonstrated in both numerical and analog models (Figures 4.3 and 4.6) (Bonnet & Crave 2003; Whipple 2009). The response of tectonically inactive, "decaying" mountain belts has been less studied, and in those contexts climate change could have a longer-term impact on erosion rates. However, few data are available for the assessment of such impacts, as the erosion rates are generally too low to be sufficiently resolved with the methods typically used to assess them on geological time scales (Section 4.3).

Climatic changes influence not only erosion rates and sediment flux from a mountain belt, but also relief of the belt. The stream-power law for fluvial incision predicts that (fluvial) relief should decrease as the climate becomes more erosive (Whipple et al. 1999): in steady state, the law can be rearranged to read:

$$S = \left(\frac{i}{K} \right)^{\frac{1}{n}} A^{\frac{m}{n}} \qquad (4.2)$$

which implies that as K increases (i.e., increasing erosional efficiency), the stream gradient required to generate the same rate of incision declines (all else being equal). Experimental models of erosion show exactly this behavior (Figure 4.6) (Bonnet & Crave 2003); both theory and experiments therefore point to a negative coupling between precipitation and relief. However, a global analysis of mountain-belt relief as a function of various tectonic and climatic parameters does not show a clear correlation between the two (Champagnac et al. 2012).

The effect of glaciation on relief is twofold. First, the strong dependence of glacial erosion rates on basal sliding velocities, and therefore ice flux, implies that glacial erosion rates should peak at the equilibrium-line altitude (ELA): the elevation that separates the accumulation and ablation regions of glaciers and at which the ice flux is necessarily maximal. Therefore, glacial erosion is very efficient at this elevation, and tends to reduce the proportion of topography above the ELA, operating as a "glacial buzzsaw" (Brozović et al. 1997; Mitchell & Montgomery 2006). As a result, the large-scale relief of glaciated mountains is limited, and large portions of glaciated landscapes are situated around the mean Quaternary ELA (Egholm et al. 2009). Second, funneling of ice masses, combined with the effect of meltwater on glacial erosion rates, focuses glacial erosion in large valleys, tending to deepen them and to increase relief (Kessler et al. 2008; Herman et al. 2011). This dual action of glacial erosion leads to a positive correlation of the small-scale relief of mountain belts with latitude (taken as a simple proxy for the importance of glacial processes), while maximum elevation decreases with latitude (Champagnac et al. 2012).

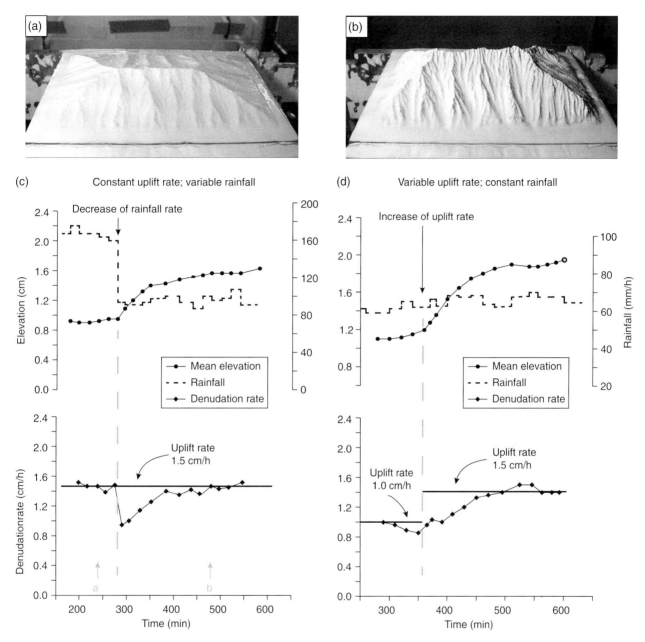

Figure 4.6 Experimental landscape response to changes in tectonic or climatic forcing. (a,b) Oblique views of an experimental landscape in a variable-rainfall experiment (experimental landscapes are 20 cm in length): (a) for a high rainfall rate of ~180 mm/h; (b) for a low rainfall rate of ~90 mm/h. (c,d) Evolution of the topography (mean elevation), rainfall, uplift and denudation rates for (c) an experiment with a constant uplift rate and a decrease in rainfall rate from ~180 to ~90 mm/h at 300 min and (d) an experiment with constant rainfall rate and an increase in uplift rate from 1.0 to 1.5 cm/h at 350 min. Both models start at topographic and erosional steady state in response to initial conditions, and are run until steady state is reattained in response to modified conditions. Times corresponding to (a) and (b) are indicated on (c). *Source:* Modified from Bonnet & Crave (2003).

4.3 Measuring (Changes in) Erosion Rates in Mountain Belts

In the previous section, I synthesized our current conceptual and theoretical understanding of the links between climate, relief and erosion rates in mountain belts. But what do we actually know about erosion rates and how they have evolved through time? Different types of data have been used to attempt to constrain the erosional history of mountain belts.

As rivers carry the erosional products of mountain belts to the sedimentary basins where they will be deposited, the sediment flux within them provides a first-order global overview of present-day erosion rates in mountain belts and their correlation with potential climatic or tectonic control parameters (Milliman & Syvitski 1992; Summerfield & Hulton 1994). In a study that was some 20 years ahead of its time, Adams (1980) used a comparison of sediment-flux measurements and geodetic uplift rates in the South Island of New Zealand to demonstrate steady state between present-day uplift and erosion. However, extrapolating measurements of modern sediment flux to erosion rates is fraught with uncertainties. First, sediment-flux data are, in general, only available for the suspended (and dissolved) sediment load; the proportion of bedload in the total flux is the subject of educated guesswork. Moreover, sediment fluxes commonly exhibit extreme variability with discharge, which is not in all cases sufficiently constrained. Finally, and perhaps most importantly, present-day sediment fluxes only provide a snapshot of modern erosion rates, which are moreover strongly influenced by anthropogenic activity, in particular deforestation, agriculture and damming of rivers (Syvitski et al. 2005).

An alternative approach to the problem is to assess variations in the deposited volume of sediments through time, either in natural or artificial lakes or offshore deltas on intermediate time and space scales (Goodbred & Kuehl 2000; Hinderer 2001) or in sedimentary basins on geological time scales. The latter studies have fairly consistently shown increasing sediment deposition rates up to the present day, on both global (Hay et al. 1988; Zhang et al. 2001) and regional (Métivier et al. 1999; Kuhlemann et al. 2002) scales. For instance, Kuhlemann et al. (2002) showed that deposited sediment volumes in sedimentary basins around the European Alps imply a two- to threefold increase in erosion rates since ~5 Ma. The cause of this inferred increase has been a source of significant speculation, considering that it occurred within a mountain belt where tectonics is waning. However, estimates of sediment volumes in sedimentary basins come with their own set of limitations and complications, among which are the problems of temporal resolution and completeness of the record. The geological reconstruction of sediment volumes can be done based on well or seismic data, or more indirectly from compiled isopach maps (i.e., maps of the thickness of sedimentary rocks of the same age). The former have the best temporal resolution but are inherently one-dimensional (1D) in nature, and extrapolation to the entire basin is nontrivial. The influential study by Zhang et al. (2001), which suggested a global increase in sedimentation rates since 2–4 Ma, was, for instance, based on 1D well data. 2D or (ideally) 3D seismic data provide for better spatial coverage, but require models for correlating seismic facies to geological units, and for

transforming seismic travel times to depths. Moreover, the vast majority of seismic data are proprietary and not readily available for scientific study. An example of the consequences of this situation is provided by the studies of sediment fluxes from Asia by Métivier et al. (1999) and Clift (2006). These two authors studied different subsets of data, made available by different companies, and came to very different conclusions: whereas Métivier et al. (1999) argued for continuously increasing fluxes since the onset of the India–Asia collision in the early Cenozoic, Clift (2006) instead suggested that a peak in sediment fluxes was reached during the middle Miocene, after which fluxes diminished, before peaking again in the Quaternary. Finally, several authors have questioned the reality of increasing sediment fluxes towards the present day (Schumer & Jerolmack 2009; Willenbring & von Blanckenburg 2010), arguing that because of the stochastic nature of sediment deposition, sediment accumulation rates depend on the measurement-time interval (known as the "Sadler effect"), which tends to be smaller for younger time periods. These authors therefore suggest that the observed increase in accumulation rates (and, therefore, inferred erosion rates) towards the present is largely an artifact of preservation.

An independent method of reconstructing erosion rates through geological time relies on the use of thermochronological techniques. Thermochronology, like geochronology, exploits naturally occurring nuclear reactions that transform unstable parent isotopes into stable daughter isotopes in minerals. In contrast to geochronology, however, thermochronology uses systems that have closure temperatures (above which the daughter isotope is mobile and removed from the mineral lattice) that are well below the formation temperature of the mineral or rock sampled. As such, thermochronology measures the timing of cooling of rock samples below the closure temperature as they are exhumed to the surface by tectonics and/or erosion. A detailed background on thermochronological systems and approaches is provided in Chapter 5. The most commonly used thermochronological systems for studies of erosion and relief development in mountain belts are based on the decay of U-isotopes through either α-emission (the (U-Th)/He system) or spontaneous fission (producing fission tracks) in apatite or zircon. Although a single system analyzed in a single sample will only provide a single cooling age that can be turned into an integrated average exhumation rate (see Chapter 5), several techniques allow the derivation of more complete cooling and exhumation histories. These approaches include applying multiple thermochronological systems on a single sample, using multiple samples with a known spatial relationship to one another (in particular, by sampling elevation transects), conducting inverse modeling of cooling paths by using additional information such as track-length measurements in fission-track analysis or $^4He/^3He$ outgassing profiles in (U-Th)/He thermochronology, or a combination of the three (see, for instance, Metcalf et al. 2009; Valla et al. 2012). An inherent complexity in thermochronology is that it provides information in a thermal reference frame (i.e., cooling histories); translating this into exhumation histories requires knowledge of the thermal structure of the upper crust, which is affected by (tectonically and/or erosionally driven) rock advection, as well as topography (e.g., Braun et al. 2006). Several models and approaches have been proposed for the consistent derivation of exhumation rates from thermochronological data sets, which take these effects into account to different degrees (e.g., Braun et al. 2006, 2012; Reiners & Brandon 2006; Safran et al. 2006; Willett & Brandon 2013; Fox et al. 2014). Using an approach that combines multiple-system thermochronology data and the geographical relationships between samples with a 1D thermal model of the crust (Fox et al. 2014), Herman et al. (2013) inverted a global thermochronology database to infer that mountain-erosion rates have increased over the last 6 My, with a peak during the Quaternary period. This finding provides support for the earlier interpretation of increased erosion towards the present day based on global sedimentation rates. Although Herman et al. (2013) observed this increase at all latitudes, it appears most pronounced in glaciated mountain ranges, pointing towards the role of glaciation in increasing mountain erosion rates.

Finally, on shorter (millennial) time scales, cosmogenic nuclide analysis is increasingly used to map erosion rates. As with thermochronological analyses, the basis of cosmogenic nuclide studies is discussed in Chapter 5, and I will not describe them in detail here. In brief, however, the approach is based on the generation of (either stable or radioactive) nuclides by the interaction of cosmic radiation with target elements in rocks, within a few meters of the Earth's surface. The method therefore allows either dating of the time of exposure of rocks at the surface or estimation of the rate at which they traveled through the surface-production layer (of a few meters' thickness), in which cosmic radiation interacts with the lithosphere (i.e., the surface erosion rate). Cosmogenic nuclide analysis can be performed either in situ on rock outcrops or on river sediments in order to obtain an estimate of catchment-averaged erosion rates (e.g., von Blanckenburg 2006; Portenga & Bierman 2011). The rates are inherently averaged over a time scale on the order of thousands of years, depending upon the erosion rate (higher erosion rates being coupled to shorter averaging times). Catchment-averaged erosion rates have been used in several studies to explore the controls on millennial-time scale erosion, both regionally (e.g.,

Wittmann et al. 2007; DiBiase et al. 2010) and globally (Portenga & Bierman 2011). Combining cosmogenic and thermochronological analyses on samples from the same catchments would allow a direct comparison of millennial and million-year time-scale erosion rates, but studies that have done so remain relatively rare (e.g., Glotzbach et al. 2013; Abrahami et al. 2016). Where it has been done, the results generally show somewhat higher millennial average erosion rates in comparison to the long-term rates, and a spatial distribution of erosion rates that can be quite dissimilar.

4.4 Reconstructing Relief Change in Mountain Belts

The preceding discussion on reconstructing erosion rates in mountain belts through geological time has been limited to either local measurements (for in situ thermochronology and cosmogenic-nuclide analysis) or catchment-averaged erosion rates (for detrital thermochronology and cosmogenic-nuclide analysis, as well as sediment flux-based approaches). It is equally important to understand spatial variations in erosion rates in mountain belts, which can be linked to spatially varying tectonic activity (rock-uplift rates) and/or to changing relief. As relief changes inherently require spatial variations in erosion rates (see Figure 4.2), approaches to reconstructing relief changes are thus based on mapping these variations in erosion rates. Changes in relief also imply that the studied landscape is in a transient state. An extensive overview of studies reporting relief changes in mountain belts has recently been provided by Champagnac et al. (2014).

The most direct approach to reconstructing paleorelief in mountain belts, and thereby relief changes, relies on the identification of "relict landscapes" or "paleosurfaces" in a transient landscape. Such relict landscapes consist of parts of the present-day landscape that have not yet responded to a change in tectonics, base level or erosional dynamics that has affected other, better connected parts of the landscape. For instance, Small & Anderson (1998) used widely occurring "summit flats" in Laramide mountain ranges of the western USA to reconstruct the paleorelief that must have existed before incision of deep valleys, and showed that relief has grown at rates of ~100 m/My over the last ~3 My in these ranges. They also showed that the isostatic response to this relief increase would amount to several tens of meters, approximately equal to the total erosion of the summit flats over this time, as constrained by cosmogenic nuclides. Similarly, Steer et al. (2012) used the well-known high-elevation, low-relief surfaces of Norway to constrain the amount of glacial fjord incision. On a larger scale, Clark et al. (2006) used relict surfaces on the south-eastern Tibetan plateau to reconstruct the paleorelief prior to incision of the major rivers and, by assuming that these relict landscapes were originally at low elevations, the uplift pattern of this region. The formation mechanism and age of many of these high-elevation, low-relief surfaces remain debated, however. While some authors have argued they are truly relict landscape remnants that predate landscape rejuvenation by focused valley incision (Clark et al. 2006; van der Beek et al. 2009; Hall et al. 2013), others have interpreted such low-relief surfaces as the result of bimodal erosion, formed by rapid valley lowering alongside lateral erosion and relief reduction on high-elevation parts of the landscape (Brozović et al. 1997; van der Beek & Bourbon 2008; Steer et al. 2012). Numerical models by Anderson (2002) and Egholm et al. (2015) show how periglacial processes can be particularly efficient in creating such high-elevation, low-relief surfaces in a time span of a few million years, and that such surfaces do not require that the landscape has been at or near base level for a prolonged period of time.

While the recognition of relict landscapes often depends on geomorphological interpretation of topographic patterns from now widely available digital topography data, direct field observations can in some cases be used to provide quantitative estimates of the timing and amount of relief change. In particular, volcanic flows fossilize paleolandscapes; mapping and dating them allows landscape changes, such as valley incision, to be tracked through time (Bishop et al. 1985; Thouret et al. 2007). Cave deposits record paleo water-table levels and therefore the elevation of valley floors in the past; dating such deposits and mapping their relationship to the present-day valley floors thus also allows valley incision to be tracked through time (Stock et al. 2004; Haeuselmann et al. 2007; Polyak et al. 2008).

A variant on the morphologic approach uses the "geophysical relief" of a catchment or mountain belt, together with the sediment-flux record, to reconstruct paleorelief. Although different definitions exist, the geophysical relief is generally defined as the relief within a spatial window of a characteristic size (typically of the order of a few kilometers). Assuming that relief has increased over time, the (compaction-corrected) sediment volumes encountered in sedimentary basins surrounding an eroding mountain belt can be preferentially stacked back on to the lowest parts of the landscape, in order to model paleorelief. Using this approach, Champagnac et al. (2007) showed that the increase in sediment flux from the Alps over the last few million years (see Section 4.3) could be entirely explained by valley deepening, while Steer et al. (2012) showed that the volume of Quaternary sediments in the basins off the coast of Norway is too large to have been derived from fjord incision alone, and required additional erosion of the high-elevation, low-relief surfaces, implying that these are young features.

Spatial variations across in-situ thermochronological ages potentially contain information about relief development, because they record the spatial variations in erosion and exhumation rates that are inherent to relief change. Moreover, they can contain information about the paleothermal structure of the crust, as near-surface crustal isotherms are not purely horizontal, but to some degree reflect the topography (Stüwe et al. 1994; Mancktelow & Grasemann 1997; Braun 2002b; Ehlers & Farley 2003; Braun et al. 2006; Safran et al. 2006). Spatial variations in thermochronological ages should relate to the crustal temperature structure (and, by inference, the relief) at the time of closure of the studied system. However, this information is generally rather subtle, and high-resolution data obtained through a dedicated sampling scheme, together with advanced numerical modeling, are required in order to extract it (Valla et al. 2010, 2011b; Braun et al. 2012). In a pioneer study, House et al. (1998) used apatite (U-Th)/He data, collected within a limited elevation range along a transect across the main valleys incising the Sierra Nevada of California, to show that the ages of valley samples were significantly older than those of interfluve samples (~70 vs. ~40 Ma), implying that these data record perturbation of the closure isotherm imposed by relief that was greater than today, several tens of millions of years ago. Braun (2002a) subsequently applied spectral analysis to quantify and better resolve the relief change (reduction, in this case) required by these data; his results corroborated the earlier, more qualitative inferences by House et al. (1998). Subsequent analyses of spatial variations in thermochronological ages in order to constrain paleorelief have relied on thermokinematic modeling of the thermal structure of the crust (Braun 2003; Braun et al. 2012). Schildgen et al. (2009) used such a model to show that spatial variations in apatite and zircon (U-Th)/He ages from the Ocaña Canyon in southern Peru required major canyon incision into the western Cordillera of the Andes since 8–11 Ma. Using a similar approach, Glotzbach et al. (2011b) showed that relief in the Mont-Blanc massif of the western European Alps should have increased significantly in the last 1 My; they attributed this increase to glacial valley incision since the mid-Pleistocene climate transition. These thermokinematic modeling studies are generally based on the assumption that the planform drainage pattern has remained stable during the episode of relief change. Although this assumption appears reasonable in many cases, there are regions – in particular, the western North American Cordillera – where thermochronology data provide evidence for significant lateral drainage-divide migration associated with Quaternary glacial erosion (Ehlers et al. 2006; Simon-Labric et al. 2014).

Modeling this situation is more complex and has been much less explored (Stüwe & Hintermüller 2000; Olen et al. 2012).

In the last 5 years, the unique potential of ^{4}He/^{3}He thermochronology to record very low-temperature (<70 °C) cooling histories, and thereby document relief development (Shuster et al. 2005), has been increasingly put to use. The advantage of this technique is that it provides an independently inferred cooling history from individual samples, thereby potentially allowing spatial variations in the onset timing and importance of relief changes to be recorded. Schildgen et al. (2010) used this technique to confirm earlier numerical modeling predictions (Schildgen et al. 2009) on canyon incision in the western Andes of Peru, but also to infer that the canyons evolved through headward incision (i.e., canyon incision started downstream and migrated upstream). Similarly, Shuster et al. (2011) showed that deep glacial valleys in Fiordland (South Island of New Zealand) were carved through headward incision since ~2 Ma, and Valla et al. (2011a) showed that the Rhone Valley in Switzerland was glacially deepened by 1.0–1.5 km during the last 1 My. On the other hand, Flowers & Farley (2012) showed that parts of the western Grand Canyon could be as old as ~70 Ma.

The short-term or present-day trends in relief development can be investigated using detrital thermochronology and/or cosmogenic nuclide analysis. Using detrital thermochronology in this manner requires prior knowledge of the spatial patterns in thermochronological ages across the studied catchments (i.e., an extensive in situ thermochronology database). The detrital data can then be used as a tracer of where in the catchment erosion is focused (Stock & Montgomery 1996; Stock et al. 2006; Vermeesch 2007; Tranel et al. 2011; Glotzbach et al. 2013). Stock et al. (2006), for instance, compared detrital and bedrock (U-Th)/He age patterns to show that nonglaciated catchments from the Sierra Nevada, California are eroding at spatially constant rates, whereas sediment from adjacent glacially influenced catchments are preferentially derived from lower elevations, implying relief increase. Although these methods are promising, they do depend on accurate knowledge not only of the distribution of thermochronological ages throughout the analyzed catchments, but also of the distribution and quality of their host minerals (Tranel et al. 2011); the generally imprecise knowledge of these factors renders the interpretation of detrital thermochronology age distributions equivocal in terms of relief development.

Detrital cosmogenic nuclide analysis can be used to map the spatial variation of erosion rates throughout a mountain belt (e.g., Wittmann et al. 2007). Delunel et al. (2010) mapped erosion rates throughout the Ecrins-Pelvoux massif in the French western Alps using this technique, and demonstrated a positive correlation between average catchment erosion rate and elevation, implying a Holocene decrease in relief at the massif scale – a trend they attributed to the increasing efficiency of frost-cracking processes at higher elevations. At a smaller (catchment) scale, local bedrock erosion rates from in-situ cosmogenic nuclide analysis can be compared to catchment-wide erosion rates from detrital cosmogenic data to infer the trend in relief change (Meyer et al. 2010). The same caveats apply to this approach as to that of comparing in situ and detrital thermochronology data sets. Moreover, temporal storage of sediments on millennial time scales within the catchments may also influence the signal, as shown by Delunel et al. (2013).

4.5 Discussion: Is There a Climatic Control on Mountain-Belt Erosion and Relief?

Despite intense research since the start of the 21st century, the question of whether climate, and climate change, exerts a significant control on erosion rates remains subject to controversy, as exemplified by a recent couple of "debate articles" in the journal *Terra Nova* (Herman & Champagnac 2016; Willenbring & Jerolmack 2016). These articles highlight how a focus on different data sets can lead to opposing conclusions: whereas Herman & Champagnac (2016) use the thermochronological record to argue for an overall increase in erosion rates (most clearly expressed in glaciated mountain belts), Willenbring & Jerolmack (2016) focus on resolution issues in both the sedimentary (the Sadler effect) and the thermochronology records to dismiss such changes, and use geochemical proxies (^{10}Be/^{9}Be, δ^7Li) to argue for stable continental weathering fluxes to the oceans over the last 10 My.

At the largest temporal scales, simple mass-budget arguments require that tectonics drive erosion rates, as pointed out by Willenbring & Jerolmack (2016): in the absence of a tectonic flux replenishing the mass lost by erosion, topography, relief and erosion rates can only diminish through time. This first-order tectonic control is clear in the distribution of present-day erosional mass fluxes (Milliman & Syvitski 1992; Summerfield & Hulton 1994). Therefore, any climatic control on erosion rates has to be transient: erosion rates may increase or decrease temporarily while the landscape (relief) adjusts to the new climatic context (Figures 4.3 and 4.6). The question therefore becomes that of the time scale at which transience may persist; that is, the "response time" of a mountain belt to changing climatic conditions.

While essentially no data exist to answer this question, theoretical arguments suggest such a response time to be at most a few million years in tectonically active environments, and to be inversely correlated to erosional efficiency (Whipple 2001; Whipple & Meade 2006). In slowly eroding, tectonically inactive environments, in contrast, transient landscapes may persist for tens to hundreds of millions of years (Baldwin et al. 2003; Bishop 2007), providing scope for long-term conservation of climatic imprints on erosion rates. However, by their nature, these regions are characterized by very slow erosion rates, and will thus not contribute significantly to global sediment fluxes. Moreover, any potential climatically controlled changes in erosion rates in such settings are likely to be below the resolution of standard thermochronological methods. From the discussion in Section 4.2, it follows that the climatic changes most capable of inducing changes in erosion rates and relief involve the onset of glaciation in mountain belts. I will therefore focus the remainder of this discussion on the effects of glaciation.

The role of glaciation in relief development is now relatively well understood. The dual effect of glacial erosion on relief outlined in Section 4.2 (i.e., increasing small-scale relief due to localized glacial valley incision coexisting with decreasing large-scale relief due to the "glacial buzzsaw" effect) leads to significant trends in these measures with latitude (Champagnac et al. 2012). Moreover, many mid- to high-latitude mountain ranges show clear indications of increasing relief during the Quaternary (Champagnac et al. 2014). Such an increase in relief has been most clearly documented – and linked to glacial valley incision since the mid-Pleistocene – in the European Alps (Champagnac et al. 2007; Haeuselmann et al. 2007; Glotzbach et al. 2011b; Valla, et al. 2011a) and the North American Rocky Mountains (Small & Anderson 1998). Interestingly, however, this effect is less clear in tectonically very active glaciated mountain ranges, such as the Himalaya or the Southern Alps of New Zealand. In the latter case, an increase in glacial relief has been demonstrated in Fiordland, where overall uplift and exhumation rates are moderate (House et al. 2002; Shuster et al. 2011), but not further north, where exhumation rates are very high (Herman & Braun 2008; Herman et al. 2010). Apparently, high rock-uplift rates in such settings efficiently counteract glacial erosion patterns, and the constant reshaping of topography during interglacial times does not allow cumulative effects of glacial erosion to develop during subsequent glaciations. This contrasts with the situation in regions characterized by more moderate rock uplift, where repeated glaciations may strengthen the imprint of glacial erosion on the landscape.

The preceding discussion leads to the question of the precise role glaciation plays in the evolution of relief and erosion rates: does it increase the overall erosion rates (i.e., the sediment flux) of a mountain belt, or does it only affect relief by redistributing erosion rates, concentrating erosion in the valleys? The analysis by Herman et al. (2013) does not answer this question, because the inversion technique employed is inherently 1D and interprets the thermochronology data in terms only of exhumation rates, not of relief change. In the European Alps, where this issue has been studied in most detail, it has been shown that thermochronology data record significant relief increases (Glotzbach et al. 2011b; Valla et al. 2011a, 2012), focused on areas where the 1D inversion of those data suggests apparent exhumation rates have increased most strongly (Fox et al. 2015). This coincidence suggests that at least part of the increase in erosion rates inferred from the in-situ thermochronology data would in fact correspond to an increase in relief, and leaves open the question of how erosion rates have evolved overall. Answering that question requires data on regional erosion rates; that is, sediment flux or detrital thermochronology records (see Section 4.3). Sediment-flux data for basins surrounding the Alps do appear to show an increase in Quaternary times (Kuhlemann et al. 2002; Leroux et al. 2016), subject of course to the importance one ascribes to the Sadler effect. Such a recent change in overall erosion rates would, however, be beyond the resolution of currently employed detrital thermochronology methods (Glotzbach et al. 2011a).

Erosion rates and relief are intrinsically coupled, however, through the scaling between them (see Section 4.2 and Figure 4.5). There may thus be an indirect coupling of glaciation and erosion rates: as relief increases through glacial erosion, erosion rates should also increase, even after the glaciers retreat (Champagnac et al. 2014). Such a "landscape-memory" or "glacial-inheritance" effect has been suggested to partially control millennial-time scale erosion rates (as measured by cosmogenic nuclides) in the European Alps (Norton et al. 2010; Glotzbach et al. 2013) and elsewhere (Abrahami et al. 2016). A potential complication arises, however, because erosion rates derived from cosmogenic nuclides may be overestimated in formerly glaciated terrain when steady-state cosmogenic nuclide concentrations have not been attained since glaciation (Wittmann et al. 2007; Glotzbach et al. 2014). The potential long-term effects of glacial inheritance on erosion rates clearly depend on the response (or recovery) time of landscapes; that is, how long it takes them to reattain steady-state relief after glaciation.

Resolving these questions will require increased spatial and temporal resolution in the record of the variability of erosion rates in (formerly) glaciated mountain ranges. High spatial resolution is required to map out in detail the spatial variations in erosion rates that are associated with relief development, and densely sampled thermochronologic data sets are needed (Valla et al. 2011b). High temporal resolution, of the order of the time span of glacial–interglacial cycles, is required to resolve the question of whether it is overall climatic cooling (intensity of glaciation) or increased climatic variability (leading to landscape transience) that has led to increased Quaternary relief and erosion rates. New thermochronological methods that are sensitive to near-surface isotherms, such as ^{4}He/^{3}He (Shuster et al. 2005) or OSL (Herman et al. 2010; King et al. 2016) thermochronometry, provide prospects for attaining the required resolution.

A final question concerns the coupling with tectonics invoked in the introduction to this chapter, both through isostatic rebound, potentially leading to the uplift of mountain peaks (the original "chicken and egg" problem of Molnar & England 1990), and through the tectonic response to changes in erosional efficiency (Dahlen & Suppe 1988; Beaumont et al. 1992; Zeitler et al. 2001). Although the mechanics of both problems are well understood, unequivocal field evidence for them has remained elusive. Uplift of mountain peaks through the isostatic response to relief increase is limited by two effects: the flexural rigidity of the lithosphere, which strongly reduces the potential isostatic response to valley incision (Montgomery 1994), and the erosion rate of the peaks themselves, which may be low, but is generally not zero. Studies that have attempted to assess the mechanism in specific field sites have concluded that the isostatic effect on mountain-peak uplift is probably small: of the order of a few tens of meters (Small & Anderson 1998; van der Beek & Bourbon 2008). Likewise, attempts have been made to demonstrate a tectonic response of critically tapered orogens to climatically induced changes in erosional efficiency in the European Alps (Willett et al. 2006) and the Chugach/St. Elias Mountains of southern Alaska (Berger et al. 2008). In both cases, it was argued that increased erosional efficiency – in the case of the Alps, due to Messinian climate change, and in the case of Alaska, due to the onset of glaciation – led to shrinkage of the range and abandonment of the outermost thrust faults. However, in both cases the chronology of thrust-fault activation and abandonment has been challenged, weakening the inference of a climatic driver (Rosenberg & Berger 2009; Meigs 2014). More recently, it has been proposed that the Alborz Mountains of northern Iran show a similar history – in this case, not directly related to climate change, but rather to a kilometer-scale base-level drop induced by desiccation of the Caspian Sea to the north

of the range (Ballato et al. 2015). The "tectonic aneurism" model, which links tectonic deformation to rapid exhumation and efficient erosion through their thermal effects (see Section 4.1), has similarly been challenged, at least in one of its original type examples: the Namche Barwa syntaxis of the eastern Himalaya. The model (in its original formulation; Zeitler et al. 2001) relies on the onset of rapid erosion by river capture and the establishment of a major through-going trunk river. However, in the case of the eastern Himalaya, establishment of such drainage (in this case, the connection of the Yarlung and Brahmaputra rivers) has recently been shown to have taken place >15 Ma (Lang & Huntington 2014; Bracciali et al. 2015), while the onset of rapid uplift appears much younger – maybe as young as ~3 Ma (Wang et al. 2014). This growing temporal gap between the inferred trigger and response casts doubt on the invoked mechanism. Thus, there is currently no consensus on the recognition of climatically driven changes in the tectonic structure of natural mountain belts; progress in this field will require detailed, high-resolution analyses of the timing of climatic, erosional and tectonic changes in mountain belts.

4.6 Conclusion

The potential role of climate in modulating erosion rates, relief and the tectonic development of mountain ranges has been a focus of interest and controversy for the last 25 years, and remains a first-order problem in the study of the coupled Earth system. On the longest time scales, the mass budget of mountain belts requires that erosion rates are ultimately driven by tectonic forcing; climatic controls on erosion rates are thus, by necessity, transient. Those climatic changes most capable of inducing changes in erosion rates and relief involve the onset of glaciation in mountain belts, and significant increases in both erosion rate and relief in response to glaciation have been reported from several mountain belts. However, it is not clear to what extent glaciation leads to an overall increase in erosion rate, and thus sediment flux from mountain belts; nor is it clear whether glaciation mostly redistributes erosion – focusing it in the valleys and thereby increasing relief, but not necessarily increasing erosional fluxes. Theoretically, climatically driven variations in erosion rates should influence the tectonic development of mountain belts through changes in both the crustal temperature and stress fields, but unequivocal field evidence for such a control has remained elusive. Resolving these issues will require increased spatial and temporal resolution in the record of the variability of erosion rates in mountain ranges worldwide.

References

Abrahami, R., van der Beek, P.A., Huyghe, P. et al. (2016) Decoupling of long-term exhumation and short-term erosion rates in the Sikkim Himalaya. *Earth and Planetary Science Letters* **433**, 76–88.

Adams, J. (1980) Contemporary uplift and erosion of the Southern Alps, New Zealand. *Geological Society of America Bulletin* **91**(1 Pt. II), 1–114.

Ahnert, F. (1970) Functional relationships between denudation, relief, and uplift in large, mid-latitude drainage basins. *American Journal of Science* **268**, 243–263.

Anderson, R.S. (2002) Modeling the tor-dotted crests, bedrock edges, and parabolic profiles of high alpine surfaces of the Wind River Range, Wyoming. *Geomorphology* **46**, 35–58.

Baldwin, J.A., Whipple, K.X. & Tucker, G.E. (2003) Implications of the shear stress river incision model for the timescale of postorogenic decay of topography. *Journal of Geophysical Research* **108**, 2158.

Ballato, P., Landgraf, A., Schildgen, T.F. et al. (2015) The growth of a mountain belt forced by base-level fall: Tectonics and surface processes during the evolution of the Alborz Mountains, N Iran. *Earth and Planetary Science Letters* **425**, 204–218.

Batt, G.E. & Braun, J. (1997) On the thermomechanical evolution of compressional orogens. *Geophysical Journal International* **128**, 364–382.

Beaumont, C., Fullsack, P. & Hamilton, J. (1992) Erosional control of active compressional orogens. In: McClay, K.R. (ed.) *Thrust Tectonics*. Philadelphia, PA: Springer, pp. 1–18.

Beerling, D.J. & Royer, D.L. (2011) Convergent Cenozoic CO_2 history. *Nature Geoscience* **4**, 418–420.

Berger, A.L., Gulick, S.P.S., Spotila, J.A. et al. (2008) Quaternary tectonic response to intensified glacial erosion in an orogenic wedge. *Nature Geoscience* **1**, 793–799.

Berner, R.A., Lasaga, A.C. & Garrels, R.M. (1983) The carbonate-silicate geochemical cycle and its effect on atmospheric carbon dioxide over the past 100 million years. *American Journal of Science* **283**, 641–683.

Bishop, P. (2007) Long-term landscape evolution: linking tectonics and surface processes. *Earth Surface Processes and Landforms* **32**, 329–365.

Bishop, P., Young, R.W. & McDougall, I. (1985) Stream profile change and longterm landscape evolution: early Miocene and modern rivers of the east Australian highland crest, central New South Wales, Australia. *Journal of Geology* **93**, 455–474.

Bonnet, S. & Crave, A. (2003) Landscape response to climate change: insights from experimental modeling and implications for tectonic versus climatic uplift of topography. *Geology* **31**, 123–126.

Bracciali, L., Najman, Y., Parrish, R.R. et al. (2015) The Brahmaputra tale of tectonics and erosion: early Miocene river capture in the Eastern Himalaya. *Earth and Planetary Science Letters* **415**, 25–37.

Braun, J. (2002a) Estimating exhumation rate and relief evolution by spectral analysis of age–elevation datasets. *Terra Nova* **14**, 210–214.

Braun, J. (2002b) Quantifying the effect of recent relief changes on age–elevation relationships. *Earth and Planetary Science Letters* **200**, 331–343.

Braun, J. (2003) Pecube: a new finite-element code to solve the 3D heat transport equation including the effects of a time-varying, finite amplitude surface topography. *Computers and Geosciences* **29**, 787–794.

Braun, J., van der Beek, P.A. & Batt, G.E. (2006) *Quantitative Thermochronology*. Cambridge: Cambridge University Press.

Braun, J., van der Beek, P.A., Valla, P. et al. (2012) Quantifying rates of landscape evolution and tectonic processes by thermochronology and numerical modeling of crustal heat transport using PECUBE. *Tectonophysics* **524–525**, 1–28.

Brozović, N., Burbank, D.W. & Meigs, A.J. (1997) Climatic limits on landscape development in the Northwestern Himalaya. *Science* **276**, 571–574.

Burbank, D.W. & Anderson, R.S. (2011) *Tectonic Geomorphology*, 2nd edn. Chichester: Wiley-Blackwell.

Burbank, D.W., Leland, J., Fielding, E. et al. (1996) Bedrock incision, rock uplift and threshold hillslopes in the northwestern Himalayas. *Nature* **379**, 505–510.

Champagnac, J.D., Molnar, P., Anderson, R.S. et al. (2007) Quaternary erosion-induced isostatic rebound in the western Alps. *Geology* **35**, 195–198.

Champagnac, J.-D., Molnar, P., Sue, C. & Herman, F. (2012) Tectonics, climate, and mountain topography. *Journal of Geophysical Research* **117**, B02403.

Champagnac, J.-D., Valla, P.G. & Herman, F. (2014) Late-Cenozoic relief evolution under evolving climate: a review. *Tectonophysics* **614**, 44–65.

Clark, M.K., Royden, L.H., Whipple, K.X. et al. (2006) Use of a regional, relict landscape to measure vertical deformation of the eastern Tibetan Plateau. *Journal of Geophysical Research* **111**, F03002.

Clift, P.D. (2006) Controls on the erosion of Cenozoic Asia and the flux of clastic sediment to the ocean. *Earth and Planetary Science Letters* **241**, 571–580.

Dadson, S.J., Hovius, N., Chen, H. et al. (2003) Links between erosion, runoff variability and seismicity in the Taiwan orogen. *Nature* **426**, 648–651.

Dahlen, F.A. (1990) Critical taper model of fold-and-thrust belts and accretionary wedges. *Annual Review of Earth and Planetary Sciences* **18**, 55–99.

Dahlen, F.A. & Suppe, J. (1988) Mechanics, growth, and erosion of mountain belts. *Geological Society of America Special Papers* **218**, 161–178.

Delunel, R., van der Beek, P.A., Carcaillet, J. et al. (2010) Frost-cracking control on catchment denudation rates: Insights from in situ produced ^{10}Be concentrations in stream sediments (Ecrins–Pelvoux massif, French Western Alps). *Earth and Planetary Science Letters* **293**, 72–83.

Delunel, R., van der Beek, P.A., Bourlès, D.L. et al. (2013) Transient sediment supply in a high-altitude Alpine environment evidenced through a ^{10}Be budget of the Etages catchment (French Western Alps). *Earth Surface Processes and Landforms* **39**, 890–899.

DiBiase, R.A., Whipple, K.X., Heimsath, A.M. & Ouimet, W.B. (2010) Landscape form and millennial erosion rates in the San Gabriel Mountains, CA. *Earth and Planetary Science Letters* **289**, 134–144.

Dietrich, W.E., Bellugi, D.G. & Sklar, L.S. (2003) Geomorphic transport laws for predicting landscape form and dynamics. In: *Prediction in Geomorphology*. Geophysical Monograph Series. Washington, DC: American Geophysical Union, pp. 103–132.

Egholm, D.L., Nielsen, S.B., Pedersen, V.K. & Lesemann, J.-E. (2009) Glacial effects limiting mountain height. *Nature* **460**, 884–887.

Egholm, D.L., Pedersen, V.K., Knudsen, M.F. & Larsen, N.K. (2012) Coupling the flow of ice, water, and sediment in a glacial landscape evolution model. *Geomorphology* **141–142**, 47–66.

Egholm, D.L., Knudsen, M.F., Larsen, N.K. et al. (2015) The periglacial engine of mountain erosion – Part 2: Modelling large-scale landscape evolution. *Earth Surface Dynamics* **3**, 463–482.

Ehlers, T.A. & Farley, K.A. (2003) Apatite (U–Th)/He thermochronometry: methods and applications to problems in tectonic and surface processes. *Earth and Planetary Science Letters* **206**, 1–14.

Ehlers, T.A., Farley, K.A., Rusmore, M.E. & Woodsworth, G.J. (2006) Apatite (U-Th)/He signal of large-magnitude accelerated glacial erosion, southwest British Columbia. *Geology* **34**, 765–768.

England, P.C. & Thompson, A.B. (1984) Pressure–temperature–time paths of regional metamorphism I. Heat transfer during the evolution of regions of thickened continental crust. *Journal of Petrology* **25**, 894–928.

Flowers, R.M. & Farley, K.A. (2012) Apatite ^{4}He/^{3}He and (U-Th)/He evidence for an ancient Grand Canyon. *Science* **338**, 1616–1619.

Fox, M., Herman, F., Willett, S.D. & May, D.A. (2014) A linear inversion method to infer exhumation rates in space and time from thermochronometric data. *Earth Surface Dynamics* **2**, 47–65.

Fox, M., Herman, F., Kissling, E. & Willett, S.D. (2015) Rapid exhumation in the Western Alps driven by slab detachment and glacial erosion. *Geology* **43**, 379–382.

Glotzbach, C., Bernet, M. & van der Beek, P.A. (2011a) Detrital thermochronology records changing source areas and steady exhumation in the Western European Alps. *Geology* **39**, 239–242.

Glotzbach, C., van der Beek, P.A. & Spiegel, C. (2011b) Episodic exhumation and relief growth in the Mont Blanc massif, Western Alps from numerical modelling of thermochronology data. *Earth and Planetary Science Letters* **304**, 417–430.

Glotzbach, C., van der Beek, P.A., Carcaillet, J. & Delunel, R. (2013) Deciphering the driving forces of erosion rates on millennial to million-year timescales in glacially impacted landscapes: an example from the Western Alps. *Journal of Geophysical Research: Earth Surface* **118**, 1491–1515.

Glotzbach, C., Röttjer, M., Hampel, A. et al. (2014) Quantifying the impact of former glaciation on catchment-wide denudation rates derived from cosmogenic ^{10}Be. *Terra Nova* **26**, 186–194.

Godon, C., Mugnier, J.L., Fallourd, R. et al. (2013) The Bossons glacier protects Europe's summit from erosion. *Earth and Planetary Science Letters* **375**, 135–147.

Goodbred, S.L. Jr. & Kuehl, S.A. (2000) Enormous Ganges–Brahmaputra sediment discharge during strengthened early Holocene monsoon. *Geology* **28**, 1083–1086.

Graveleau, F., Malavieille, J. & Dominguez, S. (2012) Experimental modelling of orogenic wedges: a review. *Tectonophysics* **538–540**, 1–66.

Haeuselmann, P., Granger, D.E. & Jeannin, P.Y. (2007) Abrupt glacial valley incision at 0.8 Ma dated from cave deposits in Switzerland. *Geology* **35**, 143–146.

Hall, A.M., Ebert, K., Kleman, J. et al. (2013) Selective glacial erosion on the Norwegian passive margin. *Geology* **41**, 1203–1206.

Hallet, B. (1979) A theoretical model of glacial abrasion. *Journal of Glaciology* **23**, 39–50.

Hallet, B. (1996) Glacial quarrying: a simple theoretical model. *Annals of Glaciology* **22**, 1–8.

Hallet, B., Hunter, L. & Bogen, J. (1996) Rates of erosion and sediment evacuation by glaciers: a review of field data and their implications. *Global and Planetary Change* **12**, 213–235.

Hay, W.W., Sloan, J.L. & Wold, C.N. (1988) Mass/age distribution and composition of sediments on the ocean floor and the global rate of sediment subduction. *Journal of Geophysical Research* **93**, 14 933–14 940.

Herman, F. & Braun, J. (2008) Evolution of the glacial landscape of the Southern Alps of New Zealand: insights from a glacial erosion model. *Journal of Geophysical Research* **113**, F02009.

Herman, F. & Champagnac, J.-D. (2016) Plio-Pleistocene increase of erosion rates in mountain belts in response to climate change. *Terra Nova* **28**, 2–10.

Herman, F., Rhodes, E.J., Braun, J. & Heiniger, L. (2010) Uniform erosion rates and relief amplitude during glacial cycles in the Southern Alps of New Zealand, as revealed from OSL-thermochronology. *Earth and Planetary Science Letters* **297**, 183–189.

Herman, F., Beaud, F., Champagnac, J.D. et al. (2011) Glacial hydrology and erosion patterns: a mechanism for carving glacial valleys. *Earth and Planetary Science Letters* **310**, 498–508.

Herman, F., Seward, D., Valla, P.G. et al. (2013) Worldwide acceleration of mountain erosion under a cooling climate. *Nature* **504**, 423–426.

Herman, F., Beyssac, O., Brughelli, M. et al. (2015) Erosion by an Alpine glacier. *Science* **350**, 193–195.

Hinderer, M. (2001) Late Quaternary denudation of the Alps, valley and lake fillings and modern river loads. *Geodinamica Acta* **14**, 231–263.

House, M.A., Wernicke, B.P. & Farley, K.A. (1998) Dating topography of the Sierra Nevada, California, using apatite (U-Th)/He ages. *Nature* **396**, 66–69.

House, M.A., Gurnis, M., Kamp, P.J.J. & Sutherland, R. (2002) Uplift in the Fiordland region, New Zealand: implications for incipient subduction. *Science* **297**, 2038–2041.

Iverson, N.R. (2012) A theory of glacial quarrying for landscape evolution models. *Geology* **40**, 679–682.

Kessler, M.A., Anderson, R.S. & Briner, J.P. (2008) Fjord insertion into continental margins driven by topographic steering of ice. *Nature Geoscience* **1**, 365–369.

King, G.E., Herman, F., Lambert, R. et al. (2016) Multi-OSL-thermochronometry of feldspar. *Quaternary Geochronology* **33**, 76–87.

Koons, P.O. (1987) Some thermal and mechanical consequences of rapid uplift: an example from the Southern Alps. *Earth and Planetary Science Letters* **86**, 307–319.

Koons, P.O., Zeitler, P.K. & Chamberlain, C.P. (2002) Mechanical links between erosion and metamorphism in Nanga Parbat, Pakistan Himalaya. *American Journal of Science* **302**, 749–773.

Koppes, M.N. & Hallet, B. (2002) Influence of rapid glacial retreat on the rate of erosion by tidewater glaciers. *Geology* **30**, 47–50.

Koppes, M.N. & Montgomery, D.R. (2009) The relative efficacy of fluvial and glacial erosion over modern to orogenic timescales. *Nature Geoscience* **2**, 644–647.

Koppes, M., Hallet, B., Rignot, E. et al. (2015) Observed latitudinal variations in erosion as a function of glacier dynamics. *Nature* **526**, 100–103.

Kuhlemann, J., Frisch, W., Székely, B. et al. (2002) Post-collisional sediment budget history of the Alps: tectonic versus climatic control. *International Journal of Earth Sciences* **91**, 818–837.

Kump, L.R., Brantley, S.L. & Arthur, M.A. (2000) Chemical weathering, atmospheric CO_2, and climate. *Annual Review of Earth and Planetary Sciences* **28**, 611–667.

Lague, D. (2014) The stream power river incision model: evidence, theory and beyond. *Earth Surface Processes and Landforms* **39**, 38–69.

Lang, K.A. & Huntington, K.W. (2014) Antecedence of the Yarlung–Siang–Brahmaputra River, eastern Himalaya. *Earth and Planetary Science Letters* **397**, 145–158.

Leroux, E., Rabineau, M., Aslanian, D. et al. (2016) High-resolution evolution of terrigenous sediment yields in the Provence Basin during the last 6 Ma: relation with climate and tectonics. *Basin Research* **29**, 305–339.

Mancktelow, N.S. & Grasemann, B. (1997) Time-dependent effects of heat advection and topography on cooling histories during erosion. *Tectonophysics* **270**, 167–195.

Meigs, A. (2014) Limited Climate Control of the Chugach/St. Elias Thrust Wedge in Southern Alaska Demonstrated by Orogenic Widening during Pliocene to Quaternary Climate Change. *Geophysical Research Abstracts* **16**, EGU2014–9549.

Metcalf, J.R., Fitzgerald, P.G., Baldwin, S.L. & Muñoz, J.A. (2009) Thermochronology of a convergent orogen: constraints on the timing of thrust faulting and subsequent exhumation of the Maladeta Pluton in the Central Pyrenean Axial Zone. *Earth and Planetary Science Letters* **287**, 488–503.

Meyer, H., Hetzel, R., Fügenshuh, B. & Strauss, H. (2010) Determining the growth rate of topographic relief using *in situ*-produced ^{10}Be: a case study in the Black Forest, Germany. *Earth and Planetary Science Letters* **290**, 391–402.

Métivier, F., Gaudemer, Y., Tapponier, P. & Klein, M. (1999) Mass accumulation rates in Asia during the Cenozoic. *Geophysical Journal International* **137**, 280–318.

Milliman, J.D. & Syvitski, J. (1992) Geomorphic/tectonic control of sediment discharge to the ocean: the importance of small mountainous rivers. *Journal of Geology* **100**, 525–544.

Misra, S. & Froelich, P.N. (2012) Lithium isotope history of Cenozoic seawater: changes in silicate weathering and reverse weathering. *Science* **335**, 818–823.

Mitchell, S.G. & Montgomery, D.R. (2006) Influence of a glacial buzzsaw on the height and morphology of the Cascade Range in central Washington State, USA. *Quaternary Research* **65**, 96–107.

Molnar, P. & England, P. (1990) Late Cenozoic uplift of mountain ranges and global climate change: chicken or egg? *Nature* **346**, 29–34.

Montgomery, D.R. (1994) Valley incision and the uplift of mountain peaks. *Journal of Geophysical Research* **99**, 13 913–13 921.

Montgomery, D.R. & Brandon, M.T. (2002) Topographic controls on erosion rates in tectonically active mountain ranges. *Earth and Planetary Science Letters* **201**, 481–489.

Norton, K.P., Abbuhl, L.M. & Schlunegger, F. (2010) Glacial conditioning as an erosional driving force in the Central Alps. *Geology* **38**, 655–658.

Olen, S.M., Ehlers, T.A. & Densmore, M.S. (2012) Limits to reconstructing paleotopography from thermochronometer data. *Journal of Geophysical Research* **117**, F01024.

Ouimet, W.B., Whipple, K.X. & Granger, D.E. (2009) Beyond threshold hillslopes: channel adjustment to base-level fall in tectonically active mountain ranges. *Geology* **37**, 579–582.

Pinet, P. & Souriau, M. (1988) Continental erosion and large-scale relief. *Tectonics* **7**, 563–582.

Polyak, V., Hill, C. & Asmerom, Y. (2008) Age and evolution of the Grand Canyon revealed by U-Pb dating of water table-type speleothems. *Science* **319**, 1377–1380.

Portenga, E.W. & Bierman, P.R. (2011) Understanding Earth's eroding surface with ^{10}Be. *GSA Today* **21**, 4–10.

Raymo, M.E. & Ruddiman, W.F. (1992) Tectonic forcing of late Cenozoic climate. *Nature* **359**, 117–122.

Reiners, P.W. & Brandon, M.T. (2006) Using thermochronology to understand orogenic erosion. *Annual Review of Earth and Planetary Sciences* **34**, 419–466.

Rosenberg, C.L. & Berger, A. (2009) On the causes and modes of exhumation and lateral growth of the Alps. *Tectonics* **28**, TC6001.

Safran, E., Blythe, A. & Dunne, T. (2006) Spatially variable exhumation rates in orogenic belts: an Andean example. *Journal of Geology* **114**, 665–681.

Schildgen, T.F., Ehlers, T.A., Whipp, D.M. Jr. et al. (2009) Quantifying canyon incision and Andean Plateau surface uplift, southwest Peru: a thermochronometer and numerical modeling approach. *Journal of Geophysical Research* **114**, F04014.

Schildgen, T.F., Balco, G. & Shuster, D.L. (2010) Canyon incision and knickpoint propagation recorded by apatite ^{4}He/^{3}He thermochronometry. *Earth and Planetary Science Letters* **293**, 377–387.

Schumer, R. & Jerolmack, D.J. (2009) Real and apparent changes in sediment deposition rates through time. *Journal of Geophysical Research* **114**, F00A06.

Shuster, D.L., Ehlers, T.A., Rusmore, M.E. & Farley, K.A. (2005) Rapid glacial erosion at 1.8 Ma revealed by ^{4}He/^{3}He thermochronometry. *Science* **310**, 1668–1670.

Shuster, D.L., Cuffey, K.M., Sanders, J.W. & Balco, G. (2011) Thermochronometry reveals headward propagation of erosion in an alpine landscape. *Science* **332**, 84–88.

Simon-Labric, T., Brocard, G.Y., Teyssier, C. et al. (2014) Low-temperature thermochronologic signature of range-divide migration and breaching in the North Cascades. *Lithosphere* **6**, 473–482.

Small, E.E. & Anderson, R.S. (1998) Pleistocene relief production in Laramide mountain ranges, western United States. *Geology* **26**, 123–126.

Steer, P., Huismans, R.S., Valla, P.G., et al. (2012) Bimodal Plio-Quaternary glacial erosion of fjords and low-relief surfaces in Scandinavia. *Nature Geoscience* **5**, 635–639.

Stock, J.D. & Montgomery, D.R. (1996) Estimating palaeorelief from detrital mineral age ranges. *Basin Research* **8**, 317–327.

Stock, G.M., Anderson, R.S. & Finkel, R.C. (2004) Pace of landscape evolution in the Sierra Nevada, California, revealed by cosmogenic dating of cave sediments. *Geology* **32**, 193–196.

Stock, G.M., Ehlers, T.A. & Farley, K.A. (2006) Where does sediment come from? Quantifying catchment erosion with detrital apatite (U-Th)/He thermochronometry. *Geology* **34**, 725–728.

Stolar, D.B., Willett, S.D. & Roe, G.H. (2006) Climatic and tectonic forcing of a critical orogen. In: *Tectonics, Climate, and Landscape Evolution*. Geological Society of America Special Paper **398**, pp. 241–250.

Stüwe, K. & Hintermüller, M. (2000) Topography and isotherms revisited: the influence of laterally migrating drainage divides. *Earth and Planetary Science Letters* **184**, 287–303.

Stüwe, K., White, L. & Brown, R. (1994) The influence of eroding topography on steady-state isotherms. Application to fission track analysis. *Earth and Planetary Science Letters* **124**, 63–74.

Summerfield, M.A. & Hulton, N.J. (1994) Natural controls of fluvial denudation rates in major world drainage basins. *Journal of Geophysical Research* **99**, 13 871–13 833.

Syvitski, J.P.M., Vörösmarty, C.J., Kettner, A. & Green, P. (2005) Impact of humans on the flux of terrestrial sediment to the global coastal ocean. *Science* **308**, 376–380.

Thomson, S.N., Brandon, M.T., Tomkin, J.H. et al. (2010) Glaciation as a destructive and constructive control on mountain building. *Nature* **467**, 313–317.

Thouret, J.C., Wörner, G., Gunnell, Y. et al. (2007) Geochronologic and stratigraphic constraints on canyon incision and Miocene uplift of the Central Andes in Peru. *Earth and Planetary Science Letters* **263**, 151–166.

Torres Acosta, V., Schildgen, T.F., Clarke, B.A. et al. (2015) Effect of vegetation cover on millennial-scale landscape denudation rates in East Africa. *Lithosphere* **7**, 408–420.

Tranel, L.M., Spotila, J.A., Kowalewski, M.J. & Waller, C.M. (2011) Spatial variation of erosion in a small, glaciated basin in the Teton Range, Wyoming, based on detrital apatite (U-Th)/He thermochronology. *Basin Research* **23**, 571–590.

Valla, P.G., Herman, F., van der Beek, P.A. & Braun, J. (2010) Inversion of thermochronological age-elevation profiles to extract independent estimates of denudation and relief history – I: Theory and conceptual model. *Earth and Planetary Science Letters* **295**, 511–522.

Valla, P.G., Shuster, D.L. & van der Beek, P.A. (2011a). Significant increase in relief of the European Alps during mid-Pleistocene glaciations. *Nature Geoscience* **4**, 688–692.

Valla, P.G., van der Beek, P.A. & Braun, J. (2011b) Rethinking low-temperature thermochronology data sampling strategies for quantification of denudation and relief histories: a case study in the French western Alps. *Earth and Planetary Science Letters* **307**, 309–322.

Valla, P.G., van der Beek, P.A., Shuster, D.L. et al. (2012) Late Neogene exhumation and relief development of the Aar and Aiguilles Rouges massifs (Swiss Alps) from low-temperature thermochronology modeling and ^{4}He/^{3}He thermochronometry. *Journal of Geophysical Research* **117**, F01004.

van der Beek, P.A. (2013) Modelling landscape evolution. In Wainwright, J. & Mulligan, M. (eds.) *Environmental Modelling: Finding Simplicity in Complexity*, 2nd edn. Chichester: John Wiley& Sons Ltd., pp. 309–332.

van der Beek, P.A. & Bourbon, P. (2008) A quantification of the glacial imprint on relief development in the French western Alps. *Geomorphology* **97**, 52–72.

van der Beek, P., Van Melle, J., Guillot, S. et al. (2009) Eocene Tibetan plateau remnants preserved in the northwest Himalaya. *Nature Geoscience* **2**, 364–368.

Vermeesch, P. (2007) Quantitative geomorphology of the White Mountains (California) using detrital apatite fission track thermochronology. *Journal of Geophysical Research* **112**, F03004.

von Blanckenburg, F. (2006) The control mechanisms of erosion and weathering at basin scale from cosmogenic nuclides in river sediment. *Earth and Planetary Science Letters* **242**, 224–239.

Wang, P., Scherler, D., Liu-Zheng, J. et al. (2014) Tectonic control of Yarlung Tsangpo Gorge revealed by a buried canyon in Southern Tibet. *Science* **346**, 978–981.

Whipple, K.X. (2001) Fluvial landscape response time: How plausible is steady-state denudation? *American Journal of Science* **301**, 313–325.

Whipple, K.X. (2004) Bedrock rivers and the geomorphology of active orogens. *Annual Review of Earth and Planetary Sciences* **32**, 151–185.

Whipple, K.X. (2009) The influence of climate on the tectonic evolution of mountain belts. *Nature Geoscience* **2**, 97–104.

Whipple, K. & Meade, B. (2006) Orogen response to changes in climatic and tectonic forcing. *Earth and Planetary Science Letters* **243**, 218–228.

Whipple, K.X. & Tucker, G.E. (1999) Dynamics of the stream-power river incision model: implications for height limits of mountain ranges, landscape response timescales, and research needs. *Journal of Geophysical Research* **104**, 17 661–17 674.

Whipple, K.X., Kirby, E. & Brocklehurst, S.H. (1999) Geomorphic limits to climate-induced increases in topographic relief. *Nature* **401**, 39–43.

Willenbring, J.K. & Jerolmack, D.J. (2016) The null hypothesis: globally steady rates of erosion, weathering fluxes and shelf sediment accumulation during Late Cenozoic mountain uplift and glaciation. *Terra Nova* **28**, 11–18.

Willenbring, J.K. & von Blanckenburg, F. (2010) Long-term stability of global erosion rates and weathering during late-Cenozoic cooling. *Nature* **465**, 211–214.

Willett, S.D. & Brandon, M.T. (2013) Some analytical methods for converting thermochronometric age to erosion rate. *Geochemistry, Geophysics, Geosystems* **14**, 209–222.

Willett, S.D., Schlunegger, F. & Picotti, V. (2006) Messinian climate change and erosional destruction of the central European Alps. *Geology* **34**, 613–616.

Wittmann, H., von Blanckenburg, F., Kruesmann, T. et al. (2007) Relation between rock uplift and denudation from cosmogenic nuclides in river sediment in the Central Alps of Switzerland. *Journal of Geophysical Research* **112**, F04010.

Zachos, J., Pagani, M., Sloan, L. et al. (2001) Trends, rhythms, and aberrations in global climate 65 Ma to present. *Science* **292**, 686–693.

Zeitler, P.K., Meltzer, A.S., Koons, P.O. et al. (2001) Erosion, Himalayan geodynamics, and the geomorphology of metamorphism. *GSA Today* **11**, 4–9.

Zhang, P., Molnar, P. & Downs, W.R. (2001) Increased sedimentation rates and grain sizes 2–4 Myr ago due to the influence of climate change on erosion rates. *Nature* **410**, 891–897.

5

Dating Mountain Building: Exhumation and Surface Uplift

Matthias Bernet[1], Verónica Torres Acosta[2] and Mauricio A. Bermúdez[3]

[1] *Institut des Sciences de la Terre (ISTerre), Grenoble Alps University, Grenoble, France*
[2] *Institut für Erd- und Umweltwissenschaften, University of Potsdam, Potsdam, Germany*
[3] *Faculty of Natural Sciences and Mathematics, University of Ibagué, Colombia*

Abstract

The formation of high mountains and deep valleys reflects the interplay between tectonic processes and climatically controlled surface processes that shape the surface of Earth over geologic time. The evolution of mountain belts can be analyzed and quantified with different isotopic dating techniques, such as low-temperature thermochronology and cosmogenic radionuclide analysis. Fission-track and (U-Th)/He dating of apatite and zircon provide information on the cooling of rocks during exhumation, whereas cosmogenic radionuclide analysis provides information on the time of exposure on Earth's surface. The data obtained from these dating techniques are used in combination with numeric modeling to determine the rates of exhumation of rocks and surface erosion during orogenesis. Two case studies, from the Merida Andes in Venezuela and the East African Rift System (EARS) in Kenya, are reviewed here to highlight the application of a multidisciplinary approach to the dating of mountain building.

Keywords: *Andes, erosion, Kenya Rift, terrestrial cosmogenic nuclides, thermal history modeling, thermochronology*

5.1 Introduction

The evolution of mountain belts influences the regional climate and biodiversity, as rising mountains function as barriers to species migration and atmospheric circulation. High mountains can alter precipitation patterns and create a large variety of ecological habitats and vegetation zones at different elevations. To understand the evolution of mountain belts, it is important to know when and at what rates the surface of the Earth was uplifted and eroded, how metamorphic rocks were exhumed and how the eroded material was transferred from the mountains to surrounding sedimentary basins.

Here, we will focus on dating mountain building by analyzing exhumation. Different isotopic dating techniques exist to estimate the rates at which rocks are exhumed towards the surface (Figure 5.1). These techniques do not provide information on surface uplift. Low-temperature thermochronology, such as fission-track or (U-Th)/He dating of apatite and zircon, is used to estimate exhumation or erosion rates over millions of years, which allows the study of the long-term

evolution of mountain belts (e.g., Reiners & Brandon 2006). Short-term erosion rates – on the millennial time scale – are determined by terrestrial cosmogenic nuclide analysis (Dunai 2010). Long- and short-term erosion rates can be compared to topography, precipitation patterns, vegetation cover and lithology in order to evaluate potential changes in erosion over geological time. To determine the amount of surface uplift and paleoelevations of mountain belts, different geomorphological, botanical, ecological and geochemical techniques can be used, as shown in Chapters 6, 7 and 27. See Chapters 2–4 for more information on the dynamics of topography formation and tectonically and climatically driven surface processes.

Exhumation-rate and surface-uplift information can be integrated into landscape evolution models together with information on, for example, present-day relief and mean elevation. Numeric modeling allows the testing of different evolution scenarios that explain the present-day topography, vegetation distribution and lithological patterns of a mountain belt. The objective of this chapter is to provide an introduction to the different techniques

Mountains, Climate and Biodiversity, First Edition. Edited by Carina Hoorn, Allison Perrigo and Alexandre Antonelli.
© 2018 John Wiley & Sons Ltd. Published 2018 by John Wiley & Sons Ltd.
Companion website: www.wiley.com/go\hoorn\mountains,climateandbiodiversity

(a)

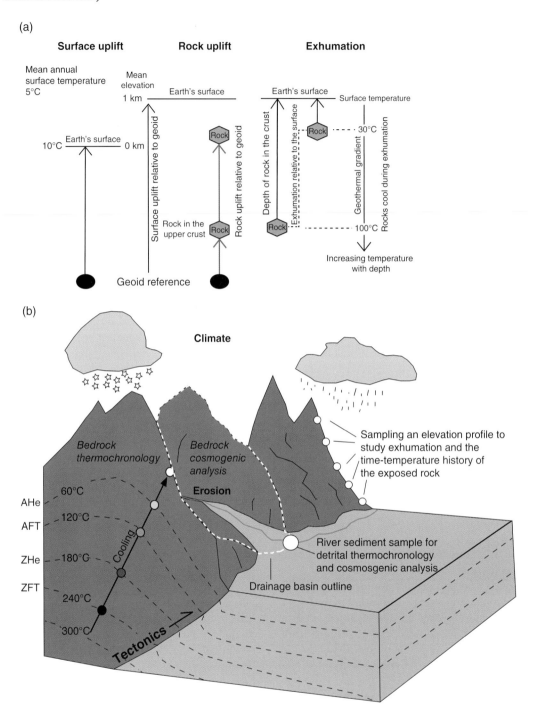

Figure 5.1 (a) Schematic presentation of the concept of surface uplift and rock uplift with respect to the centre of the Earth, and exhumation with respect to the Earth's surface. (b) Schematic concept showing the relationship between tectonics, surface processes and cooling of rocks during exhumation in a convergent orogenic setting. AFT and ZFT, apatite and zircon fission-track analysis; AHe and ZHe, apatite and zircon (U-Th)/He analysis. Isotherms (indicated by black dashed lines) relate to the average closure temperatures of the different dating techniques, depending on the cooling rate (here about 15 °C/My). See also Plate 8 in color plate section.

and underlying concepts needed to date exhumation during mountain building. Case studies from the Merida Andes in Venezuela and the East African Rift System (EARS) in Kenya are presented to demonstrate the application of such studies in different geodynamic settings.

5.2 Mountain Building

Mountainous topography is generated primarily along plate boundaries. The classic forms of mountain building and surface uplift are related to the collision of continental plates and crustal thickening (e.g., Himalaya, European Alps). However, surface uplift can also occur along plate margins where continental and oceanic plates collide (e.g., Andes), or where continental plates move away from each other, such as in continental rift systems (e.g., East African Rift). In the latter case, topography is created by surface uplift of the rift shoulders and subsidence of the rift valley between the shoulders.

In general, as soon as the Earth's surface rises above sea level, it is exposed to surface processes such as weathering and erosion (see Chapter 4). To a certain degree, erosion will intensify with increasing relief and tectonic activity (Montgomery & Brandon 2002), because the rising topography will impact the local or regional climate and the distribution of precipitation (Willett 1999). Removal of rock and soil at or near the surface of the uplifting mountains by surface processes and/or normal faulting will lead to exhumation of rocks from deeper in the crust, and to a local isostatic response to compensate for the loss of material. The rock and soil eroded from the mountains are primarily transported by rivers to surrounding sedimentary basins, which archive the erosion record of the mountains over millions of years.

5.3 Studying Long-Term Exhumation with Low-Temperature Thermochronology

Low-temperature thermochronology provides information on the timing and rate of cooling of rocks. In principle, this cooling can be post-magmatic thermal relaxation (e.g., cooling of a granitic intrusion), or it can be related to exhumation. Rocks that are exhumed will cool from upper crustal temperatures to surface temperatures (Figure 5.1). The temperature at depth depends on the mean annual surface temperature and the local geothermal gradient (commonly on the order of $30° \pm 15°C/km$). For simplicity, a linear thermal gradient is assumed for most studies.

Commonly used techniques are apatite and zircon (U–Th)/He (AHe, ZHe) and fission-track (AFT, ZFT) dating, which are sensitive to different temperature ranges. The simplifying closure temperature and partial annealing/retention zone concepts are used to interpret thermochronological data and to model cooling histories of exhumed rocks (Figure 5.1b).

5.3.1 (U-Th)/He Analysis

Apatite and zircon are minerals with naturally elevated concentrations of ^{238}U, ^{235}U and ^{232}Th. These three radioactive isotopes decay by alpha-decay to ^{206}Pb, ^{207}Pb and ^{208}Pb, respectively, with the release of one ^{4}He per disintegration step. By measuring the U, Th and He concentrations in a crystal, an apparent cooling age can be determined if the geometry and size of the analyzed crystal are known and are used to apply appropriate correction factors for potential He ejection (Farley 2000; Reiners 2005).

5.3.2 Fission-Track Analysis

For every 1 million ^{238}U isotopes that decay by alpha-decay, one ^{238}U isotope will spontaneously fission. A lighter and a heavier charged particle are formed, which travel at high velocities in opposite directions in the crystal over a combined distance of about 16.3 μm in apatite, or 11 μm in zircon. This linear damage zone is called a latent track, which can be made visible by etching. For this, the grain is mounted in epoxy or Teflon and polished to reveal an internal surface. The internal surface is chemically etched to render the fission tracks visible with an optical light microscope. A fission-track age is determined by calculating the track density over a certain surface area of the analyzed grain and determining the uranium concentration at that same location in the grain. Two different techniques are used for fission-track analysis: the external detector method and the laser ablation–inductively coupled plasma–mass spectrometry (LA-ICP-MS) method. The external detector method uses an external mica detector attached to the sample grain mount to estimate the uranium concentration of each analyzed grain. For this, the samples are irradiated with thermal neutrons in a research reactor. The thermal neutrons cause fission of ^{235}U isotopes in the crystal, and these induced fission events leave tracks in the juxtaposed mica detector. With this method, fission tracks are counted in the grains and the mica detector, and an age is calculated from the track density ratio. The LA-ICP-MS method only requires the counting of fission tracks in the grains, as the uranium concentration is determined by analyzing the grain with a laser connected to a mass spectrometer (Donelick et al. 2005; Tagami 2005).

5.3.3 Exhumation Rates

Calculating an exhumation rate from an apparent cooling age can be done using different methods, but all require some estimate of the geothermal gradient at the time of exhumation – a value that is not generally known. The simplest way of calculating an exhumation rate (ε) for a bedrock sample is:

$$\varepsilon = \left(\left(T_c - T_s\right)/G\right)/\delta t \tag{5.1}$$

where T_c is the closure temperature (see Box 5.1), T_s the surface temperature, G the geothermal gradient (°C/km) and δt the apparent cooling age (the time since cooling below T_c). If closure occurred at 240°C, T_s would be 10°C, G 30°C/km, the cooling age 15 Ma and

Box 5.1 The closure temperature concept

The closure temperature (T_c) concept can be applied to rocks with simple, monotonic cooling histories (Dodson 1973). It implies that the apatite or zircon crystals change at a certain temperature from an open to a closed system with respect to the loss of He isotopes by thermally activated volume diffusion or the loss of fission tracks by annealing. The temperature at which this closure occurs depends primarily on the rate at which the crystals are cooling during exhumation, but also on the grain size for AHe and ZHe, the chemical composition of apatites for AFT dating and the radiation damage in zircon for ZFT dating (Table 5.1). Exhumation-induced cooling rates are commonly on the order of 10–20 °C/My during orogenesis.

Table 5.1 Average closure temperatures with respect to cooling rate (°C).

Cooling rate (°C/My)	AHe	AFT	ZHe	ZFT[1]	ZFT[2]
0.1	38.4	81.9	124.0	190.6	283.0
1	51.9	98.4	161.6	210.5	321.5
10	66.6	116.4	183.1	232.1	342.4
100	82.7	136.3	206.8	255.8	365.1

Source: Data were taken from the closure program of M. Brandon. T_c given for apatite with an average chemical composition, [1]radiation-damaged and [2]non-damaged zircons.

Box 5.2 The partial annealing zone or partial retention zone concept

For slowly cooling rocks (<2 °C/My) – or rocks that have more complicated thermal histories, with different phases of cooling and reheating – the partial annealing zone (PAZ) concept for fission-track systems, or the partial retention zone (PRZ) concept for He dating, should be applied (Reiners & Brandon 2006). These concepts state that closure (or reopening during reheating) of the systems occurs over a temperature range instead of at a single temperature. As long as the crystals are held in the specific temperature ranges, some of the He may diffuse out of them and the fission tracks may shorten. Table 5.2 gives an example for a 20 My hold time in the PAZ or PRZ for the different dating techniques. The upper and lower temperature limits of these zones are given by 90 and 10% of the He retention or annealing of fission tracks, respectively. When the crystals are in these temperature ranges, not only will He be lost or tracks shortened, but new He and new tracks will be formed. For fission-track analysis, the lengths of horizontal tracks can be determined, and these represent a measure of the extent of partial annealing of fission tracks.

Table 5.2 Partial annealing and retention zones for a 20 My hold time.

	AHe	AFT	ZHe	ZFT
Upper limit (°C)	43.7	56.0	152.9	175.3
Lower limit (°C)	68.3	108.7	189.6	231.2

Source: Data were taken from the closure program of M. Brandon.

the exhumation rate 0.51 km/My. This is a first-order estimate, assuming that the exhumation rate has been linear from when the crystals cooled below the closure temperature until the present day. More sophisticated methods take into account the advection of isotherms during exhumation, as well as changes in the geothermal gradient and cooling rates (see Box 5.2) (Willett & Brandon 2013).

5.4 Studying Short-Term Erosion from Terrestrial Cosmogenic Nuclide Analysis

In the context of this chapter, we consider short-term erosion rates on a time scale of thousands to hundreds of thousands of years. We are not evaluating erosion rates obtained from sediment gauging stations in present-day rivers or sediment accumulation in artificial reservoirs such as behind dams. Rather, we will present a dating technique that directly determines surface exposure ages, and can thus be used to calculate erosion rates.

In this chapter, we focus mainly on the ^{10}Be/^{26}Al method (hereafter referred to as "^{10}Be"), but other tech-

niques are ^{21}Ne and ^{36}Cl dating. The ^{10}Be isotopes are produced by a cosmic ray-spallation nucleosynthesis reaction of ^{14}N and ^{16}O in the atmosphere and ^{16}O in silicate minerals (particularly quartz) within rocks exposed at the Earth's surface (see Box 5.3). The spallation nucleosynthesis reaction is caused by the interaction of the cosmic rays with the atoms in the elements. The ^{26}Al is derived from a spallation reaction of ^{21}Si. Isotope pairs are easier to measure accurately in mass spectrometers than absolute concentrations. ^{10}Be has a half-life of about 1.5 million years, whereas ^{26}Al has a half-life of only 717 000 years. Production of cosmogenic nuclides in the atmosphere and at the Earth's surface depends on altitude and latitude with respect to the geomagnetic equator. The higher the altitude, the higher the production, as 10 times more nuclides are produced at an elevation of 4500 m compared to sea level. The production rate also decreases in locations closer to the geomagnetic equator and increases at latitudes between 20 and 30°. Possible

---

Box 5.3 Cosmogenic radiation

The surface of the Earth is constantly bombarded with cosmic rays, which mainly consist of high-energy protons and α-particles. The flux of cosmic rays that reach the Earth's surface depends on the intensity of the galactic and solar radiation and the intensity of the Earth's magnetic field, which provides a protective shield against such rays. The weaker the magnetic shielding, the higher the intensity of cosmic radiation that reaches the surface. The intensity of solar radiation also depends on solar activity, such as solar eruptions on the surface of the sun. Whereas solar radiation and magnetic field strength are variable through time, galactic radiation – from outside of our solar system – is relatively constant.

Collision of cosmic rays with elements in the atmosphere or in the rock surface causes the formation of radioactive (^{14}C, ^{10}Be, ^{26}Al, ^{36}Cl) and nonradioactive (^{3}He, ^{21}Ne) isotopes called "terrestrial cosmogenic nuclides." The most commonly known cosmogenic nuclide is ^{14}C, which is generated by the interaction of cosmic rays with ^{14}N in the atmosphere. ^{14}C is oxidized to form CO_2, which is then stored in the oceans and in the organic material of living organisms. As long as an organism is alive, it will continuously receive new ^{14}C, replacing the decaying ^{14}C already in its system. Once the organism dies, the ^{14}C will decay with a half-life of

5730 ± 40 years back to ^{14}N by beta decay. Measurement of the ^{14}C concentration in organic material therefore allows dating of the deposition time of that material in a sedimentary layer.

Attenuation length

Attenuation length is defined as the thickness of a mass of rock necessary to attenuate (decrease the intensity) of the flux of cosmic rays. Attenuation changes with altitude and latitude because of the geomagnetic field and atmospheric effects (Gosse & Phillips 2001).

Minimum exposure age

A minimum exposure age is determined when there is little erosion during surface exposure. If erosion is considered and can be calculated, then it must be corrected by a factor:

$$f_\varepsilon = 1 + [(\varepsilon T_{exp} \rho / \Lambda_{sp})/2] \qquad (5.3.1)$$

where ρ is density of the rock (g/cm^3), Λ attenuation length and ε erosion rate.

---

shielding effects by surrounding high topography and relief have to be taken into account in order to estimate the correct production rate.

5.4.1 Bedrock-Exposure Age Dating

Depending on the density of the rock, production of ^{10}Be isotopes decreases exponentially with depth below the rock surface. The production rate is reduced to half at 40 cm for granitic rocks. ^{10}Be is produced in all silicates, although usually only quartz or olivine is analyzed. Quartz is very useful, as it is a very common mineral in the upper continental crust.

To calculate a minimum age of exposure to cosmic radiation at the Earth's surface, the ^{10}Be concentration in the quartz crystals of the exposed rocks has to be determined and the appropriate ^{10}Be production rate with respect to the topographic and latitudinal position of the rock selected (Dunai 2010). Furthermore, given the attenuation length of the incident cosmic particles, a maximum erosion rate can be calculated. In general, if the ^{10}Be concentration is low then the exposure age is young, and thus the erosion rate must be relatively fast because the quartz crystals in the rock have not accumulated a significant quantity of ^{10}Be. Conversely, if the ^{10}Be concentration is high, the exposure age is old and the erosion rate must have been slow.

5.4.2 Detrital Quartz ^{10}Be Analysis

Analysis of detrital quartz grains collected from present-day river sediments can be used to determine drainage basin-wide average erosion rates. The underlying idea is that the quartz grains were derived from different parts of the drainage area and are well mixed during fluvial transport (Figure 5.1b). Each quartz crystal will carry an individual ^{10}Be concentration, but all grains are analyzed as a single sample, yielding an average concentration. This approach is particularly suitable if the lithologies in the drainage area are rich in quartz and cover the majority or all of the drainage area, as non-quartz-bearing lithologies will not be detected by this method. As in the bedrock erosion rate calculation, for detrital quartz a low ^{10}Be concentration corresponds to relatively fast erosion rates, and vice versa.

In addition to determining drainage-basin average erosion rates, detrital quartz ^{10}Be analysis of sediment collected from incised sedimentary river terraces can be used to calculate river incision rates. As with exposure age dating, the age determined from detrital quartz grains of the terrace sediments is regarded as the time when incision started. The incision rate is then calculated from the height of the river terrace above the present-day river divided by the exposure age.

5.5 Numeric Modeling of Thermal Histories and Exhumation

Several software packages for modeling thermochronological data exist, such as *HeFTy* (Ketcham 2005) and *QTQt* (Gallagher 2012). With these programs, one can predict the thermal histories of rocks from fission-track age and track-length distributions and/or (U-Th)/He data. These programs are very useful for determining the thermal history of single-rock samples (*HeFTy*) or a suite of samples (*QTQt*), offering a range of solutions that fit the thermochronological data and independent geological constraints (Figure 5.2). Both programs use forward and inverse modeling procedures. Forward modeling predicts

(a)

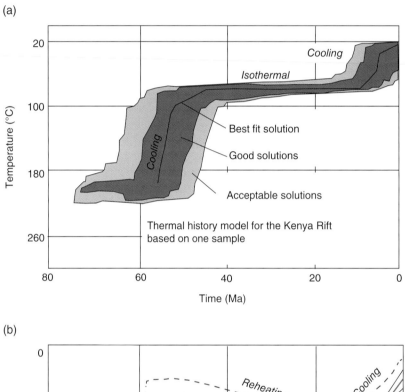

(b)

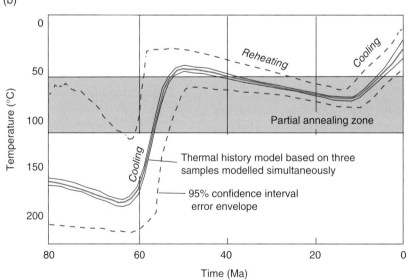

Figure 5.2 Time–temperature history models are powerful tools for interpreting thermochronological data. The examples given here are for the East African Rift. (a) *HeFTy* model showing good (dark gray) and acceptable (light gray) model solutions for the time–temperature history based on apatite fission-track data. The thermal history shows relatively fast cooling between 60 and 50 Ma, followed by slow cooling between 50 and 10 Ma and renewed fast cooling between 10 Ma and the present. (b) *QTQt* model showing t–T paths for a set of three samples, including the 95% confidence interval error envelope. The model shows rapid cooling between 60 and 50 Ma, slight reheating between 50 and 10 Ma and fast cooling from 10 Ma to the present day. *Source:* Torres Acosta et al. (2015b). Reproduced with permission of American Geophysical Union.

the anticipated distribution of thermochronological data for any thermal history, whereas inverse modeling tries to determine the thermal history that best corresponds to the input data (Ketcham 2005). However, the programs do not test what influences the thermal history or whether the model results are compatible with basic rules of heat transfer in the crust, which are important for determining long-term exhumation of mountain belts. For this, other programs are available.

Numeric modeling of long-term exhumation scenarios can be done with *FETKIN* (Almendral et al. 2015), or with the three-dimensional thermal–kinematic code *PeCUBE* (Braun 2003; Braun et al. 2012). *FETKIN*, a closed-code software, is used to combine geological data from two-dimensional geological cross-sections with thermochronological data. The program forward models thermochronological ages using a finite-element computation of temperatures and can incorporate relief change and variations in rock types (Almendral et al. 2015). By contrast, *PeCUBE* is an open-source software, and code modification allows for its adaptation for spe-cific geological conditions. In *PeCUBE*, *Pe* stands for the Péclet number and *CUBE* for the three spatial dimensions. The Péclet number is the ratio of the rate of advec-tion of a physical quantity – in this case, the heat in the crust – to the rate of diffusion of the same quantity driven by a gradient. As with *HeFTy* and *QTQt* modeling, *PeCUBE* modeling uses forward models to illustrate the reproducibility of thermochronological data from earlier model runs or of observed data from real analyses, and inverse modeling of the compiled thermochronological data to predict exhumation scenarios. Forward models are run for the family of best-fit models in order to illustrate the fit between observed and predicted ages. Figure 5.3 shows an example of two different *PeCUBE* model outputs for topography and predicted surface ZHe ages. The exhumation scenarios and relief development obtained from *PeCUBE* modeling can be combined with surface-uplift information from paleoaltimetry data (Chapters 4, 6 and 7), short-term erosion rates from cosmogenic analyses and information about vegetation cover for landscape-evolution modeling.

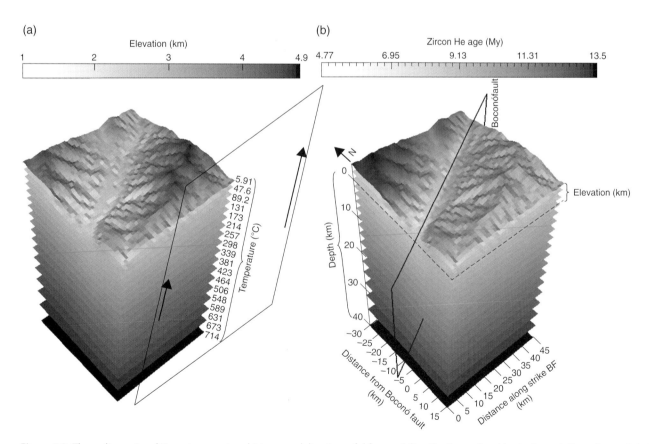

Figure 5.3 Three-dimensional time–temperature history modeling is useful for modeling the thermal and tectonic evolution of mountain belts. Shown here are *PeCUBE* models for (a) the topographic evolution (layers indicate isotherms, with the temperature values (°C) noted alongside) and (b) predicted zircon He ages at the surface of both sides of the Boconó fault. Inclusion of the fault in the middle of the study area permits the analysis of the exhumation of two tectonic blocks independently. See also Plate 9 in color plate section.

5.6 Case Study: Merida Andes of Venezuela

The Merida Andes of Venezuela are part of the Northern Andes (Figure 5.4a). The 400 km-long SW–NE-trending orogen reaches elevations of up to 4978 m in its central part, and is characterized by the SW–NE-oriented Boconó strike slip fault in its center and thrust faults on its northern and southern flanks (Figure 5.4a). The Merida Andes formed because of convergence between the Caribbean, Nazca and South American Plates and the Maracaibo continental block (Colletta et al. 1997), which had a significant impact on the evolution of the

Orinoco drainage basin and the biodiversity of northern South America (Hoorn et al. 1995, 2010).

Using fission-track and (U-Th)/He analyses and *PeCUBE* modeling, the exhumation histories of the Sierra Nevada block to the south of the Boconó fault and the El Carmen block to the north were determined (Bermúdez et al. 2011). These two blocks were exhumed following the middle Miocene at similar rates of up to 1.5 km/My but in different periods of time. The Boconó strike-slip fault must have an important oblique component to allow such rates of exhumation in combination with erosion. Because of this exhumation, crystalline rocks, which were once at tens of kilometers depth,

(a)

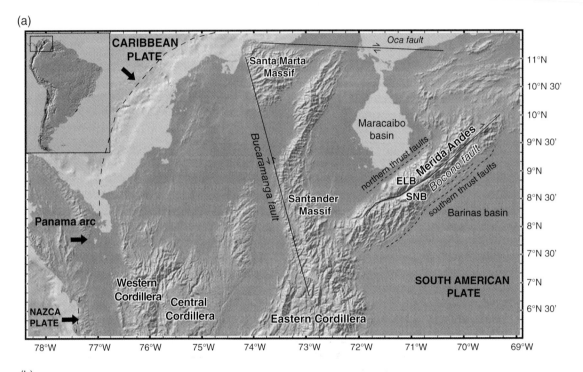

(b)

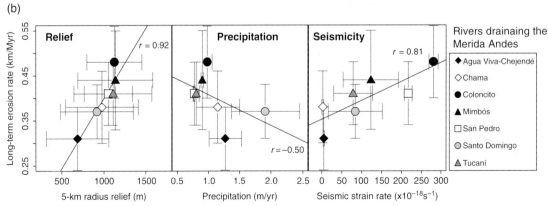

Figure 5.4 (a) Overview map of the Northern Andes in Colombia and Venezuela. ELB, El Carmen block; SNB, Sierra Nevada block. *Source:* Map made using GeoMapApp: www.geomapapp.org. (b) Correlation of long-term erosion rates derived from apatite fission-track data from river sediments in the Merida Andes against relief, precipitation and seismic strain rate determined from seismic energy release. *Source:* Bermúdez et al. (2013). Reproduced with permission of The Geological Society of America.

are now exposed at the surface. Surface uplift since the late Miocene brought these rocks to over 4000 m elevation. The exhumation does not explain the high elevation. Part of the surface uplift may be explained by compression of the tectonic plates, while part may be explained by isostatic compensation. Most of the rock removed by erosion is transported by rivers towards the flanks of the orogen and deposited in the Maracaibo and Barinas basins to the north-west and south-east of the Merida Andes, respectively (Bermúdez et al. 2015). The distribution of fission-track ages of detrital grains in sediments of rivers that drain the Merida Andes shows that average exhumation rates across the drainage basin are on the order of 0.33–0.48 km/My (Bermúdez et al. 2013). These rates are slower than the maximum rates determined in the vicinity of the Boconó fault, as the river sample presents a mixture of grains from the whole drainage area.

The average exhumation rates of the drainage basin can be used as a proxy for erosion rates and compared to different parameters. Are erosion rates controlled by lithology, relief, precipitation and/or tectonic activity? Lithologies in the core of the Merida Andes are dominated by crystalline rocks with a similar resistance to erosion, and thus lithology is not considered to be the main factor controlling erosion rates. Relief correlates positively with long-term erosion rates, but precipitation data reveal a negative correlation (Figure 5.4b). The precipitation data, however, cover a rather short time span, and climatic variability cannot be resolved. Furthermore, the total amount of precipitation is not as important as the frequency of storm events, during which erosion and sediment transport are commonly significantly higher in river systems (see Chapter 4). The seismic energy released during earthquakes in the study area can be used as a proxy to estimate the tectonic activity in the form of seismic strain rate. The available data cover only a short period (100 years), and data from a few large earthquakes may bias the calculations. Nonetheless, a positive correlation between seismic strain rate and erosion rates has been found (Figure 5.4b).

The Maracaibo and Barinas basins preserve the exhumation record of the Merida Andes (Figure 5.4a). The thermochronological data from the basin sediments reflect the episodic exhumation of individual tectonic blocks in the Merida Andes. The petrological data show that the sediments were primarily derived from the crystalline core of the orogen, and pollen analyses indicate that the sampled strata were deposited between the late Miocene and Pliocene, based on marker species. The pollen assemblages correspond to a vegetation cover normally found at 3000–4000 m elevation. Because pollen found in fluvial sediments were primarily transported in the rivers and not windblown from distant areas, the sediment sources had

mean elevations of 3000–4000 m by the late Miocene to Pliocene (Bermúdez et al. 2015).

In summary, the Merida Andes experienced rapid exhumation starting in the middle Miocene, which was controlled by tectonic activity along the Boconó fault. The Merida Andes reached present-day elevations by the late Miocene to Pliocene. This was determined using a multidisciplinary approach, in which different types of data were used to reconstruct the timing of mountain building and the surface-uplift history. The evolution of the Merida Andes is contemporaneous with the evolution of the Eastern Cordillera in Colombia, for which climatic (Mora et al. 2008, 2010) and tectonic (Parra et al. 2009) controls have also been documented, particularly over the past 3–5 million years.

5.7 Case Study: East African Rift System

The EARS is a spectacular topographic expression of ongoing rifting (Figure 5.5; see also Chapter 26) that is 4000 km long and located in the east of Africa between the Gulf of Aden to the north and Mozambique to the south. The EARS remains the archetypal Cenozoic magmatic continental rift zone, with elevated rift shoulders and deep rift basins. The area comprises elevations from sea level (the Afar Depression) to approximately 300 m above sea level along the rift floor (Suguta Valley, northern Kenya) and up to 5900 m at the highest peak (Mount Kilimanjaro in Tanzania). This typical rift morphology is generally the precursor of rift development that may lead to the formation of continental passive margins preceding oceanic opening and ocean basin formation. For the EARS, the kinematics of the area are controlled by the motion of several lithospheric plates, including the Nubia, Rovuma, Victoria, Lwandle and Somalia Plates (Figure 5.5).

The EARS consists of a western branch containing the large fresh-water lakes Tanganika, Malawi, Albert and Rukwa, and an eastern branch, which is characterized by widespread magmatism that started about 37 Ma in a region spanning between Ethiopia and Tanzania along the rift axis. The two branches, with Lake Victoria and the Tanzanian Craton in between, form the 1300 km-wide and 1200 m-high East African Plateau. This high plateau is supported by two different mechanisms: rift flank uplift by isostasy and lithospheric uplift (Wichura et al. 2011). From a climatic perspective, the EARS represents an area influenced by the convergence of the Congo Air Boundary (CAB) and traces of the African Indian Monsoon. This is a good location in which to investigate the influence of climate and tectonics on rainfall and moisture circulation through time, as well as the relationship between vegetation cover and erosion (among other processes). The Kenya Rift, located in the eastern

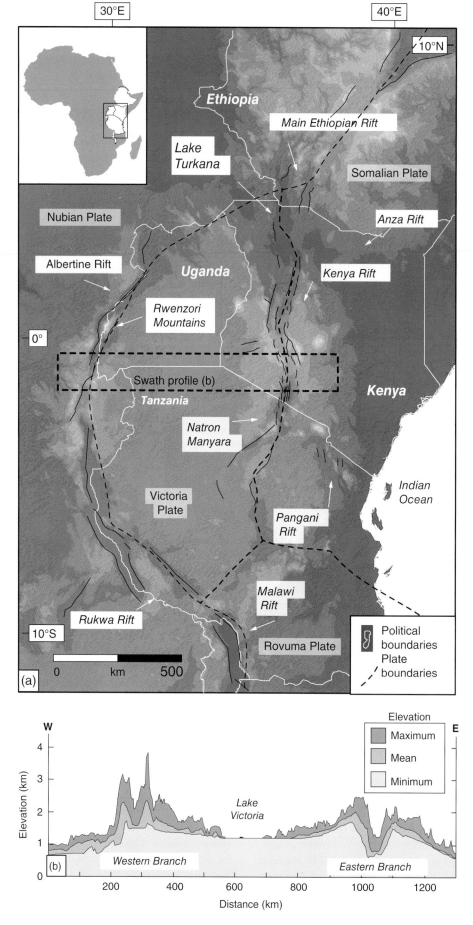

Figure 5.5 (a) Map of the East African Rift System (EARS), showing the rift shoulders and rift basin and the different lithospheric plates controlling the geodynamic evolution. The dashed gray box denotes the area shown in (b). (b) Schematic cross-section of the EARS. Swath profile from the Shuttle Radar Topography Mission (SRTM), 90 m, 100× vertical exaggeration. *Source:* Wichura et al. (2011). Reproduced with permission of The Geological Society of London.

branch of the EARS, provides information on rift-basin formation and climate change between the Ethiopian–Kenyan and Tanzanian rift basins. Evidence of early stages of extension would be needed to link the beginning of rifting with extension over time, but such evidence is either buried under younger sedimentary rocks or has been removed from the rift shoulders by erosion.

A combination of thermochronological (AFT, ZHe) and cosmogenic nuclide (^{10}Be) dating techniques with field observations, morphometric analysis of digital elevation models and satellite remote sensing data on precipitation and vegetation cover has helped to elucidate the influence of tectonics and climate on relief development and erosion on different time scales. The tectono-thermal evolution of the Kenya Rift, including the early stages of Cenozoic rifting, was determined using thermal-history modeling (see Figure 5.2) (Torres Acosta et al. 2015b). It was shown that after the initial rifting between 65 and 50 Ma, with fast cooling and rapid exhumation of the rift shoulders, a 30 My-period of stable tectonic and isothermal conditions followed. Because the rift shoulders had been uplifted to higher elevations, they were exposed to slow erosion, which caused sediments to accumulate in the rift basins during this period. The thermal-history modeling further showed that a second phase of rifting started during the Miocene, which is compatible with the rise of the Kenya Dome by 13 Ma, resulting in its extension along the present-day Kenya Rift.

With short-term erosion rates determined from ^{10}Be analysis of detrital quartz, combined with remote-sensing data, the principal factors controlling erosion along the rift escarpments and on the rift shoulders over the last thousands of years were studied. This allowed the role of climate variability on erosion rates and the importance of environmental thresholds of surface processes (e.g., the erodibility of different lithologies, the type and density of vegetation cover and relief) and landscape evolution to be documented.

Climate-driven effects on erosion during rapid early Cenozoic extension are to be expected. It is very likely that the rift-shoulder surface uplift impacted precipitation and forced erosion processes in the EARS, as this region is located at the equator and was dominated by much more humid conditions during the early Cenozoic compared to the present day. However, climate is not regarded as the main driving force on erosion over longer time scales. The long period of tectonic quiescence between about 50 and 20 Ma instead argues for very slow rates of erosion. Assuming an average geothermal gradient of 30 °C/km and an average surface temperature of 18 °C for this tropical region, the long-term exhumation rates for the Kenya Rift are on the order of 0.07–0.13 mm/y during this period.

Present-day and millennial-scale ^{10}Be-based basin-wide denudation rates in the Kenya Rift range between 0.001 and 0.13 mm/y for arid and sparsely vegetated to humid and densely vegetated regions. Given the same slope angle, areas with wetter climatic conditions and denser vegetation cover show slower denudation rates compared to areas with drier climates and sparse vegetation cover. This demonstrates that the stabilizing effect of vegetation cover exerted a fundamental influence on the steady-state slopes of escarpments and rift shoulders over the last several thousand years. Therefore, rapid shifts in climate will affect denudation rates if environmental conditions – such as vegetation cover – change significantly (Torres Acosta et al. 2015a).

5.8 Conclusion

Studying mountain building and topography evolution requires a multidisciplinary approach, which is significantly improved by using numerical modeling to test different scenarios. In many studies, the increase in exhumation rates is related to surface uplift, river incision and relief creation. Nonetheless, independent evidence for quantifying surface uplift through pollen analysis and vegetation assemblages or geochemical studies (as outlined in Chapters 4, 6 and 7) is needed. Therefore, depending on the question, an appropriate combination of analytical techniques and modeling tools should be selected. The dating techniques and erosion-rate determinations shown here are of great importance for understanding the evolution of the Earth's surface, as they allow for the quantification of geological processes. The study of erosion rates on different time scales by the combination of dating, geological, climatic, geomorphological and remote-sensing data provides a guide for potential future impacts of global change on surface processes.

References

Almendral, A., Robles, W., Parra, M. et al. (2015) FetKin: Coupling kinematic restorations and temperature to predict thrusting, exhumation histories, and thermochronometric ages. *AAPG Bulletin* **99**, 1557–1573.

Bermúdez, M.A., van der Beek, P. & Bernet, M. (2011) Asynchronous Mio-Pliocene exhumation of the central Venezuelan Andes. *Geology* **39**, 139–142.

Bermúdez, M.A., van der Beek, P. & Bernet, M. (2013) Strong tectonic and weak climatic control on exhumation rates in the Venezuelan Andes. *Lithosphere* **5**, 3–16.

Bermúdez, M.A., Hoorn, C., Bernet, M. et al. (2015) The detrital record of Late-Miocene to Pliocene surface uplift and exhumation of the Venezuelan Andes in the Maracaibo and Barinas foreland basins. *Basin Research* **29**, 370–395.

Braun, J. (2003) Pecube: a new finite element code to solve the 3D heat transport equation including the effects of a time-varying, finite amplitude surface topography. *Computers and Geosciences* **29**, 787–794.

Braun, J., van der Beek, P., Valla, P. et al. (2012) Quantifying rates of landscape evolution and tectonic processes by thermochronology and numerical modeling of crustal heat transport using PECUBE. *Tectonophysics* **524–525**, 1–28.

Colletta, B., Roure, F., De Toni, B. et al. (1997) Tectonic inheritance, crustal architecture, and contrasting structural styles in the Venezuelan Andes. *Tectonics* **16**, 777–794.

Dodson, M.H. (1973) Closure temperature in cooling geochronological and petrological systems. *Contributions to Mineralogy and Petrology* **40**, 259–274.

Donelick, R.A., O'Sullivan, P. & Ketcham, R.A. (2005) Apatite fission-track analysis. *Reviews in Minerology and Geochemistry* **58**, 49–94.

Dunai, T.J. (2010) *Cosmogenic Nuclides: Principles, Concepts and Applications in the Earth Surface Sciences.* Cambridge: Cambridge University Press.

Farley, K.A. (2000) Helium diffusion from apatite: general behavior as illustrated by Durango flourapatite. *Journal of Geophysical Research* **105**, 2903–2914.

Gallagher, K. (2012) Transdimensional inverse thermal history modelling for quantitative thermochronology. *Journal of Geophysical Research* **117**, B02408.

Gosse, J.C. & Phillips, F.M. (2001) Terestrial in situ cosmogenic nuclides: theory and application. *Quaternary Sciences Reviews* **20**, 1475–1560.

Hoorn, C., Guerrro, J., Sarmiento, G.A. & Lorente, M.A. (1995) Andean tectonics as a cause for changing drainage patterns in Miocene northern South America. *Geology* **23**, 237–240.

Hoorn, C., Wesselingh, F.P., ter Steege, H. et al. (2010) Amazonia through time: Andean uplift, climate change, landscape evolution, and biodiversity. *Science* **330**, 927–931.

Ketcham, R.A. (2005) Forward and inverse modelling of low-temperature thermochronometry data. *Reviews in Minerology and Geochemistry* **58**, 275–314.

Montgomery, D.R. & Brandon, M.T. (2002) Topographic controls on erosion rates in tectonically active mountain ranges. *Earth and Planetary Science Letters* **201**, 481–489.

Mora, A., Parra, M., Strecker, M.R. et al. (2008) Climatic forcing of asymmetric orogenic evolution in the Eastern Cordillera of Colombia. *Geological Society of America Bulletin* **120**, 930–949.

Mora, A., Baby, P., Roddaz, M. et al. (2010) Tectonic history of the Andes and sub-Andean zones: implications for the development of the Amazon drainage basin. In Hoorn, C. & Wesselingh, F.P. (eds.) *Amazonia: Landscape and Species Evolution: A Look into the Past.* Chichester: John Wiley & Sons Ltd., pp. 38–57.

Parra, M., Mora, A., Sobel, E.R. et al. (2009) Episodic orogenic front migration in the northern Andes: constraints from low-temperature thermochronology in the Eastern Cordillera, *Colombia. Tectonics* **28**.

Reiners, P.W. (2005) Zircon (U-Th)/He thermochronometry. *Reviews in Minerology and Geochemistry* **58**, 151–179.

Reiners, P.W. & Brandon, M. (2006) Using thermochronology to understand orogenic erosion. *Annual Review of Earth and Planetary Sciences* **34**, 419–466.

Tagami, T. (2005) Zircon fission-track thermochronology and applications to fault studies. *Reviews in Minerology and Geochemistry* **58**, 95–122.

Torres Acosta, V., Schildgen, T.F., Clarke, B.A. et al. (2015a) The effect of vegetation cover on millennial-scale landscape denudation rates in East Africa. *Lithosphere* **7**, 408–420.

Torres Acosta, V., Bande, A., Sobel, E.R. et al. (2015b) Cenozoic extension in the Kenya Rift from low-temperature thermochronology: links to diachronous spatiotemporal evolution of rifting in East Africa. *Tectonics* **34**, 1–20.

Wichura, H., Bousquet, R., Oberhänsli, R. et al. (2011) The Mid-Miocene East African Plateau: A Pre-Rift Topographic Model Inferred from the Emplacement of the Phonolitic Yatta Lava Flow, Kenya. *Geological Society of London, Special Publication* **357**, pp. 285–300.

Willett, S.D. (1999) Orogeny and orography: the effects of erosion on the structure of mountain belts. *Journal of Geophysical Research* **104**, 28 957–28 981.

Willett, S.D. & Brandon, M.T. (2013) Some analytical methods for converting thermochronometric age to erosion rate. *Geochemistry Geophysics Geosystems* **14**, 1–14.

6

Stable Isotope Paleoaltimetry: Paleotopography as a Key Element in the Evolution of Landscapes and Life

Andreas Mulch[1,2] and C. Page Chamberlain[3]

[1] *Senckenberg Biodiversity and Climate Research Centre (BiK-F), Frankfurt, Germany*
[2] *Institute of Geosciences, Goethe University, Frankfurt, Germany*
[3] *Earth System Sciences, Stanford University, Stanford, CA, USA*

Abstract

Reconstructing the topography of mountain ranges is a key element in the interplay of mountain building, long-term continental moisture transport, atmospheric circulation and, ultimately, the distribution of biomes and biodiversity. Stable isotope paleoaltimetry exploits systematic changes in the oxygen ($\delta^{18}O$) or hydrogen (δD) isotopic composition of precipitation along mountain slopes once the interaction of topography and atmospheric moisture transport forces rain out. When recovered from the geologic record, such changes in the $\delta^{18}O$ or δD of precipitation provide insight into the long-term evolution of landscapes and life. Stable isotope paleoaltimetry has typically focused on the geological aspects of mountain building and has produced insight into the combined tectonic and geomorphologic histories of mountain ranges, particularly when high-elevation $\delta^{18}O$ or δD data are referenced against low-elevation regions where rainfall tracks climate-modulated sea-level $\delta^{18}O$ or δD. Future challenges encompass disentangling the surface uplift component from the impact of climate and regional biome change on $\delta^{18}O$ and δD in precipitation. Studies that aim to reconcile biological and geological information on landscape and biome histories will greatly benefit from the integration of stable isotope paleoaltimetry with phylogenetic/biogeographic techniques in order to evaluate competing hypotheses with respect to the timing of surface uplift and the diversification of lineages, as well as from characterizing feedback among changes in surface elevation, atmospheric circulation and regional rainfall patterns. The accelerating pace in the development of new geochemical and phylogenetic techniques opens up new avenues in evolutionary biology and Earth surface process research, particularly if innovative tectonic and phylogenetic or biogeographic approaches are integrated into a common research framework.

Keywords: *geology, landscape evolution, biome change, atmosphere, rainfall, mountain building*

6.1 Introduction

The topography of our planet interacts with various elements of the Earth system: from a geodynamic point of view, mountains reflect the processes in the crust and mantle that facilitate orogeny, and their topography directly affects erosion rates, river transport capacity and weathering of the erosional products of mountain building (e.g., Strecker et al. 2007; Whipple 2009). At the same time, high-elevation mountain ranges and continental plateau regions control continental moisture transport and atmospheric circulation and are responsible for some of the climatic teleconnections of our planet (e.g., Ruddiman & Kutzbach 1989; Ramstein

et al. 1997; Seager et al. 2002). What do we know about the topographic history of the world's largest mountain ranges, how can such histories be recovered and what was their role in establishing present-day biodiversity?

Mountain building directly links to patterns of biomes and biodiversity at the interface of atmospheric and geodynamic processes (e.g., Antonelli et al. 2009; Finarelli & Badgley 2010; Hoorn et al. 2013; Lindner et al. 2014; Eronen et al. 2015). High-elevation, high-relief mountain ranges establish large topographic and climatic gradients extending into the alpine zone, support isolated habitats that promote allopatry and may offer local environmental conditions that serve as refugia during times of climatic or environmental change. More importantly, mountains have

Mountains, Climate and Biodiversity, First Edition. Edited by Carina Hoorn, Allison Perrigo and Alexandre Antonelli.
© 2018 John Wiley & Sons Ltd. Published 2018 by John Wiley & Sons Ltd.
Companion website: www.wiley.com\go\hoorn\mountains,climateandbiodiversity

indirect effects on the evolution of biodiversity: development of high topography can induce threshold conditions to atmospheric circulation and regional rainfall patterns that – despite the long (10^5–10^7 My) time scales of mountain building – rather rapidly alter local environmental conditions. Such changes in atmospheric circulation include the onset of rainfall seasonality (e.g., Mulch et al. 2010), onset of the Asian monsoons (Molnar et al. 2010), continental-scale rerouting of air masses (Chapter 8) (e.g., Poulsen et al. 2010) and hemispheric-scale reorganization of land–sea vapor transport (e.g., Seager et al. 2002; Takahashi & Battisti, 2007). Reconstructing the topography of our planet over geologic time is therefore a key element in understanding the evolution of landscapes and life (e.g., Antonelli et al. 2009; Hoorn et al. 2013; Lindner et al. 2014; Mulch 2016).

6.2 Oxygen and Hydrogen Isotopes in Precipitation

Stable isotopes of oxygen and hydrogen in precipitation, river water and lake water reflect meteorological processes in the hydrological cycle. Since measurements of absolute isotope ratios or of isotope abundances are analytically challenging and would render interlaboratory comparison difficult, oxygen (and, analogously, hydrogen) isotope ratios are presented with respect to a common reference, Vienna standard mean ocean water (VSMOW):

$$\delta^{18}O = \left(\frac{\left(\frac{^{18}O}{^{16}O} \right) sample}{\left(\frac{^{18}O}{^{16}O} \right) reference} - 1 \right) \cdot 1000 \, [\text{‰, VSMOW}] \quad (6.1)$$

On a global scale, the moisture flux from and to the oceans via evaporation, rainout and runoff approaches a dynamic equilibrium – a phenomenon that results in the establishment of the global meteoric water line (GMWL) (Craig 1961):

$$\delta D = 8 * \delta^{18}O + 10 \quad (6.2)$$

Regionally, the partitioning of isotopes in the hydrological cycle may deviate from this relationship, particularly in semi-arid regions where evaporation from lakes, soils or plants increases $\delta^{18}O$ (and, to a lesser extent, δD) values of atmospheric water vapor and results in spatially variable regional meteoric water lines. The key mechanisms that control $\delta^{18}O$ and δD values in nature are evaporation and precipitation (condensation), whereby repeated cycling will enhance the effects of evaporation (increasing $\delta^{18}O$ and, to a lesser extent, δD in atmospheric water vapor) and/or precipitation

(decreasing $\delta^{18}O$ and δD in rainfall). For a more detailed description of stable isotopes in the hydrological cycle, we refer to Clark and Fritz (1997).

6.3 Paleoaltimetry: Determining Surface Uplift

6.3.1 Concept of Stable Isotope Paleoaltimetry

Understanding how topography of the Earth has changed through time is a central issue in Earth sciences and biology. One commonly used substitute for past surface elevations is the timing of exhumation in active orogens. Exhumation, however, is the transport of rocks towards Earth's surface and, therefore, fundamentally different from changes in surface elevation. Exhumation rates are typically measured through low-temperature thermochronology that records cooling of minerals and rocks on their way to Earth's surface (Chapter 5) (e.g., Reiners & Ehlers 2005; Reiners 2007). The most widely used approach to the reconstruction of paleotopography (i.e., the change in surface elevation) is stable isotope paleoaltimetry, based on the systematic decrease in $\delta^{18}O$ or δD of precipitation with altitude (Figure 6.1). This phenomenon has long been recognized (Dansgaard 1964; Ambach et al. 1968; O'Neil & Silberman 1974), but targeted stable isotope paleoaltimetry applications only date back to the turn of the 21st century (e.g., Chamberlain et al. 1999; Garzione et al. 2000; Rowley et al. 2001; Poage & Chamberlain 2002; Mulch et al. 2004). Stable isotope paleoaltimetry has proven to be robust and reliable when it comes to reconstructing first-order topographic histories (for reviews on various aspects, see Blisniuk & Stern 2005; Mulch & Chamberlain 2007; Quade et al. 2007; Rowley & Garzione 2007; Mulch 2016).

The basic theoretical framework is Rayleigh-type distillation (e.g., Dansgaard 1964; Gat 1996). Ascending air parcels pass over a topographic barrier, and precipitation occurs once the relative humidity in an uplifting air parcel reaches saturation at the lifting condensation level and water vapor condenses (Figure 6.2). The $\delta^{18}O$ and δD values of such orographic precipitation (e.g., Roe 2005; Smith 2006) commonly scale with elevation. Orographic precipitation removes the heavy isotopes of oxygen (^{18}O) and hydrogen (D) from ascending water vapor, thereby decreasing $\delta^{18}O$ and δD values as a function of elevation and rainout. Assuming that changes in lifting condensation level track changes in regional topography, $\delta^{18}O$ and δD values of precipitation will thus reflect the elevation of the topographic barrier that moist air parcels have crossed.

The resulting change in $\delta^{18}O$ (or δD) with increasing elevation, $\Delta(\delta^{18}O)$, is commonly termed the "isotopic

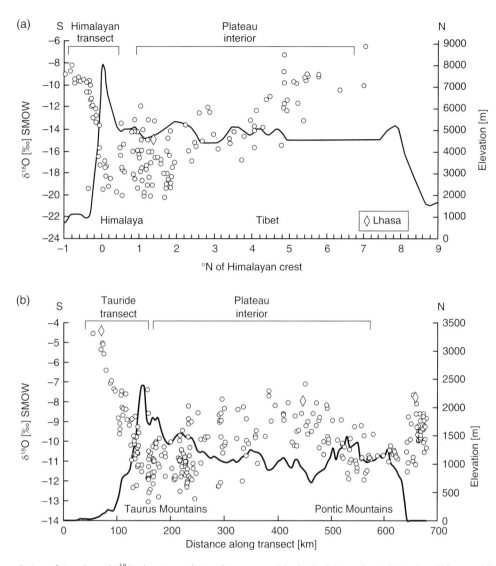

Figure 6.1 Compilation of river-based δ^{18}O-elevation relationships across: (a) a high plateau, Central Himalaya/Tibet; and (b) a low plateau, the Turkish–Anatolian Plateau. δ^{18}O values decrease systematically with elevation along the windward side of the mountain ranges (Himalayan and Tauride transects). In contrast, aridity in the lee of the mountain ranges ("plateau interior") results in higher δ^{18}O values, due to evaporation and water recycling. Circles represent data recovered from stream water and diamonds show averages from long-term station data. The solid line indicates smoothed regional topography. SMOW refers to isotope ratios normalized to standard mean ocean water. *Source:* (a) Adapted from Quade et al. (2007). (b) Schemmel et al. (2013). Reprinted with permission from the American Journal of Science.

lapse rate." The key determinant for such lapse rates is the change in water vapor content from low elevation to the site of precipitation; that is, the degree of distillation of air masses during ascent (Pierrehumbert 1999). The parameters that control distillation and isotopic fractionation are relative humidity, temperature and stratification of the atmosphere (see Rowley & Garzione 2007; Galewsky 2009). Present-day lapse rates have been determined in many mountain ranges, either through direct precipitation measurements (e.g., Ambach et al. 1968; Gonfiantini et al. 2001; Fiorella et al. 2015) or based on river and stream waters (e.g., Yonge et al. 1989; Garzione et al. 2000; Schemmel et al. 2013; Rohrmann

et al. 2014). They average −2.8‰/km in mid-latitude mountain ranges (Poage & Chamberlain 2001). River-based stable isotope lapse rates depend on catchment size, groundwater input and catchment hypsometry, and in many cases such lapse rates are in reasonable agreement with one-dimensional (1D) thermodynamic models (Figure 6.3) that predict changes in δ^{18}O and δD during orographic rainout (Rowley et al. 2001; Rowley & Garzione 2007).

Changes in isotopic lapse rates through time are difficult to quantify, and represent an important challenge for stable isotope paleoaltimetry. In particular, regions affected by high levels of evaporation may deviate from

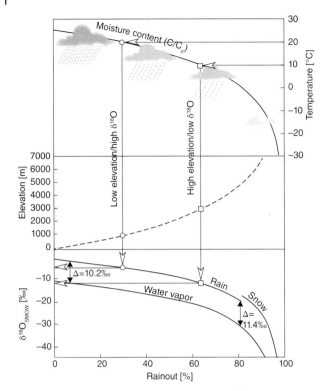

Figure 6.2 Schematic and simplified change in $\delta^{18}O$ values of precipitation following Rayleigh fractionation and cooling of air masses. Water vapor is assumed to start with $\delta^{18}O = -11‰$ at 25 °C, and cools during ascent to −25 °C. At 0 °C, oxygen isotope fractionation changes from vapor-water to snow-water fractionation. The elevation curve (dashed line) assumes a temperature lapse rate of 5 °C/km. In this example, atmospheric moisture content at 20 °C and 1000 m elevation (white circles) is still relatively high, and corresponds to higher rainfall $\delta^{18}O$ values when compared to rainfall at 10 °C and 3000 m elevation (white squares). *Source:* Adapted from Clark & Fritz (1997).

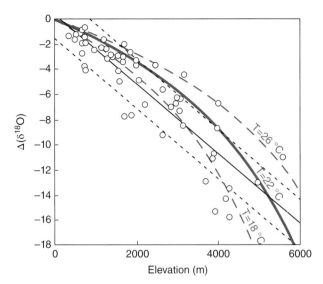

Figure 6.3 Change in $\delta^{18}O$ of river water as a function of net elevation change. Compilation is restricted to moderate elevation, mid-latitude settings and averages −2.8‰/km (solid black line; linear fit), with uncertainty estimates (dotted) after Poage & Chamberlain (2001). An alternative approach models $\delta^{18}O$ values of precipitation normalized (to sea level) as a function of elevation (curved gray line; modified after Rowley & Garzione 2007). The solid gray curve represents mean result based on Monte Carlo simulations of air parcels with starting temperature, T, and relative humidity, RH, corresponding to 22 °C and 80%, respectively. Dashed gray curves show the impact of temperature varying by ±4 °C. *Source:* Adapted from Poage & Chamberlain (2001) and Rowley & Garzione (2007).

the global average. Spatial variations in isotope lapse rates can be caused by convective precipitation regimes where ascent and cooling of air masses occurs independently of surface topography (e.g., Rohrmann et al. 2014) or by hydrologically closed basin systems where evaporation balances precipitation (i.e., there is no net water loss from the atmosphere) (e.g., Ingraham & Taylor 1991; Lechler & Niemi 2011; Mulch 2016). Similarly, topographic scenarios exist where the relationship between horizontal advection of air masses and vertical stratification of the atmosphere results in blocking of air masses and ultimately prevents orographic ascent (e.g., Galewsky 2009). Further, isotope-enabled global circulation models suggest that isotope lapse rates may change during surface uplift (Poulsen et al. 2010). Evaluating whether temporal and spatial changes in isotope lapse rates influence high-elevation $\delta^{18}O$ and δD values of precipitation on a case-by-case basis is, therefore, an important step in

paleoelevation reconstructions. When compared to the present day, however, the first 30 My of the Cenozoic were significantly warmer, resulting in shallower isotopic lapse rates (e.g., Poulsen & Jeffery 2011), typically underestimating past elevations.

6.3.2 Applications of Stable Isotope Paleoaltimetry

Systematic changes in $\delta^{18}O$ or δD values of precipitation along a topographic gradient can be recovered from the geologic record through a broad range of geologic (and sometimes biologic) materials, and recent advances in geochemical and modeling tools allow for increasingly integrative approaches to understanding the links and feedbacks among these geodynamic, atmospheric and evolutionary processes. In this section, we discuss some of the most common proxy materials. Given the rapidly expanding number of proxy materials and approaches, we would like to highlight that the use of multiple altimeters is recommended wherever possible. Not only does this bolster our confidence in the results from individual isotopic proxy materials, but more importantly, it allows us to employ methods that need to satisfy different sets of assumptions concerning, for example, mechanisms and time scales of hydrogen or oxygen incorporation into the proxy material, transport pathways of meteoric water to the site of proxy formation or post-depositional isotope exchange. It is important to note that analysis of primary (i.e., not diagenetically altered) materials is a fundamental underpinning of any stable isotope paleoaltimetry study. In particular, carbonate proxy materials can be prone to diagenesis, recrystalization and isotope exchange if burial depths and temperatures exceed critical values (e.g., DeCelles et al. 2007; Methner et al. 2016).

6.3.2.1 Oxygen Isotopes in Authigenic/Pedogenic Proxy Materials

In general, the ideal proxy materials for this type of analysis were formed close to Earth's surface in places such as ancient soils or lakes. The most commonly used proxy material is calcium carbonate ($CaCO_3$) from lacustrine and pedogenic sedimentary environments: calcium carbonate is ubiquitous in semi-arid environments and is relatively easy to analyze for both $\delta^{13}C$ and $\delta^{18}O$ (see, e.g., Chamberlain et al. 2014; Caves et al. 2015). $\delta^{13}C$ values reveal information about the carbon dynamics in soils and lakes, such as the relative abundance of C3 and C4 plants (Quade & Cerling 1995), soil productivity (Caves et al. 2014), atmosphere–water CO_2 exchange and photosynthesis/respiration of aquatic plants. $\delta^{18}O$ values of carbonate proxies mirror $\delta^{18}O$ of the source water and its evaporative ^{18}O enrichment. One disadvantage of carbonate proxy materials is the relative lack of calcium carbonate in soils from high-rainfall areas. Thus, there is a systematic bias in our understanding of terrestrial paleoclimate towards semi-arid areas (e.g., Chamberlain et al. 2014). One opportunity to overcome this bias lies in measuring $\delta^{18}O$ and δD of authigenic hydrous minerals such as smectite and kaolinite clays. Although it is more difficult to obtain pure clay samples and conduct isotopic analysis, the combined $\delta^{18}O$ and δD data provide a record of meteoric water compositions, evaporation and, in select cases, temperatures of clay formation (e.g., Chamberlain et al. 1999; Mix et al. 2016).

6.3.2.2 Stable Isotopes in Biogenic Archives

Isotopic analyses of biologic materials, such as teeth, shell material and plant remains, have been shown to reflect paleoclimate (e.g., Kohn & Dettman 2007). It is well known that $\delta^{18}O$ values of mammalian fossil teeth reflect oxygen uptake through the animal's diet, which ultimately tracks $\delta^{18}O$ values of local meteoric water (e.g., Kohn & Cerling 2002). Moreover, since there are often strong seasonal differences in $\delta^{18}O$ values of meteoric water and tooth enamel forms progressively over

months to years, it is possible to reconstruct both $\delta^{18}O$ values of drinking (and ultimately local meteoric) water and the seasonality of $\delta^{18}O$ values in the local hydrologic cycle over the growth duration of individual teeth. This information, when used in conjunction with authigenic minerals such as calcite or clay minerals that form over hundreds to thousands of years, can provide insight into both long- and short-term trends in paleoclimate. Thus, it should be possible to decipher long-term regional trends in the $\delta^{18}O$ values of water affected by mountain uplift and global climate change from water compositions in local catchments, as well as trends on shorter time scales as recorded in fossil teeth.

6.3.2.3 Hydrogen Isotopes in Volcanic Glass

Following eruption and subsequent deposition, volcanic glass from silicic ashfall deposits may incorporate water, increasing the water content of such glass from <1% to 6–8% H_2O by weight. The resulting δD values in the hydrated glass reflect a time-integrated δD signal of hydration water over the time scales of hydration (10^1–10^3 years) (e.g., Friedman et al. 1993; Mulch et al. 2008; Dettinger & Quade 2015). Laboratory isotope-exchange experiments on natural glass samples suggest that δD values may be affected under diagenetic conditions (Nolan & Bindeman 2013). There is empirical evidence, however, that once volcanic glass becomes water-saturated, only minor hydrogen (and, to an even lesser extent, oxygen) isotope exchange occurs (e.g., Friedman et al. 1993). As a consequence, δD values of hydrated volcanic glass have preservation potential over geological time scales in (semi-)arid environments (e.g., Friedman et al. 1993; Mulch et al. 2008). Paleoaltimetry studies based on δD values of hydrated volcanic glass have documented long-term persistence of elevated topography in the Sierra Nevada of California (Mulch et al. 2008; Cassel et al. 2009) and the Great Basin of western North America (Friedman et al. 1993; Cassel et al. 2012), surface uplift of the Andean plateau (Saylor & Horton 2014) and basin fill and uplift histories in the Eastern Andean Cordillera of Argentina (Canavan et al. 2014; Pingel et al. 2014, 2016).

6.3.2.4 Hydrogen Isotopes in Organic Biomarkers

Recent approaches to stable isotope paleoaltimetry include compound-specific δD measurements of leaf-wax *n*-alkanes preserved in soils or sediments. Long-chain *n*-alkanes are major components of epicuticular leaf-wax lipids produced by terrestrial vascular plants (Eglinton & Hamilton 1967) and are commonly well preserved in sediments and soils. As soil (and ultimately rain) water is the primary hydrogen source during photosynthesis, the biosynthetic products of plants, such as leaf waxes, hold the potential to track δD in the meteoric water cycle (e.g., Sachse et al. 2012). Long-chain *n*-alkanes have been documented to record the altitude effect on δD of precipitation (e.g., Jia et al. 2008; Luo et al. 2011; Nieto-Moreno et al. 2016), and an increasing number of studies have investigated the potential of *n*-alkanes for use in stable isotope paleoaltimetry (e.g., Peterse et al. 2009), with applications in the northern Sierra Nevada (Hren et al. 2010), the Tibetan Plateau (Polissar et al. 2009; Zhuang et al. 2014), the Eastern Cordillera of Colombia (Anderson et al. 2015; Rohrmann et al. 2016) and the Southern Alps (Zhuang et al. 2015). Collectively, these studies suggest that leaf-wax *n*-alkanes in ancient sediments provide a robust paleoelevation proxy. Further, in contrast to hydrated volcanic glass and pedogenic carbonates, leaf-wax *n*-alkanes permit stable isotope paleoaltimetry outside semi-arid regions of our planet, where such common proxy materials either undergo rapid weathering (volcanic glass) or are not formed (pedogenic carbonates).

6.3.2.5 Hydrogen Isotopes in Faults and Shear Zones

Hydrous silicates that crystallize from meteoric fluids in the brittle or ductile segments of extensional faults or mylonitic shear zones provide isotopic and geochronologic markers to link surface elevation to the geodynamics of orogens (e.g., Mulch et al. 2004, 2007). Despite crystallization and growth at several kilometers' depth, mylonitic hydrous silicates in such environments have been documented to reliably record δD values of meteoric fluids present at the footwall–hanging wall interface of the active detachment mylonite (e.g., Fricke et al. 1992; Mulch et al. 2004, 2007). One particular advantage of this approach is that stable isotope paleoaltimetry data can be linked to the dynamics of mountain building during the late stages of orogeny, when fault activity provides a major control on surface elevation within the orogen. The extended time scales (10^4–10^6 My) involved in the crystallization and growth of such hydrous minerals provide a robust, long-term average of meteoric water–rock interaction, characteristic for the time scales of readjustments in surface elevation. Thus, shear zone-based stable isotope paleoaltimetry not only allows access to regions of high erosion that do not contain sedimentary archives but provides a direct link between tectonics and long-term meteoric water compositions in high-elevation orogens (Figure 6.4). Using a variety of hydrous silicate proxy materials, such as muscovite and biotite, clay minerals and other hydrothermal minerals, δD-based stable isotope paleoaltimetry has been successfully applied in the North American Cordillera (Mulch et al.

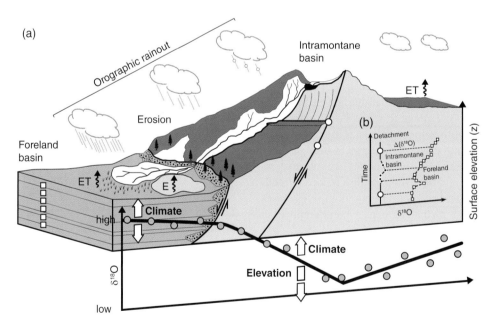

Figure 6.4 Practical approaches in stable isotope paleoaltimetry. (a) δ18O values of precipitation decrease systematically during orographic rainout and associated cooling and condensation of water vapor as a function of elevation, z. Increasing δ18O values in the lee of the mountain range result from evaporative 18O enrichment in precipitation. Over geologic time, δ18O values of advected water vapor may change due to changing climate conditions (white arrows). (b) Reconstructing differences in the oxygen isotopic composition of precipitation, Δ(δ18O), between low-elevation sites (e.g., samples from foreland basin, squares) and sites at unknown high elevation (intramontane basins or faults/detachments, circles) eliminates some of the uncertainties associated with single-site stable isotope paleoaltimetry. ET, evapotranspiration; E, evaporation. *Source:* Adapted from Campani et al. (2012) and Mulch (2016). See also Plate 12 in color plate section.

2004, 2007; Gébelin et al. 2012; Methner et al. 2015), the European Alps (Sharp et al. 2005; Campani et al. 2012), the high peaks of the Himalaya (Gébelin et al. 2013) and the eastern Mediterranean (Hetzel et al. 2013).

6.3.2.6 Clumped Isotope (Δ_{47})-based Paleoaltimetry Approaches

Clumped isotope (Δ_{47}) thermometry exploits systematic deviations from a stochastic distribution of heavy isotopes (^{13}C and ^{18}O) within the mineral lattice. This "clumping" is based on the equilibrium isotope exchange reaction:

$$Ca^{13}C^{16}O_3 + Ca^{12}C^{18}O^{16}O_2 = Ca^{12}C^{16}O_3 + Ca^{13}C^{18}O^{16}O_2 \quad (6.3)$$

among different isotopic species of carbonate molecules (isotopologues) – a temperature-controlled process that results in the increase in abundance of multiply substituted isotopologues with decreasing temperature (e.g., Eiler 2007). This temperature dependence is expressed through the Δ_{47} value (e.g., Eiler 2007, 2011). Reconstructing paleoelevation based on Δ_{47} temperatures alone, however, suffers from the same limitation as, for example, temperature-based paleobotanical approaches to paleoaltimetry: temperature lapse rates vary through time and space. It is also important to note that carbonates may be prone to diagenetic resetting of the clumped isotope information – a phenomenon that is frequently not easy to detect in sedimentary sections that have undergone various degrees of burial and post-depositional heating.

Despite these caveats, carbonate clumped isotope (Δ_{47}) (paleo)thermometry has the advantage of determining carbonate formation temperatures independent of the $\delta^{18}O$ value of ambient water from which the carbonate formed (e.g., Ghosh et al. 2006a). It thus opens opportunities to disentangle some of the confounding effects of temperature and topographic change that may affect terrestrial carbonate $\delta^{18}O$ records. Paleoelevation reconstructions using Δ_{47} measurements exploit either (i) temperature gradients, by relating proxy formation temperatures to mean annual air temperatures (MAATs), or (ii) the systematics of $\delta^{18}O$ values of meteoric water reconstructed by coupled Δ_{47} and bulk $\delta^{18}O$ analyses of carbonates. Using a variety of proxy materials, such as paleosols, lacustrine carbonate and shell material, Δ_{47} (paleo)thermometry has supported paleoaltimetry studies within the Andes (e.g., Ghosh et al. 2006b; Garzione et al. 2008, 2014; Leier et al. 2013), the Himalaya/Tibetan Plateau (e.g., Wang et al. 2013; Huntington et al. 2015) and the North American Cordillera (e.g., Huntington et al. 2010; Lechler et al. 2013; Fan et al. 2014; Methner et al. 2016).

6.4 Modeling Approaches to Determining Stable Isotopes in Precipitation Patterns

Stable isotope paleoaltimetry can benefit from approaches that include modern calibration data and modeling of past $\delta^{18}O$ and δD. In general, three classes of models have been used to estimate paleoelevation.

First, 1D thermodynamic models allow for quantification of $\delta^{18}O$ and δD values of precipitation and their relationship to elevation (Rowley et al. 2001). These models are based on Rayleigh distillation of precipitation (see, e.g., Clark & Fritz 1997), and track the temperature and relative humidity of individual air parcels during ascent and cooling. In general, such models robustly reproduce measurements of modern meteoric waters in mountainous regions (see review in Rowley 2007); more importantly, they permit the evaluation of uncertainties in isotopic lapse rates under different climate conditions.

While such models provide the theoretical underpinnings of stable isotope paleoaltimetry, they approach their limits in cases of significant vapor recycling (Rowley & Garzione 2007; Chamberlain et al. 2014). For example, a high proportion of precipitation in the continental interior of the Great Basin of the western USA is recycled through evapotranspiration (Ingraham & Taylor 1991; Lechler & Niemi 2011; Mulch 2016). The high degree of water recycling causes the moisture gradient in the atmosphere (i.e., the decrease in precipitation from the oceanic moisture source to continental interiors) to shallow and Rayleigh models, and thus cannot reliably predict the $\delta^{18}O$ and δD values of precipitation. Using $\delta^{18}O$ (or δD) values as a paleoaltimeter requires the role of water vapor recycling in continental interiors to be determined. On continental scales, a second, more robust approach is to use a 1D vapor transport model that accounts for eddy diffusion, advection (Hendricks et al. 2000) and recycling of water vapor by plants (Chamberlain et al. 2014; Winnick et al. 2014). Measuring across-continent $\delta^{18}O$ gradients along vapor transport pathways and comparing them with model outputs provides a powerful means of quantitatively reconstructing paleoelevation and assessing the importance of vapor recycling (Chamberlain et al. 2014; Winnick et al. 2014). As such, these models have been used to quantify biosphere–atmosphere interactions and grassland expansion on $\delta^{18}O$ records (Mix et al. 2013; Chamberlain et al. 2014), the controls governing the "continental effect" (Winnick et al. 2014) and the role of westerlies in controlling aridity in central Asia through time (Caves et al. 2015).

A third approach is the use of isotope-enabled global climate models (GCMs); that is, models that predict the isotopic composition of precipitation. Because $\delta^{18}O$ and δD values of precipitation are a function of how topography and climate change collectively, such GCM simulations have increased power in resolving the interactions of climate and topography, and in determining how these affect the $\delta^{18}O$ and δD values of meteoric water (e.g., Sturm et al. 2010). Therefore, such models have greatly enhanced our understanding of surface uplift in the Andes (Ehlers & Poulsen 2009) and the North American Cordillera (Feng et al. 2013). In particular, GCM simulations that track surface uplift of the Andes have challenged original interpretations of isotopic proxy data (Garzione et al. 2008; Ehlers & Poulsen 2009). GCM simulations occasionally yield unexpected results, not captured by either Rayleigh or vapor transport models. For example, it has been proposed that a Cretaceous high plateau (the "Nevadaplano") (DeCelles 2004) existed in the western USA, yet there is little stable isotopic evidence for such a plateau. Using an isotope-enabled GCM, Poulsen and Jeffery (2011) documented that $\delta^{18}O$ values of meteoric water at high elevations during a warm Cretaceous climate were likely higher than predicted by Rayleigh models, as a result of reduction of the surface to upper-atmosphere temperature gradient: reduction of this gradient reduces the oxygen isotope lapse rate, leading to higher $\delta^{18}O$ values at high altitudes. Thus, the GCM results provide insight into why $\delta^{18}O$ data may not reflect the putative high elevations that seem evident from geologic arguments.

6.5 Examples of Stable Isotope Paleoaltimetry

6.5.1 Western North America

There is robust evidence for the construction of an Eocene western North American highland and its subsequent collapse in the mid-Miocene. Stable isotope paleoaltimetry data are consistent with surface elevations that developed in a north–south fashion (Mix et al. 2011; Chamberlain et al. 2012). These highlands reached peak heights of ~4 km at ~50 Ma in British Columbia (Mulch et al. 2004) and Montana (Kent-Corson et al. 2009), with a "topographic wave" migrating southward into central and northern Nevada at ~40 Ma (Mulch et al. 2015). By the late Eocene, the Nevada highland was bordered to the west by a high proto-Sierra Nevada, with rivers from the plateau flowing over its western flanks (Mulch et al. 2006; Cassel et al. 2012), and on the east by the high Eocene Rocky Mountains. To the north, the plateau gave way to a high but narrower mountain range that formed a major drainage divide between rivers flowing to the west, draining into the Pacific, and to the east, draining into the Gulf of Mexico. These rivers cut into this Eocene highland during uplift (Dumitru et al. 2016), possibly providing corridors for animal migration from the lowlands east of the Rocky Mountains. In the Eocene, this relatively narrow topographic high extended through central Idaho into western British Columbia and was centered on the massive Challis volcanic field (Montana).

It has long been recognized that the global temperature decrease at the Eocene–Oligocene (EO) boundary had a very different impact on mammalian extinctions in Europe versus North America. Whereas major extinction accompanied the EO boundary in Eurasia, mammal communities in North America remained largely unaffected. One explanation for this difference is rooted in the North American Eocene surface elevation history (Eronen et al. 2015): due to mid- to late-Eocene (~45–40 Ma) high elevations in western North America, mammals were already adapted to a (semi-)arid and cooler climate in North America before the onset of Oligocene cooling around 34 Ma. These higher relief landscapes also provided climatic gradients along the flanks of the orogen that permitted large mammals to find their preferred niche by migration to lower-elevation regions during times of cooling. In contrast, Europe at the time was dominated by a patchwork of marine embayments with habitats near sea level that did not allow mammals to accommodate their habitat to global environmental change.

6.5.2 Central Asia, Himalaya and Tibet

Surface uplift of large mountain ranges and plateaus such as the Himalaya and Tibet or the Tien Shan is likely to have been a primary factor influencing Cenozoic climate and biodiversity both regionally and globally, providing barriers and corridors for the migration of animals and plants. Similarly, surface uplift of the Hangaay and Altai Mountains in Mongolia altered the trajectories of northern hemisphere westerlies and played a major role in the aridification of central Asia (Caves et al. 2015). Although the picture is far from complete, what is emerging is that many of these mountain ranges were high prior to the Miocene (e.g., Caves et al. 2015; Deng & Ding 2015). However, the pattern and timing of uplift is less clear than in other orogens, due to relatively poor age constraints for many of the terrestrial sedimentary sections and the complex interplay of air masses at the northern margin of the Tibetan Plateau, which impact stable isotope paleoaltimetry. At what height does a mountain become a barrier for a particular species, and are there corridors that allow dispersal of mammals? These questions are illustrated in the ability of large mammals to disperse across Tibet in the Oligocene but not in the Miocene (Deng & Ding 2015). Yet, there is excellent evidence for a high southern Tibet, segments of which have been at least 4 km high since the late Eocene (Rowley & Currie 2006). Why, then, did these high mountains not block north–south exchange of animals? Given the lack of spatially distributed paleoaltimetry data over Tibet, we are currently unable to quantify the Cenozoic relief structure of what is now the Tibetan Plateau. One could therefore envision that migration corridors may have existed for some species. By the middle Miocene, however, Mt. Everest and the high central Himalayan peaks had already attained close-to-modern elevations (Gébelin et al. 2013), which is in line with the Himalaya representing a migration barrier for large mammals since at least that time. Identifying the different elevation histories for Tibet and the central Himalaya is a focus of current research and has yet to be incorporated into models and observations on the history of biodiversity in the region.

Larger links and feedbacks exist among biological, climatic and tectonic processes of the Tibetan-Himalayan system: the rapid radiation of C4 grasses in the late Miocene may (at least in part) have been causally linked to surface-uplift exhumation and may have intensified chemical weathering of Tibet and the Himalaya (e.g., Molnar et al. 2010). Grasses that fix carbon through the C4 photosynthetic pathway are favored in seasonally dry, low-pCO_2 environments. Enhanced chemical weathering as a result of uplift, erosion and weathering may have caused a reduction of atmospheric CO_2, promoted global cooling and ultimately intensified the Indian summer monsoon (Quade et al. 1989). Such climate conditions would have favored the spread of C4 grasses, not only in those parts of Asia directly affected by the monsoonal climate, but also globally as CO_2 was scrubbed from the atmosphere. Recent analyses, however, suggest that the global drop in pCO_2 may have occurred much earlier in Earth's history, well before the expansion of C4 grasses (e.g., Beerling & Royer 2011), highlighting that the impact of Himalayan chemical weathering on global pCO_2 levels may have been superimposed on protracted global-scale climate–vegetation feedbacks. Moreover, it has been suggested based on a compilation of global stable isotope paleoclimate data that the radiation of C4 grasses may itself have altered regional hydrology by enhancing rainfall seasonality, thereby favoring the grasses' own expansion (Mix et al. 2013; Chamberlain et al. 2014).

These are but a few examples of how the uplift of the Asian mountain ranges may have affected biodiversity in Asia, and life on Earth. At this point, many of the relationships between mountain building and biology are speculative (see, e.g., Renner 2016). Despite a vast body of literature on the tectonic and uplift history of the Tibetan Plateau, it is only within the last decade that significant progress in reconstructing the elevation history of the world's largest mountain belt has been made. It is no surprise, then, that we have only started to understand the multiple interactions among lithospheric dynamics, atmospheric circulation and evolution of landscapes and biodiversity in this region. At present, we are limited by the lack of integration of paleoaltimetry data with concurrent studies in evolutionary biology, biogeography and vegetation and species distribution modeling through (deep) time.

6.6 Conclusion

With the advent of global-scale atmospheric modeling techniques capturing the impact of topography on stable isotopes in precipitation, refined paleoaltimetry approaches and rapidly expanding phylogenetic techniques in biogeography and evolutionary biology, we are increasingly able to evaluate competing hypotheses with respect to the timing of surface uplift and the development of species and biodiversity. Understanding the interplay between biodiversity and mountain building has therefore become a key challenge for the geological and biological sciences. Future advances in either of these fields will greatly benefit from: (i) identification of topographically induced teleconnections in the global climate system that affect regional precipitation patterns; and (ii) characterization of topographic thresholds to changes in atmospheric circulation and precipitation, as these are likely to be more important to the evolution of lineages than are changes in surface elevation alone. Using a broad array of isotopic and genetic tools, paleoaltimetry can develop into a truly interdisciplinary field and open new avenues for the study of the long-term evolution of landscapes and life.

References

Ambach, W., Dansgaard, W., Eisner, H. & Mollner, J. (1968) The altitude effect on the isotopic composition of precipitation and glacier ice in the Alps. *Tellus* **20**, 595–600.

Anderson, V.J., Saylor, J.E., Shanahan, T.M. & Horton, B.K. (2015) Paleoelevation records from lipid biomarkers: application to the tropical Andes. *Geological Society of America Bulletin* **127**, 1604–1616.

Antonelli, A., Nylander, J.A.A., Persson, C. & Sanmartin, I. (2009) Tracing the impact of the Andean uplift on Neotropical plant evolution. *Proceedings of the National Academy of Sciences USA* **106**, 9749–9754.

Beerling, D.J. & Royer, D.L. (2011) Convergent Cenozoic CO_2 history. *Nature Geoscience* **4**, 418–420.

Blisniuk, P.M. & Stern, L.A. (2005) Stable isotope paleoaltimetry – a critical review. *American Journal of Science* **305**, 1033–1074.

Campani, M., Mulch, A., Kempf, O. et al. (2012) Miocene paleotopography of the Central Alps. *Earth and Planetary Science Letters* **337**, 174–185.

Canavan, R.R., Carrapa, B., Clementz, M.T. et al. (2014) Early Cenozoic uplift of the Puna Plateau, Central Andes, based on stable isotope paleoaltimetry of hydrated volcanic glass. *Geology* **42**, 447–450.

Cassel, E.J., Graham, S.A. & Chamberlain, C.P. (2009) Cenozoic tectonic and topographic evolution of the northern Sierra Nevada, California, through stable isotope paleoaltimetry in volcanic glass. *Geology* **37**, 547–550.

Cassel, E.J., Graham, S.A., Chamberlain, C.P. & Henry, C.D. (2012) Early Cenozoic topography, morphology, and tectonics of the northern Sierra Nevada and western Basin and Range. *Geosphere* **8**, 229–249.

Caves, J.K., Sjostrom, D.J., Mix, H.T. et al. (2014) Aridification of Central Asia and uplift of the Altai and Hangay Mountains, Mongolia: stable isotope evidence. *American Journal of Science* **314**, 1171–1201.

Caves, J.K., Winnick, M.J., Graham, S. et al. (2015) Role of the westerlies in Central Asia over the Cenozoic. *Earth and Planetary Science Letters* **428**, 33–43.

Chamberlain, C.P., Poage, M., Craw, D. & Reynolds, R. (1999) Topographic development of the Southern Alps recorded by the isotopic composition of authigenic clay minerals, South Island, New Zealand. *Chemical Geology* **155**, 279–294.

Chamberlain, C.P., Mix, H.T., Mulch, A. et al. (2012) Cenozoic climatic and topographic evolution of the Western North America Cordillera. *American Journal of Science* **312**, 213–262.

Chamberlain, C.P., Winnick, M., Mix, H.T. et al. (2014) The impact of Neogene grassland expansion and aridification on the isotopic composition of continental precipitation. *Global Biogeochemical Cycles* **28**, 992–1004.

Clark, I. & Fritz, P. (eds.) (1997) *Environmental Isotopes in Hydrogeology*. Boca Raton, FL: CRC Press.

Craig, H. (1961) Isotopic variations in meteoric waters. *Science* **133**, 1702–1703.

Dansgaard, W. (1964) Stable isotopes in precipitation. *Tellus* **16**, 436–468.

DeCelles, P.G. (2004) Late Jurassic to Eocene evolution of the Cordilleran thrust belt and foreland basin systems, western USA. *American Journal of Science* **304**, 105–168.

DeCelles, P.G., Quade, J., Kapp, P. et al. (2007) High and dry in central Tibet during the Late Oligocene. *Earth and Planetary Science Letters* **253**, 389–401.

Deng, T. & Ding, L. (2015) Paleo-altimetry reconstructions of the Tibetan Plateau: progress and contradictions. *National Science Review* **2**, 417–437.

Dettinger, M.P. & Quade, J. (2015) Testing the analytical protocols and calibration of volcanic glass for the reconstruction of hydrogen isotopes in paleoprecipitation. *Geological Society of America Memoirs* **212**, 261–276.

Dumitru, T.A., Elder, W.P., Hourigan, J.K. et al. (2016) Four Cordilleran paleorivers that connected Sevier thrust zones in Idaho to depocenters in California, Washington, Wyoming, and, indirectly, Alaska. *Geology* **44**, 75–78.

Eglinton, G. & Hamilton, R.J. (1967) Leaf epicuticular waxes. *Science* **156**, 1322–1335.

Ehlers, T.A. & Poulsen, C.J. (2009) Influence of Andean uplift on climate and paleoaltimetry estimates. *Earth and Planetary Science Letters* **281**, 238–248.

Eiler, J.M. (2007) "Clumped-isotope" geochemistry – the study of naturally-occurring, multiply-substituted isotopologues. *Earth and Planetary Science Letters* **262**, 309–327.

Eiler, J.M. (2011) Paleoclimate reconstruction using carbonate clumped isotope thermometry. *Quaternary Science Reviews* **30**, 3575–3588.

Eronen, J.T., Janis, C.M., Chamberlain, C.P. & Mulch, A. (2015) Mountain uplift explains differences in Palaeogene patterns of mammalian evolution and extinction between North America and Europe. *Philosophical Transactions of the Royal Society of London B* **282**, 20150136.

Fan, M., Hough, B.G. & Passey, B.H. (2014) Middle to late Cenozoic cooling and high topography in the central Rocky Mountains: constraints from clumped isotope geochemistry. *Earth and Planetary Science Letters* **408**, 35–47.

Feng, R., Poulsen, C.J., Werner, M. et al. (2013) Evolution of Early Cenozoic topography, climate and stable isotopes of precipitation in the North America Cordillera. *American Journal of Science* **313**, 613–648.

Finarelli, J. A. & Badgley, C. (2010) Diversity dynamics of Miocene mammals in relation to the history of tectonism and climate. *Proceedings of the Royal Society of London B* **277**, 2721–2726.

Fiorella, R.P., Poulsen, C.J., Pillco Zolá, R.S. et al. (2015) Spatiotemporal variability of modern precipitation $\delta^{18}O$ in the Central Andes and implications for paleoclimate and paleoaltimetry estimates. *Journal of Geophysical Research – Atmospheres* **120**, 4630–4656.

Fricke, H.C., Wickham, S. & O'Neil, J.R. (1992) Oxygen and hydrogen isotope evidence for meteoric water infiltration during mylonitization and uplift in the Ruby Mountains–East Humboldt Range core complex, Nevada. *Contributions to Mineralogy and Petrology* **108**, 203–221.

Friedman, I., Gleason, J. & Warden, A. (1993) Ancient climate from deuterium content of water in volcanic glass. *Geophysical Monograph* **78**, 309–319.

Galewsky, J. (2009) Orographic precipitation isotopic ratios in stratified atmospheric flows: implications for paleoelevation studies. *Geology* **37**, 791–794.

Garzione, C.N., Quade, J., DeCelles, P.G. & English, N.B. (2000) Predicting paleoelevation of Tibet and the Himalaya from $\delta^{18}O$ vs altitude gradients in meteoric water across the Nepal Himalaya. *Earth and Planetary Science Letters* **183**, 215–229.

Garzione, C.N., Hoke, G.D., Libarkin, J.C. et al. (2008) Rise of the Andes. *Science* **320**, 1304–1307.

Garzione, C.N., Auerbach, D.J., Jin-Sook Smith, J. et al. (2014) Clumped isotope evidence for diachronous surface cooling of the Altiplano and pulsed surface uplift of the Central Andes. *Earth and Planetary Science Letters* **393**, 173–181.

Gat, J.R. (1996) Oxygen and hydrogen isotopes in the hydrologic cycle. *Annual Reviews of Earth and Planetary Sciences* **24**, 225–262.

Gébelin, A., Mulch, A., Teyssier, C. et al. (2012) Coupled basin-detachment systems as paleoaltimetry archives of the western North American Cordillera. *Earth and Planetary Science Letters* **335**, 36–47.

Gébelin, A., Mulch, A. & Teyssier, C. (2013) The Miocene elevation of Mount Everest. *Geology* **41**, 799–802.

Ghosh, P., Adkins, J., Affek, H. et al. (2006a) $^{13}C–^{18}O$ bonds in carbonate minerals: a new kind of paleothermometer. *Geochimica et Cosmochimica Acta* **70**, 1439–1456.

Ghosh, P., Garzione, C.N. & Eiler, J.M. (2006b) Rapid uplift of the Altiplano revealed through $^{13}C-^{18}O$ bonds in Paleosol carbonates. *Science* **311**, 511–515.

Gonfiantini, R., Roche, M.A., Olivry, J.C. et al. (2001) The altitude effect on the isotopic composition of tropical rains. *Chemical Geology* **181**, 147–167.

Hendricks, M.B., DePaolo, D.J. & Cohen, R.C. (2000) Space and time variation of $\delta^{18}O$ and δD in precipitation: can paleotemperature be estimated from ice cores? *Global Biogeochemical Cycles* **14**, 851–861.

Hetzel, R., Zwingmann, H., Mulch, A. et al. (2013) Spatiotemporal evolution of brittle normal faulting and fluid infiltration in detachment fault systems: a case study from the Menderes Massif, western Turkey. *Tectonics* **32**, 364–376.

Hoorn, C., Mosbrugger, V., Mulch, A. & Antonelli, A. (2013) Biodiversity from mountain building. *Nature Geoscience* **6**, 154.

Hren, M.T., Pagani, M., Erwin, D.M. & Brandon, M. (2010) Biomarker reconstruction of the early Eocene paleotopography and paleoclimate of the northern Sierra Nevada. *Geology* **38**, 7–10.

Huntington, K.W., Wernicke, B.P. & Eiler, J.M. (2010) Influence of climate change and uplift on Colorado Plateau paleotemperatures from carbonate clumped isotope thermometry. *Tectonics* **29**, TC3005.

Huntington, K.W., Saylor, J., Quade, J. & Hudson, A.M. (2015) High late Miocene–Pliocene elevation of the Zhada Basin, southwestern Tibetan Plateau, from carbonate clumped isotope thermometry. *Geological Society of America Bulletin* **127**, 181–199.

Ingraham, N.L. & Taylor B.E. (1991) Light stable isotope systematics of large-scale hydrologic regimes in California and Nevada. *Water Resources Research* **27**, 77–90.

Jia, G., Wei, K., Chen, F. & Peng, P.A. (2008) Soil n-alkane δD vs. altitude gradients along Mount Gongga, China. *Geochimica et Cosmochimica Acta* **72**, 5165–5174.

Kent-Corson, M.L., Ritts, B.D., Zhuang, G. et al. (2009) Stable isotopic constraints on the tectonic, topographic, and climatic evolution of the northern margin of the Tibetan Plateau. *Earth and Planetary Science Letters* **282**, 158–166.

Kohn, M.J. & Cerling, T.E. (2002) Stable isotope compositions of biogenic apatite. *Reviews in Mineralogy and Geochemistry* **48**, 455–488.

Kohn, M.J. & Dettman, D.L. (2007) Paleoaltimetry for stable isotope compositions of fossils. *Reviews in Mineralogy and Geochemistry* **66**, 155–171.

Lechler, A.R. & Niemi, N.A. (2011) Controls on the spatial variability of modern meteoric $\delta^{18}O$: empirical constraints from the western U.S. and East Asia and implications for stable isotope studies. *American Journal of Science* **311**, 664–700.

Lechler, A.R., Niemi, N.A., Hren, M.T. & Lohmann, K.C. (2013) Paleoelevation estimates for the northern and central proto–Basin and Range from carbonate clumped isotope thermometry. *Tectonics* **32**, 295–316.

Leier, A., McQuarrie, N., Garzione, C. & Eiler, J. (2013) Stable isotope evidence for multiple pulses of rapid surface uplift in the Central Andes, Bolivia. *Earth and Planetary Science Letters* **371**, 49–58.

Lindner, H.P., Rabosky, D.L., Antonelli, A. et al. (2014) Disentangling the influence of climatic and geological changes on species radiations. *Journal of Biogeography* **41**, 1313–1325.

Luo, P., Peng, P.A., Gleixner, G. et al. (2011) Empirical relationship between leaf wax n-alkane δD and altitude in the Wuyi, Shennongjia and Tianshan Mountains, China: implications for paleoaltimetry. *Earth and Planetary Science Letters* **301**, 285–296.

Methner, K., Mulch, A., Teyssier, C. et al. (2015) Eocene and Miocene extension, meteoric fluid infiltration, and core complex formation in the Great Basin (Raft River Mountains, Utah). *Tectonics* **34**, 680–693.

Methner, K., Fiebig, J. Wacker, U. et al. (2016) Eocene-Oligocene proto-Cascades topography revealed by clumped (Δ_{47}) and oxygen isotope ($\delta^{18}O$) geochemistry (Chumstick Basin, WA, USA). *Tectonics* **35**, 546–564.

Mix, H.T., Mulch, A., Kent-Corson, M.L. & Chamberlain, C.P. (2011) Cenozoic migration of topography in the North American Cordillera. *Geology* **39**, 87–90.

Mix, H.T., Winnick, M.J., Mulch, A. & Chamberlain, C.P. (2013) Grassland expansion as an instrument of hydrologic change in Neogene Western North America. *Earth and Planetary Science Letters* **377**, 73–83.

Mix, H., Ibarra, D., Mulch, A. et al. (2016) A hot and high Eocene Sierra Nevada. *Geological Society of America Bulletin* **128**, 531–542.

Molnar, P., Boos, W.R. & Battisti, D.S. (2010) Orographic controls on climate and paleoclimate of Asia: thermal and mechanical roles for the Tibetan Plateau. *Annual Reviews of Earth and Planetary Science* **38**, 77–102.

Mulch, A. (2016) Stable isotope paleoaltimetry and the evolution of landscapes and life. *Earth and Planetary Science Letters* **433**, 180–191.

Mulch, A. & Chamberlain, C.P. (2007) Stable isotope paleoaltimetry in orogenic belts – the silicate record in surface and crustal geological archives. *Reviews in Mineralogy and Geochemistry* **66**, 89–118.

Mulch, A., Teyssier, C., Cosca, M.A. et al. (2004) Reconstructing paleoelevation in eroded orogens. *Geology* **32**, 525–528.

Mulch, A., Graham, S.A. & Chamberlain, C.P. (2006) Hydrogen isotopes in Eocene river gravels and paleoelevation of the Sierra Nevada. *Science* **313**, 87–89.

Mulch, A., Teyssier, C., Cosca, M.A. & Chamberlain, C.P. (2007) Stable isotope paleoaltimetry of Eocene core complexes in the North American Cordillera. *Tectonics* **26**, TC4001.

Mulch, A., Sarna-Wojcicki, A.M., Perkins, M.E. & Chamberlain, C.P. (2008) A Miocene to Pleistocene climate and elevation record of the Sierra Nevada, California. *Proceedings of the National Academy of Sciences USA* **105**, 6819–6824.

Mulch, A., Uba, C.E., Strecker, M.R. et al. (2010) Late Miocene climate variability and surface elevation in the central Andes. *Earth and Planetary Science Letters* **290**, 173–182.

Mulch, A., Chamberlain, C.P., Cosca, M.A. et al. (2015) Rapid change in high-elevation precipitation patterns of western North America during the Middle Eocene Climatic Optimum (MECO). *American Journal of Science* **315**(4), 317–336.

Nieto-Moreno, V., Rohrmann, A., van der Meer, M.T.J. et al. (2016) Elevation-dependent changes in n-alkane δD and soil GDGTs across the South Central Andes. *Earth and Planetary Science Letters* **453**, 234–242.

Nolan, G.S. & Bindeman, I.N. (2013) Experimental investigation of rates and mechanisms of isotope exchange (O, H) between volcanic ash and isotopically-labeled water. *Geochimica et Cosmochimica Acta* **111**, 5–27.

O'Neil, J.R. & Silberman, M.L. (1974) Stable isotope relations in epithermal Au-Ag deposits. *Economic Geology* **69**, 902–909.

Peterse, F., van der Meer, M.T.J., Schouten, S. et al. (2009) Assessment of soil n-alkane and dD and branched tetraether membrane lipid distributions as tools for paleoelevation reconstruction. *Biogeosciences* **6**, 2799–2807.

Pierrehumbert, R.T. (1999) Huascaran $\delta^{18}O$ as an indicator of tropical climate during the Last Glacial Maximum. *Geophysical Research Letters* **26**, 1345–1348.

Pingel, H., Alonso, R.A., Mulch, A. et al. (2014) Pliocene orographic barrier uplift in the southern Central Andes. *Geology* **42**, 691–694.

Pingel, H., Mulch, A., Alonso, R.A. et al. (2016) Surface uplift and convective rainfall along the southern Central Andes (Angastaco Basin, NW Argentina). *Earth and Planetary Science Letters* **440**, 33–42.

Poage, M.A. & Chamberlain, C.P. (2001) Empirical relationships between elevation and the stable isotope composition of precipitation and surface waters: considerations for studies of paleoelevation change. *American Journal of Science* **301**, 1–15.

Poage, M.A. & Chamberlain, C.P. (2002) Stable isotopic evidence for a pre-middle Miocene rain shadow in the Western Basin and Range: implications for the paleotopography of the Sierra Nevada. *Tectonics* **21**, 16-1–16-10.

Polissar, P.J., Freeman, K.H., Rowley, D.B. et al. (2009) Paleoaltimetry of the Tibetan Plateau from D/H ratios of lipid biomarkers. *Earth and Planetary Science Letters* **287**, 64–76.

Poulsen, C.J. & Jeffery, M.L. (2011) Climate change imprinting in stable isotopic compositions of high elevation meteoric water cloaks past surface elevation of major orogens. *Geology* **39**, 595–598.

Poulsen, C.J., Ehlers, T.A. & Insel, N. (2010) Onset of convective rainfall during gradual late Miocene rise of the Central Andes. *Science* **328**, 490–493.

Quade, J. & Cerling, T.E. (1995) Expansion of C_4 grasses in the late Miocene of northern Pakistan: evidence from stable isotopes in paleosols. *Palaeogeography, Palaeoclimatology, Palaeoecology* **115**, 91–116.

Quade, J., Cerling, T.E. & Bowman, J.R. (1989) Development of Asian monsoon revealed by marked ecological shift during the latest Miocene in northern Pakistan. *Nature* **342**, 163–166.

Quade, J., Garzione, C. & Eiler, J. (2007) Paleoelevation reconstruction using pedogenic carbonates. *Reviews in Mineralogy and Geochemistry* **66**, 53–87.

Ramstein, G., Fluteau, F., Besse, J. & Joussame, S. (1997) Effect of orogeny, plate motion and land-sea distribution on Eurasian climate change over the past 30 million years. *Nature* **386**, 788–795.

Reiners, P.W. (2007) Thermochronologic approaches to paleotopography. *Reviews in Mineralogy and Geochemistry* **66**, 243–267.

Reiners, P. & Ehlers, T.A. (eds.) (2005) *Low-Temperature Thermochronology: Techniques, Interpretations, and Applications.* Reviews in Mineralogy and Geochemistry, vol. **58**. Washington, DC: Mineralogical Society of America.

Renner, S.S. (2016) Available data point to a 4-km-high Tibetan Plateau by 40 Ma, but 100 molecular-clock papers have linked supposed recent uplift to young node ages. *Journal of Biogeography* **43**, 1479–1487.

Roe, G.H. (2005) Orographic precipitation. *Annual Reviews of Earth Sciences* **33**, 645–671.

Rohrmann, A., Strecker, M.R. & Bookhagen, B. (2014) Can stable isotopes ride out the storm? The role of convection for water isotopes in models, records, and paleoaltimetry studies in the central Andes. *Earth and Planetary Science Letters* **407**, 187–195.

Rohrmann, A., Sachse, D., Mulch, A. et al. (2016) Miocene orographic uplift forces rapid hydrological change in the southern central Andes. *Nature Scientific Reports* **6**, 35678.

Rowley, D.B. (2007) Stable isotope-based paleoaltimetry: theory and validation. *Reviews in Mineralogy and Geochemistry* **66**, 23–52.

Rowley, D.B. & Currie, B.S. (2006) Palaeo-altimetry of the late Eocene to Miocene Lunpola basin, central Tibet. *Nature* **439**, 677–681.

Rowley, D.B. & Garzione, C. (2007) Stable-isotope based paleoaltimetry. *Annual Reviews of Earth and Planetary Sciences* **35**, 463–506.

Rowley, D.B., Pierrehumbert, R.T. & Currie, B. (2001) A new approach to stable isotope-based paleoaltimetry: implications for paleoaltimetry and paleohypsometry of the High Himalaya since the Late Miocene. *Earth and Planetary Science Letters* **188**, 253–268.

Ruddiman, W.F. & Kutzbach, J.E. (1989) Forcing of late Cenozoic Northern Hemisphere climate by plateau uplift in southern Asia and the American west. *Journal of Geophysical Research* **94**, 18 409–18 427.

Sachse, D., Billault, I., Bowen, G.J. et al. (2012) Molecular paleohydrology: interpreting the hydrogen-isotopic composition of lipid biomarkers from photosynthesizing organisms. *Annual Review of Earth and Planetary Sciences* **40**, 221–249.

Saylor, J.E. & Horton, B.K. (2014) Nonuniform surface uplift of the Andean plateau revealed by deuterium isotopes in Miocene volcanic glass from southern Peru. *Earth and Planetary Science Letters* **387**, 120–131.

Schemmel, F., Mikes, T., Rojay, B. & Mulch, A. (2013) The impact of topography on isotopes in precipitation across the Central Anatolian Plateau (Turkey). *American Journal of Science* **313**, 61–80.

Seager, R., Battisti, D.S., Yin, J. et al. (2002) Is the Gulf stream responsible for Europe's mild winters? *Quarterly Journal of the Royal Meteorological Society* **128**, 2563–2586.

Sharp, Z.D., Masson, H. & Lucchini, R. (2005) Stable isotope geochemistry and formation mechanisms of quartz veins; extreme paleoaltitudes of the Central Swiss Alps in the Neogene. *American Journal of Science* **305**, 187–219.

Smith, R.B. (2006) Progress on the theory of orographic precipitation. In: Willett, S.D., Hovius, N., Brandon, M. & Fisher, D.M. (eds.) *Tectonics, Climate, and Landscape Evolution.* Geological Society of America Special Paper **398**. Boulder, CO: Geological Society of America, pp. 1–16.

Strecker, M.R., Alonso, R.N. & Bookhagen, B. (2007) Tectonics and climate of the southern central Andes. *Annual Review of Earth and Planetary Sciences* **35**, 747–787.

Sturm, C., Zhang, Q. & Noone, D. (2010) An introduction to stable water isotopes in climate models: benefits of forward proxy modeling for paleoclimatology. *Climate of the Past* **6**, 115–129.

Takahashi, K. & Battisti, D.S. (2007) Processes controlling the mean tropical Pacific precipitation pattern. *Part I: The Andes and the Eastern Pacific ITCZ. Journal of Climate* **20**, 3434–3451.

Wang, Y., Xu, Y., Khawaja, S. et al. (2013) Diet and environment of a mid-Pliocene fauna from southwestern Himalaya: paleo-elevation implications. *Earth and Planetary Science Letters* **376**, 43–53.

Whipple, K.X. (2009) The influence of climate on the tectonic evolution of mountain belts. *Nature Geoscience* **2**, 97–104.

Winnick, M.J., Chamberlain, C.P., Caves, J. & Welker, J.M. (2014) Quantifying the isotopic "continental effect." *Earth and Planetary Science Letters* **406**, 123–133.

Yonge, C.J., Goldberg, L. & Krouse, H.R. (1989) An isotopic study of water bodies along a traverse of Southwestern Canada. *Journal of Hydrology* **106**, 245–255.

Zhuang, G., Brandon, M.T., Pagani, M. & Krishnan, S. (2014) Leaf wax stable isotopes from Northern Tibetan Plateau: implications for uplift and climate since 15 Ma. *Earth and Planetary Science Letters* **390**, 186–198.

Zhuang, G., Pagani, M., Chamberlin, C. et al. (2015) Altitudinal shift in stable hydrogen isotopes and microbial tetraether distribution in soils from the Southern Alps, NZ: implications for paleoclimatology and paleoaltimetry. *Organic Geochemistry* **79**, 56–64.

7

Phytopaleoaltimetry: Using Plant Fossils to Measure Past Land Surface Elevation

Robert A. Spicer

School of Environment, Earth and Ecosystem Sciences, The Open University, Milton Keynes, UK

Abstract

Plant fossils have long been used to estimate past land-surface elevations, first by reconstructing the ancient temperature regime and applying estimates of the rate at which surface temperature decreases with elevation (a lapse rate), and subsequently by using plant form to determine moist enthalpy and exploiting its relationship with a thermodynamically conserved variable of the atmosphere called "moist static energy." Surface temperatures depend on more than elevation, and vary in time and space with moisture, the albedo of the ground surface, local and regional topography, time of day, season and the nature and source of predominant air masses. Nevertheless, long-term climatic mean temperatures provide more stable values and are likely to represent the overall conditions to which plant species composition and morphology have to adapt. Although there are many kinds of lapse rates, it is important to distinguish between two kinds of lapse-rate measurements. The atmospheric or environmental lapse rate is measured in terms of a vertical column of the free atmosphere. This has a sound physical basis but it is not applicable to plants. The more empirical terrestrial lapse rate is measured along a surface transect with both vertical and horizontal components, and is based upon temperatures from meteorological stations. The terrestrial lapse rate is usually less than the atmospheric lapse rate, due to daytime heating of near-surface air along slopes. Taxon-based nearest-living-relative (NLR) methods, leaf form (physiognomy) and even plant chemistry can be used to derive surface air temperatures and, using appropriate terrestrial lapse rates, derive elevation estimates. Because moist static energy is the sum of moist enthalpy and potential energy, the difference in elevation between two locations is simply their difference in enthalpy divided by gravitational acceleration. Enthalpy is currently derived from foliar physiognomy; specifically, the Climate Leaf Analysis Multivariate Program (CLAMP) and visualizations of global CLAMP physiognomic space show that enthalpy is strongly and uniformly coded in leaf form. This makes it likely that enthalpy models derived from modern plants can be used reliably through deep time. Recent paleoelevation studies using a range of isotope methods yield identical results, within measurable uncertainties, to those obtained from the same location using CLAMP. This congruence attests to the reliability of both approaches.

Keywords: *plant fossils, enthalpy, lapse rates, elevation, climate, CLAMP*

7.1 Introduction

Attempting to understand the elevation history of Earth's surface may seem to be an esoteric quest with little relevance to society, but this appearance is deceptive. Land surface elevation is an expression of opposing buoyancy and gravitational forces within Earth; forces strongly influenced by the density and thickness of the planet's crust and underlying lithospheric mantle, as well as, in some instances, vertical components arising from mantle flow. It follows that quantifying elevation history provides a tool for testing geodynamic models and so improves our understanding of crustal movement and associated hazards (e.g., seismicity). Moreover, the interplay between uplift and erosion creates a dynamic surface topography that determines the characteristics of large-scale atmospheric phenomena such as monsoon systems, as well as more local climate regimes and their associated biodiversity and ecosystem services.

Mountains, Climate and Biodiversity, First Edition. Edited by Carina Hoorn, Allison Perrigo and Alexandre Antonelli.
© 2018 John Wiley & Sons Ltd. Published 2018 by John Wiley & Sons Ltd.
Companion website: www.wiley.com\go\hoorn\mountains,climateandbiodiversity

Elevation (Meters)

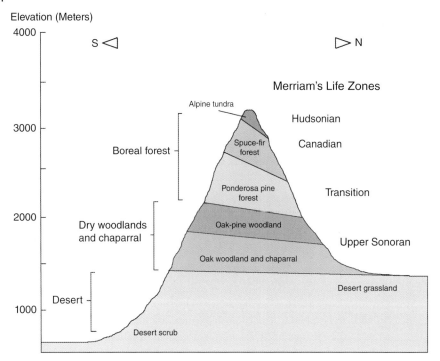

Figure 7.1 Altitude-related vegetation zones as recorded by C.H. Merriam on San Francisco Peak, AZ, USA. This zonation is specific to the southwestern USA, and elsewhere the component taxa differ. Note the aspect-related asymmetry. *Source:* Adapted from Merriam (1890).

The existence of altitude-related vegetation zones has long been known. Figure 7.1 illustrates zonation on San Francisco Peak, AZ, based on the classic work of C.H. Merriam (1890). The details of such zonation vary depending on latitude, aspect (north or south facing), slope angle, substrate, biogeographic history and prevailing wind directions, but the overriding control comes from the combination of temperature and moisture regimes at the different elevations. Because of their importance to plant growth, temperature and moisture have been the principal climate variables sought from plant fossils when attempting to determine past climates, and they also underpin plant-based techniques used to determine paleoelevation: a topic here termed "phytopaleoaltimetry."

In this chapter, the focus will be on two quite different ways of measuring ancient surface height using plant fossils. The first is to use a lapse rate, γ, which is the change in temperature, ∂T with a change in elevation, ∂Z:

$$\gamma = \delta T / \delta Z \tag{7.1}$$

The difference in elevation between locations is obtained by measuring the difference in prevailing temperature at those locations and then dividing by the lapse rate.

The second method is based on the first law of thermodynamics: the conservation of energy. Specifically, the observation that moist static energy (which is moist enthalpy, H, plus potential energy, gZ) is conserved as a parcel of air travels up and over a mountain means that the difference in enthalpy $(H_{low} - H_{high})$ between two locations divided by the acceleration due to gravity, g, gives the height difference, ΔZ, between those locations:

$$\Delta Z = \left(H_{low} - H_{high} \right) \big/ g \tag{7.2}$$

Both the lapse-rate and the enthalpy method rely on extracting reliable paleoclimate measurements from plant fossils.

7.2 Plants and Climate

To obtain paleoclimate information from plant fossils, two quite distinct approaches can be used. The first, and oldest, is taxon-based, and relies on identifying fossils in terms of their nearest living relatives (NLRs). Assuming this is done reliably, the environmental regime under which the ancient plants once lived can be derived from that tolerated today by the NLRs. Using that regime, and an accompanying one at a known elevation, paleoaltitude is calculated as a function of the environmental difference (usually temperature) experienced between the two elevations. This approach assumes that little or no evolutionary change has taken place that might alter the environmental tolerance of the NLRs from that of their presumed ancestors.

The second approach is essentially "taxon-free," in that no identification of the plant fossil is required. Instead, environmental data are obtained from the plant form (physiognomy). Again, this way of examining plant–climate relationships is not new: throughout the history of plant biology it has been recognized that plant growth forms and climate are related (e.g., Theophrastus ca. 300 BCE, based on the translation of Hort 1948; Witham 1833; Humboldt 1850; Seward 1892; Raunkiaer 1934; Givnish 1987). Plant form can be described at the whole-plant level (e.g., Holdridge 1947), including plant functional types (PFTs_) (e.g., Neilson et al. 1992; Prentice et al. 1992; Woodward & Cramer 1996), as well as at the organ level, where wood (most obviously growth rings) (e.g., Wheeler & Baas 1993; Creber & Francis 1999) and leaves (architecture and micromorphology) (e.g., Wolfe 1993; McElwain 2004; Yang et al. 2015) are known to carry strong environmental signals.

The two approaches, NLR and physiognomic, have their own advantages and disadvantages, but they are not mutually exclusive. For example, McElwain (2004) proposed using stomatal characteristics to determine the partial pressure of atmospheric carbon dioxide (pCO_2). This is a form of the physiognomic approach, in that stomata are components of leaf architecture. However, because the stomata–pCO_2 relationship is highly species-dependent (e.g., Jones et al. 1995; Bettarini et al. 1998), taxon identification is required. Some studies have also highlighted that factors other than pCO_2 affect stomatal characteristics (e.g., Whitehead & Luti 1962; Ciha & Brun 1975; Tichá 1982; Schoch et al. 1984; Awada et al. 2002; Reid et al. 2003; Konrad et al. 2008), that stomatal density can vary within a tree crown (Ceulemans et al. 1995), and, in the context of altitudinal gradients, that precipitation rather than pCO_2 can affect stomatal density (Körner et al. 1986). Even if pCO_2 is the main factor influencing stomatal characteristics, species dependency and the fact that the lifetime of a given species is often short (<1 My) compared to the time scales over which tectonic processes lead to significant surface height change mean that this approach has not been widely pursued.

Phytopaleoaltimeters also include examples based on plant chemistry, specifically deuterium/hydrogen ratios within organic molecules or biomarkers (Polissar et al. 2009; Hren et al. 2010), and spore and pollen-wall chemistry (Lomax et al. 2012). These methods, in conjunction with techniques that expose diagenetic alteration where it occurs, offer some promise, but their application to date has been limited. Techniques based on biomarkers also rely on knowing ancient lapse rates and make numerous assumptions regarding isotopic fractionation in the ancient atmosphere. Here, the focus is on phytopaleoaltimeters based not on plant chemistry, but on plant morphology.

It is worth noting that paleoaltimeters based on plant morphology are immune to the kinds of diagenetic issues that affect isotopic paleoaltimeters (for a review of isotope paleoaltimetry, see Mulch 2016). Only the form, not the chemistry, of the original plant parts needs to be preserved. Plant fossil assemblages are, however, prone to taphonomic (fossilization process) effects that can distort the characteristics of the fossil assemblage from those originally present in the once-living source vegetation (e.g., Spicer 1981; Spicer & Greer 1986; Spicer & Wolfe 1987; Spicer et al. 2005, 2011). Ideally, isotopic and phytopaleoaltimeters should be used together.

7.2.1 Taxon-Based Paleoclimate Proxies

As is evident to anyone who has travelled widely, or even just climbed up a mountain, the distribution of plant species is linked strongly to prevailing climate. This taxon-based approach began, as far as the written record is concerned, in 1086 AD. In his "Dream Pool Essays," the Chinese philosopher Shen Kuo argued that finding a plant fossil in an area where today that species does not grow indicates a change in climate since the time when the fossil was alive (Needham 1986). For its time, this was a remarkable inference, and it shows an understanding both that the geographical distribution of each plant species is determined by its climatic tolerance envelope and that the present can be used to interpret the past. In geology, this concept (uniformitarianism) was something that Western scientists only rediscovered 7 centuries later (Hutton 1788).

Shen Kuo was applying a taxon-based approach to paleoclimatology, and this approach is still used widely. Any identifiable plant part can potentially provide a clue as to the past environment, and some plant parts, such as pollen grains and spores, are extremely abundant in the fossil record because (i) they are produced in very large numbers by the parent plant and (ii) they are highly durable. This makes a taxon-based proxy useable for the vast majority of plant remains, but it assumes little or no evolutionary change in the taxon–climate relationship over time. This is problematic when taxon-based approaches are used to determine environmental change in "deep time," and recording surface height change inevitably demands techniques that are robust over time scales long enough for large-scale tectonic processes (such as mountain building) to be expressed. This means that phytopaleoaltimeters have to be immune, or at least highly robust, to evolutionary change and must be able to embrace extinct and imperfectly known species.

Correct fossil identification is at the heart of this kind of approach. Identification can be based on any part of a plant, which is a great advantage because plants rarely fossilize as whole entities and are normally preserved as isolated organs. However, not all plant parts offer the same diagnostic precision. For example, the discovery of a fossil flower, although rare, allows a plant to be compared closely with living forms, and potentially identified precisely. Fruits and seeds also offer high levels of confidence regarding identification, but leaves and pollen are usually more ambiguous in their affinity. Leaves display a great deal of phenotypic variability – useful, as we shall see, in capturing environmental signals – but this clouds species-level identification, while pollen grains, unless examined using scanning electron microscopy (SEM) (Zetter 1989; Hofmann & Zetter 2005; Ferguson et al. 2007), often can only be assigned to no better than generic level due to a limited diagnostic character pallet.

Errors in identification immediately compromise taxon-based approaches, and when a fossil represents an extinct species, it introduces further complications. The older a fossil species, the greater the likelihood that it will have had different environmental tolerances to that of its NLR. Moreover, poorly preserved material exacerbates this issue. For many leaf fossils, having a well-preserved cuticle in addition to details of leaf form and venation is highly desirable, if not essential.

Estimating the climatic tolerance for a modern species is done based on its geographic – including altitudinal – range. Because no single work can document all the occurrences of a taxon, this requires detailed records of species occurrences derived from a variety of published sources (e.g., the Palaeoflora Database: www.palaeoflora.de) (Greenwood et al. 2005; Utescher et al. 2014). In a rapidly changing modern climate, this introduces issues around when the record was made and how reliable the identification was at the time. Of course, factors other than climate also affect geographic range. These include the prior distribution of ancestral forms, disease, the presence and geographic range of animal vectors essential to plant reproduction, competition from other species and soil characteristics. The net effect of all these factors is to restrict the potential geographic and altitudinal range of a taxon, so that its observed range is almost always less than that it could occupy if climate alone controlled distribution. Moreover those nonclimatic factors that today may restrict a species' range may not have existed (or may have been different) in the past, potentially allowing that species to occupy a different environmental range.

Ideally, NLR approaches to paleoaltimetry are based not on single species, but on populations of taxa (e.g., Kershaw & Nix 1988; Kershaw 1996; Mosbrugger & Utescher 1997; Greenwood et al. 2003, 2005; Utescher et al. 2014), because while evolutionary change in climatic tolerance may be difficult to detect in a single species, the use of several species together can potentially expose anomalies. The use of numerous taxa may also improve precision, because while single taxa can have wide climatic tolerances, aggregations of taxa can only coexist where their tolerances coincide. This band – or interval – of overlapping tolerances can be used to assess the most likely climate, and thus altitude.

Despite the advantages of using multiple taxa, these approaches have attracted criticism, not just in relation to the theoretical underpinnings (Grimm & Potts 2015), but also in respect of the resolution (the width of the reconstructed interval) and the reliability (the difference between the reconstructed and observed interval for modern vegetation) with which they can correctly determine temperature (Grimm & Denk 2012).

Instead of using overlapping climate tolerances, it is possible to simply interpret a fossil assemblage in terms of a vegetation type and deduce a paleoelevation from where that type (e.g., the spruce-fir forest zone in Figure 7.1) occurs today. This approach underpinned early phytopaleoaltimetry work in the western United States (e.g., MacGinitie 1953; Axelrod 1966, 1968, 1980, 1981; Axelrod & Bailey 1976). Axelrod used the "climagraph" or nomogram developed by Bailey (1960), which characterizes vegetation types using not just the mean annual temperature (MAT), but also the concepts of "warmth" (calculated as the number of days the temperature rises above a particular value), "temperateness" (an obscure and poorly defined term), frost frequency and mean annual range of temperature (MART).

Apart from the difficulty of applying such a complex approach to the past, Axelrod & Bailey (1976) used a mean present-day terrestrial lapse rate of 5.5 °C/km to deduce the paleoelevation of their fossil floras, despite the spatial and temporal variability of such lapse rates (Section 7.4.1). Even were the appropriate paleo-lapse rates known, the precision with which paleoaltitudes can be estimated is also highly dependent on (i) the altitudinal range of the vegetation type, (ii) the propensity of the vegetation type to change over time as its constituent species evolve/migrate and (iii) similarity between past global and regional climates and those of today. This is because secular climate change will alter the altitudes of the vegetation-zone boundaries (Chapter 27). Despite these difficulties, a taxon-based approach – even one based on a single species – is sometimes the only approach that can be employed due to the nature of the fossil material (e.g., Sun et al. 2015).

7.2.2 Physiognomic Paleoclimate Proxies

Foliar physiognomic techniques assume that leaf form (both macroscopic and microscopic) represents a time-stable optimized architecture that maximizes functional efficiency within constraints imposed by the genome and the environment. In some ways, this leaf form is an "engineering solution" not unlike that of a racing car, in which the engine, brakes, tires, body shape and driver skill all operate in concert to confer competitive success (Spicer & Yang 2010). No single attribute is more "important" than the others.

This optimization results in the leaf physiognomic spectrum observed within a given patch of vegetation "coding" for the environmental conditions to which that vegetation is exposed. This coding is assumed to be time-stable because it reflects functional efficiency related to the physical laws of fluid conductance, gas diffusion, radiative balance and structural mechanics, all of which remain constant over time. Convergent evolution results in similar architectures arising under similar environmental constraints, irrespective of time and place.

Leaf form is largely an expression of "phenotypic integration" arising from highly flexible developmental pathways and pleiotropy (Schlichting 1989; Falconer & Mackay 1996; Juenger et al. 2005; Rodriguez et al. 2014). In woody dicot angiosperms, evolutionary selection has resulted in a genome capable of producing highly variable leaf phenotypes – so variable that even within a single plant leaf form is "tuned" to local microclimates within the tree crown. That is not to say leaf form can be remodeled during the life of a leaf: once a leaf has expanded, its form is largely fixed. Nevertheless, "sun leaves" high in the tree crown, exposed to intense sunlight and, at times, high wind energies, develop to be smaller, thicker and more robust, and sometimes even to have a different shape and venation, compared to "shade leaves" growing in the humid, shaded and protected subcanopy spaces (Wylie 1951; Heslop-Harrison 1964; Parkhurst & Loucks 1972; Dengler 1980; Carins Murphy et al. 2012). As a tree grows through the canopy, the proportion of sun and shade leaves it produces will change, and leaf form will alter if local conditions change due, for example, to the felling of a nearby tree. Clearly, the plant is able to sense the external environment and produce leaf forms appropriate to its conditions.

Alongside this microclimate-related plasticity within a single plant, leaf form must also be adaptive over an entire growing season, despite the fact that conditions at spring leaf expansion may well be different to those during high summer. The example usually given here is that of leaves growing in Mediterranean-type climates, where the winter and spring are cool and wet while the summers are hot and dry. If the leaves were only adaptive to conditions during the leaf-expansion phase, they would die during the summer (Spicer 2008). Thus, natural selection results in a genome that produces leaves adapted to prevailing local microclimates at expansion, as well as to the "anticipated" climate over the lifetime of a leaf. Plants with a genome incapable of doing this produce leaves poorly adapted to the summer conditions and die before they can reproduce, and so are quickly eliminated from such climates. It follows from this that, on a population basis, the spectrum of leaf physiognomic traits in a fossil assemblage likely reflects the average local environmental conditions prevailing over several years, if not decades or centuries.

On a global scale, individual leaf traits are highly correlated, and are more strongly linked to prevailing climate than to phylogenetic origin (Yang et al. 2015). Moreover, certain adaptations must have evolved several times. Adaptation to extreme cold, for example, appeared across the Cretaceous–Paleogene interval in New Zealand, disappeared later in the Paleogene (Spicer & Collinson 2014), and reappeared in high latitudes in the Quaternary. This suggests an all-pervasive selection for environment-related leaf trait spectra, independent of taxonomy, that, since the Late Cretaceous, has repeatedly achieved an optimal trade-off between investment of resources and photosynthetic return (e.g., Bloom et al. 1985; Givnish 1987). Furthermore, optimization of photosynthetic return requires the maintenance of leaf temperature within narrow limits, and there is evidence that both leaf and canopy architecture perform a homeostatic role in achieving this (Leuzinger & Körner 2007; Hellicker & Richter 2008). Of course, canopy architecture is not preserved in the fossil record, but leaves are, and in great numbers.

The highly correlated nature of leaf traits (Yang et al. 2015) has several important implications. The first is that the absence of a particular trait, or traits, has little impact on the overall climate signal derived from the assemblage, so provided that sufficient traits are properly recorded, an imperfectly preserved fossil assemblage can still yield a reliable climate signal. The second is that evolutionary changes in one trait must be accompanied by compensatory changes in others if the leaf is to remain optimized for a given climate. This means that the overall climate signal encoded in leaf form remains robust through time. This is demonstrated in the case of New Zealand, where leaf margin form shows no correlation with MAT (Kennedy 1998; Gregory-Wodzicki 2000) – unlike, for example, in eastern Asia (Wolfe 1979) or North America (Wilf 1997); however, overall New Zealand leaves carry the same coding for MAT as elsewhere (Kennedy et al. 2014; Yang et al. 2015), because collectively, other (nonmargin) traits display a spectrum of form that reflects MAT.

This trait correlation also means that no one character carries more importance than any other in regard to a given climate variable. The characters vary collectively and, as the case of New Zealand shows, MAT is coded across a spectrum of traits other than those of just the leaf margin. Because the correlation between single specific traits and specific climate variables exhibits geographical differences, it is likely that similar differences may have existed over time. It therefore follows that there is considerable danger in assuming that the importance of a given trait is time-stable, and using single (or even small numbers of) characters could yield spurious paleoclimate retrodictions. This renders single-trait analysis such as leaf-margin analysis (Wolfe 1979; Wilf 1997) or leaf-area analysis (Wilf et al. 1998) unreliable, and such approaches should be abandoned.

The use of numerous leaf traits means that several environmental variables can be returned. In CLAMP (Box 7.1), in addition to MAT, the mean temperature of the warmest and coldest months (WMMT and CMMT), length of the growing season (LGS; defined as the period of time for which the mean daily temperature is >10 °C), growing season precipitation (GSP), mean monthly growing season precipitation (MMGSP), precipitation of the three consecutive wettest months (3-WET), precipitation of the three consecutive driest months (3-DRY),

relative humidity (RH), specific humidity (SH), mean annual enthalpy (ENTH) and, in some configurations, mean annual vapor pressure deficit (VPD) are also estimated. All these climate parameters vary with elevation, but those most useful in phytopaleoaltimetry are MAT, WMMT, CMMT and ENTH.

7.3 Lapse Rates and Enthalpy

Any environmental parameter that varies with surface height can be used to estimate elevation. In phytopaleoaltimetry, this parameter is usually some measure of temperature – in which case a lapse rate is used – or of enthalpy. Enthalpy is related to moist static energy, a conserved property of the atmosphere combining both temperature and moisture. In the case of phytopaleoaltimetry, two key assumptions are involved: (i) that the environmental parameter can be accurately

---

Box 7.1 CLAMP

The most widely used and tested multivariate foliar physiognomic climate and paleoaltimetric proxy is the Climate Leaf Analysis Multivariate Program (CLAMP). First published by Wolfe (1993), and subsequently developed in numerous works (Kovach & Spicer 1995; Wolfe 1995; Herman & Spicer 1997; Spicer et al. 2009; Steart et al. 2010; Jacques et al. 2011; Teodoridis et al. 2011; Yang et al. 2011, 2015; Srivastava et al. 2012; Kennedy et al. 2014), CLAMP is calibrated using training sets of fully developed leaves collected in the field using standardized protocols from localized areas of living natural or naturalized vegetation. At each of these sites, leaves of at least 20 species of woody dicotyledonous flowering plants (angiosperms), representing the full morphological range of leaf form (e.g., including both sun and shade leaves) within each species, are scored using defined protocols for 31 macromorphological traits spanning margin form, size, apex and base form, length to width ratio and shape (Figure 7.2).

If fewer than 20 species are used, the predictive precision of the technique declines sharply (Povey et al. 1994). Summary percentage scores for each of the 31 traits across all morphotypes present in a sample form a characteristic "physiognomic trait spectrum" for that sample. Canonical correspondence analysis (CCA) (ter Braak 1986) is used to position each modern vegetation sample in multivariate space based on its characteristic physiognomic trait spectrum. Because the climate for each modern sample is known, climate trends can be determined across the cloud of samples making up modern physiognomic space. These trends are represented by climate vectors, which are positioned explicitly using CCA.

Fossil sites are similarly scored for their physiognomic trait spectra, but, lacking known climate data, they are positioned passively in physiognomic space by CCA. The position of a fossil sample along a climate vector in four-dimensional (4D) space yields the predicted paleoclimate. Details of the method and an online analysis facility are available at http://clamp.ibcas.ac.cn.

Venation and micromorphological features are not currently included in CLAMP. This is because venation patterns are used for differentiating morphotypes (groupings of leaves broadly equivalent to species but lacking any formal taxonomic assignment or nomenclature), and micromorphological features are not always preserved. By restricting the CLAMP characters to macrophysiognomic features, the technique is applicable even to poorly preserved leaf assemblages.

The most serious criticism of CLAMP, particularly in relation to its contribution to phytopaleoaltimetry, was that of Peppe et al. (2010), who contended that the categorical way leaf size is measured introduces a systematic bias. However, their critique was flawed in that they used test samples where leaf size was measured in mm^2 but failed to recalibrate CLAMP by measuring training-set leaf size the same way. By retaining categorical leaf-size measures in the calibration but using continuous leaf-size measures in the test data, they were guaranteed to get flawed results. This was exposed in Spicer & Yang (2010), who also demonstrated that on removing leaf-size data completely from the CLAMP analysis (both calibration and test samples), there was little change to CLAMP's predictive capability as regards enthalpy, and thus elevation.

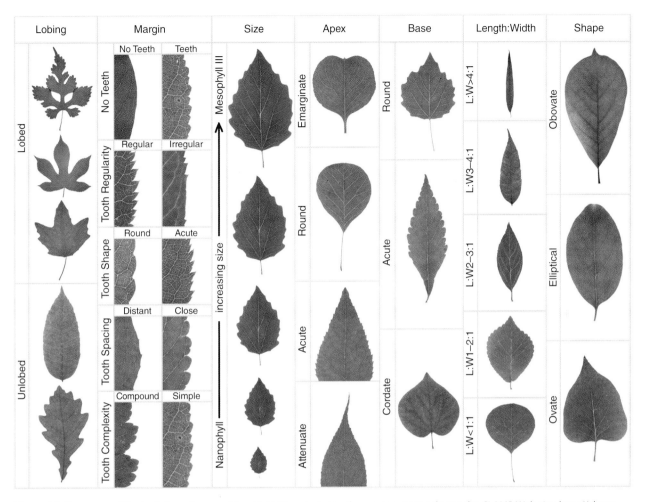

Figure 7.2 Summary of the leaf characters used in a CLAMP analysis. For the scoring protocols, see the CLAMP Web site: http://clamp. ibcas.ac.cn. See also Plate 13 in color plate section.

and precisely derived from plant fossils; and (ii) that the relationship between the chosen environmental parameter and altitude has not changed over time, or, if it has, that the paleorelationship is known and quantifiable.

7.3.1 Lapse Rates

The definitive works on lapse rates in a paleobotanical context are those of Meyer (1992, 2001, 2007). A lapse rate is any change (usually a decrease) in a variable (temperature, unless otherwise specified) with height. The global international standard lapse rate for the free atmosphere, used in meteorology and aviation, is confusingly referred to as the "environmental lapse rate" and reflects an average global temperature decrease of 6.49 °C/1000 m up to the top of the troposphere (taken as 11 km). This standard lapse rate, which here is referred to as an "atmospheric lapse rate," only applies to stationary air in a column of what is known as an International

Standard Atmosphere (ISA). An ISA has to be used because lapse rates vary in relation to atmospheric composition. In reality, the atmosphere is wet to a greater or lesser degree. For unsaturated air (RH < 100%), a dry average lapse rate of 10 °C is sometimes used, while the present-day wet atmospheric lapse rate (RH 100%) has an average value of 5.5 °C/km. In the wet tropics, it can be even lower.

Although having a sound basis in physics, standard atmospheric lapse rates are difficult to relate to plants, which are attached to land surfaces and not suspended in columns of air. Meyer (1992) refers instead to a "terrestrial lapse rate," which is measured along a surface transect with both vertical and horizontal components, and is based upon temperatures measured in standard meteorological stations sited in clearings. Usually, this empirical terrestrial lapse rate is less than the atmospheric lapse rate, because of daytime heating of near-surface air along slopes. It is only the terrestrial lapse rate that is applicable to plants.

In theory, then, the difference in elevation, ΔZ, between two locations, one high and one low, may be derived from the difference in surface temperature, T, between two locations:

$$\Delta Z = \left(T_{low} - T_{high}\right)/\gamma \qquad (7.3)$$

where $\gamma = -\delta T/\delta Z$ and is the lapse rate. Provided the elevation above sea level is known for one site, that for the other site can be derived.

Terrestrial lapse rates can be calculated from several temperature variables: MAT, WMMT, CMMT and MART. Moreover, it is possible to determine these temperature variables from plants – living or fossil – using taxon-based or physiognomic methods or organic biomarkers.

Terrestrial lapse rates are affected by a variety of conditions, including the state of the atmosphere, moisture, albedo of the ground surface, local and regional topography, time of day, season and the nature and source of predominant air masses. As such, at any one location the terrestrial lapse rate cannot be assumed to be a uniform gradient. It can be highly variable in both time and space (Figure 7.3b) due to changing meteorological conditions, including moisture content and the dynamics of adiabatic processes. Despite this variability, long-term mean temperatures represent the overall conditions to which plants have to adapt and so are likely to be reflected in terms of both species composition and morphology.

Terrestrial lapse rates vary within a local area as well as on continental scales. For example, diurnal and seasonal fluctuations are more extreme away from maritime influences, producing higher WMMTs, lower CMMTs and a higher MART in continental interiors compared to coastal and island areas. This means that in coastal and island areas, seasonal temperature variations are less than those in continental interiors, so lapse rates are different (Table 7.1).

Elevation itself is another factor. Because the ground surface is the primary source of atmospheric heating, the altitude of that surface has a large influence on the regional relationship between altitude and temperature. The classic example of this is the heating effect of the Tibetan Plateau, where strong daytime warming of the near-surface air mass results in higher near-surface mean temperatures for the plateau surface than would occur in the free atmosphere at the same altitude in adjacent locations. This has long been regarded as a contributory factor in amplifying the Asian monsoon systems (e.g., Flohn 1968; Yanai & Wu 2006), although the overall effect of Tibet on the South Asian monsoon is more complex (Boos & Kuang 2010; Molnar et al. 2010). The

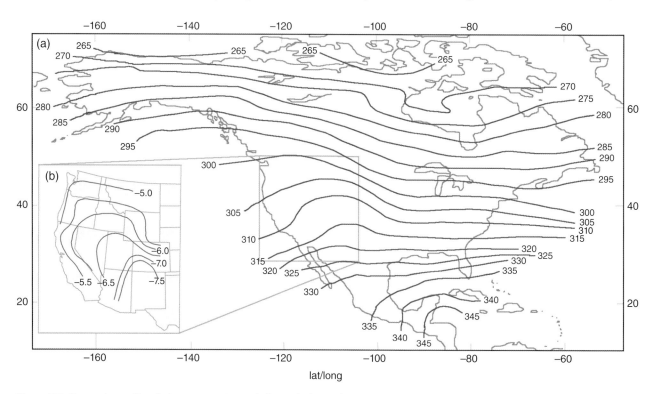

Figure 7.3 Comparison of isopleths representing enthalpy with those of terrestrial lapse rates across North America. (a) Isopleths representing the modern enthalpy field across North America (expressed as energy (kJ) per kilogram of air). (b) Lapse rates, expressed in °C/km. These are far more spatially variable than the enthalpy field, which is more zonal, varying mostly in relation to latitude and invariant with longitude. *Source:* Adapted from Forest et al. (1995) and Meyer (1992).

Table 7.1 Comparisons of terrestrial lapse rates based on mean annual temperature (MAT), warmest month mean temperature (WMMT), coldest month mean temperature (CMMT) and mean annual range of temperature (MART) for continental versus island settings. While lapse rates are usually negative (temperature cools with increasing altitude), note the positive continental MART lapse rate and the differences between the climate variables. *Source:* Data from Meyer (1992).

	MAT lapse rate (°C/km)	WMMT lapse rate (°C/km)	CMMT lapse rate (°C/km)	MART lapse rate (°C/km)
Continental (Arizona)	−6.81	−7.9	−5.0	+2.9
Island (Cyprus)	−4.92	−3.63	−6.17	−2.46

same effect is present on isolated mountains, but is not as pronounced because surface air mixes more readily with the surrounding cooler air of the open atmosphere. Of course, topographic relief also affects local wind velocity, cloud cover, solar intensity (influenced by aspect and slope) and surface roughness (Price 1981).

Prevailing wind directions also affect lapse rates because topographic barriers develop different adiabatic processes on windward and leeward slopes. Upon encountering a mountain barrier, moist air initially rises following a dry adiabat until the dew point is reached, after which it continues to rise but following a wet adiabat, determined in large part by the relative moisture content. Warming caused by the latent heat of condensation results in wet adiabatic processes being characterized by lower lapse rates. Because wet adiabatic processes predominate on the windward sides of mountains, and dry adiabatic processes on leeward sides, terrestrial lapse rates based on mean temperatures tend to be higher on leeward slopes than on windward slopes. In a paleoelevational context, this can be problematic if regional atmospheric circulation patterns are not known.

The preceding account shows that there are major complications in the application of terrestrial lapse rates on mountain slopes, but perhaps the greatest issue affecting plant fossil assemblages is that they accumulate in topographic lows (e.g., intermontane basins), and because night-time low temperatures are a component in the calculation of mean temperatures, inversions due to cold air drainage result in mean temperatures that are lower in many valleys and basins than would be expected. This often results in surface temperature rising, not cooling, with increasing elevation (Wolfe 1992).

Changes in atmospheric composition over time also affect lapse rates, meaning that for deep time, when atmospheric composition was different in ways that are poorly quantified, modern lapse rates are inappropriate. It is not only greenhouse gasses with long atmospheric residence times that are important. Greenhouse gases exert a strong influence over the temperature of the atmosphere, and warm atmospheres can hold more moisture than cold ones. It is clear from the preceding account that the amount of water vapor in the atmosphere is a critical determinant of lapse rates, and in any deep-time application of lapse rates the moisture content of the atmosphere must be known, including uncertainties.

Clearly, using lapse rates to infer past elevation is fraught with difficulties, but this has only become apparent through the work of numerous authors, especially those working in western North America, such as Axelrod (1965, 1968, 1980, 1981, 1997), Axelrod & Bailey (1976), Meyer (1986, 1992), Wolfe & Wehr (1987), Gregory & Chase (1992), Gregory (1994), Povey et al. (1994), Gregory & McIntosh (1996) and MacGinitie (1953). One of the more recent and most sophisticated uses of lapse rates is that of Gregory-Wodzicki et al. (1998), who, working in the Andes, used CLAMP to derive MAT and then attempted to correct for changes in MAT due to factors other than uplift. They made adjustments for secular climate change, lateral changes in position due to plate movement, paleogeography (e.g., proximity to paleocoastlines) and even sea-level change, although this is small compared to other methodological uncertainties.

7.3.2 Enthalpy and Moist Static Energy

Recognizing that terrestrial lapse rates are difficult to apply in paleoaltimetry, Forest et al. (1995) introduced a way of determining paleoelevation based on the principle of energy conservation. This approach, more fully explored in Forest (1996) and Forest et al. (1999), is derived from the first law of thermodynamics, and as such is time-stable.

Moist static energy, h, is the total specific energy content of air:

$$h = c'_p T + L_v q + gZ \qquad (7.4)$$

where c'_p is the specific heat capacity at a constant pressure of moist air, T is temperature (in K), L_v is the latent heat of vaporization of water, q is specific humidity, g is acceleration due to gravity (a constant) and Z is elevation. This expression excludes kinetic energy, but (except during hurricanes) this is very small (<1%) in relation to the other terms. Compared to lapse rates (Figure 7.2a), moist static energy displays far less geographic variability (Figure 7.2b). This is because it is very nearly constant and is conserved along air-parcel trajectories, being changed only by radiative heating and surface fluxes of latent and sensible heat. Moreover, the value of h in the lower (subcloud) part of the atmosphere (<~1.5 km above the Earth's surface, usually referred to as the "boundary layer") is almost the same as that in the upper troposphere, due to convection (Betts 1982; Xu & Emanuel 1989). At mid latitudes, Earth's rotation causes large-scale tropospheric airflow to move from west to east, and as a result contours of h are roughly aligned parallel to latitude (Figure 7.2b) and largely invariant with longitude (Forest et al. 1995; Forest 1996).

Moist static energy is made up of two components: enthalpy and potential energy:

$$h = H + gZ \qquad (7.5)$$

where H is enthalpy $(c'pT + Lvq)$ and gZ is potential energy. As a parcel of air rises, it gains potential energy and, because moist static energy remains the same, enthalpy decreases.

It follows, therefore, that because the value of h is conserved, the difference in elevation between two locations at the same latitude is given by:

$$\Delta Z = \left(H_{low} - H_{high} \right) \big/ g \qquad (7.6)$$

The simplicity of this equation, where the difference in enthalpy between two locations $(H_{low} - H_{high})$ divided by the acceleration due to gravity, g, gives the elevation difference, ΔZ, between those locations, offers a very attractive paleoaltimeter for obtaining either surface elevation above sea level (if H_{low} is derived from a flora bounded by, or laterally equivalent to, marine units) or relative surface height differences (if the absolute elevations of both H_{low} and H_{high} are unknown). Note that this simplicity only applies if H_{low} and H_{high} are at the same latitude, or if latitudinal differences are known and can be corrected for.

Theoretically, then, enthalpy is far superior to lapse rates for determining paleoaltitude, but as can be seen in Figure 7.3a, the zonal symmetry of enthalpy is distorted where large-scale air flow is deflected over the high topography of the western United States. Potentially, this could be a problem if the low and high locations were geographically far apart, because the ancient topography would need to be known in order to correct for disruptions to the zonal symmetry of h. There are three ways to overcome this. The first is to calculate the maximum uncertainty introduced by topography-induced zonal asymmetry and incorporate that into any elevation

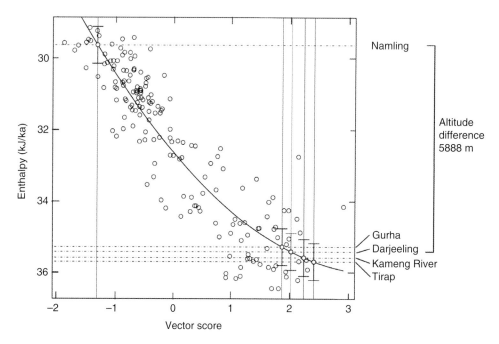

Figure 7.4 Graph showing the relationship between the relative positions of modern vegetation sites (open gray circles) along the CLAMP-derived enthalpy vector (vector score) (Box 7.1) and the observed enthalpy (measured as energy in kilojoules per kilogram of air) at those locations. The vector scores for four sea-level fossil sites of different ages (early Eocene to middle Miocene) from northern India (solid vertical gray lines) intersect at the filled circles with the enthalpy regression curve derived from the modern calibration data giving the predicted paleoenthalpy values for those sites (horizontal dotted lines). Note that all northern India sea-level sites have similar paleoenthalpies irrespective of age. In contrast, the 15 Ma Namling site from southern Tibet yields a much lower enthalpy value, meaning it must have been much higher. The figure shows the raw (uncorrected for paleoposition) enthalpy difference, which gives an elevation for Namling of 5888 m. Statistical uncertainties (±1 SD) in positioning the fossil sites are shown. The ages of the fossil sites are from Spicer et al. (2003) (Namling), Shukla et al. (2014) (Gurha), Khan et al. (2014) (Darjeeling) and Ojha et al. (2009) (Kameng River).

estimates. The second is to use climate modeling to determine H at a known height in the atmosphere above where topography has an influence and to use this, and not sea level, as the datum for measurements. The third is to again use a climate model, but one that displays sea-level –not surface – enthalpy.

In regard to the western United States, Forest et al. (1995, 1999) used records from 699 weather stations and calculated the expected error from zonal asymmetries between 30° and 60° N to be 4.5 kJ/kg, or 460 m of altitude. Lapse-rate uncertainty in altitude estimation was much larger, and increased with elevation, being 540, 660 and 920 m for altitudes of 0, 2000 and 4000 m, respectively. In a paleoaltimetric context, this lack of uniformity in lapse-rate uncertainties compounds those due to temporal changes in atmospheric composition.

To use enthalpy in paleoaltimetry, ancient enthalpy must be determined reliably and accurately from fossil floras. Leaf form reflects enthalpy well because temperature and moisture (the two components of enthalpy) are critical to plant growth. Figure 7.4 shows the relationship between observed enthalpy and leaf form using a training set of 177 modern vegetation sites from Japan, China, India, North America, Pacifica and Thailand; using this, Khan et al. (2014) calculated the paleoelevation of the 15 Ma Namling–Oiyug plant fossil site in south-central Tibet to be 5888 ± 728 m. This raw elevation was corrected for tectonically driven changes in the paleopositions of the floras involved and for translatitudinal sea-level enthalpy trends, using paleoclimate modeling (Khan et al. 2014). This reduced the estimate to 5260 ± 728 m, which is in close agreement with a surface height estimate of 5.1 km +1.4/−2.0 km based on oxygen isotopes from different minerals in fossil soils and lake sediments and on the hydrogen isotope composition of *n*-alkanes from epicuticular waxes preserved in the Namling flora (Currie et al. 2016).

If we accept the mid-Miocene elevation of the Namling–Oiyug basin floor to be 5.2 km, it is possible to estimate the terrestrial paleolapse rate for that area at that time using the CLAMP-derived MATs for both the sea-level and Namling fossil sites. CLAMP, using the PhysgAsia2 calibration, returns an MAT of 25.4 ± 2.3 °C for the Siwaliks and 8.2 ± 2.3 °C for Namling, meaning the mid-Miocene lapse rate in this region was $(25.4 - 8.2)/5.88 = 2.93$ °C/km.

If the often-used (e.g., Axelrod & Bailey 1976; Hren et al. 2010) "normal" modern terrestrial lapse rate of 5.5 °C/km had been employed to calculate the mid-Miocene elevation, the result would have been only 3.1 km: well below – and outside the uncertainties of – both the CLAMP enthalpy and isotopic estimates.

7.4 Conclusion

Phytopaleoaltimetry based on plant morphology is particularly powerful because it is immune to the diagenetic alteration that often bedevils isotope systems, but it does require (i) methodologies that are robust to evolutionary change, (ii) an environmental parameter recoverable from plant fossils that varies with altitudes in a known way over both space and time and, to avoid circularity of argument, (iii) a dating framework that is independent of biology. Methodologies that rely on species remaining evolutionarily static over time scales during which significant surface height change can take place (>1 Ma) carry with them large and unquantifiable uncertainties. Similarly, uncertainties arising from the use of terrestrial lapse rates are difficult to quantify because terrestrial lapse rates are so variable over time and space and are prone to change with poorly defined past atmospheric compositions. Note, however, that paleo-lapse rates can be determined retrospectively using CLAMP-derived MATs from fossil sites whose elevation can be determined using enthalpy. This is potentially useful where fossil leaf assemblages are present but rare and where other plant parts are available to use to derive paleo-MATs using NLR approaches.

Improvements in precision are now the main focus of CLAMP research, with new approaches being developed in order to explore and better understand the complexities of global physiognomic space (Yang et al. 2015; Li et al. 2016). It is also possible that changes in the number of leaf features and the way they are recorded will also improve precision, but precision must always be viewed alongside the topographic roughness of the ancient terrain. Phytopaleoaltimeters will always be biased towards the elevation of the depositional basin within which the plant fossils accumulate, and not that of the surrounding peaks. Uncertainties have to be viewed in this context.

References

Awada, T., Moser, L.E., Schacht, W.H. & Reece P.E. (2002) Stomatal variability of native warm-season grasses from the Nebraska Sandhills. *Canadian Journal of Plant Science* **82**, 349–355.

Axelrod, D.I. (1965) A method for determining the altitudes of Tertiary floras. *Paleobotanist* **14**, 144–171.

Axelrod, D.I. (1968) Tertiary floras and topographic history of the Snake River Basin, Idaho. *Geological Society of America Bulletin* **79**, 713–734.

Axelrod, D.I. (1980) Contributions to the Neogene paleobotany of central California. *University of California Publication in Geological Sciences* **121**, 1–212.

Axelrod, D.I. (1981) Altitudes of Tertiary forests estimated from paleotemperature. In: Liu, D.S. (ed.) *Proceedings of Symposium on Qinghai-Xizang (Tibet) Plateau, (Beijing, China); Geological and Ecological Studies of Qinghai-Xizang Plateau; vol. 1, Geology, Geological History and Origin of Qinghai-Xiang Plateau.* Beijing: Science Press and New York: Gordon and Breach, Science Publishers, Inc., pp. 131–137.

Axelrod, D.I. (1997) Paleoelevation estimates from Tertiary floras. *International Geology Review* **39**, 1124–1133.

Axelrod, D.I. & Bailey H.P. (1976) Tertiary vegetation, climate, and altitude of the Rio Grande depression, New Mexico–Colorado. *Paleobiology* **2**, 235–254.

Bailey, H.P. (1960) A method of determining the warmth and temperateness of climate. *Geografiska Annaler* **42**, 1–16.

Bettarini, I., Vaccari, F.P. & Miglietta, F. (1998) Elevated CO_2 concentrations and stomatal density: observations from 17 plant species growing in a CO_2 spring in central Italy. *Global Change Biology* **4**, 17–22.

Betts, A.K. (1982) Saturation point analysis of moist convective overturning. *Journal of the Atmospheric Sciences* **39**, 1484–1505.

Bloom, A.J., Chapin, E.S. & Mooney, H.A. (1985) Resource limitation in plants – an economic analogy. *Annual Review of Ecology and Systematics* **16**, 363–392.

Boos, W.R. & Kuang, Z. (2010) Dominant control of the South Asian monsoon by orographic insulation versus plateau heating. *Nature* **463**, 218–222.

Carins Murphy, M.R., Jordan, G.J. & Brodribb, T.J. (2012) Differential leaf expansion can enable hydraulic acclimation to sun and shade. *Plant Cell and Environment* **35**, 1407–1418.

Ceulemans, R., Van Praet, L. & Jiang, X.N. (1995) Effect of CO_2 enrichment, leaf position and clone on stomatal index and epidermal cell density in poplar (*Populus*). *New Phytolologist* **131**, 99–107.

Ciha, A.J. & Brun, W.A. (1975) Stomatal size and frequency in soybeans. *Crop Science* **15**, 309–313.

Creber, G.T. & Francis, J.E. (1999) Fossil tree-ring analysis: palaeodendrology. In: Jones, T.P. & Rowe, N.P. (eds.) *Fossil Plants and Spores: Modern Techniques.* London: Geological Society, pp. 245–250.

Currie, B.S., Polissar, P.J., Rowley, D.B. et al. (2016) Multiproxy palaeoaltimetry of the late Oligocene-Pliocene Oiyug Basin, Southern Tibet. *American Journal of Science* **316**, 401–436.

Dengler, N.G. (1980) Comparative histological basis of sun and shade leaf dimorphism in *Helianthus annuus*. *Canadian Journal of Botany* **58**, 717–730.

Falconer, D.S. & Mackay, T.F.C. (1996) *Introduction to Quantitative Genetics.* Harlow: Addison Wesley Longman.

Ferguson, D.K., Zetter, R. & Paudayal, K.N. (2007) The need for the SEM in palaeopalynology. *Comptes Rendus Palevol* **6**, 423–430.

Flohn, H. (1968) Contributions to a Meteorology of the Tibetan Highlands. Atmospheric Science Paper 130. Department of Atmospheric Sciences, Colorado State University.

Forest, C.E. (1996) Paleoaltimetry Incorporating Atmospheric Physics and Botanical Estimates of Paleoclimate. PhD Dissertation, Massachusetts Institute of Technology.

Forest, C.E., Molnar, P. & Emanuel, K.E. (1995) Palaeoaltimetry from energy conservation principles. *Nature* **343**, 249–253.

Forest, C.E., Wolfe, J.A. & Emanuel, K.A. (1999) Paleoaltimetry incorporating atmospheric physics and botanical estimates of paleoclimate. *Geological Society of America Bulletin* **111**, 497–511.

Givnish, T.J. (1987) Comparative studies of leaf form: assessing the relative roles of selective pressures and phylogenetic constraints. *New Phytologist* **106**, 131–160.

Greenwood, D.R., Moss, P., Rowett, A. et al. (2003) Plant Communities and Climate Change in Southeastern Australia during the Early Paleogene. *Geological Society of America Special Paper* **369**, pp. 365–380.

Greenwood, D.R., Archibald, S.B., Mathewes, R.W. & Moss, P.T. (2005) Fossil biotas from the Okanagan Highlands, southern British Columbia and northeastern Washington State: climates and ecosystems across an Eocene landscape. *Canadian Journal of Earth Sciences* **42**, 167–185.

Gregory, K.M. (1994) Paleoclimate and paleoelevation of the 35 Ma Florissant flora, Front Range, Colorado. *Palaeoclimates* **1**, 23–57.

Gregory, K.M. & Chase, C.G. (1992) Tectonic significance of paleobotanically estimated climate and altitude of the late Eocene erosion surface, Colorado. *Geology* **20**, 581–585.

Gregory, K.M. & McIntosh, W.C. (1996) Paleoclimate and paleoelevation of the Oligocene Pitch Pinnacle flora, Sawatch Range, Colorado. *Geological Society of America Bulletin* **108**, 545–561.

Gregory-Wodzicki, K.M. (2000) Relationships between leaf morphology and climate, Bolivia: implications for estimating paleoclimate from fossil floras. *Paleobiology* **26**, 668–688.

Gregory-Wodzicki, K.M., McIntosh, W.C. & Velasquez, K. (1998) Paleoclimate and paleoelevation of the late Miocene Jakokkota flora, Bolivian Altiplano. *Journal of South American Earth Sciences* **11**, 533–560.

Grimm, G.W. & Denk, T. (2012) Reliability and resolution of the coexistence approach – a revalidation using modern-day data. *Review of Palaeobotany and Palynology* **172**, 33–47.

Grimm, G.W. & Potts, A.J. (2015) Fallacies and fantasies: the theoretical underpinnings of the Coexistance Approach for palaeoclimate reconstruction. *Climate of the Past Discussion* **11**, 5727–5754.

Hellicker, B.R. & Richter, S. (2008) Subtropical to boreal convergence of tree-leaf temperatures. *Nature* **454**, 511–515.

Herman, A.B. & Spicer, R.A. (1997) New quantitative palaeoclimate data for the Late Cretaceous Arctic: evidence for a warm polar ocean. *Palaeogeography, Palaeoclimatology, Palaeoecology* **128**, 227–251.

Heslop-Harrison, J. (1964) Forty years of genecology. *Advances in Ecological Research* **2**, 159–247.

Hofmann, C.-C. & Zetter, R. (2005) Reconstruction of different wetland plant habitats of the Pannonian Basin System (Neogene, Eastern Austria). *Palaios* **20**, 266–279.

Holdridge, L.R. (1947) Determination of world plant formations from simple climatic data. *Science* **105**(2727), 367–368.

Hort, A. (translator) (1948) *Enquiry into Plants*, Vol. **1** by Theophrastus. Cambridge, MA: Harvard University Press.

Hren, M.T., Pagani, M., Erwin, D. & Brandon, M. (2010) Biomarker reconstruction of the early Eocene paleotopography and paleoclimate of the northern Sierra Nevada. *Geology* **38**, 7–10.

Humboldt, A. (1850) *Aspects of Nature in Different Lands and different Climates; With Scientific Elucidations.* Philadelphia, PA: Lea & Blanchard.

Hutton, J. (1788) Theory of the Earth; or an investigation of the laws observable in the composition, dissolution, and restoration of land upon the Globe. *Transactions of the Royal Society of Edinburgh* **1**(Pt. 2), 209–304.

Jacques, F.M.B., Su, T., Spicer, R.A. et al. (2011) Leaf physiognomy and climate: are monsoon systems different? *Global and Planetary Change* **76**, 56–62.

Jones, M.B., Brown, J.C., Raschi, A. & Miglietta, F. (1995) The effects on *Arbutus unedo* L. of long-term exposure to elevated CO_2. *Global Change Biology* **1**, 295–302.

Juenger, T., Pérez-Pérez, J.M. & Micol, J.L. (2005) Quantitative trait loci mapping of floral and leaf morphology traits in *Arabidopsis thaliana*: evidence for modular genetic architecture. *Evolution and Development* **7**, 259–271.

Kennedy, E.M. (1998) Cretaceous and Tertiary Megafloras from New Zealand and their Climate Signals. PhD thesis, The Open University.

Kennedy, E.M., Arens, N.C., Reichgelt, T. et al. (2014) Deriving temperature estimates from Southern Hemisphere leaves. *Palaeogeography, Palaeoclimatology, Palaeoecology* **412**, 80–90.

Kershaw, A.P. (1996) A bioclimatic analysis of Early to Middle Miocene brown coal floras, Latrobe Valley, southeastern Australia. *Australian Journal of Botany* **45**, 373–383.

Kershaw, A.P. & Nix, H.A. (1988) Quantitative palaeoclimatic estimates from pollen data using bioclimatic profiles of extant taxa. *Journal of Biogeography* **15**, 589–602.

Khan, M.A., Spicer, R.A., Bera, S. et al. (2014) Miocene to Pleistocene floras and climate of the Eastern Himalayan Siwaliks, and new palaeoelevation estimates for the Namling–Oiyug Basin, Tibet. *Global and Planetary Change* **113**, 1–10.

Konrad, W., Roth-Nebelsick, A. & Grein, M. (2008) Modelling of stomatal density response to atmospheric CO_2. *Journal of Theoretical Biology* **253**, 638–658.

Körner, C., Bannister, P. & Mark, A.F. (1986) Altitudinal variation in stomatal conductance, nitrogen content and leaf anatomy in different plant life forms in New Zealand. *Oecologia* **69**, 577–588.

Kovach, W. & Spicer, R.A. (1995) Canonical correspondence analysis of leaf physiognomy: a contribution to the development of a new palaeoclimatological tool. *Palaeoclimates: Data and Modelling* **2**, 125–138.

Leuzinger, S. & Körner, C. (2007) Tree species diversity affects canopy leaf temperatures in a mature temperate forest. *Agricultural and Forest Meteorology* **146**, 29–37.

Li, S.-F., Jacques, F.M.B., Spicer, R.A. et al. (2016) Artificial neural networks reveal a high-resolution climatic signal in leaf physiognomy. *Palaeogeography, Palaeoclimatology, Palaeoecology* **442**, 1–11.

Lomax, B.H., Fraser, W.T., Harrington, G. et al. (2012) A novel palaeoaltimetry proxy based on spore and pollen wall chemistry. *Earth and Planetary Science Letters* **353–354**, 22–28.

MacGinitie, H.D. (1953) Fossil Plants of the Florissant Beds, Colorado. Carnegie Institution of Washington Publication 599.

McElwain, J.C. (2004) Climate-independent paleoaltimetry using stomatal density in fossil leaves as a proxy for CO_2 partial pressure. *Geology* **32**, 1017–1020.

Merriam, C.H. (1890) Results of a Biological Survey of the San Francisco Mountain Region and Desert of the Little Colorado River in Arizona. US Department of Agriculture, Bureau of Biological Survey of American Fauna 3.

Meyer, H.W. (1986) An Evaluation of the Methods for Estimating Paleoaltitudes Using Tertiary Floras from the Rio Grande Rift Vicinity, New Mexico and Colorado. PhD thesis, University of California, Berkeley.

Meyer, H.W. (1992) Lapse rates and other variables applied to estimating paleoaltitudes from fossil floras. *Palaeogeography, Palaeoclimatology, Palaeoecology* **99**, 71–99.

Meyer, H.W. (2001) A review of the paleoelevation estimates from the Florissant flora, Colorado. In: Evanoff, E., Gregory-Wodzicki, K.M. & Johnson, K.R. (eds.) *Fossil Flora and Stratigraphy of the Florissant*

Formation, Colorado. Proceedings of the Denver Museum of Nature and Science, ser. 4, no. **1**. Denver, CO: Denver Museum of Nature & Science, pp. 205–216.

Meyer, H.W. (2007) A review of paleotemperature-lapse rate methods for estimating paleoelevation from fossil floras. *Reviews in Mineralogy and Geochemistry* **66**, 155–171.

Molnar, P., Boos, W.R. & Battisti, D.S. (2010) Orographic controls on climate and paleoclimate of Asia: thermal and mechanical roles for the Tibetan Plateau. *Annual Review of Earth and Planetary Sciences* **38**, 77–102.

Mosbrugger, V. & Utescher, T. (1997) The coexistence approach – a method for quantitative reconstructions of Tertiary terrestrial palaeoclimate data using plant fossils. *Palaeogeography, Palaeoclimatology, Palaeoecology* **134**, 61–86.

Mulch, A. (2016) Stable isotope paleoaltimetry and the evolution of landscapes and life. *Earth and Planetary Science Letters* **433**, 180–191.

Needham, J. (1986) *Science and Civilization in China: Volume 3, Mathematics and the Sciences of the Heavens and the Earth*. Taipei: Caves Books Ltd.

Neilson, R.P., King, G.A., DeVelice, R.L. & Lenihan, J.M. (1992) Regional and local vegetation patterns: the responses of vegetation diversity to subcontinental air masses. In: Hansen, A.J. & di Castri, F. (eds) *Landscape Boundaries*. Berlin: Springer-Verlag, pp. 129–149.

Ojha, T.P., Butler, R.F., DeCelles, P.G. & Quade, J. (2009) Magnetic polarity stratigraphy of the Neogene foreland basin deposits of Nepal. *Basin Research* **21**, 61–90.

Parkhurst, D.F. & Loucks, O.L. (1972) Optimal leaf size in relation to environment. *Journal of Ecology* **60**, 505–537.

Peppe, D.J., Royer, D.L., Wilf, P. & Kowalski, E.A. (2010) Quantification of large uncertainties in fossil leaf paleoaltimetry. *Tectonics* **29**, TC3015.

Polissar, P.J., Freeman, K.H., Rowley, D.B. et al. (2009) Paleoaltimetry of the Tibetan Plateau from D/H ratios of lipid biomarkers. *Earth and Planetary Letters* **287**, 64–76.

Povey, D.A.R., Spicer, R.A. & England, P.C. (1994) Palaeobotanical investigation of early Tertiary elevations in northeastern Nevada: initial results. *Review of Palaeobotany and Palynology* **81**, 1–10.

Prentice, I.C., Cramer, W., Harrison, S.P. et al. (1992) A global biome model based on plant physiology and dominance, soil properties and climate. *Journal of Biogeography* **19**, 117–134

Price, L.W. (1981) *Mountains and Man*. Berkeley and Los Angeles, CA: University of California Press.

Raunkiaer, C. (1934) *The Life Forms of Plants and Statistical Plant Geography*. Oxford: Oxford University Press.

Reid, C.D., Maharali, H., Johnson, H.B. et al. (2003) On the relationship between stomatal characters and atmospheric CO_2. *Geophysical Research Letters* **30**, 1983.

Rodriguez, R.E., Debernardi, J.M. & Palatnik, J.F. (2014) Morphogenesis of simple leaves: regulation of leaf size and shape. *WIREs Developmental Biology* **3**, 41–57.

Schlichting, C.D. (1989) Phenotypic integration and environmental change. *BioScience* **39**, 460–464.

Schoch, P.G., Jacques, R., Lecharny, A. & Sibi, M. (1984) Dependence of the stomatal index on environmental factors during stomatal differentiation in leaves of *Vigna sinenses* L. II. Effect of different light quality. *Journal of Experimental Botany* **35**, 1405–1409.

Seward, A.C. (1892) *Fossil Plants as Tests of Climate*. London: C.J. Clay and Sons and Cambridge University Press.

Shukla, A., Mehrotra, R.C., Spicer, R.A. et al. (2014) Cool equatorial terrestrial temperatures and the South Asian monsoon in the Early Eocene: evidence from the Gurha Mine, Rajasthan, India. *Palaeogeography, Palaeoclimatology, Palaeoecology* **412**, 187–198

Spicer, R.A. (1981) The Sorting and Deposition of Allochthonous Plant Material in a Modern Environment at Silwood Lake, Silwood Park, Berkshire, England. US Geological Survey Professional Paper 1143.

Spicer, R.A. (2008) CLAMP. In: Gornitz, V. (ed.) *Encyclopedia of Paleoclimatology and Ancient Environments*. Dordrecht: Springer, pp. 156–158.

Spicer, R.A. & Collinson, M.E. (2014) Plants and Floral Change at the Cretaceous–Paleogene Boundary: Three Decades On. *Geological Society of America Special Paper* **505**, pp. 117–132.

Spicer, R.A. & Greer, A.G. (1986) Plant taphonomy in fluvial and lacustrine systems. In: Broadhead, T. (ed.) *Land Plants*. University of Tennessee Department of Geological Sciences Studies in Geology **15**, pp. 10–26.

Spicer, R.A. & Wolfe, J.A. (1987) Taphonomy of Holocene deposits in Trinity (Clair Engle) Lake, Northern California. *Paleobiology* **13**, 227–245.

Spicer, R.A. & Yang, J. (2010) Quantification of uncertainties in fossil leaf paleoaltimetry – does leaf size matter? *Tectonics* **29**, TC6001.

Spicer, R.A., Harris, N.B.W., Widdowson, M. et al. (2003) Constant elevation of southern Tibet over the past 15 million years. *Nature* **421**, 622–624.

Spicer, R.A., Herman, A.B. & Kennedy, E.M. (2005) The sensitivity of CLAMP to taphonomic loss of foliar physiognomic characters. *Palaios* **20**, 429–438.

Spicer, R.A., Valdes, P.J., Spicer, T.E.V. et al. (2009) New development in CLAMP: calibration using global gridded meteorological data. *Palaeogeography, Palaeoclimatolgy, Palaeoecology* **283**, 91–98.

Spicer, R.A., Bera, S., De Bera, S. et al. (2011) Why do foliar physiognomic climate estimates sometimes differ from those observed? Insights from taphonomic information loss and a CLAMP case study from the Ganges Delta. *Palaeogeography, Palaeoclimatology, Palaeoecology* **299**, 39–48.

Srivastava, G., Spicer, R.A., Spicer, T.E.V. et al. (2012) Megaflora and palaeoclimate of a Late Oligocene tropical delta, Makum Coal field, Assam: evidence for the early development of the South Asia Monsoon. *Palaeogeography, Palaeoclimatology, Palaeoecology* **343**, 130–142.

Steart, D.C., Spicer, R.A. & Bamford, M.K. (2010) Is Southern Africa different? An investigation of the relationship between leaf physiognomy and climate in southern African mesic vegetation. *Review of Palaeobotany and Palynology* **162**, 607–620.

Sun, B., Wang, Y.-F., Li, C.-S. et al. (2015) Early Miocene elevation in northern Tibet estimated by palaeobotanical evidence. *Scientific Reports* **5**, 1038.

Teodoridis, V., Mazouch, P., Spicer, R.A. & Uhl, D. (2011) Refining CLAMP – investigations towards improving the climate leaf analysis multivariate program. *Palaeogeography, Palaeoclimatology, Palaeoecology* **299**, 39–48.

ter Braak, C.J.F. (1986) Canonical correspondence analysis: a new eigenvector technique for multivariate direct gradient analysis. *Ecology* **67**, 1167–1179.

Tichá, I. (1982) Photosynthetic characteristics during ontogenesis of leaves. 7. Stomata density and sizes. *Photosynthetica* **16**, 375–471.

Utescher, T., Bruch, A.A., Erdei, B. et al. (2014) The coexistence approach – theoretical background and practical considerations of using plant fossils for climate quantification. *Palaeogeography, Palaeoclimatology, Palaeoecology* **410**, 58–73.

Wheeler, E.A. & Baas, P. (1993) The potentials and limitations of dicotyledonous wood anatomy for climatic reconstructions. *Paleobiology* **19**, 487–498.

Whitehead, F.H. & Luti, R. (1962) Experimental studies of the effect of wind on plant growth and anatomy *Zea mays*. *New Phytologist* **61**, 56–58.

Wilf, P. (1997) When are leaves good thermometers? A new case for leaf margin analysis. *Palaeobiology* **23**, 373–390.

Wilf, P., Wing, S.L., Greenwood, D.R. & Greenwood, C.L. (1998) Using fossil leaves as paleoprecipitation indicators: an Eocene example. *Geology* **26**, 203–206.

Witham, W. (1833) *The Internal Structure of Fossil Vegetables Found in the Carboniferous and Oolitic Deposits of Great Britain*. Edinburgh.

Wolfe, J.A. (1979) Temperature Parameters of Humid to Mesic Forests of Eastern Asia and Relation to Forests of Other Regions of the Northern Hemisphere and Australasia. *United States Geological Survey Professional Paper* **1106**, pp. 1–37.

Wolfe, J.A. (1992) An analysis of present-day terrestrial lapse rates in the western conterminous United States and their significance to paleoaltitudinal estimates. *United States Geological Survey Bulletin* **1964**, 1–35.

Wolfe, J.A. (1993) A method of obtaining climatic parameters from leaf assemblages. *United States Geological Survey Bulletin* **2040**, 1–73.

Wolfe, J.A. (1995) Paleoclimatic estimates from Tertiary leaf assemblages. *Annual Review of Earth and Planetary Sciences* **23**, 119–142.

Wolfe, J.A. & Wehr, W. (1987) Middle Eocene dicotyledonous plants from Republic, northeastern Washington. *United States Geological Survey Bulletin* **1597**, 1–25.

Woodward, F.I. & Cramer, W. (1996) Plant functional types and climatic changes: introduction. *Journal of Vegetation Science* **7**, 306–308.

Wylie, R.B. (1951) Principles of foliar organization shown by sun-shade leaves from ten species of deciduous dicotyledonous trees. *American Journal of Botany* **38**, 355–361.

Xu, K.-M. & Emanuel, K.A. (1989) Is the tropical atmosphere conditionally unstable? *Monthly Weather Review* **117**, 1471–1479.

Yanai, M. & Wu, G.X. (2006) Effects of the Tibetan Plateau. In: Wang, B. (ed.) *The Asian Monsoon*. Berlin: Springer, pp. 513–549.

Yang, J., Spicer, R.A., Spicer, T.E.V. & Li, C-S. (2011) "CLAMP Online": a new web-based palaeoclimate tool and its application to the terrestrial Paleogene and Neogene of North America. *Palaeobiodiversity and Palaeoenvironments* **91**, 163–183.

Yang, J., Spicer, R.A., Spicer, T.E.V. et al. (2015) Leaf form-climate relationships on the global stage: an ensemble of characters. *Global Ecology and Biogeography* **10**, 1113–1125.

Zetter, R. (1989) Methodik und Bedeutung einer routinemäßig kombinierten lichtmikroskopischen und rasterelektronischen Untersuchung fossiler Mikrofloren. *Courier Forschungshefte Institut Senckenberg* **109**, 41–50.

8

Cenozoic Mountain Building and Climate Evolution

Phoebe G. Aron and Christopher J. Poulsen

Department of Earth and Environmental Sciences, University of Michigan, Ann Arbor, MI, USA

Abstract

Climate and land surfaces are intimately linked. This chapter provides an introduction to mountain and climate interactions, with a particular focus on how temperature, atmospheric circulation and precipitation change with surface uplift. Although the timing and mechanisms of the South American Andes and North American Cordillera uplift are uncertain, both orogens have a first-order control on regional climate. With surface uplift, climate models generally suggest enhanced windward precipitation, orographic deflection of prevailing winds and decreased surface temperatures. However, the details of these responses are unique to each orogen and dependent on geographic position. In South America, Andean uplift led to a shift in prevailing winds from a Pacific source to an Atlantic source, development of a low-level jet along the eastern flank, increased moisture transport from the tropics and enhanced convective rainfall along the Andean lowlands and eastern flank. In North America, topographic blocking enhanced windward precipitation and initiated monsoonal conditions along the eastern flank, modified atmospheric circulation to create a pronounced trough-and-ridge pattern over western North America and formed a rain shadow over central North America. These changes add to the mechanistic understanding of regional Cenozoic climate and environmental change.

Keywords: *surface uplift, regional climate, North American Cordillera, South American Andes*

8.1 Introduction

Mountain ranges and high topography exert a fundamental control on Earth's climate. These features steer atmospheric circulation patterns, focus precipitation and aridity and regulate surface temperatures. In this way, Cenozoic mountain building and surface uplift have played an important role in the long-term evolution of both regional and global climate change. In this chapter, we present an introduction to how mountains and climate interact in order to better understand the significance of the development of the South American Andes and the North American Cordillera and associated uplift-induced climate change.

8.2 Mountain and Climate Interactions

Climate and land surfaces are intimately linked. In particular, the high topography of mountainous regions creates unique ecosystems that drive evolution and biodiversity, supply nutrients and sediments through weathering and erosion, store water as snow and ice and dictate local and downstream weather and storms. Mountains typically have steep slopes and strong elevation gradients (e.g., biodiversity, temperature, precipitation) that create unique local and regional climate variability. In this section, we investigate how temperature, atmospheric circulation and precipitation vary in mountainous regions.

8.2.1 Temperature

Mountainous regions exhibit some of the largest elevational temperature gradients on Earth. In the Colorado Rockies, average annual temperature varies more than 10 °C with elevation on the windward (western) side of the range and up to 15 °C on the leeward (eastern) side. In South America, average annual temperatures vary more than 20 °C up the Eastern Andean Cordillera and 5 °C along the Western Andean Cordillera.

Mountains, Climate and Biodiversity, First Edition. Edited by Carina Hoorn, Allison Perrigo and Alexandre Antonelli.
© 2018 John Wiley & Sons Ltd. Published 2018 by John Wiley & Sons Ltd.
Companion website: www.wiley.com\go\hoorn\mountains,climateandbiodiversity

The rate at which the temperature of an air parcel decreases with height is the lapse rate. First-order control of parcel lapse rate is determined by adiabatic expansion and compression. As a parcel of air ascends, the surrounding atmospheric pressure decreases and the parcel expands adiabatically. Through this process, the work done to expand the parcel consumes internal energy, causing the parcel to cool. Conversely, as a parcel descends, temperature increases by compressional warming. For an ascending or descending unsaturated (dry) particle, the average dry adiabatic lapse rate is 9.8 °C/km.

The lapse rate for a rising parcel over a high mountain is typically lower than the dry adiabatic lapse rate, due to warming through latent heat release as vapor condenses during the precipitation of clouds, rain, hail and snow. The lapse rate for a saturated air parcel – the moist adiabatic lapse rate – is 6.5 °C/km. In reality, the adiabatic lapse rate of air parcels across most mountain ranges falls along a continuum between the dry and moist adiabatic lapse rates, and reflects the initial relative humidity of the air parcel and the amount of cooling that it undergoes during ascent. Latent heating of a parcel may cause it to be more buoyant than surrounding cooler air, leading to instability and upward motion, known as "moist convection." For example, the presence of moisture and propensity for moist convection accounts for low temperature lapse rates in tropical regions in comparison to those in extratropical regions.

The dry and moist adiabatic end members belie the complexity of lapse rates in mountainous regions. Over many ranges, non-adiabatic processes cause lapse rates to deviate from these end members. For example, temperature lapse rates generally follow diurnal and seasonal heating patterns, with higher values in the night-time and winter due to stable stratification of the boundary layer and lower values in the daytime and summer due to convective mixing and the breakdown of stable stratification (Pepin & Losleben 2002; Kattel et al. 2013). Air parcels are also heated and cooled through thermal emission and mixing, and can vary both spatially and temporally. On the Bolivian Altiplano, seasonal temperature lapse rates range from 4.0 to 6.5 °C/km (Gonfiantini et al. 2001), while in the central Himalaya, temperature lapse rates have an annual mean of 5.4 °C/km but vary seasonally from 4.3 to 6.1 °C/km (Kattel et al. 2013). In the Colorado Rockies, temperature lapse rates vary spatially from 1.0 to 6.3 °C/km, with greater rates at higher elevations (Pepin & Losleben 2002). In this region, steepening of the temperature lapse rate may be explained by increased snow cover, enhanced airflow over the mountains and changes in solar radiation (Pepin & Losleben 2002). In the Washington Cascades, seasonal temperature lapse rates vary more on the leeward (eastern) side of the mountains than on the windward (western) side. In the lee of the Cascades, lapse rates vary seasonally by 4 °C/km, with the smallest values (2 °C/km) in December–January and the largest values (6 °C/km) in May–July (Minder et al. 2010). On the windward side of the mountains, annual lapse rates vary from 4 to 5 °C/km (Minder et al. 2010). The greater variability of temperature lapse rates in the lee of the Cascades is likely due to pooling and damming of cold air along the eastern flank (Minder et al. 2010).

Taken together, temperature lapse rates depend on parcel humidity, temperature and mixing, and vary with topography, climatology, latitude, season, relative humidity, atmospheric circulation, cloudiness, orientation (aspect) and vegetation (Rennick 1977; Stone & Carlson 1979; Laughlin 1982). In mountainous regions, orographic orientation (aspect) affects the amount of incoming solar radiation and temperature. In the northern hemisphere, southern- and western-facing slopes receive more solar radiation and typically have warmer temperatures than shaded northern- and eastern-facing slopes. Near-surface stability and heating can also affect temperature lapse rates by influencing cloud cover. For example, under a stable cloudy sky, a reduction of daytime insolation and confinement of night-time outgoing long-wave radiation may cause moister conditions and variations in temperature lapse rates (Stone & Carlson 1979; Jeffery et al. 2012; Kattel et al. 2013). In mountainous regions, these lapse rates yield steep temperature gradients that drive pressure differences and atmospheric circulation and dictate both local and regional climate.

8.2.2 Atmospheric Circulation

Major mountain ranges redirect how and where air flows across Earth. On a planetary scale, orography induces large-scale waves that disrupt zonal (west–east) wind patterns and cause meridional (north–south) flow (Smith 1979). Major mountain ranges also affect regional-scale circulations and produce distinct climate features such as the South Asian Monsoon or the interior prairie grasslands of North America (Ruddiman & Kutzbach 1989; Broccoli & Manabe 1992; Kutzbach et al. 1993).

Topography drives large-scale circulation patterns due to the mechanical blocking effects of mountain barriers (Galewsky 2008). When winds encounter large landmasses, they flow up and over or deflect around the obstacle (Etling 1989; Whiteman 2000). In part, the path winds take depends on the kinetic energy of the system. If the flow has sufficient kinetic energy to rise against the force of gravity, it will do so (Galewsky 2008, 2009). If not, the flow is deflected around the barrier (Whiteman 2000).

Additional characteristics that determine whether airflow moves up and over or around an orogen include: (i) the length, width and height of the orographic barrier; (ii) the orientation and spacing of the mountain range; and (iii) the speed and stability of the approaching airflow, where stability is a measure of the potential for vertical motion (Justus 1985; Whiteman 2000). In the atmosphere, stable flow contains little vertical motion and is unlikely to rise, while unstable flow can more easily rise up and over a barrier. Additionally, if a mountain range is lined with valleys or channels, winds may funnel through those gaps. For example, on the eastern Altiplano, the channeling of moist flow up river valleys causes cloud formation and precipitation near river headwaters (Giovannettone & Barros 2009).

Winds tend to flow over mountain ranges when the barrier is long, the cross-barrier wind is strong and approaching flow is unstable or near neutral (Whiteman 2000). Alternatively, winds tend to be deflected around mountain ranges when barriers are narrow or high, when they are convex on the windward side, when cross-barrier winds are weak and when approaching flow is stable (Whiteman 2000). In mountainous regions, deflected winds can also form a barrier jet: a jet-like wind current that forms when stable air approaches a barrier that flows parallel to the mountain range (Parish 1982). In the northern hemisphere, barrier jets turn towards the north, while in the southern hemisphere, they turn towards the south. For example, the South American low-level jet (SALLJ), one of the dominant features of the continent's climatology, flows southward parallel to the Andes and is responsible for transporting considerable moisture from the Amazon to the La Plata Basin in southern South America (Vera et al. 2006). Moreover, the movement of winds over or around mountains is of distinct importance in understanding climate, as these barriers locate regions of precipitation and aridity and regulate surface temperatures.

When air flows up and over a mountain, buoyancy and gravitational forces act upon vertically displaced air particles, and downstream waves develop (Smith 1979; Durran 1990). These waves are known as mountain gravity waves or lee waves, and produce distinct climate patterns (Durran 2003).

Mountain waves propagate away from the orography that caused them and extend vertically and horizontally through the troposphere (Smith 1979). Formation of these waves depends on: (i) the speed, vertical temperature gradient and stability of the flow as it approaches a mountain barrier; (ii) the size and shape of the orographic barrier; and (iii) the orientation of flow relative to the barrier (Smith 1979; Durran 2003). The largest mountain waves form in the lee of an orographic barrier when approaching flow is perpendicular to the barrier and the barrier is both steep and high (Whiteman 2000; Durran 2003). Weak approaching winds form only shallow waves downwind of an orographic barrier, while moderate winds frequently overturn in the lee of mountains and form standing, or nonpropagating, downstream eddies (Smith 1979; Durran 1990; Whiteman 2000).

Orographically induced stationary waves – planetary-scale waves that remain nearly fixed in relation to the Earth's surface – have strong effects on regional climate through meridional surface winds (Smith 1979). These waves advect temperature and moisture, direct preferred paths of storm tracks and contribute to zonal circulation (Smith 1979; Nigam & DeWeaver 2003). For example, the wintertime jet stream in the western USA develops a pronounced trough and ridge pattern as it moves over the Rocky Mountains. Winds in this wave pattern advect northerly cold air over western North America, transport warm air from the Gulf of Mexico towards eastern North America and steer storm tracks across the continent.

Although this discussion has focused exclusively on planetary-scale circulation, it is important to consider small-scale orographically induced features that contribute to atmospheric circulation as well. Many of these characteristics are a result of elevation differences and temperature contrasts that induce differential along-slope heating and cooling and vertical motion (Barry 1992). The most common example, discussed briefly here, is valley circulation.

Valley winds are driven by a strong temperature control and display distinct diurnal and seasonal variation. Night-time radiative cooling causes local down-valley movement, while daytime warming causes up-valley circulation (Barry 1992, 2008). A similar seasonal pattern describes general winter and summer valley circulation, as well. Mountain valley winds are also dictated by an antitriptic wind component, which maintains a balance between the pressure gradient force and frictional force and is directed towards areas of low pressure, and a gravity wind component, which is directed downslope (Barry 1992).

8.2.3 Precipitation and Aridity

One of the primary observable effects of orographically modified circulation is continental patterns of precipitation and aridity. Air undergoing forced ascent over an orographic barrier cools adiabatically, leading to condensation of clouds and precipitation along the mountain front (Roe 2005; Hughes et al. 2009). As winds descend the leeward sides of mountains, the air undergoes compression, warms and experiences a reduction in relative humidity, leading to undersaturated conditions that produce little precipitation. This phenomenon, known as the rain shadow effect, demonstrates the potential for an orographic range to produce two distinct climates on the windward and leeward sides of a range.

The rain shadow effect describes climates where prevailing winds flow perpendicular to orographic barriers, such as in the North American Cordillera. In the Rockies, moist cool climates (Pacific coast) dominate the windward flank of the region, while interior leeward climates (Great Plains) are warmer and drier (Broccoli & Manabe 1992). However, as is the case in the western Tibetan Plateau, when orographic barriers lie parallel to prevailing winds, more nuanced mechanisms are needed to explain patterns of precipitation and climate (Broccoli & Manabe 1992).

In the absence of orography, precipitation amounts decrease gradually with distance from moisture sources (Broccoli & Manabe 1992). With orography, the magnitude of relief has a direct effect on the amount of precipitation and the development of climate features (Roe 2005). Accordingly, higher mountain ranges produce stronger windward precipitation and more intense interior aridity than lower ranges. Additionally, due to rainout, condensation decreases exponentially with height, such that regions with maximum precipitation are typically found at low elevation on the windward flank (Roe 2005). Orography in tropical regions can also act as a dynamical forcing that destabilizes the atmosphere and produces convective rainfall. For example, both rainout and convective rainfall are common phenomena in the central Andes, where westward-flowing moist air from the Amazon region ascends the eastern flanks (Garreaud et al. 2003; Insel et al. 2010; Barnes et al. 2012). The frequency and intensity of orographically induced convective rainfall vary on diurnal and seasonal time scales as a function of insolation, and are greatest during the afternoon and early evening, and in summertime (Garreaud et al. 2003).

In addition to large-scale orographic phenomena, small-scale features such as mountain gaps affect climate as well. For example, the Columbia River Gorge in the north-western United States provides a channel along which moist air flows across south-central Idaho and on to the Yellowstone Plateau (Stewart et al. 2002). As a result, regions such as Yellowstone National Park, which lies on the eastern edge of the Snake River Plain, have relatively high annual precipitation east of the Rocky Mountains. Similarly, valleys in the Andes provide a conduit for moisture to move among mountains and into regions one might otherwise assume to be dry (Bendix et al. 2006; Giovannettone & Barros 2009). As expected, convective instability and precipitation in these regions have strong diurnal patterns (Bendix et al. 2006).

In mountainous environments, contrasts in elevation induce precipitation and aridity, regulate temperatures and direct circulation patterns to develop and drive climate regimes. In the next section, we examine a brief history of Cenozoic surface uplift and paleoaltimetry in the Americas as a case study for how orography contributed to past climate change.

8.3 Paleoaltimetry Approaches

Cenozoic climate change is defined by a transition from ice-free poles and warm conditions to ice-covered poles and cold conditions (see Chapter 9). While a reduction in greenhouse gases – carbon dioxide, in particular – likely explains much of the Cenozoic global cooling, a complete mechanistic understanding of Cenozoic climate change – particularly regional changes – must consider additional climate forcings during this period. Here, we outline how progressive surface uplift of the South American Andes and North American Cordillera affected climate on regional and possibly global scales.

8.3.1 Paleoaltimetry Proxies

Elevations of continental surfaces are among the most uncertain features of past geologic times. In the absence of direct measurements, proxies are used to infer past surface elevations. Several different paleoaltimetry proxies exist, including the stable isotopic compositions of authigenic minerals and volcanic glass, the clumped isotopic composition of authigenic minerals and fossil leaf traits, all of which preserve a signal of a physical characteristic (e.g., temperature, meteoric $\delta^{18}O$, meteoric δD, moist enthalpy) that varies as a known function of elevation. For example, the stable isotopic compositions ($\delta^{18}O$ and δD) of precipitation (meteoric waters) decrease upslope as the heavy isotopologue is preferentially lost through rainout (see Chapter 6). Most of these proxies were initially founded on the assumption that lapse rates were constant through time and equivalent to modern values (Koch 1998; Poage & Chamberlain 2001; Garzione et al. 2006). This assumption has recently been challenged, and found to lead to potentially large biases – on the order of kilometers – in paleoaltimetry estimates (Rowley & Garzione 2007; Ehlers & Poulsen 2009; Poulsen et al. 2010; Insel et al. 2012; Feng et al. 2013; Fiorella et al. 2015; Feng & Poulsen 2016). In fact, lapse rates of meteoric $\delta^{18}O$ vary with topography and climate (as described in Section 8.2.1), and almost certainly have not remained constant throughout the Cenozoic (Stone & Carlson 1979; Poulsen et al. 2010; Poulsen & Jeffrey 2011; Insel et al. 2012; Feng et al. 2013). To this end, as orogens evolve, lapse rates may be affected by temperature, atmospheric CO_2 concentrations, precipitation type, moisture source and air mass mixing, among other factors (Ehlers & Poulsen 2009; Poulsen et al. 2010; Feng et al. 2013, Feng & Poulsen 2016).

8.3.2 Climate Modeling

Climate modeling has proven to be a valuable complement to proxy studies for evaluating both past elevations and the climate response to surface uplift. Terrestrial proxy data preserve a signal that reflects some combination of elevation and climate, but as demonstrated by the lapse-rate complexities discussed earlier, deconvoluting these signals can be challenging. Three-dimensional (3D) general circulation models (GCMs) can be used to directly investigate the climate response to orography and surface uplift, and can provide direction for interpreting proxy signals (e.g., Ehlers & Poulsen 2009; Poulsen et al. 2010).

Given that past continental surface elevations are not known with confidence, most GCM studies have taken a systematic approach to modifying topography, in which surface elevations are progressively adjusted from modern values (e.g., Ruddiman & Kutzbach 1989; Ehlers & Poulsen 2009). This approach is somewhat arbitrary and need not be followed, particularly when presuppositions about past surface elevations exist (e.g., Feng et al. 2013).

GCMs have been used to explore the effects of mountains and mountain uplift on climate for nearly 3 decades (e.g., Kutzbach et al. 1989; Ruddiman & Kutzbach 1989; Ruddiman et al. 1989). Early studies using atmosphere-only GCMs investigated uplift of major plateaus in western North America (Colorado Plateau, Basin and Range Province and the Rocky and Sierra Mountains) and south Asia (Tibetan Plateau and Himalaya Mountains). In addition to the diversion of the planetary wave pattern in the extratropical region of the northern hemisphere, these studies emphasized uplift-related regional changes in seasonal surface winds and precipitation due to enhanced seasonal cooling and heating during summer and winter.

Since this early seminal work, both the performance and the capabilities of GCMs have improved substantially. Although a summary of these advances is beyond the scope of this review, increases in GCM spatial resolution, improvements in the simulation of the physics of convection and cloud formation and the incorporation of water isotopes are worth mentioning for their benefits to the study of surface uplift and climate change. The first of these advances has led to significant improvements in the simulation of convection, clouds and precipitation in the vicinity of high topography. The second advance, a reduction in spatial scale and improved resolution of topographic features in GCMs, has allowed for the simulation of climate features nearer to the scale represented by the proxies. The third advance has allowed for

straightforward comparisons between simulated meteoric stable isotope tracers and stable isotope proxies.

8.4 Surface Uplift and Climate Change

8.4.1 South American Andes

The Andes are the second largest topographic feature in the world, and the only major orographic barrier in the southern hemisphere. The Andean Cordillera is 7000 km long and 100–300 km wide, with a mean height over 4000 m and many 5000–6000 m peaks (Barry 2008). The Andean Plateau, or Altiplano, which lies mainly in the Bolivian and Peruvian Andes, is approximately 400 km wide and has a mean elevation of almost 4000 m (Barry 2008). Although the mechanism and timing of Andean surface uplift are uncertain, most agree that the Andes grew as a result of convergence of the Nazca and South American Plates during the Cenozoic (Gregory-Wodzicki 2000). The Altiplano achieved its modern east–west width by 15–25 Ma, although its elevation at that time is unknown (Allmendinger et al. 1997; Isacks 1988; Gregory-Wodzicki 2000).

The two dominant – and contrasting – theories of Andean surface uplift are rapid recent plateau rise and slow and steady plateau rise (Barnes & Ehlers 2009). Geologic evidence for rapid surface uplift has been proposed from paleobotany, soil carbonate $\delta^{18}O$, volcanic glass δD and clumped isotope paleothermometry proxies, and has been interpreted to indicate pulses of Andean surface uplift of 2.5 ± 1 km from 19 to 16 Ma and from ~10 to 6 Ma (Gregory-Wodzicki 2000; Ghosh et al. 2006; Garzione et al. 2006; Saylor & Horton 2014). However, these rapid uplift estimates may be biased in neglecting climate change associated with surface uplift (Ehlers & Poulsen 2009; Poulsen et al. 2010; Insel et al. 2012) and/or in neglecting evaporative enrichment of stable isotope records (Fiorella et al. 2015). Accounting for these biases largely removes the evidence for rapid uplift and lends support to models that favor a history of slow and steady Andean surface uplift, which has been protracted over ~40 My (e.g., Barnes & Ehlers 2009; Insel et al. 2012).

Regardless of the exact timing of uplift, the rise of the Andes undoubtedly had a substantial impact on the regional climate and environment. Modern Andean climate is governed by blocked zonal flows that drive regional circulation and precipitation (Insel et al. 2010). Mechanical blocking deflects low-level Atlantic trade winds into a northerly barrier jet, the SALLJ, which flows along the eastern flank of the Andes (Campetella & Vera 2002; Vera et al. 2006). Moisture transport across the Amazon Basin, development of the SALLJ and

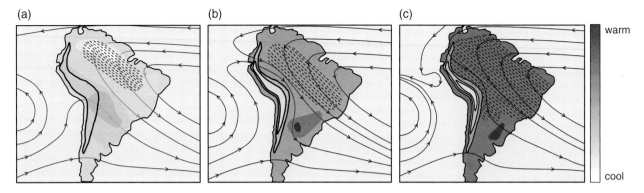

Figure 8.1 Schematic of South American climatology with (a) low, (b) medium and (c) high surface uplift. Thin vectors represent lower-level winds. Diagonal hatching indicates regions with mean annual precipitation greater than 150 cm. Dark shading indicates cooler temperatures and light shading indicates warmer temperatures. See also Plate 14 in color plate section.

orographic lifting of easterly winds combine to yield high precipitation in this region (Lenters & Cook 1995; Insel et al. 2010).

As the Andes were uplifted, paleoclimate models indicate that Altiplano temperature decreased, precipitation increased along the eastern flank and zonal winds were blocked (Figure 8.1) (Ehlers & Poulsen 2009). As elevation increased from 200 to 4800 m, temperature on the Altiplano and other high-elevation areas along the Andean Cordillera decreased more than 6 °C due to non-adiabatic cooling and increased latent and sensible heat loss (Ehlers & Poulsen 2009).

Andean growth also led to a shift in the source and direction of prevailing winds. Before the Andes uplifted, prevailing winds were westerly from the Pacific. As the Andes rose, westerly winds were blocked and zonal winds switched to an easterly Atlantic source (Ehlers & Poulsen 2009; Poulsen et al. 2010). Additionally, with uplift, mechanical blocking led to the development of a low-level jet, the SALLJ, along the eastern flank of the orogen (Insel et al. 2010).

Development of the SALLJ was integral to the evolution of uplift-induced precipitation patterns in the eastern Andes. Without the Andes, paleoclimate models indicate that the SALLJ would be absent: low-level transport of moisture would fade with distance from its source, convective processes and cumulus cloud formation would be suppressed and vertical velocities would be reduced due to the lack of a lifting mechanism (Insel et al. 2010). Additionally, due to its subtropical location, the Altiplano region would be hyper-arid without SALLJ moisture transport (Fiorella et al. 2015). At 50–75% of modern height, the SALLJ would be abruptly initiated, westerly flow would be at least partially blocked and moisture transport, humidity and precipitation would increase along the eastern flank and across the Altiplano (Ehlers & Poulsen 2009; Insel et al. 2010; Fiorella et al. 2015). As Andean elevations increase to their modern

height and the SALLJ strengthens, the eastern flank transitions from arid to humid and net evaporation decreases across the Altiplano (Ehlers & Poulsen 2009; Insel et al. 2010; Fiorella et al. 2015). Under modern conditions, moisture transport from the Amazon Basin and resulting eastern flank precipitation are so high that the South American Monsoon develops along the eastern Bolivian Andes (Garreaud 1999).

Upper-level (200 mb) flow is dominated by the Bolivian High, an upper-level anticyclone that develops during summer months. The Bolivian High affects low-level near-surface circulation (Garreaud 1999; Lenters & Cook 1999) but is not directly affected by mechanical forcing of Andean elevation (Insel et al. 2010). In its modern state, easterly upper-level flow enhances precipitation over the Andes, while westerly flow causes dry conditions (Insel et al. 2010). When the Andes are absent, the reduction of low-level moisture transport and eastward shift in the region of maximum latent heating and convergence cause the Bolivian High to shift eastward (Insel et al. 2010).

Importantly, the climate changes associated with surface uplift are nonlinear (Ehlers & Poulsen 2009; Garzione et al. 2014). For example, when the Andes were 0 and 25% of their modern height, winds were still predominantly south-westerly from the Pacific. As the Andes achieved 50 and 75% of their modern height, winds shifted to flow from the northwest but still moved over the orogen. However, at their modern height, prevailing winds flow from the east across the Amazon basin, are deflected southward due to blocking and join with the South Atlantic convergence zone (Lenters & Cook 1997; Ehlers & Poulsen 2009). This stepwise change highlights the potential for nonlinear uplift-induced climate change and the importance of uplift/climate thresholds (Insel et al. 2012).

It is also important to note that the climatic effects of Andean uplift extended far beyond the Cordillera and its flanks. As Andean elevation increased, moisture transport along the SALLJ decreased Amazonian precipitation in the eastern basin (Insel et al. 2010) but increased that in the central Amazon basin due to orographic blocking and rainout (Ehlers & Poulsen 2009). Additionally, Andean uplift contributed to the development of new ecosystems, redirected erosion and river drainage and led to the proliferation of new flora and fauna biodiversity in the Amazon basin (Hoorn et al. 2010). Along the west coast of South America, Andean uplift initiated shifts in the sea surface temperature (SST) gradient and circulation, which contributed to increased marine productivity and decreased the intensity and frequency of El Niño–Southern Oscillation (ENSO) and El Niño events (Feng & Poulsen 2014).

Taken together, uplift of the Andes contributed to a wide array of regional climate, landscape and biological effects along the Cordillera and elsewhere. A dual approach of model reconstructions and proxy records of surface uplift in the Andes explains up to 6 °C cooling during the Cenozoic, a general trend of increased precipitation along the eastern flank and Altiplano and a shift of prevailing winds from a westerly Pacific source to an easterly Atlantic one (Ehlers & Poulsen 2009). Although the timing is debated, Andean surface uplift certainly had a first-order impact on regional climate.

8.4.2 North American Cordillera

The North American Cordillera is a 6000 km-long mountainous spine that passes through most of western North America. The modern North American Cordillera stretches from Alaska to Mexico and encompasses the Rockies, Cascade Range, Sierra Nevada Mountains, Columbia Plateau and Basin and Range Province. This topography extends over 1000 km wide and has a first-order control on regional climate.

As with the Andes, the mechanisms and timing of surface uplift along the North American Cordillera are still debated, although most agree that orogenic growth was largely initiated by subduction of the Farallon Plate below the North American Plate. Evolution of the North American Cordillera occurred in three distinct tectonic phases: Sevier orogeny (140–50 Ma), a thin-skinned compressional deformation; Laramide orogeny (70–80 through 55–35 Ma), a thick-skinned compressional deformation; and Basin and Range extension (17 Ma to present), which resulted in a reduction of mean elevation.

The two dominant – and contrasting – mechanistic hypotheses of North American Cordillera evolution are Cretaceous/early Paleogene gravitational plateau collapse and Cenozoic uplift (DeCelles 2004). According to the Cretaceous gravitational collapse hypothesis, subduction of the Farallon Plate created a highland, which

later experienced gravitational strain and spread laterally through the Cenozoic (Hodges & Walker 1992; DeCelles 2004). Alternatively, the Cenozoic-uplift hypothesis proposes that the North American Cordillera formed from early-Cenozoic rapid north–south migration of high-elevation topography (Mix et al. 2011; Chamberlain et al. 2012).

Further confounding the geologic and tectonic history of the North American Cordillera, Lechler et al. (2013) note that the complex topography in this region likely reflects a combination of gravitational plateau collapse, uplift and extension processes. Additionally, reconstructions of North American paleoaltimetry are complicated by the fact that different methods yield kilometer-scale differences in elevation estimates. Stable isotope paleoaltimetry estimates suggest that the North American Cordillera plateau rose 3–4 km during the Eocene (Mix et al. 2011; Chamberlain et al. 2012; Feng et al. 2013), while fossil leaf-based paleoaltimetry estimates indicate peak elevations of 2–3 km, depending on whether temperature and/or moist enthalpy is considered (Chase et al. 1998; Wolfe et al. 1998). Despite these incongruities, recent proxy-based paleoaltimetry reconstructions and climate modeling tend to favor the Cenozoic-uplift theory (Mix et al. 2011; Chamberlain et al. 2012; Feng et al. 2013; Sewall & Fricke 2013) and development of a late-Eocene plateau (Fan et al. 2014a).

North American Cordillera paleoclimate models indicate cooler temperatures with surface uplift (Ruddiman & Kutzbach 1989; Feng et al. 2013; Sewall & Fricke 2013). With no topography, temperatures have a largely meridional gradient, with warmer temperatures in the south and cooler temperatures in the north (Sewall & Fricke 2013). With uplift, this gradient weakens and high-elevation areas cool non-adiabatically through advective heat loss (Feng et al. 2013; Sewall & Fricke 2013). Additionally, western-flank temperatures cool due to the formation of large-amplitude stationary mountain gravity waves that advect cold air from the north (Kutzbach et al. 1989; Ruddiman & Kutzbach 1989).

Modern atmospheric circulation in the North American Cordillera is predominately driven by orographically directed flows and by competition between air masses from the Pacific Ocean and the Gulf of Mexico. However, without high surface topography, climate models indicate that westerly winds would flow relatively unimpeded across North America (Kutzbach et al. 1989; Feng et al. 2013). With progressive uplift, westerlies are diverted around the Rocky Mountains (Kutzbach et al. 1989), and they develop a pronounced trough-and-ridge pattern as they are deflected over and around the orogen (Figure 8.2) (Ruddiman & Kutzbach 1989). As the Cordillera evolved, deflection intensified and a sharper trough emerged. At its modern height,

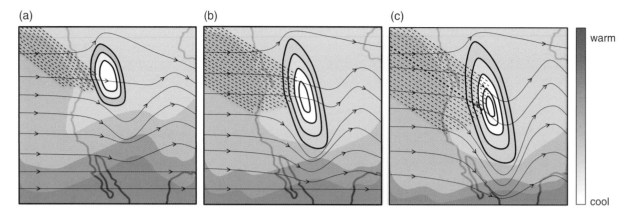

Figure 8.2 Schematic of North American climatology with (a) low, (b) medium and (c) high surface uplift. Thin vectors represent lower-level winds. Diagonal hatching indicates regions with mean annual precipitation greater than 150 cm. Dark shading indicates cooler temperatures and light shading indicates warmer temperatures. See also Plate 15 in color plate section.

southerly flow from the Gulf of Mexico transported moisture to the lee of the Colorado Rockies, where a seasonal monsoon developed (Feng et al. 2013; Sewall & Fricke 2013).

Beyond the North American Monsoon, changes in atmospheric circulation due to surface uplift led to the development of distinct seasonal precipitation patterns. As surface uplift evolved and the amplitude of the stationary wave increased, more pronounced onshore surface flow and surface heating increased windward wintertime precipitation (Figure 8.2) (Feng et al. 2013; Sewall & Fricke 2013). At modern height, forced ascent of westerly airflow over orographic masses promoted diabatic cooling, condensation and snowfall, and caused strong winter windward precipitation (Chamberlain et al. 2012; Feng et al. 2013; Sewall & Fricke 2013). Uplift caused the summertime leeward monsoon to intensify (Kutzbach et al. 1989), and at the same time induced a rain shadow that caused much of the interior continental United States to become drier (Ruddiman & Kutzbach 1989; Broccoli & Manabe 1992; Fan et al. 2014b).

As with the Andes, orography has a first-order control on regional and local climate in North America. Modern hemisphere-scale stationary waves seasonally shift the position of the jet stream and dictate temperature patterns across the continent.

8.5 Conclusion

Cenozoic surface uplift has long been associated with climate change. In this chapter, we have presented a summary of uplift-induced temperature, precipitation and atmospheric circulation changes in the Andes and North American Cordillera regions. Although the timing and mechanism of uplift are not yet constrained in either region, model results of systematic uplift generally suggest enhanced windward precipitation, orographic deflection of prevailing winds and cooler temperatures. However, there are distinct differences in the response of uplift between these regions that are related to their geographic positions. Topographic blocking at low latitudes by the Andes led to development of a low-level jet, moisture transport from the tropics and initiation of convective rainfall and cumulus cloud formation along the eastern flanks. Topographic blocking in middle latitudes by the North American Cordillera modified the stationary wave pattern creating a pronounced ridge and trough over western North America, formed a rainshadow downstream of high topography and initiated monsoonal conditions along the eastern front of the Cordillera. These changes acted in concert with other factors to produce Cenozoic climate and environmental change.

References

Allmendinger, R.W., Jordan, T.E., Kay, S. M. & Isacks, B.L. (1997) The evolution of the Altiplano-Puna plateau of the Central Andes. *Annual Review of Earth and Planetary Sciences* **25**, 139–174.

Barnes, J.B. & Ehlers, T.A. (2009) End member models for Andean Plateau uplift. *Earth Science Reviews* **97**, 105–132.

Barnes, J.B., Ehlers, T.A., Insel, N. et al. (2012) Linking orography, climate, and exhumation across the central Andes. *Geology* **40**, 1135–1138.

Barry, R.G. (1992) *Mountain Weather and Climate*, 2nd edn. London: Routledge.

Barry, R.G. (2008) *Mountain Weather and Climate*, 3rd edn. Cambridge: Cambridge University Press.

Bendix, J., Rollenbeck, R. & Reudenbach, C. (2006) Diurnal patterns of rainfall in a tropical Andean valley of southern Ecuador as seen by a vertically pointing K-band Doppler radar. *International Journal of Climatology* **26**, 829–846.

Broccoli, A.J. & Manabe, S. (1992) The effects of orography on midlatitude northern hemisphere dry Cclimates. *Journal of Climate* **5**, 1181–1201.

Campetella, C.M. & Vera, C.S. (2002) The influence of the Andes Mountains on the South American low-level flow. *Geophysical Research Letters* **29**, 1826.

Chamberlain, C.P., Mix, H.T., Mulch, A. et al. (2012) The Cenozoic climatic and topographic evolution of the western North American Cordillera. *American Journal of Science* **312**, 213–262.

Chase, C.G., Gregory-Wodzicki, K.M., Parrish-Jones, J.T. & DeCelles, P.G. (1998) Topographic history of the western Cordillera of North America and controls on climate. In: Crowley, T.J. & Burke, K. (eds.) *Tectonic Boundary Conditions for Climate Model Simulations*. New York: Oxford University Press, pp. 73–99.

DeCelles, P.G. (2004) Late Jurassic to Eocene evolution of the cordilleran thrust belt and foreland basin system, Western USA. *American Journal of Science* **304**, 105–168.

Durran, D.R. (1990) Mountain waves and downslope winds. In: Blumen, B. (ed.) *Atmospheric Processes over Complex Terrain*. Boston, MA: American Meteorological Society, pp. 59–81.

Durran, D.R. (2003) Lee waves and mountain waves. In: Holton, J., Curry, J. & Pyle, J. (eds.) *The Encyclopedia of the Atmospheric Sciences*. San Diego, CA: Academic Press, Elsevier Science, pp. 1161–1169.

Ehlers, T.A. & Poulsen, C.J. (2009) Influence of Andean uplift on climate and paleoaltimetry estimates. *Earth and Planetary Science Letters* **281**, 238–248.

Etling, D. (1989) On atmospheric vortex streets in the wake of large islands. *Meteorology and Atmospheric Physics* **41**, 157–164.

Fan, M., Hough, B.G. & Passey, B.H. (2014a) Middle to late Cenozoic cooling and high topography in the central Rocky Mountains: constraints from clumped isotope geochemistry. *Earth and Planetary Science Letters* **408**, 35–47.

Fan, M., Heller, P., Allen, S.D. & Hough, B.G. (2014b) Middle Cenozoic uplift and concomitant drying in the central Rocky Mountains and adjacent Great Plains. *Geology* **42**, 547–550.

Feng, R. & Poulsen, C.J. (2014) Andean elevation control on tropical Pacific climate and ENSO. *Paleoceanography* **29**, 795–809.

Feng, R. & Poulsen, C.J. (2016) Refinement of Eocene lapse rates, fossil-leaf altimetry, and North American Cordillera surface elevation estimates. *Earth and Planetary Science Letters* **436**, 130–141.

Feng, R., Poulsen, C.J., Werner, M. et al. (2013) Early cenozoic evolution of topography, climate, and stable isotopes in precipitation in the North American cordillera. *American Journal of Science* **313**, 613–648.

Fiorella, R.P., Poulsen, C.J., Pillco Zola, R.S. et al. (2015) Modern and long-term evaporation of central Andean surface waters suggests paleo archives underestimate Neogene elevations. *Earth and Planetary Science Letters* **432**, 59–72.

Galewsky, J. (2008) Orographic clouds in terrain-blocked flows: an idealized modeling study. *Journal of Atmospheric Sciences* **65**, 3460–3478.

Galewsky, J. (2009) Orographic precipitation isotopic ratios in stratified atmospheric flows: implications for paleoelevation studies. *Geology* **37**, 791–794.

Garreaud, R. (1999) Multiscale analysis of the summertime precipitation over the Central Andes. *Monthly Weather Review* **127**, 901–921.

Garreaud, R., Vuille, M. & Clement, A.C. (2003) The climate of the Altiplano: observed current conditions and mechanisms of past changes. *Palaeogeography, Palaeoclimatology, Palaeoecology* **194**, 5–22.

Garzione, C.N., Molnar, P., Libarkin, J.C. & MacFadden, B.J. (2006) Rapid late Miocene rise of the Bolivian Altiplano: evidence for removal of mantle lithosphere. *Earth and Planetary Science Letters* **241**, 543–556.

Garzione, C.N., Auerbach, D.J., Smith, J. et al. (2014) Clumped isotope evidence for diachronous surface cooling of the Altiplano and pulsed surface uplift of the Central Andes. *Earth and Planetary Science Letters* **393**, 173–181.

Ghosh, P., Garzione, C.N. & Eiler, J.M. (2006) Rapid uplift of the Altiplano revealed through ^{13}C-^{18}O bonds in paleosol carbonates. *Science* **311**, 511–515.

Giovannettone, J.P. & Barros, A.P. (2009) Probing regional orographic controls of precipitation and cloudiness in the Central Andes using satellite data. *Journal of Hydrometeorology* **10**, 167–182.

Gonfiantini, R., Roche, M.A., Olivry, J.C. et al. (2001) The altitude effect on the isotopic composition of tropical rains. *Chemical Geology* **181**, 147–167.

Gregory-Wodzicki, K.M. (2000) Uplift history of the Central and Northern Andes: a review. *Geological Sociecy of America Bulletin* **112**, 1091–1105.

Hodges, K.V. & Walker, J.D. (1992) Extension in the Cretaceous sevier orogen, North American Cordillera. *Geological Society of American Bulletin* **104**, 560–569.

Hoorn, C., Wesselingh, F.P., ter Steege, H. et al. (2010) Amazonia through time: Andean uplift, climate change, landscape evolution, and biodiversity. *Science* **330**, 927–931.

Hughes, M., Hall, A. & Fovell, R.G. (2009) Blocking in areas of complex topography, and its influence on rainfall distribution. *Journal of Atmospheric Sciences* **66**, 508–518.

Insel, N., Poulsen, C.J. & Ehlers, T.A. (2010) Influence of the Andes Mountains on South American moisture transport, convection, and precipitation. *Climate Dynamics* **35**, 1477–1492.

Insel, N., Poulsen, C.J., Ehlers, T.A. & Sturm, C. (2012) Response of meteoric $\delta^{18}O$ to surface uplift – implications for Cenozoic Andean Plateau growth. *Earth and Planetary Science Letters* **317–318**, 262–272.

Isacks, B.L. (1988) Uplift of the Central Andean Plateau and bending of the Bolivian orocline. *Journal of Geophysical Research* **93**, 3211–3231.

Jeffery, M.L., Poulsen, C.J. & Ehlers, T.A. (2012) Impacts of Cenozoic global cooling, surface uplift, and an inland seaway on South American paleoclimate and precipitation $\delta^{18}O$. *Bulletin of the Geological Society of America* **124**, 335–351.

Justus, C.G. (1985) Wind energy. In: Houghton, D.D. (ed.) *Handbook of Applied Meteorology*. Boston, MA: American Meteorology Soceity, pp. 915–944.

Kattel, D.B., Yao, T., Yang, K. et al. (2013) Temperature lapse rate in complex mountain terrain on the southern slope of the central Himalayas. *Theoretical and Applied Climatology* **113**, 671–682.

Koch, P.L. (1998) Isotopic reconstruction of past continental environments. *Annual Review of Earth and Planetary Sciences* **26**, 573–613.

Kutzbach, J.E., Guetter, P.J., Ruddiman, W.F. & Prell, W.L. (1989) Sensitivity of climate to late Cenozoic uplift in southern Asia and the American West: numerical experiments. *Journal of Geophysical Research* **94**, 18393–18407.

Kutzbach, J.E., Prell, W.L. & Ruddiman, W.F. (1993) Sensitivity of Eurasian climate to surface uplift of the Tibetan Plateau. *Journal of Geology* **101**, 177–190.

Laughlin, G.P. (1982) Minimum temperature and lapse rate in complex terrain: influencing factors and prediction. *Archives for Meteorology, Geophysics, and Bioclimatology Series B* **30**, 141–152.

Lechler, A.R., Neimi, N.A., Hren, M.T. & Lohmann, K.C. (2013) Paleoelevation estimates for the northern and central proto-Basin and Range from carbonate clumbed isotpoe thermometry. *Tectonics* **32**, 295–316.

Lenters, J.D. & Cook, K.H. (1995) Simulation and diagnosis of the regional summertime precipitation climatology of South America. *Journal of Climate* **8**, 2988–3005.

Lenters, J.D. & Cook, K.H. (1997) On the origin of the Bolivian high and related circulation features of the South American climate. *Journal of Atmospheric Science* **54**, 656–678.

Lenters, J.D. & Cook, K.H. (1999) Summertime precipitation variability over South America: role of the large-scale circulation. *Monthly Weather Review* **127**, 409–431.

Minder, J.R., Mote, P.W. & Lundquist, J.D. (2010) Surface temperature lapse rates over complex terrain: lessons from the Cascade Mountains. *Journal of Geophysical Research* **115**, 14122.

Mix, H.T., Mulch, A., Kent-Corson, M.L. & Chamberlain, C.P. (2011) Cenozoic migration of topography in the North American Cordillera. *Geology* **39**(1), 87–90.

Nigam, S. & DeWeaver, E. (2003) Stationary waves (orographic and thermally forced). In: Holton, J.R., Pyle, J.A. & Curry, J.A (eds.) *Encyclopedia of Atmospheric Sciences*. London: Academic Press, Elsevier Science, pp. 2121–2137.

Parish, T.R. (1982) Barrier winds along the Sierra Nevada mountains. *Journal of Applied Meteorology* **21**, 925–930.

Pepin, N. & Losleben, M. (2002) Climate change in the Colorado Rocky Mountains: free air versus surface temperature trends. *International Journal of Climatology* **22**, 311–329.

Poage, M.A. & Chamberlain, C.P. (2001) Stable isotope composition of precipitation and surface waters: considerations for studies of paleoelevation change. *American Journal of Science* **301**, 1–15.

Poulsen, C.J. & Jeffery, M.L. (2011) Climate change imprinting on stable isotopic compositions of high-elevation meteoric water cloaks past surface elevations of major orogens. *Geology* **39**, 595–598.

Poulsen, C.J., Ehlers, T.A. & Insel, N. (2010) Onset of convective rainfall during gradual late Miocene rise of the central Andes. *Science* **328**, 490–493.

Rennick, M.A. (1977) The parameterization of tropospheric lapse tates in terms of surface temperature. *Journal of the Atmospheric Sciences* **34**, 854–862.

Roe, G.H. (2005) Orographic precipitation. *Annual Review of Earth and Planetare Sciences* **33**, 645–671.

Rowley, D.B. & Garzione, C.N. (2007) Stable isotope-based paleoaltimetry. *Annual Review of Earth and Planetary Sciences* **35**, 463–508.

Ruddiman, W.F. & Kutzbach, J.E. (1989) Forcing of late Cenozoic northern hemisphere climate by plateau uplift in southern Asia and the American west. *Journal of Geophysical Research* **94**, 18409–18427.

Ruddiman, W.F., Prell, W.L. & Raymo, M.E. (1989) Late Cenozoic uplift in southern Asia and the American West: rationale for general circulation modeling experiments. *Journal of Geophysical Research* **94**, 18379–18391.

Saylor, J.E. & Horton, B.K. (2014) Nonuniform surface uplift of the Andean plateau revealed by deuterium isotopes in Miocene volcanic glass from southern Peru. *Earth and Planetary Science Letters* **387**, 120–131.

Sewall, J.O. & Fricke, H.C. (2013) Andean-scale highlands in the Late Cretaceous Cordillera of the North American western margin. *Earth and Planetary Science Letters* **362**, 88–98.

Smith, R.B. (1979) The influence of mountains on the atmosphere. *Advances in Geophysics* **21**, 87–230.

Stewart, J.Q., Whiteman, C.D., Steenburgh, W.J. & Bian, X. (2002) A climatological study of thermally driven wind systems of the US intermountain West. *Bulletin of the American Meteorological Society* **83**, 699–708.

Stone, P.H. & Carlson, J.H. (1979) Atmospheric lapse rate regimes and their parameterization. *Journal of Atmospheric Sciences* **36**, 415–423.

Vera, C., Baez, J., Douglas, M. & Emmanuel, C.B. (2006) The South American low-level jet experiment. *Bulletin of American Meteorological Society* **87**, 63–77.

Whiteman, C.D. (2000) *Mountain Meteorology*. New York: Oxford University Press.

Wolfe, J.A., Forest, C.E. & Molnar, P. (1998) Paleobotanical evidence of Eocene and Oligocene paleoaltitudes in midlatitude western North America. *Geological Society of America Bulletin* **110**, 664–678.

9

Paleoclimate

Hemmo A. Abels[1] and Martin Ziegler[2]

[1] *Department of Geosciences and Engineering, Delft University of Technology, Delft, Netherlands*
[2] *Department of Earth Sciences, Utrecht University, Utrecht, Netherlands*

Abstract

Climate is the annually averaged weather over a period of several decades. Continuously changing external factors cause climates to vary on decadal to million-year time scales. Understanding these factors, as well as the magnitude and directions of climate change, is fundamental when trying to understand its impact on the biosphere and particular parts of the Earth system, such as mountain regions. Here, we discuss climate change in its broadest sense in light of the known past variability. We briefly review the three main drivers of climate change on these time scales: Earth's surface reflectiveness, incoming solar radiation and atmospheric greenhouse gas contents. We then look at the remarkable variability that Earth's climate system experiences from million-year scale trends and ten-thousand-year scale cyclicity, to millennial-, centennial- and decadal-scale shifts. We make a long journey starting with the climate of early Earth and quickly advancing to the more recent past, about which more is known and, thus, understood. Along the way, we introduce principles and (un)certainties of climate reconstructions when using rocks and sediments, fossils and chemical parameters. This short introduction to climate change and paleoclimate is far from complete. It should give the appropriate background to understand the character and controls of paleoclimatic change and to find one's way for further study in the exciting world of paleoclimate research.

Keywords: *climate evolution, climate trends, climate shifts, climate cyclicity, drivers of climate change, ancient climates, paleoclimate*

9.1 Earth's Climate System: Lessons from the Past

Earth's climate history is a story of continuous change. The paleoclimate archives give evidence of variability at decadal to million-year time scales, chaotic as well as cyclic changes, with small- and large-scale shifts, sometimes abrupt and sometimes gradual and slow.

The climate system links all spheres of the planet: atmosphere, oceans, cryosphere and solid earth. These different spheres drive climate variability over a variety of time scales. Short-term variability – annual or even less – is related to atmospheric processes. The oceans are mainly responsible for the changes that occur on decadal to millennial time scales, and on hundred-thousand- to million-year time scales the solid earth influences climate through its impact on the concentration of atmospheric greenhouse gases and the position of the continents.

Global climate – or, in other words, Earth's average surface temperature – is ultimately related to the planet's radiative equilibrium: Earth's energy budget. The planet receives heat from the sun through electromagnetic radiation at an intensity of $1368\,W/m^2$, the solar constant. This energy is absorbed mainly at Earth's surface, warming the lower atmosphere. The warm air rises within the atmosphere and expands under decreasing pressure, causing a drop in temperature. This is why higher altitudes are cooler. The cooling, or "lapse rate," is around $10\,°C/km$ for dry air and decreases with increasing humidity due to the release of latent heat during condensation (Chapter 8).

Some of the incoming radiation is directly reflected back to space and does not warm the planet at all. Brighter surfaces, such as snow and ice, deserts and clouds are more reflective. The average reflectance of the planet is called its albedo. The absorbed energy – the part that is not reflected – gets redistributed between

Mountains, Climate and Biodiversity, First Edition. Edited by Carina Hoorn, Allison Perrigo and Alexandre Antonelli.
© 2018 John Wiley & Sons Ltd. Published 2018 by John Wiley & Sons Ltd.
Companion website: www.wiley.com\go\hoorn\mountains,climateandbiodiversity

atmosphere, oceans and land surface and is eventually lost to space by long-wave electromagnetic waves. This energy loss is described by the Stefan–Boltzmann law, which states that the rate of energy loss is proportional to the temperature. Whereas the incoming solar radiation can pass intact through the atmosphere, the outgoing long-wave radiation is absorbed during the interaction with atmospheric gases, which causes the greenhouse gas effect. The downward re-radiation of this absorbed energy yields extra heat for Earth's surface. The most important greenhouse gases are water vapor (H_2O), carbon dioxide (CO_2), methane (CH_4) and nitrous oxide (N_2O). Without the greenhouse gas effect, Earth's surface would attain its radiative equilibrium at a temperature of –19 °C and be inhabitable to current life.

At the regional to local scale, the climate state is modulated by changes in atmospheric and oceanic circulation and their associated heat transport. Atmospheric circulation does not operate in one simple large latitudinal convection cell, with rising warm air at low latitudes and cool sinking air at the poles. Instead, smaller convection cells arise as a consequence of the rotation of Earth. These cells also govern the distribution of precipitation. The picture is further complicated by temperature gradients between oceans and land surfaces, which also cause large-scale deviations in the seasonality of different locations. Ocean circulation is another important factor in the regional redistribution of heat. Over geological time scales, the distribution of land masses and development of ocean gateways can be of importance (Chapter 22). Moreover, ocean circulation impacts the biogeochemical cycling of elements in the ocean, which is crucial for the carbon cycle, as it exchanges carbon between the atmosphere and the oceans. Such processes are of fundamental importance in explaining past climate variability, such as the ice age cycles.

Over million-year time scales, Earth's average temperature is regulated by a balance in the exchange of carbon between the solid earth and the atmosphere. Volcanic outgassing adds carbon dioxide constantly from Earth's interior into the atmosphere, leading to an increase in global temperature. In turn, rock weathering, sediment burial and, ultimately, subduction of rocks remove carbon from the atmosphere, lowering the overall temperature. Long-term variations in continental drift can lead to increasing volcanic activity, which leads to a stronger greenhouse gas effect and global warming. Under warmer conditions, the weathering reactions that remove carbon dioxide from the atmosphere are quicker, keeping the whole system relatively constant. This process is also termed the "weathering thermostat" of Earth.

In general, three factors can change the rate of energy loss, and as such Earth's average temperature. These are: (i) the amount of incoming solar radiation – the solar constant; (ii) the concentration of greenhouse gases in Earth's atmosphere, in particular the content of carbon dioxide, which influences the redistribution of energy in the atmosphere; and (iii) Earth's albedo, which influences the amount of energy that is reflected back into space.

All large-scale global climate fluctuations of the past can be explained with one or a combination of these three factors. This chapter gives a brief overview of Earth's global climate history and demonstrates how the three factors can explain large-scale observations in the paleoclimate record.

The complexity of the climate system and the interactions between the different spheres and processes on a variety of time scales make paleoclimate a fascinating research area. The lessons we learn from the past, especially concerning climate tipping points and thresholds in the system, make paleoclimate research extremely relevant for the assessment of the current anthropogenic impact on Earth's climate system and biodiversity (see Box 9.1).

9.2 Early Earth's Climates

Early Earth's climates are of interest in understanding the origins of past extreme climate states. Extreme amounts of greenhouse gases caused extreme warm periods, extreme reflectiveness of Earth's surface due to snow and ice cover caused extreme cold periods and peculiar geographic conditions caused peculiar climate states with strong monsoons and strong aridity in continental interiors. All these scenarios occurred over Earth's long history, and all can help us understand the basic mechanisms that control Earth's climate.

The energy emitted by the sun was probably 25–30% lower 3–4 billion years ago (Ga) than it is today. This should have caused a very cold planet with frozen oceans. However, geological evidence points to liquid oceans and relatively high temperatures, apart from a glaciation between 2.4 and 2.1 Ga. This contradiction of warm temperatures and low solar irradiation is called the "faint young sun paradox" (Rosing et al. 2010). The paradox can only have been caused by other radically different parameters controlling Earth's temperature, such as a much higher and different greenhouse gas content in the atmosphere or a much lower surface albedo, causing higher absorption of incoming radiation. Earth's climate changed considerably after the start of photosynthetic life, as this led to the storage of atmospheric CO_2, lowered greenhouse gas warming and enhanced rock weathering.

Very cold stages occurred between 750 and 580 Ma in the Neoproterozoic era. A "snowball Earth" – with a completely or nearly completey frozen planet – developed two to possibly four times (Benn et al. 2015).

Box 9.1 The paleoclimate reconstruction toolbox

Documentary data, such as specific observations, logs and crop harvest data, are important for reconstructions of climate variability in the last millennium. Beyond this time period, alternative methods are required. Direct measurements of past change (e.g., inferred through ground temperature variations, the gas content of ice-core air bubbles, ocean sediment pore-water changes and glacier extent changes) are very limited. Therefore, paleoclimatic reconstruction methods rely to a large extent on proxy measurements involving changes in chemical, physical and biological parameters that reflect – ideally in a quantitative and well-understood manner – past change in the environment where the proxy carrier grew or existed. The field of climate proxy development has seen remarkable growth over the last several decades and continues to deliver innovative ways to reconstruct the past.

The traditional recorders of paleoclimate variability build on fossils of organisms, such as trees, corals, plankton and insects. Their evolution and alterations in their growth and population dynamics are explained as a response to changing climates.

Tree ring width networks and density chronologies are, for example, used to infer past temperature and precipitation changes based on widespread calibration with temporally overlapping instrumental data. Tree rings are wider under more favorable climate conditions. Tree ring width-based climate reconstruction has been particularly useful for reconstructing the most recent geologic past (Esper et al. 2002).

Past distributions of pollen and plankton from sediments serve as quantitative estimates of past temperature, salinity and precipitation via statistical methods calibrated against their modern distribution and associated climate parameters (Le & Shackleton 1994; Prentice & Jolly 2000).

The chemistry of several biological and physical entities reflects well-understood thermodynamic processes that can be transformed into estimates of climate parameters. Key examples include oxygen (O) isotope ratios in carbonate, used to infer past temperature, precipitation and salinity (Rohling & Cooke 1999); magnesium/calcium (Mg/Ca) and strontium/calcium (Sr/Ca) ratios in carbonate, used for temperature estimates (Elderfield & Ganssen 2000); alkenone saturation indices from marine organic molecules, used to infer past sea surface temperature (SST) (Martrat et al. 2007); and oxygen and hydrogen isotopes and combined nitrogen and argon isotope studies in ice cores, used to infer temperature and atmospheric transport (Kobashi et al. 2008).

Lastly, many sedimentary systems – marine, fluvial and aeolian deposits – change their characteristic features, such as grain size, mineralogy or elemental composition, when climate changes.

No paleoclimatic proxy method is without caveats, and knowledge of the underlying methods and processes is required when using paleoclimatic data, as are replication and verification of data by other records and methods. There is ongoing work on such methods, including improvements of our knowledge of spatial and seasonal biases of proxies. Therefore, a combination of methods is often applied, since multi-proxy series provide more rigorous estimates than a single-proxy approach.

Numerical climate models are used to simulate episodes of past climate in order to help understand the mechanisms of past climate changes. Models are key to testing sensitivity to single parameters and to quantitatively testing physical hypotheses (see Figure 9.1). They allow the linkage of cause and effect in past climate change to be investigated and help to fill the gap between the local and global scale in paleoclimate, as paleoclimatic proxy information may be sparse, patchy and seasonal.

The same climate models that are used to simulate present-day climate, or to project scenarios for the future, are also used to simulate episodes of past climate. Differences in prescribed forcing are used for the configuration of oceans and continents in the deeper past. A wide spectrum of models is available, ranging from simple conceptual ones to complex ones such as Earth system models of intermediate complexity (EMICs) and coupled general circulation models (GCMs) (Claussen et al. 2002). Since computer power remains a limiting factor, relatively "fast" coupled models are often used. Additional components that are not standard in models used for the simulation of present climate are also increasingly added for paleoclimate applications (e.g., continental ice sheet models or components that track the stable isotopes in a climate system) (LeGrande et al. 2006). Vegetation modules, as well as terrestrial and marine ecosystem modules, are increasingly included, both to capture biophysical and biogeochemical feedback to climate and to allow for validation of models against proxy paleoecological data (Otto et al. 2009). The representation of biogeochemical tracers and processes is an important new advance for paleoclimatic model simulations, as a rich body of information on the cycling of carbon and other nutrients in past climate systems has emerged (Ridgwell et al. 2007).

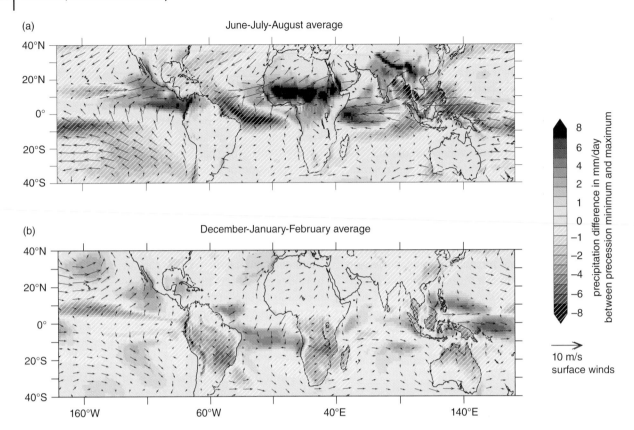

Figure 9.1 Model results from the coupled ocean–atmosphere weather model ECBilt (Bosmans 2014), simulating the impact of precession minima and maxima on precipitation. Shown are the differences in average precipitation (mm/day) and surface winds (m/s) for precession minima minus precession maxima for average northern hemisphere (a) summer and (b) winter periods. The dark-gray colors on the northern hemisphere continents in their summer and the hatched dark-gray colors on the southern hemisphere continents in their summer exemplify the intensified monsoon activity during precession minima and precession maxima, respectively. Note the large longitudinal variation in precipitation and winds within a single precession phase, highlighting the impossibility of interpreting astronomical forcing at Earth's surface directly from solar insolation changes at the top of the atmosphere. *Source:* Adapted from a version produced by Joyce Bosmans (Utrecht University, Utrecht, The Netherlands). See also Plate 16 in color plate section.

The best evidence for these states are glacial features at low latitudes. It is thought that a near-completely frozen Earth is more likely than a completely frozen one, as that would be difficult to escape from. A snowball Earth climate state would have been caused by an initial cooling, increasing snow and ice cover and consequent increasing albedo. This would have caused runaway cooling, facilitated by the distribution of land masses at the equator, where most solar radiation is received. The initial cooling might have been caused by many mechanisms, including the eruption of a supervolcano, whose plumes could have blocked solar radiation for a long time. The end of the snowball Earth states may have been caused by a gradual build-up of outgassing volcanic CO_2 in the atmosphere, as the withdrawal of greenhouse gases from the atmosphere by rock weathering was blocked by the near-complete ice cover.

The Phanerozoic eon, which we are still in, started with the Cambrian explosion of biodiversity around 540 Ma, when most of the major animal phyla originated and strong diversification of life occurred. A wide range of climates and climate extremes could be found during the Paleozoic era, between 540 and 270 Ma. The Carboniferous, towards the end of the Paleozoic, exemplifies this, as it started with very warm conditions, with global temperatures of approximately 20 °C, yet ended with clear markers of glacial and interglacial cyclicity (Fielding et al. 2008). Following this, in the early to mid Permian, a dramatic warming occurred, leading to the hothouse climates of the Mesozoic era. More and more evidence points to a 5–10-fold increase in atmospheric CO_2 content that caused this icehouse–greenhouse transition. The icehouse period of the Carboniferous is one of the first known to show periodic oscillations between glacial and interglacial states driven by astronomical climate change (see Box 9.2). This caused large sea level fluctuations, driving strong variability in coastal sediments, which

---

Box 9.2 Earth's astronomical cyclicity

The gravitational interaction of the masses in our solar system causes periodic oscillations in their orbital movements. The many bodies that play a role in these orbital movements – planets, moons and other, smaller masses – cause these variations to be quasicyclic, with a multitude of components. Nevertheless, the larger masses produce dominant orbital or astronomical cyclicity. For our Earth, key astronomical cycles stem from the variation in tilt of the rotational axis (called axial tilt or obliquity), the precession of the equinoxes (with net impact on solar insolation by the climatic precession cycle) and the eccentricity of its orbit around the sun. Very detailed astronomical solutions are calculated, which are now thought to accurately trace patterns back 10 My for obliquity and precession, 50 My for most eccentricity periods and 250 My for the 405 ky cycle of eccentricity (Laskar et al. 2004, 2011). Calculations further back in time are limited by the complexity of the solar system, with its many different masses, and by the impact of tides of Earth's solid and liquid components and the planet's variable dynamic ellipticity, which are impossible to reliably quantify back in time.

Earth's obliquity, or axial tilt, varies between 22.1 and 24.5°. This cycle has a current dominant periodicity of 41 ky and more important minor cyclic components at 54, 178 and 1.2 My. The climatic effect of a greater tilt is an amplified seasonal contrast in both hemispheres. Greater tilt causes more solar insolation at high latitudes in summertime and less at higher latitudes in wintertime. The impact of obliquity can be understood by depicting a hypothetical Earth's axis without tilt: there would be no seasons, as the sun would point to the equator year-round. Increased tilt causes increased seasonality between the hemispheres. Obliquity has a higher impact on higher latitudes, and so is thought to play a major role in global ice volume at high latitudes.

Precession of the equinoxes, or axial precession, is the semi-periodic change in the orientation of the rotational axis of Earth. This is most easily depicted as periodic "wobbling," as if Earth were a spinning top. The climatic precession cycle depicts the net impact of solar insolation on Earth, and currently has primary periods of 18.9, 22.3 and 23.6 ky. These three main periods are often averaged to a 21 ky period at present. To understand the impact of the precession cycle on insolation, the eccentricity of Earth's orbit needs first to be discussed.

With an eccentric orbit, Earth's wobbling axis causes the hemispherical summer and winter to occur alternatingly closer and farther away from the sun. The eccentricity of the orbit changes over periods of ~100 ky, 405 ky and 2.4 My. By definition, the precession minimum is reached when the northern hemisphere summer occurs closest to the sun, referred to as Earth being "in perihelion." Concurrently, the southern hemisphere summer is "in aphelion," or at its farthest from the sun. The precession maximum is reached approximately 11 ky later, when the southern hemisphere summer occurs in perihelion and the northern hemisphere summer in aphelion. The eccentricity of the orbit varies between nearly 0 and 0.05, and when the orbit is a near circle during eccentricity minima, precession is also near zero. In contrast to obliquity, precession has an impact on all latitudes and an out-of-phase impact on both hemispheres. Precession is dominant at all times, but less so in periods of major high-latitude ice volumes, as then obliquity can have a stronger impact on global climate via teleconnections.

The impact of the astronomical cycles discussed here relates to the amount of solar insolation at the top of the atmosphere. Insolation's impact on Earth's surface can be highly complex, due to the dynamic atmosphere and ocean continuously moving air and water masses vertically and horizontally. This leads to occasional highly contrasting effects of orbital parameters on local climates, even in geographically proximate sites (Bosmans et al. 2015). Straightforward interpretation of proxy records is therefore not possible, and it is more common that proof for orbital forcing in proxy records is found than that the mechanisms behind the forcing are resolved. One key tool used to help resolve the mechanism of orbital forcing in climate records is climate modeling of orbital extremes (Figure 9.1) (Bosmans 2014). This can help us understand the interaction of the ocean–atmosphere system in mitigating insolation changes into local climates, and it may explain findings in proxy records.

---

produced the landscapes now preserved in the well-known Carboniferous coal beds.

Large-scale variability of early Earth's climates was driven by the three factors controlling Earth's net temperature: solar radiation, concentration of greenhouse gases and Earth's albedo. Several extreme scenarios occurred before more stable conditions were reached in the more recent Mesozoic and Cenozoic eras, which we will discuss next.

9.3 Hothouse Climates of the Mesozoic and Paleogene

Most of the Mesozoic era and the first half of the Cenozoic, from around 250 to 40 Ma, were characterized by globally warm and ice-free conditions. These conditions are ascribed to much higher atmospheric CO_2 levels, in line with low-resolution and low-precision pCO_2 proxy data.

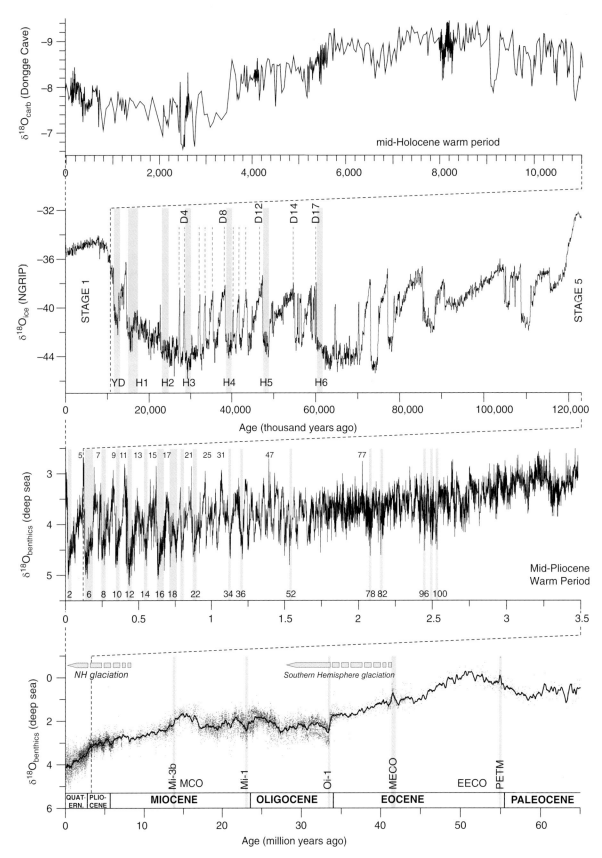

Figure 9.2 Cenozoic paleoclimate trends, rhythms and aberrations on various time scales in oxygen isotope profiles from different records, from (a) the recent past to (d) much of the Cenozoic. Generally, heavier δ¹⁸O ratio indicate glaciated and/or cooler stages. Panels (d) and (c) denote an updated version of the deep-sea benthic foraminiferal calcite stable oxygen isotope compilation by James C. Zachos (University of California, Santa Cruz, USA). Abbreviations denote specific events: PETM, Paleocene–Eocene Thermal Maximum; EECO, Early Eocene Climate Optimum; MECO, Middle Eocene Climate Optimum; Oi-1, Oligocene Isotope stage 1; Mi-1 and -3b, Miocene Isotope stages 1 and 3b; MCO, Miocene Climate Optimum. (d) Million-year scale trends. (c) Glacial–interglacial cyclicity of benthic oxygen isotopes over the last million years, fluctuating at ten- to hundred-thousand-year intervals. Along this record, marine oxygen isotope stage numbering is given with even and odd numbers for glacial and interglacial stages, respectively. (b) Stable oxygen isotope record of ice in the Northern Greenland Icecore Project (NGRIP) core from Greenland, denoting the variability that occurred between the last and the current

Several peak warming periods have been recognized, such as the Turonian stage between 93 and 89 Ma (Wilson & Norris 2001). One feature that is seen in these extreme warm periods is a low pole-to-equator or meridional temperature gradient. The cooler late Cenozoic era, after 34 Ma, had a gradient of 33 °C, while that in the warm Mesozoic and early Cenozoic was around 20 °C. Moreover, the gradient was even lower in the periods of extreme warmth occurring within the span of these warm climates. Low temperature gradients have been linked to increased ocean heat transport to the poles, but it is not thought that warmer tropics play a significant role in the polar temperature increase. Interestingly, current climate models are not capable of reproducing these periods of extreme warmth with low temperature gradients.

Oceanic anoxic events (OAEs), which occurred between 120 and 92 Ma, are important markers of Mesozoic climate (Schlanger & Jenkyns 2007). They were periods of widespread anoxia in the oceans, extreme warmth, carbon cycle disturbance and biotic crises. The ocean anoxia caused the deposition of black shales rich in organic matter. Besides these black layers, the events are recognized by a related excursion of light stable-carbon isotopes. Release of massive amounts of isotopically light carbon isotopes caused carbon isotope excursions and severe greenhouse warming during these events. No consensus has yet been formed about the common source(s) of carbon for greenhouse warming, nor for the formation mechanisms of ocean anoxia (Donnadieu et al. 2016). The light carbon is likely derived from an organic source, as light carbon is preferred during photosynthesis and is so enriched in organic matter that it decays to methane and carbon dioxide. It has been proposed that the dissociation of methane hydrates below the sea floor caused rapid influxes of methane into the ocean–atmosphere carbon pool, where methane is relatively quickly transformed into carbon dioxide. Also, carbon outgassing during the formation of continental flood basalts and ocean plateaus caused by large igneous provinces (LIPs) may have played a role. LIPs are caused by superplume core–mantle processes. This shows the large impact tectonic activity can have even on relatively short-term variability in Earth's climate.

Climates through Earth's history, such as the Paleozoic and Mesozoic climates already discussed, have been strongly influenced by astronomical forcing: climate variability caused by variation in Earth's orbit and axis due to gravitational interactions in the solar system. This causes semi-periodic changes in incoming solar irradiation, with dominant periods occurring at intervals of tens of thousands to hundreds of thousands of years, but with proven climatic impacts lasting up to millions of years. Astronomical forcing is well known as the dominant driver of glacial and interglacial climates over the last several million years (Hays et al. 1976) and has been shown to often be the principle driver of basic variability in the rock records (Hilgen et al. 2015). It drives continuous variability, for example between high and low sea or lake levels, oligotrophic and eutrophic environments, warm and cold climates and wet and dry climates. This variability is visible in rock, floral, faunal and proxy records of all ages. Interestingly, detecting astronomical climate forcing behind basic variability in a data series has proven to be much easier than understanding its driving mechanisms. Box 9.2 provides more details on this basic controlling mechanism of climate change.

The Cenozoic covers the last 66 My of Earth's history. It is the most studied of all Earth's eras, as it marks the period when the Earth we know today was formed. Cenozoic case studies are used as paleo-analogs of modern climate change, for example with early Eocene transient greenhouse warming episodes. These events occurred during a long-term warming trend that continued into the early Eocene. During this time, several sudden transient warming events, also called "hyperthermals," took place. The largest of these, the Paleocene–Eocene Thermal Maximum (PETM), is characterized by a 5–9 °C global temperature increase caused by severe ocean acidification, deep-sea benthic and terrestrial faunal turnovers and strong changes in runoff and weathering (McInerney & Wing 2011). While the floral changes on the continents were strong but mostly temporary, the mammalian faunal change was the largest after the Cretaceous–Tertiary boundary. Modern primates, artiodactyls and perissodactyls originated during the PETM. The terrestrial warming also led to a reduction of mammalian body size, due likely to the high temperatures, but possibly also to lower nutrition as a result of faster plant growth. After the PETM, several orbitally driven hyperthermal events occurred, all with various impacts on the global climate (Zachos et al. 2010). This series of events exemplifies the continuous change Earth's climate has undergone over different time scales.

9.4 The Greenhouse–Icehouse Transition of the Cenozoic

At around 50 Ma, a gradual cooling started, initiating the transition from the Eocene greenhouse world to the present-day icehouse world with high-latitude land ice in both hemispheres (Figure 9.2) (Zachos et al. 2001). The

---

Figure 9.2 (Continued) interglacials. Note the fluctuations at thousand-year time scales, such as the labelled Heinrich events (shown as H1–H6 and shaded in gray) and shorter Dansgaard–Oeschger cycles (labelled D and consecutively numbered). Stage 1 and 5 refer to MIS 1 and 5. YD, Younger Dryas cold stage. (a) A calcite stable oxygen isotope record from a stalagmite of the Dongge Cave in central China, denoting the climate variability that occurs at that spot over the course of a seemingly stable interglacial MIS 1 up to the recent past. Note the variability at hundred-year time scales.

lowering of atmospheric CO_2 (pCO_2) concentration of an unknown origin and size caused this trend. The lowering of pCO_2 is generally ascribed to tectonic processes, such as lower rates of ocean spreading leading to decreased CO_2 outgassing, higher rates of rock weathering consuming CO_2 (especially related to the uplift of the Tibetan Plateau and weathering of the Deccan Traps flood basalts) and lower rates of carbonate-rich sea-floor subduction into the mantle, which also led to CO_2 outgassing (Kent & Muttoni 2013). Other tectonic events that strongly influence the Cenozoic climate history are the opening and closing of ocean gateways connecting the main ocean basins. The existence and disappearance of these connections impact ocean circulation, which in turn impacts ocean heat transport: the global conveyor belt.

The gradual cooling that occurred from the early Eocene to the Oligocene included several cooling steps, but also warming events. The most remarkable of the latter was the Middle Eocene Climatic Optimum (MECO) at around 40 Ma. The MECO shows a warming trend, with peak warming lasting 500 ky in total before ending with a remarkably rapid cooling step (Bohaty et al. 2009). The gradual warming and rapid cooling are not easily explained by the models that exist for the early Eocene greenhouse warming events, which predict a rapid CO_2 release phase and gradual recovery by slow processes, such as rock weathering. The reason behind the rapidity of the cooling step ending the MECO is unknown. After this cooling step is when the first evidence is found for the onset of glaciations on Antarctica. From around 40 Ma onwards, the world turned cold. While several cooling and warming oscillations occurred during the subsequent overall cooling, there seems to be no return from the greenhouse–icehouse transition at the Eocene–Oligocene boundary around 34 Ma (Lear et al. 2008).

The greenhouse–icehouse transition at the Eocene–Oligocene boundary is a prominent feature in marine and terrestrial paleo-environments around the globe. The change has not always been related to the exact boundary, either because dating is poor or because the local environments had already reacted before or only reacted after the transition to the cooling temperatures or lowered sea level. The Eocene–Oligocene transition occurred in two steps according to most data sets (Lear et al. 2008). These steps were relatively rapid and were separated by hundreds of thousands of years. The first step is primarily reported as a cooling step, while the second, labelled Oi-1, is primarily identified as an ice-growth and sea level-lowering step. The terrestrial impact on North America and Asia seems to have occurred during the first, atmospheric cooling step (Xiao et al. 2010). On the other hand, European mammal associations were largely impacted by immigration from Asia – called the Grande Coupure – at the second, ice volume-increase step. This can be explained by sea level lowering during the second step, which allowed mammals to cross the Turgai Strait, a north–south seaway splitting Asia in two at the Ural Mountains, which until then had blocked migration between Asia and Europe.

Long-term cooling brought on by decreasing atmospheric CO_2 levels must have caused the gradual progression from greenhouse to icehouse states, but it does not explain the relative rapidity of the Eocene–Oligocene transition. Furthermore, additional factors have been invoked to explain the cooling, such as the isolation of Antarctica by the onset of the Antarctic Circumpolar Current (ACC) due to the opening of the Tasmanian and Drake Passages. Various models and datings of these openings disagree as to whether ocean circulation in fact caused the final greenhouse–icehouse transition at the Eocene–Oligocene boundary. However, the impact of the ACC on Antarctic glaciation is clear from the models. Furthermore, the large Antarctic continent positioned at the South Pole increases the chances for continental ice. Coupled ocean–atmosphere–ice modeling shows that a lowering of pCO_2 from 4–5 to 2–3 times pre-industrial values is sufficient to induce Antarctic glaciation (DeConto et al. 2008). The stepwise behavior leading to an icehouse state can be explained by astronomical cyclicity, which promotes glaciation during particular astronomical configurations. When long-term cooling trends occur, astronomical cycles may first extend deglaciations by causing disfavorable conditions for glaciation, and then, after threshold CO_2 values for glaciations have been reached, abruptly favor glaciations when ideal orbital configuration is reached. The stepwise character of such transition may be caused by the interaction of the multiple orbital cycles impacting Earth's climate (see Box 9.2).

The Oligocene, Miocene and Pliocene were dynamic stages with both cooling and warming phases. Much of the paleoclimate history during this interval has been accurately reconstructed from multiple ocean basins. This has been made possible, as with many of the previously mentioned paleoclimate reconstructions, by the International Ocean Discovery Program (IODP) and its predecessors. This program provides long, continuous sediment cores from ocean basins around the globe. The Oligocene and Miocene are characterized by multiple glaciation events, starting with the Oi-1. In the middle Miocene, a climatic optimum occurred, known as the Miocene Climatic Optimum (MCO). This warming period again ended with a strong cooling step – the Mid-Miocene Climate Transition (MMCT) – heralding Earth's latest descent into bipolar glaciations starting in the late Miocene or early Pliocene.

9.5 Quaternary Ice Age Cycles and Rapid Climate Change

Earth's climate transformed remarkably with the advent of the ice age cycles 2.6 Ma. During glacial maxima, ice sheets covered large parts of northern Eurasia and the northern and central parts of North America. The climate was cooler globally, and precipitation decreased almost everywhere. This had large impacts on the vegetation. Rainforest vegetation diminished during glacial periods, while deserts expanded. Globally, cold conditions were mainly a consequence of two factors. First, Earth's albedo was increased due to reflectivity of the large ice sheets. Second, the greenhouse gas concentration in the atmosphere was lower during glacials. During the Quaternary, both of these factors are linked through a positive-feedback mechanism, reinforcing each other and leading to tremendous global climate change over relatively short time periods. Due to this positive feedback, only a comparably small external forcing is required to drive the planet into or out of an ice age. Such external forcing is provided by the changes in Earth's orbit, which determine the intensity of the summer insolation at high latitudes. While in the early part of the Quaternary changes in Earth's obliquity paced ice age cycles regularly over a 41 000-year cycle, the pacing changed to 100 000-year cycles in the last 1 My, when the influence of precession increased (Hays et al. 1976).

Lower greenhouse gas concentrations during the glacial periods are linked to a number of factors. Lower ocean temperatures increased the solubility of CO_2 in seawater. On the other hand, increased salinity and a lower sea level, as well as reduced carbon storage in terrestrial biomass, had the opposite effect. Two other factors eventually led to an increased storage of carbon in the deep sea and a lower concentration in the atmosphere: one was an increase in the strength of the biological pump (Sigman et al. 2010), which is the ocean's biologically driven sequestration of carbon in the form of organic matter and carbonate – a stronger biological pump leads to a redistribution of nutrients within the water column and depletes the surface of dissolved inorganic carbon; the other was sluggish circulation and stronger stratification, which helped to sequester more carbon in the deep sea (Anderson et al. 2009).

The glacial periods were characterized by large millennial-scale variability. Such events were most pronounced in the North Atlantic, but they were are also recorded in archives in low latitudes and the southern hemisphere. The ice-core records from Greenland and Antarctica provide the best evidence for the occurrence of around 25 rapid warming events of several

degrees Celsius during the last glacial cycle (Blunier & Brook 2001). These events occurred over a few decades, or even faster. After the rapid warming, temperatures slowly returned – within a few hundred years – to baseline levels. The temperature changes in Antarctica were less abrupt, and, more importantly, they occurred in anti-phase with the temperature changes in the northern hemisphere (Barker et al. 2011). The ice-core records also document synchronous changes in atmospheric methane concentrations, with warmer conditions over Greenland coinciding with peaks in atmospheric methane (Loulergue et al. 2008). This is taken as an indicator that low-latitude wetlands, which are among the most important sources of atmospheric methane, were also affected by this millennial-scale variability. Wetlands appear to have been more extensive when the high latitudes in the northern hemisphere were warmer, presumably due to higher precipitation.

The five most extreme millennial-scale cold events are termed "Heinrich Events" (Hemming 2004). During these events, massive icebergs were discharged from the northern hemisphere ice sheets. Evidence for this comes from glacial debris that can be found in North Atlantic deep-sea sediments. The ice discharge and associated meltwater release lowered the salinity of the North Atlantic surface waters, with major consequences for the large-scale ocean circulation in this region. Modeling studies suggest that if a meltwater perturbation in the North Atlantic was large enough, it could oppose the density increase due to cooling and prevent the formation of North Atlantic Deep Water (NADW): a deep-water mass formed in the North Atlantic Ocean by cooling of southerly derived saline waters. The decreasing formation of NADW in turn reduced the flow of warm and salty waters from the Caribbean to the North Atlantic region. The North Atlantic consequently cooled further, and wintertime sea ice expanded substantially in the region. But the consequences went far beyond the North Atlantic region: paleoclimate records from tropical regions document that the Intertropical Convergence Zone migrated southwards when the North Atlantic cooled, while in the Southern Ocean, temperatures rise during the North Atlantic cooling events. This opposite temperature pattern is termed the "bipolar seesaw." A southward shift of the average position of the westerlies resulted in increased upwelling of deeper waters. The reduced stratification of the Southern Ocean ultimately facilitated the outgassing of CO_2. Large millennial-scale events are thought to have been instrumental in the termination of glacial periods. In that sense, millennial-scale and orbital-scale climate change during the Quaternary are tightly connected.

9.6 The Holocene

The Holocene epoch covers the last 11 700 years and constitutes an important benchmark against which anthropogenically induced climate change can be differentiated. Compared to the orbital- and millennial-scale changes during the last glacial cycle, global variability during the Holocene is limited and the signal-to-noise ratio is small. Global mean temperatures vary by only 0.5–1.0 °C during the Holocene. However, the Holocene should not be described as climatically stable or invariant. High-resolution records show dynamic climate variability, particularly on regional scales: for example, in low-latitude monsoon regions. Much of the regional variability can be ascribed to internal factors in the climate system, such as atmosphere–ocean and atmosphere–vegetation interactions. In addition, orbital changes, as well as variability in solar irradiance, have an impact on larger spatial scales that are hemisphere-wide or even global in nature.

Northern hemisphere summer insolation has declined over the last 10 000 years due to the influence of precession and obliquity. This has had an effect on the vegetational cover in the higher latitudes, and a particularly strong effect on monsoonal precipitation intensity. This is documented in pollen data, building on the fact that each plant species is tied to an optimal temperature range. The transition between taiga and temperate forest has shifted southward over the course of the Holocene. Similarly, the maximum northward extent of the Intertropical Convergence Zone, which is the band of highest rainfall in the tropics, has shifted progressively southward over the same time period. This shift of climate belts at a time without major northern hemisphere ice sheets occurred as a consequence of decreasing summer insolation on the northern hemisphere. Changes in the intensity of the distribution of monsoonal precipitation are well documented in stable oxygen isotope records derived from cave deposits, called speleothems. Increased precipitation during the early Holocene is also evidenced by remains of larger and more widespread lakes in North Africa.

Millennial-scale variability is also present during the Holocene, albeit with a smaller amplitude than during the last glacial period. The prominent 8.2 ka cooling event was related to the last stages of melting of the glacial ice sheets. The event was triggered by the collapse of Lake Agassiz, a large ice-dammed lake in northern North America that was filled with meltwater from the Laurentide ice sheet (Barber et al. 1999). When the ice dams collapsed under the pressure of the meltwater, the whole lake emptied into the Atlantic and formed a low-salinity layer on the ocean's surface. Similar to the larger glacial Heinrich events, this surpressed NADW formation and led to large-scale cooling in the northern hemisphere for several hundred years. The best record of this event is captured in Greenland ice cores. Smaller-amplitude millennial-scale variability during the Holocene is documented, for example, in the occurrence and distribution of ice-rafted debris, which expose the extent of icebergs in the past. Such records have been linked to solar variability. The evidence for this link comes from proxy records, which are sensitive to the brightness of the sun. Such proxies are typically radiogenic isotopes, ^{10}Be and ^{14}C, that are produced in the atmosphere at a rate that depends on the intensity of the sun. The Little Ice Age, from the 15th century through the end of the 19th century, and the Medieval Warm Period from the 9th through the 14th centuries, mark the last cycle of warmer and colder temperatures in the northern hemisphere that is linked to changes in solar radiation. These most recent centennial- to millennial-scale climate events are well documented in tree ring records, but also in marine sediment cores, which have a high sediment accumulation rate and therefore offer a high temporal resolution. The solar forcing of these events remains debated in the scientific community, since a relatively small change in solar output may apparently lead to significant climate changes. Feedback mechanisms are required to explain this response (Broecker 2001).

A large number of studies have linked climate variability during the Holocene to the decline of important human civilizations: examples are the Central American Maya, native North American cultures and Norse colonies in Greenland. Since human influence started to significantly impact global climate in the 19th century, the time period since around the 1850s is also referred to as the Anthropocene. Global surface air temperature records for the last millennium, which combine instrumental records with proxy data, show a striking pattern that is widely known as the "hockey-stick" (Mann et al. 1998). The temperature record shows a significant warming over the last 100 years, which exceeds any prior warming event in the last 1000 years. The continuous increase in the atmospheric greenhouse gas concentration will lead to further warming in the future. Some studies suggest that even the pre-industrial effects of agriculture already had an important impact on the atmospheric greenhouse gas concentration, and therefore on global climate (Ruddiman et al. 2014). The magnitude of the warming is difficult to predict, and it is the paleoclimate record that may hold crucial information about our future climate.

9.7 Conclusion

Earth's climate has been variable over its entire history on various time scales, ranging from decades to millions of years. Paleoclimate records teach us a great deal about the system response to external forcing, but it is crucial to have accurate and precise reconstructions as well as high-quality age estimates in order to really utilize the full potential of this record. The big challenge in the field in the coming years will be to further improve the accuracy and precision of paleoenvironmental reconstructions in order to ultimately achieve full integration with climate modeling.

References

Anderson, R.F., Ali, S., Bradtmiller, L.I. et al. (2009) Wind-driven upwelling in the Southern Ocean and the deglacial rise in atmospheric CO2. *Science* **323**, 1443–1448.

Barber, D.C., Dyke, A., Hillaire-Marcel, C. et al. (1999) Forcing of the cold event of 8200 years ago by catastrophic drainage of Laurentide lakes. *Nature* **400**, 344–348.

Barker, S., Knorr, G., Edwards, R.L. et al. (2011) 800 000 years of abrupt climate variability. *Science* **334**, 347–351.

Benn, D.I., Le Hir, G., Bao, H. et al. (2015) Orbitally forced ice sheet fluctuations during the Marinoan Snowball Earth glaciation. *Nature Geoscience* **8**, 704–707.

Blunier, T. & Brook, E.J. (2001) Timing of millennial-scale climate change in Antarctica and Greenland during the last glacial period. *Science* **291**, 109–112.

Bohaty, S.M., Zachos, J.C., Florindo, F. & Delaney, M.L. (2009) Coupled greenhouse warming and deep-sea acidification in the middle Eocene. *Paleoceanography* **24**, PA2207.

Bosmans, J.H.C. (2014) A Model Perspective on Orbital Forcing of Monsoons and Mediterranean Climate using EC-Earth. Utrecht Studies in Earth Sciences 055. Utrecht University Departments of Physical Geography and Earth Sciences.

Bosmans, J.H.C., Drijfhout, S.S., Tuenter, E. et al. (2015) Precession and obliquity forcing of the freshwater budget over the Mediterranean. *Quaternary Science Reviews* **123**, 16–30.

Broecker, W.S. (2001) Was the Medieval Warm Period Global? *Science* **291**, 1497–1499.

Claussen, M., Mysak, L., Weaver, A. et al. (2002) Earth system models of intermediate complexity: closing the gap in the spectrum of climate system models. *Climate Dynamics* **18**, 579–586.

DeConto, R.M., Pollard, D., Wilson, P.A. et al. (2008) Thresholds for Cenozoic bipolar glaciations. *Nature* **455**, 652–656.

Donnadieu, Y., Pucéat, E., Moiroud, M. et al. (2016) A better-ventilated ocean triggered by Late Cretaceous changes in continental configuration. *Nature Communications* **7**, n10316.

Elderfield, H. & Ganssen, G. (2000) Past temperature and d^{18}O of surface ocean waters inferred from foraminiferal Mg/Ca ratios. *Nature* **405**, 442–445.

Esper, J., Cook, E.R. & Schweingruber F.H. (2002) Low-frequency signals in long tree-ring chronologies for reconstructing past temperature variability. *Science* **295**, 2250–2253.

Fielding, C.R., Frank, T.D. & Isbell, J.L. (2008) The late Paleozoic ice age – a review of current understanding and synthesis of global climate patterns. *Geological Society of America Special Papers* **441**, 343–354.

Hays, J.D., Imbrie, J. & Shackleton, N.J. (1976) Variations in the Earth's orbit: pacemaker of the ice ages. *Science* **194**, 1121–1132.

Hemming, S.R. (2004) Heinrich events: massive late Pleistocene detritus layers of the North Atlantic and their global climate imprint. *Reviews of Geophysics* **42**, RG1005.

Hilgen, F.J., Hinnov, L.A., Aziz, H. et al. (2015) Stratigraphic Continuity and Fragmentary Sedimentation: The Success of Cyclostratigraphy as Part of Integrated Stratigraphy. *Geological Society, London, Special Publications* **404**, pp. 157–197.

Kent, D.V. & Muttoni, G. (2013) Modulation of Late Cretaceous and Cenozoic climate by variable drawdown of atmospheric CO_2 from weathering of basaltic provinces on continents drifting through the equatorial humid belt. *Climate of the Past* **9**, 525–546.

Kobashi, T., Severinghaus, J.P. & Kawamura, K. (2008) Argon and nitrogen isotopes of trapped air in the GISP2 ice core during the Holocene epoch (0–11 500 BP): methodology and implications for gas loss processes. *Geochimica et Cosmochimica Acta* **72**, 4675–4686.

Laskar, J., Robutel, P., Joutel, F. et al. (2004) A long-term numerical solution for the insolation quantities of the Earth. *Astronomy and Astrophysics* **428**, 261–285.

Laskar, J., Fienga, A., Gastineau, M. & Manche, H. (2011) La2010: a new orbital solution for the long term motion of the Earth. *Astronomy and Astrophysics* **532**, A89.

Le, J. & Shackleton, N.J. (1994) Reconstructing paleoenvironment by transfer function: model evaluation with simulated data. *Marine Micropaleontology* **24**, 187–199.

Lear, C.H., Bailey, T.R., Pearson, P.N. et al. (2008) Cooling and ice growth across the Eocene-Oligocene transition. *Geology* **36**, 251–254.

LeGrande, A.N., Schmidt, G.A., Shindell, D.T. et al. (2006) Consistent simulations of multiple proxy responses to an abrupt climate change event. *Proceedings of the National Academy of Sciences USA* **103**, 837–842.

Loulergue, L., Schilt, A., Spahni, R. et al. (2008) Orbital and millennial-scale features of atmospheric CH4 over the past 800 000 years. *Nature* **453**, 383–386.

Mann, M.E., Bradley, R.S. & Hughes, M.K. (1998) Global-scale temperature patterns and climate forcing over the past six centuries. *Nature* **392**, 779–787.

Martrat, B., Grimalt, J.O., Shackleton, N.J. et al. (2007) Four climate cycles of recurring deep and surface water destabilizations on the Iberian margin. *Science* **317**, 502–507.

McInerney, F.A. & Wing, S.L. (2011) The Paleocene-Eocene thermal maximum: a perturbation of carbon cycle, climate, and biosphere with implications for the future. *Annual Review of Earth and Planetary Sciences* **39**, 489–516.

Otto, J., Raddatz, T., Claussen, M. et al. (2009) Separation of atmosphere-ocean-vegetation feedbacks and synergies for mid-Holocene climate. *Geophysical Research Letters* **36**, L09701.

Prentice, I.C. & Jolly, D. (2000) Mid-Holocene and glacial-maximum vegetation geography of the northern continents and Africa. *Journal of Biogeography* **27**, 507–519.

Ridgwell, A., Hargreaves, J.C., Edwards, N.R. et al. (2007) Marine geochemical data assimilation in an efficient Earth System Model of global biogeochemical cycling. *Biogeosciences* **4**, 87–104.

Rohling, E.J. & Cooke, S. (1999) Stable oxygen and carbon isotopes in foraminiferal carbonate shells. In: Gupta, B.K.S. (ed.) *Modern Foraminifera*. Dordrecht: Springer Netherlands, pp. 239–258.

Rosing, M.T., Bird, D.K., Sleep, N.H. & Bjerrum, C.J. (2010) No climate paradox under the faint early sun. *Nature* **464**, 744–747.

Ruddiman, W., Vavrus, S., Kutzbach, J. & He, F. (2014) Does pre-industrial warming double the anthropogenic total? *The Anthropocene Review* **1**, 147–153.

Schlanger, S.O. & Jenkyns, H.C. (2007) Cretaceous oceanic anoxic events: causes and consequences. *Netherlands Journal of Geosciences/Geologie en Mijnbouw* **55**, 179–184.

Sigman, D.M., Hain, M.P. & Haug, G.H. (2010) The polar ocean and glacial cycles in atmospheric CO_2 concentration. *Nature* **466**, 47–55.

Wilson, P.A. & Norris, R.D. (2001) Warm tropical ocean surface and global anoxia during the mid-Cretaceous period. *Nature* **412**, 425–429.

Xiao, G.Q., Abels, H.A., Yao, Z.Q., DuPont-Nivet, G. & Hilgen, F.J. (2010) Asian aridification linked to the first step of the Eocene-Oligocene climate transition (EOT) in obliquity-dominated terrestrial records (Xining Basin, China). *Climate of the Past* **6**, 501–513.

Zachos, J., Pagani, M., Sloan, L. et al. (2001) Trends, rhythms, and aberrations in global climate 65 Ma to present. *Science* **292**, 686–693.

Zachos, J.C., McCarren, H., Murphy, B. et al. (2010) Tempo and scale of late Paleocene and early Eocene carbon isotope cycles: Implications for the origin of hyperthermals. *Earth and Planetary Science Letters* **299**, 242–249.

Part II

When Biology Meets Mountain Building

10

Mountain Geodiversity: Characteristics, Values and Climate Change

John E. Gordon

School of Geography and Sustainable Development, University of St. Andrews, St. Andrews, UK

Abstract

Mountain landscapes are exceptional for their geodiversity at a range of scales from the global to the local, reflecting the interactions between plate tectonics, geology, geomorphological processes, climate, vegetation, human activities and landscape evolution. Mountain geodiversity is an integral part of nature and has value not only for geoheritage reasons but also for the wider ecosystem services and natural resources it provides that are essential to supporting human life and biodiversity. Mountain ecosystems are particularly sensitive to climate change and human pressures. Understanding geomorphological processes and landscape evolution is fundamental to delivering sustainable management of natural resources and informing biodiversity conservation strategies and hazard risk assessment in the face of these pressures. The response of the cryosphere, in particular, to climate change will have far-reaching effects both on biotic and abiotic components of mountain ecosystems and on cultural perceptions and reactions. Mountain geodiversity merits conservation for a range of intrinsic, instrumental and relational values as part of a more integrated approach to nature conservation across the full range of IUCN Protected Area Management Categories.

Keywords: *cultural values, ecosystem services, geoconservation, geoheritage, geology, geomorphology, hazards*

> No two hills are alike...
> Jorge Luis Borges, *Utopia of a Tired Man* (1975)

10.1   Introduction

Geodiversity is a relatively new concept, broadly analogous to biodiversity, that includes the variety of rocks, minerals, fossils, landforms, sediments, soils, hydrological features and topography in an area, together with the natural processes that form and alter them (Gray 2013; Crofts & Gordon 2015). Mountains and mountain ranges are probably the most geodiverse features on Earth across a range of scales from the global to the local. They result from the complex interactions of tectonic and structural factors, lithology, geomorphological processes, climate, vegetation and human activities, acting both continuously and episodically over time scales from the diurnal to hundreds of millions of years (Chapters 2 and 4) (Geikie 1913; Barsch & Caine 1984; Gerrard 1990; Schroder & Bishop 2004; Slaymaker & Embleton-Hamann 2009;

Schroder & Price 2013). They compose one of the main relief elements of Earth's surface, ranging in form from single isolated features to ranges and chains (Barsch & Caine 1984; Summerfield 1991). Mountains occur on every continent, covering, according to different estimates, up to ~24% of the planet's land surface (Price 2015; Körner et al. 2016), and also within the oceans, spanning a range of latitudes and climate zones from polar to tropical (see inside front panel). However, "mountains are, after all, in the prodigiously protracted history of the earth's surface, merely shadows that come and go, as the subterranean energy upheaves them and as the atmospheric forces crumble them into dust" (Geikie 1896, p. 125).

Mountain environments generally have a high geomorphic sensitivity and vulnerability to natural perturbations (e.g., seismic events, volcanic activity, mass movements and floods) and human activities (e.g., land-use changes). This arises from high-energy conditions related to steep slopes, high altitudes and high relative relief, which, combined with thin soils, unconsolidated sediments and extreme or harsh climatic conditions,

Mountains, Climate and Biodiversity, First Edition. Edited by Carina Hoorn, Allison Perrigo and Alexandre Antonelli.
Companion website: www.wiley.com\go\hoorn\mountains,climateandbiodiversity

confer a susceptibility to high erosion rates and rapid, sometimes high-magnitude, geomorphological changes. Many mountains also bear a legacy of landforms inherited from geomorphological and climatic changes in the past (e.g., resulting from Pleistocene glaciations). Consequently, mountains display both spatial and temporal diversity of distinctive geomorphological process environments and landform assemblages that vary with altitude and latitude and with tectonic and denudation history (Gerrard 1990). They represent a vital chapter in the "great stone book" that contains, albeit incompletely, the collective memory of Earth's history and our common geoheritage (Ansted 1863; Macfarlane 2003). However, mountain geodiversity also has wider significance. As well as important scientific and educational interests, it has intrinsic, cultural, aesthetic, ecological and economic values and provides or contributes to a range of ecosystem services and benefits for nature and people. This chapter outlines the origins of the remarkable geodiversity of mountains, their geoheritage values, and their significance in relation to climate change.

10.2 Geodiversity and the Definition of Mountains

The difficulty of defining mountains in physical or cultural terms is widely recognized (Ives et al. 1997; Byers et al. 2013). Whatever objectifiable criteria are used, elements of geodiversity are a fundamental consideration in terms of size, relative relief, steepness, ruggedness and roughness (Gerrard 1990; Kapos et al. 2000; Meybeck et al. 2001; Körner et al. 2011). For example, Barsch & Caine (1984) considered that the definition was likely to include four characteristics of geodiversity pertaining to landforms or the processes acting on them: (i) elevation, often in absolute terms; (ii) steep or precipitous gradients; (iii) rocky terrain; and (iv) the presence of snow and ice. Some definitions exclude high plateaus such as the Tibetan Plateau, while others exclude lower, older, pre-Mesozoic mountain ranges, such as the Appalachians in the USA and the Highlands of Scotland. Developing the typology of Meybeck et al. (2001), Slaymaker & Embleton-Hamann (2009) emphasized high gradient (represented by local elevation range) and high absolute elevation above sea level as essential criteria, and defined four classes that include the latter but not the former areas (Table 10.1).

From a human perspective, mountains are cultural constructions rather than commensurable objects, with aesthetic and spiritual meanings and economic, social and political dimensions (Silva et al. 2011; Bernbaum & Price 2013; Debarbieux & Rudaz 2015). For example, Peattie (1936) considered that they should be impressive,

Table 10.1 Definition of mountain classes based on elevation and local elevation range (>300 m/5 km).

Class	Type	Elevation range (m)	Estimated area (Mkm²)
1	Low mountains	500–1000	6.3
2	Mid-elevation mountains	1000–2500	11
3	High mountains	2500–4500	3.9
4	Very high mountains	>4500	1.8

Source: Adapted from Slaymaker & Embleton-Hamann (2009).

have bulk and individuality, play a part in the imagination of the people who live in their shadow or be symbolic. Subjectively, therefore, mountains can be viewed as distinct high regions rising above the surrounding areas with physical features and a presence that have inspired humanity in many different cultures. In effect, a mountain is "a collaboration of the physical forms of the world with the imagination of humans" (Macfarlane 2003, p. 19).

10.3 Mountain Geodiversity at a Global Scale

At a global scale, mountains typically occur as long linear belts on continental margins (e.g., the Cordillera of North America and the Andes of South America), as well as in continental interiors (e.g., the Alpine–Himalayan system extending from the Iberian Peninsula to western China) (see inside front panel) (Gerrard 1990; Owen 2004; Schroder & Price 2013). They originate principally as a result of plate tectonic processes. These involve extensive deformation of the lithosphere, crustal thickening and the formation of linear orogenic belts that are subsequently uplifted into mountain ranges through a complex array of associated processes that include feedbacks between tectonics, isostasy, climate and erosional processes (Chapters 2 and 4) (Owen 2004; Owens & Slaymaker 2004; Whipple 2009; Frisch et al. 2011; Kamp & Owen 2013). Since the late Proterozoic, three major phases (with multiple components) of orogenesis have shaped the geodiversity of mountain landscapes today (Frisch et al. 2011): the Caledonian orogeny, during the late Cambrian (490 Ma) to the Middle Devonian (390 Ma); the Variscan (Hercynian)–Appalachian orogeny, during the Middle Devonian (390 Ma) to early Permian (290 Ma); and the Alpine–Himalayan orogeny, during the Cenozoic (65 Ma to present).

Various classification systems, based on criteria including geological structure, tectonic setting and

Figure 10.1 Geodiversity of mountain landscapes. (a) Sagarmatha National Park, Nepalese Himalaya: very high mountains of the Alpine–Himalayan belt characterized by high relief, active glacial and slope processes and shrinking valley glaciers with extensive rock debris cover. (b) The Cairngorm Mountains, Scotland: low mountains of the Caledonide belt, formed in a dissected Silurian granite intrusion with glacial landforms incised into a series of paleosurfaces. (c) The Blue Mountains, New South Wales, Australia: a dissected plateau on a passive margin uplifted during the Jurassic. (d) Allardyce Range, South Georgia: heavily glacierized mountains at sea level, comprising folded lower Cretaceous volcaniclastic sandstones and mudstones. (e) Zagros Mountains, Iran: differential erosion of folded Carboniferous–Miocene sedimentary rocks, forming a landscape of linear ridges and valleys and revealing a salt dome in the center of the image. *Source:* NASA, https://earthobservatory.nasa.gov/IOTD/view.php?id=6465. (f) Cerro Fitz Roy massif, southern Andes: granitic buttresses and towers formed in a Neogene igneous intrusion. (g) Tadrart Acacus, Libya: a monocline in Paleozoic sedimentary rocks uplifted and dissected during the Cenozoic. (h) Cotopaxi, Ecuador: an active, glacier-capped stratovolcano with flanks scarred by tracks of lahars and meltwater floods. (i) Mount Kenya: an extinct Plio–Pleistocene stratovolcano heavily dissected by glacial erosion and with small remnant glaciers. See also Plate 17 in color plate section.

erosional processes, highlight the diversity of mountain forms and origins (Figure 10.1) (Fairbridge 1968; Gerrard 1990). The often-quoted scheme of Fairbridge (1968) distinguishes between two main categories: primary structural, tectonic and constructional forms; and secondary erosional forms. This scheme predates the development of plate tectonic theory, which provides a more modern framework (Table 10.2). In the latter, a common distinction is between high and very high mountain belts formed during the Cenozoic at active lithospheric plate margins and mountains located in tectonically inactive plate interiors relating to pre-Cenozoic orogenesis (Summerfield 1991; Slaymaker 2004). The former (e.g., the Alpine–Himalayan and circum-Pacific systems)

(Figure 10.1a) occur in three main types of tectonic setting associated with convergent plate margins (including continent–continent collision zones and subduction zones), divergent plate margins (represented by the mid-ocean ridges) and conservative plate boundaries (characterized by transform faults) (Table 10.2). They typically display a diversity of sedimentary, igneous and metamorphic rock types (Owen 2004). These include sandstones, limestones and shales deposited under shallow water conditions, deeper-water turbidites and volcanic rocks and associated sediments. These rocks have been variously deformed, metamorphosed and intruded by igneous rocks, mainly granites. Large-scale structural features include folding, faulting and overthrusting.

Table 10.2 Classification of mountain origins by plate tectonic setting.

Plate setting	Types	Examples
Convergent plates	Oceanic to oceanic	Japanese Alps; Aleutian Arc
	Oceanic to continental	Cascade Ranges, Pacific Northwest; Andes
	Continental to continental	Alpine–Himalayan system
		Zagros Mountains, Iran
	Accreted margins	Coast Mountains, British Columbia
Divergent plates	Ocean spreading	Mid-Atlantic Ridge; Iceland
	Intercontinental rifts	Sinai Peninsula; Rwenzori Mountains
Transform plates		Southern Alps, New Zealand; Coast Ranges, California
Plate edges	Passive continental margins	Drakensberg; Western Ghats
Plate interiors	Hotspots	Grand Tetons, Wyoming; Ahaggar Mountains, Algeria
	Continental flood basalts	Deccan Plateau; Columbia Plateau
	Intracratonic uplift sites	San Rafael Swell, Utah
	Post-tectonic magmatic intrusion sites	Aïr Mountains, Nigeria

Source: Adapted from Owens & Slaymaker (2004).

The original formation and configuration of pre-Cenozoic mountain belts relate to ancient plate boundaries. They include the Caledonide system, stretching from the northern Appalachians to Scandinavia, formed by the collision of the continental masses of Laurentia, Avalonia and Baltica (Frisch et al. 2011; Woodcock & Strachan 2012). The Caledonian mountains were substantially lowered by erosion, and later fragmented, uplifted and dissected during the Cenozoic following the opening of the North Atlantic Ocean around 65 Ma (Green et al. 2013). They generally form low and mid-elevation mountains and comprise predominantly deformed and metamorphosed marine sedimentary rocks and igneous intrusions (Figure 10.1b). The Variscan (Hercynian)–Appalachian orogeny during the late Paleozoic involved a complex set of collisions of Laurasia with Gondwana, producing a system of mountains across central Europe, including the Massif Central in France, the Vosges on the border between France and Germany, the Black Forest in Germany, the Giant Mountains on the Czech–Polish border and the southern Appalachians in North America (Hatcher 2010; Schulmann et al. 2014).

Mountain belts on passive continental margins include the Western Ghats of India, the Drakensberg Mountains of South Africa and the Eastern Highlands of Australia (Figure 10.1c). The origins and evolution of these mountain belts since the initiation of their formation in the Mesozoic is complex and contested (Ollier & Pain 2000; Ollier 2004; Bishop 2007; Blenkinsop & Moore 2013). They are typically distinguished by "great escarpments" with diverse landforms related to the interplay of tectonic (flexural isostasy) and geomorphological processes over geological time scales (Ollier 2004). Many passive margins are volcanic (e.g., the margins of southern and eastern Africa and the western margin of India), often comprising large volumes of basaltic lavas (Blenkinsop & Moore 2013).

Volcanoes are a distinctive type of mountain also associated with plate tectonic activity that sometimes form isolated mountains and sometimes occur as part of a mountain chain (Gerrard 1990; Lockwood & Hazlett 2010; Huff & Owen 2013). They often occur in linear zones, principally following convergent and divergent lithospheric plate boundaries. They are typical of convergent plate margins around the Pacific, along the Pacific Ring of Fire, where they are associated with the subduction of ocean crust beneath the continental plates of North and South America, and with the formation of island arcs on the west side of the Pacific (e.g., the Aleutian, Kurile and Japanese arcs) and the South Sandwich Islands in the South Atlantic. Volcanoes associated with divergent plate margins include those of the East African Rift zone (e.g., Kilimanjaro) and those along the Mid-Atlantic Ridge and the East Pacific Rise. In the latter areas, the volcanic activity is generally submarine, except where islands rise above sea level (e.g., Iceland and Tristan da Cunha).

Stratovolcanoes (composite cones) occur along convergent, subducting plate boundaries and comprise steep cones (~30°) of interbedded lavas and pyroclastic rocks produced from higher-viscosity magmas. These explosive volcanoes include Vesuvius, Mount Pinatubo, Fujiyama, Mount Saint Helens, Popocatepetl and Cotopaxi (Figure 10.1h). Shield volcanoes are associated with divergent plate boundaries. They have generally low-angle slopes (5–10°), are composed of low-viscosity basaltic lava flows from a central vent or vents (e.g., Mauna Loa on Hawai'i and Skjalbreiður in Iceland) and contain relatively few volatiles. In some special cases, volcanic eruptions beneath Pleistocene ice sheets formed flat-topped mountains, or tuyas (e.g., in Iceland and British Columbia).

Volcanoes also occur at hot spots where uprising plumes of magma from the mantle reach the surface (Chapter 2) (Frisch et al. 2011; Bachelèry & Villeneuve 2013; Huff & Owen 2013). Continental hot spots include Yellowstone in the USA, the Massif Central and Eifel mountains in Europe and the Tibesti and Ahaggar mountains in Africa; oceanic examples include the Hawaiian and Canary islands, while Iceland is a hot spot located on a mid-ocean ridge. Where plates are moving across a hot spot, successive phases of activity have formed chains of volcanic islands (e.g., the Hawaiian–Emperor volcanic chain and the Azores, Cape Verde, Tristan da Cunha and Canary islands). Some hot spots have formed large volcanic complexes, including the massive basaltic shield volcanoes of Mauna Loa and Mauna Kea on Hawai'i. Mauna Kea is the highest mountain on Earth, rising nearly 10 km above the adjacent ocean floor.

Extensional tectonics are responsible for a further category of mountain formed by block faulting and the elevation of one crustal block relative to another (Gerrard 1990; Schroder & Price 2013). Examples include the Vosges and Black Forest in Europe and the Basin and Range Province and Sierra Nevada in North America. The Basin and Range Province is a remarkable region of narrow, approximately north–south-aligned mountain ranges and intervening arid valleys or basins in the western and south-western USA and north-western Mexico, formed by large-scale block faulting associated with crustal extension that began during the early Miocene (Nash 2013). The topography ranges in elevation from 85 m below sea level in Death Valley to 4421 m above sea level at Mount Whitney and supports a variety of desert, salt lake and montane ecosystems.

10.4 Mountain Geodiversity at Regional to Local Scales

At regional to local scales, geomorphological processes are the dominant control on mountain geodiversity. They are influenced by rock structure, lithology, climate and time. Barsch & Caine (1984) described mountain process systems in terms of sediment flux cascades (Table 10.3). High and very high mountains are dominated by the mountain cryosphere and coarse debris systems, characterized by glacial and periglacial processes together with mass movements such as landslides and

Table 10.3 Sediment cascade in mountain systems.

System	Main characteristics
Mountain cryosphere system	Characteristic of high mountains; glacier transport of debris produced by glacial erosion and rockfall
Coarse debris system	Characteristic of high mountains; transfer of coarse debris between cliffs and talus, including rockfall and rock-slope failure and subsequent post-depositional modification (e.g., rock glaciers)
Fine-grained sediment system	Mass wasting on interfluves and valley floors and consequent fluvial transfer
Geochemical system	Solutional weathering and solute transfer by water

Source: Adapted from Barsch & Caine (1984).

snow avalanches, and accompanied by high rates of fluvial sediment transfer (Gerrard 1990; Schroder & Bishop 2004). Potential energy is high, and it is the combination of processes and their intensity and rate that distinguish such mountain regions (Barsch & Caine 1984; Schroder & Bishop 2004). High relief, steep slopes, climate and vegetation contrasts and tectonic activity/seismic events generate rapid erosion rates. There is great temporal and spatial variability in processes, and episodic high-magnitude events – such as landslides and floods due to heavy rain, snowmelt or glacial lake outburst floods (GLOFs) – play an important role in shaping the landscape. The resulting spatial diversity and altitudinal zonation of landforms and processes are documented in various schematic representations (Figure 10.2a) (Brunsden & Allison 1986; Hewitt 1989; Slaymaker 1993). As differentiated by Barsch & Caine (1984), high and very high mountain types display permanent snow and ice cover and characteristic forms of glacial erosion (horns, arêtes, cirques, glacial troughs: "Alp-type" relief), with debris-covered glaciers and valley slopes characterized by rockfall, rock-slope failures, talus and rock glaciers (Figures 10.1a and 10.2). Some also include areas of rounded interfluves and flat-topped mountains ("Rocky Mountain"-type relief).

Mid-elevation and low mountains, particularly those associated with the Variscan and Caledonide orogens, are characterized by lower potential energy and lower magnitudes and rates of erosion. They generally have more rounded forms and more extensive paleosurface remnants (Calvet et al. 2015), supporting blockfields and other periglacial landforms. Quaternary glacial landforms (cirques and glacial troughs) are often incised into these surfaces, with corresponding deposits on valley floors (Figures 10.1b and 10.2). In polar and subpolar

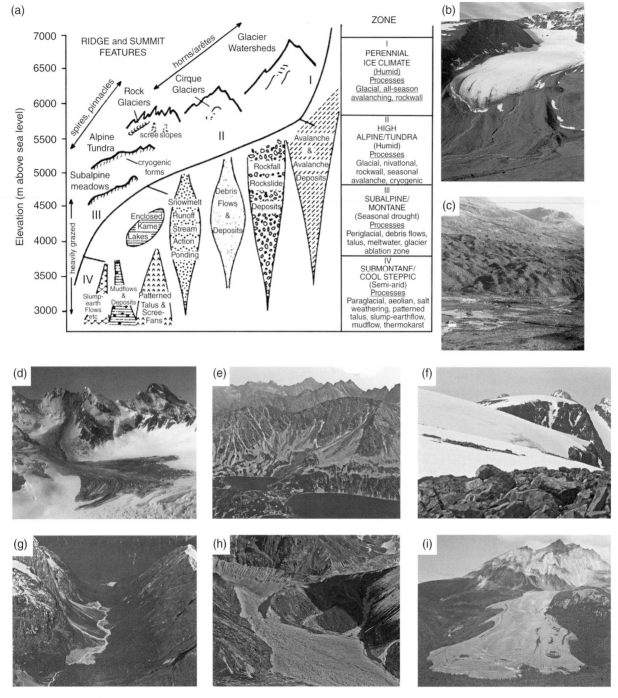

Figure 10.2 Geodiversity of mountain landforms and geomorphological processes. (a) Schematic representation of the altitudinal zonation of geomorphological processes and depositional environments, central Karakoram. *Source:* Hewitt (1989). Reproduced with permission of Schweizerbart Science Publishers. (b) Recession of cirque glacier from Little Ice Age moraines, Kebnekaise, Sweden. (c) Moraines formed by glaciers during the Younger Dryas, Scottish Highlands. (d) Landslide on Mount Dixon, Southern Alps, New Zealand, January 21, 2013. *Source:* Reproduced with permission of Arthur McBride, Alpine Guides (Aoraki) Ltd. (e) Rock weathering, talus formation and debris flows, Tatra Mountains, Poland. (f) Periglacial blockfield and small plateau icecaps, Lyngsalpene, northern Norway. (g) Glacial trough and alluvial fans, Southern Alps, New Zealand. (h) End moraine breach and glacial lake outburst flood (GLOF) deposits following the failure of the moraine dam at Tam Pokhari (Sabai Tsho) glacial lake, Hinku Valley, Nepal, September 3, 1998. (i) Rock glacier, Wrangell Mountains, Alaska. See also Plate 18 in color plate section.

areas, such as Svalbard and South Georgia, alpine forms occur at altitudes below 2500 m (Figure 10.1d).

Mountain landforms vary according to geomorphological and climate history, as well as the state of materials (rock, regolith and soil) and their sensitivity to high-energy conditions. They are landscapes in transition from past conditions: the paraglacial influence is a prime example in which the landscape is still adjusting to Quaternary glaciation (Church & Ryder 1972; Ballantyne 2002; Embleton-Hamann & Slaymaker 2012). In glaciated mountains, glacier retreat reveals unstable landscapes that, over time scales of tens to thousands of years or more, are susceptible to accelerated erosion, sediment transfer and deposition conditioned directly by glaciation and deglaciation (Ballantyne 2002). Such paraglacial adjustments from glacial to nonglacial conditions following deglaciation have extensively modified mountain slopes and valley floors. Rock slopes, drift-mantled slopes and valley-floor glacigenic deposits provide sediment sources for a range of landforms (e.g., talus accumulations, debris cones, alluvial fans, valley fills and deltas) that represent components of an interrupted sediment cascade (Figure 10.2).

However, mountain landscape development is not a simple linear or lagged response to changing climate and tectonics, but involves episodes of incomplete readjustment, both temporally and spatially, from past forcing conditions of climate or tectonics (Hewitt 2002). This has been described as a disturbance regime landscape in which disturbance occurs so frequently that the landscape is unable to readjust to contemporary processes (Hewitt 2006). Many mountain landscapes are therefore complex and heterogeneous, with diverse mosaics of landforms of different origins and ages forming a palimpsest and bearing to varying degrees the imprint of past conditions and the presence of both relict and contemporary forms and processes that reflect changes in the extent of glaciers, permafrost, periglacial activity, precipitation and vegetation (Barsch & Caine 1984; Thomas 2012; Stahr & Langenscheidt 2014; Gunnell 2015).

Various authors have applied a landsystem approach to characterize the landforms and process systems of modern and Pleistocene glacial environments, which also include lower-relief (e.g., Scotland, Labrador, Norway) and polar settings, as well as high-relief settings (Alps, Andes, Himalaya) (reviewed in Benn & Evans 2010). These models highlight the geomorphological heterogeneity of mountain landscapes, as illustrated in geomorphological maps at different scales (e.g., Kotarba 1992; Evans et al. 2006; Rączkowski et al. 2015).

In many areas, erosion has revealed a strong influence of geology and rock structure on the form of mountain landscapes, particularly where there are dipping or folded sequences of sedimentary rocks as in the Appalachians, the Jura Mountains, the Alps and the Zagros Mountains (Figure 10.1e) (Chorley et al. 1984; Selby 1985; Gerrard 1990; Schroder & Price 2013). In sedimentary rocks, the shapes of individual mountains are related to the influence of the inclination of the bedding and jointing (Cruden & Hu 1999; Cruden 2003), while weathering and erosion of igneous intrusions can produce massive rock towers and buttresses (Figure 10.1f).

Mountains have a strong influence on climate, and thus on spatial variations in tectonics and geomorphological processes (Chapters 4 and 8). Most mountains have a rain-shadow effect, with more arid leeward flanks. This is particularly well exemplified in the case of the Himalaya, where the magnitude and frequency of the processes – including the style and extent of glaciation – are greatly influenced by climate gradients and topography (Hewitt 1989; Owen 2014). In more arid areas, the landscape is often characterized by mountain fronts, pediments, pediplains and sediment accumulation in alluvial fans and intermontane basins, as well as by glacial features (Figure 10.1g) (Gerrard 1990).

Volcanoes also support a geodiverse and distinctive assemblage of landforms (Figure 10.1h,i). These include constructional forms such as lava flows and cinder cones, and a wide range of forms produced by erosional processes, including collapsed craters, large-scale mass movements, debris flows, lahars, glacial landforms and meltwater flood landforms and deposits (Gerrard 1990; Masson et al. 2002; Thouret 2004; Lockwood & Hazlett 2010; Bachelèry & Villeneuve 2013; Huff & Owen 2013).

10.5 Values of Mountain Geodiversity

Geodiversity has intrinsic, instrumental and relational (eudaimonic) values. These include scientific, educational, cultural, aesthetic, ecological and economic components that provide or contribute to a range of ecosystem services (Gordon & Barron 2013; Gray 2013; Gray et al. 2013). This is equally true for mountains as for other environments, and in the case of mountains it is particularly important because they provide direct support for 10–20% of the world's population and indirectly affect the lives of more than half of humankind (Körner et al. 2005; Slaymaker & Embleton-Hamann 2009). Geodiversity contributes in many ways to the vital range of ecosystem services that mountain areas provide (Körner et al. 2005; European Environment Agency 2010) (Table 10.4). Geodiversity mapping is an important tool for assessing the value of the abiotic components of the landscape (Chapter 11), although a site or

Table 10.4 Ecosystem services provided by mountain geodiversity.

Provisioning services

Soils and nutrients to support food and fiber production and natural medicines

Raw materials for construction and fuel

Freshwater for human consumption

Water for hydropower

Ornamental resources (rocks, fossils, minerals)

Mineral resources

Regulating services

Water regulation and storage (water towers)

Natural hazard regulation

Slope stability and erosion regulation

Cultural services

Spiritual and religious values and cultural meanings

Aesthetic value

Artistic inspiration

Landscapes and sites for recreation and nature-based tourism (skiing, rock climbing, glacier hiking, geotourism)

Influence on cultural diversity

Sense of place

Scientific knowledge and education

Supporting services

Soil formation

Biogeochemical and hydrological cycling

Rock cycling and geomorphological processes

Habitat creation and maintenance

---

area of high geoheritage significance can comprise a single feature of value, and does not require the presence of a diversity of features (Crofts & Gordon 2015).

Geodiversity has value for five main reasons.

First, elements of geodiversity, like those of biodiversity (Crofts et al. 2008; Vucetich et al. 2015), arguably have an intrinsic value (Gray 2013). This is independent of their instrumental or relational value to humanity, and they deserve to be treated with respect and preserved for future generations (Slaymaker et al. 2009).

Second, geodiversity is important for scientific research and education, enabling understanding of Earth history, materials and processes. As already noted, mountains provide fundamental evidence for tectonic processes and long-term landscape evolution since at least the Paleozoic. They support valuable paleoenvironmental and paleoclimatic archives spanning tens to thousands of years in the form of ice-core records, lake and peat bog sediments, speleothems, aeolian sediments and other

geomorphological and sedimentary evidence of Quaternary glacier fluctuations (Lowe & Walker 2015). They also provide the knowledge base for informing environmental management, particularly to help society adapt to climate change and to predict and mitigate natural hazards and their impacts (including erosion, flooding, slope failures and earthquakes) (Owens & Slaymaker 2004; Haeberli & Whiteman 2014; Davies 2015; Kargel et al. 2016). From an educational perspective, mountain geodiversity has additional value in helping to raise awareness of climate change, natural hazards and human impacts on sensitive geomorphological environments, because the effects are often immediate and striking (Reynard & Coratza 2016).

Third, as the abiotic component of ecosystems, geodiversity provides the foundation for biodiversity in mountain systems, particularly through its interactions with topography and climate. Over geological time scales, biodiversity has been strongly driven by abiotic factors and long-term landscape evolution (Antonelli et al. 2009; Benton 2009; Hoorn et al. 2010; Schnitzler et al. 2011; Gill et al. 2015). Mountain building creates the physical stage and bioclimatic gradients for species evolution, both in mountain areas and in adjacent lowlands, through nutrient supply via drainage and sediment input to soils (Hoorn et al. 2013). These links have long been recognized in geo-ecological studies and in altitudinal and latitudinal vegetation zonation systems (Humboldt & Bonpland 1805; Troll 1972). Mountains have outstanding nature conservation value. For their global extent, they incorporate a disproportionate amount of the world's protected areas and biodiversity. They include 32% of the world's protected areas and support a third of terrestrial biodiversity and nearly half of the world's biodiversity hot spots (Körner et al. 2005; Spehn et al. 2011) due to altitudinal bioclimatic gradients, geodiversity-related habitat diversity, topographic diversity and high numbers of endemic species resulting from biogeographic isolation and adaptation to particular niches (Spehn et al. 2011). This is exemplified at a global scale by centers of vascular plant diversity that coincide with highly geodiverse mountain areas in the humid tropics and subtropics (Barthlott et al. 2005). In addition, paleoenvironmental records provide evidence of ecosystem service trends, helping to inform potential future trajectories and risks to these services (Jeffers et al. 2015).

At regional and local scales, geodiversity supports habitat heterogeneity arising from the characteristics of the physical substrate, topographic effects on microclimate and disturbance regimes from continual and episodic processes. Most species – not just rare or specialized ones (e.g., those associated with limestone pavements or hot springs) – depend on the abiotic "stage" on which they exist (Hjort et al. 2015), and habitat diversity and species richness are generally greater in areas of high abiotic heterogeneity (Jačková & Romportl 2008; Anderson & Ferree 2010). Nested mosaics of habitat conditions (Haslett 1997) often result from the interplay of bedrock lithology and the physical nature of the substrate (Kruckeberg 2002; Hahm et al. 2014), soil properties and stability (Darmody et al. 2004; Holtmeier & Broll 2012; Bockheim 2015; Morrocco et al. 2016) and geomorphological processes and landforms (Jonasson et al. 2005; Kozłowska et al. 2006; Alexandrowicz & Margielewski 2010; Brazier et al. 2012), as well as the influence of snow accumulation patterns (Palacios & Sánchez-Colomer 1997), microtopography (Opedal et al. 2015) and the interactions between geomorphological processes, microclimate, water availability and vegetation (Soukupová et al.1995; Kozłowska & Rączkowska 2002).

Fourth, mountain geodiversity supports a range of aesthetic and cultural values that contribute to quality of life and human flourishing (Tobias & Drasdo 1980; Besson 2010; Ireton & Schaumann 2012; Bernbaum & Price 2013; Chan et al. 2016). Mountains have long exerted a fascination for people. Over time, they have changed from landscapes of fear and waste to places of awe, reverence and inspiration: "Climb the mountains and get their good tidings…nature's sources never fail" (Muir 1901, p. 56). Many are celebrated in mythology or have a powerful religious significance (Bernbaum 1997, 2006), including Mount Kailash in western Tibet, Mount Parnassus, Mount Olympus and Mount Athos in Greece, Croagh Patrick in Ireland and Taishan in China. Mountain geodiversity is also a potent influence on cultural heritage as a source of inspiration for art, sculpture, music, poetry and literature. In western Europe, as part of the Romantic movement of the 18th and 19th centuries, mountains were central to the aesthetic of the sublime and the development of a cultural imagery that shaped imaginative geographies (Nicolson 1959; Cosgrove & della Dora 2009). They were also celebrated in the influential nature writing of Jean-Jacques Rousseau, William Wordsworth, Henry David Thoreau and John Muir, for example. For art critic John Ruskin, mountains were "the beginning and end of all natural scenery" (Ruskin 1856, p. 353). This greatly added to their aesthetic appeal as early (geo)tourist destinations (Reynard et al. 2011; Gordon & Baker 2016; Hose 2016; Migoń 2016). In the 19th century, too, they began to attract the attention of Western science, notably in ecology (Humboldt & Bonpland 1805) and glaciology (Agassiz 1840; Forbes 1845). Today, they have become places for conducting science and valuable barometers of environmental change. In providing opportunities for tourism, recreation, outdoor activities and physical challenge, they also offer benefits to people's health and well-being. Mountains, glaciers and volcanoes all feature prominently

as modern geotourism attractions (Figure 10.3) where visitors can both learn about the shaping of the planet and enjoy the scenery and various adventure-based activities, such as glacier hiking (Wang et al. 2010; Erfurt-Cooper & Cooper 2010; Erfurt-Cooper 2014; Welling et al. 2015). Mountains also play an integral part in determining landscape character and influencing people's sense of place, cultural identity and well-being; for example, glaciers have a particular fascination in the popular imagination, although different societies relate to glaciers in different ways (Cruickshank 2005; Orlove et al. 2008; Gagné et al. 2014; Jurt et al. 2015).

Fifth, mountain geodiversity provides resources and assets for many aspects of economic development and has a profound influence on the use of land and water. The supply of fundamental materials and processes for maintaining life on Earth includes soils, fresh water and hydrological and biogeochemical cycling, which enables food and fiber production and the provision of construction materials, energy sources and minerals (European Environment Agency 2010; Price 2015). Examples of the physical and energy resources obtained from mountains include coal from the Appalachians, hydroelectric power in Norway and precious and semi-precious gemstones from Afghanistan (Gray 2013). Geodiversity also provides the basis for tourism development, in both summer and winter, including more specialized niche activities such as ecotourism and geotourism (Newsome & Dowling 2010).

Therefore, as an integral part of mountain ecosystems and their natural capital, geodiversity merits conservation for its geoheritage values, as well as for its wider values for nature and people (Crofts & Gordon 2015). Although there has been no systematic global assessment, many elements of mountain geodiversity and geoheritage are protected at international, national and local levels within the six International Union for Conservation of Nature (IUCN) Protected Area Management Categories, including World Heritage Sites and national parks (Crofts & Gordon 2015). However, in many cases, that protection is implied rather than explicit, and application of geoconservation principles that relate to sustainable management of natural systems would help to strengthen the conservation of geoheritage at the wider landscape scale (Crofts & Gordon 2015). While there has been a strong emphasis on protecting the scientific values of geoheritage, there is now increasing emphasis in many mountain areas on identifying and protecting sites for geotourism and public education (Zwoliński & Stachowiak 2012; Bollati et al. 2013; Reynard et al. 2016), notably with the Global Geoparks Network now adopted as part of the UNESCO International Geoscience and Geoparks Programme.

Figure 10.3 Breiðamerkurjökull, Vatnajökull National Park, Iceland. Many glaciers worldwide are popular (geo)tourist destinations where visitors can both enjoy the spectacular scenery and appreciate the effects of climate change, which are clearly demonstrated in glacier recession. See also Plate 19 in color plate section.

10.6 Mountain Geodiversity and Climate Change

In mountain regions, the important drivers of environmental change are hydroclimate, relief and human activity (Slaymaker & Embleton-Hamann 2009). There is wide agreement that fragile mountain ecosystems are likely to be particularly sensitive to climate change, affecting both biotic and abiotic components (Huber et al. 2005; Xu et al. 2009; La Sorte & Jetz 2010; Löffler et al. 2011; Beniston 2012; Pepin et al. 2015). Understanding the geodiversity of mountains is fundamental to assessing their vulnerability and managing hazard risk and biodiversity adaptations through natural solutions. The response of the cryosphere will have far-reaching effects due to the impact of warming on glaciers, permafrost and nival processes. 20th- and 21st-century climate change has already caused dramatic recession – and in some cases, the disappearance – of mountain glaciers, and this is projected to continue (Paul et al. 2004; Bolch et al. 2012; Cogley 2012; Rabatel et al. 2013; Shea et al. 2015). This has serious implications for crucial water

resources, especially during the dry season (Vuille et al. 2008; Immerzeel et al. 2010; Kaser et al. 2010; Sorg et al. 2012; Mark et al. 2015), and there will be impacts downstream on glacial lakes, stream communities, biodiverse wetlands and people's livelihoods (Milner et al. 2009; Xu et al. 2009; Jacobsen et al. 2012). However, increased glacier runoff in the coming decades – at least for a time – may have beneficial effects in terms of both water supply and hydropower production (Thorsteinsson et al. 2013). In some areas, glacier recession is already having a negative impact on tourism (Wang et al. 2010; Purdie 2013).

Changes in the cryosphere, combined with steep slopes and abundant sediment availability, present increased risks from geomorphic hazards for communities and human activities (Iribarren Anacona et al. 2014). Such hazards include slope instability, rock avalanches, landslides and other mass flows arising from glacier recession and melting permafrost (Gruber & Haeberli 2007; Huggel et al. 2010; Deline et al. 2012, 2014, 2015; Evans & Delaney 2014; Korup & Dunning 2015), and GLOFs from the failure of moraine dams or where large rock or ice avalanches impact on lakes (Clague & O'Connor 2014; Hewitt 2014; Quincey & Carrivick 2015). Retreating glaciers also provide a striking demonstration of climate change – explored through different cultural media (Jackson 2015) – and affect the culture, lives and social and political systems of those living nearby (Orlove et al. 2008; Carey 2010; Gagné et al. 2014; Allison 2015). Volcanoes, too, have their own particular hazards (Thouret 2010; Granados et al. 2014; McGuire 2015; Papale 2015), and there is a possible risk of glacier recession enhancing volcanic activity due to glacial unloading (Sigmundsson et al. 2010; Tuffen 2010).

Adverse impacts may also arise – particularly in low and mid-elevation mountains and on the lower slopes of high and very high mountains – through changes in land use, including agriculture and deforestation, extractive industries, infrastructure development, construction of large dams and tourism, with consequent slope instability and enhanced runoff, erosion and sediment discharge. Impacts may be enhanced through increased occurrence of extreme events such as intense rainfall and droughts or through changes in the cryosphere (Slaymaker & Embleton-Hamann 2009).

Geomorphological sensitivity and process responses to climate change will have significant impacts on mountain ecosystems and biodiversity adaptations (Brazier et al. 2012; Knight & Harrison 2013). For example, the condition of habitats and species may be affected by:

- increased rate, occurrence and intensity of flooding, landsliding, erosion, sediment supply and transfer and river channel mobility;
- changes in the spatial distributions of landforms;
- increased rates of geomorphological change that are too fast for some habitats and species to adapt to;
- reduced recovery time between extreme events;
- reduced habitat availability as morphogenetic zones move upslope;
- changes in seasonal processes (e.g., timing and duration of droughts).

Such changes may mean the loss of some features of interest or shifts in their locations, so that some biodiversity conservation targets will no longer be attainable in current protected areas (Hagerman et al. 2010). Further, geomorphological responses to perturbations are likely to be complex and nonlinear, and changes in one part of a system will likely have impacts elsewhere (Phillips 2009; Slaymaker et al. 2009; Knight & Harrison 2014). Conservation management of active systems should therefore be informed by a sound understanding of the underlying physical processes and the application of geoconservation principles (Gray et al. 2013; Crofts & Gordon 2015). This includes learning from the past – not to provide static baselines, but to help understand past

ranges of natural variability and future trajectories of change – and considering how changes in the magnitude, frequency, rate and duration of geomorphological processes will impact habitats and species.

Protected area design has largely been based on the assumption of relatively static biodiversity (Pressey et al. 2007; Hagerman et al. 2010). However, maintaining the status quo is not an option, and this requires a shift in focus from short-term preservation to planning for change in the longer term, and from preserving current species composition to maintaining the integrity of the ecological and evolutionary processes that sustains it (Bennett et al. 2009; Prober & Dunlop 2011). This underlies the concept of "conserving nature's stage," which involves the use of geodiversity as a coarse filter approach to support biodiversity conservation (Anderson & Ferree 2010; Beier et al. 2015). Such an approach requires a shift in focus from "static" biotic communities to protecting areas with a high probability of harboring high biodiversity, even if the species composition of these areas changes over time. A better understanding of geodiversity can help to inform conservation priorities in a changing world; for example, through knowledge of available habitat mosaics and the biophysical connectivity between them, and through opportunity mapping to identify macro- and microrefugia for shifting species (Elsen & Tingley 2015). Protected area design and management that incorporates the conservation of geodiverse, heterogeneous landscapes should therefore enhance biodiversity resilience and sustain key abiotic and ecological processes (Anderson et al. 2014).

10.7 Conclusion

Mountain landscapes are the ongoing products of tectonic, geomorphological and climate processes operating over different time scales. The resulting geodiversity is an integral part of nature that has value not only for geoheritage reasons, but also for the wider benefits it provides to nature and people, as recognized in IUCN Resolutions (IUCN 2008, 2012). In the face of human pressures and climate change, understanding of geodiversity and Earth system processes is fundamental to help inform the sustainable management of natural resources, biodiversity conservation and the risks from natural hazards, as well as to contextualize cultural responses to environmental change. This is implicit in several UN Sustainable Development Goals, which recognize the value of healthy natural systems (Anon. 2015; United Nations 2015). As well as acknowledging that geodiversity merits conservation for its own particular geoheritage values, a more integrated approach across the full range of IUCN Protected Area Management

Categories – and in the wider landscape, through application of geoconservation principles and geoscience knowledge – would benefit nature conservation from both a biodiversity and a geodiversity perspective.

Acknowledgments

I am grateful to Murray Gray and Lewis Owen for helpful reviews that improved the text.

References

Agassiz, L. (1840) *Études sur les Glaciers*. Neuchâtel: H. Nicolet.

Alexandrowicz, Z. & Margielewski, W. (2010) Impact of mass movements on geo- and biodiversity in the Polish Outer (Flysch) Carpathians. *Geomorphology* **123**, 290–304.

Allison, E.A. (2015) The spiritual significance of glaciers in an age of climate change. *WIREs Climate Change* **6**, 493–508.

Anderson, M.G. & Ferree, C.E. (2010) Conserving the stage: climate change and the geophysical underpinnings of species diversity. *PLoS ONE* **5**, e11554.

Anderson, M.G., Clark, M. & Sheldon, A.O. (2014) Estimating climate resilience for conservation across geophysical settings. *Conservation Biology* **28**, 959–970.

Anon. (2015) Editorial. Finite Earth. *Nature Geoscience* **8**, 735.

Ansted, D.T. (1863) *The Great Stone Book of Nature*. Philadelphia: George W. Childs.

Antonelli, A., Nylander, A.A., Persson, C. & Sanmartin, I. (2009) Tracing the impact of the Andean uplift on Neotropical plant evolution. *Proceedings of the National Academy of Sciences* **106**, 9749–9754.

Bachelèry, P. & Villeneuve, N. (2013) Hot spots and large igneous provinces. In: Owen, L.A. (ed.) *Treatise on Geomorphology*, vol. **5**. San Diego, CA: Academic Press, pp. 193–233.

Ballantyne, C.K. (2002) Paraglacial geomorphology. *Quaternary Science Reviews* **21**, 1935–2017.

Barsch, D. & Caine, N. (1984) The nature of mountain geomorphology. *Mountain Research and Development* **4**, 287–298.

Barthlott, W., Mutke, J., Rafiqpoor, M.D. et al. (2005) Global centres of vascular plant diversity. *Nova Acta Leopoldina* **92**, 61–83.

Beier, P., Hunter, M.L. & Anderson, M.G. (2015) Special section: conserving nature's stage. *Conservation Biology* **29**, 613–617.

Beniston, M. (2012) Environmental change in mountain regions. In: Matthews, J.A., Bartlein, P.J., Briffa, K.R. et al. (eds.) *The SAGE Handbook of Environmental Change*. London: SAGE Publications Ltd., pp. 262–281.

Benn, D.I. & Evans, D.J.A. (2010) *Glaciers and Glaciation*, 2nd edn. London: Arnold.

Bennett, A.F., Haslem, A., Cheal, D.C. et al. (2009) Ecological processes: a key element in strategies for nature conservation. *Ecological Management & Restoration* **10**, 192–199.

Benton, M.J. (2009) The Red Queen and the Court Jester: species diversity and the role of biotic and abiotic factors through time. *Science* **323**, 728–732.

Bernbaum, E. (1997) *Sacred Mountains of the World*. Berkeley, CA: University of California Press.

Bernbaum, E. (2006) Sacred mountains: themes and teachings. *Mountain Research and Development* **26**, 304–309

Bernbaum, E. & Price, L.W. (2013) Attitudes toward mountains. In: Price, M.F., Byers, A.C., Friend, D.A. et al. (eds.) *Mountain Geography. Physical and Human Dimensions*. Berkeley, CA: University of California Press, pp. 252–266.

Besson, F. (ed.) (2010) *Mountains Figured and Disfigured in the English-Speaking World*. Newcastle upon Tyne: Cambridge Scholars Publishing.

Bishop, P. (2007) Long-term landscape evolution: linking tectonics and surface processes. *Earth Surface Processes and Landforms* **32**, 329–365.

Blenkinsop, T. & Moore, A. (2013) Tectonic geomorphology of passive margins and continental hinterlands. In: Owen, L.A. (ed.) *Treatise on Geomorphology*, vol. **5**. San Diego, CA: Academic Press, pp. 71–92.

Bockheim, J.G. (2015) *Cryopedology*. Cham, Heidelberg: Springer.

Bolch, T., Kulkarni, A., Kääb, A. et al. (2012) The state and fate of Himalayan glaciers. *Science* **336**, 310–314.

Bollati, I., Smiraglia, C. & Pelfini, M. (2013) Assessment and selection of geomorphosites and trails in the Miage Glacier area (Western Italian Alps). *Environmental Management* **51**, 951–967.

Borges, J.L. (1975) Utopia of a tired man. *New Yorker* **51**(8), 32–33.

Brazier, V., Bruneau, P.M.C., Gordon, J.E. & Rennie, A.F. (2012) Making space for nature in a changing climate: the role of geodiversity in biodiversity conservation. *Scottish Geographical Journal* **128**, 211–233.

Brunsden, D. & Allison, R.J. (1986) Mountains and highlands. In: Fookes, P.G. & Vaughan, P.R. (eds.) *A Handbook of Engineering Geomorphology*. Glasgow: Surrey University Press, Guildford and Blackie, pp. 150–165.

Byers, A.C., Price, L.W. & Price, M.F. (2013) An introduction to mountains. In: Price, M.F., Byers, A.C., Friend, D.A. et al. (eds.) *Mountain Geography. Physical and Human Dimensions.* Berkeley, CA: University of California Press, pp. 1–10.

Calvet, M., Gunnell, Y. & Farines, B. (2015) Flat-topped mountain ranges: their global distribution and value for understanding the evolution of mountain topography. *Geomorphology* **241**, 255–291.

Carey, M. (2010) *In the Shadow of Melting Glaciers: Climate Change and Andean Society.* Oxford: Oxford University Press.

Chan, K.M.A., Balvanera, P., Benessaiah, K. et al. (2016). Why protect nature? Rethinking values and the environment. *PNAS* **113**, 1462–1465.

Clague, J.J. & O'Connor, J.E. (2014) Glacier-related outburst floods. In: Haeberli, W. & Whiteman, C. (eds.) *Snow and Ice-Related Hazards, Risks and Disasters.* Amsterdam: Elsevier, pp. 487–519.

Chorley, R.J., Schumm, S.A. & Sugden, D.E. (1984) *Geomorphology.* London: Methuen.

Church, M. & Ryder, J.M. (1972) Paraglacial sedimentation: a consideration of fluvial processes conditioned by glaciation. *Bulletin of the Geological Society of America* **83**, 3059–3072.

Cogley, G. (2012) The future of the world's glaciers. In: Henderson-Sellers, A. & McGuffie, K. (eds.) *The Future of the World's Climate*, 2nd edn. Amsterdam: Elsevier, pp. 197–222.

Cosgrove, D. & della Dora, V. (2009) *High Places. Cultural Geographies of Mountains, Ice and Science.* London and New York: I.B. Tauris.

Crofts, R. & Gordon, J.E. (2015) Geoconservation in protected areas. In: Worboys, G.L., Lockwood, M., Kothari, A. et al. (eds.) *Protected Area Governance and Management.* Canberra: ANU Press, pp. 531–568.

Crofts, R., Harmon, D. & Figgis, P. (2008) *For Life's Sake: How Protected Areas Enrich our Lives and Secure the Web of Life.* Gland: IUCN World Commission on Protected Areas.

Cruden, D.M. (2003) The shapes of cold, high mountains in sedimentary rocks. *Geomorphology* **55**, 249–261.

Cruden, D.M. & Hu, X-Q. (1999). The shapes of some mountain peaks in the Canadian Rockies. *Earth Surface Processes and Landforms* **24**, 1229–1241.

Cruickshank, J. (2005) *Do Glaciers Listen? Local Knowledge, Colonial Encounters, and Social Imagination.* Vancouver, BC: University of British Columbia Press.

Darmody, R.G., Thorn, C.E., Schlyter, P. & Dixon, J.C. (2004) Relationship of vegetation distribution to soil properties in Kärkevagge, Swedish Lapland. *Arctic, Antarctic and Alpine Research* **36**, 21–32.

Davies, T. (ed.) (2015) *Landslide Hazards, Risks, and Disasters.* Amsterdam: Elsevier.

Debarbieux, B. & Rudaz, G. (2015) *The Mountain. A Political History from the Enlightenment to the Present.* Chicago, IL and London: University of Chicago Press.

Deline, P., Gardent, M., Magnin, F. & Ravanel, L. (2012) The morphodynamics of the Mont Blanc massif in a changing cryosphere: a comprehensive review. *Geografiska Annaler* **94A**, 265–283.

Deline, P., Gruber, S., Delaloye, R. et al. (2014) Ice loss and slope stability in high-mountain regions. In: Haeberli, W. & Whiteman, C. (eds.) *Snow and Ice-Related Hazards, Risks and Disasters.* Amsterdam: Elsevier, pp. 521–561.

Deline, P., Hewitt, K., Reznichenko, N. & Shugar, D. (2015) Rock avalanches onto glaciers. In: Davies, T. (ed.) *Landslide Hazards, Risks, and Disasters.* Amsterdam: Elsevier, pp. 263–319.

Elsen, P.R. & Tingley, M.W. (2015) Global mountain topography and the fate of montane species under climate change. *Nature Climate Change* **5**, 772–776.

Embleton-Hamann, C. & Slaymaker, O. (2012) The Austrian Alps and paraglaciation. *Geografiska Annaler* **94A**, 7–16.

Erfurt-Cooper, P. (ed.) (2014) *Volcanic Tourist Destinations.* Berlin and Heidelberg: Springer-Verlag.

Erfurt-Cooper, P. & Cooper, M. (eds.) (2010) *Volcano and Geothermal Tourism. Sustainable Geo-resources for Leisure and Recreation.* London: Earthscan.

European Environment Agency (2010) *Europe's Ecological Backbone: Recognising the True Value of Our Mountains.* EEA Report No 6/2010. Copenhagen: European Environment Agency.

Evans, S.G. & Delaney, K.B. (2014) Catastrophic mass flows in the mountain glacial environment. In: Haeberli, W. & Whiteman, C. (eds.) *Snow and Ice-Related Hazards, Risks and Disasters.* Amsterdam: Elsevier, pp. 563–606.

Evans, D.J.A., Twigg, D.R. & Shand, M. (2006) Surficial geology and geomorphology of the þórisjökull plateau icefield, west-central Iceland. *Journal of Maps* **2**, 17–29.

Fairbridge, R.W. (1968) Mountain types. In: Fairbridge, R.W. (ed.) *The Encyclopedia of Geomorphology.* New York: Reinhold Book Corporation, pp. 751–761.

Forbes, J.D. (1845) *Travels Through the Alps of Savoy and Other Parts of the Pennine Chain with Observations on the Phenomena of Glaciers.* Edinburgh: A. & C. Black.

Frisch, W., Meschede, M. & Blakey, R. (2011) *Plate Tectonics. Continental Drift and Mountain Building.* Berlin and Heidelberg: Springer-Verlag.

Gagné, K., Rasmussen, M.B. & Orlove, B. (2014) Glaciers and society: attributions, perceptions, and valuations. *WIREs Climate Change* **5**, 793–808.

Geikie, A. (1896) Scottish mountains. *Scottish Mountaineering Club Journal* **4**, 113–125.

Geikie, J. (1913) *Mountains. Their Origin, Growth and Decay.* Edinburgh: Oliver & Boyd.

Gerrard, A.J. (1990) *Mountain Environments.* London: Belhaven Press.

Gill, J.L., Blois, J., Benito, B. et al. (2015) A 2.5-million-year perspective on coarse-filter strategies for conserving nature's stage. *Conservation Biology* **29**, 640–648.

Gordon, J.E. & Baker, M. (2016) Appreciating geology and the physical landscape in Scotland: from tourism of awe to experiential re-engagement. In: Hose, T.A. (ed.) *Appreciating Physical Landscapes: Three Hundred Years of Geotourism*. Special Publications **417**. London: Geological Society, pp. 25–40.

Gordon, J.E. & Barron, H.F. (2013) Geodiversity and ecosystem services in Scotland. *Scottish Journal of Geology* **49**, 41–58.

Granados, H.D., Miranda, P.J., Núñez, G.C. et al. (2014) Hazards at ice-clad volcanoes: phenomena, processes, and examples from Mexico, Colombia, Ecuador, and Chile. In: Haeberli, W. & Whiteman, C. (eds.) *Snow and Ice-Related Hazards, Risks and Disasters*. Amsterdam: Elsevier, pp. 607–646.

Gray, M. (2013) *Geodiversity: Valuing and Conserving Abiotic Nature*, 2nd edn. Chichester: Wiley-Blackwell.

Gray, M., Gordon, J.E. & Brown, E.J. (2013) Geodiversity and the ecosystem approach: the contribution of geoscience in delivering integrated environmental management. *Proceedings of the Geologists' Association* **124**, 659–673.

Green, P.F., Lidmar-Bergström, K., Japsen, P. et al. (2013) Stratigraphic Landscape Analysis, Thermochronology and the Episodic Development of Elevated, Passive Continental Margins. *Geological Survey of Denmark and Greenland Bulletin* **30**.

Gruber, S. & Haeberli, W. (2007) Permafrost in steep bedrock slopes and its temperature-related destabilization following climate change. *Journal of Geophysical Research: Earth Surface (2003–2012)* **112**, F02S18.

Gunnell, Y. (2015) Ancient landforms in dynamic landscapes: inheritance, transience and congruence in Earth-surface systems. *Geomorphology* **233**, 1–4.

Haeberli, W. & Whiteman, C. (eds.) (2014) *Snow and Ice-Related Hazards, Risks and Disasters*. Amsterdam: Elsevier.

Hagerman, S., Dowlatabad, H., Chan, K.M.A. & Satterfield, T. (2010) Integrative propositions for adapting conservation policy to the impacts of climate change. *Global Environmental Change* **20**, 351–362.

Hahm, W.J., Riebe, C.S., Lukens, C.E. & Araki, S. (2014) Bedrock composition regulates mountain ecosystems and landscape evolution. *PNAS* **111**, 3338–3343.

Haslett, J. (1997) Mountain ecology: organism responses to environmental change, an introduction. *Global Ecology and Biogeography Letters* **6**, 3–6.

Hatcher, R.D. (2010) The Appalachian orogen: a brief summary. *Geological Society of America Memoirs* **206**, 1–19.

Hewitt, K. (1989) The altitudinal organisation of Karakoram geomorphic processes and depositional environments. *Zeitschrift für Geomorphologie, Supplementband* **76**, 9–32.

Hewitt, K. (2002) Introduction: landscape assemblages and transitions in cold regions. In: Hewitt, K., Byrne, M.-L., English, M. & Young, G. (eds.) *Landscapes of Transition. Landform Assemblages and Transformations in Cold Regions*. Dordrecht: Springer, pp. 1–8.

Hewitt, K. (2006) Disturbance regime landscapes: mountain drainage systems interrupted by large rockslides. *Progress in Physical Geography* **30**, 365–393.

Hewitt, K. (2014) *Glaciers of the Karakoram Himalaya. Glacial Environments, Processes, Hazards and Resources*. Dordrecht: Springer.

Hjort, J., Gordon, J.E., Gray, M. & Hunter, M.L. (2015) Why geodiversity matters in valuing nature's stage. *Conservation Biology* **29**, 630–639.

Holtmeier, F.-K. & Broll, G. (2012) Landform influences on treeline patchiness and dynamics in a changing climate. *Physical Geography* **33**, 403–437.

Hoorn, C.F.P., Wesselingh, H., ter Steege, M.A. et al. (2010) Amazonia through time: Andean uplift, climate change, landscape evolution, and biodiversity. *Science* **330**, 927–931.

Hoorn, C., Mosbrugger, V., Mulch, A. & Antonelli, A. (2013) Biodiversity from mountain building. *Nature Geoscience* **6**, 154.

Hose, T.A. (2016) Three centuries (1670–1970) of appreciating physical landscapes. In: Hose, T.A. (ed.) *Appreciating Physical Landscapes: Three Hundred Years of Geotourism*. Special Publications **417**. London: Geological Society, pp. 1–22.

Huber, U.M., Bugmann, H.K.M. & Reasoner, M.A. (eds.) (2005) *Global Change and Mountain Regions. An Overview of Current Knowledge*. Dordrecht: Springer.

Huff, W.D. & Owen, L.A. (2013) Volcanic landforms and hazards. In: Owen, L.A. (ed.) *Treatise on Geomorphology*, vol. **5**. San Diego, CA: Academic Press, pp. 150–192.

Huggel, C., Salzmann, N., Allen, S. et al. (2010) Recent and future warm extreme events and high-mountain slope stability. *Philosophical Transactions of the Royal Society of London A* **368**, 2435–2459.

Humboldt, A. & Bonpland, A. (1805) *Essai sur la Géographie des Plantes; Accompagné d'un Tableau Physique des Régions Équinoxiales*. Paris: Chez Levrault, Schoell.

Immerzeel, W.W., van Beek, L.P.H. & Bierkens M.F.P. (2010) Climate change will affect the Asian water towers. *Science* **328**, 1382–1385.

Ireton, S. & Schaumann, C. (eds.) (2012) *Heights of Reflection. Mountains in the German Imagination from the Middle Ages to the Twenty-First Century*. Rochester, NY: Camden House.

Iribarren Anacona, P., Mackintosh, A. & Norton, K.P. (2014) Hazardous processes and events from glacier and permafrost areas: lessons from the Chilean and Argentinean Andes. *Earth Surface Processes and Landforms* **40**, 2–21.

IUCN (2008) Resolutions and Recommendations adopted at the 4th IUCN World Conservation Congress. Resolution 4 040: Conservation of geodiversity and geological heritage. Gland: IUCN. Available from: https://portals.iucn.org/library/node/44190 (accessed August 30, 2017).

IUCN (2012) Resolutions and Recommendations, World Conservation Congress, Jeju, Republic of Korea, September 6–15, 2012, WCC-2012-Res-048-EN. Valuing and Conserving Geoheritage within the IUCN Programme 2013–2016. Gland: IUCN. Available from: https://portals.iucn.org/library/node/44015 (accessed August 30, 2017).

Ives, J.D., Messerli, B. & Spiess, E. (1997) Mountains of the world – a global priority. In: Messerli, B. & Ives, J.D. (eds.) *Mountains of the World. A Global Priority.* New York and London: Parthenon Publishing Group, pp. 1–15.

Jackson, M. (2015) Glaciers and climate change: narratives of ruined futures. *WIREs Climate Change* **6**, 479–492.

Jačková, K. & Romportl, D. (2008) The relationship between geodiversity and habitat richness in Šumava National Park and Křivoklátsko Pla (Czech Republic): a quantitative analytical approach. *Journal of Landscape Ecology* **1**, 23–38.

Jacobsen, D., Milner, A.M., Brown, L.E. & Dangles, O. (2012) Biodiversity under threat in glacier-fed river systems. *Nature Climate Change* **2**, 361–364.

Jeffers, E.S., Nogué, S. & Willis, K.J. (2015) The role of palaeoecological records in assessing ecosystem services. *Quaternary Science Reviews* **112**, 17–32.

Jonasson, C., Gordon, J.E., Kociánová, M. et al. (2005) Links between geodiversity and biodiversity in European mountains: case studies from Sweden, Scotland and the Czech Republic. In: Thompson, D.B.A., Galbraith, C.A. & Price, M.F. (eds.) *The Mountains of Europe: Conservation, Management and Initiatives.* Edinburgh: The Stationery Office, pp. 57–70.

Jurt, C., Brugger, J., Dunbar, K.W. et al. (2015) Cultural values of glaciers. In: Huggel, C., Carey, M., Clague, J.J. & Kääb, A. (eds.) *The High-Mountain Cryosphere. Environmental Changes and Human Risks.* Cambridge: Cambridge University Press, pp. 90–106.

Kamp, U. & Owen, L.A. (2013) Polygenetic landscapes. In: Owen, L.A. (ed.) *Treatise on Geomorphology*, vol. **5**. San Diego, CA: Academic Press, pp. 372–393.

Kapos, V., Rhind, J., Edwards, M. et al. (2000) Developing a map of the world's mountain forests. In: Price, M.F. & Butt, N. (eds.) *Forests in Sustainable Mountain Development.* Wallingford: CAB International, pp. 4–9.

Kargel, J.S., Leonard, G.J., Shugar, D.H. et al. (2016) Geomorphic and geologic controls of geohazards induced by Nepal's 2015 Gorkha earthquake. *Science* **351**, aac8353.

Kaser, G., Großhauser, M. & Marzeion, B. (2010) Contribution potential of glaciers to water availability in different climate regimes. *PNAS* **107**, 20 223–20 227.

Knight, J. & Harrison, S. (2013) The impacts of climate change on terrestrial Earth surface systems. *Nature Climate Change* **3**, 24–29.

Knight, J. & Harrison, S. (2014) Mountain glacial and paraglacial environments under global climate change: lessons from the past, future directions and policy implications. *Geografiska Annaler* **96A**, 245–264.

Körner, C., Ohsawa, M., Spehn, E. et al. (2005) Mountain systems. In: Hassan, R., Scholes, R. & Ash, N. (eds.) *Ecosystems and Human Well-Being: Current State and Trends*, vol. **1**. Washington, DC: Island Press, pp. 681–716.

Körner, C., Spehn, E.M. & Paulsen, J. (2011) A definition of mountains and their bioclimatic belts for global comparisons of biodiversity data. *Alpine Botany* **121**, 73–78.

Körner, C., Jetz, W., Paulsen, J., Payne, D., Rudmann-Maurer, K. & Spehn, E.M. (2016) A global inventory of mountains for bio-geographical applications. *Alpine Botany* **127**, 1–15.

Korup, O. & Dunning, S. (2015) Catastrophic mass wasting in high mountains. In: Huggel, C., Carey, M., Clague, J.J. & Kääb, A. (eds.) *The High-Mountain Cryosphere. Environmental Changes and Human Risks.* Cambridge: Cambridge University Press, pp. 127–146.

Kotarba, A. (1992) Natural environment and landform dynamics of the Tatra Mountains. *Mountain Research and Development* **12**, 105–129.

Kozłowska, A. & Rączkowska, Z. (2002) Vegetation as a tool in the characterization of geomorphological forms and processes: an example from the Abisko Mountains. *Geografiska Annaler* **84A**, 233–244.

Kozłowska, A., Rączkowska, Z. & Zagajewski, B. (2006) Links between vegetation and morphodynamics of high-mountain slopes in the Tatra Mountains. *Geographia Polonica* **79**, 27–39.

Kruckeberg, A.R. (2002) *Geology and Plant Life. The Effects of Landforms and Rock Types on Plants.* Seattle, WA and London: University of Washington Press.

La Sorte, F.A. & Jetz, W. (2010) Projected range contractions of montane biodiversity under global warming. *Proceedings of the Royal Society* **277B**, 3401–3410.

Lockwood, J.P. & Hazlett, R.W. (2010) *Volcanoes. Global Perspectives.* Chichester: Wiley-Blackwell.

Löffler, J., Anschlag, K., Baker, B. et al. (2011) Mountain ecosystem response to global change. *Erdkunde* **65**,189–213.

Lowe, J.J. & Walker, M.J.C. (2015) *Reconstructing Quaternary Environments*, 3rd edn. Abingdon: Routledge.

Macfarlane, R. (2003) *Mountains of the Mind. A History of Fascination.* London: Granta Books.

Mark, B.G., Baraer, M., Fernandez, A. et al. (2015) Glaciers as water resources. In: Huggel, C., Carey, M., Clague, J.J. & Kääb, A. (eds.) *The High-Mountain Cryosphere. Environmental Changes and Human Risks.* Cambridge: Cambridge University Press, pp. 184–203.

Masson, D.G., Watts, A.B., Gee, M.J.R. et al. (2002) Slope failures on the flanks of the western Canary Islands. *Earth Science Reviews* **57**, 1–35.

McGuire, B. (2015) Implications for hazard and risk of seismic and volcanic responses to climate change in the high-mountain cryosphere. In: Huggel, C., Carey, M., Clague, J.J. & Kääb, A. (eds.) *The High-Mountain Cryosphere. Environmental Changes and Human Risks.* Cambridge: Cambridge University Press, pp. 109–126.

Meybeck, M., Green, P. & Vörösmarty, C.J. (2001) A new typology for mountains and other relief classes: an application to global continental water resources and population distribution. *Mountain Research and Development* **21**, 34–45.

Migoń, P. (2016) Rediscovering geoheritage, reinventing geotourism: 200 years of experience from the Sudetes, Central Europe. In: Hose, T.A. (ed.) *Appreciating Physical Landscapes: Three Hundred Years of Geotourism.* Special Publications **417**. London: Geological Society, pp. 215–218.

Morrocco, S.M., Ballantyne, C.K., Gordon, J.E. & Thompson, D.B.A. (2016) Assessment of terrain sensitivity on high plateaux: a novel approach based on vegetation and substrate characteristics in the Scottish Highlands. *Plant Ecology and Diversity* **9**, 219–235.

Muir, J. (1901) *Our National Parks.* Boston, MA: Houghton, Mifflin & Co.

Milner, A.M., Brown, L.E. & Hannah, D.M. (2009) Hydroecological response of river systems to shrinking glaciers. *Hydrological Processes* **23**, 62–77.

Nash, D. (2013) Tectonic geomorphology of normal fault scarps. In: Owen, L.A. (ed.) *Treatise on Geomorphology*, vol. **5**. San Diego, CA: Academic Press, pp. 234–249.

Newsome, D. & Dowling, R.K. (2010) *Geotourism. The Tourism of Geology and Landscape.* Oxford: Goodfellow.

Nicolson, M.H. (1959) *Mountain Gloom and Mountain Glory. The Development of the Aesthetics of the Infinite.* Seattle, WA and London: University of Washington Press.

Ollier, C. (2004) The evolution of mountains on passive continental margins. In: Owens, P.N. & Slaymaker, O. (eds.) *Mountain Geomorphology.* London: Arnold, pp. 59–88.

Ollier, C. & Pain, C. (2000) *The Origin of Mountains.* London: Routledge.

Opedal, Ø.H., Armbruster, W.S. & Graae, B.J. (2015) Linking small-scale topography with microclimate, plant species diversity and intra-specific trait variation in an alpine landscape. *Plant Ecology & Diversity* **8**, 305–315.

Orlove, B.S., Wiegandt, E. & Luckman, B.H. (2008) The place of glaciers in natural and cultural landscapes. In: Orlove, B., Weigand, E. & Luckman, B.H. (eds.) *Darkening Peaks. Glacier Retreat, Science and Society.* Berkeley, CA: University of California Press, pp. 3–19.

Owen, L.A. (2004) Cenozoic evolution of global mountain systems. In: Owens, P.N. & Slaymaker, O. (eds.) *Mountain Geomorphology.* London: Arnold, pp. 33–58.

Owen, L.A. (2014) Himalayan landscapes of India. In: Kale, V.S. (ed.) *Landscapes and Landforms of India.* Dordrecht: Springer Science+Business Media, pp. 41–52.

Owens, P.N. & Slaymaker, O. (2004) An introduction to mountain geomorphology. In: Owens, P.N. & Slaymaker, O. (eds.) *Mountain Geomorphology.* London: Arnold, pp. 3–29.

Palacios, D. & Sánchez-Colomer, M.G. (1997) The distribution of high mountain vegetation in relation to snow cover: Peñalara, Spain. *Catena* **30**, 1–40.

Papale, P. (ed.) (2015) *Volcanic Hazards, Risks and Disasters.* Amsterdam: Elsevier.

Paul, F., Kääb, A., Maisch, M. et al. (2004) Rapid disintegration of Alpine glaciers observed with satellite data. *Geophysical Research Letters* **31**, L21402.

Peattie, R. (1936) *Mountain Geography. A Critique and Field Study.* Cambridge, MA: Harvard University Press.

Pepin, N., Bradley, R.S., Diaz, H.F. et al. (2015) Elevation-dependent warming in mountain regions of the world. *Nature Climate Change* **5**, 424–430.

Phillips, J.D. (2009) Changes, perturbations, and responses in geomorphic systems. *Progress in Physical Geography* **33**, 17–30.

Pressey, R.L., Cabeza, M., Watts, M.E. et al. (2007) Conservation planning in a changing world. *Trends in Ecology and Evolution* **22**, 583–592.

Price, M.F. (2015) *Mountains. A Very Short Introduction.* Oxford: Oxford University Press.

Prober, S.M. & Dunlop, M. (2011) Climate change: a cause for new biodiversity conservation objectives but let's not throw the baby out with the bathwater. *Ecological Restoration & Management* **12**, 2–3.

Purdie, H. (2013) Glacier retreat and tourism: insights from New Zealand. *Mountain Research and Development* **33**, 463–472.

Quincey, D. & Carrivick, J. (2015) Glacier floods. In: Huggel, C., Carey, M., Clague, J.J. & Kääb, A. (eds.) *The High-Mountain Cryosphere. Environmental Changes and Human Risks.* Cambridge: Cambridge University Press, pp. 204–226.

Rabatel, A., Francou, B., Soruco, A. et al. (2013) Current state of glaciers in the tropical Andes: a multi-century perspective on glacier evolution and climate change. *The Cryosphere* **7**, 81–102.

Rączkowski, W., Boltiziar, M. & Rączkowska, Z. (2015) Relief. In: Dąbrowska, K. & Guzik, M. (eds.) *Atlas of the Tatra Mountains. Abiotic Nature.* Zakopane: Tatrzański Park Narodowy.

Reynard, E. & Coratza, P. (2016) The importance of mountain geomorphosites for environmental education: examples from the Italian Dolomites and the Swiss Alps. *Acta Geographica Slovenica* **56**, 291–303.

Reynard, E., Hobléa, F., Cayla, N. & Gauchon, C. (2011) Iconic sites for Alpine geology and geomorphology. Rediscovering heritage? *Revue de Géographie Alpine* **99**(1–4), 188–202.

Reynard, E., Perret, A., Bussard, J. et al. (2016) Integrated approach for the inventory and management of geomorphological heritage at the regional scale. *Geoheritage* **8**, 43–60.

Ruskin, J. (1856) *Modern Painters. Volume IV. Containing Part V. Of Mountain Beauty.* London: Smith, Elder & Co.

Schnitzler, J., Barraclough, T.G., Boatwright, J.S. et al. (2011) Causes of plant diversification in the Cape biodiversity hotspot of South Africa. *Systematic Biology* **60**, 343–357.

Schroder, J.F. & Bishop, M.P. (2004) Mountain geomorphic systems. In: Bishop, M.P. & Schroder, J.F. (eds.) *Geographic Information Science and Mountain Geomorphology.* Berlin: Springer-Verlag, pp. 33–73.

Schroder, J.F. & Price, L.W. (2013) Origins of mountains. In: Price, M.F., Byers, A.C., Friend, D.A. et al. (eds.) *Mountain Geography. Physical and Human Dimensions.* Berkeley, CA: University of California Press, pp. 11–39.

Schulmann, K., Martínez Catalán, J.R., Lardeaux, J.M. et al. (eds.) (2014) The Variscan Orogeny: extent, timescale and the formation of the European crust. Special Publications 405. London: Geological Society.

Selby, M.J. (1985) *Earth's Changing Surface. An Introduction to Geomorphology.* Oxford: Clarendon Press.

Shea, J.M., Immerzeel, W.W., Wagnon, P. et al. (2015) Modelling glacier change in the Everest region, Nepal Himalaya. *The Cryosphere* **9**, 1105–1128.

Sigmundsson, F., Pinel, V., Lund, B. et al. (2010) Climate effects on volcanism: influence on magmatic systems of loading and unloading from ice mass variations, with examples from Iceland. *Philosophical Transactions of the Royal Society of London A* **368**, 2519–2534.

Silva, C., Kastenholz, E. & Abrantes, J.L. (2011) An overview of social and cultural meanings of mountains and implications on mountain destination marketing. *Journal of Tourism* **12**, 73–90.

Slaymaker, O. (1993) Cold mountains of western Canada. In: French, H.M. & Slaymaker, O. (eds.) *Canada's Cold Environments.* Montreal, QC: McGill-Queen's University Press, pp. 171–198.

Slaymaker, O. (2004) Mountain geomorphology. In: Goudie, A. (ed.) *Encyclopedia of Geomorphology*, vol. **2**. London: Routledge, pp. 701–703.

Slaymaker, O. & Embleton-Hamann, C. (2009) Mountains. In: Slaymaker, O., Spencer, T. & Embleton-Hamann, C. (eds.) *Geomorphology and Global Environmental Change.* Cambridge: Cambridge University Press, pp. 37–70.

Slaymaker, O., Spencer, T. & Dadson, S. (2009) Landscape and landscape-scale processes as the unfilled niche in the global environmental change debate: an introduction. In: Slaymaker, O., Spencer, T. & Embleton-Hamann, C. (eds.) *Geomorphology and Global Environmental Change.* Cambridge: Cambridge University Press, pp. 1–36.

Sorg, A., Bolch, T., Stoffel, M. et al. (2012) Climate change impacts on glaciers and runoff in Tien Shan (Central Asia). *Nature Climate Change* **2**, 725–731.

Soukupová, L., Kociánová, M., Jeník, J. & Sekyra, J. (1995) Arctic-alpine tundra in the Krkonoše, the Sudetes. *Opera Corcontica* **32**, 5–88.

Spehn, E.M., Rudmann-Maurer, K. & Körner, C. (2011) Mountain biodiversity. *Plant Ecology & Diversity* **4**, 301–302.

Stahr, A. & Langenscheidt, E. (2014) *Landforms of High Mountains.* Berlin and Heidelberg: Springer-Verlag.

Summerfield, M.A. (1991) *Global Geomorphology. An Introduction to the Study of Landforms.* Harlow: Longman Scientific & Technical.

Thomas, M.F. (2012) A geomorphological approach to geodiversity – its applications to geoconservation and geotourism. *Quaestiones Geographicae* **31**, 81–89.

Thouret, J.-C. (2004) Geomorphic processes and hazards on volcanic mountains. In: Owens, P.N. & Slaymaker, O. (eds.) *Mountain Geomorphology.* London: Arnold, 242–273.

Thouret, J.-C. (2010) Volcanic hazards and risks: a geomorphological perspective. In: Alcántara-Ayala, I. & Goudie, A.S. (eds.) *Geomorphological Hazards and Disaster Prevention.* Cambridge: Cambridge University Press, pp. 13–32.

Thorsteinsson, T., Jóhannesson, T. & Snorrason, Á. (2013) Glaciers and ice caps: vulnerable water resources in a warming climate. *Current Opinion in Environmental Sustainability* **5**, 590–598.

Tobias, M.C. & Drasdo, H. (eds.) (1980) *The Mountain Spirit.* London: Victor Gollancz.

Troll, C. (1972) Geoecology and the world-wide differentiation of high-mountain ecosystems. In: Troll, C. (ed.) *Geoecology of the High-Mountain Regions of Eurasia.* Mainz: Erdwissenschaftsliche Forschung der Akademie der Wissenschaften und der Literatur and Wiesbaden: Franz Steiner Verlag GMBH, pp. 1–16.

Tuffen, H. (2010) How will melting of ice affect volcanic hazards in the twenty-first century? *Philosophical Transactions of the Royal Society of London A* **368**, 2535–2558.

United Nations (2015) Transforming Our World: the 2030 Agenda for Sustainable Development. Resolution 70/1 Adopted by the General Assembly on September 25, 2015. Available from: https://sustainabledevelopment.un.org/post2015/transformingourworld (accessed August 30, 2017).

Vucetich, J.A., Bruskotter, J.T. & Nelson, M.P. (2015) Evaluating whether nature's intrinsic value is an axiom of or anathema to conservation. *Conservation Biology* **29**, 321–332.

Vuille, M., Francou, B., Wagnon, P. et al. (2008) Climate change and tropical Andean glaciers: past, present and future. *Earth-Science Reviews* **89**, 79–96.

Wang, S., He, Y. & Song, X. (2010) Impacts of climate warming on alpine glacier tourism and adaptive measures: a case study of Baishui Glacier No. 1 in Yulong Snow Mountain, southwestern China. *Journal of Earth Science* **21**, 166–178.

Welling, J.T., Árnason, Þ. & Ólafsdottír, R. (2015) Glacier tourism: a scoping review. *Tourism Geographies* **17**, 635–672.

Whipple, K.X. (2009) The influence of climate on the tectonic evolution of mountain belts. *Nature Geoscience* **2**, 97–104.

Woodcock, N. & Strachan, R. (2012) *Geological History of Britain and Ireland*, 2nd edn. Chichester: Wiley-Blackwell.

Xu, J., Grumbine, R.E., Shrestha, A. et al. (2009) The melting Himalayas: cascading effects of climate change on water, biodiversity, and livelihoods. *Conservation Biology* **23**, 520–530.

Zwoliński, Z. & Stachowiak, J. (2012) Geodiversity map of the Tatra National Park for geotourism. *Quaestiones Geographicae* **31**, 99–107.

11

Geodiversity Mapping in Alpine Areas

*Arie C. Seijmonsbergen[1], Matheus G.G. De Jong[2], Babs Hagendoorn[1], Johannes G.B. Oostermeijer[1]
and Kenneth F. Rijsdijk[1]*

[1] *Institute for Biodiversity and Ecosystem Dynamics, University of Amsterdam, Amsterdam, Netherlands*
[2] *Research Foundation for Alpine and Subalpine Environments, Amsterdam, Netherlands*

Abstract

Geodiversity mapping has become an established tool for assessing the value of the abiotic part of the landscape. We present two methods, which have been developed in the mountains of the State of Vorarlberg (Austria), but have a wider application in alpine areas. The first is a region-wide index-based mapping of Vorarlberg that is designed with the purpose of generating, in a preliminary evaluation, an inventory of clusters of high geodiversity. The second method comprises detailed geomorphological mapping at a local scale and supports landscape planning and management by identifying potential geoconservation sites. The latter approach also enables the evaluation of the relationship between geodiversity and biotopes.

We generated a geodiversity (index) map of Vorarlberg in a GIS showing the spatial distribution of a combination of six abiotic factors (subindices) expressed as numeric values in five classes. Both conventional and unconventional data sets were used, ranging from field-based geological maps to digital data sets acquired in airborne LiDAR surveys. In this approach, high geodiversity is found more often in areas of combined complex topography and varied geological substrata. The geodiversity map can be used to rapidly assess the occurrence of clusters of high geodiversity, which subsequently can be evaluated in detail using a landform-based approach. As an example of the latter, we present the case study of a small area near the village of Au in central Vorarlberg, Austria. Using a detailed area-covering polygon-based morphogenetic map as the basis for the assessment of the Au West area, the various classes of the legend were weighted and ranked in an automated GIS procedure with four factors: scientific relevance and frequency of occurrence (primary factors) and vulnerability and disturbance (secondary factors). Three levels of importance for geoconservation potential are differentiated (low, medium and high significance) and displayed in a map on which the highly ranked units are identified as potential sites for geoconservation. Comparison of the morphogenetic types and existing biotope data in the case-study area suggests that most biotopes occur together with specific morphogenetic types. It appears that the distinction between "wet" and "dry" mass-movement processes is an important factor, together with slope steepness and material properties, for effectively characterizing the natural biotopes.

Keywords: *geomorphology, biotope, Vorarlberg, geoconservation, geodiversity index, land-surface parameters, GIS*

11.1 Geodiversity Mapping

Alpine areas are among the world's most geologically diverse and complex, as a result of the combined action of endogenic and exogenic mountain-building processes. They have endured multiple glacial and interglacial periods in the Quaternary, and consequently exhibit a wealth of landform types. The present diversity of active and inactive landforms, as well as the intensity of geomorphological processes, reflects these continually changing environmental conditions. Spatial variations in topography, soil and parent material, climatic conditions and hydrology over time also contribute to the high geodiversity of mountains.

Geodiversity, or the diversity of the geosphere, was defined by Gray (2013) as "the natural range (diversity) of geological (rocks, minerals, fossils), geomorphological (landforms, topography, physical processes), soil and hydrological features. It includes their assemblages, structures, systems and contributions to landscapes."

Mountains, Climate and Biodiversity, First Edition. Edited by Carina Hoorn, Allison Perrigo and Alexandre Antonelli.
© 2018 John Wiley & Sons Ltd. Published 2018 by John Wiley & Sons Ltd.
Companion website: www.wiley.com\go\hoorn\mountains,climateandbiodiversity

Intimately linked to geodiversity is the concept of geoconservation, which can be defined as the active management of landscapes to conserve and enhance geological and geomorphological features, processes, sites and specimens. The landscape in this definition is seen as a part of our natural heritage that is locally threatened and worthy of conservation for future generations (Burek & Prosser 2008). Geodiversity and geoconservation revolve around the notion of the value of abiotic nature. As Gray (2013) puts it, "..things of value ought to be conserved if they are threatened."

"Value," or "quality," is a broad term, and various types of value can be used to assess geodiversity: intrinsic, cultural, esthetic, economic, functional (for both physical and ecological processes) and research/educational. At a practical level (e.g., in geoconservation projects or applied studies), the values selected depend on the aims of the study and determine to a large degree the approach to assessing geodiversity. Therefore, the first question to be answered in a geoconservation project is: Which value or values are to be "captured," and for what purpose? The subsequent question is: How should we map and evaluate geodiversity? And, at the next level, we ask questions such as: Which data are available? What is the scale of the area? Are there any legislative directives? How can geodiversity maps be used to support biodiversity studies?

Questions like these have been addressed in earlier studies, for example in a pan-tropical study by Parks & Mulligan (2010), in which the functional value of geodiversity takes a central position. These authors were interested in the geosphere as an environmental resource base for biodiversity, and thus aimed to build a bridge between geodiversity and biodiversity. According to their definition, resources are properties of relevance to the development and evolution of ecosystems: specifically, energy (temperature and solar radiation), water, space and nutrients. The environmental components include climate, topography, soils and geology, which are analyzed in terms of total resource availability and their spatial and temporal heterogeneity. For this purpose, in order to compare the abiotic components of a landscape with the patterns of species distribution, quantification of geodiversity is necessary. A compound geodiversity index (GI) is generated, which is then linked to biodiversity. It is noteworthy that Parks & Mulligan (2010) use a combination of expert-derived thematic maps (such as geological maps) and proxy data (such as parameters derived from digital elevation models, DEMs) in their assessment.

Expert-derived thematic maps are increasingly available in digital format, and allow for the calculation of geological, geomorphological and soil diversity for a predefined unit or a grid of cells. An example of this is found in a study by Serrano & Ruiz-Flaño (2007), who calculated GI as the number of different physical elements in a grid cell times a coefficient of roughness, divided by the natural logarithm of the surface area or unit in square kilometers. The index values calculated for the grid cells were then ranked into classes to produce a geodiversity map. This method was tested for an area in semi-arid central Spain by Serrano & Ruiz-Flaño (2007), and Hjort & Luoto (2010) applied the same method to a boreal landscape in Finland. The need for a flexible use of GIs – as opposed to the use of a standard index for all areas – is shown by a range of studies in different environments and across different spatial scales. For example, they have been used to identify geodiversity hot spots in the Iberian Peninsula (Benito-Calvo et al. 2009) and in a Romanian highland region (Năstase et al. 2012), for geotourism in a mountainous area of Poland (Zwoliński & Stachowiak 2012), for geo- and bioconservation in the Scottish Highlands (Brazier et al. 2012) and in a rural Spanish highland (Pellitero et al. 2011; Serrano & Ruiz-Flaño 2007) and on national and regional scales in Brazil (Pereira et al. 2013).

These examples demonstrate that the different aims and the particular spatial analysis scales of applied studies, together with data availability, determine which environmental variables or sub-indices of geodiversity are appropriate, in spite of a general desire for a universal GI (Kozłowski 2004).

Geodiversity assessment studies use sets of selected fit-for-purpose indicators, or correlations among the same. For example, geosite-based approaches exist that emphasize the scientific, educational, cultural, esthetic and ecological values of geodiversity, often using weighting and ranking procedures (Brilha 2016; Reynard et al. 2016). Knowledge of the geodiversity of a landscape may help conservationists to identify regions suitable for maintaining species diversity and to design efficient corridors (Anderson & Ferree 2010). Currently, there is a trend towards using geodiversity as a surrogate for biodiversity (Hjort et al. 2012; Anderson et al. 2015). One main reason for this is that in areas without sufficient species-occurrence data, a combination of specific environmental factors can be used to estimate biodiversity, for example by using biodiversity models (Hjort et al. 2012). Recently, Anderson et al. (2014) used DEM-derived measures of landscape diversity (i.e., the diversity of topography and range of elevation at a site and in its surrounding neighborhood) to map site resilience.

DEMs ensure a transparent and repeatable calculation of topographical variables such as slope angle, aspect, solar radiation, landscape diversity and openness, which are often referred to as "land surface parameters" (LSPs). Many of these – and other, more complex quantitative indicators – are commonly used in geomorphometry, a relatively young discipline of geomorphology, to model the variation of landforms over Earth's surface (Hengl & Reuter 2009). Another new development is the inclusion of high-resolution laser altimetry (LiDAR)-based LSPs in geodiversity research (Seijmonsbergen et al. 2014). Such advances in technological developments allow for the quantification of geodiversity patterns with input data at sub-meter resolution.

In light of these developments, this chapter describes two different workflows for geodiversity mapping: an index-based geodiversity method applied on a regional scale for the state of Vorarlberg in Austria (2601 km^2) and a combined local-scale expert-driven and GIS-supported method with a focus on geomorphological mapping, implemented west of the village of Au (2.15 km^2) in central Vorarlberg. The regional approach aims to identify region-wide clusters of high geodiversity, while the local-scale approach aims to assess geoconservation potential by applying a weighting and ranking scheme for geomorphosites. The International Association of Geomorphologists Working Group defines a geomorphosite as a portion of the geosphere that presents a site of particular importance in the comprehension of Earth's history (IAG Working Group 2005). We use the term in a broader sense to include all sites, including those of no particular importance. In Section 11.4.3, the results of a local-scale assessment of geomorphosites are used to compare potential geoconservation areas to an existing inventory of biotope areas that are already protected.

11.2 Geological and Geomorphological Overview of Vorarlberg

The landscape of Vorarlberg (Figure 11.1) varies widely over relatively short distances. The elevation ranges from 396 m at Lake Constance in the north-west to 3312 m at the Piz Buin summit in the south-east. It is beyond the scope of this chapter to present an in-depth review of the highly diverse geology and geomorphology of Vorarlberg. Hence, we will give only a brief overview, largely based on Friebe (2007), Oberhauser & Rataj (1998) and Seijmonsbergen et al. (2014).

Vorarlberg is situated on the northern side of the Alps, where the Western and Eastern Alps meet. The major tectonic units of the Eastern Alps are the Silvretta Nappe (crystalline) and the Lechtal Nappe (Northern Calcareous Alps), while the Western Alps are dominated by the Helveticum and Flysch nappes, all of which have a general strike from south-west to north-east due to the collision of the African and European plates. The sedimentary rock formations of these nappes comprise a large variety of clastics, carbonates and sulfates. To the north, the rock formations of Molasse consist essentially of erosional products, deposited from the emerging Alps. The Molasse zone was partially tectonically deformed in the final stage of the mountain-building process.

Epeirogenic uplift of the Alps started about 4 million years ago (Jäckli 1985). The resulting west–east elongated and ridge-like plateau was gradually dissected by rivers, the drainage pattern of which was controlled by the structural grain of the bedrock. Broadly speaking, fluvial activity alternated with glacier activity as a function of global climate changes, especially during the last 2 million years. The action of rivers and glaciers was, and is, primarily erosive, the products being largely transported to the foreland and beyond. A subordinate amount of material is, temporarily, stored in the Alps as fluvial, deltaic and lacustrine sediments in a variety of landforms (e.g., valley fills, terraces and alluvial fans). An even smaller amount is stored in glacial deposits, such as ablation and subglacial till. Mass movement is continually active in shaping and reshaping the landscape, driven by high-precipitation events and promoted by extremes in temperature. Periglacial processes and carbonate and sulfate karst also continue to play a role in the modeling of the mountain landscape.

The natural vegetation of the lowest regions of Vorarlberg (mainly the Rhine Valley) is a mosaic of deciduous forests, reed beds and peat bogs. Moving upslope, montane coniferous–deciduous forest (500–1200 m) is succeeded by pure sub-alpine coniferous forest (comprising spruce and mountain pine) up to a tree line that varies between 1800 and 2100 m. Above the tree line, various alpine dwarf shrubland, grassland and bog communities occur. The natural vegetation has been strongly affected by different types of land use: mainly intensive agriculture in the lower regions and low- to mid-intensity agriculture (dairy farming combined with mowing and haymaking), forestry and alpine ski slopes at higher elevations. In addition to the biodiverse natural biotopes, the long land-use history (logging followed by grazing and/or mowing) has resulted in a variety of semi-natural biotopes with high biodiversity. Even though there are still many threats, biodiversity conservation has been very successful in Vorarlberg. Traditional mowing and hay-making schemes have been maintained or reinstated to conserve and restore large stretches of species-rich meadows on nutrient-poor soils, for example.

Although not densely populated (378 000 inhabitants, mainly in the valleys of the Rhine and Ill rivers and on the

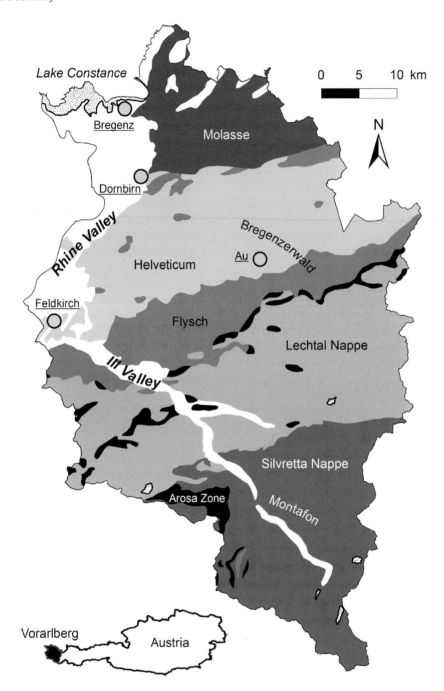

Figure 11.1 Outline of Vorarlberg, showing its main tectonic zones (in different shades of gray). The names of cities and villages are underlined. The lowest area is at Lake Constance, the highest mountains are in and around the Montafon region. The case study area is located west of the village of Au in the Bregenzerwald region. *Source:* Adapted from Friebe (2007).

border of Lake Constance), human influence on the landscape is evidenced by the dense buildings in large parts of the main valleys, the well-established infrastructure, the many reservoirs, the well-maintained natural and artificial drainage network and a highly managed forest, which functions as protection against natural hazards, especially in the steeper areas where ski resorts are also abundant.

11.3 Index-Based Geodiversity Mapping of Vorarlberg

11.3.1 General Overview

An index-based geodiversity map may be used to rapidly assess the occurrence of potential clusters of high geodiversity (hot spots) in a large area. Such a region-wide

Table 11.1 Metadata for the original input data sets used to calculate the geodiversity index (GI) for the Vorarlberg case study.

Data set name	Contains	Data type	Scale/cell size	No. of different features	No. of different variables or range	Data source
Tectonic map	Nappe units	Polygon	1:500 000	222	18 tectonic units	Oberhauser & Rataj (1998)
Geological map	Geological formations	Polygon	1:100 000	5417	109 geological formations	Geologische Bundesanstalt Wien (2007)
Fluesse50t (drainage)	Streams and large rivers	Line	1:50 000	11 460	1	Land Vorarlberg (2015)
Seen (lakes)	Lakes	Polygon	–	1048	1	Land Vorarlberg (2015)
LiDAR DEM	Elevation	Raster	5 m		391–3308 m	Land Vorarlberg (2015)
Slope map	Slope angle per cell	Raster	5 m		0–89.8°	Calculated from DEM
Solar radiation	Solar radiation per cell	Raster	25 m		$1186–1.8 \times 10^6\,WH/m^2$	Calculated from DEM

assessment can then be followed by a fine-scale evaluation of geodiversity of selected areas, using a landform-based approach. The geodiversity (index) map of Vorarlberg shows the spatial distribution of a combination of abiotic factors, expressed as numeric values for a predefined grid. We have selected tectonic diversity, geological diversity, drainage diversity, elevation diversity, slope diversity and solar radiation diversity as sub-indices. The mapping requires a number of steps, including: (i) data collection; (ii) grid definition; (iii) data pre-processing; (iv) calculation of the GI; (v) visualization of the index; and (vi) interpretation of the patterns. The main considerations behind – and the procedures for – each of these steps are addressed in the following sections.

11.3.2 Data Collection

Collecting digital map data sets for a given area usually produces a digital database of thematic maps, with differences in factors such as scale, quality, legend units, coverage and age. Tectonic and geological maps are available for Vorarlberg (Friebe 2007; Oberhauser & Rataj 1998; Geologische Bundesanstalt Wien 2007), but a state-wide inventory of soils and geomorphology does not exist. Detailed topographical information is available, however: we used a 1 m-resolution LiDAR DEM (Land Vorarlberg 2015), which was resampled in the pre-processing phase to 5 and 25 m (Table 11.1). Elevation diversity, slope-angle diversity and solar radiation diversity were derived from the DEM and were input when calculating the GI. The elevation and slope-angle maps are considered proxy data of the geomorphology at a regional scale, as they contain information on topographical variation and geometry. However, information on the genesis of landforms – glacigenic, fluvial or by

mass movements, for instance – is not included. Solar radiation (i.e., the amount of solar energy received over the period of a year) as a function of shielding and exposure induced by topography is included. Solar radiation indirectly controls soil-moisture conditions, and as such affects vegetation. Separate GIS vector layers for streams, large rivers and lakes are available in the Vorarlberg digital database and were used. Calculating the drainage network from the DEM was an alternative, but this was not employed here. The metadata of the input data sets are listed in Table 11.1, and include information on content, data type, original map scale or cell size, total number of features or range and data source.

11.3.3 Grid Definition

The optimal grid size for the generation of a GI map depends on the scale/cell size and on the quality and comprehensiveness of the input data. Hengl (2006) suggests using the scale of the input maps to calculate the coarsest (in our case, 1250 m), finest (50 m) and recommended (250 m) grid size; we decided to use a 1000 m grid in our approach, to reduce computation time. The Create-fishnet tool was used in ArcGIS 10.2 to prepare the analysis raster for the areal extent of Vorarlberg.

11.3.4 Data Pre-processing

For the region-wide analysis, we used six geodiversity sub-indices, which were subsequently classified into five classes of increasing geodiversity. The tectonic and geological maps were both rasterized at 1 m resolution using the polygon-to-raster tool in ArcGIS. The number of different tectonic and geological units per 1 × 1 km cell was counted in both maps with the variety option in the zonal-statistics

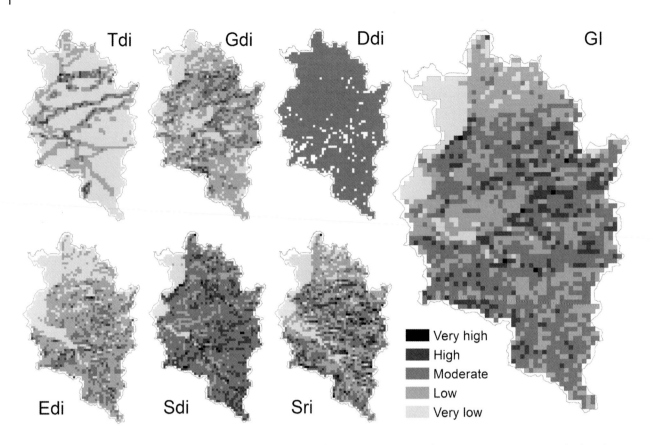

Figure 11.2 Geodiversity index (GI) map of Vorarlberg. Grid size in all maps is 1×1 km. Geodiversity for each cell follows the five-class scheme presented in the figure, ranging from very low to very high for each index, with the exception of the map for Ddi, where each cell is marked solely on presence (dark gray) or absence (white) of drainage. Tdi, tectonic diversity index; Gdi, geological diversity index; Ddi, drainage diversity index; Edi, elevation diversity index; Sdi, slope diversity index; Sri, solar radiation diversity index; GI, geodiversity index. See also Plate 22 in color plate section.

tool, generating a tectonic diversity index (Tdi) and a geological diversity index (Gdi). Given the 18 tectonic units in Vorarlberg (Table 11.1) and their generally large areal extent vis-à-vis the size of the grid cells, the Tdi range is relatively low – between 1 and 5 – and, therefore, reclassification was not deemed necessary. The maximum number of geological formations occurring in a single cell is 14; these were regrouped into five classes for the calculation of the index. The highest Gdi class (5) only covers 0.04% of the total area of Vorarlberg, while the most frequent class (2) covers 51.5% (Figure 11.2). The two hydrology data sets (see Table 11.1) were merged, and a buffer zone of 100 m was created around the drainage to reflect the potential influence of surface hydrology on the surrounding abiotic landscape. A spatial join of buffered streams to the grid in ArcGIS resulted in a drainage diversity index (Ddi) with either drainage present (1) or drainage absent (0). As there are rivers or lakes present in 93% of the cells, the Ddi sub-index is not very discriminatory. The 5 m DEM was used to construct an elevation diversity index (Edi) by calculating the standard deviation of the elevation per 1×1 km cell

with the zonal-statistics-as-table tool. The standard deviation ranges from 0.287 to 349 m; the results were reclassified into five equal size classes. In a similar way, the standard deviation of the slope angle was used to calculate a slope diversity index (Sdi). The standard deviation of the slope ranges from 0.55 to 21.8°. To reduce computational time, the 2004 solar radiation diversity index (Sri) was calculated at a lower resolution (25 m) with the ArcGIS area-solar-radiation tool. The values vary between 1186 and 1.78×10^6 WH/m^2. The standard deviation (range: 2234–440425 WH/m^2) of the solar radiation was divided into five equal size classes.

11.3.5 Calculating the Geodiversity Index

The final GI is the summation of the six diversity sub-indices, with a theoretical range of values between 5 and 26. The raster-calculator of ArcGIS was used to compute a GI raster with the following equation:

$$GI = Tdi + Gdi + Ddi + Edi + Sdi + Sri \qquad (11.1)$$

Table 11.2 Classification and areal coverage of the geodiversity classes in Vorarlberg.

Geodiversity class	Geodiversity variety per cell	Distribution (%): area/total area
1) Very low	5–8	9.27
2) Low	9–11	34.5
3) Moderate	12–14	45.7
4) High	15–17	10.2
5) Very high	18–20	0.25

In practice, the values range between 5 and 20. They were reclassified according to the class-versus-variety values set in Table 11.2.

Instead of using classified values per sub-index, the original values of the sub-indices can be used to calculate the GI, and, for example, weighting of the individual sub-indices may be considered. One should keep in mind, however, that thematic maps already contain expert-based information collected on a specific scale, and therefore they usually differ in content detail. Assigning weights to individual sub-indices should therefore be done with prudence, if at all, keeping the effect on the total GI in mind.

11.3.6 Visualization and Interpretation

The maps of the sub-indices and the resulting GI map of Vorarlberg are displayed in Figure 11.2. We emphasize that the spatial pattern can only be meaningfully interpreted with the nature and distribution pattern of the original input data in mind. The pattern of geodiversity (GI in Figure 11.2) appears to primarily reflect the diversity trends in geology (Gdi) and topography (Edi and Sdi). The very low and low geodiversity values (classes 1 and 2 in Table 11.2) of north and west Vorarlberg correspond to the valley floors of the Rhine and Ill valleys and to the relatively low topography of the Molasse zone, which are also characterized by low geological variety. High and very high GI values (classes 4 and 5 in Table 11.2) are found more frequently in areas of combined complex topography and varied geological substrata. Such areas occur in the Northern Calcareous Alps at relatively high altitudes. However, among the highest areas of Vorarlberg, the Montafon region (Figure 11.1) is characterized by low to moderate GI values, high GI values being rare as a function of the generally low to very low Gdi values. High GI scores also occur in the Hintere Bregenzerwald of central-eastern Vorarlberg, in which the Au West study area of geodiversity mapping (see Section 11.4) is situated.

It is important to realize that the function of the GI is to identify areas with potential clusters of high geodiversity; that is, those areas with a concentration of cells with high and very high GI scores.

11.4 Fine-Scale Geodiversity: The Au West Case Study

11.4.1 Area Description

The case-study area is located in the municipalities of Au and Mellau in the Bregenzerwald region of central-northern Vorarlberg (Figure 11.1), in the catchment of the Bregenzerache River and its tributary, the Argenbach. Locally, the Leuebach, Vorriedbach and Augenfallbach rivers flow to the east, into the Argenbach (Figure 11.3). The geological substratum is formed by rocks of the Helvetic Säntis Nappe. The area is the eastern part of a structurally controlled east–west-running topographic low, to the south of the Kanisfluh summit (2044 m) – with its impressive southern dip slope of Late Jurassic limestone – and to the north of the Gungern-Klippern (2066 m) mountain range – with steep face slopes of (in part) siliceous limestone, marls and shaley marls of early Cretaceous age. In contrast to the corresponding cell values of the index-based $1 \times 1\,\mathrm{km}^2$ geodiversity map (Figure 11.2), which range between 9 and 13 (low to moderate geodiversity), we show that clusters of high geoconservation potential are identified by applying the fine-scale, expert-based geomorphological approach. The discrepancy is largely due to classical geological maps focusing on bedrock geology and showing only poorly the intricacies of the glacial overburden.

11.4.2 Mapping and Assessment of Geoconservation Potential

Driven by the need for detailed information in support of landscape planning and management at the community and state levels in Vorarlberg, a method was developed by Seijmonsbergen et al. (2014) to evaluate the potential for conservation of small landforms and deposits and their surrounding areas. The method uses detailed area-covering geomorphological maps as the basis for assessment. The resulting polygon-based maps, including their attributes, depict morphogenetic classes, which are weighted and ranked in an automated GIS procedure with four factors: scientific relevance and frequency of occurrence (primary factors) and vulnerability and disturbance (secondary factors); see Table 11.3a. Three levels of importance for geoconservation potential are differentiated as rankings of low, medium and high significance (Table 11.3b). Landforms created by glacial

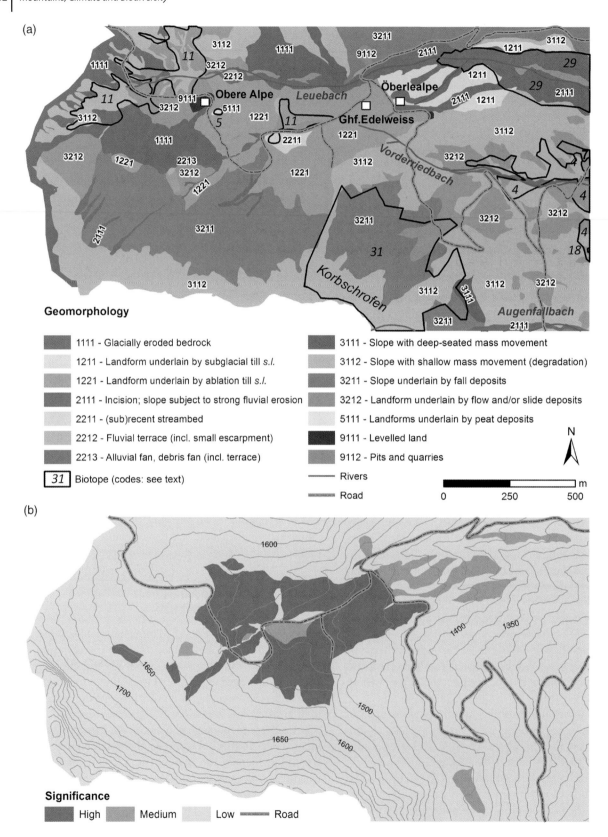

Figure 11.3 (a) Digital geomorphological map of the Au West area, displayed as a semi-transparent overlay on a DEM-derived hillshade map. The location of the biotope units is indicated. (b) Potential geoconservation map of the Au West area, shown on a background of 25 m-contour lines. Refer to Section 11.4.3 for further explanation. See also Plate 23 in color plate section.

Table 11.3 (a) The morphogenetic classification scheme used to generate the digital geomorphological map of Au West in Figure 11.3 (subset of the Seijmonsbergen et al. 2014 scheme). The standard weighting and ranking criteria (including the numerical values) used in the assessment of geoconservation potential are also shown for the morphogenetic types (see Box 11.1 for a detailed explanation). Columns include the following data for each process group: GIS code of types of geomorphosites, explanation of the GIS codes (landform and deposit types) and standard numerical values of weighting and ranking criteria used in GIS. (b) Classes of geoconservation potential. The numerical values are the summed scores of the weighting and ranking criteria for the morphogenetic types described in (a).

(a)

Process group	GIS code	Landforms and deposits	Scientific relevance (1–4–7)	Frequency of occurrence (1–3–5)	Vulnerability (1–2–3)	Disturbance (1–2–3)
Glacial	1111	Glacially eroded bedrock	1	1	1	3
	1211	Landform underlain by subglacial till *s.l.*	4	3	3	3
	1221	Landform underlain by ablation till *s.l.*	7	5	2	3
Fluvial	2111	Incision; slope subject to strong fluvial erosion	1	1	3	3
	2211	(Sub)recent streambed	1	5	3	3
	2212	Fluvial terrace (incl. small escarpment)	4	5	2	3
	2213	Alluvial fan, debris fan (incl. terrace)	4	3	1	3
Mass movement	3111	Slope with deep-seated mass movement	4	5	2	3
	3112	Slope with shallow mass movement (degradation)	1	1	3	3
	3211	Slope underlain by fall deposits	1	1	2	3
	3212	Landform underlain by flow and/or slide deposits	1	3	1	3
Organic	5111	Landforms underlain by peat deposits	7	5	3	3
Anthropogenic	9111	Levelled land	1	5	1	1
	9112	Pits and quarries	4	5	1	1

(b)

Geoconservation potential	Ranking summation
Low significance	4–8
Medium significance	9–13
High significance	14–18

Box 11.1 Brief description of the main morphogenetic mapping units in Table 11.3

Landforms and deposits of the glacial environment are formed or affected by the action of glaciers. Glacially eroded bedrock (GIS code 1111) typically has a fairly smooth surface and is relatively unaffected by postglacial processes such as weathering, erosion and mass movement. The most common deposits formed by a glacier are subglacial till *s.l.* (1211) and ablation till *s.l.* (1221). The former is an unsorted mixture of clays, silts, sands and larger rock fragments formed at the base of a glacier and compacted by the weight of it. Ablation till *s.l.* consists predominantly of sands, gravels and rock fragments, with a subordinate amount of clays and silts. It forms at the glacier margin, below the snowline, and is typically not very compacted. The main ecological difference between the types of tills is that soils developed in subglacial till are often swampy/wet and unstable due to the generally fine-grained and impermeable nature of the material, while soils in ablation till tend to be dry, permeable and stable. Fluvial activity is an important process in mountainous areas: incisions abound and many slopes are subject to strong fluvial erosion (2111). Locally, (sub)recent streambeds, fluvial terraces and small alluvial fans occur, composed of gravels and sands (2211–2213). Driven by gravity, mass movement is widespread in mountainous areas. With the disappearance of the glaciers at the end of the last ice age, many slopes, which were often oversteepened due to glacial erosion, became instable to depths tens to hundreds of meters below the land surface and collapsed (3111). Fluvial erosion also triggered – and continues to trigger – mass movements on the flanks of incisions. Surficial mass movements occur widely in mountains, resulting from the downslope movement of loose rock fragments and soil. Degradational slopes are widespread. In these, several meters of unstable wet soil have been removed to form shallow niches and gullies in an irregular topography (3112). Landforms underlain by flow and/or slide deposits (3212) form by the accumulation of relatively wet material flowing, sliding or creeping down the slopes, often coming to rest at their feet. Slopes underlain by fall deposits (3211) are the product of the accumulation of relatively dry debris which has fallen, rolled or glided from a steep cliff, usually in limestone or sandstone. Locally, peat or peaty deposits (5111) form, often in small depressions produced by mass movement and affected by high groundwater levels in a clay-rich environment. A comparison of these morphogenetic units and biotopes in the Au West area is presented in Section 11.4.3.

---

accumulation, glaciofluvial landforms, deep-seated mass movements, periglacial features and landforms of sulphate karst, among others, are ranked as highly significant in the standard application of the protocol (Table 11.3). Upgrading or downgrading the ranking of individual morphogenetic classes may be done by the application of additional assessment criteria (pedological, geological, etc.). Similarly, the standard ranking of units or associations of units may be changed on the basis of expert knowledge.

The digital geomorphological map of Au West (Figure 11.3) has been prepared in GIS using scanned and georeferenced published and unpublished classical geomorphological maps (Rupke et al. 1988) in combination with the GIS-based mapping techniques of Seijmonsbergen et al. (2014). The potential geoconservation map (Figure 11.3) was constructed by the application of the assessment protocol in Table 11.3.

11.4.3 Geomorphological Description

For a good understanding of the expert-derived geomorphological map and the subsequent assessment of geoconservation potential of the fine-scale landforms and deposits, the following explanatory notes are provided. The western part of the study area (i.e., the upper catchment of the Leuebach) can generally be described as a glacial niche with a headwall formed by the north- and east-exposed faces of Klippern. The floor of the glacial niche, which is covered in places by subglacial and ablation tills, extends from Obere Alpe (1593 m) to Öberlealpe (1473 m) along the structurally controlled low. Rockfall and debris-flow processes modify the original glacial shape of the niche (Figure 11.4).

The eastern part of the study area is dominated by erosion and mass movement, completely altering the original glacial landscape. Scree is actively produced and debris flows take place on and at the foot of the steep Korbschrofen cliff (Figure 11.3). Downslope accumulations of blocky debris are evidence of massive rockfall events. Further downslope, the irregular topography of niches and lobate features are indicative of sub-recent slope degradation, in which degradation and temporary accumulation have interacted – and are continuing to interact on a small scale – in a spasmodic way. Large slump-like features, now modified by surficial mass movement, are indicative of past deep-seated instability.

A more detailed survey of the glacial *s.l.* part of the study area reveals the presence of several types of moraines: morainic ridge, block-moraine cover and

(a)

(b)

Figure 11.4 (a) View to the south of part of the Obere Alpe glacial niche. Rockfall and debris-flow deposits cover the lower slopes of the headwall, which is part of the Gungern-Klippern mountain range. The hummocky topography around and to the left of Obere Alpe (center-right) is formed by an intricate pattern of glacially eroded bedrock and ablation tills. (b) View to the south-west of the glacial landscape in the central part of the study area. The east-sloping surface to the left of Ghf. Edelweiss is underlain by subglacial till (exposed in the flank of the Leuebach incision in the lower-central part of the photo). The steep Korbschrofen cliff (left) produces abundant scree. The central-right cliff, with an apron of talus, is the headwall of the Obere Alpe glacial niche. See also Plate 24 in color plate section.

subglacial till. A terrace with a relatively flat and gently east-sloping top, underlain by subglacial till, occurs east of Öberlealpe and Alpengasthof Edelweiss; it forms the easternmost remnant of the glacial landscape. Erosion and mass movement by the Leuebach and its small tributaries have created the terrace out of a landform that is thought to have extended much farther east. Going west to the low between the Kanisfluh and Klippern mountain ranges, the terrace grades to a surface covered by limestone erratics; remarkably, a tributary of the Leuebach flows in an unconfined manner through/below the blocky surface. Going farther west, the blocky cover becomes a hilly topography of morainic blocks, which merges with the actively building scree slope at the foot of the rock cliff of the south-eastern part of the Obere Alpe glacial niche. More morainic hills and ridges occur in the central part of the glacial niche. Their identification is not always straightforward: scree accumulation and debris-flow deposition have masked the original morphology. The subglacial till accumulation was deposited when the local Obere Alpe glacier extended in an easterly direction during the last glaciation. The cover and hills of blocks, as well as the other moraines, are interpreted to have been deposited by this local glacier while it was spasmodically receding to and in the Obere Alpe glacial niche during the final stages of deglaciation.

Table 11.4 Cross-tabulation (%) showing the occurrence of morphogenetic units in the seven main biotope units in the Au West area. For the locations of the biotope units, refer to Figure 11.3. The numerical coding of the biotope units refers to the classification presented in the municipality reports of Mellau and Au.

Biotope/morphogenetic type	Area (%)	Glacially eroded bedrock – 1111	Landform underlain by subglacial till s.l. – 1211	Landform underlain by ablation till s.l. – 1221	Incision: slope subject to strong fluvial erosion – 2111	(Sub)recent streambed – 2211	Slope with deep-seated mass movement 2212	Slope with shallow mass movement (degradation) – 3111	Slope underlain byfall deposits– 3112	Landform underlain by flow and/or slide deposits– 3211	Landforms underlain by peat deposits – 5111
Grassland and wet seepage forests – 04	9.5	–	–	–	8.2	–	–	54.6	–	37.1	–
Nutrient-poor meadows (complex) –18	<0.1	–	–	–	–	–	–	–	–	100	–
Ravine, slope and valley forests – 29	19.1	–	1.4	–	86.6	–	–	4.1	0.1	7.7	–
Montane and subalpine coniferous forests – 31	51.3	–	–	–	–	–	0.8	42.3	49.8	7.1	–
Mires and bogs –11	19.0	6.9	–	9.7	2.2	1.8	–	46.4	–	33.1	–
Lakes/ponds – 05	0.1	–	–	16.5	–	–	–	–	–	–	83.5
Reedland – 07	1.0	–	–	–	–	–	–	–	–	100	–

Source: Adapted from Gemeinde Au (2009) and Gemeinde Mellau (2014).

The ablation-till landforms (GIS code 1221), small fluvial terraces (2212) and a small pond bordered by wet and peaty deposits (5111) in the Obere Alpe glacial niche are highly significant in the standard application of the protocol. A deep-seated slump (3111) is also classified as highly significant. Although relatively well preserved and, unlike other deep-seated mass movement landforms in the eastern part of the study area, not modified by shallow mass movement, the latter is down-ranked to the level of medium significance in the protocol: it is not considered important enough within the context of the study area or that of the State of Vorarlberg to be high-ranking. No changes in the ranking of the other highly significant landforms are proposed. The final geoconservation potential is shown in Figure 11.3.

11.4.4 Comparing Geomorphological Diversity and Biotope Data

The geodiversity assessment of the Au West area paves the way for a comparison of geodiversity and biodiversity (or, more specifically, vegetation diversity): map inventories are available for both. A biotope inventory has been prepared for the State of Vorarlberg in two surveys, on two scales. The initial inventory was made between 1984 and 1989, and was revisited, updated and digitized between 2005 and 2009 (see also Broggi et al. 1991). Explanatory reports are available for all individual municipalities. Biotopes are areas with a combination of environmental conditions that host specific biological communities. In the inventory, only biotopes of conservation interest have been mapped, based on criteria such as naturalness, rarity, diversity, protected and/or endangered species and scientific relevance (Broggi et al. 1991). Using the output of the Au West analysis, a comparison of the existing biotopes on the one hand and (morphogenetic) geodiversity or geoconservation potential on the other is possible. For this study, we have done a cross-tabulation analysis of the relationship between the areal coverage of biotope units and the morphogenetic types occurring within these biotopes (Table 11.4).

In general, Table 11.4 shows that the cross-tabulation approach is promising for the quantification of geodiversity in terms of morphogenetic types within biotope units (Box 11.2). Biotope-management strategies may benefit from taking into account such relationships between biodiversity and geodiversity.

Visual inspection of the map of the geoconservation potentials and locations of biotopes (Figure 11.3) shows that there is little agreement between the grouping of morphogenetic types and biotope units. A possible explanation is that the landforms in which the biotopes predominantly occur are not unique but widespread in Vorarlberg, and, consequently, are ranked low for geoconservation purposes. In addition, such landforms are often intensely used and therefore not included in the official biotope inventory.

11.5 Conclusion

Various approaches to geodiversity mapping are possible. Common to all is that they take into account: (i) the specific goals and objectives of the study or project;

Box 11.2 Morphogenetic types and biotopes in the Au West area

Seven biotopes have been mapped within the 2.15 km^2 area covered by the morphogenetic map of Au West (see Figure 11.3 and Tables 11.3 and 11.4). The vegetation composition of these biotopes is described in detail in the municipality reports of Mellau (Gemeinde Mellau 2014) and Au (Gemeinde Au 2009) and in the associated ArcGIS database. The percentages of the surface area of the morphogenetic types falling within the boundaries of the seven biotopes are listed in Table 11.4. A few strong similarities are observed: for example, 92.1% of the biotope unit "montane and subalpine coniferous forests – 31" occurs in the morphogenetic types "slope underlain by fall deposits" (GIS code 3211) and "slope with shallow mass movement (degradation)" (3112). Such areas are characterized by steep slopes/cliffs, generally developed in resistant bedrock, with abundant talus production and large accumulations of coarse permeable debris at their feet (see Figure 11.4). Also, 86.6% of the biotope unit

"ravine, slope and valley forests – 29" occurs in the morphogenetic type "incision; slope subject to strong fluvial erosion" (2111). Streamlets, such as the Leuebach, are creating well-defined incisions characterized by steep slopes, dynamic geomorphological processes and ample water supply. These incisions are potential biotopes for the rich mixture of forests of river valleys and canyons.

The biotope units "grassland and wet seepage forests – 04," "nutrient-poor meadows (complex) – 18," "mires and bogs – 11," and "reedland – 07" seem to be present more frequently in units of the morphogenetic types "slope with shallow mass movement (degradation)" (3112) and "landform underlain by flow and/or slide deposits" (3212), although with only limited extent in the study area. These are generally low-angle slopes underlain by fine-grained weathering materials derived from marl and/or subglacial-till deposits that promote wet surface conditions due to poor drainage and the occurrence of spring zones.

(ii) the physiography of the area under consideration; (iii) the quality and scale of available data; and (iv) the required detail of mapping. We have presented the workflow of two methods, the first on a region-wide scale for inventory-mapping purposes, the second on a local scale in support of geoconservation management. The latter approach also enables the evaluation of the relationship between geodiversity and biotopes. The index-based approach is applied to the entire area of the State of Vorarlberg with the objective of identifying, in a first pass of evaluation, clusters of high geodiversity. The objective of the second approach is the identification of potential geoconservation sites (i.e., areas with high value in terms of morphogenetic type) on a local scale. Whereas the signal function is the strength of the region-wide method, the local method generates information for land-management purposes. The link between the two is clear: the quick index-based method of geodiversity assessment, although it depends on the quality, comprehensiveness, and scale of the input data, may reveal clusters of high geodiversity that subsequently can be evaluated more efficiently on a finer scale.

The index-based approach focuses on finding cells or groups of cells of geodiversity classes derived from a computer-based analysis of geological and DEM-derived data sets. On a local scale, the emphasis is on detailed geomorphological information. Expert-derived knowledge of landscape genesis is used to delineate and interpret landforms and processes, which are subsequently assessed for their degree of significance by the application of quantitative and qualitative evaluation criteria.

The fine-scale mapping approach has the potential to be supplemented by a semi-automated geomorphological mapping or classification technique, as was attempted by Anders et al. (2013) and Seijmonsbergen et al. (2014), giving the potential to transfer the method to other mountain areas (Anders et al. 2015). While the geodiversity results presented here are relevant and valid within the legislative boundaries of Vorarlberg, this does not rule out any validity within a larger frame of reference (e.g., the northern Alps). Wider application of the results, however, should be carried out with caution, taking into account possible supraregional differentiation. Alternative geosite-based approaches exist that emphasize not only the scientific, but also the educational, cultural, esthetic and ecological values of geodiversity (Brilha 2016; Reynard et al. 2016).

In all cases, geodiversity assessments provide a sound basis for geoconservation. Moreover, they are a relatively new way of analyzing the abiotic part of the environmental heterogeneity of mountains, to which biodiversity is generally thought to be closely linked (Stein et al. 2014). Comparison of morphogenetic types on a local scale with existing biotope data suggests that most biotopes occur in specific morphogenetic types. It appears that the distinction between "wet" and "dry" mass-movement processes is an important factor, together with slope steepness and material properties (e.g., coarse- vs. fine-grained), for effectively characterizing natural biotopes. Clearly, other factors, such as local land management (drainage, cattle grazing, mowing, etc.), are to be included in the analysis of biodiversity in mountain areas.

The potential of the concept of geodiversity and of DEM-derived approaches for biodiversity conservation has been demonstrated in recent studies, such as that of Anderson et al. (2015), in which eight case studies are described. These studies illustrate the mapping of ecological land units and land facets for the design of species corridors and the prioritization (from both a geodiversity and a biodiversity perspective) of conservation portfolios, all of which include particular aspects of geodiversity. Strategies for implementing geodiversity in conservation decision plans are addressed in the work of Comer et al. (2015), who emphasize the importance of support in law and policy. These recent developments acknowledge the importance of continuing research in geodiversity to our understanding of biodiversity.

Acknowledgments

The State of Vorarlberg (www.vorarlberg.at) kindly allowed us to use geographical data sets, air photo archives and LiDAR data, available from the VOGIS repository. The inatura Erlebnis Naturschau Gmbh (www.inatura.at) in Dornbirn is acknowledged for its financial and practical support of our research activities in Vorarlberg over many years. We thank the GIS-studio (www.gis-studio.nl) of the University of Amsterdam for computational support.

References

Anders, N.S., Seijmonsbergen, A.C. & Bouten, W. (2013) Geomorphological change detection using object-based feature extraction from multi-temporal LiDAR Data. *Geoscience and Remote Sensing* **10**, 1587–1591.

Anders, N.S., Seijmonsbergen, A.C. & Bouten, W. (2015) Rule set transferability for object-based feature extraction: an example for cirque mapping. *Photogrammetric Engineering and Remote Sensing* **81**, 507–514.

Anderson, M.G. & Ferree, C.E. (2010) Conserving the stage: climate change and the geophysical underpinnings of species diversity. *PLoS ONE* **5**, 1–10.

Anderson, M.G., Clark, M. & Sheldon, A.O. (2014) Estimating climate resilience for conservation across geophysical settings. *Conservation Biology* **28**, 959–970.

Anderson, M.G., Comer, P.J., Beier, P. et al. (2015) Case studies of conservation plans that incorporate geodiversity. *Conservation Biology* **29**, 680–691.

Benito-Calvo, A., Perez-Gonzalez, A., Magri, O. & Meza, P. (2009) Assessing regional geodiversity: the Iberian Peninsula. *Earth Surface Processes and Landforms* **34**, 1433–1445.

Brazier, V., Bruneau, P.M.C., Gordon, J.E. & Rennie, A.F. (2012) Making space for nature in a changing climate: the role of geodiversity in biodiversity conservation. *Scottish Geographical Journal* **128**, 211–233.

Brilha, J. (2016) Inventory and quantitative assessment of geosites and geodiversity sites: a review. *Geoheritage* **8**, 119–134.

Broggi, M.F., Grabherr, G., Alge, R. & Grabherr, G. (1991) *Biotope in Vorarlberg: Endbericht zum Biotopinventar Vorarlberg*. Natur und Landschaft in Vorarlberg 4. Dornbirn: Vorarlberger Verlagsanstalt GmbH.

Burek, C.V. & Prosser, C.D. (2008) The history of geoconservation: an introduction. In: Burek C.V. & Prosser C.D. (eds.) *The History of Geoconservation*. London: Geological Society, pp. 1–5.

Comer, P.J., Pressey, R.L., Hunter, M.L. et al. (2015) Incorporating geodiversity into conservation decisions. *Conservation Biology* **29**, 692–701.

Friebe, J.G. (2007) *Vorarlberg. Geologie der Österreichischen Bundesländer*. Wien: Geologische Bundesanstalt.

Geologische Bundesanstalt Wien (2007) Geologische Karte von Vorarlberg 1:100000.

Gemeinde Au (2009) Aktualisierung des Biotopinventars Vorarlberg, Gemeinde Au. Available from: http://www.vorarlberg.at/archiv/umweltschutz/biotopinventar/Au.pdf (accessed August 30, 2017).

Gemeinde Mellau (2014) Aktualisierung des Biotopinventars Vorarlberg, Gemeinde Mellau. Available from: http://www.vorarlberg.at/archiv/umweltschutz/biotopinventar/Mellau.pdf (accessed August 30, 2017).

Gray, M. (2013) *Geodiversity: Valuing and Conserving Abiotic Nature*, 2nd edn. Chichester: John Wiley & Sons.

Hengl, T. (2006) Finding the right pixel size. *Computers & Geosciences* **32**, 1283–1298.

Hengl, T. & Reuter, H.I. (eds.) (2009) *Geomorphometry: Concepts, Software, Applications*. Developments in Soil Science 33. Amsterdam: Elsevier.

Hjort, J. & Luoto, M. (2010) Geodiversity of high-latitude landscapes in northern Finland. *Geomorphology* **115**, 109–116.

Hjort, J., Heikkinen, R.K. & Luoto, M. (2012) Inclusion of explicit measures of geodiversity improve biodiversity models in a boreal landscape. *Biodiversity and Conservation* **21**, 3487–3506.

IAG Working Group (2005) Geomorphological Sites Research, Assessment and Improvement. Final Report 2001–2005 of the Working Group of the International Association of Geomorphologists. Available from: http://geoinfo.amu.edu.pl/iag/arch/04_Annual_report_2001-5.pdf (accessed August 30, 2017).

Jäckli, H. (1985) *Zeitmassstäbe der Erdgeschichte*. Basel, Boston, Stuttgart: Birkhäuser.

Kozłowski, S. (2004) Geodiversity. The concept and scope of geodiversity. *Przeglad Geologiczny* **52**, 833–883.

Land Vorarlberg (2015) Vorarlberg Atlas, Kartendienste des Vorarlberg Atlas. Available from: www.vorarlberg.at (accessed August 30, 2017).

Năstase, M., Cuculici, R., Muratoreanu, G. et al. (2012) A GIS-based assessment of geodiversity in the Maramures mountains natural park. A preliminary approach. European SCGIS Conference 2012. Available from: http://proc.scgis.scgisbg.org/Archive/S1-3_Nastase.pdf (accessed August 30, 2017).

Oberhauser, R. & Rataj, W. (1998) *Geologisch-Tektonische Übersichtkarte von Vorarlberg 1:200000 (mit Erläuterungen von R. Oberhauser)*. Wien: Geologische Bundesanstalt.

Parks, K.E. & Mulligan, M. (2010) On the relationship between a resource based measure of geodiversity and broad scale biodiversity patterns. *Biodiversity and Conservation* **19**, 2751–2766.

Pellitero, R., Gonzalez-Amuchastegui, M.J., Ruiz-Flaño, P. & Serrano, E. (2011) Geodiversity and geomorphosites assessment applied to a natural protected area: the Ebro and Rudron gorges Natural Park (Spain). *Geoheritage* **3**, 163–174.

Pereira, D.I., Pereira, P., Brilha, J. & Santos, L. (2013) Geodiversity Assessment of Paraná State (Brazil): An Innovative Approach. *Environmental Management* **52**, 541–552.

Reynard, E., Perret, A., Bussard, J. et al. (2016) Integrated approach for the inventory and management of geomorphological heritage at the regional scale. *Geoheritage* **8**, 43–60.

Rupke, J., Seijmonsbergen, A.C., van Westen, C.J. & Krieg, W. (1988) Erläuterungen zu den Geomorphologischen, Geotechnischen und Naturgefahrenkarten des Hinteren Bregenzerwaldes (Vlbg.Austria). Institut für Physische Geographie und Bodenkunde, Universität von Amsterdam/Vorarlberger Naturschau, Bregenz.

Seijmonsbergen, A.C., de Jong, M.G.G., de Graaff, L.W.S. & Anders, N.S. (2014) *Geodiversität von Vorarlberg und Liechtenstein*. Bern: Haupt Verlag.

Serrano, E. & Ruiz-Flaño, P. (2007) Geodiversity: a theoretical and applied concept. *Geographica Helvetica* **62**, 140–147.

Stein, A., Gerstner, K. & Kreft, H. (2014) Environmental heterogeneity as a universal driver of species richness across taxa, biomes and spatial scales. *Ecology Letters* **17**, 866–880.

Zwoliński, Z. & Stachowiak, J. (2012) Geodiversity map of the Tatra National park for geotourism. *Quastiones Geographicae* **31**, 99–106.

12

Historical Connectivity and Mountain Biodiversity

Suzette G.A. Flantua and Henry Hooghiemstra

Institute for Biodiversity and Ecosystem Dynamics, University of Amsterdam, Amsterdam, Netherlands

Abstract

The distribution of species in the present is just a snapshot in time after millions of years of change. Pleistocene climatic cycles, varying from 100 ky to sub-millennial scales, played an important role in shaping species' distributions. In mountains, these cycles pushed species rhythmically along the slopes, opening temporary dispersal pathways to new regions or dividing populations into isolated remnants. Here, we discuss the implications of the continuous connection and disconnection of populations in terms of their degree of historical connectivity. We introduce the "flickering connectivity system" and the "mountain fingerprint," which describe the temporal and spatial expression of habitat connectivity in tropical mountains as a biogeographical response to repetitive climate changes. We illustrate these concepts through paleotopographic reconstruction of alpine biome distributions in the northern Andes to exemplify the temporal and spatial dynamics that forced rapid evolutionary processes. Historical connectivity is shown to influence contemporary biodiversity on different spatial and temporal scales. We describe species richness and endemism as a consequence of historical connectivity, drawing parallels between oceanic islands and the sky islands in mountains. The continuously changing patterns of connectivity due to Pleistocene climate oscillations appear to have influenced diversification rates in evolutionarily recent time and are postulated to have been essential in shaping contemporary mountain diversity.

Keywords: *species distribution, Pleistocene, flickering connectivity system, mountain fingerprint, paleotopographic reconstruction, sky islands*

12.1   Introduction

Mountains are known for their high species richness. This richness can be the result of increases in diversity during older (mature) evolutionary radiations (e.g., Neogene) or of recent and rapid radiations (e.g., during the Plio–Pleistocene). Mountainous regions in Australia and South Africa are species-rich mostly due to mature radiations, as is typical of regions that have been climatically and geologically stable throughout the Neogene (Linder 2008). The high species richness of New Zealand, on the other hand, is the result of recent and rapid radiations (McGlone et al. 2001). Recent radiations are associated with the formation of new habitats as a result of recent (Pliocene) geotectonic activity (Linder 2008).

In the Andes, both older and very recent radiations have been identified. These mountains are extremely species-rich (45 000 plant species), with a high number of endemics (45%) (Myers et al. 2000). The geological uplift of the tropical Andes, initiated in the Oligocene, played a crucial role in the development of biodiversity in the Neotropics. As a result of this uplift, opportunities for colonization and diversification grew, leading to increased allopatric speciation owing to new high-elevation environments and greater topographic complexity. The uplift also initiated the opening of a north–south dispersal route for boreotropical lineages, and the rising orographic barrier even had an effect on the environmental conditions outside of the Andes (Chapter 8) (Luebert & Weigend 2014). Molecular phylogenies suggest that the different phases of the uplift of the Andes are echoed in the divergence times for many Andean plant groups (Luebert & Weigend 2014).

Recent diversification during the Pleistocene (the last 2.6 My) also shaped the region's species richness. For instance, most species-level variation in birds originates

Mountains, Climate and Biodiversity, First Edition. Edited by Carina Hoorn, Allison Perrigo and Alexandre Antonelli.

after the presumed onset of the Andean uplift (Smith et al. 2014), and several prominent species groups from the páramo (the alpine ecosystem of the Andes) are among the fastest radiations in the world (Sklenář et al. 2010; Madriñán et al. 2013; Nürk et al. 2013). Currently, the páramo is found in isolated formations on the mountain tops of Venezuela, Colombia, Ecuador and northern Peru, with the vegetation mainly comprising giant rosette plants, shrubs and grasses (Luteyn 1999). Triggers for rapid diversification during the Pleistocene are still debated, although the importance of new habitats (e.g., Hughes & Eastwood 2006), topography (e.g., Verboom et al. 2015) and insular environments, including islands, lakes, valleys and mountain tops (e.g., Sklenář et al. 2014), has been highlighted frequently. How Pleistocene climate change acted to further amplify diversification and create the optimum conditions for a montane "species pump" is still poorly understood.

The Pleistocene was a period of glacial–interglacial cycles that not only influenced species' distributions in temperate zones but also led to major environmental disruptions for Andean species. So how did these climatic conditions and the mountainous topography interact, leading to favorable conditions for evolutionary radiations and diversification? And why would the island-like settings of alpine ecosystems in sky-island formations have been important for explosive radiations? This chapter aims to review our current understanding of the past spatial dynamics of montane species and ecosystems, with a special focus on historical connectivity using the Andes as an example. We pay particular attention to the dynamic character of Pleistocene climates under which recent evolutionary diversifications occurred, while recognizing the importance of pre-Pleistocene diversification and intrinsic variables (see more in Chapter 18).

12.2 The Flickering Connectivity System

12.2.1 Introduction

There has been an increased appreciation of the importance of the historical processes that foster the spatial patterns of modern diversity (Chapter 14; Duncan et al. 2015), but few studies have explored the mechanistic processes from a dynamic landscape perspective. Mountain biodiversity most likely accumulated over several climatic cycles during the Pleistocene (see the simulations by Colwell & Rangel 2010), through the consecutive processes of connecting core areas, internal species mixing, disconnecting of refugia into cradles of endemism and local extinction (Rull 2005). Therefore, we propose here that rapid diversification and high

species richness can be explained and understood through a framework we call the "flickering connectivity system" (FCS). This concept describes the multi-episodic diversification of species during the last 2.6 My through cyclical phases of connectivity and isolation of mountain ecosystems. It builds on a "ghost from the past" that suggests that historical connectivity left a strong imprint on present diversity of species and that areas that have been more connected in the past have higher species richness today.

The FCS is a temporally and spatially dynamic system, where the temporal domain is set by the Pleistocene climate and the spatial terrain is the mountain topography (Figure 12.1a). The "flickering" refers to a system that moves back and forth between states with rapid or gradual change. In our case, the "flickering" refers to the climate oscillations driven by variations in Earth's orbit (Chapter 9; Hays et al. 1976).

The duration and frequency of climate states did not alternate equally throughout the Pleistocene. Only 15% of its duration consisted of warm and relatively moist interglacial conditions, as cool-to-cold and relatively dry climatic conditions (glacials) prevailed, while only ~10% involved extremely cold conditions, such as the Last Glacial Maximum (LGM). Most of the Pleistocene (75%) was characterized by slowly cooling to full glacial conditions (Figure 12.1a). During these conditions, millennial-scale climate variability of stadial (colder) to interstadial (mild) couplets occurred worldwide (Dansgaard et al. 1993; Labeyrie et al. 2007; Bogotá-Angel et al. 2011; Urrego et al. 2016). Before 1 Ma, glacial–interglacial cycles lasted ca. 40 ky, while after 1 Ma the cyclic duration slowed to ca. 100 ky. The latter cycle had a higher temperature amplitude and reached warmer and cooler temperatures than before (Figure 12.1a).

The flickering state of the Pleistocene climate caused substantial changes to plant distributions and was undoubtedly crucial in shaping contemporary biogeography (Comes & Kadereit 1998; Dynesius & Jansson 2000; Svenning et al. 2015). When rates of evolutionary adaptation are slow relative to environmental rates of change, niche conservatism prevails (Wiens 2004; Pyron et al. 2015), in which species tend to retain similar environmental niches over time and therefore follow temperature oscillations by shifting their geographic range. Species responded individualistically but similarly to changing environmental constraints, forming altitudinally restricted associations, variously called ecosystems (Golley 1993), biomes (Woodward & Cramer 1996), altitudinal belts (Tosi 1964; Van der Hammen 1974) or life zones (Holdridge et al. 1971). These associations were influenced by changing temperature, precipitation and atmospheric pCO_2 during the Pleistocene's alternating glacial and interglacial intervals.

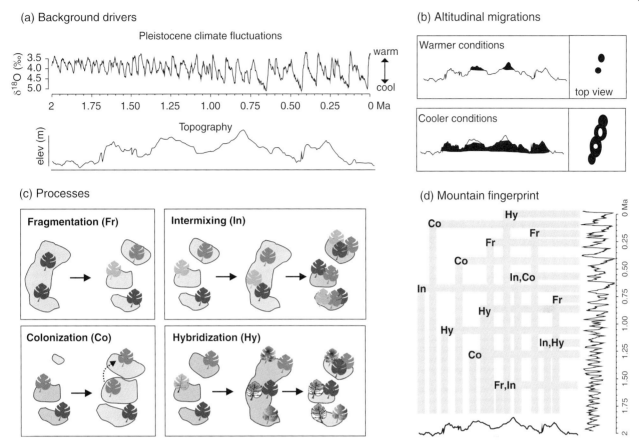

Figure 12.1 Conceptual framework of the flickering connectivity system (FCS). (a) The background drivers of speciation are the large Pleistocene climate fluctuations and highly complex montane topography. The $\delta^{18}O$ curve is based on composite stable oxygen isotope ratios from benthic foraminifera and is an indicator of global ice volume and temperature (Lisiecki & Raymo 2005). (b) Altitudinal migrations of hypothetical high-mountain biota, shown in a simple two-phase setting reflecting warmer and cooler conditions. (c) Schematic representation of the intrinsic processes of the FCS as a result of changes in connectivity: fragmentation (Fr), colonization (Co), intermixing (In) and hybridization (Hy). (d) The "mountain fingerprint" is defined by the interaction between climate and topography. It is a unique mountain identifier in which the processes of (a) occur in a spatially and temporally complex way, and therefore causes different timings and patterns of species diversification when comparing between mountains. See also Plate 25 in color plate section.

Subsequently, cold-adapted species and alpine ecosystems shifted downslope during glacial periods and upslope during interglacials (Figure 12.1b). Thus, the turbulent climate history of the Pleistocene was the pacemaker for repeated altitudinal migrations.

The mountain landscape imposed an additional set of opportunities and challenges for species' survival during the Pleistocene. The high variability of ridges, valleys, peaks and high-elevation plateaus at different elevations (Figure 12.2) creates, when pressured by climate fluctuations, a complex pattern of barriers and pathways for species to disperse around. Hence, past climate change and topographic complexity have been identified as the key extrinsic drivers of species diversification (Bouchenak-Khelladi et al. 2015), and together they form the domains of the biogeographical theatre of the Pleistocene. Therefore, the FCS builds upon these background drivers for the integration of spatial and temporal dynamics (Figure 12.1a).

12.2.2 The FCS in the Andes

Altitudinal migrations are often displayed in a simplified two-stage framework of an alpine system in a fragmented state or a highly connected state (Figure 12.1b) (e.g., Ramírez-Barahona & Eguiarte 2013). Here, we expand on this framework by showing that the climate fluctuations and the topography considered in the FCS have a very dynamic character, creating a more diverse story than is represented by a simple two-state model.

Changes in alpine ecosystems have been reconstructed using fossilized plant material such as pollen (e.g., Van der Hammen et al. 1973; Hooghiemstra & Van der Hammen 2004; McCormack et al. 2009; Brunschön &

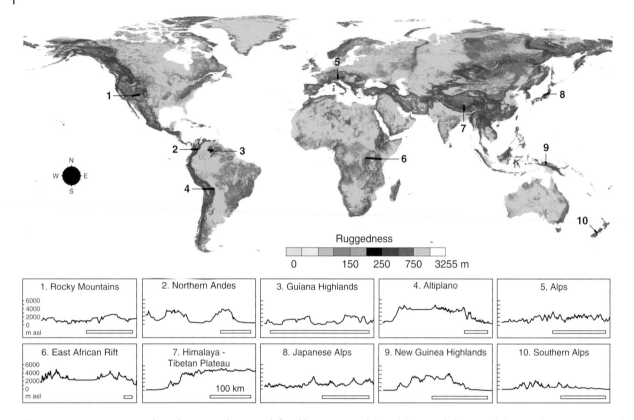

Figure 12.2 Mountain areas based on ruggedness, as defined by Körner et al. (2011) (maximal elevational distance between nine grid points of 30″ in 2.5′ pixel; for a 2.5′ pixel to be defined as "rugged" (i.e., mountainous), the difference between the lowest and highest of the nine points must exceed 200 m). Numbers indicate topographic profiles of selected mountain ranges around the world. The characteristic topography of a mountain directly relates to the potential impact and frequency of connectivity breaks caused by Pleistocene glacial cycles, and thus the expression of the flickering connectivity system. Bars below profiles indicate a 100 km distance proportional to the profile shown. *Source:* Adapted from Körner et al. (2011). Mountain ruggedness database downloaded from the Mountain Biodiversity Portal (www.mountainbiodiversity.org). See also Plate 3 in color plate section.

Behling 2010; Flantua et al. 2014; Hao et al. 2016). Paleogeographical reconstructions based on very long pollen records, such as cores from the mountain lakes Fúquene in Colombia (last 284 ky) (Bogotá-Angel et al. 2011; Groot et al. 2011) and Titicaca in Bolivia and Peru (last 370 ky) (Hanselman et al. 2011), demonstrate shifting biotic distributions during the Pleistocene. Combining paleotopographical and paleoenvironmental reconstructions using Geographic Information System (GIS) software can provide insights into past migration routes, degrees of isolation due to topographic restrictions and the persistence of core areas through time (Rull & Nogué 2007; Flantua et al. 2014; Flantua 2017).

Here, we use a long pollen record from Lake Fúquene in Colombia (Fúquene-9C) to show the spatial complexity of the repeated connection and isolation of the páramo during the past ca. 280 ky (Figure 12.3). The páramo consists of an island-like biome located on the highest mountaintops of the northern Andes; it is the most species-rich tropical alpine ecosystem (Sklenář

et al. 2014), and is considered to be a relatively young ecosystem (Plio–Pleistocene), in which numerous recent radiations have been identified (Hughes & Eastwood 2006). The time period of the last ca. 280 ky includes two full glacial–interglacial cycles and is representative of the temporal dynamics of the last million years, with a strong glacial–interglacial amplitude and a dominant 100 ky rhythm. The lower limit of the páramo, the upper forest line (UFL: the maximum elevation at which continuous forest occurs; Bakker et al. 2008), shifted altitudinally over a maximum interval of 1500 m (Van der Hammen et al. 1973; Hooghiemstra 1984; Groot et al. 2011), creating a series of very different spatial configurations of páramo distribution in the past (Figure 12.3).

The cooler periods (low UFL) alternated variably with warmer periods (high UFL) during the Pleistocene (Figure 12.3a–m). As a result, some spatial configurations occurred several times over a longer period while others occurred within a relatively short period

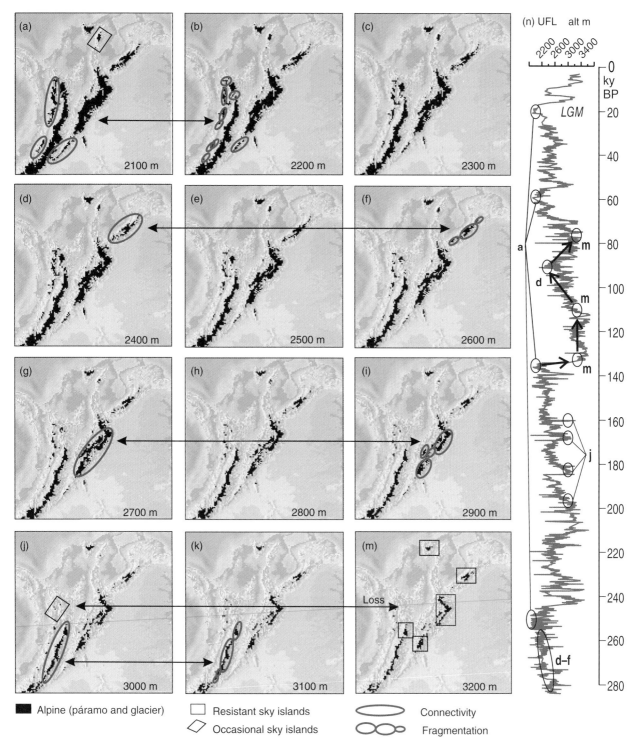

Figure 12.3 (a–m) Spatial reconstructions of tropical alpine systems (páramo and glaciers; black) in the northern Andes during the last 280 ky, showing the upper forest line (UFL) moving between elevations of 2100 and 3200 m. Each map represents a simplified reconstruction of the distribution of the alpine Andean ecosystem (the páramo) using a digital elevation model. (n) Estimated elevations of the UFL are inferred from the Fúquene-9C pollen record (Bogotá-Angel et al. 2011; Groot et al. 2011). Letters correspond to the maps. Low UFL reflects cooler periods, such as the Last Glacial Maximum (LGM), while a higher UFL reflects warmer periods (interglacial conditions, such as the present). Different regions experience alpine system connectivity and fragmentation at different moments in time. Some páramo areas persist continuously (resistant sky islands), while others appear and disappear (occasional sky islands). See also Plate 26 in color plate section.

(Figure 12.3n). Some transitions between configurations appear gradual (e.g., d–f ~260 ka in Figure 12.3n), while others are rapid (e.g., the progression a–m–d–m between 140 and 80 ka in Figure 12.3n). The contemporary páramo distribution lies above 3200 m. As can be seen from Figure 12.3, the present interglacial conditions are atypical compared to most of the Pleistocene, and extreme cold events, such as the LGM, occurred several times.

The topographic characteristics of the northern Andes allow the formation of several large sky islands and archipelagos (clusters of relatively small islands). "Sky islands" refers to a range of isolated mountain peaks separated by valleys and surrounded by a "sea" of hostile environment consisting of low-elevation habitat (Warschall 1994). By analogy with oceanic islands, the isolation of the alpine ecosystems facilitated the divergence between montane floras and faunas, creating isolated cradles of evolution.

However, sky islands such as the páramo alternated between isolation (present-day conditions) and the formation of connected islands or archipelagos with increased surface area. Lowering of the UFL facilitated páramo connectivity through the provision of ample surfaces at mid and high elevations (Flantua et al. 2014; Elsen & Tingley 2015). For instance, the extent of páramo habitat was at least three times larger during the LGM than it is under today's interglacial conditions (Hooghiemstra & Van der Hammen 2004). Thus, the cyclic climate fluctuations caused phases with significant increases in surface area and created massive opportunities for historical connectivity.

12.2.3 Why is Historical Connectivity Relevant?

Until now, most research emphasis has been put on phases of isolation, locations of refugia and the identification of the role of biogeographical barriers on patterns of genetic differentiation. However, an increasing number of studies support the idea that connectivity may have contributed more than just vicariant barriers to modern species richness and genetic variation (Rull & Nogué 2007; Edwards et al. 2012; Fjeldså et al. 2012; Smith et al. 2014; Duncan et al. 2015; Cadena et al. 2016; Kolář et al. 2016). In other words, diversification due to processes related to connectivity (e.g., colonization and episodic dispersal) could have had a bigger impact on mountain biodiversity than the isolation of populations. Here, we review how different processes intrinsic to the FCS triggered species diversification during the Pleistocene, discussing the consequences of increased and reduced connectivity and the triggers related to connectivity that stimulated diversification and species richness build-up.

Signatures of past connectivity can be identified on short (decadal) time scales and long (millennial) time scales. For instance, patches of semi-natural grasslands that are currently in an isolated state but which were connected 50–100 years ago show a higher species richness than those patches that had a lower or no historical connectivity (Lindborg & Eriksson 2004). This legacy of historical connectivity has been observed on a spatial scale of only a few kilometers for species richness (Lindborg & Eriksson 2004) and genetic diversity (Ewers et al. 2013; Münzbergová et al. 2013), but on a global scale for freshwater fish biodiversity (Dias et al. 2014) and angiosperm diversity on islands (Weigelt et al. 2016). In the global cases, it is shown that the degree of historical connectivity thousands of years ago (LGM conditions) is still reflected in contemporary patterns of biodiversity. Hence, a key mechanism in explaining contemporary spatial patterns of species richness is the degree of past connectivity.

Evidence of historical connectivity is also scattered throughout the mountainous landscape. Previous states of mountain biome connectivity are evidenced by shared species in adjacent mountain regions, such as the highest part of the páramo (superpáramo) (Luteyn 1999), that are currently highly isolated but still maintain a trans-Andean species distribution (Sklenář & Balslev 2005). Likewise, contemporary disjunct populations in neighboring mountainous regions can have more genetic resemblance than those separated by elevation. This trans-mountain pattern is common and has been observed for butterflies (Hall 2005), mammals (Patterson et al. 2012) and birds (Arctander & Fjeldså 1994; Fjeldså et al. 2012).

12.3 Components of the FCS

12.3.1 Processes and the Mountain Fingerprint

There are four processes inherent to the FCS that relate to the degree of connectivity: fragmentation, colonization (dispersal), intermixing and hybridization (Figure 12.1c). Each process plays a different and complementary role in maintaining richness or stimulating species diversification. They occur in a spatially and temporally dynamic manner, influenced by the local topography and the regional impact of climate fluctuations, and come together in what we call the "fingerprint" of a mountain (Figure 12.1d): the interaction between climate and topography, defining where and when each process of the FCS occurs. The mountain fingerprint can be considered to be a unique identifier, as mountains have different topographic profiles (Figure 12.2) and

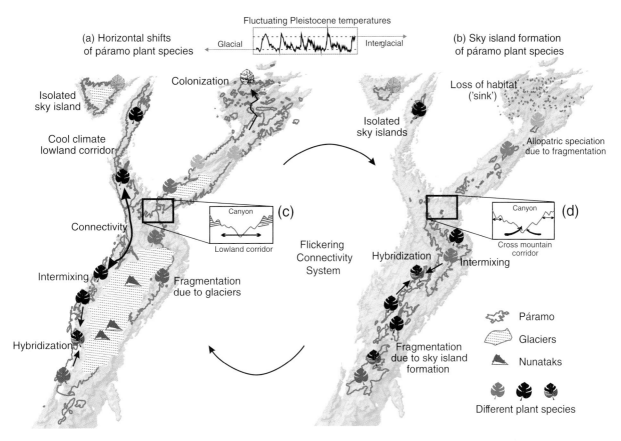

Figure 12.4 Spatial representation of the four intrinsic processes of the FCS in the Eastern Cordillera of the Colombian Andes. The potential distribution of páramo is shown during (a) cooler and (b) warmer conditions. The figure shows how different processes can occur at different locations throughout a mountain system, and as a result cause a spatially complex biogeographic pattern. The many possible intermediate configurations are shown in Figure 12.3. (c) Cool climate corridor of alpine species through a mid-elevation or lowland canyon. Glaciers are seen on the mountain tops. The arrow indicates the direction of connectivity. (d) Cross-mountain corridor between populations on either side of a mountain. Alpine species are restricted here to high elevation, and connectivity is reduced. See also Plate 27 in color plate section.

experience different degrees of climate fluctuations, and as a result the FCS is expressed in a distinctive manner for each mountain system. Figure 12.4 exemplifies the four processes of the FCS in the Eastern Cordillera of Colombia in order to highlight the mountain fingerprint in two different climate settings.

In this section, we discuss each of the different FCS processes and their implications for mountain biodiversity, looking particularly at the Andean páramo.

12.3.2 Fragmentation

Fragmentation of species distributions in mountain systems can occur in different ways. The many river valleys, intra-mountain valleys and high-elevation ridgelines form a labyrinth of potential hurdles and pathways for species to navigate. This complexity influences species

diversity. For instance, canyons influenced Andean bird (e.g., Weir 2009) and frog (e.g., Muñoz-Ortiz et al. 2015) diversity by imposing a physical barrier for the species to overcome. Similarly, mountain ridges, such as the dry forest patches in the Central Andes, capture species within geographically isolated patches, forming long-lasting isolated populations at high elevations (Pennington et al. 2010; Särkinen et al. 2012).

An additional mechanism of fragmentation in mountain ecosystems, which is often overlooked in the literature, is the glacial extent of ice caps on mountain tops and ridges during glacial periods (Osborne & Benton 1996). Mountainside populations may have differentiated due to the lack of cross-mountain gene flow, further stimulating intra-mountain endemism and diversification. For instance, periods with extensive ice caps have been linked to bird diversification events in the Southern

Alps of New Zealand (Weir et al. 2016) and to mammal diversification in various places (Chapter 14), showing the effect of recurring glacial fragmentation during the Pleistocene. Diversification rates of New Zealand birds increased fivefold due to the cyclical character of the Pleistocene climate – a pattern also observed for Neotropical birds (Weir 2006).

Small populations could have persisted on nunataks within ice shields, further contributing to the genetic variation of a region. Evidence for in situ survival exists for species in the European Alps (Stehlik et al. 2001; Schneeweiss & Schönswetter 2011). This means that allopatric speciation in alpine ecosystems such as the páramo probably occurred during both glacial (cross-mountain isolation) and interglacial (island top isolation) periods. This dual role of Andean topography, summarized in Figure 12.4, likely was a doubly reinforcing factor of Pleistocene páramo species radiations.

Restricted gene flow due to topographic barriers led to reproductive isolation and genetic drift. As a result, geographically isolated allopatric populations acquired unique and high genetic variation among populations, giving rise to high levels of endemism (Steinbauer et al. 2016). For example, refugia identified in the Alps show patterns consistent with areas known for their high levels of endemism and disjunct plant species (Schönswetter et al. 2005). In contrast to the Andean páramo, which expanded during glacial episodes (Figure 12.3), European mountain plant habitats were pushed into restricted peripheral sky islands towards the border of the European Alps (Schönswetter et al. 2005). Without a doubt, glacial refugia in the Alps played a key role in shaping biogeographical patterns of montane species endemism (Chapter 27) (Tribsch & Schönswetter 2003; Tribsch 2004).

The isolation of populations on multiple mountain tops, forming sky islands for short or extended periods during the Pleistocene, stimulated parallel radiations, population differentiation and inter-mountain endemism. A fragmented landscape is a key driver of radiations (Qian & Ricklefs 2000; Hughes & Eastwood 2006). Unsurprisingly, radiations of many plant lineages have occurred in topographically heterogeneous regions, such as the European Alps (*Gentiana*, *Globularia* and *Soldanella*: Kadereit et al. 2004), the Andes (*Lupinus*: Hughes & Eastwood 2006; *Campanulaceae*: Lagomarsino et al. 2016), the Rocky Mountains (*Penstemon*: Wolfe et al. 2006), the Himalaya (*Saussurea*: Wang et al. 2009) and the Drakensberg (*Macowania*: Bentley et al. 2014). However, fragmentation and isolation alone are not enough to explain the evolutionary dynamics during the Pleistocene or the contemporary high species richness (Benham & Witt 2016).

12.3.3 Colonization

New habitat or space becoming available for colonization (the establishment after dispersal) is a key precursor of recent and rapid radiations (Linder 2008), and is therefore an important process of the FCS. New habitat can appear due to geologically recent mountain uplift and climate fluctuations, which facilitates (re)colonization. Due to the periodic nature of Pleistocene climate variations, colonization and subsequent population expansions occurred as episodic or cyclical events. Temporary connectivity facilitated colonization of new areas, and allopatry subsequently further induced differentiation (Smith et al. 2014; Cadena et al. 2016). Species may also respond to new habitat with increased diversification rates, as has been observed for the high rates of radiation in Ericaceae (Luteyn 2002). Thus, colonization played a significant role in stimulating diversification (Pennington & Dick 2004).

Just as Weir et al. (2016) were able to relate cold conditions and glacial fragmentation to diversification events in birds, so colonization events and diversification can be linked to climate conditions becoming optimal for connectivity, as shown by Collevatti et al. (2015) in the Brazilian highlands. Also, cross-Andean movements of bird taxa were facilitated by shifting climatic conditions (Figure 12.4d), and as a result each episodic dispersal across the Andes caused the formation of independent lineages on either side of the mountains (Miller et al. 2008). Similarly, Collevatti et al. (2015) observed an increase in the number of plant lineages within the short time frame of the Pleistocene due to connectivity driven colonization. Thus, climate change was a driver of species diversification not only because it led to habitat fragmentation, but also because it connected habitats and created new habitats to be colonized.

Interestingly, mountains that are old (e.g., Cape Fold Mountains, South Africa and the Great Dividing Range, Australia) display only a few or no recent and rapid radiations (Linder 2008). Due to the fact that there have been no active geological processes during geologically recent times to create new topographies, and as the topography itself has become eroded, the mountain fingerprint (Figure 12.1d) is "flattened," and fluctuating climate conditions lack the topographic complexity pairing needed to stimulate radiations through new habitat and colonization.

New habitats are important for radiations but are not sufficient preconditions on their own (Linder 2008). Even if new habitat is potentially available through physical connectivity, differences in dispersal ability result in different biogeographic patterns among species (Dobrovolski et al. 2012; Papadopoulou & Knowles 2015a). Such differences, the degree of connectivity and the directionality of patterns of dispersal all contribute to the complex patterns of beta diversity and phylobetadiversity (phylogenetic relatedness across space) (Graham & Fine 2008). Additionally, the randomness of long-distance dispersal events, independent of propagule type, adds to the intricacy of the biogeographical picture researchers try to assemble (Pennington et al. 2006). What is clear is that within the FCS, a vast number of colonization opportunities arise, fuelling diversification.

12.3.4 Intermixing

Many sky islands formed, separated and reconnected through repeated pulses of expansion and contraction. This spatial effect is comparable to the shifting areas and connectivities of oceanic islands driven by sea level change, emphasizing the high resemblance between oceanic islands and sky islands. Interestingly, both systems are known to show rapid and recent radiations (see overview in Hughes & Atchison 2015). Furthermore, both boundaries are sharply defined, causing dispersal limitations, and are temperature-dependent: higher temperatures during Pleistocene interglacial periods correspond to a high sea level and high UFL position. During glacials, sea levels were roughly 120 m lower than they are at present (e.g., Rijsdijk et al. 2014; Spratt & Lisiecki 2016), while the UFL was between ca. 900 m (New Guinea) and ca. 1800 m (East Africa) lower (Coetzee 1967; Flenley 1979). These glacial–interglacial shifts in elevation and isolation have been widely identified as key features of the biogeographical history of insular systems, but their relationship to modern biodiversity and evolutionary radiations remains incompletely understood.

The isolation and fusion of oceanic islands caused multiple episodes of allopatric speciation and intermixing of species composition (Ali & Aitchison 2014; Gillespie & Roderick 2014). Intermixing (Figure 12.1c) facilitated by increased connectivity after isolation has been shown to be an important predictor of present diversity patterns on islands. Historical connectivity between islands facilitated intermixing of once single-island endemics into archipelago endemics, increasing the overall species richness of connected islands (Weigelt et al. 2016). Especially in clusters of islands, species diversity benefitted from parallel radiations on isolated islands during interglacial isolation, while subsequent reconnection during glacial periods had intermixing and possibly also hybridization as a consequence (Ali & Aitchison 2014). Thus, previously connected islands are characterized by higher total species richness, lower single-island endemism and higher levels of compositional similarity.

The described effect of historical connectivity on endemism has been observed for freshwater fish diversity (Dias et al. 2014) and sky islands. An example of the latter is seen in the Andes: in Ecuador, the downward shift during glacial periods connected most superpáramo, possibly causing the relatively low number of single-sky island endemics and the high number of trans-Andean species (Sklenář & Balslev 2005). In the same shift in Colombia, topographic barriers inhibited an increase of connectivity between sky islands, and as a consequence a high number of single-sky island species can now be observed. The lack of historical connectivity in Colombian superpáramo is confirmed by the extremely low genetic diversity within populations and the high genetic differentiation among them (Sklenář & Balslev 2005).

The spatial configuration of islands in the past contributes to a higher species richness today than would be expected from current settings. Small islands located between two larger islands or that were once part of a larger connected island can still display the legacy of the historically connected state, and thus disobey the expected species–area relationship. As a result of historical connectivity, small sky islands can be more species-rich than their larger counterparts, which can confound relationships between endemism and montane land area that do not account for historical effects (Adams 1985; Sklenář & Balslev 2005).

12.3.5 Hybridization

Historical connectivity increases plant species richness not only through intermixing, but also through hybridization after reconnection (Ali & Aitchison 2014; Gillespie & Roderick 2014). Hybridization (Figure 12.1c) can contribute to overall species richness through the formation of new hybrid taxa and thus new lineages alongside the parental taxa. A causal relationship between Quaternary climatic changes and the secondary contact of previously isolated populations and sympatric speciation via hybridization is observed in mountain regions such as the Alps (Comes & Kadereit 1998; Kadereit 2015). Evidence is also found for the emblematic stem-rosette plants of the Asteraceae family, for which the columnar life form of the genus *Espeletia* (also known by its local name, "frailejon") stands as an example. Together with seven other genera, *Espeletia* forms the Espeletiinae (Cuatrecasas 2013; Hooghiemstra et al. 2006), which has undergone many radiations (Madriñán et al. 2013) and frequent hybridization at numerous contact zones (Diazgranados 2012; Diazgranados & Barber 2017). Importantly, there is evidence of extensive gene flow and hybridization in páramo plant species (e.g., *Loricaria*: Kolář et al. 2016; *Espeletia*: Diazgranados 2012; Diazgranados & Barber 2017), suggesting a relationship between connectivity, hybridization and radiation.

Hybridization at contact zones (the melting pots of genetic material between former isolated populations) leads to high species richness and facilitates species radiations (Petit et al. 2003; Grant 2014). Clusters of these contact zones between connecting refugia are concentrated in mountain ridges, cross-mountain passes and mid-elevation corridors; such clusters have been identified for mountains in North America (Swenson & Howard 2005) and Europe (Hewitt 1996, 1999). For example, based on contemporary floristic evidence, it can be inferred that biogeographical boundaries in the Alps are remnants of refugia reconnection points (Schönswetter et al. 2005).

High levels of genetic diversity in refugia may reflect more stable population dynamics and larger population sizes. This diversity may also have accumulated over several climatic oscillations, not necessarily because they were stable climatically (see the páramo example, which was highly dynamic), but because they concentrated diversity from previously connected larger systems. This emphasizes the need to identify not only the locations of past refugia, but also where historical connectivity allowed exchange of genetic material between previously isolated populations, possibly enabling hybridization and consequent co-evolution.

In summary, the FCS incorporates four different processes, all of which have been identified as drivers of diversification, and which together have led to the accumulation of species in biodiverse ecosystems. Assessing the northern Andes in light of this framework shows repeating patterns of historical connectivity of sky-island archipelagos, which may have significantly contributed to the high species richness and radiations there. The reason that many sky-island complexes have high species richness and recent diversification (Hughes & Atchison 2015) is postulated to be due not only to phases of isolation but also to the opportunities provided through historical connectivity.

12.4 Perspectives on Paleogeographic Reconstructions and Historical Connectivity

Reconstructing evolutionary history and the governing processes behind diversification in complex landscapes is a difficult task, but through interdisciplinary approaches and techniques the biogeographic history can slowly be untangled. A complete review of future guidelines on paleoreconstructions goes beyond the scope of this paper, but it is clear that many opportunities and challenges lie ahead (see review on Neotropical plant evolution in Hughes et al. 2013). Here, we will briefly mention the value of spatial reconstructions in assembling the biogeographical picture of mountain areas on a global scale and the possible role historical connectivity plays through time.

Spatially explicit models that consider historical connectivity have strong explanatory power for species turnover dynamics across a region (Graham et al. 2006). Differences in plant species richness between tropical sky islands (Andes, East Africa and New Guinea; Sklenář et al. 2014) and oceanic islands (Japanese archipelago; Wepfer et al. 2016) have already been attributed to their degrees of historical connectivity. When habitat connectivity is calculated across a series of time periods under different environmental conditions, a significant relationship with contemporary biodiversity emerges (Graham et al. 2010). Assessing alpine ecosystems from different mountain regions in light of this framework can provide crucial insights into the legacy of climate and topography as summarized by the system's mountain fingerprint (Figure 12.1d).

Due to the uniqueness of each mountain's topography (Figures 12.1d and 12.2), the divergence times of radiations during the Pleistocene are mountain-, biome- and probably even sky island-specific. Expectations of concordant timing and patterns of diversification among mountain systems are unrealistic, due to the inherent heterogeneity among sky-island systems (Papadopoulou & Knowles 2015b). Genetic diversity patterns are expected to be dissimilar depending on which sky-island species survived the Pleistocene (Bidegaray-Batista et al. 2016). Therefore, it is expected that different diversification histories exist for different biomes as well (Pennington et al. 2006; Hughes et al. 2013).

Creating paleogeographic maps for different moments during the Pleistocene, as has been done for oceanic islands (Warren et al. 2010; Ali & Aitchison 2014; Rijsdijk et al. 2014), is key to understanding a region's degree of connectivity through time and space. Paleogeographic reconstruction along different altitudinal gradients also improves our understanding of the differences in diversification between lower- and upper-elevation biomes, as has been suggested by studies in the Alps (Kropf et al. 2003) and in Venezuelan tepuis (Rull & Nogué 2007; Nogué et al. 2013; see Chapter 23).

Paleo-niche reconstructions can provide additional insights into historical connectivity and increase our understanding of supposed dispersal barriers. Paleo-niche models can show where and under what climatic conditions topographic barriers were removed. For example, genetic flow has been reconstructed between sister clades on either side of a steep topographic depression between the Venezuelan and Colombian Andes (Gutiérrez et al. 2015). Furthermore, the combination of paleo-niche modeling with phylogeography is a powerful approach to reconstruct historical connectivity

(Maguire et al. 2015), and recent studies are rapidly adding to our understanding of the current biodiversity in relation to past connectivity (e.g., Sobral-Souza et al. 2015; Melville et al. 2016; Thomé et al. 2016).

12.5   Conclusion

Climate change and topography are key abiotic, extrinsic variables that influence species diversification (see Chapters 16 and 18). In this chapter, we have focused on the consequences of the reshuffling of species distributions in mountain landscapes, emphasizing how historical connectivity plays an important role in species diversification.

The present interglacial patterns of diversity and environments often serve as the implicit or explicit backdrop to most hypotheses developed to explain contemporary biodiversity. However, historical connectivity played an important role in contemporary spatial biogeographic patterns, particularly in terrestrial and oceanic systems characterized by repeated changes between high and low connectivity. The fact that different disciplines are increasingly discussing the importance of historical mechanisms in mountain diversity opens up the possibility of more interdisciplinary hypothesis-building. In this chapter, we showcased historical connectivity as an important mechanism underlying Pleistocene radiations, and posited that the combination of environmental heterogeneity, topographic complexity and Pleistocene climatic fluctuations created a "flickering connectivity system" (FCS) in which alpine sky islands

connected and disconnected at different moments in time and at different locations. In particular, the taxa from the Andean alpine biome, the páramo, have shown exceptionally recent evolutionary radiations, which we link to past extended phases of high connectivity. The dynamics of historical connectivity in archipelago formations are powerful forces in creating and explaining the spatially complex patterns of biodiversity (Gillespie & Roderick 2014).

Due to topographic complexity, the effect of climate oscillations on diversification is not expected to be spatially or temporally synchronous between or within mountains or between species. Insights derived from paleotopographic reconstructions, as exemplified here, provide the necessary platform for new hypothesis development with regard to biological evolution in sky islands and the intriguing interplay between climate and geology.

Acknowledgments

The authors are very thankful to Jack Williams, Petr Sklenář and Daniel Kissling for their helpful suggestions for improving previous drafts of this paper. We would like to thank Kenneth Rijsdijk for discussions on island biogeography, and Allison Perrigo and Carina Hoorn for their editorial work on the chapter. SGAF was supported by the research programme ALW Open Programme with project number 822.01.010, which is financed by the Netherlands Organisation for Scientific Research (NWO).

References

Adams, M.J. (1985) Speciation in the pronophiline butterflies (Satyridae) of the northern Andes. *Journal of Research on Lepidopera* **1**, 33–49.

Ali, J.R. & Aitchison, J.C. (2014) Exploring the combined role of eustasy and oceanic island thermal subsidence in shaping biodiversity on the Galápagos. *Journal of Biogeography* **41**, 1227–1241.

Arctander, P. & Fjeldså, J. (1994) Andean tapaculos of the genus *Scytalopus* (Aves, Rhinocryptidae): a study of speciation using DNA sequence data. In: Loeschcke, D.V., Jain, D.S.K. & Tomiuk, D.J. (eds.) *Conservation Genetics*. Basel: Birkhäuser EXS, pp. 205–225.

Bakker, J., Moscol Olivera, M. & Hooghiemstra, H. (2008) Holocene environmental change at the upper forest line in northern Ecuador. *The Holocene* **18**, 877–893.

Benham, P.M. & Witt, C.C. (2016) The dual role of Andean topography in primary divergence: functional and neutral variation among populations of the

hummingbird, *Metallura tyrianthina*. *BMC Evolutionary Biology* **16**, 22.

Bentley, J., Verboom, G.A. & Bergh, N.G. (2014) Erosive processes after tectonic uplift stimulate vicariant and adaptive speciation: evolution in an Afrotemperate-endemic paper daisy genus. *BMC Evolutionary Biology* **14**, 7.

Bidegaray-Batista, L., Sánchez-gracia, A., Santulli, G. et al. (2016) Imprints of multiple glacial refugia in the Pyrenees revealed by phylogeography and palaeodistribution modelling of an endemic spider. *Molecular Ecology* **25**, 2046–2064.

Bogotá-Angel, R.G., Groot, M.H.M., Hooghiemstra, H., et al. (2011) Rapid climate change from north Andean Lake Fúquene pollen records driven by obliquity: implications for a basin-wide biostratigraphic zonation for the last 284 ka. *Quaternary Science Reviews* **30**, 3321–3337.

Bouchenak-Khelladi, Y., Onstein, R.E., Xing, Y. et al. (2015) On the complexity of triggering evolutionary radiations. *New Phytologist* **207**, 313–326.

Brunschön, C. & Behling, H. (2010) Reconstruction and visualization of upper forest line and vegetation changes in the Andean depression region of southeastern Ecuador since the Last Glacial Maximum – a multi-site synthesis. *Review of Palaeobotany and Palynology* **163**, 139–152.

Cadena, C.D., Pedraza, C.A. & Brumfield, R.T. (2016) Climate, habitat associations and the potential distributions of Neotropical birds: implications for diversification across the Andes. *Revista de la Academia Colombiana de Ciencias Exactas, Físicas y Naturales* **40**, 275–287.

Collevatti, R.G., Terribile, L.C., Rabelo, S.G. & Lima-Ribeiro, M.S. (2015) Relaxed random walk model coupled with ecological niche modeling unravel the dispersal dynamics of a Neotropical savanna tree species in the deeper Quaternary. *Frontiers in Plant Science* **6**, 653.

Colwell, R.K. & Rangel, T.F. (2010) A stochastic, evolutionary model for range shifts and richness on tropical elevational gradients under Quaternary glacial cycles. *Philosophical Transactions of the Royal Society of London B: Biological Sciences* **365**, 3695–3707.

Comes, H.P. & Kadereit, J.W. (1998) The effect of Quaternary climatic changes on plant distribution and evolution. *Trends in Plant Science* **3**, 432–438.

Coetzee, J.A. (1967) Pollen analytical studies in east and southern Africa. *Paleoecology of Africa* **3**, 1–145.

Cuatrecasas, J. (2013) *A Systematic Study of the Subtribe Espeletiinae – Heliantheae, Asteraceae.* Memoirs of the New York Botanical Garden 107. New York: The New York Botanical Garden Press.

Dansgaard, W., Johnsen, S.J. & Clausen, H.B. (1993) Evidence for general instability of past climate from a 250-kyr ice-core record. *Nature* **364**, 218–220.

Dias, M.S., Oberdorff, T., Hugueny, B. et al. (2014) Global imprint of historical connectivity on freshwater fish biodiversity. *Ecology Letters* **17**, 1130–1140.

Diazgranados, M. (2012) Phylogenetic and biogeographic relationships of Frailejones (Espeletiinae, Compositae): An ongoing radiation in the tropical Andes. PhD Dissertation. Saint Louis University, St. Louis, MO.

Diazgranados, M. & Barber, J.C. (2017) Geography shapes the phylogeny of frailejones (Espeletiinae Cuatrec., Asteraceae): a remarkable example of recent rapid radiation in sky islands. *PeerJ* **5**, e2968.

Dobrovolski, R., Melo, A.S., Cassemiro, F.A.S. & Diniz-Filho, J.A.F. (2012) Climatic history and dispersal ability explain the relative importance of turnover and nestedness components of beta diversity. *Global Ecology and Biogeography* **21**, 191–197.

Duncan, S.I., Crespi, E.J., Mattheus, N.M. & Rissler, L.J. (2015) History matters more when explaining genetic diversity within the context of the core–periphery hypothesis. *Molecular Ecology* **24**, 4323–4336.

Dynesius, M. & Jansson, R. (2000) Evolutionary consequences of changes in species' geographical distributions driven by Milankovitch climate oscillations. *Proceedings of the National Academy of Sciences* **97**, 9115–9120.

Edwards, D.L., Keogh, J.S. & Knowles, L.L. (2012) Effects of vicariant barriers, habitat stability, population isolation and environmental features on species divergence in the south-western Australian coastal reptile community. *Molecular Ecology* **21**, 3809–3822.

Elsen, P.R. & Tingley, M.W. (2015) Global mountain topography and the fate of montane species under climate change. *Nature Climate Change* **5**, 772–776.

Ewers, R.M., Didham, R.K., Pearse, W.D. et al. (2013) Using landscape history to predict biodiversity patterns in fragmented landscapes. *Ecology Letters* **16**, 1221–1233.

Fjeldså, J., Bowie, R.C.K. & Rahbek, C. (2012) The role of mountain ranges in the diversification of birds. *Annual Review of Ecology, Evolution, and Systematics* **43**, 249–265.

Flantua, S.G.A. (2017) Climate change and topography as drivers of Latin American biome dynamics. PhD Dissertation. University of Amsterdam, Amsterdam.

Flantua, S.G.A., Hooghiemstra, H., Van Boxel, J.H. et al. (2014) Connectivity dynamics since the Last Glacial Maximum in the northern Andes; a pollen-driven framework to assess potential migration. In: Stevens, W.D., Montiel, O.M. & Raven, P.H. (eds.) *Monography in Systematic Botany, Vol. 128, Paleobotany and Biogeography: A Festschrift for Alan Graham in his 80th Year.* St. Louis, MO: Missouri Botanical Garden, pp. 98–123.

Flenley, J.R. (ed.) (1979) *The Equatorial Rain Forest: A Geological History.* London: Butterworths.

Gillespie, R.G. & Roderick, G.K. (2014) Evolution: geology and climate drive diversification. *Nature* **509**, 297–298.

Golley, F.B. (ed.) (1993) *A History of the Ecosystem Concept in Ecology.* New Haven, CT: Yale University Press.

Graham, C.H. & Fine, P.V.A. (2008) Phylogenetic beta diversity: linking ecological and evolutionary processes across space in time. *Ecology Letters* **11**, 1265–1277.

Graham, C.H., Moritz, C. & Williams, S.E. (2006) Habitat history improves prediction of biodiversity in rainforest fauna. *Proceedings of the National Academy of Sciences* **103**, 632–636.

Graham, C.H., VanDerWal, J., Phillips, S.J. et al. (2010) Dynamic refugia and species persistence: tracking spatial shifts in habitat through time. *Ecography* **33**, 1062–1069.

Grant, P. (2014). Adaptive radiation. In: Losos, J.B., Baum, D.A., Futuyma, D.J. et al. (eds.) *The Princeton Guide to*

Evolution. Princeton, NJ: Princeton University Press, pp. 559–566.

Groot, M.H.M., Bogotá, R.G., Lourens, L.J. et al. (2011) Ultra-high resolution pollen record from the northern Andes reveals rapid shifts in montane climates within the last two glacial cycles. *Climate of the Past* **7**, 299–316.

Gutiérrez, E.E., Maldonado, J.E. & Radosavljevic, A. (2015) The taxonomic status of Mazama bricenii and the significance of the Táchira depression for mammalian Endemism in the Cordillera de Mérida, Venezuela. *PLoS ONE* **10**, e0129113.

Hall, J.P. (2005) Montane speciation patterns in Ithomiola butterflies (Lepidoptera: Riodinidae): are they consistently moving up in the world? *Proceedings of the Royal Society B: Biological Sciences* **272**, 2457–2466.

Hanselman, J.A., Bush, M.B., Gosling, W.D. et al. (2011) A 370 000-year record of vegetation and fire history around Lake Titicaca (Bolivia/Peru). *Palaeogeography, Palaeoclimatology, Palaeoecology* **305**, 201–214.

Hao, Q., Liu, H. & Liu, X. (2016) Pollen-detected altitudinal migration of forests during the Holocene in the mountainous forest–steppe ecotone in northern China. *Palaeogeography, Palaeoclimatology, Palaeoecology* **446**, 70–77.

Hays, J.D., Imbrie, J. & Shackleton, N.J. (1976) Variations in the Earth's orbit: pacemaker of the ice ages. *Science* **194**(4270), 1121–1132.

Hewitt, G.M. (1996) Some genetic consequences of ice ages, and their role in divergence and speciation. *Biological Journal of the Linnean Society* **58**, 247–276.

Hewitt, G.M. (1999) Post-glacial re-colonization of European biota. *Biological Journal of the Linnean Society* **68**, 87–112.

Holdridge, L., Grenke, W., Hatheway, W. et al. (eds.) (1971) *Forest Environments in Tropical Life Zones: A Pilot Study*. Oxford: Pergamon Press.

Hooghiemstra, H. (ed.) (1984) Vegetational and climatic history of the high plain of Bogotá, Colombia. *Dissertaciones Botanicae* **79**, 1–368.

Hooghiemstra, H. & Van der Hammen, T. (2004) Quaternary ice-age dynamics in the Colombian Andes: developing an understanding of our legacy. *Philosophical Transactions: Biological Sciences* **359**, 173–181.

Hooghiemstra, H., Wijninga, V. & Cleef, A.M. (2006) The palaeobotanical record of Colombia: implications for biogeography and biodiversity. *Annals of the Missouri Botanical Garden* **93**, 297–324.

Hughes, C.E. & Atchison, G.W. (2015) The ubiquity of alpine plant radiations: from the Andes to the Hengduan Mountains. *New Phytologist* **207**, 275–282.

Hughes, C. & Eastwood, R. (2006) Island radiation on a continental scale: exceptional rates of plant diversification after uplift of the Andes. *Proceedings of the National Academy of Sciences* **103**, 10 334–10 339.

Hughes, C.E., Pennington, R.T. & Antonelli, A. (2013) Neotropical plant evolution: assembling the big picture. *Botanical Journal of the Linnean Society* **171**, 1–18.

Kadereit, J.W. (2015) The geography of hybrid speciation in plants. *Taxon* **64**, 673–687.

Kadereit, J.W., Griebeler, E.M. & Comes, H.P. (2004) Quaternary diversification in European alpine plants: pattern and process. *Philosophical Transactions of the Royal Society B: Biological Sciences* **359**, 265–274.

Kolář, F., Dušková, E. & Sklenář, P. (2016) Niche shifts and range expansions along cordilleras drove diversification in a high-elevation endemic plant genus in the tropical Andes. *Molecular Ecology* **25**, 4593–4610.

Körner, C., Paulsen, J. & Spehn, E.M. (2011) A definition of mountains and their bioclimatic belts for global comparisons of biodiversity data. *Alpine Botany* **121**, 73–78.

Kropf, M., Kadereit, J.W. & Comes, H.P. (2003) Differential cycles of range contraction and expansion in European high mountain plants during the Late Quaternary: insights from Pritzelago alpina (L.) O. Kuntze (Brassicaceae). *Molecular Ecology* **12**, 931–949.

Labeyrie, L., Skinner, L. & Cortijo, E. (2007) Paleoclimate reconstruction; sub-Milankovitch (DO/Heinrich) events. In: Elias, S.A. (ed.) *Encyclopedia of Quaternary Science*. Amsterdam: Elsevier, pp. 1964–1974.

Lagomarsino, L.P., Condamine, F.L., Antonelli, A. et al. (2016) The abiotic and biotic drivers of rapid diversification in Andean bellflowers (Campanulaceae). *New Phytologist* **210**, 1430–1442.

Linder, H.P. (2008) Plant species radiations: where, when, why? *Philosophical Transactions of the Royal Society B: Biological Sciences* **363**, 3097–3105.

Lindborg, R. & Eriksson, O. (2004) Historical landscape connectivity affects present plant species diversity. *Ecology* **85**, 1840–1845.

Lisiecki, L.E. & Raymo, M.E. (2005) A Pliocene-Pleistocene stack of 57 globally distributed benthic δ18O records. *Paleoceanography* **20**, PA1003.

Luebert, F. & Weigend, M. (2014) Phylogenetic insights into Andean plant diversification. *Frontiers in Ecology and Evolution* **2**, 1–17.

Luteyn, J.L. (1999) *Páramos: A Checklist of Plant Diversity, Geographical Distribution, and Botanical Literature.* Memoirs of the New York Botanical Garden 84. New York: The New York Botanical Garden Press.

Luteyn, J.L. (2002) Diversity, adaptation, and endemism in neotropical Ericaceae: biogeographical patterns in the Vaccinieae. *The Botanical Review* **68**, 55–87.

Madriñán, S., Cortés, A.J. & Richardson, J.E. (2013) Páramo is the world's fastest evolving and coolest biodiversity hotspot. *Frontiers in Genetics* **4**, 192.

Maguire, K.C., Nieto-Lugilde, D., Fitzpatrick, M.C. et al. (2015) Modeling species and community responses to past, present, and future episodes of climatic and ecological change. *Annual Review of Ecology, Evolution, and Systematics* **46**, 343–368.

McCormack, J.E., Huang, H. & Knowles, L.L. (2009) Sky islands. In: Gillespie, R.G. & Clague, D. (eds.) *Encyclopedia of Islands*. Berkeley, CA: University of Chicago Press, pp. 839–843.

McGlone, M. S., Duncan, R.P. & Heenan, P.B. (2001) Endemism, species selection and the origin and distribution of the vascular plant flora of New Zealand. *Journal of Biogeography* **28**, 199–216.

Myers, N., Mittermeier, R.A., Mittermeier, C.G. et al. (2000) Biodiversity hotspots for conservation priorities. *Nature* **403**, 853–858.

Melville, J., Haines, M.L., Hale, J. et al. (2016) Concordance in phylogeography and ecological niche modelling identify dispersal corridors for reptiles in arid Australia. *Journal of Biogeography* **43**, 1844–1855.

Miller, M.J., Bermingham, E., Klicka, J. et al. (2008) Out of Amazonia again and again: episodic crossing of the Andes promotes diversification in a lowland forest flycatcher. *Proceedings of the Royal Society of London B: Biological Sciences* **275**, 1133–1142.

Muñoz-Ortiz, A., Velásquez-Álvarez, Á.A., Guarnizo, C.E. & Crawford, A.J. (2015) Of peaks and valleys: testing the roles of orogeny and habitat heterogeneity in driving allopatry in mid-elevation frogs (Aromobatidae: Rheobates) of the northern Andes. *Journal of Biogeography* **42**, 193–205.

Münzbergová, Z., Cousins, S.A.O., Herben, T. et al. (2013) Historical habitat connectivity affects current genetic structure in a grassland species: Genetic diversity and past landscape structure. *Plant Biology* **15**, 195–202.

Nogué, S., Rull, V. & Vegas-Vilarrúbia, T. (2013). Elevational gradients in the neotropical table mountains: patterns of endemism and implications for conservation. *Diversity and Distributions* **19**, 676–687.

Nürk, N.M., Scheriau, C. & Madriñán, S. (2013) Explosive radiation in high Andean Hypericum – rates of diversification among New World lineages. *Frontiers in Genetics* **4**, 175.

Osborne, R. & Benton, R. (eds.) (1996) *The Viking Atlas of Evolution*. London: Penguin.

Papadopoulou, A. & Knowles, L.L. (2015a) Genomic tests of the species-pump hypothesis: recent island connectivity cycles drive population divergence but not speciation in Caribbean crickets across the Virgin Islands. *Evolution* **69**, 1501–1517.

Papadopoulou, A. & Knowles, L.L. (2015b) Species-specific responses to island connectivity cycles: refined models for testing phylogeographic concordance across a Mediterranean Pleistocene Aggregate Island Complex. *Molecular Ecology* **24**, 4252–4268.

Patterson, B.D., Solari, S. & Velazco, P.M. (2012) The role of the Andes in the diversification and biogeography of neotropical mammals. In: Patterson, B.D. & Costa, L.P. (eds.) *Bones, Clones and Biomes: The History of Geography of Recent Neotropical Mammals*. Chicago, IL: University of Chicago Press, pp. 351–278.

Pennington, R.T. & Dick, C.W. (2004) The role of immigrants in the assembly of the South American rainforest tree flora. *Philosophical Transactions of the Royal Society of London B: Biological Sciences* **359**, 1611–1622.

Pennington, R.T., Richardson, J.E. & Lavin, M. (2006) Insights into the historical construction of species-rich biomes from dated plant phylogenies, neutral ecological theory and phylogenetic community structure. *New Phytologist* **172**, 605–616.

Pennington, R.T., Lavin, M., Särkinen, T. et al. (2010) Contrasting plant diversification histories within the Andean biodiversity hotspot. *Proceedings of the National Academy of Sciences* **107**, 13 783–13 787.

Petit, R.J., Aguinagalde, I., de Beaulieu, J.-L. et al. (2003) Glacial refugia: hotspots but not melting pots of genetic diversity. *Science* **300**, 1563–1565.

Pyron, R.A., Costa, G.C., Patten, M.A. & Burbrink, F.T. (2015) Phylogenetic niche conservatism and the evolutionary basis of ecological speciation. *Biological Reviews* **90**, 1248–1262.

Qian, H. & Ricklefs, R.E. (2000) Large-scale processes and the Asian bias in species diversity of temperate plants. *Nature* **407**, 180–182.

Ramírez-Barahona, S. & Eguiarte, L.E. (2013) The role of glacial cycles in promoting genetic diversity in the Neotropics: the case of cloud forests during the Last Glacial Maximum. *Ecology and Evolution* **3**, 725–738.

Rijsdijk, K.F., Hengl, T., Norder, S.J. et al. (2014) Quantifying surface-area changes of volcanic islands driven by Pleistocene sea-level cycles: biogeographical implications for the Macaronesian archipelagos. *Journal of Biogeography* **41**, 1242–1254.

Rull, V. (2005) Biotic diversification in the Guayana Highlands: a proposal. *Journal of Biogeography* **32**, 921–927.

Rull, V. & Nogué, S. (2007) Potential migration routes and barriers for vascular plants of the neotropical Guyana Highlands during the Quaternary. *Journal of Biogeography* **34**, 1327–1341.

Särkinen, T., Pennington, R.T., Lavin, M. et al. (2012) Evolutionary islands in the Andes: persistence and isolation explain high endemism in Andean dry tropical forests. *Journal of Biogeography* **39**, 884–900.

Schneeweiss, G.M. & Schönswetter, P. (2011) A re-appraisal of nunatak survival in arctic-alpine phylogeography. *Molecular Ecology* **20**, 190–192.

Schönswetter, P., Stehlik, I., Holderegger, R. & Tribsch, A. (2005) Molecular evidence for glacial refugia of mountain plants in the European Alps. *Molecular Ecology* **14**, 3547–3555.

Sklenář, P. & Balslev, H. (2005) Superpáramo plant species diversity and phytogeography in Ecuador. *Flora – Morphology, Distribution, Functional Ecology of Plants* **200**, 416–433.

Sklenář, P., Dušková, E. & Balslev, H. (2010) Tropical and temperate: evolutionary history of Páramo flora. *The Botanical Review* **77**, 71–108.

Sklenář, P., Hedberg, I. & Cleef, A.M. (2014) Island biogeography of tropical alpine floras. *Journal of Biogeography* **41**, 287–297.

Smith, B.T., McCormack, J.E., Cuervo, A.M. et al. (2014) The drivers of tropical speciation. *Nature* **515**, 406–409.

Sobral-Souza, T., Lima-Ribeiro, M.S. & Solferini, V.N. (2015) Biogeography of neotropical rainforests: past connections between Amazon and Atlantic Forest detected by ecological niche modeling. *Evolutionary Ecology* **29**, 643–655.

Spratt, R.M. & Lisiecki, L.E. (2016) A Late Pleistocene sea level stack. *Climate of the Past* **12**, 1079–1092.

Steinbauer, M.J., Field, R., Grytnes, J.-A. et al. (2016) Topography-driven isolation, speciation and a global increase of endemism with elevation. *Global Ecology and Biogeography* **25**, 1097–1107.

Stehlik, I., Schneller, J.J. & Bachmann, K. (2001) Resistance or emigration: response of the high-alpine plant Eritrichium nanum (L.) Gaudin to the ice age within the Central Alps. *Molecular Ecology* **10**, 357–370.

Svenning, J.-C., Eiserhardt, W.L., Normand, S. et al. (2015) The influence of paleoclimate on present-day patterns in biodiversity and ecosystems. *Annual Review of Ecology, Evolution, and Systematics* **46**, 551–572.

Swenson, N.G. & Howard, D.J. (2005) Clustering of contact zones, hybrid zones, and phylogeographic breaks in North America. *The American Naturalist* **166**, 581–591.

Thomé, M.T.C., Sequeira, F., Brusquetti, F. et al. (2016) Recurrent connections between Amazon and Atlantic forests shaped diversity in Caatinga four-eyed frogs. *Journal of Biogeography* **43**, 1045–1056.

Tosi, J.A. (1964) Climatic control of terrestrial ecosystems: a report on the Holdridge model. *Economic Geography* **40**, 173.

Tribsch, A. (2004) Areas of endemism of vascular plants in the Eastern Alps in relation to Pleistocene glaciation. *Journal of Biogeography* **31**, 747–760.

Tribsch, A. & Schönswetter, P. (2003) Patterns of endemism and comparative phylogeography confirm palaeoenvironmental evidence for Pleistocene refugia in the Eastern Alps. *Taxon* **52**, 477–497.

Urrego, D.H., Hooghiemstra, H., Rama-Corredor, O., et al. (2016) Millennial-scale vegetation changes in the tropical Andes using ecological grouping and ordination methods. *Climate of the Past* **12**, 697–711.

Van der Hammen, T. (1974) The Pleistocene changes of vegetation and climate in tropical South America. *Journal of Biogeography* **1**, 3–26.

Van der Hammen, T., Werner, J.H. & Van Dommelen, H. (1973) Palynological record of the upheaval of the Northern Andes: a study of the Pliocene and Lower Quaternary of the Colombian Eastern Cordillera and the early evolution of its high-Andean biota. *Review of Palaeobotany and Palynology* **16**, 1–122.

Verboom, G.A., Bergh, N.G., Haiden, S.A., Hoffmann, V. & Britton, M.N. (2015) Topography as a driver of diversification in the Cape Floristic Region of South Africa. *New Phytologist* **207**, 368–376.

Wang, Y.-J., Susanna, A., von Raab-Straube, E. et al. (2009) Island-like radiation of Saussurea (Asteraceae: Cardueae) triggered by uplifts of the Qinghai–Tibetan Plateau. *Biological Journal of the Linnean Society* **97**, 893–903.

Warren, B.H., Strasberg, D., Bruggemann, J.H. et al. (2010) Why does the biota of the Madagascar region have such a strong Asiatic flavour? *Cladistics* **26**, 526–538.

Warschall, P. (1994) The Madrean sky island archipelago: a planetary overview. In DeBano, L.F., Ffolliott, P., Ortega-Rubio, A. et al. (eds.) *Biodiversity and the Management of the Madrean Archipelago: The Sky Islands of Southwestern US & Northwestern Mexico.* General Technical Report RM-GTR-264, USDA Forest Service, Rocky Mountain Forest and Range Experiment Station, Fort Collins, CO and Tucson, AZ.

Weir, J.T. (2006) Divergent timing and patterns of species accumulation in lowland and highland neotropical birds. *Evolution* **60**, 842–855.

Weir, J.T. (2009) Implications of genetic differentiation in neotropical montane forest birds. *Annals of the Missouri Botanical Garden* **96**, 410–433.

Weir, J.T., Haddrath, O., Robertson, H.A. et al. (2016) Explosive ice age diversification of kiwi. *Proceedings of the National Academy of Sciences* 201603795.

Weigelt, P., Steinbauer, M.J., Cabral, J.S. & Kreft, H. (2016) Late Quaternary climate change shapes island biodiversity. *Nature* **532**, 99–102.

Wepfer, P.H., Guénard, B. & Economo, E.P. (2016) Influences of climate and historical land connectivity on ant beta diversity in East Asia. *Journal of Biogeography* **43**, 2311–2321.

Wiens, J.J. (2004) Speciation and ecology revisited: phylogenetic niche conservatism and the origin of species. *Evolution* **58**, 193–197.

Wolfe, A.D., Randle, C.P., Datwyler, S.L. et al. (2006) Phylogeny, taxonomic affinities, and biogeography of Penstemon (Plantaginaceae) based on ITS and cpDNA sequence data. *American Journal of Botany* **93**, 1699–1713.

Woodward, F.I. & Cramer, W. (1996) Plant functional types and climatic changes: introduction. *Journal of Vegetation Science* **7**, 306–308.

13

The Environmental Heterogeneity of Mountains at a Fine Scale in a Changing World

Andrés J. Cortés[1,2,3] and Julia A. Wheeler[4,5,6]

[1] *Department of Ecology and Genetics, Uppsala University, Uppsala, Sweden*
[2] *Department of Plant Biology, Swedish Agricultural University, Uppsala, Sweden*
[3] *Department of Biological and Environmental Sciences, University of Gothenburg, Gothenburg, Sweden*
[4] *WSL Institute for Snow and Avalanche Research SLF, Davos, Switzerland*
[5] *Institute of Botany, University of Basel, Basel, Switzerland*
[6] *Department of Environmental Conservation, University of Massachusetts, Amherst, MA, USA*

Abstract

In evolutionary biology and ecological genetics, understanding the ecological, genetic and taxonomic signatures associated with local or small-scale environmental variation can provide insights into the naturally available genetic variation. The degree of genetic variation available can strongly influence how species and populations may respond to climate change. Likewise, climate change can alter the patterns of trait and genetic variation within and across species. Therefore, analyzing the relationships between traits, genomes and species is also a key step towards understanding the joint evolution of local environments and organisms under changing climatic conditions. As such, mountains represent an important and interesting research area: they are vastly heterogeneous environments that can demonstrate climatic extremes at a very fine spatial scale, which can offer a unique mosaic of highly localized environmental conditions. For example, micro-environmental conditions, such as temperature, sun and wind exposure, snow cover and water availability, can dramatically change in a matter of meters, leading to extreme differences in species distribution, species phenology, morphology and fitness at the microspatial scale. In this chapter, we explore how organisms may be affected by these sharp environmental gradients, by discussing the drivers of isolation and how isolation in turn drives diversification and adaptation patterns. We also discuss these processes in terms of how species and populations respond to changing conditions. In the long term, mountainous environments may offer safe microhabitats for many species in a warming world due to their fine-scale topographic variability, which may provide suitable habitats within only a few meters of species' current locations. Yet, such fine-scale habitat variability can also lead to locally adapted populations and species, so that individuals and populations adapted to a narrow range of conditions may respond poorly to environmental change.

Keywords: *alpine regions, evolutionary responses, genetic adaptation, genome-wide scans, heredomic scans, heritability, snowmelt timing*

13.1 The Mosaic of Environmental Heterogeneity at a Fine Scale

Understanding how organisms respond to changing environmental conditions is a major research focus in ecology and evolution. Organisms, populations or species can respond to environmental change in three ways: by migrating, persisting in current locations or going extinct (Hoffmann & Sgro 2011). Persistence of such populations in changing environments may be mediated by phenotypic plasticity, which is the range of phenotypes that a single genotype can express as a function of its environment (Nicotra et al. 2010), or by adaptation from genetic variation by increasing the frequency of existing allelic variants that can cope with the new conditions (Bridle & Vines 2007). Local adaptation to heterogeneous habitats has been documented (Gonzalo-Turpin & Hazard 2009; Savolainen et al. 2013), but the genetics of locally adapted populations are not always well understood (Savolainen et al. 2013).

Globally, some of the largest impacts of climate change are expected to occur in alpine environments, particularly

Mountains, Climate and Biodiversity, First Edition. Edited by Carina Hoorn, Allison Perrigo and Alexandre Antonelli.
© 2018 John Wiley & Sons Ltd. Published 2018 by John Wiley & Sons Ltd.
Companion website: www.wiley.com\go\hoorn\mountains,climateandbiodiversity

Figure 13.1 Mosaic of snowbeds and exposed ridges in the spring (May 2011) on Wannengrat, Switzerland. See also Plate 28 in color plate section.

near mountain summits, which are dominated by long-lived plant species and where snow cover and summer temperatures are the main drivers of vegetation composition (Körner & Basler 2010). Temperature increases over the past decades have already led to patterns of upward migration in various species (Walther et al. 2002). The alpine zone is a highly heterogeneous environment that is characterized not only by strong elevational gradients in temperature but also by local topography. Microtopographical features include depressions in which the snow accumulates and disappears very late in the summer (i.e., snowbeds) and more exposed ridges with less snow, where the snow disappears several weeks to months earlier (Figure 13.1). Similar heterogeneities or gradients that occur across the alpine zone result from wind exposure, water availability, rockiness and the presence of neighboring shrubs (Wheeler et al. 2015), among other factors. Some of these local-scale differences have been shown to cause local adaptation (e.g., in *Dryas* and in *Ranunculus*: Stanton & Galen 1997). For species that occur in heterogeneous habitats, such small-scale variation can have important implications for their response potential to changing conditions. Small-scale topographic variability may provide new locations with suitable habitats only a few meters away from current locations (Yamagishi et al. 2005; Scherrer & Körner 2011). Alternatively, such small-scale habitat variability can lead to locally adapted subpopulations (Gonzalo-Turpin & Hazard 2009), and these geno-

types, which are adapted to a more narrow range of conditions, may respond poorly to future conditions.

Therefore, in order to understand the processes involved in potential responses to changing conditions, it is important to consider not only climate differences between different altitudes but also differences between microhabitats. In this chapter, we explore the microhabitat-driven patterns (Section 13.2) and processes (Section 13.3) that are commonly observed in mountains. Although there is a focus on alpine ecosystems and sessile organisms throughout this work, the concepts are generalizable. We also summarize the methodologies used to study phenotypic and genetic variation in microhabitats on mountains (Section 13.4). To understand the interaction among fine-scale environmental (or "microhabitat") variation, diversity and the responses of populations and organisms in a changing alpine climate, we can ask the following questions:

- Are patterns of genetic differentiation and gene flow driven by small-scale environmental differences?
- Do morphological and fitness-related traits show heritable variation, and is selection currently acting on any of these traits so that they can evolve given changing conditions?
- What is the microhabitat-driven pattern of genomic divergence?
- What is the genomic architecture of ecologically relevant traits at the microhabitat level?

13.2 Drivers of Isolation at a Fine Scale

The transfer of alleles across populations is known as gene flow. When this allelic transfer is limited or interrupted, there is genetic isolation. Understanding patterns of genetic variation and gene flow across the fine-scale mosaic will help to predict the response of alpine species to climate change. As an example, under climate warming, snowmelt is expected to occur earlier in the season (Molau et al. 2005). Restricted gene flow and differentiation between subpopulations in different microhabitats (as when snowmelt timing is different) can be associated with local adaptation (Giménez-Benavides et al. 2007). In this scenario, late snowmelt-associated genotypes of long-lived species, such as the dominant shrub species, may have difficulties in persisting under warming conditions. Early snowmelt-associated genotypes, in contrast, would need to establish in new localities, and this could be difficult in long-lived species even if suitable localities are nearby. Alternatively, a lack of differentiation between subpopulations in different microhabitats and unrestricted gene flow between them could lead to genotypes capable of growing in both microhabitats, and thus persisting in situ during climate change. Apart from differentiation and gene flow, genetic variation in subpopulations in early- and late-snowmelt microhabitats could also differ due to factors such as asymmetric gene flow, and this will determine the extent to which genetic variation is lost from one of the microhabitats. In this section, we will cover the main drivers that may limit gene flow across microhabitats in mountains. In the next section, we will consider their major evolutionary consequences, specifically for adaptation and diversification in populations.

13.2.1 Mismatch in Flowering Time and Pollen Flow

Variation in the timing of flowering between subpopulations in different snow microhabitats can be a major driver of small-scale genetic structuring (Stanton et al. 1997) through the restriction of pollen-mediated gene flow (as compared to seed-mediated gene flow), regardless of whether flowering time is genetically or environmentally regulated (Stanton & Galen 1997; Jump et al. 2009; Scherrer & Körner 2011). Small-scale genetic differentiation (measured by the fixation index, F_{ST}, a measure of population differentiation due to genetic structure) has been reported in the majority of studies on snowmelt-driven genetic differentiation (Kudo 1992; Hirao &

Kudo 2004, 2008; Yamagishi et al. 2005; Shimono et al. 2009).

13.2.2 Asymmetry in Seed Dispersal

Even though there may be differentiation in populations' phenologies (the timing of periodic life cycle events, such as bud breaking, flowering, fruiting and bud setting) between microhabitats due to snowmelt timing (Figure 13.2), subpopulations growing in different microhabitats do not have to be genetically differentiated (Hülber et al. 2006, 2010). This pattern has been observed in *Empetrum hermaphroditum* (Hagerup) (Bienau et al. 2015) and *Ranunculus adoneus* (Gray) (Stanton et al. 1997), and it has perhaps been most extensively examined in *Salix herbacea* L., a clonal, dioecious, prostrate dwarf shrub common in circumpolar arctic, subarctic and alpine ecosystems (Beerling 1998). In the Swiss Alps, *S. herbacea* is an ideal species in which to study the effects of climate change, as it occurs along a relatively long elevational gradient (2100–2800 m a.s.l.) and occupies a wide range of microsite types, from rocky, early-exposure ridges to late-season snowbeds. In a study of this species by Cortés et al. (2014), although the populations growing in different microhabitats could be differentiated phenologically, they could not be differentiated genetically. F_{ST} was 0.028 ± 0.003 and 0.035 ± 0.004 for within- and between-microhabitat comparisons, P-value = 0.691. Lack of population structure was supported by a STRUCTURE analysis. This absence of population differentiation, even in microhabitats with highly different snowmelt dates, may have been mediated by high and asymmetric seed dispersal (Stanton et al. 1997; Cortés et al. 2014).

Seed dispersal can counteract isolation driven by barriers to pollen flow, such as snow, because seed dispersal occurs later in the season, when all winter snow has melted (Kudo 1992; Kudo & Hirao 2006). Gene flow via seed dispersal may result in asymmetric source/sink-like patterns driven by wind, topology and the success of seed establishment (Nathan & Muller-Landau 2000).

In the *S. herbacea* example introduced in the previous section, late-snowmelt microhabitats (snowbeds) were genetically more diverse than early-snowmelt sites (i.e., allelic richness: 8.93 ± 0.27 and 6.81 ± 0.29 for snowbeds and ridges, respectively), and gene flow, measured as the number of migrants per generation, was asymmetric towards the snowbeds (Figure 13.3). Overall, these results are consistent with snowbeds acting as sinks of genetic diversity and with seed dispersal preventing snowmelt-driven genetic isolation (Cortés et al. 2014).

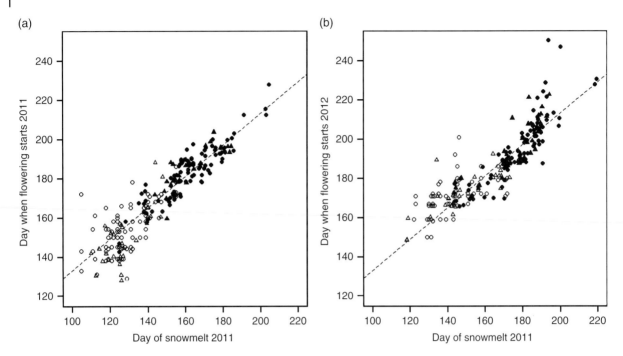

Figure 13.2 Day of snowmelt predicts when flowering starts for 274 female *Salix herbacea* patches growing on ridges (○) and snowbeds (●) and for 85 male *S. herbacea* patches growing on ridges (△) and snowbeds (▲) surveyed in (a) 2011 and (b) 2012. Dashed lines are regression lines ($R^2 = 0.827$, P-value < 0.001). *Source:* Cortés et al. (2014). Reproduced with permission of Springer.

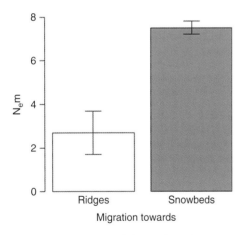

Figure 13.3 Estimates of the number of migrants per generation (N_em) between microhabitats differing in snowmelt timing (ridges and snowbeds) in *Salix herbacea* from three mountains in the Swiss Alps. *Source:* Cortés et al. (2014). Reproduced with permission of Springer.

13.3 Adaptation and Diversification at a Fine Scale

Most research on the responses of alpine species to changing snowmelt and temperature conditions has focused on species migration towards higher altitudes, where researchers can track the species' climate requirements (Lenoir et al. 2008; Eskelinen et al. 2009; Matteodo et al. 2013; Wipf et al. 2013). However, if migration potential is limited, the only way organisms can persist is by adjusting to the new environmental conditions (Jump et al. 2009; Sedlacek et al. 2014). Adjustment though phenotypic plasticity may be particularly important in long-lived species, as it can occur within the lifetime of an individual (Nicotra et al. 2010). However, plasticity may be constrained or even maladaptive if populations are exposed to novel conditions outside the range of conditions they encountered in their evolutionary history (Sedlacek et al. 2014). Alternatively, adaptation from standing variation may be promoted by increasing the frequency of existing variants that can cope with the new conditions (Bridle & Vines 2007). While adaptation is dependent on genotypes, plasticity itself depends on the environment.

Genomic divergence, which is the genetic differentiation throughout the genome, has been studied mostly among species and well-differentiated populations (Nosil & Feder 2011; Strasburg et al. 2011), but few of them have been studied at a very local scale from a genome-wide point of view. Genomes are regarded as porous, in the sense that different regions present contrasting signatures and levels of selection, isolation, drift and ancestral variation (Strasburg et al. 2011). Genome-wide patterns of divergence have recently been described in terms of metaphors about "islands" and "continents" of divergence, referring to peak- and plateau-like regions of

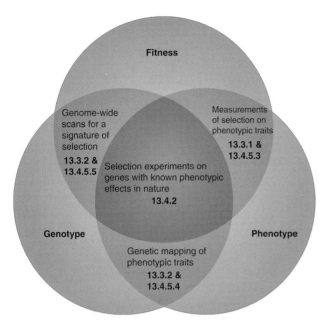

Figure 13.4 Connections between approaches to studying the genetics of ecologically relevant variation. Numbers indicate the sections in this chapter dealing with the specific concepts. *Source:* Adapted from Barrett & Hoekstra (2011).

high genetic divergence surrounded by low-divergence regions (Nosil & Feder 2011). While divergence peaks may be caused by divergent selection from novel or standing genetic variation (Roesti et al. 2014) or random drift (Keller et al. 2013), regions with low divergence may result from balancing or uniform selection, high gene flow or ancestral shared polymorphisms (Lexer et al. 2006; Jones et al. 2012).

A combined approach that explores selection gradients and association mapping of ecologically relevant traits and between-microhabitat genomic divergence allows us to understand which regions in the genome are likely to differ between populations in a mountain's different microhabitats (and therefore harbor genetic variation unique to a particular microhabitat) and how these genomic regions may relate to phenological, growth and fitness traits (Figure 13.4) (Stinchcombe & Hoekstra 2008; Barrett & Hoekstra 2011; Evans et al. 2014; Poelstra et al. 2014). Ultimately, this combined approach will allow the disentanglement of plastic and adaptive variation.

13.3.1 Selection and Evolutionary Responses

Three components are necessary in order for evolution to occur. First, there needs to be trait variation, which must be heritable and should have selection acting on it (Darwin 1874). A multivariate form of the breeder's equation (Walsh 2008) illustrates this paradigm well, and

allows for the prediction of the evolutionary response of a trait to selection over a single generation (R):

$$R = G\beta \tag{13.1}$$

where G is the variance–covariance matrix of additive genetic parameter estimates (G matrix, a proxy for heritabilities' and traits' trade-offs) and β is the vector of standardized selection gradients for the focal traits (Falconer & Mackay 1996). The evolutionary response can be calculated using selection-gradient estimates derived from fitness proxies (i.e., fitness regressed as a function of standardized trait values) and marker-based heritabilities (Lynch & Ritland 1999), as explained in Section 13.4.5.3.

Following the *S. herbacea* case study introduced in the previous section, marker-based relatedness estimates in natural populations were used to calculate heritabilities for phenological and morphological traits. For instance, there was selection towards smaller leaves and shorter thermal duration until leaf expansion, when using clonal reproduction (change in stem number) as a fitness proxy, in both ridge and snowbed microhabitats (Table 13.1). Conversely, there was selection towards longer thermal durations until flowering in both ridge and snowbed microhabitats when using sexual reproduction (proportion of flowering stems) as a fitness proxy. Selection on thermal durations until flowering diverged in the two microhabitat types when using clonal reproduction as a fitness proxy. This suggests that selection pressures on phenology may vary with ongoing climate change.

Additionally, when using the multivariate form of the breeder's equation (Equation 13.1) to estimate the potential evolutionary responses of traits in the *S. herbacea* case study, while accounting for genetic correlations among traits and selection on these traits (Sedlacek et al. 2016), the strongest predicted response was found for leaf size and the interval between snowmelt and leaf expansion ($R = -5.238$ days per generation) when using clonal reproduction as a fitness proxy. The adaptive potential found for these traits may enable *S. herbacea* to adapt to changing selection pressures. Under a climate-change scenario, with earlier snowmelt, evolutionary responses may shift towards responses that are currently observed on ridge microhabitats. Thus, longer thermal duration until flowering is expected, which might, for instance, prevent early season frost damage.

13.3.2 Adaptation and Plasticity

The persistence of populations and species given climate change may be mediated by phenotypic plasticity (Nicotra et al. 2010) or by adaptation from standing variation through an increase in the frequency

Table 13.1 Standardized selection gradients (β) across microhabitats differing in snowmelt timing (ridges and snowbeds) in *Salix herbacea* in the Swiss Alps. Linear mixed models were run separately for the two relative fitness proxies – proportion of flowering stems ($h^2 = 0.049$) and change in stem number ($h^2 = \mathbf{0.071}$) – and included the traits leaf size ($h^2 = 0.386$), interval between snowmelt and leaf expansion ($h^2 = 0.178$), thermal duration until leaf expansion ($h^2 = 0.469$) and flowering ($h^2 = 0.399$), as well as their interactions with microhabitat type (MH, microhabitat: positive interaction with ridges if $\beta > 0$ and positive interaction with snowbeds if $\beta < 0$), with the plot nested within the transect as a random effect. Estimates of narrow-sense heritability (h^2) are based on the multivariate animal model with a marker-based relatedness matrix (Lynch & Ritland 1999). Significant values (P-value < 0.5) are in bold, based on the F statistic (F) and its P-value (p). Significant h^2 values are also in bold.

Fitness	Trait (standardized)	β	df	F	p
Proportion of flowering stems	Leaf size	−0.023	67	0.032	0.86
	Interval snowmelt to leaf expansion	0.029	67	0.342	0.561
	Thermal duration until leaf expansion	−0.041	67	0.426	0.516
	Thermal duration until flowering	0.211	67	4.153	**0.046**
	Leaf size × MH	0.054	67	0.076	0.784
	Interval snowmelt to leaf expansion × MH	0.179	67	1.110	0.296
	Thermal duration until leaf expansion × MH	−0.095	67	0.376	0.542
	Thermal duration until flowering × MH	−0.075	67	0.163	0.688
Change in stem number	Leaf size	−5.232	116	−3.279	**0.011**
	Interval snowmelt to leaf expansion	−3.655	116	−2.217	0.051
	Thermal duration until leaf expansion	−3.646	116	−2.218	**0.020**
	Thermal duration until flowering	2.467	116	1.418	0.85
	Leaf size × MH	3.758	116	1.614	0.073
	Interval snowmelt to leaf expansion × MH	4.058	116	1.308	0.184
	Thermal duration until leaf expansion × MH	1.644	116	0.696	0.805
	Thermal duration until flowering × MH	−4.821	116	−2.047	**0.043**

Source: Adapted from Sedlacek et al. (2016).

of existing variants that can cope with the new conditions (Bridle & Vines 2007). Epigenetic mechanisms, which are those modulated by environmental factors that switch genes on and off and affect how cells read those genes, may also affect how plants respond to climatic variability (Bossdorf et al. 2008). In the next section, we review ways of accessing the roles of plasticity and adaptation in explaining trait variation across the fine-scale mosaic on mountains. First, however, we discuss the utility of genome-wide analysis in inferring microhabitat-driven divergent selection and the genetic basis of trait variation. We revisit the *S. herbacea* case, in which microhabitats differ in their snowmelt timing. Yet, this type of analysis also extends to other cases, such as the situation in the Andes, which concerns drought instead of snow melting (Cortés et al. 2012a,b; Blair et al. 2016).

In the *S. herbacea* case study, eight strong between-microhabitat divergence peaks and two weaker peaks were detected on seven different chromosomes (Figure 13.5). These regions coincided with regions of low single-nucleotide polymorphism (SNP; a type of molecular marker representing a change in a single base

pair) density, extensive linkage disequilibrium (a measure of dependency among loci) and negative Tajima's D values (a statistic that describes whether molecular evolution is random). This speaks for new genetic variation arising and being fixed in snowbeds and ridges separately, as compared to standing variation, which is differentially recruited between microhabitats. The same peaks persisted when the between-microhabitat F_{ST} was computed per transect, and they overlapped with "valleys" in the F_{ST} when this statistic was calculated within microhabitats and across transects, indicating that these divergent regions are not the result of genetic drift (Cortés 2015), which is the random change in the frequency of alleles. This indicates that genomic divergence can occur in the presence of gene flow and strong environmental differentiation at a very fine geographic scale. The 10 between-microhabitat divergence regions spanned a total of 219 genes, which may help us infer functional traits that diverge between microhabitats. This approach is known as "forward genetics" or "bottom-up inference," since it makes conclusions regarding unseen traits by first looking at the underlying genetic variation.

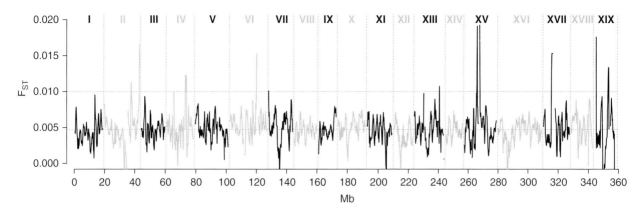

Figure 13.5 Between-microhabitat genomic divergence in *Salix herbacea*. Sliding-window analysis for the average between-microhabitat fixation index (F_{ST}). The window size is 1×106 basepairs (bps) and the step size is 200 kilobases (kb). Results of all windowed analyses are plotted against window midpoints in millions of bps (Mb). Black and gray colors highlight different chromosomes, identified by roman numerals. The lower and upper gray dashed horizontal lines indicate the genome-wide average and the threshold for the identification of outliers, respectively.

In addition to the population-genomics approach used to study microhabitat-driven divergence and identify traits that may have diverged between environments, association mapping is usually performed to explore genetically based variation in ecologically relevant traits across microhabitats (Cortés et al. in revision) – part of what is known as "reverse genetics" or "top-down inference," since it looks at the genetic basis of specific traits. In the *S. herbacea* case, a total of 57 regions comprising 66 SNP markers distributed across the genome were significantly associated with the explored traits for which heritabilities were computed (see previous section). On average, they explain 19% of the observed variation per trait. At least 10 regions included well-known candidate genes for seven of the nine surveyed traits.

13.4 Heterogeneous Microhabitats as a Field Laboratory to Study Reactions to Climate Change

Regions that are experiencing extremely high diversification rates, such as the alpine tundra ecosystems known as páramo (Madriñán et al. 2013), and convergent adaptation, such as microhabitats across mountains (Cortés 2013, 2015), are good candidates to serve as evolutionary playgrounds for today's scientists. In this section, we explore the ecological and genetic-sampling strategies and the analytical methods commonly used to cover biological variation across altitudes and microhabitats when addressing the main questions suggested at the beginning of this chapter.

13.4.1 Natural Surveys Across Microhabitats

Plot-based and transect surveys are the two main sampling strategies that allow the study of microhabitat-driven variation in mountains. Three to five different mountains spanning a range of north-east (sun/shade) exposure and covering the main elevational range are the best compromise between sampling effort and feasibility. These surveys, at a microhabitat level, follow an experimental approach known as space-for-time (SFT) substitution (Wheeler et al. 2016), in which current spatial heterogeneity is used as a proxy for predicting ecological time series (i.e., reactions to future conditions).

13.4.1.1 Plot Surveys

In this type of sampling, a few representative categorical altitudes (e.g., high and low) are chosen to cover the desired altitudinal distribution in each mountain. Depressions and ridge-like microhabitats may be chosen at each altitude based on indicators such as topology and vegetation. In each altitude/microhabitat combination, one big plot (~10 × 10 m) is designated, and several patches (~100) are sampled randomly in each plot. This sampling is best used for assessing isolation between microhabitats (e.g., ecological, trait-based or genetic isolation). Heritabilities and evolutionary responses for different traits are best computed in natural populations using this strategy (Sedlacek et al. 2016).

13.4.1.2 Transect Surveys

In this kind of sampling, the main elevational range is covered continuously. Around five to ten elevational bands along transects on each mountain, with one or more small study plots (~3 × 3 m) each, must be set up in different microhabitat sites (e.g., early-season exposure from snow and late-season exposure). In each plot, patches (at least two) are selected randomly. When compared to the plot survey, the transect survey has many smaller plots with fewer total patches, but it covers the desired altitudinal and microhabitat variation at a better resolution. This is ideal for assessing ecological responses (Wheeler et al. 2016) or genetic isolation-by-distance and for exploring trait- and microhabitat-driven genomic architecture (Cortés et al. 2014; Wheeler et al. 2016).

13.4.2 Transplant Experiments

In order to rigorously test how organisms respond to microhabitat-driven changes through phenotypic plasticity, as well as whether populations experience local adaptation (home-site advantage), reciprocal transplant experiments are needed (Figure 13.6) (Sedlacek et al. 2015). Transplant experiments are typically carried out across altitudinal gradients. However, reciprocal transplant studies explicitly examining the effects of local-scale variation (e.g., altered snowmelt timing) are scarce. Almost all reciprocal transplants have examined short-lived perennial herbs, and experiments with long-lived woody species are rare due to the difficulty in establishing clones of perennial, slow-growing species. Yet, it is important to understand how long-lived species will respond to changes in snowmelt timing, as they are a dominant functional vegetation type in alpine areas.

13.4.3 Phenotyping

Phenotyping is usually underestimated as scientists move their focus towards new developments in genotyping. Yet, it is an essential component that must be accounted for in ecological and genetic studies. Soil-temperature data loggers, nutrient probes and field observations can be used to estimate drought severity (Cortés et al. 2013), frost events (Wheeler et al. 2014), snowmelt timing (Wheeler et al. 2015), nutrient availability (Little et al. 2015) and other soil properties (Sedlacek et al. 2014). Monitoring of individuals carried out weekly during the growing season and across microhabitats during several growing seasons is the most exhaustive and informative survey method, although fewer snapshots can also be used.

13.4.4 Genotyping

Adaption and diversification are recognized as important processes that generate diversity (Cortés et al. 2011; Kelleher et al. 2012). However, their effects on genetic divergence and on the generation of morphological and ecological variation are poorly understood. Newly developed next-generation sequencing methods, together with more traditional techniques, offer the promise of major advances in the study of these interactions.

13.4.4.1 Microsatellite Genotyping

Microsatellite loci (regions of repetitive DNA that vary in the number of repeated DNA motifs, also called single-sequence repeats, SSRs) are commonly used to assess population structure (Blair et al. 2012), as shown in Section 13.2.2, and may help to estimate relatedness (Blair et al. 2013; Sedlacek et al. 2016), as shown in Section 13.3.1. Polymerase chain-reaction (PCR; a procedure used in molecular biology to replicate DNA exponentially) reactions are usually multiplexed into several runs. The PCR products are pooled afterwards and separated by capillary electrophoresis or acrylamide gels. Allele sizes are estimated in base pairs using software such as GeneMapper v.3.7 (Applied Biosystems) or by visual inspection.

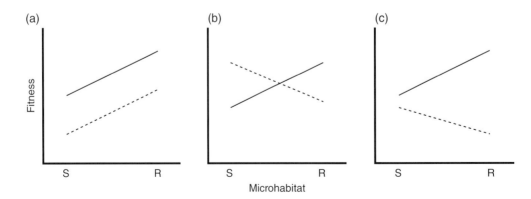

Figure 13.6 Scenarios where (a) plasticity, (b) local adaptation and (c) plasticity with a genetic basis explain trait variation across different microhabitats (snowbed, S, and ridge, R). Distinct lines (dashed: S; continuous: R) denote different genotypes reciprocally transplanted to each microhabitat. *Source:* Sedlacek et al. (2015). Licenses under CC BY 4.0, https://creativecommons.org/licenses/by/4.0/.

13.4.4.2 Genotyping-by-sequencing

Recent developments allow scientists to survey entire genomes in order to answer questions beyond the population genetics paradigm, in what is starting to be recognized as "population genomics" (as in Section 13.3.2). Genotyping-by-sequencing (GBS) is one of the cheapest and most popular methods. GBS libraries are prepared according to Elshire et al. (2011) using different enzymes for digestions. Raw Illumina DNA sequence data from the libraries can be processed through the GBS analysis pipeline as implemented in TASSEL-GBS (Glaubitz et al. 2014). Sequence tags are commonly aligned to reference genomes using the software package *BWA-aligner* (Li & Durbin 2007).

13.4.5 Common Statistical Approaches to Compare Microhabitats

13.4.5.1 Linear Models

Linear models (e.g., ANOVA, ANCOVA, linear regression) are among the most popular ways of assessing the effects of microhabitat and altitude on flowering time, as well as how temperature, humidity and snowmelt vary between microhabitats. In order to assess whether microhabitat and phenological differences trigger genetic isolation, pairwise F_{ST} values can be computed among plots or populations using, for instance, GENEPOP (Raymond & Rousset 1995). The number of alleles and the heterozygosity may be compared between microhabitats as well (as in Section 13.2.2), using linear mixed models (Venables & Ripley 2002), with the microhabitat as the fixed effect and the mountain as the random effect (covariate). This can be done in R (R Core Team), using the packages *lme4* or *lmerTest* (Bates & Sarkar 2007).

13.4.5.2 Population Structure

Population structure is usually examined using the software STRUCTURE (Pritchard et al. 2000). It is suggested that several independent runs are performed with different K (number of assumed populations) values using an admixture model and at least 100 000 iterations for the burn-in and 100 000 subsequent iterations for the Markov chain Monte Carlo (MCMC) analysis. The optimal number of subdivisions is posteriorly determined based on the rate of change of the likelihood across different K values, as described in Evanno et al. (2005). Effective population size (N_e) and pairwise migration rates (N_em) are also meaningful statistics calculated across microhabitats, as in Section 13.2.2. They can be estimated following coalescent theory and a maximum likelihood-based approach using software such as MIGRATE (Beerli & Felsenstein 1999).

13.4.5.3 Trait Heritability and Selection

When it comes to trait variation, narrow sense heritability (h^2) is estimated in natural populations using a multivariate animal model (Frentiu et al. 2008) with a marker-based relatedness matrix (Lynch & Ritland 1999). To test for selection on different traits, proxies of relative (i.e., relative to the mean across all sites or populations) clonal and fixed sexual reproductive fitness are compared against the standardized phenotypic traits using multiple regressions with linear mixed models to yield selection gradients (Lande & Arnold 1983). This analytical approach is illustrated in Section 13.3.1.

13.4.5.4 Genetic Mapping

In order to understand trait architecture in natural populations, trait-marker association studies are used (Galeano et al. 2012). Standard trait-marker association analysis can be easily implemented in FaST-LMM (Lippert et al. 2011) or BiForce (Gyenesei et al. 2012); the latter detects epistatic interactions (i.e., second-order trait-marker associations) and dominance effects.

13.4.5.5 Scans for Genome-Wide Selection

As a descriptive approach, genome-wide sliding-window analysis can be used to determine F_{ST} (as it is in Section 13.3.2) and the proportion of variable SNPs that are fixed between microhabitats, using, for instance, ARLEQUIN (Weir & Cockerham 1984; Excoffier et al. 2005). Linkage-disequilibrium (LD) and Tajima's D (Tajima 1989) are also usually computed in the same windows using the R package *PopGenome* (Pfeifer et al. 2014).

13.5 Conclusion

Fast-evolving microhabitat-driven divergence and genetically based trait variation at a larger scale may both play a role in the ability of a species to persist in diverse and variable conditions in mountainous ecosystems. Populations from different microhabitats may be isolated, and can act as sinks or sources of genetic diversity. From a genomic point of view, multiple genetic regions that diverge between microhabitats may arise at very local geographic scales even in the presence of gene flow, due to strong environmental differentiation and selection forces. Additionally, regions of high genomic divergence may be related to traits under selection that matter to the prediction of evolutionary responses. In this chapter, we have shown how small-scale environmental variability drives the responses of individuals, populations and species to changing conditions by looking at processes such as migration, adaptation and diversification at a very fine environmental scale.

Acknowledgments

We want to acknowledge our friend and colleague J. Sedlacek (RIP) for his professionalism and constancy. Thanks to S. Karrenberg for her advice and stimulating discussions. We very much appreciate C. Lexer's enthusiasm for discussing some of the ideas drafted in this chapter and for welcoming us in Fribourg, Switzerland on several occasions. We also wish to thank the rest of the SNFS-funded Sinergia Team devoted to the study of adaptation in snowmelt microhabitats – O. Bossdorf, G. Hoch, M. van Kleunen, C. Rixen and S. Wipf – for insightful discussions during the project meetings and for making our stays in Konstanz and Davos very enjoyable. B. Moreno's support while drafting this chapter is also deeply acknowledged. The invitation from the editors to contribute to this textbook on mountains, climate and biodiversity is highly appreciated, as is A. Antonelli's hospitality at his spring 2015 lab meeting in Kristineberg, Fiskebäckskil, Sweden. N. Sletvold and the staff at Norr Malma Station are also acknowledged for their support during the first author's 2015 stay in Lake Erken, Norrtälje, Sweden while drafting this chapter. Writing time by the first author was possible thanks to the grants CTS14:55 from the Carl Trygger Foundation to S. Berlin and 4.1-2016-00418 from Vetenskapsrådet (VR) to AJC.

References

Barrett, R.D.H. & Hoekstra, H.E. (2011) Molecular spandrels, tests of adaptation at the genetic level. *Nature Reviews Genetics* **12**, 767–780.

Bates, D.M. & Sarkar, D. (2007) Lme4, Linear mixed-effects models using s4 classes. R package.

Beerli, P. & Felsenstein, J. (1999) Maximum likelihood estimation of migration rates and effective population numbers in two populations using a coalescent approach. *Genetics* **152**, 763–773.

Beerling, D.J. (1998) *Salix herbacea* L. *Journal of Ecology* **86**, 872–895.

Bienau, M.J., Kröncke, M., Eiserhardt, W.L. et al. (2015) Synchronous flowering despite differences in snowmelt among habitats of *Empetrum hermaphroditum*. *Acta Oecologica* **69**, 129–136.

Blair, M.W., Soler, A. & Cortés, A.J. (2012) Diversification and population structure in common beans (*Phaseolus vulgaris* L.). *PLoS ONE* **7**, e49488.

Blair, M.W., Cortés, A.J., Penmetsa, R.V. et al. (2013) A high-throughput SNP marker system for parental polymorphism screening, and diversity analysis in common bean (*Phaseolus vulgaris* L.). *Theoretical and Applied Genetics* **126**, 535–548.

Blair M.W., Cortés A.J. & This D. (2016) Identification of an *ERECTA* gene and its drought adaptation associations with wild and cultivated common bean. *Plant Science* **242**, 250.

Bossdorf, O., Richards, C.L. & Pigliucci, M. (2008) Epigenetics for ecologists. *Ecology Letters* **11**, 106–115.

Bridle, J.R. & Vines, T.H. (2007) Limits to evolution at range margins: when and why does adaptation fail? *Trends in Ecology & Evolution* **22**, 140–147.

Cortés, A.J. (2013) On the origin of the common bean (*Phaseolus vulgaris* L.). *American Journal of Plant Sciences* **4**, 1998–2000.

Cortés, A.J. (2015) *On the Big Challenges of a Small Shrub: Ecological Genetics of Salix herbacea L.* Uppsala: Acta Universitatis Upsaliensis.

Cortés, A.J., Chavarro, M.C. & Blair, M.W. (2011) SNP marker diversity in common bean (*Phaseolus vulgaris* L.). *Theoretical and Applied Genetics* **123**, 827–845.

Cortés, A.J., Chavarro, M.C., Madriñán, S. et al. (2012a) Molecular ecology and selection in the drought-related *Asr* gene polymorphisms in wild and cultivated common bean (*Phaseolus vulgaris* L.). *BMC Genetics* **13**, 58.

Cortés, A.J., This, D., Chavarro, C. et al. (2012b) Nucleotide diversity patterns at the drought-related *DREB2* encoding genes in wild and cultivated common bean (*Phaseolus vulgaris* L.). *Theoretical and Applied Genetics* **125**, 1069–1085.

Cortés, A.J., Monserrate, F., Ramírez-Villegas, J. et al. (2013) Drought tolerance in wild plant populations, the case of common beans (*Phaseolus vulgaris* L.). *PLoS ONE* **8**, e62898.

Cortés, A.J., Waeber, S., Lexer, C. et al. (2014) Small-scale patterns in snowmelt timing affect gene flow and the distribution of genetic diversity in the alpine dwarf shrub *Salix herbacea*. *Heredity* **113**, 233–239.

Darwin, C.R. (1874) *The Descent of Man and Selection in Relation to Sex*. New York: Hurst and Company.

Elshire, R.J., Glaubitz, J.C., Sun, Q. et al. (2011) A robust, simple genotyping-by-sequencing (GBS) approach for high diversity species. *PLoS ONE* **6**, e19379.

Eskelinen, A., Stark, S. & Männistö, M. (2009) Links between plant community composition, soil organic matter quality and microbial communities in contrasting tundra habitats. *Oecologia* **161**, 113–123.

Evanno, G., Regnaut, S. & Goudet, J. (2005) Detecting the number of clusters of individuals using the software STRUCTURE: a simulation study. *Molecular Ecology* **14**, 2611–2620.

Evans, L.M., Slavov, G.T., Rodgers-Melnick, E. et al. (2014) Population genomics of *Populus trichocarpa* identifies signatures of selection and adaptive trait associations. *Nature Genetics* **46**, 1089–1096.

Excoffier, L., Laval, G. & Schneider, S. (2005) Arlequin (version 3.0), an integrated software package for population genetics data analysis. *Evolutionary Bioinformatics Online* **1**, 47–50.

Falconer, D.S. & Mackay, T.F.C. (1996) *Introduction to Quantitative Genetics*. Harlow: Longman.

Frentiu, F.D., Clegg, S.M., Chittock, J. et al. (2008) Pedigree-free animal models, the relatedness matrix reloaded. *Proceedings of the Royal Society B: Biological Sciences* **275**, 639–647.

Galeano, C.H., Cortés, A.J., Fernández, A.C. et al. (2012) Gene-based single nucleotide polymorphism markers for genetic and association mapping in common bean. *BMC Genetics* **13**, 48.

Giménez-Benavides, L., Escudero, A. & Iriondo, J.M. (2007) Local adaptation enhances seedling recruitment along an altitudinal gradient in a high mountain Mediterranean plant. *Annals of Botany* **99**, 723–734.

Glaubitz, J.C., Casstevens, T.M., Harriman, J. et al. (2014) TASSEL-GBS: a high capacity genotyping by sequencing analysis pipeline. *PLoS ONE* **9**, e90346.

Gonzalo-Turpin, H. & Hazard, L. (2009) Local adaptation occurs along altitudinal gradient despite the existence of gene flow in the alpine plant species *Festuca eskia*. *Journal of Ecology* **97**, 742–751.

Gyenesei, A., Moody, J., Semple, C.A.M. et al. (2012) High-throughput analysis of epistasis in genome-wide association studies with BiForce. *Bioinformatics* **28**, 1957–1964.

Hirao, A.S. & Kudo, G. (2004) Landscape genetics of alpine-snowbed plants, comparisons along geographic and snowmelt gradients. *Heredity* **93**, 290–298.

Hirao, A.S. & Kudo, G. (2008) The effect of segregation of flowering time on fine-scale genetic structure in an alpine-snowbed herb *Primula cuneifolia*. *Heredity* **100**, 424–430.

Hoffmann, AA. & Sgro, CM. (2011) Climate change and evolutionary adaptation. *Nature* **470**, 479–485.

Hülber, K., Gottfried, M., Pauli, H. et al. (2006) Phenological responses of snowbed species to snow removal dates in the Central Alps, implications for climate warming. *Arctic, Antarctic, and Alpine Research* **38**, 99–103.

Hülber, K., Winkler, M. & Grabherr, G. (2010) Intraseasonal climate and habitat-specific variability controls the flowering phenology of high alpine plant species. *Functional Ecology* **24**, 245–252.

Jones, F.C., Grabherr, M.G., Chan, Y.F. et al. (2012) The genomic basis of adaptive evolution in threespine sticklebacks. *Nature* **484**, 55–61.

Jump, A.S., Matyas, C. & Penuelas, J. (2009) The altitude-for-latitude disparity in the range retractions of woody species. *Trends in Ecology and Evolution* **24**, 694–701.

Kelleher, C.T., Wilkin, J., Zhuang, J. et al. (2012) SNP discovery, gene diversity, and linkage disequilibrium in wild populations of *Populus tremuloides*. *Tree Genetics & Genomes* **8**, 821–829.

Keller, I., Wagner, C.E., Greuter, L. et al. (2013) Population genomic signatures of divergent adaptation, gene flow and hybrid speciation in the rapid radiation of Lake Victoria cichlid fishes. *Molecular Ecology* **22**, 2848–2863.

Körner, C. & Basler, D. (2010) Phenology under global warming. *Science* **327**, 1461–1462.

Kudo, G. (1992) Performance and phenology of alpine herbs along a snow-melting gradient. *Ecological Research* **7**, 297–304.

Kudo, G. & Hirao, A. (2006) Habitat-specific responses in the flowering phenology and seed set of alpine plants to climate variation: implications for global-change impacts. *Population Ecology* **48**, 49–58.

Lande, R. & Arnold, S. (1983) The measurement of selection on correlated characters. *Evolution* **37**, 1210–1226.

Lenoir, J., Gegout, J.C., Marquet, P.A. et al. (2008) A significant upward shift in plant species optimum elevation during the 20th century. *Science* **320**, 1768–1771.

Lexer, C., Kremer, A. & Petit, R.J. (2006) Shared alleles in sympatric oaks: recurrent gene flow is a more parsimonious explanation than ancestral polymorphism. *Molecular Ecology* **15**, 2007–2012.

Li, H. & Durbin, R. (2007) Fast and accurate short read alignment with Burrows-Wheeler transform. *Bioinformatics* **25**, 1754–1760.

Lippert, C., Listgarten, J., Liu, Y. et al. (2011) FaST linear mixed models for genome-wide association studies. *Nature Methods* **8**, 833–835.

Little, C.J., Wheeler, J.A., Sedlacek, J. et al. (2015) Small-scale drivers: the importance of nutrient availability and snowmelt timing on performance of the alpine shrub *Salix herbacea*. *Oecologia* **180**, 1015–1024.

Lynch, M. & Ritland, K. (1999) Estimation of pairwise relatedness with molecular markers. *Genetics* **152**, 1753–1766.

Madriñán, S., Cortés, A.J. & Richardson, J.E. (2013) Páramo is the world's fastest evolving and coolest biodiversity hotspot. *Frontiers in Genetics* **4**, 192.

Molau, U., Nordenhäll, U. & Eriksen, A.B. (2005) Onset of flowering and climate variability in an alpine landscape, a 10-year study from Swedish Lapland. *American Journal of Botany* **92**, 422–431.

Matteodo, M., Wipf, S., Stöckli, V. et al. (2013) Elevation gradient of successful plant traits for colonizing alpine summits under climate change. *Environmental Research Letters* **8**, 24–43.

Nathan, R. & Muller-Landau, H.C. (2000) Spatial patterns of seed dispersal, their determinants and consequences for recruitment. *Trends in Ecology and Evolution* **15**, 278–285.

Nicotra, A.B., Atkin, O.K., Bonser, S.P. et al. (2010) Plant phenotypic plasticity in a changing climate. *Trends in Plant Science* **15**, 684–692.

Nosil, P. & Feder, J.L. (2011) Genomic divergence during speciation: causes and consequences. *Philosophical Transactions of the Royal Society B: Biological Sciences* **367**, 332–342.

Pfeifer, B., Wittelsbürger, U., Ramos-Onsins, S.E. & Lercher, M.J. (2014) PopGenome: an efficient Swiss army knife for population genomic analyses in R. *Molecular Biology and Evolution* **31**, 1929–1936.

Poelstra, J.W., Vijay, N., Bossu, C.M. et al. (2014) The genomic landscape underlying phenotypic integrity in the face of gene flow in crows. *Science* **344**, 1410–1414.

Pritchard, J.K., Stephens, M. & Donnelly, P. (2000) Inference of population structure using multilocus genotype data. *Genetics* **155**, 945–959.

Raymond, M. & Rousset, F. (1995) Genepop (version 1.2), population genetics software for exact tests and ecumenicism. *Journal of Heredity* **86**, 248–249.

Roesti, M., Gavrilets, S., Hendry, A.P. et al. (2014) The genomic signature of parallel adaptation from shared genetic variation. *Molecular Ecology* **23**, 3944–3956.

Savolainen, O., Lascoux, M. & Merilä, J. (2013) Ecological genomics of local adaptation. *Nature Reviews Genetics* **14**, 807–820.

Scherrer, D. & Körner, C. (2011) Topogaphically controlled thermal-habitat differentiation buffers alpine plant diversity against climate warming. *Journal of Biogeography* **38**, 406–416.

Sedlacek, J., Bossdorf, O., Cortés, A.J. et al. (2014) What role do plant-soil interactions play in the habitat suitability and potential range expansion of the alpine dwarf shrub *Salix herbacea*? *Basic and Applied Ecology* **15**, 305–315.

Sedlacek, J., Wheeler, J.A., Cortés, A.J. et al. (2015) The response of the alpine dwarf shrub *Salix herbacea* to altered snowmelt timing: lessons from a multi-site transplant experiment. *PloS ONE* 0122395.

Sedlacek, J., Cortés, A.J., Wheeler, J.A. et al. (2016) Evolutionary potential in the Alpine: trait heritabilities and performance variation of the dwarf willow *Salix herbacea* from different elevations and microhabitats. *Ecology and Evolution* **6**, 3940–3952.

Shimono, Y., Watanabe, M., Hirao, AS. et al. (2009) Morphological and genetic variations of *Potentilla matsumurae* (Rosaceae) between fellfield and snowbed populations. *American Journal of Botany* **96**, 728–737.

Stanton, M.L. & Galen, C. (1997) Life on the edge, adaptation versus environmentally mediated gene flow in the snow buttercup, *Ranunculus adoneus*. *The American Naturalist* **150**, 143–178.

Stanton, M.L., Galen, C. & Shore, J. (1997) Population structure along a steep environmental gradient, consequences of flowering time and habitat variation in the snow buttercup, *Ranunculus adoneus*. *Evolution* **51**, 79–94.

Stinchcombe, J.R. & Hoekstra, H.E. (2008) Combining population genomics and quantitative genetics: finding the genes underlying ecologically important traits. *Heredity* **100**, 158–170.

Strasburg, J.L., Sherman, N.A., Wright, K.M. et al. (2011) What can patterns of differentiation across plant genomes tell us about adaptation and speciation? *Philosophical Transactions of the Royal Society B: Biological Sciences* **367**, 364–373.

Tajima, F. (1989) Statistical method for testing the neutral mutation hypothesis by DNA polymorphism. *Genetics* **123**, 585–595.

Venables, W.N. & Ripley, B.D. (2002) *Modern Applied Statistics with S.*, 4th edn. Heidelberg: Springer.

Walsh, B. (2008) *Evolutionary Quantitative Genetics Handbook of Statistical Genetics*. Chichester: John Wiley & Sons Ltd., pp. 533–586.

Walther, G.R., Post, E., Convey, P. et al. (2002) Ecological responses to recent climate change. *Nature* **416**, 389–395.

Weir, B.S. & Cockerham, C. (1984) Estimating F-statistics for the analysis of population structure. *Evolution* **38**, 1358–1370.

Wheeler, J.A., Hoch, G., Cortés, A.J. et al. (2014) Increased spring freezing vulnerability for alpine shrubs under early snowmelt. *Oecologia* **175**, 219–229.

Wheeler, J.A., Schnider, F., Sedlacek, J. et al. (2015) With a little help from my friends: community facilitation increases performance in the dwarf shrub *Salix herbacea. Basic and Applied Ecology* **16**, 202–209.

Wheeler, J.A., Cortés, A.J., Sedlacek, J. et al. (2016) The snow and the willows, accelerated spring snowmelt reduces performance in the alpine shrub *Salix herbacea. Journal of Ecology* **104**, 1041–1050.

Wipf, S., Stöckli, V., Herz, K. & Rixen, C. (2013) The oldest monitoring site of the Alps revisited: accelerated increase in plant species richness on Piz Linard summit since 1835. *Plant Ecology and Diversity* **6**, 447–455.

Yamagishi, H., Allison, T.D. & Ohara, M. (2005) Effect of snowmelt timing on the genetic structure of an *Erythronium grandiflorum* population in an alpine environment. *Ecological Research* **20**, 199–204.

14

Mountains, Climate and Mammals

Catherine Badgley[1], Tara M. Smiley[2,3] and Rachel Cable[1]

[1] *Department of Ecology and Evolutionary Biology, University of Michigan, Ann Arbor, MI, USA*
[2] *Department of Earth and Environmental Sciences, University of Michigan, Ann Arbor, MI, USA*
[3] *Integrative Biology, Oregon State University, Corvallis, OR, USA*

Abstract

More than half of present-day continental mammals occur in montane regions. This concentration of diversity in regions of high topographic complexity compared to adjacent lowlands and plains constitutes the topographic diversity gradient, one of the major biogeographic patterns across continents today. Several biogeographic processes have shaped the topographic diversity gradient for mammals. Strong elevational gradients of climate, soil and vegetation in montane regions provide diverse habitats from high to low elevations; these heterogeneous habitats can accommodate high mammal diversity, especially of small mammals. Montane regions also present strong barriers – in the form of deep valleys, steep climatic gradients and abrupt changes in topography – to the distribution of populations and species. Changes in these barriers over time due to tectonic activity and erosion lead to fragmentation or mixing of populations, circumstances that can increase speciation or extinction rates in montane regions. Global or regional climate changes cause geographic ranges to shift in elevation. Climatic warming during the current interglacial has caused mammalian range shifts to higher elevations since the Last Glacial Maximum, and this process is accelerating under current global warming. The mammalian fossil record indicates that the topographic diversity gradient is strong during some time intervals and weak during others. For a 20 million-year record of North American small mammals, the interval with the greatest intensity of tectonic activity and climate change coincided with the highest origination rates and the highest diversity in the topographically complex region. This pattern suggests that landscape history has a strong influence on past and future mammal diversification.

Keywords: *biogeography, diversification, diversity gradient, topographic complexity, fossil record*

14.1 Introduction

Today, montane regions harbor more species of mammals per unit of area than adjacent lowlands. This pattern occurs at the regional scale, such as in the Rocky Mountains compared to the Great Plains in North America or in the East African Rift System (EARS) compared to the Congo Basin in Africa. At least three biophysical properties of montane regions contribute to their high species richness. First, strong elevational gradients in topography and associated climatic conditions create a diversity of habitats in terms of area, slope, soil type and depth, vegetation and seasonal cycles of temperature and precipitation. Increased environmental heterogeneity can accommodate a high diversity of species with different environmental tolerances (Coblentz &

Riitters 2004). Second, the geographic ranges of species in montane regions are often fragmented across deep valleys and extensive crests above the snowline. Over thousands of generations, isolated populations may become genetically distinct, achieve reproductive isolation and form new species (Davis et al. 2008). Third, climatic changes over decades to millions of years affect the connectivity of habitats and stimulate geographic-range shifts within montane regions and between mountains and adjacent lowlands. These elevational shifts also move species ranges into and out of montane regions, resulting in periods of higher and lower species richness in entire tectonic provinces over geologic time (e.g., Finarelli & Badgley 2010). This combination of ecological and evolutionary processes unfolding over geologically dynamic landscapes implies that montane regions may serve as both a

Mountains, Climate and Biodiversity, First Edition. Edited by Carina Hoorn, Allison Perrigo and Alexandre Antonelli.
© 2018 John Wiley & Sons Ltd. Published 2018 by John Wiley & Sons Ltd.
Companion website: www.wiley.com\go\hoorn\mountains,climateandbiodiversity

source of new species and adaptations and a long-term refuge, especially during periods of global warming (Bush et al. 2004; Badgley 2010; Merckx et al. 2015).

In this chapter, we review the biogeography of present-day mammal diversity in areas of complex topography compared to adjacent lowlands, first across continents and then in selected regions. The fossil record offers a geohistorical perspective on diversity along topographic gradients, and we present one example from the fossil record of North American mammals over the last 25 My. We describe the kinds of mammals that are known to drive this pattern via rapid speciation and high spatial turnover, as well as several biogeographic processes that shape the modern pattern of high mammal diversity in montane regions. We distinguish between an elevational diversity gradient, which refers to changes in species richness along an elevational profile, and a topographic diversity gradient, which refers to changes in diversity between a region of high topographic complexity and a region of low relief. A region of high topographic complexity has high relief (the difference in elevation between the highest and lowest altitudes) over a large area (hundreds of square kilometers). Such regions include mountain ranges, incised plateaus and basins bordered by mountain ranges. An elevational diversity gradient may occur on individual mountains or across a mountain range (e.g., Rahbek 1995; McCain 2005); a topographic diversity gradient occurs between regions on the scale of tectonic provinces (e.g., Finarelli & Badgley 2010).

The diversity of mammals in montane regions appears to be as much a response to complex topography, including multiple mountain ranges and basins, as to high elevations per se (Simpson 1964; Davis 2005; Badgley 2010). Complex topography results from the interaction of tectonic processes – which increase elevation through rock uplift within compressional regimes, or increase land area and relief through extension (lateral stretching) or rifting (see Chapters 2 and 3) – with climatic processes that cause erosion (see Chapters 4 and 8). The formation

of mountain belts (orogeny) occurs over millions to tens of millions of years (Champagnac et al. 2012). High relief, steep slopes and deep valleys arise during early to middle stages of orogeny and accentuate bioclimatic zonation. In contrast, speciation of mammals occurs on time scales of hundreds of thousands to millions of years, and substantial geographic-range shifts may occur in a few thousand years. Later stages of mountain building correspond to basin infilling, slope smoothing and more gradual bioclimatic transitions over elevational gradients. The topographic profile of mountains at all orogenic stages can be strongly influenced by glacial processes, especially at higher latitudes (Egholm et al. 2009).

The regional climate and its effects on erosion, topographic roughness and vegetation are major influences on potential mammal habitation and diversity. The altitude of the snowline, the thermal lapse rate and the precipitation gradient across a montane region determine which parts of the region have sufficient quantities of vegetation and shelter to support persistent populations of mammals. The presence of rock shelters or friable soils for burrows is critical for small mammals (<1 kg in adult body weight). In cold regions, including high elevations at low latitudes, many small mammals must be able to hibernate or enter torpor to survive daily or seasonal cold periods (Merritt 2010). In regions of low-temperature seasonality (e.g., much of the tropics), narrow climatic zones and multiple vegetation assemblages along elevation gradients contribute to dispersal limitation and species turnover in montane regions (Janzen 1967; Ghalambor et al. 2006).

The biogeographic processes that result in high mammal diversity in montane regions are the same processes that occur in all environments – speciation, extinction and geographic-range shifts via dispersal (Lomolino et al. 2010). The interaction of these processes with Earth's tectonic and climatic history shapes diversity gradients over space and time. Figure 14.1 illustrates the influences of geologic history on mammal diversity in montane regions. Tectonic and climatic processes interact to generate landscapes of

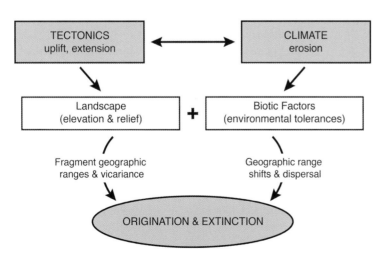

Figure 14.1 Conceptual diagram linking the tectonic and climate processes shaping regional landscapes, environmental conditions and species richness over geologic time scales. Within topographically and climatically complex regions, biogeographic processes include dispersal, geographic-range shifts and fragmentation of species ranges. These processes can promote origination and extinction, and geographic differences in origination and extinction rates in tectonically active versus passive regions can strengthen or weaken the topographic diversity gradient.

high elevation and relief. Areas of high relief feature climatically diverse habitats with a greater frequency of ecotones compared to foothills and lowlands (Coblentz & Riitters 2004). Species with environmental tolerances for montane climates and rocky slopes disperse into and colonize high elevations. Since such habitats are typically discontinuous, species occupying these regions have fragmented geographic ranges. Geologic intervals of global warming accentuate fragmentation as species ranges shift to higher elevations. Some populations may become new species as a consequence of sustained reproductive isolation. Geologic intervals of cooling cause species ranges to shift downslope, bringing fragmented populations into contact and providing opportunities for newly evolved species to disperse across the landscape. Some isolated populations at high elevations may disappear altogether, since alpine bioclimatic zones shrink during global warming and stochastic environmental or demographic factors heighten the risk of extinction for small populations (Williams & Jackson 2007; Sandel et al. 2011). With regard to diversification, the properties of mountains that distinguish them from lowlands include environmental heterogeneity resulting from high relief, numerous opportunities for fragmentation and isolation of microhabitats across complex topography and climatic changes over millions of years (Chapters 6, 12 and 13) (Brown 2001; Badgley 2010; Fjeldså et al. 2012).

14.2 Mammal Diversity Across Continents

At the continental scale, mammal diversity shows strong latitudinal and topographic gradients (Figures 14.2 and 14.3). In the New World (Figure 14.2a), the species richness (documented as the number of species per grid cell) of mammals is greatest at low latitudes. Species richness declines sharply north and south of 20–25°. The lowest values occur at high latitudes, where continental ice sheets (North America) or extensive montane and valley glaciers (South America) have dominated these landscapes for over 2 million years (Simpson 1964; Badgley & Fox 2000; Tognelli & Kelt 2004). In addition to the latitudinal diversity gradient, mammals also show strong topographic gradients in diversity. At any particular latitude in North and South America, species richness is substantially greater in montane regions than in adjacent plains or river basins. This pattern is most striking in the equatorial region, where a band of high diversity follows the Andes Mountains (Figure 14.2a). In North America, species richness rises sharply from the Great Plains to the Rocky Mountains.

Topographic gradients in elevation and relief (Figure 14.2b) reflect the history of tectonic plate movements, as well as macroclimatic gradients (Figure 14.2c) that influence temperature and precipitation, glaciation, erosion rates and the distribution of vegetation. Climatic variables show contrasting spatial gradients (Figure 14.2c). Temperature variables typically vary with latitude in response to the latitudinal gradient in the intensity of solar radiation and convective atmospheric circulation (e.g., Hadley cells). Precipitation gradients follow the major mountain ranges as a consequence of orographic processes, resulting in strong longitudinal gradients in mean annual precipitation in North and South America. The interactions of temperature and precipitation determine the dominant vegetation present in different bioclimatic regions (Woodward 1987). Consequently, the distribution of biomes shows strong latitudinal and longitudinal boundaries.

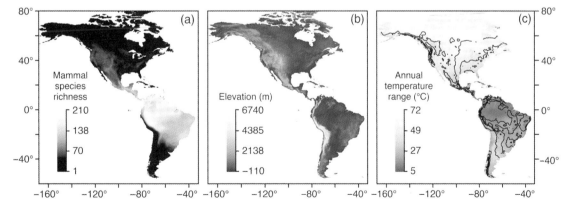

Figure 14.2 New World gradients of mammal richness, elevation and climatic variables. (a) Species richness of continental mammals, compiled at a spatial resolution of 10 km², in North and South America, based on species ranges from NatureServe. *Source:* Adapted from Patterson et al. (2007). (b) Mean elevation of grid cells at 1 km² resolution. Note that the species richness of mammals follows the topographic gradient as well as the latitudinal gradient. (c) Annual range of temperature (shading), a measure of climatic seasonality, shows a strong latitudinal gradient. Contours of mean annual precipitation follow the orientation of the major mountain ranges in North and South America. The contour interval is 500 mm for precipitation. Climatic variables and elevation are both significant predictors of mammal diversity. *Source:* Adapted from Badgley & Fox (2000) and Hijmans et al. (2005).

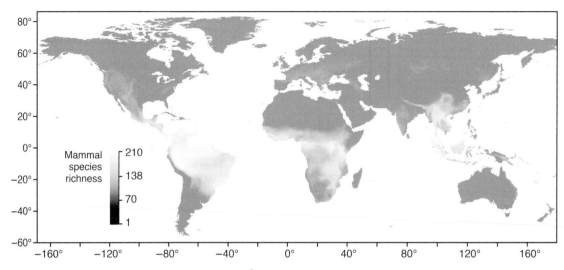

Figure 14.3 Richness of extant mammals, compiled for 10 km^2 grid cells. Overprinting the latitudinal diversity gradient is a strong topographic diversity gradient on all continents, excluding Antarctica. Mammalian species richness is elevated in the montane west of North America, along the Andes of South America, in the Alps of Europe, within the topographically complex East African Rift, over mountainous regions in India and south-eastern Asia and along the eastern coastal ranges of Australia. *Source:* Adapted from IUCN (2015). See also Plate 29 in color plate section.

Similar relationships among topography, climate and mammal diversity are found in Eurasia, Africa and Australia (Figure 14.3). In eastern Eurasia, the highest species richness of mammals occurs along the southern margin of the Himalayas, in montane Indochina and in the Malay Peninsula. In western Asia and Europe, high species richness occurs in the Caucasus Mountains, the Carpathian Mountains and the Alps. In Africa, a band of high species richness stretches across the Congo Basin and reaches its highest values along the western and eastern arms of the Rift Valley. In Australia, the highest species richness of mammals occurs in the mountain ranges and dissected plateaus that extend along the east coast (Strahan 1995). The topographic diversity gradient in mammals is most pronounced in regions of active tectonism, such as the Basin and Range Province of western North America, the Andes, the East African Rift and the southern edge of the Himalaya – regions where geologically recent topographic complexity and elevated species richness are evident. Montane regions also host high endemic diversity (number of small-range species) across the continents (Rosauer & Jetz 2015).

14.3 Topographic Diversity Gradients at the Regional Scale

A closer view of mammal diversity across topographic gradients is given in Figures 14.4–14.6, which depict transects across four present-day landscapes. Each transect spans an elevational gradient over several degrees of longitude along a single parallel. Based on detailed range maps for each region, we plotted the longitudinal span of species ranges in 1° bins of longitude. Notable changes in the number of species per bin indicate places where many range boundaries occur. For each transect, the gradients in elevation, mean annual precipitation and seasonality of precipitation illustrate some of the important topographic and climatic changes across each region.

The US state of Colorado straddles the boundary between the extensive Great Plains to the east and the Rocky Mountains to the west; these mountains form the eastern edge of a topographically complex landscape that extends westward to the Pacific Ocean. Strong elevational, orographic and diversity gradients span the transect at 40° N through northern Colorado (Figure 14.4a,b). Elevation increases from about 1000 m on the plains of eastern Colorado to over 3000 m in the Rocky Mountains (Figure 14.4a). Changes in mean annual precipitation closely match the elevation gradient, with high values over the higher peaks and low values to the west and the east; seasonality of precipitation shows almost the mirror opposite (Figure 14.4b). Shortgrass prairie occupies the plains east of the Front Range; montane conifer forest, sagebrush steppe and arid grasslands and canyons characterize the mountains and western part of the state. Species richness rises from about 50 species in eastern Colorado to over 90 species at the Rocky Mountain front, dips to 70 species within the high mountains, then rises again to over 90 species to the west on the Colorado Plateau (Figure 14.4a) (Fitzgerald et al. 1994). The highest spatial turnover occurs at the junction of the plains and the Front Range. Turnover remains high across the western part of the state, whereas it is low in eastern

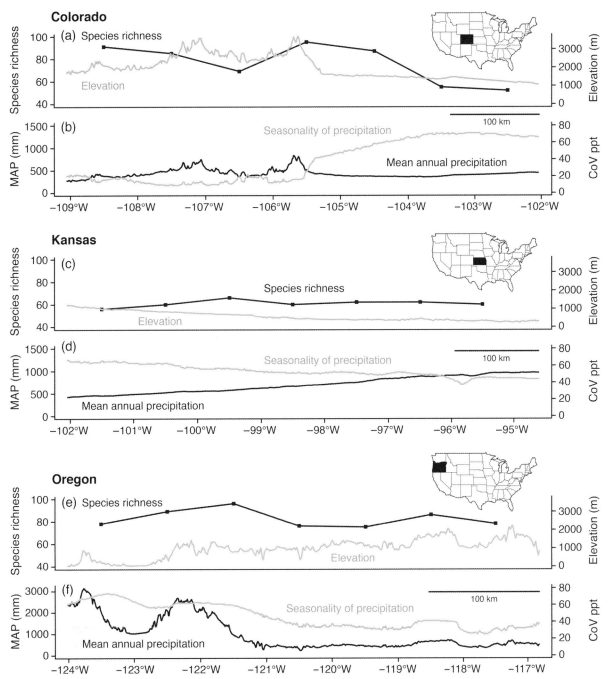

Figure 14.4 Transects illustrating topographic diversity gradients in relation to elevation and precipitation variables across three US states. Species richness is based on the distribution of species geographic ranges in 1° bins of longitude. Elevation is shown at a resolution of 1 km (lateral distance). Note that the vertical axis for mean annual precipitation changes from (b) and (d) to (f). *Source:* Data for elevation in each transect from US Geological Survey Center for Earth Resources Observation and Science (1996a). Data for mean annual precipitation (mm) and seasonality of precipitation (coefficient of variation of monthly precipitation) from the WorldClim database (Hijmans et al. 2005). (a) Transect across Colorado at 40° N. Species richness is low in eastern Colorado, where relief is low, then rises along the foothills and the front range of the Rocky Mountains around 105°W, dips slightly in the high mountains and rises again in western Colorado. *Source:* Species distributions from Fitzgerald et al. (1994). (b) Variation in mean annual precipitation (including snow) follows elevation closely, with the highest values in the high mountains. Seasonality of precipitation is almost the mirror image of precipitation, with low values in the mountains and high values over the plains of eastern Colorado. (c) Transect across Kansas at 39° N. Species richness fluctuates between 55 and 70 species per bin. Elevation increases from about 300 m in eastern Kansas to about 1200 m at the western border. *Source:* Species distributions from Bee (1981). (d) Mean annual precipitation steadily rises from west to east, the inverse of the trend in elevation; seasonality of precipitation varies little over eastern Kansas and increases slightly in the western half of the state. Gradients in species richness, elevation and precipitation variables provide a contrast with those from topographically complex regions in Colorado and Oregon. (e) Transect across Oregon at 45° N. Species richness ranges from 75 to 95 species per bin, with the highest values over the Cascade Mountains in western Oregon. The elevation profile captures the complex topography of the northern Basin and Range in the east and the Cascade Mountains, Willamette Valley and Pacific Coast Range in the west. *Source:* Species distributions from Verts & Carraway (1998). (f) Mean annual precipitation shows strong rain shadows from the Pacific coast and the Cascade Mountains in the west, with about 500 mm per year persisting over much of eastern Oregon. Seasonality of precipitation is higher in the west and declines across the eastern part of the state.

Colorado. This transect illustrates the topographic diversity gradient at the boundary between the tectonically active (western) and passive (eastern) regions of temperate North America (Badgley 2010).

The state of Kansas lies east of Colorado and is situated entirely within the Great Plains of central North America (Figure 14.4c,d). Although Kansas has the reputation of being "flatter than a pancake," the elevation increases gradually but substantially from east to west, with a gain of about 1000 m (Figure 14.4c). Mean annual precipitation is inversely proportional to elevation, whereas seasonality of precipitation increases from east to west (Figure 14.4d). Tallgrass prairie characterizes the native vegetation of eastern Kansas, while shortgrass prairie characterizes the western edge, with mixed-grass prairie in between. Species richness ranges from 55 to 67 species per bin across the state, with low levels of spatial turnover (data from Bee 1981). This transect offers a contrast in topography and diversity to the three that feature strong topographic gradients.

The state of Oregon lies in the Pacific Northwest and features complex topography throughout (Figure 14.4e,f). The transect at 45° N passes through the western edge of the northern Rocky Mountains along the eastern border, then through the Blue Mountains and the Columbia Plateau in the central region of the state. The Cascade Range, which contains active volcanoes, lies in the western third of the state, and is bordered on the west by the broad Willamette Valley. The mountain peaks are generally lower than 2500 m in elevation, but relief is on the

order of several hundred meters over short lateral distances (Figure 14.4e). The coast and west side of the Cascade Range both receive high annual precipitation and form part of a temperate rainforest, where some of the largest trees in the world grow year round. The east side of the Cascades and the interior receive much less rainfall, with grassland and sagebrush scrub extending over much of central and eastern Oregon. In contrast to Colorado and Kansas, seasonality of precipitation tracks mean annual precipitation across Oregon (Figure 14.4f). Species richness ranges between ~75 and 95 species per degree longitude, with higher values in the Blue Mountains and the Cascade Range (data from Verts & Carraway 1998).

The transect across Ecuador depicts a tropical topographic gradient across a high mountain range (Figure 14.5). This striking gradient along the equator includes the western edge of the Amazon Basin at low elevation in eastern Ecuador, the eastern Andes with peaks that reach nearly 5000 m in elevation, a high plateau bordered by the western Andes at about 3000 m, then a gradual descent towards the Pacific Ocean (Figure 14.5a). Mean annual precipitation shows a peak along the eastern Andes and another along the western foothills. Although both peaks feature mean annual precipitation of nearly 4000 mm per year, the eastern peak is accompanied by low seasonality of precipitation, whereas the western peak is highly seasonal (Figure 14.5b). Evergreen rainforest is the dominant vegetation in the eastern lowlands, with areas of savannah and scrub near

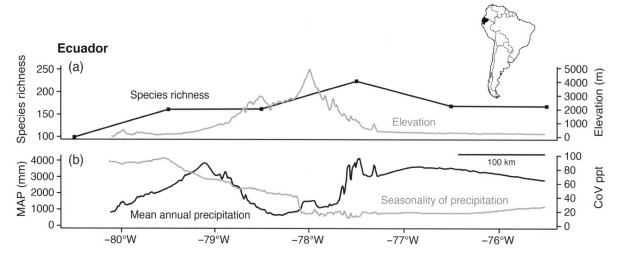

Figure 14.5 Transect across Ecuador at the equator. (a) Species richness for 1° bins of longitude, based on the distribution of geographic ranges that occur between 0 and 2° S. From east to west, species richness rises from ~160 species in the Ecuadorian Amazon Basin to ~225 along the eastern Andes, then declines towards the western Andes and the coast. Elevation, shown at a resolution of 1 km (lateral distance), is low east and west of the Andes and rises to nearly 5000 m in the eastern Andes. *Source:* Species richness adapted from Tirira (2007). Elevation data from US Geological Survey Center for Earth Resources Observation and Science (1996b). (b) The distribution of annual precipitation reflects moist air from Atlantic and Pacific sources, with peaks on the eastern and western flanks of the Andes. Mean annual precipitation is high throughout eastern Ecuador but declines steeply from the western Andes towards the Pacific coast. Seasonality of precipitation is highest in western Ecuador and declines across the mountains into the Amazon Basin. Note that the scale for all four variables differs from those in Figure 14.4. *Source:* Data for precipitation variables from the WorldClim database (Hijmans et al. 2005).

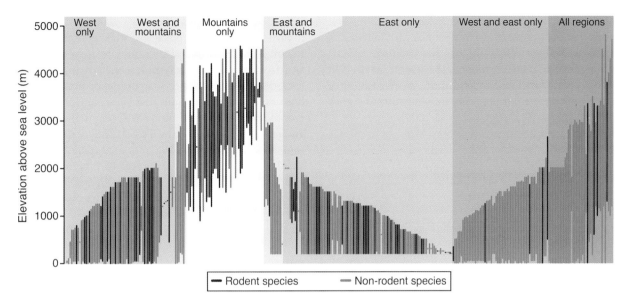

Figure 14.6 Elevational ranges for mammals in Ecuador, based on species from the transect in Figure 14.5. The range of each species is plotted as the lowest and highest reported elevational limits (Tirira 2007). Species are clustered geographically to illustrate the different geographic components of regional diversity. High regional richness is the consequence of species that uniquely occur at high elevations in combination with species that occur at lower elevations on either side of the Andes, as well as wide-ranging species.

the west coast. At higher elevations, a series of montane forest zones change to montane shrubland (páramo). The species richness of mammals ranges between 100 and 220 species per degree of longitude (data from Tirira 2007). From east to west, species richness rises from ~160 in the Amazon Basin to 220 on the eastern slopes of the Andes, then declines to ~160 in the western Andes and further to ~100 species along the west coast of Ecuador (Figure 14.5a).

The elevational ranges of Ecuadorian mammals illustrate several different geographic components of mammal diversity across topographic gradients (Figure 14.6). About 20% of Ecuadorian mammals occur west of the Andes, with elevational distributions mainly below 2000 m. Another 20% of species occur only in the mountains, with elevational ranges between 1000 and 4500 m. A few species' distributions are limited to the western side of the Andes at low to high elevations (west and mountains). About a dozen species occur only along the eastern slopes of the Andes from low to mid elevations (east and mountains). About 25% of species occur east of the Andes in the Amazon Basin. Nearly 20% of species occur both east and west of the Andes, but not at high elevations between these flanking regions. About 15% of species are found in all regions of Ecuador, some at lower elevations in river valleys and others with broad elevational ranges. The high regional diversity consists of species that occur uniquely on either side of the Andes at lower elevations, species that occur uniquely at high elevations and species with wide-ranging distributions. Notably, rodents make up more than half of the species that occur at only high elevations and about one-third of

the species that occur only west or only east of the mountains. In contrast, they make up around 10% of the species with more extensive longitudinal distributions. In addition to having more restricted geographic ranges, rodents often have lifestyles that are closely tied to properties of the substrate, which vary greatly over mountain slopes.

The four transects depicted in Figures 14.4–14.6 illustrate four general features of topographic diversity gradients for extant mammals: (i) the species richness of mammals rises sharply along major topographic boundaries between high and low elevation (Figures 14.4a and 14.5); (ii) regions of high topographic complexity show high diversity and high spatial turnover (Figures 14.4a,e, 14.5 and 14.6); (iii) regions of low topographic complexity show low diversity and low spatial turnover (Figure 14.4c); and (iv) the high diversity in montane regions consists of species restricted to high-elevation habitats (including many endemics), species that occur on either side of mountain ranges in heterogeneous habitats at lower elevations and wide-ranging species (Figure 14.6).

In contrast to transects across regions of high or low topographic complexity is a series of studies of mammal diversity in relation to elevation for individual mountains (Heaney 2001; Lomolino 2001; Rickart 2001; McCain 2004). The most common pattern is a peak in species richness at mid-elevations. The elevation at which diversity peaks is higher for taller mountains. Most of these elevational diversity gradients do not fit a mid-domain null model (the expectation that more species ranges overlap in the middle of a gradient due to spatial constraints at the end points) but correspond more closely to area effects or climatic gradients (McCain 2005). The elevation of maximum diversity declines with increasing latitude, when controlling for base elevation and mountain height. This pattern reflects the overall decline in mountain and snowline height and the greater extent of Quaternary continental and montane glaciation at higher latitudes (Körner 2007; Egholm et al. 2009).

14.4 Topographic Diversity Gradients in Deep Time

If the topographic diversity gradient is such a fundamental feature of mammal distributions over Earth's topography today, then the fossil record should also contain evidence for this gradient. However, evaluating the geologic record of high elevations is problematic for continental landscapes, because high-elevation environments are erosional rather than depositional, and thus are rarely present in the long-term stratigraphic record. Although mountains and plateaus may exist for tens of millions of years, the fossil record is preserved primarily in basins at low elevations. Caves, montane lakes and high-elevation playas are an exception to this pattern; these settings may persist for a few million years, but diminish over the long-term record.

Although it is not possible to document changes in diversity across ancient mountain ranges per se, as in Figures 14.4–14.6, it is feasible to compare diversity at lower elevations in topographically complex landscapes with diversity in plains and lowlands. Even at lower elevations preserved in the fossil record, topographically complex regions exhibit elevated diversity as a consequence of high spatial turnover across heterogeneous habitats (Davis 2005; Badgley et al. 2014). This pattern can be seen in present-day North America by comparing the species richness of mammals in the transect across Oregon at 1000 m elevation (Figure 14.4e) with the diversity of mammals in the transect across Kansas at 1000 m (Figure 14.4c). Across Oregon, species richness is greater than 70 per bin at the lowest elevations, whereas in western Kansas it is below 60 per bin at the highest elevations. The continental fossil record contains little of the high-elevation component of mammal diversity, such as the species in Figure 14.6 that occur only in mountains. But sedimentary sequences of river systems and lake margins preserve a fossil record of some of the species that inhabited foothills along with basin inhabitants at low elevations.

The North American fossil record of continental environments includes numerous widely distributed Cenozoic sequences of fluvial, lacustrine and airfall deposits from contrasting topographic landscapes (Woodburne 2004). A rich record of mammals occurs in fossil localities scattered across the tectonically active region from the Rocky Mountains to the Pacific Coast and from the tectonically quiescent Great Plains. This record allows us to contrast mammal diversity over millions of years in landscapes that were equivalent in topographic complexity to modern Oregon (Figure 14.4e) and modern Kansas (Figure 14.4c). Rodent diversity over time and space (Figure 14.7a) is especially informative, since rodents constitute over half of mammal diversity and their life habits are often closely tied to substrate properties, such as soil depth and texture (Merritt 2010). Figure 14.7b illustrates rodent diversity at the species level between 25 and 5 Ma (essentially over the Miocene) for the tectonically active and passive regions of North America. The areas sampled and the number of fossil localities in each region over time are commensurate, so these differences in diversity should not be an artifact of sampling (Finarelli & Badgley 2010).

The pattern of rodent diversity in tectonically active and passive regions of North America has three notable features (Figure 14.7b). First, diversity in the active region was lower than that in the passive region for about half of the 20 million-year record. Hence, if these records are accurate indicators of historical diversity, the topographic diversity gradient in North American mammals was absent for millions of years and present during extended time periods. Second, the strongest topographic diversity gradient occurred between 17 and 13 Ma, during an interval of global warming known as the Miocene Climatic Optimum (Zachos et al. 2001, 2008). During this interval, diversity in the active region greatly exceeded that in the passive region, reaching the highest diversity of the entire record. The highest per capita diversification rates also occurred during this middle Miocene interval (Finarelli & Badgley 2010). Using a Bayesian modeling approach, Silvestro and Schnitzler (Chapter 15) identified a peak in origination rates at this time for rodents in the active region but not in the passive region. Third, this interval featured widespread tectonic activity in the intermontane region, including eruptions of many of the Columbia Plateau flood basalts (Hooper et al. 2002), rapid extension and block faulting in the Basin and Range Province (McQuarrie & Wernicke 2005) and uplift in the northern Rocky Mountains (Fields et al. 1985).

This record offers two major insights into the topographic diversity gradient of mammals. First, this gradient is not a persistent feature over Earth history. Rather, high diversity in topographically complex regions is strongly expressed during certain periods and is absent in others. Second, the peak of diversity in the active region over the last 25 My, prior to today,

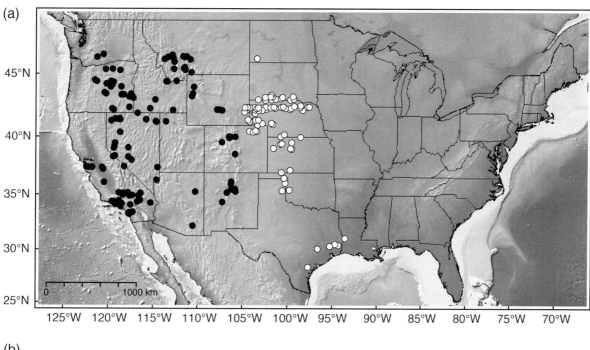

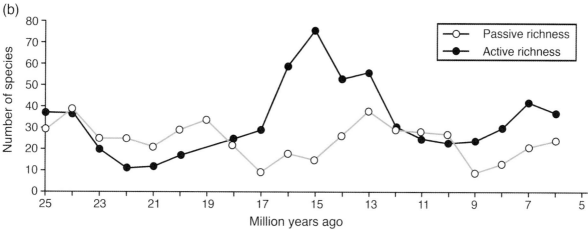

Figure 14.7 Fossil record of rodent richness from the tectonically active region of the western USA and the tectonically stable (passive) Great Plains. (a) Map of fossil localities from the western region (shaded circles) and the Great Plains (open circles) from 25 to 5 Ma. *Source:* Map made using GeoMapApp: www.geomapapp.org. (b) Rodent richness at the species level, compiled for 1 million-year intervals, from the western region (shaded circles) and the Great Plains (open circles). Rodent richness was much greater in the active region between 17 and 13 Ma, during an interval of global warming and widespread tectonic activity across western North America. *Source:* Adapted from Finarelli & Badgley (2010).

occurred when both climate change and tectonic activities were influencing the landscape – suggesting that the interaction of climate change and tectonic activity had a powerful influence on diversification (Finarelli & Badgley 2010). The modeling approach of Silvestro and Schnitzler (Chapter 15) confirmed this interpretation; in addition, they found that the rate of climate change had a greater impact on rodent diversification than did temperature per se. It is unclear how much of the diversity peak in Figure 14.7b resulted from speciation within the active region and how much resulted from immigration into the active region from the passive region or from areas beyond the documented record. It is likely that both processes contributed to the middle Miocene peak in diversity in the tectonically active region of western North America. The divergence of major lineages of heteromyid and sciurid rodents in western North America dates to the middle Miocene, according to molecular phylogenies (Hafner et al. 2007; Zelditch et al. 2015), providing evidence that this period of tectonic and climatic change coincided with cladogenesis.

14.5   Mammals that Drive the Topographic Diversity Gradient

Among the ca. 5400 species of extant mammals, most are small, weighing less than 1 kg in adult weight (Wilson & Reeder 2005). The most species-rich orders are Rodentia (rodents), Chiroptera (bats) and Soricomorpha (shrews, moles); together, these clades of primarily small mammals make up more than 70% of living mammal species and dominate latitudinal and elevational diversity gradients for mammals (Simpson 1964; Pagel et al. 1991; Patterson et al. 1998; Badgley & Fox 2000; Heaney 2001; McCain 2005). In North America, for example, the latitudinal diversity gradient is dominated by bats, whereas the topographic diversity gradient is dominated by rodents (Badgley & Fox 2000). Both groups exhibit high spatial turnover along strong environmental gradients. Within mountains and topographically complex regions, spatial turnover in diversity and taxonomic composition is greater for rodents than for other mammalian orders (Patterson et al. 1998; Badgley et al. 2014).

Small mammals show strong responses to landscape history over geologic time as well. Western North America has a robust fossil record for both large and small mammals over the last 25 My (Barnosky & Carrasco 2002; Kohn & Fremd 2008; Badgley & Finarelli 2013; Riddle et al. 2014a). The Great Basin, a topographically complex region with nearly 200 mountain ranges and intervening basins, expanded by over 200 km in longitudinal width through the geological process of extension (stretching and faulting) from 25 Ma to the present time (McQuarrie & Wernicke 2005). The fossil record of the Great Basin shows that large and small mammals responded quite differently to these changes in the landscape (Badgley et al. 2014). The species richness of large mammals declined slowly, with regional extinction of several groups, including equids (horses), rhinocerotids (rhinos) and camelids (camels). In contrast, the diversity of small mammals, including Rodentia and Lagomorpha (hares and rabbits), increased from 16 Ma to the present. Consequently, the proportion of rodent species within the total mammalian fauna increased from ~0.20 at 12 Ma to ~0.50 by 2 Ma, approaching the modern proportion of 0.63 in the Great Basin.

The elevated diversification of small mammals during the middle Miocene in western North America is also supported by molecular phylogenies and phylogeographic patterns. For example, species in the Heteromyidae (kangaroo rats and pocket mice) occur across the western USA, Mexico and Central America today. Molecular phylogenies reveal deep branching events among extant genera and species of heteromyids that occurred during episodes of mountain building over the last 20 My (Riddle 1995; Hafner et al. 2007). Spatial turnover in heteromyid species, as well as in other small-mammal clades, is high across the present-day North American landscape, with several species showing congruent patterns of geographic distribution linked to physiographic barriers (Riddle et al. 2014a). Likewise, in South America, rodents dominate high-elevation faunas and show high spatial turnover along elevational gradients (Figure 14.6). Radiations among several South American rodent groups coincided with the accelerated uplift of the Andes during the late Miocene (Hoorn et al. 2010).

Topographic, climatic and ecosystem changes influenced not only taxonomic diversity but also ecological diversity across space and time. North American rodents exhibit an impressive array of dietary and locomotor specializations that reflect adaptations to different food resources and substrates, as well as macro- and microhabitats found across topographically complex landscapes (Samuels & Van Valkenburgh 2008; Samuels 2009). The high diversity of rodents in the topographically complex region of western North America corresponds to a high frequency of small herbivores and granivores in regional and local faunas (Badgley & Fox 2000).

In contrast, some geographically and genetically distinct populations and species express little morphological or ecological divergence despite ranging across topographically complex regions. For example, a phylogenetic study of the heteromyid *Perognathus parvus* found deep divergences among populations lacking significant morphological differences, revealing the presence of a cryptic "species group" that resulted from geographic isolation among clades separated by physiographic barriers, including the Columbia Plateau and Snake River Plain (Riddle et al. 2014b). Likewise, in Sciuridae (squirrels), a decoupling of morphological disparity and taxonomic diversity among North American chipmunks and ground squirrels points to geographically mediated drivers of diversification during the tectonically active formation of western North America, rather than early rapid or steady phenotypic divergence (Zelditch et al. 2015). These examples indicate that regions of high topographic complexity can be sites of both adaptive and non-adaptive radiations in the evolutionary history of small mammals.

14.6   Biogeographic Processes in Topographically Complex Regions

The topographic diversity gradient results from the interaction of biogeographic and geomorphologic processes with Earth's climate system. Biogeographic processes, including dispersal and geographic-range shifts, speciation and extinction, have given rise to the taxonomic and ecological diversity of species whose geographic ranges are arrayed across Earth in a dynamic patchwork. The high diversity of mammals in topographically complex regions results from a high frequency of endemic species alongside a number of wide-ranging species, as in Ecuador (Figure 14.6). We present four examples in this section of biogeographic processes that contribute to the high diversity in montane regions, both in the present day and over Earth's history.

14.6.1 Geographic-Range Shifts

Species ranges expand or shift via dispersal. Dispersal is the movement of individual organisms out of their natal areas to colonize new areas (Lomolino et al. 2010). This process has contributed to changes in the topographic diversity gradient as species originating in lowland areas disperse into montane regions and species in montane regions expand into lowlands or cross drainage divides to colonize high-elevation habitats. Quaternary mammals experienced substantial shifts in their geographic ranges in response to glacial advances and retreats (Graham et al. 1996; Lyons 2003). For mammals from the contiguous 48 US states, the average shift in the centroid of the geographic range during the transitions from preglacial to glacial, glacial to Holocene and Holocene to modern time intervals was between 1200 and 1400 km (Lyons 2003), with shifts occurring along multiple geographic and climatic gradients. Species of larger adult size had larger range shifts than smaller species; however, both large and small species experienced significantly smaller range shifts in the topographically complex landscapes of the western USA than in the plains and lowlands of the eastern USA (Lyons et al. 2010). Montane species moved to lower elevations during glacial periods and to higher elevations during interglacials (Grayson 1987; Heaton 1990; Barnosky 2004).

Elevational range shifts have continued during the anthropogenic warming over the last few centuries. For example, a re-survey of small mammals (rodents, shrews and a pika) along an elevational gradient of 3000 m in Yosemite National Park, California between the early 20th and the early 21st century found that half of the species had shifted their elevational limits upward by 500 m (Moritz et al. 2008). During the 100 years of climate change between the first and second surveys, the average minimum temperature increased by ~3 °C.

14.6.2 Adaptations to High Elevation

High elevations differ in several physical properties that may require physiological or behavioral adaptations for sustained occupation. In addition to reduced land area at higher elevation, temperature and atmospheric pressure decrease with altitude (Körner 2007). The American pika (*Ochotona princeps*), a small lagomorph that inhabits montane slopes in western North America, tolerates cool temperatures well and warm temperatures poorly (Smith & Weston 1990). Some mammals that live continuously at high elevations have the ability to enter torpor (some bats) or hibernation (marmots, mountain pygmy possum) where temperature seasonality is high. Some Andean mice (*Phyllotis andium* and *P. xanthopygus*) have a greater consumption of carbohydrates than fats, compared to relatives at low elevations; carbohydrate metabolism yields more oxygen than lipid metabolism and is considered a strategy for coping with the effects of elevations over 4000 m (Schippers et al. 2012). Most mammals that inhabit high elevations are small and utilize underground microhabitats or extensive rock piles for warmth and shelter from predation. Larger mammals, such as ungulates, may migrate between higher and lower elevations according to the availability of forage. Adaptations for montane habitats contribute high-elevation species to the topographic diversity gradient.

14.6.3 Allopatric Speciation and Neo-Endemism

Populations at high elevations often occupy habitats that are discontinuous because of steep slopes, deep valleys and heterogeneous microclimates. In some regions, individual mountains or small mountain ranges are "sky islands," isolated by long distances separating high-elevation habitats. Thus, montane regions present numerous opportunities for the isolation of populations, in contrast to the distribution of the same species or sister taxa in nearby regions of low elevation and relief. Sustained isolation offers the opportunity for evolutionary divergence, which can result in genetic differentiation among populations, or in speciation. These processes contribute small-range endemic populations and species to topographic diversity gradients. For example, neoendemism and polymorphic populations in Californian mammals are concentrated in the Sierra Nevada Mountains, the coastal ranges and the Transverse and Peninsular ranges of southern California (Davis et al. 2008). Other montane regions with endemic mammals include the Ethiopian Highlands (Yalden & Largen 1992), the Trans-Mexican Volcanic Belt (Demastes et al. 2002) and the Andes (Tirira 2007). Recent divergence of species and populations in topographically complex regions is also the basis for recognizing many montane regions as areas of high conservation priority (Fjeldså et al. 2012; Rosauer & Jetz 2015).

The genetic structure of mammal populations often has a distinctive landscape signature in topographically complex regions but not in adjacent lowlands or plains, reflecting range shifts or range limits in response to topography. For example, geographic separation of distinct clades within two species of *Microdipodops* (kangaroo mice) predates Quaternary glacial cycles and instead aligns with complex topography in Nevada and increased aridification in the rain shadow of the Sierra Nevada Mountains (Hafner et al. 2008; Hafner & Upham 2011). A study of chipmunks (*Tamias*) found high phylogeographic structure and diversity in *T. amoenus* associated with specific mountain ranges from the north-western USA (Demboski & Sullivan 2003). In contrast, the phylogeographic structure of the eastern chipmunk (*T. striatus*) from the Great Plains and areas east of the Mississippi River reflects population expansion from glacial refugia in the south-eastern USA (Rowe et al. 2006).

For the American pika (*Ochotona princeps*) in western North America, Quaternary glacial cycles offered repeated opportunities for population fragmentation and coalescence. Range expansions during glacial intervals promoted genetic admixture, although montane glaciers and large pluvial lakes were barriers that enhanced isolation of populations during long glacial intervals (Galbreath et al. 2010). Phylogeographic structure in the American pika is congruent with patterns documented in co-distributed small mammals (Riddle 1996; Matocq 2002; Hafner et al. 2008; Neiswenter & Riddle 2010; Hafner & Upham 2011), suggesting that intermontane lineages share a common history of barriers to gene flow imposed by interactions between complex topography and climate.

14.6.4 Extinction of Montane Populations

Montane populations would seem to be more vulnerable to extinction than those of lowland regions because of their small areas of habitat and their isolation at high elevations. Many montane species may exist as metapopulations, with individual populations becoming extinct and dispersal leading to recolonization. Global warming should elevate extirpation rates of montane populations, as has been documented for the American pika in mountain ranges of the Great Basin (western USA) over the last century (Beever et al. 2003). Global cooling, in contrast, should reduce extinction rates in montane regions, as populations move downslope into larger areas and come into closer proximity with conspecific populations from other parts of the landscape (Brown & Kodric-Brown 1977).

For birds and plants, there is evidence that topographic complexity has buffered populations from extinction during Quaternary glacial cycles and earlier episodes of climate change (Bush et al. 2004; Fjeldså et al. 2012), due to the high heterogeneity of habitats and short distances among habitat refuges. If montane mammals have experi-

enced this influence, then reduced extinction rates would also contribute to high diversity in montane regions.

14.7 Effects of Modern Climate Change on Montane Diversity

Mammals in mountains are vulnerable to current and future climate changes. Like high latitudes, montane regions are predicted to experience some of the largest climate changes over the next 100 years, including the decline and loss of high-elevation alpine climates (Ackerly et al. 2010). Climatic warming impacts species individualistically as they shift their geographic ranges upward or poleward or adapt to new conditions, and alters biotic interactions under no-analog climates and during the formation of novel communities (Williams & Jackson 2007; Tylianakis et al. 2008; Gilman et al. 2010; Blois et al. 2013).

In western North America, upward shifts in mammalian ranges over the last century have been documented in three areas within the Sierra Nevada Mountains of California (Rowe et al. 2014) and in numerous mountain ranges of the Great Basin (Beever et al. 2003). Changes in small-mammal richness and associated changes in community abundance, structure and function in the intermontane west during anthropogenic climate warming and habitat transformation outpace natural variation and responses recorded over longer time scales in the fossil record (Grayson 2002; Blois et al. 2010; Terry et al. 2011; Terry & Rowe 2015). This combination of factors makes forecasting climate-change impacts within mountainous regions particularly challenging, yet it is critical for the protection of species, ecosystems and evolutionary processes in montane biodiversity hot spots.

14.8 Conclusion

Although mountains only cover about 10% of Earth's land surface, outside of Antarctica (Körner 2016), more than half of the world's continental mammals occur in montane regions today. This topographic diversity gradient is dominated by species of small mammals, which exhibit high spatial turnover in mountains and in topographically complex regions containing numerous mountain ranges and intervening basins. The fossil record suggests that this pattern has not characterized continental ecosystems throughout the Cenozoic. Instead, the topographic diversity gradient has been absent for millions of years and has been strongly expressed during intervals of climatic warming and tectonic activity.

A combination of ecological, evolutionary and historical processes has shaped the topographic diversity gradient of mammals. Evaluating the separate and combined influences of these processes presents an opportunity for integrating neontological and geohistorical data and perspectives (Badgley 2010; Graham et al. 2014). Protection of biodiversity in montane regions poses one of the major conservation challenges of the 21st century.

Acknowledgments

We thank Carol Abraczinskas for expert assistance with the figures, and Brett Riddle and Jan Schnitzler for constructive reviews.

References

Ackerly, D.D., Loarie, S.R., Cornwell, W.K. et al. (2010) The geography of climate change: implications for conservation biogeography. *Diversity and Distributions* **16**, 476–487.

Badgley, C. (2010) Tectonics, topography, and mammalian diversity. *Ecography* **33**, 220–231.

Badgley, C. & Fox, D.L. (2000) Ecological biogeography of North American mammals: species density and ecological structure in relation to environmental gradients. *Journal of Biogeography* **27**, 1437–1467.

Badgley, C. & Finarelli, J.A. (2013) Diversity dynamics of mammals in relation to tectonic and climatic history: comparison of three Neogene records from North America. *Paleobiology* **39**, 373–399.

Badgley, C., Smiley, T.M. & Finarelli, J.A. (2014) Great Basin mammal diversity in relation to landscape history. *Journal of Mammalogy* **95**, 1090–1106.

Barnosky, A.D. (ed.) (2004) *Biodiversity Response to Climate Change in the Middle Pleistocene.* Berkeley, CA: University of California Press.

Barnosky, A.D. & Carrasco, M.A. (2002) Effects of Oligo-Miocene global climate changes on mammalian species richness in the northwestern quarter of the USA. *Evolutionary Ecology Research* **4**, 811–841.

Bee, J.W. (1981) *Mammals in Kansas.* Lawrence, KS: University of Kansas.

Beever, E.A., Brussard, P.F. & Berger, J.B. (2003) Patterns of apparent extirpation among isolated populations of pikas (*Ochotona princeps*) in the Great Basin. *Journal of Mammalogy* **84**, 37–54.

Blois, J.L., Zarnetske, P.L., Fitzpatrick, M.C. & Finnegan, S. (2013) Climate change and the past, present, and future of biotic interactions. *Science* **341**, 499–504.

Blois, J.L., McGuire, J.L. & Hadly, E.A. (2010) Small mammal diversity loss in response to late-Pleistocene climatic change. *Nature* **465**, 771–774.

Brown, J.H. (2001) Mammals on mountainsides: elevational patterns of diversity. *Global Ecology and Biogeography* **10**, 101–109.

Brown, J.H. & Kodric-Brown, A. (1977) Turnover rates in insular biogeography: effect of immigration on extinction. *Ecology* **58**, 445–449.

Bush, M.B., Silman, M.R. & Urrego, D.H. (2004) 48 000 years of climate and forest change in a biodiversity hot spot. *Science* **303**, 827–829.

Champagnac, J.-D., Molnar, P., Sue, C. & Herman, F. (2012) Tectonics, climate, and mountain topography. *Journal of Geophysical Research* **117**, B02403.

Coblentz, D.D. & Riitters, K.H. (2004) Topographic controls on the regional-scale biodiversity of the south-western USA. *Journal of Biogeography* **31**, 1125–1138.

Davis, E.B. (2005) Mammalian beta diversity in the Great Basin, western USA: palaeontological data suggest deep origin of modern macroecological structure. *Global Ecology and Biogeography* **14**, 479–490.

Davis, E.B., Koo, M.S., Conroy, C.J. et al. (2008) The California Hotspots Project: identifying regions of rapid diversification of mammals. *Molecular Ecology* **17**, 120–138.

Demastes, J.W., Spradling, T.A., Hafner, M.S. et al. (2002) Systematics and phylogeography of pocket gophers in the genera *Cratogeomys* and *Pappogeomys*. *Molecular Phylogenetics and Evolution* **22**, 144–154.

Demboski, J.R. & Sullivan, J. (2003) Extensive mtDNA variation within the yellow-pine chipmunk, *Tamias amoenus* (Rodentia: Sciuridae), and phylogeographic inferences for northwest North America. *Molecular Phylogenetics and Evolution* **26**, 389–408.

Egholm, D.L., Nielsen, S.B., Pedersen, V.K. & Lesemann, J.-E. (2009) Glacial effects limiting mountain height. *Nature* **460**, 884–887.

Fields, R.W., Rasmussen, D.L., Tabrum, A.R. & Nichols, R. (1985) Cenozoic rocks of the intermontane basins of western Montana and eastern Idaho: a summary. In: Flores, R.M. & Kaplan, S.S. (eds.) *Cenozoic Paleogeography of the West-Central United States.* Denver, CO: Society of Economic Paleontologists and Mineralogists, pp. 9–10.

Finarelli, J.A. & Badgley, C. (2010) Diversity dynamics of Miocene mammals in relation to the history of tectonism and climate. *Proceedings of the Royal Society B: Biological Sciences* **277**, 2721–2726.

Fitzgerald, J.P., Meaney, C.A. & Armstrong, D.M. (1994) *Mammals of Colorado*. Denver, CO: Denver Museum of Natural History.

Fjeldså J., Bowie, R.C.K. & Rahbek, C. (2012) The role of mountain ranges in the diversification of birds. *Annual Review of Ecology, Evolution, and Systematics* **43**, 249–265.

Galbreath, K.E., Hafner, D.J., Zamudio, K.R. & Agnew, K. (2010) Isolation and introgression in the Intermountain West: contrasting gene genealogies reveal the complex biogeographic history of the American pika (*Ochotona princeps*). *Journal of Biogeography* **37**, 344–362.

Ghalambor, C.K., Huey, R.B., Martin, P.R. et al. (2006) Are mountain passes higher in the tropics? Janzen's hypothesis revisited. *Integrative and Comparative Biology* **46**, 5–17.

Gilman, S.E., Urban, M.C., Tewksbury, J. et al. (2010) A framework for community interactions under climate change. *Trends in Ecology & Evolution* **25**, 325–331.

Graham, C.H., Carnaval, A.C., Cadena, C.D. et al. (2014) The origin and maintenance of montane diversity: integrating evolutionary and ecological processes. *Ecography* **37**, 711–719.

Graham, R.W., Lundelius, E.L., Graham M.A. et al. (1996) Spatial response of mammals to late Quaternary environmental fluctuations. *Science* **272**, 1601–1606.

Grayson, D.K. (1987) The biogeographic history of small mammals in the Great Basin: observations on the last 20 000 years. *Journal of Mammalogy* **68**, 359–375.

Grayson, D.K. (2002). Great Basin mammals and late Quaternary climate history. *Great Basin aquatic systems history. Smithsonian Contributions to Earth Sciences* **33**, 369–386.

Hafner, J.C. & Upham, N.S. (2011) Phylogeography of the dark kangaroo mouse, *Microdipodops megacephalus*: cryptic lineages and dispersal routes in North America's Great Basin. *Journal of Biogeography* **38**, 1077–1097.

Hafner, J.C., Light, J.E., Hafner, D.J. et al. (2007) Basal clades and molecular systematics of heteromyid rodents. *Journal of Mammalogy* **88**, 1129–1145.

Hafner, J.C., Upham, N.S., Reddington, E. & Torres, C.W. (2008) Phylogeography of the pallid kangaroo mouse, *Microdipodops pallidus*: a sand-obligate endemic of the Great Basin, western North America. *Journal of Biogeography* **35**, 2102–2118.

Heaney, L.R. (2001) Small mammal diversity along elevational gradients in the Philippines: an assessment of patterns and hypotheses. *Global Ecology and Biogeography* **10**, 15–39.

Heaton, T.H. (1990) Quaternary mammals of the Great Basin: extinct giants, Pleistocene relicts, and recent immigrants. In Ross, R.M. & Allmon, W.D. (eds.) *Causes of Evolution: A Paleontological Perspective*. Chicago, IL: University of Chicago Press, pp. 422–465.

Hijmans, R.J., Cameron, S.E., Parra, J.L. et al. (2005) Very high resolution interpolated climate surfaces for global land areas. *International Journal of Climatology* **25**, 1965–1978.

Hooper, P.R., Binger, G.B. & Lees K.R. (2002) Ages of the Steens and Columbia River flood basalts and their relationship to extension-related calc-alkalic volcanism in eastern Oregon. *Geological Society of America Bulletin* **114**, 43–50.

Hoorn, C., Wesselingh, F.P., ter Steege, H. et al. (2010) Amazonia through time: Andean uplift, climate change, landscape evolution, and biodiversity. *Science* **330**, 927–931.

IUCN (2015) The IUCN Red List of Threatened Species. Version 2015-4. Available from: www.iucnredlist.org (last accessed August 30, 2017).

Janzen, D.H. (1967) Why mountain passes are higher in the tropics. *American Naturalist* **101**, 233–249.

Kohn, M.J. & Fremd, T.J. (2008) Miocene tectonics and climate forcing of biodiversity, western United States. *Geology* **36**, 783–786.

Körner, C., Jetz, W., Paulsen, J., Payne, D., Rudmann-Maurer, K. & Spehn, E.M. (2016) A global inventory of mountains for bio-geographical applications. *Alpine Botany* **127**, 1–15.

Körner, C. (2007) The use of "altitude" in ecological research. *Trends in Ecology & Evolution* **22**, 569–574.

Lomolino, M. (2001) Elevation gradients of species-density: historical and prospective views. *Global Ecology and Biogeography* **10**, 3–13.

Lomolino, M.V., Riddle, B.R. & Brown, J.H. (2010) *Biogeography*. Sunderland, MA: Sinauer Associates Inc.

Lyons, S.K. (2003) A quantitative assessment of the range shifts of Pleistocene mammals. *Journal of Mammalogy* **84**, 385–402.

Lyons, S.K., Wagner P.J. & Dzikiewicz, K. (2010) Ecological correlates of range shifts of Late Pleistocene mammals. *Philosophical Transactions of the Royal Society B: Biological Sciences* **365**, 3681–3693.

Matocq, M.D. (2002) Phylogeographical structure and regional history of the dusky-footed woodrat, *Neotoma fuscipes*. *Molecular Ecology* **11**, 229–242.

McCain, C.M. (2004) The mid-domain effect applied to elevational gradients: species richness of small mammals in Costa Rica. *Journal of Biogeography* **31**, 19–31.

McCain, C.M. (2005) Elevational gradients in diversity of small mammals. *Ecology* **86**, 366–372.

McQuarrie, N. & Wernicke, B.P. (2005) An animated tectonic reconstruction of southwestern North America since 36 Ma. *Geosphere* **1**, 147–172.

Merckx, V.S.F.T, Hendriks, K.P., Beentjes, K.K. et al. (2015) Evolution of endemism on a young tropical mountain. *Nature* **524**, 347–350.

Merritt, J.F. (2010) *The Biology of Small Mammals.* Baltimore, MD: Johns Hopkins University Press.

Moritz C., Patton, J.L., Conroy, C.J. et al. (2008) Impact of a century of climate change on small-mammal communities in Yosemite National Park, USA. *Science* **322**, 261–264.

Neiswenter, S.A. & Riddle, B.R. (2010) Diversification of the *Perognathus flavus* species group in emerging arid grasslands of western North America. *Journal of Mammalogy* **91**, 348–362.

Pagel, M.D., May, R.M. & Collie, A.R. (1991) Ecological aspects of the geographical distribution and diversity of mammalian species. *American Naturalist* **137**, 791–815.

Patterson, B.D., Stotz, D.F., Solari, S. et al. (1998) Contrasting patterns of elevational zonation for birds and mammals in the Andes of southeastern Peru. *Journal of Biogeography* **25**, 593–607.

Patterson, B.D., Ceballos, G., Sechrest, W. et al. (2007) *Digital Distribution Maps of the Mammals of the Western Hemisphere, version 3.0.* Arlington, VA: NatureServe.

Rahbek, C. (1995) The elevational gradient of species richness: a uniform pattern? *Ecography* **18**, 200–205.

Rickart, E.A. (2001) Elevational diversity gradients, biogeography and the structure of montane mammal communities in the intermountain region of North America. *Global Ecology and Biogeography* **10**, 77–100.

Riddle, B.R. (1995) Molecular biogeography in the pocket mice (*Perognathus* and *Chaetodipus*) and grasshopper mice (*Onychomys*): the late Cenozoic development of a North American aridlands rodent guild. *Journal of Mammalogy* **76**, 283–301.

Riddle, B.R. (1996) The molecular phylogeographic bridge between deep and shallow history in continental biotas. *Trends in Ecology & Evolution* **11**, 207–211.

Riddle, B.R., Jezkova T., Hornsby, A.D. & Matocq, M.D. (2014a) Assembling the modern Great Basin mammal biota: insights from molecular biogeography and the fossil record. *Journal of Mammalogy* **95**, 1107–1127.

Riddle, B.R., Jezkova, T., Eckstut, M.E. et al. (2014b) Cryptic divergence and revised species taxonomy within the Great Basin pocket mouse, *Perognathus parvus* (Peale, 1848), species group. *Journal of Mammalogy* **95**, 9–25.

Rosauer, D.F. & Jetz, W. (2015) Phylogenetic endemism in terrestrial mammals. *Global Ecology and Biogeography* **24**, 168–179.

Rowe, K.C., Heske, E.J. & Paige, K.N. (2006) Comparative phylogeography of eastern chipmunks and white-footed mice in relation to the individualistic nature of species. *Molecular Ecology* **15**, 4003–4020.

Rowe, K.C., Rowe, K.M.C., Tingley, M.W. et al. (2014) Spatially heterogeneous impact of climate change on small mammals of montane California. *Proceedings of the Royal Society B* **282**, 20141857.

Samuels, J.X. (2009) Cranial morphology and dietary habits of rodents. *Zoological Journal of the Linnean Society* **156**, 864–888.

Samuels, J.X. & Van Valkenburgh B. (2008) Skeletal indicators of locomotor adaptations in living and extinct rodents. *Journal of Morphology* **269**, 1387–1411.

Sandel, B., Arge, L., Dalsgaard, B. et al. (2011) The influence of Late Quaternary climate-change velocity on species endemism. *Science* **334**, 660–664.

Schippers, M.-P., Ramirez, O., Arana, M. et al. (2012) Increase in carbohydrate utilization in high-altitude Andean mice. *Current Biology* **22**, 2350–2354.

Simpson, G.G. (1964) Species density of North American recent mammals. *Systematic Zoology* **13**, 57–73.

Smith, A.T. & Weston, M.L. (1990) Ochotona princeps. *Mammalian Species* **352**, 1–8.

Strahan, R. (ed.) (1995) *Mammals of Australia.* Washington, DC: Smithsonian Institution Press.

Terry, R.C. & Rowe, R.J. (2015) Energy flow and functional compensation in Great Basin small mammals under natural and anthropogenic environmental change. *Proceedings of the National Academy of Sciences* **112**, 9656–9661.

Terry, R.C., Li, C.L. & Hadly, E.A. (2011) Predicting small-mammal responses to climatic warming: autecology, geographic range, and the Holocene fossil record. *Global Change Biology* **17**, 3019–3034.

Tirira, D. (2007) *Guía de campo de los mamíferos del Ecuador.* Quito: Ediciones Murciélago Blanco.

Tognelli, M.F. & Kelt, D.A. (2004) Analysis of determinants of mammalian species richness in South America using spatial autoregressive models. *Ecography* **27**, 427–436.

Tylianakis, J.M., Didham, R.K., Bascompte, J. & Wardle, D.A. (2008) Global change and species interactions in terrestrial ecosystems. *Ecology Letters* **11**, 1351–1363.

US Geological Survey Center for Earth Resources Observation and Science (1996) 30 Arc-Second DEM of North America. Available from: http://app.databasin. org/app/pages/datasetPage.jsp?id=d2198be9d2264de 19cb93fe6a380b69c (accessed via Conservation Biology Institute Data Basin, 27 February 2013).

US Geological Survey Center for Earth Resources Observation and Science (1996b) 30 Arc-Second DEM of South America. Available from: http://app.databasin. org/app/pages/datasetPage.jsp?id=d8b7e23f724d46c99 db1421623fd1b4f (accessed via Conservation Biology Institute Data Basin, 27 February 2013).

Verts, B.J. & Carraway, L.N. (1998) *Land Mammals of Oregon*. Berkeley, CA: University of California Press.

Williams, J.W. & Jackson, S.T. (2007) Novel climates, no-analog communities, and ecological surprises. *Frontiers in Ecology and the Environment* **5**, 475–482.

Wilson, D.E. & Reeder, D.M. (eds.) (2005) *Mammal Species of the World: A Taxonomic and Geographic Reference*, Vol. **12**, Baltimore, MD: Johns Hopkins University Press.

Woodburne, M.O. (2004) *Late Cretaceous and Cenozoic mammals of North America: biostratigraphy and geochronology*. New York: Columbia University Press.

Woodward, F.I. (1987) *Climate and Plant Distribution*. Cambridge: Cambridge University Press.

Yalden, D.W. & Largen, M.J. (1992) The endemic mammals of Ethiopia. *Mammal Review* **22**, 115–150.

Zachos, J., Pagani, M., Sloan, L. et al. (2001) Trends, rhythms, and aberrations in global climate 65 Ma to present. *Science* **292**, 686–693.

Zachos, J.C., Dickens, G.R. & Zeebe, R.E. (2008) An early Cenozoic perspective on greenhouse warming and carbon-cycle dynamics. *Nature* **451**, 279–283.

Zelditch, M.L., Li J., Tran L.A.P. & Swiderski, D.L. (2015) Relationships of diversity, disparity, and their evolutionary rates in squirrels (Sciuridae). *Evolution* **69**, 1284–1300.

15

Inferring Macroevolutionary Dynamics in Mountain Systems from Fossils

Daniele Silvestro[1,2] and Jan Schnitzler[3]

[1] *Department of Biological and Environmental Sciences, University of Gothenburg, Gothenburg, Sweden*
[2] *Gothenburg Global Biodiversity Centre, Gothenburg, Sweden*
[3] *Institute of Biology, Leipzig University, Leipzig, Germany*

Abstract

Mountain regions include some of the most species-rich places on Earth, with exceptionally high numbers of animals, plants and other organisms, and have been described both as cradles and as museums of biodiversity. Patterns of species diversity are governed by the complex interactions of biotic and abiotic factors, but are ultimately the result of immigration, in situ speciation and local extinction. Associations between the environment and species diversity can be studied using molecular phylogenetic trees of extant taxa and/or the fossil record to infer past changes in species diversity. In this chapter, we explore how mountain building, driven by volcanism and plate tectonics, together with climatic change, alters the regional environment and affects rates of species diversification. We first describe the expected dynamics of the key biological processes (focusing on speciation and extinction), and then illustrate ways to explore evolutionary hypotheses linking mountain building to speciation and extinction using fossil occurrence data and birth–death stochastic models. Finally, we demonstrate the applica tion of these models using the paleontological record of rodents in three adjacent regions in North America with different geological histories. Our analyses strongly support macroevolutionary models in which speciation and extinction rates vary within time intervals (based on local geological or climatic events) and as a function of global climate. Within the geographic and temporal limits of our analyses, the rate of global climate change appears to be more important in shaping rodent diversity than temperature itself. Our results highlight that geologic and climatic factors together best explain diversification dynamics, although it is often difficult to untangle the relative importance of each. With the increasing amount of data available in public databases, fossils offer a great opportunity to investigate diversification dynamics, particularly of formerly species-rich clades that today include only a small extent of their past diversity. Eventually, however, paleontological and neontological data need to be combined in order to infer macroevolutionary dynamics from both fossil and phylogenetic data.

Keywords: *speciation, extinction, fossil occurrences, diversification, rodents, PyRate*

15.1 Introduction

Mountain regions include some of the most species-rich places on Earth, with exceptionally high numbers of animals, plants and other organisms (Fjeldså & Lovett 1997; Badgley 2010; Favre et al. 2015). There are poten tially many biotic and abiotic factors behind these diver sity patterns, but ultimately the key processes that affect species diversity are immigration, in situ speciation and local extinction. The dynamics of these processes have traditionally been inferred using data from the fossil record (Simpson 1944; Stanley 1979), but advances in

molecular phylogenetics over the past several decades have created opportunities to also estimate diversifica tion rates from phylogenetic trees of extant organisms (Chapter 18) (Harvey et al. 1994; Sanderson & Donoghue 1996; Paradis 2004; Silvestro et al. 2011). Yet, fossil data often provide the most direct evidence of biodiversity changes in time and space (Silvestro et al. 2015b, 2016), allowing robust inferences of macroevolutionary dynam ics (Didier et al. 2012; Fritz et al. 2013; Schnitzler et al. 2017). Mountain ranges have been described both as cradles and as museums of biodiversity, referring to the fact that in these areas rates of speciation tend to be

Mountains, Climate and Biodiversity, First Edition. Edited by Carina Hoorn, Allison Perrigo and Alexandre Antonelli.
© 2018 John Wiley & Sons Ltd. Published 2018 by John Wiley & Sons Ltd.
Companion website: www.wiley.com\go\hoorn\mountains,climateandbiodiversity

high and that species are thought to have a higher chance of surviving changing environmental conditions (Kier et al. 2009). Accordingly, the diversification of entire clades has been linked to the uplift of major mountain chains (Hughes & Eastwood 2006). Over geological time scales, mountain systems undergo substantial alterations that affect not only their geomorphology, but also the regional and global climate (Chapters 2, 4 and 8). Associations between the environment and species diversity can be studied using molecular phylogenetic trees of extant taxa (Chapter 18) and/or the fossil record, which allow for the reconstruction of past changes in species diversity. The processes that lead to high biodiversity in mountainous regions can be complex, especially during periods with major tectonic activity and associated climatic change.

In this chapter, we explore how mountain building, driven by volcanism and plate tectonics, together with climatic change, alter the regional environment and affects species diversity. The major challenge is to understand how speciation and extinction are influenced by geological dynamics per se (e.g., mountain uplift) and by the climatic and landscape heterogeneity resulting from such geological activity. We first describe the expected dynamics of the key biological processes (focusing on speciation and extinction) and how they might be driven and altered by geological changes. We further explain how such expectations translate into patterns in the fossil record. This is crucial to formulating plausible hypotheses that can be tested against existing data. Finally, we discuss how understanding the origin of biodiversity in mountain regions depends on unraveling the dynamics of dispersal, speciation and extinction, and we show how changes in the rates of speciation and extinction may be decoupled.

15.2 Geological and Evolutionary Dynamics

Earth surface processes reflect the dynamics of lithospheric deformation through changes in (paleo)topography, erosion and weathering, which affect regional and global environments. Tectonic activity, erosion and climate are tightly linked with reciprocal effects. Mountain uplift can change climate, for example by creating rain shadows, altering air mass circulation or decreasing local temperatures. On the other hand, feedback such as a climate-driven increase of erosion rates and chemical weathering can influence the internal dynamics of actively deforming mountain ranges (Chapter 4) (Whipple 2009 and references therein; Murphy et al. 2016). In this section, we outline how Earth surface processes, climate change and biological diversification may be linked. While this will likely not cover all possible scenarios, our examples highlight some of the most common evolutionary dynam-

ics and can serve as a starting point from which to evaluate the interconnections between large-scale geological and biological processes.

As tectonic activity increases and Earth's surface is deformed and lifted, the regional environment starts to change and new habitats are created (Figure 15.1a). As a result, habitat heterogeneity increases relative to adjacent lowland areas, providing opportunities for geographic isolation and the disruption of gene flow among populations, and thus supplying an ideal context for speciation (see also Chapter 13). However, while some lineages may adapt and potentially diverge depending on the speed of these changes, the suitability of novel habitats and the adaptability of the species present, others will fail to cope with the changing conditions because their habitats are altered too rapidly or disappear entirely, or due to competitive exclusion by invading lineages. Thus, mountain uplift may lead to an increase in the rate of speciation, local extinction or both, corresponding to the rate of tectonic and environmental change (Figure 15.1c). Continued tectonic change may, however, have more profound impacts, especially if a "tipping point" is reached. Tipping points are critical thresholds at which small changes will cause a substantial shift in a system, which may not be reversible. These can include even relatively small geological changes, which have profound impacts on regional and/or global climates and can result in rapid environmental changes over short periods of time, substantially altering the conditions necessary for species to survive. For example, Mulch et al. (2008) suggested that surface uplift associated with tectonic activities in the Eastern Cordillera could have led to a change in atmospheric circulation through the deflection of the South American low-level jet along the eastern flanks of the central Andes, resulting in markedly different regional rainfall patterns. If such changes are too fast for species to adapt, these events will lead to a peak in the extinction rate and the disappearance of many lineages (Figure 15.1c). Such tipping-point events are, however, comparatively rare, and biodiversity changes in mountain regions are likely to result from complex interactions of evolutionary processes responding to both gradual and rapid environmental changes.

After numerous lineages go locally extinct, the rate of extinction will eventually decline again (Figure 15.1c), while at the same time providing the opportunity for colonization by other species, which benefit from a growing amount of unoccupied environmental space (niche space). These may be either lineages that have survived in the region or lineages migrating from similar environments elsewhere with some degree of pre-adaptation. These lineages will encounter a large amount of available environmental space, while competition for resources will be low, potentially leading to ecological differentiation and a rapid increase in local diversity

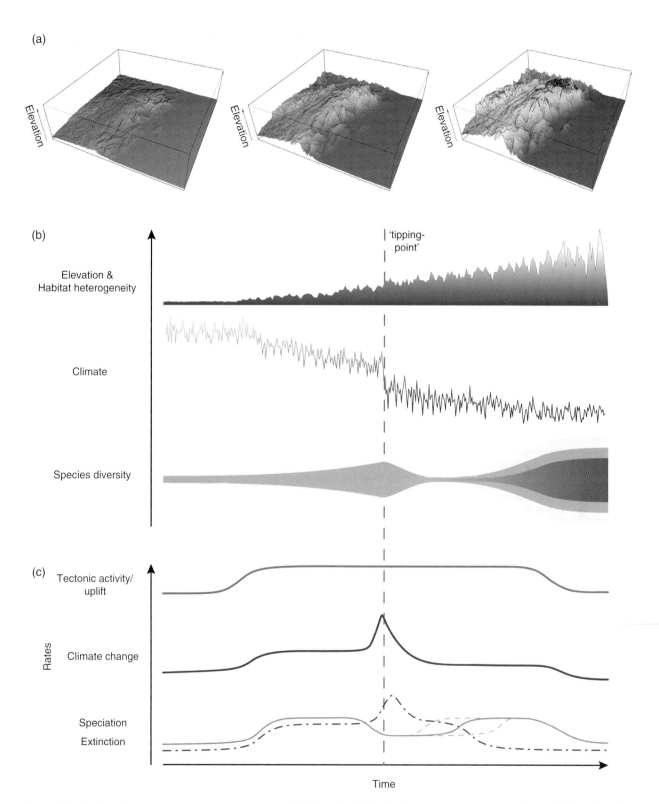

Figure 15.1 Relationships between mountains, climate and biodiversity. (a) Uplift of a mountain range through time. (b) Associated changes in elevation, habitat heterogeneity, climate and species diversity. (c) Rates of change corresponding to the different processes involved, namely the rate of tectonic activity/uplift, the rate of climatic change and the rates of speciation (gray line) and local extinction (dashed line). The gradual change of the relief increases topographic complexity, creates novel habitats and affects regional climatic conditions. Such changes, even if moderate, will likely affect the rates of speciation (afforded by adaptation to the novel conditions and divergence) and local extinction (if lineages fail to adapt). Continued tectonic change may result in a state shift ("tipping point"), causing profound and rapid climatic changes. A corresponding peak in species extinction is followed by an increase in immigration and in situ speciation, for example by pre-adapted lineages from other regions. As clades diversify, they fill ecological niche space, and the rate of speciation slows again. See also Plate 30 in color plate section.

(adaptive radiation). This process may occur shortly after environmental changes (especially if adaptation and diversification within lineages already present predominates), or following a dispersal event and the successful establishment of a lineage – in some cases, even millions of years later (Figure 15.1c). Thus, changes in the rates of speciation and extinction may become decoupled, with extinction more immediately associated with rapid environmental changes, while new radiations are dependent on successful immigration events and may start at any point (indicated by the dashed lines in Figure 15.1c). If this takes place during a phase of global (or large-scale) climate change, species diversification and extinction will likely respond to both local tectonic activity and global change. As species richness increases and environmental space is filled, biotic interactions may again become stronger. With an increasing number of species, resources will become more limited and opportunities for ecological divergence will decrease, negatively affecting the rate of speciation (diversity dependence).

15.3   Case Study: Rodent Diversification in North America

In this section, we illustrate some of the ways to explore the evolutionary hypotheses linking mountain building to species diversification and extinction using fossil occurrence data and birth–death stochastic models. We take, as an empirical example, the paleontological record of rodents in three adjacent regions in North America with different geological histories, following a study by Badgley & Finarelli (2013). This study shows that the highest rates of speciation coincided with intervals of intensified tectonic activity in the montane regions and that rates of speciation and extinction changed in relation to the regional climate, in particular around the Miocene Climatic Optimum (MCO; 15–17 Ma). Thus, Badgley & Finarelli (2013) suggested that topographic complexity and climatic changes have driven rodent diversification. Here, we transfer these conclusions into an explicit model-testing framework to show how statistical tools can be used to discriminate the fit of alternative diversification scenarios, and to evaluate the strength of the correlation between species diversification and potential external drivers. We examine the impact of geologic events, as well as local and global climate change, on rates of speciation and extinction in rodents, and compare these effects among regions using a Bayesian probabilistic framework (implemented in the program PyRate; Silvestro et al. 2014a,b), which allows us to explicitly model different evolutionary hypotheses and rank alternative diversification scenarios by their relative probabilities (Box 15.1).

For a detailed description of rodent species diversity patterns in North America, see Chapter 14.

15.3.1   The Three Regions

Region 1 spans the Cascade Range in the US states of Oregon and Washington (Figure 15.3). Today, it is a montane region with a mean elevation around 1350 m, resulting from a 40 My-history of intense tectonic and volcanic activity. Volcanism associated with the Yellowstone hot spot affected the south-eastern part of the region between 16 and 13 Ma (Pierce & Morgan 1992). The Cascades in northern Washington were characterized by high elevation and relief during the Miocene, causing a rain shadow towards the east from around 16 Ma. The southern Cascades later increased in elevation and relief, between 12 and 8 Ma, intensifying the rain shadow around 8 Ma (Mitchell & Montgomery 2006). At the same time, mesic forest covering the region during most of the Miocene was replaced in the east by vegetation adapted to dry summers (Leopold & Denton 1987). Today, the region is characterized by temperate rainforests west of the Cascades, montane conifer forest in the mountains and semi-desert vegetation east of the Cascades (Verts & Carraway 1998).

Region 2 encompasses the northern Rocky Mountains in Idaho, Montana and Wyoming (Figure 15.3). Similar to Region 1, this area experienced intense tectonic activity during the Cenozoic. Extensive volcanism took place from the early Cenozoic to the Eocene (56 Ma) and intensified again in the late Miocene (10 Ma). This was associated with the migration of the Yellowstone hot spot to its present position at the center of Region 2 (Pierce & Morgan 1992). The climate experienced aridification during the Neogene (23.0–2.58 Ma), interrupted only by a phase of warm and humid climate around the MCO (18–16 Ma). Following the MCO, uplift and erosion caused increased relief in the region and produced a rain shadow to the east that lasted until the end of the Miocene (5.3 Ma) (Elliott et al. 2003). Today, the region is characterized by a high mean elevation (2100 m) and relief, and by low temperatures (Badgley & Fox 2000).

Region 3 is part of the northern Great Plains, an area that has been tectonically quiescent throughout the Cenozoic, but which has been indirectly affected by the tectonic activity in adjacent Region 2 (Figure 15.3) (Swinehart et al. 1985). With the renewed uplift of the Rocky Mountains ca. 19 Ma, Region 3 experienced increased erosion and the formation of river systems. Incision and relief intensified again around 5 Ma. The climate was warm and increased in aridity in the middle Miocene (18–12 Ma), then cooled and increased in humidity during 12–5 Ma (Passey et al. 2002). From the late Oligocene (26 Ma), the vegetation was increasingly

---

Box 15.1 Estimating diversification dynamics from fossil data: the methodological background

Here, we explain the conceptual and methodological background of the statistical analyses presented in this chapter. PyRate uses a hierarchical Bayesian framework to incorporate different sources of uncertainties in its analysis and to provide posterior estimates of the parameters of interest, along with their 95% credible intervals (95% CIs). Parameter estimation is obtained through a Markov chain Monte Carlo (MCMC) simulation to sample rates from their joint posterior distribution.

Preservation

The preservation process is modeled as a nonhomogeneous Poisson process (Liow et al. 2010; Silvestro et al. 2014b), in which the preservation rate is estimated from the data and represents the expected number of fossil occurrences/lineage per My. While a single mean preservation rate (q) is estimated from the entire data set to avoid overparameterization, rate heterogeneity across species is incorporated in the analysis using a Gamma model analogous to those commonly implemented in molecular substitution models (Yang 1994; Silvestro et al. 2014b). Under the Gamma model, preservation rates are assumed to vary among lineages according to a Gamma distribution, with mean q and shape parameter α, both of which are estimated from the data. For any given value of q, the parameter α determines the amount of rate heterogeneity across species in a data set (Table 15.1).

Species diversification

Species diversification is modeled as a birth–death stochastic process (see also Chapter 18), in which speciations and extinctions occur as random events in continuous time. The temporal distribution of speciation and extinction times (s

and e, respectively, as inferred based on the preservation process; Figure 15.2) allows us to estimate the birth–death rates and to quantify the expected number of speciation/extinction events per lineage per My. The basic time-homogeneous birth–death model can be modified to incorporate temporal rate variation (see Section 15.4 and Table 15.1). In the case of birth–death models with a correlation to paleoenvironmental changes, variations in speciation and extinction rates are determined by exponential correlations with the covariate. For a given environmental value θ_t at time t, the resulting speciation rate is:

$$\lambda_t = \lambda_0 \, exp(G_\lambda \, \theta_t) \qquad (15.1)$$

where G_λ indicates the estimated strength and sign of the correlation (Table 15.1). An analogous equation is used to calculate μ_t from μ_0 and G_μ. Prior to the analyses, the values of the environmental curve are adjusted so that $\theta_0 = 0$; that is, the curve equals zero at the present. Thus, the baseline rates λ_0 and μ_0 represent the speciation and extinction rates at the present, regardless of the estimated values of G_λ and G_μ (Silvestro et al. 2015a).

Model testing

The statistical support for alternative birth–death models (e.g., with a different number of rate shifts or paleoenvironmental covariates) can be estimated by their respective marginal likelihoods. The marginal likelihood of each model is quantified through MCMC using an algorithm named "path sampling" (Kass & Raftery 1995; Lartillot & Philippe 2006) and measures how well a model fits the data (higher values indicate stronger support relative to alternative models). We compared the marginal likelihoods using Bayes factors (Kass & Raftery 1995) and calculated the relative probabilities of all models (e.g., Silvestro et al. 2014c).

---

characterized by open-habitat grasses (Strömberg 2011), and from the Miocene to the Quaternary the fraction of C_4 vegetation steadily increased, reaching 80% of the biomass (Fox & Koch 2003).

15.3.2 Rodent Fossil Occurrences

We compiled three data sets combining fossil occurrences from the Paleobiology Database (www.paleobiodb.org) and NeoMap (www.ucmp.berkeley.edu/neomap/), both accessed in January 2016. We selected all fossil occurrences attributed to the mammalian order Rodentia sampled within the regions, following the geographic boundaries defined by Badgley & Finarelli (2013). As in their study, we considered species with an

observed lifespan of <1 My to be singletons and removed them from the analysis. This procedure is commonly applied in fossil analyses to remove potential biases (Foote 2003), especially when singletons represent a substantial proportion of taxa.

The data set from Region 1 included 479 fossil occurrences and 77 species, after removing 70 species as singletons. The oldest occurrence was dated to ca. 32 Ma. The fossil record in Region 2 comprised 282 fossil occurrences and 69 species, after excluding 118 species as singletons. The oldest occurrence was dated to ca. 56 Ma. Finally, the record in Region 3 comprised 1504 fossil occurrences and 177 species, after excluding 173 species as singletons. The oldest occurrence was dated to ca. 37 Ma.

Table 15.1 List of the main parameters inferred within the PyRate analytical framework.

	Notation	Name	Unit/Interpretation
Parameters of the preservation process	q	Mean preservation rate	Expected number of fossil occurrences per lineage per My
	α	Rate heterogeneity among lineages	Shape parameter of the gamma distribution modeling rate heterogeneity. Small values (e.g., $\alpha < 5$) indicate a high degree of heterogeneity. Large values ($\alpha > 15$) indicate roughly homogeneous rates
	s, e	Times of speciation (or immigration) and extinction for each lineage	Lifespan of each lineage inferred from fossil occurrences and preservation processes
Parameters of a constant birth–death process	λ	Speciation or origination rate	Expected number of origination events per lineage per My*
	μ	Extinction rate	Expected number of extinction events per lineage per My
Parameters of a birth–death process with predefined rate shifts	$\boldsymbol{\lambda} = [\lambda_1, ..., \lambda_S]$	Speciation or origination rates within each time frame	As above
	$\boldsymbol{\mu} = [\mu_1, ..., \mu_S]$	Extinction rates within each time frame	As above
Parameters of a birth–death process with paleoenvironmental covariate	λ_0	Baseline speciation rate	Speciation rate at the present
	μ_0	Baseline extinction rate	Extinction rate at the present
	G_λ	Correlation parameter (speciation)	G_λ quantifies the exponential correlation between speciation rate and a covariate $G_\lambda \approx 0$: no correlation $G_\lambda > 0$: positive correlation $G_\lambda < 0$: negative correlation
	G_μ	Correlation parameter (extinction)	As above

Source: Adapted from Pires et al. (2015).

*When looking at the fossil record within a limited geographic region, origination events may include both in situ speciation and immigration of lineages (Pires et al. 2015).

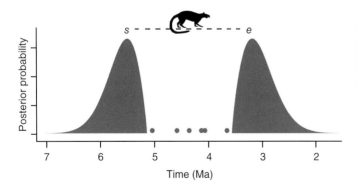

Figure 15.2 Example showing the estimation of speciation and extinction times (*s, e*) for a species, while modeling the preservation process. Circles indicate dated fossil occurrences; curves represent the resulting posterior distributions of *s* and *e*. *Source:* Silhouette from http://phylopic.org. Licensed under CC-BY SA 3.0.

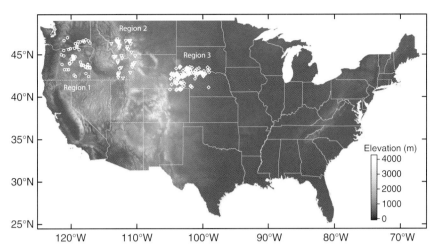

Figure 15.3 Map of North American rodent fossil localities. Localities in Regions 1–3 from 56 to <0.0002 Ma were used for the diversification analysis (circles, Region 1; triangles, Region 2; diamonds, Region 3). The base map shows mean elevation per 1 km² grid cell. *Source:* Data from www.worldclim.com.

15.4 PyRate Analytical Framework

We analyzed the fossil data sets using the open-source program PyRate (available from https://github.com/dsilvestro/PyRate) (Silvestro et al. 2014a). This program performs a joint estimation of the preservation and the diversification processes in continuous time and provides a robust probabilistic framework in which to assess the statistical fit of different speciation and extinction models. Incorporating the preservation process when inferring speciation and extinction rates from fossil data is crucial, because it provides the means to account for the inevitable incompleteness of the paleontological record. For the sake of this research, "preservation" comprises fossilization, sampling and taxonomic identification (see Box 15.1 for details). Based on the preservation process, PyRate estimates the times of speciation and extinction for each lineage (Box 15.1, Figure 15.2). This procedure is important because the oldest and most recent fossil occurrences of a species are likely to underestimate the true extent of its lifespan (Liow & Stenseth 2007).

Species diversification is modeled by a birth–death process in which the appearance of new species is governed by a speciation rate (λ) and their disappearance is determined by an extinction rate (μ). Speciation and extinction rates are estimated from the temporal distribution of lineages and their lifespans, as inferred from the preservation process and fossil occurrences (Box 15.1, Figure 15.4). We emphasize that, when analyzing the diversity dynamics within a region, species appearances and disappearances may also be linked to the immigration or emigration from and to other regions. In this case, speciation and extinction rates should be interpreted as rates of in situ speciation plus immigration and of local extinction, respectively (Pires et al. 2015). The simplest birth–death model assumes that

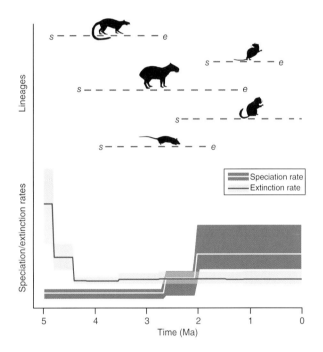

Figure 15.4 Estimation of speciation and extinction rates from a set of species, with origination and extinction times inferred from fossil occurrences (see Figure 15.2). Speciation and extinction rates (white and black curves, respectively) may vary through time, depending on the birth–death model used in the analysis. Shaded areas indicate 95% credible intervals. *Source:* Silhouettes from http://phylopic.org. Licensed under CC BY- SA 3.0.

speciation and extinction rates are constant through time. In the following analyses, we incorporated rate variation through time in three ways: (i) by introducing times of rate shifts (Silvestro et al. 2014b); (ii) by estimating the correlation between a curve (approximating a climatic variable through time) and the birth–death rates (Silvestro et al. 2015a); and (iii) by combining rate shifts and an environmental covariate.

15.4.1 Analysis Settings

We compared the fit of six birth–death models (BD1–6) in order to assess the impact of local and global environmental changes on rodent diversification in the three regions under discussion. The first model (BD1) assumed constant speciation and extinction rates through time. In the second (BD2), speciation and extinction rates varied through time following rate-shift events (BD2). When assuming a birth–death model with rate shifts, PyRate estimates speciation and extinction rates within each predefined time interval (Silvestro et al. 2015b) (Box 15.1, Table 15.1). While it is possible to infer the number and temporal placement of rate shifts from the data (Silvestro et al. 2014b), for the purpose of comparing alternative hypotheses we decided to use fixed configurations of shifts defined on the basis of historical events that might have induced environmental changes in each region (e.g., volcanism, mountain uplift, local climate change) (Section 15.3) (Badgley & Finarelli 2013). Thus, we defined a number of time intervals for each data set based on the main geologic and climatic events taking place in the respective geographic regions. In Region 1 we defined four time intervals with shift times at 16, 12 and 8 Ma; in Region 2, we set shift times at 34, 18, 16, 10 and 5.3 Ma; in Region 3, we set shift times at 26, 18, 12 and 5 Ma.

Furthermore, we tested two models (BD3 and BD4) linking the speciation and extinction dynamics of rodents with proxies of global climate change. Here, speciation and extinction rates vary in time through an exponential correlation with a curve approximating a paleoenvironmental variable (Silvestro et al. 2015a). This model is similar to the birth–death models described in a phylogenetic framework by Condamine et al. (2013) (Chapter 18) and involve the estimation of parameters G_λ and G_μ, quantifying the correlation between changes in birth–death rates and the paleoenvironmental curve (Box 15.1, Table 15.1). Thus, for instance, $G_\lambda > 0$ indicates a positive correlation between the paleoclimatic value and speciation rates and $G_\lambda < 0$ indicates a negative correlation. The correlation is considered statistically significant when 0 is not included within the 95% CI of parameter G. We considered two aspects of global climate change: the global temperature and the speed of temperature change. Estimates of the global paleotemperature are derived from stable isotope proxies (Zachos et al. 2008) and provide an estimate of global climate trends (Figure 15.5). We used this curve (spanning the past 50 My) and piecewise linear regressions to quantify the pace of temperature change through time. After testing different numbers of break points using the R package *segmented* (Muggeo 2008), we determined that five break points capture the main trends in

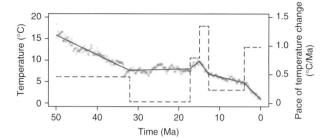

Figure 15.5 Approximations of global climate change through time. Gray circles represent paleotemperatures inferred from stable isotopes. The solid line shows the result of a piecewise regression. The absolute slope of the piecewise regression is interpreted as a proxy for the rapidity of climate change (dashed line). *Source:* Adapted from Zachos et al. (2008).

paleotemperature (Figure 15.5). The estimated temporal placements of the break points were 3.97, 12.66, 14.94, 17.23 and 32.14 Ma. We used the absolute values of the piecewise slopes (Figure 15.5) as a measure of the speed of temperature change through time. We then fitted time-varying birth–death models with speciation and extinction rates correlating with paleotemperature (BD3) and pace of temperature change (BD4).

Finally, we tested birth–death models that integrated both local (times of rate shift, based on local geologic and climatic events) and global (paleoclimate proxies) factors. We kept the region-specific times of rate shifts as defined in model BD2, but added the possibility for correlations with paleotemperature (BD5) and the pace of temperature change (BD6). We therefore estimated within each time interval t (delimited by shift times), a set of independent baseline rates $\lambda_0(t)$ and $\mu_0(t)$ and correlation parameters $G_\lambda(t)$ and $G_\mu(t)$. We ran six analyses for each region and estimated the marginal likelihoods of the BD1–6 birth–death models through path sampling. After selecting the best fitting model, we plotted the estimated speciation and extinction rates through time (RTT plot).

15.5 Preservation Rates and Model Selection

15.5.1 Preservation Rates in the Three Regions

The estimated preservation rates (expected number of fossil occurrences per lineage per My) differed among regions and among lineages within each region. The median preservation rates were lower in montane regions (0.79 in Region 1 and 0.57 in Region 2) and highest in the Great Plains (1.21 in Region 3). We estimated the lowest preservation rate in the northern Rocky Mountains, potentially because intense tectonic activity and erosion had a negative effect on fossilization and preservation potential. The Gamma model also inferred a considerable amount of preservation-rate heterogeneity across lineages within each data set. The posterior preservation rates ranged between 0.23 and 1.82 in the Cascades, between 0.24 and 1.09 in the northern Rocky Mountains and between 0.25 and 3.30 in the Great Plains.

15.5.2 Model Selection Across Alternative Birth–Death Scenarios

The estimated marginal likelihoods for the six birth–death models in each region are provided in Table 15.2. In Region 1, accounting for rate shifts based on local

tectonic and climate events (model BD2) did not improve model fit compared with a constant birth–death rate. Better marginal likelihoods (by 4–6 log units) were obtained when considering global temperature or its pace of temporal variation (BD3–4). Model fit improved substantially (>20 log units) only when including both local events and global climate (models BD5–6). The best model overall included rate shifts and the rapidity of climate change (BD6). In Region 2, none of the models accounting for either local events or global climate (BD2–4) received support over a constant-rate birth–death model. The integration of both, however, yielded a significantly better fit (12–20 log units), and Bayes factors favored model BD6 against all other alternatives. In

Table 15.2 Results of model testing. The statistical fit of six birth–death models for each region was evaluated by estimating the respective marginal likelihoods. Log Bayes factors were calculated between the best fitting model (BD6 in all regions) and all other models. An alternative way to look at the statistical support obtained by each model is to compute its relative probability, reported in the last column.

Region	Model name	Model interpretation	No. birth–death parameters	Marginal likelihood	2 log BF	Support in favor of best model	Relative probability
1. Cascade Range	BD1	Constant rates	2	−490.961	61.514	Very strong	0
	BD2	Correlation with local geologic/climatic events	8	−491.8558	63.304	Very strong	0
	BD3	Correlation with global climate	4	−484.435	48.462	Very strong	0
	BD4	Correlation with rapidity of climate change	4	−486.098	51.788	Very strong	0
	BD5	Correlations with local events and global climate	16	−466.516	12.624	Very strong	0.002
	BD6	Correlations with local events and rapidity of climate change	16	−460.204	0	–	0.998
2. Northern Rocky Mountains	BD1	Constant rates	2	−430.084	37.624	Very strong	0
	BD2	Correlation with local geologic/climatic events	12	−433.88773	45.232	Very strong	0
	BD3	Correlation with global climate	4	−431.845	41.146	Very strong	0
	BD4	Correlation with rapidity of climate change	4	−433.837	45.13	Very strong	0
	BD5	Correlations with local events and global climate	24	−419.044	15.544	Very strong	0.001
	BD6	Correlations with local events and rapidity of climate change	24	−411.272	0	–	0.999
3. Great Plains	BD1	Constant rates	2	−1115.448	81.522	Very strong	0
	BD2	Correlation with local geologic/climatic events	10	−1086.1074	22.84	Very strong	0
	BD3	Correlation with global climate	4	−1108.123	66.872	Very strong	0
	BD4	Correlation with rapidity of climate change	4	−1114.614	79.854	Very strong	0
	BD5	Correlations with local events and global climate	20	−1075.366	1.358	Weak	0.332
	BD6	Correlations with local events and rapidity of climate change	20	−1074.687	0	–	0.655

Region 3, the introduction of rate shifts based on local events (BD2) substantially improved the model fit compared to BD1, whereas models that only incorporated global temperature proxies did not receive strong statistical support. As also documented for Regions 1 and 2, however, the combination of both local and global events in the birth–death model yielded a significantly better fit. Model BD6 was only weakly supported over BD5, suggesting that global temperature and its rate of change can similarly explain rate changes in the diversification of rodents.

In summary, all three data sets strongly supported models in which speciation and extinction rates were allowed to change within time intervals (defined for each region based on local geological or climatic events) and as a function of global climate. Within the geographic and temporal limits of our analyses, the rapidity of global climate change appears to be more important in shaping rodent diversity than temperature itself. Our results also highlight that geologic and climatic factors together best explain diversification dynamics, although it is often difficult to untangle the relative importance of each.

15.6   Rodent Diversification in Active Montane Regions and Quiescent Plains

The estimated rates of speciation and extinction through time (Figure 15.6) are in the same range of values across regions, suggesting that, while regional differences may induce different temporal dynamics of diversification, the overall magnitude of these processes is comparable in adjacent regions. Despite this similarity in the overall pace of diversification, however, there are some important differences between the two montane and tectonically active regions and the quiescent lowland region. First, changes in both speciation and extinction rates appear to be more sudden and of greater magnitude in Regions 1 and 2, and more gradual in Region 3. This difference in the rate dynamics seems to reflect the complex history of landscape evolution in the Cascades and Rocky Mountains. Second, the MCO coincides with a short period of fast species diversification in Regions 1 and 2, matched by low extinction rates. In contrast, diversification in the northern Great Plains did not vary significantly during this time interval. Finally, in montane areas, the highest extinction rates are recorded from the most recent time interval, and result from a very strong rate increase, which takes place during the Pliocene. In Region 3, extinction slowly intensified through the whole Miocene and Pliocene, and the extinction rate at the present remains lower than in the adjacent montane areas.

The estimated diversification dynamics within each area are detailed in the this section.

In Region 1, we infer a positive and significant correlation between the pace of global temperature change and speciation rate in the oldest time frame (32–16 Ma). This interval includes an early period of global cooling in the late Eocene, followed by a phase of stable climate during the Oligocene and early Miocene. The end of this time frame includes the MCO, which is characterized by a rapid change in global temperature between 18 and 16 Ma. Speciation rates were higher during times of rapid climate change (i.e., close to the oldest occurrences ca. 32 Ma and during the MCO, ca. 18–16 Ma) and lower in between (Figure 15.6), as indicated by the positive correlation parameter, $G_\lambda = 1.51$ (95% CI: 0.56–2.39). The speciation rates inferred during the MCO are the highest in 35 My in Region 1 ($\lambda = 0.39$, 95% CI: 0.17–0.64) and are three times higher than in the early Miocene ($\lambda = 0.12$, 95% CI: 0.08–0.16). The correlation parameters were not significantly different from zero in the other time frames, indicating that variation in speciation rates following 16 Ma can be explained by the rate shifts describing local geologic and climatic events. During this period, speciation was highest ($\lambda = 0.17$, 95% CI: 0.10–0.25) during a phase of intense uplift and increased relief in the Cascades (12–8 Ma), whereas it drops significantly towards the present ($\lambda = 0.06$, 95% CI: 0.02–0.10). The extinction dynamics are fairly uniform throughout most of the time span examined here, with estimated correlation parameters (G_μ) not significantly different from zero. Extinction rates increased from ca. $\mu = 0.08$ in the first two time intervals (32–12 Ma) to $\mu = 0.12$ (95% CI: 0.06–0.18) between 12 and 8 Ma. Thus, while the MCO event did not leave a visible signature in terms of extinction in rodents, a slightly higher extinction rate is associated with a phase of rapid uplift (12–8 Ma). The most dramatic change in extinction occurred around the late Pliocene (3 Ma), when rates increased from $\mu = 0.05$ (95% CI: 0.02–0.09) to $\mu = 0.76$ (95% CI: 0.51–1.00). During this phase, extinction rates strongly correlate with global climatic trends ($G_\mu = 3.85$, 95% CI: 2.59–4.84), indicating that their increase can be explained by the rapid cooling taking place during the Pliocene and Pleistocene.

The speciation rate of rodents in Region 2 was comparatively high ($\lambda = 0.20$, 95% CI: 0.13–0.27) during the Eocene, a time interval with intense volcanic activity in the area, whereas it dropped to values around $\lambda = 0.09$ to 0.12 during the Oligocene and early Miocene. The highest speciation rate ($\lambda = 0.79$, 95% CI: 0.26–1.42) is estimated during the MCO (18–16 Ma), which, in the northern Rocky Mountains, coincides with a short phase with a warm and humid climate. The speciation rate in this interval is negatively correlated with the pace of global temperature change ($G_\lambda = -1.80$, 95% CI: −3.29

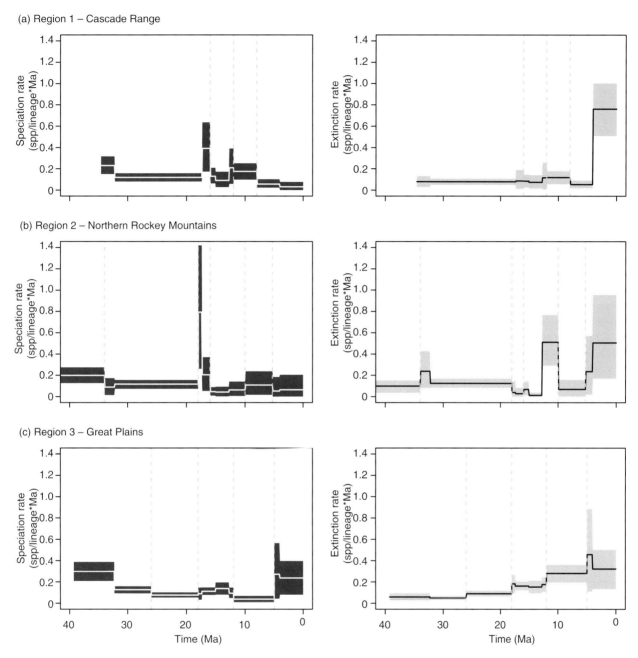

Figure 15.6 Diversification of rodents in three US regions: (a) Cascade Range; (b) northern Rocky Mountains; (c) Great Plains. The white and black lines indicate the posterior speciation and extinction rates through time, respectively. Birth–death rates were estimated from the best fitting birth–death model (BD6 in all regions), and therefore account for both local geologic events and global climate change. Shaded areas show the 95% CI around the estimates. Dashed lines indicate the time intervals identified by local tectonic or climatic changes affecting each region.

to −0.23). The MCO interval is characterized by an initial phase of global warming followed by a more rapid phase of global cooling (Figure 15.5). Therefore, a negative correlation between speciation rate and the pace of temperature change in this interval indicates that speciation rates were higher during the warming phase than during the cooling phase. Besides this rate variation between 18–16 Ma, the peak speciation rate during the

MCO identified in this analysis represents a sevenfold increase compared to the speciation rate during the previous time interval. Speciation rates dropped drastically during the aridification that followed the MCO, and showed no significant correlation with global climate change. Extinction rates show a fourfold decrease during the MCO, from $\mu = 0.12$ (95% CI: 0.08–0.17) to $\mu = 0.03$ (95% CI: 0–0.08). Thus, rodents diversified very rapidly in the northern Rocky Mountains, under a net diversification rate of about 0.76. Extinction rates increased again in the time interval that followed (16–10 Ma), and correlated negatively with global climate change ($G\mu$ = -3.82, 95% CI: –5.93 to –1.83). Thus, extinction increased from $\mu = 0.01$ (95% CI: 0–0.04) to $\mu = 0.51$ (95% CI: 0.29–0.77) after the rapid cooling that marks the end of the MCO (15 Ma), in a phase when both temperature and precipitation continued to gradually decline. After a temporary phase of lower extinction between 10 and 5 Ma, extinction rates increased again ($\mu = 0.51$, 95% CI: 0.17–0.96) in the most recent time interval (5–0 Ma), possibly as a consequence of rapid global cooling following the Miocene/Pliocene boundary (5.3 Ma) (Cerling et al. 1997) and increased volcanic activity linked with the migration of the Yellowstone hot spot towards the center of the region.

In Region 3, the first time interval included in our analyses extends from the Middle Eocene to the late Oligocene. During this time, the global climate went from a phase of cooling (until ca. 32 Ma) to a period of relative stability, which lasted until the onset of the MCO. The speciation rate of rodents in the northern plains was higher during the early phase ($\lambda = 0.30$, 95% CI: 0.21–0.39), while it underwent a threefold decrease around 32 Ma as a result of a positive and significant correlation with the change in temperature ($G_\lambda = 1.95$, 95% CI: 1.00–2.89). Extinction was low during the entire time interval, indicating that this was a phase of positive species diversification. Speciation rates of rodents in Region 3 remained low and stable ($\lambda = 0.04$–0.11) until the late Miocene, indicating that local and global climate (and vegetation) changes did not have a detectable effect on the pace of species diversification. In contrast, extinction rates appear to have slowly but steadily increased starting around 26 Ma and continuing through the Miocene, coinciding with a gradual change in vegetation in the Great Plains, which became increasingly dominated by open-habitat grasses (Strömberg 2011). The intensification of relief and erosion in Region 3, connected with river systems from the adjacent Rocky Mountains, coincides with a burst in speciation rates, which shifted from $\lambda = 0.04$ (95% CI: 0.01–0.07) to $\lambda = 0.24$ (95% CI: 0.08–0.40) in the most recent time interval. This is paired with comparatively high extinction rates, suggesting that rodents experienced high level of species turnover during the Pliocene and Pleistocene.

Taken together, these results suggest that both local events and global climate change contribute to the temporal dynamics of speciation and extinction rates. Additionally, major tectonic activities can affect adjacent areas. However, in mountain regions, the magnitude of these effects is typically larger, both in promoting diver-

sification through high speciation and in hindering it through episodes of high extinction.

15.7 Conclusion

We have shown how fossil information can provide insights into past biodiversity dynamics, and we have demonstrated that quantitative methods allow us to assess changes in rates of speciation and extinction as a function of geomorphological and climatic changes. Concerted changes in speciation and extinction rates, as shown for the Great Plains during the Pliocene and Pleistocene (Figure 15.6c), highlight the impact of Earth surface processes on evolutionary dynamics (see also Figure 15.1c). However, the processes of speciation and extinction need to be considered independently, as they may show distinct relationships with environmental changes. The decoupling of speciation and extinction processes is evident in our empirical analyses, where peaks in origination and extinction clearly follow different temporal dynamics. In the mountain regions (Regions 1 and 2), the MCO is associated with a burst in speciation, but unchanged (or even slightly decreased) extinction rates (Figure 15.6a,b). Conversely, the strong mid-late Miocene increase in extinction rate in the northern Rocky Mountains is not matched in speciation rates (Figure 15.6b). Finally, the mountain regions have experienced extinction bursts during the Pliocene and Pleistocene, whereas speciation rates remained low during this time (Figure 15.6a,b). Thus, while net diversification rates (speciation minus extinction) are a useful measure by which to quantify biodiversity changes (and in some cases, the only reliable one) (Nee 2006), disentangling the contributions of speciation and extinction is crucial to understanding which factors affect these changes.

The diversification dynamics described here do not explicitly distinguish between in situ speciation and immigration, nor between the extinction of a species and its local disappearance from an area. Recent methodological advances have shown that it is possible to estimate rates of dispersal between geographic areas based on the fossil record (Silvestro et al. 2016). However, the joint estimation of area-specific speciation, dispersal and extinction rates remains difficult. Future methodological developments and an improved understanding of the phylogenetic relationships among extinct and extant lineages will allow for better resolution in our estimates.

In our empirical examples, we found that preservation rates (which combine fossilization and sampling processes) are, on average, slightly lower in montane and tectonically active regions than in geologically stable lowlands. Despite the lower preservation rates, the size of the empirical data sets analyzed here was still sufficient to perform diversification rate analyses. However, for other taxa and geographic areas, this might not be the case. For instance, large areas in the western Alps either show no fossiliferous rock bodies due to erosion and strong metamorphism that erased potential paleontological evidence, or include fossils of organisms whose existence is unlinked with the montane environment (e.g., marine species). With the increasing amount of data that are available in public databases, fossils offer a great opportunity to investigate diversification dynamics, particularly of formerly species-rich clades that today comprise only a small extent of their past diversity. Eventually, however, paleontological and neontological data need to be combined in order to infer macroevolutionary dynamics from both fossil and phylogenetic data.

References

Badgley, C. (2010) Tectonics, topography, and mammalian diversity. *Ecography* **33**, 220–231.

Badgley, C. & Finarelli, J.A. (2013) Diversity dynamics of mammals in relation to tectonic and climatic history: comparison of three Neogene records from North America. *Paleobiology* **39**, 373–399.

Badgley, C. & Fox, D.L. (2000) Ecological biogeography of North American mammals: species density and ecological structure in relation to environmental gradients. *Journal of Biogeography* **27**, 1437–1467.

Cerling, T.E., Harris, J.M., MacFadden, B.J. et al. (1997) Global vegetation change through the Miocene/Pliocene boundary. *Nature* **389**, 153–158.

Condamine, F.L., Rolland, J. & Morlon, H. (2013) Macroevolutionary perspectives to environmental change. *Ecology Letters* **16**, 72–85.

Didier, G., Royer-Carenzi, M. & Laurin, M. (2012) The reconstructed evolutionary process with the fossil record. *Journal of Theoretical Biology* **315**, 26–37.

Elliott, W., Douglas, B. & Suttner, L. (2003) Structural control on Quaternary and Tertiary sedimentation in the Harrison Basin, Madison county, Montana. *The Mountain Geologist* **40**, 1–18.

Favre, A., Päckert, M., Pauls, S.U. et al. (2015) The role of the uplift of the Qinghai-Tibetan Plateau for the evolution of Tibetan biotas. *Biological Reviews* **90**, 236–253.

Fjeldså, J. & Lovett, J.C. (1997) Geographical patterns of old and young species in African forest biota: the significance of specific montane areas as evolutionary centres. *Biodiversity and Conservation* **6**, 325–346.

Foote, M. (2003) Origination and extinction through the Phanerozoic: a new approach. *Journal of Geology* **111**, 125–148.

Fox, D.L. & Koch, P.L. (2003) Tertiary history of C4 biomass in the Great Plains, USA. *Geology* **31**, 809–812.

Fritz, S.A., Schnitzler, J., Eronen, J.T. et al. (2013) Diversity in time and space: wanted dead and alive. *Trends in Ecology & Evolution* **28**, 509–516.

Harvey, P.H., May, R.M. & Nee, S. (1994) Phylogenies without fossils. *Evolution* **48**, 523–529.

Hughes, C. & Eastwood, R. (2006) Island radiation on a continental scale: Exceptional rates of plant diversification after uplift of the Andes. *Proceedings of the National Academy of Sciences USA* **103**, 10 334–10 339.

Kass, R.E. & Raftery, A.E. (1995) Bayes factors. *Journal of the American Statistical Association* **90**, 773–795.

Kier, G., Kreft, H., Lee, T.M. et al. (2009) A global assessment of endemism and species richness across island and mainland regions. *Proceedings of the National Academy of Sciences USA* **106**, 9322–9327.

Lartillot, N. & Philippe, H. (2006) Computing Bayes factors using thermodynamic integration. *Systematic Biology* **55**, 195–207.

Leopold, E. & Denton, M. (1987) Comparative age of grassland and steppe east and west of the northern rocky mountain. *Annals of the Missouri Botanical Garden* **74**, 841–867.

Liow, L.H. & Stenseth, N.C. (2007) The rise and fall of species: implications for macroevolutionary and macroecological studies. *Proceedings of the Royal Society of London B: Biological Sciences* **274**, 2745–2752.

Liow, L., Skaug, H., Ergon, T. & Schweder, T. (2010) Global occurrence trajectories of microfossils: environmental volatility and the rises and falls of individual species. *Paleobiology* **36**, 224–252.

Mitchell, S.G. & Montgomery, D.R. (2006) Polygenetic topography of the Cascade Range, Washington State, USA. *American Journal of Science* **306**, 736–768.

Muggeo, V.M.R. (2008) Segmented: an R package to fit regression models with broken-line relationships. *R News* **8**, 20–25.

Mulch, A., Sarna-Wojcicki, A.M., Perkins, M.E. & Chamberlain, C.P. (2008) A Miocene to Pleistocene climate and elevation record of the Sierra Nevada (California). *Proceedings of the National Academy of Sciences USA* **105**, 6819–6824.

Murphy, B.P., Johnson, J.P.L., Gasparini, N.M. & Sklar, L.S. (2016) Chemical weathering as a mechanism for the climatic control of bedrock river incision. *Nature* **532**, 223–227.

Nee, S. (2006) Birth-death models in macroevolution. *Annual Review of Ecology Evolution and Systematics* **37**, 1–17.

Paradis, E. (2004) Can extinction rates be estimated without fossils? *Journal of Theoretical Biology* **229**, 19–30.

Passey, B.H., Cerling, T.E., Perkins, M.E. et al. (2002) Environmental change in the Great Plains: an isotopic record from fossil horses. *Journal of Geology* **110**, 123–140.

Pierce, K.L. & Morgan, L.A. (1992) The track of the Yellowstone hot spot: volcanism, faulting, and uplift. *Geological Society of America Memoirs* **179**, 1–54.

Pires, M.M., Silvestro, D. & Quental, T.B. (2015) Continental faunal exchange and the asymmetrical radiation of carnivores. *Proceedings of the Royal Society of London B: Biological Sciences* **282**, 20151952.

Sanderson, M.J. & Donoghue, M.J. (1996) Reconstructing shifts in diversification rates on phylogenetic trees. *Trends in Ecology & Evolution* **11**, 15–20.

Schnitzler, J., Theis, C., Polly, P.D. & Eronen, J.T. (2017) Fossils matter – understanding modes and rates of trait evolution in Musteloidea (Carnivora). *Evolutionary Ecology Research* **18**, 187–200.

Silvestro, D., Schnitzler, J. & Zizka, G. (2011) A Bayesian framework to estimate diversification rates and their variation through time and space. *BMC Evolutionary Biology* **11**, 311.

Silvestro, D., Salamin, N. & Schnitzler, J. (2014a) PyRate: a new program to estimate speciation and extinction rates from incomplete fossil data. *Methods in Ecology and Evolution* **5**, 1126–1131.

Silvestro, D., Schnitzler, J., Liow, L.H. et al. (2014b) Bayesian estimation of speciation and extinction from incomplete fossil occurrence data. *Systematic Biology* **63**, 349–367.

Silvestro, D., Zizka, G. & Schulte, K. (2014c) Disentangling the effects of key innovations on the diversification of Bromelioideae (Bromeliaceae). *Evolution* **68**, 163–175.

Silvestro, D., Antonelli, A., Salamin, N. & Quental, T.B. (2015a) The role of clade competition in the diversification of North American canids. *Proceedings of the National Academy of Sciences USA* **112**, 8684–8689.

Silvestro, D., Cascales-Miñana, B., Bacon, C.D. & Antonelli, A. (2015b) Revisiting the origin and diversification of vascular plants through a comprehensive Bayesian analysis of the fossil record. *New Phytologist* **207**, 425–436.

Silvestro, D., Zizka, A., Bacon, C.D. et al. (2016) Fossil Biogeography: a new model to infer dispersal, extinction and sampling from paleontological data. *Philosophical Transactions of the Royal Society B: Biological Sciences* **371**, 20150225.

Simpson, G.G. (1944) *Tempo and Mode in Evolution.* New York: Columbia University Press.

Stanley, S.M. (1979) *Macroevolution: Pattern and Process.* Baltimore, MD: John Hopkins University Press.

Strömberg, C.A.E. (2011) Evolution of grasses and grassland ecosystems. *Annual Review of Earth and Planetary Sciences* **39**, 517–544.

Swinehart, J.B., Souders, V.L., DeGraw, H.M. & Diffendal, R.F. (1985) Cenozoic paleogeography of western Nebraska. In: Flores, R.M. & Kaplan, S.S. (eds.) *Cenozoic Paleogeography of the West-Central United States.* Denver, CO: Society of Economic Paleontologists and Mineralogists, pp. 209–229.

Verts, B. & Carraway, L.N. (1998) *Land Mammals of Oregon.* Berkeley, CA: University of California Press.

Whipple, K.X. (2009) The influence of climate on the tectonic evolution of mountain belts. *Nature Geoscience* **2**, 97–104.

Yang, Z. (1994) Maximum likelihood phylogenetic estimation from DNA sequences with variable rates over sites: approximate methods. *Journal of Molecular Evolution* **39**, 306–314.

Zachos, J.C., Dickens, G.R. & Zeebe, R.E. (2008) An early Cenozoic perspective on greenhouse warming and carbon-cycle dynamics. *Nature* **451**, 279–283.

16

The Interplay between Geological History and Ecology in Mountains

Catherine H. Graham[1,2], Mauricio Parra[3], Andrés Mora[4] and Camilo Higuera[4]

[1] *Stony Brook University, Stony Brook, NY, USA*
[2] *Swiss Federal Research Institute WSL, Birmensdorf, Switzerland*
[3] *Institute of Energy and Environment, University of Sao Paulo, Brazil*
[4] *Ecopetrol, Exploration Division, Bogota, Colombia*

Abstract

Mountain biodiversity is strongly influenced by the geologic and climatic history of a given mountain region. While this phenomenon has been long acknowledged, the rich history of mountains documented by geologists and climatologists is often poorly integrated with ecological and evolutionary hypotheses about the origin and maintenance of diversity. Here, we detail two main approaches for such integration: a broad correlative approach, where characteristics such as uplift age, area or heterogeneity from multiple mountains are related to biological diversity, and a detailed study of the climatic and uplift history of a single mountain system and its influence on diversity patterns, using the Colombian Andes as an example. When applying either approach, ecologists and evolutionary biologists should consider the different temporal scales of ecological/evolutionary and geologic processes. For instance, many ecological and evolutionary processes occur at scales of one to thousands of years, while geologic change occurs over millions of years. Hence, a mountain considered geologically dynamic might be quite stable from an ecological perspective. In the Andes, we found that some areas considered to be geologically dynamic – a result of both proximity to a tectonic boundary resulting in rapid uplift and constant erosion caused by high precipitation – had high biological diversity. However, other regions that did not have these characteristics also had high diversity, implicating other mechanisms in the generation and maintenance of diversity. Based on our results, we suggest that while generalizations about how mountain history influences biological diversity are beginning to emerge, expected relationships are not always found, suggesting that much more can be learned by a careful integration of geology and climate utilizing ecological and evolutionary approaches.

Keywords: *Andes mountains, Colombia, erosion, stability, topographic complexity*

16.1 Introduction

Mountains represent some of the most dynamic landscapes on Earth, and as such, they provide a template on which ecological and evolutionary processes act to produce regions of often very high diversity. The dynamic nature of mountain orogeny means that characteristics such as area, productivity and topographic complexity, which are often hypothesized to influence ecological and evolutionary processes, vary through time. For instance, some mountains, such as the Andes of South America, are massive ($\sim$3.37 million km^2) and uplifted rapidly (locally more than 1 mm/y) and relatively recently (final uplift by the late Pliocene to the present), while others, such as the Appalachian Mountains, are smaller ($\sim$1.9 million km^2) and ancient (uplifting approximately 480 Ma as part of the plate collisions that formed the supercontinent Pangaea). While the link between mountain dynamics and species richness has long been recognized (Simpson 1980), there are several challenges to studying the interplay between the geological and climatic history of mountains and the ecological and evolutionary processes that generate and maintain diversity within them. Perhaps the most obvious challenge is that most ecological and evolutionary processes occur over short time scales as compared to montane geologic

Mountains, Climate and Biodiversity, First Edition. Edited by Carina Hoorn, Allison Perrigo and Alexandre Antonelli.
© 2018 John Wiley & Sons Ltd. Published 2018 by John Wiley & Sons Ltd.
Companion website: www.wiley.com\go\hoorn\mountains,climateandbiodiversity

history. Most ecological processes occur at scales of one to hundreds of years, while evolutionary processes occur across tens to thousands of years and species life spans range between 1 and 10 million years (Lawton & May 2005). Geologic or climate changes occur over millions of years, with even rapid changes acting at much longer time scales than most ecological and even some evolutionary events. Nonetheless, correlations between diversity and mountain characteristics are often observed, suggesting that such an interplay has an important role in generating and maintaining diversity.

There are two main approaches to evaluating the interplay between geologic history and ecology in mountains: (i) correlative approaches to identify mountain characteristics assumed to influence diversity and biodiversity (Rahbek 1997; Fjeldså et al. 1999; Körner 2000; Graham et al. 2014; Craw et al. 2016); and (ii) detailed studies of the climatic and uplift history of a single mountain system and its influence on diversity patterns (Graham et al. 2006; Carnaval et al. 2009; da Silva et al. 2014; Merckx et al. 2015). The former usually includes a relatively large number of mountain systems and correlates current mountain characteristics with species or phylogenetic diversity. Factors such as net primary production or energy, topographic complexity and mountain age are hypothesized to influence the ecological and evolutionary processes that generate and maintain biodiversity (Currie et al. 2004; Mittelbach et al. 2007). The latter approach attempts to uncover the geological and climatic mechanisms associated with mountain height, topographic complexity, connectivity, age and rock and soil characteristics, and then relate these directly to specific patterns of species or phylogenetic diversity. In this chapter, we review the predominant ecological and evolutionary hypotheses associated with these correlates and suggest how geological and climatic history can inform them. We then focus on a specific mountain region, the Northern Andes of South America, to provide a more in-depth view of mountain-building dynamics and spatial patterns of biodiversity.

16.2 Overview of Mountain Formation and Resulting Geologic and Climatic Complexity

The most dynamic changes to Earth's topography occur along the boundaries of tectonic plates, where mantle convection caused by gravitational forces results in the formation of massive mountain chains such as the Andes, the Alps and the Himalaya (Molnar 1988; Oncken et al. 2006). There are two facets to mountain building: surface uplift and erosion. Surface uplift, which is the change in

position of Earth's surface with respect to the geoid, is the result of uplift (displacement of rocks with respect to the geoid) and exhumation (upwards displacement of rocks with respect to the surface) (England & Molnar 1990). Erosion occurs when rocks are weathered and eroded once they reach the surface (Chapter 4). It constantly removes material beginning when a mountain grows above mean sea level, becoming most efficient once it reaches a height of 2 km (Mora et al. 2008). Thermochronometric studies have shown that in the tropical Andes, erosion can destroy approximately 4 km of relief in 1 My (Mora et al. 2008; Bermúdez et al. 2011).

This dynamic of uplift and erosion can have a strong impact on ecology and biodiversity. Mountains that are high and continuous act as biotic corridors, (Chapter 12). For instance, deeply incised canyons resulting from erosion often contain distinct environmental conditions that influence the formation of unique habitats. In addition, mountains undergoing aggressive erosion generate massive amounts of sediments and nutrients, which can accumulate both in certain mountain regions and in adjacent low-elevation areas.

16.3 Geologic and Climatic Factors Influencing Montane Diversity

Several geological and climatic factors, such as mountain age, topographic complexity and environmental productivity (driven by temperature, rainfall and nutrients), have been hypothesized to influence patterns of biodiversity on mountains (Jetz et al. 2004; Davies et al. 2007; Mittelbach et al. 2007; Weir & Price 2011; Fine 2015). These factors have been studied using correlative approaches, where patterns of diversity – usually species, but increasingly genetic/phylogenetic, trait or even interaction (i.e., food web complexity, specialization) diversity – are correlated with static factors representing a snapshot of the current geological conditions (Graham et al. 2006; Davies et al. 2007; Dalsgaard et al. 2011). While this snapshot approach can be informative, it is also limited, and often requires assumptions about the importance of geologic history. Information on geologic history provides additional information by which to evaluate ecological and evolutionary hypotheses, but it has yet to be fully integrated into these fields (Carnaval et al. 2009; Hoorn et al. 2013; Favre et al. 2015; Mastretta-Yanes et al. 2015). In this section, we first review the main ecological and evolutionary hypotheses used to explain correlations between biodiversity and montane geological history. We then describe how information on geological and climate history may provide new insight into these hypotheses. We do not attempt to evaluate an exhaustive list of correlates, but instead focus on those that are most prevalent in the mountain systems where we think the integration of ecological and evolutionary hypotheses with geological history will be most fruitful.

16.3.1 Topographic Complexity and Associated Habitat Heterogeneity

All other factors being equal, mountains with greater topographic complexity will have more species than mountains with little topographic complexity. This is because greater topographic relief, often measured as the degree of change in the range of elevations in a grid, generally yields greater habitat heterogeneity. This hypothesis should be consistent across latitudes, but may be stronger in the tropics because of increased altitudinal zonation (Janzen 1967). From an ecological perspective, heterogeneity should provide more niches, which provide opportunities for more species to coexist (Macarthur & Levins 1967; Ricklefs 1977; Davies et al. 2007; Dyer et al. 2007). Further, topographic complexity should result in increased buffering for species as climate shifts, because species will be more likely to track their climatic niches (Bush 2002; Colwell et al. 2008). Habitat heterogeneity would thus be expected to be associated with lower extinction rates. However, if species have exceptionally small ranges, the opposite may instead be the case.

Topographic complexity should also result in high speciation rates by providing increased opportunities for isolation and divergent selection (Ricklefs 2004; Antonelli & Sanmartin 2011; Stein et al. 2014). At macroecological scales, the complexity of topographic relief positively correlates with species richness in mountains (Rahbek & Graves 2001). Moreover, isolation, combined with the environmental variation along mountain gradients, should further increase rates of speciation because of adaptation to a given habitat type (Lynch et al. 1986; Kozak & Wiens 2010; Antonelli & Sanmartin 2011). For instance, Graham et al. (2004) combined geographic and environmental information to show that sister taxa of Ecuadorian frogs (mostly montane) tended to exist in disjunct areas with different environmental conditions – predominately precipitation and temperature. In a similar vein, phylogeograhic analyses have combined morphological and genetic information with measures of geographic connectivity and environmental conditions to show that both isolation and ecological speciation are important mechanisms leading to within-species divergence (Guarnizo et al. 2009; Thomassen et al. 2010).

The tempo of changes in topographic and habitat complexity resulting from geological and climatic dynamics should also be considered when evaluating their influence on speciation rates. The rate of speciation can vary from tens to hundreds of thousands of years (Hendry et al. 2007) and is likely influenced by species' ability to adapt, their life histories and morphological traits and the topographic and climatic context in which they occur. A slow change in topography or habitat (stability) may influence the rate of species production by providing sufficient time for speciation before disturbance wipes out incipient species (Fjeldså et al. 1999). However, while instability erodes incipient differentiation, high stability may lead to evolutionary stasis. Thus, one might expect that diversification rates would peak at intermediate levels of stability.

Two models hypothesize increased generation of species with such intermediate levels of stability, which we refer to as the "classical refuge or stability hypothesis" and the "vanishing refuge model" (Vanzolini & Williams 1981). The classical refuge hypothesis emphasizes that vicariance resulting from topographic isolation combined with stabilizing selection on the physiological (abiotic) niche, and associated with niche conservatism, should drive speciation (Janzen 1967; Wiens & Graham 2005; Futuyma 2010; Kozak & Wiens 2010). In contrast, the vanishing refuge model emphasizes adaptive evolution (i.e., an ecological niche shift) into new environments as refugia (i.e., historic environmental conditions for a species) vanish in climatically and geologically unstable regions (Vanzolini & Williams 1981; Damasceno et al. 2014). More generally, we can determine whether there is an increase in net diversification with "intermediate" levels of stability, as measured by rates of change of geological topographic change associated with mountain building (uplift and exhumation) and climatic variation. Further, we can identify whether there is a signature of niche filling (slowing diversification with filling of morphological or functional space) associated with different rates of topographic change.

In this regard, it is important to define what geologists consider fast topographic change or fast topographic destruction. Rates of relief destruction or growth of about 1 km/My are considered to be quite fast, and only occur in the youngest orogenic belts undergoing active tectonism and rapid erosion (Thiede et al. 2005; Mora et al. 2008). Species evolve over tens to thousands of years but can then persist for one to several million years before going extinct (Lawton & May 2005). This suggests that even the most aggressively deforming mountain chains, with fast erosion rates and rapid topographic growth, may represent intermediate levels of stability, facilitating ecological processes and speciation (Kattan et al. 2004; Robin et al. 2015). An example of such a dynamic environment is the Cocuy region of Colombia, which has been growing rapidly as judged by denudation rates (1 km/My) but relatively slowly as compared with speciation (Mora et al. 2015).

16.3.2 Climate and Net Primary Productivity

Montane species richness is often positively correlated with temperature (until critical thermal limits are met), precipitation, water-energy balance and, more generally, energy availability (usually estimated as net primary productivity, NPP) (Evans et al. 2005). The hypothesis behind this correlation is that in productive environments, there are more individuals and little variation in the fraction of NPP (resources) secured by each species, resulting in more species existing in such environments (Wright 1983; Hawkins et al. 2003; Currie et al. 2004; Evans et al. 2005; McCain 2009). As environments become harsher, there should be fewer species, because harsher environments support fewer individuals and extinction may be more likely when population sizes are small.

A second hypothesis behind the positive correlation between richness and climate parameters is the environmental limits hypothesis, which posits that extreme climatic conditions, such as very cool temperatures, act as filters, which do not permit the establishment of all species (Wright 1983; Currie et al. 2004). Phylogenetic information can be used to evaluate the extent to which functional traits permitting persistence in extreme environmental conditions are labile and over what time period such traits might have evolved. In the case of mountain building, if a clade originated in a benign lowland environment and niches are generally conserved (which can be tested by evaluating how species functional traits and climatic niches evolve through time), then only a few lineages will have evolved the niche characteristics that allow them to persist and diversify in newly emerging high elevations, potentially leading to lower species richness in these environments (Ricklefs 2007; Wiens et al. 2007).

From a geological perspective, it is clear that climate (temperature and rainfall) usually changes much more rapidly than the topography and morphology of mountains (Chapter 9). For example, Dansgaard–Oeschger cycles, measured from ice cores from Greenland, suggest there may have been periods of rapid warming during the last glacial (Alley 2000). Also, it is well known that Milankovich cycles (cyclic changes in Earth's orbital parameters) dramatically influenced climatic fluctuations, resulting in periods of glaciation or eustatic sea-level rise, which globally altered rainfall and temperature (Chapter 9). These cycles operate over periods of tens of thousands of years and may interact with geological factors to influence ecosystems and species evolution.

16.3.3 Age and Stability

Mountains vary in age, which influences the time over which evolution can occur. The evolutionary time hypothesis states that species richness in a region will be lower when there is limited time for either colonization or speciation (Rohde 1992; Willig et al. 2003). The time-for-speciation effect is a result of how long members of a group have been present in a region, and is considered an important determinant of species richness because, assuming limited extinction, it permits the slow build-up of diversity (Stephens & Wiens 2003; Mittelbach et al. 2007). In some cases, the time-for-speciation effect appears to be a more important determinant than climate or topographic complexity (Smith et al. 2007). In many studies invoking the time-for-speciation effect as a major driver of diversity, time is estimated simply based on the branch length of the oldest lineage on a given mountain, which assumes that the ancestors of any present taxon must have persisted on that mountain. The geological and climatic factors that may have permitted old taxa to persist in a given place are generally not considered.

Mountain stability, which can be either climatic or geological but is most often characterized based on climate, has been correlated with species richness and endemism both within and among mountain areas. Stability can influence species richness by minimizing environmentally driven extinctions, resulting in a build-up of species through time (Colinvaux et al. 1996; Jansson 2003; Araujo et al. 2008). This relationship is often particularly high for endemic species (i.e., those that originated in a given mountain area and have a limited geographic distribution), because once endemic species go extinct, they are simply gone, while widespread species can recolonize from adjacent regions after instances of local extinction (Graham et al. 2006). High instability can cause extinctions if lineages are confronted with conditions outside of their fundamental niche and are unable to either move to areas with suitable conditions or adapt to new conditions. For instance, work in the Australian Wet Tropics, a montane region in eastern Australia, showed that the richness of endemic species was highest on mountains that were hypothesized to have been climatically stable through several historical climatic shifts, including the Last Glacial Maximum (LGM) (Graham et al. 2006). Phylogeographic information from the same region confirmed that areas where lineages had persisted for longer periods had been climatically stable (Hugall et al. 2002; Moritz et al. 2005, 2012; Moussalli et al. 2009; Bell et al. 2012). In fact, when climate conditions were tracked across time and species' ability to track these conditions was considered in a dynamic refugia model, patterns of endemism were better explained than when a single mean measure of stability was used (Carnaval et al. 2009; Graham et al. 2010b).

It appears that even the geologically fastest growing mountains may be old and stable with regard to ecology. While young from a geologic perspective, mountains may be quite old in relation to speciation rates and the ages of their associated species. Speciation generally occurs on a time scale of hundreds of thousands of years, and species persist for between 1 and 10 million years. Therefore, it is not only the age of mountains that influences ecology and speciation but, consistent with our discussion on topographic complexity, also the dynamics of mountain building.

16.4   Case Study: The Northern Andes

A rapid and paroxysmal surface uplift (the creation of topography and relief higher than 2 km) of the Eastern Cordillera since the late Miocene (11 Ma to the present) is hypothesized to have influenced the generation of the high species richness that we see today (Gentry 1982; Hoorn et al. 2010). This paroxysmal uplift may have acted as a trigger for a variety of mechanisms promoting biodiversity, including: adaptive radiations caused by habitat heterogeneity; isolation and allopatric speciation; and increased rainfall and nutrient deposition, resulting in greater primary productivity (Antonelli & Sanmartin 2011). Empirical evidence that the uplift triggered an increase in biodiversity has been observed across multiple taxonomic groups (Weir 2006; Ribas et al. 2007; Sedano & Burns 2010; Hutter et al. 2013; De-Silva et al. 2016), although the exact mechanisms promoting this increase remain somewhat elusive (Mutke et al. 2014). One difficulty in identifying these mechanisms may be that the scale of most analyses tends to be relatively large (e.g., the entire Andes chain or the Northern Andes). A more detailed analysis focused on variation within a mountain chain might provide additional insight into how the geologic history associated with topographic complexity and primary productivity influenced the build-up of biodiversity (Figure 16.1a).

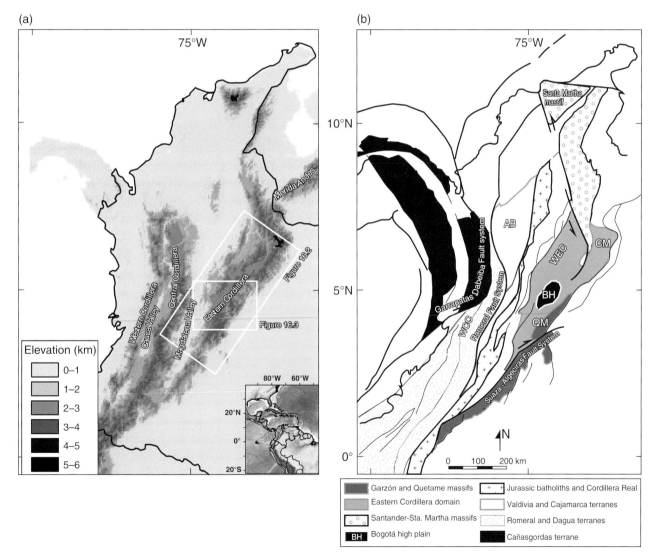

Figure 16.1 (a) SRTM-90 m digital elevation model of the Northern Andes in Colombia and Ecuador. (b) Tectonic setting showing the main tectonic provinces. The black lines are the boundaries of these provinces, represented by faults or fault systems where the thickest lines concentrate more strain. AB: Antioquia Batholith; BH: Bogota high plain; CM: Cocuy Massif; QM: Quetame Massif; WCC: Western side of Central Cordillera; WEC: Western side of Eastern Cordillera.

Table 16.1 Geological, climatic and biodiversity characteristics of different regions in the Colombian Andes of South America. Local relief is based on Figure 16.3 and is calculated with a 3 km moving window using SRTM-90 m topographic data. Precipitation is from radar-based precipitation measurements, based on the Tropical Rainfall Measuring Mission (TRMM) data (see Figure 16.3). Species richness is calculated by combing species distribution models from 5400 plant and animal species.

Location	Tectonic deformation	Topography	Precipitation	Relative richness
Quetame Massif (QM)	Suaza–Algeciras fault system	Variable	High	High
Cocuy Massif (CM)	Pajarito fault system	Variable	High	Lower
Cauca Valley (CV) and western side of Central Cordillera (WCC)	Garrapatas fault	Variable	High	High
Western side of Eastern Cordillera (WEC)	Rock resistance to erosion; no significant fault	Variable	Medium	High
Antioquia Batholith (AB)	Deep crustal controls on exhumed plateau	Flat	High	Highest
Bogota High Plain	Deep crustal controls on exhumed plateau	Flat	Low	Lowest
Santa Marta Massif (SM)	Santa Marta and Oca faults	Variable	High	Lower

In this section, we describe the rates of topographic growth and relief construction in the Colombian Andes. To explore the factors that result in high topographic variation and precipitation, we first provide a detailed evaluation of a single region (the Quetame Massif: Figure 16.1a). We then identify regions with similar geologic characteristics to the Quetame, such as recent rapid uplift and high precipitation, as well as areas with different characteristics, including slow or intermediate rates of uplift and different levels of precipitation (Table 16.1 and Figure 16.1b). We chose these regions to represent a range of geologic histories and climatic conditions so that we could better explore how this variation might influence biodiversity.

We suggest that erosion is an important paroxysmal mechanism and is very closely associated with zones containing strong deformation of Earth's crust materials, and in some cases with particularly high localized rainfall or possible Pleistocene glaciation (see also Chapter 4). By dating Quaternary (<2.6 Ma) rapid erosion in the Andes, we can determine when and where the topography became jagged and complex, creating deeply incised river canyons, which likely resulted in more heterogeneous ecosystems in and adjacent to these canyons. Finally, using richness maps based on 5400 species for Colombia, we test the prediction that these regions of high topographic complexity also have relatively high rainfall and species richness.

16.4.1 Mechanism Influencing Topographic Complexity

The Quetame Massif on the eastern side of the Colombian Eastern Cordillera is probably the best studied example in the Northern Andes of the main factors that result in high topographic variation and precipitation (Mora et al. 2008,

2015; Parra et al. 2009; Ramirez-Arias et al. 2012). We use this region to demonstrate the relative importance of geologic and climate forcing in generating high topographic complexity and precipitation, which are related to NPP and are hypothesized to generate and maintain biodiversity. To demonstrate the importance of focused strain in generating variable topography, we use a geophysical relief map from the Quetame Massif (Figure 16.2). This map, which illustrates the mean elevation difference between a smooth surface connecting the highest points in the landscape and the current topography, provides a detailed picture of the topographic variation because it can uncover both shallow surface processes and deeper crustal anisotropies. Young (Pliocene) apatite fission track ages, jagged topography and prominent geophysical relief all suggest that the area of the Quetame Massif is related to the termination of the strike-slip zone of the Suaza–Algeciras fault (Figure 16.1b).

Data from the Quetame Massif also suggest that over the last 3 My, erosion rates in the Colombian Eastern Cordillera were very high (Figure 16.3). This conclusion is based on the very young (<3 My) low-temperature thermochronometric ages, which spatially coincide with very high topographic variation (Figure 16.4b) and rainfall (Figure 16.4c). From this evidence, Mora et al. (2008) suggested that climatic forcing may be the dominant factor in the topographic evolution over at least the last 5 My. However, other evidence from the Quetame Massif, as well as the Cocuy Massif to the north and the Venezuelan Andes, indicates that climatic forcing of mountain building and topographic variation requires a significantly focused strain (tectonic forces deforming geological terrains) localized in areas where climatic forcing is hypothesized to enhance deformation and topographic change (Bermúdez et al. 2011; Bande et al. 2012;

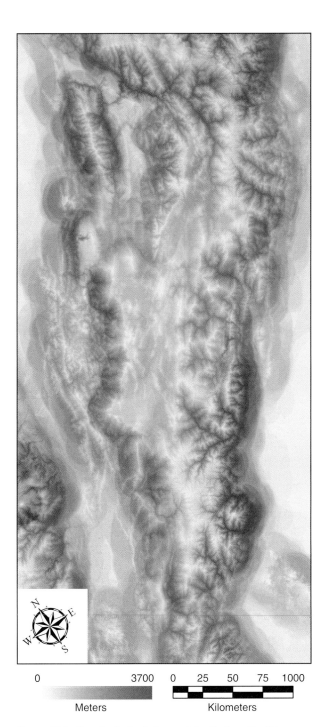

0   3700   0   25   50   75 1000

Meters   Kilometers

Figure 16.2 Geophysical relief map of the Eastern Cordillera, showing the elevation difference between a smooth surface connecting the highest points in the landscape and the current topography. See Figure 16.1 for location.

Ramirez-Arias et al. 2012; Mora et al. 2015). If this is the case, then areas with high precipitation will only display high long-term erosion rates (as measured by thermochronology) if they are located close to important weaknesses or key boundaries in the Earth crust.

16.4.2 Topographic Complexity, Precipitation and Biodiversity in the Northern Andes

As already described, precipitation in tectonically active regions of the tropics is the key factor that causes erosion, so precipitation is expected to be directly related to variable relief. Therefore, to further test our hypothesis that highly variable relief combined with significant rainfall resulted in particularly high diversity, we compare maps of four variables (geology, rainfall, local relief and species richness; Figure 16.4) for seven regions in Colombia (Figure 16.1 and Table 16.1). First, we evaluate whether the relationship between focused tectonic deformation and variable topography is general across the Northern Andes. Second, we evaluate whether these are also areas of high rainfall. Third, we identify areas with lower tectonic deformation and less variable topography. Finally, we evaluate whether these two factors are consistently associated with high species richness.

The species-richness map was generated using species-distribution modeling based on climate and species-occurrence data. Worldclim climate data describing means and standard deviations of temperature and precipitation generated from climate stations (Hijmans et al. 2005) were used for the species-distributional modeling. Maps based only on climate often predict that a species' range extends beyond known occurrence data, especially in the Colombian Andes, where the three different mountain chains have somewhat similar climates (Graham et al. 2010a). Therefore, to obtain an accurate map of richness, individual species distribution models are often modified by experts to minimize prediction to a region where a species has not been observed (Chapter 21). Our richness map was compiled from distribution models of 5800 plant and animal species in Colombia (Olaya-Rodriguez et al. unpublished data).

From a visual spatial comparison of the map of geological regions (Figure 16.1b) and the local relief map (Figure 16.4b), we can infer that several areas, including the Quetame Massif, Cauca Valley, Cocuy and Santa Marta, correspond with key geological boundaries, such as the Garrapatas fault between the Dagua-Romeral terrain and the Cañasgordas terrain (Figure 16.1b) (Cediel et al. 2003), which have extensive earthquake activity. Based on our discussion of the mechanisms influencing topographic variation and erosion in the Quetame Massif, we suggested that in all four areas where tectonic forces deformed Earth's crust there would be significant topographic variation, high erosion and high precipitation. However, of these four areas, only the Quetame Massif, the Cauca Valley and the western side of the Central Cordillera have high biodiversity.

Why the Cocuy and Santa Marta show a different pattern is hard to determine with the available data, but

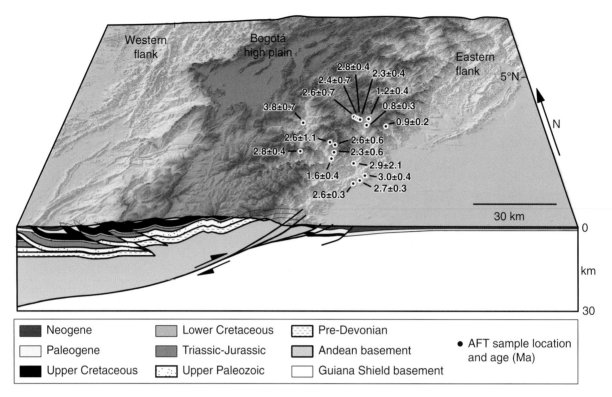

Figure 16.3 Three-dimensional (3D) model depicting the topography (SRTM-90) and geological structure of the Eastern Cordillera at ~4° N latitude. Numbers indicate apatite fission track ages (including 1σ error), which broadly correspond to the times when rocks cropping out at the surface were buried at a temperature of >120 °C and thus illustrate the time span required for the erosion of 3–4 km of rocks. See Figure 16.1 for location. *Source:* Mora et al. (2008). Reproduced with permission of The Geological Society of America. See also Plate 31 in color plate section.

perhaps the lower species richness can be attributed to the relatively small size and high geographic isolation of these two areas (Rosenzweig 1995). Further, we evaluated total species richness, whereas an evaluation of endemism might provide a somewhat different picture; Santa Marta, in particular, has a number of species that are endemic to this single mountain massif (Kattan et al. 2004).

The western side of the Eastern Cordillera presents a different scenario, because the tectonic forces are likely less strongly focused there, and precipitation is lower. The area of relatively high biodiversity along the western side of the Eastern Cordillera is adjacent to the cliffs located along the western boundary of the Bogota Plateau. The plateau consists of erosion-resistant sandy Cretaceous rock units that are part of the Guadalupe Group. In this case, it appears that a breakdown in topography is related to the resistance of the units to erosion. In sum, the topographic variability presumably provided opportunities for the generation and maintenance of biodiversity.

Finally, we can evaluate biodiversity in areas without significant faults. These, in theory, should not coincide with areas of high local topographic relief and precipitation. One such area is the Antioquia Batholith, a prominent high plateau incised by the Medellin and Porce Rivers. Local relief and rainfall are not especially high in this area. In addition, exhumation ages from

thermochronology are generally older than 25 Ma, and the denudation rates in this area peaked between 75 and 40 Ma (Restrepo-Moreno et al. 2009; Villagómez & Spikings 2013). Therefore, it appears the exhumation rate has been very low for at least the last 20 My. If our hypothesis is correct, and high biodiversity occurs in areas of high topographic variability and precipitation, then we would expect low biodiversity. Yet, there is very high biodiversity in the Antioquia Batholith. Interestingly, the outcropping rocks in this region are acid rocks of the Antioquia Batholith covered by Quaternary volcanic rocks that come from active volcanoes to the south. In general, the quality and presence of nutrients in soils is high when parent rocks are igneous, especially when they have volcanic deposits. Indeed, the Antioquia Batholith is recognized for its extremely rich soils (Hermelin 1992). We suggest that these stable, nutrient-rich soils resulted in high NPP, which may have influenced the very high biodiversity in the Antioquia Batholith region (Evans et al. 2005).

The last case is the Bogota High Plain, which is a high-elevation, low-relief high plain associated with homogeneous tectonic uplift (Carrillo et al. 2016) and where biodiversity is low. It does not behave like the Antioquia Batholith, which is also a high-elevation, low-relief plateau, potentially because it is isolated from the orographic

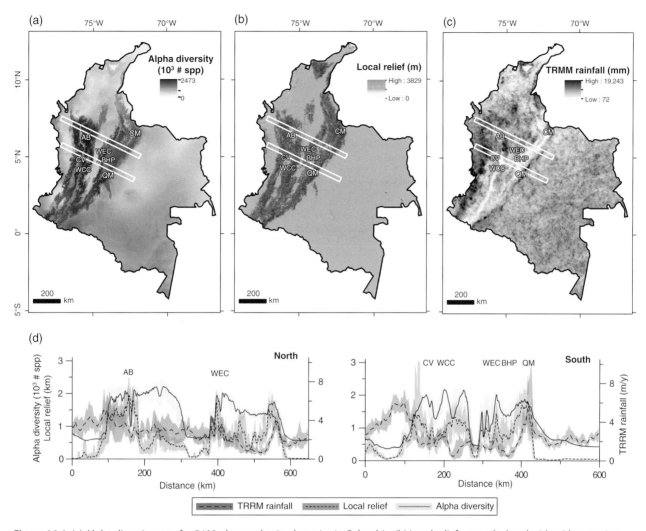

Figure 16.4 (a) Alpha diversity map for 5400 plant and animal species in Colombia. (b) Local relief map calculated with a 3 km moving window using SRTM-90 m topographic data. (c) Radar-based precipitation, from Tropical Rainfall Measuring Mission (TRMM) data. (d) Data extracted from 30 km-wide swath profiles along northern and central Colombia (indicated by white rectangles in (a), (b) and (c)) for alpha diversity, local relief and TRMM rainfall. Abbreviations as indicated in Table 16.1. See also Plate 32 in color plate section. AB: Antioquia Batholith; BHP: Bogota high plain; CM: Cocuy Massif; CV: Cauca Valley; QM: Quetame Massif; WCC: Western side of Central Cordillera; WEC: Western side of Eastern Cordillera.

precipitation (Mora et al. 2008), and therefore reduced rainfall and poor sediment evacuation by the rivers make it a flat region with comparatively low biodiversity.

16.5 Conclusion

In this chapter, we reviewed the potential roles of geologic and climatic factors, including topographic complexity, precipitation, productivity, age and isolation, in generating biodiversity patterns in mountains. We then explored how geologic and climatic factors might have influenced animal diversity in the Northern Andes of Colombia. We tested the prediction that biodiversity hot spots coincide with areas of significant local topographic relief, which were generated by focused tectonism (higher strain or deformation of Earth's crust

materials) and high precipitation. We found support for this prediction in several regions. However, the western side of the Eastern Cordillera and the Antioquia Batholith, in particular, have high species richness and are located far from major crustal boundaries. How the geologic history of mountains influences biological diversity is beginning to emerge; however, expected relationships were not always found, suggesting that much more can be learned by a careful integration of geological, ecological and evolutionary approaches. Properly and precisely dating the geological processes and landscape evolution will thus be instrumental to our understanding biodiversity and species evolution. The Northern Andes represents an ideal area for more detailed analyses, because the topographic variability is geologically young and hosts high but variable species richness.

References

Alley, R.B. (2000) Ice-core evidence of abrupt climate changes. *Proceedings of the National Academy of Sciences* **97**, 1331–1334.

Antonelli, A. & Sanmartin, I. (2011) Why are there so many plant species in the Neotropics? *Taxon* **60**, 403–414.

Araujo, M.B., Nogues-Bravo, D., Diniz-Filho, J.A.F. et al. (2008) Quaternary climate changes explain diversity among reptiles and amphibians. *Ecography* **31**, 8–15.

Bande, A., Horton, B.K., Ramírez, J.C. et al. (2012) Clastic deposition, provenance, and sequence of Andean thrusting in the frontal Eastern Cordillera and Llanos foreland basin of Colombia. *Geological Society of America Bulletin* **124**, 59–76.

Bell, R.C., MacKenzie, J.B., Hickerson, M.J. et al. (2012) Comparative multi-locus phylogeography confirms multiple vicariance events in co-distributed rainforest frogs. *Proceedings of the Royal Society B: Biological Sciences* **279**, 991–999.

Bermúdez, M.A., van der Beek, P. & Bernet, M. (2011) Asynchronous Miocene–Pliocene exhumation of the central Venezuelan Andes. *Geology* **39**, 139–142.

Bush, M.B. (2002) Distributional change and conservation on the Andean flank: a palaeoecological perspective. *Global Ecology and Biogeography* **11**, 463–473.

Carnaval, A.C., Hickerson, M.J., Haddad, C.F.B. et al. (2009) Stability predicts genetic diversity in the Brazilian Atlantic Forest Hotspot. *Science* **323**, 785–789.

Carrillo, E., Mora, A., Ketcham, R.A. et al. (2016) Movement vectors and deformation mechanisms in kinematic restorations: a case study from the Colombian Eastern Cordillera. *Interpretation* **4**, T31–T48.

Cediel, F., Shaw, R. & Cáceres, C. (2003) Tectonic assembly of the northern Andean block. In: Bartolini, C., Buffler, R.T. & Blickwede, J. (eds.) *The Circum-Gulf of Mexico and the Caribbean: Hydrocarbon Habitats, Basin Formation and Plate Tectonics*. American Association of Petrolum Geologists Memoir 79. Tulsa, OK: American Association of Petroleum Geologists, pp. 815–848.

Colinvaux, P.A., DeOliveira, P.E., Moreno, J.E. et al. (1996) A long pollen record from lowland Amazonia: forest and cooling in glacial times. *Science* **274**, 85–88.

Colwell, R.K., Brehm, G., Cardelus, C.L. et al. (2008) Global warming, elevational range shifts, and lowland biotic attrition in the wet tropics. *Science* **322**, 258–261.

Craw, D., Upton, P., Burridge, C.P. et al. (2016) Rapid biological speciation driven by tectonic evolution in New Zealand. *Nature Geoscience* **9**, 140–144.

Currie, D.J., Mittelbach, G.G., Cornell, H.V. et al. (2004) Predictions and tests of climate-based hypotheses of broad-scale variation in taxonomic richness. *Ecology Letters* **7**, 1121–1134.

Dalsgaard, B., Magard, E., Fjeldså, J. et al. (2011) Specialization in plant-hummingbird networks is associated with species richness, contemporary precipitation and quaternary climate-change velocity. *PLoS ONE* **6**, e25891.

Damasceno, R., Strangas, M.L., Carnaval, A.C. et al. (2014) Revisiting the vanishing refuge model of diversification. *Frontiers in Genetics* **5**, 353.

da Silva, F., Almeida-Neto, M. & Arena, M.V.N. (2014) Amphibian beta diversity in the Brazilian Atlantic forest: contrasting the roles of historical events and contemporary conditions at different spatial scales. *PLoS ONE* **9**, e109642.

Davies, R.G., Orme, C.D.L., Storch, D. et al. (2007) Topography, energy and the global distribution of bird species richness. *Proceedings of the Royal Society B: Biological Sciences* **274**, 1189–1197.

De-Silva, D.L., Elias, M., Willmott, K. et al. (2016) Diversification of clearwing butterflies with the rise of the Andes. *Journal of Biogeography* **43**, 44–58.

Dyer, L.A., Singer, M.S., Lill, J.T. et al. (2007) Host specificity of Lepidoptera in tropical and temperate forests. *Nature* **448**, 696–699.

England, P. & Molnar, P. (1990) Surface uplift, uplift of rocks, and exhumation of rocks. *Geology* **18**, 1173–1177.

Evans, K.L., Warren, P.H. & Gaston, K.J. (2005) Species-energy relationships at the macroecological scale: a review of the mechanisms. *Biological Reviews* **80**, 1–25.

Favre, A., Päckert, M., Pauls, S.U. et al. (2015) The role of the uplift of the Qinghai-Tibetan Plateau for the evolution of Tibetan biotas. *Biological Reviews* **90**, 236–253.

Fine, P.V.A. (2015) Ecological and evolutionary drivers of geographic variation in species diversity. In: Futuyma, D.J. (ed.) *Annual Review of Ecology, Evolution, and Systematics* **46**, 369–392.

Fjeldså, J., Lambin, E. & Mertens, B. (1999) Correlation between endemism and local ecoclimatic stability documented by comparing Andean bird distributions and remotely sensed land surface data. *Ecography* **22**, 63–78.

Futuyma, D.J. (2010) Evolutionary constraint and ecological consequences. *Evolution* **64**, 1865–1884.

Gentry, A.H. (1982) Neotropical floristic diversity - phytogeographical connections between Central and South America, pleistocene climatic fluctuations or an accident of the Andean orogeny. *Annals of the Missouri Botanical Garden* **69**, 557–593.

Graham, C.H., Ron, S.R., Santos, J.C. et al. (2004) Integrating phylogenetics and environmental niche models to explore speciation mechanisms in dendrobatid frogs. *Evolution* **58**, 1781–1793.

Graham, C.H., Moritz, C. & Williams, S.E. (2006) Habitat history improves prediction of biodiversity in rainforest fauna. *Proceedings of the National Academy of Sciences USA* **103**, 632–636.

Graham, C.H., Silva, N. & Velasquez-Tibata, J. (2010a) Evaluating the potential causes of range limits of birds of the Colombian Andes. *Journal of Biogeography* **37**, 1863–1875.

Graham, C.H., Van Der Wal, J., Phillips, S.J. et al. (2010b) Dynamic refugia and species persistence: tracking spatial shifts in habitat through time. *Ecography* **33**, 1062–1069.

Graham, C.H., Carnaval, A.C. Cadena, C.D. et al. (2014) The origin and maintenance of montane diversity: integrating evolutionary and ecological processes. *Ecography* **37**, 711–719.

Guarnizo, C.E., Amezquita, A. & Bermingham, E. (2009) The relative roles of vicariance versus elevational gradients in the genetic differentiation of the high Andean tree frog, *Dendropsophus labialis*. *Molecular Phylogenetics and Evolution* **50**, 84–92.

Hawkins, B.A., Field, R., Cornell, H.V. et al. (2003) Energy, water, and broad-scale geographic patterns of species richness. *Ecology* **84**, 3105–3117.

Hendry, A.P., Nosil, P. & Rieseberg, L.H. (2007) The speed of ecological speciation. *Functional Ecology* **21**, 455–464.

Hermelin, M. (1992) Los suelos del oriente antioqueño un recurso no renovable. *Bulletin de l'Institut Français d'Etudes Andines* **21**, 25–36.

Hijmans, R.J., Cameron, S.E., Parra, J.L. et al. (2005) Very high resolution interpolated climate surfaces for global land areas. *International Journal of Climatology* **25**, 1965–1978.

Hoorn, C., Wesselingh, F.P., ter Steege, H. et al. (2010) Amazonia through time: Andean uplift, climate change, landscape evolution, and biodiversity. *Science* **330**, 927–931.

Hoorn, C., Mosbrugger, V., Mulch, A. & Antonelli, A. (2013) Biodiversity from mountain building. *Nature Geoscience* **6**, 154–154.

Hugall, A., Moritz, C., Moussalli, A. & Stanisic, J. (2002) Reconciling paleodistribution models and comparative phylogeography in the Wet Tropics rainforest land snail Gnarosophia bellendenkerensis (Brazier 1875). *Proceedings of the National Academy of Sciences USA* **99**, 6112–6117.

Hutter, C.R., Guayasamin, J.M. & Wiens, J.J. (2013) Explaining Andean megadiversity: the evolutionary and ecological causes of glassfrog elevational richness patterns. *Ecology Letters* **16**, 1135–1144.

Jansson, R. (2003) Global patterns in endemism explained by past climatic change. *Proceedings of the Royal Society B: Biological Sciences* **270**, 583–590.

Janzen, D.H. (1967) Why mountain passes are higher in the tropics. *American Naturalist* **101**, 233–249.

Jetz, W., Rahbek, C. & Colwell, R.K. (2004) The coincidence of rarity and richness and the potential signature of history in centres of endemism. *Ecology Letters* **7**, 1180–1191.

Kattan, G.H., Franco, P., Rojas, V. & Morales, G. (2004) Biological diversification in a complex region: a spatial analysis of faunistic diversity and biogeography of the Andes of Colombia. *Journal of Biogeography* **31**, 1829–1839.

Körner, C. (2000) Why are there global gradients in species richness? Mountains might hold the answer. *Trends in Ecology & Evolution* **15**, 513–514.

Kozak, K.H., & Wiens, J.J. (2010) Niche conservatism drives elevational diversity patterns in Appalachian salamanders. *American Naturalist* **176**, 40–54.

Lawton, J.H. & May, R.M. (2005) *Extinction Rates*. Oxford: Oxford University Press.

Lynch, J.D., Vuilleumier, F. & Monasterio, M. (1986) Origins of the high Andean herpetological fauna. In: Vuilleumier, F. & Monasterio, M. (eds.) *High Altitude Tropical Biology*. New York: Oxford University Press, pp. 478–499.

Macarthur, R. & Levins, R. (1967) Limiting similarity convergence and divergence of coexisting species. *American Naturalist* **101**, 377–385.

Mastretta-Yanes, A., Moreno-Letelier, A., Pinero, D. et al. (2015) Biodiversity in the Mexican highlands and the interaction of geology, geography and climate within the Trans-Mexican Volcanic Belt. *Journal of Biogeography* **42**, 1586–1600.

McCain, C.M. (2009) Global analysis of bird elevational diversity. *Global Ecology and Biogeography* **18**, 346–360.

Merckx, V., Hendriks, K.P., Beentjes, K.K. et al. (2015) Evolution of endemismon a young tropical mountain. *Nature* **524**, 347–350.

Mittelbach, G.G., Schemske, D.W., Cornell, H.V. et al. (2007) Evolution and the latitudinal diversity gradient: speciation, extinction and biogeography. *Ecology Letters* **10**, 315–331.

Molnar, P. (1988) A review of geophysical constraints on the deep structure of the Tibetan Plateau, the Himalaya and the Karakoram, and their tectonic implications. *Philosophical Transactions of the Royal Society of London A* **326**, 33–88.

Mora, A., Parra, M., Strecker, M.R. et al. (2008) Climatic forcing of asymmetric orogenic evolution in the Eastern Cordillera of Colombia. *Geological Society of America Bulletin* **120**, 930–949.

Mora, A., Casallas, W., Ketcham, R.A. et al. (2015) Kinematic restoration of contractional basement structures using thermokinematic models: a key tool for petroleum system modeling. *AAPG Bulletin* **99**, 1575–1598.

Moritz, C., Hoskin, C., Graham, C.H. et al. (2005) Historical biogeography, diversity and conservation of Australia's tropical rainforest herpetofauna. In: Purvis, A., Gittleman, J.L. & Brooks, T. (eds.) *Phylogeny and Conservation*. Cambridge: Cambridge University Press, pp. 243–264.

Moritz, C., Langham, G., Kearney, M. et al. (2012) Integrating phylogeography and physiology reveals divergence of thermal traits between central and peripheral lineages of tropical rainforest lizards. *Philosophical Transactions of the Royal Society B: Biological Sciences* **367**, 1680–1687.

Moussalli, A., Moritz, C., Williams, S.E. & Carnaval, A.C. (2009) Variable responses of skinks to a common history of rainforest fluctuation: concordance between phylogeography and palaeo-distribution models. *Molecular Ecology* **18**, 483–499.

Mutke, J., Jacobs, R., Meyers, K. et al. (2014) Diversity patterns of selected Andean plant groups correspond to topography and habitat dynamics, not orogeny. *Frontiers in Genetics* **5**, 351.

Oncken, O., Hindle, D., Kley, J. et al. (2006) Deformation of the Central Andean Upper Plate System – facts, fiction, and constraints for plateau models. In: Oncken, O., Chong, G., Franz, G. et al. (eds.) *The Andes: Active Subduction Orogeny*. Berlin: Springer, pp. 3–27.

Parra, M., Mora, A., Sobel, E.R. et al. (2009) Episodic orogenic front migration in the northern Andes: constraints from low-temperature thermochronology in the Eastern Cordillera, Colombia. *Tectonics* **28**.

Rahbek, C. (1997) The relationship among area, elevation, and regional species richness in neotropical birds. *American Naturalist* **149**, 875–902.

Rahbek, C. & Graves, G.R. (2001) Multiscale assessment of patterns of avian species richness. *Proceedings of the National Academy of Sciences USA* **98**, 4534–4539.

Ramirez-Arias, J.C., Mora, A., Rubiano, J. et al. (2012) The asymmetric evolution of the Colombian Eastern Cordillera. Tectonic inheritance or climatic forcing? New evidence from thermochronology and sedimentology. *Journal of South American Earth Sciences* **39**, 112–137.

Restrepo-Moreno, S.A., Foster, D.A., Stockli, D.F. & Parra-Sánchez, L.N. (2009) Long-term erosion and exhumation of the "Altiplano Antioqueño," Northern Andes (Colombia) from apatite (U–Th)/He thermochronology. *Earth and Planetary Science Letters* **278**, 1–12.

Ribas, C.C., Moyle, R.G., Miyaki, C.Y. & Cracraft, J. (2007) The assembly of montane biotas: linking Andean tectonics and climatic oscillations to independent regimes of diversification in Pionus parrots. *Proceedings of the Royal Society B: Biological Sciences* **274**, 2399–2408.

Ricklefs, R.E. (1977) Environmental heterogeneity and plant species diversity – hypothesis. *American Naturalist* **111**, 376–381.

Ricklefs, R.E. (2004) A comprehensive framework for global patterns in biodiversity. *Ecology Letters* **7**, 1–15.

Ricklefs, R.E. (2007) History and diversity: explorations at the intersection of ecology and evolution. *American Naturalist* **170**, S56–S70.

Robin, V.V., Vishnudas, C.K., Gupta, P. & Ramakrishnan, U. (2015) Deep and wide valleys drive nested phylogeographic patterns across a montane bird community. *Proceedings of the Royal Society B: Biological Sciences* **282**, 20150861.

Rohde, K. (1992) Latitudinal gradients in species-diversity – the search for the primary cause. *Oikos* **65**, 514–527.

Rosenzweig, M.L. (1995) *Species Diversity in Space and Time*. Cambridge: Cambridge University Press.

Sedano, R.E. & Burns, K.J. (2010) Are the Northern Andes a species pump for Neotropical birds? Phylogenetics and biogeography of a clade of Neotropical tanagers (Aves: Thraupini). *Journal of Biogeography* **37**, 325–343.

Simpson, G.G. (1980) *Spledid Isolation: The Curious History of South American Mammals*. New Haven, CT: Yale University Press.

Smith, S.A., de Oca, A.N.M, Reeder, T.W. & Wiens, J.J. (2007) A phylogenetic perspective on elevational species richness patterns in Middle American treefrogs: why so few species in lowland tropical rainforests? *Evolution* **61**, 1188–1207.

Stein, A., Gerstner, K. & Kreft, H. (2014) Environmental heterogeneity as a universal driver of species richness across taxa, biomes and spatial scales. *Ecology Letters* **17**, 866–880.

Stephens, P.R. & Wiens, J.J. (2003) Explaining species richness from continents to communities: the time-for-speciation effect in emydid turtles. *American Naturalist* **161**, 112–128.

Thiede, R.C., Arrowsmith, J.R., Bookhagen, B. et al. (2005) From tectonically to erosionally controlled development of the Himalayan orogen. *Geology* **33**, 689–692.

Thomassen, H.A., Buermann, W., Mila, B. et al. (2010) Modeling environmentally associated morphological and genetic variation in a rainforest bird, and its application to conservation prioritization. *Evolutionary Applications* **3**, 1–16.

Vanzolini, P.E. & Williams. E.E. (1981) The vanishing refuge: a mechanism for ecogeographic speciation. *Papeis Avulsos de Zoologia (Sao Paulo)* **34**, 251–255.

Villagómez, D. & Spikings, R. (2013) Thermochronology and tectonics of the Central and Western Cordilleras of Colombia: Early Cretaceous–Tertiary evolution of the Northern Andes. *Lithos* **160**, 228–249.

Weir, J.T. (2006) Divergent timing and patterns of species accumulation in lowland and highland neotropical birds. *Evolution* **60**, 842–855.

Weir, J.T. & Price, M. (2011) Andean uplift promotes lowland speciation through vicariance and dispersal in Dendrocincla woodcreepers. *Molecular Ecology* **20**, 4550–4563.

Wiens, J.J. & Graham, C.H. (2005) Niche conservatism: integrating evolution, ecology, and conservation biology. *Annual Review of Ecology Evolution and Systematics* **36**, 519–539.

Wiens, J.J., Parra-Olea, G., Garcia-Paris, M. & Wake, D.B. (2007) Phylogenetic history underlies elevational biodiversity patterns in tropical salamanders. *Proceedings of the Royal Society B: Biological Sciences* **274**, 919–928.

Willig, M.R., Kaufman, D.M. & Stevens, R.D. (2003) Latitudinal gradients of biodiversity: pattern, process, scale, and synthesis. *Annual Review of Ecology Evolution and Systematics* **34**, 273–309.

Wright, D.H. (1983) Species-energy theory – an extension of the species-area theory. *Oikos* **41**, 496–506.

17

Mountains and the Diversity of Birds

Jon Fjeldså

Center for Macroecology, Evolution and Climate at the Natural History Museum of Denmark, University of Copenhagen, Copenhagen, Denmark

Abstract

Avian diversity varies greatly across the montane regions of the world. High-latitude mountains are bleak and harsh environments and marginal habitats for birds, but some tropical montane regions have a rich endemic avifauna and may be cradles of diversification within their respective regions. This chapter provides a quantitative overview of the diversity of avian species for all the montane areas of the world, and aims to explain the immense variation across regions. Some mountain tracts at low latitudes harbor species with no near relatives, in addition to many species representing recent radiations. Areas with this combination of old and young endemics are generally located near thermally stable ocean currents, suggesting that oceanic influence and stable water delivery may have played a key role in the accumulation of species diversity over time. Such climatic stability may also allow bird populations to adapt and specialize in response to the environmental pressure generated by local conditions within the montane habitat mosaics.

Keywords: *biodiversity hot spots, endemism, global patterns, latitude, montane regions, passerines*

17.1   Introduction

Mountains have been described as bleak environments where the diversity of birds is often constrained by extreme weather (e.g., Martin & Wiebe 2004). Several decades ago, when montane avifaunas were mainly studied in Europe and North America, the prevailing view was that mountains were fauna sinks with a limited capacity to support the evolution of greater diversity: they were mainly believed to be colonized, upslope, from the larger source pools of species in the surrounding lowlands, or by arctic birds, which, during the Pleistocene, were constrained to refugia in mountains at lower latitudes, where isolated populations survived after the ice sheets withdrew (Vuilleumier & Monasterio 1986).

With a more global outlook today, it has become evident that certain mountain regions are exceedingly rich in birds, and may in fact have been cradles of diversity (e.g., Weir 2006; Fjeldså & Bowie 2008; Lei et al. 2015; Liu et al. 2016). In fact, 74% of the non-insular endemic bird areas of the world are located in montane regions (Stattersfield et al. 1998). Montane areas are also widely recognized as focal areas or hot spots for bird conservation. The high diversity reflects the complexity of montane landscapes, with isolated "sky islands" and barren top ridges alternating with deep and sheltered valleys, and with habitat mosaics created by variation in slope and succession stages following frequent land slips.

Regardless of whether mountains were initially colonized from the Arctic or uphill from the adjacent lowlands, it has long been acknowledged that speciation can also proceed by isolation of local populations (vicariance) within a mountain region, or by founding events between sky islands. Given the number of potential dispersal barriers in montane areas, sedentary birds may easily become isolated locally; the ornithological literature provides an abundance of case studies (e.g., Vuilleumier & Monasterio 1986; García-Moreno & Fjeldså 2000; Päckert et al. 2015). However, most of these cases are from a few mountain ranges at low latitudes.

Little has been done yet to pursue a global analysis that can place the many idiosyncratic patterns in context and provide a more complete picture of the processes through

Mountains, Climate and Biodiversity, First Edition. Edited by Carina Hoorn, Allison Perrigo and Alexandre Antonelli.
© 2018 John Wiley & Sons Ltd. Published 2018 by John Wiley & Sons Ltd.
Companion website: www.wiley.com\go\hoorn\mountains,climateandbiodiversity

which the patterns evolved and were maintained. In this chapter, I will present a global overview of how the diversity of birds varies within and among montane regions. I will demonstrate some fundamental differences in species richness and diversification between montane regions at different geographical latitudes, and highlight some of the evolutionary and environmental factors that may have driven the diversification of birds in these regions.

17.2 Methods

17.2.1 How the Montane Regions of the World are Defined

In order to detect general patterns, the world's mountain areas were subdivided into a manageable number of regions, defined by contemporary and historical isolation between them. This required a set of rules to define mountain regions.

Mountains are rugged landscapes that are uplifted above the surrounding landscape to an extent that affects local climate, vegetation and the life histories of birds. In spite of the key role of climate, climatic thresholds are practically impossible to apply at a global scale because of the complex ways in which climate varies among montane regions and over time. The time component comprises changes between intense heat during the day and freezing conditions at night (often referred to as day-summer and night-winter conditions) in high tropical mountains, as well as the effects of Pleistocene climatic cycles. To avoid dealing with such temporal variation, it is more feasible to define mountains from landscape structures that persist over the time span during which species evolve (Fjeldså et al. 2012). A fixed elevational threshold is not useful, as many birds of rugged montane landscapes descend to sea level at the highest latitudes. Many areas in the interior of the continents are elevated but lack other characteristic features of mountains. Many lowland areas may, on the other hand, have scattered peaks projecting slightly above a given elevational threshold. Such places will mainly represent marginal (sink) habitats for birds from the surrounding habitat matrix, and should not be recognized as montane areas.

Based on these considerations, I have defined montane areas as follows. The pixelized maps produced by the Mountain Research Institute (Kapos et al. 2000; Körner et al. 2011) were used as a baseline for defining broader polygones including sloping areas >1000 m and excluding plateaus and isolated intermontane basins. The polygons circumscribing a montane region include fairly broad valleys intersecting the highlands, as well as outliers, which may be under some climatic influence from the larger highland. Near tropical coasts, I have recognized some mountains of <1000 m elevation, as these have compressed vegetation zones and a strong mist effect because of humidity from the ocean (Foster 2001). All boundary zones of larger montane regions were scrutinized using various data layers in Google Earth in order to determine local elevations, steepness gradients and surface structure. However, isolated peaks located >50 km from a larger highland were generally ignored, as they are considered too small to generate a highland climate.

Adjoining mountain ranges were defined as different montane polygons if they represented different biogeographic units (following Udvardy 1975 and Holt et al. 2013 for major subdivisions, and Stattersfield et al. 1998 for endemic bird areas). However, it was problematic to tease apart endemic bird areas that form interlocking ecological and elevational zones, which are like pieces of a jigsaw puzzle (notably in the Andes and Sino-Himalayan mountains, i.e., the mountain region comprising southwestern China in Sichuan and Yunnan provinces, and adjacent northern Myanmar and the Himalaya). Here, most of the regional montane endemics would be shared among adjacent zones, and with a rather coarse resolution of the distributional data it was not feasible to recognize them as separate units.

Altogether, 130 montane regions of very different shapes and extents were recognized, 38 of them located on relatively large islands and archipelagos (Figure 17.1). For each region, the area was calculated according to the Mollweide projection.

17.2.2 Recording the Avian Diversity in Montane Regions

Digital species distribution maps with one-degree resolution (Rahbek et al. 2012) were used. They represent conservative estimates of breeding ranges for all birds, and have been expert-validated and used in numerous other studies (e.g., Holt et al. 2013). A total of 90% of all nonpasserine and 92% of passerine bird species have some local overlap with a montane polygon. Not all of these species reside in montane landscapes, and it was therefore deemed necessary to inspect every species distribution and make a judgment based on the amount of montane habitat in the grid cells and on ecological information (including personal familiarity with a great portion of the species) regarding whether a species was likely to be resident inside a montane polygon. Quantitative graphics were based on the 6220 species of passerine birds only, as these have a fairly uniform (small) size and have successfully colonized all terrestrial habitats and land areas (Ericson et al. 2003; Fjeldså 2013). For each montane polygon, I recorded: (i) the number of resident passerine species; (ii) the number of endemic

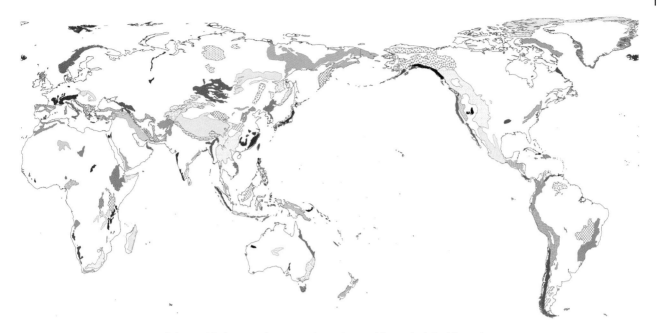

Figure 17.1 Montane regions of the world, showing the 130 region units used for analysis in this study.

species, defined as those with two-thirds or more of their range inside the polygon; and (iii) the number of small-ranged species, defined as species recorded in 1–10 grid cells, with at least half of these inside the polygon.

17.3 The Avifauna of Montane Environments

Birds are constrained in their ability to survive in environments where they will experience long periods of frost and absence of food. Birds breeding in mountains at high latitudes therefore leave these locations outside the growing season and undertake seasonal migrations, which often bring them to tropical winter quarters. Compared with many other organisms, whose life cycles allow them a seasonal diapause, the birds of northern mountains are dispersive, and therefore tend to become widely distributed. However, several avian groups have evolved mechanisms to tolerate the night-winter conditions in high mountains at low latitudes, and such birds could diversify and specialize to local conditions within montane regions.

Guilds of tropical montane birds have been described by Dorst & Vuilleumier (1986). They suggest a certain degree of convergently similar guilds among the montane avifaunas of South America, Africa and the Tibet area, and note a particularly large number of similarities among the faunas at the highest elevations.

Only a few nonpasserine birds can be classified as genuine highland birds. In general, large birds have large geographical distributions that encompass lowlands as well as highlands, as viable populations need to be able to track scarce and fluctuating resources. Genuine highland species are mainly found among wildfowl (Phasianidae), which can sustain themselves during even long-lasting periods of frost by browsing on scrubby vegetation or digging out tubers. Some raptors favor mountain landscapes, as the thermals over the mountain slopes aid their soaring flight high above the terrain in search of prey. These and many other nonpasserine birds use inaccessible cliffs as safe nest sites. Waterbirds are generally widespread and dispersive, and often rather opportunistic in their response to the changes in water levels in their favored shallow lakes and marshes. Waterbirds find little food in the deep, cold and nutrient-poor lakes in glacier-eroded high mountains, but wetlands on the large mountain plateaus in Asia and the Andes can hold large populations of waterfowl (Anatidae), flamingos (Phoenicopteridae) and shorebirds (Charadrii). A number of waterbird species are specifically adapted to live in these environments, where they forage in the shore meadows or on the large amounts of zooplankton that exist in the absence of fish populations (Fjeldså 1985; Hurlbert et al. 1986).

Most of the genuine highland birds are small birds, as these can best take advantage of the limited food resources and maintain viable populations even on isolated mountains of limited extent (see Ferenc et al. 2016). Among nonpasserine groups, only hummingbirds (Trochilidae) can be characterized as truly successful colonizers of highland environments. A monophyletic clade (with groups known as coquettes, brilliants, mountain gems, bees and emeralds, making up 77% of all hummingbirds) radiated mainly in the montane regions along the western axis of the New World continents (Chapter 18).

The passerine birds (Passeriformes), which make up 61% of all avian species, have successfully invaded montane environments all over the world (Dorst & Vuilleumier 1986). Although species diversity generally declines towards the highest elevations (Rahbek 1995), certain bird groups, which have adapted to tolerate climatic harshness, maintain high diversities in the bamboo thickets and densely tangled, gnarled and lichen-covered dwarf forests on the borderline towards the barren top ridges. In South America, these include in particular synallaxine ovenbirds (Furnariidae), antpittas (Grallaridae), fluvicoline flycatchers (Tyrannidae), tanagers, and sparrows (Thraupidae, Passerellidae). In Eurasia and Africa, there are other groups of grain-eating finches and sparrows, and a large diversity of insectivorous birds, such as leaf warblers (Phylloscopidae), flycatchers and chats (Muscicapidae). A large proportion of the highland birds use flower nectar as fuel, or can use seasonal fruiting for premigratory fattening. In the alpine zone, all continents have some grain-eating finches and ground insect-eaters.

17.3.1 The Global Species Diversity Pattern

The following sections are based only on the passerine bird data. Their global species richness pattern corresponds well to the well-known diversity maps for all birds (Jetz et al. 2012), mammals (Schipper et al. 2008) and plants (Kreft & Jetz 2007), and may provide a rather generalized picture of how biodiversity varies around the world. According to these maps, biodiversity peaks in the tropical rainforest regions and gradually decreases towards colder, or more arid, climates. To interpret this pattern, one needs to acknowledge that it is strongly dominated by co-distributed, widespread species (Rahbek et al. 2007). The upper quartile (25%) of most widespread avian species make up 82.6% of the total number of records in the global bird distributional database (Rahbek et al. 2012). The next quartile represents 13.2%, the next again 3.6% and the quartile of the least widespread species contributes only 0.6% of the data points. Thus, the majority of species contribute little to the overall pattern, but interestingly, those with the smallest distributions give rise to a distinct spatial pattern with a remarkable local turnover rate in species composition in tropical montane regions (Figure 17.2b).

Species breeding at high latitudes are generally dispersive and widespread, which gives rise to a very uniform species richness across highlands and surrounding lowlands. At lower latitudes, the mountain regions stand out as having elevated species richness. Peak values (up to 498 resident passerine species per grid cell) are found where the widespread species of the Amazon Basin meet the montane avifauna in the Andean foothills, giving a high species turnover (β-diversity) within grid cells. The peak concentrations of small-ranged species are found in

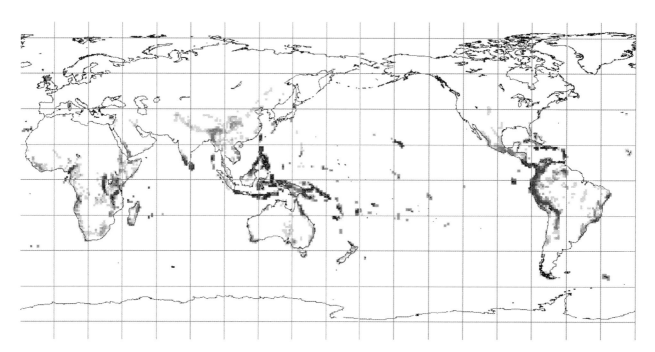

Figure 17.2 Global variation in diversity of small-ranged passerine birds, made up from data for the 25% of the world's 6220 species with the smallest geographical ranges (1–11 grid cells; lightest gray indicates a single species and darkest gray (in Ecuador) encompasses 40 species).

the tropical Andes region and the Costa Rica/Panamanian Mountains, and in other marked peaks in African sky islands and on larger and mountainous tropical islands. This means that, over time, very localized aggregations of small-ranged species can only be maintained in the fairly stable climates near the equator, and mainly in mountain ranges located along the margins of the continental plates or along the headwaters of large tropical floodplains (Fjeldså et al. 2012).

17.3.2 Correlations with Environmental Factors

Ecological hypotheses assume that, regardless of where on Earth an organismal group originated, it will have had ample time to diversify and redistribute itself in a way that corresponds to rates of metabolism in different ecosystems (Brown 2014; Pigot et al. 2016). Correlation analyses comparing bird distribution data and environmental parameters generally confirm that species richness is explained well by water and energy availability or by net primary productivity (e.g., Rahbek et al. 2007; Hawkins et al. 2011). The peak diversity of birds along the western margin of the Amazon Basin is consistent with hot and humid conditions, and with the way in which the humid air from the Intertropical Convergence Zone has been captured for millions of years by the concave shape of the Andean Cordillera. However, climate parameters alone cannot account for the extraordinary local accumulation of small-ranged species (Rahbek et al. 2007). This is better explained by topography, which acts in concert with mesoscale climatic variation in montane areas (Ruggiero & Hawkins 2008). The role of topography becomes even more prominent at a more coarse-grained spatial resolution, where sister species replacing one another in different mountains become included in the same grid cells. This reflects how a long history of allopatric speciation has gradually given rise to a geographical patchwork of closely related species (e.g., Sánchez-González et al. 2014). In the fairly stable tropical climate, it may take millions of years before full ecological compatibility evolves and allows sister species to coexist in the same location. This process has been particularly important along the 4000 km of continuous montane habitat in the tropical Andes (Fjeldså & Irestedt 2009).

It has also been suggested that small local populations could easily survive in montane regions, as they could track Pleistocene climate change by relocating short distances up or down the mountain slopes (in contrast to lowland birds, which would have to move greater distances) (Loarie et al. 2009). The distribution of small-ranged species has been found to correlate well with a "climate change velocity index," generated by draping a global climate change model onto a topographic model

(Sandel et al. 2011). However, a positive correlation does not necessarily explain causality. For instance, it does not explain the absence of small-ranged species in high mountains outside the tropics (Figure 17.2b). Already in the Himalaya (at 30° N), there are very few cases of local speciation, as phylogeographic data suggest that species richness has instead been built up by populations colonizing from southern China and adjacent Indochina (Johansson et al. 2007). Fjeldså et al. (2012) suggested that extraordinary accumulation of small-ranged species may first of all be determined by predictable water supply from thermally stable parts of the oceans, and how this interacts with topography.

17.3.3 Variation in Species Richness and Endemism Across Montane Regions

If we assume that species richness in montane regions evolved by local isolation within each region, we should expect the largest montane polygons to have the most species. This should above all apply to the most small-ranged species. Figure 17.3 presents the variation in species richness and endemism in relation to the area of montane polygons and across latitudinal zones.

High-latitude montane regions may harbor a moderate species diversity, which varies little with available area (Figure 17.3a). Some of the montane regions in North America and Asia have one or two endemic or range-restricted species – mainly accentors (*Prunella*) and rosefinches (*Carpodacus* and *Leucosticte*) – and isolated austral areas such as the Patagonian uplands, Tasmania and the South Island of New Zealand have slightly more, but most of the high-latitude montane regions have no endemics at all (Figures 17.3b,c). Considering the uniform species richness across northern adjacent montane and non-montane regions, it is apparent that high-latitude montane avifaunas represent an effective sampling of the communities inhabiting the surrounding matrix habitats.

At 20–40° latitude, the species counts per highland become more area-dependent. Still, many highlands (notably in arid regions) are species-poor, with no endemics. Significant numbers of endemics and smaller-ranged species are found in some of the larger highlands (the Madrean Cordilleras of Mexico, the Brazilian Atlantic Mountains, the Southern Andes, the South African Highlands, the Himalaya and the montane areas of south-western China).

In the tropics, the species counts become even more area-dependent, but still with clear deviations from the trend, since islands and arid mountains generally have few species. Most highlands have some endemic and smaller-ranged species, and some large mountain tracts – which mainly correspond to rift zones and subduction zones along tectonic plate margins – have very

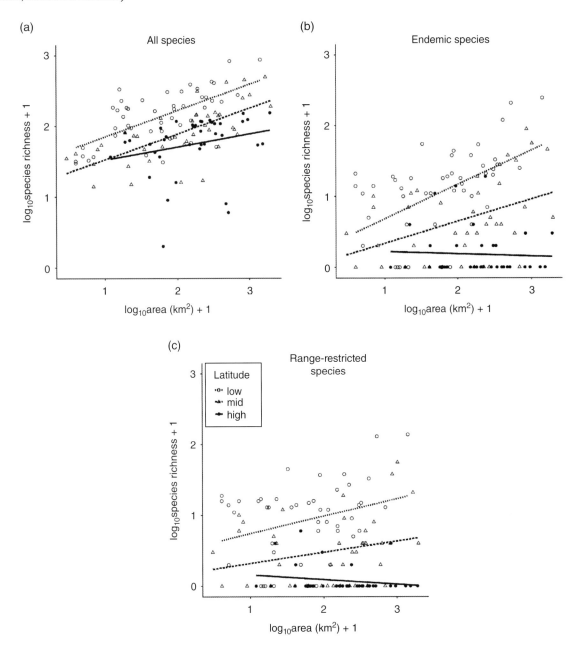

Figure 17.3 Variation in avian diversity with area of montane regions. (a) Variation in total number of passerine species. (b) Number of endemic species. (c) Number of species with restricted geographical distributions (1–11 grid cells). Black circles and solid regression lines represent data from the 39 montane regions in Figure 17.1 located mainly at >40° latitude; triangles and dashed regression lines represent data from the 43 montane regions located mainly at 20–40° latitude; and open circles and stippled lines represent data from the 48 montane regions located at <20° latitude.

high numbers, suggesting high rates of diversification in certain clades of highland-adapted birds. The highest species diversity is found in the Northern and Central Andes, with 828 and 865 species, respectively, of which 206 and 255 are endemic and 129 and 137 smaller-ranged. Other outstanding areas are the moderate-sized Costa Rica/Panamanian Highland (232 species, 54 endemic, 44 smaller-ranged), the Tanzania–Malawi Mountains (337 species, 53 endemic, 36 smaller-ranged) and the New Guinea Highlands (219 species, 118 endemics, 32 smaller-ranged). In a large-scale study of commensalistic hummingbird-plant networks, the small Costa Rica/Panamanian Highland proved to have the highest scores for specialized and small-ranged species, closely followed by study sites in the tropical Andes (Sonne et al. 2015).

Although no explicit analyses have explored the role of mountain ages, it seems that high diversity and endemism are more strongly associated with ruggedness, as found in young mountain ranges, than with time since uplift. Ancient erosional remnants (such as the mountains of the old cratonic shields) have few endemic species (or their original endemics have either died out or had time to disperse to the surrounding habitat matrix).

17.3.4 Speciation in Tropical Montane Regions

Montane avifaunas might evolve in three different ways. (i) Birds of the surrounding lowlands could move into the valleys and gradually adapt to highland conditions. In this way, montane avifauna become subsamples of the broader regional source pool, and gradually some of the colonists might diverge to become genuine highland species. (ii) Birds that are well adapted to highland conditions could disperse and establish populations in other mountain regions, or could be subdivided by vicariance within a large mountain range. (iii) Highland birds could colonize surrounding lowlands.

Most phylogeographic studies suggest that evolutionary shifts from tropical lowlands into cold highland environments are rather rare (Hughes & Eastwood 2006; Merckx et al. 2015). With the large molecular phylogenies that are now available, it has been demonstrated that most of the species inhabiting high elevations in the tropical Andes are members of distinct montane lineages, many of which originated in the southern cone of the continent (e.g., fluvicoline flycatchers, tapaculos and synallaxine ovenbirds: Fjeldså & Irestedt 2009; Borregaard et al. 2014), while others were rooted in the mountainous western Nearctic and diversified as they dispersed along the mountainous western axis of Central and South America (especially tanagers and finches: Barker et al. 2015; Kennedy et al. 2015). Distribution patterns within the Andes are complex, with several local aggregates of more or less narrowly endemic species (Fjeldså et al. 1999). Graves (1988) pointed out that the highest degree of differentiation is found in lineages inhabiting the narrow transition zone between the montane cloud forest and the barren upper montane zone. This suggests a role for random fragmentation of populations inhabiting this extremely narrow habitat band.

For Africa, special focus has been on the East African mountains, and notably the 600 km chain of 13 discreet sky islands in Tanzania's Eastern Arc Mountains. Many species groups are distributed more or less continuously across the whole range, but breaks in gene flow and divergence ages of the different populations are not congruent across the lineages. Speciation therefore appears to be uncoupled from tectonic history (Fjeldså & Bowie 2008). In Asia, researchers are analyzing the large radiations that took place in the Sino-Himalayan region, as well as speciation in glacial refugia in the Central Asian highland (e.g., Qu et al. 2010; Lei et al. 2015; Päckert et al. 2015).

17.4   The Effect of Latitude

The most remarkable aspect of global variation in the diversity of birds is probably the contrast between those regions where species had to move around in response to seasonal and long-term environmental changes and those where species could remain in the same place, perhaps for millions of years, and adapt to specific local conditions.

Birds that colonize montane environments at high latitudes may benefit from a boom of food during summer, but they will have to leave at the end of the season. They are forced to move in order to track changing conditions, both on a seasonal basis and on larger time scales (Jansson & Dynesius 2002). Although some species will only descend to the nearest lowlands, a large proportion make long seasonal migrations, often to tropical winter quarters (Newton 2008, ch. 13). This requires a high dispersal potential, and these species will therefore be able to disperse between mountain regions. Range dynamics and secondary introgression often erased the divergence that took place among different Pleistocene refugial populations. This means that incipient speciation was reversed and high genetic diversity is instead maintained over large geographical areas (e.g., Zink et al. 2008; Hogner et al. 2012), limiting the amount of speciation.

In contrast, tropical mountains are characterized by relative constancy in incoming radiations. This leads to year-round forage for many birds, including nectar for highland hummingbirds and sunbirds.

As pointed out by Janzen (1967), seasonal temperature amplitudes exceed the elevational temperature gradient at high latitudes, but not in the tropics, making topographic barriers less important there. Diversification in tropical mountains could therefore proceed fairly easily, as minor shifts in climate could lead to the isolation of populations inhabiting separate locations. Once isolated, small local populations would gradually diverge to become different species. It appears that montane areas act as cradles of diversity where climate is stable enough to allow populations to be maintained in the same location for sufficient time to become distinct species.

17.5   The Role of Niche Conservatism

Cadena et al. (2012) demonstrated that recently diverged sister species among Neotropical birds tend to replace one another in the same elevational zone on different slopes. It may take significant amounts of evolutionary time before related species diverge in their environmental preferences so as to segregate into different ecological zones on the same mountain slope (García-Moreno & Fjeldså 2000). Cadena et al. (2012) found that sister species in North America more often had different elevational niches.

In agreement with this, a study of the thermal traits and basal metabolic rates of 255 birds and 297 mammal species in a phylogenetic context found a general tendency for tropical species to be highly conservative in their thermal tolerance, while groups that evolved outside the tropics were more flexible (Khaliq et al. 2015). Winger et al. (2014) analyzed distributions and migratory habits in relation to the phylogeny of the 850 species of the New World nine-primaried songbirds (Emberizoidea), and found that several of the ancestral North American lineages had shifted their winter distributions from the southern Nearctic to the Neotropics, and then established breeding populations in the south, which gave rise to rapid expansion along the Andes. This was the case with the large radiation of tanagers (Thraupidae), for instance, which, after their early establishment in the northern Andes, rapidly expanded and diversified across all South American ecoregions (Barker et al. 2015; Kennedy et al. 2015). Phylogeographic studies have revealed several other cases where migratory northern birds established themselves and diversified in tropical montane areas (e.g., Moyle et al. 2014; Rolland et al. 2015).

Thermal niche conservatism seems mainly to be a constraint affecting ancient tropical rainforest groups, which rarely evolved tolerance of the night-winter conditions in the highlands. Other groups, which over time adapted to tolerate cold and seasonality at high latitudes, are more flexible: they can successfully colonize tropical highlands, and can also descend to tropical lowlands (Kennedy et al. 2015).

17.6   How did Species Diversity Build Up in Tropical Mountain Regions?

It is widely assumed that competition for resources limits the number of species, on local as well as regional spatial scales. Thus, a slowing of speciation rates over evolutionary time has been documented for many groups (Rabosky & Hurlbert 2015). However, some large phylogenies suggest continued high diversification, so it has

been argued that diversification can also be unbounded and dynamic (Harmon & Harrison 2015). When discussing this controversy, it is important to note that larger phylogenies will include a multitude of lineages that, at different times in their evolutionary history, have moved into new geographical regions. This sometimes provides new opportunities for intensive diversification; for instance, in archipelagos or montane regions (Chapter 15) (Kennedy et al. 2015). In the New World sparrows (Passerellidae), the highest species richness in one-degree grid cells is found in the boreal woodland biome, and is caused by the broad geographical overlap of a dozen widespread species. In the tropical Andes, in contrast, three recently colonizing lineages gave rise to twice as many species, but most of them replace one another on different mountain slopes (Sánchez-González et al. 2014). The rapid proliferation in the Andes could therefore proceed without raising the species diversity much at the local scale, where competition takes place.

The species richness in the Eastern Arc Mountains of Tanzania follows the species–area relationship among the different sky islands, but endemism does not. Three sky islands (the East Usambara, Uluguru and Udzungwa mountains, which are well spaced within the mountain chain) have excessive levels of endemism for their relative areas (Burgess et al. 2007), and palaeoecological data suggest an exceptional degree of climatic stability for these mountains (e.g., Marchant et al. 2007). Here, and in many other tropical montane areas, locations with many phylogenetically young species also harbor some ancient (relict) species, suggesting low rates of extinction in these places (Fjeldså & Bowie 2008; Fjeldså et al. 2012). One cannot know how many local mountaintop populations have gone extinct in the past, but careful census works in tropical mountains suggest that species inhabiting the upper montane forests today tend to have dense and therefore viable populations (Ferenc et al. 2016). Such studies suggest that tropical mountains can continue to accumulate new species, which offers hope for conservation work in these places.

The greatest concentrations of small-ranged avian species are associated with cloud forests or other mist-dependent vegetation, which are mainly found within 300 km of tropical coasts (Bruijnzeel et al. 2010). A few montane areas far from the coasts also have high endemism (Figure 17.1b). The eastern slope of the tropical Andes, the Albertine Rift mountains and the Sino-Himalayan mountains form the boundary of large forest-rich fluvial basins with high evapotranspiration. These areas are characterized by cloud formation and rainfall, and the removal of water vapor from the atmosphere means a drop in air pressure that sucks denser and more humid air from the ocean into the interior of the continent. The forest cover and biological diversity of tropical lowlands therefore becomes a self-perpetuating system, although it ultimately relies on the stability of oceanic currents. Makarieva & Gorshkov (2007) described this mechanism for the Amazon area, and documented how deforestation is constraining the flow of humid air that until now sustained the high forest biodiversity in the Brazilian Atlantic Mountains. The accumulation of small-ranged species over time mainly took place in mountain ranges near tropical coasts, or in tropical archipelagos, where large and mountainous islands could serve important roles as sources (termed "module and network hubs") that acted to maintain species diversity within the larger island region (Carstensen et al. 2011).

17.7 The Next Challenge: Does Geology also Play a Role?

The most prominent aggregation of small-ranged birds is found in mountains representing zones of active tectonic subduction along the edges of continental plates, suggesting that geology could also play a role. Birds may not be directly affected by the bedrock or soil type, but plants are, and the vegetation and food chains may therefore affect the avifauna. This would be particularly relevant in stable low-latitude climates, where birds may adapt to specific local environments for millions of years.

Zones of geological subduction are characterized by rough topographic relief, but are also geochemically complex. This is particularly pronounced in the Ring of Fire along the periphery of the Pacific Ocean, which is characterized by andesitic volcanic rocks and serpentines and highly metalliferous rocks that originated in Earth's mantle. Such rocks are alkaline, very rich in magnesium, and lacking in phosphorus in a form that plants can utilize (Batjes 2011). Botanists have known for a long time that these substrates constitute strong environmental filters (e.g., Macnair 1997) and that the exclusion of many plant groups from the "toxic" soils has allowed explosive speciation in other plant groups that can

tolerate such conditions (e.g., Hughes & Eastwood 2006). What is less known is how this affects the metabolism of the botanical communities, or the associated commensalistic networks. The vegetation on ultramafic soils often has a distinct, sclerophyllic structure. High phenol and lignin contents in the leaves are assumed to protect against grazing and to secure leaf longevity, but they also affect the cycling of nutrients in the thick leaf litter that accumulates (Kitayama et al. 2004). It could also be important that some plants, constrained by nutrient availability, cannot use the products of photosynthesis for growth and instead use them to attract reliable pollinators and fruit dispersers, or simply excrete them as sap or sweet exudates (Orians & Milewski 2007).

In a study in Bolivia, Dehling et al. (2014) detected a deep dichotomy between commensalistic networks of plants and frugivorous birds inhabiting the sedimentary environments in the lowlands and on Andean slopes above 1000 m elevation. Lowland and highland birds often represent distinct phylogenetic lineages (see McGuire et al. 2014 and Borregaard et al. 2014 for hummingbirds), which could be physiologically adapted to deal with different botanical communities and kinds of carbohydrates. Similar dichotomies seem to exist across Wallace's Line in Indonesia, with different frugivore and nectarivore groups dominating in the ultramafic and sedimentary environments to its east and west, respectively. However, nobody has explored the possible functional links between geochemistry and avian communities in these regions as of yet. This will be a challenge for future research, which needs to consider metabolism at the community level and in relation to phylogenetic lineages.

Acknowledgments

Bjørn Hermansen, Louis A. Hansen, Mikkel Willemoes and Michael Krabbe Borregaard are thanked for assistance with data and graphical presentations, and the Danish National Research Foundation for grant DNRF96 to the Center for Macroecology, Evolution and Climate.

References

Barker, F.K., Burns, K.J., Klicka, J. et al. (2015) New insights into New World biogeography: an integrated view from the phylogeny of blackbirds, cardinals, sparrows, tanagers, warblers, and allies. *Auk* **132**, 333–348.

Batjes, N. (2011) Global Distribution of Soil Phosphorus Potential. Report 2011/06. Wageningen: ISRIC World Soil Information.

Borregaard, M.K., Rahbek, C., Fjeldså, J. et al. (2014) Node-based analysis of species distributions. *Methods in Ecology and Evolution* **5**, 1225–1235.

Brown, J.H. (2014) Why are there so many species in the tropics? *Journal of Biogeography* **41**, 8–22.

Bruijnzeel, L.A., Scatena, F.N. & Hamilton, L.S. (2010) *Tropical Montane Cloud Forests.* Cambridge: Cambridge University Press.

Burgess, N.D., Butynski, T.M., Cordeiro, N.J. et al. (2007) The biological importance of the Eastern Arc Mountains of Tanzania and Kenya. *Biological Conservation* **134**, 209–231.

Cadena, C.D., Kozak, K.H., Gómez, J.P. et al. (2012) Latitude, elevational climatic zonation and speciation in New World vertebrates. *Proceedings of the Royal Society B* **279**, 194–201.

Carstensen, D.W., Dalsgaard, B., Svenning, J.-C. et al. (2011) Biogeographical modules and island roles: a comparison of Wallacea and the West Indies. *Journal of Biogeography* **39**, 739–749.

Dehling, S.M., Fritz, S.A., Töpfer, T. et al. (2014) Functional and phylogenetic diversity and assemblage structure of frugivorous birds along an elevational gradient in the tropical Andes. *Ecography* **37**, 1047–1055.

Dorst, J. & Vuilleumier, F. (1986) Convergence in bird communities at high altitudes in the tropics (especially the Andes and Africa) and at temperate latitudes (Tibet). In: Vuilleumier, F. & Monasterio, M. (eds.) *High Altitude Tropical Biogeography*. New York: Oxford University Press, pp. 120–149.

Ericson, P.G.P., Irestedt, J. & Johansson, U.S. (2003) Evolution, biogeography, and patterns of diversification in passerine birds. *Journal of Avian Biology* **34**, 3–15.

Ferenc, M., Fjeldså, J., Sedláček, O. et al. (2016) Abundance–area relationships in bird assemblages along an Afrotropical elevational gradient: montane forest species compensate for less space available. *Oecologia* **181**, 225–233.

Fjeldså, J. (1985) Origin, evolution and status of the avifauna of Andean wetlands. *Ornithological Monographs* **36**, 85–112.

Fjeldså, J. (2013) The global diversification of songbirds (Oscines) and the build-up of the Sino-Himalayan diversity hotspot. *Chinese Birds* **4**, 132–143.

Fjeldså, J. & Bowie, R.C.K. (2008) New perspectives on Africa's ancient forest avifauna. *African Journal of Ecology* **46**, 235–247.

Fjeldså, J. & Irestedt, M. (2009) Diversification of the South American avifauna: patterns and implications for conservation in the Andes. *Annals of Missouri Botanical Garden* **96**, 398–409.

Fjeldså, J., Lambin, E. & Mertens, B. (1999) Correlation between endemism and local ecoclimatic stability documented by comparing Andean bird distributions and remotely sensed land surface data. *Ecography* **22**, 63–78.

Fjeldså, J., Bowie, R.C.K. & Rahbek, C. (2012) The role of mountain ranges in the diversification of birds. *Annual Review of Ecology, Evolution, and Systematics* **43**, 244–265.

Foster, P. (2001) The potential negative impacts of global climate change on tropical montane cloud forests. *Earth Science Reviews* **55**, 73–106.

García-Moreno, J. & Fjeldså, J. (2000) Chronology and mode of speciation in the Andean avifauna. *Bonner Zoological Monographs* **46**, 25–46.

Graves, G.R. (1988) Linearity of geographic range and its possible effect on the population structure of Andean birds. *Auk* **105**, 47–52.

Harmon, L.J. & Harrison, S. (2015) Species diversity is dynamic and unbounded at local and continental scales. *American Naturalist* **185**, 584–593.

Hawkins, B.A., McCain, C.M., Davies, T.J. et al. (2011) Different evolutionary histories underlie congruent species richness gradients of birds and mammals. *Journal of Biogeography* **39**, 825–841.

Hogner, S., Laskemoen, T., Lifjeld, J.T. et al. (2012) Deep sympatric mitochondrial divergence without reproductive isolation in the common redstart *Phoenicurus phoenicurus*. *Ecology and Evolution* **2**, 2974–2988.

Holt, B.G., Lessard, J.-P., Borregaard, M.K. et al. (2013) An update of Wallace's zoogeographic regions of the World. *Science* **339**, 74–78.

Hughes, C. & Eastwood, R. (2006) Island radiation on a continental scale: exceptional rates of plant diversification after uplift of the Andes. *Proceedings of the National Academy of Sciences USA* **103**, 10 334–10 339.

Hurlbert, S.H., Loayza, W. & Moreno, T. (1986) Fish-flamingo-plankton interactions in the Peruvian Andes. *Limnology and Oceanography* **31**, 457–468.

Jansson, R. & Dynesius, M. (2002) The fate of clades in a world of recurrent climatic change: Milankovitch oscillations and evolution. *Annual Review of Ecology and Systematics* **33**, 741–777.

Janzen, D.H. (1967) Why mountain passes are higher in the tropics. *American Naturalist* **101**, 233–249.

Jetz, W., Thomas, F.H., Joy, J.B. et al. (2012) The global diversity of birds in space and time. *Nature* **491**, 444–448.

Johansson, U.S., Alström, P., Olsson, U. et al. (2007) Build-up of the Himalayan avifauna through immigration: a biogeographical analysis of the *Phylloscopus* and *Seicercus* warblers. *Evolution* **61**, 324–333.

Kapos, V., Rhind, J., Edwards, M. et al. (2000) Developing a map of the world's mountain forests. In: Price, M.F. & Butt, N. (eds.) *Forests in Sustainable Mountain Development: A State of Knowledge Report for 2000*. Wallingford: CABI.

Kennedy, J.D., Price, T.D., Fjeldså, J. & Rahbek, C. (2015) Historical limits on species co-occurrence determines variation in clade richness among New World passerine birds. *Journal of Biogeography* **41**, 1746–1757.

Khaliq, I., Fritz, S.A., Prinzinger, R. et al. (2015) Global variation in thermal physiology of birds and mammals: evidence for phylogenetic niche conservatism only in the tropics. *Journal of Biogeography* **42**, 2187–2196.

Kitayama, K., Suzuki, S., Hori, et al. (2004) On the relationships between leaf-litter lignin and net primary productivity in tropical rain forests. *Oecologia* **140**, 335–339.

Körner, C., Paulsen, J. & Spehn, E.J. (2011) A definition of mountains and their bioclimatic belts for global comparisons of biodiversity data. *Alpine Botany* **121**, 73–78.

Kreft, H. & Jetz, W. (2007) Global patterns and determinants of vascular plant diversity. *Proceedings of National Academy of Sciences USA* **104**, 5925–5930.

Lei, F., Qu, Y., Song, G. et al. (2015) The potential drivers in forming avian biodiversity hotspots in the East Himalaya Mountains of Southwest China. *Integrative Zoology* **10**, 171–181.

Liu, Y., Hu, J., Li, S-H. et al. (2016) Sino-Himalayan mountains act as cradles of diversity and immigration centers in the diversification of parrotbills (Paradoxornithidae). *Journal of Biogeography* **43**, 1488–1501.

Loarie, S.R., Duffy, P.B., Hamilton, H. et al. (2009) The velocity of climate change. *Nature* **462**, 1052–1055.

Macnair, M.R. (1997) The evolution of plants in metal-contaminated environments. In: Bijlsma, R. & Loeschcke, V. (eds.) *Environmental Stress, Adaptation and Evolution.* Basel: Birkhäuser, pp. 3–24.

Makarieva, A.M. & Gorshkov, V.G. (2007) Biotic pump of atmospheric moisture as driver of the hydrological cycle on land. *Hydrology and Earth System Sciences* **11**, 1013–1033.

Marchant, R., Mumbi, C., Behera, S. & Yamagata, T. (2007) The Indian Ocean dipole – the unsung driver of climate variability in East Africa. *African Journal of Ecology* **45**, 4–16.

Martin, K. & Wiebe, K.L. (2004) Coping mechanisms of alpine and arctic breeding birds: extreme weather and limitations to reproductive resilience. *Integrative and Comparative Biology* **44**, 177–185.

McGuire, J.A., Witt, C.C., Remsen, J.V. Jr. et al. (2014) Molecular phylogenetics and the diversification of hummingbirds. *Current Biology* **24**, 910–916.

Merckx, V.S.F.T., Hendriks, K.P., Beentjes, K.K. et al. (2015) Evolution of endemism on a young tropical mountain. *Nature* **524**, 347–350.

Moyle, R.G., Hosner, P.A., Jones, A.W. & Outlaw, D.C. (2014) Phylogeny and biogeography of *Ficedula* flycatchers (Aves: Muscicapidae): novel results from fresh source material. *Molecular Phylogenetics and Evolution* **82**, 87–94.

Newton, I. (2008) *The Migration Ecology of Birds.* Amsterdam: Elsevier.

Orians, G.H. & Milewski, A.V. (2007) Ecology of Australia: the effects of nutrient-poor soils and intense fires. *Biological Reviews* **82**, 393–423.

Päckert, M., Martens, J., Sun, Y.-H. & Tietze, D.T. (2015) Evolutionary history of passerine birds (Aves: Passeriformes) from the Qhinghai–Tibetan plateau: from a pre-Quaternary perspective to an integrative biodiversity assessment. *Journal of Ornithology* **156**(S1), 355–365.

Pigot, A.L., Tobias, J.A. & Jetz, W. (2016) Energetic constraints on species coexistence in birds. *PLoS Biology* **14**, e1002407.

Qu, Y., Lei, F., Zhang, R. & Lu, X. (2010) Comparative phylogeography of five avian species: implications for Pleistocene evolutionary history in the Quinghai–Tibetan plateau. *Molecular Ecology* **19**, 338–351.

Rabosky, D.L. & Hurlbert, A.H. (2015) Species richness at continental scales is dominated by ecological limits. *American Naturalist* **185**, 572–583.

Rahbek, C. (1995) The elevational gradient of species richness: a uniform pattern? *Ecography* **18**, 200–205.

Rahbek, C., Gotelli, N.J., Colwell, R.K. et al. (2007) Predicting continental-scale patterns of bird species richness with spatially explicit models. *Proceeding of the Royal Society of London B* **274**, 165–174.

Rahbek, C., Hansen, L.A. & Fjeldså, J. (2012) *One degree resolution database of the global distribution of birds.* Denmark: University of Copenhagen, Zoological Museum.

Rolland, J., Jiguet, F., Jønsson, K.A. et al. (2015) Settling down of seasonal migrants promotes bird diversification. *Proceedings of the Royal Society B* **281**, 20140473.

Ruggiero, A. & Hawkins, B.A. (2008) Why do mountains support so many species of birds? *Ecography* **31**, 306–315.

Sánchez-González, L.A., García-Moreno, J., Navarro-Sigüenza, A.G. et al. (2014) Diversification in a Neotropical montane bird: the *Atlapetes* brush-finches. *Zoologica Scripta* **44**, 135–152.

Sandel, B., Arge, L., Dalsgaard, B. et al. (2011) The influence of late quaternary climate-change velocity on species endemism. *Science* **334**, 660–664.

Schipper, J., Chanson, J.S., Chiozza, et al. (2008) The status of the World's land and marine mammals: diversity, threat, and knowledge. *Science* **322**, 225–230.

Sonne, J., González, A.M.M., Maruyama, P.K. et al. (2015) High proportion of smaller-ranged hummingbird species coincides with ecological specialization across the Americas. *Proceeding of the Royal Society B* **283**, 20152512.

Stattersfield, A.J., Crosby, M.J., Long, A.J. & Wege, D.C. (1998) *Endemic Bird Areas of the World.* Cambridge: BirdLife International.

Udvardy, M.D.F. (1975) A Classification of the Biogeographical Provinces of the World. IUCN Occasional Paper 18. Morges: IUCN.

Vuilleumier, F. & Monasterio, M. (eds.) (1986) *High Altitude Tropical Biogeography.* Oxford: Oxford University Press.

Weir, J.T. (2006) Divergent timing and patterns of species accumulation in lowland and highland neotropical birds. *Evolution* **60**, 842–855.

Winger, B.M., Barker, F.K. & Ree, R.H. (2014) Temperate origins of long-distance seasonal migration in New World songbirds. *Proceedings of National Academy of Sciences USA* **111**, 12 115–12 120.

Zink, R.M., Pavlova, A., Drovetski, S.W. & Rohwer, S. (2008) Mitochondrial phylogeographies of five widespread Eurasian bird species. *Journal of Ornithology* **149**, 399–413.

18

Teasing Apart Mountain Uplift, Climate Change and Biotic Drivers of Species Diversification

Fabien L. Condamine[1,2], Alexandre Antonelli[2,3,4], Laura P. Lagomarsino[2,3,5], Carina Hoorn[6] and Lee Hsiang Liow[7,8]

[1] *CNRS, Institut des Sciences de l'Evolution de Montpellier, University of Montpellier, Montpellier, France*
[2] *Gothenburg Global Biodiversity Centre, Gothenburg, Sweden*
[3] *Department of Biological and Environmental Sciences, University of Gothenburg, Gothenburg, Sweden*
[4] *Gothenburg Botanical Garden, Gothenburg, Sweden*
[5] *Department of Biology, University of Missouri St. Louis, St. Louis, MO, USA*
[6] *Institute for Biodiversity and Ecosystem Dynamics, University of Amsterdam, Amsterdam, Netherlands*
[7] *Natural History Museum, University of Oslo, Oslo, Norway*
[8] *Centre for Ecological and Evolutionary Synthesis (CEES), Department of Biosciences, University of Oslo, Oslo, Norway*

Abstract

Identifying the causes of species diversification and extinction remains a major challenge. Such biodiversity dynamics can be influenced by two major classes of factors: (i) biotic (intrinsic to the species, such as biological traits or species interactions), as in the Red Queen scenario; and (ii) abiotic (extrinsic to the species, such as climatic and geological events), as in the Court Jester scenario. Both classes are likely at play in most montane systems, where the interaction between mountain building and climate change may generate species diversity in a variety of ways; for example, via increased environmental heterogeneity, the generation of local habitats, the immigration of species or the formation of island-like ecological opportunities. Teasing apart the relative contributions of abiotic and biotic processes is challenging, because both may occur simultaneously and interact with each other, and a statistical framework that enables the separation of their relative contributions is still lacking. Here, we review the origin and evolution of biodiversity within a unified phylogenetic framework that explicitly disentangles the influences of mountain orogeny, climate change and ecological interactions. Relying on recently developed birth–death models, we build a model-testing approach that compares various diversification scenarios. Our approach includes a series of biologically realistic models to estimate speciation and extinction rates using a phylogeny, while assessing the relationship between diversification in the focal clade with an environmental variable, with growing species diversity within the focal clade or with the diversity of interacting clades. We illustrate the usefulness of this approach on two clades of Andean hummingbirds. We find that hummingbird speciation is positively correlated with temperature throughout their history. In contrast, speciation is negatively correlated with paleo-elevation, indicating that hummingbirds diversified faster in the early stages of the Andean orogeny. The analytical framework and empirical examples presented here demonstrate the power of combining phylogenetic and Earth-science models to untangle the complex interplay of geology, climate and ecology in generating biodiversity.

Keywords: *abiotic factors, biotic factors, birth–death models, diversification, macroevolution, past environments*

18.1 Seeking the Causes of Species Diversification and Extinction

Biodiversity is determined by temporal variations in speciation (the formation of new species) and extinction (the full demise of species), which lead clades (branches in the tree of life) to wax and wane. In addition to diversification (the net result of speciation and extinction, specifically speciation minus extinction), immigration also contributes to increased or decreased diversity on local to regional scales (Figure 18.1a). Ecologists and evolutionary biologists have long endeavored to identify which factors control the assembly of ecological communities and the regulation of long-term biodiversity

Mountains, Climate and Biodiversity, First Edition. Edited by Carina Hoorn, Allison Perrigo and Alexandre Antonelli.
© 2018 John Wiley & Sons Ltd. Published 2018 by John Wiley & Sons Ltd.
Companion website: www.wiley.com\go\hoorn\mountains,climateandbiodiversity

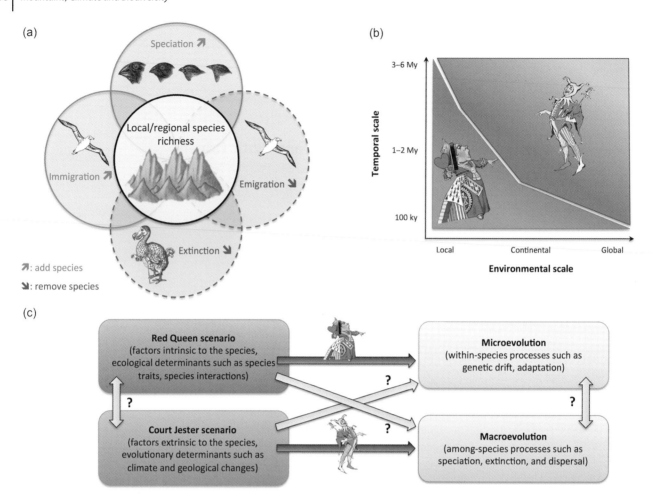

Figure 18.1 Paradigm of the Red Queen and the Court Jester as explanations for macroevolutionary dynamics of biodiversity. (a) Biodiversity dynamics are ultimately driven by speciation (adding species to a system), extinction (removing species) and dispersal (either adding or removing species). (b) The Red Queen is often regarded as a driver of evolutionary changes at fine taxonomic (within-species) and environmental (local) scales, and the Court Jester is seen as a driver of changes at larger taxonomic (genus, family) and environmental (continental, global) scales. (c) The links between these evolutionary processes are a major source of division between population genetics and paleontology/phylogenetics. While the two models are thought to operate over different geographic and temporal scales, their interactions are poorly understood (as denoted by the question marks).

dynamics (Benton 2015; Donoghue & Sanderson 2015). Proposed drivers governing and maintaining the evolution of biodiversity are numerous, but can be grouped into two classes: biotic (intrinsic) and abiotic (extrinsic) (Van Valen 1973; Barnosky 2001; Benton 2009).

In macroevolution (evolution that takes place over millions of years), explanations for the generation and maintenance of biodiversity can also be divided into two broad classes. The Red Queen scenario (RQS), conceptually based on the theories of Darwin and Wallace, and empirically rooted in observations from the fossil record (Van Valen 1973), posits that ecological interactions among species exert a large proportion of control on diversification rates. On the other hand, the Court Jester scenario (CJS), built mainly on geological evidence, including the fossil record (Barnosky 2001), posits

that longer-term changes in the physical environment – including climate fluctuations, sea-level changes and plate tectonics – are more important in shaping major diversity patterns, including sudden mass-extinction events (Figure 18.1b). In a simplified context, the RQS and CJS are thought to operate over different spatial and temporal scales: biotic factors such as species interactions shape ecosystems locally and over short time spans (thousands of years), whereas abiotic factors such as climate and tectonic events shape larger-scale patterns regionally and globally on million-year scales (Figure 18.1b) (Barnosky 2001). Although we use this dichotomy to aid discussion in this chapter, it is sometimes counterproductive (see later sections for examples). For instance, Van Valen (1973) proposed that taxa must constantly evolve to keep up with changes in their surroundings (*including* the physical environment and other taxa) or go extinct. In addition, interactions between environmental conditions and species' intrinsic characteristics (known as autecological traits) may act as key innovations, promoting diversification, yet they cannot, in general, be shoehorned into either of the original models. In this chapter, we broadly classify geological processes leading to mountain formation and associated climate changes (regional and local) as abiotic, and autecological traits, community composition and interspecific interactions as biotic, and recognize that these factors can interact to control diversification.

A growing number of species radiations (i.e., the relatively fast and substantial net diversification of clades) have been documented across disparate geographical areas and diverse habitats (Hunt et al. 2007; Meredith et al. 2011; Jetz et al. 2012; Magallón et al. 2015). However, relatively few studies conclude that diversification rate shifts can be attributed to both biotic and abiotic factors (but see Drummond et al. 2012; Jønsson et al. 2012; Bouchenak-Khelladi et al. 2015; Donoghue & Sanderson 2015). While synergy between these broad classes of factors has been documented (Ezard et al. 2011; Wagner et al. 2012), the underlying mechanisms and consequences of such interactions remain unknown in most cases, especially for mountain biota (Drummond et al. 2012; Hughes & Atchison 2015). An approach that simultaneously contrasts potential determinants of diversification and reconstructs the ancestral characteristics and biogeographic history of the focal clade offers a unique opportunity to shed light on the timing and underlying causes of species radiations (Condamine et al. 2012; Voje et al. 2015). This is important in understanding why some lineages diversify extensively while others do not, and consequently why some regions and habitats today contain many more species than do others. Further, a joint statistical analysis allows us to discern the individual impact of specific factors (e.g., biotic interactions, trait evolution, niche opportunity or changes in the physical environment) on each of the diversification components, speciation and extinction.

We lack definite answers to many fundamental questions regarding species radiations, such as: When and how do they start? Are there limits to global or regional species richness as imposed by carrying capacity, or as set by geographic area and/or niche heterogeneity (Rabosky 2013)? Are these limits to richness ever reached, and do we expect ecological systems (such as an isolated mountain, an island or a whole continent) to reach and maintain equilibria over geological time scales (see Harmon & Harrison 2015 and Rabosky & Hurlbert 2015 for a debate)? And, as specifically addressed by this chapter, do different montane clades share similar diversification histories as they diversify in response

to common abiotic factors with respect to mountain building, as predicted by the CJS? Or do clades, rather, have idiosyncratic patterns that are better explained by species' traits or by their biotic interactions (e.g., with prey, predators, competitors, pollinators or hosts), lending support to the RQS?

Macroevolutionary studies that disentangle which ecological and evolutionary processes underly patterns of diversity across space, time and clades are increasingly common, due to the increased availability of statistical tools for the analysis of molecular and fossil data. The accessibility of time-calibrated molecular phylogenies – branching trees that represent the evolutionary relationships among species – has stimulated the development of analytical tools that detect changes in diversification rates as a function of time, diversity and character state(s). These methods allow for the quantitative exploration of factors that are potentially linked to speciation and/or extinction of lineages (Morlon 2014). The ongoing development of phylogeny-based methods has provided new opportunities to address questions about the mechanisms that shape diversity patterns, shedding light on the processes explaining biodiversity at global (Condamine et al. 2012; Rolland et al. 2014), regional (Drummond et al. 2012; Hutter et al. 2013) and local (Schnitzler et al. 2011) scales. While we focus on phylogeny-based approaches in this chapter, these are beginning to converge with fossil-based approaches in their conceptual development and inferences (Morlon et al. 2011; Silvestro et al. 2014). Here, we review key diversification methods and build a framework to tease apart the effects of mountain building, climate change and species diversity on the diversification dynamics of Andean hummingbirds, embracing the challenge laid out in Voje et al. (2015) to simultaneously study the contributions of both biotic and abiotic drivers to evolutionary rates.

18.2 Defining the Abiotic and Biotic Drivers of Diversification: A Real Dichotomy?

In the RQS, biotic forces (including, but not limited to, interspecific interactions) are the main factors involved in shaping diversity dynamics over time (Voje et al. 2015). Biotic factors that may control macroevolutionary rates include competition with ecologically similar clade(s), predation, mutualism, symbiosis and trait evolution (Jablonski 2008). A prediction of the RQS is that the evolutionary arms race of change and counter-response prompts interacting species to drive one another to adapt and ultimately speciate (Weber & Agrawal 2014). One example is bats developing successively longer tongues in order to reach nectar in flowers with successively longer corolla tubes (Muchhala & Thomson 2009). Thus, we would expect that species richness in the RQS depends primarily on biotic interactions or intrinsic factors, such as life-history traits (Jablonski 2008). All things being equal, over larger temporal and spatial scales, we would expect (i) speciation rates to be greater in areas where species interactions are strongest (e.g., in a tropical rainforest) (Schemske et al. 2009) and (ii) net diversification rates to slow over time as communities become filled with species (e.g., as an island is successively colonized). These two expectations are not mutually exclusive, and can concurrently impact the diversification of a given group. Nonetheless, diversity can positively or negatively affect diversification rates. In the first case, speciation is positively linked to diversity, such that the more diversity increases, the more speciation increases (diversity begets diversity) (Janz et al. 2006). In the second case, speciation is negatively linked to diversity, such that the more diversity increases over time, the lower the speciation rate (ecological limits) (Rabosky 2013). A slowing of diversification could derive either from increasing extinction rates or from decreasing speciation and immigration rates, as the number of ways to compete for space and/or resources declines (Wagner et al. 2012; Mahler et al. 2013). In extreme cases of the RQS, we can expect both diversification rates and trait evolution rates to decrease as species richness increases (Rabosky 2013).

The CJS, on the other hand, assumes that changes in speciation and extinction rarely happen (i.e., they remain constant), except in response to sudden and major changes in the physical environment (Barnosky 2001; Benton 2009). In this case, we would expect that diversity dynamics depend on abiotic factors: changes in climate (cooling or warming), geography (appearance or disappearance of areas), geology (mountain building) and sea level (increase or decrease). Many paleontological studies have found that changes in the physical environment have fostered the diversification of clades or have contributed to their extinction (Mayhew et al. 2008; Peters 2005, 2008; Hannisdal & Peters 2011). For instance, Peters (2008) showed that the turnover of marine-shelf habitats and correlated environmental changes have been consistent determinants of extinction, extinction selectivity and the shifting composition of the marine biota during the Phanerozoic. Over longer temporal and larger spatial scales, according to the CJS, we expect (i) clade-specific speciation and/or extinction rates to be greater in physically dynamic areas, such as rising mountains and changing landscapes (Hoorn et al. 2013; Hughes & Atchison 2015) and (ii) that diversification rates will co-vary with environmental change (Mayhew et al. 2008; Condamine et al. 2013, 2015).

Traditionally, ecologists who study extant organisms have mainly thought in an RQS framework, and paleontologists in a CJS framework (Benton 2009). Support for either the RQS or the CJS is thought to depend on the time scale considered (Barnosky 2001; Benton 2009, 2010). Ecological interactions drive the local-scale success (or failure) of individuals, populations and species over microevolutionary scales (hundreds to thousands of years), but long-term tectonic and climatic processes erase the signature of these processes on macroevolutionary time scales (millions of years) (Figure 18.1c). However, long-term ecological interactions have recently been shown to play a more important and persistent role than previously thought (Liow et al. 2015; Silvestro et al. 2015).

It is probable that the importance of biotic and abiotic factors varies through time (Ezard et al. 2011; Condamine et al. 2012; Cantalapiedra et al. 2014), a view coined the "multilevel mixed model" (Benton 2009). Marx and Uhen's (2010) study of Cetacea provides an example of this model. To explain the dynamics of cetacean diversity, Marx and Uhen analyzed both abiotic and biotic factors as potential drivers of the past cetacean diversity by considering the effect of paleotemperature and phytoplankton (diatom and coccolithopore) diversities in the fossil record. They fitted models in which the past diversity of cetaceans can be correlated to each variable individually or to multiple variables simultaneously, and then used model comparison to assess which best explained the genus-level diversity through time. Diversity in this group is explained by a model that combines both phytoplankton diversity and climate fluctuations through time (as opposed to a model in which diversification is driven by one factor only), suggesting that diversification is driven by both temperature and food supply (Marx & Uhen 2010). Models with only change in diatom diversity and models with only climate change were significantly outperformed (Marx & Uhen 2010, Table 1).

18.3 Phylogenetic Approaches to Study Diversification

The limited (or completely absent) fossil record for many clades has encouraged the development of alternative approaches to the study of diversification that do not rely on fossil evidence (Raup et al. 1973; Nee et al. 1994). Phylogenetic trees of extant species can be inferred using molecular and/or morphological data. Statistical models can estimate a phylogeny's underlying diversification parameters (i.e., origination or speciation, denoted λ, and death or extinction, denoted μ), and ultimately make inferences about the clade's diversification dynamics

(Morlon 2014). Phylogenies have thus become an essential tool for studying macroevolutionary dynamics (reviewed in Morlon 2014), and such emphasis is further bolstered by the rapidly increasing availability of large dated phylogenies (Hunt et al. 2007; Meredith et al. 2011; Jetz et al. 2012; Pyron & Wiens 2013; Magallón et al. 2015).

Many models have been developed to estimate speciation and extinction rates, commonly called "birth–death models." The difference between speciation and extinction gives the net diversification rate:

$$r = \lambda - \mu \tag{18.1}$$

The simplest model is a constant-rate speciation, also called the Yule model or pure-birth model (no extinction is included). Adding non-zero extinction, the constant-rate birth–death model is obtained. Then, one can adjust each diversification parameter to vary according to time, trait or environment, and so estimate speciation and extinction rates and their variation through time (Morlon et al. 2011; Stadler 2011; Rabosky et al. 2014). Extinction rates estimated using molecular phylogenies may be congruent with those stemming from the fossil record (Morlon et al. 2011), but extinction estimates determined only from molecular phylogenies are often associated with large margins of error (confidence intervals).

Past species diversity and its variation through time can be inferred by estimating speciation and extinction rates over time (Morlon et al. 2011). If temporal variation is detected, additional tests can be applied to determine which processes are at play (Morlon 2014). Although the use of these models is in its infancy, they provide statistical tools for testing mechanistic models of evolution such as the two macroevolutionary hypotheses described earlier, the RQS and the CJS.

The effect of the abiotic environment on diversification has been primarily examined via likelihood models that incorporate discrete shifts in diversification rates over the course of a clade's history. For instance, birth–death likelihood methods have been used to test the hypothesis that a shift in speciation rate occurred at specific Cenozoic climatic events (the "shift approach": Condamine et al. 2012). In these studies, climatic events were modeled as punctuated events (25 Ma in the case of the late Oligocene warming event), and a two-rate model with one shift coinciding with the climatic event was tested against a one-rate model (corresponding to the null hypothesis of no rate shift). While these analyses were performed with a likelihood expression that assumed no extinction, the expression including extinction is now available in the R package *TreePar* (Stadler 2011). This approach is not restricted to a single rate shift and could thus be used to test support for multiple shifts in speciation or extinction rates over a time series and their concordance with temperature

shifts (Stadler 2011). This framework allows for testing of specific hypotheses about diversification changes in response to major periods of global warming (such as the Paleocene–Eocene Thermal Optimum ca. 56 Ma and the Mid-Miocene Climatic Optimum ca. 15 Ma) and global cooling (such as at the Eocene–Oligocene transition 33.9 Ma and the onset of northern hemisphere glaciations in the Pleistocene ca. 2.6 Ma).

The study of ecological interactions as drivers of macroevolutionary change has relied on two classes of birth–death models: diversity-dependent models (Phillimore & Price 2008; Etienne et al. 2012) and trait-dependent models (Maddison et al. 2007; FitzJohn 2012). These approaches identify patterns that fit historical ideas of evolution, such as rapid morphological evolution allowing a clade to colonize new areas/resources (adaptive radiation through natural selection) (Mahler et al. 2013), or factors that are associated with changes in diversification rates (Jablonski 2008), such as body size, reproductive modes, color polymorphism or trophic strategies.

While a molecular phylogenetic perspective is relevant to understanding the determinants and causes of biodiversity dynamics of extant taxa (Condamine et al. 2012; Jønsson et al. 2012; Drummond et al. 2012), other macroevolutionary models that incorporate only the fossil record are particularly valuable for understanding the evolution of extinct clades or groups with a substantial fossil record. A recently developed Bayesian model jointly estimates taxon-specific rates of speciation and extinction and their variation through time based on fossil data (Silvestro et al. 2014). By applying this method, Silvestro et al. (2015) found that felids came into competition with two canid subfamilies in North America, driving their local extinction and subsequent replacement. A different approach, based on capture–recapture models developed in statistical ecology, uses Markov models to simultaneously estimate speciation, extinction and sampling probabilities through observations of fossils (Liow & Nichols 2010). Using this method to explore climate and sea level as potential drivers of diversification histories, it was shown that bivalves suppressed brachiopod diversification (Liow et al. 2015).

18.4 A Unified Framework to Tease Apart the Drivers of Diversification

Understanding the historical processes that contributed to current patterns of species richness and distribution requires a reconciliation and integration of the RQS and CJS in a common analytical framework (Voje et al. 2015). In this section, we present a model selection framework that compares potential determinants of

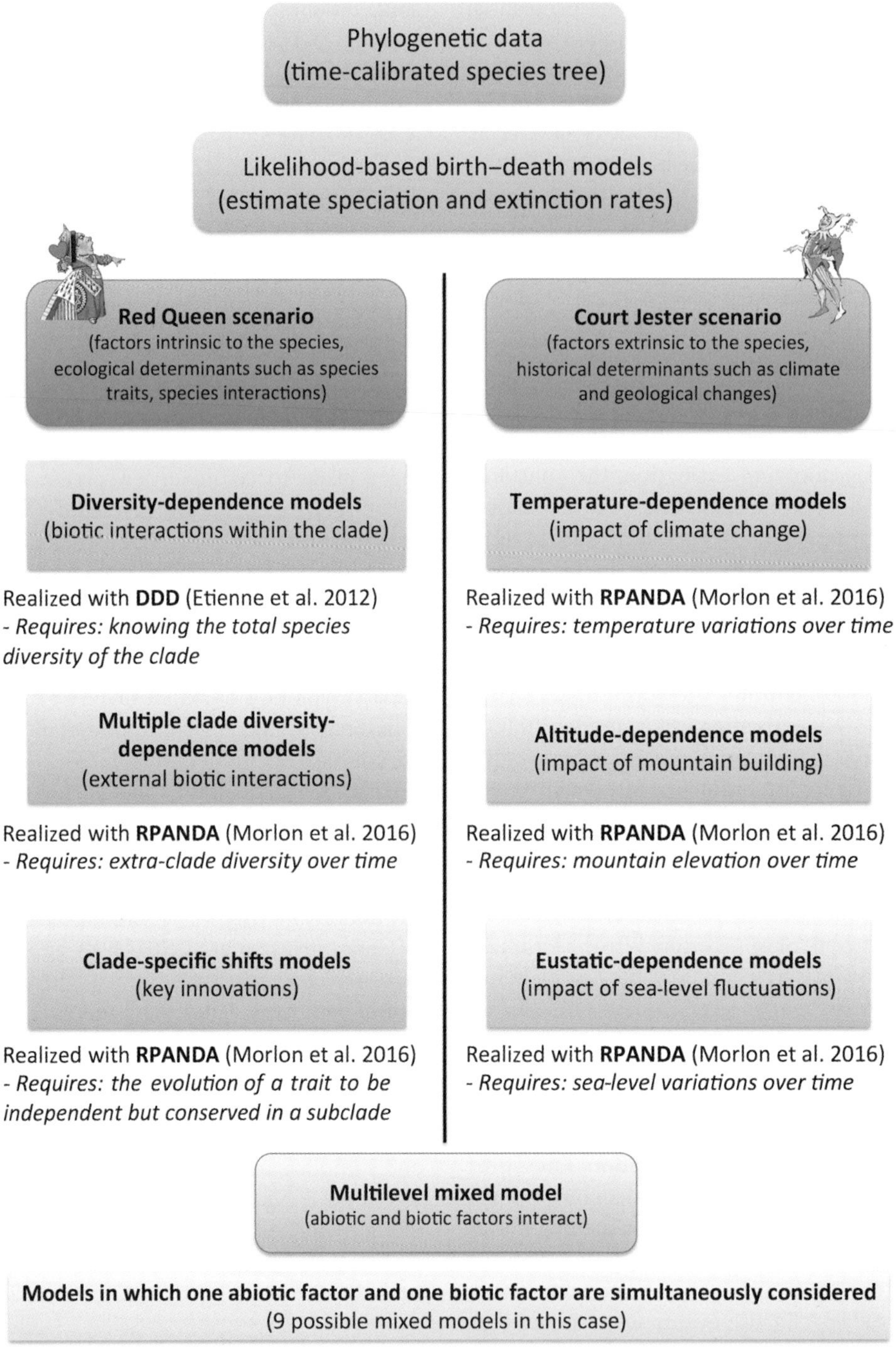

Figure 18.2 General framework for testing the Red Queen scenario (RQS) and Court Jester scenario (CJS) with phylogenetic data and birth–death models. The left panel highlights diversification models that are appropriate for testing the RQS, while the right panel illustrates models that are appropriate for testing the CJS. The models are not comprehensive, and more elaborate ones can be built, depending on the data at hand.

species diversification and extinction over macroevolutionary time scales by estimating the relative roles of abiotic (extrinsic) and biotic (intrinsic) factors and their influences on species richness patterns (Figure 18.2). The full statistical framework is available at GitHub (https://github.com/FabienCondamine/Diversification_analyses). After introducing our approach in general terms, we will apply it specifically to the Andean biota to illustrate how it may help us understand biotic evolution in mountain systems.

18.4.1 Modeling the Impact of Past Environments on Diversity Dynamics

Many studies have suggested a prominent role of environmental changes in diversification, following the CJS (Peters 2005, 2008; Erwin 2009). Such studies have mostly relied on descriptive ad hoc comparisons of dated phylogenies to paleoenvironmental curves, or used the shift approach (discussed earlier). However, these methods are not always appropriate, because climatic events are generally not instantaneous. For example, the Oligocene warming (25 Ma) lasted 3 My, and warming during the late Permian (260–252 Ma) lasted even longer. More ambiguities may appear when the shift approach is used over a collection of trees, randomly taken from a posterior distribution of a Bayesian dating analysis, due to the dating and phylogenetic uncertainties. In addition, climate change is often linked to geological events, such as the uplift of the Andes (Sepulchre et al. 2010) and of the Qinghai–Tibetan Plateau (Chapter 29) (Favre et al. 2015). Finally, these approaches do not quantify the potential effects of other environmental variables on diversification rates.

However, recently developed methods now allow us to quantify such effects. One approach developed by Condamine et al. (2013) – hereafter termed the "paleoenvironment-dependent diversification model" (PDDM) – builds on time-dependent diversification models (Nee et al. 1994; Morlon et al. 2011) to allow speciation and extinction rates to depend not only on time but also on an external variable (which may vary through time). This methodology, implemented in the R package *RPANDA* (Morlon et al. 2016), assumes that clades evolve under a birth–death process. Speciation (λ) and extinction (μ) rates can vary through time, and both can be influenced by one or several environmental variables (E) that also vary through time, such that $E_1(t), E_2(t), ..., E_k(t)$. For simplicity, we will call E temperature in this discussion. Note that while the observations of E are discrete, we use a smoothing function to allow the modeling of diversification rates in continuous time.

Thus:

$$\tilde{\lambda}(t) = \lambda\left(t, E_1(t), E_2(t), ..., E_k(t)\right) \quad (18.2)$$

and

$$\tilde{\mu}(t) = \mu\left(t, E_1(t), E_2(t), ..., E_k(t)\right) \quad (18.3)$$

stand for speciation and extinction rates influenced by temperature and time, respectively. We consider the phylogeny of n species sampled from the present, and allow for the possibility that some extant species are not included in the sample by assuming that each extant species was sampled with probability $f \le 1$. Time is measured from the present to the past such that $t_1 > t_2 > ... > t_n$ denotes branching times in the phylogeny, where t_1 is the stem age and t_2 the crown age of the clade. The probability density of observing such a phylogeny, conditioned on the presence of at least one descendant in the sample, is based on Morlon et al. (2011). The approach can be used to derive likelihoods for any functional form of λ and μ. For example, λ may be an exponential function of temperature, such that:

$$\tilde{\lambda}(t) = \lambda_0 e^{\alpha T(t)} \quad (18.4)$$

or a linear function, such that:

$$\tilde{\lambda}(t) = \lambda_0 + \alpha T(t) \quad (18.5)$$

where λ_0 and α are the two parameters to estimate. In the case of exponential dependency on temperature, the estimation of a positive α would indicate that higher temperatures increase speciation rates, whereas a negative α would indicate that higher temperatures decrease speciation rates.

In this section, temperature is treated as a potential determinant of speciation and/or extinction rates, such that the inferred time variation in rates matches the time variation in temperatures. However, in other studies, it might be more relevant to consider the effects of other environmental variables on species diversification, such as sea-level fluctuations, atmospheric carbon concentration (a proxy for climate change), the area of available habitats through time, coastal lengths, sea depth or orogeny (the deformation of continental plates into mountains). Despite their importance as descriptors of the physical environment, there is much less theory about the effect of the aforementioned variables on diversity dynamics relative to temperature (Peters 2005, 2008; Hannisdal & Peters 2011). However, we are beginning to apply some of these alternative variables to macroevolutionary models. A recent study using this approach investigated the role of past sea-level variation on an island-dwelling swallowtail butterfly clade in Indonesia (Condamine et al. 2015) using the raw sea-level data of Miller et al. (2005). It was inferred that extinction is positively linked to higher sea level through time, suggesting that butterfly species went extinct during periods of high sea level.

18.4.2 Modeling the Impact of Biotic Factors and Interactions on Diversity Dynamics

The study of the role of biotic drivers has arguably been the most productive field in macroevolutionary biology (reviewed in Rabosky 2013). Here, trait-dependent diversification, as inferred in the State-Dependent Speciation and Extinction class of models (e.g., BiSSE) (Maddison et al. 2007), has played a prominent role (e.g., Rolland et al. 2014). However, these models have recently received criticisms stemming from their mathematical foundations and false positive results (Rabosky & Goldberg 2015). Furthermore, these trait-dependent models are not directly comparable in a common framework with other approaches, and they cannot quantify how a biotic variable (within-clade diversity, extra-clade diversity) influences diversification rates. We mention this class of models because they have been widely used to study biotic factors in diversification, such as intrinsic species traits (Ng & Smith 2014), but due to the criticisms, we will focus on other models here. To circumvent these issues, alternative models are emerging (Rabosky & Huang 2016).

The decrease in diversification rates over time is a very common phylogenetic pattern that is generally attributed to radiations reaching their ecological limits (Jønsson et al. 2012; Mahler et al. 2013). A birth–death method termed "diversity-dependent diversification" (DDD) explicitly links speciation and extinction rates to the number of species in the clade (Etienne et al. 2012). Like the *RPANDA* approach, DDD is an extension of the constant-rate birth–death model. Speciation and extinction rates depend exponentially or linearly on the number of species, and eventually become equal at the carrying capacity, K, which is interpreted as the maximum number of species a clade can have in a given environment. The R package *DDD* implements this method and can compute the likelihood of a phylogeny under a wide range of diversity-dependent birth–death models of diversification (see later) in order to estimate the diversification parameters and carrying capacity (Etienne et al. 2012). A DDD model that estimates a carrying capacity close to the extant number of species in a clade suggests that the clade is at, or close to, its ecological limits (Rabosky 2013). These models are thus appropriate for testing biotic interactions within a focal clade (e.g., competition for resources).

Using this framework, it would be possible to test for biotic interactions among clades – a scenario termed "multiple-clade diversity-dependent diversification" (Silvestro et al. 2015). At the moment, no comparable model exists in phylogenetics. It is nonetheless possible to use the PDDM framework (Condamine et al. 2013) to address a similar question, by replacing the paleoenvironmental curve (e.g., temperature) with a diversity or relative abundance curve for an interacting clade. For instance, if one wants to test the role of angiosperms as direct competitors of gymnosperms, the model can be adjusted to use the floral composition through time (angiosperms began from zero in the Early Cretaceous and reached 80% of the floral composition by the end of the Late Cretaceous) (Benton 2010). Another application would be to study the effect of the appearance and increased abundance of grasslands in the mid-Miocene on grass-specialist insects.

18.4.3 Which Model Best Explains Diversity Dynamics: A Model Comparison Procedure

Likelihood approaches involve computing the probability of observed data given a model and a set of parameters, such as speciation and extinction rates. Here, "model" refers to a mathematical statement of how proposed factors might influence the parameters of interest. The maximum likelihood approach then optimizes the estimates of the parameters given a model and a phylogeny. Because there is often a "forest" of equally probable phylogenetic trees based on sampled taxa, molecular sequences and calibration points for molecular dating, we recommend running the models over a set of multiple trees (e.g., Bayesian posterior trees from a dating analysis) in order to infer the confidence interval of each parameter based on the uncertainty of tree topology and age estimates.

In addition to estimating diversification parameters within a given model, we are interested in comparing models incorporating different variables. For example, does mountain formation explain the diversification dynamics of mountain-adapted species better than species richness through time? Or have mountain-adapted species diversified at a constant rate, not dependent on any external factor? In a likelihood framework, this can be done easily, provided that consistent conditioning and normalization factors have been used for the likelihood computation (Stadler 2013). Once normalized, likelihoods, and hence models, can be compared. Likelihood ratio tests can be used for comparisons of nested models; that is, the most complex model in the set of models can be transformed into all the simpler models by imposing constraints on its parameters. When the models to be tested are not nested, other methods can be used to compare them, including the corrected Akaike Information Criterion (AICc) and Bayes factors (Burnham & Anderson 2002). The AICc is useful in comparing the probabilities of observing branching times as explained by various individual paleoenvironmental variables (Condamine et al. 2015). A series of models can be designed to quantify the effect of various environmental variables, in isolation or in combination, on diversification. For instance, one can compare the AICc scores for the best-fit models between multiple environmental variables (e.g., temperature, sea level) to determine which has the strongest effect on diversification.

18.5 Case Study: The Andean Radiation of Hummingbirds

To illustrate how this analytical framework can be applied to investigate the diversification history of montane organisms, we explore the RQS and CJS using hummingbirds. In particular, we focus on the "Andean clade" of South America *sensu* McGuire et al. (2014), comprising the Brilliants (51 extant species) and Coquettes (61 extant species). As more than 80% of this clade's current species diversity lives in the Andes, and some species occur as high as 5200 m (hillstars, genus *Oreotrochilus*), a reasonable hypothesis is that Andean orogeny was a major trigger of diversification (date phylogeny in Figure 18.3a). However, there is a debate over whether mountain uplift indeed promoted diversification of the mega-diverse tropical Andes (Drummond et al. 2012; Hoorn et al. 2013; Hutter et al. 2013; Hughes & Atchison 2015). Besides mountain building per se, climate change may have been another important factor impacting biodiversity. Lagomarsino et al. (2016) were the first to tackle this issue, by testing the influence of mountain uplift, climate change and biotic interactions on the diversification of South American lobelioids (Campanulaceae). To date, no study has investigated this hypothesis with respect to the evolution of an animal clade.

The diversification of Andean hummingbirds could also be the result of intra-clade biotic interactions, possibly manifested as diversity dependence, where different species compete for similar resources, including nectar and invertebrate prey. Alternatively, their diversification may be driven by extra-clade interactions, including the mutualistic interactions between hummingbirds and the plants they pollinate or predator–prey interactions. However, it has been difficult to pinpoint clades that have co-evolved with hummingbirds. For instance, Abrahamczyk and Renner (2015) provide absolute time frames for the build-up of hummingbird/plant mutualism in North America and temperate South America. They conclude that time frames differ greatly between the interacting clades, and while plant groups successively entered the hummingbird-pollination adaptive zone, this was not associated with rapid speciation (Abrahamczyk & Renner 2015). On the other hand, there is clear evidence for ecological interactions *within* the clade: hummingbirds are specialized nectar-feeders whose bill morphology often closely matches the floral morphology of their food plants. Thus, within-clade interactions likely influence species survival, and may promote diversification in resource use that leads to speciation.

Using our analytical framework, we aim to tease apart the contribution of mountain formation, climate change and biotic drivers on the diversification of Andean hummingbirds. We obtained the dated phylogeny from McGuire et al. (2014), and first applied two null models – the Yule and constant-rate birth–death models – which serve as a reference point for model comparison. We then tested models reflecting the RQS by using diversity-dependent models (Etienne et al. 2012) to investigate whether lineages diversified rapidly early in their evolutionary history and subsequently reached equilibrium or "ecological limits" to their richness (Rabosky 2013). The DDD function *dd_ML* was used, and set such that comparison with the other models is possible. We tested three diversification models: (i) speciation declines linearly with diversity and no extinction (DDL); (ii) speciation declines linearly with diversity and extinction is non-zero but constant (DDL + E); and (iii) speciation declines linearly and extinction increases linearly with diversity (DDL + EL). In the latter, the parameter r gives the ratio of linear dependencies in speciation and extinction rates. The initial carrying capacity was set to the current species diversity, and the final carrying capacity, K, was estimated according to the models and parameters. To ensure sufficient convergence of the likelihood estimates, four independent analyses were performed for each model, each with different initial values for the diversification parameters (λ = 0.1, 0.2, 0.3, 0.4 and μ = 0.0, 0.02, 0.05, 0.1). We also applied environment-dependent models consistent with the CJS using both global temperature (derived from the Cenozoic deep-sea oxygen isotope record; Figure 18.3b; Zachos et al. 2008) and Andean elevation (using a compilation of past elevation records of the Andes; Figure 18.3c; Lagomarsino et al. 2016). For each series, we designed four models using the method of Condamine et al. (2013): (i) speciation varies with the paleoenvironmental variable and no extinction (λ varies with T°); (ii) speciation varies with the paleoenvironmental variable and constant extinction (λ varies with T°, constant μ); (iii) constant speciation, and extinction varies with the paleoenvironmental variable (constant λ, μ varies with T°); and (iv) both speciation and extinction vary with the paleoenvironmental variable (both λ and μ vary with T°). The same rationale applies for the paleoelevation of the Andes. Using the same data (i.e., branching times) and conditioning for non-extinction of the descendants of the root lineage, all models were compared using the AICc. The best-fitting model was selected using Akaike weights.

The results of our analyses are summarized in Table 18.1 and Figure 18.3. Adding the extinction parameter did not improve the fit of the models despite there being non-zero extinction rates when this parameter was included (Table 18.1a). Of the null models, the Yule model is a better fit than the constant-rate birth–death model, and this is thus used as the reference for further comparison. Of the

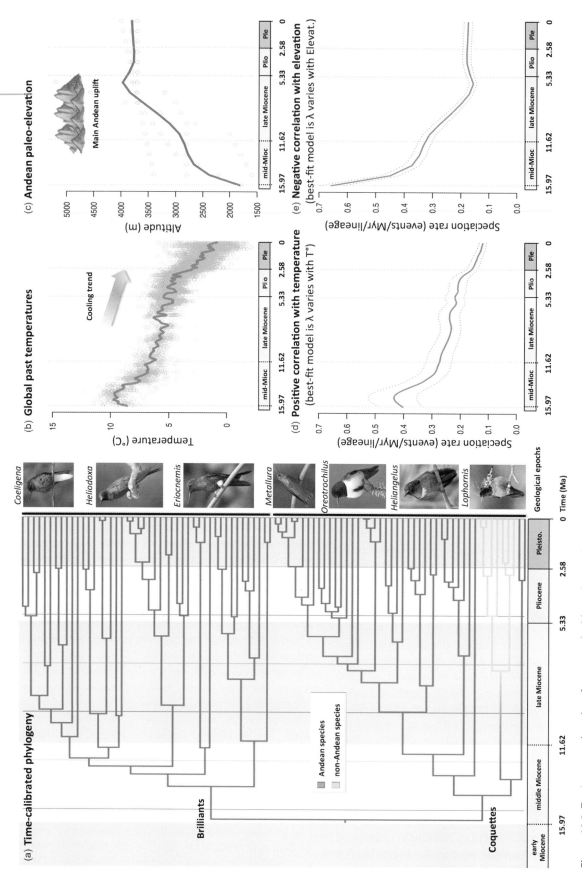

Figure 18.3 Teasing apart the roles of mountain building, climate change and biotic drivers in the diversification of Andean hummingbirds. (a) The dated phylogeny highlights the Andean and non-Andean species of the two hummingbird clades (Brilliants and Coquettes, with extant exemplars on the right). Two past environmental variables that may have influenced the hummingbirds' diversification are temperature (b) and elevation (c). The best-fitting diversification models varying with temperature (d) and elevation (e) are shown, as determined by the analytical framework described in the text (see Table 18.1 for all tested models). *Source:* Adapted from McGuire 2014. See also Plate 33 in color plate section.

Table 18.1 Summary of diversification analyses on the Andean hummingbird clade. (a) Eleven models were fitted using the analytical framework presented. The models are categorized according to the three model types tested: null models, models testing the Red Queen and models testing the Court Jester. For each model, we reported values corresponding to the fit of maximum likelihood approaches. (b) Model comparison between the best models of each series. DDL, speciation declines linearly with diversity and no extinction; DDL + E, speciation declines linearly with diversity and extinction is non-zero but constant; DDL + EL, speciation declines linearly with diversity and extinction increases linearly with diversity; NP, number of parameters; logL, log likelihood; AICc, corrected Akaike Information Criterion; λ, speciation rate; α, rate of variation of the speciation according to the paleoenvironmental variable; μ, extinction rate; β, rate of variation of the extinction according to the paleoenvironmental variable; K, estimated carrying capacity for the diversity-dependence models; and r, ratio of linear dependencies in speciation and extinction rates. The units of λ and μ are events/lineage/My; α and β are the rate of change over time (no unit).

(a)

Macroevolutionary scenario	Model	NP	logL	AICc	λ	α	μ	β	K	r
Null models	Yule model	1	−246.41	494.87	0.1929	–	–	–	–	–
	Constant birth-death	2	−246.41	496.96	0.1929	–	7E−08	–	–	–
Red Queen	DDL	2	−243.86	491.85	0.2828	–	–	–	193	–
	DDL + E	3	−242.69	491.65	0.5556	–	0.147	–	118	–
	DDL + EL	4	−242.69	493.84	0.5553	–	0.147	–	118	0.0005
Court Jester	λ varies with T°	2	−243.08	490.29	0.1009	0.1523	–	–	–	–
	λ varies with T°, constant μ	3	−242.304	490.88	0.1177	0.2342	0.2226	–	–	–
	constant λ, μ varies with T°	3	−246.44	499.15	0.1928	–	0.0	0.0252	–	–
	both λ and μ vary w. T°	4	−242.27	493.00	0.1101	0.260	0.1882	0.0688	–	–
	λ varies with elevation	2	−243.413	490.96	2.2007	−0.0007	–	–	–	–
	λ varies with elevation, constant μ	3	−243.30	492.87	4.2548	−0.0008	0.0658	–	–	–
	constant λ, μ varies with elevation	3	−246.44	499.15	0.1928	–	0.0699	−0.0355	–	–
	both λ and μ vary with elevation	4	−243.41	495.29	2.1961	−0.0007	2.0836	−0.0417	–	–

(b)

Model	Name	NP	AICc	ΔAIC	Akaike weights
Best null model	Yule model	1	494.87	4.58	0.044
Best diversity-dependence	DDL	2	491.85	1.56	0.201
Best temperature-dependence	λ varies with T°	2	490.29	0	0.44
Best elevation-dependence	λ varies with elevation	2	490.96	0.67	0.314

Red Queen models, the DDL was a better fit than the DDD models with extinction (DDL + E or DDE + EL). The DDL model estimated a carrying capacity at 193 species (compared with 112 extant species), suggesting that the Andean clade diversity is not currently saturated. The analyses estimate that the clade currently comprises about 60% of the carrying capacity in the Andes, suggesting that within-clade interactions are at play. Of the Court Jester models, those termed "λ varies with T°" and "λ varies with elevation" are the best fitting in their corresponding series (Table 18.1a). We found a positive correlation between paleotemperatures and speciation: hummingbirds diversified faster during periods of warmer temperatures (Figure 18.3d). In addition, we detected a negative correlation between paleoelevation and speciation: hummingbirds diversified faster in the early stages of the Andean orogeny (Figure 18.3e). Overall, these results indicate a slowdown of speciation over time, attributed either to the RQS or to the CJS. To tease apart the roles of mountain building, climate change and biotic interactions, we relied on the Akaike weights. We rejected the hypothesis of constant speciation in favor of more complex models that take into account the role of biotic interactions or the effect of changing environments (Table 18.1b). Furthermore, the CJS received slightly more support than the RQS, but it is less obvious which model incorporating extrinsic data best explains the Andean hummingbirds' radiation. This result may suggest that both Andean uplift and climate change foster the diversification, but it may also be attributed to a lack of statistical power, or to another factor we did not assess. Interestingly, a phylogeographic (microevolutionary) study on *Metallura tyrianthina*, a species belonging to the "Andean clade," indicates that geographic isolation rather than climatic dissimilarity explains the greatest proportion of genetic variance in this species (Benham & Witt 2016), in agreement with the fact that Andean topography causes both isolation and climatic variation, underscoring its dual role in biotic diversification.

In this empirical study, we have not embraced the full potential of the analytical framework. Following the general analytical workflow described in Figure 18.2, our analyses could be further extended. For instance, we did not take into account the role of other biological features such as migratory behavior, dispersal ability or sexual selection. Depending on the biological system, this analytical framework would allow researchers to test the role of the RQS, CJS and other hypotheses in governing macroevolutionary changes.

18.6 Limitations and Perspectives

Distinguishing between the RQS and CJS requires excellent data sets that adequately sample large geographical and taxonomic scales through both long and short time intervals (Marx & Uhen 2010; Drummond et al. 2012; Condamine et al. 2015; Liow et al. 2015; Silvestro et al. 2015). Unraveling the processes governing the tempo and mode of lineage diversification ideally requires densely sampled species trees that encompass at least 80% of the total standing diversity within the clade of interest (Cusimano & Renner 2010; Davis et al. 2013). To obtain reliable estimates of diversification rates, this sampling should ideally capture the range of ecological, geographic and morphological diversity. This requires accurate taxonomy within the clades of interest, with clear documentation of the numbers of identifiable species in the region(s) of interest. These types of basic information are increasingly hard to come by, as taxonomists remain underfunded. Additionally, knowledge of species ecology, including species interactions and species traits, may greatly enhance the design of macroevolutionary analyses. Fortunately, phylogenies are increasingly improving, even in species-rich groups, and biological data – including on morphology, ecology and distribution – are accumulating at an unprecedented rate.

We have detailed here an analytical framework based on phylogenetic approaches to diversification in order to tease apart the drivers of diversification. However, this framework is certainly not definitive, and it has its own limitations. First, one must keep in mind that these models, as with many macroevolutionary models, suffer from the fundamental dilemma of the extrapolation of correlations to causations. For instance, mountain uplift or paleoelevation change per se was probably not the direct driver of diversification; rather, diversification was driven by the many indirect consequences of that uplift (Lagomarsino et al. 2016). In the context of mountains, these models cannot test the role of allopatry, but we propose that macroevolutionary studies need to combine diversification (birth–death) models with parametric biogeography models in order to uncover the extent to which speciation and extinction vary, and according to what factors. This can also be used to estimate whether speciation occurred via allopatry (vicariance) or sympatry (ecological speciation). It is worth mentioning that some models exist to test the role of diversity dependence in the context of allopatry (Pigot et al. 2010; Valente et al. 2015).

Moreover, continued progress will increase the resolution of our understanding of the interactions between abiotic and biotic factors in diversity dynamics. More elaborate applications of the approach, considering other paleoenvironmental data either alone or in combination, might provide valuable models for understanding the drivers of diversification. Developments towards this end have been made based on trait-dependent models (Maddison et al. 2007). Cantalapiedra et al. (2014) developed a climate-based trait-dependent diversification model in which rates of speciation vary not only among traits, but also as a function of the same global temperature profile that we incorporated into our analysis. Their approach is similar to that of Condamine et al. (2013), except that it uses a smoothing spline through the raw isotopic data, and scales this so that the response is allowed to vary from 0 to 1. This model may be used to test whether a particular trait had an effect under past climate change. It has been successfully used with mammals, showing that ruminant clades with three feeding modes (browsers, grazers and mixed feeders) diversified differently during warm and cold periods (Cantalapiedra et al. 2014).

Some challenges remain if we want to improve our understanding of the relative contributions of the RQS and CJS. For instance, one of the persistant questions is what role competition plays in ecologically similar clades that are phylogenetically unrelated. It is difficult to model the interaction of external clades with the focal clade and its effect on speciation and/or extinction using current models. Yet, clade competition has been shown to occur over long time scales (Liow et al. 2015; Silvestro et al. 2015). Notably, a Bayesian model (Silvestro et al. 2015) is available for use with fossil data in order to assess the effect of competition on diversification, in which speciation and extinction rates are linearly correlated with the diversity trajectory of another clade. Under competitive interaction scenarios, increasing species diversity has the effect of suppressing speciation rates and/or increasing extinction rates. Although in phylogenetics diversity dependence is typically tested within a single clade, the fossil-based model allows for competition among species that are not closely related but which share similar ecological niches. At the moment, no comparable model exists in phylogenetics.

18.7 Conclusion

The study of macroevolution using phylogenies has never been more exciting and promising than today. The successful development of powerful analytical tools in recent years, in conjunction with the rapid and massive increase in the availability of biological data (including molecular phylogenies, fossils, georeferrenced occurrences and ecological traits), will allow us to disentangle complex diversification histories. We still face important limitations in data availability and methodological shortcomings, but by acknowledging them we can better target our joint efforts as a scientific community. Together, we will gain a much better understanding of how historical and biotic triggers are intertwined, and of the rich, deep past of Earth's stupendous biological diversity. We hope that our approach (and analytical framework) will help movement in that direction, and that it will provide interesting perspectives for future investigations of other model groups. Applied to an Andean bird radiation, we were able to tease apart the relative contributions of biotic (intra-clade interaction) versus abiotic (climate and mountain orogeny) factors. We found that the Andean hummingbird biodiversity is better explained by changing temperature and the rise of the Andes.

Acknowledgments

We are thankful to Colin Hughes, Luis Valente and an anonymous reviewer, who all provided constructive and insightful comments on the manuscript. We are grateful to Allison Perrigo for additional astute comments and edits, as well as for assistance in this work. Financial support was provided by a Marie Curie International Outgoing Fellowship (project 627684 BIOMME) to F.L.C. A.A. is supported by grants from the Swedish Research Council (B0569601), the European Research Council under the European Union's Seventh Framework Programme (FP/2007-2013, ERC Grant Agreement n. 331024), the Swedish Foundation for Strategic Research and a Wallenberg Academy Fellowship. This is contribution ISEM 2016-099 of the Institut des Sciences de l'Evolution de Montpellier (UMR 5554 – CNRS).

References

Abrahamczyk, S. & Renner, S.S. (2015) The temporal build-up of hummingbird/plant mutualisms in North America and temperate South America. *BMC Evolutionary Biology* **15**, 104.

Barnosky, A.D. (2001) Distinguishing the effects of the Red Queen and Court Jester on Miocene mammal evolution in the northern Rocky Mountains. *Journal of Vertebrate Paleontology* **21**, 172–185.

Benham, P.M. & Witt, C.C. (2016) The dual role of Andean topography in primary divergence: functional and neutral variation among populations of the hummingbird, *Metallura tyrianthina*. *BMC Evolutionary Biology* **16**, 22.

Benton, M.J. (2009) The red queen and the court jester: species diversity and the role of biotic and abiotic factors through time. *Science* **323**, 728–32.

Benton, M.J. (2010) The origins of modern biodiversity on land. *Philosophical Transactions of the Royal Society of London: Biological Sciences* **365**, 3667–3679.

Benton, M.J. (2015) Exploring macroevolution using modern and fossil data. *Proceedings of the Royal Society of London: Biological Sciences* **282**, 20150569.

Bouchenak-Khelladi, Y., Onstein, R.E., Xing, Y. et al. (2015) On the complexity of triggering evolutionary radiations. *New Phytologist* **207**, 313–326.

Burnham, K.P. & Anderson, D.R. (2002) *Model Selection and Multi-Model Inference: A Practical Information-Theoretic Approach*, 2nd edn. New York: Springer-Verlag.

Cantalapiedra, J.L., FitzJohn, R.G., Kuhn, T.S. et al. (2014) Dietary innovations spurred the diversification of ruminants during the Caenozoic. *Proceedings of the Royal Society of London: Biological Sciences* **281**, 20132746.

Condamine, F.L., Sperling, F.A.H., Wahlberg, N. et al. (2012) What causes latitudinal gradients in species diversity? Evolutionary processes and ecological constraints on swallowtail biodiversity. *Ecology Letters* **15**, 267–277.

Condamine, F.L., Rolland, J. & Morlon, H. (2013) Macroevolutionary perspectives to environmental change. *Ecology Letters* **16**, 72–85.

Condamine, F.L., Toussaint, E.F.A., Clamens, A.-L. et al. (2015) Deciphering the evolution of birdwing butterflies 150 years after Alfred Russel Wallace. *Scientific Reports* **5**, 11 860.

Cusimano, N. & Renner, S.S. (2010) Slowdowns in diversification rates from real phylogenies may be not real. *Systematic Biology* **59**, 458–464.

Davis, M.P., Midford, P.E. & Maddison, W.P. (2013) Exploring power and parameter estimation of the BiSSE method for analysing species diversification. *BMC Evolutionary Biology* **13**, 38.

Donoghue, M.J. & Sanderson, M.J. (2015) Confluence, synnovation, and depauperons in plant diversification. *New Phytologist* **207**, 260–274.

Drummond, C.S., Eastwood, R.J., Miotto, S.T. & Hughes, C.E. (2012) Multiple continental radiations and correlates of diversification in *Lupinus* (Leguminosae): testing for key innovation with incomplete taxon sampling. *Systematic Biology* **61**, 443–460.

Erwin, D.H. (2009) Climate as a driver of evolutionary change. *Current Biology* **19**, 575–583.

Etienne, R.S., Haegeman, B., Stadler, T. et al. (2012) Diversity-dependence brings molecular phylogenies closer to agreement with the fossil record. *Proceedings of the Royal Society of London: Biological Sciences* **279**, 1300–1309.

Ezard, T.H.G., Aze, T., Pearson, P.N. & Purvis, A. (2011) Interplay between changing climate and species' ecology drives macroevolutionary dynamics. *Science* **332**, 349–351.

Favre, A., Päckert, M., Pauls, S.U. et al. (2015) The role of the uplift of the Qinghai–Tibetan Plateau for the evolution of Tibetan biotas. *Biological Reviews* **90**, 236–253.

FitzJohn, R.G. (2012) Diversitree: comparative phylogenetic analyses of diversification in R. *Methods in Ecology and Evolution* **3**, 1084–1092.

Hannisdal, B. & Peters, S.E. (2011) Phanerozoic Earth system evolution and marine biodiversity. *Science* **334**, 1121–1124.

Harmon, L.J. & Harrison, S. (2015) Species diversity is dynamic and unbounded at local and continental scales. *American Naturalist* **185**, 584–593.

Hoorn, C., Mosbrugger, V., Mulch, A. & Antonelli, A. (2013) Biodiversity from mountain building. *Nature Geosciences* **6**, 154.

Hughes, C.E. & Atchison, G.W. (2015) The ubiquity of alpine plant radiations: from the Andes to the Hengduan Mountains. *New Phytologist* **207**, 275–282.

Hunt, T., Bergsten, J., Levkaničova, Z. et al. (2007) A comprehensive phylogeny of beetles reveals the evolutionary origins of a superradiation. *Science* **318**, 1913–1916.

Hutter, C.R., Guayasamin, J.M. & Wiens, J.J. (2013) Explaining Andean megadiversity: the evolutionary and ecological causes of glassfrog elevational richness patterns. *Ecology Letters* **16**, 1135–1144.

Jablonski, D. (2008) Biotic interactions and macroevolution: extensions and mismatches across scales and levels. *Evolution* **62**, 715–739.

Janz, N., Nylin, S. & Wahlberg, N. (2006) Diversity begets diversity: host expansions and the diversification of plant-feeding insects. *BMC Evolutionary Biology* **6**, 4.

Jetz, W., Thomas, G.H., Joy, J.B. et al. (2012) The global diversity of birds in space and time. *Nature* **491**, 444–448.

Jønsson, K.A., Fabre, P.-H., Fritz, S.A. et al. (2012) Ecological and evolutionary determinants for the adaptive radiation of the Madagascan vangas. *Proceedings of the National Academy of Sciences USA* **109**, 6620–6625.

Lagomarsino, L.P., Condamine, F.L., Antonelli, A. et al. (2016) The abiotic and biotic drivers of rapid diversification in Andean bellflowers (Campanulaceae). *New Phytologist* **210**, 1430–1442.

Liow, L.H. & Nichols, J.D. (2010) Estimating rates and probabilities of origination and extinction using taxonomic occurrence data: capture–recapture approaches. In: Alroy, J. & Hunt, G. (eds.) *Quantitative Paleobiology Short Course*. Boulder, CO: Paleontological Society, pp. 81–94.

Liow, L.H., Reitan, T. & Harnik, P.G. (2015) Ecological interactions on macroevolutionary time scales: clams and brachiopods are more than ships that pass in the night. *Ecology Letters* **18**, 1030–1039.

Maddison, W.P., Midford, P.E. & Otto, S.P. (2007) Estimating a binary character's effect on speciation and extinction. *Systematic Biology* **56**, 701–710.

Magallón, S., Gómez-Acevedo, S., Sánchez-Reyes, L.L. & Hernández-Hernández, T. (2015) A metacalibrated time-tree documents the early rise of flowering plant phylogenetic diversity. *New Phytologist* **207**, 437–453.

Mahler, D.L., Ingram, T., Revell, L.J. & Losos, J.B. (2013) Exceptional convergence on the macroevolutionary landscape in island lizard radiations. *Science* **341**, 292–295.

Marx, F.G. & Uhen, M.D. (2010) Climate, critters, and cetaceans: Cenozoic drivers of the evolution of modern whales. *Science* **327**, 993–996.

Mayhew, P.J., Jenkins, G.N. & Benton, T.G. (2008) A long-term association between global temperature and biodiversity, origination and extinction in the fossil record. *Proceedings of the Royal Society of London: Biological Sciences* **275**, 47–53.

McGuire, J.A., Witt, C.C., Remsen, J.V. et al. (2014) Molecular phylogenetics and the diversification of hummingbirds. *Current Biology* **24**, 910–916.

Meredith, R.W., Janečka, J.E., Gatesy, J. et al. (2011) Impacts of the Cretaceous terrestrial revolution and KPg extinction on mammal diversification. *Science* **334**, 521–524.

Miller, K.G., Kominz, M.A., Browning, J.V. et al. (2005) The Phanerozoic record of global sea-level change. *Science* **312**, 1293–1298.

Morlon, H. (2014) Phylogenetic approaches for studying diversification. *Ecology Letters* **17**, 508–525.

Morlon, H., Parsons, T.L. & Plotkin, J. (2011) Reconciling molecular phylogenies with the fossil record. *Proceedings of the National Academy of Sciences USA* **108**, 16 327–16 332.

Morlon, H., Lewitus, E., Condamine, F.L. et al. (2016) RPANDA: an R package for macroevolutionary analyses on phylogenetic trees. *Methods in Ecology and Evolution* **7**, 589–597.

Muchhala, N. & Thomson, J.D. (2009) Going to great lengths: selection for long corolla tubes in an extremely specialized bat-flower mutualism. *Proceedings of the Royal Society of London: Biological Sciences* **276**, 2147–2152.

Nee, S., May, R.M. & Harvey, P.H. (1994) The reconstructed evolutionary process. *Philosophical Transactions of the Royal Society of London: Biological Sciences* **344**, 305–311.

Ng, J. & Smith, S.D. (2014) How traits shape trees: new approaches for detecting character state-dependent lineage diversification. *Journal of Evolutionary Biology* **27**, 2035–2045.

Peters, S.E. (2005) Geologic constraints on the macroevolutionary history of marine animals. *Proceedings of the National Academy of Sciences USA* **102**, 12 326–12 331.

Peters, S.E. (2008) Environmental determinants of extinction selectivity in the fossil record. *Nature* **454**, 626–629.

Phillimore, A.B. & Price, T.D. (2008) Density-dependent cladogenesis in birds. *PLoS Biology* **6**, e71.

Pigot, A.L., Phillimore, A.B., Owens, I. & Orme, C.D.L. (2010) The shape and temporal dynamics of phylogenetic trees arising from geographic speciation. *Systematic Biology* **59**, 660–673.

Pyron, R.A. & Wiens, J.J. (2013) Large-scale phylogenetic analyses reveal the causes of high tropical amphibian diversity. *Proceedings of the Royal Society of London: Biological Sciences* **280**, 20131622.

Rabosky, D.L. (2013) Diversity-dependence, ecological speciation, and the role of competition in macroevolution. *Annual Review of Ecology, Evolution, and Systematics* **44**, 481–502.

Rabosky, D.L. & Goldberg, E.E. (2015) Model inadequacy and mistaken inferences of trait-dependent speciation. *Systematic Biology* **64**, 340–355.

Rabosky, D.L. & Huang, H. (2016) A robust semi-parametric test for detecting trait-dependent diversification. *Systematic Biology* **65**, 181–193.

Rabosky, D.L. & Hurlbert, A.H. (2015) Species richness at continental scales is dominated by ecological limits. *American Naturalist* **185**, 572–583.

Rabosky, D.L., Donnellan, S.C., Grundler, M. & Lovette, I.J. (2014) Analysis and visualization of complex macroevolutionary dynamics: an example from Australian scincid lizards. *Systematic Biology* **63**, 610–627.

Raup, D.M., Gould, S.J., Schopf, T.J. & Simberloff, D.S. (1973) Stochastic models of phylogeny and the evolution of diversity. *Journal of Geology* **81**, 525–542.

Rolland, J., Condamine, F.L., Jiguet, F. & Morlon, H. (2014) Faster speciation and reduced extinction in the tropics contribute to the mammalian latitudinal diversity gradient. *PLoS Biology* **12**, e1001775.

Schemske, D.W., Mittelbach, G.G., Cornell, H.V. et al. (2009) Is there a latitudinal gradient in the importance of biotic interactions? *Annual Review of Ecology, Evolution, and Systematics* **40**, 245–269.

Schnitzler, J., Barraclough, T.G., Boatwright, J.S. et al. (2011) Causes of plant diversification in the Cape biodiversity hotspot of South Africa. *Systematic Biology* **60,** 343–357.

Sepulchre, P., Sloan, L.C. & Fluteau, F. (2010) Modelling the response of Amazonian climate to the uplift of the Andean mountain range. In: Hoorn, C. & Wesselingh, F. (eds.) *Amazonia: Landscape and Species Evolution: A Look into the Past*. Oxford: Blackwell, pp. 211–222.

Silvestro, D., Schnitzler, J., Liow, L.H. et al. (2014) Bayesian estimation of speciation and extinction from incomplete fossil occurrence data. *Systematic Biology* **63**, 349–367.

Silvestro, D., Antonelli, A., Salamin, N. & Quental, T.B. (2015) The role of clade competition in the diversification of North American canids. *Proceedings of the National Academy of Sciences USA* **112**, 8684–8689.

Stadler, T. (2011) Mammalian phylogeny reveals recent diversification rate shifts. *Proceedings of the National Academy of Sciences USA* **108**, 6187–6192.

Stadler, T. (2013) How can we improve accuracy of macroevolutionary rate estimates? *Systematic Biology* **62**, 321–339.

Valente, L.M., Phillimore, A.B. & Etienne, R.S. (2015) Equilibrium and non-equilibrium dynamics simultaneously operate in the Galápagos Islands. *Ecology Letters* **18**, 844–852.

Van Valen, L.M. (1973) A new evolutionary law. *Evolutionary Theory* **1**, 1–30.

Voje, K.L., Holen, Ø.H., Liow, L.H. & Stenseth, N.C. (2015) The role of biotic forces in driving macroevolution: beyond the Red Queen. *Proceedings of the Royal Society of London: Biological Sciences* **282**, 20150186.

Wagner, C.E., Harmon, L.J. & Seehausen, O. (2012) Ecological opportunity and sexual selection together predict adaptive radiation. *Nature* **487**, 366–369.

Weber, M.G. & Agrawal, A.A. (2014) Defense mutualisms enhance plant diversification. *Proceedings of the National Academy of Sciences USA* **111**, 16442–16447.

Zachos, J.C., Dickens, G.R. & Zeebe, R.E. (2008) An early Cenozoic perspective on greenhouse warming and carbon-cycle dynamics. *Nature* **451**, 279–283.

19

Upland and Lowland Fishes: A Test of the River Capture Hypothesis

James S. Albert[1], Jack M. Craig[1], Victor A. Tagliacollo[2] and Paulo Petry[3,4]

[1] Department of Biology, University of Louisiana at Lafayette, Lafayette, LA, USA
[2] Universidade Federal do Tocantins, Palmas, Tocantins, Brazil
[3] The Nature Conservancy, Arlington, VA, USA
[4] Museum of Comparative Zoology, Harvard University, Cambridge, MA, USA

Abstract

Continental freshwaters are among the most species-dense ecosystems on Earth, with ~6% of all described species compressed into ~0.8% of the world's surface area and ~0.01% of the total water supply. This uneven concentration of biodiversity is pronounced in vertebrates, where >16 700 freshwater fishes represent ~26% of all vertebrate species. However, unlike many terrestrial taxa, which are most diverse at mid-elevations and in foothills, freshwater fishes exhibit greatest species richness in lowland river basins, below the fall line (~250 m elevation). Here, we explore the role of river capture as a landscape evolution process generating species richness in freshwater ecosystems. River capture is a geomorphological process that moves the location of watershed divides between adjacent basins, separating and merging portions of drainage networks and their resident biotas. River capture promotes speciation and extinction by subdividing species geographic ranges, and also inhibits these processes by facilitating dispersal (geographic range expansion). Upland and lowland fish assemblages represent an excellent system in which to test several predictions of the river capture hypothesis (RCH). There are: (i) more and larger (in km^2) river-capture events in lowland than upland basins, because most continental surfaces are drained by flat (low-topographic relief) lowlands, even though erosion rates per unit area are higher in rugged (high-relief) uplands; (ii) higher rates of speciation and dispersal, lower rates of extinction and therefore higher total species richness, in lowland fish assemblages; and (iii) more endemic species, and higher regional endemism, in upland basins. We constructed river-capture curves (RCCs), plotting geological age (in Ma) against size (km^2), for 40 empirical basin-capture events: the first such compilation at a global scale. Our empirical RCCs show: strong positive correlations between area and age of river-capture events for lowlands and uplands, analyzed both together and separately; that larger capture events are less common than, and overwrite the geographic signal of, smaller capture events; a steeper slope, indicating a higher rate of river capture, for lowland than upland RCCs; and strong positive correlations among paleospecies richness estimates, geographic area and geological age for lowlands and uplands, analyzed both together and separately. In combination, these results support the hypothesis that river capture accelerates evolutionary diversification and contributes to the high diversity of freshwater fishes and other taxa in lowland river basins.

Keywords: *biodiversity, biogeography, drainage basin, landscape evolution, speciation*

19.1 Introduction

19.1.1 Species Richness and Topography

Many groups of terrestrial organisms exhibit high diversity in and around large mountain ranges (Merriam & Stejneger 1890; Pianka 1966; Terborgh 1977; Rahbek 1997; McCain & Grytnes 2010). In many terrestrial taxa, the highest levels of alpha diversity are recorded in foothills of geologically young ranges, especially at low latitudes (e.g., in the Andes, East Africa, Himalaya and New Guinea: Brehm et al. 2005; Barthlott et al. 2007; Schipper et al. 2008; Kier et al. 2009; Jetz et al. 2012) and where there are high rates of uplift and erosion (Schumm 1963; Montgomery & Brandon 2002; Whipple & Tucker 2002). Numerous explanations have been offered for these mid-elevation biodiversity peaks, focusing on

Mountains, Climate and Biodiversity, First Edition. Edited by Carina Hoorn, Allison Perrigo and Alexandre Antonelli.
© 2018 John Wiley & Sons Ltd. Published 2018 by John Wiley & Sons Ltd.
Companion website: www.wiley.com/go/hoorn\mountains,climateandbiodiversity

ecological and physiological variables (Wake et al. 1992; Brown 2001) and on neutral processes like the species–area relationship (Sanders 2002; McCain 2007) and the mid-domain effect (McCain 2004). According to macro-evolutionary theory, the number of species in a biogeographic region arises from the contributions of speciation and dispersal, which increase, and extinction, which decreases, the species richness of regional biotas (Stanley 1979; Vermeij 1987; Jablonski & Roy 2003). Whatever the exact combination of underlying processes, the unique geographic conditions of montane settings apparently contribute to the accumulation of elevated terrestrial species richness though geological time.

Fishes differ from most continentally distributed taxa in exhibiting a monotonic decline in species richness with elevation, such that the highest diversity is usually observed near (although not quite at) the river mouth (Ward 1998; Hoeinghaus et al. 2004; Muneepeerakul et al. 2008). In this regard, freshwater fish diversity closely matches estimates of total aquatic habitat area in a river basin, a regularity known as the species–discharge relationship (Davies & Walker 2013; McGarvey & Terra 2015). Distinct upland and lowland fish species assemblages exist on all continents that have freshwater ichthyofaunas (Matthews & Robison 1988; Ibarra & Stewart 1989; Jowett & Richardson 1996; Dudgeon 2000; Skelton 2001). Fishes that inhabit rivers at high elevations (>1000 m) are represented by many different families and genera, but by relatively few species (Hutchinson 1939; Daget et al. 1984; He et al. 2001; Schaefer 2011). In brief, mountains appear to provide a good substrate for fish speciation, but also for extinction. Mountain and hill-stream fishes have evolved multiple times independently in each of the world's major mountainous regions (Table 19.1), each clade acquiring a characteristic suite of morphological and ecophysiological specializations that facilitates life in high-elevation and torrential waters (rheophily) (Lujan & Conway 2015).

In sharp contrast to montane ichthyofaunas, fishes of lowland river basins exhibit high diversity at all taxonomic levels, on all continents except arid Australia and frozen Antarctica (Abell et al. 2008; Levêque et al. 2008). More than 15 100 species of obligate freshwater fishes are known, representing about 49% of all fishes, or about 23% of all vertebrate species on Earth (Betancur-R et al. 2015). Yet, all these freshwater species are compressed into less than 0.01% of Earth's water supply (Gleick 1993; Shiklomanov & Rodda 2004). The very tiny fraction of Earth's water that flows in rivers and streams (0.0002%) therefore supports an intensely dense number of independent evolutionary lineages (Lundberg et al. 2000). Many hypotheses have been proposed to explain the exceptional diversity of continental fishes (see Vega &

Wiens 2012). Some focus on ecological factors of modern aquatic environments and watersheds (Levêque et al. 2008; Dias et al. 2013), others on mechanisms affecting evolutionary diversification at continental scales and across geological time frames (Albert et al. 2011a; Matthews 2012; Bloom et al. 2013). Many hypotheses focus on the particular role of dispersal limitation among populations isolated in different tributaries (Smith 1978; Fagan et al. 2009; Albert & Carvalho 2011).

19.1.2 Barrier Displacement

Landscape evolution can strongly affect the evolutionary diversification of freshwater fishes and other riverine taxa (Lundberg et al. 1998; Smith et al. 2010; Albert et al. 2011b; López-Fernández & Albert 2011). Certain landscape evolution processes create and remove barriers to biotic dispersal and gene flow, both in aquatic taxa (Craw et al. 2008, 2016; Gutiérrez-Pinto et al. 2012; Tagliacollo et al. 2015a) and in terrestrial taxa that inhabit rivers, floodplains and riparian woodlands (Junk et al. 1989; Stanford & Ward 1993; Robinson et al. 2002). Landscape evolution processes that displace geographic barriers, like orogenic uplift and river capture, operate continually and perennially on all continental platforms (Gilchrist & Summerfield 1991; Bishop 2007).

Barrier displacement occurs when an erosionally dissected cordillera is uplifted through climatically defined elevation zones (Antonelli et al. 2009; Badgley 2010; Gates et al. 2012; Toussaint et al. 2014; Hughes & Atchison 2015; Badgley et al. 2016). For example, Neogene uplift of the Northern Andes has been hypothesized to have fragmented species ranges and promote the speciation and extinction in many taxa (Remsen 1984; Doan 2003; Kattan et al. 2004; Hall 2005; Albert et al. 2006; Roberts et al. 2006; Antonelli et al. 2009, 2010). Barrier displacement also occurs during river capture (i.e., stream piracy), when a portion of one river basin is diverted into a different drainage, thereby moving the watershed between the two basins and so altering connectivity patterns among portions of river networks through time (Figure 19.1) (Bishop 1995; Brookfield 1998; Albert & Crampton 2010).

19.1.3 Big Rivers

Big, ancient rivers exert strong controls on the geomorphology of continental land surfaces (Potter 1978; Miall 2006) and on the biodiversity of organisms adapted to riverine habitats (Lundberg et al. 1998; Gascon et al. 2000; Aleixo 2004). Evidence of continental erosion from big rivers can be found in some of Earth's earliest preserved rocks, in Greenland: metamorphic sandstones

Table 19.1 Some well-known groups of montane fishes.

Order	Clade	Common name	No. montane species	Mountain range(s)	Continent
Salmoniformes	*Oncorhynchus*	Pacific trouts	10	Rockies	North America
	Galaxias	Galaxiids	24	Southern Alps	New Zealand
Cypriniformes	*Cyprinella*	Satinfin shiners	20	Appalachians and Ozarks	North America
	Notropis	Eastern shiners	44	Appalachians and Ozarks	North America
	Balitoridae	Hillstream loaches	99	Himalaya–Borneo	S and SE Asia
	Schizothoracinae[a]	Snow minnows	63	Himalaya–Tibet	Asia
Characiformes	*Creagruttus*	Creagrutus	70	Andes	South America
	Parodon	Scrapetooths	4	Andes	S and SE Asia
Siluriformes	Astroblepidae[a]	Climbing catfishes	75	Andes	South America
	Akysidae	Hillstream catfishes	57	Himalaya–Tenasserim	Southeast Asia
	Amphiliidae	Loach catfishes	76	East African ranges	Africa
	Chaetostoma	Bristlemouth catfishes	48	Andes	South America
	Sisoridae	Sisorids	235	Turkey–Bornoa	S and SE Asia
	Trichomycteridae	Pencil catfishes	49	Andes	South America
Cyprinodontiformes	*Orestias*[a]	Carache	45	Andes	South America
Perciformes	*Etheostoma*	North American Darters	105	Appalachians and Ozarks	North America
	Percina	Log Perch	27	Appalachians and Ozarks	North America
Total	17		1051		

[a] Demonstrably nonmonophyletic mountain assemblages.
Note: this list includes many but not all fish taxa at high elevations.

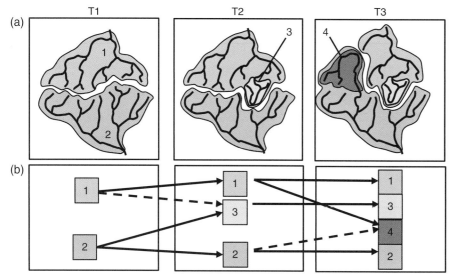

Figure 19.1 Landscape evolution processes that result in barrier displacement. (a) River capture. Regions shaded in gray are river basins; watersheds are dispersal barriers. (b) Changing patterns of area connectivity through time. Solid arrows indicate area identity or vicariance events; dashed arrows indicate area identity or geodispersal events. *Note:* River capture results in both separation and merging of portions of areas 1–4. T1–T3, time steps 1–3.

that formed ca. 3.7 billion years ago (Fedo et al. 2001; de Wit & Furnes 2013). Certain fluvial processes occur only in big rivers, generating broad valleys with high width–depth ratios, extensive low-gradient floodplains and large floodplain lakes that serve as sinks for fine-grained sediments and organic deposition. The sediment-rich floodplains of large tropical rivers represent important centers of diversity and diversification for freshwater fishes and many other continentally distributed taxa (Zedler & Kercher 2005; Winemiller et al. 2016).

All the world's continental-scale river basins formed through the accumulation of many smaller basins during a lengthy history of river capture (Figure 19.1a) (Ashworth & Lewin 2012). These capture events unfolded over time frames of millions of years, and the ages of big river basins range over several orders of magnitude. Some of the world's biggest river basins are quite ancient: the Michigan–Mississippi basin draining central North America can be traced to the early break-up of Pangaea in the Lower Triassic (250 Ma), meaning it has existed for about 6% of the planet's life span (Potter 1978). However, most of the world's big rivers drain Neogene orogens (e.g., Amazon, Brahmaputra, Congo, Yangtze) that assumed modern configurations only in the last ca. 20 My.

19.1.4 The River Capture Hypothesis

River capture has been implicated in the formation of high freshwater diversity in many regions and taxa (Sepkoski & Rex 1974; Smith 1981; Livingstone et al. 1982; Smith et al. 1982; Yap 2002; Smith & Bermingham 2005; Spencer et al. 2008; Hughes et al. 2009; Griffiths

2010; Albert & Carvalho 2011; Bossu et al. 2013; Carrea et al. 2014; Roxo et al. 2014). The "river capture hypothesis" (RCH) posits that river capture events at appropriate spatial scales increase the probability of speciation in riverine taxa following basin isolation and decrease the probability of extinction following species range expansion. River capture affects diversification in riverine taxa by forming new geographic barriers (i.e., vicariance) and removing old ones (i.e., dispersal), thereby separating and merging portions of river basin drainage networks (Waters & Wallis 2000; Burridge et al. 2006; Waters et al. 2015). Although framed here as a macroevolutionary theory, the population-level processes underlying the RCH may operate over substantially different time scales, with dispersal occurring over ecological time scales (10^2–10^3 years) and speciation and extinction over evolutionary ones (10^4–10^7 years) (Tagliacollo et al. 2015a,b). The RCH as presented here does not model these temporal heterogeneities.

The RCH makes specific predictions regarding all three processes of macroevolutionary diversification calling for higher rates of speciation and dispersal and lower rates of extinction in areas with higher rates of river capture (Albert et al. 2011a). It predicts more species on low-relief (flat) lowlands, and more endemic species on high-relief (rough) uplands. It also predicts higher rates of speciation and extinction (i.e., species turnover) in areas more centrally connected within a drainage network, in which there are higher rates of lateral tributary transfer (Smith et al. 2010; Altermatt et al. 2013). In the geological provinces of the world, low-relief areas include structural basins and areas of extended crust, whereas high-relief areas include upland shields, platforms and orogens.

19.1.5 Geomorphological Processes of River Capture

Many geomorphological processes can affect the age–area relationship of capture events. The main processes controlling river capture are erosion and avulsion at smaller scales (moving basin drainage networks towards equilibrium) and tectonic and climatic factors at larger ones (moving basins away from equilibrium) (Bishop 2007; Goren et al. 2014). Further, surface processes such as denudational isostasy (i.e., erosionally induced rebound) contribute to mountain uplift (Gilchrist & Summerfield 1991), and river processes such as watershed dynamics control drainage area competition (Tucker & Bras 1998).

Hack's law is an empirically derived relationship between the length of a river and the area of its basin (Hack 1957):

$$L = cA^h \qquad (19.1)$$

where L is the longitudinal length of the mainstem from headwaters to mouth, A is the total basin area, c is a proportionality constant and h is a scaling exponent. Empirical values of h decrease with basin size, from about 0.6 in basins on the order of $10^4\,km^2$ to 0.5 in basins on the order of $10^5\,km^2$ (Muller 1973). The decreasing values of h suggest that different processes dominate basin evolution at different scales. In fact, erosion dominates basin evolution at relatively smaller scales and tectonics at larger ones (Rigon et al. 1996). In an erosionally dominated landscape, larger basins draining the continental interior are expected to become larger still, at the expense of smaller peripheral ones (Maritan et al. 1996; Willett et al. 2014).

Therefore, an equilibrium expectation of river capture is that younger drainage networks (i.e., those with a shorter time since the last basin-wide reconfiguration due to orogenies or other factors) will exhibit a radial (centripetal) drainage pattern away from centrally located uplands and a more equitable distribution of basin areas (e.g., Madagascar) (Roberts et al. 2012). By contrast, older drainage networks should be characterized by a centrifugal drainage pattern with marginally located uplands, and by a less equitable distribution of basin areas. The RCH therefore predicts that species richness will increase towards the geographic periphery (towards the river mouths) in landscapes with younger drainage networks, and towards the center in landscapes with older ones.

19.1.6 Erosion Rates and River Capture

Comparisons among upland and lowland fish assemblages represent an excellent system in which to test the RCH. Erosion (or denudation) rates vary widely by climate zone, rock type and tectonic setting. Measured erosion rates are higher in uplands, where the steepest 10% of the continental surfaces contribute >50% of the total sediments delivered to the seas (Larsen et al. 2014; Warrick et al. 2014). Yet, the great majority of continental surfaces have low topographic relief, and much of the global sediment flux occurs across lowland basins (~81% from landscape slopes <6.0°, and ~40% from slopes <0.6°) (Willenbring et al. 2013). Therefore, despite lower surface erosion rates, lowland basins presumably have a greater number of and larger-scale river capture events than upland basins. They may also have higher river capture rates because their watershed boundaries are made of soft alluvium rather than rocky mountain outcrops (Portenga & Bierman 2011). Further, due to the well-known species–area relationship, many more species are expected to be affected by river capture events in lowlands, regardless of the rates of watershed displacement (Albert et al. 2011a).

Little is actually known about rates of rivers capture. To date, there have been no empirical studies reporting river capture rates at regional or continental scales, where the effects would most strongly affect net rates of diversification in resident aquatic biotas. There have also been no studies comparing rates of river capture in upland and lowland portions of the same watershed. Empirical river capture studies are largely restricted to relatively small spatial scales (<$10^3\,km^2$), well below those expected to be relevant in fish species diversification (Wilkinson et al. 2006, 2010; Kisel & Barraclough 2010).

In natural rivers, stream gradient generally decreases downstream, meaning upstream portions of a river have a higher gradient or slope (Figure 19.2). The relationship between stream gradient, G, and basin area, A, is a well-documented power function:

$$G = AK^Z \qquad (19.2)$$

where K is a scaling constant and $Z < 0$ is a measure of the topographic relief of the landscape surface (Strahler 1952; Flint 1974; Whipple & Tucker 2002). Using computer simulations, Howard (1971) calculated rates of river capture for areas with different levels of topographic relief, and found the strength, Z, of this relationship to be more pronounced in high-gradient mountain and upland rivers with greater relief (more negative Z values) than in low-gradient rivers of continental lowlands (Figure 19.3).

Howard (1971) was the first to report a greater rate (higher number per unit time) of river capture events in rugged uplands with high relief, as compared to flat lowlands with low relief (Figure 19.4c). Subsequent modeling studies have confirmed these expectations for higher rates of river capture per unit area in uplands but more total (and larger) capture events in lowlands

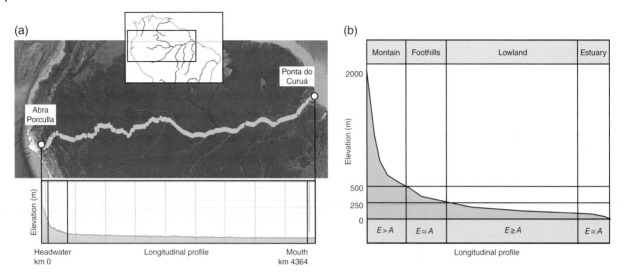

Figure 19.2 River elevation profiles. (a) Longitudinal profile (gray line) of Amazon thalweg (valley line) from a headwater at Abra Porculla at 2145 m in the Peruvian Andes (5°50′24″ S, 79°21′12″ W) to the mouth at Ponta do Curuá at 0 m (0°41′16″ S, 50°11′17″ W). Inset depicts map location in northern South America. *Source:* Map image from Google Earth; elevation profile traced using "path" tool. (b) Schematic of a longitudinal profile illustrating alternating zones of fluvial processes that affect river capture. *E*, erosion rates from sediment denudation; *A*, avulsion rates from sediment accumulation. Mountains and lowlands are largely (~98%) non-floodplain surfaces where $E > A$ and tributary capture predominates, and where $E \approx A$ on floodplains. Foothills of large mountains and estuaries of large rivers are largely alluvial fans, where tributary capture predominates when the fan is retreating ($E > A$) and river-mouth capture predominates when the fan is accreting ($E < A$).

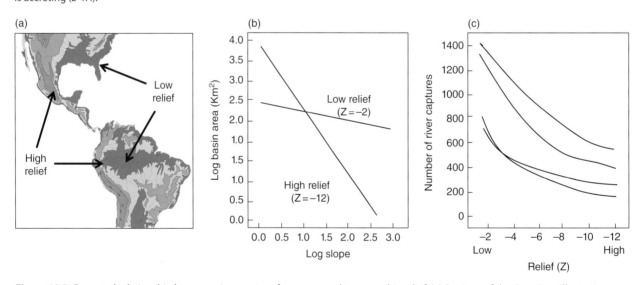

Figure 19.3 Expected relationship between river capture frequency and topographic relief. (a) Regions of the Americas, illustrating areas with high (Mesa Central and Andean piedmont) and low (Coastal plains and Central Amazon) relief. Gray shading indicates elevation levels. (b) Relationship between river gradient (slope) and basin (drainage) area, from Equation 19.2. *Note:* Upland rivers have higher relief (less negative *Z* values) than lowland rivers. (c) River capture as a function of topographic relief. *Note:* Higher rate (number per unit time) of capture events in flat lowlands (low relief) as compared with rugged uplands (high relief). Multiple curves result from differing initial conditions. *Source:* Adapted from Howard (1971).

(Willgoose et al. 1991; Howard 1994; Tucker & Hancock 2010; Goren et al. 2014).

19.1.7 Low-Frequency, High-Impact Events

Evolutionary biologists have long struggled with the fact that rare or even unique (low-frequency) events can have highly important (high-impact) effects on the history of life on Earth (Simpson 1970; Monod 1974; Sepkoski 1989). Some events with outsized influences occur only once, or extremely rarely, frustrating efforts to induce generalizations; for example, the Cambrian explosion of animal body plans (Marshall 2006), overseas dispersal events and the formation of insular biotas (De Queiroz

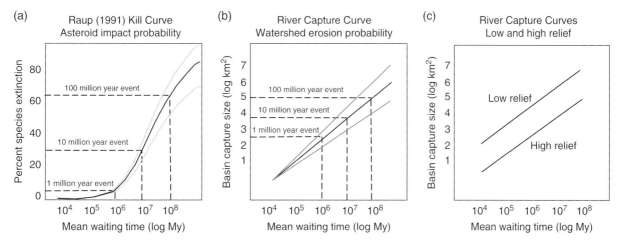

Figure 19.4 Hypothetical frequency distributions, illustrating the continuous relationship between the rarity and impact of historical events. (a) The asteroid impact Kill Curve, plotting log mean waiting time in years against the percentage of global species that go extinct. Gray curves indicate expected 95% confidence intervals (95% CIs). *Source:* Adapted from Raup (1991). (b) The river capture curve (RCC), plotting mean waiting time (log My) against size (log area) of basin capture, bounded by 95% CIs. The RCC predicts larger basin-capture events to be rarer. *Note:* Larger basin captures overwrite the signals of smaller events via basin-wide reorganization of the drainage net. (c) Hypothetical RCCs for regions with different topographic reliefs. River capture events at a given spatial scale are more frequent (occur more often) in flat lowland (low-relief) than in rugged upland (high-relief) basins.

2005; Tagliacollo et al. 2015b) and the end-Cretaceous (KPg) mass extinctions resulting from an asteroid impact (Alvarez et al. 1980). In an effort to provide a single explanatory framework for multiple distinct mass extinction events during the Phanerozoic, David Raup (1991) developed a "kill curve" from the stratigraphic ranges of >17 000 animal genera in Sepkoski's (1989) compendium, plotting log mean waiting time in years against the percentage of global species that go extinct (Figure 19.4a). Although this kill curve was derived empirically from the fossil record, it models how the biosphere responds stochastically to the orbital dynamics of Earth and other bodies in the solar system. Raup's kill curve treats natural phenomena such as asteroid impacts on Earth as probabilistic – not deterministic or chaotic – events drawn from a regular distribution in time.

19.2 Methods: Developing a River Capture Curve

In this section, we propose a river capture curve (RCC) that treats river capture events as drawn from a stochastic distribution, plotting log mean waiting time in years against size (area) of basin capture (Figure 19.4b). We construct empirically based RCCs for upland and lowland basins worldwide, with the goal of testing the hypothesis that the higher diversity of fishes in lowlands results, at least in part, from higher rates of river capture. One expectation of the RCH is that larger basin-capture events, with greater areal extent in km^2, are rarer than

smaller events. A corollary of this is the expectation that larger events overwrite signals of multiple smaller events via basin-wide reorganization of the drainage network. A second prediction is that the quantitative relationship between mean waiting time and basin capture area (i.e., the RCC slope, *b*) varies with topographic relief, such that areas with low relief (flatlands in lowlands basins) experience higher rates of river capture than rugged uplands (Figure 19.4c).

Empirical data on mega-river capture events were compiled from the primary literature (Table 19.2). Studies were selected for our analysis if they included estimates of basin areas – or maps illustrating basins before and after the capture event – and the geological age of the capture event. Recent advances in analytical and geochronological techniques have enabled a precise determination of dates and rates of geomorphological processes, including apatite fission track and (U–Th)/He analysis for rates of rock uplift and denudational exhumation and cosmogenic isotope analysis (Be) for ages of bedrock and sedimentary surfaces and/or the rates of denudation of these surfaces (Chapter 5) (Bishop 2007). Dates reported as epochs or stages were converted to Ma by taking the median age of the chronostratigraphic unit reported. The area of the captured basin was estimated as the difference in the areas of the encroaching basin before and after capture, rounded to the nearest 100 km^2, as measured on published maps using ImageJ (Abràmoff et al. 2004). Both age and area estimates are poorly constrained for most capture events, and must be treated as accurate to within approximately one- to twofold error.

Table 19.2 Literature references for data used in river capture curve (RCC) analyses. Capture age and area estimates from geology references. Species estimates as fraction of native total in modern basin from fish species references. Additional fish biogeographic studies from biogeography references. Data arranged alphabetically by continent and capture event.

Capture event	Continent	Geology	Fish species	Biogeography
Congo: U. Zambezi	Africa	Goudie (2005)	Brosse et al. (2013)	
Gilf: Quena (L. Nile)	Africa	Goudie (2005)	Brosse et al. (2013)	
Okavango: U. Cunene	Africa	Buch (1997); Goudie (2005)	Hogan (2012)	
Senegal: Niger Delta	Africa	Goudie (2005)	Brosse et al. (2013)	
Indus: Ganges	Asia	Clift et al. (2006)	Brosse et al. (2013)	
Jinsha: Yalong	Asia	Yang et al. (2008); Xu et al. (2011)	Chen (2013)	
L Yangtze: U. Yangtze	Asia	Clift et al. (2006); Zheng et al. (2013)	Yu et al. (2004)	He & Chen (2006)
L. Lena: U. Lena	Asia	Alekseev & Drouchits (2004)	Brosse et al. (2013)	
Lohit: Parlung	Asia	Cina et al. (2009)	Bakalial et al. (2014)	
Qaidam: Xinjiang	Asia	Wang et al. (2006)	Brosse et al. (2013)	
Salween: U. Mekong	Asia	Clark et al. (2004)	Brosse et al. (2013)	He & Chen (2006)
Selenga (Yenisei): Amur	Asia	Korzhuev (1979)	Brosse et al. (2013)	Froufe et al. (2005)
Siang: Yalu	Asia	Cina et al. (2009); Robinson et al. (2014)	Brosse et al. (2013)	He & Chen (2006)
Yalu: Subansiri	Asia	Cina et al. (2009); Bracciali et al. (2015)	Bakalial et al. (2014)	
Danube: Paleo lake Pannon	Europe	Fielitz & Seghedi (2005)	Brosse et al. (2013)	Banarescu (1998); Harzhauser & Mandic (2008)
L.: U. Duero	Europe	Antón et al. (2014)	Brosse et al. (2013)	
Rhine: Aare	Europe	Yanites et al. (2013)	Brosse et al. (2013)	
Sorbas: Alias	Europe	Stokes et al. (2002)	Stokes et al. (2002)	
Haast: Burke & Clark	New Zealand	Cooper & Beck (2009)	Brosse et al. (2013)	
Taieri: Kye Burn	New Zealand	Waters et al. (2015)	Waters et al. (2015)	Waters et al. (2015)
Wairu: Pelorus	New Zealand	Burridge et al. (2006)	Burridge et al. (2006)	Burridge et al. (2006)
Columbia: Snake	North America	Smith et al. (2000)	Beecher et al. 1988	Smith et al. (2002)
Fraser: modern delta	North America	Andrews et al. (2012)	Brosse et al. (2013)	
Guadalupe: Blanco	North America	Woodruff (1977)	Brosse et al. (2013)	
Old-Kentucky: Teays	North America	Bossu et al. (2013)	Smith et al. (2002)	
Sabine: Red	North America	Manning (1990)	Brosse et al. (2013)	
Branco: Uraricoera	South America	Crawford et al. (1985)	Albert et al. (2011b)	Lujan & Armbuster (2011)
Chapare: U. Madeira: Paraguay	South America	Sempere (1995)	Albert et al. (2011b)	Lundberg et al. (1998)

E. Amazon: W. Amazon	South America	Sempere (1995); Gibbs & Barron (1993)	Albert et al. (2011b)	Lundberg et al. (1998)
El Baul: L. Orinoco: M. Orinoco	South America	Sempere (1995)	Albert et al. (2011b)	Lundberg et al. (1998); Albert et al. (2011b)
Fitzcarrald: L & U Madeira	South America	Espurt et al. (2007); Mora et al. (2010)	Albert et al. (2011b)	Carvalho & Albert (2011)
L. Jacui: Uruguay trib.	South America	Lisbôa & Castro (1998)	Albert et al. (2011b)	
L. Negro shift	South America	Almeida-Filho & Miranda (2007)	Albert et al. (2011b)	
L. Tocantins shift	South America	Rossetti & Valeriano (2007)	Albert et al. (2011b)	Albert et al. (2011b)
Maracaibo: Orinoco	South America	Lundberg et al. (1998)	Albert et al. (2006)	Albert et al. (2006)
Michicola: Paraguay: U. Madeira	South America	Lundberg et al. (1998)	Albert et al. (2011b)	Tagliacollo et al. (2015b)
Paraguay: U. Amazon tribs.	South America	Ribeiro et al. (2013)	Albert et al. (2011b)	Carvalho & Albert (2011)
Ribeira de Iguape: U. Tiete	South America	Ribeiro et al. (2006)	Albert et al. (2011b)	Roxo et al. (2014)
S. Francisco: U. Parnaiba	South America	Saadi (1995); Karner & Driscoll (1999)	Brosse et al. (2013)	Pereira et al. (2013)
U. Paraiba do Sur: U. Tiete	South America	Lundberg et al. (1998)	Albert et al. (2011b)	

Basin capture events were scored as taking place in a lowland or upland setting, in a geologically active or stable setting and in a tributary or distributary of the main channel. In upland basins, erosion generally exceeds avulsion, landscapes are generally denuded of sediments (except in growing alluvial fans) and the slope is generally more than ~10° (Figure 19.2). In lowland basins, erosion also exceeds avulsion in non-floodplain areas, sediments accumulate on floodplains (except some sinking estuaries) and the slope is generally much less than 10° (Willenbring et al. 2013). Upland and lowland basins are often delimited by the fall line, a geomorphologic discontinuity at around 200–300 m elevation in most basins worldwide, lying between uplands of relatively hard crystalline basement rock and lowlands of softer sedimentary rock and unconsolidated sediments (McGee 1888).

We focused our study on mega-river capture events, where the area exposed to drainage network reorganization is greater than $1000\,km^2$. This cut-off is arbitrary in terms of hydrological and geomorphological processes, but represents an area that can plausibly affect rates of net diversification in freshwater fishes (Wilkinson et al. 2006). RCCs were compared for all mega-capture events pooled, and for a subset of capture events that occurred on stable geological platforms, where there is a higher likelihood of estimating the elevation of a capture event at the time of capture. Regions in tectonically active areas like Eastern Africa and the Himalaya were rapidly uplifted during the Neogene, and current watersheds between captured basins are likely to have a higher elevation than they did at the time of the actual capture event. RCCs of upland and lowland areas were statistically compared using an analysis of covariance in R.

We estimated the species richness of the area (in km^2) exposed to the river capture (paleobasin) as the percentage of the species richness observed in the modern encroaching basin, as:

$$S_P = S_M \times \left(\frac{A_P}{A_M} \right) \tag{19.3}$$

where S is species richness, A is area (in km^2), P is paleobasin size (before capture) and M is modern basin size (after capture). For example, the species richness of the Paleo-Western Amazon (S_{PW}), which was captured by the Eastern Amazon from the former Proto-Orinoco-Amazon, is calculated as: $S_{PW} = 910$ species $\times 0.84 = 763$ species, because there are 910 fish species in the modern Western Amazon, and the Paleo-Western Amazon occupied 0.84 of the areal extent of the modern Amazon. Due to the possibly large and unconstrained errors in published and inferred estimates of S_P and A_P, we did not attempt to correct species density for the exponent of the species–area relationship (Albert et al. 2011b).

19.3 Results

We compiled data for 40 river capture events extracted from the geological literature that satisfied our criteria for inclusion in this study, including 38 events at or above the $1000\,km^2$ cut-off for being characterized as a mega-capture event (Table 19.3). These events were located on all continents except Antarctica and Australia, and included all of the largest 10 – and 18 of the largest 20 (excepting the Japurá and St. Lawrence) – river basins on Earth by water discharge (Figure 19.5). They ranged in size over more than two orders of magnitude, from about 600 to >2 million km^2, and in age by over two orders of magnitude, from 0.1 to 45.0 Ma. Of these events, 19 (48%) occurred in areas that are uplands, at least on the modern landscape, 21 (53%) in lowlands, 16 (40%) in geologically stable landscapes and 24 (60%) in tectonically active landscapes. The oldest event in our data set is the middle Eocene (ca. 45 Ma) capture of the Upper São Francisco by the Lower São Francisco basin from the Paraíba basin (Karner & Driscoll 1999). Only three events are known from the Paleogene (66.0–23.1 Ma), from either upland or lowland areas. The majority (21 events; 53%) of documented capture events occurred during the Neogene (23.0–2.7 Ma), and 17 (43%) occurred in the Quaternary (2.6–0 Ma).

One of the main findings of this study is a strong positive correlation ($R^2 = 0.46$, $P < 0.0001$) between the area and the age of river capture events (Figure 19.6). This relationship, referred to as an RCC, is observed for all 40 basins, whether examined together (Figure 19.6a) or separated by lowland and upland basins (Figure 19.6b). The RCC for lowland basins has a steeper slope (b value) than that for upland basins, as predicted from the results of numerical simulations (Howard 1971; Goren et al. 2014). The y-intercepts of the lowland and upland RCCs do not differ significantly in the analysis of all 40 basins pooled, but do in the analysis of 16 basins in tectonically stable areas (Figure 19.6c). We interpret this latter result as evidence for the poor accuracy of modern elevations as estimates for paleoelevations in tectonically active areas.

The RCC has important implications for understanding continental aquatic biodiversity, because species richness is so intimately associated with geographic area (MacArthur & Wilson 1967; Warren et al. 2015) and geological age (Goldberg et al. 2005; Harmon & Harrison 2015). Estimates of species richness, geographic area and geological age are all highly intercorrelated (Figure 19.6b,c) for all basins pooled, as well as when examined separately by elevation. In fact, RCC area–age relationships of upland and lowland basins (Figure 19.6a) closely resemble the species–age relationships of these basins (Figure 19.6c), and both pairs of curves are significantly different ($P < 0.0001$) in analyses of geologically stable terrains.

Table 19.3 Mega-river capture events used to generate river capture curves (RCCs). Geographic locations illustrated in Figure 19.5. Encroaching basin listed first for each capture event. Estimated fish species richness of paleobasins estimated from capture area as percentage of modern basin area. Events classified as lowland or upland with reference to the 250 m elevation contour, as active or stable and as tributary or distributary, based on conditions of modern landscapes. References in Table 19.2. L, Lower; M, Middle; U, Upper; E, East; W. West.

Capture event	Continent	Basin capture Area (km²)	Age (My)	Paleobasin species	Capture area as % modern	Modern Area (km²)	Fish species	Geometry	Min. watershed elev. (m)	Lowland/ upland	Active/ stable
Congo: U. Zambezi	Africa	272 000	5.3	35	0.20	1 390 000	180	Tributary	1205	Upland	Active
Gilf: Quena (L. Nile)	Africa	268 500	6.0	11	0.08	3 400 000	135	Distributary	185	Lowland	Active
Okavango: U. Cunene	Africa	16 000	9.7	9	0.15	110 000	62	Tributary	1128	Upland	Stable
Senegal: Niger Delta	Africa	70 000	6.0	12	0.10	736 000	127	Tributary	274	Upland	Stable
Indus: Ganges	Asia	50 000	5.0	14	0.05	1 016 124	282	Tributary	230	Lowland	Active
Jinsha: Yalong	Asia	3600	1.0	3	0.02	187 260	142	Tributary	552	Upland	Active
L Yangtze: U. Yangtse	Asia	1 387 800	23.0	290	0.77	1 808 600	378	Tributary	65	Lowland	Active
L. Lena: U. Lena	Asia	2 036 000	2.6	35	0.88	2 306 743	40	Distributary	107	Lowland	Stable
Lohit: Parlung	Asia	159 734	10.0	204	1.00	159 734	204	Tributary	130	Lowland	Active
Qaidam: Xinjiang	Asia	2000	3.0	NA	0.02	120 000	12	Tributary	700	Upland	Active
Salween: U. Mekong	Asia	150 000	3.4	134	0.19	774 000	691	Tributary	4500	Upland	Active
Selenga (Yenisei): Amur	Asia	447 000	1.8	6	0.17	2 554 388	37	Tributary	1255	Upland	Active
Siang: Yalu	Asia	48 331	4.0	95	1.00	48 331	95	Tributary	155	Lowland	Active
Yalu: Subansiri	Asia	32 640	10.0	96	1.00	32 640	204	Distributary	1563	Upland	Active
Danube: Paleo lake Pannon	Europe	290 000	5.8	33	0.36	800 000	90	Tributary	67	Lowland	Active
L.: U. Duero	Europe	50 000	2.6	11	0.51	98 258	22	Tributary	600	Upland	Active
Rhine: Aare	Europe	45 000	4.2	9	0.20	225 000	45	Tributary	639	Upland	Active
Sorbas: Alias	Europe	1600	0.1	1	0.03	52 600	26	Tributary	72	Lowland	Stable
Haast: Burke & Clark	New Zealand	800	1.0	NA	0.97	821	NA	Tributary	106	Lowland	Active
Taieri: Kye Burn	New Zealand	7200	1.0	NA	0.65	11 000	NA	Tributary	660	Upland	Active
Wairu: Pelorus	New Zealand	600	0.1	NA	0.01	50 000	NA	Tributary	50	Lowland	Active
Columbia: Snake	North America	107 510	2.6	32	1.00	107 510	32	Tributary	105	Lowland	Active
Fraser: modern delta	North America	1000	1.1	NA	0.00	234 000	39	Distributary	300	Upland	Active
Guadalupe: Blanco	North America	1100	0.7	5	0.07	15 540	71	Tributary	307	Upland	Stable
Old-Kentucky: Teays	North America	188 400	2.6	199	1.00	188 400	199	Tributary	220	Lowland	Stable

(Continued)

Table 19.3 (Continued)

Capture event	Continent	Basin capture Area (km²)	Age (My)	Paleobasin species	Capture area as % modern	Modern Area (km²)	Fish species	Geometry	Min. watershed elev. (m)	Lowland/ upland	Active/ stable
Sabine: Red	North America	102800	7.0	62	0.60	170000	102	Tributary	152	Lowland	Stable
Branco: Uraricoera	South America	85400	2.6	61	0.14	605000	430	Tributary	82	Lowland	Stable
Chapare: U. Madeira: Paraguay	South America	63000	15.0	77	0.17	378000	463	Tributary	230	Lowland	Active
E. Amazon: W. Amazon	South America	1600000	10.0	763	0.84	1909000	910	Tributary	95	Lowland	Active
El Baul: L. Orinoco: M. Orinoco	South America	138000	5.0	315	1.00	138000	315	Distributary	8	Lowland	Stable
Fitzcarrald: L & U Madeira	South America	300000	4.0	367	0.79	378000	463	Tributary	100	Lowland	Active
L. Jacui: Uruguay trib.	South America	24100	1.8	22	0.15	166000	150	Tributary	124	Lowland	Stable
L. Negro shift	South America	9800	1.0	13	0.02	496000	668	Distributary	30	Lowland	Stable
L. Tocantins shift	South America	30000	1.3	14	0.04	717000	346	Distributary	11	Lowland	Stable
Maracaibo: Orinoco	South America	88700	8.0	127	1.00	88700	127	Tributary	420	Upland	Active
Michicola: Paraguay: U. Madeira	South America	63000	33.0	77	0.17	378000	463	Tributary	345	Upland	Active
Paraguay: U. Amazon tribs.	South America	75200	5.3	51	0.15	493000	332	Tributary	155	Lowland	Stable
Ribeira de Iguape: U. Tiete	South America	10200	2.6	44	0.40	25700	110	Tributary	865	Upland	Stable
S. Francisco: U. Parnaiba	South America	475000	45.0	145	0.80	593000	181	Tributary	518	Upland	Stable
U. Paraiba do Sur: U. Tiete	South America	18000	3.6	30	0.31	57700	97	Tributary	480	Upland	Stable
Min.		600	0.1	1	0.00	821	12		8		
Max.		2036000	45	763	1.00	3400000	910		4500		
Avg.		218000	6.4	97	0.43	575501	224		470		
Count		40									

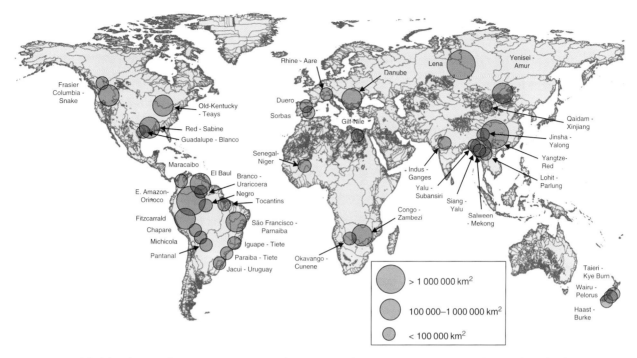

Figure 19.5 Global distribution of river capture events used to construct the river capture curve (*n* = 40). Figure based on literature references in Table 19.2 and data on capture event size, age, and species richness in Table 19.3. Base map from HydroSHEDS depicting 36 429 basins (>50 km^2) encompassing >116 million km^2, or about 85.1% of Earth's total land surface area, excluding Antarctica. *Source:* Adapted from HydroSHEDS (Lehner et al. 2006).

19.4 Discussion

19.4.1 Empirical RCCs are Consistent with Several RCH Predictions

The RCCs derived here (Figure 19.6) are consistent with several predictions of the RCH (Figure 19.4), including more and larger (in km^2) river capture events in lowland versus upland basins, higher species richness values in lowland versus upland fish assemblages and more endemic species with higher regional endemism in upland versus lowland basins. At a pattern level, the geological age, areal extent and species richness of these river capture events are all significantly correlated. Endemic species – those with restricted range distributions – constitute a majority of species in most biotas, and this is true for the freshwater fish faunas of all continents (Albert et al. 2011b; Reyjol et al. 2007). Yet, the total number and proportion of endemic fish species are greater in the upper and middle reaches of river basins; upland fish species tend to be more ecologically restricted and have lower vagility than do lowland species (Banarescu 1998). Data gathered for this study are not sufficient to test the RCH prediction that fish species richness is higher towards the margins of geological platforms with younger drainage networks and towards the centers of platforms with older drainage networks.

The RCC presented here is a neutral model, with the simplified assumptions of no meaningful differences among basins in the mechanisms or rates of fluvial dynamics, or among species responses to macroevolutionary processes of speciation, extinction or dispersal. No doubt, many additional factors are involved in the diversification of continental freshwater taxa. Real systems are rarely if ever neutral; taxa with different ecological requirements and dispersal capacities have different evolutionary responses to the same river capture event (Rosindell et al. 2011) and watershed evolution dynamics vary by geological and climatic setting (Bishop 1995). Indeed, each river capture is a unique historical event, with many local and idiosyncratic factors that can influence the precise quantitative relationship between age, area and biodiversity levels. Contributing factors include the frequency or magnitude of climatic and tectonic events that push basin properties away from expected equilibria; the size, latitude, elevation and tectonic setting of the capture event; and the proximity of the basin to regional centers of species diversity. Drainage networks closer to erosional equilibrium (sensu Sinha & Parker 1996; Carretier & Lucazeau 2005), with larger sizes, similar latitudes and elevations and stable tectonic settings, are expected to more closely match the expectations of the RCC and to exhibit smaller residual values. Rates of river capture

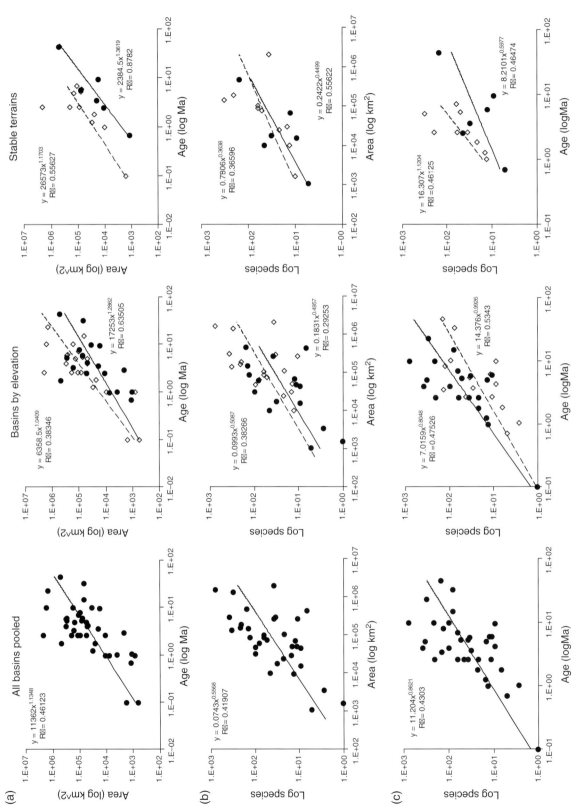

Figure 19.6 Correlations among geographic area, geological age and estimated paleobasin species richness of mega-river capture events. Data from Table 19.2. (a) River capture curves (RCCs) of paleobasin area versus paleobasin age. (b) Paleobasin species richness (calculated as percentage of modern species-area relationship) versus paleobasin area. (c) Paleobasin species richness (as in (b)) versus capture age. Left: Area and age data for 40 capture events; paleospecies richness for 35 paleobasins (as in (b)). Middle: Data plotted separately for uplands (closed circles; n = 18) and lowlands (open diamonds; n = 22). Right: Data for 16 paleobasins in geologically stable terrains (6 uplands, 10 lowlands).

may also be non-neutral through geological time. For example, erosion rates were more highly variable in temperate-zone drainage during the Pleistocene than in previous epochs.

To test for some of these non-neutral effects, we examined RCCs for all capture events pooled (Figure 19.6a), separately as lowlands and uplands (Figure 19.6b) and in tectonically stable areas (Figure 19.6c). RCCs for lowlands and uplands have similar slopes ($b = 1.1$ and 1.3, respectively) and do not differ significantly ($P = 0.14$). However, the lowland curve has a higher y-intercept value, as predicted by Howard's (1971) simulations. This partial separation in the predicted direction merits attention, given the very rough estimates available in the literature for the geographic area and geological age of river capture events.

19.4.2   Veil Line

The RCCs recovered from this study suggest the possibility of a veil line for discerning paleoriver captures from the information on modern landscapes. In ecological sampling theory, a veil line is the point at which rare items cease to be sampled (Preston 1948; Hubbell 2001). No capture event is recorded from before the middle Eocene (ca. 45 Ma), although several Cretaceous and Paleogene events are inferred from sediment records in eastern North America (Milici 1968), southern Africa (Dingle & Hendry 1984; Haddon & McCarthy 2005) and Australia (Bishop 1982; Beek & Braun 1999). Most of the well-documented capture events in the geological record, and all of the largest ones (>1 million km^2), occurred during the Neogene or Quaternary. These observations suggest a limit on our ability to reconstruct river captures from the configuration of modern landforms and drainage networks. Such a veil line means that few or no small and ancient capture events are expected to be observed, as these are likely to have been overwritten by subsequent capture events. Only the larger of the older capture events tend to survive to the present.

19.5   Conclusion

River capture is a ubiquitous and perennial landscape evolution process that merges and subdivides portions of river drainage networks and the populations and species that inhabit these waterways. River capture affects all three terms of macroevolutionary diversification, leaving a strong and characteristic genetic and biogeographic signature on riverine taxa (Smith 1978; Hughes 2007; Fagan et al. 2009). River capture is potent in generating species richness in freshwater and riparian taxa, because it results in both allopatric speciation and geodispersal, the later facilitating biotic dispersal and gene flow, thereby reducing extinction risk. The net effect of river capture is a rapid accumulation of species lineages in the regional pool (i.e., river basins). The RCCs reported here are derived from the first compilation of mega-river capture events at a global scale. These empirical results are a good fit to expectations of river capture based on numerical simulations, both for the RCC in general and under the different erosional regimes that control river capture in upland and lowland settings in particular.

Acknowledgments

This chapter is dedicated to Dr. Gerald R. Smith, for his pioneering work on the effects of river capture on fish diversification. We thank Alexandre Antonelli, Jon Armbruster, Maxwell Bernt, Tiago Carvalho, Prosanta Chakrabarty, William Crampton, Scott Duke-Sylvester, Kory Evans, Lesley Kim, Michael Goulding, Nathan Lujan, John Lundberg, Hernán López-Fernández, Nathan Lovejoy, Kevin Padian, Donald Schoolmaster, Roberto Reis, Isabel Sanmartín, Peter Van der Sleen, Brandon Waltz and Kirk Winemiller for stimulating discussions. We are indebted to Drs. Carina Hoorn and Alexandre Antonelli for the invitation to contribute this chapter. This work was supported by United States National Science Foundation DEB 0614334, 0741450 and 1354511 to J.S.A.

References

Abell, R., Thieme, M. L., Revenga, C. et al. (2008) Freshwater ecoregions of the world: a new map of biogeographic units for freshwater biodiversity conservation. *BioScience* **58**, 403–414.

Abràmoff, M.D., Magalhães, P.J. & Ram, S.J. (2004) Image processing with ImageJ. *Biophotonics International* **11**, 36–42.

Albert, J.S. & Carvalho, T.P. (2011) *Neogene Assembly of Modern Faunas. Historical Biogeography of Neotropical Freshwater Fishes.* Berkeley, CA: University of California Press, pp. 119–136.

Albert, J.S. & Crampton W.G.R. (2010) The geography and ecology of diversification in Neotropical freshwaters. *Nature Knowledge* **1**, 13–19.

Albert, J.S., Lovejoy, N.R. & Crampton, W.G.R. (2006) Miocene tectonism and the separation of cis-and trans-Andean river basins: evidence from Neotropical fishes. *Journal of South American Earth Sciences* **21**, 14–27.

Albert, J.S., Petry, P. & Reis, R.E. (2011a) Major biogeographic and phylogenetic patterns. In: Albert, J.S. & Reis, R. (eds.) *Historical Biogeography of Neotropical Freshwater Fishes*. Berkeley, CA: University of California Press, pp. 21–59.

Albert, J.S., Carvalho, T., Petry, P. et al. (2011b) Aquatic biodiversity in the Amazon: habitat specialization and geographic isolation promote species richness. *Animals* **1**, 205–241.

Aleixo, A. (2004) Historical diversification of a terra-firme forest bird superspecies: a phylogeographic perspective on the role of different hypotheses of Amazonian diversification. *Evolution* **58**, 1303–1317.

Alekseev, M.N. & Drouchits, V.A. (2004) Quaternary fluvial sediments in the Russian Arctic and Subarctic: late Cenozoic development of the Lena River system, northeastern Siberia. *Proceedings of the Geologists' Association* **115**(4), 339–346.

Almeida-Filho, R. & Miranda, F.P. (2007) Mega capture of the Rio Negro and formation of the Anavilhanas Archipelago, Central Amazônia, Brazil: evidences in an SRTM digital elevation model. *Remote Sensing of Environment* **110**(3), 387–392.

Altermatt, F., Seymour, M. & Martinez, N. (2013) River network properties shape α-diversity and community similarity patterns of aquatic insect communities across major drainage basins. *Journal of Biogeography* **40**, 2249–2260.

Alvarez, L.W., Alvarez, W., Asaro, F. & Michel, H.V. (1980) Extraterrestrial cause for the Cretaceous-Tertiary extinction. *Science* **208**, 1095–1108.

Andrews, G.D., Russell, J.K., Brown, S.R. & Enkin, R.J. (2012) Pleistocene reversal of the Fraser River, British Columbia. *Geology* **40**(2), 111–114.

Antón, L., De Vicente, G., Muñoz-Martín, A. & Stokes, M. (2014) Using river long profiles and geomorphic indices to evaluate the geomorphological signature of continental scale drainage capture, *Duero basin (NW Iberia) Geomorphology* **206**, 250–261.

Antonelli, A., Nylander, J.A., Persson, C. & Sanmartín, I. (2009) Tracing the impact of the Andean uplift on Neotropical plant evolution. *Proceedings of the National Academy of Sciences* **106**, 9749–9754.

Antonelli, A., Quijada-Mascareñas, A., Crawford, A.J. et al. (2010) Molecular studies and phylogeography of Amazonian tetrapods and their relation to geological and climatic models. In: Hoorn, C. & Wesselingh, F.P. (eds.) *Amazonia: Landscape and Species Evolution: A Look into the Past*. Chichester: John Wiley & Sons, pp. 386–404.

Ashworth, P.J. & Lewin, J. (2012) How do big rivers come to be different? *Earth-Science Reviews* **114**, 84–107.

Badgley, C. (2010) Tectonics, topography, and mammalian diversity. *Ecography* **33**, 220–231.

Badgley, C., Domingo, M.S., Barry, J.C. et al. (2016) Continental gateways and the dynamics of mammalian faunas. *Comptes Rendus Palevol* **15**, 763–779.

Bakalial, B., Biswas, S.P., Borah, S. et al. (2014) Checklist of fishes of Lower Subansiri River drainage, northeast India. *Annals of Biological Research* **5**(2), 55–67.

Banarescu, P.M. (1998) On the relations between hydrography and the ranges of freshwater fish species and subspecies. *Italian Journal of Zoology* **65**, 87–93.

Barthlott, W., Hostert, A., Kier, G. et al. (2007) Geographic patterns of vascular plant diversity at continental to global scales. *Erdkunde* **61**, 305–315.

Beek, O. & Braun, J. (1999) Controls on post-mid-Cretaceous landscape evolution in the southeastern highlands of Australia: insights from numerical surface process models. *Journal of Geophysical Research: Solid Earth* **104**, 4945–4966.

Betancur-R, R., Ortí, G. & Pyron, R.A. (2015) Fossil-based comparative analyses reveal ancient marine ancestry erased by extinction in ray-finned fishes. *Ecology Letters* **18**, 441–450.

Bishop, P. (1982) Stability or change: a review of ideas on ancient drainage in eastern New South Wales. *The Australian Geographer* **15**, 219–230.

Bishop, P. (1995) Drainage rearrangement by river capture, beheading and diversion. *Progress in Physical Geography* **19**, 449–473.

Bishop, P. (2007) Long-term landscape evolution: linking tectonics and surface processes. *Earth Surface Processes and Landforms* **32**, 329–365.

Bloom, D.D., Weir, J.T., Piller, K.R. & Lovejoy, N.R. (2013) Do freshwater fishes diversify faster than marine fishes? A test using state-dependent diversification analyses and molecular phylogenetics of new world silversides (Atherinopsidae). *Evolution* **67**, 2040–2057.

Bossu, C.M., Beaulieu, J.M., Ceas, P.A. & Near, T.J. (2013) Explicit tests of palaeodrainage connections of southeastern North America and the historical biogeography of Orangethroat Darters (Percidae: *Etheostoma*: *Ceasia*). *Molecular Ecology* **22**, 5397–5417.

Bracciali, L., Najman, Y., Parrish, R.R. et al. (2015) The Brahmaputra tale of tectonics and erosion: Early Miocene river capture in the Eastern Himalaya. *Earth and Planetary Science Letters* **415**, 25–37.

Brehm, G., Pitkin, L.M., Hilt, N. & Fiedler, K. (2005) Montane Andean rain forests are a global diversity hotspot of geometrid moths. *Journal of Biogeography* **32**, 1621–1627.

Brookfield, M.E. (1998) The evolution of the great river systems of southern Asia during the Cenozoic India–Asia collision: rivers draining southwards. *Geomorphology* **22**, 285–312.

Brosse, S., Beauchard, O., Blanchet, S. et al. (2013) Fish-SPRICH: a database of freshwater fish species richness throughout the World. *Hydrobiologia* **700**(1), 343–349.

Brown, J.H. (2001) Mammals on mountainsides: elevational patterns of diversity. *Global Ecology and Biogeography* **10**, 101–109.

Buch, M.W. (1997) Etosha Pan – the third largest lake in the world. *Madoqua* **20**, 49–64.

Burridge, C.P., Craw, D. & Waters, J.M. (2006) River capture, range expansion, and cladogenesis: the genetic signature of freshwater vicariance. *Evolution* **60**, 1038–1049.

Carrea, C., Anderson, L.V., Craw, D. et al. (2014) The significance of past interdrainage connectivity for studies of diversity, distribution and movement of freshwater-limited taxa within a catchment. *Journal of Biogeography* **41**, 536–547.

Carretier, S. & Lucazeau, F. (2005) How does alluvial sedimentation at range fronts modify the erosional dynamics of mountain catchments? *Basin Research* **17**, 361–381.

Carvalho, T.P. & Albert, J.S. (2011) The Amazon–Paraguay divide. In: Albert, J.S. & Reis, R. (eds.) *Historical Biogeography of Neotropical Freshwater Fishes*. Berkeley, CA: University of California Press, pp. 193–202.

Chen, X.Y. (2013) Checklist of fishes of Yunnan. *Zoological Research* **34**(4), 281–343.

Cina, S.E., Yin, A., Grove, M. et al. (2009) Gangdese arc detritus within the eastern Himalayan Neogene foreland basin: implications for the Neogene evolution of the Yalu–Brahmaputra River system. *Earth and Planetary Science Letters* **285**(1), 150–162.

Clark, M.K., Schoenbohm, L.M., Royden, L.H. et al. (2004) Surface uplift, tectonics, and erosion of eastern Tibet from large-scale drainage patterns. *Tectonics* **23**, 1006–1029.

Clift, P.D., Blusztajn, J. & Nguyen, A.D. (2006) Large-scale drainage capture and surface uplift in eastern Tibet–SW China before 24 Ma inferred from sediments of the Hanoi Basin, Vietnam. *Geophysical Research Letters* **33**, L19403.

Cooper, A.F. & Beck, R.J. (2009) River capture and Main Divide migration in the Haast River catchment assessed from the fluvial distribution of sodalite-bearing dike rocks and fenites. *New Zealand Journal of Geology and Geophysics* **52**(1), 27–36.

Craw, D., Burridge, C.P., Upton, P. et al. (2008) Evolution of biological dispersal corridors through a tectonically active mountain range in New Zealand. *Journal of Biogeography* **35**, 1790–1802.

Craw, D., Upton, P., Burridge, C.P. et al. (2016) Rapid biological speciation driven by tectonic evolution in New Zealand. *Nature Geoscience* **9**, 140–144.

Crawford, F.D., Szelewski, C.E & Alvey, G.D. (1985) Geology and exploration in the Takutu graben of Guyana Brazil. *Journal of Petroleum Geology* **8**(1), 5–36.

Daget, J., Gosse, J.P. & Thys van den Audenaerde, D.F. (1984) *Check-List of the Freshwater Fishes of Africa (CLOFFA)*. Marseille: Institut de Recherche pour le Developpement.

Davies, B.R. & Walker, K.F. (eds.) (2013) *The Ecology of River Systems*, Vol. **60**. Berlin: Springer Science and Business Media.

De Queiroz, A. (2005) The resurrection of oceanic dispersal in historical biogeography. *Trends in Ecology and Evolution* **20**, 68–73.

de Wit, M. & Furnes, H. (2013) Earth's oldest preserved unconformity – prospect of a beginning in the tectono-sedimentary continental cycle? *Gondwana Research* **23**, 429–435.

Dias, M.S., Cornu, J.F., Oberdorff, T. et al. (2013) Natural fragmentation in river networks as a driver of speciation for freshwater fishes. *Ecography* **36**, 683–689.

Dingle, R.V. & Hendry, Q.B. (1984) Late Mesozoic and Tertiary sediment supply to the eastern Cape Basin (SE Atlantic) and palaeo-drainage systems in southwestern Africa. *Marine Geology* **56**, 13–26.

Doan, T.M. (2003) A south-to-north biogeographic hypothesis for Andean speciation: evidence from the lizard genus *Proctoporus* (Reptilia, Gymnophthalmidae). *Journal of Biogeography* **30**, 361–374.

Dudgeon, D. (2000) The ecology of tropical Asian rivers and streams in relation to biodiversity conservation. *Annual Review of Ecology and Systematics* **31**, 239–263.

Espurt, N., Baby, P., Brusset, S. et al. (2007) How does the Nazca Ridge subduction influence the modern Amazonian foreland basin? *Geology* **35**(6), 515–518.

Fagan, W.F., Grant, E.H.C., Lynch, H. & Unmack, P.J. (2009) Riverine landscapes: ecology for an alternative geometry. In Cantrell, S., Cosner, C. & Ruan, S. (eds.) *Spatial Ecology*. Boca Raton, FL: CRC Press, pp. 85–100.

Fedo, C.M., Myers, J.S. & Appel, P.W. (2001) Depositional setting and paleogeographic implications of earth's oldest supracrustal rocks, the >3.7 Ga Isua Greenstone belt, West Greenland. *Sedimentary Geology* **141**, 61–77.

Fielitz, W. & Seghedi, I. (2005) Late Miocene–Quaternary volcanism, tectonics and drainage system evolution in the East Carpathians, Romania. *Tectonophysics* **410**(1), 111–136.

Flint, J.J. (1974) Stream gradient as a function of order, magnitude, and discharge. *Water Resources Research* **10**, 969–973.

Froufe, E., Alekseyev, S., Knizhin, I. & Weiss, S. (2005) Comparative mtDNA sequence (control region, ATPase 6 and NADH-1) divergence in *Hucho taimen* (Pallas) across four Siberian river basins. *Journal of Fish Biology* **67**, 1040–1053.

Gascon, C., Malcolm, J.R., Patton, J.L. et al. (2000) Riverine barriers and the geographic distribution of Amazonian species. *Proceedings of the National Academy of Sciences* **97**, 13672–13677.

Gates, T.A., Prieto-Márquez, A. & Zanno L.E. (2012) Mountain building triggered Late Cretaceous North American megaherbivore dinosaur radiation. *PLoS ONE* **7**, e42135.

Gibbs, A. & Barron C. (1993) *The Geology of the Guiana Shield*. New York: Oxford University Press.

Gilchrist, A.R. & Summerfield, M.A. (1991) Denudation, isostasy and landscape evolution. *Earth Surface Processes and Landforms* **16**, 555–562.

Gleick, P.H. (1993) *Water in Crisis: A Guide to the World's Fresh Water Resources*. Oxford: Oxford University Press.

Goldberg, E.E., Roy, K., Lande, R. & Jablonski, D. (2005.) Diversity, endemism, and age distributions in macroevolutionary sources and sinks. *The American Naturalist* **165**, 623–633.

Goren, L., Willett, S. D., Herman, F. & Braun, J. (2014) Coupled numerical–analytical approach to landscape evolution modeling. *Earth Surface Processes and Landforms* **39**, 522–545.

Goudie, A.S. (2005) The drainage of Africa since the Cretaceous. *Geomorphology* **67**, 437–456.

Griffiths, D. (2010) Pattern and process in the distribution of North American freshwater fish. *Biological Journal of the Linnaean Society* **100**, 46–61.

Gutiérrez-Pinto, N., Cuervo, A.M., Miranda, J. et al. (2012) Non-monophyly and deep genetic differentiation across low-elevation barriers in a Neotropical montane bird (*Basileuterus tristriatus*; Aves: Parulidae). *Molecular Phylogenetics and Evolution* **64**, 156–165.

Hack, J. (1957) Studies of Longitudinal Stream Profiles in Virginia and Maryland. US Geological Survey Professional Paper 294-B. Reston, VA: US Geological Survey.

Haddon, I.G. & McCarthy, T.S. (2005) The Mesozoic–Cenozoic interior sag basins of Central Africa: the Late-Cretaceous–Cenozoic Kalahari and Okavango basins. *Journal of African Earth Sciences* **43**, 316–333.

Hall, J.P. (2005) Montane speciation patterns in *Ithomiola* butterflies (Lepidoptera: Riodinidae): are they consistently moving up in the world? *Proceedings of the Royal Society of London B: Biological Sciences* **272**, 2457–2466.

Harzhauser, M. & Mandic, O. (2008) Neogene lake systems of Central and South-Eastern Europe: faunal diversity, gradients and interrelations. *Palaeogeography, Palaeoclimatology, Palaeoecology* **260**(3), 417–434.

Harmon L.J. & Harrison S. (2015) Species diversity is dynamic and unbounded at local and continental scales. *The American Naturalist* **185**, 584–593.

He, D. & Chen, Y. (2006) Biogeography and molecular phylogeny of the genus *Schizothorax* (Teleostei: Cyprinidae) in China inferred from cytochrome b sequences. *Journal of Biogeography* **33**(8), 1448–1460.

He, S., Cao, W. & Chen, Y. (2001) The uplift of Qinghai-Xizang (Tibet) Plateau and the vicariance speciation of glyptosternoid fishes (Siluriformes: Sisoridae). *Science in China Series C: Life Sciences* **44**, 644–651.

Hoeinghaus, D.J., Winemiller, K.O. & Taphorn, D.C. (2004) Compositional change in fish assemblages along the Andean piedmont–Llanos floodplain gradient of the río Portuguesa, Venezuela. *Neotropical Ichthyology* **2**, 85–92.

Hogan, C.M. (2012) Kunene River. *Encyclopedia of Earth*. Available from: http://www.eoearth.org/view/article/174385 (last accessed August 30, 2017).

Howard, A.D. (1971) Simulation model of stream capture. *Geological Society of America Bulletin* **82**, 1355–1376.

Howard, A.D. (1994) A detachment-limited model of drainage basin evolution. *Water Resources Research* **30**, 2261–2285.

Hubbell, S.P. (2001) *The Unified Neutral Theory of Biodiversity and Biogeography (MPB-32)*. Princeton, NJ: Princeton University Press.

Hughes, J.M. (2007) Constraints on recovery: using molecular methods to study connectivity of aquatic biota in rivers and streams. *Freshwater Biology* **52**, 616–631.

Hughes, C.E. & Atchison G.W. (2015) The ubiquity of alpine plant radiations: from the Andes to the Hengduan Mountains. *New Phytologist* **207**, 275–282.

Hughes, J.M., Schmidt, D.J. & Finn, D.S. (2009) Genes in streams: using DNA to understand the movement of freshwater fauna and their riverine habitat. *BioScience* **59**, 573–583.

Hutchinson, G.E. (1939) Ecological observations on the fishes of Kashmir and Indian Tibet. *Ecological Monographs* **9**, 145–182.

Ibarra, M. & Stewart, D. J. (1989) Longitudinal zonation of sandy beach fishes in the Napo River basin, eastern Ecuador. *Copeia* **1989**, 364–381.

Jablonski, D. & Roy, K. (2003) Geographical range and speciation in fossil and living molluscs. *Proceedings of the Royal Society of London B: Biological Sciences* **270**, 401–406.

Jetz, W., Thomas, G.H., Joy, J.B. et al. (2012) The global diversity of birds in space and time. *Nature* **491**, 444–448.

Jowett, I.G. & Richardson, J. (1996) Distribution and abundance of freshwater fish in New Zealand rivers. *New Zealand Journal of Marine and Freshwater Research* **30**, 239–255.

Junk, W.J., Bayley, P.B. & Sparks, R.E. (1989) The flood pulse concept in river-floodplain systems. *Canadian Special Publication of Fisheries and Aquatic Sciences* **106**, 110–127.

Karner, G.D. & Driscoll, N.W. (1999) Tectonic and stratigraphic development of the West African and eastern Brazilian Margins: insights from quantitative basin modeling. *Geological Society of London Special Publications* **153**, 11–40.

Kattan, G.H., Franco, P., Rojas, V. & Morales, G. (2004) Biological diversification in a complex region: a spatial analysis of faunistic diversity and biogeography of the Andes of Colombia. *Journal of Biogeography* **31**, 1829–1839.

Kier, G., Kreft, H., Lee, T.M. et al. (2009) A global assessment of endemism and species richness across island and mainland regions. *Proceedings of the National Academy of Sciences* **106**, 9322–9327.

Kisel, Y. & Barraclough, T. G. (2010) Speciation has a spatial scale that depends on levels of gene flow. *The American Naturalist* **175**, 316–334.

Korzhuev, S.S. (1979) Migration of the divides. In: Gerasimov, I.P. & Korzhuev, S.S. (eds.) *Morfostrukturnyi analiz gidroseti SSSR (Morphostructural Analysis of the Hydrological Network of the USSR).* Moscow: "Nauka," pp. 256–261. [In Russian.]

Larsen, I.J., Montgomery, D.R. & Greenberg, H.M. (2014) The contribution of mountains to global denudation. *Geology* **42**, 527–530.

Lehner, B., Verdin, K. & Jarvis, A. (2006) HydroSHEDS Technical Documentation, version 1.0. Washington, DC: World Wildlife Fund US, pp. 1–27.

Levêque, C., Oberdorff, T., Paugy, D. et al. (2008) Global diversity of fish (Pisces) in freshwater. *Hydrobiologia* **595**, 545–567.

Lisbôa, N.A. & Castro, J.H.W. (1998) Captura do sistema fluvial Camaquã pelo sistema fluvial Jacuí-São Gabriel, RS. *Anais do IX Simpósio Brasileiro de Sensoriamento Remoto*, 415–424.

Livingstone, D.A., Rowland, M. & Bailey, P.E. (1982) On the size of African riverine fish faunas. *American Zoologist* **22**, 361–369.

López-Fernández, H. & Albert, J.S. (2011) Paleogene radiations. In: Albert, J.S. & Reis R.E. (eds.) *Historical Biogeography of Neotropical Freshwater Fishes.* Berkeley, CA: University of California Press, pp. 105–118.

Lujan, N.K. & Armbruster, J.W. (2011) The Guiana Shield. In: Albert, J. & Reis, R. (eds.) *Historical Biogeography of Neotropical Freshwater Fishes.* Berkely, CA: University of California Press, pp. 211–224.

Lujan, N.K. & Conway, K.W. (2015) Life in the fast lane: a review of rheophily in freshwater fishes. In: Riesch, R., Tobler, M. & Plath, M. (eds.) *Extremophile Fishes.* New York: Springer, pp. 107–136

Lundberg, J.G., Marshall, L.G., Guerrero, J. et al. (1998) The stage for Neotropical fish diversification: a history of tropical South American rivers. In: Malabarba, L.R., Reis, R.E., Vari, R.P. et al. (eds.) *Phylogenetics and Classification of Neotropical Fishes.* Porto Alegre: Edipucrs, pp. 13–48.

Lundberg, J.G., Kottelat, M., Smith, G.R. et al. (2000) So many fishes, so little time: an overview of recent ichthyological discovery in continental waters. *Annals of the Missouri Botanical Garden* **87**, 26–62.

MacArthur R. & Wilson E.O. (1967) *The Theory of Island Biogeography.* Princeton, NJ: Princeton University Press.

Manning, E. (1990) The Early Miocene Sabine River. *Gulf Coast Association of Geological Societies Transactions, v.* **40**.

Maritan, A., Rinaldo, A., Rigon, R. et al. (1996) Scaling laws for river networks. *Physical Review E* **53**, 1510.

Marshall, C.R. (2006) Explaining the Cambrian "explosion" of animals. *Annual Review of Earth and Planetary Sciences* **34**, 355–384.

Matthews, W.J. (2012) *Patterns in Freshwater Fish Ecology.* New York: Springer Science & Business Media.

Matthews, W.J. & Robison, H.W. (1988) The distribution of the fishes of Arkansas: a multivariate analysis. *Copeia* **1988**, 358–374.

McCain, C.M. (2004) The mid-domain effect applied to elevational gradients: species richness of small mammals in Costa Rica. *Journal of Biogeography* **31**, 19–31.

McCain, C.M. (2007) Area and mammalian elevational diversity. *Ecology* **88**, 76–86.

McCain, C.M. & Grytnes, J.A. (2010) Elevational gradients in species richness. In: *Encyclopedia of Life Sciences.* Chichester: John Wiley & Sons, Ltd.

McGarvey, D.J. & Terra, B.D.F. (2015) Using river discharge to model and deconstruct the latitudinal diversity gradient for fishes of the Western Hemisphere. *Journal of Biogeography* **43**, 1436–1449.

McGee, W.J. (1888) The geology of the head of Chesapeake Bay. In: Geological Survey 7th Annual Report, pp. 537–646.

Merriam, C.H. & Stejneger, L. (1890) Results of a biological survey of the San Francisco Mountain region and desert of the Little Colorado, Arizona. *North American Fauna* **3**, 1–136.

Miall, A.D. (2006) How do we identify big rivers? And how big is big? *Sedimentary Geology* **186**, 39–50.

Milici, R. (1968) Mesozoic and Cenozoic physiographic development of the lower Tennessee River: in terms of the dynamic equilibrium concept. *Journal of Geology* **76**, 472–479.

Monod, J. (1974) On chance and necessity. In Ayala, F. (ed.) *Studies in the Philosophy of Biology: Reduction and Related Problems.* London: Macmillan Education, pp. 357–375.

Montgomery, D.R. & Brandon, M.T. (2002) Topographic controls on erosion rates in tectonically active mountain ranges. *Earth and Planetary Science Letters* **201**, 481–489.

Mora, A., Baby, P., Roddaz, M. et al. (2010) Tectonic history of the Andes and sub-Andean zones: implications for the development of the Amazon drainage basin. In: Hoorn, C. (ed.) *Amazonia, Landscape and Species Evolution: A Look into the Past.* Chichester: John Wiley & Sons, Ltd., pp. 38–60.

Muller, J.E. (1973) Re-evaluation of the relationship of master streams and drainage basins: reply. *Geological Society of America Bulletin* **84**, 3127–3130.

Muneepeerakul, R., Bertuzzo, E., Lynch, H.J. et al. (2008) Neutral metacommunity models predict fish diversity patterns in Mississippi–Missouri basin. *Nature* **453**, 220–222.

Pereira, T. L., Santos, U., Schaefer, C. E. et al. (2013) Dispersal and vicariance of *Hoplias malabaricus* (Bloch, 1794) (Teleostei, Erythrinidae) populations of the Brazilian continental margin. *Journal of Biogeography* **40**(5), 905–914.

Pianka, E.R. (1966) Latitudinal gradients in species diversity: a review of concepts. *American Naturalist* **100**, 33–46.

Portenga, E.W. & Bierman, P.R. (2011) Understanding Earth's eroding surface with 10 Be. *GSA Today* **21**, 4–10.

Potter, P.E. (1978) Significance and origin of big rivers. *Journal of Geology* **86**, 13–33.

Preston, F.W. (1948) The commonness, and rarity, of species. *Ecology* **29**, 254–283.

Rahbek, C. (1997) The relationship among area, elevation, and regional species richness in neotropical birds. *American Naturalist* **149**, 875–902.

Raup, D.M. (1991) A kill curve for Phanerozoic marine species. *Paleobiology* **17**, 37–48.

Remsen, J.V. (1984) High incidence of "leapfrog" pattern of geographic variation in Andean birds: implications for the speciation process. *Science* **224**, 171–173.

Reyjol, Y., Hugueny, B., Pont, D. et al. (2007) Patterns in species richness and endemism of European freshwater fish. *Global Ecology and Biogeography* **16**, 65–75.

Ribeiro, A.C., Lima, F.C., Riccomini, C. et al. (2006) Fishes of the Atlantic Rainforest of Boracéia: testimonies of the Quaternary fault reactivation within a Neoproterozoic tectonic province in Southeastern Brazil. *Ichthyological Exploration of Freshwaters* **17**(2), 157.

Ribeiro, A.C., Jacob, R.M., Silva, R.R.S.R. et al. (2013) Distributions and phylogeographic data of rheophilic freshwater fishes provide evidences on the geographic extension of a Central-Brazilian Amazonian palaeoplateau in the area of the present day Pantanal Wetland. *Neotropical Ichthyology* **11**(2), 319–326.

Rigon, R., Rodriguez-Iturbe, I., Maritan, A. et al. (1996) On Hack's law. *Water Resources Research* **32**, 3367–3374.

Roberts, J.L., Brown, J.L., von May, R. et al. (2006) Genetic divergence and speciation in lowland and montane Peruvian poison frogs. *Molecular Phylogenetics and Evolution* **41**, 149–164.

Roberts, G.G., Paul, J.D., White, N. & Winterbourne, J. (2012) Temporal and spatial evolution of dynamic support from river profiles: a framework for Madagascar. *Geochemistry, Geophysics, Geosystems* **13**, Q04004.

Robinson, C.T., Tockner, K. & Ward, J.V. (2002) The fauna of dynamic riverine landscapes. *Freshwater Biology* **47**, 661–677.

Robinson, R.A., Brezina, C.A., Parrish, R.R. et al. (2014) Large rivers and orogens: the evolution of the Yarlung Tsangpo–Irrawaddy system and the eastern Himalayan syntaxis. *Gondwana Research* **26**(1), 112–121.

Rosindell, J., Hubbell, S.P. & Etienne, R.S. (2011) The unified neutral theory of biodiversity and biogeography at age ten. *Trends in Ecology & Evolution* **26**, 340–348.

Rossetti, D.F. & Valeriano, M.M. (2007) Evolution of the lowest Amazon basin modeled from the integration of geological and SRTM topographic data. *Catena* **70**(2), 253–265.

Roxo, F.F., Albert, J.S., Silva, G.S. et al. (2014) Molecular phylogeny and biogeographic history of the armored Neotropical catfish subfamilies Hypoptopomatinae, Neoplecostominae and Otothyrinae (Siluriformes: Loricariidae). *PLoS ONE* **9**, e105564.

Saadi, A. (1995) A geomorfologia da Serra do Espinhaço em Minas Gerais e suas margens. *Geonomos* **3**, 41–63.

Sanders, N.J. (2002) Elevational gradients in ant species richness: area, geometry, and Rapoport's rule. *Ecography* **25**, 25–32.

Schaefer, S. (2011) The Andes: riding the tectonic uplift. In Albert, J.S. & Reis, R.E. (eds.) *Historical Biogeography of Neotropical Freshwater Fishes.* Berkeley, CA: University of California Press, pp. 259–278.

Schipper, J., Chanson, J.S., Chiozza, F. et al. (2008) The status of the world's land and marine mammals: diversity, threat, and knowledge. *Science* **322**, 225–230.

Schumm, S.A. (1963) *The Disparity between Present Rates of Denudation and Progeny.* Washington, DC: US Government Printing Office.

Sempere, T. (1995) Phanerozoic evolution of Bolivia and adjacent regions. In: Tankard, A.J. Suarez Soruco, R. & Welsink, H.J. (eds.). *Petroleum Basins of South America.* AAPG Memoir 62, pp. 207–230.

Sepkoski, J.J. (1989) Periodicity in extinction and the problem of catastrophism in the history of life. *Journal of the Geological Society* **146**, 7–19.

Sepkoski, J.J. & Rex, M.A. (1974) Distribution of freshwater mussels: coastal rivers as biogeographic islands. *Systematic Biology* **23**, 165–188.

Shiklomanov, I.A. & Rodda, J.C. (2004) *World Water Resources at the Beginning of the Twenty-First Century.* Cambridge: Cambridge University Press.

Simpson, G.G. (1970) Uniformitarianism. An inquiry into principle, theory, and method in geohistory and biohistory. In: Hecht, M.K. (ed.) *Essays in Evolution and Henetics in Honor of Theodosius Dobzhansky*. New York: Springer, pp. 43–96.

Sinha, S.K. & Parker, G. (1996) Causes of concavity in longitudinal profiles of rivers. *Water Resources Research* **32**, 1417–1428.

Skelton, P.H. (2001) *A Complete Guide to the Freshwater Fishes of Southern Africa*. Cape Town: Struik.

Smith, G.R. (1978) Biogeography of intermountain fishes. *Great Basin Naturalist Memoirs*, 17–42.

Smith, G.R. (1981) Late Cenozoic freshwater fishes of North America. *Annual Review of Ecology and Systematics* **12**, 163–193.

Smith, S.A. & Bermingham, E. (2005) The biogeography of lower Mesoamerican freshwater fishes. *Journal of Biogeography* **32**, 1835–1854.

Smith, G.R., Swirydczuk, K., Kimmel, P.G. & Wilkinson, B.H. (1982) Fish biostratigraphy of late Miocene to Pleistocene sediments of the western Snake River Plain, Idaho. *Cenozoic Geology of Idaho: Idaho Bureau of Mines and Geology Bulletin* **26**, 519–541.

Smith, G.R., Morgan, E. & Gustafson, E. (2000) Fishes of the Mio-Pliocene Ringold Formation, Washington: pliocene capture of the Snake river by the Columbia River. *University of Michigan Papers on Paleontology* **32**, 1–47.

Smith, G.R., Dowling, T.E., Gobalet, K.W. et al. (2002) Biogeography and timing of evolutionary events among Great Basin fishes. *Great Basin Aquatic Systems History* **33**, 175–234.

Smith, G.R., Badgley, C., Eiting, T.P. & Larson, P.S. (2010) Species diversity gradients in relation to geological history in North American freshwater fishes. *Evolutionary Ecology Research* **12**, 693–726.

Spencer, J.E., Smith, G.R. & Dowling, T.E. (2008) Middle to late Cenozoic geology, hydrography, and fish evolution in the American Southwest. *Geological Society of America Special Papers* **439**, 279–299.

Stanford, J.A. & Ward, J.V. (1993) An ecosystem perspective of alluvial rivers: connectivity and the hyporheic corridor. *Journal of the North American Benthological Society* **12**, 48–60.

Stanley, S.M. (1979) *Macroevolution: Pattern and Process*. Baltimore, MD: Johns Hopkins University Press.

Stokes, M., Mather, A.E. & Harvey, A.M. (2002) Quantification of river-capture-induced base-level changes and landscape development, Sorbas Basin, SE Spain. *Geological Society, London, Special Publications* **191**(1), 23–35.

Strahler, A.N. (1952) Hypsometric (area-altitude) analysis of erosional topography. *Geological Society of America Bulletin* **63**, 1117–1142.

Tagliacollo, V.A., Roxo, F.F., Duke-Sylvester, S.M. et al. (2015a) Biogeographical signature of river capture events in Amazonian lowlands. *Journal of Biogeography* **42**, 2349–2362.

Tagliacollo, V.A., Duke-Sylvester, S.M., Matamoros, W.A. et al. (2015b) Coordinated Dispersal and Pre-Isthmian Assembly of the Central American Ichthyofauna. *Systematic Biology* **66**, 183–196.

Terborgh, J. (1977) Bird species diversity on an Andean elevational gradient. *Ecology* **58**, 1007–1019.

Toussaint, E.F., Hall, R., Monaghan, M.T. et al. (2014) The towering orogeny of New Guinea as a trigger for arthropod megadiversity. *Nature Communications* **5**, 4001.

Tucker, G.E. & Bras, R.L. (1998) Hillslope processes, drainage density, and landscape morphology. *Water Resources Research* **34**, 2751–2764.

Tucker, G.E. & Hancock, G.R. (2010) Modeling landscape evolution. *Earth Surface Processes and Landforms* **35**, 28–50.

Vega, G.C. & Wiens, J.J. (2012) Why are there so few fish in the sea? *Proceedings of the Royal Society of London B: Biological Sciences* **279**, 2323–2329.

Vermeij, G.J. (1987) The dispersal barrier in the tropical Pacific: implications for molluscan speciation and extinction. *Evolution* **41**, 1046–1058.

Wake, D.B., Papenfuss, T.J. & Lynch, J.F. (1992) Distribution of salamanders along elevational transects in Mexico and Guatemala. *Tulane Studies in Zoology and Botany* **1**, 303–319.

Wang, E., Xu, F.Y., Zhou, J.X. et al. (2006) Eastward migration of the Qaidam basin and its implications for Cenozoic evolution of the Altyn Tagh fault and associated river systems. *Geological Society of America Bulletin* **118**(3–4), 349–365.

Ward, J. (1998) Riverine landscapes: biodiversity patterns, disturbance regimes, and aquatic conservation. *Biological Conservation* **83**, 269–278.

Warren, B.H., Simberloff, D., Ricklefs, R.E. et al. 2015. Islands as model systems in ecology and evolution: prospects fifty years after MacArthur-Wilson. *Ecology Letters* **18**, 200–217.

Warrick, J.A., Milliman, J.D., Walling, D.E. et al. (2014) Earth is (mostly) flat: apportionment of the flux of continental sediment over millennial time scales: comment. *Geology* **42**, e316.

Waters, J.M. & Wallis, G.P. (2000) Across the southern Alps by river capture? Freshwater fish phylogeography in South Island, New Zealand. *Molecular Ecology* **9**, 1577–1582.

Waters, J.M., Wallis, G.P., Burridge, C.P. & Craw, D. (2015) Geology shapes biogeography: Quaternary river-capture explains New Zealand's biologically "composite" Taieri River. *Quaternary Science Reviews* **120**, 47–56.

Whipple, K.X. & Tucker, G.E. (2002). Implications of sediment-flux-dependent river incision models for landscape evolution. *Journal of Geophysical Research: Solid Earth* **107**, ETG 3-1–ETG 3–20.

Wilkinson, M.J., Marshall, L.G. & Lundberg, J.G. (2006) River behavior on megafans and potential influences on diversification and distribution of aquatic organisms. *Journal of South American Earth Sciences* **21**, 151–172.

Wilkinson, M.J., Marshall, L.G., Lundberg, J.G. & Kreslavsky, M.H. (2010) Megafan environments in northern South America and their impact on Amazon Neogene aquatic ecosystems. In: Hoorn, C. & Wesselingh, F.P. (eds.) *Amazonia: Landscape and Species Evolution: A Look into the Past.* Oxford: John Wiley & Sons, pp. 162–184.

Willenbring, J.K., Codilean, A.T. & McElroy, B. (2013) Earth is (mostly) flat: apportionment of the flux of continental sediment over millennial time scales. *Geology* **41**(3), 343–346.

Willett, S.D., McCoy, S.W., Perron, J.T. et al. (2014) Dynamic reorganization of river basins. *Science* **343**, 1248765.

Willgoose, G., Bras, R.L. & Rodriguez-Iturbe, I. (1991) A coupled channel network growth and hillslope evolution model: 1. Theory. *Water Resources Research* **27**, 1671–1684.

Winemiller, K.O., McIntyre, P.B., Castello, L. et al. (2016) Balancing hydropower and biodiversity in the Amazon, Congo, and Mekong. *Science* **351**, 128–129.

Woodruff, C.M. Jr. (1977) Stream piracy near the Balcones fault zone, central Texas. *Journal of Geology* **85**(4), 483–490.

Xu, Q., Yang, D. & Zhaoshuai, G. (2011) The response of geomorphological process of the diamicton in Jinpingzi in Jinsha River to neotectonics and climate change of Eastern Tibet. *Acta Sedimentologica Sinica* **5**, 14.

Yang D., Han, Z., Ge, Q. et al. (2008) Geomorphic process of the formation and incision of the section from Shigu to Yibin of the Jinshajiang river. *Quaternary Sciences* **4**, 007.

Yanites, B.J., Ehlers, T.A., Becker, J.K. et al. (2013) High magnitude and rapid incision from river capture: Rhine River, Switzerland. *Journal of Geophysical Research: Earth Surface* **118**(2), 1060–1084.

Yap, S.Y. (2002) On the distributional patterns of Southeast-East Asian freshwater fish and their history. *Journal of Biogeography* **29**, 1187–1199.

Yu, X., Luo, T. & Zhou, H. (2004) Large-scale patterns in species diversity of fishes in the Yangtze River Basin. *Chinese Biodiversity* **13**(6), 473–495.

Zedler, J.B. & Kercher, S. (2005) Wetland resources: status, trends, ecosystem services, and restorability. *Annual Review of Environment and Resources* **30**, 39–74.

Zheng, H., Clift, P.D., Wang, P. et al. (2013) Pre-Miocene birth of the Yangtze River. *Proceedings of the National Academy of Sciences* **110**(19), 7556–7561.

20

Different Ways of Defining Diversity, and How to Apply Them in Montane Systems

Hanna Tuomisto

Department of Biology, University of Turku, Turku, Finland

Abstract

Comparing diversity measurements from different areas is far from straightforward, especially when results from different studies need to be combined in order to obtain a general understanding of diversity patterns. Problems may also arise for conceptual reasons: different studies may have understood the word "diversity" in different ways, so that even if they use the same terms, they are actually talking about conceptually different things. This is especially the case with beta diversity, as this term has been used to refer to dozens of different phenomena. Problems may also be of a more practical nature: even studies that have measured the same kind of diversity may have done so using different sampling approaches. Sampling decisions (such as how the organism group of interest is delimited and observed, how much sampling effort is invested and how that sampling effort is distributed in space and time) have a huge impact on perceived species diversity and gradients in it. Topographical complexity makes montane environments highly heterogeneous, which gives rise to high habitat diversity. Habitats can differ in direct and indirect environmental variables, such as temperature, air moisture, soil properties and elevation. There is also variation in species diversity and species composition at different spatial scales. Quantifying this diversity and finding out whether it is systematically related to the environmental variables is a big challenge, especially as species diversity also varies along latitudinal gradients and among biogeographical regions. Many sampling decisions have a direct effect on observed diversity. Ensuring strictly comparable data is difficult, because it is never possible to standardize all important factors, so a choice needs to be made and attention focused on the ones that are most important for the questions at hand. The purpose of this chapter is to provide an overview of the considerations affecting that choice for studies in montane ecosystems.

Keywords: *alpha diversity, beta diversity, effective number of species, gamma diversity, sample representativeness, sampling, richness, turnover*

20.1 Introduction

"Diversity" is a term that has a diversity of meanings. In the ecological literature, the traditional idea behind diversity is usually related to the question, "How many types?" (e.g., "How many species?"). Intuitively, a community with many species has higher species diversity than a community with few species, and a landscape with many habitats has higher habitat diversity than a landscape with few habitats. In contrast, indices of functional and phylogenetic diversity are often more related to the question, "How much divergence in measurable characters?" Functional diversity is intuitively higher in a community where plant height ranges from a few centimeters to 20 m (as in forests) than in a herbaceous community where all species are less than 50 cm tall, even if species diversity is the same. Similarly, phylogenetic diversity is higher in a community that contains species whose common ancestor lived a long time ago (as with pines and birches) and lower in a community consisting of more closely related species (such as different species of birch).

The approach of counting the number of types (such as species) gives a very different measure of diversity than the approach of quantifying the degree of difference in some quantitative variable (such as height). Numerous diversity indices have been developed for both approaches, and each of these quantifies something different. For example, species diversity has been quantified both with a simple count of the observed species and with more nuanced measures such as the effective number of species or various kinds of entropies, probabilities and species accumulation rates. All of these measures are somehow related to the intuitive concept of diversity, but in very different ways.

Because the numerical values obtained with different indices can have very different ranges and behaviors, comparing the values across indices easily becomes misleading. Some indices have a minimum value of 1, others a minimum value of 0 and some are constrained between 0 and 1. This may not be immediately obvious when all of them are referred to as "diversity," so a lot of care is needed when reading the ecological literature in order to avoid comparing apples with oranges. Therefore, a more standardized and accurate use of diversity-related terminology has been called for (Jost 2006, 2007; Moreno & Rodríguez 2010, 2011; Tuomisto 2010a,c, 2011; Veech & Crist 2010; Jurasinski & Koch 2011).

Quantifying diversity requires that units of observation have been classified into types (such as species), and generally also that information on the abundances of these types exists. For simplicity, the present text mostly assumes that the measure of abundance is the number of individuals. However, any other measure of abundance could be used instead, such as biomass or surface cover. More than one measure of abundance may be appropriate in any particular case, but different measures are not equivalent, so the results may change if the unit of abundance is changed. Similarly, the focus here is mostly on species diversity, although the individuals (or other units of abundance) can equally well be classified into genera (to obtain genus diversity), families (to obtain family diversity), operational taxonomic units (OTUs), guilds, functional types or any other categories of interest. In each case, the ecological interpretation of the result will be different, but the mathematical calculations applied to the data are the same.

Much of the discussion on how to define "diversity" has actually focused on how to accurately estimate diversity for a community of interest. For example, it has often been recommended that diversity is best measured using diversity indices whose values stabilize at small sample sizes (Lande 1996; Magurran 2004; Beck & Schwanghart 2010). However, it can also be argued that the conceptual problem of defining diversity should be kept separate from the practical problem of sampling it (Tuomisto 2010c). In this vein, the present chapter first defines the relevant concepts, assuming that an appropriate data set exists, and only thereafter starts to worry about all the biases that might make the data set unsuitable for answering the diversity-related ecological questions of interest.

20.2 Quantifying Diversity

20.2.1 Species Richness, *R*

Although species richness is often used as a measure of species diversity, the two are conceptually different. Species richness is quantified by simply counting the number of species present in a data set. Therefore, species richness can be quantified on the basis of presence/absence data or even presence-only data, whereas the definition of species diversity is based on taking both species richness and species abundance distribution into account. This is done because two communities can have the same species richness (i.e., equally long species lists), but nevertheless they are intuitively considered to differ in diversity if species abundances in one of them are more equal than in the other.

To make the distinction explicit, the abbreviation R is here used for richness and the abbreviation ^qD for diversity of order q (where q defines how the abundances are taken into account; this will be explained in the next section). Because every individual that is added to a survey can potentially represent a species that has not yet been encountered, species richness is very sensitive to sampling effort. For this reason, standardized sampling is crucial to avoid creating spurious trends in species richness.

Many studies discussing elevational species diversity gradients have actually focused on species richness only (e.g., Patterson et al. 1998; Kattan & Franco 2004; Krömer et al. 2005; Kluge et al. 2006; McCain 2010; Rodríguez-Castañeda et al. 2010; Acharya et al. 2011), but a few recent ones have presented data on true diversity, ^qD (Veijalainen et al. 2014; Tello et al. 2015).

20.2.2 Effective Number of Species (True Diversity), ^qD

Richness, R, can be quantified even if species abundances are not known, because it simply equals the length of a species list. It can also be thought of as the number of named slots needed to place all observed individuals (or other entities of interest) in a slot with an appropriate species name (first step in Figure 20.1). Diversity, ^qD, is a more complicated concept, because it takes into account species abundances. A community in which all species are equally abundant has higher species diversity than a community in which species richness is the same but some species are more abundant than others. This idea can be quantified in many different ways, but the simplest and most intuitive approach is the "effective number of species," ^qD (final step in Figure 20.1; explained in detail later). This measure (which is also known as true diversity or the Hill number) has many useful properties, and

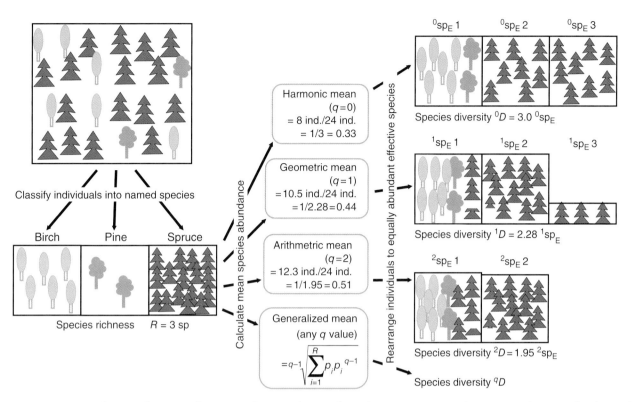

Figure 20.1 Quantification of species richness, R, and species diversity, ^qD, within a tree inventory plot. Species richness can be obtained in a single step, by classifying all the trees that belong to the inventory into a species (visualized with a slot labeled with a species name). The number of slots needed equals the actual (observed) number of species; that is, species richness. To quantify species diversity, two additional steps are necessary. First, the mean of the proportional species abundance values, p_i, over all species, i, is calculated using the weighted generalized mean with exponent $q-1$ (the numeric results are shown for three q values that correspond to familiar means, and the general equation is shown at the bottom). Second, the individuals are reallocated to new slots such that the abundance within each new slot equals the mean abundance of the named species. The number of new slots needed equals the effective number of species; that is, species diversity. The measurement unit is actual species, sp, in the case of richness, R, and effective species, ^qsp_E, in the case of diversity, ^qD. *Source:* Adapted from Tuomisto (2011).

seems to be becoming the trade standard (Hill 1973; Routledge 1979; Jost 2006; Gauthier et al. 2010; Tuomisto 2010a,c, 2011; Dauby & Hardy 2012; Culmsee & Leuschner 2013; Chao et al. 2014; Broms et al. 2015).

The concept of diversity (effective number of species, ^qD) can be derived from the concept of richness (actual number of species, R) as follows. If all R species are equally abundant in the data set, each species must have $1/R$th of all the individuals (or whatever unit of abundance is used). Then, the mean of the proportional abundances of the species also equals $1/R$. In other words, by taking the reciprocal of the actual number of species, we find out what the mean proportional abundance of the species would be if all of them were equally abundant. In reality, species are almost never equally abundant, but the weighted mean of their proportional abundances can always be expressed with a ratio of the form $1/^qD$. The denominator, ^qD, is the effective number of species; that is, species diversity (Hill 1973; Routledge 1979; Tuomisto 2010a,c, 2011). The value of ^qD can range from 1 to R. Any meaningful data set must have at least one species (both effective and

actual), but when species abundances are unequal, the effective number of species can be considerably smaller than the actual number of species. If all species really are equally abundant, then there are as many effective as actual species, and ^qD obtains the same value as R.

This definition of diversity can be expressed as follows (Hill 1973; Tuomisto 2010a):

$$^qD = 1/^q\overline{p}_i \qquad (20.1)$$

Here, ^qD is diversity of order q, and $^q\overline{p}_i$ is the weighted mean of the proportional species abundances. The mean that is applied here is the weighted generalized mean with exponent $q-1$. Special cases of this mean are the arithmetic mean (obtained when $q=2$), the geometric mean ($q=1$) and the harmonic mean ($q=0$). The general form of the equation for calculating a mean with any value of q is:

$$^q\overline{p}_i = {}^{q-1}\!\!\sqrt{\sum_{i=1}^{R} p_i p_i^{q-1}} \qquad (20.2)$$

Within the summation of the equation, the second p_i is the species proportional abundance for which the weighted mean is calculated, and the first p_i is the one used as the weight.

Weighting (= multiplying) p_i by itself has the effect that each species contributes to the value of the mean in proportion to its abundance: abundant species have more weight and rare species less. As a result, each individual has the same weight no matter which species it belongs to (without weighting, each species would have the same weight no matter how many individuals it had, so individuals of rare species would get more weight than individuals of abundant species).

At the same time, the second p_i is raised to the power $q-1$, which modifies the species weighting. Increasing the value of q shifts the mean towards the p_i value of the most abundant species in the data set, and decreasing the value of q shifts it towards that of the least abundant. This is why the arithmetic mean gives higher values than the geometric mean. When $q=1$, each species affects the value of the mean exactly in proportion to its abundance. When $q>1$, abundance differences are exaggerated, and when $q<1$, abundance differences are downplayed. At $q=0$, abundances cancel out in the equation and the mean proportional species abundance obtains the value $1/R$ even when all species are not equally abundant.

Because the zeroth root is not defined, the geometric mean is calculated by multiplying the weighted p_i values rather than summing them:

$$^1\overline{p}_i = \prod_{i=1}^{R} p_i{}^{p_i} \qquad (20.3)$$

Here, the exponents are the weights.

Diversity ^qD is then calculated by replacing $^q\overline{p}_i$ in Equation 20.1 with either its formula in Equation 20.3 (for $q=1$) or its formula in Equation 20.2 (for all other values of q). ^0D (diversity at $q=0$, also known as diversity of order zero) has the same value as richness R, but increasing q causes ^qD to decrease (Figure 20.2). The exact manner in which ^qD changes with q depends on the rank abundance distribution of the species in the data set. Some studies on montane systems have already taken advantage of visualizing this relationship, as it can reveal interesting details about community structure (Kindt et al. 2006; Veijalainen et al. 2014). In general, the more even the abundances of the species, the smaller the influence of q on the value of ^qD.

All values of q are equally valid, and the resulting diversities can be interpreted in a conceptually similar way. However, it is important to be explicit about which mean (i.e., which value of q) has been used in a particular case: results based on different means emphasize species abundance differences in different ways, which needs to

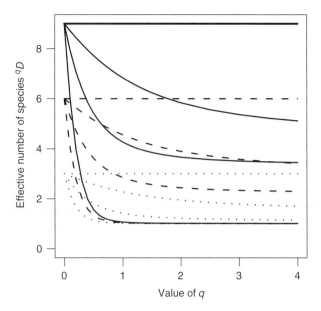

Figure 20.2 Dependency of the effective number of species (species diversity ^qD) in a sampling unit on the value of the parameter q, which defines the mean used when calculating mean proportional species abundance (=$1/^qD$). Each line corresponds to one sampling unit, and the same line type is used for all sampling units with the same species richness. In the sampling units with three species (dotted lines), the species abundance distributions are, from top to bottom, 1:1:1, 16:6:2, 100:10:1 and 1000:1:1. The sampling units with higher species richness have been obtained by replicating all species either two times (dashed lines; e.g., to obtain the species abundance distribution 100:100:10:10:1:1) or three times (solid lines), except that only one species in any sampling unit has abundance 1000, while all the others have abundance 1. The species abundance distribution 16:6:2 is the same as in the example shown in Figure 20.1. The more uneven the species abundance distribution, the faster the fall of the curve with increasing value of q.

be taken into account when their values are compared. The closer the value of q to 0, the more the obtained ^qD value reflects the number of species. Conversely, the larger the value of q, the more ^qD reflects the proportional abundance of the most abundant species.

20.2.3 Diversity Indices Related to ^qD

The most popular diversity index by far in the ecological literature has been the Shannon entropy H', which equals $\log_b(^1D)$. The logarithm can have any base b, although 2-based and natural logarithms have been the most popular ones. The log base determines the measurement unit, and hence the range of possible index values, so only values obtained with the same log base can be meaningfully compared. In practice, the Shannon entropy (also known as Shannon index) quantifies the degree of uncertainty in the species identity of an individual that is picked at random from the data set.

Through misunderstandings, the Shannon entropy has also been referred to as the Shannon–Weaver index and the Shannon–Wiener index, and a misspelling of the latter has led to the misnomer Shannon–Weiner index (Spellerberg & Fedor 2003). The index has also been called the Shannon diversity, but this is confusing, because the term "Shannon diversity" has (arguably with better justification) also been used for the exponential of the Shannon entropy ($b^{H'} = {}^1D$), which is a measure of true diversity (= effective number of species) rather than of entropy (= uncertainty) (Jost 2006; Tuomisto 2010a).

Another well-known diversity index is the Simpson index, of which there are a few different variants. The original Simpson index equals the weighted arithmetic mean of the proportional species abundances, ${}^2\bar{p}_i$. This is a counterintuitive index of diversity, because its value actually decreases when the number of species increases, so a transformation is generally applied to obtain an index whose value increases with species richness. A popular transformation is to subtract the value of the Simpson index from 1, which yields the Gini–Simpson index, $1 - {}^2\bar{p}_i$. This quantifies the probability that two individuals picked at random from the data set (with replacement of the first before the second is picked) represent different species (the original Simpson index quantifies the probability that they represent the same species). Another logical transformation is the reciprocal of the original Simpson index:

$$1/{}^2\bar{p}_i = 1/(1/{}^2D) = {}^2D \qquad (20.4)$$

This is known as the inverse Simpson index, which is a true diversity (= effective number of species) with $q = 2$.

Both richness and diversity follow the important replication principle: if each of the original species is replicated to form r different species, each of which has the same absolute abundance as the original, then both the richness and the diversity of the data set increase to r times their original values (Hill 1973; Jost 2006, 2007). This can be seen in Figure 20.2: in the six- and nine-species sampling units that were obtained by replicating a three-species sampling unit, the diversities at any value of q are exactly two- or threefold those in the three-species unit. This corresponds to an intuitive sense of how diversity is expected to behave. The Shannon and Simpson indices do not follow the replication principle, so their values are easily misinterpreted (Jost 2006, 2007; Jost et al. 2010).

20.2.4 Alpha, Beta and Gamma Diversity

When Robert H. Whittaker (1960) coined the terms "alpha," "beta," and "gamma" diversity, his intention was to understand the total species diversity in a landscape (gamma diversity) as the combined result of two different phenomena: the average species diversity within localities (alpha diversity) and the compositional heterogeneity among localities (beta diversity). The idea can be expressed mathematically by partitioning gamma diversity into two multiplicative components: $\gamma = \alpha \times \beta$. This, in turn, led Whittaker to quantify beta diversity using the equation $\beta = \gamma/\alpha$. When alpha diversity is interpreted as the mean concentration of species diversity per locality and gamma diversity as the total species diversity in all localities collectively, then beta diversity gets a simple and intuitive interpretation as the diversity of localities (Figure 20.3). In other words, beta diversity answers the question: If the total species diversity in the data set (γ) were evenly distributed among subunits such that each subunit received the same species diversity as the original localities had on average (α), how many subunits would be obtained (β)? By definition, the obtained hypothetical subunits are compositionally non-overlapping, so they have been called "compositional units" (Routledge 1977; Tuomisto 2010a,c).

The product of α (mean species density: species per compositional unit) and β (number of compositional units) gives the total number of species in the data set γ, in the same way that the product of mean density (kg/dm^3) and total volume (dm^3) gives the total weight of an object (kg) (Tuomisto 2010a,c, 2013).

The partitioning of γ into α and β components applies equally well to richness as to diversity. To make this distinction explicit, the richness components can be denoted R_α, R_β and R_γ, and the corresponding diversity components ${}^qD_\alpha$, ${}^qD_\beta$ and ${}^qD_\gamma$. Then, gamma richness, R_γ, is the actual number of (named) species in the data set and gamma diversity, ${}^qD_\gamma$, is the effective number of (hypothetical equally abundant) species. To be explicit, these measurement units can be abbreviated sp and ${}^q\text{sp}_E$, respectively. Both R_γ and ${}^qD_\gamma$ can be calculated independently of whether the data set is divided into subunits. In contrast, the alpha and beta components are defined only when the data set consists of explicit subunits, which may or may not be the same sampling units that were used during the field inventory. Partitioning of gamma diversity in this classical way has been relatively rare in the ecological literature, but some examples for montane diversity can be found (Veijalainen et al. 2014; Tello et al. 2015). Most studies discussing the three components of diversity actually address some measure of species turnover (discussed in the next section) rather than beta diversity in the strict sense (${}^qD_\beta$).

It is possible to divide a single data set into subunits in many different ways, and each different subdivision corresponds to different values of α and β. Dividing the data set into a larger number of subunits decreases the value of α and increases the value of β. Both components have

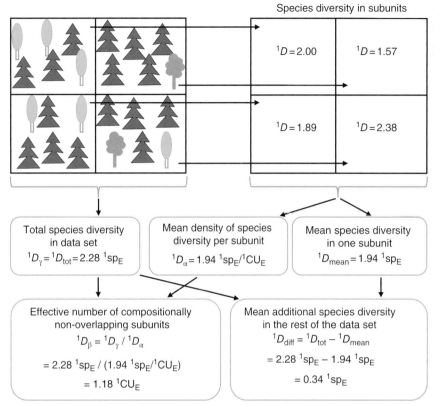

Species diversity in subunits

$^1D = 2.00$ $^1D = 1.57$

$^1D = 1.89$ $^1D = 2.38$

Figure 20.3 Quantification of beta diversity within the inventory plot introduced in Figure 20.1 when the plot consists of four subunits. Numerical values are based on $q = 1$, so all means are geometric means. The effective number of compositionally non-overlapping subunits (beta diversity) is smaller than the number of actual (original) subunits, because the latter share species. Measurement units are effective species (^1sp_E) for species diversity and effective compositional unit (^1CU_E) for beta diversity.

Total species diversity in data set
$^1D_\gamma = {}^1D_{tot} = 2.28 \; {}^1sp_E$

Mean density of species diversity per subunit
$^1D_\alpha = 1.94 \; {}^1sp_E/{}^1CU_E$

Mean species diversity in one subunit
$^1D_{mean} = 1.94 \; {}^1sp_E$

Effective number of compositionally non-overlapping subunits
$^1D_\beta = {}^1D_\gamma \,/\, {}^1D_\alpha$
$= 2.28 \; {}^1sp_E \,/\, (1.94 \; {}^1sp_E/{}^1CU_E)$
$= 1.18 \; {}^1CU_E$

Mean additional species diversity in the rest of the data set
$^1D_{diff} = {}^1D_{tot} - {}^1D_{mean}$
$= 2.28 \; {}^1sp_E - 1.94 \; {}^1sp_E$
$= 0.34 \; {}^1sp_E$

a minimum value of 1: for the analyses to be meaningful, a data set must have at least one subunit, and each subunit must have at least one species. From this follows that both α and β have an upper limit equaling γ. In addition, β is constrained by the number of inventoried subunits (Tuomisto 2010c).

20.2.5 Species Turnover and Related Measures

If all observed species are not present in all inventoried subunits, there will be species turnover (i.e., change in species composition) among subunits. The total amount of species turnover in a data set can be defined as the number of species that are swapped to new ones (turned over) as one starts from one compositional unit and goes through all the other ones in the data set (Tuomisto 2010a). This can be quantified as $\beta_{At} = \gamma - \alpha_t$, which is derived from an additive partitioning of gamma diversity or gamma richness (hence the subscript A): $\gamma = \alpha_t + \beta_{At}$. The additive alpha component α_t expresses how many species there are, on average, in one sampling unit (or compositional unit), and the additive beta component β_{At} expresses how many additional species there are in the rest of the data set. The sum of the two gives the total (effective) number of species in the data set, in the same way that the total weight of an object can be expressed as the sum of the weights of two parts into which it has been divided.

The additive beta component has commonly been referred to as "additive beta diversity" or just "beta diversity" (Lande 1996; Veech et al. 2002; Kiflawi & Spencer 2004; Ricotta 2008; De Bello et al. 2010). However, using the term "beta diversity" in this context is confusing, because β_{At} differs from the original multiplicative beta diversity (which can be given the subscript M for clarity: β_M) both conceptually and in numerical value (Figure 20.3). The subscript t (for turnover) highlights the fact that the alpha and beta components in the additive equation must have the same measurement unit as γ (i.e., species), unlike in the multiplicative equation, where they cannot have the same unit (instead, the unit of alpha is species/compositional unit and the unit of beta is compositional unit) (Tuomisto 2010a, 2013).

To reduce the risk of confusion, $\beta_{At} = \gamma - \alpha_t$ can also be expressed with the equation $R_{diff} = R_{tot} - R_{mean}$ when the interest is in species richness (presence/absence data) or with the equation $^qD_{diff} = {}^qD_{tot} - {}^qD_{mean}$ when the interest is in species diversity (abundance data) (Tuomisto 2013). To obtain a turnover measure that is independent of the species richness of the system, β_{At} can be divided either by γ or by α_t. The former has been called proportional species turnover, β_P, and the latter Whittaker's species turnover, β_W, in order to explicitly differentiate them from beta diversity in the strict sense, β_M (Tuomisto 2010a). All three turnover measures have been used to document heterogeneity in montane systems (Pineda & Halffter 2004; Mandl et al. 2010; Culmsee & Leuschner 2013; Karger et al. 2015; Tello et al. 2015).

Since β_{At} quantifies a number of species, its value is directly constrained by γ: β_{At} cannot be larger than $\gamma - 1$, where 1 is the smallest value that α_t can take. Since β_M quantifies the number of compositional units that do not share any species (but have at least one species each), its value is primarily constrained by the number of subunits N. However, if a data set has fewer species than subunits, then some subunits must share species and β_M becomes limited by γ.

Apart from $\beta_M = \gamma/\alpha$, Whittaker (1960) used several other measures related to compositional heterogeneity, and other researchers have developed dozens more. These correspond to over 30 distinct (and very dissimilar) phenomena, and quite some confusion has resulted from the fact that all of them have been referred to as "beta diversity" in one ecological publication or another (discussed in Vellend 2001; Koleff et al. 2003; Jost 2007; Jurasinski et al. 2009; Tuomisto 2010a,b, 2011; Anderson et al. 2011).

All "beta diversity indices" are somehow related to compositional heterogeneity, but in very different ways. In fact, many of them quantify such different phenomena that their values are not even correlated. Apart from true beta diversity, β_M, and the species turnover measures already discussed, the most popular of these measures have been different compositional dissimilarity indices (which quantify the degree of compositional difference between two or among many sampling units) and various summary indices that can be derived from the pairwise values (such as their mean, the amount of dispersion around an ordination centroid, the length of the first ordination axis or their rate of change along an external gradient) (Tuomisto 2010b, 2013). Because different beta diversity indices measure very different things, their values are generally not comparable. This makes the ecological literature on "beta diversity" much less informative (and more easily misleading) than it could be if the central terms were used more consistently.

20.3 Documenting Diversity Patterns

20.3.1 Target Organism Group

When reporting results on diversity, it is important to be explicit about which organisms were included in the inventory. For example, observed "plant species richness" will be very different depending on whether "plants" includes all land plants, all angiosperms or trees only, and in tree inventories it also makes a big difference what minimum stem diameter value is chosen to define a tree.

In seasonal climates, the timing of the inventories in relation to the growing season of plants and possible diurnal or annual migrations of animals may be crucial in determining what the target group actually is. For example, only a small proportion of the bird species that breed in an area may be encountered in an inventory carried out during the local winter. On the other hand, inventories carried out simultaneously in the wintering grounds may include a high proportion of species that never breed in the area.

Possible target groups differ greatly in the size of individuals, population density per unit area, detectability of the individuals and existing number of species, and all of these variables have a great impact both on the observed diversity in one site and on the diversity differences and other patterns that may emerge among sites. If some of these properties change systematically along an environmental gradient of interest, this needs to be taken into account when interpreting possible trends in species diversity. For example, if trees get smaller and their stem density increases towards higher elevations, this can be expected to have an effect on the diversity patterns as observed with inventory plots of uniform surface area.

20.3.2 Taxonomic Classification

Both richness and diversity depend on how individuals (or other observed entities) are classified into types (often taxa, such as species, genera or families). Obviously, data sets should only be compared if they have used the same classification: comparing species richness in one area with genus richness in another is more likely to be misleading than useful. It is maybe less obvious that specifying the taxonomic level is not enough, as the numbers of accepted taxa also vary among classification systems that have been used at different times or by different taxonomists. For example, in fern systematic, there has been a tendency to split families and genera into several smaller ones, which has resulted in an increase in family and genus richness over time even when species richness remains the same (e.g., Almeida et al. 2016).

There may also be biases that directly affect observed richness gradients, because the taxonomical traditions may tend more towards lumping in some areas and splitting in others. For example, the European biota is already very well known, so it has been possible to separate species on the basis of subtle morphological or behavioral differences, especially in well-known groups such as birds and vascular plants. In the tropics, the level of knowledge may not yet be sufficient to recognize such characteristics as taxonomically significant, so many new species can be expected to emerge through a splitting of the old ones as they become better studied (Ruokolainen et al. 2002; Tobias et al. 2010). If the species concepts applied in temperate areas are generally narrower than those in the tropics, tropical species richness will be underestimated in relation to temperate species richness. Similar biases may affect assessments of elevational species richness gradients if the biotas at some elevation belts are taxonomically better resolved than those at others (e.g., due to differences in accessibility).

20.3.3 Abundance Measure

The calculation of diversity involves species abundances, so the obtained values and patterns in them may be very different if different abundance measures are used. Possible measures include (but are not limited to) the number of individuals, biomass, basal area, volume and the percentage of surface cover. Obviously, if the size of individuals varies (as it generally does), the number of individuals and biomass can be expected to give different diversity values even when calculated for the same data set. For example, canopy trees typically have much larger biomass but fewer stems than do subcanopy trees, so using the number of stems as the abundance measure gives more weight to the understory and using biomass gives more weight to the canopy when quantifying overall tree species diversity in a forest plot. When comparing diversity values across sites, it is therefore important that the abundance measure is the same for all sites, and that it has been chosen in such a way that it is relevant for the ecological questions of interest.

20.3.4 Sampling Effort

The more you search, the more you find. Therefore, raw species counts are only comparable across sites or organism groups if sampling effort has been standardized in a way that is both appropriate for the organisms sampled and relevant for the questions at hand (Gotelli & Colwell 2001; Tuomisto 2010b). How this should be done is a contentious issue, because many different methods are available and justifiable. However, they are not interchangeable, so different results may be obtained depending on which aspect of sampling effort is standardized (i.e., fixed).

Since each new individual that is added to a data set may represent a new species, the most straightforward approach to standardizing sampling effort is to fix the number of individuals that are observed and identified to species (or other categories of interest) at each site. Alternatively, sampling effort may be defined by way of the surface area of an inventory plot (common with plants), the observation time dedicated to a given site (common with birds) or trap operating time (common with insects). Even in these cases, sampling effort is generally defined such that the number of individuals is of a reasonable order of magnitude given the aims of the study. For example, trees with diameter at breast height (DBH) exceeding 10 cm are often sampled using a plot of 1 ha, whereas for smaller trees (2.5 cm DBH), it has been more common to use 0.1 ha. Many studies of understory plants have used plots of 400 m², while bryophyte inventories may be based on plots as small as 0.04 m² (van der Maarel 2005).

In each case, the observed patterns in richness and diversity are contingent on the measure of sampling effort that is chosen and on the level to which it has been fixed. When sampling is started, every individual represents a species that has not yet been observed, but the more individuals the data set already contains, the less likely it is that the next individual represents a new species, so the species accumulation rate decreases (Figure 20.4a). There can be big differences among sites in how fast the species accumulation rate slows with increasing number of individuals sampled. Therefore, observed diversity patterns may be different for 100-individual samples than for 1000-individual ones.

The exact shape of the species accumulation curve depends on how many species the relevant metacommunity (or species pool) contains and what its species abundance distribution is. The more species there are, the more individuals are needed to sample them all (Figure 20.4). In addition, the larger the proportional abundances of the most abundant species, the larger the number of individuals that need to be observed for the rarer species to be included in the sample at all.

The proportional abundance of a very abundant species can be relatively accurately estimated already from a small sample, whereas a rare species may appear absent until the sample is very large. Therefore, the accumulation of diversity, ^{q}D, with sampling effort tends to slow down faster than the accumulation of richness, R, which means that a representative estimate of diversity can be obtained with a smaller number of individuals than a representative estimate of richness. This is especially the case when diversity is calculated with a large value of q, because increasing the value of q gives more weight to the proportional abundance of the most abundant species and less weight to the number of species (Figure 20.2) (Routledge 1980; Tuomisto 2010a; Dauby & Hardy 2012).

When interpreting the results, it also needs to be taken into account that only one aspect of sampling effort can be fixed at a time, and variation in the others may partly explain the observed patterns in diversity. For example, differences in diversity among inventory plots of a fixed surface area may reflect differences in the density of individuals within them, especially if the number of individuals per plot is small relative to the relevant species pool size (for an example from New Caledonian mountains, see Ibanez et al. 2016).

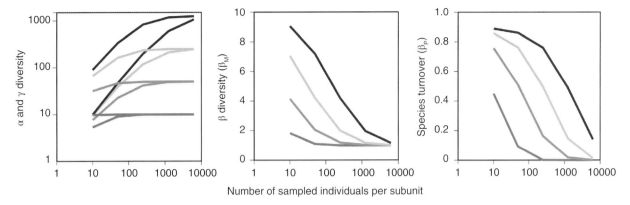

Figure 20.4 The dependency of (a) α and γ, (b) β and (c) proportional species turnover on species pool size and within-subunit number of individuals in 10-subunit data sets. Each gray shade corresponds to a different species pool size: from bottom to top, 10, 50, 250 and 1250 species. Each line shows the mean of 1000 replicates where 10 subunits of either 10, 50, 250, 1250 or 6250 individuals were randomly drawn from a log-normal species-abundance distribution. In (a), the lower line for a given species pool size shows α and the upper line, γ. With a sufficiently large number of individuals, α and γ converge; all subunits were drawn from the same species pool, so any compositional differences among them are caused by undersampling. In (b), the number of compositionally distinct sampling units ($\beta_M = \gamma/\alpha$) decreases with increasing number of individuals sampled. In (c), the proportion of species in the entire set of 10 subunits that does not fit within a single subunit ($\beta_P = 1 - \alpha/\gamma$) decreases with increasing number of individuals sampled. *Source:* Tuomisto & Ruokolainen (2012). Reproduced with permission of The American Association for the Advancement of Science.

Legacy data (i.e., data available for reuse by researchers other than the original collector) often have the problem that sampling effort has not been uniform across sites. A similar problem emerges in insect data that are obtained by passive methods (e.g., pitfall traps or Malaise traps), as their efficiency is very sensitive to uncontrollable factors such as weather conditions. However, if the data have recorded species abundances (and not only species presences), it is possible to standardize sampling effort after the fact through rarefaction to a fixed number of individuals (Chiarucci et al. 2009; Chao et al. 2013). This has been applied in montane studies when assessing elevational species richness or diversity patterns (Culmsee & Leuschner 2013; Ibanez et al. 2016).

The drawback of rarefaction is that it leads to loss of data. The lower the chosen number of individuals, the less representative each sampling unit becomes of the community surrounding it. The higher the chosen number of individuals, the more sampling units get discarded because they contain too few individuals. A more complete picture of relative diversities can be obtained by comparing rarefaction curves (Cayuela et al. 2015). Such curves can also provide useful insight when choosing an appropriate target number of individuals when rarefied data are needed.

20.3.5 Local Representativeness

Often, the main interest is in comparing the species richness or diversity values of entire communities, but only a tiny fraction of the individuals belonging to those communities are included in the available data set. Since any one of the unseen individuals could represent a new species, the community species richness is generally (and often considerably) underestimated by the observed richness, and an undersampling problem emerges. For example, a 1 ha tree plot gives an accurate picture of the tree species density at the resolution of 1 ha, because this is what it actually measures, but the same plot may not be an adequate representation of tree species diversity in the entire forest. The undersampling problem is more acute in species-rich communities than in species-poor ones, so if sampling units have few individuals, they are likely to underestimate species richness differences among communities and may, therefore, fail to observe existing diversity gradients.

Downward bias in estimating alpha diversity easily leads to upward bias in estimating beta diversity and any related measures of species turnover. For example, a 100-stem sample of trees can be quite adequate to quantify the species richness of a community with 10 species (such as a boreal forest) but is woefully inadequate to quantify the species richness of a community with 1000 species (such as a tropical rainforest) (Figure 20.4a). In the latter case, no more than 10% of the species that are actually present in the community can be included in the sample, and even if two species-rich communities shared a large proportion of their species, it is unlikely that the same shared species would happen to be sampled in both communities (Figure 20.4b,c) (Wolda 1981; Scheiner 1990; Colwell & Coddington 1994; Plotkin & Muller-Landau 2002; Chao et al. 2006; Tuomisto 2010b).

The degree to which richness or diversity is underestimated depends on sample representativeness, and the standardization of this may require the application of different sampling efforts at different sites. In vegetation science, representative samples have traditionally been obtained by increasing the inventory plot size until it is estimated that almost all of the species in the community are included in the plot (van der Maarel 2005). In animal studies, sampling may be continued until no new species are found in a predefined amount of additional observation time. Another proposed approach is to continue sampling until the species accumulation curve has reached an asymptote. However, this may be difficult to define objectively, and it is questionable whether species accumulation curves even have asymptotes (Condit et al. 1996; Williamson et al. 2001).

Different kinds of sampling and data analysis methods can be used to improve the accuracy of the species richness estimate, just as in other situations where results are extrapolated beyond actual observations (e.g., Colwell & Coddington 1994; Chao & Tsay 1998; Plotkin & Muller-Landau 2002; Chao & Shen 2004; Cardoso et al. 2009; Chao et al. 2009, 2013; Beck & Schwanghart 2010; Colwell et al. 2012; Dauby & Hardy 2012). An approach that has recently gained in popularity is to standardize sample cover; that is, the (estimated) proportion of individuals in the community that belong to species already represented in the sample (Chao & Jost 2012).

A general challenge when interpreting patterns in observed species richness and diversity is that the results can be affected by many factors, only some of which are quantifiable. Natural communities do not usually have obvious boundaries, but increasing the sampling effort often causes the sample to become more heterogeneous. This may be due to the inclusion of species from more than one community, which can prevent the sample from becoming fully representative of the first community. In addition, a community is already a sample from the larger metacommunity (or species pool), so differences in observed species diversity, beta diversity and species turnover may arise due to differences in metacommunity diversity, community assembly processes or local community size, all of which are beyond experimental control. In montane systems, such problems are especially acute, because topographic heterogeneity is often related to considerable habitat heterogeneity. This can cause local communities to be associated with partly different species pools even if they are not separated by long distances.

20.3.6 Larger Subunits Versus More of Them

The issues related to sampling effort and representativeness are equally relevant for the total species diversity in a data set (gamma diversity) and for the mean species diversity per subunit (alpha diversity). When subunits are considered, another aspect of sampling effort becomes important: the partitioning of total sampling effort between making individual subunits more comprehensive versus inventorying more subunits. This partitioning constrains the relative magnitudes of the alpha and beta components of diversity.

The number of individuals identified per subunit sets an upper limit to alpha diversity, and the number of subunits inventoried sets an upper limit to beta diversity (β_M). The larger the proportion of sampling effort that is dedicated to each subunit, the larger the alpha component. The extreme is when all sampling effort goes to a single subunit, in which case α must equal γ and β_M must equal 1. Increasing the number of subunits increases the upper limit of β_M. Therefore, to obtain meaningful comparisons of the different components of diversity among regions or organism groups, the relevant aspect of sampling effort (or representativeness) needs to be standardized: total sampling effort for γ, within-subunit sampling effort for α and number of sampling units for β_M (and for all species turnover and other measures related to β_M). If needed, standardization can be achieved by rarefaction (Chiarucci et al. 2009).

It is especially noteworthy that since α quantifies diversity for fewer individuals than γ does, α is more severely limited by sampling effort than γ is. Therefore, sample representativeness is smaller for α than it is for γ. In particular, when subunits contain fewer individuals than there are species in the local community, α will have a ceiling set by the number of individuals per sample, and an increase in γ beyond this constraint is only possible if β_M increases (Figure 20.4). A gradient in sample representativeness may, therefore, express itself as a gradient in beta diversity, which detaches the latter from the biological processes it might otherwise be an indication of (Colwell & Hurtt 1994; Kraft et al. 2011; Tuomisto & Ruokolainen 2012; Tello et al. 2015).

When two communities have very similar species compositions, their pairwise species turnover is small, and hence can be overestimated by a much wider margin than when the communities are very different. Severe undersampling therefore causes pairwise species turnover and compositional heterogeneity measures to converge towards high values for all sampling unit pairs, unless the shared species are much more abundant than the non-shared ones (Jobe 2008; Cardoso et al. 2009). This makes unraveling ecological or spatial trends in beta diversity more difficult for species-rich than for species-poor communities or target organisms (Jones et al. 2008).

20.3.7 Spatial and Environmental Representativeness

Species distributions are usually spatially structured, and often their structure is related to dispersal limitation or environmental heterogeneity, or a combination of both. This increases compositional heterogeneity among localities (and hence beta diversity and related measures), and may also give rise to gradients in species richness and diversity. Typically, mountains are highly heterogeneous landscapes: elevation can change rapidly over short distances, and many environmental variables co-vary with elevation. In addition, topography can trigger local heterogeneity independent of elevation: sunny slopes can be much warmer and shaded slopes and closed valleys much colder than the average for their elevation, and leeward and windward slopes can have very different moisture regimes.

High-elevation sites on different mountaintops are likely to be separated by environmentally very different lower elevations, and hence to experience a higher degree of isolation and more differentiated species pools (including a larger proportion of endemics) than lowland sites separated by the same geographical distance would. In addition, the species pools relevant for different elevations may consist of different numbers of species, which can cause sample representativeness to vary with elevation even if sampling effort is the same.

Because of the high degree of local heterogeneity, any study that aims to obtain a comprehensive view of species diversity patterns and their determinants in mountains needs to pay special attention to the spatial and environmental representativeness of the sample. This implies a sufficient number of sampling units that are both standardized among areas and distributed within areas in a representative manner.

A geographically biased sample can be expected to underestimate both gamma diversity and beta diversity of a landscape more than a random sample of the same sampling effort would. In practice, it is very common for sampling to be geographically biased, due to limited accessibility in some parts of the area of interest. In mountains, this problem is especially acute, because steep topography can make access to many areas practically impossible. The more heterogeneous the environment, the more problematic this can become, because the number of sampling units needed to obtain a representative sample increases rapidly with the number of relevant environmental variables. Ideally, sampling could be stratified according to the variation in the most important environmental variables (Austin & Heyligers 1989), but this may not be possible due to

lack or inaccuracy of local-scale environmental data (Soria-Auza et al. 2010).

20.4 Final Notes Related to Diversity in Montane Systems

Species diversity gradients in mountains are the product of a multitude of different ecological and evolutionary mechanisms (Graham et al. 2014). Clarifying the relative roles of these mechanisms is not easy, and even some of the main patterns are still controversial. For example, some studies have reported species richness to decrease with elevation, others have found a mid-elevation peak, and some have reported different trends in different mountain systems (Rahbek 1995; Krömer et al. 2005; Kluge et al. 2006; Kessler et al. 2011). It is an interesting – but, as yet, unresolved – question to what degree such contrasting results reflect differences among organism groups (e.g., inherent physiological properties or evolutionary histories), differences among mountain systems (e.g., maximum elevation, topographic complexity, how climatic conditions co-vary with elevation) or differences among climatic zones or biogeographical regions.

Some of the uncertainties in unraveling general patterns are no doubt due to differences in sampling effort and other aspects of sampling strategy. As discussed earlier, representativeness can easily vary among studies, organism groups and geographical regions, and results may vary according to whether the focus is on quantifying the total number of species per elevational belt or on local species density within elevational belts (Patterson et al. 1998; Krömer et al. 2005). In either case, the results may also be affected by the surface area covered by each elevational belt, as this can have an effect on the relevant species pool size (Rahbek 1997; Karger et al. 2015; Pouteau et al. 2015).

The availability of digital species occurrence data through portals such as GBIF provides intriguing prospects for broad-scale and even global mapping of diversity along elevational and other gradients. However, it remains to be seen to what degree problems related to uneven sampling effort and the erratic quality of species identifications and geographical location data in the available records can be solved (Rowe 2005; Maldonado et al. 2015).

References

Acharya, B.K., Sanders, N.J., Vijayan, L. & Chettri, B. (2011) Elevational gradients in bird diversity in the eastern Himalaya: an evaluation of distribution patterns and their underlying mechanisms. *PLoS ONE* **6**, e29097.

Almeida, T.E., Hennequin, S., Schneider, H. et al. (2016) Towards a phylogenetic generic classification of Thelypteridaceae: additional sampling suggests alterations of neotropical taxa and further study of paleotropical genera. *Molecular Phylogenetics and Evolution* **94**, 688–700.

Anderson, M.J., Crist, T.O., Chase, J.M. et al. (2011) Navigating the multiple meanings of β diversity: a roadmap for the practicing ecologist. *Ecology Letters* **14**, 19–28.

Austin, M.P. & Heyligers, P.C. (1989) Vegetation survey design for conservation: Gradsect sampling of forests in North-eastern New South Wales. *Biological Conservation* **50**, 13–32.

Beck, J. & Schwanghart, W. (2010) Comparing measures of species diversity from incomplete inventories: an update: *Methods in Ecology and Evolution* **1**, 38–44.

Broms, K.M., Hooten, M.B. & Fitzpatrick, R.M. (2015) Accounting for imperfect detection in Hill numbers for biodiversity studies. *Methods in Ecology and Evolution* **6**, 99–108.

Cardoso, P., Borges, P.A.V. & Veech, J.A. (2009) Testing the performance of beta diversity measures based on incidence data: the robustness to undersampling. *Diversity and Distributions* **15**, 1081–1090.

Cayuela, L., Gotelli, N.J. & Colwell, R.K. (2015) Ecological and biogeographic null hypotheses for comparing rarefaction curves. *Ecological Monographs* **85**, 437–455.

Chao, A. & Jost, L. (2012) Coverage-based rarefaction and extrapolation: standardizing samples by completeness rather than size. *Ecology* **93**, 2533–2547.

Chao, A. & Shen, T.-J. (2004) Nonparametric prediction in species sampling. *Journal of Agricultural, Biological, and Environmental Statistics* **9**, 253–269.

Chao, A. & Tsay, P.K. (1998) A sample coverage approach to multiple-system estimation with application to census undercount. *Journal of the American Statistical Association* **93**, 283–293.

Chao, A., Chazdon, R.L., Colwell, R.K. & Shen, T.-J. (2006) Abundance-based similarity indices and their estimation when there are unseen species in samples. *Biometrics* **62**, 361–371.

Chao, A., Colwell, R.K., Lin, C.-W. & Gotelli, N.J. (2009) Sufficient sampling for asymptotic minimum species richness estimators. *Ecology* **90**, 1125–1133.

Chao, A., Gotelli, N.J., Hsieh, T.C. et al. (2013) Rarefaction and extrapolation with Hill numbers: a framework for sampling and estimation in species diversity studies. *Ecological Monographs* **84**, 45–67.

Chao, A., Chiu, C.-H. & Jost, L. (2014) Unifying species diversity, phylogenetic diversity, functional diversity, and related similarity and differentiation measures through Hill numbers. *Annual Review of Ecology, Evolution, and Systematics* **45**, 297–324.

Chiarucci, A., Bacaro, G., Rocchini, D. et al. (2009) Spatially constrained rarefaction: incorporating the autocorrelated structure of biological communities into sample-based rarefaction. *Community Ecology* **10**, 209–214.

Colwell, R.K. & Coddington, J.A. (1994) Estimating terrestrial biodiversity through extrapolation. *Philosophical Transactions of the Royal Society of London B: Biological Sciences* **345**, 101–118.

Colwell, R.K. & Hurtt, G.C. (1994) Nonbiological gradients in species richness and a spurious Rapoport effect. *The American Naturalist* **144**, 570–595.

Colwell, R.K., Chao, A., Gotelli, N.J. et al. (2012) Models and estimators linking individual-based and sample-based rarefaction, extrapolation and comparison of assemblages. *Journal of Plant Ecology* **5**, 3–21.

Condit, R., Hubbell, S.P., Lafrankie, J.V. et al. (1996) Species-area and species-individual relationships for tropical trees: a comparison of three 50-ha plots. *Journal of Ecology* **84**, 549–562.

Culmsee, H. & Leuschner, C. (2013) Consistent patterns of elevational change in tree taxonomic and phylogenetic diversity across Malesian mountain forests. *Journal of Biogeography* **40**, 1997–2010.

Dauby, G. & Hardy, O.J. (2012) Sampled-based estimation of diversity sensu stricto by transforming Hurlbert diversities into effective number of species. *Ecography* **35**, 661–672.

De Bello, F., Lavergne, S., Meynard, C.N. et al. (2010) The partitioning of diversity: showing Theseus a way out of the labyrinth. *Journal of Vegetation Science* **21**, 992–1000.

Gauthier, O., Sarrazin, J. & Desbruyères, D. (2010) Measure and mis-measure of species diversity in deep-sea chemosynthetic communities. *Marine Ecology Progress Series* **402**, 285–302.

Gotelli, N.J. & Colwell, R.K. (2001) Quantifying biodiversity: procedures and pitfalls in the measurement and comparison of species richness. *Ecology Letters* **4**, 379–391.

Graham, C.H., Carnaval, A.C., Cadena, C.D. et al. (2014) The origin and maintenance of montane diversity: integrating evolutionary and ecological processes. *Ecography* **37**, 711–719.

Hill, M.O. (1973) Diversity and evenness: a unifying notation and its consequences. *Ecology* **54**, 427–432.

Ibanez, T., Grytnes, J.-A. & Birnbaum, P. (2016) Rarefaction and elevational richness pattern: a case study in a high tropical island (New Caledonia, SW Pacific). *Journal of Vegetation Science* **27**, 441–451.

Jobe, R.T. (2008) Estimating landscape-scale species richness: reconciling frequency- and turnover-based approaches. *Ecology* **89**, 174–182.

Jones, M.M., Tuomisto, H. & Olivas, P.C. (2008) Differences in the degree of environmental control on large and small tropical plants: just a sampling effect? *Journal of Ecology* **96**, 367–377.

Jost, L. (2006) Entropy and diversity. *Oikos* **113**, 363–375.

Jost, L. (2007) Partitioning diversity into independent alpha and beta components. *Ecology* **88**, 2427–2439.

Jost, L., DeVries, P., Walla, T. et al. (2010) Partitioning diversity for conservation analyses. *Diversity and Distributions* **16**, 65–76.

Jurasinski, G. & Koch, M. (2011) Commentary: do we have a consistent terminology for species diversity? We are on the way. *Oecologia* **167**, 893–902.

Jurasinski, G., Retzer, V. & Beierkuhnlein, C. (2009) Inventory, differentiation, and proportional diversity: a consistent terminology for quantifying species diversity. *Oecologia* **159**, 15–26.

Karger, D.N., Tuomisto, H., Amoroso, V.B. et al. (2015) The importance of species pool size for community composition. *Ecography* **38**, 1243–1253.

Kattan, G.H. & Franco, P. (2004) Bird diversity along elevational gradients in the Andes of Colombia: area and mass effects. *Global Ecology and Biogeography* **13**, 451–458.

Kessler, M., Kluge, J., Hemp, A. & Ohlemüller, R. (2011) A global comparative analysis of elevational species richness patterns of ferns. *Global Ecology and Biogeography* **20**, 868–880.

Kiflawi, M. & Spencer, M. (2004) Confidence intervals and hypothesis testing for beta diversity. *Ecology* **85**, 2895–2900.

Kindt, R., Van Damme, P. & Simons, A.J. (2006) Tree diversity in western Kenya: using profiles to characterise richness and evenness. *Biodiversity & Conservation* **15**, 1253–1270.

Kluge, J., Kessler, M. & Dunn, R.R. (2006) What drives elevational patterns of diversity? A test of geometric constraints, climate and species pool effects for pteridophytes on an elevational gradient in Costa Rica. *Global Ecology and Biogeography* **15**, 358–371.

Koleff, P., Lennon, J.J. & Gaston, K.J. (2003) Are there latitudinal gradients in species turnover? *Global Ecology and Biogeography* **12**, 483–498.

Kraft, N.J.B., Comita, L.S., Chase, J.M. et al. (2011) Disentangling the drivers of diversity along latitudinal and elevational gradients. *Science* **333**, 1755–1758.

Krömer, T., Kessler, M., Robbert Gradstein, S. & Acebey, A. (2005) Diversity patterns of vascular epiphytes along an elevational gradient in the Andes. *Journal of Biogeography* **32**, 1799–1809.

Lande, R. (1996) Statistics and partitioning of species diversity, and similarity among multiple communities. *Oikos* **76**, 5–13.

Magurran, A.E. (2004) *Measuring Biological Diversity*. Chichester: John Wiley & Sons, Ltd.

Maldonado, C., Molina, C.I., Zizka, A. et al. (2015) Estimating species diversity and distribution in the era of Big Data: to what extent can we trust public databases? *Global Ecology and Biogeography* **24**, 973–984.

Mandl, N., Lehnert, M., Kessler, M. & Gradstein, S.R. (2010) A comparison of alpha and beta diversity patterns of ferns, bryophytes and macrolichens in tropical montane forests of southern Ecuador. *Biodiversity and Conservation* **19**, 2359–2369.

McCain, C.M. (2010) Global analysis of reptile elevational diversity. *Global Ecology and Biogeography* **19**, 541–553.

Moreno, C.E. & Rodríguez, P. (2010) A consistent terminology for quantifying species diversity? *Oecologia* **163**, 279–282.

Moreno, C.E. & Rodríguez, P. (2011) Commentary: do we have a consistent terminology for species diversity? Back to basics and toward a unifying framework. *Oecologia* **167**, 889–892.

Patterson, B.D., Stotz, D.F., Solari, S. et al. (1998) Contrasting patterns of elevational zonation for birds and mammals in the Andes of southeastern Peru. *Journal of Biogeography* **25**, 593–607.

Pineda, E. & Halffter, G. (2004) Species diversity and habitat fragmentation: frogs in a tropical montane landscape in Mexico. *Biological Conservation* **117**, 499–508.

Plotkin, J.B. & Muller-Landau, H.C. (2002) Sampling the species composition of a landscape. *Ecology* **83**, 3344–3356.

Pouteau, R., Bayle, É., Blanchard, É. et al. (2015) Accounting for the indirect area effect in stacked species distribution models to map species richness in a montane biodiversity hotspot. *Diversity and Distributions* **21**, 1329–1338.

Rahbek, C. (1995) The elevational gradient of species richness: a uniform pattern? *Ecography* **18**, 200–205.

Rahbek, C. (1997) The relationship among area, elevation, and regional species richness in neotropical birds. *The American Naturalist* **149**, 875–902.

Ricotta, C. (2008) Computing additive-diversity from presence and absence scores: a critique and alternative parameters. *Theoretical Population Biology* **73**, 244–249.

Rodríguez-Castañeda, G., Dyer, L.A., Brehm, G. et al. (2010) Tropical forests are not flat: how mountains affect herbivore diversity. *Ecology Letters* **13**, 1348–1357.

Routledge, R.D. (1977) On Whittaker's components of diversity. *Ecology* **58**, 1120–1127.

Routledge, R.D. (1979) Diversity indices: which ones are admissible? *Journal of Theoretical Biology* **76**, 503–515.

Routledge, R.D. (1980) Bias in estimating the diversity of large, uncensused communities. *Ecology* **61**, 276–281.

Rowe, R.J. (2005) Elevational gradient analyses and the use of historical museum specimens: a cautionary tale. *Journal of Biogeography* **32**, 1883–1897.

Ruokolainen, K., Tuomisto, H., Vormisto, J. & Pitman, N. (2002) Two biases in estimating range sizes of Amazonian plant species. *Journal of Tropical Ecology* **18**, 935–942.

Scheiner, S.M. (1990) Affinity analysis: effects of sampling. *Vegetatio* **86**, 175–181.

Soria-Auza, R.W., Kessler, M., Bach, K. et al. (2010) Impact of the quality of climate models for modelling species occurrences in countries with poor climatic documentation: a case study from Bolivia. *Ecological Modelling* **221**, 1221–1229.

Spellerberg, I.F. & Fedor, P.J. (2003) A tribute to Claude Shannon (1916–2001) and a plea for more rigorous use of species richness, species diversity and the "Shannon–Wiener" Index. *Global Ecology and Biogeography* **12**, 177–179.

Tello, J.S., Myers, J.A., Macía, M.J. et al. (2015) Elevational gradients in β-diversity reflect variation in the strength of local community assembly mechanisms across spatial scales. *PLoS ONE* **10**, e0121458.

Tobias, J.A., Seddon, N., Spottiswoode, C.N. et al. (2010) Quantitative criteria for species delimitation. *Ibis* **152**, 724–746.

Tuomisto, H. (2010a) A diversity of beta diversities: straightening up a concept gone awry. Part 1. Defining beta diversity as a function of alpha and gamma diversity. *Ecography* **33**, 2–22.

Tuomisto, H. (2010b) A diversity of beta diversities: straightening up a concept gone awry. Part 2. Quantifying beta diversity and related phenomena. *Ecography* **33**, 23–45.

Tuomisto, H. (2010c) A consistent terminology for quantifying species diversity? Yes, it does exist. *Oecologia* **164**, 853–860.

Tuomisto, H. (2011) Commentary: do we have a consistent terminology for species diversity? Yes, if we choose to use it. *Oecologia* **167**, 903–911.

Tuomisto, H. (2013) Defining, measuring, and partitioning species diversity. In: Levin, S. (ed.) *Encyclopaedia of Biodiversity*. Philadelphia, PA: Elsevier, pp. 434–446.

Tuomisto, H. & Ruokolainen, K. (2012) Comment on "Disentangling the drivers of diversity along latitudinal and elevational gradients." *Science* **335**, 1573.

van der Maarel, E. (2005) Vegetation ecology – an overview. In: van der Maarel, E. (ed.) *Vegetation Ecology*. Oxford: Blackwell Publishing, pp. 1–51.

Veech, J.A. & Crist, T.O. (2010) Toward a unified view of diversity partitioning. *Ecology* **91**, 1988–1992.

Veech, J.A., Summerville, K.S., Crist, T.O. & Gering, J.C. (2002) The additive partitioning of species diversity: recent revival of an old idea. *Oikos* **99**, 3–9.

Veijalainen, A., Sääksjärvi, I.E., Tuomisto, H. et al. (2014) Altitudinal trends in species richness and diversity of Mesoamerican parasitoid wasps (Hymenoptera: Ichneumonidae). *Insect Conservation and Diversity* **7**, 496–507.

Vellend, M. (2001) Do commonly used indices of β-diversity measure species turnover? *Journal of Vegetation Science* **12**, 545–552.

Whittaker, R.H. (1960) Vegetation of the Siskiyou Mountains, Oregon and California. *Ecological Monographs* **30**, 279–338.

Williamson, M., Gaston, K.J. & Lonsdale, W.M. (2001) The species–area relationship does not have an asymptote! *Journal of Biogeography* **28**, 827–830.

Wolda, H. (1981) Similarity indices, sample size and diversity. *Oecologia* **50**, 296–302.

21

A Modeling Framework to Estimate and Project Species Distributions in Space and Time

Niels Raes[1] and Jesús Aguirre-Gutiérrez[1,2,3]

[1] *Naturalis Biodiversity Center, Leiden, Netherlands*
[2] *Institute for Biodiversity and Ecosystem Dynamics, University of Amsterdam, Amsterdam, Netherlands*
[3] *Environmental Change Institute, School of Geography and the Environment, University of Oxford, Oxford, UK*

Abstract

Over the past decade, species distribution models (SDMs) have become an indispensable item in the ecologist's toolbox. SDMs, also known as ecological niche models, bioclimatic models or habitat suitability models, characterize the multivariate ecological space delimiting species' distributions and project this subset of ecological space back on to geography, resulting in a map of habitat suitability. Although SDMs build on correlations, they offer an important capacity to elucidate the altitudinal zoning on mountains, to forecast the effects of climate change on the risk of plant invasions onto mountains and to predict the potential of mountains as climate refugia, among many other ecological applications. There are more than 600 million digitized and georeferenced collection records currently available through the Global Biodiversity Information Facility (GBIF) portal, which, together with large amounts of spatial data on past, present and future climatic conditions, quantitative soil conditions and high-resolution topographic data, will further increase the popularity and applications of SDMs. In this chapter, we describe the principles and data requirements of SDMs, and provide an introduction to estimating species ranges from collection records and projecting these estimates in space and time.

Keywords: *ecological niche model, species distribution model, ensemble model, climate change, non-analog conditions, range shift*

21.1 Species Niches and Their Reciprocal Spatial Distributions

A species' niche consists of three components, and largely builds on Hutchinson's quantification of a species' niche as n-dimensional space that reflects suitable values of n biologically important and independent variables (e.g., temperature and precipitation) (Hutchinson 1957; Colwell & Rangel 2009; Blonder et al. 2014). Hutchinson's key innovation was the separation of the physical distribution of a species characterized by its geographical coordinates from the local values of n environmental conditions at a given time. The definition of a species' niche by n environmental attributes allows reciprocal projections between a species' niche and its present, past and future geographical distributions. Importantly, Hutchinson's niche expresses the effects of species interactions, but also the constraints of dispersal limitation (Colwell & Rangel 2009), known as a species' realized niche. Building on Hutchinson's niche concept,

Soberón & Peterson (2005) developed the biotic, abiotic, movement (BAM) framework. The first and most important component of a species' niche is represented by the abiotic conditions within which a species population can establish and maintain itself, given its intrinsic physiological limits (Hutchinson 1957; Boulangeat et al. 2012). The second component consists of dispersal or movement limitations, which may prevent species from reaching sites with suitable abiotic conditions (e.g., a mountain range with suitable abiotic conditions separated by a vast lowland region, ocean, ocean strait or large river) (Bateman et al. 2013; Vasudev et al. 2015). The third component represents biotic interactions such as specific plant–pollinator interactions, the presence of pathogens or mutualistic relationships between plants and fungi or soil microbes. A species is present where all three niche components overlap, in what is known as its "realized niche" (Soberón & Nakamura 2009). It is from the realized niche that species presence records are collected, subsequently to be used in species

Mountains, Climate and Biodiversity, First Edition. Edited by Carina Hoorn, Allison Perrigo and Alexandre Antonelli.
© 2018 John Wiley & Sons Ltd. Published 2018 by John Wiley & Sons Ltd.
Companion website: www.wiley.com\go\hoorn\mountains,climateandbiodiversity

distribution models (SDMs). The extent to which the three niche components overlap is often unknown, and this is a caveat of SDMs that should be taken in consideration. Advances are being made towards including biotic interactions (Boulangeat et al. 2012; Giannini et al. 2013; Thuiller et al. 2015) and dispersal limitations (Engler et al. 2012; Miller & Holloway 2015) in SDMs. The majority of SDM studies, however, estimate the spatial distribution of suitable abiotic niche conditions based on a species' realized niche. Despite this caveat, abiotic conditions, both at present and historically, govern at least the broadest outlines of the distribution of species and biomes (Thomas 2010; Boucher-Lalonde et al. 2016; Lee-Yaw et al. 2016). SDMs have successfully been used to forecast the effects of climate change on species' distributions (Thuiller et al. 2011), to identify historical refugia (Waltari et al. 2007) and map past distribution ranges (Raes et al. 2014), to predict the potential geographical ranges of invasive species (Broennimann et al. 2007) and to overcome (at least partly) the Wallacean shortfall (Hortal et al. 2015) or lack of knowledge on geographical distributions of species (Vollering et al. 2016), among many other applications (Araújo & Peterson 2012).

21.2 Species Presence Data

Without data on species occurrences from which to infer niche dimensions, it would be impossible to develop an SDM. These records are obtained from survey data and digitized herbarium and natural history museum specimens, which represent verifiable presences. The largest data portal with collection records is arguably the Global Biodiversity Information Facility data portal (www.gbif.org). For South America, the speciesLink data portal (www.splink.cria.org.br) is an additional source. Absence records are far more difficult to obtain, as "the absence of presence does not equal the presence of absence." Some SDM algorithms use presence-only data, while others require pseudo-absences or a background sample as replacement for true absence records (see Table 21.1).

Given that most species are rare, the number of records that are available to model the distributions of many species is limited. This poses a potential problem, as no relationship between species occurrence and abiotic conditions can be inferred based on only a few records. Various authors have used the subjective number of five spatially unique records as the absolute minimum requirement

Table 21.1 SDM algorithms. The most widely used are indicated with bold text.

SDM	Description	"Absence" data	References
ANNs	Artificial neural networks	Pseudo-absence	Hilbert & Ostendorf (2001)
BIOCLIM	Bioclimatic envelope – rectilinear	Presence only	Busby (1991)
BRTs	Boosted regression trees	Pseudo-absence	Elith et al. (2008)
CART	Classification and regression trees	Pseudo-absence	Breiman et al. (1984); De'ath & Fabricius (2000)
DOMAIN	Proximity to presences in multidimensional predictor space measured by the Gower metric	Presence only	Carpenter et al. (1993)
ENFA	Ecological niche factor analysis	Background sample	Hirzel et al. (2002)
GARP	Genetic algorithm for rule set prediction	Pseudo-absence	Stockwell & Peters (1999)
GAMs	**Generalized additive models**	**Pseudo-absence**	**Hastie & Tibshirani (1986); Yee & Mitchell (1991)**
GBMs	Generalized boosted models	Pseudo-absence	Ridgeway (1999)
GDM	Generalized dissimilarity modeling	Pseudo-absence	Ferrier et al. (2007)
GLMs	**Generalized linear models**	**Pseudo-absence**	**McCullagh & Nelder (1989); Venables & Ripley (2002)**
HABITAT	Bioclimatic envelope – convex hull	Presence only	Walker & Cocks (1991)
Mahalanobis distance	Multidimensional distance to the mean value for each predictor across presence localities	Presence only	Rotenberry et al. (2006); Calenge et al. (2008)
MARS	Multivariate adaptive regression splines	Pseudo-absence	Elith & Leathwick (2007)
Maxent	**Maximum entropy**	**Background sample**	**Phillips et al. (2006)**
MDA	Mixture discriminant analysis	Pseudo-absence	Hastie et al. (1994)
RFs	Random forests	Pseudo-absence	Breiman (2001)
SVMs	Support vector machines	Presence-only	Guo et al. (2005)

(Pearson et al. 2007; Raes et al. 2014). A recent study has shown that the minimum required number of presence records depends on the prevalence, or proportional presence area, relative to the study region (van Proosdij et al. 2016). Prevalence values should range between 0.1 and 0.9 in order to obtain reliable results. Taking these considerations into account, and using virtual species distributions with a stringent accuracy test, the results of van Proosdij et al. (2016) indicate that at least 10 spatially unique presence records are required to calibrate an SDM. Additionally, these authors provide a methodology to arrive at an accurate estimate of the minimum required number of presence records for a given study region.

Although the region under study might cover only part of a species' range, it is important to include *all* available presence records for that species to calibrate an SDM, in order to avoid modeling partial or truncated niches (Raes 2012; Hannemann et al. 2016). Partial niche models tend to underestimate the probability of occurrence at the edges of "niche space" covered by the artificially delimited study region, and to overestimate it at the centre (Raes 2012). Furthermore, caution should be taken when modeling the distribution of invasive species. Lack of biotic interactions (e.g., pathogens, predators) or niche shifts in the invaded range can lead to the inclusion of presence records with abiotic conditions that do not exist in the native range, and hence potentially result in overprediction of the native range (Broennimann et al. 2007).

Another issue of concern is taxonomic synonyms. Institutes that contribute data to the global data portals may not always use the same taxonomy, or they may file records under synonymous names. Synonyms can be resolved using the Taxonomic Name Resolution Service (TNRS) (Boyle et al. 2013), while the Encyclopedia of Life (www.eol.org) provides synonyms for a wide taxonomic range of organisms. Moreover, specimens can also be stored under false taxonomic names as a result of misidentification (Goodwin et al. 2015), or as a result of belonging to as of yet undescribed taxa ("the Linnean shortfall") (Hortal et al. 2015).

Finally, geographical coordinates should be checked against specimen locality descriptions. Too often, latitudinal and longitudinal coordinates are reversed or centroid country coordinates are linked to specimens, among other potential sources of errors (Maldonado et al. 2015; Töpel et al. 2017).

21.3 Abiotic Spatial Data

21.3.1 Bioclimatic Variables

SDM algorithms are powerful tools that identify correlations between species presence records and abiotic – or, in fact, any spatially explicit – variables.

Therefore, in order to obtain meaningful SDM results, it is important to select abiotic variables that relate to the ecological niche of the species. For the terrestrial realm, abiotic climatic conditions, such as temperature and precipitation, account for the majority of the spatial variation in the probability-of-occurrence estimation of a species (Boucher-Lalonde et al. 2012, 2014; Lee-Yaw et al. 2016). The widely used Bioclim data set consists of 19 bioclimatic variables derived from monthly minimum and maximum temperatures and monthly precipitation data (Hijmans et al. 2005). Bioclimatic variables represent biological limits such as "minimum temperature of the coldest month" or "precipitation of the driest quarter." Bioclimatic data sets are available at different spatial resolutions, ranging between 0.5 degree ($\sim$3000 km^2 at the equator) and 30 arc-seconds ($\sim$1 km^2), and can be downloaded from www.worldclim.org (Hijmans et al. 2005), www.climond.org (Kriticos et al. 2012), www.ccafs-climate.org and www.ecoclimate.org (Lima-Ribeiro et al. 2015).

21.3.2 Altitude and Derived Variables

In addition to bioclimatic variables, altitude can be used as an abiotic variable. Altitude seems relevant when modeling species distributions in montane regions. However, it is very often highly negatively correlated with the annual mean temperature, as temperature decreases with increasing altitude (Körner 2007). If the goal of an SDM is to predict the impact of future climate change on species distributions, or to project the model on to past climatic conditions, it is strongly advised not to use altitude as a variable: altitude is static, whereas global climate models (GCMs) predict increasing future temperatures, resulting in upslope range shifts of species.

Related to altitude is "topographic heterogeneity." The NASA Shuttle Radar Topographic Mission (SRTM) has delivered a digital elevation model at 3 arc-seconds, or 90 m spatial resolution, at the equator. When 90 m SRTM data are aggregated to resolutions between 1 km^2 and 5 arc-minutes ($\sim$9.3$\times$9.3 km), the standard deviation (SD) around the mean is a measure of topographic heterogeneity. Altitudinal plains are represented by raster cells with low topographic heterogeneity values, and rugged mountainous terrains have high topographic heterogeneity values.

Additionally, slope and aspect can be derived from altitudinal data. Aspect describes the direction in which a slope faces, and relates to the degree of solar exposure. It should be noted that various other variables can be derived from altitudinal measurements and that the inclusion of topographic heterogeneity, slope and aspect variables in SDMs may be recommended instead of the inclusion of altitude.

21.3.3 Quantitative Soil Property Variables

A third category of abiotic variables is made up of quantitative soil variables such as pH, water holding capacity and organic carbon content. These abiotic variables have recently become available through various portals, such as the Harmonized World Soil Database (HWSD) (FAO/IIASA/ISRIC/ISSCAS/JRC 2012), SoilGrids1km (Hengl et al. 2014) and the European Soil DataBase (ESDB). The quantitative soil information is derived from interpolated US Food and Agricultural Organization (FAO) soil profile data, and is also available as categorical variables.

SDM algorithms that use regression modeling transform categorical data into presence/absence dummy variables, with one dummy variable for each category in the data layer (Franklin 2009). Thus, for regression models, the use of categorical variables may be unwanted – especially when many environmental variables are used and few species presence records are available, which may result in overfitted SDMs. Decision-tree algorithms may be a better option when handling categorical variables.

21.3.4 Land-Cover Data and Satellite Imagery

Land-cover data and satellite imagery with global coverage can be useful, but should be used with caution. Most land-cover data are interpretations of satellite images and/or aerial photos. The earliest satellite images, from Landsat 1, were taken in 1973, but many species collection records predate that year; a specimen collected in 1960 can easily be associated with agricultural land based on satellite imagery postdating 1973. Furthermore, while land cover, the Normalized Difference Vegetation Index (NDVI) and the Enhanced Vegetation Index (EVI) data may be useful for modeling animal distributions, we advise against the use of these sources for plant distributions, as this can be classified as circular reasoning. Furthermore, when the intention is to predict future (or past) species distributions under different climate change scenarios, it should be kept in mind that no future land-cover, NDVI or EVI data are readily available, although advances are being made (Martinuzzi et al. 2015). Land-cover data are useful, however, for correcting the predicted distributions of species for remaining natural vegetation cover by removing all areas classified as "urban" and "agricultural land" from the predicted distribution range.

21.3.5 Selecting Uncorrelated Abiotic Variables

Most SDM algorithms require uncorrelated predictor variables, in order to avoid problems with collinearity (Dormann et al. 2013). Once ecologically relevant predictors are identified, these can be tested for correlations with a Pearson's r-correlation test, or with a Spearman's rank correlation test in the case of non-normally distributed variables. As a rule of thumb, Pearson's $|r| > 0.7$ or Spearman's $|rho| > 0.7$ is an appropriate indicator for when collinearity begins to severely distort model estimations and subsequent predictions (Dormann et al. 2013).

Another measure of variable correlation or collinearity is the Variance Inflation Factor (VIF). A VIF value of >10 is often used to indicate high collinearity (O'brien 2007). From sets of correlated variables, the one with the highest ecological relevance should be kept to develop the SDM. Once all correlated predictors are removed, the correlation table should not have values above 0.7, or VIF values should not exceed 10.

21.3.6 Future and Past Bioclimatic Data

When the aim is to predict the impacts of future climate change on the distributions of species, data from global climate/circulation models (GCMs) are required. The latest report from the Intergovernmental Panel on Climate Change (IPCC) uses four different scenarios for global development, known as representative concentration pathways (RCPs), which lead to increased global average temperatures of between 2 and 4 °C (IPCC 2013). At local scales, the predicted increase in temperature can be much higher or lower, however. No fewer than 61 different GCMs, developed by 20 different institutes, have contributed to the latest IPCC-AR5 report (IPCC 2013). Details of the different GCMs can be found in the Climate Model Intercomparison Project – phase 5 (CMIP5) portal (Taylor et al. 2012). Given the complexity of GCMs, the spatial resolutions of the data are coarse, typically ranging between 1.0 and 2.75°.

To predict the future distributions of species, data from GCMs need to be downscaled to the desired spatial resolution. Two different methods are widely used: the Delta method (GCM portal) and the bias-corrected method (www.worldclim.org). The Delta method calculates the difference (anomaly) between predicted future values and recorded present values at the coarse spatial resolution of the GCM. These anomalies are then interpolated to the desired high spatial resolution used for modeling. Finally, the interpolated values are added to the present high-resolution values in order to maintain the high-resolution climate differences related to, for example, topographic differences. The bias-corrected method calculates the difference between the predicted future GCM values and the predicted present GCM values at the coarse resolution of the GCM. Not all GCMs correctly predict present values as derived from weather stations. The anomalies between predicted present and predicted future values are then interpolated to the desired spatial resolution and added to the present data, which are interpolated from weather stations. This procedure corrects for biases in GCM predictions concerning present climatic conditions.

It might be equally interesting to assess past distributions of species; for example, to identify glacial refugia (Waltari et al. 2007) or to predict the vegetation types that covered exposed sea beds during glacial periods (Raes et al. 2014). Varela et al. (2015) provide a detailed summary of the available paleoclimatic data.

21.4   Species Distribution Models

The applications of SDMs are twofold. They can be used (i) to predict habitat suitability for areas where species collection records are lacking (Wallacean shortfall) and (ii) to describe a species' ecology based on its occurrence records and the abiotic conditions at those localities. SDMs combine data from species presence/absence records – taxonomically synonymized and georeferenced – with a selection of ecologically relevant and uncorrelated predictor variables (see Figure 21.1).

Over the past 2 decades, many different algorithms have been developed, compared and scrutinized in comparative tests (Elith et al. 2006; Aguirre-Gutiérrez et al. 2013; Qiao et al. 2015). Three main classes of modeling algorithms can be distinguished based on their requirements with respect to absence records (Table 21.1). The first class requires presence records only. The second requires absences, or pseudo-absences if true absences are lacking. Pseudo-absences are randomly drawn absences from the study area, taken from any locality where no presence was recorded. The third class do not require any absence data, but use a background sample defined as randomly drawn localities from the entire study area, including presence localities. Depending on the SDM algorithm, the "distributions" are either assumed to be parametric (normal, binomial, Poisson distribution) or are more relaxed in their assumptions (semi-parametric or nonparametric). We do not intend to be exhaustive here, nor to provide detailed descriptions of the different SDM algorithms. For that purpose, we refer to the textbooks of Franklin (2009), Peterson et al. (2011) and Guisan et al. (2017) and the references in Table 21.1. Presently, Maxent, GLM and GAM are the most widely used algorithms (Merow et al. 2013; Qiao et al. 2015).

Several of the SDM algorithms listed in Table 21.1 are implemented in applications (software) with a graphical user interface (GUI), notably Maxent. Most of them can be operated directly through R (R Development Core Team 2014). Several R-libraries have been developed especially for species distribution modeling, including "dismo" (Hijmans et al. 2015), "biomod2" (Thuiller et al.

2014) and "SSDM" (Schmitt et al. 2016). The R-vignette (manual) "Species Distribution Modeling with R" (Hijmans & Elith 2016) is highly recommended and covers the entire modeling process for various algorithms using the R framework.

21.4.1   Measures of SDM Accuracy and the Null-Model Test

Testing the accuracy of SDMs is challenging because independent test data are generally lacking. As a solution, presence records are often partitioned into a training and a testing data set. Either single partitions, multiple random partitions or *k*-fold partitions are used (e.g., 75% for training and 25% for testing) to develop SDMs and assess their predictive power on the test data. When the number of presence records is small, a jackknife (or "leave-one-out") procedure can be used (Pearson et al. 2007). For each run, one record is left out of the training data set and is used to measure the predictive accuracy. Both *k*-fold partitioning and the jackknife procedure result in a distribution of accuracy values that can be interpreted as the sensitivity of the SDM to different partitions of the data.

Most measures of SDM accuracy depend on a binary confusion matrix (Fielding & Bell 1997). A confusion matrix is a 2×2 contingency table that captures (i) the number of presences correctly predicted as present ("sensitivity"), (ii) the number of absences falsely predicted as present ("false positives" or "commission error"), (iii) the number of presences falsely predicted as absent ("false negatives" or "omission error") and (iv) the number of absences correctly predicted as absent ("specificity"). To calculate the different fractions of the confusion matrix, the continuous SDM output should first be converted into a discrete presence/absence prediction based on a threshold value. For an overview of different threshold rules, we refer to the work of Liu et al. (2013). We advocate the use of the "10 percentile training presence threshold." This is a conservative threshold that excludes 10% of the presence records with the lowest probability of occurrence from the predicted presence range and does not rely on absences (which are replaced by pseudo-absences). This threshold accounts for taxonomic misidentifications and georeferencing errors.

From the available threshold-dependent measures of SDM accuracy, Cohen's kappa statistic and true skill statistic (TSS) are widely used (Allouche et al. 2006). Arguably the most widespread, and one of the few threshold independent measures of SDM accuracy, is the area under the curve (AUC) of the receiver operating characteristic (ROC) plot (Hanley & McNeil 1982). The major advantage of the AUC value, in addition to its threshold independence, is that it is relatively insensitive

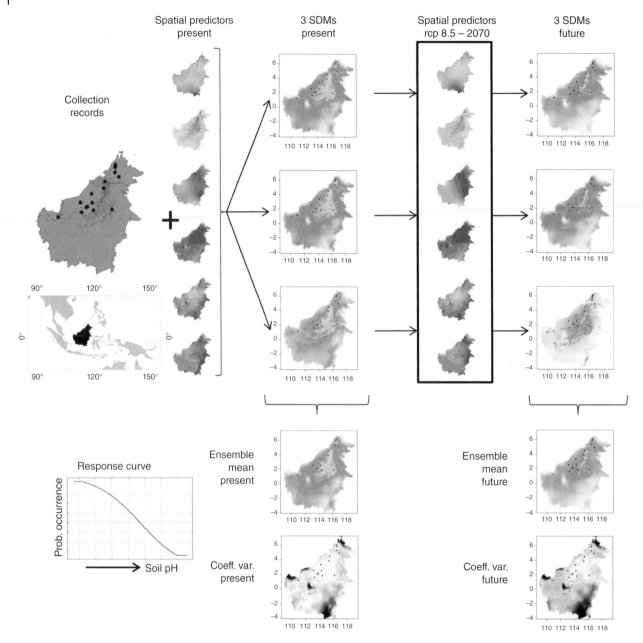

Figure 21.1 Species distribution model (SDM) workflow for *Vaccinium phillyreoides* occurring on Borneo. Collection records (dots) and uncorrelated spatial predictors of present conditions are used to create three different SDMs using different algorithms (Table 21.1); white indicates high probability of occurrence. An ensemble (mean) of the three SDMs shows where the models agree and mapping of the coefficient of variation identifies areas where predictions are least consistent (dark gray). The SDMs are then projected to future climatic conditions (here, scenario RCP8.5, the most pessimistic climate change scenario, where greenhouse gas emissions continue increasing after the year 2100), resulting in three individual future projections. These are assembled in an ensemble mean forecast. The lower left corner shows a response curve of probability of occurrence decreasing with increasing Soil pH. See also Plate 34 in color plate section.

to prevalence (McPherson et al. 2004). The AUC value is a measure of the area under the curve of sensitivity (proportion of correctly predicted presences) plotted against 1-specificity (proportion of correctly predicted absences) for the range of all possible thresholds, and hence is threshold-independent. The AUC value is interpreted as the chance that a randomly drawn presence record has a higher predicted probability of occurrence value than a randomly drawn absence. For SDMs developed with presence and absence data, AUC values >0.7 are generally accepted as useful models, and an AUC value of 1 indicates perfect model fit (Swets et al. 2000).

A major drawback of all measures of SDM accuracy is that they rely on true absences, which are lacking in most cases and are replaced by pseudo-absences or a background sample. Under such conditions, part of the pseudo-absences or background samples are randomly drawn from the species presence area proportional to the species' true presence distribution (prevalence). The maximum AUC value under these conditions is not 1, but $1 - a/2$, where a stands in for the species' true prevalence, which is typically not known (Phillips et al. 2006; Raes & ter Steege 2007). For example, for a species with a prevalence of 0.4, the maximum AUC value is 0.8 ($1 - 0.4/2$). Therefore, measures of SDM accuracy that rely on standard threshold values (e.g., AUC > 0.7) that are calculated using pseudo-absences or a background sample instead of true absences are flawed. It should be noted, however, that when the aim is to compare the performance of different SDM algorithms on the same input data, the SDM with the highest AUC value is the most accurate.

Recognition of this caveat led Raes and ter Steege (2007) to develop a null-model that tests whether SDM accuracy values significantly deviate from random expectation. The procedure is straightforward, and uses a random sample of pseudo-presence records from the study area with the same number of records as was used for the real SDM. This is replicated 999 times. These 999 random sets of pseudo-presence records are modeled in a similar way as the real species, using the same SDM algorithm and abiotic spatial data. The 999 measures of SDM accuracy, plus the one measure of accuracy for the real species, are subsequently ranked from high to low. If the real species' measure of SDM accuracy ranks among the top 5%, then the chance that a random set of presence points can produce an equally good model is less than 5%; hence, significantly better than random expectation. The test can be further improved by drawing the pseudo-presence records from a target group background sample (Phillips et al. 2009). The target group background sample represents all presence records in the study area from species of the same group (e.g., same genus) to which the species being modeled belongs. This procedure also corrects for collection biases (Phillips et al. 2009).

21.4.2 SDM Complexity

Many of the SDM algorithms listed in Table 21.1 can fit very complex relationships between species presence records and spatial predictors. Complex SDMs often have very high model accuracy values but limited predictive power, as a result of model overparameterization or overfitting (Merow et al. 2014). Model overparameterization refers to the inclusion of too many predictor variables relative to the number of presences (and absences), or the inclusion of predictors that do not relate to the ecology of the study species. An overfit SDM is fitted to noise in the presence data, and fails to capture the species' response to environmental gradients. Ecological niche theory suggests that species' response curves are (at least for fundamental niches) often unimodal (Dolédec et al. 2000; Austin 2005, 2007), and hence quadratic responses to environmental gradients may be most appropriate (Figure 21.2) (Merow et al. 2013). When only part of a unimodal response is captured by the study area, a linear response might be sufficient. Threshold responses are appealing when physiological tolerance limits exist, such as a freezing intolerance resulting in predicted presence for areas where the temperature in the coldest month is above 0 °C. Any other modeling rules should only be included based on ecologically motivated reasoning.

21.4.3 Ensemble Models

Different SDM algorithms (Table 21.1), given their statistical assumptions and ways to handle absence data, result in different outputs from the same input data (Figure 21.1). The variation in output can occur not only when comparing different SDM algorithms, but also when comparing models from a single algorithm across multiple cross-validation runs and with different model parameterization. The between and within modeling variability in SDM outputs has been widely documented (Araújo & New 2007; Elith & Graham 2009), and has led to the development of ensemble models (EMs). EMs are rooted in the

Figure 21.2 A linear (light gray) and a unimodal or quadratic response curve (dark gray), covering the present and future/past non-analog range of values (abiotic conditions that are not present in the model training data) of an ecological gradient. Dotted lines represent extrapolation to non-analog future/past conditions. Horizontal lines represent the clamped values (future probability values are set constant at the value of the present range edge). The vertical line represents no extrapolation and the edge of the present range of values.

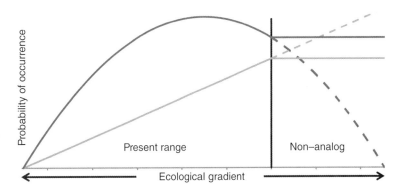

general idea described by Bates and Granger (1969) that, "Given that unknown conditions cannot be exactly predicted, then an ensemble of predictions may render a smaller error than any single prediction." Ensemble modeling can be summarized as a technique that captures the uncertainty in model predictions generated by different SDM algorithms, multiple cross-validation runs or different model parameterizations using a single SDM algorithm. Moreover, EMs may also render more consistent predictions when projecting models from different SDM algorithms to future climate scenarios.

A growing number of studies have compared the outputs of single SDMs against an EM. Aguirre-Gutiérrez et al. (2013) showed that EMs were among the best performing models, consistent across spatial scales, for different prevalence classes (i.e., widely to narrowly distributed species) and for rare to common species. Buisson et al. (2010) showed, using EMs of fish distributions in France, that most of the variation in future climate projections could be attributed to different SDM algorithms, followed by differences in GCMs. Similar conclusions were drawn by Diniz-Filho et al. (2009) for projections of bird SDMs onto future climate projections in South America. Next to differences in the outputs of different SDM algorithms and projections to different global climate change scenarios, differences in ensemble rules also yield different EM outputs.

Currently, the R-libraries "biomod2" (Thuiller et al. 2014), "BiodiversityR" (Kindt & Coe 2005) and "SSDM" (Schmitt et al. 2016) facilitate a semi-automated construction of EMs. First, individual SDMs using different algorithms are constructed and tested for their predictive power: only SDMs above a predefined threshold may be retained. Second, the ensemble rule should be selected. Among the choices are the mean, median and weighted mean. The latter applies a weight to the different SDMs according to the results of individual model evaluations. In this way, better performing SDMs drive the outcome of the final EM. Although it is tempting to present only the final projected EM, this output should be accompanied by a representation of the measure of uncertainty (Thuiller 2014). The uncertainty measure can be obtained by computing, for example, the coefficient of variation of single SDM predictions used to construct the EM (Figure 21.1). The measure of uncertainty indicates where single SDMs differ in their predictions.

21.4.4   The Ecology of Species as Derived from SDMs

Not all selected and uncorrelated abiotic variables contribute equally to an SDM. Often, just two or three variables largely determine the ecology of a species (Aguirre-Gutiérrez et al. 2015). Several methods have been developed to estimate the importance of each environmental variable to the final SDM. The first is a randomization procedure where the values of the variable under investigation are randomly permutated. Subsequently, the Pearson correlation between the predictions of the original SDM and the SDM with one permutated variable is determined. If the correlation is high (i.e., little difference between the two predictions), the permutated variable is considered unimportant for the SDM (Thuiller et al. 2009). This procedure is implemented in the R-library "biomod2" and can be used for all SDM algorithms (Thuiller et al. 2014). Maxent offers a similar permutation methodology, but here the relative drop in AUC value is used to determine the importance of the permutated variable (Phillips et al. 2006). Another method is a jackknife, or leave-one-out, analysis. Each environmental variable is left out of the SDM, and the relative drop in predictive power can be used as a measure of the variable's importance. Alternatively, an SDM can be fitted on a single environmental variable, and the predictive power can be used as a performance indicator for this variable alone.

Once the variables that determine a species' distribution are identified, the responses to these environmental gradients can be analyzed. For that purpose, Elith et al. (2005) developed the "evaluation strip," which allows the user to plot the response curves of an SDM to a single environmental gradient. The evaluation strip consists of generated environmental data, where the range of values of each variable in turn is systematically varied over its range, while all other variables are held constant at their mean (or minimum or maximum). Plotting of the predicted probability of occurrence values on the evaluation strip (response curves) shows how the model responds to increasing values of each variable independently (Figures 21.1 and 21.2). The evaluation strip is included in Maxent and in the R-library "biomod2" (Thuiller et al. 2014). Maxent has one additional type of response curve that is based on the predicted probability of occurrence of an SDM developed with a single environmental predictor, which can be readily plotted.

21.5   Projecting SDMs in Time and Space

When the aim of an SDM is to predict the impact of future climate change on species distributions (time), or to assess the invasive potential of a species in a certain area (space), the created SDM is projected to the future abiotic (often climatic) conditions, or to the new area of interest. In both situations, the data sets can include abiotic conditions that are not present in the training data set, known as "novel" or "non-analog" conditions (Figure 21.2) (Williams & Jackson 2007). Projection of SDMs to novel conditions requires extrapolation of the species' responses along ecological gradients. This is especially troublesome if this involves linear responses (e.g., if the probability of occurrence increases with increasing temperature, in which case extrapolation results in continuous increasing probabilities beyond the present range of temperatures) (Figure 21.2).

Multivariate environmental similarity surfaces (MESS) measure the similarity of any given point to a reference set of points, and allow an analysis of the extent to which future values exceed the present range of values (Elith et al. 2010). Negative values indicate the maximum extension of the present range of values, expressed as a percentage; for example, if present temperatures range from 10–20 °C and the future value at a locality is 25 °C, then the MESS value is $(20-25)/(20-10) \times 100 = -50$. A future temperature of 5 °C results in the same negative MESS value. Future values within the present range have MESS values between 0 and 100. The most negative MESS value across all environmental variables reported for each locality is plotted, resulting in a MESS map. Predicted SDM probabilities of occurrence in areas with highly negative MESS values should be treated with caution. The MESS analysis is implemented in Maxent and in the R-library "dismo" (Hijmans et al. 2015).

Several SDM algorithms allow regulation of the degree of extrapolation to non-analog future values. The first option is not to extrapolate, effectively enforcing a value for the probability of occurrence of 0 at localities with future values exceeding the present range of values (vertical line in Figure 21.2). A second option is to "clamp" the probability of occurrence values beyond the present range of values. This sets the future probability values constant at the value of the present range edge (horizontal lines in Figure 21.2). The third

option is to extrapolate the probability values to non-analog conditions (dashed lines in Figure 21.2).

A final word is dedicated to the dispersal capacity of species to track suitable future abiotic conditions. If a species is a poor disperser, it may not be able to migrate fast enough to keep up with shifting environmental conditions. The R-library "MigClim" (Engler et al. 2012) allows the user to integrate dispersal constraints on future SDM projections. Key to the calibration, however, is to select a realistic dispersal kernel (a probability density function of distance of dispersal), which is difficult because the kernel shape is determined by many variables. Furthermore, it is hard to include rare long-distance dispersal events. Therefore, most future SDM projections use either no dispersal, full dispersal or both.

21.6 Conclusion

SDMs are powerful statistical models that relate species presences to climatic and landscape features in order to predict their distributions and the potential impacts of future and past climatic conditions. We would like to stress that ecological knowledge cannot be disregarded and should be taken into account when selecting the predictors used to model species distributions. A model is only as good as the data used to train it. Although SDM outputs often look beautiful, it is of utmost importance to determine whether they reflect reality.

Acknowledgements

We thank Rafael O. Wüest and Jose A.F. Diniz-Filho for useful comment on the manuscript, and Vitor Freitas for sharing R-scripts.

References

Aguirre-Gutiérrez, J., Carvalheiro, L.G., Polce, C. et al. (2013) Fit-for-purpose: species distribution model performance depends on evaluation criteria – dutch hoverflies as a case study. *PLoS ONE* **8**, e63708.

Aguirre-Gutiérrez, J. & Serna-Chavez, H.M., Villalobos-Arambula, A.R. et al. (2015) Similar but not equivalent: ecological niche comparison across closely-related Mexican white pines. *Diversity and Distributions* **21**, 245–257.

Allouche, O., Tsoar, A. & Kadmon, R. (2006) Assessing the accuracy of species distribution models: prevalence, kappa and the true skill statistic (TSS). *Journal of Applied Ecology* **43**, 1223–1232.

Araújo, M.B. & New, M. (2007) Ensemble forecasting of species distributions. *Trends in Ecology & Evolution* **22**, 42–47.

Araújo, M.B. & Peterson, A.T. (2012) Uses and misuses of bioclimatic envelope modeling. *Ecology* **93**, 1527–1539.

Austin, M.P. (2005) Vegetation and environment: discontinuities and continuities. In: van der Maarel, E. (ed.) *Vegetation Ecology*. Hoboken, NJ: Blackwell Science, pp. 52–84.

Austin, M.P. (2007) Species distribution models and ecological theory: a critical assessment and some possible new approaches. *Ecological Modelling* **200**, 1–19.

Bateman, B.L., Murphy, H.T., Reside, A.E. et al. (2013) Appropriateness of full-, partial- and no-dispersal scenarios in climate change impact modelling. *Diversity and Distributions* **19**, 1224–1234.

Bates, J.M. & Granger, C.W. (1969) The combination of forecasts. *Or* **20**, 451–468.

Blonder, B., Lamanna, C., Violle, C. et al. (2014) The n-dimensional hypervolume. *Global Ecology and Biogeography* **23**, 595–609.

Boucher-Lalonde, V., Morin, A. & Currie, D.J. (2012) How are tree species distributed in climatic space? A simple and general pattern. *Global Ecology and Biogeography* **21**, 1157–1166.

Boucher-Lalonde, V., Morin, A. & Currie, D.J. (2014) A consistent occupancy–climate relationship across birds and mammals of the Americas. *Oikos* **123**, 1029–1036.

Boucher-Lalonde, V., Morin, A. & Currie, D.J. (2016) Can the richness–climate relationship be explained by systematic variations in how individual species' ranges relate to climate? *Global Ecology and Biogeography* **25**, 527–539.

Boulangeat, I., Gravel, D. & Thuiller, W. (2012) Accounting for dispersal and biotic interactions to disentangle the drivers of species distributions and their abundances. *Ecology Letters* **15**, 584–593.

Boyle, B., Hopkins, N., Lu, Z. et al. (2013) The taxonomic name resolution service: an online tool for automated standardization of plant names. *BMC Bioinformatics* **14**, 16.

Breiman, L. (2001) Random forests. *Machine Learning* **45**, 5–32.

Breiman, L., Friedman, J.H., Olshen, R.A. et al. (1984) *Classification and Regression Trees*. Pacific Grove, CA: Wadsworth.

Broennimann, O., Treier, U.A., Muller-Scharer, H. et al. (2007) Evidence of climatic niche shift during biological invasion. *Ecology Letters* **10**, 701–709.

Buisson, L., Thuiller, W., Casajus, N. et al. (2010) Uncertainty in ensemble forecasting of species distribution. *Global Change Biology* **16**, 1145–1157.

Busby, J.R. (1991) BIOCLIM – a bioclimate analysis and prediction system. In: Margules, R. & Austin, M.P. (eds.) *Nature Conservation: Cost Effective Biological Surveys and Data Analysis*. Canberra: CSIRO, pp. 64–68.

Calenge, C., Darmon, G., Basille, M. et al. (2008) The factorial decomposition of the Mahalanobis distances in habitat selection studies. *Ecology* **89**, 555–566.

Carpenter, G., Gillison, A.N. & Winter, J. (1993) DOMAIN: a flexible modelling procedure for mapping potential distributions of plants and animals. *Biodiversity and Conservation* **2**, 667–680.

Colwell, R.K. & Rangel, T.F. (2009) Hutchinson's duality: the once and future niche. *Proceedings of the National Academy of Sciences* **106**, 19 651–19 658.

De'ath, G. & Fabricius, K.E. (2000) Classification and regression trees: a powerful yet simple technique for eological data analysis. *Ecology* **81**, 3178–3192.

Diniz-Filho, J.A.F., Bini, L.M., Rangel, T.F. et al. (2009) Partitioning and mapping uncertainties in ensembles of forecasts of species turnover under climate change. *Ecography* **32**, 897–906.

Dolédec, S., Chessel, D. & Gimaret-Carpentier, C. (2000) Niche separation in community analysis: a new method. *Ecology* **81**, 2914–2927.

Elith, J. & Graham, C.H. (2009) Do they? How do they? WHY do they differ? On finding reasons for differing performances of species distribution models. *Ecography* **32**, 66–77.

Elith, J. & Leathwick, J. (2007) Predicting species distributions from museum and herbarium records using multiresponse models fitted with multivariate adaptive regression splines. *Diversity and Distributions* **13**, 265–275.

Elith, J., Ferrier, S., Huettmann, F. et al. (2005) The evaluation strip: a new and robust method for plotting predicted responses from species distribution models. *Ecological Modelling* **186**, 280–289.

Elith, J., Graham, C.H., Anderson, R.P. et al. (2006) Novel methods improve prediction of species' distributions from occurrence data. *Ecography* **29**, 129–151.

Elith, J., Leathwick, J.R. & Hastie, T. (2008) A working guide to boosted regression trees. *Journal of Animal Ecology* **77**, 802–813.

Elith, J., Kearney, M. & Phillips, S. (2010) The art of modelling range-shifting species. *Methods in Ecology and Evolution* **1**, 330–342.

Dormann, C.F., Elith, J., Bacher, S. et al. (2013) Collinearity: a review of methods to deal with it and a simulation study evaluating their performance. *Ecography* **36**, 27–46.

Engler, R., Hordijk, W. & Guisan, A. (2012) The MIGCLIM R package – seamless integration of dispersal constraints into projections of species distribution models. *Ecography* **35**, 872–878.

FAO/IIASA/ISRIC/ISSCAS/JRC (2012) Harmonized World Soil Database, version 1.2. Available from: http://webarchive.iiasa.ac.at/Research/LUC/External-World-soil-database/HTML/ (last accessed August 30, 2017).

Ferrier, S., Manion, G., Elith, J. et al. (2007) Using generalized dissimilarity modelling to analyse and predict patterns of beta diversity in regional biodiversity assessment. *Diversity and Distributions* **13**, 252–264.

Fielding, A.H. & Bell, J.F. (1997) A review of methods for the assessment of prediction errors in conservation presence/absence models. *Environmental Conservation* **24**, 38–49.

Franklin, J. (2009) *Mapping Species Distributions: Spatial Inference and Prediction*. Cambridge: Cambridge University Press.

Giannini, T.C., Chapman, D.S., Saraiva, A.M. et al. (2013) Improving species distribution models using biotic interactions: a case study of parasites, pollinators and plants. *Ecography* **36**, 649–656.

Goodwin, Z.A., Harris, D.J., Filer, D. et al. (2015) Widespread mistaken identity in tropical plant collections. *Current Biology* **25**, R1066–R1067.

Guisan, A., Thuiller, W. & Zimmermann, N.E. (2017) *Habitat suitability and distribution models; with applications in R.* Cambridge: Cambridge University Press.

Guo, Q., Kelly, M. & Graham, C.H. (2005) Support vector machines for predicting distribution of Sudden Oak Death in California. *Ecological Modelling* **182**, 75–90.

Hanley, J.A. & McNeil, B.J. (1982) The meaning and use of the area under a receiver operating characteristic (ROC) curve. *Radiology* **143**, 29–36.

Hannemann, H., Willis, K.J. & Macias-Fauria, M. (2016) The devil is in the detail: unstable response functions in species distribution models challenge bulk ensemble modelling. *Global Ecology and Biogeography* **25**, 26–35.

Hastie, T. & Tibshirani, R. (1986) Generalized additive models. *Statistical Science* **1**, 297–310.

Hastie, T., Tibshirani, R. & Buja, A. (1994) Flexible discriminant analysis by optimal scoring. *Journal of the American Statistical Association* **89**, 1255–1270.

Hengl, T., de Jesus, J.M., MacMillan, R.A. et al. (2014) SoilGrids1km – global soil information based on automated mapping. *PLoS ONE* **9**, e105992.

Hijmans, R. & Elith, J. (2013). Species distribution modeling with R. *Encyclopedia of Biodiversity* **6**, 10.1016/B978-0-12-384719-5.00318-X.

Hijmans, R.J., Cameron, S.E., Parra, J.L. et al. (2005) Very high resolution interpolated climate surfaces for global land areas. *International Journal of Climatology* **25**, 1965–1978.

Hijmans, R.J., Phillips, S., Leathwick, J. et al. (2015) Dismo: Species Distribution Modeling. R package version 1.0–12.

Hilbert, D.W. & Ostendorf, B. (2001) The utility of artificial neural networks for modelling the distribution of vegetation in past, present and future climates. *Ecological Modelling* **146**, 311–327.

Hirzel, A.H., Hauser, J., Chessel, D. et al. (2002) Ecological-niche factor analysis: how to compute habitat-suitability maps without absence data. *Ecology* **83**, 2027–2036.

Hortal, J., Bello, F.d., Diniz-Filho, J.A.F. et al. (2015) Seven shortfalls that beset large-scale knowledge of biodiversity. *Annual Review of Ecology, Evolution, and Systematics* **46**, 523–549.

Hutchinson, G.E. (1957) Concluding remarks. Cold Spring Harbor Symposia on Quantitative Biology, pp. 415–427.

IPCC (2013) *Climate Change 2013: The Physical Science Basis. Contribution of Working Group I to the Fifth Assessment Report of the Intergovernmental Panel on Climate Change.* Cambridge: Cambridge University Press.

Kindt, R. & Coe, R. (2005) *Tree Diversity Analysis. A Manual and Software for Common Statistical Methods for Ecological and Biodiversity Studies.* Nairobi: World Agroforestry Centre (ICRAF).

Körner, C. (2007) The use of "altitude" in ecological research. *Trends in Ecology & Evolution* **22**, 569–574.

Kriticos, D.J., Webber, B.L., Leriche, A. et al. (2012) CliMond: global high-resolution historical and future scenario climate surfaces for bioclimatic modelling. *Methods in Ecology and Evolution* **3**, 53–64.

Lee-Yaw, J.A., Kharouba, H.M., Bontrager, M. et al. (2016) A synthesis of transplant experiments and ecological niche models suggests that range limits are often niche limits. *Ecology Letters* **19**, 710–722.

Lima-Ribeiro, M.S., Varela, S., González-Hernández, J. et al. (2015) EcoClimate: a database of climate data from multiple models for past, present, and future for macroecologists and biogeographers. *Biodiversity Informatics* **10**, 1–21.

Liu, C., White, M. & Newell, G. (2013) Selecting thresholds for the prediction of species occurrence with presence-only data. *Journal of Biogeography* **40**, 778–789.

Maldonado, C., Molina, C.I., Zizka, A. et al. (2015) Estimating species diversity and distribution in the era of Big Data: to what extent can we trust public databases? *Global Ecology and Biogeography* **24**, 973–984.

Martinuzzi, S., Radeloff, V.C., Joppa, L.N. et al. (2015) Scenarios of future land use change around United States' protected areas. *Biological Conservation* **184**, 446–455.

McCullagh, P. & Nelder, J.A. (1989) *Generalized Linear Models.* Boca Raton, FL: CRC Press.

McPherson, J.M., Jetz, W. & Rogers, D.J. (2004) The effects of species' range sizes on the accuracy of distribution models: ecological phenomenon or statistical artefact? *Journal of Applied Ecology* **41**, 811–823.

Merow, C., Smith, M.J. & Silander, J.A. (2013) A practical guide to MaxEnt for modeling species' distributions: what it does, and why inputs and settings matter. *Ecography* **36**, 1058–1069.

Merow, C., Smith, M.J., Edwards, T.C. et al. (2014) What do we gain from simplicity versus complexity in species distribution models? *Ecography* **37**, 1267–1281.

Miller, J.A. & Holloway, P. (2015) Incorporating movement in species distribution models. *Progress in Physical Geography* **39**, 837–849.

O'brien, R.M. (2007) A caution regarding rules of thumb for variance inflation factors. *Quality & Quantity* **41**, 673–690.

Pearson, R.G., Raxworthy, C.J., Nakamura, M. et al. (2007) Predicting species distributions from small numbers of occurrence records: a test case using cryptic geckos in Madagascar. *Journal of Biogeography* **34**, 102–117.

Peterson, A.T., Soberón, J., Pearson, R.G. et al. (2011) *Ecological Niches and Geographic Distributions.* Princeton, NJ: Princeton University Press.

Phillips, S.J., Anderson, R.P. & Schapire, R.E. (2006) Maximum entropy modeling of species geographic distributions. *Ecological Modelling* **190**, 231–259.

Phillips, S.J., Dudík, M., Elith, J. et al. (2009) Sample selection bias and presence-only distribution models: implications for background and pseudo-absence data. *Ecological Applications* **19**, 181–197.

Qiao, H., Soberón, J. & Peterson, T.A. (2015) No silver bullets in correlative ecological niche modeling: Insights from testing among many potential algorithms for niche estimation. *Methods in Ecology and Evolution* **6**, 1126–1136.

R Development Core Team (2014) *R: A Language and Environment for Statistical Computing.* Vienna: R Foundation for Statistical Computing.

Raes, N. (2012) Partial versus full species distribution models. *Natureza & Conservação* **10**, 127–138.

Raes, N., Cannon, C.H., Hijmans, R.J. et al. (2014) Historical distribution of Sundaland's Dipterocarp rainforests at Quaternary glacial maxima. *Proceedings of the National Academy of Sciences* **111**, 16790–16795.

Raes, N. & ter Steege, H. (2007) A null-model for significance testing of presence-only species distribution models. *Ecography* **30**, 727–736.

Ridgeway, G. (1999) The state of boosting. *Computing Science and Statistics* **31**, 172–181.

Rotenberry, J.T., Preston, K.L. & Knick, S.T. (2006) GIS-based niche modeling fro mapping species' habitat. *Ecology* **87**, 1458–1464.

Schmitt, S., Pouteau, R., Justeau, D. et al. (2016) SSDM: Stacked Species Distribution Modelling. R package version 0.1.1.

Soberón, J. & Nakamura, M. (2009) Niches and distributional areas: concepts, methods, and assumptions. *Proceedings of the National Academy of Sciences* **106**, 19644–19650.

Soberón, J. & Peterson, A.T. (2005) Interpretation of models of fundamental ecological niches and species' distributional areas. *Biodiversity Informatics* **2**, 1–10.

Stockwell, D. & Peters, D. (1999) The GARP modelling system: problems and solutions to automated spatial prediction. *International Journal of Geographical Information Science* **13**, 143–158.

Swets, J.A., Dawes, R.M. & Monahan, J. (2000) Better decisions through science. *Scientific American* **283**, 82–87.

Taylor, K.E., Stouffer, R.J. & Meehl, G.A. (2012) An overview of CMIP5 and the experiment design. *Bulletin of the American Meteorological Society* **93**, 485–498.

Thomas, C.D. (2010) Climate, climate change and range boundaries. *Diversity and Distributions* **16**, 488–495.

Thuiller, W. (2014) Editorial commentary on "BIOMOD – optimizing predictions of species distributions and projecting potential future shifts under global change." *Global Change Biology* **20**, 3591–3592.

Thuiller, W., Lafourcade, B., Engler, R. et al. (2009) BIOMOD – a platform for ensemble forecasting of species distributions. *Ecography* **32**, 369–373.

Thuiller, W., Lavergne, S., Roquet, C. et al. (2011) Consequences of climate change on the tree of life in Europe. *Nature* **470**, 531–534.

Thuiller, W., Georges, D. & Engler, R. (2014) Biomod2: ensemble platform for species distribution modeling. R package version 3.1-64.

Thuiller, W., Pollock, L.J., Gueguen, M. et al. (2015) From species distributions to meta-communities. *Ecology Letters* **18**, 1321–1328.

Töpel, M., Zizka, A., Calió, M.F. et al. (2017) SpeciesGeoCoder: Fast Categorization of Species Occurrences for Analyses of Biodiversity, Biogeography, Ecology, and Evolution. *Systematic Biology* **66**, 145–151.

van Proosdij, A.S.J., Sosef, M.S.M., Wieringa, J.J. et al. (2016) Minimum required number of specimen records to develop accurate species distribution models. *Ecography* **39**, 542–552.

Varela, S., Lima-Ribeiro, M.S. & Terribile, L.C. (2015) A short guide to the climatic variables of the Last Glacial Maximum for biogeographers. *PLoS ONE* **10**, e0129037.

Vasudev, D., Fletcher, R.J., Goswami, V.R. et al. (2015) From dispersal constraints to landscape connectivity: lessons from species distribution modeling. *Ecography* **38**, 967–978.

Venables, W.N. & Ripley, B.D. (2002) *Modern Applied Statistics with S-PLUS*, 4th edn. New York: Springer.

Vollering, J., Schuiteman, A., de Vogel, E. et al. (2016) Phytogeography of New Guinean orchids: patterns of species richness and turnover. *Journal of Biogeography* **43**, 204–214.

Walker, P.A. & Cocks, K.D. (1991) HABITAT: a procedure for modelling a disjoint environmental envelope for a plant or animal species. *Global Ecology and Biogeography Letters* **1**, 108–118.

Waltari, E., Hijmans, R.J., Peterson, A.T. et al. (2007) Locating pleistocene refugia: comparing phylogeographic and ecological niche model predictions. *PLoS ONE* **2**, e563.

Williams, J.W. & Jackson, S.T. (2007) Novel climates, no-analog communities, and ecological surprises. *Frontiers in Ecology and the Environment* **5**, 475–482.

Yee, T.W. & Mitchell, N.D. (1991) Generalized additive models in plant ecology. *Journal of Vegetation Science* **2**, 587–602.

Part III

Mountains and Biota of the World

22

Evolution of the Isthmus of Panama: Biological, Paleoceanographic and Paleoclimatological Implications

Carlos Jaramillo

Smithsonian Tropical Research Institute, Panama, Republic of Panama

Abstract

The rise of the Isthmus of Panama was the product of small-scale geological processes that, nonetheless, had worldwide repercussions. The building of the Panamanian landscape can be summarized in four phases: (i) the emergence of a large, late Eocene island across central Panama and the Azuero Peninsula; (ii) the early Miocene large-scale generation of terrestrial landscapes in Central America, which connected central Panama with North America; (iii) the full closure of the Central American Seaway (CAS) at 10 Ma, which interrupted the exchange of deep waters between the Caribbean and the Pacific, and generated most of the landscape across the Isthmus (exchange of shallow waters continued until 3.5 Ma, albeit intermittently); and (iv) the persistence of a terrestrial landscape across the Isthmus over the past 3.5 My. Four major events have been linked to the rise of the Isthmus: (i) the onset of thermohaline circulation (TCH); (ii) the onset of northern hemisphere glaciation (NHG); (iii) the birth of the Caribbean Sea; and (iv) the Great American Biotic Interchange (GABI). The available evidence indicates a strong link between the closure of the CAS and the onset of Atlantic meridional overturning circulation (AMOC, a precursor of THC), but at 10.0 rather than 3.5 Ma, as had been assumed. There is no evidence of a connection between the full emergence of the Isthmus at 3.5 Ma and the onset of the NHG. There is, however, strong evidence that the full emergence of the Isthmus at 3.5 Ma changed the oceanography of the Caribbean Sea to its modern conditions, although the role of other variables influencing Pleistocene Caribbean Sea conditions, including the changes in the Pleistocene climate and the cessation of the freshwater flow of several South American rivers into the Caribbean, still needs to be evaluated. The GABI is more complex than is often assumed, and it seems that variables other than a continuous terrestrial Isthmus have controlled the direction, timing and speed of migrations in both directions.

Keywords: *biogeography, phylogeny, oceanography, biological invasion, biome, neotropics*

22.1 Introduction

Southern Caribbean geology is a very complex product of three colliding tectonic plates and their interactions since the Late Cretaceous (e.g., Wadge & Burke 1983; Adamek et al. 1988; Mann & Corrigan 1990; Pindell & Barret 1990; Kolarsky et al. 1995; Mann & Kolarsky 1995; Coates et al. 2004; Rockwell et al. 2010; Montes et al. 2012a; Barat et al. 2014). These geological processes drive both the generation and the movement of terrestrial landmasses, which in turn control the distribution, migration and, ultimately, the evolution of terrestrial biotas. One of most interesting geological events in the southern Caribbean has been the rise of the Isthmus of Panama, also simply called the Isthmus. Even though the

geological build-up of Panama occurred over a small spatial scale, it produced major paleoceanographic, climatic, biogeographic and evolutionary changes that have attracted the attention of researchers for a long time.

Over 2000 publications (compiled in Bacon et al. 2015a) have cited an age of 4.2–3.5 Ma for the rise of the Isthmus. This range was derived from a number of geological (Coates et al. 1992, 2003, 2004, 2005; Coates & Obando 1996) and paleoceanographic (Haug & Tiedemann 1998; Haug et al. 2001) studies. Four major events have been linked to the rise to the Isthmus: (i) the onset of thermohaline circulation (THC); (iii) the onset of northern hemisphere glaciation (NHG); (iii) the formation of the Caribbean Sea; and (iv) the onset of the Great American Biotic

Mountains, Climate and Biodiversity, First Edition. Edited by Carina Hoorn, Allison Perrigo and Alexandre Antonelli.
© 2018 John Wiley & Sons Ltd. Published 2018 by John Wiley & Sons Ltd.
Companion website: www.wiley.com/go/hoorn/mountains,climateandbiodiversity

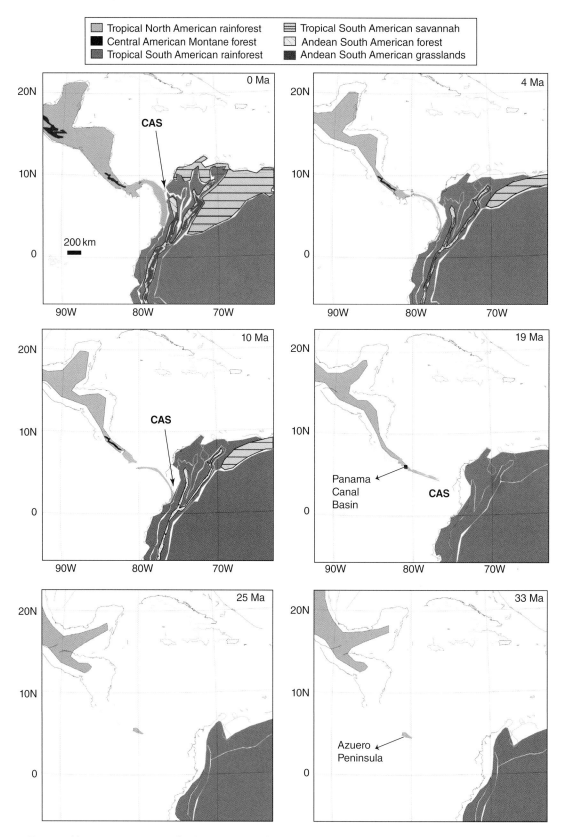

Figure 22.1 Terrestrial biome reconstruction for the past 36 My of the Isthmus of Panama. The reconstruction is an orthographic projection based on the plate tectonic model of GPlates 1.5.0, using the plate reconstruction of Seton et al. (2012). Terrestrial biomes include the tropical rainforest from South America, the tropical rainforest of North America (corresponding to the Central American rainforest), the Central American montane forests (forest >2000 m elevation), the Andean South American forest (forest >2000 m elevation) and the Andean South American grasslands (grasslands above the tree line in the Andes of South America). The first development of terrestrial landscape occurred during the late Eocene, as an island spanning from the present site of central Panama to

Interchange (GABI). Some of these links have been heavily criticized, however (Molnar 2008).

In late 2007, Panama started an 8-year construction effort to expand the Panama Canal, during which numerous new outcrops were exposed, providing an unparalleled opportunity for additional studies on the timing and dynamics of the Isthmus' formation. This set of new data has revised our understanding of the rise of the Isthmus and its consequences, which will be explored in this chapter.

22.2 A Brief History of the Isthmus Landscape Construction

The Isthmus is predominantly the product of three events of arc magmatism build-up on the trailing edge of the Caribbean Plate during the late Cretaceous–Cenozoic, together with the successive accretion of Pacific-born sea-mounts. These arcs are deformed by collisions with both the South American Plate and oceanic plateaus (e.g., Wadge & Burke 1983; Mann & Corrigan 1990; Coates et al. 2004; Rockwell et al. 2010; Buchs et al. 2011; Wegner et al. 2011; Montes et al. 2012a,b).

During the early–middle Eocene (50–43 Ma) (Montes et al. 2012b; Ramirez et al. 2016), a large exhumation pulse occurred, which was probably the result of the accretion of intraplate Pacific volcanoes to the southern boundary of the Caribbean plate (Buchs et al. 2011). This accretion generated the first terrestrial land in Panama during the late Eocene: a large island that extended from the present-day Azuero Peninsula to central Panama (Figures 22.1 and 22.2). There are extensive fossil wood remains in the Tonosi Formation of the Azuero Peninsula, dated as late Eocene (Baumgartner-Mora et al. 2008), that include large Arecaceae and Humiriaceae trees (Stern & Eyde 1963), as well as fruits and seeds of Arecaceae, Vitaceae, Humiriacae, Anacardiaceae and Lamiales (Herrera et al. 2012b). The landscape extended into central Panama, where deposits of the late Eocene

Gatuncillo Formations have yielded a rich palynoflora that indicates a tropical rainforest biome (Graham 1984, 1985, 1994).

A second major phase of terrestrial landscape development along the Isthmus occurred during the early Miocene (~21 Ma), probably derived from an extensive exhumation pulse during the late Oligocene–early Miocene (25–22 Ma) produced by the onset of the tectonic collision between South America and Panama (Farris et al. 2011). There are several deposits in central Panama, including the Cascadas, Cucarachas and Pedro Miguel formations, that indicate the presence of a terrestrial landscape since ~21 Ma (Figures 22.1 and 22.2) (Montes et al. 2012a,b). The North American-derived mammal fauna of the Cascadas and Cucaracha formations indicates a continuous landscape from central Panama to North America across Central America (Kirby & MacFadden 2005; MacFadden 2006b, 2009; Kirby et al. 2008; MacFadden et al. 2010, 2014) made up of biomes that were dominated by tropical rainforest (Jaramillo et al. 2014). The width of the Central American Seaway (CAS) at this time has been estimated at ~200 km (Montes et al. 2012b).

A third major phase occurred during the late Miocene (~12–10 Ma) (Figures 22.1 and 22.2). There was an extensive exhumation pulse during the late–middle Miocene to the late Miocene (15–10 Ma) (Coates et al. 2004; Montes et al. 2012b). This pulse was produced by the collision of the Panamanian Volcanic Arc with South America, which resulted in the closure of the CAS (Montes et al. 2015). The widespread effects of this collision can be clearly seen in north-eastern Colombia (Vargas & Mann 2013). It was also concomitant with the onset of subduction in Costa Rica and eastern Panama of the oceanic crust that originally formed above the Galápagos mantle plume, resulting in the production of new continental crust and widespread exhumation in Central America (Gazel et al. 2015). This drastic change in the configuration of the Panamanian landscape has also been recognized by previous studies (Duque 1990; Duque-Caro 1990a,b; Coates et al. 1992, 2003, 2004). Most of the landscape was dominated by a tropical

Figure 22.1 (Continued) the Azuero Peninsula. A second major build-up of the terrestrial landscape occurred during the early Miocene. A third occurred during the late Miocene. The Central American Seaway (CAS; defined as the ocean gap along the tectonic boundary between the South American plate and the Panama microplate) was closed by 10 Ma. From 10.0 to 3.5 Ma, there were intermittent Caribbean–Pacific connections through pathways other than the CAS. At 3.5 Ma, there was a complete closure of the Isthmus. A movie of the GPlates landscape reconstruction can be found in online Appendix 22.1 (www.wiley.com\go\hoorn\mountains, climateandbiodiversity). *Source:* Terrestrial biomes adapted from Jaramillo & Cardenas (2013), the exhumation evolution of the Isthmus of Panama from the Montes models (Farris et al. 2011; Montes et al. 2012a,b, 2015), the terrestrial plant fossil record of the region from Stern & Eyde (1963), Jaramillo et al. (2006, 2014), Herrera et al. (2010, 2012a,b, 2014a,b), Jaramillo & Cardenas (2013) and Rodriguez-Reyes et al. (2014), and the terrestrial vertebrate record from (Whitmore & Stewart 1965; Slaughter 1981; MacFadden & Higgins 2004; MacFadden 2006a,b, 2009, 2010; Cadena et al. 2012; Head et al. 2012; MacFadden et al. 2012; Rincon et al. 2012, 2013; Hastings et al. 2013). See also Plate 35 in color plate section.

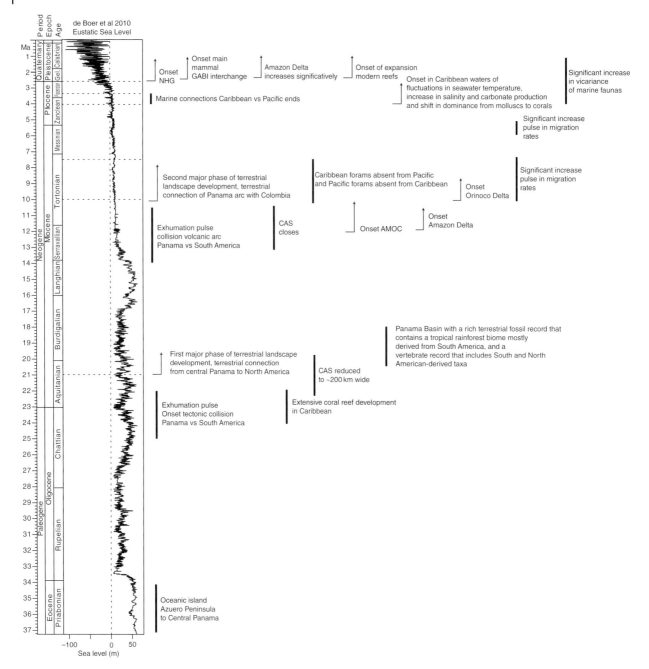

Figure 22.2 Summary of the main events discussed in the text. *Source:* Adapted from Walker et al. (2012) and de Boer et al. (2010).

rainforest biome (Jaramillo et al. 2014), and there is no evidence of extensive savannahs.

From 10.0 to 3.5 Ma, the Caribbean Sea and Pacific Ocean were still connected by shallow water through other passages than the CAS (Figures 22.1 and 22.2) (Coates et al. 2004), although these connections were intermittent, as seen in the 10.0–7.5 Ma interval when Caribbean foraminifera faunas are absent from Pacific deposits in the Atrato Basin (Duque-Caro 1990b) and Pacific faunas are mostly absent from Caribbean

deposits (Collins et al. 1996). The evolution of Tonnoid gastropods also indicates intermittent marine connections during this interval (Beu 2001). Most of the landscape was dominated by a tropical rainforest biome (Jaramillo et al. 2014).

All marine connections between the Caribbean and Pacific finally ended in the interval between 4.2 and 3.5 Ma (Haug & Tiedemann 1998), and since 3.5 Ma there has been a continuous terrestrial landscape across the Isthmus of Panama (Figures 22.1 and 22.2).

22.3   Thermohaline Circulation

Numerous paleoclimatic and paleoceanographic studies have addressed the implications of the rise of the Isthmus, from both modeling and empirical perspectives. More than 24 different modeling exercises have explored the consequences of the rise (Maier-Raimer et al. 1990; Mikolajewicz et al. 1993; Mikolajewicz & Crowley 1997; Murdock et al. 1997; Nisancioglu et al. 2003; Nof & van Gorder 2003; Prange & Schulz 2004; van der Heydt & Dijkstra 2005, 2006; Klocker et al. 2005; Schneider & Schmittner 2006; Steph et al. 2006a,b, 2010; Lunt et al. 2008a,b; Vizcaíno et al. 2010; Butzin et al. 2011; Zhang et al. 2012; Fedorov et al. 2013, 2015; Fyke et al. 2015; Brierley & Fedorov 2016), most of which are summarized in Sepulchre et al. (2014). Although models differ in the assumed width and depth of the CAS, almost all of them produce similar results. Once the CAS is closed, the salinity of the Atlantic Ocean increases and Atlantic Meridional Overturning Circulation (AMOC) is greatly enhanced. Sepulchre et al. (2014) found that the CAS depth is critical for paleoceanographic purposes. Once the flow of deep and intermediate waters (>200 – 500 m depth) of the Pacific Ocean into the Caribbean is interrupted, AMOC is enhanced. Additional reduction of shallow water flow across the Isthmus does not greatly affect Atlantic Ocean circulation.

Empirical evidence indicates that AMOC is much older than 3.5 Ma, and that it has become significantly stronger since the late Miocene, 12–10 Ma (Poore et al. 2006). Bell et al. (2015) showed that the North Atlantic Deep Water, a key component of AMOC, was vigorous prior to 4.7 Ma, and that during the 4.2–3.5 Ma interval there were no major changes in its strength. Therefore, the modeling/empirical results contradict the assumption of a 3.5 Ma closure of the CAS.

New geological results indicate that the CAS started to close by the late–middle Miocene (~14 Ma), and was fully closed by the late Miocene (10 Ma) (Figures 22.1 and 22.2) (Montes et al. 2012a, 2015). This full closure of the CAS (as it is defined here; see Glossary) has been previously recognized (Coates et al. 2004). The closure ceased most of the flow of deep and intermediate waters from the Pacific into the Caribbean along the CAS, as registered by the neodymium record of the Caribbean (Sepulchre et al. 2014). This 12–10 Ma major uplift of the Isthmus has also been previously recognized, in multiple studies (Bandy & Casey 1973; McDougall 1985; Duque-Caro 1990b; McNeill et al. 2000; Harmon 2005). Exchange of shallow waters continued across the Isthmus along pathways other than the CAS until complete closure at 3.5 Ma, but this exchange was not sufficient to alter AMOC significantly.

This new geological evidence allows for the reconciliation of the empirical evidence of the link between AMOC and the closure of the CAS, as most modeling exercises have predicted – not at 3.5 Ma, as has been traditionally accepted, but rather at 12–10 Ma.

22.4   Northern Hemisphere Glaciation

It has been argued that the closure of the CAS at 4.2–3.5 Ma resulted in the onset of NHG at 2.7 Ma (Haug & Tiedemann 1998; Haug et al. 2004). However, this hypothesis has been heavily criticized based on both the empirical data available and climate modeling (Molnar 2008; Fedorov et al. 2013). Some models indicate that closing the CAS does not have a significant effect either on Arctic ice volume (Vizcaíno et al. 2010) or on any of the major climatic changes that occurred at the end of the Pliocene, 2.7 Ma (Fedorov et al. 2013, 2015; Brierley & Fedorov 2016). The climate of the last 2.7 My (Quaternary) is characterized by the combination of four major elements, including a steep temperature gradient from the tropical zone to the poles, a temperature gradient from the western to the eastern equatorial Pacific, fluctuating CO_2 levels reaching below 200 ppm and permanent ice caps on both poles (Fedorov et al. 2013). The coincidence of these four features at ~2.7 Ma represents a major shift in Earth's climate, the causes of which are still a major unresolved question and the object of intensive research (Fedorov et al. 2015). It seems, however, that neither the closure of the CAS at 12–10 Ma nor the full closure of the Isthmus at 4.2–3.5 Ma had a major influence in the onset of NHG or the climate of the Quaternary.

22.5   The Caribbean Sea

The modern Caribbean Sea is characterized by high salinities, warm waters experiencing low temperature seasonality, low productivity and extensive reef development, in contrast to the western Pacific of Panama, which is characterized by low salinities, interannual and seasonal variations in temperature, high productivity and relatively low reef development (O'Dea et al. 2007). Abundant empirical evidence indicates that several changes occurred in the Caribbean during the 4.2–3.5 Ma interval, including a decrease in the fluctuations of Caribbean seawater temperature, an increase in salinity, an increase in carbonate production, a shift in dominance in benthic ecosystems from molluscs to reef corals and a more pronounced difference in taxonomic composition between shallow marine faunas from the Caribbean versus the Pacific Ocean (Keigwin 1982; Coates et al. 1992, 2003, 2004, 2005; Collins et al. 1996; Haug & Tiedemann 1998; Collins & Coates 1999; McNeill et al. 2000; Haug et al. 2001; Kirby & Jackson 2004; Steph et al. 2006b; Jain & Collins 2007; O'Dea et al. 2007; Groeneveld et al. 2008; Jackson & O'Dea 2013). All of these changes have been interpreted as indicating the final closure of the Isthmus and the cessation of shallow-water exchange between the Pacific and Caribbean water masses. This interpretation has not been challenged by any of the more recent studies on the geology of the Isthmus (Farris et al. 2011; Montes et al. 2012a,b, 2015), although a recent paleoceanographic study has proposed that the differences in the Caribbean versus eastern tropical Pacific salinity could be consistent with the transition from permanent El Niño-like conditions of the early Pliocene to La Niña-like modern conditions (Mestas-Nuñez & Molnar 2014).

However, two additional variables that could affect the climate and ocean of the Caribbean region still need to be investigated. The first is the reorganization of South American fluvial drainages, a byproduct of the Andean uplift. Modern freshwater fluvial discharge into the southern Caribbean is ~700 km^3/y (Restrepo et al. 2014), equivalent to about 16% of the Orinoco (830 km^3/y) and Amazon (3420 km^3/y) fluvial discharges combined. During the Miocene, however, most of the tropical South American drainages flowed into the southern Caribbean. Following the rise of the Andes, rivers shifted into their present configuration (Hoorn 1994; Hoorn et al. 1995, 2010a,b). The formation of the Amazon delta had two main stages: 9.4–2.4 Ma, when sedimentation rates were relatively low (0.05–0.30 m/ka), and 2.4 Ma to modern times, when sedimentation rates have been very high (1.22 m/ka) (Figueiredo et al. 2009; Hoorn et al. 2017). The Orinoco (or proto-Orinoco) delta, located in western Venezuela and eastern Colombia, drained into the southern Caribbean. At ~10 Ma, it shifted to its modern delta, which drains into the Atlantic (Diaz de Gamero 1996). What were the consequences in the southern Caribbean Sea of this reduction of freshwater discharge? Preliminary studies indicate that the Caribbean Colombia Basin would have had a lower salinity before the shifting of South America's major drainages into the Atlantic (P. Sepulchre, pers. comm.).

The second variable that needs to be investigated is the extensive coral reef development across the Caribbean during the late Oligocene, 24–22 Ma (e.g., Johnson 2008; Johnson et al. 2009). At this time, the CAS was deep and ~200 km wide (Farris et al. 2011; Montes et al. 2012b). Why did reefs develop in the Caribbean if there was massive water exchange across the CAS between the Pacific and Caribbean? Why didn't they develop there during the late Miocene (~10.0–7.5 Ma), when there was a temporary disconnection between Caribbean and Pacific

waters (Duque-Caro 1990b; Collins et al. 1996)? This suggests that there are factors other than the rise of the Isthmus, such as slope and upwelling intensity, that controlled the development of reefs in the Caribbean (e.g., van der Heydt & Dijkstra 2005; Maier et al. 2007; Klaus et al. 2011). Indeed, modern reefs have developed extensively only since the onset of the NHG, as the sharp drops in sea level produced a change of low-angle ramps into steep-dipping shelf margins, thereby shifting coral growth toward shallower environments conducive to the large, fast-growing species that dominate reefs worldwide today (Budd & Johnson 1999; Klaus et al. 2012; Renema et al. 2016).

22.6 The Great American Biotic Interchange

The rise of the Isthmus is one of the major biogeographical events of the Cenozoic – a massive experiment in biological invasions – and as such it has been the focus of intensive research over many decades. It is a vicariant event from a marine perspective, as two oceans that were formerly connected – the Pacific and the Caribbean – became separated; however, it is a migration event from a terrestrial perspective, as lands formerly separated – South America and Central/North America – became connected.

Hundreds of biogeographic and phylogenetic studies have used 3.5 Ma as the a priori date for this event (Bacon et al. 2015a), although several genetic studies of taxa with low dispersal capabilities have reported evidence of earlier exchanges, including of bees (Roubik & Camargo 2011), tree frogs (Pinto-Sanchez et al. 2012), salamanders (Elmer et al. 2013), freshwater *Poecilia* fishes (Alda et al. 2013) and *Amazilia* hummingbirds (Ornelas et al. 2014). Is 3.5 Ma a safe assumption? We have explored this question by examining the DNA record of extant taxa and the fossil record. A meta-analysis of the molecular phylogenetic data of 426 taxa across many clades, both terrestrial and marine, indicates a large increase in the rate of migrations/vicariance events at around 10 Ma (Figure 22.2). The analysis shows four significant migration rate shifts, at 41.1 (46.2–35.9), 23.7 (26.2–19.9), 8.7 (10.0–7.2) and 5.2 (6.0–5.1) Ma (Bacon et al. 2015a). The latter two dates represent drastic rate increases, and all predate the inferred migration at 3.5 Ma. Even restricting the analysis to marine vicariance, there is a significant pulse at ~9 Ma (Bacon et al. 2015b); this predates a rate increase at 2.06 (4.3–1.0) Ma (Bacon et al. 2015a) that is indeed associated with the final closure of the Isthmus at 3.5 Ma. The overall results indicate that the GABI occurred over a much longer time period than previously proposed, that it comprised several distinct migrational pulses and that the onset of significant migration was at ~10 Ma.

The fossil record indicates a similar pattern to that shown by the molecular analysis. The early Miocene plant record (20–18 Ma) from Panama (Jaramillo et al. 2014; Rodriguez-Reyes et al. 2014) is dominated by South American-derived lineages and indicates an earlier crossing of the CAS by many South American plant families (e.g., Humiriaceae, Annonaceae, Euphorbiaceae), implying that plants were one the first groups crossing the CAS (Cody et al. 2010). These results derived from the plant fossil record were also suggested by Graham in his multiple studies of Panamanian floras (Graham 1988a,b, 1991, 1992, 1999, 2010, 2011). Other recent studies of fossils have found earlier migrations (~19 Ma) of vertebrates from South America into central Panama across the CAS, including turtles (Cadena et al. 2012), snakes (Head et al. 2012) and crocodiles (Hastings et al. 2013; Scheyer et al. 2013).

Most of what has been written about the GABI in the paleontological literature is derived from the mammal record (e.g., Woodburne 2010). Consequently, the mammal-derived GABI has been accepted as the de facto pattern for all other organisms, even though mammals represent only a small fraction (~0.02%) of the high biodiversity of the Americas (1418 mammal species, versus more than half a million plants, invertebrates and other vertebrates). There are two major ideas about the GABI that are ingrained in the literature, both scientific and popular: first, the onset was due to the closure of the Isthmus, and second, the North American lineages replaced most of the South American lineages in South America.

Let us discuss the first idea first: the closure driving the GABI. Both the fossil and genetic records show that mammal exchange started at 10 Ma and accelerated greatly at ~2.5 Ma (Webb 1976, 2006; Woodburne 2010; Forasiepi et al. 2014; Leite et al. 2014; Bacon et al. 2015a; Carrillo et al. 2015), rather than at 3.5 Ma. This difference in timing between the final closure of the Isthmus and the onset of the massive GABI mammal migrations has been used to suggest that other factors than a land connection were major drivers of the GABI acceleration: mainly habitat change due to the onset of the NHG and concomitant reductions in precipitation across the Americas (Webb 1976, 1978, 2006; Molnar 2008; Smith et al. 2012; Leigh et al. 2014; Bacon et al. 2016), and lower sea levels during glacial intervals (Woodburne 2010). Recent new findings in Guajira, northern Colombia, of late Pliocene age (~2.7 Ma) – the closest paleontological site in time and space to the Isthmus of Panama – indicate that most vertebrate faunas, including mammals, are still strongly dominated by South American clades (Forasiepi et al. 2014; Jaramillo et al. 2015; Moreno et al. 2015; Suarez et al. 2016), suggesting that the GABI acceleration occurred during the Pleistocene in tropical South America, rather than during the late

Pliocene. These new localities support the hypothesis that factors other than the final closure of the Isthmus accelerated the GABI among mammals.

The second major idea – the replacement of South American lineages by North American ones – involves two different hypotheses. The first is that North American taxa had a competitive advantage over South American fauna and therefore, when they migrated into South America, they displaced the South American taxa (Simpson 1983). The second suggests that major habitat changes drove many South American lineages to extinction prior to the arrival of North American lineages (Vrba 1992). These contrasting hypotheses are difficult to test using the fossil record, as that would require fine stratigraphic resolution and a large spatial distribution of the sampling. Nevertheless, the few studies addressing the point indicate that at least when comparing carnivorans (North American-derived) with Sparassodonta (South American-derived), factors other than competitive displacement caused the extinction of the latter (Prevosti et al. 2013).

It is interesting to see how the proportion of South American- versus North American-derived taxa in the fossil mammal fauna of Panama compares with that of the extant fauna. The early Miocene (20–18 Ma) mammal record of central Panama includes a wide variety of groups, including horses, camels, peccaries, bear-dogs, anthracotheriums, rhinocerids, geomyoid rodents, dogs, oreodonts and protoceratids (Whitmore & Stewart 1965; Slaughter 1981; MacFadden & Higgins 2004; MacFadden 2006a,b, 2009; MacFadden et al. 2010, 2012; Rincon et al. 2012, 2013), all of which are derived from North American lineages. In contrast, only two are South American-derived taxa (a bat and a monkey: Bloch et al. 2016), representing 5% of the early Miocene assemblage (from a total of 40 fossil taxa). In the extant mammal assemblage of Panama, however, 57% (118 of 208 species) are South American-derived lineages. Therefore, from a Panamanian perspective (i.e., looking from the southern tip of North America), the prevalent direction of mammal migration was south to north, rather than the traditional view of north to south (57% of modern taxa in Panama are South American, versus 5% during the Miocene).

Coates & Stallard (2013) compared the Caribbean region during the Neogene with the modern Indonesian–Australia tectonic configuration. Because the collision of Australia with Indonesia has not fully closed the marine gap between them, the geographic situation could be thought of as an analog to the Isthmus of Panama ~11 Ma, when the CAS was still open (Coates & Stallard 2013). The CAS would be analogous to the Timor Trough–Ceram Sea: the deep strait that follows the southern tectonic boundary of Wallacea (Figure 22.3) (Hall 1998). The Timor Trough is about 100 km wide (Figure 22.1) – similar to the CAS at 16–15 Ma (Montes et al. 2012a). This tectonic contact point also demarks a biogeographic boundary called

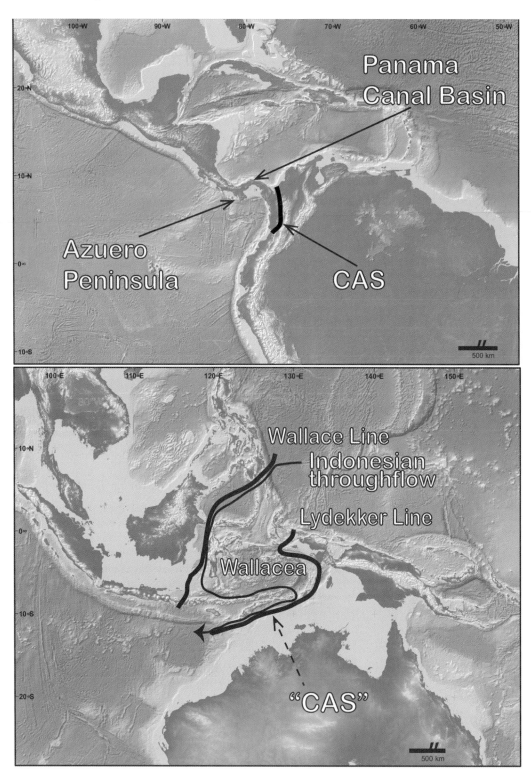

Figure 22.3 Topographic map illustrating Indo-Australia versus the Caribbean and Isthmus of Panama (Amante & Eakins 2009). Both maps are at the same scale. The region between Wallace and Lydekker's Line is called Wallacea, and corresponds to a complex plate boundary between the Indo-Australian plate to the south and the Eurasia Plate to the north (Hall 1998). The Central American Seaway (CAS) corresponds to the Uramita suture on the western Cordillera of Colombia (Montes et al. 2015), and it is an analog to the Timor Trough, the deep strait that follows the southern boundary of Wallacea, which also corresponds to Lydekker's Line. Most ocean water flow through the region (Indonesian throughflow) goes via the Timor Trough (Godfrey 1996), but it is small enough that there is already a distinctive difference between the Indian Ocean and the western equatorial Pacific (Cane & Molnar 2001). Note that Wallacea is ~1000 km wide – as wide as half of Central America from Panama to Nicaragua. Wallacea is a biogeographical region of transition between Australian and Asian biotas (Lohman et al. 2011), and has several mammal lineages derived from Australia that have been able to cross Lydekker's Line.

Lydekker's Line (Lohman et al. 2011). Most of the water output into the Indian Ocean, the "Indonesian throughflow," arrives through the Timor Trough (Godfrey 1996). As Coates & Stallard (2013) point out, the modern setting of Indonesia–Australia provides a valid analog to the Panamanian configuration just prior to the closure of the CAS at 12–11 Ma.

The comparison provides several insights. (i) Even though the Timor Trough has not been fully closed and the waters of the Pacific and Indian oceans are still connected by the Indonesian throughflow, the total output is not sufficient to prevent an oceanographic distinction between the Indian and Pacific oceans. In other words, a complete collision and a fully formed land bridge between Australia and Indonesia are not necessary to produce an oceanographic effect on the Pacific and Indian oceans. Such a "closing" of the Indonesian seaway has been proposed as one of the causes for major climatic changes around 4–3 Ma (Cane & Molnar 2001). This comparison underscores the major effects that closing the CAS 10 Ma would have had on the Atlantic–Pacific connection as it has been discussed in this chapter, relative to the minor effect produced by the final closure of the Isthmus 3.5 Ma. (ii) The Wallace Line is located ~1000 km west of Lydekker's Line (Lohman et al. 2011). Wallacea, the region in between, represents a transition zone of faunas and floras derived from both Asia and Australia (Lohman et al. 2011). This transition region is equivalent in area to half of Central America, from Panama to Nicaragua, and consists of many islands that have not been connected recently by dry land to either of the large continental masses. However, in this zone there are several mammal lineages with an Australian affinity that are east of the Timor Trough and have been able to reach the islands of eastern Wallacea by crossing sea passages. Why, then, was there not an active mammal migration between Panama and South America until 2.7 Ma? This question raises the possibility that factors other than continuous land across the Isthmus drove mammal migration during the GABI.

22.7 Unresolved Questions

The rise of the Isthmus is probably one of the best cases in which the interactions among geology, climate and biota can be observed. In spite of the intensive research in the region, several questions remain:

1) There were sloths, South American capromyid rodents and primates in Cuba during the early Miocene (MacPhee et al. 2003). Why, then, were there primates in Panama in the early Miocene, but no sloths or South American rodents? One hypothesis is that a continuous terrestrial bridge or a continuous chain of islands (known as GAARlandia) connected the Antilles with South America during the Oligocene (Iturralde-Vinent & MacPhee 1999); in other words, Cuba was more connected to South America than Panama during the Oligocene/early Miocene. This hypothesis is still highly controversial (e.g., Graham 2003; Hedges 2006).

2) What controls the rate of migration across the Isthmus? Why were mammals so delayed in migrating in either direction compared to other groups?

3) What were the roles of biomes and sea-level change over the past 2.7 My in the migration process? It has been argued that migration was facilitated by a massive expansion of savannahs in both Central and South America. Were there extensive savannahs during the glacial intervals in Central America?

4) It has been assumed that the extensive development of reefs across the Caribbean was due to the final closure of the Isthmus. If that is the case, why were Caribbean reefs widespread during the late Oligocene (Johnson et al. 2009) if the CAS was a wide and deep marine strait connecting the Atlantic and Pacific oceans at this time? Why didn't reef development occur during the late Miocene (~10.0–7.5 Ma) when there was a separation of Caribbean and Pacific waters (Duque-Caro 1990b; Collins et al. 1996)? Perhaps Caribbean reef development is driven by other factors than the closure of the Isthmus, such as tectonism through the generation of new suitable habitats for reef development.

5) How did the closure of the Isthmus affect the position and migration of the Intertropical Convergence Zone (ITZC)? Novel research on the ITZC suggests that a strong AMOC pushed it north of the equator (Schneider et al. 2014).

6) What effect, if any, did the closure of the Isthmus have on the shallowing of the thermocline of the eastern equatorial Pacific?

Answering these questions will require intercontinental comparisons and additional fieldwork. The continent–continent collision of Asia with Australia is an excellent analog for the Panamanian arc–South America collision during the late Miocene (Coates & Stallard 2013). We could learn about the process and mechanism of migration by comparing the two events across multiples clades, both terrestrial and marine. It is also imperative to find many more late Miocene and Pliocene terrestrial deposits in eastern Colombia, Panama and Costa Rica. There is a large gap of information during this time period that is probably critical to understanding the GABI and the role of the Isthmus in mass migration. Several geological formations in eastern and western Costa Rica, eastern Panama and eastern Colombia have great potential to fill this void.

The geological building of the terrestrial landscape that represents the Isthmus of Panama developed across a very small geographical area, but nonetheless it produced changes on a global scale. There is no better example of the relationship between biological evolution and the development of landforms.

Recently O'Dea et al. (2016) proposed that findings regarding a Miocene closure of CAS are unsupported and provided a new age for the formation of the Isthmus at 2.8 Ma. Both conclusions have been rejected (Jaramillo et al. 2017; Molnar 2017).

Acknowledgments

This research was made possible through the funding of the Smithsonian Institution, Senacyt, the Autoridad del Canal de Panamá (ACP), the Mark Tupper Fellowship, Ricardo Perez S.A, the National Science Foundation grant EAR 0824299 and OISE, EAR, DRL 0966884, the Anders Foundation, 1923 Fund, Gregory D. and Jennifer Walston Johnson and the National Geographic Society. Thanks to Camilo Montes, German Bayona, Aldo Rincon, Bruce McFadden, Austin Hendy, Fabiany Herrera, Edwin Cadena, Douglas Jones, John Bloch, David Farris, Natalia Hoyos and the paleontology-geology team at STRI, University of Florida, Corporación Geológica Ares and Universidad de Los Andes. Thanks for the reviews of P. Mann, W. Renema, C. Hoorn and A. Antonelli.

References

Adamek, S., Frohlich, C. & Pennington-Wayne, D. (1988) Seismicity of the Caribbean-Nazca boundary: constraints on microplate tectonics of the Panama region. *Journal of Geophysical Research* **93**, 2053–2075.

Alda, F., Reina, R.G., Doadrio, I. & Bermingham, E. (2013) Phylogeny and biogeography of the Poecilia sphenops species complex (Actinopterygii, Poeciliidae) in Central America. *Molecular Phylogenetics and Evolution* **66**, 1011–1026.

Amante, C. & Eakins, B.W. (2009) ETOPO1 1 Arc-Minute Global Relief Model: Procedures, Data Sources and Analysis, NGDC-24, NOAA.

Bacon, C.D., Silvestro, D., Jaramillo, C.A. et al. (2015a) Biological evidence supports an early and complex emergence of the Isthmus of Panama. *Proceedings of the National Academy of Sciences* **112**, 6110–6115.

Bacon, C.D., Silvestro, D., Jaramillo, C.A. et al. (2015b) Reply to Lessios and Marko et al.: early and progressive migration across the Isthmus of Panama is robust to missing data and biases. *Proceedings of the National Academy of Sciences* **112**, E5767–E5768.

Bacon, C.D., Molnar, P., Antonelli, A. et al. (2016) Quaternary glaciation and the Great American Biotic Interchange. *Geology* **44**, 375–378.

Bandy, O.L. & Casey, R.E. (1973) Reflector horizons and paleobathymetric history, eastern Panama. *Geological Society of America Bulletin* **84**, 3081–3086.

Barat, F., Mercier, B., Sosson, M. et al. (2014) Transition from the Farallon Plate subduction to the collision between South and Central America: Geological evolution of the Panama Isthmus. *Tectonophysics* **622**, 145–167.

Baumgartner-Mora, C., Baumgartner, P.O., Buchs, D. et al. (2008) Paleocene to Oligocene foraminifera from the Azuero Peninsula (Panama): the timing of the seamount formation and accretion to the mid-American margin. Available from: https://serval.unil.ch/resource/serval:BIB_C2D31C7C6363.P001/REF (last accessed August 30, 2017).

Bell, D.B., Jung, S.J.A., Kroon, D. et al. (2015) Atlantic deep-water response to the Early Pliocene shoaling of the Central American Seaway. *Scientific Reports* **5**, 12252.

Beu, A. (2001) Gradual Miocene to Pleistocene uplift of the Central American Isthmus: evidence from tropical American Tonnoidean gastropods. *Journal of Paleontology* **75**, 706–720.

Bloch, J., Woodruff, E.D., Wood, A.R. et al. (2016) First North American fossil monkey and early Miocene tropical biotic interchange. *Nature* **533**, 243–246.

Brierley, C.M. & Fedorov, A.V. (2016) Comparing the impacts of Miocene–Pliocene changes in inter-ocean gateways on climate: Central American Seaway, Bering Strait, and Indonesia. *Earth and Planetary Science Letters* **444**, 116–130.

Buchs, D.M., Arculus, R.J., Baumgartner, P.O. & Ulianov, A. (2011) Oceanic intraplate volcanoes exposed: example from seamounts accreted in Panama. *Geology* **39**, 335–338.

Budd, A.F. & Johnson, K.G. (1999) Origination preceding extinction during Late Cenozoic turnover of Caribbean reefs. *Paleobiology* **25**, 188–200.

Butzin, M., Lohmann, G. & Bickert, T. (2011) Miocene ocean circulation inferred from marine carbon cycle modeling combined with benthic isotope records. *Paleoceanography* **26**, PA11203.

Cadena, E., Bourque, J., Rincon, A. et al. (2012) New turtles (Chelonia) from the Late Eocene through Late Miocene of the Panama Canal Basin. *Journal of Paleontology* **86**, 539–557.

Cane, M.A. & Molnar, P. (2001) Closing of the Indonesian seaway as a precursor to east African aridification around 3 ± 4 million years ago. *Nature* **411**, 157–162.

Carrillo, J.D., Forasiepi, A., Jaramillo, C. & Sánchez-Villagra, M.R. (2015) Neotropical mammal diversity and the Great American Biotic Interchange: spatial and temporal variation in South America's fossil record. *Frontiers in Genetics* **5**, 451.

Coates, A. & Obando, J. (1996) The geologic evolution of the Central American Isthmus. In: Jackson, J., Budd, A. & Coates, A. (eds.) *Evolution and Environment in Tropical America*. Chicago, IL: University of Chicago Press, pp. 21–56.

Coates, A. & Stallard, R.F. (2013) How old is the Isthmus of Panama? *Bulletin of Marine Science* **89**, 801–813.

Coates, A., Jackson, J.B.C., Collins, A.G. et al. (1992) Closure of the Ishtmus of Panama: the near-shore marine record of Costa Rica and western Panama. *Geological Society of America Bulletin* **104**, 814–828.

Coates, A.G., Aubry, M.P., Berggren, W.A. & Collins, L.S. (2003) Early Neogene history of the Central American arc from Bocas del Toro, western Panama. *Geological Society of America Bulletin* **115**, 271–287.

Coates, A.G., Collins, L.S., Aubry, M.P. & Berggren, W.A. (2004) The geology of the Darien, Panama, and the late Miocene-Pliocene collision of the Panama Arc with northwestern South America. *Geological Society of America Bulletin* **116**, 1327–1344.

Coates, A.G., McNeill, D.F., Aubry, M.P. et al. (2005) An introduction to the geology of the Bocas del Toro Archipelago, Panama. *Caribbean Journal of Science* **41**, 374–391.

Cody, S., Richardson, J.E., Rull, V. et al. (2010) The Great American Biotic Interchange revisited. *Ecography* **33**, 326–332.

Collins, L.S. & Coates, A.G. (1999) A Paleobiotic survey of Caribbean faunas from the Neogene of the Isthmus of Panama. *Bulletins of American Paleontology* **357**, 1–351.

Collins, L., Coates, A., Berggren, W. et al. (1996) The late Miocene Panama isthmian strait. *Geology* **24**, 687–690.

de Boer, B., van de Wal, R.S.W., Bintanja, R. et al. (2010) Cenozoic global ice-volume and temperature simulations with 1-D ice-sheet models forced by benthic $\delta 18O$ records. *Annals of Glaciology* **51**, 23–33.

Diaz de Gamero, M.L. (1996) The changing course of the Orinoco River during the Neogene: a review. *Palaeogeography, Palaeoclimatology, Palaeoecology* **123**, 385–402.

Duque, H. (1990) The Choco Block in the Northwestern corner of South America: structural, tectonostratigraphic and paleogeographic implications. *Journal of South America Earth Sciences* **3**, 71–84.

Duque-Caro, H. (ed.) (1990a) *Major Neogene Events in Panamaic South America*. Tokyo: Tokyo University Press.

Duque-Caro, H. (1990b) Neogene stratigraphy, paleoceanography and paleobiogeography in northwest South America and evolution of the Panama Seaway. *Palaeogeography, Palaeoclimatology, Palaeoecology* **77**, 203–234.

Elmer, K.R., Bonett, R.M., Wake, D.B. & Lougheed, S. (2013) Early Miocene origin and cryptic diversification of South American salamanders. *BMC Evolutionary Biology* **13**, 59.

Farris, D. W., Jaramillo, C., Bayona, G. et al. (2011) Fracturing of the Panamanian Isthmus during initial collision with South America. *Geology* **39**, 1007–1010.

Fedorov, A.V., Brierley, C.M., Lawrence, K.T. et al. (2013) Patterns and mechanisms of early Pliocene warmth. *Nature* **496**, 43–49.

Fedorov, A.V., Burls, N.J., Lawrence, K.T. & Peterson, L.C. (2015) Tightly linked zonal and meridional sea surface temperature gradients over the past five million years. *Nature Geoscience* **8**, 975–980.

Figueiredo, J., Hoorn, C., van der Ven, P. & Soares, E. (2009) Late Miocene onset of the Amazon River and the Amazon deep-sea fan: evidence from the Foz do Amazonas Basin. *Geology* **37**, 619–622.

Forasiepi, A.M., Soibelzon, L.H., Suarez, C. et al. (2014) Carnivorans at the Great American Biotic Interchange: new discoveries from the northern Neotropics. *Naturwissenschaften* **101**, 965–974.

Fyke, J.G., D'Orgeville, M. & Weaver, A.J. (2015) Drake Passage and Central American Seaway controls on the distribution of the oceanic carbon reservoir. *Global and Planetary Change* **128**, 75–82.

Gazel, E., Hayes, J.L., Hoernle, K. et al. (2015) Continental crust generated in oceanic arcs. *Nature Geosciences* **8**, 321–327.

Godfrey, J.S. (1996) The effect of the Indonesian throughflow on ocean circulation and heat exchange with the atmosphere: a review. *Journal of Geophysical Research* **101**, 217–237.

Graham, A. (1984) Lisanthius pollen from the Eocene of Panama. *Annals of the Missouri Botanical Garden* **71**, 987–993.

Graham, A. (1985) Studies in Neotropical paleobotany. IV. The Eocene communities of Panama. *Annals Missouri Botanical Garden* **63**, 787–842.

Graham, A. (1988a) Studies in Neotropical paleobotany. VI. The lower Miocene communities of Panama – the Cucaracha formation. *Annals Missouri Botanical Garden* **75**, 1467–1479.

Graham, A. (1988b) Studies in Neotropical paleobotany. V. The lower Miocene communities of Panama – the Culebra formation. *Annals Missouri Botanical Garden* **75**, 1440–1466.

Graham, A. (1991) Studies in Neotropical paleobotany. X. The Pliocene communities of Panama – composition, numerical representation, and paleocommunity paleoenvironmental reconstructions. *Annals Missouri Botanical Garden* **78**, 465–475.

Graham, A. (1992) Utilization of the isthmian land bridge during the Cenozoic – paleobotanical evidence for timing, and the selective influence of altitudes and climate. *Review of Palaeobotany and Palynology* **72**, 119–128.

Graham, A. (1994) Neotropical Eocene coastal floras and $\delta^{18}O/\delta^{16}O$-estimated warmer vs. cooler equatorial waters. *American Journal of Botany* **81**, 301–306.

Graham, A. (1999) *Late Cretaceous and Cenozoic History of North American Vegetation*. New York: Oxford University Press.

Graham, A. (2003) Geohistory models and Cenozoic paleoenvironments of the Caribbean region. *Systematic Botany* **28**, 378–386.

Graham, A. (2010) *Late Cretaceous and Cenozoic History of Latin American Vegetation and Terrestrial Environments*. St. Louis, MO: Missouri Botanical Garden Press.

Graham, A. (2011) The age and diversification of terrestrial New World ecosystems through Cretaceous and Cenozoic time. *American Journal of Botany* **98**, 336–351.

Groeneveld, J., Nurnberg, D., Tiedemann, R. et al. (2008) Foraminiferal Mg/Ca increase in the Caribbean during the Pliocene: Western Atlantic Warm Pool formation, salinity influence, or diagenetic overprint? *Geochemistry Geophysics Geosystems* **9**, Q01P23.

Hall, R. (1998) The plate tectonics of Cenozoic SE Asia and the distribution of land and sea. In: Hall, R. & Holloway, J.D. (eds.) *Biogeography and Geological Evolution of SE Asia*. Leiden: Backhuys Publishers, pp. 99–131.

Harmon, R.S. (2005) Geological development of Panama. In: Harmon, R.S. (ed.) *The Rio Chagres, Panama: A Multidisciplinary Profile of a Tropical Watershed*. Philadelphia, PA: Springer, pp. 45–62.

Hastings, A., Bloch, J., Jaramillo, C. et al. (2013) Systematics and biogeography of crocodylians from the Miocene of Panama. *Journal of Vertebrate Paleontology* **33**, 239–263.

Haug, G.H. & Tiedemann, R. (1998) Effect of the formation of the Isthmus of Panama on Atlantic Ocean thermohaline circulation. *Nature* **393**, 673–676.

Haug, G.H., Tiedemann, R., Zahn, R. & Ravelo, A.C. (2001) Role of Panama uplift on oceanic freshwater balance. *Geology* **29**, 207–210.

Haug, G., Tiedemann, R. & Keigwin, L. (2004) How the Isthmus of Panama put ice in the Arctic. *Oceanus* **42**, 94–97.

Head, J., Rincon, A., Suarez, C. et al. (2012) Fossil evidence for earliest Neogene American faunal interchange: boa (Serpentes, Boinae) from the early Miocene of Panama. *Journal of Vertebrate Paleontology* **32**, 1328–1334.

Hedges, S.B. (2006) Paleogeography of the Antilles and the origin of West Indian terrestrial vertebrates. *Annals of the Missouri Botanical Garden* **93**, 231–244.

Herrera, F., Manchester, S., Jaramillo, C. et al. (2010) Phytogeographic history and phylogeny of the Humiriaceae. *International Journal of Plant Sciences* **171**, 392–408.

Herrera, F., Manchester, S., Carvalho, M. et al. (2012a) *Permineralized Fruits and Seeds from the Early Middle Miocene Cucaracha Formation of Panama*. Columbus, OH: Botanical Society of America.

Herrera, F., Manchester, S. & Jaramillo, C. (2012b) Permineralized fruits from the late Eocene of Panama give clues of the composition of forests established early in the uplift of Central America. *Review of Palaeobotany and Palynology* **175**, 10–24.

Herrera, F., Manchester, S., Vélez-Juarbe, J. & Jaramillo, C. (2014a) Phytogeographic history of the Humiriaceae (Part 2). *International Journal of Plant Science* **175**, 828–840.

Herrera, F., Manchester, S.R., Koll, R. & Jaramillo, C. (2014b) Fruits of Oreomunnea (Juglandaceae) in the Early Miocene of Panama. In: Stevens, W.D., Montiel, O.M. & Raven, P. (eds.) *Paleobotany and Biogeography: A Festschrift for Alan Graham in His 80th Year*. St. Louis, MO: Missouri Botanical Garden Press, pp. 124–133.

Hoorn, C. (1994) An environmental reconstruction of the palaeo-Amazon River system (Middle–Late Miocene, NW Amazonia). *Palaeogeography, Palaeoclimatology, Palaeoecology* **112**, 187–238.

Hoorn, C., Guerrero, J., Sarmiento, G.A. & Lorente, M.A. (1995) Andean tectonics as a cause for changing drainage patterns in Miocene northern South America. *Geology* **23**, 237–240.

Hoorn, C., Roddaz, M., Dino, R. et al. (2010a) The Amazonian craton and its influence on past fluvial systems (Mesozoic–Cenozoic, Amazonia). In: Hoorn, M.C. & Wesselingh, F.P. (eds.) *Amazonia, Landscape and Species Evolution*. Oxford: Wiley-Blackwell Publishing.

Hoorn, C., Wesselingh, F., ter Steege, H. et al. (2010b) Amazonia through time: Andean uplift, climate change, landscape evolution and biodiversity. *Science* **331**, 399–400.

Hoorn, C., Bogotá-A, G.R., Romero-Baez, M. et al. (2017) The Amazon at sea: Onset and stages of the Amazon River from a marine record, with special reference to Neogene plant turnover in the drainage basin. *Global and Planetary Change* **153**, 51–65.

Iturralde-Vinent, M.A. & MacPhee, R.D.E. (1999) Paleogeography of the Caribbean region: implications for Cenozoic biogeography. *Bulletin of the American Museum of Natural History* **238**, 1–95.

Jackson, J. & O'Dea, A. (2013) Timing of the oceanographic and biological isolation of the Caribbean Sea from the Tropical Eastern Pacific Ocean. *Bulletin of Marine Science* **89**, 779–800.

Jain, S. & Collins, A.G. (2007) Trends in Caribbean Paleoproductivity related to the Neogene closure of the Central American Seaway. *Marine Micropaleontology* **63**, 57–74.

Jaramillo, C., Montes, C., Cardona, A., Silvestro, D., Antonelli, A. & Bacon, C. (2017) Comment (1) on "Formation of the Isthmus of Panama" by O'Dea et al. *Science Advances* **3**, e1602321.

Jaramillo, C. & Cardenas, A. (2013) Global warming and Neotropical rainforests: a historical perspective. *Annual Reviews of Earth and Planetary Sciences* **41**, 741–766.

Jaramillo, C., Rueda, M. & Mora, G. (2006) Cenozoic plant diversity in the Neotropics. *Science* **311**, 1893–1896.

Jaramillo, C., Moreno, E., Ramirez, V. et al. (2014) Palynological record of the last 20 million years in Panama. In: Stevens, W.D., Montiel, O.M. & Raven, P. (eds.) *Paleobotany and Biogeography: A Festschrift for Alan Graham in His 80th Year*. St. Louis, MO: Missouri Botanical Garden Press, pp. 134–253.

Jaramillo, C., Moreno, F., Hendy, F. et al. (2015) Preface: La Guajira, Colombia: a new window into the Cenozoic neotropical biodiversity and the Great American Biotic Interchange. *Swiss Journal of Palaeontology* **134**, 1–4.

Johnson, K.G. (2008) Caribbean reef development was independent of coral diversity over 28 million years. *Science* **319**, 1521–1523.

Johnson, K.G., Sanchez-Villagra, M. & Aguilera, O.A. (2009) The Oligocene-Miocene transition on coral reefs in the Falcon Basin (NW Venezuela). *PALAIOS* **24**, 59–69.

Keigwin, L.D. (1982) Isotope paleoceanography of the Caribbean and east Pacific: role of Panama uplift in late Neogene time. *Science* **217**, 350–353.

Kirby, M.X. & Jackson, J.B.C. (2004) Extinction of a fast-growing oyster and changing ocean circulation in Pliocene tropical America. *Geology* **32**, 1025–1028.

Kirby, M. & MacFadden, B.J. (2005) Was southern Central America an archipelago or a peninsula in the middle Miocene? A test using land-mammal body size. *Palaeogeography, Palaeoclimatology, Palaeoecology* **228**, 193–202.

Kirby, M.X., Jones, S.J. & MacFadden, B.J. (2008) Lower Miocene stratigraphy along the Panama Canal and its bearing on the Central American peninsula. *PLoS ONE* **3**, e2791.

Klaus, J.S., Lutz, B.P., McNeill, D.F. et al. (2011) Rise and fall of Pliocene free-living corals in the Caribbean. *Geology* **39**, 375–378.

Klaus, J.S., McNeill, D.F., Budd, A.F. & Coates, A. (2012) Neogene reef coral assemblages of the Bocas del Toro region, Panama: the rise of *Acropora palmata*. *Coral Reefs* **31**, 191–203.

Klocker, A., Prange, M. & Schulz, M. (2005) Testing the influence of the Central American Seaway on orbitally forced Northern Hemisphere glaciation. *Geophysical Research Letters* **32**, L03703.

Kolarsky, R.A., Mann, P. & Monechi, S. (1995) Stratigraphic development of southwestern Panama as determined from integration of marine seismic data and onshore geology. *Geological Society of America Special Papers* **295**, 159–200.

Leigh, E.G., O'Dea, A. & Vermeij, G. J. (2014) Historical biogeography of the Isthmus of Panama. *Biological Reviews* **89**, 148–172.

Leite, R.N., Kolokotronis, S.-O., Almeida, F.C. et al. (2014) In the wake of invasion: tracing the historical biogeography of the South American Cricetid radiation (Rodentia, Sigmodontinae). *PLoS ONE* **9**, e110081.

Lohman, D.J., de Bruyn, M., Page, T. et al. (2011) Biogeography of the Indo-Australian Archipelado. *Annual Review of Ecology and Systematics* **42**, 205–226.

Lunt, D.J., Foster, G.L., Haywood, A.M. & Stone, E.J. (2008a) Late Pliocene Greenland glaciation controlled by a decline in atmospheric CO_2 levels. *Nature* **454**, 1102–1106.

Lunt, D.J., Valdes, P.J., Haywood, A.M. & Rutt, I.C. (2008b) Closure of the Panama Seaway during the Pliocene: implications for climate and Northern Hemisphere glaciation. *Climate Dynamics* **30**, 1–18.

MacFadden, B.J. (2006a) Extinct mammalian biodiversity of the ancient New World tropics. *Trends in Ecology & Evolution* **21**, 157–165.

MacFadden, B.J. (2006b) North American Miocene land mammals from Panama. *Journal of Vertebrate Paleontology* **26**, 720–734.

MacFadden, B.J. (2009) Three-toes browing horse Anchiterium (Echidae) from the Miocene of Panamá. *Journal of Paleontology* **83**, 489–492.

MacFadden, B.J. (2010) Extinct peccary "Cynorca" occidentale (Tayassuidae, Tayassuinae) from the Miocene of Panama and correlations to North America. *Journal of Paleontology* **84**, 288–289.

MacFadden, B.J. & Higgins, P. (2004) Ancient ecology of 15-million-year-old browsing mammals within C3 plant communities from Panama. *Oecologia* **140**, 169–182.

MacFadden, B.J., Kirby, M.X., Rincon, A. et al. (2010) Extinct Peccary "Cynorca" occidentale (Tayasuidae) from the Miocene of Panama and Correlations to North America. *Journal of Vertebrate Paleontology* **84**, 288–298.

MacFadden, B.J., Foster, D.A., Rincón, A.F. et al. (2012) The New World Tropics as a cradle of biodiversity during the Early Miocene: calibration of the centenario fauna from Panama. *Geological Society of America Abstracts with Programs* **44**, 163.

MacFadden, B.J., Bloch, J., Evans, H.F. et al. (2014) Temporal calibration and biochronology of the Centenario fauna, Early Miocene of Panamá. *Journal of Geology* **122**, 113–135.

MacPhee, R.D.E., Iturralde-Vinent, M.A. & Gaffney, E.S. (2003) Domo de Zaza, an Early Miocene vertebrate locality in South-Central Cuba, with notes on the tectonic evolution of Puerto Rico and the Mona Passage. *American Museum Novitates* **3394**, 1–42.

Maier, K.L., Klaus, J.S., McNeill, D.F. & Budd, A. (2007) A late Miocene low-nutrient window for Caribbean reef formation? *Coral Reefs* **26**, 635–639.

Maier-Raimer, E., Mikolajewicz, G. & Crowley, T. (1990) Ocean general circulation model sensitivity experiment with an open Central American isthmus. *Paleoceanography* **5**, 349–366.

Mann, P. & Corrigan, J. (1990) Model for late Neogene deformation in Panama. *Geology* **18**, 558–562.

Mann, P. & Kolarsky, R.A. (1995) East Panama deformed belt: structure, age, and neotectonic significance. In: Mann, P. (ed.) *Geologic and Tectonic Development of the Caribbean Plate Boundary in Southern Central America*. Boulder, CO: Geological Society of America, pp. 111–130.

McDougall, K. (1985) Miocene to Pleistocene Benthic Foraminifers and Paleoceanography of the Middle America Slope. Deep Sea Drilling Project Leg 841, Initial Reports DSDP 84, 363–418.

McNeill, D.F., Coates, A., Budd, A. & Borne, P.F. (2000) Integrated paleontologic and paleomagnetic stratigraphy of the upper Neogene deposits around Limon, Costa Rica: a coastal emergence record of the Central American Isthmus. *Geological Society of America Bulletin* **112**, 963–981.

Mestas-Nuñez, A.M. & Molnar, P. (2014) A mechanism for freshening the Caribbean Sea in pre-Ice Age time. *Paleoceanography* **29**, 508–517.

Mikolajewicz, G. & Crowley, J.L. (1997) Response of a coupled ocean/energy balance model to restricted flow through the central American isthmus. *Paleoceanography* **12**, 429–441.

Mikolajewicz, U., Maier-Raimer, E., Crowley, T.J. & Kim, K.Y. (1993) Effect of Drake and Panamanian gateways on the circulation of an ocean model. *Paleoceanography* **8**, 409–426.

Molnar, P. (2017) Comment (2) on "Formation of the Isthmus of Panama" by O'Dea et al. *Science Advances* **3**, e1602320.

Molnar, P. (2008) Closing of the Central American Seaway and the Ice Age: a critical review. *Paleoceanography* **23**, 1-A2201.

Montes, C., Bayona, G., Cardona, A. et al. (2012a) Arc-continent collision and orocline formation: closing of the Central American Seaway. *Journal of Geophysical Research* **117**, B04105.

Montes, C., Cardona, A., McFadden, R.R. et al. (2012b) Evidence for middle Eocene and younger emergence in Central Panama: implications for Ishtmus closure. *Geological Society of America Bulletin* **124**, 780–799.

Montes, C., Cardona, A., Jaramillo, C. et al. (2015) Middle Miocene closure of the Central American Seaway. *Science* **348**, 226–229.

Moreno, J.F., Hendy, A.J.W., Quiroz, L. et al. (2015) An overview and revised stratigraphy of early Miocene–Pliocene deposits in Cocinetas Basin, La Guajira, Colombia. *Swiss Journal of Palaeontology* **134**, 5–43.

Murdock, T.Q., Weaver, A.J. & Fanning, A.F. (1997) Paleoclimatic response of the closing of the Isthmus of Panama in a coupled ocean–atmosphere model. *Geophysical Research Letters* **24**, 253–256.

Nisancioglu, K.H., Raymo, M.E. & Stone, S.M. (2003) Reorganization of Miocene deep water circulation in response to the shoaling of the Central American Seaway. *Paleoceanography* **18**, PAO00767.

Nof, D. & van Gorder, S. (2003) Did an open Panama Isthmus correspond to an invasion of Pacific water into the Atlantic? *Journal of Physical Oceanography* **33**, 1324–1336.

O'Dea, A., Lessios, H.A., Coates, A.G. et al. (2016) Formation of the Isthmus of Panama. *Science Advances* **2**, e1600883.

O'Dea, A., Jackson, J.B.C., Fortunato, H. et al. (2007) Environmental change preceded Caribbean extinction by 2 million years. *Proceedings of the National Academy of Sciences* **104**, 5501–5506.

Ornelas, J.F., Gonzales, C., Espinosa, A. et al. (2014) In and out of Mesoamerica: temporal divergence of Amazilia hummingbirds pre-dates the orthodox account of the completion of the Isthmus of Panama. *Journal of Biogeography* **41**, 168–181.

Pindell, J.L. & Barret, S.F. (1990) Geological evolution of the Caribbean region: a plate tectonic perspective. In: Dengo, G. & Case, J. (eds.) *The Caribbean Region*. Boulder, CO: Geological Society of America, pp. 405–432.

Pinto-Sanchez, N., Ibañez, R., Madriñan, S. et al. (2012) The Great American Biotic Interchange in frogs: multiple and early colonization of Central America by the South American genus Pristimantis (Anura: Craugastoridae). *Molecular Phylogenetics and Evolution* **62**, 954–972.

Poore, H.R., Samworth, R., White, N.J. et al. (2006) Neogene overflow of Northern Component Water at the

Greenland–Scotland Ridge. *Geochemistry Geophysics Geosystems* **7**, Q06010.

Prange, M. & Schulz, M. (2004) A coastal upwelling seesaw in the Atlantic Ocean as a result of the closure of the Central American Seaway. *Geophysical Research Letters* **31**, L17207.

Prevosti, F.J., Forasiepi, A. & Zimicz, N. (2013) The evolution of the Cenozoic terrestrial mammalian predator guild in South America: competition or replacement? *Journal of Mammalian Evolution* **20**, 3–21.

Ramirez, D.A., Foster, D.A., Min, K. et al. (2016) Exhumation of the Panama basement complex and basins: implications for the closure of the Central American seaway. *Geochemistry Geophysics Geosystems* **17**, 1758–1777.

Renema, W., Pandolfi, J.M., Kiessling, W. et al. (2016) Are coral reefs victims of their own past success? *Science Advances* **2**, e1500850.

Restrepo, J.C., Ortiz, J.C., Pierini, J. et al. (2014) Freshwater discharge into the Caribbean Sea from the rivers of Northwestern South America (Colombia): magnitude, variability and recent changes. *Journal of Hydrology* **509**, 266–281.

Rincon, A., Bloch, J.I., Suarez, C. et al. (2012) New Floridatragulines (Mammalia, Camelidae) from the Early Miocene Las Cascadas Formation, Panama. *Journal of Vertebrate Paleontology* **32**, 456–475.

Rincon, A., Bloch, J.I., MacFadden, B.J. & Jaramillo, C. (2013) First Central American record of Anthracotheriidae (Mammalia, Bothriodontinae) from the early Miocene of Panama. *Journal of Vertebrate Paleontology* **33**, 421–433.

Rockwell, T.K., Bennett, R.A., Gath, E. & Fransechi, P. (2010) Unhinging an indenter: a new tectonic model for the internal deformation of Panama. *Tectonics* **29**, TC4027.

Rodriguez-Reyes, O., Falcon-Lang, H.J., Gasson, P. et al. (2014) Fossil woods (Malvaceae) from the lower Miocene (early to mid-Burdigalian) part of the Cucaracha Formation of Panama (Central America) and their biogeographic implications. *Review of Palaeobotany and Palynology* **209**, 11–34.

Roubik, D.W. & Camargo, J.M. (2011) The Panama microplate, island studies and relictual species of Melipona (Melikerria) (Hymenoptera: Apidae: Meliponini). *Systematic Entomology* **37**, 189–199.

Scheyer, T.M., Aguilera, O.A., Delfino, M. et al. (2013) Crocodylian diversity peak and extinction in the late Cenozoic of the northern Neotropics. *Nature Communications* **4**, 1907.

Schneider, B. & Schmittner, A. (2006) Simulating the impact of the Panamanian seaway closure on ocean circulation, marine productivity and nutrient cycling. *Earth and Planetary Science Letters* **246**, 367–380.

Schneider, T., Bischoff, T. & Haug, G.H. (2014) Migrations and dynamics of the intertropical convergence zone. *Nature* **513**, 45–53.

Sepulchre, P., Arsouze, T., Donnadieu, Y. et al. (2014) Consequences of shoaling of the Central American Seaway determined from modeling Nd isotope. *Paleoceanography* **29**, 176–189.

Seton, M., Müller, R.D., Zahirovic, S. et al. (2012) Global continental and ocean basin reconstructions since 200 Ma. *Earth-Science Reviews* **113**, 212–270.

Simpson, G.G. (1983) *Splendid Isolation: The Curious History of South American Mammals*. New Haven, CT: Yale University Press.

Slaughter, B.H. (1981) A new genus of geomyoid rodent from the Miocene of Texas and Panama. *Journal of Vertebrate Paleontology* **1**, 111–115.

Smith, B.T., Amei, A. & Klicka, J. (2012) Evaluating the role of contracting and expanding rainforest in initiating cycles of speciation across the Isthmus of Panama. *Proceedings of the Royal Society B: Biological Sciences* **279**, 3520–3526.

Steph, S., Tiedemann, R., Groeneveld, J. et al. (2006a) Pliocene changes in tropical East Pacific upper ocean stratification: response to tropical gateways? *Proceedings of the Ocean Drilling Program, Scientific Results* **202**, 1–51.

Steph, S., Tiedemann, R., Prange, M. et al. (2006b) Changes in Caribbean surface hydrography during the Pliocene shoaling of the Central American Seaway. *Paleoceanography* **21**, PA4221.

Steph, S., Tiedemann, R., Prange, M. et al. (2010) Early Pliocene increase in thermohaline overturning: a precondition for the development of the modern equatorial Pacific cold tongue. *Paleoceanography* **25**, PA2202.

Stern, W.L. & Eyde, R.H. (1963) Fossil forests of Ocú, Panama. *Science* **140**, 1214.

Suarez, C., Forasiepi, A.M., Goin, F.J. & Jaramillo, C. (2016) Insights into the Neotropics prior to the Great American Biotic Interchange: new evidence of mammalian predators from the Miocene of Northern Colombia. *Journal of Vertebrate Paleontology* **36**, e1029581.

van der Heydt, A. & Dijkstra, H.A. (2005) Flow reorganizations in the Panama Seaway: a cause for the demise of Miocene corals? *Geophysical Research Letters* **32**, L02609.

van der Heydt, A. & Dijkstra, H.A. (2006) Effect of ocean gateways on the global ocean circulation in the late Oligocene and early Miocene. *Paleoceanography* **21**, PA1011.

Vargas, C.A. & Mann, P. (2013) Tearing and breaking off of subducted slabs as the result of collision of the Panama Arc-indenter with northwestern South America. *Bulletin of the Seismological Society of America* **103**, 2025–2046.

Vizcaíno, M., Rupper, S. & Chiang, J.C.H. (2010) Permanent El Niño and the onset of Northern Hemisphere glaciations: mechanism and comparison with other hypotheses. *Paleoceanography* **25**, PA2205.

Vrba, E.S. (1992) Mammals as a key to evolutionary theory. *Journal of Mammalogy* **73**, 1–28.

Wadge, G. & Burke, K. (1983) Neogene Caribbean Plate rotation and associated Central American tectonic evolution. *Tectonics* **2**, 633–643.

Walker, J.D., Geissman, J.W., Bowring, S.A. & Babcock, L.E. (2012) *Geologic Time Scale v. 4.0.* Boulder, CO: Geological Society of America.

Webb, S.D. (1976) Mammalian faunal dynamics of the Great American Interchange. *Paleobiology* **2**, 220–234.

Webb, S.D. (1978) A history of savanna vertebrates in the New World. Part II: South America and the great interchange. *Annual Review of Ecology and Systematics* **9**, 393–426.

Webb, S.D. (2006) The great American biotic interchange: patterns and processes. *Annals of the Missouri Botanical Garden* **93**, 245–257.

Wegner, W., Wörner, G., Harmon, M.E. & Jicha, B.R. (2011) Magmatic history and evolution of the Central American land bridge in Panama since the Cretaceous times. *Geological Society of America Bulletin* **123**, 703–724.

Whitmore, F.C. & Stewart, R.H. (1965) Miocene mammals and Central American Seaways. *Science* **148**, 180–185.

Woodburne, M.O. (2010) The Great American Biotic Interchange: dispersals, tectonics, climate, sea level and holding pens. *Journal of Mammalian Evolution* **17**, 245–264.

Zhang, X., Prange, M., Steph, S. et al. (2012) Changes in equatorial Pacific thermocline depth in response to Panamanian seaway closure: insights from a multi-model study. *Earth and Planetary Science Letters* **317**, 76–84.

23

The Tepuis of the Guiana Highlands

Otto Huber[1], Ghillean T. Prance[2], Salomon B. Kroonenberg[3,4] and Alexandre Antonelli[5,6,7]

[1] *Instituto Experimental Jardín Botánico de Caracas (UCV), Caracas, Venezuela*
[2] *Herbarium, Royal Botanic Gardens, Kew, Richmond, UK*
[3] *Department of Geosciences, Delft University of Technology, Delft, Netherlands*
[4] *Department of Geosciences, Anton de Kom University of Suriname, Tammenga, Suriname*
[5] *Gothenburg Global Biodiversity Centre, Gothenburg, Sweden*
[6] *Department of Biological and Environmental Sciences, University of Gothenburg, Gothenburg, Sweden*
[7] *Gothenburg Botanical Garden, Gothenburg, Sweden*

Abstract

The interior of the Guiana region in north-eastern South America was one of the last mountain areas to be explored in the Americas; the flat and rocky summit of Roraima was ascended for the first time only in 1884. This is one of more than 60 such table mountains, called *tepuis* by the local indigenous inhabitants. Most of these mountains and mountain ranges, distributed over the still largely inaccessible interfluvium between the Amazon and Orinoco rivers, have now been identified and mapped. Some have been partially explored scientifically, but the majority are still virtually unexplored. The geology, hydrology, climate, biodiversity, contemporary and evolutionary biology and exploration of the region are discussed in this chapter. Four different age groups of sandstone plateaus occur in the region: the Roraima tepuis, the Neblina tepuis, the Tunuí sandstone plateaus and the Chiribiquete–Araracuara sandstone plateaus. The latter are located in Colombia, while the first three are mainly found in Venezuela and adjacent Brazil and Guyana. The relatively isolated mountain summits of the Guiana tepuis above ~1500 m form the Pantepui Province of the Guiana biogeographical region, one of the most biologically interesting bioregions of the Neotropics. This province harbors an assemblage of highly characteristic biota: the flora consists of almost 130 plant families with 336 (including 61 endemic) genera and over 2100 (with almost 1300 endemic) species, which are distributed in several floristic areas. The animal life – of which birds, mammals, frogs and lizards are the best known – is characterized by similarly high levels of endemicity, in spite of relatively low population densities, probably due to the low nutrient contents of the dominant plant groups. Recent genetic studies show that diversification of this rich and endemic biota, once thought to have taken place in nearly complete isolation, is in fact characterized by sparse but continuous biotic interchange with other regions and ecosystems in South America. In this sense, the Guiana Highlands have functioned both as a source and a sink of diversity to the rest of the continent. Although some of the lineages colonizing the tepuis were pre-adapted to the mountain environment, most had to develop features to cope with the novel ecological conditions encountered. Phylogenetic analyses further show that current diversity may be underestimated, with both well-differentiated and cryptic species awaiting discovery.

Keywords: *Neotropics, Orinoco Basin, Roraima Formation, table mountains, geology, biogeography, ecology, exploration*

23.1 Introduction

The Guiana Shield is the vast rainforest-clad basement area in northern South America between the Orinoco and Amazon rivers. Large parts of this area are overlain by impressive sandstone plateaus (*tepuis*), collectively referred to as the Guiana Highlands. The largest plateau – the Pakaraima Mountains – covers an area of 73 000 km², but outlier plateaus occur in Colombia, Venezuela, Brazil, Guyana and Suriname, over an area of more than 1 350 000 km². Mount Roraima – the peak where Venezuela, Guyana and Brazil meet – is the highest point of the Pakaraima Mountains, reaching 2810 m, while the Pico da Neblina – 2994 m and entirely

Mountains, Climate and Biodiversity, First Edition. Edited by Carina Hoorn, Allison Perrigo and Alexandre Antonelli.
© 2018 John Wiley & Sons Ltd. Published 2018 by John Wiley & Sons Ltd.
Companion website: www.wiley.com\go\hoorn\mountains,climateandbiodiversity

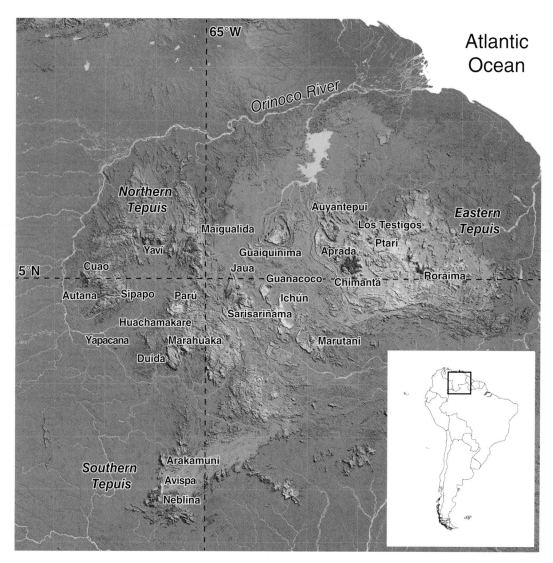

Figure 23.1 Map showing the distribution and locations of the main mountain systems in the Pantepui Province of the Guiana Highlands. *Source:* Courtesy of Charles Brewer-Carías. See also Plate 36 in color plate section.

in Brazil – is the highest point in the whole of extra-Andean South America. At least 60 individual plateaus have been distinguished in the Venezuelan part alone (see Figure 23.1 and Table 23.1) (Huber 1995a). The table mountains are separated from one another by deep and wide valleys covered by dense tropical rainforest and alternating shrub and grassland areas. Many plateaus are bordered by insurmountable vertical cliffs up to 1500 m high, and streams originating on the plateaus produce impressive waterfalls. These include the Angel Falls in Venezuela, the highest waterfall in the world (~980 m) (Figure 23.2a) and the Kaieteur Falls in Guyana (251 m), the world's widest single-drop falls. Outcrop patterns, valley systems and escarpments are often structurally controlled by lineaments, faults or other discontinuities, as appears clearly in the combined geological map of

southern Venezuela and the Shuttle Radar Topography Mission (SRTM) data (Hackley et al. 2005). Deep vertical caves, formed by the dissolution of sandstone, reach down to over 300 m (Brewer-Carías 1975), and some of them are connected deep down in the plateau through horizontal galleries in particularly susceptible sandstone beds (Aubrecht et al. 2008, 2012).

To the south-east of Roraima, in adjacent north-western Guyana, four tepuis with small summit plateaus were visited for the first time during recent scientific explorations (Kelloff et al. 2011). They are (from north to south): Waukauyengtipu (1570 m), Mt. Maringma (2134 m), Mt. Ayanganna (2080 m) and Mt. Wokomong (1680 m). In addition to the widely visible Uei-tepui or Serra do Sol (2100 m), shared by Venezuela and Brazil along the watershed, three low- to mid-elevation sandstone

Table 23.1 High mountains and tepuis of Pantepui in the Venezuelan Guiana.

Drainage	Mountain massif	Tepui/mountain unit	Max. elevation (m, approx.)	Summit area (km^2)
Caroní basin	Eastern Tepui chain	Uei-tepui	2150	2.5
		Roraima-tepui	2723	34.38
		Kukenán (Mpatauí)-tepui	2650	20.63
		Yuruaní-tepui	2400	4.38
		Wadakapiapué-tepui	2000	<0.01
		Karaurín-tepui	2500	1.88
		Ilú / Tramen-tepui	2700	5.63
	Ptari massif	Ptari-tepui	2400	1.25
		Carrao-tepui	2200	1.25
		Sororopán-tepui	2050	N.A.
	Los Testigos massif	(undifferentiated)	1900–2400	12
		Kamarkawarai-tepui	2400	5
		Tereke-yurén-tepui	1900	0.63
		Murisipán-tepui	2350	5
		Aparamán-tepui	2100	1.25
	Auyan massif	Auyán-tepui	2450	666.9
		Cerro La Luna	1650	0.2
		Cerro El Sol	1750	0.6
		Uaipán-tepui	1950	2.5
	Aprada massif	Aprada-tepui	2500	4.37
		Araopán-tepui	2450	1.25
	Chimantá massif	(undifferentiated, incl. 11 tepuis)	2200–2650	615
		Angasima-tepui	2250	2
		Upuigma-tepui	2100	0.63
Paragua basin	Paragua uplands	Sierra Marutaní (Pia-Zoi)	1500	740
		Cerro Guanacoco	1500	526.25
		Cerro Guaiquinima	1650	1096.26
Caura basin	Jaua massif	Cerro Sarisariñama	2350	546.88
		Cerro Jaua	2250	625.62
	Maigualida massif	Sierra de Maigualida	2400	440
		Serranía de Uasadi	1300–1800	N.A.
Ventuari basin				
Manapiare basin	–	Cerro Ualipano	1800	N.A.
	Yaví massif	Cerro Yaví	2300	5.62
	Yutajé massif	Serranía Yutajé	2140	95.63
		Cerro Coro Coro	2400	179.38
		Cerro Guanay	2080	165
		Cerro Camani	1800	1.88
	Sipapo uplands	Cerro Ovaña (Ouana)	1800	N.A.
	Parú massif	Cerro Parú (A'roko) (incl. Cerro Asisa)	2200 1700	724.38
Lower Ventuari basin *Upper Ventuari basin*		Cerro Euaja	2000	205.62

(Continued)

Table 23.1 (Continued)

Drainage	Mountain massif	Tepui/mountain unit	Max. elevation (m, approx.)	Summit area (km²)
Middle Orinoco basin	Cuao-Sipapo massif	Cerro Autana	1300	1.88
		Cerro Cuao	2000	80
		Cerro Sipapo	1800	56
Upper Orinoco basin	Duida-Marahuaka massif	Cerro Huachamakari	1900	8.75
		Cerro Marahuaka	2800	121
		Cerro Duida	2358	1089
Casiquiare – Rio Negro basin	Tapirapecó massif	Cerro Aratityope	1700	<0.01
		Sierra Unturán	1600	N.A.
		Cerro Tamacuari	2340	<0.01
		Serranía Tapirapecó	2000	N.A.
	Arakamuni – Avispa uplands	Cerro Arakamuni	1600	238
		Cerro Avispa	1600	
	Imeri massif	Sierra de la Neblina	2994	235
		Pirapucú (Brazil)	N.A.	N.A.

N.A., not ascertained.

massifs occur in adjacent Brazil: the isolated plateau of Serra Tepequém (ca. 1050 m, severely impacted by diamond miners using dynamite) in Roraima State and the Serra das Sururucús and Serra Aracá, neither of which exceeds 1500 m in elevation, in Amazonas State. The easternmost remnant is the Tafelberg in central Suriname, at 1026 m.

The indigenous Pemón inhabitants living at the base of these flat-topped mountains in eastern Guiana call them "*tepui*" ("*tipu*" in adjacent Guyana, but "*jidi*" amongst the Makiritare in the upper Orinoco region), a term possibly indicating the home of their gods. In Pemón cosmovision, these are seen as scary emperor-gods, and for that reason, the local populations have never dared to penetrate into their homeland. The summits of these many table mountains can thus be considered one of the only pristine places left in the tropical world, relatively untouched by direct human impact. Although almost all tepui summits have been visited by humans over the last century, previously on foot and more recently by helicopter, the tepui environment as such can still be considered to be in its original state. However, slight impacts are noticeable in certain parts of Roraima Tepui, the only officially accessible tepui, which is visited annually by 5000–8000 tourists.

23.2 Geology

The sandstones that form the tepuis lack fossils, but early researchers regarded the age of the plateau sandstones as Cretaceous. However, the first radiometric data that became available in the 1960s established the age as Paleoproterozoic, the era between 2500 and 1600 Ma (Snelling 1963). Modern geochronological work (Santos et al. 2003) shows that there are at least four different age groups of sandstone plateaus, as detailed later (Figure 23.3), although not all may fit the biogeographical Pantepui concept. Santos et al. (2003) also give a comprehensive overview of the history of geological research on the tepuis.

23.2.1 Roraima Tepuis

The sedimentary rocks of the Pakaraima Mountains are designated as part of the Roraima Supergroup (Reis et al. 1990; Cox et al. 1993; Gibbs & Barron 1993; Mendoza 2005) and reach a total thickness of over 2900 m (Reis et al. 2003). The Roraima Supergroup also crops out in the Jaua, Cácaro, Guaiquinima, Ichún and Urutaní plateaus in southern Venezuela, as well as in the Uafaranda and Tepequém outliers in northern Brazil, the Makari outlier in Guyana and the Tafelberg in Suriname (Figure 23.3). Geologists have distinguished up to 10 subdivisions of the Roraima Supergroup (Keats 1976; Ghosh 1985; Reis et al. 1990, 2003; Sidder & Mendoza 1991; Gibbs & Barron 1993; Santos et al. 2003; Beyer et al. 2015). Sandstones predominate (Figure 23.2b), but there are conglomeratic and shaley intercalations as well. Sedimentary structures indicate a predominantly braided fluvial to deltaic origin, with varying paleocurrents. Several horizons are diamond-bearing. A salient characteristic is the presence of various levels of jasper-like fine-grained indurated volcanic ash-fall tuffs. The Roraima Supergroup is intruded by conspicuous dolerite dykes and sills, a basalt-like igneous rock.

Figure 23.2 (a) View of Cañón del Diablo (Devil's Valley), with Salto Angel (Angel Falls) on the left, falling from Auyantepui in Venezuela. *Source:* Courtesy of Charles Brewer-Carías. (b) Roraima sandstone geology in the Chimantá massif, Venezuela. *Source:* Gérard Vigo, https://commons.wikimedia.org/wiki/File:Roraima_Rocks7.JPG. Licensed under CC BY-SA 3.0. (c) Mt. Roraima and the Eastern Tepui chain. *Source:* Courtesy of Charles Brewer-Carías. (d) *Bonnetia crassa* Gleason, endemic treelet of the northern Pantepui province. (e) *Oreophrynella quelchii* Boulenger, endemic toad of the summit of Mt. Roraima. *Source:* Courtesy of Javier Mesa. See also Plate 37 in color plate section.

Formerly, the age of the Roraima sandstones was the subject of considerable controversy, as the underlying basement, the intercalated volcanic ash horizons and the cross-cutting dolerite dykes gave conflicting results (e.g., Priem et al. 1973). Also, Precambrian paleosols at the base of the formation at Tafelberg, Suriname were earlier erroneously identified as intrusive contacts (Kroonenberg 1978). Recently, Santos et al. (2003) dated the intercalated ash-fall tuffs at 1873 ± 3 Ma, the underlying basement at around 1990–1921 Ma and older and intruding

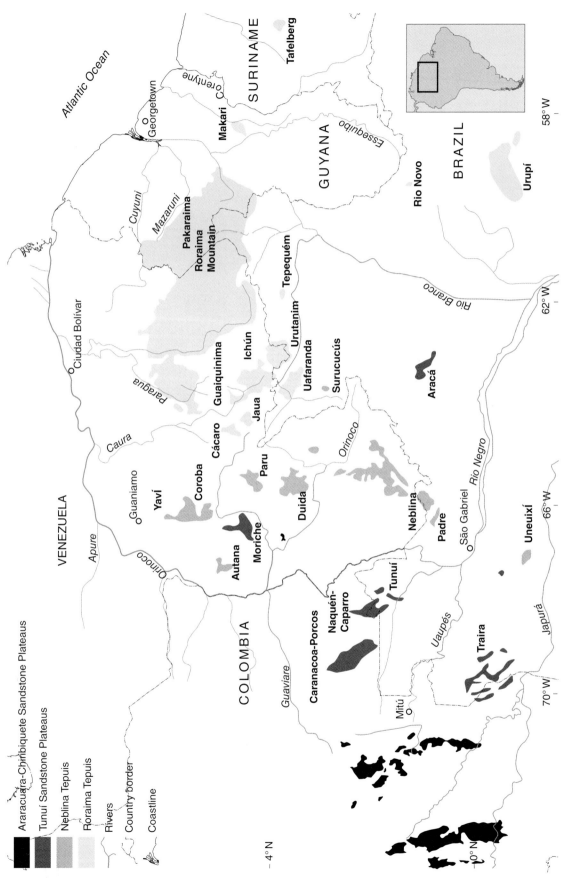

Figure 23.3 Distribution of tepuis and sandstone plateaus in the Guiana Shield. *Source:* Adapted from Santos *et al.* (2003).

dolerite dykes at around 1787 ± 14 Ma. Zircon sand grains in sandstone from the Pakaraima Mountains show ages around 1950 Ma and older (Santos et al. 2003; Beyer et al. 2015). This suggests that the sediments were deposited in the foreland basin of the 2260–2090 Ma Trans-Amazonian Orogenic Belt, the major granite-greenstone belt that stretches 1500 km along the northern border of the Guiana Shield from Venezuela to French Guiana and continues into West Africa in a pre-drift setting of the Gondwana supercontinent (Kroonenberg et al. 2016). Paleocurrent analyses support a northern provenance for the sediments as well (Reis et al. 1990). The prominent mesas in the uppermost part of the Pakaraima Mountains are underlain by the coarse fluvial Matauí Formation; this probably does not belong to the Roraima Supergroup sensu stricto, because of an unconformity at its base, a different dip angle and the absence of dolerite dykes and ash-fall tuffs (Reid 1972; Santos et al. 2003).

23.2.2 Neblina Tepuis

The second group of sandstone plateaus is situated farther to the west, in southern Venezuela. These rocks cannot belong to the Roraima Supergroup because they have been deposited on top of basement rocks that are younger than that group. This applies to the Yaví, Coroba, Parú, Duida, Autana and Avispa tepuis in Venezuela and the Pico da Neblina in Brazil. The sandstones reflect periods of prolonged basement erosion and deposition in a basin that partly overlaps with the older Roraima Supergroup basin (Santos et al. 2003). Because of the absence of volcanic tuff intercalations and dolerite intrusions, their exact age cannot be established. The youngest zircon sand grains in this sandstone are 1878 Ma (Santos et al. 2003). They show some evidence of metamorphism, dated around 1334 Ma (Santos et al. 2003). It cannot be excluded that the uppermost rock unit of the Pakaraima Mountains, the Matauí Formation, belongs to the Neblina group as well (Santos et al. 2003).

23.2.3 Tunuí Sandstone Plateaus

A third group of sandstone plateaus is found in southeastern Colombia and adjacent parts of Brazil: the Tunuí Group (Gibbs & Barron 1993; Kroonenberg & de Roever 2010; Kroonenberg & Reeves 2012 and references therein). These include the Naquén (800 m), Caracanoa, Machado, La Pedrera, La Libertad and Taraira in Colombia and the Tunuí and Caparro ranges in Brazil. They are sandstones, siltstones, mudstones and some conglomerates, deposited in meandering and braided river systems. Many are gold-bearing. In contrast to the Neblina plateaus, these rocks are strongly tilted and folded, and show low-grade metamorphism. No direct

age data are available from these sandstones either. The age of the basement is locally as young as 1500 My, the youngest zircon sand grains are 1720 My and the age of the metamorphism is between 1000 and 1300 My. Unfortunately, older literature refers to these rocks also as the "Roraima Formation" (Galvis et al. 1979). This is no longer valid. Deformation and metamorphism may be the result of the Grenvillian orogeny at the western border of the Guiana Shield between 1200 and 1000 Ma (Kroonenberg 1982; Santos et al. 2003; Ibáñez-Mejía et al. 2011; Kroonenberg & Reeves 2012; Kroonenberg et al. 2016), which led to the amalgamation of the supercontinent Rodinia. The metamorphism in the Neblina group of tepuis may also be due to this event.

23.2.4 Chiribiquete and Araracuara Sandstone Plateaus

Finally, there is a series of impressive horizontal sandstone plateaus in Colombian Amazonia – the Araracuara (240 m) and Chiribiquete (840 m) mesetas – that are not Precambrian but Ordovician, as borne out by the presence of acritarchs and other fossils (Bogotá-Ruiz 1982; Théry et al. 1985; Estrada & Fuertes 1993; Kroonenberg & Reeves 2012). They represent a shortlived and shallow Ordovician transgression from the Solimões Basin on to the Precambrian Guiana Shield.

23.2.5 Uplift and Erosion

All these sandstone formations were originally deposited in basins that constituted the lowest part of the Paleoproterozoic and Ordovician landscapes. Since then, they have suffered relief inversion as a result of uplift and differential erosion, due to their greater resistance to denudation compared to the crystalline rocks on which they rest. Parallel escarpment retreat is the main denudational process in the tepuis. At least part of the uplift is probably related to the Mesozoic break-up of Gondwana, possibly as rift shoulders of the Atlantic margin and of the 6000 m-deep Jurassic Takutu rift that divides the Guiana Shield in two parts (Givnish et al. 2000; Santos et al. 2003; Kroonenberg & de Roever 2010). Tectonic uplift is still continuing, as is evident from flights of terraces along the main rivers, drainage anomalies and stream captures (Khobzi et al. 1980; Sacek 2014).

23.3 Hydrology

The mountains of the Guiana Shield are drained by an extensive river system belonging to three continental hydrologic regions – the Orinoco, Cuyuní and Amazonas basins – which all deliver their huge water masses into the western Atlantic Ocean. The southern Orinoco drainage basin is by far the most extensive in the Venezuelan Guiana, since it covers approximately 79% of the surface, compared to 12% for the Amazonas basin and 9% for the Cuyuní. An additional, although negligible, drainage area into the Amazon basin is from the narrow belt formed by the slopes descending from the southern frontier of the Venezuelan watershed in the Pakaraima, Parima, Tapirapecó and Neblina mountains.

The sandstone tepuis with high and extensive accumulations of peats drain into extremely acidic (pH < 5), oligotrophic and transparent black-waters, which are characteristic of some larger drainage basins in the Amazon region, such as the Río Negro in the western Guiana and the Río Caroní in the eastern Guiana (Zinck & Huber 2011). On the other hand, rivers coming from mountains consisting primarily of granitic or other igneous-metamorphic rocks are normally white-water courses, recognizable by their opaque grayish to whitish color, usually caused by a heavy load of dissolved sediments of low or neutral acidity (pH 5–8). The Rio Branco in Brazil is a typical white-water river of the southern Guiana Province. The third water type is clear-water, intermediate between the former two and draining mainly densely forested mid-elevation hill- and uplands (e.g., Río Ventuari from the Sierra Maigualida or Río Padamo from the Sierra Parima).

23.4 Climate

The overwhelming majority of the Guiana region lies between sea level and 1500 m; that is, within the tropical macro- and mesothermic belts, where average annual temperature ranges between ca. 26 °C in the lowlands and 16 °C in the uplands. Based on empirical observations and scattered measurements (e.g., Huber 1976), the tepui summits between ca. 1800 and 2800 m are exposed to a mesic- (18–15 °C) or microthermic (ca. 15–10 °C) air temperature regime, depending on elevation and exposure.

The rainfall regime in the Guiana lowlands oscillates between less than 1000 mm/y in north-eastern Guiana on the lower Orinoco and probably up to 5000 mm/y in the Upper Rio Negro region of Brazil and southern Venezuela. In the Guiana Highlands, there is a short dry period, usually between January and March/early April, in which rains do not occur daily, or just in short showers. The rainy season, in contrast, is more intense than in the surrounding low- and uplands, and average annual rainfall amounts to probably between 2000 and 4000 mm. No climatic data are available from the extensive rocky

wall habitats. For more detailed information on climate on the summit of Roraima, Auyántepui and Chimantá, see Adlassnig et al. (2010), Galán (1992) and Huber (1976, 1995a), respectively.

One aspect worth mentioning here is the probable absence of frosts on top of most tepuis (except the summit of Pico da Neblina). It appears that no ice formation has ever been documented on the summit of Roraima (2810 m), which is permanently windswept by the cold trade winds and is one of the highest mountains in eastern Venezuela. Therefore, it can be assumed that the plants and animals living in this upper life zone of the Guiana Highlands do not possess special physiological mechanisms for frost resistance, unlike most alpine plants of Europe, Asia, the Andes and North America.

The virtual total lack of weather records is the main limitation to discussing the climate of the interior Guiana region, and especially of its montane environment. The only data available to the public, although with notable gaps and methodological deficiencies, are from a few larger settlements with airports, almost always located near a larger river, and often not reflecting the macro- or mesoclimate of a given landscape or geographical entity. In the late 1960s, a modern network of meteorological stations was installed in Venezuela along the main watercourses and in selected regional centers by CODESUR, a massive Venezuelan governmental development program for the southern region. However, even then, no permanent rainfall gauges or temperature stations were installed on top of any mountain in the Guiana region.

These limitations in climatic and environmental data preclude a proper assessment of the relationship between abiotic variables and species-richness patterns, and make it difficult to predict species' potential responses to climate change. For such studies, data must be mostly derived from modeling approaches and data interpolation, sometimes aided by remote sensing techniques (Waltari et al. 2014).

23.5 Guiana Orography

The Guiana Highlands, reaching almost 3000 m on just one mountain summit, are much lower than comparable large mountain systems in South America (Andes), Africa (East African volcanic chain) and tropical Asia (e.g., Mt. Wilhelm in New Guinea), all of which exceed at least 4000 m in height. They also have considerable orographic and biogeographic zonation, although, in the case of the tepuis, that zonation is not as recognizable, especially in those mountains with sheer walls on all sides. Because of the impressive cliffs crowning

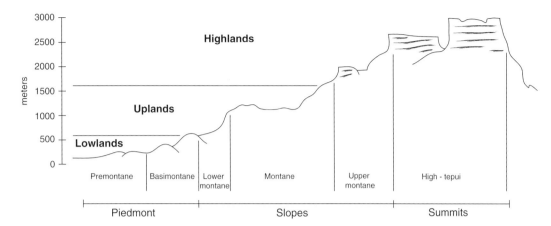

Figure 23.4 Schematic altitudinal transect of Pantepui.

(e.g., the mighty Roraima massif: Figure 23.2c), it has been generally assumed that the classic subdivision of altitudinal life zones – such as the one documented by Humboldt for the Chimborazo volcano in Ecuador (Humboldt & Bonpland 1807) – cannot be applied to mountains lacking a continuously inclined slope. This view implies that a sustained exchange of plant and animal species between the lower slope and the mountain summit has been minimal for a long time, and that, therefore, the summit biota were physically isolated from the lower-lying slope and lowland biota. This so-called "Lost World" effect was vividly described in Conan Doyle's homonymous novel (Doyle 1912).

However, this is only true in a few tepuis, such as Roraima, Marahuaka and Autana. Only after the exploration of more tepuis, and especially of the larger massifs in the central and southern sections of the Venezuelan Guiana Highlands, did it became clear that such isolated summit plateaus, like that of Roraima, are the exception (only 14), and that the great majority of the 60+ tepuis and mountains of the Guiana Highlands have a diverse and complex orography, enabling biotic interchange between lowland, upland and highland biota during both the past and the present (Figure 23.4, and the evolutionary overview in Section 23.8).

Today, we know that the uppermost, usually nonforested tepui life zone forms a fragmented altitudinal belt at submontane to montane elevations, similar but not corresponding to the alpine belt above the forest line of other cordilleras of the world, where the climatic gradient is much more influential than in the tropical mid-elevation mountains of the Guiana Highlands. This specific tepui belt, called "Pantepui" in the biogeographical sense, is not restricted to a definite altitudinal level, but can be found, although with variable taxonomic composition, at all levels between 1200 and 3000 m, and so far no trends for particular topographic or environmental conditions have been documented.

23.6 Phytogeographical Provinces in the Guiana Shield

Based on the great number of botanical collections made during the past 100 years in many areas of the Guiana Highlands (see Howard & Boom 1980; Steyermark et al. 1995–2005), a more detailed phytogeographic definition, subdivision and floristic characterization of the Pantepui than that given by Mayr & Phelps (1967) was needed. In spite of the lack of sufficient botanical exploration in some areas and at all elevations (especially in southern Venezuela), a proposal was presented (Huber 1988, 1994, 1995b; Berry et al. 1995), of which a slightly modified version is given here.

The phytogeographic Guiana Region (*Región Guayana* in Spanish, *Região Guiana* in Portuguese) is composed of the following four provinces (Figure 23.5):

1) *Eastern Guiana Province*, in the coastal lowlands and uplands of the Atlantic foreland, from the Orinoco delta to Amapá state in Brazil.
2) *Central Guiana Province*, covering the core area of the Guiana region: the lands where the main mountain systems and tepuis are located. The major part of this province is in southern Venezuela, as well as north-western Guyana, south-western Suriname, the northern frontier region of Brazil with Venezuela and the Chiribiquete Mountains in Colombia. The main characteristic of this province is its extensive forest ecosystems, found in the lowlands and uplands (up to ca. 1000–1500 m) and covering the lower and medium-elevation slopes of the tepuis, usually up to the rocky walls or to the submontane cloud forest belt. Some non-forest ecosystems, such as savannahs, saxicolous (rock-dwelling) vegetation and mid-elevation shrublands also occur there.
3) *Pantepui Province*, including all upper and high-mountain ecosystems of the Guiana region, located

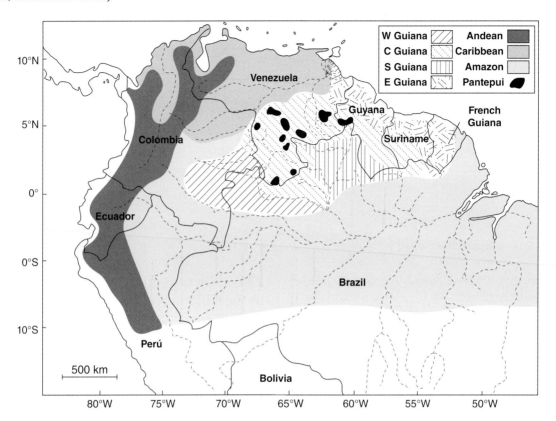

Figure 23.5 Phytogeography of the Guiana Region, showing its four provinces and the four floristic regions of northern South America. Major rivers are indicated by dotted lines, political by solid lines. Regions (Andean, Caribbean and Amazon) are indicated with solid gray shading, and the provinces of the Guiana Region (W, Western; C, Central; S, Southern; E, Eastern) with patterned shading, except the Pantepui Province, which is indicated in solid black. *Source:* Adapted from Berry et al. (1995).

above the continuous submontane forest belt of the Central Guiana Province. In biogeography, the term "Pantepui" refers to the high-mountain ecosystems of the Guiana Region in equatorial South America at elevations from 1300 to 3000 m, found predominantly on Precambrian sandstone table mountains or occasionally on plateaus of different lithology (e.g., intrusive diabase, granite). When present, it also includes a saxicolous biota growing on the rocky walls and cliffs. The typical ecosystems of Pantepui are dominated by broadleaved meadows and sclerophyllous woodlands growing on peat, pioneering saxicolous plant communities and small patches of low forests growing in depressions and crevasses (see Brewer-Carías 1978; George 1988; Huber 1992). The most characteristic plant families of Pantepui are the herbaceous Rapateaceae, the woody Bonnetiaceae (Figure 23.2d) and the pitcher-plant family Sarraceniaceae, with its endemic pantepuian genus *Heliamphora*. According to recent studies, Pantepui, as defined in this broad ecological and phytogeographical way, harbors nearly 1300 endemic plant species in over 60 endemic genera, growing in a highly fragmented area of less than 8000 km^2

(tepui summits and a transition zone at mid-elevations between 1300 and 1800 m); these figures differ from those cited by Berry & Riina (2005), due mainly to substantial changes in modern taxonomy and a wider geographical delimitation.

4) *South and western Guiana Province*, comprising mainly the low- and uplands of the Brazilian Guiana, extending along the slopes descending from the Pirapucú–Neblina–Tapirapecó–Parima–Pakaraima mountain chain (and the watershed between Venezuela and Brazil) towards the river valley of the Amazon, as well as the eastern Colombian lowlands and mid-elevation sierras of Naquén, Araracuara and Chiribiquete. The predominant vegetation types of this province are lowland rainforests, frequently found growing on white sand soils derived from the adjacent Roraima sandstone massifs and called *Caatinga Amazônica* in northern Brazil. Another characteristic ecosystem is the granite outcrops (*lajas* or inselbergs), with their specialized saxicolous lowland flora, also rich in endemics. This province is in direct contact with the vast lowland forests of the Amazonian biome, with a transition zone that may be difficult to delimit biogeographically.

23.7 Animal Life in the Pantepui Region

As with the plants, the avifauna of the Pantepui region encompasses considerable endemism. Of the 628 species recorded, 41 are thought to be endemic. Birds are more vagile than plants, and although some species are endemic to the region, they tend to be more widespread within Pantepui. Examples include three species of manakins (*Pipra cornuta*, *P. pipra* and the rather widespread olive manakin *Chloropipo uniformis*); the duida grassfinch (*Emberizoides duidae*), endemic to Duida; the tepui tinamou (*Crypturellus ptaritepui*), a bird of the moist forests between 1330 and 1800 m that is only recorded on Chimantá-tepui, Sororopán tepui and – as its name implies – Ptari-tepui; and the roraiman nightjar (*Setopagis whitelyi*), known only from Roraima, Ptari-tepui, Jaua, Urutani and Duida. A wider-ranging endemic is the tepui swift (*Streptoprocne phelpsi*), usually seen flying along vertical cliffs. Two species of hummingbirds are widespread, but confined to the tepui region: the rufous-breasted sabrewing (*Campylopterus hyperythrus*) and the buff-breasted sabrewing (*C. duidae*). The roraiman antbird (*Myrmelastes saturatus*) is confined to Mount Roraima and its environs, but the white-throated foliage-gleaner (*Syndactyla roraimeae*) is much more widespread. Other notable tepui widespread endemics include the tepui parrotlet (*Nannopsittaca panychlora*), the tepui goldenthroat (*Polytmus milleri*), the velvet-browed brilliant (*Heliodoxa xanthogonys*), the brown-breasted antpitta (*Myrmothera simplex*) and the Roraima barbtail (*Roraima adusta*) (see Mayr & Phelps 1967; Hilty 2003).

Only a few of the 186 known mammal species from the region are endemic. Examples include the Roraima mouse (*Podoxymys roraimae*), in a genus of its own, which is known only from the Venezuelan side of Mount Roraima; Tyler's mouse opossum (*Marmosa tyleriana*), known from only four montane localities; and another marsupial, *Marmosops pakaraimae*, described by Voss et al. (2013) from the Pakaraima Highlands of the eastern subregion of Pantepui. Among the amphibians and reptiles found in the Pantepui, 68% are endemic (e.g., the black frog genus *Oreophrynella*: Figure 23.2e), and most are confined to one or a few tepuis (Gorzula & Señaris 1998). Reptilian endemics include the gymnothalamid lizard *Riolama inopinata* from Murisipán-tepui, which is so different from other members of the family that a new subfamily was erected to accommodate it (Kok et al. 2015).

23.8 Evolution of the Pantepui Biota

As previously described, the Pantepui mountain system has a considerably older geological history as compared to other mountain systems in South America. In addition, the region has been geologically stable through hundreds of millions of years, with the exception of low levels of uplift and the long-term erosion that formed the tepuis from the original bedrock. Hence, the evolutionary history of the lineages occurring here should have little or nothing to do with mountain uplift and other major landscape changes, as has been documented for other parts of Amazonia and the tropical Andes (Hoorn et al. 2010; Wesselingh et al. 2010; Lagomarsino et al. 2016).

The evolutionary history of the biodiversity on the tepui mountains has attracted the attention of prominent scientists, including the Schomburgk brothers (Schomburgk 1848), Mayr & Phelps (1967), Maguire (1979) and Steyermark (1986). The key questions addressed to date refer, inter alia, to whether or not the tepui biota comprises ancient relicts of once widespread Gondwanan taxa (as proposed in Doyle's *The Lost World*, which includes dinosaurs: Doyle 1912); whether disjunct taxa atop the tepuis obtained their current distribution through vicariance, caused by bedrock erosion or dispersal; whether the tepuis have been a source or sink of biodiversity to neighboring biomes; and whether the tepui diversity mainly derives from lowland ancestors or from species in other mountain regions pre-adapted to similar environmental conditions. Although many of these questions remain poorly addressed, some patterns have emerged, mainly thanks to recent studies based on molecular sequence divergence of extant species.

It is now clear that the tepui biota has been much more dynamic and connected to other regions and habitats than was originally thought. Some of its elements appear to have originated in situ, such as the species-rich plant family Bromeliaceae, which probably originated some 100 Ma in the Guiana region and from there dispersed to other Neotropical regions (Givnish et al. 2011). However, this deep origin does not imply that the single species in the family naturally occurring in West Africa achieved this distribution through the break-up of Gondwana: rather, this represents a much more recent (Miocene) long-distance dispersal event, a pattern mirrored in the family Rapateaceae (Givnish et al. 2000).

Tepui diversity has its origins in multiple regions and habitats, as has been documented for other mountain systems around the world (Antonelli 2015; Merckx et al. 2015). Among liverworts, up to one-third of all species are thought to derive from cold-adapted Andean ancestors, whereas only ca. 10% of the species originally come from lowland Amazonia (Désamoré et al. 2010). In contrast, at least one of the relatively few Pantepui endemic mammal species, the opossum *Monodelphis reigi*, descends from Amazonian ancestors, colonizing the tepuis only in the early Pleistocene (Pavan et al. 2016). In the case of the Roraima mouse *Podoxymys roraimae*, this species has its closest relatives in the Brazilian savannahs (Leite et al. 2015).

It remains unclear whether there have been any general mechanisms facilitating biotic interchange among the various tepuis. There is scarce pollen evidence (stretching ca. 6500 years) for vertical vegetation shifts (Rull 2004), and there are climatic models (Rull & Nogué 2007) that partly support a model of biotic interchange among tepuis during glacial periods (Rull 2005). There is molecular evidence for genetic interchange and speciation long post-dating mountain formation by erosion, such as in the tree frog genus *Tepuihyla*, whose extant species only diversified in the Pliocene (Salerno et al. 2012), and to which newly formed cryptic species may exist (Salerno et al. 2015). Clearly, much further molecular work is required in order to shed light on the evolution of the rich and endemic Pantepui biota.

23.9 Conclusion

Undoubtedly, the scientific exploration of Pantepui has made impressive progress in the course of the last century. Interdisciplinary research is beginning to be published on some important characteristics of the Pantepui environment and ecology (e.g., Zinck & Huber 2011). But still too many areas of this wide montane tropical archipelago are waiting for more detailed biological and geological exploration. This is necessary both to fill out the as yet very uneven floristic and faunistic inventory and also to help us better understand the evolutionary and geological history of this "Lost World."

As long as these remote montane areas continue to be relatively difficult to reach and all of them remain legally protected in six national parks, twenty-six natural monuments, one biosphere reserve and one IUCN World Natural Heritage Site, it is hoped that the Guiana tepuis will avoid future threats – such as unsustainable touristic development and gold and diamond mining. In the future, it is important to monitor the reaction of these unique high-mountain ecosystems to the impact of climate changes, especially because there is no escape to higher life zones for the biota.

Acknowledgments

OH thanks an anonymous supporter for funds to work on this chapter. OH and GTP thank the various systematists who have helped us to use an updated taxonomy, and Charles Brewer Carías and Javier Mesa for the use of their color images. AA is supported by grants from the Swedish Research Council (B0569601), the European Research Council under the European Union's Seventh Framework Programme (FP/2007–2013, ERC Grant Agreement n. 331024), the Swedish Foundation for Strategic Research and a Wallenberg Academy Fellowship.

References

Adlassnig, W. Pranji, K., Mayer, E. et al. (2010) The abiotic environment of Heliamphora nutans (Sarraceniaceae): pedological and microclimatic observations on Roraima Tepui. *Brazilian Archives of Biology and Technology* **53**, 425–430.

Antonelli, A. (2015) Multiple origins of mountain life. *Nature* **524**, 300–301.

Aubrecht, R., Lánczos, T., Šmída, B. et al. (2008) Venezuelan sandstone caves: a new view on their genesis, hydrogeology and speleothems. *Geologia Croatica* **61**, 345–362.

Aubrecht, R., Barrio-Amorós, C.L., Breure, A.S.H. et al. (2012) *Venezuelan Tepuis: Their Caves and Biota*. Acta Geologica Slovaca – Monograph. Bratislava: Comenius University.

Berry, P.E. & Riina, R. (2005) Insights into the diversity of the Pantepui Flora and the biogeographic complexity of the Guayana Shield. *Biologiske Skrifter* **55**, 145–167.

Berry, P.E., Huber, O. & Holst, B.K. (1995) Floristic analysis and phytogeography. In: Steyermark, J.A., Berry, P.E. & Holst, B.K. (eds.) *Flora of the Venezuelan Guayana 1: Introduction*. St. Louis, MO: Missouri Botanical Garden and Portland, OR: Timber Press.

Beyer, S.R., Hiatt, E.E., Kyser, K. et al. (2015) Stratigraphy, diagenesis and geological evolution of the Paleoproterozoic Roraima Basin, Guyana: links to tectonic events on the Amazon Craton and assessment for uranium mineralization potential. *Precambrian Research* **267**, 227–249.

Bogotá-Ruiz, J. (1982) Estratigrafía del Paleozóico inferior en el área amazónica de Colombia. *Geología Norandina* **6**, 29–38.

Brewer-Carías, C. (1975) *Sarisariñama*. Caracas: Ediciones Fundación Explora.

Brewer-Carías, C. (1978) *La vegetación del mundo perdido*. Caracas: Fundación Eugenio Mendoza.

Cox, D.C., Wynn, J.C., Skidder, G.B. & Page, N.J. (1993) Geology of the Venezuelan Guayana Shield. In: USGS and CVG Técnica Minera CA. *Geology and Mineral Resource Assessment of the Venezuelan Guayana Shield*. USGS Bulletin 2062. Reston, VA: USGS, pp. 9–15.

Désamoré, A., Vanderpoorten, A., Laenen, B. et al. (2010) Biogeography of the Lost World (Pantepui region, northeastern South America): insights from bryophytes. *Phytotaxa* **9**, 254–265.

Doyle, A.C. (1912) *The Lost World*. London: Hodder & Stoughton.

Estrada, J. & Fuertes, J. (1993) Estudios botánicos en la Guayana Colombiana IV. *Notas sobre la vegetación y la flora de la Sierra de Chiribiquete. Revista de la Academia Colombiana de Ciencias* **18**, 484–497.

Galán, C. (1992) El clima. In: Huber, O. (ed.) *El Macizo del Chimantá, Escudo de Guayana, Venezuela. Un ensayo ecológico tepuyano*. Caracas: Oscar Todtmann Editores.

Galvis, J., Huguett, A. & Ruge, P. (1979) Geología de la Amazonia Colombiana. *Boletín Geológico INGEOMINAS* **22**, 3–86.

George, U. (1988) *Inseln in der Zeit. Venezuela – Expeditionen zu den letzten weißen Flecken der Erde*. Hamburg: GEO Verlag.

Ghosh, S.K. (1985) Geology of the Roraima Group and its implications. In: *Memoria Simposium Amazonico, 1st, Venezuela, 1981: Caracas, Venezuela*. Dirección General Sectorial de Minas y Geología, Publicación Especial 10, pp. 22–30.

Gibbs, A.K. & Barron, C.N. (1993) *Geology of the Guiana Shield*. Oxford: Oxford University Press.

Givnish, T.J., Evans, T.M., Zjhra, M.L. et al. (2000) Molecular evolution, adaptive radiation, and geographic diversification in the amphiatlantic family Rapateaceae: evidence from ndhF sequences and morphology. *Evolution* **54**, 1915–1937.

Givnish, T.J., Barfuss, M.H., Van Ee, B. et al. (2011) Phylogeny, adaptive radiation, and historical biogeography in Bromeliaceae: insights from an eight-locus plastid phylogeny. *American Journal of Botany* **98**, 872–895.

Gorzula, S. & Señaris, J.C. (1998) Contribution to the Herpetofauna of the Venezuelan Guayana I. Database. *Scientia Guaianæ* 8. Caracas.

Hackley, P.C., Urbani, F., Karlsen, A.W. & Garrity, C.P. (2005) Geologic Shaded Relief Map of Venezuela. USGS Open File Report 2005, 1038.

Hilty, S.L. (2003) *Birds of Venezuela*, 2nd edn. Princeton, NJ: Princeton University Press.

Hoorn, C., Wesselingh, F.P., ter Steege, H. et al. (2010). Amazonia through time: Andean uplift, climate change, landscape evolution, and biodiversity. *Science* **330**, 927–931.

Howard, R.A. & Boom, B.M. (1990) Bassett Maguire – an annotated biography. In: Buck, W.R., Boom, B.M. & Howard, R.A. (eds.) *The Bassett Maguire Festschrift. A Tribute to the Man and his Deeds*. Memoirs of the New York Botanical Garden 64, pp. 1–28.

Huber, O. (1976) Observaciones climatológicas sobre la región del Auyan-tepui (Edo. Bolívar). *Boletín de la Sociedad Venezolana de Ciencias Naturales* **32**, 509–525.

Huber, O. (1988) Guayana highlands versus Guayana lowlands, a reappraisal. *Taxon* **36**, 595–614.

Huber, O. (1992) La vegetación. In: Huber, O. (ed.) *El Macizo del Chimantá, Escudo de Guayana, Venezuela. Un ensayo ecológico tepuyano*. Caracas: Oscar Todtmann Editores.

Huber, O. (1994) Recent advances in the phytogeography of the Guayana region, South America. *Mémoires de la Société de Biogéographie* (3. Série) **4**, 53–63.

Huber, O. (1995a) Geographical and physical features. In: Steyermark, J.A., Berry, P.E. & Holst, B.K. (eds.) *Flora of the Venezuelan Guayana 1: Introduction*. St. Louis, MO: Missouri Botanical Garden and Portland, OR: Timber Press.

Huber, O. (1995b) Vegetation. In: Steyermark, J.A., Berry, P.E. & Holst, B.K. (eds.). *Flora of the Venezuelan Guayana 1: Introduction*. St. Louis, MO: Missouri Botanical Garden and Portland, OR: Timber Press.

Humboldt, A. & Bonpland, A. (1807) *Essai sur la Géographie des Plantes*.

Ibáñez-Mejía, M., Ruiz, J., Valencia, V.A. et al. (2011) The Putumayo Orogen of Amazonia and its implications for Rodinia reconstructions: new U–Pb geochronological insights into the Proterozoic tectonic evolution of northwestern South America. *Precambrian Research* **191**, 58– 77.

Keats, W. (1976) *The Roraima Formation in Guyana: A Revised Stratigraphy and a Proposed Environment of Deposition*. Memoria 2ndo Congreso Latinoamericano de Geología, Caracas 1973, pp. 901–940.

Kelloff, C.L., Alexander, S.N., Funk, V.A. & Clarke, H.D. (2011) *Smithsonian Plant Collections, Guyana: 1995–2004, H. David Clarke*. Smithsonian Contributions to Botany 97, pp. 1–307.

Khobzi, J., Kroonenberg, S.B., Faivre, P. & Weeda, A. (1980) *Aspectos geomorfológicos de la Amazonia y Orinoquia colombianas*. Revista CIAF, Bogotá 5, pp. 97–126.

Kok, P.J.R. (2015) A new species of the Pantepui endemic genus Riolama (Squamata: Gymnophthalmidae) from the summit of Murisipán-tepui, with the erection of a new gymnophthalmid subfamily. *Zoological Journal of the Linnean Society* **174**, 500–518.

Kroonenberg, S.B. (1978) Precambrian paleosols at the base of the Roraima Formation in Surinam. *Geologie en Mijnbouw* **57**, 445–450.

Kroonenberg, S.B. (1982) A Grenvillian granulite belt in the Colombian Andes and its relation to the Guiana Shield. *Geologie & Mijnbouw* **61**, 325–333.

Kroonenberg, S.B. & de Roever, E.W.F. (2010) Geological evolution of the Amazonian craton. In: Hoorn, C. & Wesselingh, F.P. (eds.) *Amazonia, Landscape and Species Evolution*. Chichester: John Wiley & Sons Ltd., pp. 9–28.

Kroonenberg, S.B. & Reeves, C.V. (2012) Geology and petroleum potential, Vaupés-Amazonas Basin, Colombia. In: Cediel, F. (ed.) *Petroleum Geology of Colombia, 15*. Medellín: Universidad EAFIT, pp. 1–92.

Kroonenberg, S.B., de Roever, E.W.F., Fraga, L.M. et al. (2016) Paleoproterozoic evolution of the Guiana Shield in Suriname: a revised model. *Netherlands Journal of Geosciences – Geologie en Mijnbouw* **95**, 491–522.

Lagomarsino, L.P., Condamine, F.L., Antonelli, A. et al. (2016) The abiotic and biotic drivers of rapid diversification in Andean bellflowers (Campanulaceae). *New Phytologist* **210**, 1430–1442.

Leite, Y.L., Kok, P.J. & Weksler, M. (2015) Evolutionary affinities of the "Lost World" mouse suggest a late Pliocene connection between the Guiana and Brazilian shields. *Journal of Biogeography* **42**, 706–715.

Maguire, B. (1979) Guayana, region of the Roraima Sandstone Formation. In: Larsen, K. & Holm-Nielsen, L.B. (eds.) *Tropical Botany*. London: Academic Press, pp. 223–238.

Mayr, E. & Phelps, W.H. Jr. (1967) The origin of the bird fauna of the South Venezuelan Highlands. *Bulletin of the American Museum of Natural History* **136**, 269–328.

Mendoza, V.S. (2005) *Geología de Venezuela. Tomo I: Escudo de Guayana, Andes Venezolanos y Sistema Montañoso del Caribe*. Ciudad Bolívar, Venezuela: Universidad de Oriente.

Merckx, V.S.F.T., Hendriks, K.P., Beentjes, K.K. et al. (2015) Evolution of endemism on a young tropical mountain. *Nature* **524**, 347–350.

Pavan, S.E., Jansa, S.A. & Voss, R.S. (2016) Spatiotemporal diversification of a low-vagility Neotropical vertebrate clade (short-tailed opossums, Didelphidae: Monodelphis). *Journal of Biogeography* **43**, 1299–1309.

Priem, H.N.A., Boelrijk, N.A.I.M., Hebeda, E.H. et al. (1973) Age of the Precambrian Roraima Formation in northeastern South America: evidence from isotopic dating of Roraima pyroclastic volcanic rocks in Suriname. *Geological Society of America Bulletin* **84**, 1677–1684.

Reid, A.R. (1972) Stratigraphy of the type area of the Roraima Group, Venezuela. Memoria, Conferencia Geológica Interguianas, 9th, Puerto Ordaz, Venezuela, Caracas. Ministerio de Minas y Hidrocarburos, Boletín de Geologia, Publicación Especial 6, pp. 343–353.

Reis, N.J., Pinheiro, S. da S., Costi H.T. & Costa, J.B.S. (1990) A Cobertura sedimentar Proterozóica Média do Supergrupo Roraima no Norte do Estado de Roraima, Brasil: Atribuições aos seus Sistemas Deposicionais e Esquema Evolutivo da sua Borda Meridional. *Anais Congresso Brasileiro de Geologia* **36**, 66–81.

Reis, N.J., Fraga, L.M., de Faria, M.S.G. & Almeida, M.E. (2003) Geologia do Estado de Roraima, Brasil. *Géologie de la France* **2**, 121–134.

Rull, V. (2004) Is the "Lost World" really lost? Palaeoecological insights into the origin of the peculiar flora of the Guayana Highlands. *Naturwissenschaften* **91**, 139–142.

Rull, V. (2005) Biotic diversification in the Guayana Highlands: a proposal. *Journal of Biogeography* **32**, 921–927.

Rull, V. & Nogué, S. (2007) Potential migration routes and barriers for vascular plants of the Neotropical Guyana Highlands during the Quaternary. *Journal of Biogeography* **34**, 1327–1341.

Sacek, V. (2014) Drainage reversal of the Amazon River due to coupling of surface and lithospheric processes. *Earth and Planetary Science Letters* **401**, 301–312.

Salerno, P., Ron, S.R., Señaris, J.C. et al. (2012) Ancient tepui summits harbor young rather than old lineages of endemic frogs. *Evolution* **66**, 3000–3013.

Salerno, P., Señaris, J., Rojas-Runjaic, F. & Cannatella, D. (2015) Recent evolutionary history of Lost World endemics: population genetics, species delimitation, and phylogeography of sky-island treefrogs. *Molecular Phylogenetics and Evolution* **82**, 314–323.

Santos, J.O.S., Potter, P.E., Reis, N.J. et al. (2003) Age, source and regional stratigraphy of the Roraima Supergroup and Roraima-like outliers in northern South America based on U-Pb geochronology. *Geological Society of America Bulletin* **115**, 331–348.

Schomburgk, M.R. (1848) *Reisen in Britisch-Guiana in den Jahren 1840–1844*, **3** volumes. Leipzig: Verlagsbuchhandlung von J. J. Weber.

Sidder, G.B. & Mendoza, V. (1991) Geology of the Venezuelan Guayana Shield and its Relation to the Entire Guayana Shield. United States Department of the Interior US Geological Survey Open-File Report, pp. 91–141.

Snelling, N.J. (1963) Age of the Roraima Formation, British Guiana. *Nature* **98**, 1079.

Steyermark, J.A. (1986) Speciation and endemism in the flora of the Venezuelan tepuis. In: Vuilleumier, F. & Monasterio, M. (eds.) *High Altitude Tropical Biogeography*. New York: Oxford University Press, pp. 317–373.

Steyermark, J.A., Berry, P.E. & Holst, B.K. (eds.) (1995–2005) *Flora of the Venezuelan Guayana*, **9** volumes. St. Louis, MO: Missouri Botanical Garden Press (vols. 1–9) and Portland, OR: Timber Press (vols. 1–2).

Théry, J.M., Peniguel, T. & Haye, G. (1985) Descubrimiento de acritarcos del Arenigiano cerca a Araracuara (Caquetá-Colombia). Ensayo de reinterpretación de esta región de la saliente del Vaupés. *Geología Norandina* **9**, 3–14.

Voss, R.S., Lim, B.K., Díax-Nieto, J.F. & Jansa, S.A. (2013) A new species of Marmosops (Marsupialia, Didelphidae) from the Pakaraima Highlands of Guyana, with remarks on the origin of the endemic Pantepui mammal fauna. *American Museum Novitates* **3778**, 1–27.

Waltari, E., Schroeder, R., McDonald, K. et al. (2014) Bioclimatic variables derived from remote sensing: assessment and application for species distribution modelling. *Methods in Ecology and Evolution* **5**, 1033–1042.

Wesselingh, F.P., Hoorn, C., Kroonenberg, S.B. et al. (2010) On the origin of Amazonian landscapes and biodiversity: a synthesis. In: Hoorn, C. & Wesselingh, F.P. *Amazonia, Landscape and Species Evolution.* New York: Blackwell, pp. 421–431.

Zinck, J.A. & Huber, O. (eds.) (2011) *Peatlands of the Western Guayana Highlands, Venezuela. Properties and Palaeogeographic Significance.* Ecological Studies 217. Berlin-Heidelberg: Springer.

24

Ice-Bound Antarctica: Biotic Consequences of the Shift from a Temperate to a Polar Climate

Peter Convey[1], Vanessa C. Bowman[1], Steven L. Chown[2], Jane E. Francis[1], Ceridwen Fraser[3], John L. Smellie[4], Bryan Storey[5] and Aleks Terauds[6]

[1] British Antarctic Survey, Cambridge, UK
[2] School of Biological Sciences, Monash University, Melbourne, Australia
[3] Fenner School of Environment and Society, Australian National University, Canberra, Australia
[4] Department of Geology, University of Leicester, Leicester, UK
[5] Gateway Antarctica, University of Canterbury, Christchurch, New Zealand
[6] Australian Antarctic Division, Kingston, Tasmania, Australia

Abstract

The Antarctica of today is a forbidding place: isolated from the other landmasses of the planet, with more than 99.6% of its surface covered in permanent ice, which is on average more than 2 km thick. Organisms that live there face chronic and extreme environmental stresses, and generally belong to cryptic and microscopic groups that the casual observer often fails even to notice. No wonder the early human explorers a little over a century ago considered it a barren and desolate place! It has not always been like this, and Antarctica's fossil record tells us that it once hosted lush temperate and even subtropical forests, dinosaurs, early mammals and diverse groups of other biota. The main Antarctic landmass has been located at high southern paleolatitudes for around 130 My. However, while transient alpine (mountaintop) glaciations are likely to have been a feature of high Antarctic mountains even during the "greenhouse Earth" periods of the geological past, the first ice sheets did not form until at least 34 Ma, in a period of declining global atmospheric CO_2 concentrations and global cooling. Antarctica originally formed part of the southern supercontinent called Gondwana. As that broke up, the other southern continents drifted northwards, finally leaving the isolated Antarctica that we see today. The opening of the Drake Passage and Tasman Gateway led to the eventual physical isolation of the continent and – once this was surrounded by deep ocean water – allowed the formation of the Antarctic Circumpolar Current in the Southern Ocean at 25–20 Ma. While this accelerated the formation of continent-wide ice sheets, tundra ecosystems and cool temperate forests are thought to have persisted until at least 14 Ma on parts of East Antarctica, and possibly as late as 5 Ma on the Antarctic Peninsula. Montane systems are a fundamental element of Antarctica's architecture, with different parts of the continent undergoing several episodes of intense volcanism and mountain building over geological time. Vast mountain ranges – the Transantarctic Mountains, the Ellsworth Mountains, the Antarctic Peninsula, the enigmatic ice-buried Gamburtsev Mountains and other ranges across the continent – characterize the Antarctica of today. Indeed, arguably most of Antarctica's contemporary biology is found in what would elsewhere be characterized as montane or alpine ecosystems, from the coasts to the high mountain summits. The contemporary biology of Antarctica, despite its apparently low diversity and cryptic nature, has revealed remarkable and important insights over the last decade or so, particularly as the rapidly developing tools of molecular biology have been brought to bear on it. These show that, across all areas of the continent, the life that is present today is not the result of sporadic and opportunistic recent colonization processes following complete glacial wipe-out, but rather in large part represents the last vestiges of the ancient fauna, flora and microbiota that Antarctica inherited as it became free from Gondwana. Over the millions and tens of millions of years of its isolation, this biota has evolved and radiated, and we are only now appreciating the complex signal of regional endemism and dispersal contained in its contemporary biodiversity, and the implications this has for understanding the geological and glaciological history of the continent.

Keywords: *adaptation, evolution and radiation, extinction, glaciation, montane ecosystems, refugia, terrestrial biota, volcanism*

Mountains, Climate and Biodiversity, First Edition. Edited by Carina Hoorn, Allison Perrigo and Alexandre Antonelli.
© 2018 John Wiley & Sons Ltd. Published 2018 by John Wiley & Sons Ltd.
Companion website: www.wiley.com\go\hoorn\mountains,climateandbiodiversity

24.1 Introduction

It is only just over a century since the first human landings on the Antarctic continent, and the initiation of the "Heroic Age" of exploration. Antarctica (Figure 24.1) was the last continent to be discovered and explored, and even now areas exist that have barely or never been visited by humans. Roughly twice the size of Australia,

Antarctica is a continent of extremes, inspiring awe and fascination. It is the coldest, windiest and highest continent, and, with an average depth of around 2 km, it holds the majority of the planet's ice. Antarctica's extremes place it at the end of many global environmental gradients, providing a focus for ecological, physiological and evolutionary studies (Peck et al. 2006; Convey et al. 2014).

Figure 24.1 Overview map of Antarctica, indicating major regions of the continent and surrounding Southern Ocean mentioned in the text. Gray shading indicates permanent floating ice shelves.

Recent studies have estimated that the continental area exposed from the otherwise permanent cover of snow or ice ranges from $45\,886\,km^2$ (0.32%) (Terauds & Lee 2016) to $21\,745\,km^2$ (0.18%) (Burton-Johnson et al. 2016). In consequence, most contemporary terrestrial habitats are small and, in effect, islands. Nevertheless, a surprising and diverse range of terrestrial ecosystems and organisms is found (e.g., Convey 2013), especially among the microbiota (López-Bueno et al. 2009; Cary et al. 2010; Aguirre de Cárcer et al. 2015; Cavicchioli 2015), and the complexity of the continent's biogeography and biological history has only recently come to be appreciated (Convey et al. 2008; Terauds et al. 2012; Terauds & Lee 2016).

With ice covering the continent to and beyond the geographical coastline (Bentley 1991), most of Antarctica's areas of ice-free habitat are rocky mountain peaks (nunataks) and ranges (e.g., the Transantarctic Mountains) exposed above the surface (Fretwell et al. 2013). The entire Antarctic Peninsula, as well as the island archipelagos of the Scotia Arc, is geologically continuous with the Andean mountain chain of South America (Barker et al. 1991).

Much of Antarctica is climatically a frigid desert. But in addition to true desert ecosystems, such as the McMurdo Dry Valleys and elsewhere in the Transantarctic Mountains, terrestrial ecosystems include the nunataks, cliffs and montane habitats of the continental interior, the moister fellfields that characterize the Antarctic Peninsula and Scotia Arc (as well as small "oases" along the continental coastline) and the considerably more developed, vegetated and often marine vertebrate-fertilized sub-Antarctic islands. Many of these ecosystems are often apparently biologically barren to those not familiar with them, and their biodiversity is composed of organisms that are small to microscopic (Cavicchioli 2015; Chown et al. 2015).

24.2 Early Geological History of Antarctica

Rocks in Antarctica tell a tale of a continent that has experienced a wide range of climates, has seen a variety of life forms and has hosted mountain ranges that are now completely eroded away (Tingey 1991; Torsvik et al. 2008). At the same time, reconstructions of atmospheric gas compositions paint a picture of long-term and sometimes drastic global climate variation (Retallack 2001; Joachimski et al. 2012; Sun et al. 2012). The rocks of East Antarctica were part of a huge supercontinent that moved across the globe before breaking up and positioning Antarctica as an isolated continent at the South Pole (see illustrations available through the Palaeomap project, www.scotese.com). The rocks of West Antarctica

have a variety of origins, and most of them are relatively young additions to the continent (Dalziel & Elliott 1982). Those of the Antarctic Peninsula are related to rocks of Patagonian South America (Barker et al. 1991).

From 550 to 450 Ma, during the Cambrian and Ordovician periods, the Pacific Oceanic crust subducted beneath Gondwana along what is now the edge of East Antarctica, resulting in a mountain-building episode known as the Ross Orogeny (Stump 1995). This most likely ended when a mass of continental crust on the subducting plate collided with East Antarctica and stopped the subduction process. The Ross Orogeny brought deep-level rocks to the surface, including granite, the solidified roots of volcanoes and metamorphic rocks formed deep in Earth's crust during subduction. Once subduction had stopped, erosion of these mountains produced an extensive surface, known as the Kukri Peneplain, which cuts through the igneous and metamorphic roots of the Ross Orogeny.

Over the next 300 million years of Gondwana's history, a 2.5 km-thick sequence of sedimentary rocks known as the Beacon Supergroup (Barrett 1991) was deposited on the Kukri Peneplain in Antarctica (Figure 24.2). During that time, Gondwana experienced climatic variations that affected the whole globe and which are reflected in the fossil record and the rock compositions. In addition, Gondwana crossed different climatic regions as it moved from the equator towards the South Pole, which led to a constantly evolving flora and fauna (Torsvik et al. 2008).

The oldest rocks at the base of the Beacon Supergroup show that the East Antarctic segment of Gondwana supported large meandering rivers and was partly covered in lakes and shallow seas during the Devonian period (Barrett 1991). This period saw rapid evolution of fishes across the world, and the rivers of Antarctica were no exception. Fossils of various fishes, including primitive lungfish and armor-plated placoderms, are preserved in late Devonian strata of the southern Victoria Land sector of the Transantarctic Mountains. The Beacon Supergroup rocks also suggest that about 300 Ma, during the Carboniferous period, the climate changed and the high latitudes of Gondwana became partially covered by a large ice sheet. This ice sheet advanced and retreated, depositing glacial debris in the Transantarctic Mountains as thick layers of sediment known as tillite. As the ice moved, boulders frozen within it cut furrows and grooves in the underlying rock surface that are still visible today, and which record the passage of the ice sheet. After the Carboniferous glaciation, the climate warmed gradually, and plants that thrived in extensive cool-temperate swamps evolved quickly during the Permian period. This swamp vegetation eventually turned into coal deposits, now seen as thick seams at several places in the Transantarctic Mountains.

Figure 24.2 (a) North-east face of Finger Mountain, Victoria Land, showing three key elements of the stratigraphy: 1) the Devonian Beacon quartz sandstone (horizontal massive light-colored rocks) grading into thin late Devonian fish-bearing sandstone and siltstone; 2) a thin white feldspathic sandstone, marking an unconformity in which the Carboniferous System is missing, and which forms the base of the Permian Glossopteris-bearing coal measures; and 3) a largely sand-filled channel cut into Devonian sandstone by pre-Permian glaciers. The dark brown horizontal layer (lower middle) is a 200 m-thick sill of Ferrar Dolerite, which connects with an inclined sheet of dolerite reaching the summit from the right. *Source:* Courtesy of J. Smellie. (b) Fossilized leaves of *Glossopteris* from Permian sediments, Allan Hills, Transantarctic Mountains. The large leaf is approximately 14 cm long. *Source:* Courtesy of J. Francis. (c) Large piece of fossilized tree trunk, latest Cretaceous age, Seymour Island, Antarctic Peninsula. The log is approximately 60 cm in height. *Source:* Courtesy of J. Francis. See also Plate 38 in color plate section.

Approximately 250 Ma, at the end of the Permian period, up to 96% of all marine species and 70% of all land vertebrate species died out, in Earth's most extreme extinction event (Hallam & Wignall 1997). Many plants were also affected, including *Glossopteris*. The initial triggers of the extinction were probably gradual environmental changes such as dropping sea levels, the development of anoxic conditions in the oceans and increased aridity on land. Late Permian volcanism in the Siberian traps is closely associated with the timing of this extinction event (Burgess & Bowring 2015). A later catastrophic event, such as a large meteor impact (Kaiho et al. 2001), which could have led to a sudden release of methane hydrates from the sea floor, may also have acted on this already stressed system and made the extinction particularly devastating. Nonetheless, certain groups, including some conifers, ferns and ginkgos, survived the extinction. They became part of the next burst in evolution,

during the warm and wet Triassic period, which produced many plants that are widely represented in South America, Australia and New Zealand today. This new flora was characterized by *Dicroidium* (a seed fern with forked fronds), podocarps and horsetails. As the Triassic progressed, the climate became increasingly hot and dry, and many plants evolved drought-resistant adaptations.

24.3 Antarctica and Gondwana: the Break-up of a Supercontinent

When Gondwana eventually broke apart, the continents and landmasses of Antarctica, South America, India, Africa, Australia and New Zealand were formed (Torsvik et al. 2008). The elements that would become Antarctica had a temperate or subtropical climate even though they were already located at relatively high paleolatitude. Gondwana started to break into smaller fragments about 183 Ma, during the Jurassic period. The break-up was marked by intensive basaltic volcanism that lasted less than a million years (Storey 1995). The volcanic activity spanned Gondwana from what is now southern Africa, through East Antarctica along the Transantarctic Mountains and into Tasmania and New Zealand, a distance of over 4000 km. Most volcanoes and vents that formed during this period have since been eroded. However, just below Earth's surface, a network of vertical sheets (dykes) and horizontal layers (sills) of basaltic igneous rock formed, which originally supplied magma to the volcanoes. This structure is now solidified and preserved, and has been uplifted to Earth's surface and exposed by erosion. It is known as the Ferrar Dolerite (Elliot 1992), and the wider basaltic volcanic activity across East Antarctica is known as the Ferrar Large Igneous Province (Storey & Kyle 1997). Volcanic rocks in southern Africa known as the Karoo Province (Cox 1988) are similar to those found in Dronning Maud Land in East Antarctica. The Ferrar and Karoo igneous rocks may have been derived from an abnormally large hot part of Earth's mantle: a plume (Storey & Kyle 1997). As the plume-related thermal anomaly migrated out from its center below South Africa and Dronning Maud Land, it triggered more volcanic "flare-ups" throughout the Antarctic Peninsula and southern South America (Pankhurst et al. 2000). Plumes appear to be associated with unusually large-volume volcanic events, and can also be linked with continental rifting (Storey & Kyle 1997).

The break-up of Gondwana started either at the same time as or just after the outburst of Ferrar volcanic activity in the Jurassic period, and led to a thinning of the continental crust and to the sea encroaching between West Gondwana (South America and Africa) and East Gondwana (East Antarctica, Australia, India and New Zealand) (Lawver et al. 1998). As rifting progressed, new tectonic plates formed in what are now East Africa and the Weddell Sea, ocean crust formed, and sea floor spreading developed. The second stage of the break-up happened about 130 Ma (early Cretaceous), when the South American plate separated from a combined African–Indian plate, which in turn was splitting from the combined Antarctic–Australian plate (Lawver et al. 1998). From 100 to 90 Ma (Late Cretaceous), New Zealand and Australia started to separate from Antarctica, and by approximately 32 Ma the break-up of Gondwana was almost complete. The final stage was the separation of the tip of South America from the Antarctica Peninsula, which opened up the Drake Passage (Whitehead et al. 2006; Livermore et al. 2007).

As Gondwana was breaking apart, the Pacific Ocean floor was continually being subducted beneath the supercontinent in the regions that are now New Zealand, the Antarctic Peninsula and the west coast of South America. This resulted in the formation of a volcanic mountain chain and in sedimentary material being scraped off the subducting Pacific plate and accreted (added) to the western margin of Gondwana. This subduction was the first stage in the emergence of the Antarctic Peninsula (Storey & Garrett 1985). The eroded remnants of the volcanoes and the accreted sediment remain today. The subduction process has now ceased along most of the Peninsula region of the Antarctic plate, because the spreading ridge between the Antarctic and Nazca plates collided with the subduction zone and brought the process to a halt (Larter et al. 2002). However, in a small region in the northern part of the Antarctic Peninsula, the subduction that was initiated during the break-up of Gondwana continues today.

In addition to the formation of new tectonic plates and the large southern hemisphere continents, some small fragments (microplates) were also created, particularly in the South Atlantic region, and their movements help explain some apparently anomalous geological features of Antarctica (Dalziel & Elliot 1982; Grunow et al. 1987), as well as some distinctive features of Antarctic terrestrial biogeography (e.g., Convey & McInnes 2005; Allegrucci et al. 2012). The Ellsworth Mountains of West Antarctica consist of Ross Orogeny and Gondwanan rocks like those of the Transantarctic Mountains, rather than the younger volcanic rocks typical of the rest of West Antarctica. These rocks form the Ellsworth microplate, which was part of East Antarctica but has now moved some distance away from the Transantarctic Mountains and is oriented more or less at a right angle to them (Grunow et al. 1987). The Haag Nunataks are a microplate of 1200 million-year-old metamorphic rocks from East Antarctica, which now form part of West Antarctica (Dalziel & Elliot 1982).

Displaced microplates are also a significant component of the final separation of the South American plate from the Antarctic Peninsula and the creation of the Drake Passage (Barker et al. 1991). The South Georgia block (now part of the South American plate) moved from a location much closer to southern South America to its present position, and the South Orkney block (on the Antarctic plate) moved away from the northern tip of the Peninsula to form the Powell Basin. The South Shetland Islands (on the Shetland plate) are now separated from the Peninsula by Bransfield Strait. A further consequence of the break-up was the initiation of the West Antarctic Rift System (WARS) (Behrendt et al. 1991). The rifting is thought to be part of a process that initially led to the separation of New Zealand from Antarctica, but which continues today with the separation of East and West Antarctica.

The continental rifting represented by the WARS took place within the Antarctic plate itself, between East and West Antarctica (Behrendt et al. 1991; Elliot 2013). On one side of the rift system, in the McMurdo Volcanic Province in the western Ross Sea, volcanic activity began about 40 Ma (Kyle 1990; Rocchi et al. 2002). This area hosts Antarctica's most active volcano, Mount Erebus, and several extinct volcanoes, including Mount Terror and Mount Bird on Ross Island, and Mount Discovery and Mount Morning on the mainland. Although Mount Erebus does not sit on a plate boundary, volcanic activity exists due to tension and extension within the Antarctic plate resulting from the break-up of Gondwana. As the Antarctic plate extended, the continental crust gradually thinned to the point that hot magma from the mantle rose and produced volcanic activity. Associated volcanism occurred along the Transantarctic Mountains as far north as Victoria Land. Today, Mount Melbourne has warm ground, and dating of an ash layer exposed in ice on the volcano's flank suggests an eruption less than 200 years ago. Fumaroles are also present on Mount Rittmann.

The Transantarctic Mountains (ten Brink et al. 1993) are also a product of the crustal tension that formed the WARS. Traditionally, it is thought that during the initial part of continental extension, Antarctica's surface bulged upwards, creating an elevated landscape over a wide area before the eventual rift. The Marie Byrd Land sector of West Antarctica gradually separated from East Antarctica, leaving behind the Transantarctic Mountains, which have steep escarpments on the side of the rift. There is no corresponding escarpment on the West Antarctica side of the rift, because West Antarctica is made of several smaller landmasses, allowing the extension to be distributed over a much wider area within West Antarctica itself. In recent years, the origin of the Transantarctic Mountains and the tectonics of the Ross Sea have been explored further. Numerical modeling suggests the Transantarctic Mountains may have been the marginal remnants of a high plateau that existed until the mid-Cretaceous (~110 Ma), which then subsided and rifted (e.g., Bialas et al. 2007). In contrast, a vast intracratonic basin across West Antarctica is invoked during the late Triassic to late Cretaceous, uplifting the Transantarctic Mountains as a consequence of rifting (Lisker & Läufer 2013).

The terrestrial fauna and flora were not driven extinct during these transitions, with fossil evidence showing communities typical of southern temperate rainforest (Francis & Poole 2002; Bowman et al. 2014). The final stages of Antarctica's isolation, involving the opening of the Drake Passage and the Tasman Gateway (45–30 Ma), eventually permitted the creation of deep-water oceanic circulation surrounding the continent and the formation of the Antarctic Circumpolar Current (ACC) (Pfuhl & McCave 2005; Scher & Martin 2006). The formation of the ACC and the extent of the Southern Ocean isolated the Antarctic continent from the other southern continents. In concert, global cooling driven by decreasing atmospheric CO_2 concentrations led to the East Antarctic Ice Sheet becoming extensive and stable, and to the evolution of isolated and polar-adapted Antarctic organisms and communities, both on land and in the sea.

Today's Antarctica is therefore built of several distinct geological elements whose juxtaposition is a result of post-Gondwanan tectonic activity. These include East Antarctica, West Antarctica (including the Antarctic Peninsula), the Scotia Arc archipelagos (the South Sandwich, South Orkney and South Shetland islands) and South Georgia. Other remote sub-Antarctic islands, while sharing some biological linkages with the more extreme southern regions, have never formed part of another major continental landmass, being either volcanic islands of various ages (Prince Edward Islands, Heard and McDonald Islands, Îles Kerguelen and Crozet) or uplifted oceanic crust (Macquarie Island).

24.4 Volcanism

Antarctica has experienced a range of types of volcanic activity, briefly mentioned in the previous section. They range from one of the best examples globally of an oceanic island arc (the South Sandwich Islands: Figure 24.3a; Leat et al. 2003) to numerous volcanoes of mainly Mesozoic–Cenozoic age related to subduction situated along the Pacific margin (Weaver et al. 1982). The continent contains voluminous products of Jurassic magmatism caused by continental break-up – mainly basaltic intrusions in the Transantarctic Mountains (called the Ferrar Supergroup: Figure 24.2), but also small remnants of large-volume basalt lavas (Kirkpatrick Basalt Formation) and geographically widespread explosively erupted rocks in the Antarctic Peninsula (Elliot 1992; Pankhurst et al. 2000; Storey et al. 2013). Antarctica is also host to one of the world's largest extension-related volcanic provinces, situated mainly within the Neogene (c. <34 Ma) WARS (Figure 24.3b) (LeMasurier & Thomson 1990). Active volcanoes are present in the WARS, attesting to continuing volcanic activity. These include several under the ice, identified remotely by geophysical methods, and there is local evidence for high and probably long-lived subglacial heat fluxes (Lough et al. 2013; Patrick & Smellie 2013; Fisher et al. 2015). Because they were erupting during the period in which the Antarctic ice sheet was initiated and developed (after 34 Ma), many of the volcanoes show abundant conspicuous evidence for interactions with the expanding sheet (Smellie et al. 2009, 2011).

Volcanoes are thus widespread, and are found across the entire continent, although they are apparently rare in East

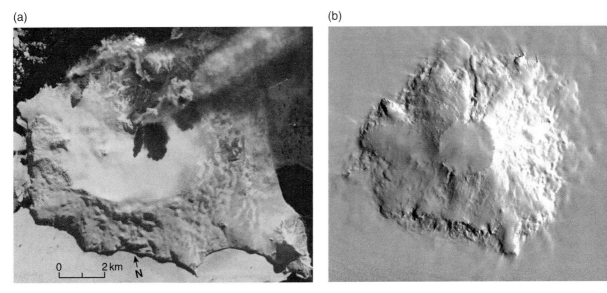

Figure 24.3 (a) Landsat image showing a volcanic eruption on Montagu Island, South Sandwich Islands, in 2005. Although extensively ice-covered and thought to be dead, with no historical records of activity, the volcano unexpectedly burst into life in a long-lived eruption between 2001 and 2007. It thus probably supports currently undetected subglacial geothermal areas possibly capable of sustaining life during glacials. *Source:* Patrick & Smellie (2013). Reproduced with permission of Antarctic Science Ltd. (b) GoogleEarth satellite image showing Mt. Takahe, a large stratovolcano in Marie Byrd Land, lying within the WARS. Although ice-covered and seemingly dead, this volcano is the source of numerous ash layers recovered in regional boreholes, representing eruptions extending back hundreds of thousands of years. It is thus merely quiescent, and, although no surface heat has currently been detected, probably supports subglacial heat under the extensive ice cover. See also Plate 39 in color plate section.

Antarctica, apart from on the western margin of the Ross Sea. Volcanoes can accurately record climate-related environmental changes. The practice of using Antarctica's Cenozoic volcanoes as a proxy for past ice on the continent, including assessing its relative stability, is now well developed (Smellie et al. 2009, 2011, 2014), and volcanoes are being used to document ice-poor or even ice-free interglacial conditions for different geological periods (J. Smellie, unpubl. data). Furthermore, the volcanic heat produced by subglacial eruptions has the potential to affect ice-sheet behavior, by increasing the amount of basal melting and potentially destabilizing the overlying ice. Subglacial volcanoes may thus influence future global change by causing the delivery of increased volumes of ice to the sea and raising global sea levels, although the reality of this threat has not been assessed by modeling studies. Knowledge of Antarctica's volcanism is therefore crucial for understanding the continent more completely, not only in terms of its geological development, but also in terms of its paleoenvironmental, paleoclimatic and biological evolution (Francis et al. 2008; Fraser et al. 2014).

Probably the most influential volcanic episode in terms of modern biological diversity in the region is the one that occurred after 34 Ma, which is associated with the growth and development of the Antarctic Ice Sheet. Antarctica hosts the largest and longest-lived glacio-volcanic province in the world, extending several thousand kilometers from the Antarctic Peninsula, through Marie Byrd Land,

into Victoria Land in East Antarctica (LeMasurier & Thomson 1990; Smellie & Edwards 2016). This province contains a uniquely important record of Antarctic glaciation (Haywood et al. 2009; Smellie et al. 2014) and consists of numerous large volcanoes, together with several volcanic fields formed of multiple small, short-lived volcanic centers. The eruptive lifetimes of the large volcanoes typically lasted for 1–2 My. Conversely, although overall eruptive activity in the individual volcanic fields may have persisted for similar lengths of time as in the large volcanic centers, the small volcanoes in the volcanic fields individually probably erupted for only a few weeks or months (possibly years) and then died. It is the ability of the larger volcanic complexes to sustain geothermal activity for substantial periods of time (many tens of thousands of years, at least) that makes them important for potentially providing habitable refugia in some parts of Antarctica, overlapping in time and capable of sustaining life through multiple glacial cycles (Fraser et al. 2014).

24.5 How Antarctica Became An Ice-bound Continent

The first transient alpine glaciation events may have taken place in Antarctica as early as the late Cretaceous (Bowman et al. 2013), then the mid-Eocene (42 Ma) (e.g., Birkenmajer et al. 2005; Tripati et al. 2005).

Continent-wide ice sheets first formed around the Eocene–Oligocene transition (34 Ma) (e.g., Barrett 1996; Coxall et al. 2005; Rocchi et al. 2006). The geographical separation of Antarctica from the other southern continents is not thought to have been the main driver of ice-sheet formation, with stronger influences coming from changing atmospheric CO_2 concentrations, orbital forcing and ice–climate feedbacks (DeConto & Pollard 2003). Evidence for the formation of permanent ice sheets in East Antarctica and the Antarctic Peninsula suggests that they were probably present by c. 15–12 Ma (Shevenell et al. 2004; Hillenbrand & Ehrmann 2005). It is less clear when the West Antarctic Ice Sheet formed, but grounded West Antarctic ice may have been present in the Ross Sea embayment by c. 16 Ma (Barrett 2003; Rocchi et al. 2006).

There have subsequently been multiple fluctuations in ice-sheet extent and thickness across Antarctica through the Oligocene, Miocene and Pliocene, with the finest temporal detail available from ice-coring studies covering the Pleistocene era (EPICA 2004), while geological sediment-coring studies provide a physical and biotic record of orbitally driven glacial advance and retreat over various periods extending back to 34 Ma (ANDRILL, Cape Roberts: Barrett 2007; McKay et al. 2009; Naish et al. 2001, 2009; Galeotti et al. 2016; Levy et al. 2016). Reconstructing the detail of Antarctica's glacial history, particularly at physical and temporal scales relevant to interpreting biological and refugial distributions, remains a challenge, as physical evidence pertaining to previous glaciations is largely wiped clear by subsequent ice advances, and there are relatively few locations across Antarctica where clear and distinct evidence of multiple cycles is preserved (but see, e.g., Hodgson et al. 2012). Studies of volcanoes that interacted with past ice sheets are now quantifying past ice thicknesses and other critical parameters (Smellie et al. 2009, 2014). More recent modeling studies (e.g., Pollard & DeConto 2009; DeConto & Pollard 2016), made possible through improved knowledge of bedrock topography and glacial processes, have suggested considerable dynamism within climatic cycles, for instance including the presence of seaways, and by implication biological connections, across the base of the Antarctic Peninsula (see also Barnes & Hillenbrand 2010). At present, these modeling studies do not provide the resolution to allow precise identification of potential terrestrial biological refugia, but they do raise the tantalizing possibility of such refugia being considerably more widespread than is indicated by most glaciological modeling and geomorphological studies – as is required by recent and rapid advances in classical and molecular biological understanding of Antarctic diversity (see discussion in Convey et al. 2008, 2009; Pugh & Convey 2008).

24.6 Antarctica's Fossil Biota

The fossil record in Antarctica is extensive, particularly for ancient terrestrial environments. Restricting our consideration here largely to the Cretaceous onwards, when Antarctica already occupied high paleolatitudes, fossil plants indicate that, from about 100 Ma, the continent was covered with temperate rainforest vegetation much like that found in Patagonia, Tasmania and New Zealand today (Francis et al. 2008; Bowman et al. 2014). These cool to warm temperate rainforests were characterized by podocarps and southern beeches (*Nothofagus*), along with higher altitude araucarian forests and even higher elevation heathland (Bowman et al. 2014). The forests were inhabited by dinosaurs until their demise at the end of the Cretaceous (Stillwell & Long 2011), when they were replaced by early mammals during the Paleocene (Reguero & Gelfo 2014; Gelfo et al. 2015). During the early Eocene, globally warm climates that reached high latitudes allowed more subtropical vegetation to flourish on Antarctica, and along the Wilkes Land coast of East Antarctica, tropical forests were characterized by groups such as palms and tropical mallows (Bombacoideae) (55–48 Ma) (Pross et al. 2012).

Interpreting this fossil record has at times proven controversial. An important example of this, comprehensively summarized by Barrett (2013), arose from the discovery in the 1980s of marine Pliocene diatoms in the Sirius group high in the Transantarctic Mountains. Their presence implied instability of the East Antarctic Ice Sheet, with a collapse as recently as 3 Ma (Webb et al. 1984). This idea contrasted strongly with previous views that an Antarctic Ice Sheet had persisted at least since the middle Miocene. Subsequent to this proposal of instability, improvements in knowledge of ice-sheet behavior and fossil and modern diatom assemblages suggested a different explanation: marine diatoms from the Southern Ocean and Antarctic margin are commonly found in the aeolian biota, and, therefore, the rare Pliocene marine diatoms of the Sirius group could be windblown contaminants. A stable East Antarctic Ice Sheet has persisted since at least 14 Ma (Barrett 2013; Lewis & Ashworth 2015). However, "stability" is a relative concept, and there are indications that parts of the margins of East Antarctica, at least, may have been quite dynamic, with substantial contractions and expansions in the great subglacial embayments such as the Lambert Graben (Pollard & DeConto 2009).

Other elements of the fossil record are perhaps less controversial, but no less extraordinary. Several major reviews have described the fossil biota of Antarctica, which covers much of life's known diversity, extending over most of the geological record (comprehensive and accessible overviews are provided by Stillwell & Long 2011; Cantrill & Poole 2012). In keeping with global turnover of groups, many of these – such as the early amphibian *Lystrosaurus* and a suite of dinosaurs – have long since vanished from the continent. So, too, have the lush forests that once characterized much of it (Francis et al. 2008; Stillwell & Long 2011).

As the global climate cooled and ice sheets began to expand over the Antarctic landscape from the mid–late Eocene onwards, temperate forest vegetation no longer survived. Instead, the fossil record shows that vegetation became dominated by *Nothofagus*, tolerant of cool climates. Remains of the last Antarctic "forests" can be found as small twigs, leaves and pollen within glacial sediments of the Meyer Desert Formation of the Sirius Group at around 1500 m altitude in the Beardmore region of the Transantarctic Mountains. The exact age of these deposits has yet to be confirmed, but they are Neogene (Miocene or Pliocene) in origin (Francis & Hill 1996; Barrett 2013) and represent periods of glacial retreat, which allowed small dwarf trees and other plants to colonize ground once covered by glaciers. Similar ecosystems may have survived as late as 5 Ma close to sea level on the Antarctic Peninsula (Anderson et al. 2011). This tundra-like vegetation suggests summer temperatures of ca. +5 °C for at least several months and cold (−15 to −22 °C) winter temperatures (Francis & Hill 1996; Ashworth & Kuschel 2003; Ashworth & Thompson 2003; Ashworth & Cantrill 2004; Lewis et al. 2008). A community of insects, including flies and weevils, lived among the plants.

These fossils, along with more recent molecular and traditional biogeographical evidence (Convey et al. 2009; Chown & Convey 2016), suggest that some refugia must have existed on the continent despite significant ice sheets at various times (see also Haywood et al. 2009). Indeed, the juxtaposition of fossil and molecular data on modern groups such as flies, springtails and mites provides an important alternative line of evidence for understanding the behavior of ice sheets (with the necessary caveats – see the Sirius debate discussed earlier) (Convey et al. 2008, 2009). It also shows that the Antarctic terrestrial biota is not a largely contemporary one, derived by colonization since the Last Glacial Maximum (LGM), but is much more biogeographically complex (Terauds et al. 2012), with both more ancient and more recent events contributing to local and regional diversity (Convey et al. 2009; Chown et al. 2015).

24.7 Antarctica's Contemporary Biota

Two terrestrial biogeographic zones are generally recognized within the continent in the Antarctic literature – the maritime and continental Antarctic – with a further sub-Antarctic zone encompassing the remote high-latitude oceanic islands close to the Antarctic Polar Front. The ecosystems and climatic characteristics in each zone are distinctively different (Convey 2013). While recent research has identified increasingly strong evidence supporting far more complex and ancient patterns of regionalization within Antarctica (Terauds et al. 2012; Terauds & Lee 2016), this broad division still has much practical utility. The maritime Antarctic includes the western coastal regions of the Antarctic Peninsula, the Scotia Arc archipelagos (South Shetland, South Orkney and South Sandwich islands) and the isolated Bouvetøya and Peter I Øya. The continental Antarctic incorporates all of East Antarctica, the Balleny Islands, Scott Island, the eastern side and base of the Antarctic Peninsula and the remainder of West Antarctica.

Today there are no native terrestrial mammals, reptiles or amphibians in Antarctica, and true terrestrial vertebrates are limited in the sub-Antarctic to an endemic passerine (South Georgia), freshwater ducks (South Georgia, Îles Kerguelen) and scavenging sheathbills (Marion Island, South Georgia), with the latter also occurring in the maritime Antarctic. The abundant marine vertebrates that rest and breed on land exert a strong influence on some terrestrial habitats through deposition of guano and other biological debris, and by physical trampling. Unlike most other regions globally, Antarctic terrestrial faunas therefore predominantly or completely consist of invertebrates (Convey 2013; Chown & Convey 2016). Antarctic faunas are dominated by communities of micro-arthropods (e.g., mites and springtails) and micro-invertebrates (nematodes, tardigrades, rotifers). Higher insects are rare, with the only native species being two flies that occur on the Antarctic Peninsula. The diversity of higher or flowering plants is also very low, with only two species in the maritime Antarctic and none at all in the continental Antarctic. Floras across both continental biogeographic zones are otherwise dominated by bryophytes (mosses and liverworts) and lichens.

As already noted, the available terrestrial habitat for Antarctic biodiversity is small, with permanently ice- and snow-free areas making up less than 0.4% of the continental land mass, although the extent and potential influence of snow-surface, ice and under-ice ecosystems is increasingly being recognized (Stibal et al. 2012). Ice-free areas range from the extensive McMurdo Dry Valleys in the Ross Sea region to the isolated nunataks that are present and patchily distributed over much of East Antarctica. The insular nature of these areas mean connectivity is limited, which has a profound effect on biodiversity (Chown & Convey 2007). While the species richness of higher plants and insects is low, plants like mosses and lichens are relatively well represented (Peat et al. 2007), as are invertebrates like springtails, nematodes, tardigrades and mites (Convey 2013; Velasco-Castrillon et al. 2014) and microbial groups including viruses (Cary et al. 2010; Aguirre de Cárcer et al. 2015; Cavicchioli 2015; Chong et al. 2015). Studies have particularly highlighted the diversity of microbial life in the terrestrial and freshwater environments of Antarctica (Vincent 1988; Chown et al. 2015), with the application of high-throughput sequencing and metagenomic techniques clearly showing that it is much higher than previously thought (Cary et al. 2010; Fierer et al. 2012; Cavicchioli 2015). The environmental conditions that all of these species face across much of the continent are often described as some of the harshest on the planet, and to survive them many have developed unique physiological adaptations, including the ability to survive prolonged desiccation and/or freezing.

Liquid water is one of the fundamental environmental drivers of the evolution and existence of Antarctic terrestrial biodiversity (Convey et al. 2014). The presence of free water is largely mediated by the intensity of solar radiation, which also has direct effects on plants and microbial primary producers as a driver of photosynthetically active radiation (PAR). Solar radiation is directly affected by physical aspects of the landscape (e.g., altitude and aspect) and by other climate factors such as cloud cover. At smaller spatial scales, the physical characteristics of the landscape, including soil structure and chemistry, are extremely influential on the existence and type of biological life, including the distribution and abundance of microbes. Soil geochemical gradients can exist over very small scales, and their analysis can provide insights into the potential of ice-free areas to act as refugia during glacial cycles (Lyons et al. 2016). Perhaps counterintuitively, increasing age of exposure of some terrestrial substrate can lead to the development of conditions less compatible with the survival of some biota (e.g., increasing salinity), giving a negative correlation of diversity with age (Czechowski et al. 2016).

Although abiotic processes play a critical role in shaping the patterns of Antarctic terrestrial biodiversity, biological processes also play a fundamental one. For example, mosses provide habitat for a range of invertebrates that could not otherwise survive. In turn, some mosses thrive in areas where there are (or were) penguin colonies, which provide an influx of nutrients to an environment otherwise characterized by nutrient scarcity (Wasley et al. 2012). At larger spatial scales, processes over geological time scales, including shifting tectonic plates, glaciation events and areas of geothermal activity, have all played important roles in the current distribution of species (Convey et al. 2009; Fraser et al. 2014). In the case of geothermal areas, species richness has been found to be correlated with distance from such

features, with higher numbers of species typically found closer to sites of geothermal activity (Fraser et al. 2014).

Given the glaciological history of Antarctica, it is not surprising that, until relatively recently, Antarctic biodiversity was thought to be a result of sporadic and opportunistic recent colonization processes. However, it has become clear that across all areas of the continent, what we find today are in many cases the last vestiges of the ancient fauna, flora and microbiota that Antarctica inherited as it became free from Gondwana. Over the tens of millions of years of the continent's isolation, this biota has evolved and radiated, and we are only now appreciating the complex signal of regional endemism and evolution in isolation contained in the contemporary biodiversity, and the challenges this creates for understanding the geological and glaciological history of the continent across scientific disciplines.

Evidence for the persistence of terrestrial biota in refugia over thousands or even millions of years is provided by both geological and biological signals. Cosmogenic dating, which uses isotopic ratios to estimate historic ice thickness over geological time scales, suggests that some areas may have been ice-free for millions of years (Fink et al. 2006). Biological data, including fossil evidence, indicate the persistence of some taxa over similar time periods (Convey et al. 2008, 2009).

Biogeographical patterns on a continental scale also appear linked to the historical processes shaping the distribution of Antarctic terrestrial biodiversity. The most obvious of these biogeographical differences can be most clearly seen between the Antarctic Peninsula and the rest of the continent (Chown & Convey 2007 and references therein). More recent studies have shown that terrestrial Antarctica can be divided into 16 biologically distinct regions (Antarctic Conservation Biogeographic Regions, ACBRs) (Terauds et al. 2012; Terauds & Lee 2016), which, in addition to reinforcing the biological differences between the Antarctic Peninsula and the rest of the continent, show that the continent itself can also be delineated into biologically distinct areas (Figure 24.4).

24.8 The Role of Volcanism and Montane Ecosystems in Supporting Antarctica's Unique Biota

The biological evidence for long-term survival of diverse terrestrial organisms on the Antarctic continent, illustrated particularly by high levels of endemism and genetic diversity, conflicts with geomorphological and modeling evidence of glacial ice covering all land in most regions of the continent either at the LGM or during

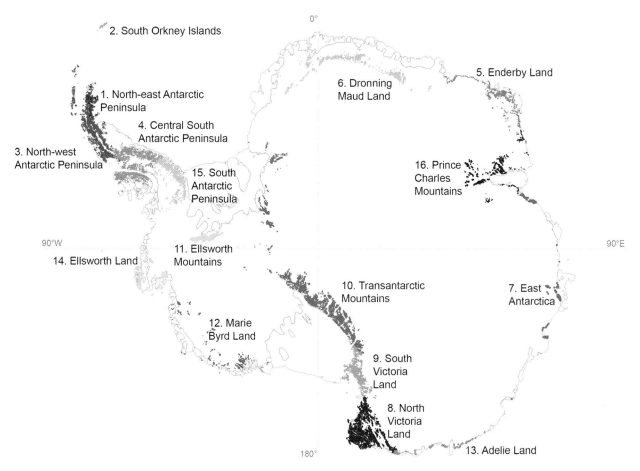

Figure 24.4 The 16 currently recognized Antarctic Conservation Biogeographic Regions (ACBRs). *Source:* Terauds & Lee (2016). Reproduced with permission of John Wiley & Sons.

previous maxima (Convey et al. 2008, 2009; Fraser et al. 2012, 2014). Some ice-free areas must have remained as refugia for a wide range of terrestrial plants, animals and microbes during glacial periods. Two major hypotheses have been proposed about the form refugia may have taken for at least some of the biota of Antarctica (Hawes 2015). The first is the typical ice-free areas found in coastal regions, nunataks and ablation valleys today (Figures 24.5 and 24.6); indeed, there is evidence from cosmogenic data that many mountain peaks and ridges remained exposed at the LGM, and that parts of the McMurdo Dry Valleys may have remained ice-free for more than 10 My, although at present there is no such evidence for the coastal regions, which host the majority of Antarctica's contemporary biodiversity across most ACBRs. The second hypothesis is that refugia consisted of ice-free land maintained by geothermal activity. Both forms of refugia may have played important roles during repeated glacial maxima. There has also been a suggestion that montane ridges and spurs kept clear of snow, and supplied with condensation water by winds, may have provided another form (Green et al. 2011).

24.8.1 Nunataks and Ablation Valleys

Nunataks are rocky outcrops that protrude through glacial and snow cover, and are often the summits and high ridges of mountains (Figure 24.5). Ablation valleys are areas effectively kept clear of lying snow by a combination of low precipitation and high rates of direct evaporation (ablation) of any snow that does fall. The latter are particularly well represented in Victoria Land and elsewhere in the Transantarctic Mountains (Figure 24.6), but they are also found at smaller scales elsewhere on the Antarctic continent, for instance on Alexander Island near the southern Antarctic Peninsula. Isolated nunataks and far more extensive mountain ranges occur throughout Antarctica, including both high-altitude inland areas and low-altitude and even coastal exposures. They support organisms such as cyanobacteria, algae and fungi (Broady & Weinstein 1998; Brinkmann et al. 2007), nematodes, rotifers, tardigrades (Sohlenius et al. 2004; Convey & McInnes 2005), mites, lichens and mosses (Ryan & Watkins 1989; Lee et al. 2013). Nunatak communities are typically extremely simple, comprising few species, and

Figure 24.5 Nunataks near Sanae IV base in Queen Maud Land, illustrating the physically isolated and "island-like" nature of many terrestrial habitats in Antarctica. *Source:* Courtesy of Santjie du Toit.

often appear to have few affinities with lowland, coastal ecosystems. As a result, although nunataks may well have supported specific diversity through past glacial periods in Antarctica, they cannot have been the only or even the primary form of glacial refugium (Convey et al. 2008).

24.8.2   Geothermal Areas

Antarctica has a large number of volcanoes that are either currently active or have been active in recent geological time (i.e., the last million years) (Figure 24.7) (LeMasurier & Thomson 1990). Additionally, some areas of the continent have unusually high levels of geothermal heat through radiogenic decay of granites (Carson & Pittard 2012; Carson et al. 2014). Geothermal activity can cause overlying ice and snow to melt, producing ice-free terrain that can support life. Spatial analyses of broad-scale patterns of species richness in Antarctica indicate that the diversity of mosses and lichens decreases with distance from geothermal areas in the parts of the continent where

current geothermal activity occurs, supporting the hypothesis that such areas are important for the persistence of terrestrial life in these regions (Fraser et al. 2014).

Geothermal sources of heat can not only keep areas completely free of ice, but also hollow out caves beneath snow and ice (Lyon & Giggenbach 1974). Many of the subglacial cave systems on Mount Erebus, Victoria Land are warm (some are tens of degrees Celsius warmer than the external temperature) and are thought to be highly interconnected, possibly even harboring liquid water in the form of lakes or running water. These caves are home to diverse fungal (Connell & Staudigel 2013) and microbial (Tebo et al. 2015) life, but whether they can support plants and invertebrates has yet to be determined. Likewise, how the biota persists over millions of years has not been determined. Volcanoes such as Mount Erebus and Mount Melbourne are very young in geological terms, but have still had long lifetimes (probably 1–2 My). If they did not have earlier precursors, they could not have hosted refugia that preserved biota over

Figure 24.6 Davis Valley (83° S) in the Pensacola Mountains at the eastern extremity of the Transantarctic Mountains, an example of an ablation area kept free of snow by a combination of very low atmospheric humidity and precipitation rates. Davis Valley also contains some of the lowest-diversity terrestrial ecosystems known on the planet (Hodgson et al. 2010). *Source:* Courtesy of P. Convey.

substantive geological time scales. Both centers are flanked by older, eroded edifices that overlap in age and extend the age of local volcanism cumulatively to a few million years, but not to the tens of millions of years that are required by biological studies. However, as already noted, an explanation may lie in the presence of larger, longer-lived volcanic fields across Antarctica, several of which were constructed of multiple volcanic centers with very long aggregate lifetimes (several millions to a few tens of millions of years): volcanoes in the fields grew and died on many occasions, but enough active centers may have overlapped in time for biota to migrate and thus survive during the harsh glacial periods.

24.9 Conclusion

Antarctica was a key element of the supercontinent Gondwana, whose ancient biota is now revealed in fossils and coal deposits. Clearly, the continent's current extremes have not always pertained over geological time, and at various stages Antarctica has hosted widespread temperate, subtropical and even tropical ecosystems. Its

Paleogene biota provided a foundation for subsequent developments. While Antarctica remained connected to other continents (in particular South America and the Australian region), there was the possibility of northward and southward migrations of biota in response to climatic changes. This was ended by Antarctica's geographic isolation in the final stages of break-up of Gondwana and its initial glaciation. Successive glacial fluctuations then served to progressively reduce the remaining Antarctic biota, culminating in its almost total exclusion and the transition to very reduced tundra-type ecosystems around 14 Ma, and in the development of Antarctica's modern ecosystems.

The Antarctica of today is exceptional among Earth's continents in the extent of its ice cover, the extremes of its physical conditions and the degree to which montane and alpine ecosystems dominate its ecology and biodiversity. With over 99.6% of the continent's area covered by ice, the majority of terrestrial exposures are in the form of mountain "islands" of varying size and isolation that protrude through this blanket of ice or fringe its coast. Even the largest ice-free area – the McMurdo Dry Valleys – is montane in character, but experiences the

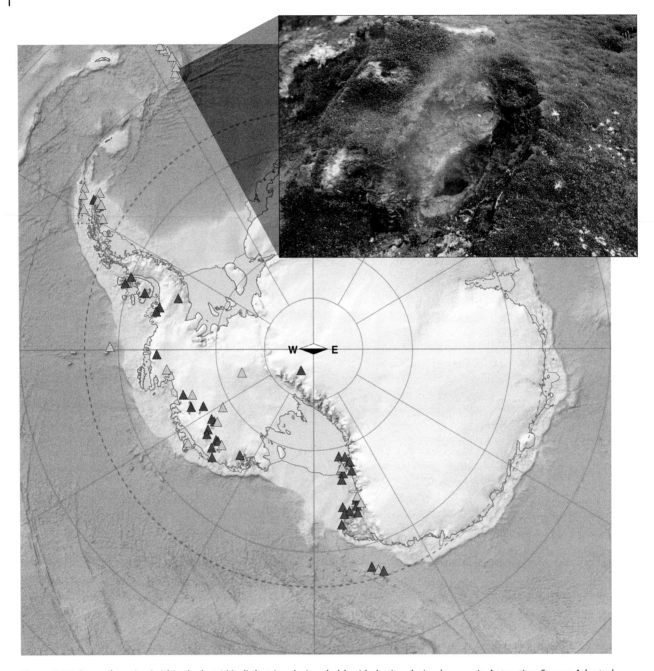

Figure 24.7 Recently active (within the last 1 My, light triangles) and older (dark triangles) volcanoes in Antarctica. *Source:* Adapted from Dalziel & Lawver (2001). Inset: lush moss grows around a fumarole on Bellingshausen Island in the South Sandwich Islands. *Source:* Courtesy of P. Convey.

additional stresses of frigid desert conditions. The contemporary terrestrial biota, dominated by lichens, bryophytes, micro-invertebrates and microbes, has provided a focus in studies of adaptation to some of the most extreme environmental conditions currently experienced on Earth. In recent years, biogeographic studies have led to a paradigm shift demonstrating the antiquity and continuous presence over multimillion-year time scales of much of this contemporary biota, providing a new and exciting challenge across multiple research disciplines to integrate biological, geological and glaciological understanding of the continent's history.

Acknowledgments

We are grateful to the volume editors for advice, and to Professor Peter Barrett and Professor David Cantrill for detailed and constructive comments and discussion on an initial version of this chapter.

References

Aguirre de Cárcer, D., López-Bueno, A., Pearce, D.A. & Alcamí, A. (2015) Biodiversity and distribution of polar freshwater DNA viruses. *Science Advances* e1400127.

Allegrucci, G., Carchini, G., Convey, P. & Sbordoni, V. (2012) Evolutionary geographic relationships among chironomid midges from maritime Antarctic and sub-Antarctic islands. *Biological Journal of the Linnean Society* **106**, 258–274.

Anderson, J.B., Warny, S., Askin, R.A. et al. (2011) Progressive Cenozoic cooling and the demise of Antarctica's last refugium. *Proceedings of the National Academy of Sciences* **108**, 11 356–11 360.

Ashworth, A.C. & Cantrill, D.J. (2004) Neogene vegetation of the Meyer Desert Formation (Sirius Group) Transantarctic Mountains, Antarctica. *Palaeogeography, Palaeoclimatology, Palaeoecology* **213**, 65–82.

Ashworth, A.C. & Kuschel, G. (2003) Fossil weevils (Coleoptera: Curculionidae) from latitude 85°S Antarctica. *Palaeogeography, Palaeoclimatology, Palaeoecology* **191**, 191–202.

Ashworth, A.C. & Thompson, F.C. (2003) A fly in the biogeographic ointment. *Nature* **423**, 135–136.

Barker, P.F., Dalziel, I.W.D. & Storey, B.C. (1991) Tectonic development of the Scotia Arc region. In: Tingey, R.J. (ed.) *The Geology of Antarctica*. Oxford: Clarendon, pp. 215–246.

Barnes, D.K.A. & Hillenbrand, C.-D. (2010) Faunal evidence for a late Quaternary trans-Antarctic seaway. *Global Change Biology* **16**, 3297–3303.

Barrett, P.J. (1991) The Devonian to Jurassic Beacon Supergroup of the Transantarctic Mountains and correlatives in other parts of the world. In: Tingey, R.J. (ed.) *The Geology of Antarctica*. Oxford: Clarendon, pp. 120–152.

Barrett, P.J. (1996) Antarctic paleoenvironments through Cenozoic time – a review. *Terra Antarctica* **3**, 103–119.

Barrett, P.J. (2003) Were West Antarctic Ice Sheet grounding events in the Ross Sea a consequence of East Antarctic Ice Sheet expansion during the middle Miocene? *Earth and Planetary Science Letters* **216**, 93–107.

Barrett, P.J. (2007) Cenozoic climate and sea-level history from glacimarine strata off the Victoria Land Coast, Cape Roberts Project, Antarctica. In: Hambrey, M.J., Christoffersen, P., Glasser, N.F. & Hubbart, B. (eds.) *Glacial Processes and Products*. International Association of Sedimentologists Special Publication 39, pp. 259–287.

Barrett, P.J. (2013) Resolving views on Antarctic Neogene glacial history – the Sirius debate. *Earth and Environmental Science Transactions of the Royal Society of Edinburgh* **104**, 31–53.

Behrendt, J.C., LeMasurier, W.E., Cooper, A.K. et al. (1991) Geophysical studies of the West Antarctic Rift System. *Tectonics* **10**, 1257–1273.

Bentley, C.R. (1991) Configuration and structure of the subglacial crust. In: Tingey, R.J. (ed.) *The Geology of Antarctica*. Oxford: Clarendon, pp. 120–152.

Bialas, R.W., Buck, W.R., Studinger, M. & Fitzgerald, P. (2007) Plateau collapse model for the Transantarctic Mountains–West Antarctic Rift System: insights from numerical experiments. *Geology* **35**, 687–690.

Birkenmajer, K., Gaździcki, A., Krajewski, K.P. et al. (2005) First Cenozoic glaciers in West Antarctica. *Polish Polar Research* **26**, 3–12.

Bowman, V.C., Francis, J.E. & Riding, J.B. (2013) Late Cretaceous winter sea ice in Antarctica? *Geology* **41**, 1227–1230.

Bowman, V.C., Francis, J.E., Askin, R.A. et al. (2014) Latest Cretaceous-earliest Paleogene vegetation and climate change at the high southern latitudes: playnological evidence from Seymour Island, Antarctic Peninsula. *Paleogeography, Paleoclimatology, Paleocology* **408**, 26–47.

Brinkmann, M., Pearce, D.A., Convey, P. & Ott, S. (2007) The cyanobacterial community of polygon soils at an inland Antarctic nunatak. *Polar Biology* **30**, 1505–1511.

Broady, P.A. & Weinstein, R.N. (1998) Algae, lichens and fungi in La Gorce Mountains, Antarctica. *Antarctic Science* **10**, 376–385.

Burgess, S.D. & Bowring. S.A. (2015) High-precision geochronology confirms voluminous magmatism before, during, and after Earth's most severe extinction. *Science Advances* **1**, e1500470.

Burton-Johnson, A., Black, M., Fretwell, P.T. & Kaluza-Gilbert, J. (2016) An automated methodology for differentiating rock from snow, clouds and sea in Antarctica from Landsat 8 imagery: a new rock outcrop map and area estimation for the entire Antarctic continent. *The Cryosphere* **10**, 1665–1677.

Cantrill, D.J. & Poole, I. (2012) *The Vegetation of Antarctica Through Geological Time*. Cambridge: Cambridge University Press.

Carson, C.J. & Pittard, M. (2012) A Reconnaissance Crustal Heat Production Assessment of the Australian Antarctic Territory (AAT). Geoscience Australia Record 2012/63. Canberra.

Carson, C.J., McLaren, S., Roberts, J.L. et al. (2014) Hot rocks in a cold place: high sub-glacial heat flow in East Antarctica. *Journal of the Geological Society* **171**, 9–12.

Cary, S.C., McDonald, I.R., Barrett, J.E. & Cowan, D.A. (2010) On the rocks: the microbiology of Antarctic Dry Valley soils. *Nature Reviews Microbiology* **8**, 129–138.

Cavicchioli, R. (2015) Microbial ecology of Antarctic aquatic systems. *Nature Reviews Microbiology* **13**, 691–706.

Chong, C.-W., Pearce, D.A. & Convey, P. (2015) Emerging spatial patterns in Antarctic prokaryotes. *Frontiers in Microbiology* **6**, 1058.

Chown, S.L. & Convey, P. (2007) Spatial and temporal variability across life's hierarchies in the terrestrial Antarctic. *Philosophical Transactions of the Royal Society of London, Series B* **362**, 2307–2331.

Chown, S.L. & Convey, P. (2016) Antarctic entomology. *Annual Reviews of Entomology* **61**, 119–137.

Chown, S.L., Clarke, A., Fraser, C.I. et al. (2015) The changing form of Antarctic biodiversity. *Nature* **522**, 431–438.

Connell, L. & Staudigel, H. (2013) Fungal diversity in a dark oligotrophic volcanic ecosystem (DOVE) on Mount Erebus, Antarctica. *Biology* **2**, 798–809.

Convey, P. (2013) Antarctic ecosystems. In Levin, S.A. (ed.) *Encyclopaedia of Biodiversity*, Vol. **1**, 2nd edn. San Diego, CA: Elsevier, pp. 179–188.

Convey, P. & McInnes, S.J. (2005) Exceptional tardigrade-dominated ecosystems in Ellsworth Land, Antarctica. *Ecology* **86**, 519–527.

Convey, P., Gibson, J.A.E., Hillenbrand, C.D. et al. (2008) Antarctic terrestrial life – challenging the history of the frozen continent? *Biological Reviews* **83**, 103–117.

Convey, P., Stevens, M.I., Hodgson, D.A. et al. (2009) Exploring biological constraints on the glacial history of Antarctica. *Quaternary Science Reviews* **28**, 3035–3048.

Convey, P., Chown, S.L., Clarke, A. et al. (2014) The spatial structure of Antarctic biodiversity. *Ecological Monographs* **84**, 203–244.

Cox, K.G. (1988) The Karoo Province. In: MacDougall, J.D. (ed.) *Continental Flood Basalts*. Dordrecht: Kluwer Academic Publishers, pp. 239–271.

Coxall, H.K., Wilson, P.A., Päike, H. et al. (2005) Rapid stepwise onset of Antarctic glaciation and deeper calcite compensation in the Pacific Ocean. *Nature* **433**, 53–57.

Czechowski, P., White, D., Clarke, L. et al. (2016) Age-related environmental gradients influence invertebrate distribution in the Prince Charles Mountains, East Antarctica. *Royal Society Open Science* **3**, 160296.

Dalziel, I.W.D. & Elliot, D.H. (1982) West Antarctica: problem child of Antarctica. *Tectonics* **1**, 3–19.

Dalziel, I.W.D. & Lawver, L.A. (2001) The lithospheric setting of the West Antarctic Ice Sheet. In: Alley, R.B. & Bindschadle, R.A. (eds.) *The West Antarctic Ice Sheet: Behavior and Environment*. Chichester: John Wiley & Sons Ltd., pp. 29–44.

DeConto, R.M. & Pollard, D. (2003) Rapid Cenozoic glaciation of Antarctica induced by declining atmospheric CO2. *Nature* **421**, 245–249.

DeConto, R.M. & Pollard, D. (2016) Contribution of Antarctica to past and future sea-level rise. *Nature* **531**, 591–597.

Elliot, D.H. (1992) *Jurassic Magmatism and Tectonism Associated with Gondwanaland Break-up: An Antarctic Perspective*. Geological Society, London, Special Publications 68, 165–184.

Elliot, D.H. (2013) The geological and tectonic evolution of the Transantarctic Mountains: a review. In: Hambrey, M.J., Barker, P.F., Barrett, P.J. et al. (eds.) *Antarctic Palaeoenvironments and Earth-Surface Processes*. Geological Society, London, Special Publications 381, pp. 7–35.

EPICA (2004) Eight glacial cycles from an Antarctic ice core. *Nature* **429**, 623–628.

Fierer, N., Leff, J.W., Adams, B.J. et al. (2012) Cross-biome metagenomic analyses of soil microbial communities and their functional attributes. *Proceedings of the National Academy of Sciences* **109**, 21 390–21 395.

Fink, D., McKelvey, B., Hambrey, M.J. et al. (2006) Pleistocene deglaciation chronology of the Amery Oasis and Radok Lake, northern Prince Charles Mountains, Antarctica. *Earth and Planetary Science Letters* **243**, 229–243.

Fisher, A.T., Mankoff, K.D., Tulaczyk, S.M. et al. (2015) High geothermal heat flux measured below the West Antarctic Ice Sheet. *Science Advances* **1**, e1500093.

Francis, J.E. & Hill, R.S. (1996) Fossil plants from the Pliocene Sirius Group, Transantarctic Mountains: evidence for climate from growth rings and fossil leaves. *Palaios* **11**, 389–396.

Francis, J.E. & Poole, I. (2002) Cretaceous and early Tertiary climates of Antarctica: evidence from fossil wood. *Palaeogeography, Palaeoclimatology, Palaeoecology* **182**, 47–64.

Francis, J.E., Ashworth, A., Cantrill, D.J. et al. (2008) 100 million years of Antarctic climate evolution: evidence from fossil plants. In: Cooper, A.K., Raymond, C.R. & the 10th ISAES Editorial Team (eds.) *Antarctica: A Keystone in a Changing World*. Washington, DC: The National Academies Press, pp. 19–27.

Fraser, C.I., Nikula, R., Ruzzante, D.E. & Waters, J.M. (2012) Poleward bound: biological impacts of Southern Hemisphere glaciation. *Trends in Ecology & Evolution* **27**, 462–471.

Fraser, C.I., Terauds, A., Smellie, J. et al. (2014) Geothermal activity helps life survive glacial cycles. *Proceedings of the National Academy of Sciences USA* **111**, 5634–5639.

Fretwell, P., Pritchard, H.D., Vaughan, D.G. et al. (2013) Bedmap2: improved ice bed, surface and thickness datasets for Antarctica. *The Cryosphere* **7**, 375–393.

Galeotti, S., DeConto, R., Naish, T. et al. (2016) Antarctic Ice Sheet variability across the Eocene-Oligocene boundary climate transition. *Science* **352**, 76–80.

Gelfo, J.N., Mörs, T., Lorente, M. et al. (2015) The oldest mammals from Antarctica, early Eocene of the La Meseta Formation, Seymour Island. *Palaeontology* **58**, 101–110.

Green, T.G.A., Sancho, L.G., Türk, R. et al. (2011) High diversity of lichens at 84° S, Queen Maud Mountains, suggests preglacial survival of species in the Ross Sea region, Antarctica. *Polar Biology* **34**, 1211–1220.

Grunow, A.M., Kent, D.V. & Dalziel, I.W.D. (1987) Evolution of the Weddell Sea basin: New palaeomagnetic constraints. *Earth and Planetary Science Letters* **86**, 16–26

Hallam, A. & Wignall, P.B. (1997) *Mass Extinctions and their Aftermath*. Oxford: Oxford University Press.

Hawes, T.C. (2015) Antarctica's geological arks of life. *Journal of Biogeography* **42**, 207–208.

Haywood, A.M., Smellie, J.L., Ashworth, A.C. et al. (2009) Middle Miocene to Pliocene history of Antarctica and the Southern Ocean. In: Siegert, M.J. & Florindo, F. (eds.) *Developments in Earth & Environmental Sciences 8, Antarctic Climate Evolution*. London: Elsevier, pp. 401–463.

Hillenbrand, C.-D. & Ehrmann, W. (2005) Late Neogene to Quaternary environmental changes in the Antarctic Peninsula region: evidence from drift sediments. *Global and Planetary Change* **45**, 165–191.

Hodgson, D.A, Convey, P., Verleyen, E. et al. (2010) Observations on the limnology and biology of the Dufek Massif, Transantarctic Mountains 82° south. *Polar Science* **4**, 197–214.

Hodgson, D.A., Bentley, M.J., Schnabel, C. et al. (2012) Glacial geomorphology and cosmogenic ^{10}Be and ^{26}Al exposure ages in the northern Dufek Massif, Weddell Sea embayment, Antarctica. *Antarctic Science* **24**, 377–394.

Joachimski, M.M., Lai, X., Shen, S. et al. (2012) Climate warming in the latest Permian and the Permian–Triassic mass extinction. *Geology* **40**, 195–198.

Kaiho, K., Kajiwara, Y., Nakano, T. et al. (2001) End-Permian catastrophe by a bolide impact: evidence of a gigantic release of sulfur from the mantle. *Geology* **29**, 815–818.

Kyle, P.R. (1990) McMurdo Volcanic Group Western Ross Embayment. Introduction. In: LeMasurier, W. & Thompson, J. (eds.) *Volcanism of the Antarctic Plate and Southern Ocean*. Antarctic Research Series, American Geophysical Union 48, pp. 18–25.

Larter, R.D., Cunnigham, A.P. & Barker, P.F. (2002) Tectonic evolution of the Pacific margin of Antarctica. 1. Late Cretaceous reconstructions. *Journal of Geophysical Research* **107**, 23–45.

Lawver, L.A., Gahagan, L.M. & Dalziel, I.W.D. (1998) A tight fit-Early Mesozoic Gondwana, a plate reconstruction perspective. *Memoires of the National Institute of Polar Research* **53**, 214–229.

Leat, P.T., Smellie, J.L., Millar, I.L. & Larter, R.D. (2003) Magmatism in the South Sandwich arc. In: Larter, R.D. & Leat P.T. (eds.) *Intra-Oceanic Subduction Systems: Tectonic and Magmatic Processes*. Geological Society, London, Special Publication 219, pp. 285–313.

Lee, J.E., Le Roux, P.C., Meiklejohn, K. & Chown, S.L. (2013) Species distribution modelling in low-interaction environments: insights from a terrestrial Antarctic system. *Austral Ecology* **38**, 279–288.

LeMasurier, W.E. & Thomson, J.W. (1990) *Volcanoes of the Antarctic Plate and Southern Oceans*. Antarctic Research Series. Washington, DC: American Geophysical Union.

Levy, R., Harwood, D., Florindo, F. et al. (2016) Antarctic ice sheet sensitivity to atmospheric CO_2 variations in the early to mid-Miocene. *Proceedings of the National Academy of Sciences USA* **113**, 3453–3458.

Lewis, A.R. & Ashworth, A.C. (2015) An early to middle Miocene record of ice-sheet and landscape evolution from the Friis Hills, Antarctica. *Geological Society of America Bulletin* **128**, 719–713.

Lewis, A.R., Marchant, D.R., Ashworth, A.C. et al. (2008) Mid-Miocene cooling and the extinction of tundra in continental Antarctica. *Proceedings of National Academy of Sciences* **105**, 10676–10680.

Lisker, F. & Läufer, A.L. (2013) The Mesozoic Victoria Basin: vanished link between Antarctica and Australia. *Geology* **41**, 1043–1046.

Livermore, R., Hillenbrand, C.-D., Meredith, M. & Eagles, G. (2007) Drake Passage and Cenozoic climate: an open and shut case? *Geochemistry Geophysics Geosystems* **8**, Q01005.

López-Bueno, A., Tamames, J., Velázquez, D. et al. (2009) High diversity of the viral community from an Antarctic lake. *Science* **326**, 858–861.

Lough, A.C., Wiens, D.A., Barcheck, C.G. et al. (2013) Seismic detection of an active subglacial magmatic complex in Marie Byrd Land, Antarctica. *Nature Geoscience* **6**, 1031–1035.

Lyon, G.L. & Giggenbach, W.F. (1974) Geothermal activity in Victoria Land, Antarctica. *New Zealand Journal of Geology and Geophysics* **17**, 511–521.

Lyons, W.B., Dueurling, K., Welch, K.A. et al. (2016) The soil geochemistry in the Beardmore Glacier region, Antarctica: implications for terrestrial ecosystem history. *Scientific Reports* **6**, 26189.

McKay, R., Browne, G., Carter, L. et al. (2009) The stratigraphic signature of the late Cenozoic Antarctic Ice Sheets in the Ross Embayment. *GSA Bulletin* **121**, 1537–1561.

Naish, T.R., Woolfe, K.J., Barrett, P.J. et al. (2001) Orbitally induced oscillations in the East Antarctic ice sheet at the Oligocene/Miocene boundary. *Nature* **413**, 719–723.

Naish, T., Powell, R., Levy, R. et al. (2009) Obliquity-paced Pliocene West Antarctic Ice Sheet Oscillations. *Nature* **458**, 322–328.

Pankhurst, R.J., Riley, T.R., Fanning, C.M. & Kelley, S.P. (2000) Episodic silicic volcanism in Patagonia and the Antarctic Peninsula: chronology of magmatism associated with the break-up of Gondwana. *Journal of Petrology* **41**, 605–625.

Patrick, M.R. & Smellie, J.L. (2013) A spaceborne inventory of volcanic activity in Antarctica and southern oceans, 2000–10. *Antarctic Science* **25**, 475–500.

Peat, H.J., Clarke, A. & Convey, P. (2007) Diversity and biogeography of the Antarctic flora. *Journal of Biogeography* **34**, 132–146.

Peck, L.S., Convey, P. & Barnes, D.K.A. (2006) Environmental constraints on life histories in Antarctic ecosystems: tempos, timings and predictability. *Biological Reviews* **81**, 75–109.

Pfuhl, H.A. & McCave, I.N. (2005) Evidence for late Oligocene establishment of the Antarctic Circumpolar Current. *Earth and Planetary Science Letters* **235**, 715–728.

Pollard, D. & DeConto, R.M. (2009) Modelling West Antarctic ice sheet growth and collapse through the past five million years. *Nature* **458**, 329–332.

Pross, J., Contreras, L., Bijl, P.K. et al. (2012) Persistent near-tropical warmth on the Antarctic continent during the early Eocene Epoch. *Nature* **488**, 73–77.

Pugh, P.J.A. & Convey, P. (2008) Surviving out in the cold: Antarctic endemic invertebrates and their refugia. *Journal of Biogeography* **35**, 2176–2186.

Reguero, M.A. & Gelfo, J.N. (2014) Final Gondwana breakup: the Paleogene South American native ungulates and the demise of the South America–Antarctica land connection. *Global and Planetary Change* **123**, 400–413.

Retallack, G.J. (2001) A 300-million-year record of atmospheric carbon dioxide from plant cuticles. *Nature* **411**, 287–290.

Rocchi, S., Armienti, P., D'Orazio, M. et al. (2002) Cenozoic magmatism in the western Ross Embayment: role of mantle plume versus plate dynamics in the development of the West Antarctic Rift System. *Journal of Geophysical Research* **107**, 2195.

Rocchi, S., Lemasurier, W.E. & Di Vincenzo, G. (2006) Oligocene to Holocene erosion in Marie Byrd Land, West Antarctica, inferred from exhumation of the Dorrel Rock intrusive complex and from volcano morphologies. *Geological Society of America Bulletin* **118**, 991–1005.

Ryan, P.G. & Watkins, B.P. (1989) The influence of physical factors and ornithogenic products on plant and arthropod abundance at an inland nunatak group in Antarctica. *Polar Biology* **10**, 151–160.

Scher, H.D. & Martin, E.E. (2006) Timing and climatic consequences of the opening of the Drake Passage. *Science* **312**, 428–430.

Shevenell, A.E., Kennett, J.P. & Lea, D.W. (2004) Middle Miocene Southern Ocean cooling and Antarctic cryosphere expansion. *Science* **305**, 1766–1770.

Smellie, J.L. & Edwards, B.E. (2016) *Glaciovolcanism on Earth and Mars. Products, Processes and Palaeoenvironmental Significance.* Cambridge: Cambridge University Press.

Smellie, J.L., Haywood, A.M., Hillenbrand, C.-D. et al. (2009) Nature of the Antarctic Peninsula Ice Sheet during the Pliocene: geological evidence and modelling results compared. *Earth-Science Reviews* **94**, 79–94.

Smellie, J.L., Rocchi, S., Gemelli, M. et al. (2011) Late Miocene East Antarctic ice sheet characteristics deduced from terrestrial glaciovolcanic sequences in northern Victoria Land, Antarctica. *Palaeogeography, Palaeoclimatology, Palaeoecology* **307**, 129–149.

Smellie, J.L., Rocchi, S., Wilch, T.I. et al. (2014) Glaciovolcanic evidence for a polythermal Neogene East Antarctic Ice Sheet. *Geology* **42**, 39–41.

Sohlenius, B., Boström, S. & Ingemar Jönsson, K. (2004) Occurrence of nematodes, tardigrades and rotifers on ice-free areas in East Antarctica. *Pedobiologia* **48**, 395–408.

Stibal, M., Šabacká, M. & Žárský, J. (2012) Biological processes on glacier and ice sheet surfaces. *Nature Geoscience* **5**, 771–774.

Stillwell, J.D. & Long, J.A. (2011) *Frozen in Time. Prehistoric Life in Antarctica.* Collingwood: CSIRO Publishing.

Storey, B.C. (1995) The role of mantle plumes in continental breakup: case histories from Gondwanaland. *Nature* **377**, 301–306.

Storey, B.C. & Garrett, S.W. (1985) Crustal growth of the Antarctic Peninsula by accretion, magmatism and extension. *Geological Magazine* **122**, 5–14.

Storey, B.C. & Kyle, P.R. (1997) An active mantle mechanism for Gondwana breakup. *South African Journal of Geology* **100**, 283–290.

Storey, B.C., Vaughan, A.P.M. & Riley, T.R. (2013) The links between large igneous provinces, continental break-up and environmental change: evidence reviewed from Antarctica. *Earth and Environmental Science Transactions of the Royal Society of Edinburgh* **104**, 1–14.

Stump, E. (1995) *The Ross Orogen of the Transantarctic Mountains.* Cambridge: Cambridge University Press.

Sun, Y., Joachimski, M.M., Wignall, P.B. et al. (2012) Lethally hot temperatures during the early Triassic greenhouse. *Science* **338**, 336–370.

Tebo, B.M., Davis, R.E., Anitori, R.P. et al. (2015) Microbial communities in dark oligotrophic volcanic ice cave ecosystems of Mt. Erebus, Antarctica. *Frontiers in Microbiology* **6**, 179.

ten Brink, U.S., Bannister, S., Beaudoin, B.C. & Stern, T.A. (1993) Geophysical investigations of the tectonic boundary between East and West Antarctica. *Science* **261**, 45–50.

Terauds, A. & Lee, J.R. (2016) Antarctic biogeography revisited: updating the Antarctic Conservation Biogeographic Regions. *Diversity and Distributions* **22**, 836–840.

Terauds, A., Chown, S.L., Morgan, F. et al. (2012) Conservation biogeography of the Antarctic. *Diversity and Distributions* **18**, 726–741.

Tingey, R.J. (ed.) (1991) *The Geology of Antarctica*. Oxford: Clarendon.

Torsvik, T.H., Gaina, C. & Redfield, T.F. (2008) Antarctica and global paleogeography: from Rodinia, through Gondwanaland and Pangea, to the birth of the Southern Ocean and the opening of gateways. In: Cooper, A.K., Barrett, P.J., Stagg, H. et al. (eds.) *Antarctica: A Keystone in a Changing World. Proceedings of the 10th International symposium on Antarctic Earth Sciences.* Washington DC: The National Academies Press.

Tripati, A., Backman, J., Elderfield, H. & Ferretti, P. (2005) Eocene bipolar glaciation associated with global carbon cycle changes. *Nature* **436**, 341–346.

Velasco-Castrillon, A., Gibson, J.A.E. & Stevens, M.I. (2014) A review of current Antarctic limno-terrestrial microfauna. *Polar Biology* **37**, 1517–1531.

Vincent, W.F. (1988) *Microbial Ecosystems of Antarctica*. Cambridge: Cambridge University Press.

Wasley, J., Robinson, S.A., Turnbull, J.D. et al. (2012) Bryophyte species composition over moisture gradients in the Windmill Islands, East Antarctica: development of a baseline for monitoring climate change impacts. *Biodiversity* **13**, 257–264.

Weaver, S.D., Saunders, A.D. & Tarney, J. (1982) Mesozoic-Cenozoic volcanism in the South Shetland Islands and the Antarctic Peninsula: geochemical nature and plate tectonic significance. In: Craddock, C. (ed.) *Antarctic Geosciences*. Madison, WI: University of Wisconsin Press, pp. 263–273.

Webb, P.-N., Harwood, D.M., McKelvey, B.C. et al. (1984) Cenozoic marine sedimentation and ice-volume variation on the East Antarctic craton. *Geology* **12**, 287–291.

Whitehead, J.M., Quilty, P.G., McKelvey, B.C. & O'Brien, P.E. (2006) A review of the Cenozoic stratigraphy and glacial history of the Lambert Graben – Prydz Bay region, East Antarctica. *Antarctic Science* **18**, 83–99.

25

The Biogeography, Origin and Characteristics of the Vascular Plant Flora and Vegetation of the New Zealand Mountains

Matt S. McGlone[1], Peter Heenan[1], Timothy Millar[2] and Ellen Cieraad[3]

[1] Landcare Research, Lincoln, New Zealand
[2] Plant & Food Research, Christchurch, New Zealand
[3] Institute of Environmental Sciences (CML), Leiden University, Leiden, Netherlands

Abstract

We discuss the origin of the New Zealand montane-alpine environment, its plant communities and alpine plant traits, the evolution of alpine lineages and the biogeography of the region in a global context. The extensive, mountainous topography of New Zealand is recent: tall mountains formed only in the late Pliocene, and permanent snow and ice by the mid-Pleistocene. Treelines are at low elevations, reflecting the highly oceanic climate and cool growing seasons. Around 800 vascular plant taxa grow above the treeline, many in species-rich radiations of morphologically diverse, closely related taxa. Plants of the New Zealand alpine region are similar to elsewhere, but there is a strong representation of tussock grasses and sedges, specialized scree plants and many (70+) cushion species. The alpine flora has a predominance of white, green or dull, small, unspecialized flowers (over 80%), most likely because pollination is largely carried out by short-tongue bees. Around 20% of the alpine flora is gender-dimorphic, reflecting its prevalence in the flora as a whole. The biogeographic origins of the alpine genera are: 42% Australasian, 13% from the greater southern hemisphere and 45% from the northern hemisphere. The recent formation of the alpine environment and New Zealand's oceanic setting mean that many species are derived from ancestors that arrived via trans-oceanic dispersal.

Keywords: *treeline, plant radiation, dispersal, climate change, floral biology, plant form, gender dimorphism*

25.1 New Zealand Mountain Environments

Mountainous terrain (land over 600 m in elevation) forms ~31% of the New Zealand mainland (i.e., North, South and Stewart islands) and creates a near-continuous south-west–north-east barrier to the persistent westerly wind flow in the southern two-thirds of the archipelago (Figure 25.1).

The Southern Alps, the highest part of this mountain chain, extend over 500 km down the west coast of the central South Island, with numerous peaks over 3000 m and many glaciers and permanent ice fields. The North Island axial mountain chain is lower, with few alpine areas. For the most part, the mountains are formed from graywacke and schists, with limited areas of limestone, granite and andesitic rocks. About 24 500 km² (9%) of the mainland lies above treeline, mainly in the Southern Alps (Figure 25.1). Uplift rates are high throughout the rapidly eroding central Southern Alps, peaking at 10 mm per year, but decrease rapidly towards the stable eastern regions of broad-topped low mountains. The easily shattered graywacke and schist rocks, steep terrain and rapid uplift of the axial mountains have resulted in large areas of scree, particularly in the drier, leeward eastern ranges, where intense freeze–thaw activity, peaking at 2300 m, feeds active scree that may extend down to 500 m (Hales & Roering 2005, 2009). To the west, river down-cutting, rockfall and landslides dominate erosion processes. Streams and rivers draining the mountains deliver large amounts of sediment to the floodplains (Hicks et al. 2011), creating wide, shingle-filled braided rivers.

Mountainous areas receive over 1000 mm of precipitation a year, and the Southern Alps between 3000 and 10 000 mm, with the highest totals on the windward,

Mountains, Climate and Biodiversity, First Edition. Edited by Carina Hoorn, Allison Perrigo and Alexandre Antonelli.
© 2018 John Wiley & Sons Ltd. Published 2018 by John Wiley & Sons Ltd.
Companion website: www.wiley.com\go\hoorn\mountains,climateandbiodiversity

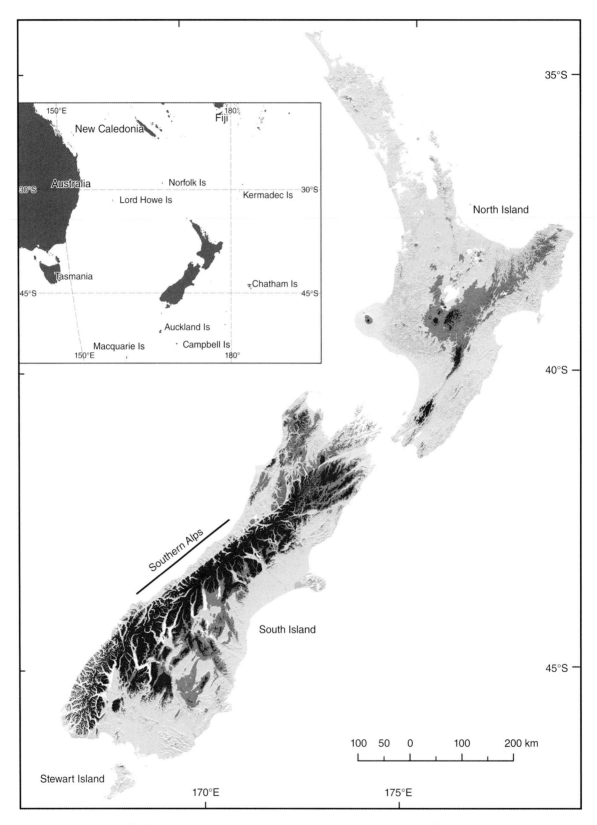

Figure 25.1 New Zealand mainland alpine and montane regions. Dark shading indicates the area above the alpine treeline; gray shading indicates the area above 600 m elevation. Inset: The South-West Pacific and the New Zealand archipelago.

western flanks. Warm surrounding oceans moderate temperature extremes in winter, while cloudy, mild, wet summers create a long, variable growing season. Frost and snow are possible in any month. Winter snow cover is highly variable, but snowlines average 1000 m in the south and 1400 m in the north-east of the Southern Alps; the summer snowline retreats to 1500 m in the south and 2200 m in the north (Fitzharris et al. 1999; Clare et al. 2002). Snow rarely lies for more than a few days below the treeline, and only exceptionally longer than a month or two at treeline. On clear days in summer, soil temperatures can be high and evaporation intense, but drought is rare. Bogs, tarns and seepages are plentiful throughout the alpine zone (Johnson & Gerbeaux 2004).

25.2 Origin of the New Zealand Mountain Landscape

Although Zealandia – the ancestral continent that separated from Antarctica and Australia ca. 85 Ma with the opening of the Tasman Sea – had a mountainous landscape, erosion, rising sea levels and the foundering of the landmass in the early Cenozoic reduced it to a low-lying scatter of islands by the Oligocene (Lee et al. 2001a). South-west to north-east propagation of the Australian–Pacific plate boundary through the New Zealand region at 45 Ma created the present tectonic framework. Uplift and lateral movement along this major boundary beginning ca. 23 Ma drove the emergence of the current mountainous landscape.

In the early Miocene, the New Zealand archipelago consisted of low-lying islands surrounded by subtropical shallow seas. Volcanic islands in the north and east of the region formed the tallest mountains. The first significant mountain ranges in the South Island arose in the middle Miocene (Nathan et al. 1986). Rapid uplift along the Alpine Fault system (which runs the length of the Southern Alps) began ca. 8 Ma, but initially mountain relief was limited as the soft overlying sedimentary rocks eroded away rapidly. These rising ranges intercepted the prevailing westerly wind flow, increasing rainfall in the west but creating drier rain-shadow climates in the east from the late Miocene onwards (Chamberlain et al. 1999). An abrupt reorganization of the plate tectonic regime occurred ca. 5 Ma (Batt & Braun 1999). Rapid uplift of erosion-resistant graywacke began ca. 4 Ma in the far south, and propagated northwards, with tall mountainous relief forming as late as 1.3 Ma in the north-west of the South Island. Mountains of the north-eastern South Island, the axial ranges of the North Island and the Volcanic Plateau formed later, and ranges in the southern North Island began to form only about 500 000 years ago.

Profound cooling of the Antarctic continent and the Southern Ocean began around 43 Ma. However, the New Zealand landmass shifted equatorward over this period, warming to a subtropical maximum in the middle Miocene. While montane to subalpine habitat increased from the middle Miocene, it is unlikely that alpine environments formed until the late Pliocene cooling. From the beginning of the Pleistocene 2.6 Ma, cold glacials alternated with warm interstadials, resulting in periods of cool subalpine and alpine environments along the axial mountains. In the mid-Pleistocene, mountains became sufficiently tall for permanent ice fields to persist through warm interglacials (Heenan & McGlone 2013).

25.3 Vegetation of the New Zealand Mountains

25.3.1 Montane Forests

The species-rich, multistoried evergreen lowland forest gives way to montane or cool temperate evergreen rainforest between 300 and 800 m elevation, depending on latitude (McGlone et al. 2016). Small-leaved Nothofagaceae dominates this montane forest. In the absence of Nothofagaceae, a diverse montane conifer–angiosperm rainforest prevails, typically with tall podocarp trees emergent over small-leaved angiosperm canopies. There are few deciduous trees or shrubs, and they are largely confined to limited areas of disturbed or fertile soils along waterways (McGlone et al. 2004). Tree ferns, lianas and vascular epiphytes are abundant in the lower montane zone, but rapidly decline in abundance and diversity with elevation (Jimenez-Castillo et al. 2007).

25.3.2 Subalpine Forest

The subalpine forest zone, defined by the absence of lianas and vascular epiphytes, begins at elevations of ca. 1300 m in the north and 250 m or less in the south. It includes two distinctive vegetation types: where Nothofagaceae are the dominant canopy trees, an abrupt treeline forms with trees 3–6 m tall, mainly of *Fuscospora cliffortioides* and/or *Lophozonia menziesii*, edged by a dense shrubland, where genera such as *Coprosma*, *Dracophyllum*, *Myrsine*, *Olearia*, *Phyllocladus* and *Veronica* are prominent, giving way with increasing elevation to *Chionochloa* tall tussock grassland, which includes large-leaved forbs. In some areas, forest directly abuts tall tussocks. In the absence of Nothofagaceae (mainly in the central western Southern Alps), forest gradually transitions to grassland.

Subalpine forest (often including genera such as *Libocedrus*, *Podocarpus*, *Weinmania* and *Metrosideros*) gives way to a diverse low forest-scrub of small trees, including angiosperm genera *Myrsine*, *Griselinia*, *Pseudopanax*, the podocarps *Podocarpus*, *Phyllocladus* and *Halocarpus*, and then to shrubland of asterad species (*Brachyglottis* and *Olearia*), ericads (*Archeria* and *Dracophyllum*), *Coprosma* and *Veronica*. In high-rainfall cloudy regions, the subalpine trees become stunted and gnarled with many leaders, and tufted growth forms occur (Wardle 1978)

25.3.3 Treeline

Treeline varies from ca. 1500 m in the North Island to around 900 m in the far southern South Island; in the central South Island, eastern abrupt treelines are about 200 m higher than the gradual treelines on western mountains. While these elevations are much lower than at similar latitudes on continents, the growing season mean temperature ranges from 6.6 to 7.8 °C (Cieraad et al. 2014), similar to other temperate regions (Körner & Paulsen 2004). However, New Zealand treeline winters are relatively mild (minimum temperatures rarely below −10 °C), and growing seasons are 6–9 months long (Benecke et al. 1981; Cieraad et al. 2012, 2014). Local topographic features may depress treeline elevations by several hundred meters (Case & Hale 2015).

New Zealand treelines resemble those of the tropics (Troll 1973). In New Guinea, for example, species-poor alpine forests with emergent conifers, tree ferns and sclerophyllous evergreen angiosperms border alpine tussock grasslands and herbfields (Wardle 1973), much as they do in New Zealand ever-wet mountains. About 45% of the genera found exclusively in the montane/alpine zone in New Guinea are also present in New Zealand (van Balgooy 1976).

25.3.4 Alpine Communities

Absence of woody plants more than 3 m in height defines the alpine zone. The forest-to-alpine sequence begins with a shrub/tall tussock/large-leaved forb community (low alpine), followed by a low grassland transition to the high-alpine cushionfields, herbfield and fell field, and eventually to bare rock, shingle and ice (Wardle 1991, 2008; Mark et al. 2000). Where snow cover is prolonged, forbs dominate, whereas grassland prevails where snow vanishes early. New Zealand alpine communities are similar to those elsewhere (Mark 2012), but physiognomically most closely resemble those of the tropics.

Cushion plants are abundant in New Zealand, with ~72 species – about 6% of the global total (Aubert et al.

2014). Their relative dominance is comparable to regions such as the Sierra Nevada in California (Rundel 2011), alpine Iran (Noroozi et al. 2008) and Tierra del Fuego (Mark et al. 2001). Some cushion plants achieve exceptional sizes; *Raoulia eximia*, one of the "vegetable sheep," grows more than 2 m wide and over 30 cm thick. Cushion plants are characteristic of cool, windswept and often high-radiation dry environments such as exposed ridgelines and slopes, and nearly three-quarters of New Zealand species reach the high-alpine zone. However, some are plants of peat bogs and seepages (e.g., *Centrolepis*, *Gaimardia*, *Oreobolus*, *Donatia*) (Johnson & Gerbeaux 2004). Cushion bogs are prominent also in Tasmania, the Andes, Tierra del Fuego, New Guinea and Hawaii, but not elsewhere (Godley 1978).

Wetlands are abundant in the high-rainfall mountains (Johnson & Gerbeaux 2004; McGlone 2009), ranging from small alpine lakes and pools (tarns) to montane and alpine bogs (Froggatt & Rogers 1990), sloping fens and patterned wetlands (Mark et al. 1995). Wetland species make up over 20% of the total alpine plant flora, and tend to have wide altitude and latitude ranges.

Scree mantles many upper mountain slopes with a shifting surface of shingle with little vegetation cover and high insolation and temperatures at the surface, but a damp, cool, deep soil beneath. There are ~50 scree species, and some, such as the penwiper plant (*Notothlaspi rosulatum*), have deep tap roots and compact rosettes of thick, gray leaves. The characteristic cryptic gray colorations may conceal scree species from herbivorous insects such as moths, butterflies and grasshoppers (Strauss et al. 2015). Species of rocky places, such as *Pachycladon* (Brassicaceae), have deep tap roots and leaf pubescence to counter the stress of a high-radiation environment (Mershon et al. 2015).

25.3.5 Alpine Species Numbers

Just over 800 (~31%) of the nearly 2600 indigenous New Zealand vascular plant species and subspecies (de Lange et al. 2013) occur in the alpine zone. About 300, or nearly 12% of the total vascular flora, are "obligate" alpine species that never – or only occasionally – occur below treeline.

A large number (64%) of the alpine species extend into open, unstable or frost-prone habitats below treeline, some ranging from lowland or coastal to high-alpine settings (e.g., *Blechnum penna-marina*, *Leucopogon fraseri*, *Poa colensoi*). Alpine vascular plant species richness shows a striking unimodal elevational pattern, with 91% of species present between 1200 and 1400 m (Figure 25.2).

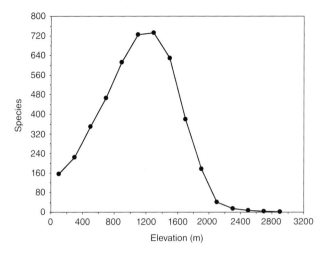

Figure 25.2 Species richness of indigenous alpine vascular plant species in New Zealand, reported in 200 m altitudinal increments.

25.4   Alpine Plant Traits

25.4.1   Growth Forms

Alpine growth forms and adaptations in New Zealand are similar to elsewhere (Halloy 1990; Halloy & Mark 1996). With altitude, the vegetation becomes dwarfed or appressed, the proportion of graminoids and erect shrub species falls and the relative percentage of perennial herb and cushion plant species increases (Figure 25.3).

25.4.2   Reproductive Traits

In contrast to growth forms, the reproductive traits of the New Zealand alpine species differ as a whole from those of most other alpine regions.

25.4.2.1   Floral Morphology and Color

Small white or green simple flowers dominate in the New Zealand alpine. Of the 560 petal-bearing species, 82% are mainly white or green, 13% are yellow and only ~6% are red, purple, blue or orange. Over 40% of the noncapitate, petal-bearing species have small (maximum dimension ≤5 mm) flowers. Asteraceae (capitate flowers) make up nearly half of the larger flowers(maximum dimension ≥10.1 mm) and are mostly (70%) white. While small white flowers are common in other alpine regions – for instance, in the Snowy Mountains of Australia (Inouye & Pyke 1988) and in Greece (Makrodimos et al. 2008) – New Zealand seems to be an extreme. Genera with brightly colored flowers elsewhere (e.g., *Epilobium*, *Gentianella* and *Veronica*) tend towards

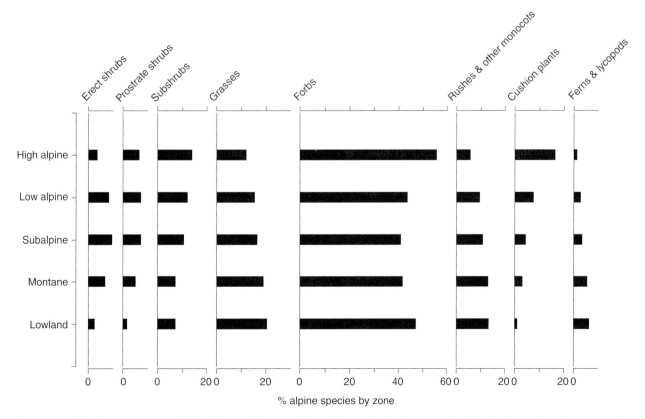

Figure 25.3 Relative percentages of eight different alpine growth forms in New Zealand, presented by altitudinal zone. Note that many alpine species also occur below treeline. Cushion plants are also included within other growth-form species counts, so percentages do not sum to 100.

white or pastel versions of their ancestral coloration. New Zealand alpine flowers appear to be strongly scented (Mark 2012).

The abundance of white or dull-colored small unspecialized flowers in New Zealand has been attributed to the lack of social bees and long-tongued pollinators and a predominance of undiscriminating flower-visiting flies and moths (Godley 1979; Lloyd 1985; Pickering & Stock 2003). New Zealand Colletidae bees, like bees elsewhere, prefer blue but readily train to flowers that provide rewards, rather than consistently favoring "bee" colors (Arnold et al. 2009). The claim that white is a generalized default flower coloration is no longer supported (Campbell et al. 2010; McGimpsey & Lord 2015). Moreover, as white is the most common flower color in most temperate European and Mediterranean habitats (Chittka & Raine 2006), absence of the long-tongued pollinator guild cannot be responsible for the frequency of white flowers (Campbell et al. 2012). In the New Zealand alpine, even when syrphid flies make more visits, bees are vastly more effective at pollen transfer (Bischoff et al. 2013). The small, unspecialized flowers of New Zealand are therefore most likely the result of a "small-bee syndrome" (Newstrom & Robertson 2005), although why they are predominately white or dull-colored remains unclear.

25.4.2.2   Gender Dimorphism

In New Zealand, around 13% of species and 23% of genera of vascular plants are gender dimorphic, mostly dioecious (male and female plants) or gynodioecious (female and hermaphrodite plants), compared with the global average of ~6% (Webb et al. 1999; Renner 2014). About 20% of New Zealand alpine species are gender dimorphic and 21% of alpine genera have gender-dimorphic species. Woody and/or fleshy fruited genera are the most likely to have gender-dimorphic species. However, the closely related dry-fruited herbaceous Apiaceae genera *Aciphylla* (49 spp.), *Anisotome* (22 spp.) and *Gingidia* (5 spp.) make up ~50% of the gender-dimorphic alpine total. A striking correlation occurs with flower size: nearly 50% of alpine New Zealand species with small (≤5 mm) flowers are gender dimorphic, while less than 5% of species with large (≥10 mm) flowers are. The small separation of anther and stigma in the minute flowers typical of many species may increase the risk of inbreeding and thus favor gender dimorphism (Barrett & Hough 2013).

25.4.2.3   Fruit and Seed Dispersal

Only 10% of the alpine flora has fleshy fruits – mostly shrubs and prostrate shrubs. The remainder are largely wind-dispersed, aside from a handful of species with spines or hooks (e.g., *Acaena*) (Lee et al. 2001b). Many

alpine species tend towards white, variably pale or blue-colored fruits, a probable adaption to fruit dispersal by lizards (Lord & Marshall 2001). However, more high-altitude *Coprosma* species have red-colored fruit than do low-altitude species (Lee et al. 1988). Mast seeding occurs in the subalpine Nothofagaceae, and in some large grasses and giant forbs such as *Chionochloa* grasses, *Phormium cookianum* and *Aciphylla* (Kelly 1994; Schauber et al. 2002; Brookes & Jesson 2007).

25.5   Alpine Radiations and Endemism

Because of their fragmented topography and unique habitats, alpine regions globally offer ample opportunities for allopatric speciation, and many alpine plant lineages are impressively species-rich, often with >100 species (Hughes & Atchison 2015). New Zealand alpine genera are moderately large (Figure 25.4). A number result from radiations within or into the alpine zone: for instance, *Myosotis* (Meudt et al. 2015), *Ranunculus* (Hörandl & Emadzade 2011), *Chionochloa* (Pirie et al. 2010), *Ourisia* (Meudt 2007), *Pachycladon* (Mandakova et al. 2010) and *Veronica* (Meudt & Bayly 2008).

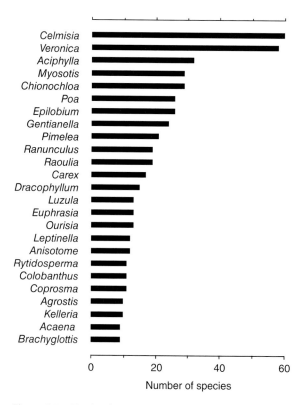

Figure 25.4 Total indigenous alpine species richness of the 25 most speciose alpine vascular plant genera in New Zealand.

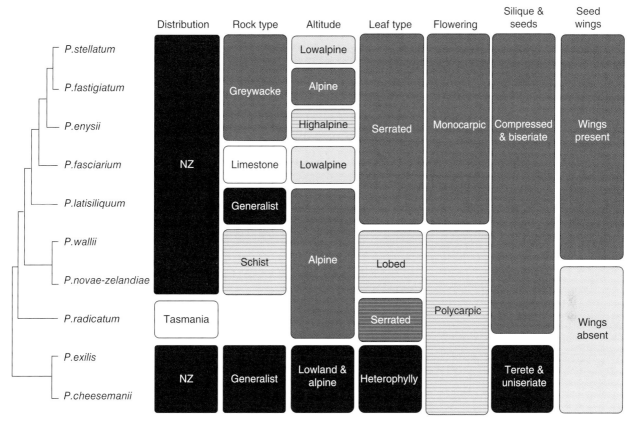

Figure 25.5 Morphological, niche-diversification and distribution characteristics in *Pachycladon* (Brassicaceae), plotted on to a phylogeny showing species relationships within the genus.

New alpine habitat and long-lived lifestyles accelerate growth-form evolution and radiation (Hughes & Atchison 2015). Globally, numerous alpine lineages have shifted from annual to perennial life histories and from herbaceous to woody or cushion growth forms, and this holds true in New Zealand: the large *Veronica* radiation consists entirely of shrubs (80% of the alpine species) and subshrubs derived from a northern hemisphere herbaceous ancestor; woody caudexes have developed in many radiating New Zealand genera (e.g., *Pachycladon*, *Celmisia*, *Euphrasia*); and over half of the 23 largest genera have cushion species.

Habitat and dispersal ability explain local endemism and broad distributional patterns of alpine species in New Zealand (McGlone et al. 2001). Widespread species tend to have broad altitudinal ranges and habitat preferences, and good dispersal (spores, dust seeds, fleshy fruits, etc.) (Fenner et al. 2001; Lee et al. 2001b). Repeated glaciation forced species distributions down and away from the mountains or isolated them on nunataks (Heenan & Mitchell 2003); subsequent ice retreat and reforestation then reversed the movement. These cycles of expansion and contraction have enhanced speciation opportunities for plants with low dispersal potential (Winkworth et al. 2005), while encouraging wide distributions of those with good dispersal. *Pachycladon* shows that these radiation events are due not only to allopatric speciation, but also to significant niche-shifts and gene-flow changes (Figure 25.5) (Joly et al. 2014).

The far south and the north-west of the South Island are particularly rich in alpine endemics (Mark 2012). The mountains in these regions tend to be low but deeply divided and older than the rapidly growing, heavily glaciated central Southern Alps, and thus may have provided more opportunities for speciation (McGlone et al. 2001). New spatial analyses (Figure 25.6) show that the areas of most pronounced alpine endemism are on relatively low mountains outside or at the fringe of the position of the glacial ice sheet at the Last Glacial Maximum (LGM), confirming the general pattern. Isolated mountain ranges in the North Island also tend to have high endemism. On the other hand, the central Southern Alps have less endemism than expected, as they were covered by glacial ice sheets.

Figure 25.6 Species-corrected weighted (CWE) endemism for 579 alpine species comprising 50 117 georeferenced samples. CWE statistically analyses and corrects for species range sizes to provide a quantitative estimate of endemism (Laffan et al. 2016). Increasing CWE is shown by grid cells transitioning from light to dark gray. Grid cells with higher endemism than expected are noted with a plus (+), while those with lower endemism than expected are marked with a dot (·). Gray overlay: Last Glacial Maximum (LGM) ice distribution. *Source:* Provided by David Barrell (2011), Geological and Nuclear Sciences, Dunedin, New Zealand. See also Plate 42 in color plate section.

25.6 Biogeographic Relationships of the Alpine Flora

The New Zealand alpine flora is drawn largely from families widespread and common in all other alpine regions, but is more distinctive at the generic level. The genera fall into four broad distributional patterns (Table 25.1): genera that are (i) endemic to the New Zealand region (inclusive of the subantarctic islands immediately to the south, including Macquarie); (ii) Australasian endemics (including Australia, New Guinea and the mountains of Malesia); (iii) Southern (ranges extending to South America and southern Africa); and (iv) Bihemispheric (genera present in both the northern and southern hemispheres).

Bihemispheric genera (e.g., *Veronica*, *Epilobium*, *Myosotis*, *Gentianella*, *Poa*) make up 41.5% of the alpine total and are almost entirely derived from the northern hemisphere. The northern hemisphere thus has been the largest source of plant genera for the New Zealand alpine region, in stark contrast with the <6% of the genera of lowland woody plants shared between the two areas (McGlone et al. 2016).

Australasian genera are the second largest group, and are dominated by New Zealand-centered examples (e.g., *Celmisia*, *Chionochloa*, *Kelleria*, *Aciphylla*, *Anisotome*, *Coprosma*). New Zealand endemics and those with a broader southern hemisphere distribution form the smallest groups, and contribute the fewest species.

New Zealand has been a source for other austral alpine areas (Winkworth et al. 2005), including Australia, New Guinea and South America: examples are *Myosotis* (Meudt et al. 2015), *Pachycladon* (Heenan & Mitchell 2003), *Chionochloa* (Pirie et al. 2010) and *Veronica* [= *Chionohebe*] (Meudt & Bayly 2008). While few genera in the alpine vascular flora are endemic to the New Zealand region, ~93% of the species are (Mark 2012). Non-endemic species are mostly shared with Australia: 18% of the species in the Kosciuszko Alpine Flora in the Australian Alps are conspecific with New Zealand species (Costin et al. 2000), and of the 64 non-endemic species in the New Zealand alpine region, 55 also occur in Tasmania (Kirkpatrick 1997).

25.7 Origins of the Vascular Montane and Alpine Flora

Nearly all extant plant lineages appear to have arrived after the separation of Australia and New Zealand, and many much more recently. Molecular dates for the divergences of 72 austral plant taxa with disjunct distributions between Australasia and southern South America or between Australia and New Zealand (Winkworth et al. 2015) show that "cold-climate" disjunctions (96% of which are herbs and ferns) have a mean age of 8 My, versus 17 My for "temperate-climate" disjunctions (60% of which are trees and shrubs).

25.7.1 Montane–subalpine Flora

The montane–subalpine tree and shrub flora has two sources: recent autochthonous evolution from old elements, and radiations of more recent arrivals. Woody genera (and most of the families) are nearly entirely absent from the northern temperate zone, and only 12% are shared with southern South America (Ezcurra et al. 2008). The dominant tree genera in montane–subalpine regions – *Fuscospora*, *Lophozonia*, *Halocarpus*, *Phyllocladus*, *Podocarpus*, *Libocedrus*, *Metrosideros*, *Elaeocarpus*, *Weinmannia*, *Quintinia*, *Melicytus*, *Coprosma* and *Pseudopanax* – all date back to the Oligocene or earlier (Mildenhall 1980). On the other hand, the shrubby subalpine–alpine species, and in particular members of the *Melicytus*, *Coprosma* and *Dracophyllum* radiations, are all Pliocene or younger in age (Anderson et al. 2001; Bremer & Eriksson 2009; Mitchell et al. 2009; Wagstaff et al. 2010; Cantley et al. 2014). Many of the subalpine–alpine small tree and

Table 25.1 Global distribution patterns of New Zealand alpine genera.

Genus distribution	Genera		Species in genera		Alpine species (per genus)
	Number	% alpine total	Number	% alpine total	
Endemic	15	10.2	58	7.2	3.9
Australasian only	48	32.7	278	34.7	5.8
Southern hemisphere	23	15.6	102	12.7	4.4
Bihemispheric	61	41.5	364	45.4	6.0
Total	147	100	802	100	5.5

Note: The New Zealand alpine flora is largely (93%) endemic at the species level.

shrub species are derived from late-arriving herbaceous or suffrutescent (herbs with woody stocks) ancestors, including *Veronica*, *Olearia* and *Carmichaelia*. However, even long-present groups underwent radiation: for example, the ancient podocarp *Phyllocladus* consists in New Zealand of five closely related species that cannot have diverged earlier than 6.3 Ma, and the subalpine shrub ecotype *P. alpinus* is derived from a lowland tree ecotype (Wagstaff 2004).

25.7.2 Alpine Flora

The origin of the alpine flora has been controversial, even though the biogeographical pattern has long been clear. In this section, we discuss the four main scenarios.

25.7.2.1 Persistence of Cool-Adapted Taxa from the Early Cretaceous Onwards

Peter Wardle (1968) suggested that some isolated genera (e.g., *Lophozonia*, *Stilbocarpa*, *Hectorella*, *Oreostylidium*, *Donatia*) are ancient taxa adapted to the cooler landscapes of the Cretaceous and Paleocene when the New Zealand landmass lay close to the Antarctic Circle. Peter Raven (1973) supported this view, adding *Coprosma*, *Dracophyllum*, *Caltha*, *Gunnera*, *Acaena*, *Geum*, *Pittosporum*, *Uncinia* and hydrocotyloid Apiaceae to the list. According to Wardle, these ancient genera persisted through the warm–temperate to subtropical environments of the Palaeogene and Miocene in cloudy, wet locations in stunted vegetation on low-nutrient or poorly drained soils. Molecular phylogenies do not support ancient origins for any of these putative Tertiary holdouts: none extend back beyond the Oligocene, and some evolved as recently as the late Miocene to Pliocene. For example, the three genera in the Stylidiaceae mentioned by Wardle (*Oreostylidium*, *Forstera* and *Phyllachne*) and the species-rich *Dracophyllum* clade could not have been present before 6 Ma (Wagstaff & Wege 2002; Wagstaff et al. 2010).

25.7.2.2 Transoceanic Dispersal from Antarctica and High-Latitude Southern Ocean Islands

Antarctica and southern extensions of the Zealandia landmass have been suggested as a source of cold-adapted plants well into the Tertiary, the subantarctic islands acting as stepping stones (Wagstaff & Hennion 2007; Winkworth et al. 2015). While it was once thought that Antarctica supported subalpine habitats well into the Pliocene, it is now clear that it was fully glaciated from ca. 14 Ma (Barrett 2013). Molecular phylogenies do not support the subantarctic islands as a significant donor of plant lineages – rather, the reverse. Unusual plant taxa unique to the subantarctic islands (the large-leaved *Pleurophyllum*

spp. and *Stilbocarpa* spp. of Antipodes, Campbell and Auckland islands to the south of New Zealand and *Pringlea antiscorbutica* of the Kerguelen and adjacent archipelagos) are mainland derivatives: *Pleurophyllum* from the mainland genus *Olearia* (Wagstaff et al. 2011), *Stilbocarpa* from mainland *Schizeilema* (Mitchell et al. 1999) and *Pringlea* nested within a southern South American radiation (Bartish et al. 2012).

25.7.2.3 Derivation from Lowland to Montane Lineages

While Cretaceous holdouts in Antarctica are not significant sources of alpine lineages, genera from the Neogene lowland to lower montane flora have donated to the alpine flora. These include: *Podocarpus* (1 sp.), *Phyllocladus* (1 sp.) and *Lepidothamnus* (1 sp.) (Podocarpaceae); *Dracophyllum* (15 spp.), *Pittosporum* (4 spp.), *Coprosma* (11 spp.) and *Myrsine* (2 spp.) – all of which have overseas subtropical or tropical representatives.

25.7.2.4 Transoceanic Dispersal, Especially from Australia, Southern South America and the Northern Hemisphere

Peter Raven (1973) argued that the mountains of the Asian mainland and Andean South America were sources of alpine plants for the rapidly uplifting mountains of New Zealand, Australia, New Guinea and Malaysia in the late Pliocene and Pleistocene. He suggested the mountains of Malaysia, New Guinea, and Australia acted as pathways for Asian lineages that gave rise to New Zealand *Aciphylla*, *Anisotome*, Brassicaceae, *Carex*, *Celmisia*, *Colobanthus*, *Epilobium*, *Veronica*, *Myosotis*, *Poa*, *Ranunculus* and *Scleranthus*. Molecular phylogenies confirm Asian ancestry for a large number of these, but there is little evidence for mountain stepping stones: New Guinea and Australian alpine–subalpine species are mostly nested within New Zealand radiations; for example, *Myosotis* (Meudt et al. 2015), *Chionochloa* (Pirie et al. 2010) and *Veronica* spp. (Albach et al. 2005). South America contributed *Caltha* (=*Psychrophila*), *Gentianella* and *Ourisia* (Winkworth et al. 2005).

There are 23 alpine species, mostly herbs, that are disjunct between Tasmania and New Zealand (Jordan 2001), a result of alpine–alpine dispersal. Some are clearly Australian in origin (e.g., ericads *Pentachondra pumila* and *Montitega dealbata*). *Plantago* has dispersed at least three times from Australia to New Zealand, and two of these are likely to have been alpine–alpine dispersals (Tay et al. 2010). The wetland herb *Tetrachondra* is disjunct between southern South America (*T. patagonia*: alpine) and New Zealand (*T. hamiltoni*: lowland to subalpine), and Pleistocene dispersal is likely (Wagstaff et al. 2000).

25.7.3 Alpine Immigration Pathways within New Zealand

The major pathway to the alpine region for species of both autochthonous and transoceanic origins has been via lowland to montane environments. Heenan & McGlone (2013) argue that lower-altitude open habitats such as riverbeds, cliffs, bogs and unstable ground provided entry points into a mostly forested landscape during the late Miocene–Pliocene period of mountain building (Figure 25.7).

Many species (~65%) present in the alpine zone extend below treeline, and most large alpine genera include lowland–lower montane species. For example, the alpine *Ranunculus* in New Zealand are derived from the cosmopolitan water buttercup clade (sect. Batrachium) (Hörandl & Emadzade 2011) and probably entered via a lowland wetland pathway. The snowgrasses (*Chionochloa*) are from a southern African danthoid ancestor via Australia (Linder & Barker 2005) and probably entered via lowland–montane grassland habitat (Antonelli et al. 2011; McGlone et al. 2014), and thence to the alpine zone. Some alpine radiations have given rise to lowland–montane species. For example, lowland shrub and tree

species of *Veronica* are nested within a largely alpine clade, and *Ourisia* – which represents a single dispersal from an alpine lineage in the Andes – gave rise to New Zealand lowland–montane species. As noted in the previous subsection, there has been a limited amount of alpine–alpine species transfer, mainly from Australia.

25.8 Conclusion

The montane and alpine flora and vegetation of New Zealand bear many floristic and vegetational similarities to those of the tropics. In particular, the diverse evergreen montane and treeline communities and low-alpine shrub and tussock strongly resemble those of the mountains of New Guinea. Montane vegetation bears almost no resemblance to that of the temperate northern hemisphere, as it lacks pines, firs and spruces and is largely without deciduous trees. However, the alpine flora has strong floristic biogeographic connections to the northern hemisphere, a consequence of major transoceanic dispersal in the last few million years. As with other cordilleras and mountainous island archipelagos, spec-

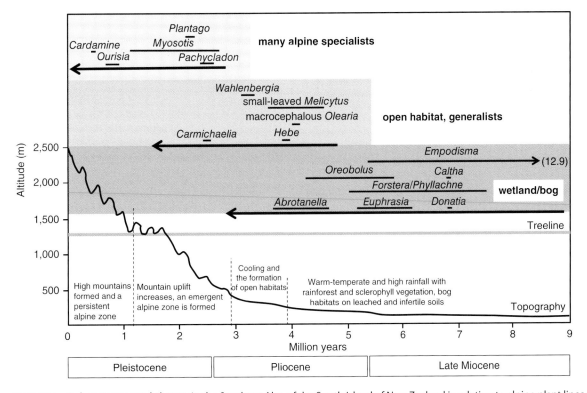

Figure 25.7 Historical environmental change in the Southern Alps of the South Island of New Zealand in relation to alpine plant lineage ages. Bars underneath genera indicate the molecular age estimate for their arrival in New Zealand. Shaded bands indicate broad groupings of species according to habitat preferences. The dark line indicates the rising elevation of the landmass relative to treeline as a consequence of mountain building and falling temperatures in the course of the late Miocene to Pleistocene. A number of genera that were later to evolve alpine members were first present on open or wet habitats below treeline before true alpine environments arose. *Source:* Heenan & McGlone (2013). Reproduced with permission of New Zealand Ecological Society.

tacular, morphologically diverse radiations have rapidly evolved. Harsh alpine environments impose similar selective forces across the globe, and the range of alpine growth forms in New Zealand is familiar in nearly all alpine areas. The truly distinctive features of the New Zealand alpine zone lie in reproductive biology. New Zealand alpine flowers are small and largely white, green or pale, and have a high level of gender dimorphism. Absence of long-tongued bees and butterflies and the presence of short-tongued small bees as the most efficient pollinators may be responsible for this striking pattern.

Acknowledgments

We acknowledge the pioneering work by Sir Alan Mark and the late Peter Wardle, which has done so much to advance understanding and enjoyment of the alpine and montane environment. We also thank two anonymous reviewers for their helpful comments.

References

Albach, D.C., Utteridge, T. & Wagstaff, S.J. (2005) Origin of Veroniceae (Plantaginaceae, formerly Scrophulariaceae) on New Guinea. *Systematic Botany* **30**, 412–423.

Anderson, C.L., Rova, J.H. & Andersson, L. (2001) Molecular phylogeny of the tribe Anthospermeae (Rubiaceae): systematic and biogeographic implications. *Australian Systematic Botany* **14**, 231–244.

Antonelli, A., Humphreys, A.M., Lee, W.G. & Linder, H.P. (2011) Absence of mammals and the evolution of New Zealand grasses. *Proceedings of the Royal Society B: Biological Sciences* **278**, 695–701.

Arnold, S.E.J., Savolainen, V. & Chittka, L. (2009) Flower colours along an alpine altitude gradient, seen through the eyes of fly and bee pollinators. *Arthropod–Plant Interactions* **3**, 27–43.

Aubert, S., Boucher, F., Lavergne, S. et al. (2014) 1914–2014: a revised worldwide catalogue of cushion plants 100 years after Hauri and Schroter. *Alpine Botany* **124**, 59–70.

Barrell, D. (2011) Quaternary glaciers of New Zealand. In: Ehlers, J., Gibbard, P. & Hughes, P. (eds.) *Quaternary Glaciations: Extent and Chronology – A Closer Look.* Amsterdam: Elsevier, pp. 1047–1064.

Barrett, P.J. (2013) Resolving views on Antarctic Neogene glacial history – the Sirius debate. *Earth and Environmental Science Transactions of the Royal Society of Edinburgh* **104**, 31–53.

Barrett, S.C.H. & Hough, J. (2013) Sexual dimorphism in flowering plants. *Journal of Experimental Botany* **64**, 67–82.

Bartish, I.V., Aïnouche, A., Jia, D. et al. (2012) Phylogeny and colonization history of *Pringlea antiscorbutica* (Brassicaceae), an emblematic endemic from the South Indian Ocean Province. *Molecular Phylogenetics and Evolution* **65**, 748–756.

Batt, G.E. & Braun, J. (1999) The tectonic evolution of the Southern Alps, New Zealand: insights from fully

thermally coupled dynamical modelling. *Geophysical Journal International* **136**, 403–420.

Benecke, U., Schulze, E.-D., Matyssek, R. & Havranek, W.M. (1981) Environmental control of CO_2 assimilation and leaf conductance in *Larix decidua* Mill. *Oecologia* **50**, 54–61.

Bischoff, M., Campbell, D.R., Lord, J.M. & Robertson, A.W. (2013) The relative importance of solitary bees and syrphid flies as pollinators of two outcrossing plant species in the New Zealand alpine. *Austral Ecology* **38**, 169–176.

Bremer, B. & Eriksson, T. (2009) Time tree of Rubiaceae: phylogeny and dating the family, subfamilies, and tribes. *International Journal of Plant Sciences* **170**, 766–793.

Brookes, R.H. & Jesson, L.K. (2007) No evidence for simultaneous pollen and resource limitation in Aciphylla squarrosa: a long-lived, masting herb. *Austral Ecology* **32**, 370–377.

Campbell, D.R., Bischoff, M., Lord, J.M. & Robertson, A.W. (2010) Flower color influences insect visitation in alpine New Zealand. *Ecology* **91**, 2638–2649.

Campbell, D.R., Bischoff, M., Lord, J.M. & Robertson, A.W. (2012) Where have all the blue flowers gone: pollinator responses and selection on flower colour in New Zealand *Wahlenbergia* albomarginata. *Journal of Evolutionary Biology* **25**, 352–364.

Cantley, J.T., Swenson, N.G., Markey, A. & Keeley, S.C. (2014) Biogeographic insights on Pacific *Coprosma* (Rubiaceae) indicate two colonizations to the Hawaiian Islands. *Botanical Journal of the Linnean Society* **174**, 412–424.

Case, B.S. & Hale, R.J. (2015) Using novel metrics to assess biogeographic patterns of abrupt treelines in relation to abiotic influences. *Progress in Physical Geography* **39**, 310–336.

Chamberlain, C.P., Poage, M.A., Craw, D. & Reynolds, R.C. (1999) Topographic development of the Southern Alps recorded by the isotopic composition of authigenic clay minerals, South Island, New Zealand. *Chemical Geology* **155**, 279–294.

Chittka, L. & Raine, N.E. (2006) Recognition of flowers by pollinators. *Current Opinion in Plant Biology* **9**, 428–431.

Cieraad, E., McGlone, M., Barbour, M.M. & Huntley, B. (2012) Seasonal frost tolerance of trees in the New Zealand treeline ecotone. *Arctic Antarctic and Alpine Research* **44**, 332–342.

Cieraad, E., McGlone, M.S. & Huntley, B. (2014) Southern Hemisphere temperate tree lines are not climatically depressed. *Journal of Biogeography* **41**, 1456–1466.

Clare, G.R., Fitzharris, B.B., Chinn, T.J.H. & Salinger, M.J. (2002) Interannual variation in end-of-summer snowlines of the Southern Alps of New Zealand, and relationships with Southern Hemisphere atmospheric circulation and sea surface temperature patterns. *International Journal of Climatology* **22**, 107–120.

Costin, A.B., Gray, M., Totterdell, C. & Wimbush, D. (2000) *Kosciuszko Alpine Flora*. Canberra: CSIRO Publishing.

de Lange, P.J., Rolfe, J.R., Champion, P.D. et al. (2013) *Conservation Status of New Zealand Indigenous Vascular Plants, 2012*. New Zealand Threat Classification Series 3, p. 70.

Ezcurra, C., Baccala, N. & Wardle, P. (2008) Floristic relationships among vegetation types of new zealand and the southern andes: Similarities and biogeographic implications. *Annals of Botany* **101**, 1401–1412.

Fenner, M., Lee, W.G. & Pinn, E.H. (2001) Reproductive features of *Celmisia* species (Asteraceae) in relation to altitude and geographical range in New Zealand. *Biological Journal of the Linnean Society* **74**, 51–58.

Fitzharris, B., Lawson, W. & Owens, I. (1999) Research on glaciers and snow in New Zealand. *Progress in Physical Geography* **23**, 469–500.

Froggatt, P.C. & Rogers, G.M. (1990) Tephrostratigraphy of high altitude peat bogs along the axial ranges, North Island, New Zealand. *New Zealand Journal of Geology and Geophysics* **33**, 111–124.

Godley, E.J. (1978) Cushion bogs. In: Troll, C. & Lauer, W. (eds.) *Geoecological Relations between the Southern Temperate Zone and the Tropical Mountains*. Wiesbaden: Franz Steiner Verlag GMBH, pp. 141–158.

Godley, E.J. (1979) Flower biology in New Zealand. *New Zealand Journal of Botany* **17**, 441–466.

Hales, T.C. & Roering, J.J. (2005) Climate-controlled variations in scree production, Southern Alps, New Zealand. *Geology* **33**, 701–704.

Hales, T.C. & Roering, J.J. (2009) A frost "buzzsaw" mechanism for erosion of the eastern Southern Alps, New Zealand. *Geomorphology* **107**, 241–253.

Halloy, S. (1990) A morphological classification of plants, with special reference to the New Zealand alpine flora. *Journal of Vegetation Science* **1**, 291–304.

Halloy, S.R.P. & Mark, A.F. (1996) Comparative leaf morphology spectra of plant communities in New Zealand, the Andes and the European Alps. *Journal of the Royal Society of New Zealand* **26**, 41–78.

Heenan, P.B. & McGlone, M.S. (2013) Evolution of New Zealand alpine and open-habitat plant species during the late Cenozoic. *New Zealand Journal of Ecology* **37**, 105–113.

Heenan, P.B. & Mitchell, A.D. (2003) Phylogeny, biogeography and adaptive radiation of *Pachycladon* (Brassicaceae) in the mountains of South Island, New Zealand. *Journal of Biogeography* **30**, 1737–1749.

Hicks, D.M., Shankar, U., McKerchar, A.I. et al. (2011) Suspended sediment yields from New Zealand rivers. *Journal of Hydrology, New Zealand* **50**, 81–142.

Hörandl, E. & Emadzade, K. (2011) The evolution and biogeography of alpine species in *Ranunculus* (Ranunculaceae): a global comparison. *Taxon* **60**, 415–426.

Hughes, C.E. & Atchison, G.W. (2015) The ubiquity of alpine plant radiations: from the Andes to the Hengduan Mountains. *New Phytologist* **207**, 275–282.

Inouye, D.W. & Pyke, G.H. (1988) Pollination in the Snowy Mountains of Australia: comparisions with montane Colorado, USA. *Australian Journal of Ecology* **13**, 191–210.

Jimenez-Castillo, M., Wiser, S.K. & Lusk, C.H. (2007) Elevational parallels of latitudinal variation in the proportion of lianas in woody floras. *Journal of Biogeography* **34**, 163–168.

Johnson, P. & Gerbeaux, P. (2004) *Wetland Types in New Zealand*. Wellington: Department of Conservation.

Joly, S., Heenan, P.B. & Lockhart, P.J. (2014) Species radiation by niche shifts in New Zealand's rockcresses (*Pachycladon*, Brassicaceae). *Systematic Biology* **63**, 192–202.

Jordan, G.J. (2001) An investigation of long-distance dispersal based on species native to both Tasmania and New Zealand. *Australian Journal of Botany* **49**, 333–340.

Kelly, D. (1994) The evolutionary ecology of mast seeding. *Trends in Ecology & Evolution* **9**, 465–470.

Kirkpatrick, J.B. (1997) *Alpine Tasmania: An Illustrated Guide to the Flora and Vegetation*. Oxford: Oxford University Press.

Körner, C. & Paulsen, J. (2004) A world-wide study of high altitude treeline temperatures. *Journal of Biogeography* **31**, 713–732.

Laffan, S.W., Rosauer, D.F., Di Virgilio, G. et al. 2016. Range-weighted metrics of species and phylogenetic turnover can better resolve biogeographic transition zones. *Methods in Ecology and Evolution* **7**, 580–588.

Lee, W.G., Wilson, J.B. & Johnson, P.N. (1988) Fruit colour in relation to the ecology and habit of *Coprosma* (Rubiaceae) species in New Zealand. *Oikos* **53**, 325–331.

Lee, D.E., Lee, W.G. & Mortimer, N. (2001a) Where and why have all the flowers gone? Depletion and turnover in the New Zealand Cenozoic angiosperm flora in relation to palaeogeography and climate. *Australian Journal of Botany* **49**, 341–356.

Lee, W.G., Macmillan, B.H., Partridge, T.R. et al. (2001b) Fruit features in relation to the ecology and distribution of *Acaena* (Rosaceae) species in New Zealand. *New Zealand Journal of Ecology* **25**, 17–27.

Linder, H.P. & Barker, N.P. (2005) From Nees to now – changing questions in the systematics of the grass subfamily Danthonioideae. *Nova Acta Leopoldina Neue Folge* **92**, 29–44.

Lloyd, D.G. (1985) Progress in understanding the natural history of New Zealand plants. *New Zealand Journal of Botany* **23**, 707–722.

Lord, J.M. & Marshall, J. (2001) Correlations between growth form, habitat, and fruit colour in the New Zealand flora, with reference to frugivory by lizards. *New Zealand Journal of Botany* **39**, 567–576.

Mark, A.F. (2012) *Above the Treeline: A Nature Guide to Alpine New Zealand*. Nelson: Craig Potton Publishing.

Mark, A.F., Johnson, P.N., Dickinson, K.J.M. & McGlone, M.S. (1995) Southern hemisphere patterned mires, with emphasis on southern New Zealand. *Journal of the Royal Society of New Zealand* **25**, 23–54.

Mark, A.F., Dickinson, K.J.M. & Hofstede, R.G.M. (2000) Alpine vegetation, plant distribution, life forms, and environments in a perhumid New Zealand region: oceanic and tropical high mountain affinities. *Arctic, Antarctic, and Alpine Research* **32**, 240–254.

Mark, A.F., Dickinson, K.J.M., Allen, J. et al. (2001) Vegetation patterns, plant distribution and life forms across the alpine zone in southern Tierra del Fuego, Argentina. *Austral Ecology* **26**, 423–440.

Makrodimos, N., Blionis, G.J., Krigas, N. & Vokou, D. (2008) Flower morphology, phenology and visitor patterns in an alpine community on Mt Olympos, Greece. *Flora* **203**, 449–468.

Mandakova, T., Heenan, P.B. & Lysak, M.A. (2010) Island species radiation and karyotypic stasis in *Pachycladon* allopolyploids. *BMC Evolutionary Biology* **10**, 367.

McGimpsey, V.J. & Lord, J.M. (2015) In a world of white, flower colour matters: a white-purple transition signals lack of reward in an alpine *Euphrasia*. *Austral Ecology* **40**, 701–708.

McGlone, M.S. (2009) Postglacial history of New Zealand wetlands and implications for their conservation. *New Zealand Journal of Ecology* **33**, 1–23.

McGlone, M.S., Duncan, R.P. & Heenan, P.B. (2001) Endemism, species selection and the origin and distribution of the vascular plant flora of New Zealand. *Journal of Biogeography* **28**, 199–216.

McGlone, M.S., Dungan, R.J., Hall, G.M.J. & Allen, R.B. (2004) Winter leaf loss in the New Zealand woody flora. *New Zealand Journal of Botany* **42**, 1–19.

McGlone, M.S., Perry, G.L.W., Houliston, G.J. & Connor, H.E. (2014) Fire, grazing and the evolution of New Zealand grasses. *New Zealand Journal of Ecology* **38**, 1–11.

McGlone, M.S., Buitenwerf, R. & Richardson, S.J. (2016) The formation of the oceanic temperate forests of New Zealand. *New Zealand Journal of Botany* **54**, 128–155.

Mershon, J.P., Becker, M. & Bickford, C.P. (2015) Linkage between trichome morphology and leaf optical properties in New Zealand alpine *Pachycladon* (Brassicaceae). *New Zealand Journal of Botany* **53**, 175–182.

Meudt, H.M. (2007) Phylogeography of *Ourisia* (Plantaginaceae): an integrated study of a New Zealand alpine radiation. *New Zealand Journal of Botany* **45**, 296–297.

Meudt, H.M. & Bayly, M.J. (2008) Phylogeographic patterns in the Australasian genus *Chionohebe* (*Veronica* s.l., Plantaginaceae) based on AFLP and chloroplast DNA sequences. *Molecular Phylogenetics and Evolution* **47**, 319–338.

Meudt, H.M., Prebble, J.M. & Lehnebach, C.A. (2015) Native New Zealand forget-me-nots (*Myosotis*, Boraginaceae) comprise a Pleistocene species radiation with very low genetic divergence. *Plant Systematics and Evolution* **301**, 1455–1471.

Mildenhall, D.C. (1980) New Zealand Late Cretaceous and Cenozoic plant biogeography – a contribution. *Palaeogeography, Palaeoclimatology, Palaeoecology* **31**, 197–233.

Mitchell, A.D., Meurk, C.D. & Wagstaff, S.J. (1999) Evolution of *Stilbocarpa*, a megaherb from New Zealand's sub-antarctic islands. *New Zealand Journal of Botany* **37**, 205–211.

Mitchell, A.D., Heenan, P.B., Murray, B.G. et al. (2009) Evolution of the south-western Pacific genus *Melicytus* (Violaceae): evidence from DNA sequence data, cytology and sex expression. *Australian Systematic Botany* **22**, 143–157.

Nathan, S., Anderson, H.J., Cook, R.A. et al. (1986) *Cretaceous and Cenozoic Sedimentary Basins of the West Coast Region, South Island, New Zealand*. Wellington: Department of Scientific and Industrial Research.

Newstrom, L. & Robertson, A. (2005) Progress in understanding pollination systems in New Zealand. *New Zealand Journal of Botany* **43**, 1–59.

Noroozi, J., Akhani, H. & Breckle, S.-W. (2008) Biodiversity and phytogeography of the alpine flora of Iran. *Biodiversity and Conservation* **17**, 493–521.

Pickering, C.M. & Stock, M. (2003) Insect colour preference compared to flower colours in the Australian Alps. *Nordic Journal of Botany* **23**, 217–223.

Pirie, M.D., Lloyd, K.M., Lee, W.G. & Linder, H.P. (2010) Diversification of *Chionochloa* (Poaceae) and biogeography of the New Zealand Southern Alps. *Journal of Biogeography* **37**, 379–392.

Raven, P.H. (1973) Evolution of subalpine and alpine plant groups in New Zealand. *New Zealand Journal of Botany* **11**, 177–200.

Renner, S.S. (2014) The relative and absolute frequencies of angiosperm sexual systems: dioecy, monoecy, gynodioecy, and an updated online database. *American Journal of Botany* **101**, 1588–1596.

Rundel, P.W. (2011) The diversity and biogeography of the alpine flora of the Sierra Nevada, California. *Madrono* **58**, 153–184.

Schauber, E.M., Kelly, D., Turchin, P. et al. (2002) Masting by eighteen New Zealand plant species: the role of temperature as a synchronizing cue. *Ecology* **83**, 1214–1225.

Strauss, S.Y., Cacho, N.I., Schwartz, M.W. et al. (2015) Apparency revisited. *Entomologia Experimentalis et Applicata* **157**, 74–85.

Tay, M.L., Meudt, H.M., Garnock-Jones, P.J. & Ritchie, P.A. (2010) DNA sequences from three genomes reveal multiple long-distance dispersals and non-monophyly of sections in Australasian *Plantago* (Plantaginaceae). *Australian Systematic Botany* **23**, 47–68.

Troll, C. (1973) The upper timberlines in different climatic zones. *Arctic and Alpine Research* **5**, 3–18.

van Balgooy, M.M.J. (1976) Phytogeography. In: Paijmans, K. (ed.) *New Guinea Vegetation*. Amsterdam: Elsevier Scientific Publishing, pp. 1–22.

Wagstaff, S.J. (2004) Evolution and biogeography of the austral genus *Phyllocladus* (Podocarpaceae). *Journal of Biogeography* **31**, 1569–1577.

Wagstaff, S.J. & Hennion, F. (2007) Evolution and biogeography of *Lyallia* and *Hectorella* (Portulacaceae), geographically isolated sisters from the Southern Hemisphere. *Antarctic Science* **19**, 417–426.

Wagstaff, S.J. & Wege, J. (2002) Patterns of diversification in New Zealand Stylidiaceae. *American Journal of Botany* **89**, 865–874.

Wagstaff, S.J., Martinsson, K. & Swenson, U. (2000) Divergence estimates of *Tetrachondra hamiltonii* and *T. patagonica* (Tetrachondraceae) and their implications for austral biogeography. *New Zealand Journal of Botany* **38**, 587–596.

Wagstaff, S.J., Dawson, M.I., Venter, S. et al. (2010) Origin, diversification, and classification of the Australasian genus *Dracophyllum* (Richeeae, Ericaceae) 1. *Annals of the Missouri Botanical Garden* **97**, 235–258.

Wagstaff, S.J., Breitwieser, I. & Ito, M. (2011) Evolution and biogeography of *Pleurophyllum* (Astereae, Asteraceae), a small genus of megaherbs endemic to the subantarctic islands. *American Journal of Botany* **98**, 62–75.

Wardle, P. (1968) Evidence for an indigenous pre-Quaternary element in the mountain flora of New Zealand. *New Zealand Journal of Botany* **6**, 120–125.

Wardle, P. (1973) New Guinea: our tropical counterpart. *Tuatara* **20**, 113–124.

Wardle, P. (1978) Ecological and geographical significance of some New Zealand growth forms. In: Troll, C. & Lauer, W. (eds.) *Geoecological Relations between the Southern Temperate Zone and the Tropical Mountains*. Wiesbaden: Franz Steiner Verlag GMBH, pp. 531–536.

Wardle, P. (1991) *Vegetation of New Zealand*. Cambridge: Cambridge University Press.

Wardle, P. (2008) New Zealand forest to alpine transitions in global context. *Arctic Antarctic and Alpine Research* **40**, 240–249.

Webb, C., Lloyd, D.G. & Delph, L.F. (1999) Gender dimorphism in indigenous New Zealand seed plants. *New Zealand Journal of Botany* **37**, 119–130.

Winkworth, R.C., Wagstaff, S.J., Glenny, D. & Lockhart, P.J. (2005) Evolution of the New Zealand mountain flora: origins, diversification and dispersal. *Organisms Diversity & Evolution* **5**, 237–247.

Winkworth, R.C., Hennion, F., Prinzing, A. & Wagstaff, S.J. (2015) Explaining the disjunct distributions of austral plants: the roles of Antarctic and direct dispersal routes. *Journal of Biogeography* **42**, 1197–1209.

26

The East African Rift System: Tectonics, Climate and Biodiversity

Uwe Ring[1], Christian Albrecht[2,3] and Friedemann Schrenk[4]

[1] *Institutionen för geologiska vetenskaper, Stockholm University, Stockholm, Sweden*
[2] *Department of Animal Ecology and Systematics, Justus Liebig University, Giessen, Germany*
[3] *Department of Biology, Mbarara University of Science and Technology, Mbarara, Uganda*
[4] *Senckenberg Research Institute and Institute for Ecology, Evolution and Diversity, Goethe University, Frankfurt, Germany*

Abstract

The East African Rift System (EARS) is one of the most prominent rift systems on Earth and transects the high-elevation East African Plateau. The EARS is famous for its tectonics and geology, and has also been suggested to be the "cradle of mankind," making it a natural laboratory for interdisciplinary research straddling the Earth and Life sciences. Rifting commenced as a result of mantle plume activity under East Africa. Two distinct rift branches are observed: an older, volcanically active Eastern Branch and a younger, much less volcanic Western Branch. The Eastern Branch is generally characterized by high elevation, whereas the Western Branch comprises a number of deep rift lakes (e.g., Lake Tanganyika, Lake Malaŵi). The onset of topographic uplift in the EARS is poorly dated, but it preceded graben development, which commenced at ca. 24 Ma in the Ethiopian Rift, at ca. 12 Ma in Kenya and at ca. 10 Ma in the Western Branch. Pronounced uplift of the East African Plateau since ca. 10 Ma might be connected to climate change in East Africa and to floral and faunal – including human – evolution. But linking global climate changes to biodiversity in East Africa is difficult, as long-term global climate records come from marine cores off the African coast, while faunal and floral evidence from the late Miocene onwards comes from local rift valley sediments. East Africa experienced cooling starting at 15.6–12.5 Ma that heralded profound faunal changes at 8–5 Ma, when modern taxa began to take shape and the hominin lineage split from the chimpanzee lineage. The Pliocene can be divided into a period of long-term environmental stability between 5.3 and 3.3 Ma and a period of cooler and more arid conditions after ~2.8 Ma. The end of the Pliocene warm period coincides with the earliest known stone tools at ca. 3.3 Ma and the earliest known fossils of the hominin genus *Homo* at 2.8 Ma.

Keywords: *continental graben, structural architecture, volcanism, topography, fossils, hominins, climate change*

26.1 The East African Rift System

The East African Rift System (EARS) is more than 3000 km long and transects the high-elevation Ethiopian and Kenyan plateaus (Figure 26.1). This spectacular structure is a biodiversity hot spot, and according to Leakey (1973) the "cradle of mankind." The latter proposition highlights a conceivable link between tectonics, climate, biodiversity and hominin evolution.

The EARS is a diffuse rift system separating the Somalian Plate in the east from the Nubian Plate in the west (Calais et al. 2003). It is subdivided into an Eastern and a Western Branch (Figure 26.1). The Eastern Branch extends from Afar in northern Ethiopia to the Manyara Basin in northern Tanzania. Most of the Eastern Branch dissects the high-elevation Ethiopian and Kenyan plateaus (herein collectively referred to as the "East African Plateau"), which are separated by the Turkana depression. Lake Turkana formed in this depression, and thus has an unusual tectonic position in the high-elevation Eastern Branch. It is the only large elongated rift lake in the Eastern Branch (6405 km^2 surface area at an elevation of 360 m) (Figure 26.2). In contrast, the rift lakes in Ethiopia (e.g., Lake Zway in the Awash valley, 485 km^2 surface area) and central Kenya (e.g., Lake Naivasha, 139 km^2 surface area) are situated at 1636 m and 1884 m

Mountains, Climate and Biodiversity, First Edition. Edited by Carina Hoorn, Allison Perrigo and Alexandre Antonelli.

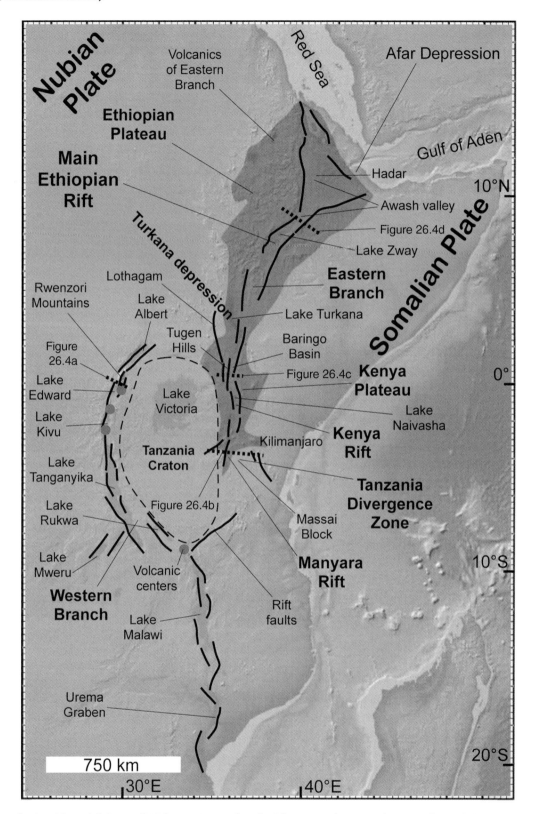

Figure 26.1 The East African Rift System (EARS), superimposed on the Ethiopian and Kenyan plateaus, collectively known as the East African Plateau (www.geomapapp.org). The EARS comprises series of individual graben that link up to form the Western and Eastern branches. The rift floors of the Eastern Branch, with vast volcanic rock accumulations (in dark gray shading), have high elevations, and only the Turkana graben of this branch has a lower elevation. Localities mentioned in the text are indicated, and dotted lines for cross-sections depicted in Figure 26.4 are shown. See also Plate 43 in color plate section.

altitude, respectively, and are relatively small. The Eastern Branch is magmatically very active, with large shield volcanoes like Kilimanjaro, Mt. Kenya and Mt. Elgon on the rift shoulders.

The Western Branch extends from Lake Albert in Uganda to the Urema Graben in Mozambique (Figure 26.1) and is magmatically much less active than the Eastern Branch. Basin floors have lower elevations with a number of long and deep rift lakes, indicating greater absolute basin subsidence. The largest lakes are Lake Albert (5300 km^2 surface area), Lake Tanganyika (32 000 km^2 surface area) and Lake Malaŵi (29 600 km^2 surface area), all with surface elevations of ~500–800 m (Figure 26.2). The lake floor of Lake Tanganyika is ~700 m below sea level, and together with Lake Baikal and the Dead Sea these represent the deepest continental points.

Collectively, the rift lakes of the EARS are the largest and most species-rich freshwater feature in the world (Salzburger et al. 2014). These lakes vary in age, however: many of the small lakes (100–200 km^2 surface area) of the Eastern Branch formed since the middle Pleistocene, while Lake Tanganyika in the Western Branch formed beginning in the middle/late Miocene (Cohen et al. 1993). Lake Victoria, located in between the two rift branches, is much younger, and resulted from river reversal and ponding due to pronounced rift shoulder uplift at ~0.4 Ma (Ebinger 1989). It can be described as a lake induced by rift tectonics, although it is not a rift lake.

The major differences in geography, elevation and hydrography between the EARS lakes are largely related to the tectonic development. An exception is Lake Turkana, which formed in an unusual tectonic position at the crossroads between the Eastern Branch of the EARS and the Cretaceous north-west–south-east oriented Sudan–Anza Rift system, and has geological characteristics of a "typical" Western Branch rift lake in the more arid, generally high-altitude Eastern Branch.

An important question is what role these hydrographical and topographical differences play in local rift mesoclimate, biodiversity and hominin evolution. Davies et al. (1985) demonstrated that the present-day topography of the Kenyan Plateau has a strong modifying effect on regional precipitation patterns. As a result, the Eastern Branch is relatively dry, and its large savannahs today host extraordinary wildlife. In contrast, the Western Branch has mountain forests and a rich lake biodiversity. Both the lake systems and the "sky islands" (isolated mountaintops) have produced an unusual degree of endemism among the incredible overall biodiversity. Examples include the Rwenzori Mountains (Eggermont et al. 2009) and rift lakes such as Lake Tanganyika and Lake Malaŵi (Salzburger et al. 2014).

26.2 Continental Rift Zones

Continental rift zones are sites of lithospheric extension (Figure 26.3) (Chapter 2). Extension is achieved through normal faulting that thins the crust. Where the brittle crust breaks, continental graben (tectonically controlled depressions) develop. Dense lithospheric mantle rocks rise upward to replace the thinning crust, enhancing subsidence in the fault-bounded basins. The lithospheric mantle also extends and thins, and is replaced by hotter and thus less dense asthenosphere, thereby transferring heat to the lithosphere beneath the extending region, reducing rock density and subsequently causing regional, time-dependent uplift over tens of millions of years. Mantle upwelling may be enhanced by plumes and/or small-scale mantle convection induced by steep thickness gradients at the transition between thinned and nonthinned lithosphere (Ebinger et al. 2013).

Continental rift zones are typically made up of a series of asymmetric graben. From their very inception, rift zones show regular along-axis structural segmentation into graben bounded on one or both sides by large offset border faults (Ebinger 2012). The border faults are flanked by broad uplifts that may rise 2–4 km above the surrounding regional elevations (Figure 26.1). Border fault length, rift flank uplift and basin dimension increase with increasing strength of the lithosphere (Weissel & Karner 1989). This is why young rift segments in strong lithosphere are characterized by long, skinny graben with deep rift lakes bounded by impressive tall escarpments.

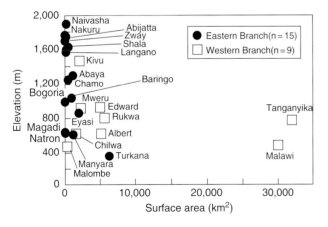

Figure 26.2 Elevation versus size (surface area) for the main East African Rift System (EARS) rift lakes. Lakes in the Western Branch (indicated by squares) are relatively larger and at lower elevations than those in the Eastern Branch (indicated by filled circles). Note that the data point for Lake Turkana plots in the same general area as those of the Western Branch lakes.

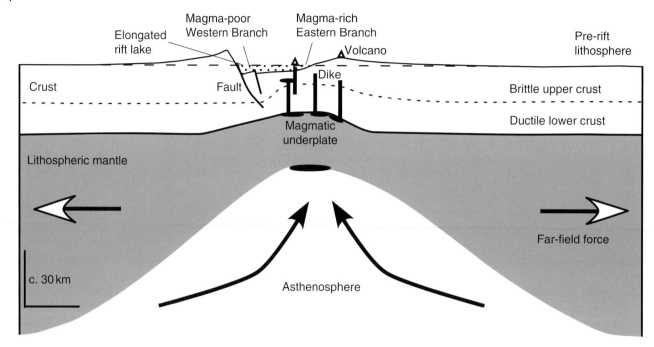

Figure 26.3 Simplified sketch showing changes in lithospheric structure during rifting. The pre-rift lithosphere at either end of the cross-section is modified by faulting and magmatism in the rift. Extension thins the lithosphere, causing subsidence, and asthenospheric inflow beneath the rift replaces the denser lithospheric mantle, causing long-term uplift. Note the elevated transition zone between the brittle upper crust and ductile lower crust in the rift zone, caused by a thermal perturbation, especially in the magma-rich part of the rifts, where mafic melt that accumulated beneath the crust (i.e., underplated) forms dikes (vertical black lines) and intrusions (black ellipses). Note that "underplating" refers to melting that forms "ponds" beneath the crust. See Chapter 2 for a simplified depiction of this process. *Source:* Adapted from Ebinger (2012).

26.3 Tectonic Development of the East African Rift System

26.3.1 African Superswell

Africa sits above a major mantle upwelling, the African Superswell, and is breaking apart along the EARS. The African plume is rising from the core–mantle boundary (~2900 km deep) and flowing to the north-east beneath East Africa (Bagley & Nyblade 2013). Geophysical and geochemical data demonstrate that mantle upwelling and magmatism developed above an anomalously hot asthenosphere (Hart et al. 1989). This might be the reason why Africa has the highest mean elevation of all continents, and why especially the Eastern Branch, with its vast volcanic rock accumulations, has such elevated rift floors.

The EARS comprises several discrete and diachronous rift sectors. Field data have shown that the Eastern and Western branches skirt around the Tanzania craton, an Archean/Proterozoic continental nucleus (Ring 1994). Tomographic sections show that the deep root of the Tanzania craton plays a major role in guiding the upwelling hot mantle emanating from the top of the

African Superswell in the upper few hundred kilometers of the mantle (Ebinger & Sleep 1998). The earliest basaltic volcanism in the EARS occurred between 45 and 39 Ma in south-western Ethiopia and northernmost Kenya (Morley et al. 1992). The widespread distribution of the volcanics suggests heating along the asthenosphere–lithosphere boundary (~100 km deep) long (~20 My) before any regional extension and surface expression of rifting (Ebinger et al. 2013).

Rift faulting started to form at ~24 Ma in Ethiopia, with major extension since ~11 Ma; at ~12 Ma in Kenya; and at ~10 Ma in the Western Branch (Ebinger 1989; Ebinger et al. 2000; Corti 2009) – highlighting that the ~12–10 Ma time window was important for the tectonic development of the EARS. At ~0.4 Ma, a major tectonic reorganization occurred, and extension became more oblique, resulting in strike-slip faulting and local uplift (Strecker & Bosworth 1991; Ring et al. 1992). Rift-flank uplift at ~0.4 Ma was responsible for the formation of Lake Victoria (Ebinger 1989).

Rifting has progressed to incipient sea-floor spreading in the Afar depression (Corti 2009). Further south, the Eastern Branch is superposed on the broad Ethiopian and Kenyan plateaus. The Turkana depression between the two plateaus marks a failed Mesozoic rift system, allowing for the possibility that the plateaus are part of one large zone of uplift extending from southern Africa to the Red Sea.

26.3.2 Magmatism of Rift Sectors

Partial melting, which is chiefly controlled by the temperature and volatile content of the asthenosphere, as well as the degree of decompression, can be used as a proxy for the maturity of a rift sector (White et al. 1987; Ebinger et al. 2013). We follow Ebinger (2012) and describe the differences from the most juvenile, primitive stage in the Western Branch and the southern tip of the Eastern Branch to the most evolved stage in the Ethiopian Rift at the northern end of the Eastern Branch. The differences in volcanism, uplift and subsidence of the two EARS branches reflect the way the mantle plume was channeled underneath East Africa. The young rift sectors have not been significantly affected by plume-related heating, and the entire lithosphere (tectonic plate) is strong and has not been thinned to any significant extent. The farther north one goes in the Eastern Branch, the longer the rift sectors have been affected by plume-related heating, leading to more pronounced igneous activity, a viscous lower crust and more advanced rifting, causing weak lithosphere.

The individual rift basins of the Western Branch are long (~100–150 km) and narrow (~50–70 km). Volcanism occurs in four isolated centers (Figure 26.1), and commenced at 10 ± 2 Ma, largely coeval with rift faulting (Ring & Betzler 1995). Overall, the morphologic and magmatic evolution of the Western Branch suggests a relatively strong plate that has been very modestly thinned, and border faults penetrate the entire lower crust, consistent with deep seismicity (Figure 26.4a).

The Eastern Branch shows a striking progression of magmatism (and rift evolution) from south to north. The most juvenile rift sector is the Tanzanian Divergence Zone, which is ~300–400 km wide and consists of three separate graben (Figure 26.4b). As in the Western Branch, individual rift basins are half graben-bounded by a faulted rift escarpment on one side and a flexural warp on the other (Foster et al. 1997). Each basin is ~100 km long and ~50 km wide (Ebinger et al. 1997), and contains thin (<3 km) sequences of volcanics and sediments. Most rift basins in the Tanzanian Divergence Zone are younger than ~1 Ma (Ring et al. 2005).

The Kenya Rift, farther north, is distinctly older and more mature than the Tanzanian Divergence Zone, and forms a narrow (~50 km-wide) rift sector. The basins are still asymmetric half-graben, and extension is of the order of 10 km, with the lithosphere thinning up to ~90 km (Figure 26.4c) (Mechie et al. 1997). Volcanism

commenced at ~23 Ma in broad downwarps that subsequently became the sites of half-graben. Between 14 and 11 Ma, a period of intense volcanism filled and overflowed the rift basins. Another period of pronounced volcanism occurred between 5 and 2 Ma, again overspilling the rift basins (Morley et al. 1992). A stunning feature of young volcanism is the large off-rift shield volcanoes, such as Mt. Elgon, Mt. Kenya and Kilimanjaro in the central Kenya Rift (Figure 26.1). All data indicate a higher degree of lithospheric extension in the central Kenya Rift when compared to the Western Branch and the Tanzanian Divergence Zone.

A review of the various rift segments shows that with increasing maturity of the rift sectors comes thinning of the lithosphere and increasing decompressional melt (Figure 26.4), creating a weaker lithosphere. In the EARS, these features are best expressed in the Main Ethiopian Rift, which links into the Afar Depression with the spreading centers of the Red Sea and Gulf of Aden.

Mohr (1983) estimated that the vast Ethiopian Plateau amounts to ~300 000 km³ of volcanic rocks that erupted between 32 and 21 Ma, with a short-lived period of extensive and aerially widespread flood basalts at 31–30 Ma (Corti 2009). Initial basin formation and minor extension occurred from 24 to 11 Ma, followed by extension in the Main Ethiopian Rift at ~11 Ma, possibly related to the onset of sea-floor spreading in the Gulf of Aden and the initiation of extension in Kenya (Corti 2009). The region of active extension narrowed with time to a ~20 km-wide zone near the center of the rift since ~2 Ma. This period of rifting is probably related to sea-floor spreading in the Red Sea and a rearrangement in global plate motions at 5–3 Ma (Calais et al. 2003).

Collectively, these findings indicate that the crust beneath the rifted regions in Ethiopia has been extensively modified by magmatic processes through the addition of mafic rocks in the mid to lower crust (Figure 26.4d). The drastic thinning of the lithospheric mantle is probably a combined effect of lithospheric extension and plume-related thermal erosion of the lithospheric mantle. Beneath Afar, the mantle structure is much akin to that of mid-ocean ridge systems, and magmatism and faulting are focused on the central rift axis (Corti 2009).

The EARS is not only an archetypal rift system but also a typical example of an active rift. In active rifts, rising mantle plumes cause regional updoming and thermal erosion of the lithospheric mantle, and these processes are the major drivers initiating plate divergence. Plume-related processes also drive topography, primarily through heat-controlled density changes; a fascinating corollary of this is that the plume-driven topography may have been an important cause of climate change in East Africa, and thus influenced the development and evolution of mankind.

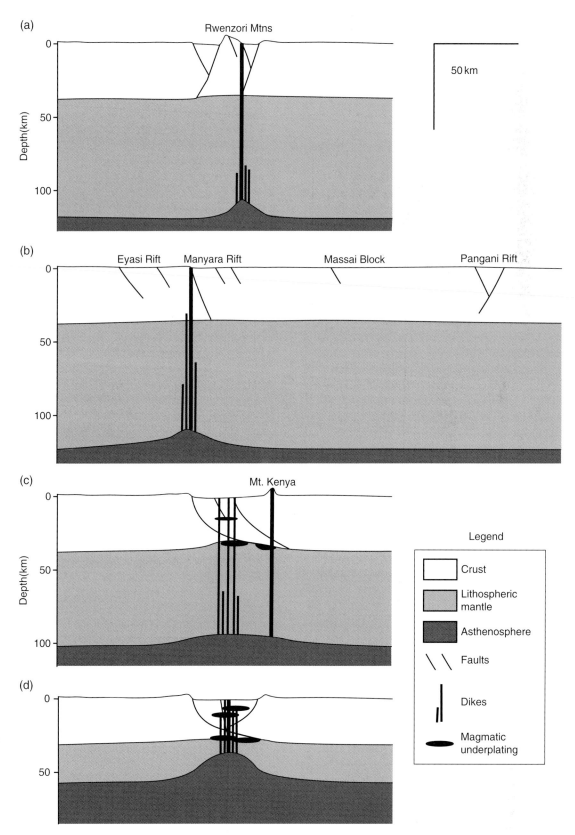

Figure 26.4 Simplified lithospheric cross-sections through the Western and Eastern Rift branches, showing different styles of rifting, lithospheric thinning and magmatism; for the localities of the cross-sections, see the dotted lines in Figure 26.1. (a) Albertine Rift of the Western Branch. The lithosphere is hardly thinned and the lower crust is not viscously deforming. The chemistry of the magmatic rocks is primitive and the volume of magmatism is very limited. Pronounced rift-flank uplift created the >5000 m-high Rwenzori Mountains. (b) Tanzania Divergence Zone of the Eastern Branch. Note the wide area affected by rifting, caused by the strong lithosphere of the Massai Block. A lithospheric structure similar to (a) indicates the early stages of rifting. (c) Central Kenya Rift. The lithosphere is notably attenuated and thinned, resulting in a viscous lower crust in which the high-angle upper crustal normal faults flatten out. Note off-axis volcanism, which forms large volcanic edifices like Mt. Kenya, on the rift shoulder. (d) Main Ethiopian Rift. The most advanced stage of continental rifting, resulting in strongly thinned lithosphere. The style of rifting becomes more symmetric, and rifting is strongly assisted by pronounced magmatism.

26.4 Global Climate Change and East Africa

26.4.1 Global Climate in the Neogene

The Neogene was a period of long-term global cooling and increased climate variability. Miller et al. (2011) proposed a major cooling step at ~15.5–12.5 Ma, heralding the first glaciations in the northern hemisphere at 10–6 Ma (Jansen et al. 1990), and another one at 2.95–2.52 Ma, causing further major northern hemisphere glaciation. Global cooling and an associated aridification trend in continental Africa occurred at 3.5–3.35, 2.5–2.0 and 1.8–1.6 Ma (Wynn 2004). Between these steps was the Pliocene warm (wet) period, between 5.3 and 3.3 Ma (Tiedemann et al. 1994), with a weak east–west zonal Walker sea circulation (i.e., easterly trade winds that move water and air warmed by the sun towards the west). During the Pliocene, global surface temperature was ~3 °C warmer than at present, and atmospheric CO_2 concentrations were ~30% higher (Ravelo et al. 2004). Tropical and subtropical climates during this time significantly influenced the evolution of Pliocene biodiversity in East Africa.

The most significant global climate event of the Pliocene was the onset of major northern hemisphere glaciation and a shift in the dominant period of climate oscillation, both of which commenced by ~2.8 Ma (Chapter 9) (Martinez-Boti et al. 2015). Wet phases became less frequent after 2.8 Ma. Changes reflect a shift from orbital precession (23 ka) to obliquity (41 ka). The development of strong Walker circulation took place in two steps, neither of which is temporally linked with the onset of northern hemisphere glaciation. The first step, at 4.5–4.0 Ma, was marked by altered surface-water gradients and ocean circulation, possibly connected to restriction of the Panamanian and Indonesian seaways (Cane & Molnar 2001; Molnar 2008; Pfister et al. 2014). The second step, at 2.0–1.5 Ma, established strong Walker circulation, a steeper sea-surface east–west temperature gradient across the Pacific, and the modern tropical climate system (Ravelo et al. 2004). Overall, while the onset of significant northern hemisphere glaciation occurred as subtropical conditions began to cool, changes in the Pliocene tropical climate system were to a certain degree independent (Ravelo et al. 2004).

26.4.2 Possible Drivers of Climate Change in East Africa Prior to Hominin Evolution

Sepulchre et al. (2006) speculated that "massive eastern African topographic uplift" led to a pronounced reorganization of atmospheric circulation and caused increased aridification in East Africa. However, the development of pronounced topography of the rift shoulders, used as evidence in their study, has not been dated, and paleoelevation is largely unconstrained. Current work strongly suggests that pronounced rift-shoulder uplift occurred much too recently to explain late Mio/Pliocene climate shifts (Foster & Gleadow 1993; Ring 2008; Bauer et al. 2010, 2013, 2016).

A more significant factor that may have influenced East African climate in the late Miocene appears to be the topography exerted by the development of the African Superswell. Moucha & Forte (2011) developed a numerical simulation based on flow pattern changes in the mantle that allows for the reconstruction of the amplitude and timing of topographic uplift above the African Superswell through time. Their results suggest that geographically extensive (~$5 \times 10^6 \, km^2$) topographic uplift in excess of 500 m of elevation started to develop by 15 Ma and became pronounced at 10 Ma. Wichura et al. (2010) documented extensive topography in the Kenya Rift by 13.5 Ma, and Gani et al. (2007) on the Ethiopian Plateau between 10 and 6 Ma. Major surface uplift of the East African plateau is probably causally connected with extensive volcanism in the Kenya Rift by 14–11 Ma and the surge of rift faulting in Kenya and Ethiopia at 12–11 Ma.

These findings provide evidence that topography in East Africa coincided with climate changes in the region and may have influenced a shift towards colder and more arid conditions by 15–10 Ma. However, further evidence for the timing of topography is needed to better evaluate tectonic drivers of climate change in Africa in the Miocene. Furthermore, a deeper understanding of the causes of climate shifts in Africa also needs to consider more global drivers, such as the Asian monsoon.

The combination of isotopic shifts, expansion of grassland at the expense of forests, and suggestions of increased aridification concurs with a shift in South Asia near 10 Ma towards today's monsoonal climate (Kroon et al. 1991; Molnar et al. 2010), largely concurrently with the increased abundance of C_4 plants (tropical grasses, sedges) in East Africa (Cerling 2014). If monsoon winds did strengthen, sea-surface temperatures would have dropped if the thermocline had been deeper at ~10 Ma than it is today, as Philander & Fedorov (2003) suggested. Colder sea-surface temperature in the Indian Ocean off the East Africa coast would have deflected moisture transport away from East Africa towards Southeast Asia, which collectively may have cooled East Africa and caused a shift towards arid conditions. A final element in the global climate mix might be the glaciers that started to grow in the northern hemisphere by 10–6 Ma.

26.4.3 More Recent Drivers of East African Climate Change

The next profound change in climate took place in the mid/late Pliocene. Geo- and biochemical data sets, as well as climate models, have shown that East African aridification is primarily controlled by Indian Ocean sea-surface temperature (Cane & Molnar 2001) and that precessional variations in C_3 (mostly trees) and C_4 plants have been controlled by changes in monsoonal precipitation driven by changes in low-latitude insolation (Philander & Fedorov 2003). These findings imply that East African climate change since ~2.8 Ma has largely been governed by ocean–atmosphere interactions in the low latitudes. Consequently, Cane & Molnar (2001) argued that changes in surface-ocean circulation, controlled by the final closing of the Indonesian seaway at ~4–3 Ma, were responsible for the aridification of East Africa. According to their model, the northward motion of Australia and the associated northward displacement of New Guinea switched the source of flow through Indonesia from warm South Pacific to relatively cold North Pacific waters. This, in turn, decreased sea-surface temperatures in the Indian Ocean, leading to reduced rainfall over East Africa (Cane & Molnar 2001). The Cane and Molnar hypothesis can be seen as a novel alternative to the commonly held opinion that the onset of significant northern hemisphere glaciation was the main driver for aridification in East Africa (deMenocal 2004).

The second-order climate changes in the Pleistocene, namely the wet phases at 1.9–1.7 Ma and 1.1–0.9 Ma, coincide with the intensification of the Walker circulation (1.9–1.7 Ma) and the mid-Pleistocene revolution (1.0–0.7 Ma) (Trauth et al. 2007). High-latitude forcing is required to compress the Intertropical Convergence Zone so that East Africa becomes locally sensitive to precessional forcing, resulting in rapid shifts from wet to dry conditions (Maslin et al. 2015).

Taken together, the evidence suggests that there was no unique cause-and-effect scenario for the evolution of hominins. Instead, there may have been a combination of many geological events – such as the changing East African topography, the closing of the Indonesian throughflow, the Asian monsoon and northern hemisphere glaciation – and local rift valley climate regimes.

26.5 Biodiversity in the East African Rift Lakes

The record of lake and floodplain environments in the rift basins provides important information on the former aquatic and terrestrial habitats in the EARS. Fossils further add to our knowledge of past environments in the region. In addition, the sedimentary record of the rift lakes reflects the interplay between tectonics and climate. Tectonics also controls sedimentary patterns and explains why some lake sediments are exposed and others are not. The latter point is especially important when it comes to interpreting fossil biodiversity, which is usually hampered by taphonomic biases (Behrensmeyer et al. 2007). In this section, we look at the response of fauna (including hominins) and flora in the EARS throughout the Neogene.

26.5.1 Lacustrine Fauna

The five largest EARS lakes (total surface area $\sim 146\,000\,km^2$) contain a combined ~1800 species, of which ~95% are endemic to a single lake (Salzburger et al. 2014). In contrast, the five North American Great Lakes (combined surface area $\sim 244\,000\,km^2$) have 176 species, of which only 3% are endemic to a single lake. This disparity is largely a result of the tectonic history of the region. Furthermore, in the EARS, there is a striking correlation between biodiversity and lake size (Figure 26.5), again highlighting stark differences between Eastern and Western branch lakes. Ancient lakes, like Lake Tanganyika, are hot spots of diversity and endemicity.

In the EARS, very few species occur in more than one of the lakes. For instance, not a single cichlid species is naturally shared between Lakes Malaŵi and Tanganyika. The rift lakes, with their high levels of endemicity and their restricted geographical occurrence, are quasi eco-insular systems (Salzburger et al. 2014). They are to aquatic organisms what oceanic islands are to terrestrial biota. Ocean islands (like the EARS as a whole) are mantle plume-related, and terrestrial biodiversity shows an age progression in relation to plate motion over the plume, as is also found in many studies from the Galapagos, Hawaii and other plume-related ocean islands (Merten 2014; Harpp et al. 2014).

The lakes are quasi-replicate systems, in which closely related faunas evolved through iterative evolution. In Lakes Tanganyika and Malaŵi, a whole set of similar cichlid ecotypes have evolved both independently and in parallel (Kocher et al. 1993). Biodiversity in the EARS lakes is primarily controlled by the age of the lake, topography, subsidence history, climate and river in- and outflow (Van Bocxlaer et al. 2008). The taxonomic diversity, endemism and morphological disparity are primarily the product of evolutionary radiations (reviewed in Salzburger et al. 2014). A major factor behind the extraordinary levels of genetic diversity and morphological disparity is the long-term stability of the large EARS lakes (Figure 26.5).

Salzburger et al. (2014) showed the transience of freshwater connections and drainage patterns as a function of

Figure 26.5 Lake size and ecosystem stability versus species diversity in the EARS lakes, showing a positive correlation. Number of species was calculated by summing the known diversity of fish, ostracodes and mollusks as a general approximation for all diversity. Size and stability were calculated as an index of area (in km^2) and maximum depth (m). The metric indicates that the large Western Branch lakes, and also Lake Victoria, are stratified and are biodiversity hot spots. The high biodiversity in Lake Victoria is possibly due to invasion history, intrinsic biological factors and the saucer-like (nonrift) basin morphology. Holomictic, completely mixed; meromictic, permanently stratified. The dotted line with a question mark indicates the unknown size of Lake Obweruka. *Source:* Adapted from Salzburger et al. (2014).

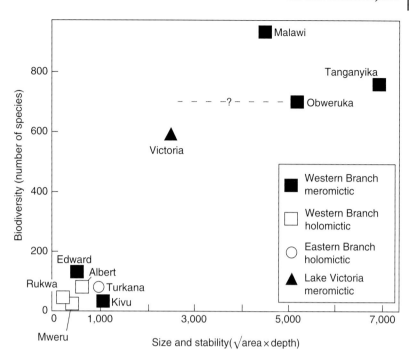

rift tectonics (Figure 26.6). Biodiversity in more stable lakes (e.g., Tanganyika and Malaŵi) contrasts with that in lakes of more limited ecological stability (Eastern Branch lakes or Lake Kivu) (Hauffe et al. 2014). The faunas of stable lakes are often characterized by species flocks that are endemic to the lake, whereas less stable lakes contain relatively more taxa that are inherited from large-scale biogeographical connections (Schultheiß et al. 2009).

A combination of tectonic and topographic changes with climatically driven cycles of aridity and humidity has periodically changed the dispersal rates of both terrestrial and aquatic organisms (Danley et al. 2012). For example, during the Quaternary, Lake Rukwa had multiple highstands and spilled over into Lake Tanganyika, causing exchange of faunal taxa that were previously endemic to a single lake (Cohen et al. 2013).

In the Western Branch, an extensive rift lake, Lake Obweruka, formed by 8 Ma (Pickford et al. 1993) and occupied an area from the northern end of present Lake Albert to the southern end of Lake Edward (Figure 26.6). Lake Obweruka persisted for ~5.5 My and was the site of remarkable evolutionary activity, especially among mollusks and their chief predators, fishes.

The regional surroundings of Lake Obweruka consisted of humid biotopes, including tropical forest and typical forest–savannah mosaics (Pickford et al. 1993). Strong enrichment of iron in thick lateritic soils, vertical and lateral transport of dissolved iron by groundwater and precipitation of iron at the groundwater–lakewater chemocline indicates a wet tropical climate when Lake

Obweruka formed (Roller et al. 2010), suggesting that the lake was meromictic (i.e., had layers of water that did not intermix) from 8 Ma (Van Damme & Pickford 1995). Overall, there was a change from a semi-arid climate before ~8 Ma to a wet climate with times of high seasonality thereafter. This highlights the importance of local rift climate, as, in general, there was a shift towards increased aridity in East Africa after 8 Ma. At ~2.5 Ma, the uplift of the Rwenzori Mountains commenced, and progressively split Lake Obweruka into Lakes Albert and Edward (Pickford et al. 1993). Some mollusks and fish initially survived the reorganization of the lakes, but eventually their diversity collapsed (Van Damme & Pickford 1995). Brachert et al. (2010) demonstrated through $\delta^{18}O$ signatures of hippopotamus teeth that the commencing uplift of the Rwenzori Mountains at ~2.5 Ma did not result in a growing rain shadow. The missing rain shadow is interpreted to be due to the overriding effect of evaporation on $\delta^{18}O$ responding to aridification of the basin floor by a valley air-circulation system through relative deepening of the valley. This, again, shows that mesoscale climate in rift valleys is an important factor.

The development of the Lake Victoria cichlid superflock after the Last Glacial Maximum (LGM) at ~12.5 ka shows hydrological connections between Lake Victoria and Lakes Albert, Edward and Kivu in the Western Branch (Elmer et al. 2009). Presently, there is much more limited hydrological connectivity between the Western Branch lakes and Lake Victoria via the Nile River (Schultheiß et al. 2014).

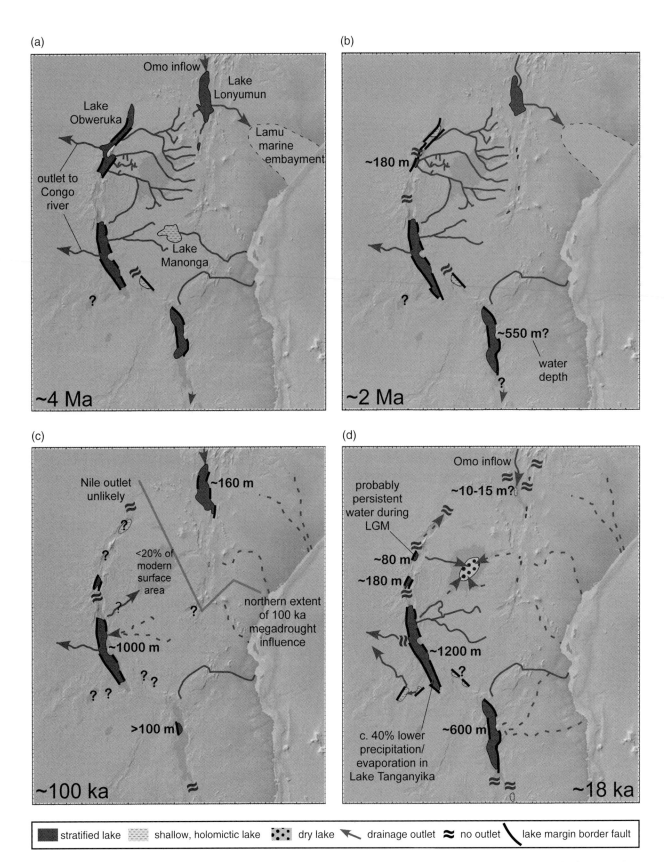

Figure 26.6 Paleogeographic maps of the EARS lakes, showing hydrographic configurations of the major lakes and rivers for four periods over the past 4 My: (a) ~4 Ma, (b) ~2 Ma, (c) ~100 ka and (d) ~18 ka. Solid lines indicate major perennial rivers; dashed lines show intermittent rivers. The main drainage until ~18 ka (Last Glacial Maximum, LGM) was to the west, into the Western Branch depocenters and ultimately the Atlantic Ocean. The Lamu marine embayment was controlled by a Cretaceous rift, which also controls the Turkana depression. Note the dramatic changes in hydrography at 18 ka. *Source:* Adapted from Salzburger et al. (2014). See also Plate 44 in color plate section.

26.5.2 Terrestrial Fauna and Flora Successions in Relation to Lake Development

We present the terrestrial fauna (mainly mammals) and flora together because they are intimately related, as a number of animals feed on vegetation. A fundamental aspect of the floral evolution in the wider EARS, especially in conjunction with hominin evolution, is the relationship between C_3 and C_4 plants (Cerling et al. 2013).

26.5.2.1 Mid Miocene

During this period, a gradual replacement of subtropical forested environments by more seasonal, open country savannah-mosaic habitats took place (Bonnefille 2010). One of the best-exposed mid-Miocene strata in the EARS is at Tugen Hills in the Baringo Basin of the Kenya Rift (Figure 26.1). The sediments there yield primate fossils at ~15.5 Ma and ~12.5 Ma that retain the primitive characters typical of their predecessors (Hill et al. 1991, 2002). In the younger part of the sedimentary sequence, crocodiles, Hyaenidae, Felidae, pigs, rodents, bovids, giraffe, Proboscidae (elephants) and Rhinocerotidae also occur. $\delta^{13}C$ values of tooth enamel of Rhinocerotidae and Proboscidae suggest a minor dietary component of C_4 grasses at ~15 Ma (Morgan et al. 1994). Environments in the Baringo Basin were mostly tropical moist or wet forest in the tropical lower montane or premontane forest category during the 15.5–12.5 Ma period (Jacobs & Kabuye 1987). Rodent assemblages from the early to middle Miocene indicate faunal stability during this period in the Baringo Basin (Winkler 2002).

26.5.2.2 Late Miocene

The next critical interval in biodiversity evolution is the 8–5 Ma period, when a dramatic change in African biota took place. This period is characterized by a major faunal transition that postdated widespread volcanism in Kenya at 14–11 Ma, major rift faulting in Kenya and Ethiopia at ~12–11 Ma and the formation of the Western Branch at ~10 Ma. The latter had a profound impact on East African lacustrine evolution, as at least two major long-lived lakes (Tanganyika and Obweruka) formed in the late Miocene.

Shrinkage of the equatorial forests coincided with an expansion of C_4 plants (Cerling et al. 1997). In Africa, this resulted in the emergence of the mammal elements that would dominate the late Cenozoic, including hippos, giant pigs, grazing antelopes, true giraffes and elephants and hominins. Faunal assemblages during the 8–5 Ma period are transitional, reflecting the replacement of mid-Miocene communities containing creodonts (carnivorous mammals), climacoceras (large mammals), caprins (goat-like animals) and boselaphines (antelopes) with more modern tribes (Kingston et al. 2002). By ~5 Ma, faunas were distinctly different to mid-Miocene ones in East Africa, as well as in Asia (Hill 1999). These profound changes are linked to two main factors: (i) migratory exchange with Eurasia and (ii) gradual replacement of subtropical environments by more seasonal open country savannah-mosaic habitats, considered characteristic of the latest Miocene (Leakey et al. 1996; Cerling 2014).

In Tugen Hills, sediments dated between ~7.5 and 6.2 Ma show the earliest change towards the modern fauna. New families such as elephantids, modern rhinoceros, hippopotamids, giraffids, bovids and the first leporids in sub-Saharan Africa appear (Kingston et al. 2002). Associated with the faunas are fossil wood fragments representing six taxa (Kingston et al. 2002). Living representatives of the fossil taxa show fairly limited environmental tolerance and suggest that the fossil forest grew in a lowland or upland forest with floral affinities to West and Central Africa. Maximum tree heights were >50 m, a height typical for wet or moist forest communities (Menaut et al. 1995), as drier African forests and montane forests have tree heights <35–40 m (Richards 1996). At Lothagam, paleosols document a period of increased aridity between ~6.7 and 5.0 Ma, and depositional stasis at ~6.5–5.2 Ma is indicated by two very well developed luvisols (Wynn 2004). Vegetation throughout the interval appears to have been a mosaic of floodplain savannahs dissected by gallery woodland.

Nonetheless, fossil remains are rare in the 8–5 Ma interval of the EARS, and the tempo and mode of change during this important transitional period are therefore poorly known (Kingston et al. 2002). However, molecular data can help to fill this gap, and phylogenies of snakes (Menegon et al. 2014) and fishes (Dorn et al. 2014) suggest that tectonic processes may have triggered speciation events in these groups at this time.

26.5.2.3 Pliocene

The Pliocene warm (wet) period had a pronounced impact on lacustrine conditions, and more large lakes formed by 4.5–4.0 Ma (Malaŵi and Lonyumun). However, there is no known tectonic driver for climate change during this period. Nonetheless, the formation of Lakes Lonyumun and Malaŵi and the important lake stage of Obweruka at 4.5–4.0 Ma coincided with the first step in the development of strong Walker circulation. If this step was indeed related to processes restricting and finally closing the Indonesian and Panamanian seaways (see earlier and Chapter 22), then global tectonic processes can be linked to major shifts in hydrography in East Africa in the early Pliocene.

The faunal and climatic proxies for the Pliocene warm period show wet conditions coinciding with extensive highstands of rift lakes and the existence of paleolakes Obweruka in the Albertine Basin, and in Lonyumun in the Turkana Basin (Figure 26.6). During this time, the fauna of the northern Malaŵi Rift of the Western Branch was dominated by typical modern African equids, suids, hippopotamids, giraffids, bovids, elephantids and rhinoceratids (Sandrock et al. 2007). The high percentage of terrestrial species (~90%) suggests a tropical semi-arid bushland or tropical grassland environment. Towards the end of the Pliocene, the faunas indicate a strong trend towards arid grassland that persisted into the Pleistocene (Sandrock et al. 2007).

Lüdecke et al. (2016) presented a long-term Plio–Pleistocene $\delta^{13}C$ record from pedogenic carbonate and suidae (pig) teeth of the northern Malaŵi Rift. The sediments contain fossils of *Homo rudolfensis* (Schrenk et al. 1993) and *Paranthropus boisei* (Kullmer et al. 1999). Consistent $\delta^{13}C$ values of –9‰ indicate a C_3-dominated closed environment with regional patches of C_4 grasslands. The overall fraction of woody cover of 60–70% reflects more forest canopy in the Malaŵi Rift than in the Eastern Branch. The appearance of C_4 grasses is considered to be a possible driver of evolutionary faunal shifts (Cerling et al. 2013). However, despite the different climatic and ecosystem evolutions in the north (e.g., Kenya) and south (Malaŵi), similar hominins and suids occurred in both landscapes (Bromage et al. 1995), pointing to habitat flexibility and/or microhabitat tracking, which may also indicate their nutritional versatility (Lüdecke et al. 2016).

26.5.2.4 Pleistocene

East Africa became cooler and more arid after ~2.8 Ma. Vrba (1988, 1999) proposed a major turnover pulse in EARS mammals at ~2.8 Ma, linked to the origin of hominin genera *Paranthropus* and *Homo* and the extinction of *Australopithecus*. Despite the fact that some causes of faunal turnover during the late Pliocene seem reasonably well established (Reed 2008), a number of technical difficulties in recognizing faunal patterns have been proposed, related for example to the resolution of time scales (Faith & Behrensmeyer 2013), stratigraphic gaps and sampling errors (Frost 2007). Nevertheless, in recent studies, faunal turnover events were established to have occurred around 2.8 Ma at Afar (DiMaggio et al. 2015) and around 2.0–1.75 Ma on a larger African scale (Bibi & Kissling 2015).

Vrba (1999) also proposed a turnover pulse among bovids at 2.8–2.5 Ma. Suids do not seem to show an increased amount of turnover at this time (White 1995).

This discrepancy may be the result of the difference in trophic ecology between these groups: the bovids may have been relatively specific in their habitat requirements, dividing the ecosystem into several niches, while the suids may have been broader in theirs. Suids are eurybiomic (tolerating variable habitats), and therefore would not have been affected by a possible 2.8–2.5 Ma turnover event (Frost 2007). In mammals, herbivory (i.e., plant eating) generally leads to fast speciation rates during periods of climatic fluctuations (Hernandez Fernandez & Vrba 2005), whereas omnivory (eating both plants and animals) results in reduced diversification rates. The origins of *Paranthropus* and *Homo* at around 2.5 Ma are directly linked to changes in diet availability (Bromage & Schrenk 1999).

All possible events of increased turnover at 3.5, 3.0, 2.8–2.5 and 2.0–1.75 Ma occur at periods when there are also possible stratigraphic gaps, even in the larger data sets. Forest fragmentation may have been important in speciation and diversification. Thus, while climate change may have been a significant factor in faunal evolution, it seems that global cooling between 2.8 and 2.5 Ma alone did not necessarily cause a faunal and hominin turnover pulse.

26.6 The Advent of Hominins

26.6.1 Hominins and their Ecological Context

Hominin fossils are exceedingly rare in the geological record. Therefore, detailed aspects of their evolution, as reflected in patterns of distribution and diversification, are not well resolved. Moreover, the climatic and environmental factors that drove the chimpanzee/hominin split and the consequent evolution of hominins remain poorly known. A challenging problem in hominin research is therefore to relate tectonics and both global and regional climate changes to shifts in paleoenvironments and species occurrences, dispersal and evolution (Behrensmeyer 2006; Potts 2007; Levin 2015).

All but a very few hominin fossils in the EARS are from the Eastern Branch (Bonnefille 2010), with the northern part of the Malaŵi Rift (Schrenk et al. 1993) and the Albertine Rift (Crevecoeur et al. 2014) being the only localities so far where hominin remains have been discovered in the Western Branch (Figure 26.7). Widespread lake formation led to more diverse environments, which potentially influenced the evolution of hominins by altering their habitat. However, it is questionable whether the EARS basins were indeed the only sites where our ancestors evolved. The discovery of abundant fossils in the EARS may be taphonomically biased, as it is basically

Figure 26.7 Locations of African early hominin sites in relation to the distribution of EARS lakes. Hominin sites outside of the EARS are known mainly from the Lake Chad Basin (Mio–Pliocene sites at Toros Menalla, Koro Toro, Yayo), from cave deposits in South Africa (Plio–Pleistocene) and from north-western African sites (Pleistocene). Light gray dots on the continent indicate topography; dark gray lines show major rivers and lakes.

in the rift-basin "badlands" where sediments of the right age occur, with reasonable potential for preservation and discovery of hominin fossils (Ring 2014).

Hill (1995) suggested that the human ancestor must have been an "anthropomorphic ape" that lived during the Miocene, probably before 7.0–6.5 Ma. When and where did the major evolutionary transition to hominids take place? Most likely not only in the EARS, but also in early wooded savannah ecosystems, developing all around the diminishing African rainforest towards the end of the Miocene (Schrenk et al. 2004; Bonnefille 2010). As Hill (1995) noted, only 0.1% of Africa is represented by fossil localities, providing perspective on the scale of knowledge versus the huge area of diverse habitats where early hominins evolved. However sparse this record might seem, most of the available evidence indicates that environmental changes in Africa, driven by tectonics and climate, have had a profound impact on early human evolution, and probably paced these transitions in our early history.

Vertebrate faunas provide important evidence for the ecological context of hominin evolution over a wide range of scales, from site-specific analysis of taxa directly associated with hominin fossils to faunal trends indicating long-term environmental change that could have affected human evolution (Behrensmeyer et al. 2007; Macho 2013; Potts 2013). Site-specific faunal information from fossil records can address paleoecological questions at local (Cooke 2007), regional (Bobe et al. 2007; Sandrock et al. 2007) and continental (Frost 2007) scales. However, a major issue is the discontinuity between local stratigraphic records and their relation with global-scale processes (Bibi & Kiessling 2015).

An additional problem in this context is that most available long-term climate records are based on marine-core data from basins proximal to Africa (Tiedemann et al. 1994; deMenocal 2004; Feakins & deMenocal 2010), while most of the late Miocene and Pliocene hominins and other fauna occur in continental deposits of the EARS and further west in the Chad Basin. In response to this issue, researchers increasingly attempt to obtain climate proxy records directly from outcropping hominin-bearing deposits (Kingston et al. 2002; Joordens et al. 2013; Cerling 2014). To bypass the problematic characteristics of outcrops (such as diagenesis and stratigraphic discontinuity), cores have been drilled into lakes and ancient lake sediments to allow the capture of relatively pristine and continuous climate records that can be correlated to fossil records (Cohen et al. 2016).

26.6.2 Chronology of Hominin Findings

The first African hominin remains were found in 1921 in Zimbabwe (Woodward 1921) and 1924 in South Africa (Dart 1925) – outside the EARS. Thousands of hominin fragments, comprising roughly 10 genera and 20 species, have been discovered at more than 50 African sites (Figure 26.7). The majority of specimens originate from South African cave fillings. Whereas caves have the advantage of a higher preservation potential, EARS sites are generally more accurately dated. Two of the major breakthroughs in human evolution in Africa were the origin of bipedal locomotion between 8 and 6 Ma and the onset of cultural evolution at ~3.3 Ma.

The oldest hominin species found so far, *Sahelanthropus tchadensis*, is from the Chad Basin, about 2500 km west of the EARS (~7 Ma) (Figure 26.7) (Brunet et al. 2002). This age broadly fits with the chimpanzee–hominin divergence, estimated at 8–6 Ma (Patterson et al. 2006; Steiper & Young 2006; Langergraber et al. 2012). Other

late-Miocene hominins are *Orrorin tugenensis*, from the Kenya Rift (6.0–5.7 Ma) (Pickford & Senut 2001), and *Ardipithecus kadabba*, from Afar (5.8–5.2 Ma) (Haile-Selassie 2001). The oldest Pliocene (4.4 Ma) hominin, *Ardipithecus ramidus*, is also from Afar (White et al. 2009).

Fossil remains belonging to the genus *Australopithecus*, whose attribution to the hominin lineage is unambiguous, are known from ~4.2 Ma onwards. They derive mainly from the Eastern Branch and South Africa (Wood & Lonergan 2008), with the exception of *Australopithecus bahrelghazali* (Brunet et al. 1995; Strait 2013) from northern Chad. The youngest Pliocene hominin fossil at 2.8–2.75 Ma is from the northern Ethiopian Rift, and so far represents the earliest appearance of the genus *Homo* (Villmoare et al. 2015). The earliest record of stone tools just postdates the end of the Pliocene warm period: Harmand et al. (2015) reported stone tools from Lake Turkana dated at 3.31–3.21 Ma, discovered in the same sediments where fossils from the hominin species *Kenyanthropus platyops* were found (Leakey et al. 2001). This is broadly coeval with animal bones from Ethiopia that bear stone-inflicted cut marks, dated at ~3.39 Ma (McPherron et al. 2010).

Many of the early-Pleistocene EARS hominins attributed to the genus *Homo* (Leakey et al. 2012; Spoor et al. 2015) co-occur with hominins of the genus *Australopithecus* (e.g., *Australopithecus sediba* in South Africa), as well as *Paranthropus* (Berger et al. 2010). Most of the younger (<0.3 Ma) hominin species are found outside the EARS and outside Africa (e.g., in Georgia: Lordkipanidze et al. 2013). This is interpreted to reflect that hominins dispersed "out of Africa" between 2.0 and 0.1 Ma (Mithen & Reed 2002).

26.6.3 Hominins, Climate and the Changing Landscape

Paleoanthropologists and geologists often argue that climate changes causing a transformation from tropical forest to savannah-type habitats (change from C_4 to C_3 environments) were the main drivers of early hominin evolution. A review of Cenozoic vegetation and climate by Bonnefille (2010) concluded that an expansion of savannah/grassland at ~10 Ma in East Africa took place after the 15.5–12.5 Ma cooling event. Another pronounced change in vegetation took place at 6.3–6.0 Ma and was marked by a decrease in tree cover across all of tropical Africa.

The earliest hominins found in the Kenyan and Ethiopian Rift, and also in Chad, inhabited mixed C_3/C_4 environments, probably grassland with patches of woodland (Cerling 1992). There is no record of closed-canopy forest at localities where early hominins were found.

Current interpretations of the paleoenvironment suggest that *Orrorin* in Kenya lived in open woodland habitats with dense woodland or forest in the vicinity, possibly along lake margins. *Sahelanthropus* likely dwelled in a mosaic of environments, ranging from gallery forest at the edge of a lake to savannah woodland to open grassland, although there are indications that there was a dominance of shrub/bushland and grassy woodland habitats within the Chadian lake basin. *Ardipithecus kadabba* in Ethiopia lived in riparian woodland and floodplain grassland along water margins (Su 2014). WoldeGabriel et al. (1994) suggested that after 4.4 Ma, hominins started to inhabit environments with more open vegetation. By contrast, a mosaic of open and wooded habitat is reconstructed by Leakey et al. (1996) for the Miocene/Pliocene boundary in the Turkana Basin just south of the Ethiopian border. The early Pliocene *Ardipithecus ramidus* is found in closed woodland habitats with possible patches of forest (White et al. 2009; Jolly-Saad & Bonnefille 2014 – but see Cerling et al. 2013 for a different view) and associated with bushland and grassland habitats in the northern Ethiopian Rift (Levin et al. 2004). Combined, these habitat interpretations of the African late Miocene and early Pliocene suggest that the beginnings of our lineage did not occur in open, semi-arid to arid habitat conditions, but rather in more closed and/or wet habitats. However, a definitive conclusion is difficult to draw at this time, given the lack of detailed paleoecological reconstruction for the paleohabitat where *Orrorin* is found, the possibility that *Sahelanthropus* was found in more open habitats, the discordance in interpretation of the Ethiopian data set and the general paucity of late-Miocene/Pliocene hominin-bearing sites in Africa.

During the late Pliocene, at ~3.5 Ma, hominin diet, especially of *Kenyanthropus*, began to include higher amounts of C_4 food resources (Cerling et al. 2013). By ~2.8 Ma, dryer conditions prevailed in the African landscape (Vrba 1999; deMenocal 2004). The behavioral flexibility of early hominins was likely triggered by abiotic changes (Macho 2013). Extensive open habitats with more aridity-tolerant vegetation developed. The resulting selective pressures apparently led to increased survival of megadont varieties capable of feeding on tougher fruit and open woodland/open savannah food items. This is true for early hominins as well as numerous large eastern and southern terrestrial African vertebrate lineages at ~2.5 Ma (Turner & Wood 1993), and resulted in the phyletic splitting of *Australopithecus afarensis* into *Paranthropus* and *Homo* lineages after ~2.8 Ma (Bromage & Schrenk 1999; Strait et al. 2015). It appears likely that our ancestors had a eurybiomic lifestyle.

Work by Lüdecke et al. (2016) highlights major differences in ecosystems between the Eastern Branch and the Malaŵi Rift of the Western Branch in the Plio–Pleistocene. Brachert et al. (2010) demonstrated that the large water bodies entrenched between strongly uplifted rift shoulders of the Western Branch had an impact on local rift climate by creating a valley air-circulation system through relative deepening of the lake basin. Bobe et al. (2007) showed that aridity trends in the Turkana and Hadar areas demonstrated different responses at different times, evidence that local geography and tectonics played an important role in mediating environmental change. This strongly suggests that mesoscale climate in rift sectors may be more important, and not necessarily directly linked to global climate. These factors may also explain local climate proxy variations and different habitat reconstructions for the Miocene/Pliocene boundary in the Ethiopian/Kenya border region.

The evolution of the genus *Homo* in the early Pleistocene can only be understood by integrating fossil, archaeological and environmental data on seasonal, decadal and evolutionary time scales. In East Africa, climatic and environmental fluctuations were intensified by tectonic activity, increased landscape fragmentation and amplifier lakes (high-precipitation, high-evaporation lakes formed between graben that respond rapidly to moderate climate shifts) (Trauth et al. 2010). Several scenarios and hypotheses have been put forward to explain and test processes of selection and speciation in Plio–Pleistocene hominin evolution in the context of climate change. Among other things, these include the turnover pulse hypothesis (Vrba 1999) and the pulsed climate variability hypothesis (Maslin et al. 2015). The variability selection hypothesis (Potts 2007, 2013; Potts & Faith 2015), the most comprehensive approach to early hominin evolutionary patterns to date, refers to adaptive change in response to environmental variation. Whereas short-term variability takes into account habitat variations, long-term variability selection assumes that certain adaptations have evolved as a result of large environmentally caused inconsistencies in the selective conditions.

This review highlights that, whereas the influence of climate changes can be documented in many instances, the complex interactions of various drivers of hominin and faunal evolutions in space and time are not yet necessarily statistically solid. Behrensmeyer et al. (2007) summarized the taphonomic biases of habitat reconstruction, especially in the deposition context. Mid/late-Miocene to Pleistocene sediments are only exposed in a few rift sectors, and hardly any deposits of this age occur outside the EARS. This probably leads to a strong bias in reconstructing general hominin habitats. The small sample size, spatially concentrated fossil localities (i.e., lots of fossil material from the Kenyan and Ethiopian rift sectors and hardly any from the Western Branch) and preservation biases make correlations between climate and hominin evolution speculative.

26.7 Conclusion

The ~10 Ma time window is of great importance for overall East African climate, biology and tectonics. At this time, the Western Branch started to form, and seafloor spreading in the Gulf of Aden caused a surge of extensional deformation in the Ethiopian and Kenya Rift. Rift structure and architecture are probably less important for regional East African climate change. The latter appears controlled by plume-related density changes and resulting surface uplift, as well as the strengthening of the Indian monsoon, which made East Africa drier.

Regional surface uplift also largely controlled the development of the EARS lakes and the pronounced differences in lake size, depth and biodiversity between the Eastern and Western Branch lakes (Figures 26.2 and 26.6). Rift structure is important for local climate effects. Studies by Bobe et al. (2007) and Brachert et al. (2010) show that rift mesoclimate may represent an underestimated aspect in previous paleoclimate reconstructions from rift-valley data.

Can climate change, driving and controlling hominin evolution, be linked to stages in the evolution of the EARS? As outlined in this chapter, numerical models of uplift above the African Superswell can broadly be linked with pronounced changes in EARS tectonics and biodiversity evolution. Significant tectonic reorganization at ~0.4 Ma coincides with the formation of Lake Victoria. Because the Lake Victoria basin is tectonically connected to rift-shoulder uplift of the two EARS branches, its formation suggests enhanced uplift by 0.4 Ma. Rift-shoulder uplift at 0.4 Ma was also important for uplift and exposure of rift sediments, creating the potential for the discovery of fossils. The occurrence of fossils over the continent and within the rift is influenced by preservational biases. In the rift, hominin fossils are largely found in the more arid Eastern Branch. In contrast, lake biodiversity is much more pronounced in the Western Branch, with its great subsidence and wetter climate. Lake Turkana is an exception, as it is tectonically controlled by the position of a Cretaceous rift, causing a depression. The tectonic differences between the Western and Eastern branches reflect different plate strengths, largely controlled by mantle-plume activity.

Acknowledgments

Thanks to Carina Hoorn for inviting us to put this review together and for continued editorial help. We also appreciate helpful comments and reviews by Josephine Joordens, Henry Wichura, Thomas Lehmann and Bert Van Bocxlaer, and editorial assistance by Allison Perrigo.

References

Bagley, B. & Nyblade, A.A. (2013) Seismic anisotropy in eastern Africa, mantle flow, and the African superplume. *Geophysical Research Letters* **40**, 1500–1505.

Bauer, F.U., Glasmacher, U.A., Ring, U. et al. (2010) Thermal and exhumation history of the central Rwenzori Mts, western branch of the East African Rift System, Uganda. *International Journal of Earth Science* **99**, 1575–1597.

Bauer, F.U., Glasmacher, U.A., Ring, U. et al. (2013) Tracing the exhumation history of the Rwenzori Mountains, Albertine Rift, Uganda, using low-temperature thermochronology. *Tectonophysics* **599**, 8–28.

Bauer, F.U., Glasmacher, U.A., Ring, U. et al. (2016) Long-term cooling history of the Albertine Rift: new evidence from the western rift shoulder, D.R. Congo. *International Journal of Earth Sciences* **105**, 1707–1728.

Behrensmeyer, A.K. (2006) Climate change and human evolution. *Science* **311**, 476–478.

Behrensmeyer, A.K., Bobe, R. & Alemseged, Z. (2007) Approaches to the analysis of faunal change during the East African Pliocene. In: Bobe, R., Alemseged, Z. & Behrensmeyer, A.K. (eds.) *Hominin Environments in the East African Pliocene: An Assessment of the Faunal Evidence.* Dordrecht: Springer, pp. 1–24.

Berger L., de Ruiter, D.J., Churchill, S.E. et al. (2010) *Australopithecus sediba*: a new species of *Homo*-like Australopith from South Africa. *Science* **328**, 195–204.

Bibi, F. & Kiessling, W. (2015) Continuous evolutionary change in Plio-Pleistocene mammals of eastern Africa. *Proceedings of the National Academy of Sciences USA* **112**, 10623–10628.

Bobe, R., Behrensmeyer, A.K., Eck, G.G. & Harris, J.M. (2007) Patterns of abundance and diversity in late Cenozoic bovids from the Turkana and Hadar Basins, Kenya and Ethiopia. In: Bobe, R., Alemseged, Z. & Behrensmeyer, A.K. (eds.) *Hominin Environments in the East African Pliocene: An Assessment of the Faunal Evidence.* Dordrecht: Springer, pp. 129–157.

Bonnefille, R. (2010) Cenozoic vegetation, climate change and hominid evolution in tropical Africa. *Global and Planetary Change* **72**, 390–411.

Brachert, T., Brügmann, G., Mertz, D. et al. (2010) Stable isotope variation in tooth enamel from Neogene Hippopotamids: monitor of meso and global climate and rift dynamics on the Albertine Rift, Uganda. *International Journal of Earth Sciences* **99**, 1663–1675.

Bromage, T. & Schrenk, F. (eds.) (1999) *African Biogeography, Climate Change and Early Hominid Evolution*. New York: Oxford University Press.

Bromage, T., Schrenk, F. & Juwayeyi, Y. (1995) Palaeobiogeography of the Malawi Rift: age and vertebrate palaeontology of the Chiwondo Beds, northern Malawi. *Journal of Human Evolution* **28**, 37–57.

Brunet, M, Beauvilan, A., Coppens, Y. et al. (1995) The first australopithecine 2500 kilometers west of the Rift Valley (Chad). *Nature* **378**, 273–275.

Brunet, M., Guy, F., Pilbeam, D. et al. (2002) A new hominin from the upper Miocene of Chad, Central Africa. *Nature* **418**, 145–151.

Calais, E., DeMets, C.J. & Nocquet, C.J. (2003) Evidence for a post-3.16 Ma change in Nubia–Eurasia–North America plate motions? *Earth Planetary Science Letters* **216**, 81–92.

Cane, M.A. & Molnar, P. (2001) Closing of the Indonesian seaway as a precursor to east African aridification around 3–4 million years ago. *Nature* **411**, 157–162.

Cerling, T.E. (1992) Development of grasslands and savannas in East Africa during the Neogene. *Palaeogeography, Palaeoclimatology, Palaeoecology* **97**, 241–247.

Cerling T.E. (2014) Stable isotope evidence for hominin environments in Africa. In: Cerling, T.E. (ed.) *Treatise on Geochemistry 14: Archaeology and Anthropology*. Amsterdam: Elsevier, pp. 157–167.

Ccrling, T.E., Harris, J.M., MacFadden, B.A. et al. (1997) Global vegetation change through the Miocene-Pliocene boundary. *Nature* **389**, 153–158.

Cerling, T.E., Manthi, F.E., Mbua, E.N. et al. (2013) Woody cover and hominin environments in the past 6 million years. *Proceedings of the National Academy of Sciences USA* **110**, 10501–10506.

Cohen, A.S., Soreghan, M.J. & Scholz, C.A. (1993) Estimating the age of formation of lakes: an example from Lake Tanganyika, East African Rift system. *Geology* **21**, 511–514.

Cohen, A.S., Van Bocxlaer, B., Todd, J.A. et al. (2013) Quaternary ostracodes and molluscs from the Rukwa Basin (Tanzania) and their evolutionary and palaeobiogeographic implications. *Palaeogeography, Palaeoclimatology, Palaeoecology* **392**, 79–97.

Cohen, A.S., Campisano, C., Arrowsmith, R. et al. (2016) The Hominin Sites and Paleolakes Drilling Project: inferring the environmental context of human evolution from eastern African rift lake deposits. *Scientific Drilling* **21**, 1.

Cooke, H.B.S. (2007) Stratigraphic variation in Suidae from the Shungura Formation and some coeval deposits. In: Bobe, R., Alemseged, Z. & Behrensmeyer, A.K. (eds.) *Hominin Environments in the East African Pliocene: An Assessment of the Faunal Evidence*. Dordrecht: Springer, pp. 211–241.

Corti, G. (2009) Continental rift evolution: from rift initiation to incipient break-up in the Main Ethiopian Rift, East Africa. *Earth Science Reviews* **96**, 1–53.

Crevecoeur, I., Skinner, M.M., Bailey, S.E. et al. (2014) First early hominin from Central Africa (Ishango, Democratic Republic of Congo). *PLoS ONE* **9**, 846–852.

Danley, P.D., Husemann, M., Ding, B. et al. (2012) The impact of the geological history and paleoclimate on the diversification of East African cichlids. *International Journal of Evolution Biology* **57**, 48–51.

Dart, R.D (1925) *Australopithecus africanus*: the man-ape of South Africa. *Nature* **115**, 195–199.

Davies, T.D., Vincent, C.E. & Beresford, A.K.C. (1985) Rainfall in west-central Kenya. *Journal of Climatology* **5**, 17–33.

deMenocal, P.B. (2004) African climate change and faunal evolution during the Pliocene–Pleistocene. *Earth and Planetary Science Letters* **220**, 3–24.

DiMaggio, D.N., Campisano, C., Rowen, J. et al. (2015) Late Pliocene fossiliferous sedimentary record and the environmental context of early *Homo* from Afar, Ethiopia. *Science* **347**, 1355–1358.

Dorn, A., Musilová, Z., Platzer, M. et al. (2014) The strange case of East African annual fishes: aridification correlates with diversification for a savannah aquatic group? *Evolutionary Biology* **14**, 210–222.

Ebinger, C.J. (1989) Tectonic development of the western branch of the East African Rift System. *Geological Society of America Bulletin* **101**, 885–903.

Ebinger C.J. (2012) Evolution of the Cenozoic East African rift system: cratons, plumes, and continental breakup. In: Roberts, D.G. & Bally A.W. (eds.) *Phanerozoic Rift Systems and Sedimentary Basins*. Burlington, MA: Elsevier, pp. 133–162.

Ebinger, C. & Sleep, N.H. (1998) Cenozoic magmatism in central and east Africa resulting from impact of one large plume. *Nature* **395**, 788–791.

Ebinger, C., Djomani, Y. P., Mbede, E. et al. (1997) Rifting Archean lithosphere: the Eyasi–Manyara–Natron rifts, East Africa. *Journal of the Geological Society, London* **154**, 947–960.

Ebinger, C.J., Yemane, T., Harding, D.J. et al. (2000) Rift deflection, migration, and propagation: linkage of the Ethiopian and Eastern rifts, Africa. *Geological Society of America Bulletin* **112**, 163–176.

Ebinger, C.J. van Wijk, J. & Keir, D. (2013) The time scales of continental rifting: Implications for global processes. *Geological Society of America Special Paper* **500**, 1–13.

Eggermont, H., Van Damme, K. & Russell, J.M. (2009) Rwenzori Mountains (Mountains of the Moon): headwaters of the White Nile. In: Dumont, H.J. (ed.) *The Nile*. Berlin: Springer, pp. 243–261.

Elmer, K.R., Reggio, C., Wirth, T. et al. (2009) Pleistocene desiccation in East Africa bottlenecked but did not extirpate the adaptive radiation of Lake Victoria haplochromine cichlid fishes. *Proceedings of the National Academy of Sciences USA* **106**, 13404–13409.

Faith, J.T. & Behrensmeyer, A.K. (2013) Climate change and faunal turnover: testing the mechanics of the turnover-pulse hypothesis with South African fossil data. *Paleobiology* **39**, 609–627.

Feakins, S.J. & deMenocal, P.B. (2010) Global and African regional climate during the Cenozoic. In: Werdelin, F. & Sanders, E.J. (eds.) *Cenozoic Mammals of Africa.* Berkeley, CA: University of California Press, pp. 45–55.

Foster, D.A. & Gleadow, A.J.W. (1993) Episodic denudation in east Africa: a legacy of intracontinental tectonism. *Geophysical Research Letters* **20**, 2395–2398.

Foster, A., Ebinger, C., Mbede, E. & Rex, D. (1997) Tectonic development of the northern Taiizaiiian sector of the East African Rift System. *Journal of the Geological Society, London* **154**, 689–700.

Frost, S.R. (2007) African Pliocene and Pleistocene cercopithecid evolution and global climatic change. In: Bobe, R., Alemseged, Z. & Behrensmeyer, A.K. (eds.) *Hominin Environments in the East African Pliocene: An Assessment of the Faunal Evidence.* Dordrecht: Springer, pp. 312–345.

Gani N.D., Gani M.R. & Abdelsalam M.G. (2007) Blue Nile incision on the Ethiopian Plateau: pulsed plateau growth, Pliocene uplift, and hominin evolution. *GSA Today* **17**, 4–11.

Haile-Selassie, Y. (2001) Late Miocene hominids from the Middle Awash, Ethiopia. *Nature* **412**, 178–181.

Harmand, S., Lewis, J.E., Roche, H. et al. (2015) 3.3-million-year-old stone tools from Lomekwi 3, West Turkana, Kenya. *Nature* **521**, 310–313.

Harpp K.S., Wirth K.R., Teasdale R. et al. (2014) *Plume–Ridge Interaction in the Galápagos: Perspectives from Wolf, Darwin, and Genovesa Islands.* Geophysical Monograph Series: Galapagos: A Natural Laboratory for the Earth Sciences. Washington, DC: American Geophysical Union, pp. 187–211.

Hart, W.K., WoldeGabriel, G., Walter, R.C. & Mertzmann, S.A. (1989) Basaltic volcanism in Ethiopia: constraints on continental rifting and mantle interactions. *Journal of Geophysical Research* **94**, 7731–7748.

Hauffe, T., Schultheiß, R., Van Bocxlaer, B. et al. (2014) Environmental heterogeneity predicts species richness of freshwater mollusks in sub-Saharan Africa. *International Journal of Earth Sciences* **103**, 1–16.

Hernandez Fernandez, M. & Vrba, E.S. (2005) Macroevolutionary processes and biomic specialization: testing the resource-use hypothesis. *Evolutionary Ecology* **19**, 199–219.

Hill, A. (1995) Fauna and environmental change in the Neogene of East Africa: evidence from the Tugen Hills sequence, Baringo District, Kenya. In: Vrba, E.S., Denton, G.H., Partridge, T.C. & Burckle, L.H. (eds.) *Paleoclimate and Evolution, With Emphasis on Human Origins.* New Haven, CT: Yale University Press, pp. 178–193.

Hill, A. (1999) The Baringo Basin, Kenya: from Bill Bishop to BPRP. In Andrews, P. & Banham, P. (eds.) *Late Cenozoic Environments and Hominid Evolution: A Tribute to Bill Bishop.* London: Geological Society of London, pp. 85–97.

Hill, A., Behrensmeyer, A.K., Brown, B. et al. (1991) Kipsaramon: a lower Miocene hominoid site in the Tugen Hills, Baringo District, Kenya. *Journal of Human Evolution* **20**, 67–75.

Hill, A., Leakey, M., Kingston, J. & Ward, S. (2002) New cercopithcoids and a hominoid from 12.5 Ma in the Tugen Hill succession, Kenya. *Journal of Human Evolution* **42**, 75–93.

Jacobs, B.F. & Kabuye, C.H.S. (1987) A middle Miocene (12.2 Ma) forest in the East African Rift Valley, Kenya. *Journal of Human Evolution* **6**, 147–155.

Jansen, E., Sjoholm, J., Bleil, U. & Erichsen, J.A. (1990) Neogene and Pleistocene glaciations in the northern hemisphere and late Miocene-Pliocene global ice volume fluctuations: evidence from the Norwegian Sea. In: Bleil, U. & Thiede, J. (eds.) *Geological History of the Polar Oceans: Arctic versus Antarctic.* Amsterdam: Kluwer, pp. 677–705.

Jolly-Saad, M.C. & Bonnefille, R (2014) Lower Pliocene fossil wood from the Middle Awash Valley, Ethiopia. *Palaeontographica* **289**, 43–73.

Joordens, J.C.A., Dupont-Nivet, G., Feibel, C.S. et al. (2013) Improved age control on early *Homo* fossils from the upper Burgi Member at Koobi Fora, Kenya. *Journal of Human Evolution* **65**, 731–745.

Kingston, J.D., Jacobs, B.F., Hill, A. & Deino, A. (2002) Stratigraphy, age and environments of the late Miocene Mpesida Beds, Tugen Hills, Kenya. *Journal of Human Evolution* **42**, 95–116

Kocher, T.D., Conroy, J.A., McKaye, K.R. & Stauffer, J.R. (1993) Similar morphologies of cichlid fish in Lakes Tanganyika and Malawi are due to convergence. *Molecular Phylogenetic Evolution* **2**, 158–165.

Kroon, D., Steens, T. & Troelstra, S.R. (1991) Onset of monsoonal related upwelling in the western Arabian Sea as revealed by planktonic foraminifers. *Proceedings of the Ocean Drilling Program, Scientific Results* **117**, 257–263.

Kullmer, O., Sandrock, O., Abel, R. et al. (1999) The first Paranthropus from the Malawi Rift. *Journal of Human Evolution* **37**, 121–127.

Langergraber, K.E., Prüfer, K., Rowney, C. et al. (2012) Generation times in wild chimpanzees and gorillas suggest earlier divergence times in great ape and human evolution. *Proceedings of the National Academy of Sciences USA* **109**, 15 716–15 721.

Leakey, R.E.F. (1973) Evidence for an advanced plio-pleistocene hominid from East Rudolf, Kenya. *Nature* **242**, 447–450.

Leakey, M.G., Feibel, C.S., Bernor, R.L. et al. (1996) Lothagam: a record of faunal change in the Late Miocene of East Africa. *Journal of Vertebrate Paleontology* **16**, 556–570.

Leakey, M.G., Spoor, F., Brown, F.H. et al. (2001) New hominin genus from eastern Africa shows diverse middle Pliocene lineages. *Nature* **410**, 433–440.

Leakey, M.G., Spoor, F., Dean, M.C. et al. (2012) New fossils from Koobi Fora in northern Kenya confirm taxonomic diversity in early *Homo*. *Nature* **488**, 201–204.

Levin, N.E. (2015) Environment and climate of early human evolution. *Annual Review of Earth Planetary Science* **43**, 405–429.

Levin, N.E., Quade, J., Simpson, S.W. et al. (2004) Isotopic evidence for Plio-Pleistocene environmental change at Gona, Ethiopia. *Earth and Planetary Science Letters* **219**, 93–110.

Lordkipanidze, D., Ponce de León, M.S., Margvelashvili, A. et al. (2013) A complete skull from Dmanisi, Georgia, and the evolutionary biology of early *Homo*. *Science* **342**, 326–331.

Lüdecke, T., Schrenk, F., Thiemeyer, H. et al. (2016) Persistent C3 vegetation accompanied by Plio-Pleistocene hominin evolution in the Malawi Rift (Chiwondo Beds, Malawi). *Journal of Human Evolution* **90**, 163–175.

Macho, G.A (2013) An ecological and behavioural approach to hominin evolution during the Pliocene. *Quaternary Science Reviews* **96**, 23–31.

Martinez-Boti, M.A., Foster, G.L., Chalk, T.B. et al. (2015) Plio-Pleistocene climate sensitivity evaluated using high-resolution CO2 records. *Nature* **518**, 49–54.

Maslin, M.A., Shultz, S. & Trauth, M.H. (2015) A synthesis of the theories and concepts of early human evolution. *Philosophical Transactions of the Royal Society* **370**, 2014–2064.

McPherron, S.P., Alemseged Z., Marean C.W. et al. (2010) Evidence for stone-tool-assisted consumption of animal tissues before 3.39 million years ago at Dikika, Ethiopia. *Nature* **466**, 857–860.

Mechie J., Keller, G.R., Prodehl, C. et al. (1997) A model for the structure, composition and evolution of the Kenya rift. *Tectonophysics* **278**, 95–119.

Menaut, J.C., LePage, M., Abbadie, L. (1995) Savannas, woodlands and dry forests in Africa. In: Bullock, S.,

Mooney H.A. & Medina, E. (eds.) *Seasonally Dry Tropical Forests*. Cambridge: Cambridge University Press, pp. 64–92.

Menegon, M., Loader, S.P., Marsden, S.J. et al. (2014) The genus Atheris (Serpentes: Viperidae) in East Africa: phylogeny and the role of rifting and climate in shaping the current pattern of species diversity. *Molecular Phylogenetics and Evolution* **79**, 12–22

Merten, G. (2014) Plate tectonics, evolution, and the survival of species: a modern day hotspot. In: Harpp, K.S., Mittelstaedt, E. & d'Ozouville, N. (eds.) *The Galapagos: A Natural Laboratory for the Earth Sciences*. Geophysical Monograph Series 445. Chichester: John Wiley & Sons, Ltd., pp. 119–144.

Miller, K.G., Mountain, G.S., Wright, J.D. & Browning, J.V. (2011) A 180-million-year record of sea level and ice volume variations from continental margin and deep-sea isotopic records. *Oceanography* **24**, 40–53.

Mithen, S. & Reed, M. (2002) Stepping out: a computer simulation of hominid dispersal from Africa. *Journal of Human Evolution* **43**, 433–462.

Mohr, P. (1983) Volcanotectonic aspects of the Ethiopian Rift evolution. *Bulletin Centre Recherches Elf Aquitaine Exploration Production* **7**, 175–189.

Molnar, P. (2008) Closing of the Central American Seaway and the Ice Age: a critical review. *Paleocenaography* **23**, 220–231.

Molnar, P., Boos, W.R. & Battisti, D.S. (2010) Orographic controls on climate and paleoclimate of Asia: thermal and mechanical roles for the Tibetan Plateau. *Annual Reviews of Earth Planetary Science* **38**, 77–102.

Morgan, M.E., Kingston, L.D. & Marinoa, B.D. (1994) Carbon isotopic evidence for the emergence of C4 plants in the Neogene from Pakistan and Kenya. *Nature* **367**, 162–165.

Morley, C.K., Westcott, W.A., Stone, D.M. et al. (1992) Tectonic evolution of the northern Kenyan Rift. *Journal of the Geological Society London* **149**, 333–348.

Moucha, R. & Forte, A.M. (2011) Changes in African topography driven by mantle convection. *Nature Geoscience* **4**, 707–712.

Patterson N., Richter D.J., Gnerre S. et al. (2006) Genetic evidence for complex speciation of humans and chimpanzees. *Nature* **441**, 1103–1108.

Pfister, P.L., Stocker, T.F., Rempfer, J. & Ritz, S.P. (2014) Influence of the Central American Seaway and Drake Passage on ocean circulation and neodymium isotopes: a model study. *Paleoceanography* **29**, 1214–1237.

Philander, S.G. & Fedorov, A.V. (2003) Role of tropics in changing the response to Milankovich forcing some three million years ago. *Paleoceanography* **18**, 1045–1055.

Pickford M. & Senut, B. (2001) The geological and faunal context of Late Miocene hominid remains from Lukeino,

Kenya. *Comptes Rendus de l'Académie des Sciences – Series IIA – Earth and Planetary Science* **332**, 145–152

Pickford M., Senut, B. & Adoto, D. (1993) Geology and Palaeobiology of the Albertine Rift Valley Uganda–Zaire. CIFEG Orleans, Occasional Publication, Volume I: Geology, 1993/24.

Potts, R. (2007) Environmental hypotheses of Pliocene human evolution. In: Bobe, R., Alemseged, Z. & Behrensmeyer, A.K. (eds.) *Hominin Environments in the East African Pliocene: An Assessment of the Faunal Evidence*. Dordrecht: Springer, pp. 48–73.

Potts, R. (2013) Hominin evolution in settings of strong environmental variability. *Quaternary Science Reviews* **73**, 1–13

Potts, R. & Faith, J.T. (2015) Alternating high and low climate variability: the context of natural selection and speciation in Plio-Pleistocene hominin evolution. *Journal of Human Evolution* **87**, 5–20

Ravelo, C., Andreasen, D., Lyle, M. et al. (2004) Regional climate shifts caused by gradual global cooling in the Pliocene epoch. *Nature* **429**, 263–267.

Reed, K.E. (2008) Paleoecological patterns at the Hadar hominin site. Afar Regional State, Ethiopia. *Journal of Human Evolution* **54**, 743–768.

Richards, P.W. (1996) *The Tropical Rain Forest*. Cambridge: Cambridge University Press.

Ring, U. (1994) The influence of preexisting crustal anisotropies on the evolution of the Cenozoic Malawi rift (East African rift system). *Tectonics* **13**, 313–326.

Ring U. (2008) Extreme uplift of the Rwenzori Mountains, Uganda, East African Rift: structural framework and possible role of glaciations. *Tectonics* **27**, TC4018.

Ring, U. (2014) The East African Rift System. *Austrian Journal of Earth Sciences* **107**, 132–146.

Ring, U. & Betzler, C. (1995) Geology of the Malawi Rift: kinematic and tectonosedimentary background to the Chiwondo beds. *Journal of Human Evolution* **28**, 1–22.

Ring, U., Betzler, C. & Delvaux, D. (1992) Normal versus strike-slip faulting during rift development in East Africa: the Malawi rift. *Geology* **20**, 1015–1018.

Ring, U., Schwartz, H., Bromage, T.G. & Sanaane, C. (2005) Kinematic and sedimentologic evolution of the Manyara Rift in northern Tanzania, East Africa. *Geological Magazine* **142**, 355–368.

Roller, S., Hornung, H., Hinderer, M. & Ssemmanda, I. (2010) Middle Miocene to Pleistocene sedimentary record of rift evolution in the southern Albert Rift (Uganda). *International Journal of Earth Sciences* **99**, 560–577.

Salzburger, W., van Bocxlaer, B. & Cohen, A.S. (2014) Ecology and evolution of the African Great Lakes and their faunas. *Annual Review of Ecology, Evolution, Systematics* **45**, 519–545.

Sandrock, O., Kullmer, O., Schrenk, F. et al. (2007) Fauna, taphonomy and ecology of the Plio- Pleistocene Chiwondo Beds, Northern Malawi. In: Bobe, R., Alemseged, Z. & Behrensmeyer, A.K. (eds.) *Hominin Environments in the East African Pliocene: An Assessment of the Faunal Evidence*. Dordrecht: Springer, pp. 315–332.

Schrenk, F., Bromage, T., Betzler, C. et al. (1993) Oldest *Homo* and Pliocene biogeography of the Malawi Rift. *Nature* **365**, 833–836.

Schrenk, F., Kullmer, O. & Sandrock, O. (2004) An open source perspective of earliest hominid origins. *Collegium Anthropologicum* **28**, 113–120.

Schultheiß, R., Van Bocxlaer, B., Wilke, T. & Albrecht, C. (2009) Old fossils–young species: evolutionary history of an endemic gastropod assemblage in Lake Malawi. *Proceedings of the Royal Society* **276**, 2837–2846.

Schultheiß, R., van Bocxlaer, B., Riedel, F. et al. (2014) Disjunct distributions of freshwater snails testify to a central role of the Congo system in shaping biogeographical patterns in Africa. *Evolutionary Biology* **14**, 42–53.

Sepulchre, P., Ramstein, G., Fluteau, F. et al. (2006) Tectonic uplift and Eastern Africa aridification. *Science* **313**, 1419–1423.

Spoor, F., Gunz, P., Neubauer, S. et al. (2015) Reconstructed *Homo habilis* type OH 7 suggests deep-rooted species diversity in early *Homo*. *Nature* **519**, 83–86.

Steiper, M.E. & Young, N.M. (2006) Primate molecular divergence dates. *Molecular Phylogenetics and Evolution* **41**, 384–394.

Strait, D.S. (2013) The biogeographic implications of early hominin phylogeny. In: Reed, K.E., Fleagle, J.G. & Leakey, R.E. (eds.) *The Palaeobiology of Australopithecus. Vertebrate Paleobiology and Paleoanthropology*. Dordrecht: Springer, pp. 183–191.

Strait, D., Grine, F. & Feagle, J. (2015) Analyzing hominin hominin phylogeny: cladistic approach. In: Henke, W. & Tattersall, I. (eds.) *Handbook of Paleoanthropology*. Dordrecht: Springer, pp. 1989–2014

Strecker, M. & Bosworth, W. (1991) Quaternary stress-field change in the Gregory Rift, Kenya. *EOS* **72**, 1721–1722.

Su, D.F. (2014) Taphonomy and paleoecology of KSD-VP-1/1. In: Haile-Selassie, Y. & Su, D.F. (eds.) *The Paleobiology of Early Australopithecus afarensis: Comparative Description of KSD-VP-1/1 from Woranso-Mille, Ethiopia*. Dordrecht: Springer, pp. 109–121.

Tiedemann, R., Sarnthein, M. & Shackleton, N.J. (1994) Astronomical timescale for the Pliocene Atlantic δ18O and dust flux records of ODP Site 659. *Paleoceanography* **9**, 619–638.

Trauth, M.H., Maslin, M.A., Deino, A.L. et al. (2007) High- and low-latitude forcing of Plio-Pleistocene East African climate and human evolution. *Journal of Human Evolution* **53**, 475–486.

Trauth, M.H., Maslin, M.A., Deino, A.L. et al. (2010) Human evolution in a variable environment: the amplifier lakes of Eastern Africa. *Quaternary Science Reviews* **29**, 2981–2988.

Turner, A. & Wood, B. (1993) Comparative palaeontological context for the evolution of the early hominid masticatory system. *Journal of Human Evolution* **24**, 301–318

Van Bocxlaer, B., Van Damme, D. & Feibel, C.S. (2008) Gradual versus punctuated equilibrium evolution in the Turkana Basin molluscs: evolutionary events or biological invasions? *Evolution* **62**, 511–520.

Van Damme, D. & Pickford, M. (1995) The late Cenozoic ampullariidae (mollusca, gastropoda) of the Albertine Rift Valley (Uganda-Zaire). *Hydrobiologia* **316**, 1–3.

Villmoare, B., Kimbel, W.H., Seyoum, C. et al. (2015) Early *Homo* at 2.8 Ma from Ledi-Geraru, Afar, Ethiopia. *Science* **347** 1352–1355.

Vrba, E.S. (1988) Late Pliocene climatic events and hominid evolution. In: Grine, F.E. (ed.) *Evolutionary History of the "Robust" Australopithecines*. New York: Aldine, pp. 405–426.

Vrba, E.S. (1999) Habitat theory in relation to the evolution in African biota and hominids. In: Bromage, T. & Schrenk, F. (eds.) *African Biogeography, Climate Change and Early Hominid Evolution*. New York: Oxford University Press, pp. 19–34.

Weissel, J.K. & Karner, G.D. (1989) Flexural uplift of rift flanks due to mechanical unloading of the lithosphere under extension. *Journal of Geophysical Research* **94**, 13 919–13 950.

White, T.D. (1995) African omnivores: global climatic change and Plio-Pleistocene hominids and suids. In: Vrba, E.S. (ed.) *Paleoclimate and Evolution with an Emphasis on Human Origins*. New Haven, CT: Yale University Press, pp. 112–131.

White, R.S., Spence, G.D., Fowler, S.R. et al. (1987) Magmatism at a rifted continental margin. *Nature* **330**, 439–444.

White, T.D., Asfaw, B., Beyene, Y. et al. (2009) *Ardipithecus ramidus* and the paleobiology of early hominids. *Science* **326**, 75–86.

Wichura, H., Bousquet R., Oberhänsli R. et al. (2010) Evidence for middle Miocene uplift of the East African Plateau. *Geology* **38**, 543–546.

Winkler, A. (2002) Neogene paleobiogeography and East African paleoenvironments: contributions from the Tugen Hills rodents and lagomorphs. *Journal of Human Evolution* **42**, 237–256.

WoldeGabriel, G., Yemane, T., White, T. et al. (1994) Age of volcanism and fossils in the Burji-Soyoma area, Amaro Horst, southern Main Ethiopian Rift. *Journal of African Earth Sciences* **13**, 437–447.

Wood, B.A. & Lonergan, N. (2008) The hominin fossil record: taxa, grades and clades. *Journal of Anatomy* **212**, 354–376.

Woodward, A.S. (1921) A new cave man from Rhodesia, South Africa. *Nature* **108**, 371–372.

Wynn, J.G. (2004) Influence of Plio-Pleistocene aridification on human evolution: evidence from paleosols of the Turkana Basin, Kenya. *American Journal of Physical Anthropology* **123**, 106–118.

27

The Alps: A Geological, Climatic and Human Perspective on Vegetation History and Modern Plant Diversity

Séverine Fauquette[1], Jean-Pierre Suc[2], Frédéric Médail[3], Serge D. Muller[1], Gonzalo Jiménez-Moreno[4], Adele Bertini[5], Edoardo Martinetto[6], Speranta-Maria Popescu[7], Zhuo Zheng[8] and Jacques-Louis de Beaulieu[3]

[1] *Institut des Sciences de l'Evolution, University of Montpellier, Montpellier, France*
[2] *Institut des Sciences de la Terre Paris, Sorbonne University, Paris, France*
[3] *Institut Méditerranéen de Biodiversité et d'Ecologie marine et continentale, Aix-Marseille University, Aix-en-Provence, France*
[4] *Departamento de Estratigrafía y Paleontología, University of Granada, Granada, Spain*
[5] *Department of Earth Sciences, University of Florence, Florence, Italy*
[6] *Department of Earth Sciences, University of Torino, Torino, Italy*
[7] *GeoBioStratData.Consulting, Rillieux la Pape, France*
[8] *Department of Earth Sciences, Sun Yat-sen University, Guangzhou, China*

Abstract

Mountain ecosystems, and in particular European Alpine ecosystems, have a rich biodiversity, as they represent complex associations controlled by elevation, soils and rocks, and climatic conditions following latitude, longitude, slope orientation and aspect. Vegetation belts are organized with respect to altitude according to a concomitant decrease in temperature and increase in precipitation, and are defined by their dominant plant elements. Plant diversity within the Alps also results from these mountains' location at the transition point between cold-temperate and warm-temperate climates, as well as from their historical biogeography. In order to explain how past environmental changes have shaped the modern plant diversity and the organization of vegetation in altitudinal belts, a temporal dimension may be provided by paleovegetation data. Abundant micro- (pollen) and macro-remains (leaves, fruits, seeds) show that European Cenozoic vegetation exhibits a similar latitudinal and altitudinal organization to the vegetation belts observed today in south-eastern China, where most of the taxa – missing in Europe since the late Neogene – now occur. Late Eocene to Pliocene pollen floras of the south-western and Eastern Alps are characterized by the presence of megatherm plants and an abundance of mega-mesotherm plants. These are typical of moist evergreen low-altitude forests, and are characterized by the presence of *Cathaya*, a conifer now restricted to subtropical China at mid to high elevations. Pollen data indicate the presence of *Abies/Picea* forests at high altitudes since at least the Oligocene. Since then, several thermophilous taxa have declined, and some have disappeared from the Alps due to natural or human-related processes that are responsible for the modern plant diversity pattern there. These processes include the uplift of the mountain range, which began ca. 35 Ma, late Neogene global cooling, Pleistocene glacials/interglacials and more recent human impact.

Keywords: *paleo-phytogeography, pollen, mountain uplift, paleoaltitude, Cenozoic, global cooling, climatic cycles*

27.1 Introduction

Mountain vegetation is characterized by the altitudinal stratification of plants related to temperature and precipitation. Vegetation zones change from low to high latitude or from low to high altitude in the same way (Ozenda 1989): the replacement of taxa occurs as vegetation types adapt to colder conditions.

Mountain plant ecosystems generally have high diversity, as they represent complex assemblages influenced by elevation, climatic conditions – which follow latitude and longitude – and variable amounts of rainfall. Moreover, the physiographic complexity of mountains creates environments that can vary over short distances due to, for instance, exposure to sun or slope angle and aspect.

Mountains, Climate and Biodiversity, First Edition. Edited by Carina Hoorn, Allison Perrigo and Alexandre Antonelli.
© 2018 John Wiley & Sons Ltd. Published 2018 by John Wiley & Sons Ltd.
Companion website: www.wiley.com\go\hoorn\mountains,climateandbiodiversity

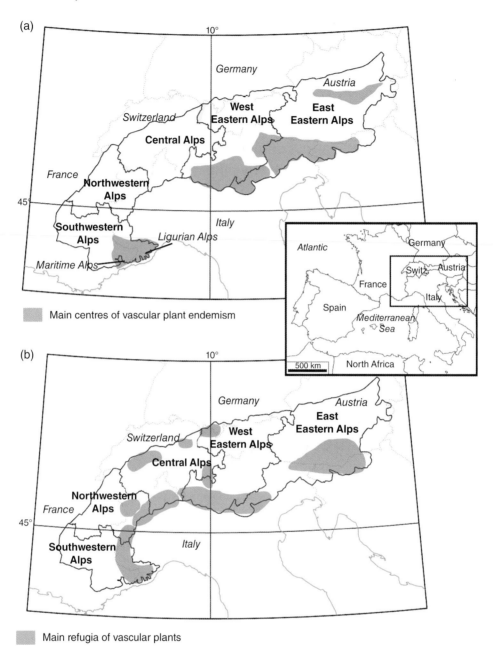

Figure 27.1 Maps of the five main phytogeographic zones of the Alps. *Source:* Adapted from Aeschimann et al. (2011). In gray: (a) The main centers of vascular plant endemism, defined by Noble & Diadema (2011), and (b) The main refugia of vascular plants in the Alps, defined by Schönswetter et al. (2005).

In the Alps, vegetation organization is also complex because of the large size of the mountains and their position at the crossroads of many floristic provinces (Ozenda 2009). Moreover, the Alpine arc is oriented north–south in France but west–east for the rest of the chain (Figure 27.1a). The French section of the chain is thus characterized by important differences between the southern and the northern parts, with the Southern European Alps being drier, as they are under Mediterranean influences.

Plant diversity within the Alps mainly results from their location at the transition point between cold-temperate and warm-temperate climates, their landscape heterogeneity, the local climate variability that arises from complex topographies and historical biogeography. A detailed knowledge of the past is thus essential to an understanding of the processes that have controlled changes in plant diversity up to the present.

This chapter, which focuses primarily on the Western Alps, starts with a description of the present flora and vegetation patterns in their physiographic and geological context. It then shows how the intense tectonic activity in this region over the last 40 My has influenced the flora and vegetation. Finally, it shows the role of Pleistocene climatic cycles and human impact on the mountains and their flora.

27.2 Present Flora and Vegetation Patterns in the Physiographic, Climatic and Geological Context of the Alps

27.2.1 Plant Distribution and Endemism Patterns

The European Alps are one of the most important hot spots of plant biodiversity in Europe (Ozenda 1985, 2002), with a high number of endemic taxa. According to Aeschimann et al. (2004), the Alps encompass 3983 indigenous taxa of vascular plants, which is more than one-third of the total European flora. The average taxonomic richness of the Alps is estimated at about 2200 taxa per $10\,000\,km^2$, and despite their relatively restricted area, the Southwestern Alps are home to more than 70% of the entire Alpine flora. In the Alps, 501 endemic taxa are registered, making up 4% of the European flora. This corresponds to an endemism rate of 12.6% within the area's native flora.

The distribution of endemics (Figure 27.1a) is mainly concentrated in the southernmost Western Alps and the southern part of the East Eastern and West Eastern Alps (Noble & Diadema 2011). At the junction of the Mediterranean Basin and the Alps, the Maritime and Ligurian Alps constitute a zone of geographic overlap between major biotic assemblages (Comes & Kadereit 2003), an important hot spot of plant biodiversity (Médail & Quézel 1997) and also a major refugium (Figure 27.1b) (Médail & Diadema 2009). This region, with 25% of the endemic plants of the whole range, represents one of the most important biogeographical areas in Europe, due to the high concentration of paleoendemic and endemic plants (Casazza et al. 2008; Noble & Diadema 2011). Recent molecular investigations of endemic plants from the Maritime and Ligurian Alps show that vicariance events are the most important factor explaining the distribution and genetic patterns of plant populations there (Casazza et al. 2013). The Maritime Alps do not act as a strong barrier to the migration of plants, and moderate gene flow seems to create an admixture between some plant populations, even when separated by deep valleys (Diadema et al. 2005). The persistence of several endemics throughout various glacial periods appears to be linked to the

capacity of mountains to provide a wide diversity of microhabitats within a reduced space.

In the Alps and Carpathian Mountains, species richness and genetic diversity do not co-vary for mountain vascular plants (Taberlet et al. 2012); rather, genetic diversity is associated with the glacial and postglacial history. In the Alps, genetic rarity (the presence of rare alleles or restricted haplotypes) is correlated with both species rarity (i.e., rare markers tend to be located in the same areas as rare species) and species endemism (i.e., rare makers in widespread species tend to be situated in the same areas as endemic species). Rare markers are often fixed in long-term isolated refugial populations, and these crucial areas correspond to distinct glacial refuges, in which the highest endemism levels also occur (Schönswetter et al. 2005).

27.2.2 Vegetation

Vegetation belts (also called altitudinal zones) in the Alps are well characterized by bioclimatic factors and their principal vegetation types, especially forests. Vegetation belts are organized altitudinally according to steep ecological gradients, with a concomitant decrease in temperature and increase in precipitation, extending from the lowlands to the nival zone. In the Central Alps, five vegetation belts are distinguished: the collinean, mountain, sub-alpine, alpine and nival belts (Table 27.1).

In the Southwestern Alps, vegetation is similarly organized in altitudinal belts on both flanks of the mountains, but with more pronounced Mediterranean influences occurring on the southern slopes, from the lowest thermo-Mediterranean to the oro-Mediterranean belt, which corresponds to the alpine belt of the Central Alps (Table 27.1). These southern mountains are heterogeneous from a biogeographical viewpoint, with the southern side clearly influenced by Mediterranean elements and the northern side composed of medio-European vegetation (Barbero & Quézel 1975). The altitudinal limits of these vegetation belts vary along a latitudinal gradient, but they are ~700 m wide in vertical elevation (Ozenda 1989). For collinean to sub-alpine species, especially trees, temperature often represents the limiting factor for survival at the upper limit and moisture at the lower limit.

These vegetation belts are well delimited by modern pollen data collected from moss polsters along two transects across the Southwestern Alps (Figure 27.2). In order to reconstruct Cenozoic vegetation and climate information from pollen data, Fauquette et al. (2015) established the relationship between modern pollen data and vegetation to serve as a baseline. To create this reference data set, pollen grains were grouped according to vegetation belts. However, *Pinus* was separated because

Table 27.1 Structure of the vegetation belts of the Alps in relation to the mean minimum temperatures of the coldest month of the year (*m*), and their correspondence with two concepts of the Mediterranean vegetation belts.

m (°C)	Alpine vegetation belts		Mediterranean vegetation belts	
	Ozenda (1985)	Main vegetation types	Ozenda (2002)	Quézel & Médail (2003)
−10				
−9	Nival Alpine	Bryophytes and lichens Grasslands with Poaceae and Cyperaceae; low heathlands with *Salix*, *Juniperus* and *Vaccinium*	Alti-Mediterranean	Oro-Mediterranean
−7				
−6	Sub-alpine	Coniferous forests with *Picea abies*, *Larix decidua*, *Pinus mugo* and *Pinus cembra*	Mountain-Mediterranean	Mountain-Mediterranean
−4				
−3	Mountain	Forests with *Fagus sylvatica* and/or *Abies alba*; mesophilous forests with *Pinus sylvestris*	Supra-Mediterranean	Supra-Mediterranean
−1				
0				
1	Collinean	Deciduous forests *with Quercus pubescens*, *Q. robur* or *Q. petraea*; xeric forests with *Pinus sylvestris*; mesophilous forests with *Tilia*, *Fraxinus*, *Acer* and *Prunus*	Medio-Mediterranean	Meso-Mediterranean
			Per-Mediterranean	Thermo-Mediterranean
9				
10		Mediterranean vegetation in the Southwestern Alps (southern slopes)		
≥10				Infra-Mediterranean

its pollen is not identifiable at the species level and may be representative of several vegetation belts. Pollen grains of *Larix* are scarce; samples containing this genus were collected within its forests or from sites immediately below such forests. Pollen grains of Cupressaceae are grouped together because infra-family identification (genus identification) is impossible, even though their species grow at different altitudes in the Southwestern Alps: *Cupressus sempervirens*, *Juniperus oxycedrus* and *J. phoenicea* in the two lowermost Mediterranean vegetation belts; *J. thurifera* in the higher supra-Mediterranean belt and in the lower part of mountain-Mediterranean belt; and *J. communis* and *J. sabina* in the mountain-Mediterranean belt. Most of the herbaceous pollen grains cannot be distinguished at the genus level, whereas such plants can inhabit different altitudes; their abundance may indicate

open and/or undergrowth vegetation. Finally, alpine tundra-like plants (*Aconogonum alpinum*, *Gentiana*, *Centaurea uniflora*) were identified in only a few sites.

These west–east transects show that there is a good correlation between pollen data and observed vegetation distributions along altitudinal and longitudinal gradients, and that the vegetation distribution follows a clear asymmetric topography (Figure 27.2): replacement of pollen groups as elevation increases appears progressive on the western flank, but much more abrupt in the east.

Pollen data may therefore be used to reconstruct Paleogene to Quaternary paleovegetation. Moreover, for these periods, information from pollen may be complemented by macrofossils when they are available in the same region. However, these fossils are produced from plants that lived not more than a few kilometers from the

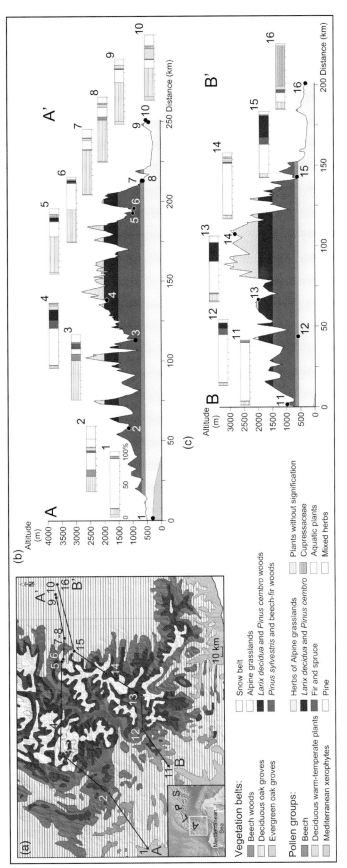

Figure 27.2 Relationship between vegetation belts positioned on a topographic profile, following studies by Noirfalise et al. (1987) and Ozenda (2002), and modern pollen data collected from moss polsters along two transects. (a) Simplified vegetation map of the Southwestern Alps (from Noirfalise et al. 1987), on which the transects are shown (dots correspond to the moss polsters). Pollen localities and altitudes are indicated below. (b) Transect AA′: 1, Nyons (300 m); 2, Luz la Croix haute (1100 m); 3, Le Freney (800 m); 4, Col du Lautaret (1850 m); 5,6, Pinerolese (Lago Villaretto di Roure and Balma di Roure, 980 and 890 m, respectively); 7,8, Cumiana Giaveno (612 and 605 m, respectively); 9,10, Baldissero Torinese (Val Samfrà and Superga road, 383 and 537 m, respectively). (c) Transect BB′: 11, Taulanne (995 m); 12, Sisteron (540 m); 13, Saint-Léger (2000 m); 14, Agnel (2800 m); 15, Pinerolese, Inverso Pinasca (560 m); 16, Moncucco Torinese, Borelli (315 m). *Source:* Adapted from Fauquette et al. (2015). See also Plate 45 in color plate section.

fossilization site, and rarely record those growing in the higher belts. To get around this problem, Bertini & Martinetto (2011) used the presence of pollen and the concomitant absence of macrofossils in several deposits of the same age to pinpoint those plants that probably grew at higher elevations (e.g., *Cedrus, Picea, Tsuga*).

27.3 Vegetation History of the Alps Since the Late Eocene

The Alps resulted from the convergence of the European and African plates, trapping the Apulian microplate between them, which led to the closure of the Tethys Ocean by the Eocene, followed by a collision between the European and Adriatic plates (e.g., Rosenbaum & Lister 2005). This complex collision history led to a nappe stack involving basement and cover rocks of the European margin, of the intervening ocean and micro-continents, and of the Adriatic margin.

The intense tectonic activity in this region over the last 40 My has influenced the vegetation and flora, with the appearance of high mountains and new marine basins, and consequent landscape fragmentation, environmental heterogeneities and new climatic patterns.

27.3.1 Vegetation History Before Human Impact

Abundant pollen records (e.g., Suc 1989; Suc et al. 1995; Jiménez-Moreno et al. 2005, 2010; Bertini 2010) show that European Neogene vegetation followed belts that were aligned similarly in latitude and altitude to the belts seen today in south-eastern China (Wang 1961), where most of the taxa that disappeared from Europe by the late Neogene are still extant.

The study by Fauquette et al. (2015) shows that in the Southwestern Alps, late Eocene to Pliocene pollen floras (Figures 27.3 and 27.4) are characterized by prevalent mega-mesotherm plants (see the definitions of these plant groups in the legend of Figure 27.4), reflecting sub-tropical conditions, and sometimes high frequencies of *Cathaya*, a conifer restricted today to subtropical China at mid to high elevations, below the *Abies* and/or *Picea* belts (Wang 1961). Pollen records suggest that the past representatives of *Cathaya* in Europe probably occupied a similar ecological niche (e.g., Suc 1989; Combourieu-Nebout et al. 2000, 2015; Biltekin et al. 2015). Megatherm plants supporting an interpretation of tropical conditions were present in all of the Eocene to mid-Miocene samples (Figure 27.4). The maximum frequency of such species occurred during the early Langhian (~15 Ma) (Jiménez-Moreno et al. 2005), corresponding to the

mid-Miocene Climatic Optimum (MCO). The pollen record of mega-mesotherm plants decreases from the early Zanclean (ca. 5 Ma) to the Piacenzian (3.6–2.6 Ma), and major disappearance events have occurred since the earliest Pleistocene (2.6 Ma), due to the onset of northern hemisphere glaciations (Suc 1984; Bertini 2010). Climate gradients linked to the latitude, altitude and physiography of pollen sites explain the different timings of the extinctions of *Taxodium*-type, *Cathaya* and *Tsuga* species, among others (e.g., Suc et al. 1995; Bertini 2010).

On the whole, microtherm trees support the presence of boreal conditions (*Abies, Picea*) and meso-microtherm trees (particularly *Cathaya*) occur in significantly higher frequencies on the eastern side of the Southwestern Alps than on the western side (Figure 27.4). This indicates that the present-day asymmetric topography of the Southwestern Alps, with a relatively gentle western flank and steeper eastern flank, was established early, and has been in existence since at least the early Miocene, and possibly since the Oligocene or late Eocene. Therefore, the high topography and asymmetric morphology of this part of the Alps have been maintained throughout the past ~30 My.

In the East Eastern Alps, early to late Miocene pollen data (Figures 27.3 and 27.4) show a very diverse flora (Jiménez-Moreno et al. 2005, 2008). The Burdigalian and Langhian (20.4–13.6 Ma) vegetation was dominated by thermophilous elements such as evergreen trees, typical of a present-day low-altitude evergreen rainforest (e.g., south-eastern China), reflecting the MCO. During the Serravallian and Tortonian (13.6–7.2 Ma), important vegetation changes are observed: *Avicennia*, which previously populated the coastal areas up to these high latitudes, vanished, and several other megatherm elements (*Buxus bahamensis*-type, *Alchornea*, Melastomataceae), typical of broad-leaved evergreen forests, became rarer, before finally disappearing from Central Europe. These changes can be linked to cooling and correlated with global and regional climatic changes. Later, this vegetation was progressively replaced by communities rich in mesotherm plants. Percentages of mid- and high-altitude conifers increased considerably during the Langhian and, later, during the Serravallian and Tortonian, indicating their relocation to a lower altitude, nearer to the sedimentary basin, which is correlated with significant cooling phases. These coniferous belts developed again at higher altitudes during warmer phases, as signaled by their reduced frequency in younger pollen records.

Subsequently, several thermophilous plants declined in abundance, and some of them disappeared from the Alps due to global cooling and progressive amplification of the glacial–interglacial cycles (see further details in Chapter 9). The data from the Stirone site (Figure 27.4) (Bertini 2001) at the foot of the Apennines provide a good example of this phenomenon. From their onset,

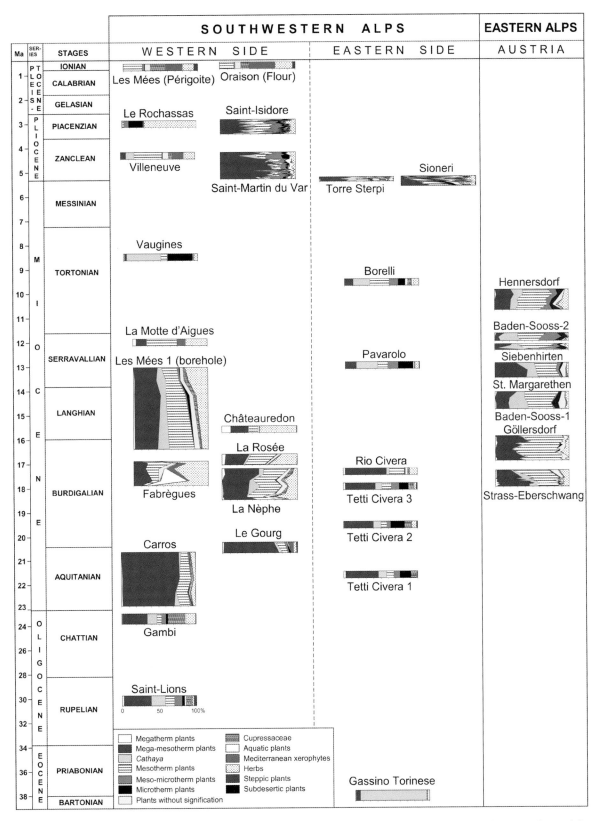

Figure 27.3 Synthetic pollen diagrams from late Eocene to Pleistocene localities of the Southwestern and Eastern Alps, and their positions in a chronostratigraphic frame. Note the change of time scale at the Paleogene–Neogene boundary. Taxa have been arranged into 13 different groups (*Pinus* excluded) based on ecological criteria to visualize the main changes in the paleovegetation cover. Plants without signification indicates families (e.g., Rosaceae, Ranunculaceae, Rutaceae) in which pollen identification is difficult at the genus level and includes cosmopolitan taxa and/or taxa with wide ecological requirements, plus unidentified pollen grains. Some groups are classified according to thermic requirements with respect to the Nix (1982) classification: equatorial and tropical forests are inhabited by megatherm plants (mean annual temperature, MAT >24 °C), subtropical forests by mega-mesotherm plants (MAT 20–24 °C), temperate deciduous forests by mesotherm plants (MAT 14–20 °C), boreal coniferous forests successively by meso-microtherm (MAT 12–14 °C) and microtherm (MAT <12 °C) plants and tundra by microtherm plants. *Source:* Adapted from Fauquette et al. (2015) and Jiménez-Moreno et al. (2008). See also Plate 46 in color plate section.

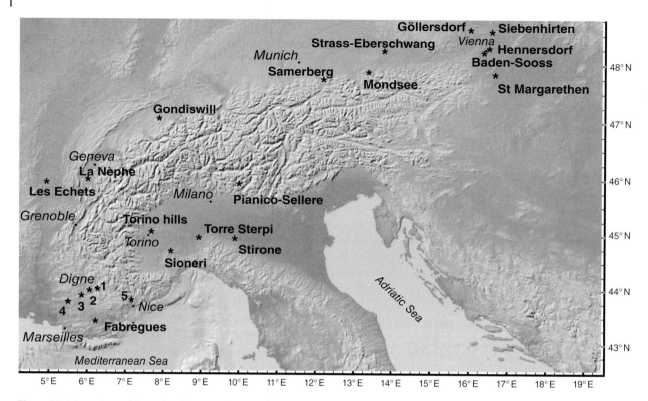

Figure 27.4 Locations of the palynological sites mentioned in the text and in Figure 27.3. The sites of the Torino Hills are: Gassino Torinese, Tetti Civera (samples 1, 2 and 3 in Figure 27.3) and Rio Civera, Pavarolo, Borelli. The numbered sites in Southern France are: 1, Saint Lions, Gambi, Châteauredon, Le Rochassas; 2, La Rosée, Les Mées 1, Les Mées-Périgoite; 3, Villeneuve, Oraison (Flour); 4, La Motte d'Aigues, Vaugines; and 5, Le Gourg, Carros, Saint Martin du Var, Saint-Isidore

glacial–interglacial cycles are marked by changes in both vegetation and reconstructed climate compared to the early Pliocene (Fauquette & Bertini 2003). During interglacials, the climate was almost as warm as in the early Pliocene, and the vegetation remained a mega-mesotherm to mesotherm plant forest. During glacials, mega-mesotherm taxa disappeared (e.g., *Distylium*, *Myrica*) or declined in abundance and distribution (*Taxodium*-type, *Sequoia*-type). Temperatures were markedly lower than during interglacials – nearly equal to modern values – and not suitable for these thermophilous taxa.

The known macrofloras mostly come from the northern and eastern borders of the Alps for the Eocene to mid-Miocene interval (Mai 1995), and from the southern Alps for the Messinian and Pliocene (Bertini & Martinetto 2011; Martinetto et al. 2015). In general, these data confirm the picture provided by pollen with respect to the flora and vegetation of the lowland Alpine fringes. The most interesting information added by macrofossils is connected with the marked climatic cooling that has occurred since ~2.6 Ma: several plants now extinct in Europe entered the macrofossil record for the first time during this period (Martinetto et al. 2015). The cones of *Picea florschuetzii*, which are very similar to a modern spruce that grows in the Himalaya (*P. smithiana*), or of *Tsuga chairugii*, whose relatives

today live only in America and East Asia, are frequent in a few early Pleistocene sites of northern Italy (Ravazzi 2003). The presence of these taxa indicates either the descent of the Alpine high-vegetation belts or immigration from Northern Europe/Siberia. Reliable records of plant species typical of the modern Alpine high-altitude belt (e.g., *Larix* cf. *decidua* and *Picea abies*) are first found only in the late early Pleistocene (Ravazzi et al. 2005). Records of diagnostic species of the alpine belt are restricted to the late Pleistocene glacials and most commonly occur in the lowlands to the north of the Alps (e.g., *Salix herbacea*) (Alsos et al. 2009). This appears to be a good example of a taphonomic window, where alpine taxa are only preserved when they descend close to potential depositional environments, which are restricted to lowlands or basin floors.

Very few pollen data are available from the Pleistocene. Pollen data from Pianico-Sellere (Bergamo, Italy; Figure 27.4), attributed to early–middle Pleistocene, represent vegetation dynamics similar to those of the late Pleistocene interglacials, although "Tertiary relicts" (*Pterocarya*, *Tsuga*, *Rhododendron ponticum*) are still present (Martinetto 2009). Other pollen successions in Switzerland and Bavaria attributed to the mid-Pleistocene (Welten 1982; Grüger 1983) illustrate, as elsewhere in Europe, a moderate expansion of *Pterocarya* during the

Holstein interglacial. Several pollen records from France, Switzerland, Bavaria and Austria (respectively, Les Echets (de Beaulieu & Reille 1984), Gondiswill (Wegmüller 1992), Samerberg (Grüger 1979) and Mondsee (Drescher-Schneider 2000); Figure 27.4) document the vegetation dynamics of the last interglacial (Eemian), highlighting the significant role played by *Carpinus betulus* and the quasi-absence of *Fagus*.

27.3.2 The Onset of Modern Vegetation: An Example from the Western Alps

Numerous pollen records are available in the western Alps for the Holocene, which allow for the depiction of the general trend of vegetation changes during this period.

The Last Glacial Maximum (LGM), between 24 and 17 ka (calibrated ages), constitutes the state of maximal retreat of temperate vegetation types, to the benefit of alpine-boreal plant communities. Temperate taxa were mostly restricted to southern refuges, mainly located in the Iberian, Italian and Balkan peninsulas and around the Black Sea, but some could also have survived within cryptic northern refugia (e.g., Nieto Feliner 2011). Most of the temperate plant recolonizations occurring thereafter originated from southern regions (e.g., Brewer et al. 2002; Muller et al. 2007).

The Lateglacial period (17.0–11.3 ka) was marked by an ecological succession arising from an initial climate warming: the cold steppe vegetation dominated by *Artemisia* and Amaranthaceae was replaced by *Juniperus* scrub, then *Betula* woods and finally *Pinus* forests (de Beaulieu & Reille 1983). The Younger Dryas climatic event, which allowed a widespread expansion of steppe communities throughout Europe, led to a general lowering of the treeline to ~300 m.

The subsequent temperature increase at the beginning of the Holocene induced a second phase of successional dynamics, involving temperate tree communities that replaced *Pinus* forests. The first trees to establish were *Corylus* and deciduous *Quercus*. These communities constituted, from sea level to about 1800 m, diversified forests, including *Fraxinus*, *Ulmus* and *Tilia*. At higher altitudes, these forests were replaced by coniferous ones.

Simultaneously, around 10.1 ka, *Abies alba* penetrated into the southern French Alps, migrating from its nearest refugium in the northern Adriatic region (Muller et al. 2007). Fir forests expanded during the Atlantic period (8.9–5.4 ka), in both the southern (de Beaulieu & Reille 1983) and the northern French Alps (David 1995). This range expansion was slower in the dryer internal massifs, and ended about 6.5 ka, when fir penetrated into the extensive pine woods of the most arid region of the Alps, the Briançonnais (Muller et al. 2007). Throughout its Alpine range, this highly competitive species found suitable conditions at middle altitudes,

progressively forcing pines and *Larix decidua* to migrate to higher altitudes and deciduous broad-leaved trees to migrate to lower ones.

Between 6.5 and 2.8 ka, *Pinus cembra* formed dense forests between 2000 and more than 2300 m (Muller et al. 2006). This occurred at the same time as an increase in the upper treeline, which exceeded 2900 m in the Queyras Valley (Talon et al. 1998). During the Subboreal period (5.4–2.8 ka), fir and deciduous oak declined as Neolithic human populations progressively colonized mountain territories (Walsh et al. 2014). The anthropogenic pressure was intense during the last 2.8 ka, since the beginning of the Subatlantic period, and the forest exploitation of the Bronze and Iron ages focused on some useful species, such as fir, which was intensively exploited to supply coastal shipyards (Nakagawa et al. 2000). Such human activity induced major vegetation changes from the shores to the mountains: at low altitudes, evergreen pioneer formations of *Quercus ilex* and *Pinus halepensis* replaced the deciduous forests of *Q. pubescens*, forming the anthropogenic meso-Mediterranean belt. At mid-altitudes, above the remnant deciduous oak forests of the supra-Mediterranean belt and within the mountain-Mediterranean belt, *Abies alba* forests declined to give way to *Fagus sylvatica*, while in the sub-alpine belt, *Pinus cembra* declined and *Abies* disappeared from the internal valleys (Muller et al. 2007), which allowed larch and mountain pine to disperse to lower elevations.

At the time of the Roman Empire, several cultivated trees were introduced (e.g., *Castanea*, *Juglans*), and the industrial forest exploitation (Nakagawa et al. 2000) allowed for the immigration of *Picea abies* into the newly opened vegetation zones. This species spread into the sub-alpine belt of the Northern Alps, where it found a suitably wet climate. The period between AD 500 and 1000 was marked by a forest recovery, followed by a new deforestation trend from feudal times onwards, which finally ended during the Industrial Revolution of the 19th century. Thereafter, the abandonment of agriculture in remote valleys resulted in a re-establishment of mountain forests and the significant establishment of firs and oaks in secondary pioneer forests during the 20th century.

27.4 Climate and Paleoaltitude Reconstructions of the Alps Since the Late Eocene

27.4.1 Methods Based on Pollen Data

Pollen grains offer a large number of morphological traits (such as aperture, structure and sculpture) that can be analyzed by light microscopy (Punt et al. 2007) and scanning electronic microscopy (SEM) for surface features (Halbritter et al. 2007). These features can then be used to identify the pollen to species level. The implementation of these techniques in the 1970s led to vastly improved identification rates of the Neogene pollen flora in the Mediterranean region, based on systematic comparative observations of fossil pollen grains with those of living plants (Suc 1976; Suc & Bessedik 1981). This botanical approach afforded the identification of more than 100 new taxa, including 14 robust identifications at the species level in the study area (*Avicennia officinalis*, *A. alba*, *Hamamelis mollis*, *Parrotia persica*, *Parrotiopsis jacquemontiana*, *Distylium chinense*, *Microtropis fallax*, *Juglans cathayensis*, *Nyssa sinensis*, *Rhoiptelea chiliantha*, *Carpinus orientalis*, *Rhus cotinus*, *Lygeum spartum* and *Cedrus atlantica*) (Suc et al. 2004). This achievement significantly enriched the understanding of the regional Neogene flora and contributed to reliable climate reconstructions.

The climate reconstructions assembled from the Neogene data (Fauquette et al. 1999; Jiménez-Moreno et al. 2008; Fauquette et al. 2015) are based on the "climatic amplitude method," which was developed by Fauquette et al. (1998) specifically to quantify the climate of periods with no modern analogs for the pollen spectra. In this method, the past climate is estimated by transposing the climatic requirements of the maximum number of modern taxa to the fossil data. The climatic amplitude method takes into account pollen percentages, and allows, despite some biases, for a greater refinement and precision of climatic estimates, because taxa with very low pollen percentages are not always significant. The most probable climate for a set of taxa is estimated as the climatic interval suitable for the highest number of taxa and a "most likely value" corresponding to a weighted mean.

In this method, *Pinus* pollen grains are excluded from the pollen sum, as they are generally over-represented in marine sediments due to a prolific production and over-abundance resulting from long-distance air and water transport (Heusser 1988; Beaudouin et al. 2007). High-latitude/altitude taxa are also excluded from the reconstruction process to avoid a cold bias linked to pollen transport from higher elevations. The excluded taxa are defined on the basis of their occurrence in modern vegetation zones: vegetation types described for the studied time period are found today in south-eastern China. There, the vertical distribution of vegetation is characterized from the base to the top of the massifs by evergreen broad-leaved forest, mixed evergreen and deciduous broad-leaved forest, *Picea*/*Tsuga* forest, *Abies* forest and high mountain meadows (Wang 1961; Hou 1983). This stratification allows us to exclude the modern high-altitude taxa from the climatic reconstruction in order to infer the low- to middle-low-altitude climate (Fauquette et al. 1998). This procedure also allows us to discriminate between climatic variations and elevation changes.

Estimates based on this method show that the latitudinal temperature gradient in western Europe was ~0.48 °C/degree latitude during the mid-Miocene (Fauquette et al. 2007) and ~0.46 °C/degree latitude during the early Oligocene (Fauquette et al. 2015), and that a modern gradient was established during the Tortonian.

The method used to quantify paleoaltitudes, developed by Fauquette et al. (1999), is based on pollen floras, which provide a regional view of the vegetation from the lowest to the highest altitudinal belts feeding into the drainage basin from which the sediments were sourced, and the mean annual temperature (MAT) reconstructed at sea level, which is transposed into a modern latitude. The latitude difference between the Cenozoic site and the modern latitude indicates the shift in altitude of the vegetation belts; in particular, the *Abies*/*Picea* belt, which is the best indicator of elevated topography in the fossil pollen floras. This method uses a standard modern relationship established by Ozenda (1989, 2002), as the data do not allow the estimation of past terrestrial lapse rates. Ozenda's relationship assumes that, in the Alps, modern vegetation belts shift by 110 m in altitude per degree of latitude. It should be noted that terrestrial lapse rates are highly variable geographically in relation to altitude, slope aspect, prevailing wind directions, continentality and atmospheric composition, among other factors, and they are difficult to apply where the past atmospheric composition differed from the present (see Chapter 7). Meyer (1992) estimated modern terrestrial lapse rates for many areas of the world, and found that they vary between 0.36 and 0.81 °C/100 m for MAT. However, because there is no information concerning paleoterrestrial lapse rates in the Alps, Fauquette et al. (2015) hypothesized that the ratio between altitudinal and latitudinal temperature gradients was similar to the present ratio; taking into account the range of modern terrestrial lapse rates, the relationship between the latitudinal and altitudinal gradients would vary, giving different values for the shift of vegetation belts per degree of latitude (Table 27.2). If this range of values were used, the estimates of paleoaltitude would differ as well. However, Table 27.2 highlights that the estimates of the terrestrial lapse rate for each studied period based on the standard modern relationship are included in Meyer's estimated range.

Moreover, paleoelevation estimates correspond to the most probable paleoelevations based on the "most likely values" of MAT and to ranges of paleoelevation based on the minimum and maximum MAT. This partially compensates for the use of the modern altitudinal/latitudinal ratio.

In the case of sites with unknown elevation (Suc & Fauquette 2012), which were all younger than 10 Ma in Fauquette et al. (2015), the latitudinal temperature

Table 27.2 Shifts in altitude of vegetation belts in relation to different terrestrial lapse rates and to latitudinal gradients estimated for different epochs (Fauquette et al. 2015). The minimum (0.36 °C/100 m) and maximum (0.8 °C/100 m) lapse rates are those reconstructed by Meyer (1992) for many areas of the world. The lines in bold corresponds to the estimates of terrestrial lapse rate for each studied period based on the standard modern relationship.

	Latitudinal gradient (°C/°lat.)	Terrestrial lapse rate (°C/100 m)	Shift in altitude/°lat.
Standard modern	0.6	0.55	1 °lat. → 110 m
Pliocene	0.6	0.36	1 °lat. → 167 m
		0.55	**1 °lat. → 110 m**
		0.81	1 °lat. → 74 m
Middle Miocene	0.48	0.36	1 °lat. → 134 m
		0.44	**1 °lat. → 110 m**
		0.55	1 °lat. → 87 m
		0.81	1 °lat. → 60 m
Early Oligocene	0.46	0.36	1 °lat. → 128 m
		0.42	**1 °lat. → 110 m**
		0.55	1 °lat. → 84 m
		0.81	1 °lat. → 57 m

gradient was similar to the modern one, and the authors chose to keep the terrestrial lapse rate equivalent to the modern rate, taking into account that the position and continentality of the Alps were the same as today.

The record in the fossil pollen flora of microtherm trees such as *Abies* and *Picea* indicates elevated topography close to the sampled sites. Nevertheless, pollen data do not permit altitudinal range estimation for this vegetation belt, nor a determination of whether alpine herbaceous and perpetual-snow belts existed above it, due to difficulties in differentiating between the pollen of alpine herbaceous elements and that of plants growing at lower elevations. As a consequence, these pollen-based paleoelevations are minimum estimates, corresponding to the lower limit of the highest forested belt.

We must emphasize the consistency of our previous estimates: the paleoaltitude we inferred for the Mercantour Massif (Southwestern Alps) is consistent with a geomorphological reconstruction (Fauquette et al. 1999); the paleoaltitude of the Ganos Mountain in the Dardanelles Strait (Melinte-Dobrinescu et al. 2009) is in agreement with the tectonic hypothesis of Armijo et al. (1999); and the paleoaltitude of the Canigou Mount in the Eastern Pyrenees (Suc & Fauquette 2012) is supported by a paleogeographic validation (Clauzon et al. 2015).

27.4.2 Resulting Paleoaltitudes

Based on this method, the study by Fauquette et al. (2015) shows that the topography of the internal zone of the Southwestern Alps reached elevations of around 1800 m as early as the Oligocene (ca. 30 Ma). In contrast, the external massifs, including the Mercantour Massif, uplifted later, during the late Neogene, while the post-Serravallian (ca. 12–13 Ma) emplacement of the Digne thrust nappe was reinforced. Pollen data show that the present-day topographic asymmetry of the Southwestern Alps, with a relatively gentle western flank and a much steeper eastern one, dates back to at least the early Miocene, and possibly to the Oligocene or late Eocene.

In the East Eastern Alps (Jiménez-Moreno et al. 2008), paleoaltitudes during the Burdigalian were not high enough for *Abies* and *Picea* to form forests, but were sufficient to allow *Cathaya* to survive. Therefore, the relief was less than 1800 m. On the other hand, Serravallian and Tortonian pollen records, rich in conifers typical of high elevations, indicate an intense uplift during the mid-Miocene. The abundance increase in high-elevation trees is due to uplift and not to climatic changes, as the pollen records are still dominated by thermophilous plants. Paleoaltitudes were higher than 1600 m at that time.

27.5 How Do Regional Geological Evolution, Global Climatic Changes and Human Pressure Affect Alpine Plant Diversity and Vegetation?

Due to the formation of the Alps since the late Paleogene (40–35 Ma) under warm-temperate to subtropical conditions, boreal-type environments could only develop at the higher elevations. Since the Oligocene, high elevations (~1800 m) have allowed meso-microtherm (*Cedrus*, *Cathaya*) and microtherm (*Abies*, *Picea*) trees to establish.

The development of plant populations on mountains arises from two major sources (Ozenda 1985, 2009): an autochthonous (in situ) population, formed of plants already present at the beginning of the uplift, and an allochthonous component, arising from migration from other cold regions.

The autochthonous part formed from a subtropical to tropical flora. Most of these taxa, vulnerable to cold, became extinct later due to glaciations and competition, but a few, such as *Juniperus thurifera* and *Berardia subacaulis*, survived in the Western Alps (Ozenda 2009). Plants of temperate origin are represented by two groups: Holarctic and Mediterranean plants (Ozenda 1985).

The allochthonous part is made up of plants that migrated from the Mediterranean and/or central Asian mountains and Arctic regions (Ozenda 1995, 2009; Kadereit et al. 2008). Exchanges with Arctic and Northern European floras during glacials/interglacials is evidenced by the presence of "arctic-alpine" plants. However, the Alpine flora shows less pronounced affinities to the northern (Scandinavia, North America) than to the southern (Mediterranean Basin, Central and Eastern Asia) floras (Ozenda 1995, 2009; Comes & Kadereit 2003). The European Alpine system and major Central Asian mountains (Altai, Tibetan Plateau and Himalaya) are connected by a more or less continuous system of mountain ranges. This implies both a northern (boreal, sub-arctic, arctic) and a southern (southern mountains) connection between the European Alpine system and the Central Asian high mountains.

Pleistocene glaciations played a major role in the changing vegetation of the Alps: (i) the pre-existing thermophilous flora almost completely disappeared; (ii) plants of temperate origin only survived in refugia, leading to a disjunct distribution of some plants; and (iii) exchanges between the Alps and the other mountain ranges and/or northern regions were amplified. Pleistocene warming phases led to the expansion of the thermophilous floras, forcing microtherm plants to move towards colder areas to the north or to higher elevations (Ozenda 1985, 2009), which resulted in disjunct distributions of arctic-alpine plants.

The postglacial period of the French Alps is the period in which the different vegetation belts recognized today were established (Table 27.1). Around 10 ka, only two forested belts existed: a lower one, occupied by deciduous oak forest, and an upper one, characterized by coniferous forests (mainly *Pinus sylvestris* and *P. uncinata*). The boundary between these two belts was not abrupt, but formed a transition zone with mixed formations of oaks and pines. This intermediate belt only transformed into a belt in its own right with the immigration of fir, from ca. 9 ka. The fir forests, mixed with numerous deciduous trees, forced oak forests downslope and pine forests upslope. The late onset of this belt is clearly due to the migratory dynamics of firs that arrived in the French Alps from their Adriatic and Apennine refuges. Between 7.5 and 5.0 ka, the sub-alpine belt, which was made up of open woodlands of *Pinus uncinata* mixed with *Larix* and *Pinus cembra*, differed significantly from its modern physiognomy. The modern forest structure of this belt was only achieved around 5 ka, with the development of dense forests of *P. cembra* up to ~2300 m. The onset of a real forested sub-alpine belt appears to be related to a general aridification, evidenced by low lake levels (Digerfeldt et al. 1997).

The last belt to appear in the Southwestern Alps was the lower meso-Mediterranean belt, at around 5–4 ka, characterized by the establishment of typical Mediterranean plants at low altitudes, in particular *Quercus ilex* and *Pinus halepensis*. The presence of Neolithic human populations supports the anthropogenic origin of this belt, which corresponds to the replacement of deciduous forests by pioneer disturbance-adapted plant formations. Nevertheless, recent multiproxy attempts to quantify Holocene climate fluctuations have shown that a succession of dry episodes from ca. 5 ka may have also triggered the expansion of sclerophyllous Mediterranean taxa (Magny et al. 2002). Unlike the other vegetation belts, the meso-Mediterranean belt appears to be a partly human-triggered subdivision of the former (supra-)Mediterranean belt, naturally dominated by deciduous oak forests. At higher altitudes, human activity also had strong effects, such as lowering the treeline by at least 300 m (Carnetti et al. 2004) and creating sub-alpine meadows. Moreover, increased human population sizes in mountains led to the fragmentation of forest ecosystems and thus to erosion events. In the second part of the 20th century, a policy of intensive reforestation started.

27.6 Conclusion

The flora of the Alps encompasses more than one-third of the European flora. Endemic taxa, mainly concentrated in the south-westernmost Alps and the south-western part of the Eastern Alps, represent 4% of the European flora. Maritime and Ligurian Alps constitute an important hot spot of plant species richness and form a major refugium where endemic plants represent 25% of all alpine endemics.

In the Alps, vegetation belts are organized by altitude, governed by decreases in temperature and increases in precipitation at higher elevations. These vegetation belts are well recognized in modern pollen data, established on two transects along the slopes of the Southwestern Alps. The modern pollen data also reflect the asymmetric west–east elevation of the massif: the replacement of pollen groups as elevation increases appears progressive on the western flank but much more abrupt on the eastern one.

The formation of the Alps dates back to ca. 35 Ma and has influenced both the flora and the vegetation of the area, due to the appearance of high mountains and new marine basins, and consequent landscape fragmentation, environmental heterogeneities and new climatic patterns. As early as the Oligocene, high relief allowed boreal-type environments to develop at the higher elevations, and meso-microtherm (*Cedrus*, *Cathaya*) and microtherm trees (*Abies*, *Picea*) established early in the history of Alpine vegetation. During the Pleistocene, due to glaciations, the earlier, more thermophilous flora became extinct, while arctic-alpine plants developed on the massifs. Holarctic plants only survived in refugia, leading to disjunct distributions. However, warm phases also led to disjunct distributions of the arctic-alpine plants, which moved towards colder areas to the north or to higher elevations. Today, only some "Tertiary relicts," such as *Juniperus thurifera* and *Berardia subacaulis*, still survive in the Alps.

The onset of the modern vegetation patterns started in the early postglacial period, around 10 ka. The different vegetation belts recognized today gradually developed following local climate and soil conditions and the migratory dynamics of the species. The presence of human populations in the Alps since the Neolithic triggered the expansion of some vegetation types, in particular typical Mediterranean plant formations at low altitude, and the lowering of the treeline, which led to the development of the sub-alpine meadows. Finally, the recent abandonment of land cultivation has resulted in the return of mountain forests.

Acknowledgments

The authors are deeply indebted to the late Professor Pierre Quézel for his teaching, advice and unfailing support for their work on the history of the Mediterranean flora and vegetation. A part of this study has been realized in the framework of the project "Erosion and Relief Development in the Western Alps," funded by the Agence Nationale de la Recherche. The authors thank the two reviewers, R.A. Spicer and J. Carrión, who helped us to improve the chapter. This is ISEM contribution no. 2016-115.

References

Aeschimann, D., Lauber, K., Moser, D.M. & Theurillat, J.-P. (2004) *Flora Alpina*. Belin & Zanichelli: Haupt.

Aeschimann, D., Rasolofo, N. & Theurillat, J.-P. (2011) Analyse de la flore des Alpes. 1: historique et biodiversité. *Candollea* **66**, 27–55.

Alsos, I.G., Alm, T., Normand, S. & Brochmann, C. (2009) Past and future range shifts and loss of diversity in dwarf willow (*Salix herbacea* L.) inferred from genetics, fossils and modelling. *Global Ecology and Biogeography* **18**, 223–239.

Armijo, R., Meyer, B., Hubert, A. & Barka, A. (1999) Westward propagation of the North Anatolian fault into the northern Aegean: timing and kinematics. *Geology* **27**, 267–270.

Barbero, M. & Quézel, P. (1975) Végétation culminale du mont Ventoux; sa signification dans une interprétation phytogéographique des Préalpes méridionales. *Ecologia Mediterranea* **1**, 3–33.

Beaudouin, C., Suc, J.-P., Escarguel, G. et al. (2007) The significance of pollen signal in present-day marine terrigenous sediments: the example of the Gulf of Lions (western Mediterranean Sea). *Geobios* **40**, 159–172.

Bertini, A. (2001) Pliocene climatic cycles and altitudinal forest development from 2.7 Ma in the Northern Apennines (Italy): evidence from the pollen record of the Stirone section (~5.1 to ~2.2 Ma). *Geobios* **34**, 253–265.

Bertini, A. (2010) Pliocene to Pleistocene palynoflora and vegetation in Italy: state of the art. *Quaternary International* **225**, 5–24.

Bertini, A. & Martinetto, E. (2011) Reconstruction of vegetation transects for the Messinian/Piacenzian of Italy by means of comparative analysis of pollen, leaf and carpological records. *Palaeogeography, Palaeoclimatology, Palaeoecology* **304**, 230–246.

Biltekin, D., Popescu, S.-M., Suc, J.-P. et al. (2015) Anatolia: a long-time plant refuge area documented by pollen records over the last 23 million years. *Review of Palaeobotany and Palynology* **215**, 1–22.

Brewer, S., Cheddadi, R., de Beaulieu, J.-L. et al. (2002) The spread of deciduous *Quercus* throughout Europe since the last glacial period Forest. *Ecology and Management* **156**, 27–48.

Carnetti, A.L., Theurillat, J.-P., Thinon, M. et al. (2004) Past uppermost tree limit in the Central European Alps (Switzerland) based on soil and soil charcoal. *The Holocene* **14**, 393–405.

Casazza, G., Zappa, E., Mariotti, M.G. et al. (2008) Ecological and historical factors affecting distribution pattern and richness of endemic plant species: the case of Maritime and Ligurian Alps hotspot. *Diversity and Distributions* **14**, 47–58.

Casazza, G., Grassi, F., Zecca, G., et al. (2013) Phylogeography of *Primula allionii* (Primulaceae), a narrow endemic of the Maritime Alps. *Botanical Journal of the Linnean Society* **173**, 637–653.

Clauzon, G., Le Strat, P., Duvail, C. et al. (2015) The Roussillon Basin (S. France): a case-study to distinguish local and regional events between 6 and 3 Ma. *Marine and Petroleum Geology* **66**, 18–40.

Combourieu-Nebout, N., Fauquette, S. & Quézel, P. (2000) What was the late Pliocene Mediterranean climate like: a preliminary quantification from vegetation. *Bulletin de la Société géologique de France* **171**, 271–277.

Combourieu-Nebout, N., Bertini, A., Russo-Ermolli, E. et al. (2015) Climate changes in the central Mediterranean and Italian vegetation dynamics since the Pliocene. *Review of Palaeobotany and Palynology* **218**, 127–147.

Comes, H.P. & Kadereit, J.W. (2003) Spatial and temporal patterns in the evolution of the flora of the European Alpine system. *Taxon* **52**, 451–462.

David, F. (1995) Vegetation dynamics in the northern French Alps. *Historical Biology* **9**, 269–295.

de Beaulieu, J.-L. & Reille, M. (1983) Paléoenvironnement tardiglaciaire et holocène des lacs de Pelléautier et Siguret (Hautes-Alpes, France). L'histoire de la végétation d'après les analyses polliniques. *Ecologia Mediterranea* **9**, 19–36.

de Beaulieu, J.-L. & Reille, M. (1984) A long Upper Pleistocene pollen record from Les Echets, near Lyon, France. *Boreas* **13**, 111–132.

Diadema, K., Bretagnolle, F., Affre, L. et al. (2005) Geographic structure of molecular variation of *Gentiana ligustica* (Gentianaceae) in the Maritime and Ligurian regional hotspot, inferred from ITS sequences. *Taxon* **54**, 887–894.

Digerfeldt, G., de Beaulieu, J.-L., Guiot, J. & Mouthon, J. (1997) Reconstruction and paleoclimatic interpretation of Holocene lake-level changes in Lac de Saint-Léger, Haute-Provence, southeast France. *Palaeogeography, Palaeoclimatology, Palaeoecology* **136**, 231–258.

Drescher-Schneider, R. (2000) The Riss-Wurm interglacial from West to East in the Alps: an overview of the vegetational succession and climatic development. *Geologie en Mijnbouw* **79**, 233–239.

Fauquette, S. & Bertini, A. (2003) Quantification of the northern Italy Pliocene climate from pollen data – evidence for a very peculiar climate pattern. *Boreas* **32**, 361–369.

Fauquette, S., Guiot, J. & Suc, J.-P. (1998) A method for climatic reconstruction of the Mediterranean Pliocene using pollen data. *Palaeogeography, Palaeoclimatology, Palaeoecology* **144**, 183–201.

Fauquette, S., Clauzon, G., Suc, J.-P. & Zheng, Z. (1999) A new approach for paleoaltitude estimates based on pollen records: example of the Mercantour Massif (southeastern France) at the earliest Pliocene. *Earth and Planetary Science Letters* **170**, 35–47.

Fauquette, S., Suc, J.-P., Jiménez-Moreno, G. et al. (2007) Latitudinal climatic gradients in western European and Mediterranean regions from the Mid-Miocene (~15 Ma) to the Mid-Pliocene (~3.5 Ma) as quantified from pollen data. In: Williams, M., Haywood, A., Gregory, J. & Schmidt, D. (eds.) *Deep Time Perspectives on Climate Change: Marrying the Signal from Computer Models and Biological Proxies.* The Micropaleontological Society, Special Publications. London: The Geological Society of London, pp. 481–502.

Fauquette, S., Bernet, M., Suc, J.-P. et al. (2015) Quantifying the Eocene to Pleistocene topographic evolution of the southwestern Alps, France and Italy. *Earth and Planetary Science Letters* **412**, 220–234.

Grüger, E. (1979) Spätriß, Riß/Würm und Frühwürm am Samerberg in Oberbayern – ein vegetationsgeschichtlicher Beitrag zur Gliederung des Jungpleistozäns. *Geologica Bavarica* **80**, 5–64.

Grüger, E. (1983) Untersuchungen zur Gliederung und Vegetationsgeschichte des Mittelpleistozäns am Samerberg in Oberbayern. *Geologica Bavarica* **84**, 21–40.

Halbritter, H., Weber, M., Zetter, R. et al. (2007) *PalDat – Illustrated Handbook on Pollen Terminology.* Vienna: Springer.

Heusser, L. (1988) Pollen distribution in marine sediments on the continental margin of Northern California. *Marine Geology* **80**, 131–147.

Hou, H.-Y. (1983) Vegetation of China with reference to its geographical distribution. *Annals of Missouri Botanical Garden* **70**, 509–548.

Jiménez-Moreno, G., Rodríguez-Tovar, F.-J., Pardo-Igúzquiza, E. et al. (2005) High-resolution palynological analysis in late early-middle Miocene core from the Pannonian Basin, Hungary: climatic changes, astronomical forcing and eustatic fluctuations in the Central Paratethys. *Palaeogeography, Palaeoclimatology, Palaeoecology* **216**, 73–97.

Jiménez-Moreno, G., Fauquette, S. & Suc, J.-P. (2008) Vegetation, climate and paleoaltitude reconstructions of the Eastern Alps during the Miocene based on pollen records from Austria, Central Europe. *Journal of Biogeography* **35**, 1638–1649.

Jiménez-Moreno, G., Suc, J.-P. & Fauquette, S. (2010) Miocene to Pliocene vegetation reconstruction and climate estimates in the Iberian Peninsula from pollen data. *Review of Palaebotany and Palynology* **162**, 403–415.

Kadereit, J.W., Licht, W. & Uhink, C.H. (2008) Asian relationships of the flora of the European Alps. *Plant Ecology and Diversity* **1**(2), 171–179.

Magny, M., Miramont, C. & Sivan, O. (2002) Assessment of the impact of climate and anthropogenic factors on Holocene Mediterranean vegetation in Europe on the basis of palaeohydrological records. *Palaeogeography, Palaeoclimatology, Palaeoecology* **186**, 47–59.

Mai, D.H. (1995) *Tertiäre Vegetationsgeschichte Europas.* Jena: Gustav Fischer.

Martinetto, E. (2009) Palaeoenvironmental significance of plant macrofossils from the Pianico Formation, Middle Pleistocene of Lombardy, North Italy. *Quaternary International* **204**, 20–30.

Martinetto, E., Monegato, G., Irace, A. et al. (2015) Pliocene and Early Pleistocene carpological records of terrestrial plants from the southern border of the Po Plain (northern Italy). *Review of Palaeobotany and Palynology* **218**, 148–166.

Médail, F. & Diadema, K. (2009) Glacial refugia influence plant diversity patterns in the Mediterranean Basin. *Journal of Biogeography* **36**, 1333–1345.

Médail, F. & Quézel, P. (1997) Hot-Spots analysis for conservation of plant biodiversity in the Mediterranean Basin. *Annals of the Missouri Botanical Garden* **84**, 112–127.

Melinte-Dobrinescu, M.C., Suc, J.-P., Clauzon, G. et al. (2009) The Messinian Salinity Crisis in the Dardanelles region: chronostratigraphic constraints. *Palaeogeography, Palaeoclimatology, Palaeoecology* **278**, 24–39.

Meyer, H.W. (1992) Lapse rates and other variables applied to estimating paleoaltitudes from fossil floras. *Palaeogeography, Palaeoclimatology, Palaeoecology* **99**, 71–99.

Muller, S.D., Nakagawa, T., de Beaulieu, J.-L. et al. (2006) Paléostructures de végétation à la limite supérieure des forêts, dans les Alpes françaises internes. *Comptes Rendus Biologies* **329**, 502–511.

Muller, S.D., Nakagawa, T., de Beaulieu, J.-L. et al. (2007) Postglacial migration of silver fir (*Abies alba* Mill.) in the southwestern Alps. *Journal of Biogeography* **34**, 876–899.

Nakagawa, T., de Beaulieu, J.-L. & Kitagawa, H. (2000) Pollen-derived history of timber exploitation from the Roman period onwards in the Romanche valley, central French Alps. *Vegetation History and Archaeobotany* **9**, 85–89.

Nieto Feliner, G. (2011) Southern European glacial refugia: a tale of tales. *Taxon* **60**, 365–372.

Nix, H. (1982) Environmental determinants of biogeography and evolution in Terra Australis. In: Barker, W.R. & Greenslade, P.J.M. (eds.) *Evolution of the Flora and fauna of Arid Australia*. Frewville: Peacock Publishing, pp. 47–66

Noble, V. & Diadema, K. (2011) *La flore des Alpes-Maritimes et de la Principauté de Monaco. Originalité et diversité.* Turriers: Conservatoire botanique national méditerranéen de Porquerolles & Naturalia Publications.

Noirfalise, A., Dahl, E., Ozenda, P. & Quézel, P. (1987) Carte de la végétation naturelle des états membres des communautés européennes et du conseil de l'Europe. Luxembourg: Office des publications officielles des Communautés européennes.

Ozenda, P. (1985) *La végétation de la chaîne alpine dans l'espace montagnard européen*. Paris: Masson.

Ozenda, P. (1989) Le déplacement vertical des étages de végétation en fonction de la latitude: un modèle simple et ses limites. *Bulletin de la Société Géologique de France* **8**, 535–540.

Ozenda, P. (1995) L'endémisme au niveau de l'ensemble du Système alpin. *Acta Botanica Gallica* **142**, 753–762.

Ozenda, P. (2002) *Perspectives pour une géobiologie des montagnes*. Lausanne: Presses polytechniques et universitaires romandes.

Ozenda, P. (2009) On the genesis of the plant population in the Alps: new or critical aspects. *Comptes Rendus Biologies* **332**, 1092–1103.

Punt, W., Hoen, P.P., Blackmore, S. et al. (2007) Glossary of pollen and spore terminology. *Review of Palaeobotany and Palynology* **143**, 1–81.

Quézel, P. & Médail, F. (2003) *Ecologie et biogéographie des forêts méditerranéennes*. Paris: Elsevier.

Ravazzi, C. (2003) *Gli antichi bacini lacustri e i fossili di Leffe, Ranica e Piànico-Sèllere (Prealpi Lombarde).* Bergamo: CNR Special Publications Quaderni di Geodinamica Alpina e Quaternaria.

Ravazzi, C., Pini, R., Breda, M. et al. (2005) The lacustrine deposits of Fornaci di Ranica (late Early Pleistocene, Italian Pre-Alps): stratigraphy, palaeoenvironment and geological evolution. *Quaternary International* **131**, 35–58.

Rosenbaum, G. & Lister, G.S. (2005) The Western Alps from the Jurassic to Oligocene: spatio-temporal constraints and evolutionary reconstructions. *Earth-Science Reviews* **69**, 281–306.

Schönswetter, P., Stehlik, I., Holderegger, R. & Tribsch, A. (2005) Molecular evidence for glacial refugia of mountain plants in the European Alps. *Molecular Ecology* **14**, 3547–3555.

Suc, J.-P. (1976) Quelques taxons-guides dans l'étude paléoclimatique du Pliocène et du Pléistocène inférieur du Languedoc (France). *Revue de Micropaléontologie* **18**, 246–255.

Suc, J.-P. (1984) Origin and evolution of the Mediterranean vegetation and climate in Europe. *Nature* **307**, 429–432.

Suc, J.-P. (1989) Distribution latitudinale et étagement des associations végétales au Cénozoïque supérieur dans l'aire ouest-méditerranéenne. *Bulletin de la Société géologique de France* **3**, 541–550.

Suc, J.-P. & Bessedik, M. (1981) A methodology for Neogene palynostratigraphy. International Symposium on Concepts and Methods in Paleontology, Barcelona, pp. 205–208.

Suc, J.-P. & Fauquette, S. (2012) The use of pollen floras as a tool to estimate palaeoaltitude of mountains: the Eastern Pyrenees in the Late Neogene, a case study. *Palaeogeography, Palaeoclimatology, Palaeoecology* **321–322**, 41–54.

Suc, J.-P., Bertini, A., Combourieu-Nebout, N. et al. (1995) Structure of West Mediterranean and climate since 5.3 Ma. *Acta Zoologica Cracovia* **38**, 3–16.

Suc, J.-P., Fauquette, S. & Popescu, S.-M. (2004) L'investigation palynologique du Cénozoïque passe par les herbiers. In: Aupic, C., Labat, J. & Pignal, M. (eds.) *Les herbiers: un outil d'avenir. Tradition et modernité.* Nancy: Association française pour la Conservation des Espèces Végétales, pp. 67–87.

Taberlet, P., Zimmermann, N.E., Englisch, T. et al. (2012) Genetic diversity in widespread species is not congruent with species richness in alpine plant communities. *Ecology Letters* **15**, 1439–1448.

Talon, B., Carcaillet, C. & Thinon, M. (1998) Etudes pédoanthracologiques des variations de la limite supérieure des arbres au cours de l'Holocène dans les Alpes françaises. *Géographie physique et Quaternaire* **52**, 195–208.

Walsh, K., Court-Picon, M., de Beaulieu, J.-L. et al. (2014) A historical ecology of the Ecrins (Southern French Alps): archaeology and palaeoecology of the Mesolithic to the Medieval period. *Quaternary International* **353**, 52–73.

Wang, C.W. (1961) *The Forests of China with a Survey of Grassland and Desert Vegetation.* Maria Moors Cabot Foundation 5. Cambridge, MA: Harvard University Press.

Wegmüller, S. (1992) *Vegetationsgeschichtliche und stratigraphische Untersuchungen am Schieferkohlen des nördlichen Alpenvorlandes,* vol. **102**. Basel: Birkhauser.

Welten, M. (1982) Stand der palynologischen Quartörforschung am schweizerischen Nordalpenrand. *Geographica Helvetica* **2**, 75–83.

28

Cenozoic Evolution of Geobiodiversity in the Tibeto-Himalayan Region

Volker Mosbrugger[1], Adrien Favre[2,3], Alexandra N. Muellner-Riehl[2,4], Martin Päckert[5] and Andreas Mulch[6,7]

[1] *Senckenberg Research Institute and Natural History Museum, Frankfurt, Germany*
[2] *Institute of Biology, Leipzig University, Leipzig, Germany*
[3] *Institute for Ecology, Evolution and Diversity, Goethe University, Frankfurt, Germany*
[4] *German Centre for Integrative Biodiversity Research (iDiv), Halle-Jena-Leipzig, Leipzig, Germany*
[5] *Senckenberg Natural History Collections, Museum für Tierkunde, Dresden, Germany*
[6] *Senckenberg Biodiversity and Climate Research Centre (BiK-F), Frankfurt, Germany*
[7] *Institute of Geosciences, Goethe University, Frankfurt, Germany*

Abstract

The Tibeto-Himalayan region, also known as the "Third Pole," is a unique, vast and fascinating area of the Earth. Similar to the North and South Poles, the Tibeto-Himalayan region plays a major role in global climate dynamics, including hemispheric-scale climate connections. In particular, the Tibeto-Himalayan highlands influence the Indian and East Asian monsoons. The region, with its humid and high-relief mountains in the south and south-east, is characterized by a complex biological history and encompasses areas of remarkably high biodiversity. Here, we provide a short summary of the evolution of geo-biodiversity in the Tibeto-Himalayan region since its formation in the Eocene. We use a systemic approach considering both the local abiotic and biotic dynamics, thus integrating the available geological, climatic and biological information. Our focus is thus on the evolution of present-day biodiversity patterns, for which we provide examples across a wide range of organisms. Based on dated molecular phylogenies of extant species, it appears that already during the early uplift phase in the Eocene to mid-Miocene, alpine lineages had emerged in the Tibeto-Himalayan region that later contributed to existing patterns of biodiversity. These data further show three putative phases of high net diversification. The first peak of diversification of modern families and genera occurred during the mid-Miocene (20–15 Ma), the second peak towards the Miocene–Pliocene boundary and the third between 2 Ma and 150 ka. To explain this pattern, we propose the "mountain-geobiodiversity hypothesis." It postulates three boundary conditions that are required to maximize the impact of mountain formation and surface uplift on regional biodiversity patterns in mountainous regions: (i) the presence of lowland, montane and alpine zones; (ii) climatic oscillations that enable mountains to act as species pumps; and (iii) high-relief terrain with environmental gradients that minimize migration distances during climate oscillations, providing refugia and geographical barriers that allow for the survival of many species, as well as for allopatric speciation. We propose that the combination of these three boundary conditions is key for the origination of montane biodiversity hotspots.

Keywords: *Third Pole, diversification, mountain-geobiodiversity hypothesis, montane biodiversity hotspots, biodiversity patterns*

28.1 Introduction: Tropical and Subtropical Mountains and Biodiversity

Maps of global biodiversity distribution (e.g., Mutke & Barthlott 2005; Barthlott et al. 2014) unequivocally show the presence of latitudinal biodiversity gradients: for many taxa, the tropics display the highest diversity, when compared to northern and southern temperate zones. A number of different hypotheses to explain this pattern have been formulated (e.g., Kreft & Jetz 2007; Donoghue 2008). Within the tropical and subtropical zone, mountain ranges appear to be particularly species-rich. Although this pattern is evident from all biodiversity distribution maps, it has received very little attention, and

Mountains, Climate and Biodiversity, First Edition. Edited by Carina Hoorn, Allison Perrigo and Alexandre Antonelli.
© 2018 John Wiley & Sons Ltd. Published 2018 by John Wiley & Sons Ltd.
Companion website: www.wiley.com\go\hoorn\mountains,climateandbiodiversity

explanations are still largely lacking (e.g., Körner 2004; Martinelli 2007; Hoorn et al. 2013). The establishment and persistence of high biodiversity levels in tropical and subtropical mountain systems seem to result from a number of interacting, non-mutually exclusive factors.

28.1.1 Habitat Diversity

High geological diversity fosters high biodiversity (Chapter 11). Mountains are characterized by an elevation-dependent ecological zonation, which typically allows the distinction of lowland, montane, alpine and nival zones. However, the presence of these four basic zones depends not only on surface elevation, but also on latitude and climate. This ecologic–climatic gradient in tropical and subtropical mountains is locally much steeper than the corresponding global latitudinal trend. In combination with a high-relief terrain, and a multitude of lithologies, slopes and expositions, mountains provide a high variety of (micro)habitats, allowing for a correspondingly high local biodiversity (species diversity). It thus appears that because mountains have a particularly high geodiversity (i.e., diversity of abiotic environmental parameters) (Mutke & Barthlott 2005), they also facilitate a particularly high biodiversity, thereby promoting the evolution of more complex biological interactions (e.g., symbioses, pathogens, etc.), which in turn may foster speciation.

28.1.2 Speciation in Mountains

Mountains provide an ideal set-up for speciation. Either they behave like islands, where ecozones of higher altitudinal belts are separated by lowland ecozones (e.g., Sklenář et al. 2014), or the variety of (micro)habitats typical for mountains (see Section 28.1.1) coincides with an abundance of geographical barriers (ridges, valleys, rivers or glaciers), favoring allopatric speciation, or both.

28.1.3 Biome Shifts

In mountainous areas, cooling or warming phases in Earth's history may lead to compression or expansion of biomes, as well as to changes in the altitudinal zonation of mountain ecosystems (Figure 28.1; see Chapter 12). This pattern has several important consequences.

The steeper the topographic gradient, the smaller the spatial distance organisms have to migrate in order to cope with climate change. Consequently, rapid climate changes are expected to pose a smaller challenge to mountain versus lowland organisms (Sandel et al. 2011), at least for those occupying mid elevations (but see Sections 28.3.2 and 28.3.3). In combination with high geological diversity, short distances between favorable habitats increase the chance of successful dispersal and establishment. As such, mountains provide excellent refugia during climate change.

Extreme climate cooling, for example during the Pleistocene transitions from interglacials to glacials, will induce a downward movement of the mountainous ecozones and may even lead to the disappearance or complete change of the previous lowland ecozone (Figure 28.1); alternatively, the nival zone may (re)appear in mountains that had none because of their modest elevations, opening new niches to be colonized locally.

Similarly, in the case of extreme warming, the nival and even the alpine zone may disappear (Figure 28.1), and new lowland ecozones may develop if the mountains existed in a very cold climate before the warming (e.g., the European lowland vegetation during the interglacials).

Surface uplift or extensional collapse and denudation of mountains may alter the topographic structure and affect the distribution of elevational and temperature zones.

Under conditions of climatic oscillations at millennial time scales or longer, mountains may act as "species pumps," producing repeated pulses of speciation followed by species expansions (Chapter 12) (Haffer 2008;

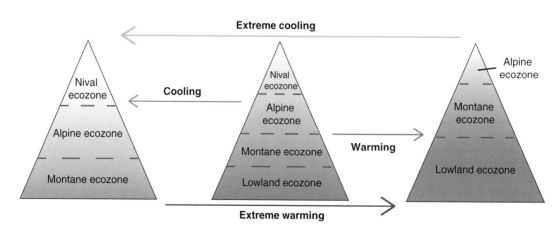

Figure 28.1 Schematic illustration of montane ecozone changes due to climatic warming or cooling.

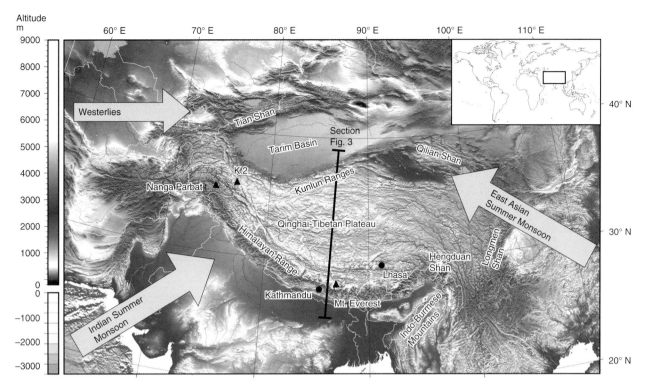

Figure 28.2 Map of the Tibeto-Himalayan region (THR), illustrating three major atmospheric circulation patterns: westerlies, the Indian summer monsoon and the East Asian summer monsoon. The black line marks the position of the profile in Figure 28.3. Triangles indicate the positions of prominent mountain peaks, circles indicate cities.

Adams 2009). Variations in connectivity through time may thus cyclically modify the intensity of allopatric speciation.

28.1.4 Understanding Origins of Montane Biodiversity

Therefore, depending on the dominant processes, mountainous regions containing high levels of biodiversity may represent one of the following conditions (cf. Bellwood et al. 2012): they may be (i) centers of origin, where many taxa have evolved and survived; (ii) centers of overlap, where the distribution areas of many taxa overlap; (iii) centers of accumulation, to which many taxa have migrated; (iv) centers of survival, where many taxa have survived a major diversity loss in comparison to neighboring areas; or (v) a combination of these center types.

Thus, it is easily understandable that mountains – and tropical to subtropical mountains, in particular – are especially species-rich. Correspondingly, various authors have associated mountain building directly with the formation of biodiversity (cf. Fjeldså et al. 2012; Hoorn et al. 2013; Favre et al. 2015). Here, we take a closer look at this linkage, focusing on the Tibeto-Himalayan region (THR), including the Qinghai–Tibetan Plateau, the Himalaya and the biodiversity hotspot known as the "Mountains

of Southwest China" (incl. the Hengduan Shan) (Figure 28.2). We consider a time frame corresponding to the most prominent Cenozoic surface uplift, linking the evolution of present-day biodiversity to changes in climate and the formation and uplift of the Qinghai-Tibetan Plateau, the Himalaya, and the Hengduan Shan.

28.2 Evolution of Geodiversity in the Tibeto-Himalayan Region

28.2.1 Geography

The Qinghai–Tibetan Plateau, as part of the THR, is the world's highest and largest plateau, with a mean elevation exceeding 4500 m and a surface area of 2.3 million km^2. Surface elevation ranges from 1500 m in the north and 500 m at the base of the Himalaya to 8848 m at Mt. Everest/Qomolangma. With an average elevation of 4500 m, the Qinghai–Tibetan Plateau has a crucial role in the climate system, as it exerts a substantial influence on atmospheric circulation. The interaction between the wind systems caused by the seasonal shift of the Intertropical Convergence Zone (ITCZ) and the warm air masses ascending from the Qinghai–Tibetan Plateau during summer months, both in combination with the

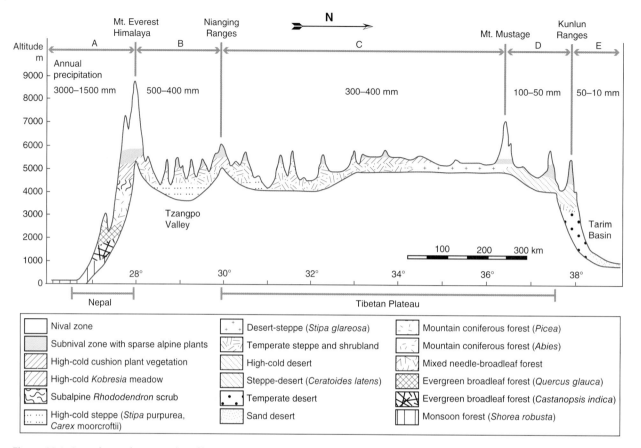

Figure 28.3 A north–south-oriented profile across the Tibetan Plateau along longitude 85° E (see location in Figure 28.2). The zones are (a) the tropical monsoon forest zone; (b) the Tsangpo Valley xeric shrubland-steppe plateau zone; (c) the Chiangtang high-cold steppe plateau zone; (d) the northern Chiangtang high cold desert plateau zone; and (e) the Tarim temperate desert region. *Source:* Chang (1981). Reproduced with permission of International Mountain Society.

Coriolis effect and moisture enrichment, lead to intense summer monsoon rainfall reaching the Himalaya from the west and south-west (Figure 28.2). During the winter months, the system reverses to (north-)easterly directions and rainfall reaches the oceans.

Because the Himalayan mountain chain acts as a topographic barrier, annual precipitation is unevenly distributed, ranging from 3000 mm in Darjeeling (Indian Himalaya) to less than ca. 100–300 mm within the plateau center. The surrounding areas of the Qinghai–Tibetan Plateau therefore generally offer more attractive habitats for organisms than the colder and drier plateau interior, and encompass parts of three different biodiversity hotspots (see Section 28.3.1) (Marchese 2015). Due to the large elevation gradients, high elevation and resulting variations in precipitation and temperature across the entire THR, several biomes co-occur within this region (e.g., alpine tundra, montane forests, subtropical dry forests and rainforests) (Figure 28.3) (cf. Chang 1981). Today, the upper tree line in the Qinghai–Tibetan Plateau area is found between ca. 3700 and 4000 m, with rare and isolated forest patches

(*Juniperus*) occurring as high as 4900 m in south-eastern Tibet (Miehe et al. 2007).

Animal species dependent on forested habitats are generally absent from the highest parts of the Qinghai–Tibetan Plateau region, including considerable amounts of the actual plateau interior. Only a few vertebrate and vascular plant species are able to survive under the unfavorable conditions of the high plateau, characterized by extremely low precipitation and short vegetation periods, with soils frozen for a considerable amount of the year. Parts of the Qinghai–Tibetan Plateau harbor permafrost, representing a total estimated area of between ca. 1.3 and 1.6 million km^2, making it the largest expanse of high-elevation permafrost in the world (Jin et al. 2007). The existing permafrost on the Qinghai–Tibetan Plateau was formed during the last two major glaciations, and survived the intervening warmer periods. In contrast, the southern and south-eastern fringes of the Qinghai–Tibetan Plateau (i.e., the Himalaya and the Hengduan Shan region in south-western China) profit from more favorable temperature and precipitation regimes across both dry and rainy seasons, with less seasonality and ample water supply for vascular plant growth, providing a variety of habitats for vertebrates and other animal groups.

28.2.2 Climate, Monsoon Evolution and Central Asian Aridification

Together with the northern hemisphere westerlies, the Indian and East Asian monsoons represent the key controls on Central Asian precipitation patterns. Increased insolation within the continental interior results in enhanced land–sea atmospheric pressure gradients that force convergence of oceanic air masses over East and Southeast Asia and ultimately produce summer monsoonal rainfall. This situation is reversed when the cooling of continents enhances continental aridity with seaward moisture transport (winter monsoon). This monsoonal circulation is primarily modulated by global climate variations responding to $p\mathrm{CO_2}$ fluctuations and astronomical forcing (e.g., Cheng et al. 2016). The extended high topography over the THR enhances the land–sea thermal contrast sufficiently to seasonally deflect the ITCZ far north of the equator (Broccoli & Manabe 1987). Today, the Qinghai–Tibetan Plateau and adjacent high mountain ranges create barriers to west–east subtropical tropospheric flow of air masses, favor orographic rainfall and ultimately enhance summer monsoonal conditions. These perturbations to atmospheric circulation affect not only regional climate patterns, but also possibly the global distribution of heat and moisture (see reviews in Rodwell & Hoskins 1996; Molnar et al. 2010; Huber & Goldner 2012; Lippert et al. 2014).

The Qinghai–Tibetan Plateau and the Himalaya also prevent cool, dry, subtropical air from Central Asia from mixing with warm, humid, tropical air from the Indian Ocean, and thus may simultaneously accentuate the Indian monsoon and the subtropical aridity that characterizes the climate of western China and surrounding regions (Boos & Kuang 2010). Understanding the history of topography and rainfall is therefore key when relating geodynamic, climatic and evolutionary processes over geological time. In this context, it seems reasonable to assume that once the Qinghai–Tibetan Plateau and the Himalayan ranges had reached a critical size and elevation, they increasingly acted as an orographic barrier to atmospheric circulation in Asia, and consequently may have contributed to gradual or punctuated precipitation and temperature changes in the region. These could have included differences in overall moisture transport to the plateau interior, and to Central Asia in general, or changes in rainfall seasonality.

The age and timing of the onset of the Asian monsoons, their intensity through time and their relationship with global climate variations, as well as with Himalayan and Tibetan surface uplift, is still poorly constrained for large parts of their Cenozoic history. Evidence from fossil flora (Molnar et al. 2010; Quan et al. 2012; Shukla et al. 2014), windblown dust provenance and oxygen isotope-based paleoclimate records, as well as climate simulations (Huber & Goldner 2012; Licht et al. 2014, 2016), suggest that a pronounced monsoonal regime (distinct wet/dry seasonality) most probably existed in the Eocene at the time of the India–Asia collision. These studies show that the monsoonal circulations form a resilient system that has existed at least since the Eocene and has been modulated by a combination of various factors, including THR surface uplift (and possibly the retreat of the Tarim Sea) (Bosboom et al. 2014), changes in global atmospheric $p\mathrm{CO_2}$ (Licht et al. 2014) and global climate shifts including the Eocene–Oligocene transition (Dupont-Nivet et al. 2007) and general global cooling (Jiang & Ding 2008).

Aeolian loess-like deposits, which record arid winter monsoonal storms north of the Tibetan Plateau, have been used as a proxy for monsoonal intensity through time, enhanced by orographic effects (Guo et al. 2002; Licht et al. 2016). These loess-like deposits suggest that aridity increased through the Eocene–Oligocene transition in relation to global cooling and associated retreat of the Tarim Sea (Dupont-Nivet et al. 2007; Bosboom et al. 2014). Through Miocene to Pliocene times, these loess deposits, together with other proxies, indicate a continued increase in aridity that has been related to global cooling, as well as to regional surface uplift in Central Asia, enhancement of monsoonal circulation and shielding of monsoonal and westerly moisture (Guo et al. 2002; Caves et al. 2015, 2016). Miocene–Pliocene intensification of monsoonal strength has also been reported from various seasonal rainfall proxy records in the Himalayan foreland (e.g., Quade et al. 1995, 2011; Dettman et al. 2001; Wan et al. 2007). These records of monsoonal intensification have been traditionally associated with enhanced orography (e.g., Clift et al. 2008), although the role of Mio–Pliocene global cooling is being recognized as equally – if not, more – important (e.g., Gupta et al. 2003).

Given these multiple interactions, the evolution of the Asian monsoons through geological time and their relationship to either global climate or Himalayan and Tibetan surface uplift is still a matter of debate. Global climate variations can be very precisely constrained through time and compared to (admittedly much coarser) surface uplift reconstructions of the THR. With increasing temporal resolution, it will be possible to assess the relative contribution of global climate to observed paleoenvironmental variations and evolutionary patterns. In contrast, the geological history controlling regional surface uplift is still rather poorly constrained and remains controversial. Some general patterns, however, can be discerned from data sets, which we will briefly review later in the chapter (see Sections 28.3 and 28.4) in order to estimate the potential relationships between the formation of orographic barriers, atmospheric circulation in Asia and – ultimately – biodiversity in the region in response to gradual or punctuated changes in topography, precipitation and temperature.

28.2.3 Surface Uplift History

Reconstructing the surface uplift history of the THR has received increased attention over the past decade, mainly due to the development of new paleoaltimetry techniques. Even though we are far from establishing detailed regional surface elevation histories, there is increasing evidence that the significant topographic features of the plateau developed at least as early as the timing of the India–Eurasia collision (ca. 55–40 Ma, Eocene) (Yin & Harrison 2000; Tapponnier et al. 2001; Rowley & Currie 2006; Deng & Ding 2016) and high Himalayan peaks developed no later than the early to mid-Miocene (Gébelin et al. 2013).

The exact timing of the India–Asia collision, its along-strike variation and its relationship to surface uplift is a focus of ongoing research, and simple convergence–uplift relationships have been challenged in the past (Tapponnier et al. 2001; Rowley & Currie 2006). In fact, a proto-Tibetan Plateau had probably already developed before the collision in Cretaceous–Paleogene times (Murphy et al. 1997), which characterized the physiography of the southern margin of Asia from at least 45 Ma (Eocene) (Rowley & Currie 2006). As with the Andean Altiplano, significant topography likely developed at subtropical latitudes above an oceanic subduction zone prior to continent–continent collision, with crustal thickening already promoting the rise of surface elevation above sea level since ca. 110 Ma (Cretaceous) (Lippert et al. 2014). Nevertheless, the collision onset at ca. 55–50 Ma, corresponding to a nearly twofold decrease in the relative motion rates between the Indian and Asian plates, is well supported by both geological observations (e.g., sediment provenance, biostratigraphy) (Najman et al. 2010) and paleomagnetic/geophysical data (Lippert et al. 2014). The minimum age estimate for the latest paleomagnetically determined collision is 49.4 ± 4.5 Ma, and available evidence points towards a scenario in which the collision between continental units of Indian affinity with the southern margin of Asia was ongoing by 52 Ma at ~$21 \pm 4°$N (Lippert et al. 2014).

One view holds that following collision, surface uplift was likely to have been diachronous, progressing from south to north (e.g., Tapponnier et al. 2001; Mulch & Chamberlain 2006; Rowley & Currie 2006), and the center of the Qinghai–Tibetan Plateau may have reached an average elevation of ~4000 m no later than ca. 35 Ma (late Eocene) (Rowley & Currie 2006), whereas more northern localities might have been only half as high at

that time (Cyr et al. 2005; see also Deng & Ding 2016). Nevertheless, thermochronologic results indicate fast exhumation in northern and central Tibet as early as 55 Ma, suggesting high elevations (e.g., Wang et al. 2008; Clark et al. 2010). This is supported by palynological assemblages from the Xining Basin (north-eastern Qinghai–Tibetan Plateau), showing the presence of high-elevation vegetation suggesting the occurrence of regional surface uplift as early as 38 Ma (late Eocene) (Dupont-Nivet et al. 2008). In summary, current stable isotope paleoaltimetry data consistently indicate that high-elevation landscapes of at least 4000 m were already present during the Eocene at 55–35 Ma (Rowley & Currie 2006; Ding et al. 2014; Deng & Ding 2016).

Starting in early Miocene through Pliocene times (ca. 25–5 Ma), a generalized expansion of the Qinghai–Tibetan Plateau is suggested by thermochronology-constrained fast exhumation, likely accompanied by continued surface uplift of high mountain ranges, such as the higher Himalaya in the south, the Pamir and Tian Shan in the north and west, the Qilian Shan and Nan Shan in the north-east and the Longmen Shan in the east (e.g., Li & Fang 1999; Zheng et al. 2000; Tapponnier et al. 2001; Mulch & Chamberlain 2006; Clift et al. 2008). This is supported by stable isotope paleoaltimetry data from various areas, in particular the central Himalaya, reaching an elevation comparable to present no later than the middle Miocene (ca. 19–17 Ma) (Gébelin et al. 2013). As part of the continued expansion of high-elevation regions during the Miocene and into the Pliocene, the "Mountains of Southwest China" biodiversity hotspot (Marchese 2015) is of particular interest for the present study. Paleobotanical and paleoclimatic data (Sun et al. 2011) suggest that the Hengduan Shan may have an even younger final phase of uplift, probably dating back only to the Pliocene, and reaching peak elevation before the late Pliocene.

The THR constitutes one of the fastest-growing mountain systems in the world, and we still lack a comprehensive understanding of the long-term climate–topography feedback. Future studies will have to account for the multiple links that exist among crustal and mantle convergence and climate–tectonics interactions, including rapid, localized exhumation, as well as regional surface uplift. The relationship with evolutionary biological processes further requires a better understanding of Miocene–recent landscape changes.

28.3 Evolution of Present-Day Biodiversity in the THR

28.3.1 Spatial Patterns of Biodiversity

When Myers et al. (2000) united the Himalaya with the Indo-Burma hotspot, they ranked this region among the

Table 28.1 Species richness (number of species) of plants, mammals and birds in forest and high-alpine ecosystems along the southern fringe and the alpine central region of the Qinghai–Tibetan Plateau. Species numbers have ranges where more than one ecoregion is considered. The ecoregions included here are (with ecoregion numbers according to Wikramanayake at el. (2002) following in parentheses): Himalayan subtropical broadleaf and pine forests (ecoregions 25 and 31), Eastern and Western Himalayan broadleaf forests (ecoregions 26 and 27), Eastern and Western Himalayan sub-alpine conifer forests (ecoregions 28 and 29), Northwestern, Western and Eastern Himalayan alpine shrub and meadows (ecoregions 37, 38 and 39) and Central Tibetan Plateau alpine steppe (ecoregion 41). N.A., not applicable, as Wikramanayake at el. (2002) did not define a separate ecoregion for central Himalayan forests.

Himalayan subtropical broadleaf and pine forests	Plant species 1000–1500			Mammal species 97–121			Bird species 343–480		
	West	Central	East	West	Central	East	West	Central	East
Temperate and sub-alpine forests (Himalayan broadleaf forests and sub-alpine conifer forests)	1000–1200	N.A.	1500–2000	58–76	N.A.	89–128	285–315	N.A.	202–490
Alpine shrub and meadows and Central Tibetan Plateau alpine steppe	1500–2000	1000	7000	36–50	34	63	127–139	122	115

Source: Adapted from Wikramanayake et al. (2002).

"eight hottest hotspots" (out of 25) worldwide in terms of plant and vertebrate endemism. Today, the Himalaya and the adjacent "Mountains of Southwest China" represent two (out of 36) separate hotspots (Marchese 2015), each of which harbors an impressive faunal and floral diversity. Furthermore, the alpine shrub and meadow landscapes in the Eastern Himalaya and the Hengduan Shan host one of the richest subnival floral communities worldwide (Wikramanayake et al. 2002; Luo et al. 2016). The Qinghai–Tibetan Plateau, including its forest margins, is thus counted among the most species-rich regions of the northern hemisphere.

The high positive effect of precipitation and temperature and the negative effect of temperature range are the strongest predictors of vertebrate diversity in the THR (Antonelli et al., unpublished data). Cooling age, long and short erosion, relief and number of soil types additionally show a positive correlation with vertebrate species richness. In summary, terrestrial vertebrate diversity is highest at the southern flanks and south-east of the Qinghai–Tibetan Plateau (i.e., in the Himalaya and the Hengduan Shan) and is lowest on the Qinghai–Tibetan Plateau itself (Table 28.1) (Aliabadian et al. 2008; Fjeldså et al. 2012), which may be explained by the unfavorable climate and the predominantly semi-arid conditions at high elevations. The THR harbors more than 15 000 vascular plant species (Mittermeier et al. 2004), with the highest diversity throughout the Sino-Himalayan mountain forests (4000–5000 species per km^2); this diversity decreases with increasing elevation and latitude towards the center of the Qinghai–Tibetan Plateau (Wen et al. 2014; Zhang et al. 2016).

There is a clear trend of increasing biodiversity from west to east across the Sino-Himalayan mountain region (Wikramanayake et al. 2002; Yan et al. 2013), which correlates well with the observed trend of increasing precipitation towards the east. Compared to the drier Western Himalaya, the Eastern Himalaya harbors about twice as many passerine bird species in the temperate and sub-alpine forests (Price et al. 2011, 2014) and over three times more alpine shrub and meadow plant species (Table 28.1) (Wikramanayake et al. 2002). Therefore, the eastern fringe of the Qinghai–Tibetan Plateau is claimed to constitute the "evolutionary front" of China, with a high proportion of endemic seed plants, probably resulting from the uninterrupted uplift of the region since the late Neogene (López-Pujol et al. 2011). Increasing aridity of the Qinghai–Tibetan Plateau during the Miocene led to drastic shifts in the distribution of plant communities (Sun & Wang 2005; Zhang & Sun 2011; Miao et al. 2012) and major faunal turnovers (Meng & McKenna 1998; Barry et al. 2002; Badgley et al. 2008). One driving factor that probably shaped the diversity gradient in birds is arthropod abundance as a limiting resource of diversity in the north-west (Ghosh-Harihar & Price 2014; Price et al. 2014). The Himalaya also deviates from the expected relationship between mammal species richness and phylogenetic diversity, which suggests that this region has unique patterns of community diversity and assembly that are different from those in the surrounding regions (Tamma & Ramakrishnan 2015).

In general, there seems to be limited evidence of a mid-domain effect of species richness (peaking at middle elevations) and of the elevational extension of Rapoport's rule (an increase in the altitudinal range of species with increasing elevation; see Stevens 1992) in plants (Bhattarai & Vetaas 2006; Grau et al. 2007; Tang et al. 2014) and montane vertebrates (McCain & Bracy Knight

2013). In Himalayan woody plants, only temperate species richness follows the mid-domain expectation (Oomen & Shanker 2005), and in birds there is mixed support for it (Acharya et al. 2011). Among reptiles, lizards show a linear rate of diversity decline along the gradient, whereas snakes show a nonlinear pattern with a peak between 500 and 1000 m (Chettri et al. 2010).

Biogeographical scenarios inferred from dated phylogenies and fossil evidence hypothesize that the southeastern flanks of the Qinghai–Tibetan Plateau were a "cradle of evolution"; that is, a center of diversification where new species have continuously evolved. A Tibetan/ Himalayan cradle of evolution has been hypothesized for plants (Zhang et al. 2014b; Ren et al. 2015), birds (Päckert et al. 2015a), mammals (Deng et al. 2011; Wang et al. 2014; Pisano et al. 2015) and ground beetles (Schmidt et al. 2012). In contrast, certain regions of the Qinghai– Tibetan Plateau have often been considered a "museum of evolution," where relict lineages survived over long periods of time and contributed to a regional fauna and flora with a high percentage of very ancient species. Recent studies have suggested that evolutionary histories of diversification may exhibit characteristics of both the cradle and the museum models (McKenna & Farell 2006; López-Pujol et al. 2011; Moreau & Bell 2013; Tamma & Ramakrishnan 2015). However, recent phylogenetic studies have somewhat reanimated the "Out of Tibet" hypothesis that was advocated by Weigold (2005) for some Palearctic bird genera. For plants, Wen et al. (2014) identified the Qinghai–Tibetan Plateau as both a source (center of origin) and a sink (center of accumulation) for different organisms with particular biogeographical relationships to other Palearctic regions. For example, the Qinghai–Tibetan Plateau acted as a source area for the worldwide diversity in *Gentiana* (Favre et al. 2016), but *Saxifraga* dispersed to the Qinghai–Tibetan Plateau from North America and then diversified in the THR (Ebersbach et al. 2016). Also, in birds, the Himalaya apparently served as a source area for montane specialists that subsequently colonized other Palearctic regions (Drovetski et al. 2013; Tietze et al. 2013; Voelker et al. 2015).

28.3.2 Temporal Patterns of Species Diversification

Renner (2016) criticized the fact that the vast majority of phylogenetic papers linked a postulated recent Qinghai– Tibetan Plateau uplift to young (post-Miocene) node ages despite firm evidence from geological, paleoclimatic and fossil data that the major Qinghai–Tibetan Plateau uplifting processes had already occurred during the early Oligocene. In fact, the persistent paradigm of a "young Tibet," reinforced by the strong focus of evolutionary biologists on Quaternary processes shaping biodiversity patterns on the Qinghai–Tibetan Plateau, greatly neglects the earlier phases of evolution, particularly in relation to the early surface uplift during Eocene to Miocene times.

28.3.2.1 Pre-Miocene Events

Although most diversification events derived from molecular estimates of modern taxa in the THR date back to times when at least the core plateau region was already high, there are rare but striking examples of the pre-Miocene emergence of lineages that gave rise to near-exclusive Tibetan families or genera. Martens (2015) even discussed a few rare "pre-Himalayan Tethys elements," including representatives of the Tibeto-Himalayan millipede, dragonfly and beetle fauna, whose evolutionary origin is assumed to date back to when the Tethys Sea started to close. As one of the oldest Tibetan vertebrate faunal elements, Asian hynobiid salamanders were suggested to have diversified in the Paleocene (ca. 62 Ma), and phylogenetic analyses of this group allowed for discussions on the geography and climate during the early phases of Qinghai–Tibetan Plateau uplift (Zhang et al. 2006). Further Eocene and Oligocene splits of Tibetan lineages from their closest relatives have been identified in plants (Ebersbach et al. 2016; Favre et al. 2016), mammals (Lanier & Olson 2009; Pisano et al. 2015) and birds (Baker et al. 2007; Päckert et al. 2016) (Table 28.2). It is unclear whether monotypic families among these ancient Tibetan faunal elements did not diversify to a notable degree or if they represent relict lineages from formerly diverse clades (Päckert et al. 2015a). One well-documented example of climate-triggered extinction can be found in a formerly speciose mammal group, the Tibetan pikas (Ochotonidae).

28.3.2.2 Miocene Diversification

The onset of a burst of terrestrial faunal and floral diversification in Asia occurred from the early to mid-Miocene (Table 28.2). This included plants (Zhang et al. 2014a), passerine birds (Päckert et al. 2012) and mammals; for example, pikas (Lanier & Olson 2009) and pantherine cats (Tseng et al. 2014). Similarly, the fauna of freshwater ecosystems on the Qinghai–Tibetan Plateau experienced early lineage diversification in high-elevation lakes and among the main drainage systems of the core plateau region (Table 28.2) (Che et al. 2010; Wang et al. 2013). In passerine birds, this phase of diversification coincided with a major phylogenetic lineage split between exclusively alpine species and species associated with shrubs in the Qinghai–Tibetan Plateau region (Drovetski et al. 2013; Tietze et al. 2013). As a consequence, diversification of body size and foraging strategies of passerines had already evolved before niche filling caused a slowdown of diversification towards the present – particularly of morphological evolution

Table 28.2 Divergence time estimates for key events during the evolutionary history of terrestrial vertebrates and plants in the Tibeto-Himalayan region (THR).

Period and epoch	Age (Ma)	Group	Taxon	Key evolutionary event(s)	References
Palaeogene					
Paleocene	~62	Amphibians	Salamanders, Hynobiidae	Early emergence and diversification of Qinghai–Tibetan Plateau endemic alpine lineages	Zhang et al. (2006)
Eocene	~40	Mammals	Rodents, Dipodoidea	Early diversification of Qinghai–Tibetan Plateau lineages	Pisano et al. (2015)
	~44–50	Birds	Charadriiformes	Early diversification from a Qinghai–Tibetan Plateau center of origin	Baker et al. (2007)
	~37	Mammals	Pikas, Ochotonidae	Mid-Eocene emergence of Qinghai–Tibetan Plateau endemic ibisbill (*Ibidorhyncha*)	Lanier & Olsson (2009)
Oligocene	~45–25	Plants	*Gentiana*, Gentianaceae	Split from other Lagomorphs	Favre et al. (2016)
	~35–25	Plants	*Saxifraga*, Saxifragaceae	Split from other related genera, followed by a slow, but continuous increase of diversification rates	Ebersbach et al. (2016)
	~26	Birds	Passerines, Urocynchramidae	Colonisation of the Qinghai–Tibetan Plateau by *S.* section *Ciliatae*	Päckert et al. (2016)
				Split from weavers and waxbills	
Miocene				Bursts of diversification from Asian centers of origin; increased emergence of alpine lineages, establishment of endemism on the Qinghai–Tibetan Plateau and along its margins, ecological segregation among high alpine and forest fauna	
	25–20	Plants	*Gentiana*, Gentianaceae	Single dispersal "Out of Tibet" (to Europe)	Favre et al. (2015)
	25–20	Amphibians	Frogs, Paini	Diversification of four circum-Qinghai–Tibetan Plateau clades, emergence of core Qinghai–Tibetan Plateau *Nanorana*	Che et al. (2010)
	23.5	Fishes	Tibetan loaches, *Triplophysa*	Basal divergence and successive adaptive radiation of 33 extant species of high-elevation lakes	Wang et al. (2016)
	~20–10	Birds, fishes, plants	Taxon-wide	Burst of faunal and floral diversity from Asian centers of origin	Päckert et al. (2012, 2015a), Wang et al. (2013), Zhang et al. (2014a)
	~20	Plants	Tamaricaceae	Himalayan origin and successive circum Qinghai–Tibetan Plateau/West Palearctic radiation of *Myricaria*	Zhang et al. (2014b)
	~16	Mammals	Pantherine cats, *Panthera*	Split of modern pantherine cats from the clouded leopard (*Neofelis nebulosa*)	Tseng et al. (2014)
	20–14	Plants	*Isodon*, Lamiaceae	Single dispersal "Out of Tibet" (to Africa)	Yu et al. (2014)
	13.7	Fishes	Cyprinidae, *Percocypris*	Basal diversification among lineages from the major drainage systems of the south-eastern Qinghai–Tibetan Plateau margin	Wang et al. (2013)
	12–11	Mammals	Gazelles, *Procapra*	Basal diversification of Antilopini with Qinghai–Tibetan Plateau endemic *Procapra* as earliest offshoot	Hassanin et al. (2012)

(*Continued*)

Table 28.2 (Continued)

Period and epoch	Age (Ma)	Group	Taxon	Key evolutionary event(s)	References
	~12	Mammals	Pikas, *Ochotona*	Diversification into two Qinghai–Tibetan Plateau clades and one northern clade	Lanier & Olsson (2009)
	~10	Plants	*Rheum*, Polygonaceae	First burst of diversification	Sun et al. (2012)
	10–7	Birds	Passeriformes	Split of alpine Qinghai–Tibetan Plateau endemic species	Päckert et al. (2015a,b), Qu et al. (2013), Tietze et al. (2013)
	9.7	Plants	Delphinieae, Ranunculaceae	Burst of diversification in Asian *Delphinium*	Jabbour & Renner (2012)
	7.8–5.3			Burst of diversification in Asian *Aconitum* and *Lycoctotum*	
	7.1	Birds	Passeriformes	Mean split age among Himalayan sister species (forest assemblages)	Price et al. (2014)
	9.5–7.0	Mammals	Grazing herbivores	Faunal turnover at the southern Qinghai–Tibetan Plateau margins	Barry et al. (2002), Badgley et al. (2008)
	~9	Plants	*Rhodiola*, Crassulaceae	Burst of diversification in Qinghai–Tibetan Plateau *Rhodiola*	Zhang et al. (2014a)
	~8	Plants	*Saxifraga* sect. *Ciliatae*, Saxifragaceae	Burst of diversification in Qinghai–Tibetan Plateau *S.* sect. *Ciliatae*	Ebersbach et al. (2016)
	9–7	Mammals	Pantherine cats, *Panthera*	Origin and diversification of high-alpine Qinghai–Tibetan Plateau endemics (snow leopard, *P. uncia*)	Tseng et al. (2014)
	10–6	Plants	*Isodon*, Lamiaceae	Bursts of diversification in Asian *Isodon*	Yu et al. (2014)
Pliocene				Elevational niche diversification in Qinghai–Tibetan Plateau marginal forests, onset of alpine in situ speciation processes on the high-alpine plateau, "Out of Tibet" radiations of many plant genera	
	~7–1	Plants	*Spiraea*, Rosaceae	Bursts of diversification in Qinghai–Tibetan Plateau *Spiraea*	Khan et al. (2016)
	~6–1	Plants	*Gentiana*, Gentianaceae	Massive dispersal "Out of Tibet" (to subcosmopolitan distribution)	Favre et al. (2016)
	~6–1	Plants	*Isodon*, Lamiaceae	Multiple dispersal "Out of Tibet" (to neighboring Asia)	Yu et al. (2014)
	5.3	Fishes	Glyptosternoid catfish	Basal divergence among lineages from different drainage systems, followed by morphological and physiological adaptation to high-elevation environments	Ma et al. (2015)
	~5.0–2.5	Birds	Passeriformes	Faunal interchange between Qinghai–Tibetan Plateau, Central Asia and North Palearctic	Päckert et al. (2012)
	~5.0–2.5	Birds	Passeriformes	Diversification along the Himalayan elevational gradient	Price et al. (2014)
	~5–3	Plants	*Rhodiola*, Crassulaceae	Dispersal "Out of Tibet"	Zhang et al. (2014a)
	3.8–3.1	Birds	Passeriformes, snow sparrows, rosefinches	Parallel in situ speciation into six and three endemic species	Tietze et al. (2013), Lei et al. (2014)
	5.0–3.7	Reptiles	Lizards, *Phrynocephalus vlangalii, P. erythrurus*	Circum-Qinghai–Tibetan Plateau diversification of seven and two highland desert lineages	Jin et al. (2008), Jin & Liu (2010)

Age (Ma)	Group	Taxon	Pattern/process	Reference
5.0–3.6	Mammals	Tibetan Arctic fox, *Vulpes qiuzhudingi*	Himalayan origin and dispersal "Out of Tibet" of the Arctic fox	Wang et al. (2014)
4.6	Mammals	*Hipparion* sp.	Three-toed horses occupying high-alpine niches of the Zanda Basin	Deng et al. (2011, 2012)
3.7		*Coelodonta thibetana*	Split of Tibetan *C. thibetana* from other wooly rhinos	Sun et al. (2012)
4.2–3.6	Plants	*Rheum*, Polygonaceae	Second burst of diversification	Favre et al. (2016)
~4	Plants	*Gentiana* section *Cruciata*	Simultaneous bursts of diversification in two Asian (incl. Qinghai–Tibetan Plateau) Gentianinae lineages following niche shifts (from alpine to sub-alpine)	Matuszak et al. (2016)
~5–4		*Tripterospermum*, Gentianaceae		
3.6–3.7	Fishes	Cyprinidae, schizothoracine fishes	Basal divergence among lineages from different core Qinghai–Tibetan Plateau drainage systems	He & Feng (2007)
~2–present	Plants	*Saxifraga*, Saxifragaceae	Multiple dispersal "Out of Tibet"	Ebersbach et al. (2016)
Pleistocene			Intraspecific diversification in glacial refuges, formation of extant circum-Qinghai–Tibetan Plateau distribution patterns: vicariance, east–west disjunctions in the Himalaya and north–south disjunctions along the eastern Qinghai–Tibetan Plateau margin (Hengduan Shan and adjacent central and northern Chinese mountain systems)	
~2.5–present	Birds	Leaf warblers, Phylloscopidae	East–west disjunctions in the Himalaya	Päckert et al. (2012, 2015a)
0.16–0.36	Birds	Snow sparrows, ground tit, twite, redstarts	Intraspecific diversification on the alpine core Qinghai–Tibetan Plateau	Lei et al. (2014)
2.2–2.8	Birds	Old World buntings, *Emberiza*	North–south disjunctions along the eastern Qinghai–Tibetan Plateau margin	Päckert et al. (2015b)
0.04–0.4	Birds	Pheasants, Himalayan snowcock	Intraspecific diversification on the alpine core Qinghai–Tibetan Plateau	An et al. (2015)
3.7–1.4	Amphibians	Frogs, *Nanorana parkeri*	Intraspecific east–west disjunction	Liu et al. (2015)
1.2	Mammals	Rodents, *Eospalax baileyi*	Divergence of plateau-edge and interior plateau populations	Tang et al. (2010)
2.1–2.2	Mammals	Rodents, *Apodemus draco*	Early divergence among populations from deep river gorges at the south-east Qinghai–Tibetan Plateau margin and Mekong-Salween divide	Fan et al. (2012)
0.8–1.5		*A. ilex*		Liu et al. (2012)
0.4–0.8				Liu et al. (2012)
0.3–0.5	Mammals	Insectivores, mole shrew, *Anourosorex squamipes*	North–south disjunctions along the eastern Qinghai–Tibetan Plateau margin	He et al. (2016)
2–present	Plants	*Anisodus tanguticus*, Solanaceae	Intraspecific diversification in the Himalaya and Hengduan Shan	Wan et al. (2015)
~2.5–1.5	Mammals	Pikas, *Ochotona*	Circum-Qinghai–Tibetan Plateau speciation (sister species and intraspecific lineages; north–south split in *O. pallasi*)	Lanier & Olsson (2009)

(e.g., rates of body-size diversification, as shown for Himalayan assemblages: Kennedy et al. 2012; Price et al. 2014). In the course of adaptive evolution, high-alpine vertebrate species on the core Qinghai–Tibetan Plateau acquired anatomical adaptations to ground-dwelling life (birds) and physiological adaptations to extreme environments (e.g., oxidative stress at high elevations: birds and lizards) (Jin et al. 2008; Qu et al. 2013; Zhang et al. 2015).

Modern alpine plant genera started to diversify in the Qinghai–Tibetan Plateau region during the late Miocene (10–5 Ma) (Ebersbach et al. 2016; Favre et al. 2016). In some taxa, species diversification rates increased along with the acquisition of perennial growth as a possible adaptation to cold, high elevations (Jabbour & Renner 2012). At lower elevations, the fossil ungulate faunal record documents a late Miocene faunal turnover (caused by vegetation changes in forests and grasslands at the Himalayan foothills between 9.5 and 7 Ma), when the southern Qinghai–Tibetan Plateau margins received a considerable influx of grazing herbivores, migrating northwards from the Indian subcontinent and adjacent Southeast Asia (Barry et al. 2002; Badgley et al. 2008). However, this time period also included one severe alpine fauna extinction event. After a peak in the middle Miocene, pika diversity drastically decreased up until the Pliocene–Pleistocene boundary. Only one genus survived, *Ochotona*, although many of its species went extinct, presumably as a consequence of the expansion of C4 plants in the late Miocene (Lopez-Martinez 2008; Erbajeva et al. 2011; Ge et al. 2013).

28.3.2.3 Pliocene Radiations

As global temperatures decreased in the Pliocene, a new phase of diversification started on the Qinghai–Tibetan Plateau and along its margins. For example, ancestors of extant boreal bird species from the upper forest belt (just below the timberline) occupied the newly emerging niches and separated from their subtropical relatives (Päckert et al. 2012). In fact, the most recent lineage divergences documented for Asian passerine birds have been associated with the elevational gradient; moreover, elevation-dependent disparity measures show a steep increase from 5 Ma towards a peak at the Pliocene–Pleistocene boundary (Price et al. 2014). In plants, elevational niche shifts (from open/alpine to forested environments) have been linked to late Miocene/Pliocene diversification in *Gentiana* (Favre et al. 2016) and *Tripterospermum* (Matuszak et al. 2016). Together, this shows that elevational niche diversification within the forest belt may have played an important role in the generation of modern Sino-Himalayan biodiversity.

Pliocene passerine radiations are characterized by a boreal faunal interchange along the southern and southwestern Qinghai–Tibetan Plateau margins with Central Asia and with the Palearctic (Päckert et al. 2012). Postulated forest corridors (Kitamura 1955) may have enhanced boreal faunal interchange between Southeast Asian mountain systems and the northern Palearctic. Such radiations may have been supported by the development of higher topographic complexity in Central Asia, including the high elevations of Mongolia. Along with global cooling and desertification, the central Loess Plateau underwent two major Neogene vegetation shifts towards an increased diversification of desert plants before 3.7 Ma (Wang et al. 2006; Meng et al. 2015). In succession, endemic animals of the dry grassland and desert steppe occupied the newly emerging niches (e.g., lizards) (Jin et al. 2008; Jin & Liu 2010). On the plateau itself, the Pliocene onset of in situ speciation was documented in birds (Lei et al. 2014) and pikas (Lanier & Olson 2009), and cold-adapted mammals started to diversify from there (Deng et al. 2011; Wang et al. 2014).

28.3.2.4 Pleistocene Range Fragmentation and Vicariance

The most recent diversification phase is associated with species range contractions and expansions during glacial cycles on the Qinghai–Tibetan Plateau and along its margins (Table 28.2 and Chapter 12). Allopatric speciation shaped extant species distributions, including zones of secondary contact and hybridization (Himalaya: Martens & Eck 1995; Palearctic: Aliabadian et al. 2005). Typically, east–west phylogeographic disjunctions are observed in many species pairs along the southern Qinghai–Tibetan Plateau margins (Päckert et al. 2012, 2015a; Martens 2015) and north–south disjunctions are seen in the Hengduan Shan and adjacent northern Chinese mountain ranges (Päckert et al. 2015b; Song et al. 2016). Strikingly, Pleistocene speciation processes were most commonly observed in higher-elevation species assemblages: in the upper forest belt below timberline, in the sub-alpine shrubs and alpine rocks and grasslands and to a lesser degree in the temperate and subtropical species assemblages at the lower elevations (Päckert et al. 2012).

Glacial refugia on the plateau were recently reconstructed using a combination of population genetics and species distribution modeling under past climatic conditions (Lei et al. 2014; Lu et al. 2016), and there is evidence from these studies that extant patterns of intraspecific diversity were substantially shaped by Quaternary climate rather than by orogenesis (Wan et al. 2015). These findings strengthen Renner's (2016) criticism that recent (post-Miocene) divergence of lineages endemic to the Qinghai–Tibetan Plateau was not associated with any recent geological uplift processes. It seems that strong intraspecific phylogeographical structuring is a rather common phenomenon at the eastern Qinghai–Tibetan

Plateau margin (in the Hengduan Shan, and further south in the Indo-Burmese mountains), where forested mountain ranges or single summits served as glacial sky island refugia (He & Jiang 2014). Strikingly, the strongest phylogenetic structuring corresponding to two or more glacial refugia was found in sedentary species. This structuring is more common in the few reptile and amphibian species that are adapted to the high Qinghai–Tibetan Plateau elevations, such as snakes (Hofmann et al. 2014), lizards (Jin & Liu 2010) and frogs (Liu et al. 2015), than in highly mobile species like birds (Yang et al. 2009; Qu et al. 2013; Lei et al. 2014), pikas and rodents (Ci et al. 2009; Tang et al. 2010; Lin et al. 2014; He et al. 2016).

28.4 The Mountain-Geobiodiversity Hypothesis

Considering the spatial and temporal patterns reported in this chapter, it becomes evident that there are complex relationships among the formation of the THR, climate changes and biotic evolution (e.g., Hoorn et al. 2013; Mulch 2016). Importantly, there seems to be a time lag between the formation/uplift of the older mountains of the THR (Qinghai-Tibetan Plateau, Himalaya) and species radiations (Favre et al. 2015; Päckert et al. 2015a; Deng & Ding 2016; Renner 2016). The former had already started by the Eocene, and reached its present-day elevation and topography in the Oligocene to mid-Miocene at the latest. However, rates of evolution (as derived from divergence times of modern taxa) seem to have peaked in the Miocene and Pliocene in the Himalaya and on the Qinghai–Tibetan Plateau, and again between 2 Ma and 150 ka.

Favre et al. (2015) proposed that this time lag might reflect the time needed for mountain habitats to establish and for species to occupy them. Another possibility is that species diversification did happen shortly after mountain uplift, and that diversity levels in the Eocene and Oligocene were just as high as today (see e.g., Jaramillo et al. 2006; Hoorn et al. 2010 for an analogous situation in Amazonia), but that molecular phylogenies based on extant taxa cannot properly detect subsequent extinction and the temporal dynamics of species richness. To further test these two alternatives, better integration between molecular phylogenies and fossil-based inferences (e.g., Morlon et al. 2011) is necessary.

If, indeed, there was a substantial lag between surface uplift in the Eocene and Oligocene and major speciation postdating the mid-Miocene, factors other than surface elevation (and associated climate and relief conditions) must have contributed to the diversity patterns observed today. We therefore formulate a testable mountain-geobiodiversity hypothesis and suggest that three principles may be required for mountain formation and surface uplift to have a substantial impact on species diversity.

28.4.1 Steep Zonation

In addition to a lowland and a mountainous zone, an alpine zone must also have relatively steep altitudinal zonation. The existence of an alpine zone depends on elevation, latitude and global climate. In lower latitudes and/or under warmer and wetter global climates, alpine zones require a higher elevation, and vice versa. Other factors, such as bedrock and soil types, nutrient levels and erosion may also play a role.

28.4.2 Climate Oscillations

Climate oscillations are required and must be sufficiently strong to stimulate speciation. This is in accordance with the predictions of the turnover-pulse hypothesis that, during evolutionary history, lineage turnovers should occur in pulses correlated with global or regional climate change (Vrba 1992; Simões et al. 2016).

28.4.3 High-Relief Terrain

A high-relief terrain, which typically implies high geodiversity (i.e., diversity of abiotic environmental parameters), is required such that migration distances caused by climate oscillations remain relatively small and a sufficient number of refugia and potential geographical barriers exist to allow both for survival of many species despite the climatic oscillations and for allopatric speciation.

28.4.4 Testing the Mountain-Geobiodiversity Hypothesis

Can a combination of these conditions explain the postulated delay between the onset of surface uplift and the response of Qinghai–Tibetan Plateau lineages? From the description of the spatial and temporal pattern of mountain uplift and biodiversity distribution, it is obvious that already around 40 Ma at least some parts of the THR exceeded an elevation of 4000 m. Moreover, it is clear that some areas with high elevation and alpine conditions existed since the late Eocene (Dupont-Nivet et al. 2008; Wang et al. 2008). In order to test this hypothesis, it remains to be explored whether the frequency and magnitude of early Cenozoic climate oscillations were large enough to impact speciation, or whether equitable climates over extended periods countered the presence of transient climate gradients along high-relief landscapes.

With successive global cooling since the early Eocene thermal maximum (ca. 55 Ma) and the formation of permanent polar icecaps, climatic oscillations have become increasingly important. This is particularly evident since the mid-Miocene climatic optimum, showing oscillations with a predominant obliquity cyclicity (40 ky cycle) (Zachos et al. 2001, 2008). Starting in the Pleistocene, around 2.6 Ma, these oscillations intensified, leading to true glacials and interglacials, with important periodicities ranging from eccentricity (100 ky) to precession (19 ky); the 100 ky cycle has been dominant since about 800 ka (Zachos et al. 2001). This successive Cenozoic cooling with increasing climate cyclicity since the early Eocene thermal maximum should have led to a steepening of the altitudinal zonation and to an increasing stimulation of the mountainous species pump. At least in the Pleistocene, the nival zone should have appeared so that the ecological gradient covered the entire range from the lowland to the nival zone.

28.5   Conclusion

We suggest that the evolution of present-day biodiversity on and around the Qinghai–Tibetan Plateau can be largely explained by the proposed mountain-geobiodiversity hypothesis. In this hypothesis, an increase in local geodiversity, in combination with rapid (millennial-scale) climatic oscillations, is expected to lead to an increase in biodiversity. The effect of mountain formation and regional surface uplift is to (i) increase the altitudinal gradient and thus local geodiversity; (ii) allow for barriers that promote allopatric speciation; and (iii) provide refugia to prevent extinctions during times of climatic change (Chapter 12) (see also Hoorn et al. 2013; Condamine et al. 2016). We therefore hypothesize that the evolution of biodiversity in the THR may be explained by the factors outlined in Figure 28.4.

Early uplift of the Qinghai–Tibetan Plateau (Eocene to mid-Miocene) did not lead to a major radiation of (modern) lineages, because the regional climate was still too warm and too wet to allow for steep elevational zonation, and climatic oscillations were too weak to act as a species pump.

The earliest, rare emergences of modern Qinghai–Tibetan Plateau endemic lineages and Asian vertebrate radiations in the Qinghai–Tibetan Plateau region occurred at the Eocene–Oligocene boundary. The first peak in the rate of radiation of modern lineages during the mid-Miocene at 20–15 Ma was associated with ecological segregation and the successive diversification of body size and feeding ecology between forest and alpine faunal elements. The second diversification rate peak, before the Miocene–Pliocene boundary, was caused by the cooler and (in particular) drier regional climate, which induced steeper elevational zonation. In combination with more pronounced wet–dry oscillations, the first species pump at 7 Ma was initiated. This early radiation coincided with aridification, the expansion of C4 plants and the intensification of the Asian monsoons.

Between 2 Ma and 150 ka, the third peak in modern lineage diversification occurred, which was linked with increasingly cooler climate and more intense climatic oscillations. Presumably, the regional relief was steeper compared to previous times because of more intense weathering and denudation during glacial–interglacial cycles.

Thus, according to the mountain-geobiodiversity hypothesis, the Cenozoic evolution of the diversification rate in the Qinghai–Tibetan Plateau region, with its three peaks, can be explained by the temporal evolution of (i) steep ecological gradients; (ii) rapid climatic oscillations; and (iii) a high-relief terrain. A high diversification rate requires all three factors.

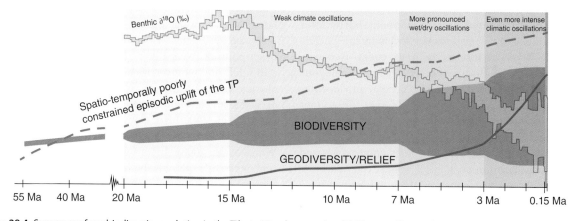

Figure 28.4 Summary of geobiodiversity evolution in the Tibeto-Himalayan region (THR), according to the mountain-geobiodiversity hypothesis. TP, Tibetan Plateau. All temporally parallel trends are displayed schematically and without mutual vertical scaling. The oxygen isotope curve is based on Zachos et al. (2001, 2008), the uplift curve on Wang et al. (2011). The biodiversity evolution curve shows the merged data from Table 28.2. The geodiversity curve is displayed according to the mountain-geobiodiversity hypothesis, developed in the discussion.

Acknowledgments

We thank Dr. Sybille Roller, Frankfurt, for her creative help with the figures, in particular Figure 28.4. We are particularly grateful to our reviewers, Alexandre Antonelli, Guillaume Dupont-Nivet and Susanne Renner, for their valuable comments and suggestions. This work was supported by the German Science Foundation (DFG-PAK 807/1) and the German Ministry for Education and Research (BMBF "CAME").

References

Acharya, B.K., Sanders, N.J., Vijayan, L. & Chettri, B. (2011) Elevational gradients in bird diversity in the Eastern Himalaya: an evaluation of distribution patterns and their underlying mechanisms. *PLoS ONE* **6**, e29097.

Adams, J. (2009) *Species Richness: Patterns and Diversity of Life*. Berlin-Heidelberg: Springer.

Aliabadian, M., Roselaar, C.S., Nijman, V. et al. (2005) Identifying contact zone hotspots of passerine birds in the Palaearctic region. *Biology Letters* **1**, 21–23.

Aliabadian, M., Sluys, R., Roselaar, C.S. & Nijman, V. (2008) Species diversity and endemism: testing the mid-domain effect on species richness patterns of songbirds in the Palearctic region. *Contributions to Zoology/ Bijdragen tot de dierkunde* **77**, 99–108.

An, B., Zhang, L. & Wang, L. (2015) Refugia persistence of Qinghai-Tibetan Plateau by the cold-tolerant bird *Tetraogallus tibetanus* (Galliformes: Phasianidae). *PLoS ONE* **10**, e0121118.

Badgley, C., Barry, J.C., Morgan, M.E. et al. (2008) Ecological changes in Miocene mammalian record show impact of prolonged climatic forcing. *Proceedings of the National Academy of Sciences USA* **105**, 12145–12149.

Baker, A.J., Pereira, S.L. & Paton, T.A. (2007) Phylogenetic relationships and divergence times of Charadriiformes genera: multigene evidence for the Cretaceous origin of at least 14 clades of shorebirds. *Biology Letters* **3**, 205–209.

Barry, J.C., Morgan, M.E., Flynn, L.J. et al. (2002) Faunal and environmental change in the late Miocene Siwaliks of northern Pakistan. *Paleobiology* **28**, 1–71.

Barthlott, W., Erdelen, W.R. & Rafiqpoor, M.D. (2014) Biodiversity and technical innovations: bionics. In: Lanzerath, D. & Friele, M.B. (eds.) *Concepts and Values in Biodiversity*. Routledge Studies in Biodiversity Politics and Management. London: Routledge, pp. 300–315.

Bellwood, D.R., Renema, W. & Rosen, B.R. (2012) Biodiversity hotspots, evolution and coral reef biogeography: a review. In: Gower, D.J., Johnson, K., Richardson, J. et al. (eds.) *Biotic Evolution and Environmental Change in Southeast Asia*. The Systematics Association Special, vol. **82**. Cambridge: Cambridge University Press, pp. 216–245.

Bhattarai, K.R. & Vetaas, O.R. (2006) Can Rapoport's rule explain tree species richness along the Himalayan elevation gradient, Nepal? *Diversity and Distribution* **12**, 373–378.

Boos, W.R. & Kuang, Z. (2010) Dominant control of the South Asian monsoon by orographic insulation versus plateau heating. *Nature* **463**, 218–222.

Bosboom, R., Dupont-Nivet, G., Grothe, A. et al. (2014) Linking Tarim Basin sea retreat (west China) and Asian aridification in the late Eocene. *Basin Research* **26**, 621–640.

Broccoli, A.J. & Manabe, S. (1987) The influence of continental ice, atmospheric CO_2, and land albedo on the climate of the last glacial maximum. *Climate Dynamics* **1**, 87–99.

Caves, J.K., Winnick, M.J., Graham, S.A. et al. (2015) Role of the westerlies in Central Asia climate over the Cenozoic. *Earth and Planetary Science Letters* **428**, 33–43.

Caves J.K., Moragne D.Y., Ibarra D.E. et al. (2016) The Neogene de-greening of Central Asia. *Geology* **44**, 887–890.

Chang, D.H.S. (1981) The vegetation zonation of the Tibetan Plateau. *Mountain Research and Development* **1**, 29.

Che, J., Zhou, W.-W., Hu, J.-S. et al. (2010) Spiny frogs (Paini) illuminate the history of the Himalayan region and Southeast Asia. *Proceedings of the National Academy of Sciences USA* **107**, 13765–13770.

Cheng, H., Edwards, R.L., Sinha, A. et al. (2016) The Asian monsoon over the past 640000 years and ice age terminations. *Nature* **534**, 640–646.

Chettri, B., Bhupathy, S. & Acharya, B.K. (2010) Distribution pattern of reptiles along an eastern Himalayan elevation gradient, India. *Acta Oecologica* **36**, 16–22.

Ci, H.X., Lin, G.H., Cai, Z.Y. et al. (2009) Population history of the plateau pika endemic to the Qinghai-Tibetan Plateau based on mtDNA sequence data. *Journal of Zoology* **279**, 396–403.

Clark, M.K., Farley, K.A., Zheng, D. et al. (2010) Early Cenozoic faulting of the northern Tibetan Plateau margin from apatite (U–Th)/He ages. *Earth and Planetary Science Letters* **296**, 78–88.

Clift, P.D., Hodges, K.V., Heslop, D. et al. (2008) Correlation of Himalayan exhumation rates and Asian monsoon intensity. *Nature Geoscience* **1**, 875–880.

Condamine, F.B., Leslie, A.B. & Antonelli, A. (2016) Ancient islands acted as refugia and pumps for conifer diversity. *Cladistics* **2016**, 1–24.

Cyr, A.J., Currie, B.S. & Rowley, D.B. (2005) Geochemical evaluation of Fenghuoshan group lacustrine carbonates, North-Central Tibet: implications for the paleoaltimetry of the Eocene Tibetan Plateau. *Journal of Geology* **113**, 517–533.

Deng, T. & Ding, L. (2016). Paleoaltimetry reconstructions of the Tibetan Plateau: progress and contradictions. *National Science Review* **2**, 417–437.

Deng, T., Wang, X., Fortelius, M. et al. (2011) Out of Tibet: Pliocene woolly rhino suggests high-plateau origin of Ice Age megaherbivores. *Science* **333**, 1285–1288.

Deng, T., Li, Q., Tseng, Z.J. et al. (2012) Locomotive implication of a Pliocene three-toed horse skeleton from Tibet and its paleo-altimetry significance. *Proceedings of the National Academy of Sciences USA* **109**, 7374–7378.

Dettman, D.L., Kohn, M.J., Quade, J. et al. (2001) Seasonal stable isotope evidence for a strong Asian monsoon throughout the past 10.7 my. *Geology* **29**, 31–34.

Ding, L., Xu, Q., Yue, Y. et al. (2014) The Andean-type Gangdese Mountains: Paleoelevation record from the Paleocene–Eocene Linzhou Basin. *Earth and Planetary Science Letters* **392**, 250–264.

Donoghue, M.J. (2008) A phylogenetic perspective on the distribution of plant diversity. *Proceedings of the National Academy of Sciences USA* **105**, 11 549–11 555.

Drovetski, S.V., Semenov, G., Drovetskaya, S.S. et al. (2013) Geographic mode of speciation in a mountain specialist Avian family endemic to the Palearctic. *Ecology and Evolution* **3**, 1518–1528.

Dupont-Nivet, G., Krijgsman, W., Langereis, C.G. et al. (2007) Tibetan plateau aridification linked to global cooling at the Eocene-Oligocene transition. *Nature* **445**, 635–638.

Dupont-Nivet, G., Hoorn, C. & Konert, M. (2008) Tibetan uplift prior to the Eocene-Oligocene climate transition: evidence from pollen analysis of the Xining Basin. *Geology* **36**, 987.

Ebersbach, J., Muellner-Riehl, A.N., Michalak, I. et al. (2016) In and out of the Qinghai–Tibet Plateau: divergence time estimation and historical biogeography of the large arctic-alpine genus *Saxifraga* L. *Journal of Biogeography* **44**, 900–910.

Erbajeva, M.A., Mead, J.I., Alexeeva, N.V. et al. (2011) Taxonomic diversity of Late Cenozoic Asian and North American ochotonids (an overview). *Paleontologia Electronica* **14**, 1–9.

Fan, Z., Liu, S., Liu, Y. et al. (2012) Phylogeography of the South China field mouse (*Apodemus draco*) on the Southeastern Tibetan Plateau reveals high genetic diversity and glacial refugia. *PLoS ONE* **7**, e38184.

Favre, A., Päckert, M., Pauls, S.U. et al. (2015) The role of the uplift of the Qinghai-Tibetan Plateau for the evolution of Tibetan biotas. *Biological Reviews* **90**, 236–253.

Favre, A., Michalak, I., Chen, C.H. et al. (2016) Out-of-Tibet: the spatio-temporal evolution of *Gentiana* (Gentianaceae). *Journal of Biogeography* **43**, 1967–1978.

Fjeldså, J., Bowie, R.C. & Rahbek, C. (2012) The role of mountain ranges in the diversification of birds. *Annual Review of Ecology, Evolution, and Systematics* **43**, 249–265.

Ge, D., Wen, Z., Xia, L. et al. (2013) Evolutionary history of lagomorphs in response to global environmental change. *PLoS ONE* **8**, e59668.

Gébelin, A., Mulch, A., Teyssier, C. et al. (2013) The Miocene elevation of Mount Everest. *Geology* **41**, 799–802.

Ghosh-Harihar, M. & Price, T.D. (2014) A test for community saturation along the Himalayan bird diversity gradient, based on within-species geographical variation. *Journal of Animal Ecology* **83**, 628–638.

Grau, O., Grytnes, J.-A. & Birks, H.J.B. (2007) A comparison of altitudinal species richness patterns of bryophytes with other plant groups in Nepal, Central Himalaya. *Journal of Biogeography* **34**, 1907–1915.

Guo, Z. T., Ruddiman, W.F., Hao, Q.Z. et al. (2002) Onset of Asian desertification by 22 Myr ago inferred from loess deposits in China. *Nature* **416**, 159–163.

Gupta, A. K., Anderson, D.M. & Overpeck, J.T. (2003) Abrupt changes in the Asian southwest monsoon during the Holocene and their links to the North Atlantic Ocean. *Nature* **421**, 354–357.

Haffer, J. (2008) Hypotheses to explain the origin of species in Amazonia. *Brazilian Journal of Biology* **68**, 917–947.

Hassanin, A., Delsuc, F., Ropiquet, A. et al. (2012) Pattern and timing of diversification of Cetartiodactyla (Mammalia, Laurasiatheria), as revealed by a comprehensive analysis of mitochondrial genomes. *Comptes Rendus Biologies* **335**, 32–50.

He, D.K. & Feng, C.Y. (2007) Molecular phylogeny and biogeography of the highly specialized grade schizothoracine fishes (Teleostei: Cyprinidae) inferred from cytochrome b sequences. *Chinese Science Bulletin* **52**, 777–788.

He, K. & Jiang, X. (2014) Sky islands of southwest China. I: An overview of phylogeographic patterns. *Chinese Science Bulletin* **59**, 585–597.

He, K., Hu, N.Q, Chen, X. et al. (2016) Interglacial refugia preserved high genetic diversity of the Chinese mole shrew in the mountains of southwest China. *Heredity* **116**, 23–32.

Hofmann, S., Kraus, S., Dorge, T. et al. (2014) Effects of Pleistocene climatic fluctuations on the phylogeography, demography and population structure of a high-elevation snake species, *Thermophis baileyi*, on the Tibetan Plateau. *Journal of Biogeography* **41**, 2162–2172.

Hoorn, C., Wesselingh, F.P., ter Steege, H. et al. (2010). Amazonia through time: Andean uplift, climate change, landscape evolution, and biodiversity. *Science* **330**, 927–931.

Hoorn, C., Mosbrugger, V., Mulch, A. & Antonelli, A. (2013) Biodiversity from mountain building. *Nature Geoscience* **6**, 154.

Huber, M. & Goldner, A. (2012) Eocene monsoons. *Journal of Asian Earth Sciences* **44**, 3–23.

Jabbour, F. & Renner, S.S. (2012) A phylogeny of Delphinieae (Ranunculaceae) shows that *Aconitum* is nested within *Delphinium* and that Late Miocene transitions to long life cycles in the Himalayas and Southwest China coincide with bursts in diversification. *Molecular Phylogenetics and Evolution* **62**, 928–942.

Jaramillo, C. Rueda, M.J. & Moira, G. (2006) Cenozoic plant diversity in the Neotropics. *Science* **311**, 1893–1896.

Jiang, H. & Ding, Z. (2008) A 20 Ma pollen record of East-Asian summer monsoon evolution from Guyuan, Ningxia, China. *Palaeogeography, Palaeoclimatology, Palaeoecology* **265**, 30–38.

Jin, Y.-T. & Liu, N.-F. (2010) Phylogeography of *Phrynocephalus erythrurus* from the Qiangtang Plateau of the Tibetan Plateau. *Molecular Phylogenetics and Evolution* **54**, 933–940.

Jin, H.J., Chang, X.L. & Wang, S.L. (2007) Evolution of permafrost on the Qinghai-Xizang (Tibet) Plateau since the end of the late Pleistocene. *Journal of Geophysical Research* **112**, 1–23.

Jin, Y.-T., Brown, R.P. & Liu, N.-F. (2008) Cladogenesis and phylogeography of the lizard *Phrynocephalus vlangalii* (Agamidae) on the Tibetan plateau. *Molecular Ecology* **17**, 1971–1982.

Kennedy, J.D., Weir, J.T., Hooper, D.M. et al. (2012) Ecological limits on diversification of the Himalayan core Corvoidea. *Evolution; International Journal of Organic Evolution* **66**, 2599–2613.

Khan, G. Zhang, F.Q., Gao, Q.B. et al. (2016) Phylogenetic analyses of *Spiraea* (Rosaceae) distributed in the Qinghai-Tibetan Plateau and adjacent regions: insights from molecular data. *Plant Systematics and Evolution* **302**, 11–21.

Kitamura, S. (1955) Flowering plants and ferns. In: Kihara, H. (ed.) *Fauna and Flora of Nepal Himalaya*. Kyoto: Fauna and Flora Research Society.

Körner, C. (2004) Mountain biodiversity, its causes and function. *Ambio* **13**, 11–17.

Kreft, H. & Jetz, W. (2007) Global patterns and determinants of vascular plant diversity. *Proceedings of the National Academy of Sciences USA* **104**, 5925–5930.

Lanier, H.C. & Olson, L.E. (2009) Inferring divergence times within pikas (*Ochotona* spp.) using mtDNA and relaxed molecular dating techniques. *Molecular Phylogenetics and Evolution* **53**, 1–12.

Lei, F., Qu, Y. & Song, G. (2014) Species diversification and phylogeographical patterns of birds in response to the uplift of the Qinghai-Tibet Plateau and Quaternary glaciations. *Current Zoology* **60**, 149–161.

Li, J. & Fang, X. (1999) Uplift of the Tibetan Plateau and environmental changes. *Chinese Science Bulletin* **44**, 2117–2124.

Licht, A., van Cappelle, M., Abels, H.A. et al. (2014) Asian monsoons in a late Eocene greenhouse world. *Nature* **513**, 501–506.

Licht, A., Dupont-Nivet, G., Pullen, A. et al. (2016) Resilience of the Asian atmospheric circulation shown by Paleogene dust provenance. *Nature Communications* **7**, 12390.

Lin, G., Zhao, F., Chen, H. et al. (2014) Comparative phylogeography of the plateau zokor (*Eospalax baileyi*) and its host-associated flea (*Neopsylla paranoma*) in the Qinghai-Tibet Plateau. *BMC Evolutionary Biology* **14**, 180.

Lippert, P.C., van Hinsbergen, D.J.J. & Dupont-Nivet, G. (2014) Early Cretaceous to present latitude of the central proto-Tibetan Plateau: a paleomagnetic synthesis with implications for Cenozoic tectonics, paleogeography and climate of Asia. In: Nie, J., Horton, B.K. & Hoke, G.D. (eds.) *Toward an Improved Understanding of Uplift Mechanisms and the Elevation History of the Tibetan Plateau*. The Geological Society of America Special Paper 507, pp. 1–21.

Liu, Q., Chen, P., He, K. et al. (2012) Phylogeographic Study of *Apodemus ilex* (Rodentia: Muridae) in Southwest China. *PLoS ONE* **7**, e31453.

Liu, J., Wang, C., Fu, D. et al. (2015) Phylogeography of *Nanorana parkeri* (Anura: Ranidae) and multiple refugia on the Tibetan Plateau revealed by mitochondrial and nuclear DNA. *Scientific Reports* **5**, 9857.

Lopez-Martinez, N. (2008) The Lagomorph fossil record of and the origin of the European rabbit. In: Alves, P.C., Ferrand, N. & Hackländer, K. (eds.) *Lagomorph Biology: Evolution, Ecology and Conservation*. Berlin-Heidelberg: Springer Verlag, pp. 27–46.

López-Pujol, J., Zhang, F.-M., Sun, H.-Q., Ying, T.S. & Ge, S. (2011) Mountains of southern China as "plant museums" and "plant cradles": evolutionary and conservation insights. *Mountain Research and Development* **31**, 261–269.

Lu, Q., Zhu, J., Yu, D. & Xu. X. (2016) Genetic and geographical structure of boreal plants in their southern range: phylogeography of *Hippuris vulgaris* in China. *BMC Evolutionary Biology* **16**, 34.

Luo, D., Yue, J.P., Sung, W.G. et al. (2016) Evolutionary history of the subnival flora of the Himalaya-Hengduan Mountains: first insights from comparative phylogeography of four perennial herbs. *Journal of Biogeography* **43**, 31–43.

Ma, X., Kang, J., Chen, W.T. et al. (2015) Biogeographic history and high-elevation adaptations inferred from the mitochondrial genome of Glyptosternoid fishes (Sisoridae, Siluriformes) from the southeastern Tibetan Plateau. *BMC Evolutionary Ecology* **15**, 233.

Marchese, C. (2015) Biodiversity hotspots: a shortcut for a more complicated concept. *Global Ecology and Conservation* **3**, 297–309.

Martens, J. (2015) Fauna – Himalayan patterns of diversity. In: Miehe, G. & Pendry, C. (eds.) *Nepal: An Introduction to the Natural History, Ecology and Human Environment of the Himalayas.* Edinburgh: Royal Botanic Garden, pp. 211–249.

Martens, J. & Eck, S. (1995) *Towards an Ornithology of the Himalayas: Systematics, Ecology and Vocalizations of Nepal Birds.* Bonn: Zoologisches Forschungsinstitut und Museum Alexander Koenig.

Martinelli, G. (2007) Mountain biodiversity in Brazil. *Brazilian Journal of Botany* **30**, 587–597.

Matuszak, S., Favre, A., Schnitzler, J. & Muellner-Riehl, A.N. (2016) Key innovations and climatic niche divergence as drivers of diversification in subtropical Gentianinae in southeastern and eastern Asia. *American Journal of Botany* **103**, 1–13.

McCain, C.M. & Bracy Knight, K. (2013) Elevational Rapoport's rule is not pervasive on mountains. *Global Ecology and Biogeography* **22**, 750–759.

McKenna, D.D. & Farell, B.D. (2006) Tropical forests are both evolutionary cradles and museums of leaf beetle diversity. *Proceedings of the National Academy of Sciences USA* **103**, 10947–10951.

Meng, J. & McKenna, M.C. (1998) Faunal turnovers of Palaeogene mammals from the Mongolian plateau. *Nature* **394**, 364–367.

Meng, H.H., Gao, X.Y., Huang, J.F. & Zhang, M.L. (2015) Plant phylogeography in arid Northwest China: retrospectives and perspectives. *Journal of Systematics and Evolution* **53**, 33–46.

Miao, Y., Herrmann, M., Wu, F. et al. (2012) What controlled Mid–Late Miocene long-term aridification in Central Asia? – Global cooling or Tibetan Plateau uplift: a review. *Earth-Science Reviews* **112**, 155–172.

Miehe, G., Miehe, S., Vogel, J. et al. (2007) Highest treeline in the northern hemisphere found in Southern Tibet. *Mountain Research and Development* **27**, 169–173.

Mittermeier, R.A., Seligmann, P.A. & Ford, H., eds. (2004). Hotspots revisited. CEMEX, Mexico City.

Molnar, P., Boos, W.R. & Battisti, D.S. (2010) Orographic controls on climate and paleoclimate of Asia: thermal and mechanical roles for the Tibetan Plateau. *Annual Review of Earth and Planetary Sciences* **38**, 77–102.

Moreau, C.S. & Bell, C. (2013) Testing the museum versus cradle tropical biological diversity hypothesis: Phylogeny, diversification, and ancestral biogeographic range evolution of the ants. *Evolution* **67**, 2240–2257.

Morlon, H., Parsons, T.L. & Plotkin, J.B. (2011) Reconciling molecular phylogenies with the fossil record. *Proceedings of the National Academy of Sciences USA* **108**, 16327–16332.

Mulch, A. (2016) Stable isotope paleoaltimetry and the evolution of landscapes and life. *Earth and Planetary Science Letters* **433**, 180–191.

Mulch, A. & Chamberlain, C.P. (2006) Earth science: the rise and growth of Tibet. *Nature* **439**, 670–671.

Murphy, M.A., Yin, A., Harrison, T.M. et al. (1997) Did the Indo-Asian collision alone create the Tibetan plateau? *Geology* **25**, 719–722.

Mutke, J. & Barthlott, W. (2005) Patterns of vascular plant diversity at continental to global scales. *Biologiske skrifter* **55**, 521–531.

Myers, N., Mittermeier, R.A., Mittermeier, C.G. et al. (2000) Biodiversity hotspots for conservation priorities. *Nature* **403**, 853–858.

Najman, Y., Appel, E., Boudagher-Fadel, M. et al. (2010) Timing of India-Asia collision: geological, biostratigraphic, and palaeomagnetic constraints. *Journal of Geophysical Research* **115**, B12416.

Oomen, M.A. & Shanker, K. (2005) Elevational species richness patterns emerge from multiple local mechanisms in Himalayan woody plants. *Ecology* **86**, 3039–3047.

Päckert, M., Martens, J., Sun, Y.-H. et al. (2012) Horizontal and elevational phylogeographic patterns of Himalayan and Southeast Asian forest passerines (Aves: Passeriformes). *Journal of Biogeography* **39**, 556–573.

Päckert, M., Martens, J., Sun, Y.-H. & Tietze, D.T. (2015a) Evolutionary history of passerine birds (Aves Passeriformes) from the Qinghai-Tibetan plateau: from a pre-Quarternary perspective to an integrative biodiversity assessment. *Journal of Ornithology* **156**, 355–365.

Päckert, M., Sun, Y.H., Strutzenberger, P. et al. (2015b) Phylogenetic relationships of endemic bunting species (Aves, Passeriformes, Emberizidae, *Emberiza koslowi*) from the eastern Qinghai-Tibet Plateau. *Vertebrate Zoology* **65**, 135–150.

Päckert, M., Martens, J., Sun, Y.-H. & Strutzenberger, P. (2016) The phylogenetic relationships of Przevalski's Finch *Urocynchramus pylzowi*, the most ancient Tibetan endemic passerine known to date. *Ibis* **158**, 530–540.

Pisano, J., Condamine, F.L., Lebedev, V. et al. (2015) Out of Himalaya: the impact of past Asian environmental changes on the evolutionary and biogeographical history of Dipodoidea (Rodentia). *Journal of Biogeography* **42**, 856–870.

Price, T.D., Mohan, D., Tietze, D.T. et al. (2011) Determinants of northerly range limits along the Himalayan bird diversity gradient. *The American Naturalist* **178**, 97–108.

Price, T.D., Hooper, D.M., Buchanan, C.D. et al. (2014) Niche filling slows the diversification of Himalayan songbirds. *Nature* **509**, 222–225.

Qu, Y., Zhao, H., Han, N. et al. (2013) Ground tit genome reveals avian adaptation to living at high altitudes in the Tibetan plateau. *Nature Communications* **4**, 2071.

Quade, J., Cater, J.M.L., Ojha, T.P. et al. (1995) Dramatic carbon and oxygen isotopic shift in paleosols from Nepal and late Miocene environmental change across the northern Indian sub-continent. *Geological Society of America Bulletin* **107**, 1381–1397.

Quade, J., Breecker, D.O., Daëron, M. & Eiler, J. (2011) The paleoaltimetry of Tibet: an isotopic perspective. *American Journal of Science* **311**, 77–115.

Quan, C., Liu, Y.-S. & Utescher, T. (2012) Eocene monsoon prevalence over China: a paleobotanical perspective. *Palaeogeography, Palaeoclimatology, Palaeoecology* **365**, 302–311.

Ren, G., Conti, E. & Salamin, N. (2015) Phylogeny and biogeography of *Primula* sect. *Armerina*: implications for plant evolution under climate change and the uplift of the Qinghai-Tibet Plateau. *BMC Evolutionary Biology* **15**, 161.

Renner, S.S. (2016) Available data point to a 4-km-high Tibetan Plateau by 40 Ma, but 100 molecular-clock papers have linked supposed recent uplift to young node ages. *Journal of Biogeography* **43**, 1479–1487.

Rodwell, M.J. & Hoskins, B.J. (1996) Monsoons and the dynamics of deserts. *Quarterly Journal of the Royal Meteorological Society* **122**, 1385–1404.

Rowley, D.B. & Currie, B.S. (2006) Palaeo-altimetry of the late Eocene to Miocene Lunpola basin, central Tibet. *Nature* **439**, 677–681.

Sandel B., Arge, L., Dalsgaard, B. et al. (2011) The influence of Late Quaternary climate-change velocity on species endemism. *Science* **334**, 660–664.

Schmidt, J., Opgenoorth, L., Holl, S. & Bastrop R. (2012) Into the Himalayan exile: the phylogeography of the ground beetle *Ethira* clade supports the Tibetan origin of forest-dwelling Himalayan species groups. *PloS ONE* **7**, e45482.

Shukla, A., Mehrotra, R.C., Spicer, R.A. et al. (2014) Cool equatorial terrestrial temperatures and the South Asian monsoon in the Early Eocene: evidence from the Gurha Mine, Rajasthan, India. *Palaeogeography, Palaeoclimatology, Palaeoecology* **412**, 187–198.

Simões, M., Breitkreuz, L., Alvarado, M. et al. (2016) The evolving theory of evolutionary radiations. *Trends in Ecology and Evolution* **31**, 27–34.

Sklenář, P., Hedberg, I. & Cleef, A.M. (2014) Island biogeography of tropical alpine floras. *Journal of Biogeography* **41**, 287–297.

Song, G., Zhang, R., Qu, Y. et al. (2016) A zoogeographical boundary between the Palaearctic and Sino-Japanese realms documented by consistent north/south phylogeographical divergences in three woodland birds in eastern China. *Journal of Biogeography* **43**, 2099–2112.

Stevens, G.C. (1992) The elevational gradient in altitudinal range: an extension of Rapoport's latitudinal rule to altitude. *The American Naturalist* **140**, 893–911.

Sun, X. & Wang, P. (2005) How old is the Asian monsoon system? – Palaeobotanical records from China. *Palaeogeography, Palaeoclimatology, Palaeoecology* **222**, 181–222.

Sun, B.-N., Wu, J.-Y., Liu, Y.-S. et al. (2011) Reconstructing Neogene vegetation and climates to infer tectonic uplift in western Yunnan, China. *Palaeogeography, Palaeoclimatology, Palaeoecology* **304**, 328–336.

Sun, Y.S., Wang, A.L., Wan, D.S. et al. (2012) Rapid radiation of *Rheum* (Polygonaceae) and parallel evolution of morphological traits. *Molecular Phylogenetics and Evolution* **63**, 150–158.

Tamma, K. & Ramakrishnan, U. (2015) Higher speciation and lower extinction rates influence mammal diversity gradients in Asia. *BMC Evolutionary Biology* **15**, 11.

Tang, L.-Z., Wang, L.-Y., Cai, Z.-Y. et al. (2010) Allopatric divergence and phylogeographic structure of the plateau zokor (*Eospalax baileyi*), a fossorial rodent endemic to the Qinghai-Tibetan Plateau. *Journal of Biogeography* **37**, 657–668.

Tang, L., Li, T., Li, D. & Meng, X. (2014) Elevational patterns of plant richness in the Taibai Mountain, China. *The Scientific World Journal* **2014**, 309053.

Tapponnier, P., Zhiqin, X., Roger, F. et al. (2001) Oblique stepwise rise and growth of the Tibet plateau. *Science* **294**, 1671–1677.

Tietze, D.T., Päckert, M., Martens, J. et al. (2013) Complete phylogeny and historical biogeography of true rosefinches (Aves: *Carpodacus*). *Zoological Journal of the Linnean Society* **169**, 215–234.

Tseng, Z.J., Wang, X., Slater, G.J. et al. (2014) Himalayan fossils of the oldest known pantherine establish ancient origin of big cats. *Proceedings of the Royal Society B: Biological Sciences* **281**, 20132686.

Voelker, G., Semenov, G., Fadeev, I.V. et al. (2015) The biogeographic history of *Phoenicurus* redstarts reveals an allopatric mode of speciation and an out-of-Himalayas colonization pattern. *Systematics and Biodiversity* **13**, 296–305.

Vrba, E.S. (1992) Mammals as a key to evolutionary theory. *Journal of Mammalogy* **73**, 1–28.

Wan, S., Li, A., Clift, P.D. & Stuut, J.-B.W. (2007) Development of the East Asian monsoon: mineralogical and sedimentologic records in the northern South China Sea since 20 Ma. *Palaeogeography, Palaeoclimatology, Palaeoecology* **254**, 561–582.

Wan, D.S., Feng, J.J., Jiang, D.C. et al. (2015) The Quaternary evolutionary history, potential distribution dynamics, and conservation implications for a Qinghai-Tibet Plateau endemic herbaceous perennial, *Anisodus tanguticus* (Solanaceae). *Ecology and Evolution* **6**, 1977–1995.

Wang, L., Lü, H.Y., Wu, N.Q. et al. (2006) Palynological evidence for Late Miocene–Pliocene vegetation evolution recorded in the red clay sequence of the central Chinese Loess Plateau and implication for palaeoenvironmental change. *Palaeogeography, Palaeoclimatology, Palaeoecology* **241**, 118–128.

Wang, C., Zhao, X., Liu, Z. et al. (2008) Constraints on the early uplift history of the Tibetan Plateau. *Proceedings of the National Academy of Sciences USA* **105**, 4987–4992.

Wang, G.C., Cao, K., Zhang K.X. et al. (2011) Spatio-temporal framework of tectonic uplift stages of the Tibetan Plateau in Cenozoic. *Science China Earth Sciences* **54**, 29–44.

Wang, M., Yang, J.-X. & Chen, X.-Y. (2013) Molecular phylogeny and biogeography of *Percocypris* (Cyprinidae, Teleostei). *PloS ONE* **8**, e61827.

Wang, X., Tseng, Z.J., Li, Q. et al. (2014) From "third pole" to north pole: a Himalayan origin for the arctic fox. *Proceedings of the Royal Society B* **281**, 20140893.

Wang, Y., Shen, Y., Feng et al. (2016) Mitogenomic perspectives on the origin of Tibetan loaches and their adaptation to high altitude. *Scientific Reports* **6**, 29690.

Weigold, H. (2005) *Die Biogeographie Tibets und seiner Vorländer*. Hohenstein-Ernstthal: Verein Sächsischer Ornithologen.

Wen, J., Zhang, J.-Q., Nie, Z.-L. et al. (2014) Evolutionary diversifications of plants on the Qinghai-Tibetan Plateau. *Frontiers in Genetics* **5**, 4.

Wikramanayake, E.D., Dinerstein, E. & Loucks, C.J. (2002) *Terrestrial Ecoregions of the Indo-Pacific: A Conservation Assessment*. Washington, DC: Island Press.

Yan, Y., Yang, X. & Tang, Z. (2013) Patterns of species diversity and phylogenetic structure of vascular plants on the Qinghai-Tibetan Plateau. *Ecology and Evolution* **3**, 4584–4595.

Yang, S., Dong, H. & Lei, F. (2009) Phylogeography of regional fauna on the Tibetan Plateau: a review. *Progress in Natural Science* **19**, 789–799.

Yin, A. & Harrison, T.M. (2000) Geologic evolution of the Himalayan-Tibetan orogen. *Annual Review of Earth and Planetary Sciences* **28**, 211–280.

Yu, X.Q., Maki, M., Drew, B.T. et al. (2014) Phylogeny and historical biogeography of *Isodon* (Lamiaceae): rapid radiation in south-west China and Miocene overland dispersal into Africa. *Molecular Phylogenetics and Evolution* **77**, 183–194.

Zachos, J., Pagani, M., Sloan, L. et al. (2001) Trends, rhythms, and aberrations in global climate 65 Ma to present. *Science* **292**, 686–693.

Zachos, J.C., Dickens, G.R. & Zeebe, R.E. (2008) An early Cenozoic perspective on greenhouse warming and carbon-cycle dynamics. *Nature* **451**, 279–283.

Zhang, Z. & Sun, J. (2011) Palynological evidence for Neogene environmental change in the foreland basin of the southern Tianshan range, northwestern China. *Global and Planetary Change* **75**, 56–66.

Zhang, P., Chen, Y.-Q., Zhou, H. et al. (2006) Phylogeny, evolution, and biogeography of Asiatic Salamanders (Hynobiidae). *Proceedings of the National Academy of Sciences USA* **103**, 7360–7365.

Zhang, J.Q., Meng, S.Y., Allen, G.A. et al. (2014a) Rapid radiation and dispersal out of the Qinghai-Tibetan Plateau of an alpine plant lineage *Rhodiola* (Crassulaceae). *Molecular Phylogenetics and Evolution* **77**, 147–158.

Zhang, M.-L., Meng, H.-H., Zhang, H.-X. et al. (2014b) Himalayan origin and evolution of *Myricaria* (Tamaricaeae) in the Neogene. *PloS ONE* **9**, e97582.

Zhang, Y., Liang, S., He, J. et al. (2015) Oxidative stress and antioxidant status in a lizard *Phrynocephalus vlangalii* at different altitudes or acclimated to hypoxia. *Comparative Biochemistry and Physiology Part A: Molecular & Integrative Physiology* **190**, 9–14.

Zhang, D.C., Ye, J.X. & Sun, H. (2016) Quantitative approaches to identify floristic units and centres of species endemism in the Qinghai-Tibetan Plateau, south-western China. *Journal of Biogeography* **43**, 2465–2476.

Zheng, H., Powell, C.M., An, Z. et al. (2000) Pliocene uplift of the northern Tibetan Plateau. *Geology* **28**, 715–718.

29

Neogene Paleoenvironmental Changes and their Role in Plant Diversity in Yunnan, South-Western China

Zhe-Kun Zhou[1,2], Tao Su[1] and Yong-Jiang Huang[2]

[1] *Key Laboratory of Tropical Forest Ecology, Xishuangbanna Tropical Botanical Garden, Chinese Academy of Sciences, Mengla, China*
[2] *Key Laboratory for Plant Diversity and Biogeography of East Asia, Kunming Institute of Botany, Chinese Academy of Sciences, Kunming, China*

Abstract

The Chinese province of Yunnan, bordered by the Tibeto-Himalayan Region, has long been renowned for its high plant diversity. Understanding how this diversity arose is a focus in various research areas, such as botany, paleobotany, molecular phylogenetics and environmental biology. Paleoclimatic studies indicate that during the Neogene, Yunnan was warmer and wetter than today, with a weaker Asian monsoon climate. As temperature declined, precipitation decreased and the monsoon intensified plant diversity: some plant taxa disappeared (e.g., *Cedrus*, *Sequoia*), while numerous others flourished and diversified (e.g., *Pedicularis*, *Rhododendron*). These changes in plant diversity may also be associated with the topographic deformations that occurred as a result of the south-eastern extrusion of the Tibeto-Himalayan Region, including the uplift of mountains and the formation of deep valleys.

Keywords: *extinction, Himalaya, Hengduan Mountains, monsoon climate, topographic deformation*

29.1 Geomorphology, Tectonic History and Modern Plant Diversity of Yunnan

Yunnan, situated in south-western China, is a topographically complex area with large altitude and climate gradients (Figure 29.1). It is located within a transitional zone in terms of altitude: it extends from below 500 m in the south-east to above 6000 m in the north-west, which constitutes the south-eastern border of the Tibeto-Himalayan region (THR). Tectonically, Yunnan lies in the conjunctional region between the Eurasian and Indian plates, and therefore its tectonic history has been largely influenced by the collision of the two (Royden et al. 2008). As a result of the continuous uplift of the THR, which began no later than the early Miocene, Yunnan has experienced dramatic tectonic activities. During the Neogene, many coal-bearing basins were formed among the mountains (Ge & Li 1999) and a series of faults were created (Schoenbohm et al. 2006). Both clockwise rotation and slip occurred in some parts of Yunnan as a result of the collision. For example, a clockwise rotation of about 30° and a southward slip of about 800 km occurred for the Simao Terrane (Chen et al. 1995). The eastward expansion

of the THR led to asynchronous uplifts in different parts of Yunnan. The northern part uplifted earlier than the southern (Jacques et al. 2014), and western Yunnan experienced continuous uplift during the post-Pliocene (Su et al. 2013a). This tectonic activity shaped the modern landscape in Yunnan: the north has much higher altitudes than the south. Despite the presumed pre-Miocene origin of the Yangtze River, the largest river in China (Zheng et al. 2013), the Neogene was still a vital period, during which the modern drainage systems evolved. Besides mountain uplift, water erosion of the plateau also played a role in the formation of huge mountains and deep valleys, especially in western and north-western Yunnan. Largely associated with its complex geology and geomorphology, Yunnan is home to a diverse number of climates occurring over a relatively small area. These range from cold and alpine climates in the north-west to warm and humid seasonal tropical climates in the south (Wang 2006).

Its unique position, complex topography and diverse climates together make Yunnan one of the world's biodiversity hotspots (Ruth et al. 2008; López-Pujol et al. 2011). Even though it accounts for only 4% of the Chinese territory, according to Wu (1979), it is now home to more

Mountains, Climate and Biodiversity, First Edition. Edited by Carina Hoorn, Allison Perrigo and Alexandre Antonelli.
© 2018 John Wiley & Sons Ltd. Published 2018 by John Wiley & Sons Ltd.
Companion website: www.wiley.com\go\hoorn\mountains,climateandbiodiversity

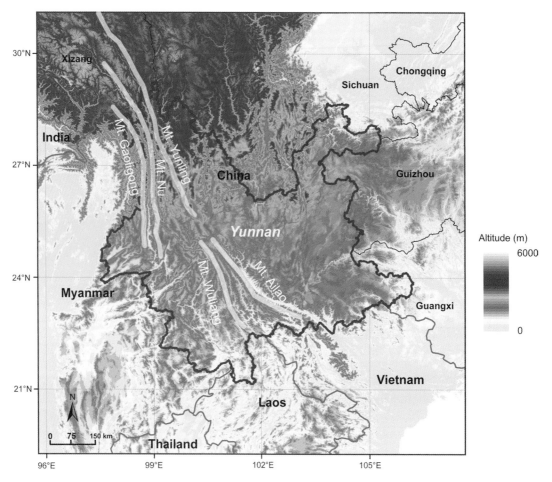

Figure 29.1 Geographic Information System (GIS) map showing the geographical position and topography of Yunnan. *Source:* Created with ArcGIS 9.0 (Environmental Systems Research Institute, Inc.). See also Plate 47 in color plate section.

than 16 000 species of vascular plants, grouped into over 3000 genera and 433 families (6% of the world's total vascular plant species). Of these, nearly half are endemic to Yunnan. Huge mountains, especially the famous Hengduan Mountains (Wu 1988; Ruth et al. 2008), display particularly rich plant biodiversity as a consequence of the number of ecosystems they foster. Many plant groups, such as *Rhododendron* and *Pedicularis*, with a considerable level of species richness in Yunnan are also present in adjacent regions (e.g., in Nepal), but with much lower species diversity. Almost all of the ecosystem types of China, including tropical seasonal rainforests, evergreen broadleaved forests, sclerophyllous evergreen broad-leaved forests, deciduous broad-leaved forests, both warm and cold temperate coniferous forests, dry and warm valley scrubs and sub-alpine and alpine scrubs can be found in Yunnan (Wu & Zhu 1987). Moreover, Yunnan possesses a mixture of different floristic components: Indo-Chinese and Chinese-Japanese elements frequently commingle in this region. This may have led to the increased complexity of the floristic composition in the region.

29.2   Neogene Climates of Yunnan

Plant fossils are important proxies for quantitative reconstructions of paleoclimates (Greenwood 2005). Yunnan bears numerous sites harboring abundant plant fossils from the Neogene in the form of leaves, fruits, seeds, wood and pollen/spores (Ge & Li 1999). These fossil remains have been used to reconstruct the Neogene climates of the region (e.g., Kou et al. 2006; Xia et al. 2009; Xie et al. 2012) by three quantitative approaches: the coexistence approach, leaf-margin analysis and the Climate Leaf Analysis Multivariate Program (CLAMP). The coexistence approach is based on the assumption that the fossil taxa have similar climate requirements to those of their closest living relatives (Mosbrugger & Utescher 1997; Utescher et al. 2014). The climate space in which the highest number of closest living relatives of a given fossil flora can coexist is considered to represent its paleoclimatic condition. This approach is valid with all types of plant organs, as long as their taxonomic statuses can be determined reliably (Mosbrugger &

Utescher 1997). Leaf-margin analysis is based on the positive correlation between the percentage of woody species with entire-margin leaves in a given flora and the mean annual temperature (MAT) (Bailey & Sinnott 1915; Greenwood 2005). The Neogene temperature estimates for Yunnan inferred by leaf-margin analysis are based primarily on an equation established using data drawn from Chinese forests by Su et al. (2010). CLAMP uses canonical correspondence analysis to calculate the relationship between 31 leaf physiognomic characters and 11 climate parameters (Wolfe 1993; Wolfe & Spicer 1999). This approach is based on the hypothesis that there is selection for those features of leaves that confer the maximum functional advantage under various environmental conditions (Wolfe & Spicer 1999). The reliability of CLAMP-based climate reconstructions of Neogene Yunnan can be improved when using the data set derived from Chinese forests (Jacques et al. 2011b).

In Yunnan, the climates of different Neogene subepochs (early Miocene, middle Miocene, late Miocene and late Pliocene) have been studied quantitatively. Reconstructed data allow for a fairly detailed interpretation of temperature changes in Yunnan throughout the late Cenozoic. Early Miocene temperatures have only been looked at for the north-western part of the region, and the calculated temperatures indicate that north-western Yunnan was warmer during the early Miocene than it is now (Sun et al. 2011). The late Miocene of northern Yunnan has also been shown to be warmer than it is today: temperature estimates from north-central Yunnan are noticeably higher than modern values (Xu et al. 2004; Xing et al. 2012). In contrast to northern Yunnan, Miocene temperatures in southern Yunnan were similar to today. Temperature estimates for south-central Yunnan during the middle Miocene (Zhang et al. 2012), as well as for south-eastern and south-western Yunnan during the late Miocene (Zhao et al. 2004; Xia et al. 2009; Li et al. 2015), match temperatures in those regions in modern times. Apparently, temperatures in southern Yunnan have changed only moderately. The temperature decline measured in northern Yunnan seems to be in line with a global drop (Zachos et al. 2001), but differs from that recorded in middle- to high-latitude regions (e.g., Europe), where temperature has declined more significantly (Utescher et al. 2000; Mosbrugger et al. 2005; Uhl et al. 2007). This supports previous interpretations that the Cenozoic cooling was more pronounced in higher latitudes of the northern hemisphere than in lower latitudes (Uhl et al. 2007; Utescher et al. 2011). Although the post-Miocene temperature drop in northern Yunnan may be related to the global cooling, it might be regionally specific and thus cannot be explained merely from a global perspective. Severe mountain uplift caused by the south-eastern extrusion of the THR

(Clark & Royden 2000; Zhang & Ding 2003; Clark et al. 2005; Shen et al. 2005; Westaway 2009) may have also played a major role. Paleoelevation reconstruction indicates that northern Yunnan experienced significant uplift before the late Miocene, while southern Yunnan remained at about the same level (Jacques et al. 2014). Increasingly higher altitudes in northern Yunnan may have led to the temperature decline, while the relatively stable elevation in southern Yunnan may have resulted in the almost unchanged temperature.

Latitudinal temperature gradients in Yunnan remained negligible during the Miocene, but grew more evident in the late Pliocene. At that time, northern Yunnan, particularly the north-western section, was markedly cooler, while southern Yunnan was warmer than other parts of the region (Huang et al. 2015a). This gradient temperature pattern has been largely attributed to differences in altitude: northern Yunnan was probably already higher than southern Yunnan by the late Pliocene (Huang et al. 2015a). Moreover, the post-Pliocene decline of summer temperatures was pronounced, while winter temperatures remained almost unchanged, in contrast to observations in other territories. The different patterns in summer and winter cooling trends since the late Pliocene may be explained by increases in summer precipitation and simultaneous decreases in winter precipitation (Su et al. 2013a), as rainfall commonly decreases Earth's surface temperature (Déry & Wood 2005). The reduced impact of cold winter air from the north due to the eastward extension of the THR may have played a role as well (Huang et al. 2015a).

Besides the cooling trend, Yunnan has also been affected by the post-Miocene aridification. Paleoclimatic reconstructions indicate that conditions were humid in Yunnan during the Neogene but that a drying trend has occurred since then (Xu et al. 2008; Xia et al. 2009; Xing et al. 2012; Yao et al. 2012). With respect to precipitation changes, Yunnan can be divided into two regions, with the Ailao-Yunling Mountains roughly serving as the boundary (Jacques et al. 2011a). The area east of the Ailao-Yunling Mountains has experienced decreased precipitation (Xia et al. 2009; Xing et al. 2012), while the area to the west has generally undergone increased precipitation (e.g., Xu et al. 2004; Zhang et al. 2012). This provides a historical background that partly explains the rainfall pattern observed in Yunnan today, where eastern Yunnan receives higher levels of precipitation than western Yunnan.

The post-Miocene changes in precipitation may allow us to deduce the development of the Asian monsoon in Yunnan, which is characterized by warm and humid summers but cool and dry winters. In East Asia, precipitation changes are closely linked to the Asian monsoon system, which consists of the Indian summer monsoon and East Asian summer monsoon. In summer, the former mainly influences southern Asian countries, while the latter mainly affects eastern Asia, with the Ailao-Yunling Mountains roughly forming the boundary zone (Jacques et al. 2011a). The Asian monsoon is one of the typical climates in Yunnan that started at least as far back as the early Miocene (Sun & Wang 2005; Guo et al. 2008), and possibly even before that (Licht et al. 2014). It is characterized by humid summers but dry winters, and is associated with the uplift of the THR (Harris 2006). Since fluctuations of the monsoon climate greatly impact both social activities and the economy, especially in East Asia, the intensification of the Asian monsoon has long attracted attention (Sun & Wang 2005). A number of variables, known as the monsoon strength index (MSI), have been proposed to precisely describe monsoon intensity (e.g., Liu & Yin 2002; van Dam 2006; Xing et al. 2012). The contrast between summer and winter precipitations appears to be an appropriate means by which to define monsoon strength (Xing et al. 2012), and has been used in a recent study by Li et al. (2015). Precipitation estimates for Yunnan from both the Miocene and the Pliocene indicate higher precipitation in summer but lower precipitation in winter (Xu et al. 2008; Jacques et al. 2011a; Xing et al. 2012; Su et al. 2013a; Huang et al. 2015a), suggesting the prevalence of the Asian monsoon in Yunnan during the Neogene. The Asian monsoon in Yunnan has intensified since the late Miocene (Li et al. 2015), which has been widely accredited to the uplift of the THR (e.g., An et al. 2001; Zheng et al. 2004; Harris 2006; Su et al. 2013a) and the onset and expansion of the Arctic glaciation (Su et al. 2013a).

29.3 Plant Diversity Changes in Response to Monsoon Intensification in Yunnan

The evolutionary history of modern vascular plant diversity in Yunnan can be traced by looking at the fossil record. Yunnan holds one of the richest Neogene floras in China (Li 1995). Situated at the south-eastern margin of the THR, it experienced intense tectonic activity during the Neogene (Schoenbohm et al. 2006), which created more than 200 Neogene coal-bearing basins in the west of the region (Ge & Li 1999). A number of fossil floras have been recovered from Neogene strata of these isolated basins (WGCPC 1978; Li 1995). Over the last 2 decades, many paleobotanical studies focusing on Neogene floras in Yunnan have been conducted (Li et al. 2008; Tao 2000). These have significantly improved our understanding of the history of plant diversity following environmental changes in the region.

Presently, the monsoon system dominates the climate of Yunnan, which is characterized by seasonal rainfall – specifically, wet summers and dry winters. As already mentioned, paleoclimatic reconstructions indicate that the monsoon climate gradually intensified during the Neogene, and weaker rainfall seasonality occurred during that period (Xia et al. 2009; Jacques et al. 2011a; Xing et al. 2012; Su et al. 2013a; Huang et al. 2015a). Monsoon intensification played an important role in the history of plant diversity in Yunnan. Seasonal drought is thought to have profoundly influenced seed germination and leaf growth, and is therefore considered the main factor in shaping plant diversity.

Many plants adapted to paleoclimatic changes and were shielded by the complex topography during the Pleistocene glaciations. These plant taxa diversified in Yunnan and nearby areas, making this one of the principal centers of plant diversity in the modern world. Increasing fossil evidence suggests that many taxa, having their modern diversification centers in Yunnan, already occurred in this region during the Neogene. Examples include *Pinus* (Xing et al. 2010; Zhang et al. 2015a), *Quercus* (Hu et al. 2014; Xing et al. 2013) and *Rosa* (Su et al. 2016). It is difficult to explore the evolutionary history of these taxa through fossils from earlier stages, simply due to the paucity of pre-Neogene floras known from Yunnan. *Pinus kesiya* Royle ex Gord and *P. yunnanensis* Franch. are two species with close phylogenetic affinity confirmed by molecular analysis (Eckert & Hall 2006). Both are dominant elements in forests of Yunnan, with the modern distribution of *P. kesiya* more southerly than that of *P. yunnanensis*. Generally, precipitation in southern Yunnan is higher than in the rest of the province. Late Miocene female cones of *P. prekesiya* Y.W Xing, Y.S. Liu et Z.K. Zhou, which were found at Xianfeng, northern Yunnan, show high morphological similarity to both *P. kesiya* and *P. yunnanensis*. Therefore, *P. prekesiya* may be the ancestor of *P. kesiya* and *P. yunnanensis*. According to paleoclimatic reconstructions, central Yunnan experienced a wetter climate during the late Miocene than today (Xing et al. 2012). The intensification of the monsoon climate may have led to the divergence of *P. prekesiya* into *P. kesiya* and *P. yunnanensis*. Recent molecular studies also reveal that many plant groups in south-western China, such as *Incarvillea* (Chen et al. 2012) and *Pedicularis* (Yu et al. 2015), diversified in response to environmental changes during the Neogene.

Even though Yunnan is rich in plant species today, not all plants adapted and diversified under the monsoon climate. Some plant taxa, such as *Cedrus*, disappeared from this region, according to the fossil record (Figure 29.2). At present, there are only four living species of *Cedrus*, which are naturally distributed in the western Himalaya

Figure 29.2 Selected fossil taxa, with their latest occurrence in Yunnan. *Metasequoia* sp. from the middle Miocene of Zhenyuan, southern Yunnan; *Sequoia maguanensis* J.W. Zhang et Z.K. Zhou from the late Miocene of Wenshan and Maguan, south-eastern Yunnan; *Cedrelospermum asiaticum* L.B. Jia, Y.J. Huang et Z.K. Zhou from the late Miocene of Maguan, south-eastern Yunnan; and *Cedrus angusta* T. Su, Z.K. Zhou et Y.S. Liu from the late Pliocene of Yongping, western Yunnan. All three of these genera are now absent from Yunnan.

and Mediterranean regions (Maheshwari & Biswas 1970). Although *Cedrus* is widely cultivated in Yunnan, its natural distribution does not extend into this region (Fu et al. 1999). Recently, fossil seed scales of a species described as *C. angusta* T. Su, Z.K. Zhou et Y.S. Liu were reported from the late Pliocene (ca. 3 Ma) of western Yunnan, suggesting the past existence of the genus in this region. Pollen evidence from Neogene and Quaternary strata further supports the post-Pliocene disappearance of *Cedrus* from Yunnan (Su et al. 2013b). Comparisons of climatic conditions between Yunnan and the natural ranges of *Cedrus* indicate that the former has much drier winters (Su et al. 2013b). Interestingly, *Cedrus* seeds are recalcitrant, and germinate immediately upon their maturity (Kumar et al. 2011). It can therefore be hypothesized that dry winters and early springs may have caused the extinction of *Cedrus* from Yunnan by preventing the community establishment of young shoots (Su et al. 2013b).

An analogous example is *Sequoia* (coast redwood) of the Cupressaceae family (Figure 29.2). This genus has only one living species, *S. sempervirens* (D. Don) Endlicher, native to California and Oregon along the Pacific coast of the USA. It has a rich fossil record from Cenozoic strata in the northern hemisphere (Taylor et al. 2008). Many leaves and cones of *S. maguanensis* J.W. Zhang et Z.K. Zhou have been recovered from the late Miocene Lühe flora in central Yunnan (Ma et al. 2005) and the late Miocene Wenshan flora in south-eastern Yunnan (Zhang et al. 2015b). These fossils are morphologically similar to *S. sempervirens*, indicating a

close affinity to their modern relative. Climate comparison shows that winter is much wetter in the distribution area of *S. sempervirens* in the USA than it is in present-day Yunnan (Zhang et al. 2015b). Similar to *Cedrus*, seeds of *S. sempervirens* are recalcitrant and mature during late autumn. Therefore, it is reasonable to hypothesize that the disappearance of *Sequoia* in Yunnan was likely caused by the strengthening of drought in the dry season since the Neogene, which has been associated with the intensification of the Asian monsoon. The relationship between tree susceptibility and seasonal drought has also been demonstrated by a recent controlled experiment with saplings of *S. sempervirens* (Quirk et al. 2013). There are several other taxa that do not exist in Yunnan today that have been found as fossils: for example, *Cedrelospermum* (Jia et al. 2015) and *Metasequoia* (unpublished data). Both taxa have an excellent fossil record in the northern hemisphere (Taylor et al. 2008). *Metasequoia* has only one living species, *M. glyptostroboides* Hu et Cheng, restricted to a limited area in central China, whereas *Cedrelospermum* is globally extinct. Their disappearance from Yunnan can also be linked to the intensification of the Asian monsoon climate.

The monsoon climate has influenced not only plant diversity throughout geological time, but also the spatial distribution pattern of modern biodiversity. South-western China is well known for the diversity of its relict plants, which are now considered to be local endemics, but are well represented in the fossil record in other regions of the world (Manchester et al. 2009). Spatial distribution patterns of relict plants in China indicate that they exhibit the highest diversity in areas with relatively wet autumns and springs (Huang et al. 2015b). High mountains and deep valleys in Yunnan have a positive effect on precipitation in some geographically limited areas during winter, where the climate is different from adjacent monsoonal regions and can provide a suitable environment for the survival of relict plants. Fossil records from Europe also indicate the importance of rainfall for the survival of many East Asian relict taxa during the Neogene and Quaternary (Martinetto et al. 2015). It is possible to observe the different adaptations of seed physiology to the dry season among genera. *Quercus* exhibits a high level of diversity in modern Yunnan, with about 63 species (Wu 1979). Among them, *Quercus sichourensis* (Hu) C.C. Huang & Y.T. Chang is an endemic species that now has fewer than 10 wild individuals in south-eastern Yunnan and south-western Guizhou provinces (Xia et al. 2008). The fact that the seed-drying rate of *Q. sichourensis* is significantly higher than that of other species in *Quercus*, such as *Q. annulata*, *Q. fabri* and *Q. franchetii* (Xia et al. 2012), can largely explain the narrower distribution of *Q. sichourensis* (Wu 1979). Therefore, the drought in dry season associated with the Asian monsoon could have a great influence on seed germination physiology, and thus drives the differences in tree recruitment among *Quercus* species.

In addition to the effects of the intensification of the monsoon climate since the Neogene, other factors may also have contributed to modern plant diversity in Yunnan. The most important of these is the uplift of the THR. Even though models of the uplift are still debated, continued uplift of the south-eastern THR during the Neogene has been verified by several lines of evidence, such as thermochronology, mapping, sedimentology and paleobotany (Schoenbohm et al. 2006). The complex topography caused by the mountain uplift has created high environmental diversity (Xu et al. 2014), allowing species with different ecological niches to coexist. It is common for different species or varieties of a genus to live in a small restricted area in Yunnan. This makes Yunnan one of the diversity centers for numerous plant groups, such as *Rhododendron* (Ma et al. 2016) and *Pedicularis* (Yu et al. 2015). Moreover, due to its complex topography, Yunnan plays a pivotal role as a refugium during climatic perturbations over geological time. A recent molecular study on *Hippophae tibetana* (Schlechtendal) Servettaz. revealed that more than one refugium has survived from the eastern or western region of the THR since the Pleistocene glaciations (Jia et al. 2011). In addition, the uplift of the THR has created many high mountains isolated by deep valleys. It is difficult for both wind- and insect-dispersed plants to exchange genes or hybridize among populations,

because the mountains are geographical barriers to the transport of pollen grains/spores by wind or pollinators (Cun & Wang 2010). Consequently, this reproductive isolation among plant populations may accelerate the differentiation and speciation of wind-dispersed plants such as *Pinus* (Mao et al. 2008), as well as of insect-dispersed plants such as *Meconopsis* (Yang et al. 2012) and *Rhododendron* (Ma et al. 2016).

Overall, even though some taxa have disappeared from this region, many others have adapted and diversified in response to paleoenvironmental changes, especially the monsoon climate and its characteristic seasonal drought. More evidence combining paleobotanical, paleoclimatological, ecological, plant physiological and molecular studies is urgently needed in order to explore the mechanisms driving plant diversity following environmental changes in Yunnan, a fascinating hotspot of modern biodiversity.

29.4 Conclusion

Impacted by the uplift of the THR, Yunnan Province in south-western China has experienced surface uplift, topographic deformation, temperature decline and aridification since the early Miocene. On one hand, these environmental changes have promoted the diversification of numerous plant groups, and the complex landscape, featuring huge mountains and deep valleys, has created a variety of climatic conditions that may allow plant species with different climatic tolerances to coexist in a limited geographical area. On the other, the climatic fluctuations may have resulted in a number of species extinctions. These include the severe aridification associated with the intensification of the Asian monsoon, particularly during the dry season, which may have influenced seed germination and young leaf development, and thus led to the disappearance of several woody genera from Yunnan, such as *Cedrus* and *Sequoia*. To conclude, the pronounced environmental changes caused mainly by the uplift of the THR have driven numerous speciation events – as well as a few extinctions of plants – during the Late Cenozoic, and eventually led to the establishment of modern plant diversity in Yunnan.

Acknowledgments

This work was supported by a grant from the National Natural Science Foundation of China (U1502231).

References

An, Z., Kutzbach, J.E., Prell, W.L. & Porter, S.C. (2001) Evolution of Asian monsoons and phased uplift of the Himalaya-Tibetan plateau since Late Miocene times. *Nature* **411**, 62–66.

Bailey, I.W. & Sinnott, E.W. (1915) A botanical index of Cretaceous and Tertiary climates. *Science* **41**, 831–834.

Chen, H.H., Dobson, J., Heller, F. & Hao, J. (1995) Paleomagnetic evidence for clockwise rotation of the Simao region since the Cretaceous, a consequence of India-Asia collision. *Earth and Planetary Science Letters* **134**, 203–217.

Chen, S-T., Xing, Y-W., Su, T. et al. (2012) Phylogeographic analysis reveals significant spatial genetic structure of *Incarvillea sinensis* as a product of mountain building. *BMC Plant Biology* **12**, 58.

Clark, M.K. & Royden, L.H. (2000) Topographic ooze: building the eastern margin of Tibet by lower crustal flow. *Geology* **28**, 703–706.

Clark, M.K., House, M.A., Royden, L.H. et al. (2005) Late Cenozoic uplift of southeastern Tibet. *Geology* **33**, 525–528.

Cun, Y.-Z. & Wang, X.-Q. (2010) Plant recolonization in the Himalaya from the southeastern Qinghai-Tibetan Plateau: geographical isolation contributed to high population differentiation. *Molecular Phylogenetics and Evolution* **56**, 972–982.

Déry, S. & Wood, E.F. (2005) Observed twentieth century land surface air temperature and precipitation covariability. *Geophysical Research Letters* **32**, L21414.

Eckert, A.J. & Hall, B.D. (2006) Phylogeny, historical biogeography, and patterns of diversification for *Pinus* (Pinaceae): phylogenetic tests of fossil-based hypotheses. *Molecular Phylogenetics and Evolution* **40**, 166–182.

Fu, L.-G., Li, N., Elias, T.S. & Mill, R.R. (1999) *Flora of China*, vol. **4**. Beijing: Science Press and St. Louis, MO: Missouri Botanical Garden.

Ge, H.-R. & Li, D.-Y. (1999) *Cenozoic Coal-Bearing Basins and Coal-Forming Regularity in West Yunnan*. Kunming: Yunnan Science and Technology Press.

Greenwood, D.R. (2005) Leaf form and the reconstruction of past climates. *New Phytologist* **166**, 355–357.

Guo, Z.T., Sun, B., Zhang, Z.S. et al. (2008) A major reorganization of Asian climate by the early Miocene. *Climate of the Past* **4**, 153–174.

Harris, N. (2006) The elevation history of the Tibetan Plateau and its implications for the Asian monsoon. *Palaeogeography, Palaeoclimatology, Palaeoecology* **241**, 4–15.

Hu, Q., Xing, Y., Hu, J. et al. (2014) Evolution of stomatal and trichome density of the Quercus delavayi complex since the late Miocene. *Chinese Science Bulletin* **59**, 310–319.

Huang, Y.-J., Chen, W.-Y., Jacques, F.M.B. et al. (2015a) Late Pliocene temperatures and their spatial variation at the southeastern border of the Qinghai-Tibet Plateau. *Journal of Asian Earth Sciences* **111**, 44–53.

Huang, Y., Jacques, F.M.B., Su, T. et al. (2015b) Distribution of Cenozoic plant relicts in China explained by drought in dry season. *Scientific Reports* **5**, 14212.

Jacques, F.M.B., Guo, S.-X., Su, T., et al. (2011a) Quantitative reconstruction of the Late Miocene monsoon climates of southwest China: a case study of the Lincang flora from Yunnan Province. *Palaeogeology, Palaeoclimatology, Palaeoecology* **304**, 318–327.

Jacques, F.M.B., Su, T., Spicer, R.A., et al. (2011b) Leaf physiognomy and climate: are monsoon systems different? *Global and Planetary Change* **76**, 56–62.

Jacques, F.M.B., Su, T., Spicer, R.A. et al. (2014) Late Miocene southwestern Chinese floristic diversity shaped by the southeastern uplift of the Tibetan Plateau. *Palaeogeography, Palaeoclimatology, Palaeoecology* **411**, 208–215.

Jia, D.-R., Liu, T.-L., Wang, L.-Y. et al. (2011) Evolutionary history of an alpine shrub *Hippophae tibetana* (Elaeagnaceae): allopatric divergence and regional expansion. *Biological Journal of the Linnean Society* **102**, 37–50.

Jia, L.-B., Manchester, S.R., Su, T. et al. (2015) First occurrence of *Cedrelospermum* (Ulmaceae) in Asia and its biogeographic implications. *Journal of Plant Research* **128**, 747–761.

Kou, X.-Y., Ferguson, D.K., Xu, J.-X. et al. (2006) The reconstruction of palaeovegetation and palaeoclimate in the Late Pliocene of west Yunnan, China. *Climatic Change* **77**, 431–448.

Kumar, R., Singh, C., Malik, S. et al. (2011) Effect of storage conditions on germinability of Himalayan Cedar (*Cedrus deodara* Roxb., G. Don) seeds. *Indian Forester* **137**, 1099–1102.

Li, X.-X. (1995) *Fossil Floras of China through the Geological Ages*. Guangzhou: Guangdong Science and Technology Press.

Li, C.-S., Yi, T.-M. & Yao, Y.-F. (2008) *Vegetation Succession and Environment Change in China*, vol. **1**. Nanjing: Jiangsu Science and Technology Publishing House.

Li, S.-F., Mao, L.-M., Spicer, R.A. et al. (2015) Late Miocene vegetation dynamics under monsoonal climate in southwestern China. *Palaeogeography, Palaeoclimatology, Palaeoecology* **425**, 14–40.

Licht, A., van Cappelle, M., Abels, H.A. et al. (2014) Asian monsoons in a late Eocene greenhouse world. *Nature* **513**, 501–506.

Liu, X.D. & Yin, Z.Y. (2002) Sensitivity of East Asian monsoon climate to the uplift of the Tibetan Plateau. *Palaeogeography, Palaeoclimatology, Palaeoecology* **183**, 223–245.

López-Pujol, J., Zhang, F-M., Sun, H-Q. et al. (2011) Centres of plant endemism in China: places for survival or for speciation? *Journal of Biogeography* **38**, 1267–1280.

Ma, Q.-W., Li, F.-L. & Li, C.-S. (2005) The coast redwoods (*Sequoia*, Taxodiaceae) from the Eocene of Heilongjiang and the Miocene of Yunnan, China. *Review of Palaeobotany and Palynology* **135**, 117–129.

Ma, Y.-P., Xie, W.-J., Sun, W.-B. & Marczewski, T. (2016) Strong reproductive isolation despite occasional hybridization between a widely distributed and a narrow endemic *Rhododendron* species. *Scientific Reports* **6**, 19146.

Maheshwari, P. & Biswas, C. (1970) *Cedrus*. Botanical Monograph No. 5. New Delhi: Council of Scientific and Industrial Research, Rafi Marg.

Manchester, S.R., Chen, Z.-D., Lu, A.-M. & Uemura, K. (2009) Eastern Asian endemic seed plant genera and their paleogeographic history throughout the Northern Hemisphere. *Journal of Systematics and Evolution* **47**, 1–42.

Mao, J.-F., Li, Y. & Wang, X.-R. (2008) Empirical assessment of the reproductive fitness components of the hybrid pine *Pinus densata* on the Tibetan Plateau. *Evolutionary Ecology* **23**, 447–462.

Martinetto, E., Momohara, A., Bizzarri, R. et al. (2015) Late persistence and deterministic extinction of "humid thermophilous plant taxa of East Asian affinity" (HUTEA) in southern Europe. *Palaeogeography, Palaeoclimatology, Palaeoecology* doi: 10.1016/j. palaeo.2015.08.015.

Mosbrugger, V. & Utescher, T. (1997) The coexistence approach – a method for quantitative reconstructions of Tertiary terrestrial palaeoclimate data using plant fossils. *Palaeogeography, Palaeoclimatology, Palaeoecology* **134**, 61–86.

Mosbrugger, V., Utescher, T. & Dilcher, D.L. (2005) Cenozoic continental climatic evolution of Central Europe. *Proceedings of the National Academy of Sciences* **102**, 14964–14969.

Quirk, J., McDowell, N.G., Leake, J.R. et al. (2013) Increased susceptibility to drought-induced mortality in *Sequoia sempervirens* (Cupressaceae) trees under Cenozoic atmospheric carbon dioxide starvation. *American Journal of Botany* **100**, 582–591.

Royden, L.H., Burchfiel, B.C. & van der Hilst, R.D. (2008) The geological evolution of the Tibetan Plateau. *Science* **321**, 1054–1058.

Ruth, S., Renee, M., Li, H. et al. (2008) Spatial patterns of plant diversity and communities in Alpine ecosystems of the Hengduan Mountains, northwest Yunnan, China. *Journal of Plant Ecology* **1**, 117–136.

Schoenbohm, L.M., Burchfiel, B.C. & Chen, L. (2006) Propagation of surface uplift, lower crustal flow, and Cenozoic tectonics of the southeast margin of the Tibetan Plateau. *Geology* **34**, 813–816.

Shen, Z.-K., Lü, J., Wang, M. & Bürgmann, R. (2005) Contemporary crustal deformation around the southeast borderland of the Tibetan Plateau. *Journal of Geophysical Research* **110**, B11409.

Su, T., Xing, Y.-W., Liu, Y.-S. et al. (2010) Leaf margin analysis: a new equation from humid to mesic forests in China. *Palaios* **25**, 234–238.

Su, T., Jacques, F.M.B., Spicer, R.A. et al. (2013a) Post-Pliocene establishment of the present monsoonal climate in SW China: evidence from the late Pliocene Longmen megaflora. *Climate of the Past* **9**, 1911–1920.

Su, T., Liu, Y.-S., Jacques, F.M.B. et al. (2013b) The intensification of the East Asian winter monsoon contributed to the disappearance of *Cedrus* (Pinaceae) in southwestern China. *Quaternary Research* **80**, 316–325.

Su, T., Huang, Y.-J., Meng, J. et al. (2016) A Miocene leaf fossil record of *Rosa* (*R. fortuita* nov. sp.) from its modern diversity center in SW China. *Palaeoworld* **25**, 104–115.

Sun, X. & Wang, P. (2005) How old is the Asian monsoon system? – Palaeobotanical records from China. *Palaeogeography, Palaeoclimatology, Palaeoecology* **222**, 181–222.

Sun, B.-N., Wu, J.-Y., Liu, Y.-S. et al. (2011) Reconstructing Neogene vegetation and climates to infer tectonic uplift in western Yunnan, China. *Palaeogeography, Palaeoclimatology, Palaeoecology* **304**, 328–336.

Tao, J.-R. (2000) *The Evolution of the Late Cretaceous-Cenozoic Flora in China*. Beijing: Science Press.

Taylor, T.N., Taylor, E.L. & Krings, M. (2008) *Paleobotany: The Biology and Evolution of Fossil Plants*, 2nd edn. New York: Academic Press.

Uhl, D., Klotz, S., Traiser, C. et al. (2007) Cenozoic paleotemperatures and leaf physiognomy – a European perspective. *Palaeogeography, Palaeoclimatology, Palaeoecology* **248**, 24–31.

Utescher, T., Mosbruger, V. & Ashraf, A.R. (2000) Terrestrial climate evolution in Northwest Germany over the last 25 million years. *Palaios* **15**, 430–449.

Utescher, T., Bruch, A.A., Micheels, A. et al. (2011) Cenozoic climate gradients in Eurasia – a palaeo-perspective on future climate change? *Palaeogeography, Palaeoclimatology, Palaeoecology* **304**, 351–358.

Utescher, T., Bruch, A.A., Erdei, B. et al. (2014) The Coexistence Approach – theoretical background and practical considerations of using plant fossils for climate quantification. *Palaeogeography, Palaeoclimatology, Palaeoecology* **410**, 58–73.

van Dam, J.A. (2006) Geographic and temporal patterns in the late Neogene (12–3 Ma) aridification of Europe: the use of small mammals as paleoprecipitation proxies. *Palaeogeography, Palaeoclimatology, Palaeoecology* **238**, 190–218.

Wang, Y. (2006) *Yunnan Mountain Climate*. Kunming: Yunnan Science and Technology Press.

Westaway, R. (2009) Active crustal deformation beyond the SE margin of the Tibetan Plateau: constraints from the evolution of fluvial systems. *Global and Planetary Change* **68**, 395–417.

Wolfe, J.A. (1993) Method of obtaining climatic parameters from leaf assemblages. *US Geological Survey Bulletin* **2040**, 1–71.

Wolfe, J.A. & Spicer, R.A. (1999) Fossil leaf character states: multivariate analysis. In: Jones, T.P. & Rowe, N.P. (eds.) *Fossil Plants and Spores: Modern Techniques.* London: Geological Society, pp. 233–239.

WGCPC (1978) *Cenozoic Plants from China*. Fossil Plants of China, vol. **3**. Beijing: Science Press.

Wu, Z.-Y. (1979) *Flora of Yunnan*, vol. **2**. Beijing: Science Press.

Wu Z.-Y. (1988) The Hengduan Mountain flora and her significance. *Journal of Japanese Botany* **63**, 297–311.

Wu, Z.-Y. & Zhu, Y.-C. (1987) *Vegetation of Yunnan*. Beijing: Science Press.

Xia, K., Zhou, Z.-K., Chen, W.-Y. & Sun, W.-B. (2008) Rescuing the Sichou oak *Quercus sichourensis* in China. *Oryx* **42**, 3–4.

Xia, K., Su, T., Liu Y.-S. et al. (2009) Quantitative climate reconstructions of the late Miocene Xiaolongtan megaflora from Yunnan, southwest China. *Palaeogeography, Palaeoclimatology, Palaeoecology* **276**, 80–86.

Xia, K., Daws, M.I., Stuppy, W. et al. (2012) Rates of water loss and uptake in recalcitrant fruits of *Quercus* species are determined by pericarp anatomy. *PLoS ONE* **7**, e47368.

Xie, S., Sun, B., Wu, J. et al. (2012) Palaeoclimatic estimates for the late Pliocene based on leaf physiognomy from western Yunnan, China. *Turkish Journal of Earth Sciences* **21**, 251–261.

Xing, Y.-W., Liu, Y.-S., Su, T. et al. (2010) *Pinus prekesiya* sp. nov. from the upper Miocene of Yunnan, southwestern China and its biogeographical implications. *Review of Palaeobotany and Palynology* **160**, 1–9.

Xing, Y., Utescher, T., Jacques, F.M.B. et al. (2012) Paleoclimatic estimation reveals a weak winter monsoon in southwestern China during the late Miocene: evidence from plant macrofossils. *Palaeogeography, Palaeoclimatology, Palaeoecology* **358–360**, 19–26.

Xing, Y.-W., Hu, J.-J., Jacques, F.M.B. et al. (2013) A new *Quercus* species from the upper Miocene of southwestern China and its ecological significance. *Review of Palaeobotany and Palynology* **193**, 99–109.

Xu, J.-X., Ferguson, D.K., Li, C.-S. et al. (2004) Climatic and ecological implications of late Pliocene palynoflora from Longling, Yunnan, China. *Quaternary International* **117**, 91–103.

Xu, J.-X., Ferguson, D.K., Li, C.-S. & Wang, Y.-F. (2008) Late Miocene vegetation and climate of the Lühe region in Yuanmou, southwestern China. *Review of Palaeobotany and Palynology* **148**, 36–59.

Xu, B., Li, Z.-M. & Sun, H. (2014) Plant diversity and floristic characters of the alpine subnival belt flora in the Hengduan Mountains, SW China. *Journal of Systematics and Evolution* **52**, 271–279.

Yang, F.-S., Qin, A.-L., Li, Y-F. & Wang, X-Q. (2012) Great genetic differentiation among populations of *Meconopsis integrifolia* and its implication for plant speciation in the Qinghai-Tibetan Plateau. *PLoS ONE* **7**, e37196.

Yao, Y.-F., Bruch, A.A., Cheng, Y.-M. et al. (2012) Monsoon versus uplift in southwestern China – late Pliocene climate in Yuanmou Basin, Yunnan. *PloS ONE* **7**, e37760.

Yu, W.-B., Liu, M.-L., Wang, H. et al. (2015) Towards a comprehensive phylogeny of the large temperate genus *Pedicularis* (Orobanchaceae), with a emphasis on species from the Himalaya-Hengduan Mountains. *BMC Plant Biology* **15**, 176.

Zachos, J., Pagani, M., Sloan, L. et al. (2001) Trends, rhythms, and aberrations in global climate 65 Ma to present. *Science* **292**, 686–693.

Zhang, J.-J. & Ding, L. (2003) East west extension in Tibetan Plateau and its significance to tectonic evolution. *Chinese Journal of Geology* **38**, 179–189.

Zhang, Q.-Q., Ferguson, D.K., Mosbrugger, V. et al. (2012) Vegetation and climatic changes of SW China in response to the uplift of Tibetan Plateau. *Palaeogeography, Palaeoclimatology, Palaeoecology* **363–364**, 23–36.

Zhang, J.-W., D'Rozario, A., Adams, J.M. et al. (2015a) The occurrence of *Pinus massoniana* Lambert (Pinaceae) from the upper Miocene of Yunnan, SW China and its implications for paleogeography and paleoclimate. *Review of Palaeobotany and Palynology* **215**, 57–67.

Zhang, J.-W., D'Rozario, A., Adams, J.M. et al. (2015b) *Sequoia maguanensis*, a new Miocene relative of the coast redwood, *Sequoia sempervirens*, from China: implications for paleogeography and paleoclimate. *American Journal of Botany* **102**, 103–118.

Zhao, L.-C., Wang, Y.-F., Liu, C.-J. & Li, C.-S. (2004) Climatic implications of fruit and seed assemblage from Miocene of Yunnan, southwestern China. *Quaternary International* **117**, 81–89.

Zheng, H., Powell, C.M., Rea, D.K. et al. (2004) Late Miocene and mid-Pliocene enhancement of the East Asian monsoon as viewed from the land and sea. *Global and Planetary Change* **41**, 147–155.

Zheng, H., Clift, P.D., Wang, P. et al. (2013) Pre-Miocene birth of the Yangtze River. *Proceedings of the National Academy of Sciences* **110**, 7556–7561.

30

Influence of Mountain Formation on Floral Diversification in Japan, Based on Macrofossil Evidence

Arata Momohara

Graduate School of Horticulture, Chiba University, Matsudo, Chiba, Japan

Abstract

This chapter reviews the development of the geomorphology and flora of Japan since the Paleogene and discusses the influence of mountains on floral diversity. From the late Eocene to the Oligocene, cool-temperate genera comprising the present mountain vegetation migrated from higher latitudes and established in Japan, which was on the margin of the Asian continent. During the mid-Miocene climate optimum (MCO), the Japanese islands were small islands in the Pacific Ocean where volcanic mountains provided habitats for boreal plants and conifers at higher altitudes. Since the late Miocene, crust compression has uplifted mountain ranges and differentiated the inland climate. The increasing formation of mountain ranges since the late Pliocene, in association with climate changes, has influenced the habitats and migration routes of plants, ultimately resulting in the extinction of plants that were dominant in the Neogene.

Keywords: *Cenozoic, plant extinction, tectonic movement, volcanic mountain*

30.1 Introduction

The Japanese islands have a rich flora, comprising 5685 vascular plant species, including 1719 endemic to the area, which covers 377 972 km^2 (Kato & Ebihara 2011). Its high diversity and endemism are a result not only of the Quaternary climate history but also of the development of the present geomorphology during the Neogene and Quaternary. Mountains played an important role in the diversification of the island's biota, as evidenced by the distribution of endemic plants, which are concentrated in alpine regions in addition to remote locations such as the Ogasawara and Ryukyu islands (Kato & Ebihara 2011).

Mountains occupy 75% of the total area of the Japanese islands (GIAJ 1990), and those 2000–3000 m in height constitute the backbone range of the country (Figure 30.1). The mountainous landscape is ascribed to the country's position along an active margin of the Eurasian continental plate, where thrust tectonism has shaped the steep geomorphology. Subduction of the oceanic plate has caused a volcanic frontal zone along the island arc that has formed high-altitude volcanoes, represented by Mt. Fuji (3776 m high) (Figure 30.2) and several 3000 m examples in the Hida Mountains. The Neogene and Quaternary sedimentary basins are well developed, and include marine deposits containing index fossils useful for the correlation of the sediments. Active volcanism has supplied pyroclastic rocks that can be used for the absolute dating and correlation of sediments within and among basins. This chapter reviews the development of mountain geomorphology and flora since the Paleogene as a basis for discussing the influences of mountains on biodiversity in the Japanese islands.

30.2 Distribution and Characteristics of Mountain Vegetation and Flora in Japan

In Japan, where annual precipitation (700–4700 mm) is sufficient for plant growth, temperature is the major factor determining the altitudinal and latitudinal distribution of vegetation (Figure 30.1). The lowlands and hilly zones south of 38°N in north-eastern Honshu are classified as a warm-temperate zone, where the mean annual temperature (MAT)

Mountains, Climate and Biodiversity, First Edition. Edited by Carina Hoorn, Allison Perrigo and Alexandre Antonelli.
© 2018 John Wiley & Sons Ltd. Published 2018 by John Wiley & Sons Ltd.
Companion website: www.wiley.com\go\hoorn\mountains,climateandbiodiversity

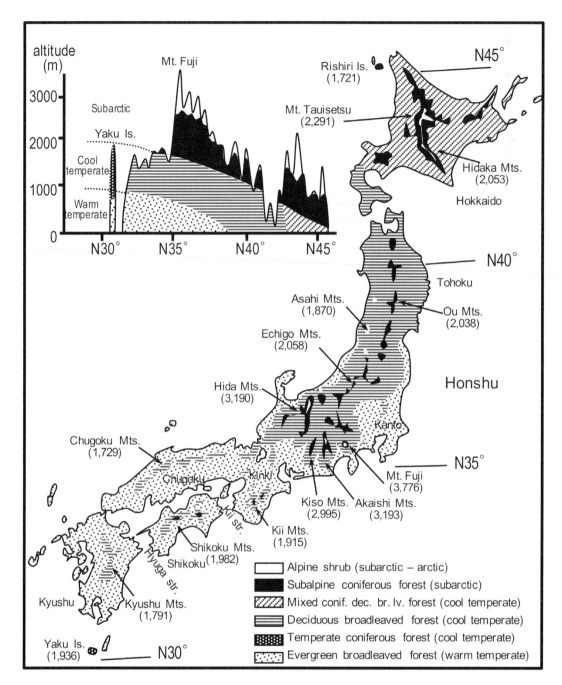

Figure 30.1 Altitudinal and latitudinal distribution of forest formations north of 30°N in Japan. Parentheses indicate the altitude (m) of the highest mountain peak or range. *Source:* Adapted from Yoshioka (1973) and Horikoshi & Aoki (1985).

is 12–20 °C. The vegetation in this zone is evergreen broadleaved forest dominated by evergreen Fagaceae (*Quercus* subgen. *Cyclobalanopsis* and *Castanopsis*) and Lauraceae (*Machilus, Cinnamomum,* etc.), except in inland basins, where deciduous broadleaved forest has established under colder winter temperatures. The subtropical zone (MAT >20 °C), distributed in the south of Okinawa Island (26° N), is represented by tree ferns, strangler *Ficus* and mangroves (Kira 1991).

The mountain zone in central and western Japan and the lowland zone in northern Japan represent a cool-temperate zone (MAT ~5–12 °C). Deciduous broadleaved forest dominated by beech (*Fagus crenata*) is distributed south of the Oshima Peninsula in south-west Hokkaido, and mixed coniferous and deciduous broadleaved tree forest dominated by *Quercus crispula, Abies sachalinensis* and *Picea jezoensis* is found in the north. On Yaku Island, beech

Figure 30.2 Mt. Fuji (3776 m) and its lateral volcanoes (on the right side, 1424–1566 m), which erupted in AD 864–866, with basaltic lava flow covering the north-western foot. See also Plate 48 in color plate section.

is absent and coniferous forest of *Cryptomeria japonica*, *Abies firma* and *Tsuga sieboldii* is dominant. Such coniferous forest has also developed around the boundary between the warm and cool temperate zones on the Pacific side of the mountains in western Japan (Yoshioka 1973).

The sub-alpine (or sub-arctic) zone in Honshu (MAT < ~5 °C) is covered by evergreen coniferous forest, in which the dominant trees are *Abies veitchii*, *A. mariesii* and *Tsuga diversifolia*. The uppermost mountain zone in central and northern Japan is covered by shrubs of *Pinus pumila* and Ericaceae, together with alpine plants that share a common origin with the flora of the circumpolar arctic zone.

Another climatic factor controlling forest physiognomy is the heavier snowfall on mountains facing the Sea of Japan. The cold and dry winter monsoon absorbs moisture from the Sea of Japan, where the Tsushima Warm Current prevails. This collides with the mountains, resulting in snow accumulation in the mountain and sub-alpine zones. Its physical impact diminishes the diversity of cool temperate trees and has ultimately established forests of pure beech in the mountains in the western part of the Tohoku District, including the Asahi Mountains (Figure 30.1).

30.3 Mountains in the Development of Geomorphology in Japan

The dominant volcanic rocks from the Cenozoic indicate the persistence of a volcanic frontal zone along the length of Japan (Figure 30.3) and the existence of volcanic mountains. Mountain ranges in Japan began to uplift in the late Miocene, when the crust of the Japanese islands began to be compressed. Evidence for this is provided by volcanic and sedimentary rocks dated by the fission-track (see Chapter 5) and potassium–argon (K/A) methods, as well as by marine index fossils such as foraminifera, diatoms and calcareous nanoplankton. Until the beginning of the late Oligocene rifting, the crust of the present-day Japanese islands was attached to the continental margin – as evidenced by the correlation of the geotectonic belt (Yamakita & Otoh 2000) – except for eastern Hokkaido and the Izu Peninsula, which migrated on the oceanic plate and accreted to the Japanese islands in the early Miocene and the Pleistocene, respectively (Figure 30.3a).

Rifting along the continental margin began ca. 30 Ma and moved the crusts of Japan to their present position, accompanied by very active volcanism, before ceasing

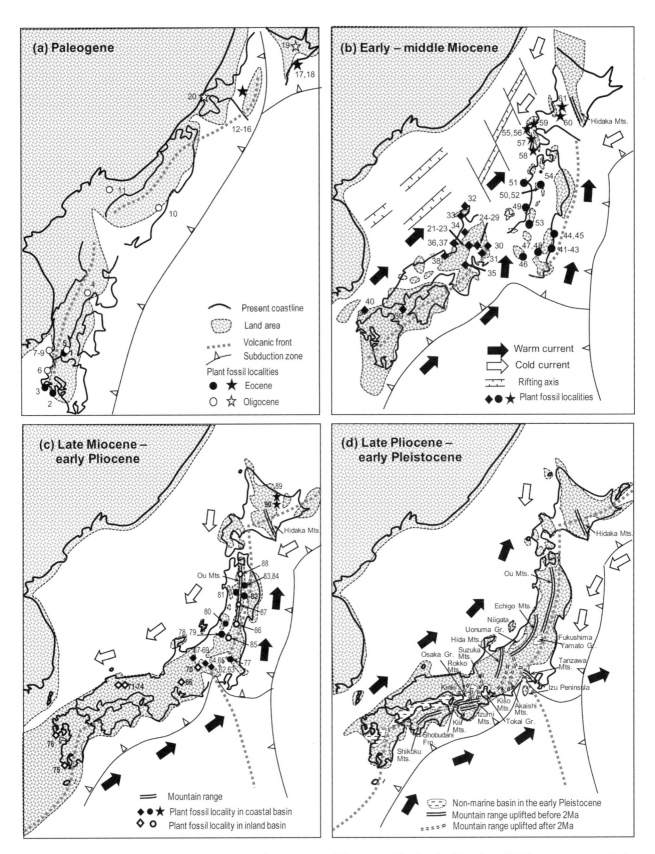

Figure 30.3 Topographical changes in the Japanese islands since the Paleogene, with plant fossil localities. (a) Oligocene geomorphology and Paleogene fossil localities. Black circle: Eocene localities in western Japan. 1, Ube (Huzioka & Takahashi 1970); 2, Takashima (Matsuo 1967); 3, Sakito (Matsuo 1970). Black star: Eocene localities in Hokkaido (Tanai 1970, 1991). 12–16, Ishikari (12, Noborikawa; 13, Yubari; 14, Bibai; 15, Ikushunbetsu; 16, Ashibetsu); 17, 18, Kushiro (17, Harutori; 18, Shakubetsu). White circle: Oligocene localities in central and western Japan. 4, Kobe (Kobatake 1983); 5, Shimokatakura (Matsushita et al. 1994); 6, Ainoura (Tanai 1961); 7, Ouchiyama-kami (Uemura et al. 1999); 8, Noda (Tanai & Uemura 1991); 9, Daibo (Huzioka 1974); 10, Shichiku (Huzioka 1964); 11, Seki (Huzioka 1964). White star: Oligocene localities in Hokkaido. 19, Wakamatsuzawa (Tanai & Uemura 1983; Tanai 1991); 20, Garo (Tanai & Suzuki 1972). (b) Late early Miocene geomorphology and early and middle Miocene fossil localities. Diamond: Fossil localities in western Japan. 21, Nakamura (Ina et al. 2007a); 22, Hachiya; 23, Hiramaki; 24, Agi; 25, Inkyoyama; 26, Yamanouchi; 27, Toyama; 28, Shukunohora; 29, Oidawara; 30, Tomikusa; 31, Shibaishitoge (22–31 all Ina 1992); 32, Noroshi (Ishida 1970); 33, Noto-Nakajima (Matsuo 1963); 34, Tatsunokuchi (Kitanaka & Fuji 1988);

at ca. 14 Ma (Figure 30.3b) (Jolivet et al. 1994). During this movement, the eastern Japan crust rotated anticlockwise between 20 and 15 Ma, and the western Japan crust rotated clockwise between 17 and 15 Ma. The dominant tension field in the crust, which was stretched by the rifting, depressed the marginal sea surrounding the small islands of Japan. Along the collision zone in central Hokkaido, the Hidaka Mountains were formed by fast uplift in the late early Miocene (Arita et al. 2001).

In the late Miocene, compression of oceanic plates prevailed, as evidenced by the increasing reverse-fault in the crust. This movement increased the area of the Japanese islands and uplifted mountain ranges (Figure 30.3c). The Ou Mountains, which are a backbone range of north Honshu, began to uplift, and intramontane basins subsided during the thrusting phase between 12 and 9 Ma, and again since 6.5 Ma (Nakajima et al. 2006).

Mountain ranges in central Honshu, with peaks of 2000–3200 m high, began to uplift in the late Pliocene (Figures 30.3d and 30.4). The expansion of a pebbly fan delta in the south Niigata sedimentary basin indicates uplift of the Echigo Mountains since 2.6 Ma (Kazaoka et al. 1986). The Hida Mountains, which have upheaved and have produced pebbles since the late Miocene, began to uplift actively in a stepwise manner between 2.5 and 1.5 Ma, and again since 0.8 Ma, as evidenced by the increased supply of gravel and pyroclastic rocks (Oikawa 2003). The collision of the present-day Izu Peninsula on the Philippine Plate with Honshu has promoted the active uplift of mountains since the late early Pleistocene. The Akaishi and Kiso mountains began to uplift from 1.4 and 0.7 Ma, respectively (Suganuma et al. 2003).

In western Japan, inland basins with fluvial plains developed eastward and westward from the Tokai District to central Kyushu (Figure 30.3d) from the late Pliocene onwards. Sedimentary basins were formed by a northward compression of the Philippine Plate, along with mountains in the north (Chugoku Mountains) and south (Kii, Izumi and Shikoku mountains), running parallel with the basin (Huzita 1980). Since the late early Pleistocene (ca. 1.2 Ma), the prevailing westward compression formed mountains running south and north, which dissected the basins. The corresponding axes of subsidence have depressed the Kii and Hyuga straits and isolated the Shikoku and Kyushu islands in the interglacial stages since the late early Pleistocene (Huzita 1980).

The western part of Japan became disconnected from the continent in the late Pliocene. The Tsushima Warm Current began to flow into the Sea of Japan at 3.2 Ma, and became frequent from 1.6 Ma (Figure 30.4) (Kitamura & Kimoto 2006). Comparative studies of proboscidean fauna between Japan and China indicate that the land bridges between western Japan and the continent, reflected in faunal immigration, were limited to three glacial stages since the late Pliocene at 1.2, 0.63 and 0.43 Ma (Taruno 2010).

30.4 Development of Mountain Flora Since the Paleogene

30.4.1 Evidence of Mountain Flora in Fossil Records

Cenozoic fossil assemblages often include combinations of thermophilous taxa, distributed in subtropical to warm-temperate regions, and boreal taxa, limited to cool-temperate to sub-arctic regions. This admixture has sometimes been ascribed to intermingled growth of both climatic elements within the same vegetation under a less seasonal climate (e.g., Miki 1948). However, evidence of volcanic activity and tectonic movements around the sedimentary basins confirms the existence of mountains. Boreal taxa at higher altitudes were transported to locations where fossils were deposited and mixed with thermophilous taxa. Compared with the

---

Figure 30.3 (Continued) 35, Igami (Ina et al. 2007b); 36, Itoo (Yabe 2008b); 37, Komegawaki (Uemura & Yasuno 1991); 38, Yosa (Onoe 1978); 39, Heigun (Uemura 2000); 40, Chojabaru (Hayashi 1975). Circle: Fossil localities in eastern Honshu. 41, Lower Kunugidaira; 42, Upper Kunugidaira; 43, Nakayama; 44, Shiote; 45, Ouchi (41–45 all Yabe 2008a); 46, Itsukaichi (Uemura et al. 2001); 47, Asakawa (Horiuchi & Takimoto 2001); 48, Kitataki (Horiuchi 1996); 49, Nishitagawa (Huzioka 1964); 50; Aniai (Huzioka 1964); 51, Daijima (Kano et al. 2011); 52, Utto (Huzioka 1963); 53, Oguni (Onoe 1974); 54, Fujikura (Tanai & Uemura 1988). Star: Fossil localities in Hokkaido. 55, Kudo; 56, Wakamatsu; 57, Kaminokuni; 58, Yoshioka; 59, Abura (55–59 all Tanai & Suzuki 1963, 1972); 60, Takinoue (Tanai & Uemura 1988); 61, Sakipenpetsu (Tanai 1971). (c) Late Miocene geomorphology and late Miocene and early Pliocene fossil localities. Black diamond: Fossil localities in coastal basins of western and central Japan. 62, Middle Yagii (Horiuchi 1996); 63, Upper Yagii (Horiuchi 1996); 64, Lower Itahana; 65, Upper Itahana; 66, Seto; 67, Sashikiri; 68, Chausuyama; 69, Ohoka (64–70 all Ozaki 1991). White diamond: Fossil localities in inland basins in western and central Japan. 70, Kabutoiwa; 71, Tatsumi-Toge (Ozaki 1981); 72, Mitoku (Tanai & Onoe 1961); 73, Onbara (Tanai & Onoe 1961); 74, Ningyo-Toge (Tanai & Onoe 1961); 75, Nakayama (Ina & Ishikawa 1982); 76, Mogi (Tanai 1976). Black circle: Fossil localities in coastal basins in eastern Honshu. 77, Kume (Takimoto et al. 1998); 78, Nishihaga; 79, Koyanaizu; 80, Takamine; 81, Miyata; 82, Hishinai; 83, Gomyojin; 84, Gosho (78–84 all Uemura 1988). White circle: Fossil localities in inland basins in eastern Honshu. 85, Kuromori; 86, Shirakawa; 87, Sanzugawa; 88, Tayama (85–88 all Uemura 1988). Black star: Fossil localities in Hokkaido. 89, Syanabuchi (Tanai & Suzuki 1965); 90, Rubeshibe (Tanai & Suzuki 1965). (d) Geomorphology of Japan in the early Pleistocene. *Source:* Land area, Kano et al. (1991); volcanic front, Kano et al. (2007); crust in the Oligocene, Yamakita & Otoh (2000); crust, rifting axis, subduction zone in the Neogene, Jolivet et al. (1994) and Honza et al. (2004); land area and sea currents in the Neogene, Kano et al. (2007) and Ogasawara (1994).

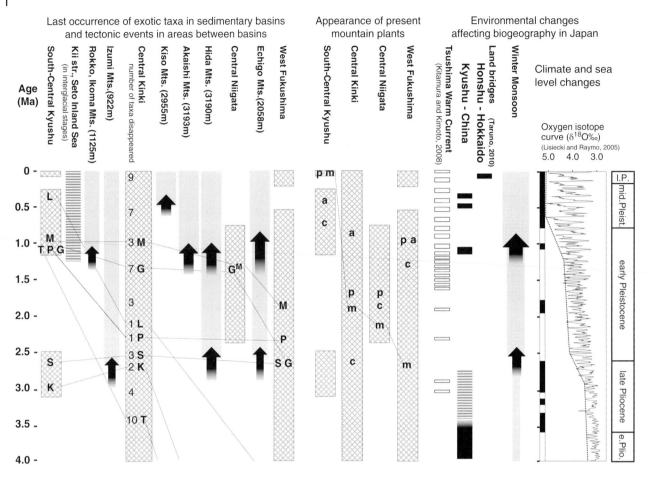

Figure 30.4 The last occurrences of major exotic taxa and the first appearances of taxa in present-day mountain vegetation in four sedimentary basins associated with the formation of the mountains surrounding the basins, with major environmental events. Rectangles with a checked pattern indicate the stratigraphic range of the sediments of each basin. Horizons of last occurrence are indicated by the following abbreviations: G, *Glyptostrobus*; K, *Keteleeria*; L, *Liquidambar*; M, *Metasequoia*; P, *Pseudolarix*; S, *Sequoia*; T, *Taiwania*. Horizons of appearance: a, *Abies veitchii*; c, *Cryptomeria japonica*; m, *Menyanthes trifoliata*; p, *Picea jezoensis*. *Source:* Adapted from Momohara (2016) and Lisiecki & Raymo (2005).

present-day altitudinal forest zonation, Japanese paleo-botanists have assumed the existence of high mountains surrounding sedimentary basins, and they have reconstructed altitudinal distributions of the fossil taxa in these mountains (Tanai 1961, 1967; Ishida 1970; Uemura 1988; Ina 1992).

30.4.2 Paleogene

Most Paleogene floras in Japan (Figure 30.3a) are correlated with stages later than the late Lutetian in the middle Eocene (since ca. 44 Ma) (Ozaki et al. 1996). Climatic change since this stage is reconstructed based on leaf-margin characters, which are calculated by finding the percentage of tree taxa with entire margined leaves among the total broadleaved tree taxa in a local flora. This characteristic exhibits a positive correlation with MAT (Chapter 7) (Wolfe 1979). Cooling trends with temperature depressions at 41 and 37–34 Ma are reconstructed in

the coal-field sequence in Hokkaido (Figure 30.5) (Tanai 1992; Ozaki et al. 1996). Among the five stages that Tanai (1990) defined in the Eocene floral sequence in Hokkaido, the first (no. 12 in Figure 30.5), third (no. 14) and fourth (no. 15) are characterized by rich evergreen broadleaved tree taxa and tropical genera such as Sterculiaceae, Icacinaceae, Menyspermaceae, *Sabalites* and *Musophyllum*. The Ube (no. 1) and Takashima (no. 2) floras in western Japan are associated with the third stage in Hokkaido and exhibit the most tropical characteristics (Tanai 1992). In contrast, floras of the second (no. 13) and fifth (nos. 16–18) stages in Hokkaido are dominated by temperate deciduous broadleaved trees and conifers, while evergreen broadleaved trees decreased and tropical genera disappeared during these stages.

Global cooling, culminating in the Oligocene, promoted the migration of boreal and cool-temperate taxa from the polar region to the East Asian mid-latitude region. The latest Eocene and earliest Oligocene floras in the Kobe

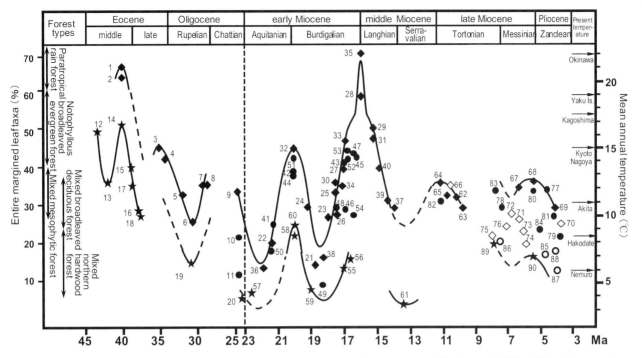

Figure 30.5 Paleotemperature changes reconstructed from the percentage of entire-margined leaf taxa in leaf fossil floras between the middle Eocene and early Pliocene in Japan. Fossil records (4, 5, 7, 21–31, 34–48, 62–70, 75, 77, 85, 86) were added to the original figures (based on Tanai 1991), and their ages were revised based on an up-to-date radiometric and marine micropaleontological correlation. Forest types are based on Wolfe (1979). Data used for leaf floras cover more than 30 broadleaved taxa, except for nos. 20, 38 and 68. Mean annual temperatures (MATs) in Paleogene floras at nos. 2, 3, 14, 16 and 18 are estimated based on the composition of fossil floras (Tanai 1991). Locality number and symbol are the same as those indicated in Figure 30.3b,c. Black diamond, localities in western and central Japan; black circle, localities in eastern Honshu; black star, localities in Hokkaido; white symbols, Neogene inland localities.

Group (no. 4 in Figure 30.5), western Honshu, are dominated by temperate deciduous broadleaved trees such as *Fagus, Betula, Castanea, Zelkova* and *Liquidambar,* and include *Thuja,* which has modern relatives distributed in the sub-alpine zone in Japan. Evergreen broadleaved Fagaceae and Lauraceae trees and subtropical taxa such as Palmae are mixed in this flora. Because of the dominance of the temperate taxa common in the Miocene floras, the age of this flora was correlated with the Miocene (Tanai 1961; Kobatake 1983) before Ozaki (1992) and Ozaki et al. (1996) dated the groups to between 37 and 30 Ma. The late early Oligocene Wakamatsuzawa flora (no. 19 in Figure 30.5) in eastern Hokkaido is composed mainly of temperate taxa, represented by conifers and deciduous broadleaved trees, that are comparable to Miocene species in Japan (Tanai & Uemura 1983). The coniferous genera that make up modern mountain vegetation in Japan, such as *Abies, Picea, Tsuga* and *Sciadopitys,* first appeared in this flora.

30.4.3 Early and Middle Miocene

The Japanese islands consisted of small islands in the Pacific Ocean during this stage (Figure 30.3b), and were strongly influenced by marine climate changes (Ogasawara

et al. 2003). The percentage of species with entire-margined leaves in the early Miocene floras in Honshu varied with temperature fluctuations between warmer (ca. 21–19 and 17–15 Ma) and cooler (23–21 and 19–17 Ma) stages (Figure 30.3b). Fossil floras in the cooler stages are dominated by temperate deciduous trees, including Betulaceae and Aceraceae, together with conifers. The floristic composition is similar to that in the Oligocene, but consists of plants representing extant species in the mountain and sub-alpine zones of central Japan, including *Abies homolepis, A. mariesii, Picea bicolor, Pinus pentaphylla, Thuja standishii, Thujopsis dolabrata* and *Alnus maximowicziana.*

Assemblages in alternative warm stages in Honshu are characterized by their high species diversity, with abundant evergreen broadleaved trees and warm-temperate genera. They include many evergreen Fagaceae, such as *Cyclobalanopsis, Castanopsis* and *Pasania,* and Lauraceae, including *Cinnamomum, Actinodaphne, Machilus* and *Neolitsea,* although deciduous trees such as *Comptonia, Zelkova* and *Quercus* sect. *Cerris* are dominant in number. The mixed evergreen and deciduous broadleaved forest was distributed up to north-eastern Honshu and changed to mixed coniferous and deciduous broadleaved forest in southern Hokkaido, where it was influenced by cold sea currents (Tanai & Uemura 1988).

Assemblages during warm phases occasionally exhibit an admixture of warm-temperate and boreal plants, which is ascribed to the transportation of higher-altitude plants to a lower place. Based on the early Miocene Noroshi flora (no. 32 in Figure 30.5), dated 20 Ma (Kano et al. 2002), Ishida (1970) reconstructed the altitudinal zonation of paleovegetation. Forests dominated by evergreen broadleaved trees, including evergreen Fagaceae, Lauraceae and Camelliaceae, were distributed on the floodplain and slope below 100 m, with riverside vegetation including *Glyptostrobus* and *Livistonia* (Palmae). Forest types occurring between 100 and 800 m were made up of valley forest species, including *Zelkova, Pterocarya, Acer, Taiwania* and *Cunninghamia*, while the slope and ridge forest was occupied by conifers, such as *Libocedrus, Keteleeria* and *Sequoia*. The montane zone between 800 and 1200 m was a forest comprising mixed coniferous and deciduous broadleaved trees, including *Picea, Pinus* subgen. *Haploxylon, Thuja, Betula* and *Acer*. *Picea* pollen accounts for about 20% of tree pollen in sediments at the macrofossil localities (Yamanoi 1989), indicating the dominance of cool-temperate to sub-alpine conifers in the montane zone.

The tropical marine condition was reconstructed based on the occurrence of tropical mollusks around ca. 16 Ma, the boundary of the early and middle Miocene (Ogasawara et al. 2003). The occurrence of pollen from mangroves such as *Excoecaria, Rhizophora* and *Sonneratia*, which grow south of the Ryukyu Islands, indicates that mangrove forests were widely distributed in central and western Honshu (Yamanoi & Tsuda 1986). The warmest condition in the Neogene is suggested by the highest percentages of entire-margined leaf species in the Shukunohora (no. 26, 61%) (Ina 1992) and Igami Sandstone (no. 35, 72%) (Ina et al. 2007b) floras in central Honshu (Figure 30.5). The Igami Sandstone flora is dominated by evergreen broadleaved tree species (31% of the total taxa), but includes seeds of *Picea*, which was distributed possibly at the highest altitude among the components of the flora.

30.4.4 Late Miocene to Early Pliocene

Neogene floras after the mid-Miocene Climatic Optimum (MCO) in Honshu were dominated by deciduous broadleaved trees. The diversity of evergreen broadleaved trees that require mild winter conditions and the percentage of entire-margined leaf taxa are dependent on the topographic setting (Figure 30.5). Leaf assemblages in coastal basins in central (nos. 67–69 in Figure 30.3c) and north-eastern (nos. 77–84) Honshu have a higher entire-margined rate and include more evergreen broadleaved trees than contemporaneous assemblages in

inland basins of western Japan (nos. 70–76) and intramontane basins in the Ou Mountains (nos. 85–88).

The compositions of the Mio–Pliocene fossil floras are constrained by the geomorphology in the hinterland of the sedimentary basins. Based on the assemblage composition in sedimentary basins in north-eastern Honshu, Uemura (1988) reconstructed the altitudinal and geomorphological organization of forest types (Figure 30.6). The most dominant forest species found in the sedimentary basins was *Fagus stuxbergii*, which is a fossil species related to modern *Fagus hayatae* in China. The beech forest included diverse deciduous broadleaved trees, including *Ulmus, Betula, Carpinus, Quercus, Sorbus* and *Acer*. Littoral forest included evergreen broadleaved trees with many exotic elements. Basins with wide fluvial lowlands included wetland forest vegetation of *Alnus, Populus, Salix* and *Glyptostrobus*. However, conifers such as *Picea, Abies, Tsuga, Thujopsis* and *Thuja*, and boreal broadleaved trees such as *Betula* and *Sorbus*, are limited in assemblages in volcanic intramontane basins in the Ou Mountains, which exhibit mixed coniferous and deciduous broadleaved forest at high altitudes.

The sedimentological changes in fossil assemblages in an early late Miocene basin (ca. 9 Ma) in central Japan indicate an uplift of mountains in its hinterland, from where higher-altitude plants were transported (Momohara & Saito 2001). Fine sediments of the Tokiguchi Porcelain Clay Formation indicate deposition in a fluvial backmarsh system in small collapsed basins. The fruit and seed fossil assemblages are dominated by components of swamp forest such as *Sequoia* and *Glyptostrobus*, with plants that possibly inhabited mesic locations surrounding the backmarsh, including *Pinus trifolia, Keteleeria, Fagus stuxbergii, Quercus* sect. *Cerris* and *Carpinus*. The fine sediment was covered by sub-rounded gravels (including Nohi Rhyolite) distributed in areas far from the sedimentary basin, which indicates that expansion of the drainage basin and active tectonism resulted in the uplift of mountains. Warm-temperate elements that were common in porcelain clay formations continued to occur, but cool-temperate conifers, such as *Picea, Pinus* subgen. *Haploxylon* and *Chamaecyparis*, appeared in the gravel formation. This indicates that these conifers were transported with gravels from higher-altitude areas in the mountains where they grew (Momohara & Saito 2001).

30.4.5 Late Pliocene to Pleistocene

Floral changes since the late Pliocene represent the extinction of plants that were major elements in the Neogene fossil assemblages and an expansion of elements that are dominant in modern cool-temperate and sub-alpine forests in Japan (Momohara 1994). The floral

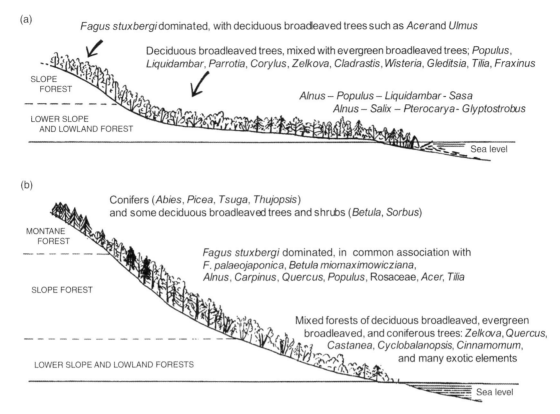

Figure 30.6 Forest distribution in relation to the geomorphology and altitude of the hinterland of the Mio–Pliocene basins in north-eastern Honshu. (a) Sedimentary basin with a flood plain but no montane zone around the Takamine site (no. 80). (b) Sedimentary basins with slope and montane forest along the Ou Mountains around the Miyata (no. 81) and Sanzugawa (no. 87) sites. *Source:* Uemura (1988). Reproduced with permission of the National Museum of Nature and Science, Tokyo.

change was stepwise, and the events were concentrated in transition periods of climatic fluctuation and/or in a downward shift of the glacial climate (Momohara 2016). These events in central Kinki District are characterized by the last occurrence of exotic taxa in the late Pliocene (at 3.35, 2.9 and 2.7–2.5 Ma) and late Quaternary (at 1.2, 0.9 and 0.5 Ma and in the late Pleistocene) (Figure 30.4) (Momohara 2016).

The middle Pliocene flora (between 3.5 and 3.3 Ma) in the central Kinki District is characterized by a richness of taxa that later became extinct in Japan. The abundance of thermophilous taxa and rare cool-temperate conifers indicates that this was the warmest climate in the Plio–Pleistocene (Momohara 2016). After the first conspicuous decrease in exotic elements (10 taxa) at 3.35 Ma, a compositional change in flora, including the extinction of nine taxa, occurred in a stepwise manner from 2.9 to 2.5 Ma (Figure 30.4) (Momohara 2016). While exotic and thermophilous elements decreased during this stage, cool-temperate trees, such as *Picea* sect. *Picea*, *Thuja*, *Thujopsis dolabrata*, *Pinus* subgen. *Haploxylon*, *Cryptomeria japonica* and *Betula maximowicziana*, increased in the south Osaka Basin (Momohara 1992). The floral change is correlated with stages of marked

climatic deterioration, with the development of the northern hemisphere ice sheet shown in marine isotope curves (Figure 30.4) (Lisiecki & Raymo 2005). However, thermophilous plants were occasionally accompanied by cool-temperate trees. Thus, Momohara (1992) ascribed the increase of cool-temperate trees to an uplift of the Izumi Mountains along the southern margin of the basin, which is evidenced by the disappearance of chert that had been supplied from areas south of the mountains.

Extinction events were also concentrated in a period during the mid-Pleistocene climatic transition that marked the onset of the 100 000-year glacial cycle and a shift towards a colder glacial regime since ca. 1.2 Ma. During this stage, a total of 17 taxa, including *Metasequoia*, *Glyptostrobus* and *Davidia*, became extinct from the central Kinki District. Sub-arctic conifers, such as *Abies veitchii*, *Larix kaempheri*, *Picea jezonensis*, *Pinus koraiensis* and *Tsuga diversifolia*, appeared and increased during the late early to middle Pleistocene, indicating a colder glacial climate than in the earlier stages.

The extinction events in west Fukushima, central Niigata and the central and southern parts of Kyushu were concentrated near the end of the late Pliocene and the late early Pleistocene, as in the Kinki District (Figure 30.4) (Momohara 2016). However, the local extinction of these plants occurred earlier in northern and inland basins. The geographical trends indicate that milder climate conditions sustained populations of exotic plants to a later stage in southern and maritime basins.

The appearance of elements dominant in present-day Japanese vegetation is less regular than the trends in local extinctions, except for *Menyanthes*, which appeared earlier in the northern sedimentary basin (Figure 30.4). The occurrence of *Menyanthes*, which is common in cool-temperate bogs, indicates that cool-temperate climate conditions prevailed earlier in the north and in inland basins. However, fossils of cool-temperate and sub-arctic trees are generally present in the sandy deposits of river channels that include plants transported from higher altitudes. Thus, their first occurrence in each basin was dependent on the probability of their transport from habitats at higher altitudes in mountains surrounding the basin.

30.5 Discussion

30.5.1 Mountains as a Habitat for Boreal Plants and Conifers

Latitudinal and altitudinal zonation of vegetation in Japan has existed since the Eocene (Tanai 1961, 1967). Boreal elements similar to the floras in Hokkaido occupied a higher altitude than warm-temperate elements in western Honshu. Paleotemperature has fluctuated since the middle Eocene, as indicated by the percentage change of entire-margined leaf taxa in leaf floras (Figure 30.5). Boreal elements migrated from northern regions during cool stages and established their habitats in mountains during warm stages in southern regions.

During the rifting stage, when the tension field prevailed, volcanic mountains played an important role as a habitat for boreal plants and conifers. The coexistence of boreal conifers such as *Picea* and *Thuja* with subtropical plants such as palms and mangroves in the mid-Miocene Climatic Optimum (MCO) phase is evidence of high-altitude mountains (Ishida 1970). Based on the present-day lowermost distribution limit of *Picea* (2000 m) in Taiwan (Li 1980), where mangrove plants such as *Excoecaria* and *Sonneratia* are distributed along the coast, high mountains are assumed to have been present behind the mid-Miocene basin. However, the dominance of temperate trees in the leaf floras indicates that only the coastal zone had winter temperatures warm enough for mangroves to grow, as suggested by Uemura (1990). This subtropical marine climate in this stage has been ascribed to a strong northward compulsion of warm currents produced by the closing of the Indonesian seaway (Ogasawara et al. 2003).

Edaphic control by active volcanism during this stage may have allowed for the expansion of sub-alpine conifers to lower altitudes. A corresponding modern example is the evergreen coniferous forest dominated by cool-temperate taxa of Pinaceous and Cupressaceous conifers that developed on basalt (900–1500 m) in the north-west foothills of Mt. Fuji (Figure 30.7). Sub-alpine conifers such as *Picea jezoensis* var. *hondoensis*, *Tsuga diversifolia* and *Abies veitchii* occur in the upper part (1200–1500 m) of the forest. The altitudinal zone from 800 to 1600 m in central Japan is dominated by deciduous broadleaved forest, and sub-alpine coniferous forest is dominant in areas higher than 1600 m (Hayashi 1952). This forest is assumed to be an edaphic climax community on poorly developed soil from lava flows that formed in AD 864–866 (Ohtsuka et al. 2008). Temperate coniferous forest is also common on steep slopes and ridges in the mountains on the Pacific side of western Japan, including Yaku Island (Figure 30.1). In these areas, higher annual precipitation (3000–4700 mm) provided from the Pacific Ocean flushes out and leaches soil. Both active volcanism and higher precipitation sustained by warm currents possibly released conifers and boreal plants from competition against thermophilous broadleaved trees at relatively lower altitudes and increased the beta diversity of vegetation in volcanic mountains during the mid-MCO.

30.5.2 Influence of Mountain Ranges on Local Climate and Floras

The expansion of land areas during the compression stage since the late Miocene may have intensified the inland climate, which is characterized by a greater annual range of temperature and less precipitation than coastal areas. Less diverse evergreen broadleaved trees and lower paleotemperatures in inland basins in western Japan may have been caused by the unique topographic setting. Rotation of the western Japan crust in the early Miocene formed a narrow contiguity between its western part and the continent, preventing warm currents from flowing into the Sea of Japan. The dominant cold current in the Sea of Japan likely increased the coldness and dryness in western Japan, and this has been accompanied by an intensification of the Asian Monsoon since 8 Ma (Filippelli 1997). On the other hand, the warm currents prevailing in the Pacific Ocean maintained an oceanic climate in north-eastern Japan, including the small islands.

Mountain ranges thrust up by crust compression differentiated the local climate more effectively than did discontinuous volcanic mountains. The uplift of mountain ranges since the late Pliocene culminated in the late Quaternary and differentiated the climates between the two sides of the mountains. A comparison of the composition of macrofossil assemblages between sedimentary basins in Niigata Prefecture and those in central Kinki District indicates differentiation of the flora in the mountains between the Sea of Japan and the Pacific Ocean sides since the late early Pleistocene. *Stewartia monadelpha* and *Chamaecyparis obtusa*, which are endemic to the present-day Pacific Ocean-side mountains, disappeared from the Niigata sedimentary basin at about 1.3 Ma, while occurrences of *Cryptomeria*, which is common in the Sea of Japan-side mountains, increased in their place (Momohara 2016).

The heavy snowfalls along the Sea of Japan side possibly began in the late early Pleistocene. The frequent inflow of the Tsushima Warm Current into the Sea of Japan began in the interglacial stages around ca. 1.7 Ma (Kitamura & Kimoto 2006), and the winter monsoon became strong ca. 1.3 Ma (Sun et al. 2010). The Echigo Mountains were uplifted actively in the late early Pleistocene, as evidenced by the increasing gravel size (Kazaoka et al. 1986), and this promoted heavy snowfall on the north-western slopes behind the locations where fossils accumulated.

30.5.3 Mountain Ranges as Barriers for Plant Migration

Increased mountain building in the Quaternary provided continuous corridors of migration for boreal and alpine plants during the glacial stages. On the other hand, mountains constituted barriers that hindered the migration of thermophilous plants growing at lower altitudes and wetland plants inhabiting fluvial plains, such as *Metasequoia* and *Glyptostrobus*. The uplift of mountains surrounding the sedimentary basins is almost concurrent with the stages of plant extinction accompanying the shift to a colder glacial climate (Figure 30.4) (Momohara 2016). From the latest early Pleistocene, the uplift of the alpine zone in central Honshu became conspicuous, while in western Japan, mountains dissected basins, and the formation of straits isolated the Shikoku and Kyushu islands in interglacial stages (Figure 30.4). Such geomorphological changes limited the distribution of lowland inhabitants and prevented their expansion (Momohara 2016).

Drought and cold due to winter monsoons increased in a stepwise manner since the late Pliocene (Figure 30.4) (Sun et al. 2010), and have altered the climate in migration routes and prevented plants from establishing

Figure 30.7 Evergreen coniferous forest in Aokigahara (900–1500 m) at the north-western foot of Mt. Fuji: an edaphic climax forest that developed on a lava flow formed in AD 864–866. The mountains in the far background are the southern part of the Akaishi Mountains (2500–3000 m). See also Plate 49 in color plate section.

populations in new locations. Dry winter monsoon conditions prevail in the present-day mountain areas of central Japan, with an altitudinal temperature lapse rate greater than 0.6 °C per 100 m (JMA 2011). In contrast, the mountains in central China, which are inhabited by taxa that are locally extinct in Japan, are under a more humid climate regime, and the altitudinal lapse rate is less than 0.45 °C per 100 m (Fang & Yoda 1988). The mild winter climate results in a wide altitudinal range of mixed evergreen and deciduous broadleaved forests (mixed mesophytic forest) in mountain zones, whereas such a mixed forest zone is very limited at high altitudes in Japan (Tang & Ohsawa 1997). The decreasing winter temperatures in higher altitudes would have hindered migration to the north and inland over mountain ranges, even as exotic taxa surviving in southern sedimentary basins had the potential to expand their distribution under improving climatic conditions during interglacial stages (Momohara 2016). The process of reducing the distribution of exotic taxa to southern and/or maritime areas in Japan finalized the present-day distributions in continental China, which has vast potential areas of glacial refugia extending to the south (Chapter 29).

30.6 Conclusion

Late Oligocene to early middle Miocene rifting moved the Japanese islands from the continental margin to their present-day position in the Pacific Ocean. These islands formed volcanic mountains that provided habitats for boreal plants and conifers that had migrated from higher latitudes during the late Paleogene.

The expansion of land areas and the uplift of mountain ranges caused by crust compression, along with the development of a winter monsoon, have differentiated the local climate and floras of Japan since the late Miocene.

Active uplift of mountain ranges, which culminated in the late Quaternary, provided corridors of migration for boreal and alpine plants in the glacial stages, while mountain ranges hindered the migration of plants that inhabited lowlands and caused their extinction.

References

Arita, K., Ganzawa, Y. & Itaya, T. (2001) Tectonics and uplift process of the Hidaka Mountains, Hokkaido, Japan inferred from thermochronology. *Bulletin of the Earthquake Research Institute, the University of Tokyo* **76**, 93–104.

Fang, J. & Yoda, K. (1988) Climate and vegetation in China (I): changes in the altitudinal lapse rate of temperature and distribution of sea level temperature. *Ecological Research* **3**, 37–51.

Filippelli, G.M. (1997) Intensification of the Asian monsoon and a chemical weathering event in the late Miocene – early Pliocene implications for late Neogene climate change. *Geology* **25**, 27–30.

GIAJ (1990) The National Atlas of Japan, revised edition. Geospacial Information Authority of Japan. CD-Rom.

Hayashi, Y. (1952) *The Natural Distribution of Important Trees, Indigenous to Japan*. Conifer Report 2. Bulletin of the Government Forest Experiment Station 55.

Hayashi, T. (1975) *Fossil from Chojabaru Iki Island, Japan*. Nagasaki: Island Science Institute.

Honza, E., Tokuyama, H. & Soh, W. (2004) Formation of the Japan and Kuril basins in the late Tertiary. In: Clift, P., Kuhnt, W., Wang, P. & Hayes, D. (eds.) *Continent–Ocean Interactions within East Asian Marginal Seas*. New York: John Wiley & Sons Ltd., pp. 87–108.

Horikoshi, M. & Aoki, J. (1985) *Biota in Japan*. Tokyo: Iwanami-shoten.

Horiuchi, J. (1996) Neogene floras of the Kanto District. *Science Reports of the Institute of Geoscience, University of Tsukuba, Section B* **17**, 109–208.

Horiuchi, J. & Takimoto, H. (2001) Plant mega-fossils from the late Early to early Middle Miocene Asakawa Formation at Inuboe Pass, Ibaraki Prefecture, Kanto District, Japan. *Bulletin of Ibaraki Nature Museum* **4**, 1–32.

Huzioka, K. (1963) The Utto flora of northern Honshu. In: Tanai, T. et al. (eds.) *Tertiary Floras of Japan*, Miocene Floras. Tokyo: The Collaborating Association to Commemorate the 80th Anniversary of the Geological Survey of Japan, pp. 153–216.

Huzioka, K. (1964) The Aniai flora of Akita Prefecture, and the Aniai-type floras in Honshu, Japan. *Journal of the Mining College, Akita University, Series A* **3**, 1–105.

Huzioka, K. (1974) The Miocene Daibo flora from the western end of Honshu, Japan. *Journal of the Mining College, Akita University, Series A* **5**, 85–108.

Huzioka, K. & Takahashi, E. (1970) The Eocene flora of the Ube coal-field, southwest Honshu, Japan. *Journal of the Mining College, Akita University, Series A* **4**, 1–88.

Huzita, K. (1980) Role of the Median Tectonic Line in the Quaternary tectonics of the Japanese Islands. *Memoirs of the Geological Society of Japan* **18**, 129–153.

Ina, H. (1992) Miocene vegetational and climatic history of the eastern part of the Setouchi Geologic Province, Japan. *Journal of Earth and Planetary Sciences, Nagoya University* **39**, 47–82.

Ina, H. & Ishikawa, T. (1982) Late Miocene flora from the west part of Satsuma Peninsula, Kagoshima Prefecture, Japan. *Bulletin of the Mizunami Fossil Museum* **9**, 35–58.

Ina, H., Saito, T., Kawase, M. & Wang, W. (2007a) Fossil leaves, fruits and pollen of Liquidambar (Hamamelidaceae) from the Lower Miocene Nakamura Formation in Gifu Prefecture, central Japan. *Journal of the Geological Society of Japan* **113**, 542–545.

Ina, H., Shibata, K. Ichihara, T. & Ujihara, A. (2007b) An early Middle Miocene flora from the Igami Sandstone Member of the Yamagasu Group in the eastern Nara Prefecture, Japan. *Science Report of the Toyohashi Museum of Natural History* **17**, 1–6.

Ishida, S. (1970) The Noroshi flora of Noto Peninsula, Central Japan. *Memoirs of the Faculty of Science, Kyoto University, Series of Geology and Mineralogy* **37**, 1–112.

JMA (2011) Monthly climatological normals (1981–2010). Japan Meteorological Agency. Available from: http://www.data.jma.go.jp/obd/stats/etrn/index.php (last accessed August 30, 2017).

Jolivet, L., Tamaki, K. & Fournier, M. (1994) Japan Sea, opening history and mechanism: a synthesis. *Journal of Geophysical Research* **99**, 22 237–22 259.

Kano, K., Kato, H., Yanagisawa, S. & Yoshida, S. (eds.) (1991) Stratigraphy and geologic history of the Cenozoic of Japan. *Report of Geological Survey of Japan* **274**, 1–114.

Kano, K., Yoshikawa, T., Yanagisawa, Y. et al. (2002) An unconformity in the early Miocene syn-rifting succession, northern Noto Peninsula, Japan: evidence for short-term uplifting precedent to the rapid opening of the Japan Sea. *The Island Arc* **11**, 170–184.

Kano, K., Uto, K. & Ohguchi, T. (2007) Stratigraphic review of Eocene to Oligocene succession along the eastern Japan Sea: implication for early opening of the Japan Sea. *Journal of Asian Earth Sciences* **30**, 20–32.

Kano, K., Ohguchi, T., Kobayashi, N. & Sato., Y. (2011) Upper Cretaceous, Upper Eocene, and Lower Miocene. In: Kano, K., Ohguchi, T., Yanagisawa, Y. et al. (eds.) *Geology of the Toga and Funakawa District.* Quadrangle Series, 1 : 50 000. Tsukuba: Geological Survey of Japan, AIST, pp. 34–55.

Kato, M. & Ebihara, A. (eds.) (2011) *Endemic Plants of Japan.* Tokyo: Tokai University Press.

Kazaoka, O., Tateishi, M. & Kobayashi, I. (1986) Stratigraphy and facies of the Uonuma Group in the Uonuma district, Niigata Prefecture, central Japan. *Journal of the Geological Society of Japan* **92**, 829–853.

Kitanaka, T. & Fuji, N. (1988) Neogene Daijima-type "Tatsunokuchi fossil flora" in Kaga, Ishikawa Prefecture, central Japan. *Bulletin of the Faculty of Education, Kanazawa University, Natural Science* **37**, 97–117.

Kitamura, A. & Kimoto, K. (2006) History of the inflow of the warm Tsushima Current into the Sea of Japan between 3.5 and 0.8 Ma. *Palaeogeography, Palaeoclimatology, Palaeoecology* **236**, 355–366.

Kira, T. (1991) Forest ecosystems of East and Southeast Asia in a global perspective. *Ecological Research* **6**, 185–200.

Kobatake, N. (1983) Plant fossils from the Kobe Group. In: Fuzita, K. & Kasama, T. (eds.) *Geology of the Kobe District.* Quadrangle Series, 1 : 50 000. Tsukuba: Geological Survey of Japan, pp. 24–32.

Li, H.L. (1980) Pinaceae. In: Flora of Taiwan Editorial Committee (ed.) *Flora of Taiwan*, Vol. **1**, 2nd edn. Taipei: Epoch Publishing, pp. 514–529.

Lisiecki L.E. & Raymo, M.E. (2005) A Pliocene–Pleistocene stack of 57 globally distributed benthic $\delta^{18}O$ records. *Paleoceanography* **20**, PA1003.

Matsuo, H. (1963) The Notonakajima flora of Noto Peninsula. In: Tanai, T. et al. (eds.) *Tertiary Floras of Japan*, Miocene Floras. Tokyo: The Collaborating Association to Commemorate the 80th Anniversary of the Geological Survey of Japan, pp. 219–243.

Matsuo, H. (1967) Palaeogene floras of Northwest Kyushu. Part I: Takashima flora. *Annals of Science, College of Liberal Arts, Kanazawa University* **4**, 15–90.

Matsuo, H. (1970) Palaeogene floras of Northwestern Kyushu. Part II: The Sakito flora. *Annals of Science, College of Liberal Arts, Kanazawa University* **7**, 13–62.

Matsushita, S., Matsuo, S. & Ishida, S. (1994) The Shimokatakura flora of Ube City, Yamaguchi Prefecture, southwest Honshu, Japan. *Bulletin of Mine City Museum* **10**, 1–49.

Miki, S. (1948) Floral remains in Kinki and adjacent districts since the Pliocene with description of 8 new species. *Mineralogy and Geology* **2**, 105–144.

Momohara, A. (1992) Late Pliocene plant biostratigraphy of the lowermost part of the Osaka Group, southwest Japan, with reference to extinction of plants. *The Quaternary Research (Tokyo)* **31**, 76–88.

Momohara, A. (1994) Floral and paleoenvironmental history from the late Pliocene to middle Pleistocene in and around central Japan. *Palaeogeography, Palaeoclimatology, Palaeoecology* **108**, 281–293.

Momohara, A. (2016) Stages of major floral change in Japan based on macrofossil evidence and their connection to climate and geomorphological changes since the Pliocene. *Quaternary International* **397**, 92–105.

Momohara, A. & Saito, T. (2001) Change of paleovegetation caused by topographic change in and around a sedimentary basin of the Upper Miocene Tokiguchi Porcelain Clay Formation, central Japan. *Geoscience Reports of Shimane University* **20**, 49–58.

Nakajima, T., Danhara, T., Iwano, H. & Chinzei, K. (2006) Uplift of the Ou Backbone Range in Northeast Japan at around 10 Ma and its implication for tectonic evolution of the eastern margin of Asia. *Palaeogeography, Paleoclimatology, Paleoecology* **241**, 28–48.

Ogasawara, K. (1994) Neogene paleogeography and marine climate of the Japanese Islands based on shallow-marine molluscs. *Palaeogeography, Palaeoclimatology, Palaeoecology* 108, 335–351.

Ogasawara, K., Ugai, H. & Kurihara, Y. (2003) Short-Term Early Miocene Climatic Fluctuations in the Japanese Islands. RCPNS 46 – 8th International Congress on Pacific Neogene Stratigraphy, February, 2003, Chiang Mai, 181–190.

Oikawa, T. (2003) The spatial and temporal relationship between uplifting and magmatism in the Hida Mountain Range, Central Japan. *The Quaternary Research (Tokyo)* 42, 141–156.

Ohtsuka, T., Yokosawa, T. & Ohtake, M. (2008) Community structure and dynamics of coniferous forest on Aokigahara lava flow, Mt. Fuji, Japan. *Vegetation Science* 25, 95–107.

Onoe, T. (1974) A Middle Miocene flora from Oguni-machi, Yamagata Prefecture, Japan. *Report of Geological Survey of Japan* 253, 1–66.

Onoe, T. (1978) New knowledge on Miocene floras in the northern part of Kinki District, central Japan. *Bulletin of the Geological Survey of Japan* 29, 53–58.

Ozaki, K. (1981) On the paleoenvironments of the Late Miocene Tatsumitoge flora. *Science Report of Yokohama National University, Section II* 28, 47–75.

Ozaki, K. (1991) Late Miocene and Pliocene floras in central Honshu, Japan. *Bulletin of the Kanagawa Prefectural Museum*. Natural Science, Special Issue, 1–188.

Ozaki, M. (1992) Paleogene floral and climatic changes of Japan. *Bulletin of the Geological Survey of Japan* 43, 68–85.

Ozaki, M., Matsuura, H. & Sato, Y. (1996) Geologic age of the Kobe Group. *Journal of the Geological Society of Japan* 102, 73–83.

Suganuma, Y., Suzuki, T., Yamazaki, H. & Kikuchi, T. (2003) Chrono-stratigraphy of the Ina Group, Central Japan, based on correlation of volcanic ash layers with Pleistocene widespread tephras. *The Quaternary Research (Tokyo)* 42, 321–334.

Sun, Y.B., An, Z.S., Clemens, S. et al. (2010) Seven million years of wind and precipitation variability on the Chinese Loess Plateau. *Earth and Planetary Science Letters* 297, 525–535.

Takimoto, H., Horiuchi, J., Sugaya, M. & Hosogai, T. (1998) Plant megafossils from the Pliocene Kume Formation in the Osato area, Ibaraki Prefecture, Central Japan. *Bulletin of Ibaraki Nature Museum* 1, 47–68.

Tanai, T. (1961) Neogene floral change in Japan. Journal of the Faculty of Science, Hokkaido University, Ser. 4. *Geology and Mineralogy* 11, 119–398.

Tanai, T. (1967) Miocene floras and climate in East Asia. *Abhandlungen des Zentralen Geologischen Instituts* 10, 195–205.

Tanai, T. (1970) The Oligocene floras from the Kushiro coal field, Hokkaido, Japan. *Journal of the Faculty of Science, Hokkaido University, Ser. 4, Geology and Mineralogy* 14, 383–514.

Tanai, T. (1971) The Miocene Sakipenpetsu flora from Ashibetsu Area, central Hokkaido, Japan. *Memoirs of the National Science Museum (Tokyo)* 4, 127–172.

Tanai, T. (1976) The revision of the Pliocene Mogi flora, described by Nathorst (1883) and Florin (1921). *Journal of the Faculty of Science, Hokkaido University, Ser. 4, Geology and Mineralogy* 17, 277–346

Tanai, T. (1990) Euphorbiaceae and Icacinaceae from the Paleogene of Hokkaido, Japan. *Bulletin of the Natural Science Museum (Tokyo), Ser. C* 18, 91–118.

Tanai, T. (1991) Tertiary climate and vegetation changes in the Northern Hemisphere. *Journal of Geography* 100, 951–966.

Tanai, T. (1992) Tertiary vegetational history of East Asia. *Bulletin of the Mizunami Fossil Museum* 19, 125–163.

Tanai, T. & Onoe, T. (1961) A Mio-Pliocene flora from the Ningyo-toge area on the border between Tottori and Okayama Prefectures, Japan. *Report of the Geological Survey of Japan* 187, 1–63.

Tanai, T. & Suzuki, N. (1963) Miocene floras of southwestern Hokkaido, Japan. In: Tanai, T. et al. (eds) *Tertiary Floras of Japan*, Miocene Floras. Tokyo: The Collaborating Association to Commemorate the 80th Anniversary of the Geological Survey of Japan, pp. 9–149.

Tanai, T. & Suzuki, N. (1965) Late Tertiary floras from northeastern Hokkaido, Japan. *Palaeontological Society of Japan, Special Papers* 10, 1–117.

Tanai, T. & Suzuki, N. (1972) Additions to the Miocene floras of southwestern Hokkaido, Japan. *Journal of the Faculty of Science, Hokkaido University, Ser. 4, Geology and Mineralogy* 15, 281–359.

Tanai, T. & Uemura, K. (1983) Engelhardia fruits from the Tertiary of Japan. *Journal of the Faculty of Science, Hokkaido University, Ser. 4, Geology and Mineralogy* 20, 249–260.

Tanai, T. & Uemura, K. (1988) Daijima-type floras (Miocene) in southwestern Hokkaido and the northern part of Honshu, Japan. *Memoirs of the National Science Museum (Tokyo)* 21, 7–16.

Tanai, T. & Uemura, K. (1991) The Oligocene Noda flora from the Yuya-wan area of the Western end of Honshu, Japan. Part 2. *Bulletin of the National Science Museum (Tokyo) Ser. C* 17, 81–90.

Tang, C.Q. & Ohsawa, M. (1997) Zonal transition of evergreen, deciduous, and coniferous forests along the altitudinal gradient on a humid subtropical mountain, Mt. Emei, Sichuan, China. *Plant Ecology* 133, 63–78.

Taruno, H. (2010) The stages of land bridge formation between the Japanese Islands and the continent on the basis of faunal succession. *The Quaternary Research (Tokyo)* 49, 309–314.

Uemura, K. (1988) *Late Miocene floras in northeast Honshu, Japan*. Tokyo: National Science Museum, 1–197.

Uemura, K. (1990) Tertiary conifers in time and space. *Japanese Journal of Historical Botany* **5**, 27–38.

Uemura, K. (2000) Middle Miocene plants from Heigun-to Island, Yamaguchi Prefecture, Southwestern Japan. *Memoirs of the National Science Museum (Tokyo)* **32**, 39–54.

Uemura, K. & Yasuno, T. (1991) *Miocene Plants from the Komegawaki Formation, Fukui Prefecture, Central Japan*. Prof. S. Miura Memorial Volume, 43–54.

Uemura, K., Doi, E. & Takahashi, F. (1999) Plant megafossil assemblage from the Kiwado Formation (Oligocene) from Ouchiyama-kami in Yamaguchi Pref., western Honshu, Japan. *Bulletin of Mine City Museum* **15**, 1–59.

Uemura, K., Tanokura K. & Hamano, I. (2001) The Early Miocene flora from Itsukaichi in the western part of Tokyo Prefecture, Japan. *Memoirs of the National Science Museum (Tokyo)* **37**, 53–70.

Wolfe, J. A. (1979) Temperature parameters of humid to mesic forests of Eastern Asia and relation to forests of other regions of the Northern Hemisphere and Australasia. *USGS Professional Paper* **1106**, 1–36.

Yamakita, S. & Otoh, S. (2000) Cretaceous rearrangement processes of pre-Cretaceous geologic units of the Japanese Islands by MTL-Kurosegawa left-lateral strike-slip fault system. *Memoirs of the Geological Society of Japan* **56**, 23–38.

Yamanoi, T. (1989) *Palyno-Flora of the Middle Miocene Sediments in Noto Peninsula, Central Japan*. Prof. Hidekuni Matsuo Memorial Volume, pp. 5–13.

Yamanoi, T. & Tsuda, K. (1986) On the conditions of paleo-mangrove forest in the Kurosedani Formation (Middle Miocene), central Japan. *Memoirs of the National Science Museum (Tokyo)* **19**, 55–66.

Yabe, A. (2008a) Early Miocene terrestrial climate inferred from plant megafossil assemblages of the Joban and Soma areas, Northeast Honshu, Japan. *Bulletin of the Geological Survey of Japan* **59**, 397–413.

Yabe, A. (2008b) Plant megafossil assemblage from the Lower Miocene Ito-o Formation, Fukui Prefccture, Central Japan. *Memoir of the Fukui Prefectural Dinosaur Museum* **7**, 1–24.

Yoshioka, K. (1973) *Phytogeography*. Tokyo: Kyoritsu.

31

The Complex History of Mountain Building and the Establishment of Mountain Biota in Southeast Asia and Eastern Indonesia

Robert J. Morley

Palynova UK and Royal Holloway, University of London, London, UK

Abstract

The Cenozoic history of uplift and denudation of mountain regions across Southeast Asia and Eastern Indonesia is complex, paralleling the intricate nature of the geological history of the region. Establishing the timing of uplift and erosion of each upland region requires state-of-the-art geological evaluation, and this is still ongoing. The uplift scenario for many of the present and former upland areas has been established using a combination of tectonic reconstruction, evaluation of sedimentation rate changes that reflect rates of denudation and utilization of the pollen records of montane plants. Montane vegetation changes gradually from lower montane forest at 1000 m (which bears similarities to the more complex lowland forests), to simple subalpine woodland above about 2600 m and on the tallest mountains, to grassland above the treeline. The altitudes at which the main transitions occur are controlled mainly by climate and soils. They will thus shift position with changing global climate, as established from Quaternary studies over the last glacial cycle. For the deep-time Cenozoic, records of changing montane floras are mostly derived from data generated during the course of petroleum exploration programs, and many of these data remain unpublished. The Cenozoic pollen record reflects three different patterns with respect to mountain building across the region. First, the pollen record of southern conifers from different areas of Indonesia shows the Pliocene and Pleistocene uplift of the Lengguru fold belt and the uplift of the Bird's Head in West Papua, Timor and Sulawesi – and also Sumatra and Java – and provides examples of dispersal of Gondwanan elements into the Sunda region or Sundaland (the region that includes the islands of Borneo, Sumatra and Java, the Malay Peninsula and the surrounding continental shelves). Second, for equatorial Borneo, Ericaceae pollen is the most reliable indicator of former upland vegetation, and reflects the mid and late Miocene uplift of the Borneo Central Ranges. Third, pollen of the Laurasian conifer genera *Picea*, *Abies* and *Tsuga* from the northern Sunda Shelf, Borneo and Indochina indicate the former occurrence of seasonal climate uplands in those areas. With the increasing availability of molecular data sets from montane biota, it is becoming possible to construct detailed scenarios for the establishment, dispersal times and patterns of evolution of diverse montane taxa across the region. Additional molecular evaluations, when integrated with geological histories, will further clarify the manner in which montane communities have become established and will also provide an independent indication of the timing of uplift.

Keywords: *plate tectonics, vegetation dynamics, montane forest, palynology, plant dispersal, migration tracks, geogenomics*

31.1 Introduction

Montane biota yield surprising insights with respect to climate and tectonic change. In 1861, Alfred Russell Wallace scaled the dormant volcano Gunung (Mount) Pangrango (3019 m) in West Java (Wallace 1869). As the volcano was clothed with dense rainforest to the crater rim, he was astonished to find a grassy hollow at the summit, rich in alpine herbs, including Laurasian temperate genera such as *Ranunculus*, *Gentiana* and *Primula*: a temperate island in a tropical sea. He speculated that these plants had colonized the mountain during a glacial period, when the equatorial climate was cooler and montane plants could migrate along lower-altitude ridges. He reached this conclusion 100 years before the dynamics of equatorial montane vegetation began to be studied from pollen records in East Africa in the early 1960s (e.g., Van Zinderen Bakker 1962). Subsequently, in the 1930s, the

Mountains, Climate and Biodiversity, First Edition. Edited by Carina Hoorn, Allison Perrigo and Alexandre Antonelli.
© 2018 John Wiley & Sons Ltd. Published 2018 by John Wiley & Sons Ltd.
Companion website: www.wiley.com/go/hoorn\mountains,climateandbiodiversity

visionary botanist Cornelius van Steenis, working on Gunung Pangrango and other mountains across Indonesia and Malaysia, recognized that mountain plants dispersed into the region along three trackways, one from the Himalaya, one from East Asia and one from Australia (van Steenis 1934a,b, 1936). These trackways identified plate tectonic collisions, which were not unraveled for another 60 years.

The Southeast Asia region is geologically complex, a region currently undergoing active tectonic collision, with many microplates jostling for position within an area that is gradually shrinking as the Australian Plate slowly moves northward, and the Pacific Plate westward. Sometimes, rocks are squeezed and pushed to high altitudes, creating mountain ranges, whereas elsewhere plates may subside, often rapidly, creating deep troughs and basins. Despite the overall compression, some areas have exhibited extension, with the South China Sea opening during the mid-Cenozoic, and the Andaman Sea during the late Miocene.

Mountain ranges form as a result of plate collisions and rifting. The Eocene collision of India with Asia resulted in the formation of mountain ranges in Indochina during the early Cenozoic, and the collision of the Australian and Pacific plates with the Sunda region from the latest Oligocene onward and the closure of the proto-South China Sea resulted in uplift in Borneo. The Australian Plate then continued northward, resulting in the uplift of Sulawesi and New Guinea. At the same time, subduction along the Sumatra and Java trenches resulted in the formation of the volcanic arcs that presently dominate these islands.

The pattern of uplift and erosion of the mountain ranges of the region has been greatly clarified through the tectonic reconstructions – coupled with paleogeographic models – of Hall (1998, 2002, 2009, 2012a,b). Patterns of changing rates of sedimentation are also very valuable, and can provide information on the precise timing of uplift – and of subsequent erosion – based on changing sedimentation rates. Regional sedimentation rates for the Makassar Straits have recently been published by Morley (2014), allowing the timing of initiation of uplift of mountain ranges bordering the Straits to be proposed from sudden increases in rates of sedimentation across the region.

The formation of the broad Southeast Asia region and its mountain ranges (Figure 31.1) can be visualized in six phases. First, the Indian Plate collided with Asia, with its northerly drift slowing at about 55 Ma at the beginning of the Eocene (Tapponnier et al. 1982, 1986), but with the final collision interpreted variously from the middle Eocene (Bouihol et al. 2013) to the Eocene/Oligocene boundary

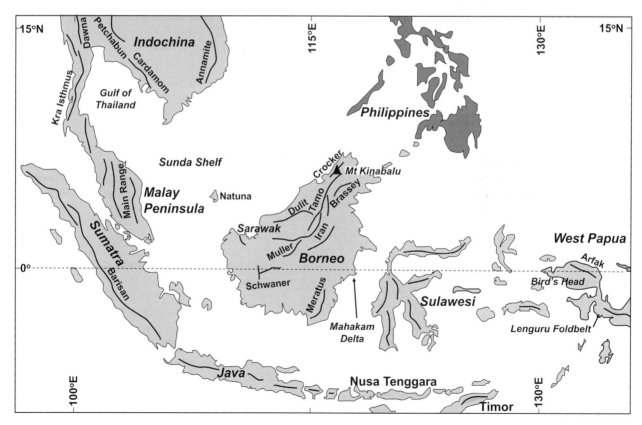

Figure 31.1 Map indicating the positions of mountain ranges mentioned in this chapter.

(Aitchison et al. 2007). The collision of the Indian Plate with Asia resulted in the eastward extrusion of much of Indochina, but probably not to the extent originally envisaged (Hall et al. 2009). However, the process of easterly extrusion of Indochina is thought to have resulted in the mid-Cenozoic uplift of the Sino-Burmese ranges, the mid-Thailand uplands, the Amman Mountains in Vietnam and probably the Main Range of the Malay Peninsula. The second phase involved the opening of the South China Sea, with initial rifting in the late Eocene, which continued into the Oligocene. The initial rifting extended across Sundaland, with rifts across the Natuna region, northward through the Malay Basin and the Gulf of Thailand and south to Sumatra and Java, with a probable triple junction in the vicinity of the Penyu Basin (Ngah et al. 1996; Morley & Morley 2013). These rifts are thought to have been bordered by substantial uplands. The third phase involved the collision of the Australian Plate with Sundaland at the end of the Oligocene (Hall 2002). This did not initially have a big impact on elevation, but resulted in the closure of the Indonesian Throughflow and the establishment of the Asian monsoon (Morley 2012), which is distinct from the Indian monsoon (Wang et al. 2003). The fourth phase involved the closure of the proto-South China Sea and the mid-Miocene elevation of Borneo, the uplift of the Meratus Mountains of south-east Borneo and the initial elevation of the Barisan Range of Sumatra, with the emplacement of the Kinabalu Granite – implying very high altitudes for Kinabalu and the Crocker Range – at about 8 Ma. The fifth phase involved the uplift of islands of Eastern Indonesia, including Sulawesi and the Bird's Head, from the late Miocene to Pleistocene, with the ongoing northward movement of the Australian Plate. The sixth phase involved the establishment of the island of Java during the late Pliocene and Pleistocene.

The pattern of development of Southeast Asia's montane biota is thus also likely to be complex. The taxa that occupy each montane area will be determined by the presence of dispersal pathways, dictated mainly by plate tectonics, time of uplift and the maximum elevation reached, along with soils, climate and geographical isolation. Chance dispersal also plays an important role.

The main upland areas of the region are shown in Figure 31.1. The Philippines are not included in this review.

31.2 Present Montane Vegetation of Southeast Asia

Montane vegetation across the area displays a broadly similar pattern of physiognomic change with increasing altitude, with complex multiple-story mesophyll (leaf size range 4500–18 225 mm^2) rainforest at low altitudes giving way to single-story notophyll (2025–4500 mm^2) lower-montane forests at intermediate altitudes and then to upper-montane microphyll (225–2025 mm^2) forests grading to dwarf sub-alpine woodland and then grassland above the treeline (Figure 31.2). This succession is paralleled by a gradual reduction in floristic diversity.

The changes observed show many parallels to the latitudinal ecotone (Andrade Marin 1945; van Steenis 1972). However, there are some clear differences: because of the lack of seasons in the equatorial tropics, the minimum temperature limits for tree survival on Southeast Asian mountains occur in the opposite order compared to the latitudinal gradient (Ohsawa 1993), with the boundary limiting minimum heat requirements for tree growth occurring well below the level at which the coldest winter temperatures are too low for tree survival (Figure 31.3). Because of this, altitudinal boundaries do not reduce gradually with increasing latitude, but within the tropical zone show consistencies from area to area. Ashton (2014) suggests that the ecotone between lowland and lower-montane forests coincides with a change from lowland organic-poor yellow–red soils to more humic-rich mull soils at around 1000 m, whereas the boundary between

Figure 31.2 Altitudinal forest formation series in the Malay Peninsula. (a) Lowland evergreen rainforest. (b) Lower-montane rainforest (upper dipterocarp forest facies). (c) Lower-montane "oak-laurel" forest. (d) Upper-montane "Ericaceous" forest. *Source:* Morley (2000). Reproduced with permission of John Wiley & Sons.

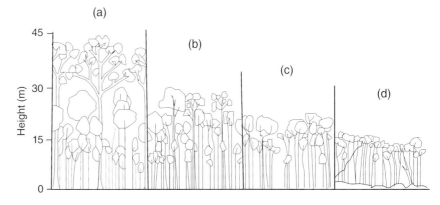

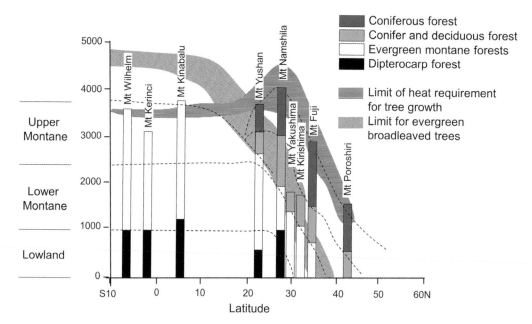

Figure 31.3 Altitudinal and latitudinal representation of forest zones in Southeast and East Asia. *Source:* Ohsawa (1993). Reproduced with permission of IAVS – the International Association of Vegetation Science.

lower- and upper-montane forest coincides with diurnal fog persistently penetrating the canopy.

It has been suggested that the position of altitudinal breaks in montane vegetation varies according to the mountain mass-elevation (or Massenerhebung) effect (Grubb 1971), but Ashton (2014) emphasizes that this effect applies mainly to the mossy facies of upper-montane forests, which are drawn downward on coastal mountains in areas of high humidity. The boundary between upper-montane and sub-alpine woodlands is thought to be correlated with a change in mean ambient temperature, but it perhaps also correlates with minimum night temperatures and periodic frosts. The treeline naturally occurs at about 3800 m in the tropics, above which altitude there is insufficient heat for tree growth (Ohsawa 1993), as the mean temperature during the growing season falls below 7 °C (Hoch & Korner 2012).

Across Sundaland and Indochina, lower-montane forests are dominated by species of Fagaceae (*Lithocarpus, Castanopsis* and *Quercus*) and Lauraceae; hence the name, "oak-laurel" forest. Additional common elements include *Altingia excelsa, Schima wallichii* and gymnosperms such as *Podocarpus* and *Agathis* spp. The "upper dipterocarp forest" of Symington (1943) is a facies of the lower-altitude lower-montane forest and is widespread across the Sunda region. Similarly, in eastern Malaysia, lower-montane forests are typified by the oak genera *Lithocarpus* and *Castanopsis*. Seasonally dry lower-montane forests in Indochina are characterised by *Pinus merkusii* (which extends to its only southern hemisphere locality in central Sumatra), and the physiognomically similar *Casuarina junghuhniana* occurs in the same

setting in East Java and Nusa Tenggara. Upper-montane forests across the region are typified by members of the family Ericaceae, together with oaks, Lauraceae, Myrtaceae and *Myrica* spp., but taxa of Gondwanan origin are also well represented, and include podocarps such as *Dacrycarpus* and *Phyllocladus*, Cunoniaceae, Winteraceae and others. The floristic composition of trees in sub-alpine woodland is similar to that in upper-montane forests, but temperate herbaceous genera are also common, such as *Ranunculus* and *Primula*, as seen by Wallace on Gunung Pangrango. In New Guinea, many herbaceous taxa of Gondwanan origin are common, including *Astelia* and *Coprosma*, and some of these – such as *Thelymitra* and *Gahnia* – extend to the Sunda region.

Many higher-altitude plants produce seeds that are easily dispersed by wind or birds, and many are wind-pollinated, facilitating dispersal between isolated peaks. In higher-latitude areas of Indochina, upper-montane forests are replaced by temperate forests, with the genera *Picea, Abies* and *Tsuga* becoming dominant.

31.3 Late Quaternary Vegetation Dynamics

The late Quaternary dynamics of montane vegetation have been established from numerous palynological studies on lake deposits, mainly from New Guinea and Sumatra (Flenley 1979, 1984), but also from Java and Sulawesi. From the palynological study of lake cores, it is possible to determine the history of the vegetation that formerly grew around a specific lake, and from the nature of the vegetation changes, to determine the extent of climate changes that affected the area for the duration of the core. For cores that penetrate the last glacial period, by comparing records from such multiple cores, regional trends in temperature and moisture change over time can be determined. Data from lake cores (which penetrate the last glacial) from 950 m a.s.l. in Sumatra to 2200 m a.s.l. in New Guinea (Figure 31.4) have been plotted against altitude by Morley & Flenley (1987), who show that while the treeline was depressed by up to 1600 m at the time of the Last Glacial Maximum (LGM), the depression of vegetation zones at lower altitudes was less, with the lower/upper-montane boundary being depressed by just 500 m.

The movements of the lowland to lower-montane and lower/upper-montane forest boundaries and the position of the forest limits are usually interpreted to reflect global cooling during the last glacial, with cooling of perhaps 5–6 °C at the LGM commonly suggested. However, the different extent of movement

for different vegetational boundaries implies an increased adiabatic lapse rate during the LGM compared to today: a conclusion that has been found unacceptable by climatologists (Kutzbach & Guetter 1986). Flenley (1996) emphasized that the difference in altitudinal shift need not be due to changes in lapse rates with time, but could result from the different degree of development of upper-montane forests during glacial versus interglacial periods, with upper-montane forests being virtually absent, or present only as isolated fragments, near the forest limit at the time of glacial maxima. This is borne out for Borneo from a late Quaternary core from the ocean floor offshore from Mahakam Delta (Morley et al. 2004; Morley 2010), which shows that during the last glacial, lower-montane forest in the Mahakam catchment greatly expanded prior to 16 ka. This is based on the increased abundance of *Lithocarpus*-type pollen, suggesting cooler global temperatures at the time of the LGM. However, the pollen records of upper-montane elements, such as *Phyllocladus* and

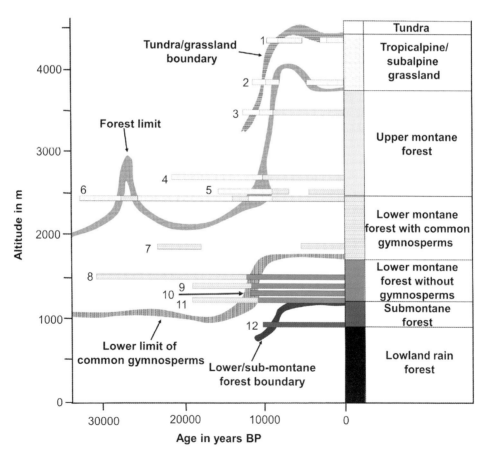

Figure 31.4 Late Quaternary vegetational history of the New Guinea highlands, Sumatra and West Java, based on palynological evidence (modified from Morley & Flenley 1987). Horizontal numbered lines reflect pollen profiles, plotted at their representative altitudes, with length reflecting time in radiocarbon years. 1, Summit pool, Mt. Wilhelm, New Guinea; 2, Brass Tarn, Mt. Wilhelm, New Guinea; 3, Ibuka Swamp, Mt. Wilhelm, New Guinea; 4, Kumanimambuno Mire, Wabag area, New Guinea; 5, Wabag area, New Guinea; 6, Sirunki, Mt. Hagen, New Guinea; 7, Draepi swamp, Mt. Hagen area, New Guinea; 8, Danau Di Atas, Sumatra; 9, Pea Sim Sim, Sumatra; 10, Rawa Sippingan, Sumatra; 11, Bayongrong, West Java; 12, Danau Padang, Sumatra. *Source:* Morley (2000). Reproduced with permission of John Wiley & Sons.

Dacrycarpus, occurred only to a limited extent during the late Pleistocene but are much more prominent during the Holocene, suggesting a Holocene expansion of upper-montane forests on Bornean mountains. Flenley (1996) suggested that the development of upper-montane forests may have been driven by increased ultraviolet radiation during the LGM, but this suggestion has met with little favor, and currently the factors determining the differential movement of such vegetation on the mountains of this region remain unexplained.

31.4 The Mountain Ranges of Southeast Asia and New Guinea, and Their Uplift History

The uplift history of the main montane areas across Southeast Asia is complex (Figure 31.5). This section discusses each in turn.

31.4.1 Indochinese Ranges and Former Uplands of the Sunda Shelf

The Indochinese ranges consist of three NNW–SSE trending mountain belts: the Annamite Cordillera, which follows the border between Vietnam and Thailand; the Dawna Range, which forms the boundary between Thailand and Burma; and the Petchabun Range, which runs through Central Thailand, terminating with the Cardamom Mountains at their southern end to the south of the Tonle Sap in Cambodia. These ranges are very old relative to most other Southeast Asian mountain chains, and probably date from the collision of the Indian Plate and the extrusion of Indochina during the early Cenozoic or even earlier. The Dawna Range is weakly connected to the Main Range of the Malay Peninsula via the Kra Isthmus, whereas the Phetchabun/Cardamom Range was probably aligned with the uplands that followed previously elevated areas such as the Con Son Rise (Morley 2014), Natuna Arch and the former uplands of Western Sarawak.

During the mid-Cenozoic, these mountain ranges were probably much more highly elevated than at present, being sufficiently high to yield a temperate montane flora. This is indicated by pollen evidence, since pollen of the frost-tolerant conifers *Abies*, *Tsuga* and *Picea* occur commonly in sediments from widespread Oligocene and early Miocene localities in Thailand (Watanasak 1990; Songtham et al. 2003), and also by the common occurrence of macrofossils of temperate conifer taxa from intermontane basins in the same areas (Grote 2013). The presence of pollen from these conifers is noteworthy, since today none of them occur on low-latitude mountains. It is thought that this reflects seasonally dry montane climates, with the development of frosts during the dry season. However, this suggestion is purely speculative and requires further investigation.

Conifer pollen is also associated with common pollen of temperate angiosperms, especially *Alnus*, *Altingia*, *Carya*, Ericaceae, *Pterocarya* and *Tilia*. It is likely that it was produced from seasonally cool temperate forests growing at altitudes of 1500 m or more, and that the lowland vegetation would have been deciduous forest, as is the case in northern Thailand today. The presence of seasonally dry lowland and montane climates is suggested also from the occurrence of common *Pinus* pollen associated with the other montane taxa. Prior to anthropogenic influences, *Pinus* was a common element of mid-altitude savannah vegetation and of some dipterocarp communities in Thailand (Ashton 2014). *Altingia*-type pollen is common in Miocene sediments from the northern Gulf of Thailand, and likely derived from lower-montane evergreen rainforests, as the present-day altitudinal range for *Altingia* is between 1300 and 2000 m. The vegetation zonation on the uplands bordering giant rift lakes during the Oligocene and early Miocene of the Gulf of Thailand (Shoup et al. 2013) would most likely have been very similar to that of northern Thailand today, and probably comparable to the vegetation growing on mountains such as Doi Chiangdao (Ashton 2014). This suggests that – especially in the Oligocene – seasonally dry and cool climates extended farther south than today (Morley 2012).

The occurrences of elevated uplands in the Sunda Shelf region is suggested from the presence of common *Picea*, *Tsuga* and *Abies* pollen in petroleum exploration wells studied from the West Natuna and Malay basins (Morley 2014). A clear provenance can be seen from the Con Son Rise (Figure 31.6), which would have been an upland area at the time (Shoup et al. 2013).

31.4.2 Main Range of the Malay Peninsula

The Main Range is essentially the southern expression of the Dawna Range of Indochina, with which it is connected via the Kra Isthmus. A similar uplift history is suggested, and the current topography is the remnant of a much older and probably higher mountain range that would have been present throughout the early Cenozoic. Based on data from Malay Basin petroleum exploration wells that contain common conifer pollen, higher elevations may have borne seasonally cool/dry montane forests with common Laurasian conifers during the Oligocene (Figure 31.6), at least in the northern part of

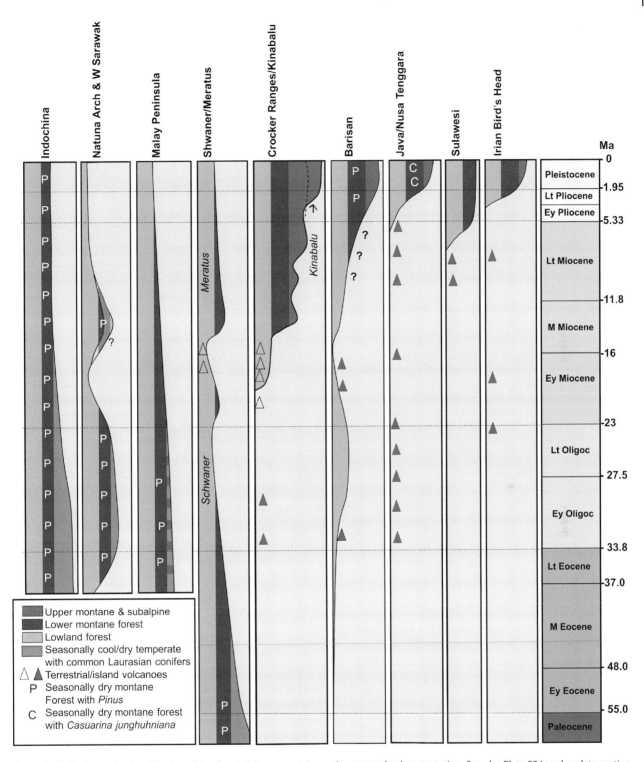

Figure 31.5 Timing and paleoaltitudes of Southeast Asian mountains and suggested paleovegetation. See also Plate 52 in color plate section.

the range. Samples analyzed for palynology from the Batu Arang succession from near Kuala Lumpur, of possible late Eocene or early Oligocene age (Morley, unpublished), suggest deposition in an intermontane basin. This has yielded rich angiosperm tree pollen assemblages but only rare conifer pollen, suggesting a wet climate setting. The southern part of the Main Range may therefore have been characterized by a more typically everwet montane climate during the Oligocene.

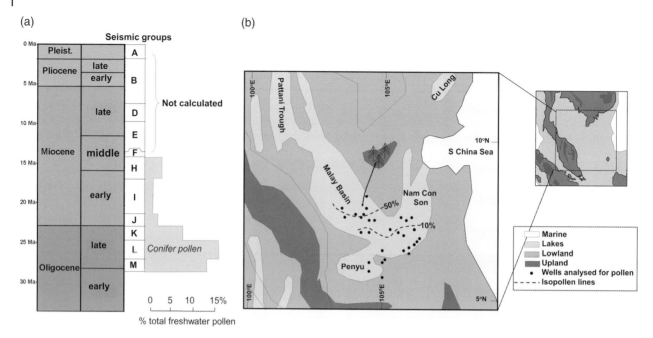

Figure 31.6 (a) Averaged percentage abundance of Laurasian conifers in Malay Basin and West Natuna petroleum exploration wells during the Oligocene and early Miocene. (b) Paleogeographic map for the late Oligocene of Sunda Shelf (ca. 24 Ma). Pollen of Laurasian conifers from exploration wells (black circles) is presented in terms of total freshwater pollen, calculated for each Malay Basin seismic group (stratigraphic packages applied to Malay Basin geology where traditional use of formational terminology fails) and their lateral equivalents in the West Natuna Basin (Morley 2014). Isopollen lines demonstrate that pollen is transported from upland in the region of the Con Son Rise to the north, and the black arrow shows the direction of transportation. The area of the inset differs from the area of the map, because this region has compressed since the Oligocene due to the rotation of Borneo and the extrusion of Indochina. *Source:* Morley (2014). Reproduced with permission of Indonesian Petroleum Association. Basemap from de Bruyn et al. (2012). See also Plate 53 in color plate section.

31.4.3 Uplift of Borneo

Borneo's uplift is complex, but the history of the elevation of its uplands is beginning to become understood. Bornean uplands can be divided into four main areas: Schwaner, western Sarawak (linked to the Natuna Arch), Meratus and the Kinabalu–Central Bornean ranges.

31.4.3.1 Schwaner Mountains

The Schwaner Mountains are the oldest uplands on Borneo, occurring in the south-west. They formed during the mid-Cretaceous with the emplacement of the Schwaner granites (Davies et al. 2015) and have been in place throughout the Cenozoic. Their erosive products are extensively present in the latest Cretaceous to Paleocene Kayan Group in Sarawak (Breitfeld et al. 2015). The older Kayan sediments contain common Laurasian bisaccate pollen derived from Pinaceae (Muller 1968; Morley 1998). This suggests elevations of at least 1500 m (probably more than 2000 m) for the Schwaner at this time, and also a seasonally dry/cool climate. The Schwaner Mountains would have occurred at or to the south of the equator during the Paleocene (Hall 2012b), and would have thus borne the only substantial establishment of Pinaceae south of the equator during the entire history of the family (although very localized occurrences of *Pinus*

merkusii do occur today just south of the equator in Sumatra). Schwaner-derived sediments are also present in the Tanjung Formation in south-eastern Borneo (Witts et al. 2012), but these contain hardly any montane-derived pollen (just very rare Ericaceae grains), suggesting that by the middle Eocene, the Schwaner were substantially denuded with minimal elevations.

Davies et al. (2015) identified a phase of tectonic shearing at around 25 Ma, indicated by the character of Schwaner metamorphism, and suggested that it could reflect further exhumation. This could relate to an increase in sedimentation, as seen in the Mahakam Delta (Morley 2014). However, palynological studies of Mahakam sediments thought to be derived from Schwaner do not suggest derivation from a high-altitude source, since montane pollen is missing from sediments of this period (although data are limited to a few profiles from onshore Mahakam). The Schwaner Mountain Range at this time would probably have been a low range when most of Borneo was either a shallow sea or characterized by low-lying terrain.

31.4.3.2 Western Sarawak and the Natuna Arch

The western Sarawak uplands essentially formed part of the Sunda Shelf uplands. They incorporate Paleocene and Eocene sediments (Breitfeld et al. 2015) and relate to a mid-Cenozoic period of uplift, probably during the late Eocene and Oligocene. Their presence is suggested by the occurrence of common Laurasian conifer pollen, including *Abies*, *Picea* and *Tsuga*, in Oligocene sediments from offshore Sarawak (Muller 1966, 1972) and by unpublished records from the East Natuna Basin. Montane forests in this area would also have included common temperate angiosperms, such as *Alnus*, *Altingia*, *Carpinus*, Ericaceae, *Pterocarya* and *Tilia*, suggesting that the upland flora was of considerable diversity. Toward the northern end of the Natuna Arch, *Ephedra* and *Cedrus* pollen are regularly found, and this may suggest a strongly seasonal element.

This range underwent erosion during the early Miocene, at which time montane conifer pollen becomes rare, but following renewed uplift across the Sunda region due to inversion in the middle Miocene (Ginger et al. 1993), temperate elements returned, although in reduced numbers. The most common indicator is *Alnus* pollen, suggesting that some elevations along the Natuna Arch were sufficient for upper-montane forests to return to the area.

31.4.3.3 Meratus Mountains

In the south-eastern part of Borneo, the Meratus Mountains form a low range aligned north to south. The origin of the Meratus has been a geological puzzle for many years, but they have recently been shown to have undergone rapid uplift in the middle Miocene, after about 14 Ma (Witts et al. 2012). This area presently yields pockets of tree taxa not otherwise seen south of the central range, and is likely to have acted as a refuge for such taxa during the late Neogene and Pleistocene.

Late Quaternary studies (Morley 1982, 2012; Kershaw et al. 2001) suggest that during glacial periods, the climate of lowland south-east Borneo was too seasonal to support evergreen rainforests. However, it is likely that the Meratus Mountains formed a wet enclave that allowed montane rainforest species to survive in this area, at the same time as seasonal climate vegetation was probably prevalent in the surrounding lowlands. Such a scenario would explain the preservation of additional diversity of everwet rainforest taxa in this area, as suggested by Slik et al. (2009).

31.4.3.4 Mount Kinabalu and the Central Borneo Ranges

The Central Borneo Ranges consist of a concentric series of massive ridges, curving in a giant arc from Sabah to central Sarawak, and form the divide between Indonesian Kalimantan and East Malaysia, sometimes informally termed the "Crocker Range." The Central Borneo Ranges consist of extensive mountains and include the Kapuas Hulu, Muller, Iran, Tamo, Dulit and Brassey ranges (Figure 31.1), radiating out from the center of the island. The Crocker Range is the most accessible of these in Sabah. Mount Kinabalu occurs at the northern end of the Crocker Range and consists of a massive granite batholith. It is the highest mountain in Southeast Asia, just surpassing 4000 m in altitude. The granites of Kinabalu were shown by Cottam et al. (2010) to have been emplaced as a laccolith during a period of less than 800 ky between 8 and 7 Ma. The granite, which outcrops today, would have been deeply buried at that time, emphasizing the extensive uplift, and also that the sedimentary overburden has now been removed by erosion. Hall (pers. comm. in Merckx et al. 2015) suggested that Kinabalu then underwent further uplift to its current altitude, probably in the Pliocene. It bore an icecap on its summit during Pleistocene glacial maxima (Stauffer 1968).

The timing and manner of uplift of the Central Borneo Ranges can be inferred from changes in the rate of sedimentation (and hence erosion) in nearby basins and deltas, and from the pollen record from petroleum exploration wells in the surrounding area. The sedimentation records of the Mahakam Delta (Morley 2014; Marshall et al. 2015), the Tarakan Basin deltas (Morley 2014) and Sabah deep-water fans provide a record of their uplift. The Mahakam Delta is the oldest of Borneo's deltas, and first began to prograde after about 20 Ma, suggesting the beginning of uplift in Borneo. The rate of Mahakam sedimentation then increased dramatically after about 15 Ma, suggesting increased uplift following that time, most probably in the central part of the island. The Crocker Range of Sabah probably first became elevated a little later, after 13 or 12 Ma. During the late Miocene, sedimentation rates in the Mahakam Delta and offshore Sabah show parallel fluctuations (Morley 2014), which suggest that uplift was episodal.

It is much more difficult to use the pollen record to infer uplift in Borneo compared to the uplands of Indochina and the Sunda Shelf. This is because Laurasian conifers did not extend to the equatorial zone during the Neogene (none adapted to equatorial everwet climates), and the ecologically equivalent Podocarpaceae did not disperse to Borneo until the Pliocene (Morley 2010). In Quaternary sediments, *Lithocarpus* pollen provides the best montane indicator, but in older Cenozoic rocks it is difficult to differentiate *Lithocarpus* from morphologically similar *Rhizophora* pollen (Morley 2014). One pollen type from montane plants that can be readily differentiated is that of Ericaceae, which has characteristic tetrads. This type is quite rare in deltaic sediments, but if unpublished records from all of the palynologically studied localities from the circum-Borneo region are grouped together, the record of Ericaceae pollen provides a good indicator of uplift in the area (Morley 2014).

There are no Neogene Ericaceae pollen records from the circum-Borneo region before 15 Ma, but after that time Ericaceae tetrads occur, initially in low numbers, then increasing sharply after 8 Ma, suggesting an expansion of Ericaceae-bearing vegetation. This suggests, in turn, a sudden increase in elevation, with upper-montane or sub-alpine vegetation becoming more prominent. This is the time at which the Kinabalu granite was emplaced, as already noted, which probably resulted in the (geologically) sudden rapid uplift of Kinabalu. The Ericaceae pollen record thus suggests initial uplift after 15 Ma, followed by further uplift, with significant elevations – probably above 2000 m – after 8 Ma. Muller (1972) noted the same trend for Ericaceae pollen in the Brunei area, but did not publish details of changes in abundance through time.

Many studied petroleum exploration wells from the circum-Sabah and Kutai areas contain high proportions of *Eugeissona* pollen, especially during the late Miocene and latest middle Miocene, when sedimentation rates were particularly high, suggesting periods of uplift and increased erosion (Morley, in prep.). *Eugeissona* is an understory palm that can be very common along Bornean foothills in steeply incised valleys below 1000 m, and today some species become particularly common following selective logging, suggesting that this genus is often a pioneer following disturbance (Dransfield, pers. comm.). These *Eugeissona* abundance maxima, first noted as "palm pollen maxima" by Muller (1972), are thought to relate to periods of rapid uplift and river incision. These assemblages occur with minimal numbers of pollen of obvious montane forest taxa, and suggest that for much of the late Miocene, sediment was eroded along incised valleys, with much of the topography taking the form of deeply incised foothills with valley bottoms within the range of 500–1000 m.

Merckx et al. (2015) provide molecular evidence for the uplift of Kinabalu (see later) and suggest that the mountain rose to its present altitude after 6 Ma, but that adjacent upland areas were probably never elevated to the same level.

The appearance of regular *Dacrycarpus* and *Phyllocladus* pollen in late Pliocene to Pleistocene circum-Borneo sediments, also noted by Muller (1966), reflects the dispersal of these taxa following uplift in West Papua and other islands of eastern Indonesia.

31.4.4 Barisan Range of Sumatra

The history of uplands along the western margin of Sumatra begins in the Oligocene, at which time inter-montane basins formed in areas such as Ombilin in central Sumatra. The occurrence of upland areas was limited during the early Miocene, but renewed subduction along the Sumatra Trench during the middle Miocene resulted in the initial stage of uplift of the Barisan Range, which includes numerous active volcanoes (especially Gunung Kerinci, at 3805 m the highest mountain in Southeast Asia after Kinabalu). The time of initiation of uplift is about the same as the uplift of the Meratus Mountains and could suggest a common broad causal factor. The current volcanoes are all of Quaternary age. There is little palynological evidence from Sumatra itself to suggest the development of the range. However, the main rivers that flowed out of Sumatra during the Plio–Pleistocene glacio-eustatic lowstands had their deltas in the East Natuna Basin, offshore Western Sarawak and delta sediments from there have yielded scattered *Dacrycarpus* pollen from the basal Pleistocene onward, suggesting that *Dacrycarpus* probably dispersed to Sumatra a little later than it did to Borneo.

The occurrence of pockets of *Pinus merkusii* in the northern part of the Barisan Range, reaching just south of the equator near Gunung Kerinci (the only natural southern hemisphere stand of *Pinus*), reflects previously more seasonally dry montane climates, which would have been more extensive during glacials, as indicated from palynological studies of marine cores from offshore Sumatra by van der Kaars et al. (2010).

31.4.5 Javanese Volcanoes

The present volcanoes of Java are all of Quaternary age, and the island itself became established only during the late Pliocene in the west and the Pleistocene in the east, as a result of increased subduction along the Java Trench (Lunt 2013). However, the area of Java would have comprised a long arc of insular volcanoes through much of its history, and these would have more or less merged to form a continuous land area at times, especially during the early Miocene. The Javanese mountains currently bear everwet montane rainforest in the west and seasonally dry montane forests in the east, where *Casuarina junghuhniana* is dominant.

Palynological studies through the last glacial from Bandung Lake in West Java (van der Kaars & Dam 1995) show that seasonal montane vegetation extended to that area during glacial maxima. A pollen and phytolith record from a "Java Man" site from Perning, East Java, of early Pleistocene age, provides evidence that nearby volcanoes bore vegetation with *Casuarina* and pooid grasses at intermediate altitudes, suggesting seasonally dry lower-montane forests and regular occurrences of *Dacrycarpus imbricatus* and other podocarps, presumably from wetter upper-montane forest (Morley 2010).

Van Steenis (1934a, 1972) observed in Java that some montane taxa with intermediate altitudinal ranges occur on high peaks but not on lower peaks with summits within their altitudinal range. He proposed that montane taxa have a zone of "permanent establishment," from which they can migrate either upslope or downslope with changing global temperatures, and that if a summit occurs below the "permanent" zone of a taxon, it will not be present on that peak, although it may occur at the same altitude on neighboring higher peaks. He called this the "elevation effect" – not to be confused with the mountain mass-elevation or Massenerhebung effect.

31.4.6 Sulawesi

Hall (2009, 2012a) suggested the development of uplands on the island of Sulawesi during the Pliocene. Unpublished data on sedimentation rates from the Makassar Straits adjacent to west Sulawesi show a rapid rise in sedimentation from about 7 Ma onward, and this suggests latest Miocene initiation of uplift.

Today, upland areas in north-east Sulawesi bear evergreen montane rainforest, but a palynological study from Lake Tondano at 680 m (Dam et al. 2001) indicates seasonally dry conditions at the time of the LGM, with predominantly open forest in the surrounding lower-montane forest zone.

31.4.7 Bird's Head of West Papua and Timor

The Bird's Head region of West Papua has undergone dramatic tectonic modification within the last few million years, indicated by the occurrence of strongly dipping latest Miocene rocks on the western side of the Lengguru Fold Belt, which forms the "Bird's Neck." In petroleum exploration wells drilled in this area, the first appearance of montane pollen – indicated by the occurrence of regular *Dacrycarpus* and *Nothofagus* pollen – is during the mid-Pliocene at about 4.6 Ma, from which time it remains common throughout the later Pliocene and Pleistocene. This pollen was probably sourced from the Arfak Mountains, which currently reach 2955 m. It is likely that other areas, such as Misool, were elevated even later – since the earliest Pleistocene – because in that area, *Dacrycarpus* pollen first appears later, at about 1.9 Ma. Pollen increases in abundance after 1.5 Ma, suggesting a further phase of uplift.

Further east, in New Guinea, *Dacrycarpus* pollen appears a little earlier, during the late Miocene (Morley 2000). However, the mid-Pliocene appearance in the Bird's Head, coupled with the first appearance of this pollen at the same time in offshore Sulawesi (van der Kaars 1991) and in Borneo (Muller 1966; Morley 2010), suggests that following its arrival in the Bird's Head, *Dacrycarpus* quickly dispersed westward, facilitated by the presence of other sufficiently high mountains that had formed by this time in intervening areas, such as Sulawesi.

Montane podocarp pollen has also been recorded from the Pliocene of Timor, by Nguyen et al. (2013), who suggest late Pliocene uplift to 2000 m. However, this may be an overestimation of the elevation, as *Podocarpus* and *Dacrycarpus* are common in seasonal montane forest above 1000 m in Timor (Monk et al. 1997).

The pollen record from West Papua and Timor demonstrates the remarkably young age of this mountain range, and emphasizes that any montane species that are endemic to the Bird's Head have differentiated over a very short time period.

31.5 Dispersal and Evolution

The manner in which montane plants dispersed to high mountain peaks puzzled both Wallace (1869), during his visit to Gunung Pangrango, and van Steenis (1936), while investigating montane floras during the 1930s. Van Steenis needed to make judgments to explain plant dispersal prior to the understanding of plate tectonics, and proposed former "land bridges" in his famous, but now often discredited, "land bridge theory" of botany (van Steenis 1962). However, 40 years later, most of his "land bridges" were demonstrated to coincide very closely to dispersal routes formed as a result of plate tectonic collisions (Morley 2003).

By comparing plant distributions, van Steenis recognized three dispersal patterns relating to microtherm elements in the mountain flora (Figure 31.7). One group, which dispersed along his "Sumatran Track," had the majority of its records from the volcanoes of Sumatra and Java, but originated in the Himalaya. *Primula imperialis* is a good example of this group (Figure 31.8). A second group dispersed along the Luzon–Formosa Track, with most occurrences in the Philippines and Wallacea. A third group followed the New Guinea Track, along which taxa dispersed from Australia or New Zealand via New Guinea. The terrestrial orchid *Microtis*, with a center of diversity and presumed origin in Australia, dispersed along both of these tracks to Japan (Figure 31.8). Many of the localities for such microtherm elements are volcanoes, which are ephemeral with respect to their highest altitudes. Current ranges may therefore relate to the former occurrence of older volcanoes that are now eroded stumps, or that have already undergone collapse.

The Southeast Asia region provides a rich laboratory within which many details of the processes of evolution can be studied. This applies especially to montane biota, which are diverse across the region, and have many endemics. Molecular studies are already providing a rich resource for such research, with results being brought together in meta-studies

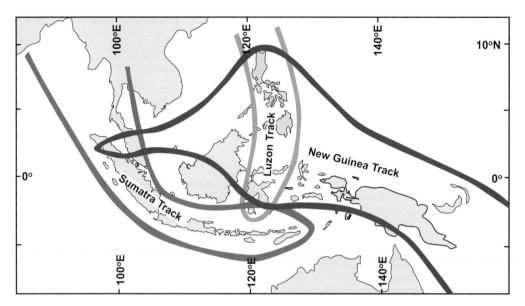

Figure 31.7 Migration tracks suggested by van Steenis. *Source:* van Steenis (1934a). Reproduced with permission of Reinwardtia.

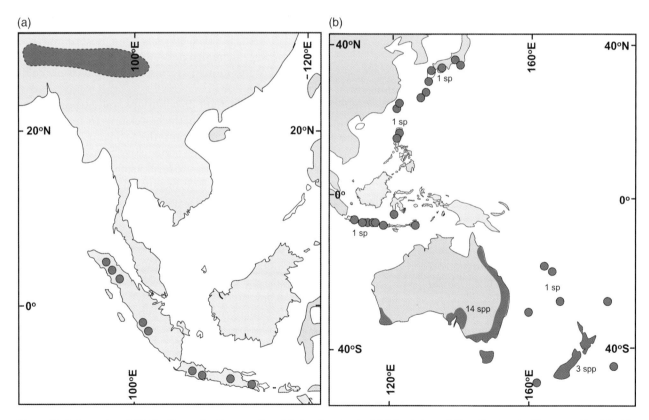

Figure 31.8 Examples of distributions reflecting van Steenis' migration tracks. (a) *Primula imperialis*, reflecting the "Sumatran Track," with its main distribution in the Himalaya and scattered occurrences on mountaintops in Sumatra and Java. (b) The orchid genus *Microtis*, with a diversity center in Australia, but with one species (*M. unifolia*) dispersing via the New Guinea and Luzon tracks to Japan. *Source:* van Steenis (1972). Reproduced with permission of EJ Brill Publishers.

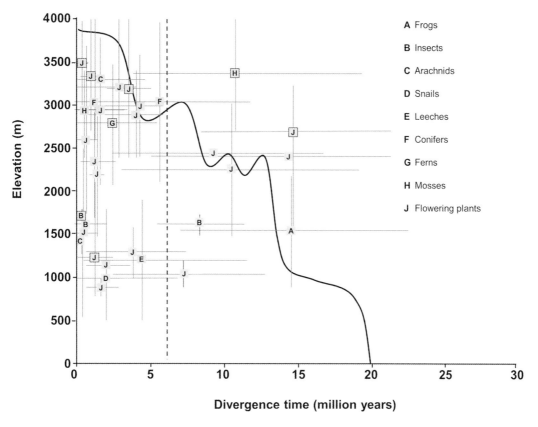

Figure 31.9 Elevations and ages for Mount Kinabalu endemic species, based on molecular analyses. Elevations (mid-points, minima and maxima) and dates of origination are derived from molecular dating for endemic species. Eccentric species are indicated with a black box around the data point. Vertical dashed line indicates the earliest possible date for Kinabalu to have reached its current elevation (Merckx et al. 2015); solid line indicates the suggested altitude of Kinabalu and the surrounding area, based on geological and palynological data (extracted from Figure 31.5). *Source:* Adapted from Merckx et al. (2015).

that integrate molecular, distributional, geological and fossil pollen data (de Bruyn et al. 2012).

The origin of the remarkable diversity of the Mount Kinabalu flora has long attracted attention (Christenden & Holttum 1934; van Steenis 1964). It has recently been investigated using a molecular approach by Merckx et al. (2015), who sequenced DNA from a diverse group of organisms, including plants, fungi, frogs, snails and insects, rather than a single taxonomic group. These authors compared results with collections from adjacent areas, both within and outside Borneo. By using a wide range of taxa, errors inherent in the study of a single taxonomic group were minimized. The analysis indicated that most of the endemic taxa examined were quite young, starting to speciate after 6 Ma, which was after, or at the same time as, the uplift of the mountain they inhabited. Merckx et al. (2015) concluded that organisms reached the mountain by two routes. Those from the highest elevations in particular were most closely related to taxa from outside Borneo, and were good dispersers, with small seeds or spores. Their dispersal route would most likely have followed one or more of van Steenis' dispersal tracks. These taxa were termed "eccentric" species. A significantly larger number were derived from local ancestors, and were termed "centric" species, living

at lower altitudes on the same island and thus originating by parapatry. The study also emphasized the role of niche conservatism, with organisms maintaining their environmental preferences over time: the long-distance dispersers stayed in their preferred altitudinal zone, to which they were pre-adapted, while those migrating upslope tended to stay in the same broadly defined vegetation zone.

Figure 31.9 superimposes the altitudinal development of Kinabalu suggested from geology and palynology (this study) over the time of origin and altitudinal distribution of the sequenced taxa. It is noteworthy that the lower-altitude montane taxa appeared between 15 and 10 Ma, tying closely with the predicted elevation for the region at that time.

Two additional studies, based on single taxa, also help to demonstrate how the phylogeography of montane plant taxa can be clarified through the integration of molecular data with historical pollen/macrofossil records and geological history. *Rhododendron* sect *Vireya* (Ericaceae) is a common and spectacular component of upper-montane forests and sub-alpine woodlands. Molecular data provide a phylogeny, and the Ericaceae pollen record provides a "time envelope" within which the section diversified. *Lithocarpus*, on the other hand, is one of the most important trees of lower-montane forests, the pollen of which is difficult to identify with certainty (so it has a poor pollen record), but the phylogeography can be understood from the integration of molecular and geological data. The record in *Lithocarpus* of endemism hints at the antiquity of mountain ranges.

A molecular evolutionary study of *Rhododendron* sect *Vireya* was undertaken by Brown et al. (2006), and subsequently evaluated using the program LAGRANGE by Webb & Ree (2012). The age of the clade was suggested to be younger than 46–32 Ma, based on fossil dates for *Rhododendron* subgenus

Rhododendron, whereas Richardson (pers. comm. in Webb & Ree 2012) suggested an age of about 13 Ma for the crown group *Vireya*, based on a specific nuclear marker. Pollen of *Vireya* cannot be differentiated from other Ericaceae, but with a good fossil record for Ericaceae pollen from Borneo, and with no records prior to 15 Ma, a realistic "maximum age" picture of the development of *Vireya* can be proposed (Figure 31.10). The age of subsequent clades can then be used to provide ages for the timing of uplift of uplands across the region, although it must be emphasized that such suggestions are very tentative if based on a single taxon.

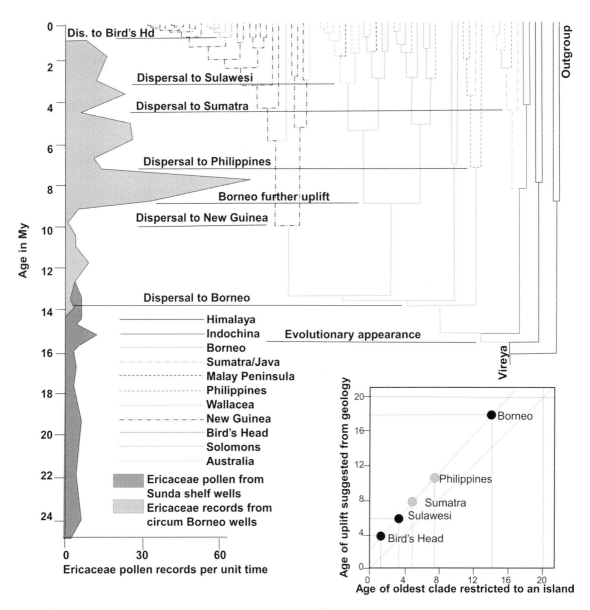

Figure 31.10 Timing of entry of *Vireya* clades across insular Southeast Asia. The inset shows the relationship between the known timing of the uplift and appearance of *Vireya* spp. (dark gray circles) and the predicted timing of the uplift of uplands for which there is incomplete geological evidence based on the first appearance of *Vireya* (light gray circles). *Source:* Adapted from Webb & Ree (2012). See also Plate 54 in color plate section.

The *Euvireya* clade is characteristic of Sundaland, and if this clade first became established after 15 Ma, as is suggested from the pollen record, then the maximum age for the *Vireya* clade elements centered in the Himalaya and Indochina is about 16 Ma. The group would have become established in mainland New Guinea by about 10 Ma, tying in with the appearance of *Dacrycarpus* pollen there (Morley 2010); in Sulawesi by about 5 Ma, just after the initiation of uplift based on sedimentation rates (see earlier); and on the Bird's Head around 1 Ma, just after its uplift, based on the pollen record. The conclusion to be reached from this is that following the initial appearance of Sect *Vireya* in

Indochina during the middle Miocene, it subsequently dispersed to each upland area as soon as it became available for colonization, and then diversified. The molecular ages for Filipino (7 Ma) and Sumatran (5 Ma) *Vireya* might then indicate minimum ages for uplift in those areas.

For the stone oak genus *Lithocarpus*, Cannon & Manos (2002) suggest establishment across Indochina and Borneo in three stages (Figure 31.11). Initially, the ancestral group spread throughout both regions. Next, this widespread population fragmented and underwent a severe bottleneck in central Sundaland. This and subsequent fragmentation caused the Bornean lineage to

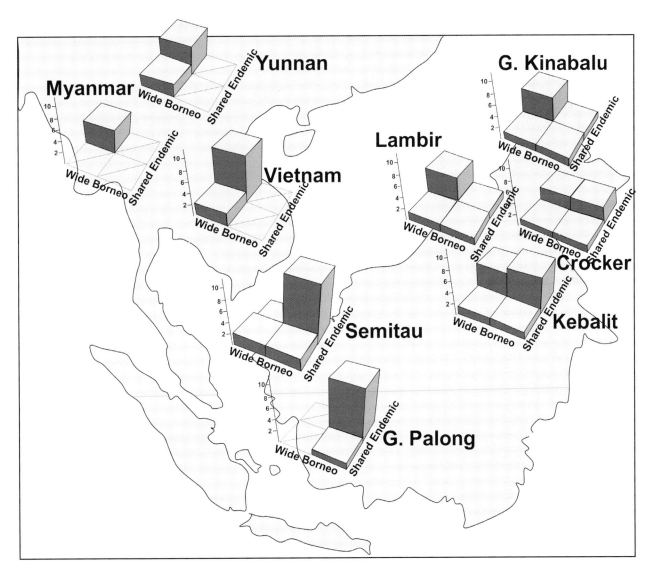

Figure 31.11 *Lithocarpus* diversity across Southeast Asia, showing the geographical distribution of shared versus endemic types from each of the major clades. For each location, the vertical axis represents the frequency of taxa for each category. Taxa from the "widespread" clade are shown in the two columns on the left and taxa from the "Bornean" clade in the two on the right. The front row represents shared taxa, while the back row represents endemic taxa. *Source:* Cannon & Manos (2002). Reproduced with permission of John Wiley & Sons.

diversify in isolation. Finally, the populations containing the Bornean lineage migrated into central and northern Borneo from western Borneo. This scenario would fit well with *Lithocarpus* dispersing between Indochina and Borneo when these areas were connected via uplands stretching intermittently from the Cardamom Mountains in Cambodia, through the Con Son Rise and Natuna and into Western Sarawak, perhaps during the Oligocene. At this time, Borneo was of more reduced extent than at present, with upland areas only in the west (West Sarawak and Schwaner Mountains). Isolation in Borneo may have come about following denudation of the Con Son and Natuna ranges, with subsequent initial diversification taking place in Western Sarawak and the Schwaner Range. With the uplift of the central Bornean Range starting during the middle Miocene (about 15 Ma), stone oaks probably dispersed to central Borneo, and subsequently to Sabah, with the uplift of the Crocker Range beginning about 13 Ma, and the rapid uplift of Kinabalu from about 8 Ma. The rate of speciation in each area would suggest that Bornean endemics each take about 2 My to evolve.

Sundaic montane birds show biogeographic patterns that have long intrigued avian biogeographers (Banks 1937; Harrisson 1956; Smythies 1957). Distributions are now thought to reflect phylogeography and climate history; this has been summarized in a recent review of molecular data by Sheldon et al. (2015).

Taxa considered "ancient" among montane birds, such as the endemic Bornean ferruginous partridge (*Caloperdix oculus*) and the more widespread crimson-headed partridge (*Haematortyx sanguiniceps*), are restricted in Borneo to the Central Ranges and Sabah, and are absent from older uplands such as the Schwaner. This suggests that their arrival in Borneo, and the appearance of subsequent clades, postdates the middle Miocene uplift of the Central Bornean ranges.

Some montane bird taxa demonstrate Himalayan origins, such as the Sundaic laughing thrush (*Garrulax palliatus*), the Bornean endemic chestnut-crested yuhina (*Yuhina everetti*) and the more widespread yellow-breasted warbler (*Seicerus montis*). Others have been derived from Australasia, such as the Bornean whistler (*Pachycephala hypoxantha*). These patterns echo the migration trackways of van Steenis (1936), and thus some montane birds display similar dispersal routes to plants.

Sheldon et al. (2015) suggest that many Bornean montane endemics have arisen by allopatry, rather than parapatry, in that sister taxa tend to occur on mountains on other islands, such as the Philippines, Sumatra or Java, and that when closely related congeners are elevationally parapatric, the lowland species is more likely to be a recent invader, restricting the montane species through competition. This is best demonstrated by the distribution of the mountain black-eye (*Chlorocharis emeliae*), which has a "sky island" distribution, with populations on the higher peaks in north-west Borneo and on isolated mountains in western Sarawak (Gawin et al. 2014). Based on a detailed molecular evaluation, Sheldon et al. (2015) suggest that this pattern has arisen as a result of vicariance during dryer periods in the middle or early Pleistocene (when mountains may have afforded wet refugia), with dispersal occurring during wetter times. So far, palynological and plant geographical and isotope studies have provided little evidence for late Quaternary drier climates in areas such as Sarawak (Cannon et al. 2009; Wurster et al. 2010; Morley 2012), but the occurrence of periods with drier lowland climates during earlier Quaternary glacials cannot be ruled out. Detailed palynological studies of the thick Plio–Pleistocene sediments penetrated by hydrocarbon exploration wells from offshore Sarawak may clarify this.

31.6 Conclusion

The Southeast Asia region and Eastern Indonesia include some of the world's youngest mountain ranges, as well as older eroded roots of ranges dating back to the late Cretaceous. The time and extent of uplift of several of the uplands in the region have been determined from timings of tectonic events, the time of emplacement of igneous rocks, timing of metamorphism, sedimentation rate patterns and the record of pollen from upland vegetation and minerals from sediments produced through the erosion of those uplands. However, molecular studies utilizing sequencing of mountain plants and animals to determine times of origin of endemic taxa are also beginning to play a part in shaping our understanding of the timing and extent of uplift. Timing the uplift of many of the upland areas across the region based on geologic criteria will always be problematic, but from the few pioneering studies so far, it is clear that comparative studies of the time of origin of floral and faunal elements on uplands may eventually provide a valuable means of determining the timing of uplift for each of the uplands within the region.

Molecular studies are of value not only in dating the time of origin of biota on young mountains, but also in identifying ancient mountain roots that have acted as refugia for montane genera over long time periods. The genus *Lithocarpus* displays this pattern, the diversity of its endemics reflecting the age of dispersal into different upland areas within the region of Borneo and Indochina.

Further studies on *Nepenthes*, *Ranunculus*, *Begonia*, *Impatiens*, Asteraceae, Caprifoliaceae, Labiatae, Ericaceae, Orchidaceae and others are bound to provide rich pickings with respect to establishing the origins of sub-alpine plant communities, but the study of taxa from upper- and lower-montane forests restricted to intermediate altitudes will be more challenging. Utilizing the niche conservatism demonstrated for Kinabalu endemics by Merckx et al. (2015), by comparing time of origin of diverse endemics from specific vegetation zones across the region, the time of establishment of lower-montane, then upper-montane and finally sub-alpine communities could allow the timing and rate of uplift of montane areas and the timing of montane ecosystem establishment across the region to be determined.

The Southeast Asia region is geologically complex, and its biota ranks among the most diverse globally (Slik et al. 2015). This diversity applies especially to the flora and fauna of mountain areas, and every effort should be made to better understand how it has come to be and what measures can be put in place to ensure the long-term conservation of mountain biota across the region.

Acknowledgments

The timing of mountain building was discussed variously with Tony Swiecicki, Robert Hall and Tim Breitfeld, but the suggestions made here are the author's personal perspective. Chuck Cannon checked the section on *Lithocarpus*.

References

Aitchison, J.C., Ali, J.R. & Davis, A.M. (2007) When and where did Asia and India collide? *Journal of Geophysical Research* **112**, BO5423.

Andrade Marin, L. (1945) *Cuadro Sinoptico de Climatological Ecuatoriana*. Quito.

Ashton, P.S. (2014) *On the Forests of Tropical Asia. Lest the Memory Fade*. London: Kew Publishing, Royal Botanic Gardens Kew and Cambridge, MA: Arnold Arboretum, Harvard University.

Banks, E. (1937) Birds from the highlands of Sarawak. *Sarawak Museum Journal* **4**.

Bouihol, P., Jagoutz, O., Hanchar, J.M. & Dudas, F.O. (2013) Dating the India-Eurasia collision through arc magmatic records. *Earth and Planetary Science Letters* **355**, 163–175.

Breitfeld, H.T., Galin, T., Hall, R. et al. (2015) Proto-South China Sea and South China Sea Early History: A View from Sarawak Proc AAPG workshop on the South China Sea, Kota Kinabalu.

Brown, G.K., Nelson, G. & Ladiges, P.Y. (2006) Historical biogeography of Rhododendron sect Vireya and the Malesian Archipelago. *Journal of Biogeography* **33**, 1929–44.

Cannon, C.H. & Manos, P.S. (2002) Phylogeography of the Southeast Asian stone oaks (*Lithocarpus*). *Journal of Biogeography* **30**, 211–226.

Cannon, C.H, Morley, R.J. & Bush, A.B.G. (2009) The current refugial rainforests of Sundaland are unrepresentative of their biogeographic past and highly vulnerable to disturbance. *Proceedings of the National Academy of Sciences USA* **106**, 11 188–11 193.

Christenden, C. & Holttum, R.E. (1934) The ferns of Mount Kinabalu. *Gardens Bulletin Straits Settlements* **VII**, 191–231.

Cottam, M.A, Hall, R., Sperber, C. & Armstrong, R. (2010) Pulsed emplacement of the Mount Kinabalu Granite, North Borneo. *Journal of the Geological Society of London* **167**, 49–60.

Dam, R.A.C., Fluin, J., Suparan, P. & van der Kaars, S. (2001) Palaeoenvironmental developments in the Lake Tonada area, (N Sulawesi, Indonesia) since 33 000 yr B.P. *Palaeogeography, Palaeoclimatology, Palaeoecology* **171**, 147–183.

Davies, L., Hall, R. & Forster, M. (2015) Age and Character of Basement Rocks in SW Borneo: New Insights from Ar-Ar Dating of Pinoh Metamorphic Group Rocks. AAPG Workshop on the South China Sea, Kota Kinabalu.

de Bruyn, M., Stelbrink, B., Morley, R.J. et al. (2012) Borneo and Indochina are major evolutionary hotspots for Southeast Asian biodiversity. *Systematic Biology* **63**, 879–901.

Flenley, J.R. (1979) *The Equatorial Rain Forest: A Geological History*. London: Butterworths.

Flenley, J.R. (1984) Late Quaternary changes of vegetation and climate in the Malesian mountains. *Erdwissenschaftliche Forschung* **18**, 261–267.

Flenley, J.R. (1996) Problems of the Quaternary on mountains of the Sunda-Sahul region. *Quaternary Science Reviews* **15**, 549–555.

Gawin, D.F, Rahman, M.A, Ramji, M.F.S. et al. (2014) Patterns of avian diversification in Borneo: the case of the endemic mountain black-eye (*Chlorocharis emiliae*). *The Auk* **131**, 86–99.

Ginger, D.C., Ardjakusumah, W.O., Hedley, R.J. & Pothecary, J. (1993) Inversion History of the West Natuna Basin: Examples from the Cumi-Cumi PSC. Proceedings of the 22nd Indonesian Petroleum Association Convention Jakarta 1993, pp. 637–657.

Grote, P. (2013) *Migration and Extinction of Conifers during the Cenozoic: Evidence from the Fossil Record in Thailand.* Berlin: SAGE Proceedings.

Grubb, P.J. (1971) Interpretation of the Massenerhebung' effect on tropical mountains. *Nature* **229**, 39–40.

Hall, R. (1998) The plate tectonics of Cenozoic SE Asia and the distribution of land and sea. In: Hall, R. & Holloway, J. *Biogeography and Geological Evolution in SE Asia.* Amsterdam: Bukhuys Publishers, pp. 99–131.

Hall, R. (2002) Cenozoic geological and plate tectonic evolution of SE Asia and the SW Pacific: computer-based reconstructions, model and animations. *Journal of Asian Earth Sciences* **20**, 353–431.

Hall, R. (2009) Southeast Asia's changing palaeogeography. *Blumea* **54**, 148–161.

Hall, R. (2012a) Sundaland and Wallacea, geology, plate tectonics and palaeogeography. In: Gower, D., Johnson, K., Richardson, J. et al. (eds.) *Biotic Evolution and Environmental Change in Southeast Asia.* Cambridge: Cambridge University Press, pp. 32–78.

Hall, R. (2012b) Late Jurassic to Cenozoic reconstructions of the Indonesian region and the Indian Ocean. *Tectonophysics* **570–571**, 1–41.

Hall, R., Clements, B. & Smythe, H.R. (2009) Sundaland: Basement Character, Structure and Plate Tectonic Development. Proceedings, Indonesian Petroleum Association Thirty-Third Annual Convention & Exhibition, May 2009. Paper IPA09-G-134.

Harrisson, T. (1956) A new mountain black-eye (Chlorocharis) from North Borneo. *The Sarawak Museum Journal* **7**, 518–521.

Hoch, G. & Korner, C. (2012) Global patterns of mobile carbon stores in trees at the high-elevation tree line. *Global Ecology and Biogeography* **21**, 861–871.

Kershaw, A.P., Penny, D., van der Kaars, S. et al. (2001) Vegetation and climate in lowland Southeast Asia at the last glacial maximum. In: Metcalfe, I., Smith, J., Morwood, M. & Davidson, I. (eds.) *Faunal and Floral Migrations and Evolution in SE Asia-Australasia.* Lisse: Balkema, pp. 227–338.

Kutzbach, J.E. & Guetter, P.J. (1986) The influence of changing orbital parameters and surface boundary conditions on climate simulators for the past 18 000 years. *Journal of Atmospheric Science* **43**, 1726–1759.

Lunt, P. (2013) *The Sedimentary Geology of Java.* Jakarta: Indonesian Petroleum Association.

Marshall, N., Novak, V., Cibaj, I. et al. (2015) Dating Borneo's deltaic deluge: middle Miocene progradation of the Mahakam Delta. *Palaios* **30**, 7–25.

Merckx, V.S.F.T, Hendriks, K.P., Beentjes, K.K. et al. (2015) Evolution of endemism on a young tropical mountain. *Nature* **524**, 347–350.

Monk, K.A., de Fretes, Y. & Reskodihardjo-Lilley, G. (1997) *The Ecology of Nusa Tenggaara and Malukyu.* The Ecology of Indonesia Series V. Hong Kong: Periplus.

Morley, R.J. (1982) A palaeoecological interpretation of a 10 000 year pollen record from Danau Padang, Central Sumatra, Indonesia. *Journal of Biogeography* **9**, 151–190.

Morley, R.J. (1998) Palynological evidence for Tertiary plant dispersals in the SE Asia region in relation to plate tectonics and climate. In: Hall, R. & Holloway J. (eds.) *Biogeography and Geological Evolution of SE Asia.* Amsterdam: Bakhuys Publishers, pp. 177–200.

Morley, R.J. (2000) *Origin and Evolution of Tropical Rain Forests.* Chichester: John Wiley & Sons Ltd.

Morley, R. J. (2003) Interplate dispersal routes for megathermal angiosperms. *Perspectives in Plant Ecology, Evolution and Systematics* **6**, 5–20.

Morley, R.J. (2010) Palaeoecology of tropical podocarps. In: Turner, B.L. & Cernusak, L.M. (eds.) *Ecology of the Podocarpaceae in Tropical Forests.* Smithsonian Institution Contributions to Botany 95. Washington, DC: Smithsonian Institution Scholarly Press, pp. 21–41.

Morley, R.J. (2012) A review of the Cenozoic palaeoclimate history of Southeast Asia. In: Gower, D., Johnson, K., Richardson, J. et al. (eds.) *Biotic Evolution and Environmental Change in Southeast Asia.* Cambridge: Cambridge University Press, pp. 79–114.

Morley, R.J. (2014) Rifting and Mountain Building across Sundaland, a Palynological and Sequence Biostratigraphic Perspective. Proceedings, Indonesian Petroleum Association Thirty-Eighth Annual Convention & Exhibition IPA14-G-011.

Morley, R.J. & Flenley, J.R. (1987) Late Cainozoic vegetational and environmental changes in the Malay Archipelago. In: Whitmore, T.C. (ed.) *Biogeographical Evolution of the Malay Archipelago.* Oxford Monographs on Biogeography 4. Oxford: Oxford Scientific Publications, pp. 50–59.

Morley, R.J. & Morley, H.P. (2013) Mid Cenozoic freshwater wetlands of the Sunda region. *Journal of Limnology* **72**, 18–35.

Morley, R.J., Morley, H.P. & Wonders, A.A. (2004) Biostratigraphy of Modern (Holocene and Late Pleistocene) Sediment Cores from Makassar Straits, Deepwater and Frontier Exploration in Asia & Australasia Proceedings, Jakarta, December 2004.

Muller, J. (1966) Montane pollen from the Tertiary of NW Borneo. *Blumea* **14**, 231–235.

Muller, J. (1968) Palynology of the Pedawan and plateau sandstone formations (Cretaceous – Eocene) in Sarawak, Malaysia. *Micropalaeontology* **14**, 1–37.

Muller, J. (1972) Palynological evidence for change in geomorphology, climate and vegetation in the Mio-Pliocene of Malesia. In: Ashton, P.S. & Ashton, M. (eds.)

The Quaternary Era in Malesia. Geography Department, University of Hull, Misc. Ser. 13, pp. 6–34.

Ngah, K., Madon, M. & Tjia, H.D. (1996) The role of pre-Tertiary fracture in the formation and development of the Malay and Penyu Basins. In: Hall, R. & Blundell, D.J. (eds.) *Tectonic Evolution of Southeast Asia*. Geological Society of London Special Publication 106, pp. 281–290.

Nguyen, N., Duffy, B., Shulmeister, J. & Quigley, M. (2013) Rapid Pliocene uplift of Timor. *Geology* **41**, 179–182.

Ohsawa, M. (1993) Latitudinal pattern of mountain vegetation zonation in southern and eastern Asia. *Journal of Vegetation Science* **4**, 13–18.

Robbins, R.G. & Wyatt-Smith, J. (1964) Dry-land forest formations and forest types in the Malay Peninsula. *Malayan Forester* **27**, 188–217.

Sheldon, F.H., Lim, H.C. & Moyle, R.G. (2015) Return to the Malay Archipelago: the biogeography of Sundaic rainforest birds. *Journal of Ornithology* **15**(Suppl. 1), 91–113.

Smythies, B.E. (1957) An annotated checklist of the birds of Borneo. *The Sarawak Museum Journal* **7**, 523–818.

Slik, J.W.F., Raes, N., Aiba, S.I. et al. (2009) Environmental correlates for tropical tree diversity and distribution patterns in Borneo. *Diversity and Distributions* **15**, 523–532.

Slik, J.W.F., Arroyo-Rodríguez, V., Aiba, S.I. et al. (2015) An estimate of the number of tropical tree species. *Proceedings of the National Academy of Sciences USA* **112**, 7472–7477.

Shoup, R.C., Morley, R.J., Swiecicki, T. & Clark, S. (2013) Tectono-stratigraphic framework and Tertiary paleogeography of Southeast Asia: Gulf of Thailand to South Vietnam Shelf. Search and Discovery Article #30246. Posted September 24, 2012.

Songtham, W., Ratanasthien, B., Mildenhall, D.C. et al. (2003) Oligocene-Miocene climatic changes in northern Thailand resulting from extrusion tectonics of Southeast Asian landmass. *Science Asia* **29**, 221–233.

Stauffer, P.H. (1968) Glaciation on Mount Kinabalu. *Bulletin of the Geological Society of Malaysia* **1**, 63.

Symington, C.F. (1943) Malayan Forest Records No. 16. Kuala Lumpur: Forest Research Institute Malaysia.

Tapponnier, P., Peltzer, G., LeDain, A. et al. (1982) Propogating extrusion tectonics in Asia; new insights with simple experiments with plasticine. *Geology* **10**, 611–616.

Tapponnier, P., Peltzer, G. & Armijo, R. (1986) On the mechanics of the collision between India and Asia. In: Coward, M.P. & Ries, A.C. (eds.) *Collision Tectonics*. Geological Society Special Publication 19, pp. 115–157.

van der Kaars, W.A. (1991) Palynological aspects of site 767 in the Celebes Sea. *Proceedings of the Ocean Drilling Program, Scientific Results* **124**, 369–374.

van der Kaars, W.A. & Dam, M.A.C. (1995) A 135 000-year old record of vegetation and climatic change from the Bandung area, West Java, Indonesia. *Palaeogeography, Palaeoclimatology, Palaeoecology* **117**, 55–71.

van der Kaars, W.A, Bassinot, F., De Deckker, P. & Guichard, F. (2010) Changes in monsoon and ocean circulation and the vegetation cover of southwest Sumatra through the last 83 000 years: the record from marine core BAR94-42. *Palaeogeography, Palaeoclimatology, Palaeoecology* **296**, 52–78.

van Steenis, C.G.G.J. (1934a) On the origin of the Malaysian mountain flora, Part 1. *Bulletin du Jardin Botanique de Buitenzorg III* **13**, 135–262.

van Steenis, C.G.G.J. (1934b) On the origin of the Malaysian mountain flora, Part 2. *Bulletin du Jardin Botanique de Buitenzorg III* **13**, 289–417.

van Steenis, C.G.G.J. (1936) On the origin of the Malaysian mountain flora, Part 3. Analysis of floristic relationships (1st installment). *Bulletin du Jardin Botanique de Buitenzorg III* **14**, 36–72.

van Steenis, C.G.G.J. (1962) The land-bridge theory in botany. *Blumea* **11**, 235–372.

van Steenis, C.G.G.J. (1964) Plant geography of the mountain flora of Mount Kinabalu. *Proceedings of the Royal Society of London* **16**, 7–38.

van Steenis, C.G.G.J. (1972) *The Mountain Flora of Java*. Leiden: E.J. Brill.

van Zinderen Bakker, E.M. (1962). A late-glacial and post-glacial climatic correlation between East Africa and Europe. *Nature* **194**, 201–203.

Wallace, A.R. (1869) *The Malay Archipelago*. London: Macmillan.

Wang, B., Clemens, S.C. & Liu, P. (2003) Contrasting the Indian and East Asian monsoons: implications on geologic timescales. *Marine Geology* **201**, 5–21.

Watanasak, M. (1990) Mid Tertiary palynostratigraphy of Thailand. *Journal of Southeast Asian Earth Sciences* **4**, 203–218.

Webb, C.O. & Ree, R. (2012) Historical biogeography inference in Malesia. In: Gower, D., Johnson, K., Richardson, J. et al. (eds.) *Biotic Evolution and Environmental Change in Southeast Asia*. Cambridge: Cambridge University Press, pp. 191–215.

Witts, D., Hall, R., Nichols, G. & Morley, R.J. (2012) A new depositional and provenance model for the Tanjung formation, Barito Basin, SE Kalimantan, Indonesia. *Journal of Asian Earth Sciences* **56**, 77–104.

Wurster, C.M., Bird, M.I., Bull, I.D. et al. (2010). Forest contraction in north equatorial Southeast Asia during the Last Glacial Period. *Proceedings of the National Academy of Sciences* **107**, 15 508–15 511.

Index

Page numbers in *italic* refer to figures; **bold** to tables

a

abiotic data, species distribution
models 311–313
abiotic nature, geoconservation 156
abiotic processes
and biodiversity 3
drivers of diversification 259–260,
261, 263
mountain geodiversity 144
ablation valleys, Antarctica
365–366, *367*
abundance measures, definitions
296, 302
ACBR (Antarctic Conservation
Biogeographic Region) *365*
ACC (Antarctic Circumpolar
Current) 130
adaptation
Antarctica 358, 364, 368
to climate change 190–193, *191*,
192, *193*, *194*
mammals 211
adaptive radiation *see* radiation
advection processes 53
Africa *see also* East African Rift
System
biodiversity 171
mammals 204
mountain building dating *74*, 77,
78, 79
African Superswell 394–395
Airy isostasy 20, 21, *21*, *23*, 27,
27, 28
Akaike Information Criterion
(AIC) 264
Alai Mountains, Kyrgyzstan *499*
albedo effect 6, *7*
Alborz Mountains, Iran 63
Allardyce Range, South Georgia *139*

allopatric speciation, mammals
211–212
alpha diversity, definitions
299–300, 304
alpine flora, New Zealand 378–380,
379, *380*, 384
see also New Zealand plants
Alps, European *3*, 9, 413–415, *414*,
424–425
anthropogenic climate change
187–188, *188*, 189, *190*, 424–425
crust thickening processes 29
erosion and mountain uplift 60,
61, 62
flora/fauna 204, 415–418
landscape evolution 423–424
mountain areas based on
ruggedness *4*, *174*
paleoaltitude 421–423
pollen analysis 415–416, 418,
419, 420–423
study sites *420*
vegetation belts in relation to
temperature **416**
vegetation belts in relation to
topography *417*
vegetation history 418–421
Vorarlberg 157–161, *158*, **159**,
160, **161**
Altiplano, Andean. *see* Andean
Altiplano
altitude
in relation to mammals *205*,
207, 211
species distribution models 311
altitude-related vegetation zones. *see*
biomes
Amazon Basin 42–45, *43*, *44*,
45, 113

American pika (*Ochotona princeps*)
211, 212
AMOC (Atlantic Meridional
Overturning Circulation) 327
ancient climates. *see* paleoclimate
research
Andean Altiplano *3*, *501*
crust thickening processes 20, 28
mountain areas based on
ruggedness *4*, *174*
mountain-climate
interactions 115–117, *116*
Andes *3*, *495*, *496*, *497*
crust thickening processes 20, 23,
24, 28, 30
digital elevation model *235*
dynamic topography 42
environmental heterogeneity 192
erosion and mountain uplift 60
geology, effects on ecology 231,
232, 235–239
historical connectivity 171–176,
177, 178, 179
hummingbird radiation study
265–268, **267**
mammals 203, *206*, 206–207, 210
mountain areas based on
ruggedness *4*, *174*
mountain building dating study
76–77, *76*
mountain-climate interactions
111, 113–117, *116*
mountain geodiversity *139*
orographic barrier effect 6
spatial reconstructions *175*
subduction case study 42–45, *43*,
44, *45*
surface-uplift analysis 88
topography *177*

Mountains, Climate and Biodiversity, First Edition. Edited by Carina Hoorn, Allison Perrigo and Alexandre Antonelli.
© 2018 John Wiley & Sons Ltd. Published 2018 by John Wiley & Sons Ltd.
Companion website: www.wiley.com\go\hoorn\mountains,climateandbiodiversity

Antarctica   9, 355–357, *356, 365,*
    367–368
    ablation valleys   365–366, *367*
    crust thickening processes   29
    Davis Valley, Pensacola
        Mountains   *367*
    Finger Mountain, Victoria
        Land   *358*
    fossil record   362–363
    geological history   357–358
    geothermal areas   366–367
    glaciation   361–362
    Gondwana landmass, breaking up
        of   358–360
    montane ecosystems   364–367
    nunataks   365–366, *366*
    terrestrial biota   363–364
    volcanism   360–361, *361,*
        364–367, *368*
Antarctic Circumpolar Current
    (ACC)   130
anthropogenic pressures. *see also*
    climate change, anthropogenic
    and biodiversity   xxi–xxii, 4
    European Alps   423–424
    Vorarlberg, Austria   158
apatite, mountain building
    dating   71, 72
Apennine Mountains, Italy   *20,* 24
Appalachian Mountains, USA   231
Araracuara Sandstone Plateau,
    Guiana   *344,* 345
area under the curve (AUC),
    receiver operating
    plots   313–315
Arenal Volcano, Costa Rica   *497*
aridity, mountain-climate interactions
    113–114
    see also precipitation
Asia. *see* Southeast Asia
asteroid impact, Cretaceous mass
    extinction   279, *279*
asthenosphere, crust thickening
    processes   *18,* 19–23, 25–27,
    30–31
astronomical cyclicity, paleoclimate
    research   127
astronomical forcing, paleoclimate
    research   *126,* 129, 133
Atlantic Meridional Overturning
    Circulation (AMOC)   327
atmosphere-geosphere-biosphere-
    hydrosphere interactions   6–7, *7*
atmospheric carbon dioxide, fossil
    analysis   97
atmospheric circulation patterns
    mountain-climate interactions
        112–113
    paleoclimate research   124

Tibeto-Himalayan region   *430,*
    431–432
atmospheric modeling, stable
    isotope paleoaltimetry
    81–85, 88, 89
astronomical cyclicity, paleoclimate
    research   127
    attenuation length, cosmogenic
        nuclide analysis   73
Australia   171, 204
Austria. *see* Vorarlberg, Austria
authigenic proxy materials   85, 114
Au West case study, Austria
    161–167
    biotope units   *162,* **166**
    morphogenetic classification
        scheme **163**
    Obere Alpe glacial niche
        photograph   *165*
AUC. *see* area under the curve
avifauna. *see* birds

b
Barisan Range, Sumatra   484
barrier displacement, fish   274, *276*
Basin and Range Province, USA   *20,*
    29, 30, 31
Bayesian probabilistic framework. *see*
    PyRate analytical framework
bedrock-exposure age dating   73
beta diversity, definitions   299–300,
    300, 301
big river systems, fish   274–276
bioclimatic variables, species
    distribution models   311,
    312–313
    see also climate change
biodiversity   xxi, 1–2, 4, 9–10,
    81–82, 217
    see also diversity (definitions);
        diversification; geodiversity;
        historical connectivity
    Andes **236**
    birds   249–250, *250*
    East African Rift System
        398–402, *399*
    fish   273–274, 277, 282, **283–284,**
        285, *285, 286*
    and geodiversity   144, 156
    and geologic history   231–235,
        237–239, *239*
    and human populations   xxi, 4
    macroevolutionary
        dynamics   *219, 258*
    and mountain physiography   5–7
    numbers of species of vascular
        plants   xx, *5*
    Tibeto-Himalayan region
        429–441, **435**

and topography   273–274
    Yunnan, China   449–450
biodiversity hot spots
    birds   245
    East African Rift System   391
    European Alps   415
    Skadar Lake, Montenegro   *501*
    Tibeto-Himalayan region   432,
        434–435
biogenic archives, stable isotope
    paleoaltimetry   85
biogeography   7
    European Alps   413, *414*
    Guiana Highlands   348
    Isthmus of Panama   323, 328, 329,
        330
    mammals   210–212
    New Zealand plants   383
    river capture hypothesis   274,
        280–281
biological invasion, Isthmus of
    Panama   328
biological pump   131
biomes   xix–xxi
    Antarctica   364–367
    Chimborazo Volcano,
        Ecuador   *2*
    European Alps   415–418
    Guiana Highlands   *347,*
        347–348
    historical connectivity   172
    Isthmus of Panama   *324*
    Japan   459–461, *467, 468*
    mammals   203
    New Zealand   377–378, 384
    in relation to terrestrial lapse rates/
        latitudinal gradients   **423**
    in relation to temperature   **416**
    in relation to topography   *417*
    San Francisco Peak   96, *96*
    Southeast Asia   *477,* 477–478,
        478, 487
    stable isotope paleoaltimetry   81
    Tibeto-Himalayan region
        430–433, 441
biosphere–atmosphere interactions,
    stable isotope
    paleoaltimetry   87
biosphere-atmosphere-geosphere-
    hydrosphere interactions
    6–7, *7*
biotic, abiotic, movement (BAM)
    framework   309
biotic processes
    Andean hummingbird
        study   *266*
    and biodiversity   3
    drivers of diversification   259–260,
        263–264

biotopes
  East African Rift System   399
  Vorarlberg, Austria   157, *162*, 164, **166**, 167–168
birds, biodiversity in mountains   9, 245–246
  Andean hummingbird study   265–268, **267**
  definitions/concepts   246
  environmental correlations   249
  global patterns   245–246, *248*, 248–249
  Guiana Highlands   349–350
  latitudinal effects   249–252, *250*
  montane fauna   247–251, *248*, *250*
  niche conservatism   252
  role of geology   253
  speciation in tropical regions   251
  study methodology   246–247, *247*
  Tibeto-Himalayan region   **437–439**
  variations in biodiversity and endemism   249–251, *250*
Bird's Head region, West Papua   485
birth–death models
  diversification drivers   261, *262*, 264, 265, 268
  rodents   223–225, **225**
Blue Mountains, Australia   *139*
boreal plants, Japan   468, *469*
Borneo   482–484, 487, *487*
bottom-up inference, environmental heterogeneity   193
Brazilian Shield, crust thickening   23–24, *24*
breeder's equation   191
Breiðamerkurjökull, Vatnajökull National Park, Iceland   *146*

C
C$_4$ vegetation   221, 440, 442
  East African Rift System   395, 398, 401, 402, 404, 405
  stable isotope paleoaltimetry   85, 88
Cairngorm Mountains, Scotland   *139*
calcium carbonate, stable isotope paleoaltimetry   85
calcium, paleoclimate research   125
Caledonian Mountains   140
Cambrian explosion, biodiversity   126
carbon dioxide, plant fossils   97
Caribbean Sea   327–328
Carpathian Mountains   204
CAS. *see* Central American Seaway
Cascade Range, US
  mammals   206
  rodents   220–228, *222*, *227*

Caucasus Mountains, Asia   204
Cedrus spp., Yunnan, China   452–453, *453*
Cenozoic era   87
  dating techniques, mountain building   79
  erosion and mountain uplift   51, 52, *52*, 58
  European Alps   415, 422
  Japanese floral diversification   461, 463
  paleoclimate research   129–130
  stable isotope paleoaltimetry   84, 88, 89
Central American Seaway (CAS)   *324*, 325, 326, *326*, 327, 328, 329, *330*
  see also Isthmus of Panama
Cerro Fitz Roy massif, Andes   *139*
Chimborazo Volcano, Ecuador   *2*
China, biomes   xx
  see also Yunnan
chipmunks (*Tamias* spp.)   212
Chiribiquete Sandstone Plateau, Guiana   *344*, 345
Chiroptera (bats)   210
CJS. *see* court jester scenario
CLAMP (Climate Leaf Analysis Multivariate Program)   100, *101*, *104*, 105, 450–451
classical refuge hypothesis   233, 234–235
climate change, anthropogenic environmental heterogeneity   187–190, *190*, 193–195
  effect on mammals   212
  paleoclimate research   132
climate change, natural   1–2, 9
  see also flickering connectivity; mountain-climate interactions; paleoclimate research; precipitation; temperature
  Andean hummingbird study   265–268, *266*
  Andes   **236**
  Antarctica   361–362
  biodiversity   xxi–xxii, 232–235
  East African Rift System   397–398, 404–405
  European Alps   **416**, 421–423
  geologic history and ecology   232, 234
  Guiana Highlands   346
  historical connectivity   172, *173*, 176–178, *177*, 180, 181
  Japan   459–461, 468–469
  mammals   201, 203

mountain-climate interactions   111–114
mountain geodiversity   146–147
  and mountain uplift   51, 55–56, 61–63
  New Zealand   377, 383
  phytopaleoaltimetry   95
  plant fossils   96–100
  and primary productivity   234
  rodents   *219*, *224*, 225, **225**, 226, *227*
  species distribution models   311, 312–313
  Tibeto-Himalayan region   *431*, 431–434, 441
  Yunnan, China   450–452
Climate Leaf Analysis Multivariate Program (CLAMP)   100, *101*, *104*, 105, 450–451
climate modeling   115, 312
  see also global climate models
climatic cycles   172, 246, 362, 415
climatic gradients   81
clock analogy, geologic time   *502*
closure temperature (Tc) concept   72
clumped isotope thermometry   87
Cohen's kappa statistic   313
Colombia. *see* Eastern Cordillera; Guiana Highlands
colonisation. *see* dispersal/ colonisation
Colorado, USA
  crust thickening processes   26
  mammals   204–206, *205*
  mountain-climate interactions   111, 112
conduction processes, earth's mantle   25, 30
conifers, Japan   468, *469*
Congo Basin, Africa   201, 204
connectivity. *see* historical connectivity
continental drift   7–8
  see also plate tectonics
continental graben, East African Rift System   393
continental rift zones, East African Rift System   393, *394*
convection processes, mantle   25, 37, 38, 39, 41–42 *see also* dynamic topography
Cordillera Blanca, Peru   *24*, 30
Coriolis effect   432
cosmogenic nuclide analysis   55, 59, 61, 62, 69, 72–73, 79
Costa Rica, Arenal Volcano   *497*
Cotopaxi, Ecuador   *139*

coupled ocean–atmosphere weather modeling, paleoclimate research *126*, 130

court jester scenario (CJS), diversification drivers *258*, 258–259, 261, *262*, 263, **267**, 268, 269

Cretaceous era
asteroid impact, mass extinctions 279, *279*
plant fossils 99
stable isotope paleoaltimetry 88

critically tapered wedges, mountain uplift processes 52–53, *54*, 56

crustal thickening
dynamic topography 38
mountain building 23, 23–25, *24*, 26, *27*, 70
phytopaleoaltimetry 95

crustal thinning 29–30

cultural values 138, 143–145, **144**, 156

d

dating techniques, mountain building. *see* cosmogenic nuclide analysis; thermochronology

Davis Valley, Antarctica *367*

DDD. *see* diversity-dependent diversification

Deccan Traps, India *20*, 26

detrital quartz, cosmogenic nuclide analysis 73

Dicroidium (seed fern) 358

digital elevation models (DEMs) 156, 159, *162*, **163**

dispersal/colonisation of organisms 8
historical connectivity 171, *173*, 176, *177*, 178–179
New Zealand 380, *381*, 384–385
Southeast Asia 476, 485–490, *486*
stable isotope paleoaltimetry 88

diversification 9, 257–259, 268–269
see also Japan (floral diversification); radiation; speciation
abiotic and biotic processes 259–260
Andean hummingbird study 265–268, *266*, **267**
anthropogenic climate change 190–193, *191*, **192**, *193*
mammals 203, 208, 209, 210
model comparisons 264, 265, 268
phylogenetic approaches 260–261, *262*, 268

red queen/court jester paradigms *258*, 258–259, 261, *262*, **267**, 268, 269

rodents 218, *219*, 220, 221, 223–229, *227*

theoretical frameworks 261–264, *262*

Tibeto-Himalayan region **437–439**, 440–441

diversity, definitions of 9, 295–296, 305
abundance measures 296, 302
alpha, beta and gamma diversity 299–300, *300*
effective number of species 296–299, *298*
local representativeness 303–304
number vs. size of subunits 304
sampling effort 302–303
spatial/environmental representativeness 304–305
species richness 296, *297*
species turnover 300–301, *303*
target organism group 301
taxonomic classification 301–302

diversity-dependent diversification (DDD) 264, **267**, 268

diversity gradients, mammals. *see* topographic diversity gradients

diversity indexes 298–299, 301

Dolomites, Italy *498*

Dream Pool Essays (Shen Kuo) 97

dynamic topography 8, 22, 37–38, *38*, 46
convection processes 38, *39*
definitions/concepts 38–39
modeled mantle flow 41–42
mountains 42
Peruvian case study 42–45, *43*, *44*, *45*
phytopaleoaltimetry 95
residual topography *39*, *40*, 40–41
role of geological record 42

e

Earth, planetary structure *18*, 19
see also mantle

Earth system models of intermediate complexity (EMICs) 125

East African Rift System (EARS) *3*, 9, 391–393, *392*, 406
biodiversity 398–402, *399*
climate change 397–398
continental rift zones 393, *394*
dynamic topography 42
exhumation-rates/surface-uplift analysis case study 77–79, *78*
hominins 402–405, *403*

lakes 391, 393, *393*, 398–402, *399*, *400*
mammals 201
mountain areas based on ruggedness *4*, *174*
paleogeographic maps *400*
plate tectonics 394–395, *396*
terrestrial biota 401–402
volcanism 395, *396*

Eastern Cordillera, Colombia
crust thickening processes 24, 25
geologic history in relation to ecology 235, 235–239, **236**, *239*
geophysical relief map *237*
three dimensional model *238*

Eastern Indonesia. *see* Southeast Asia

East Pacific Rise, plate tectonics *19*, 22

ecology. *see also* geologic history in relation to ecology
Andes 231–232
Guiana Highlands 348, 350
species distribution models *315*, 316, 317

ecosystem services, mountain geodiversity 143–144, **144**

ecotones, mammals 203

Ecrins-Pelvoux Massif, French Alps 61

Ecuador, mammals *206*, 206–207, *207*

effective number of species, definitions 296–299, *298*

EH. *see* environmental heterogeneity

elevation above sea level, plant fossil analysis 96, *96*, 98, 100, 102–105, *104*

elevational gradient (EG), and biodiversity 6

El Niño–Southern Oscillation (ENSO) 117

EMICs (Earth system models of intermediate complexity) 125

Empetrum hermaphroditum (crowberry) 189

endemism 6
birds 9, 249–251, *250*
East African Rift System 398
European Alps 415
historical connectivity 171, 172, 177, 178, 179
mammals 211–212
mountain geodiversity 144
New Zealand plants 380–381, *382*, **383**
Southeast Asia *487*, *489*

energy budgets, crust thickening processes 25, 25–28, *26*, 30

enhanced vegetation index (EVI), species distribution models 312
ensemble model, species distribution models 315–316
ENSO (El Niño–Southern Oscillation) 117
enthalpy, plant fossil analysis 100, *102*, 103–105, *104*
environmental drivers of diversification 263
environmental heterogeneity (EH) 6, 8, 195
  adaptation/diversification 190–193, *191*, *193*
  adaptation/diversification at fine scales **192**
  adaptation/plasticity 191–193, *193*, *194*
  and biodiversity 233
  climate change 187–190, *190*, 193–196
  drivers of genetic isolation 189–190
  fine scale mosaic of 187–188
  link smallscale genetics of ecologically relevant variation *193*
  mammals 201, 203
  number of migrants per generation *190*
  reactions to climate change 193–195, *194*
  selection/evolutionary responses 191, **192**
  snowmelt timing 189, 190, *190*, **192**
  statistical approaches 195
environmental limits hypothesis 234
environmental representativeness, definitions 304–305
Eocene, Japanese floral diversification 464–465, *465*
Eocene–Oligocene (EO) transition 261
  Antarctica 362
  paleoclimate research 130
  Southeast Asia 481
  stable isotope paleoaltimetry 88
  Tibeto-Himalayan region 433, 442
equilibria, isostatic. *see* isostasy
equilibrium-line altitude (ELA) 56
erosion 51–53, *52*, *53*, 63
  climatic controls 61–63
  critically tapered wedges 52–53, *54*, 56
  crust thickening processes 17, 28–29

feedback loops 52–53, *55*
  geologic history in relation to ecology 232
  Guiana Highlands 345
  measurement of change 57–59, *57*
  models/concepts 53–56
  and mountain building 71
  reconstruction of relief changes 59–61
  river capture hypothesis 277–278, *279*
  sediment flux 55, *55*, 56, 58–60, 62
Ethiopian Plateau 18, *20*, *26*
Eurasia
  biomes xx
  mammals 204
European Alps. *see* Alps, European
European brown bear (*Ursus arctos*) 4
evaluation strips, species distribution models 316
evapotranspiration
  influence of biota on the environment 6, *7*
  stable isotope paleoaltimetry 87
Everest, Mount 38
evolutionary history
  Japan *464*
  Southeast Asia 485–490, *488*
  Tibeto-Himalayan region **437–439**
evolutionary processes 231–232, 234
  see also geologic history in relation to ecology
evolutionary responses, environmental heterogeneity 191
evolution, biotic
  Antarctica 357, 358, 360, 361, 364
  and mountain building 2–3, 8, 17
  New Zealand plants 383–385
  plate tectonics and mountain building 8
exhumation-rates 8, 69–71, 79
  closure temperature concept 72
  cosmogenic nuclide analysis 69, 72–73, 79
  East African Rift System 77–79, *78*
  Merida Andes of Venezuela study 76–77, *76*
  schematic representation *70*
  thermochronology 69, 71–72, 74–75, *75*, 82
  time-temperature history models *74*, *75*, *77*, 79

exploration, scientific
  Antarctica 355
  Guiana Highlands 340, 347, 350
extinction xx, xxi, xxii, 1, 8, 9, 10
  see also rodent macroevolutionary dynamics
  Antarctica 358
  diversification drivers 257–259, *258*, 260, 261
  Eocene–Oligocene boundary 88
  Japan 466, 467–470
  low-frequency, high-impact events 279, *279*
  mammals 212
  river capture hypothesis 287
  Yunnan, China 453, 454

f

faint young sun paradox 124
faults/shear zones 86, 86–87
feedback loops
  erosion and mountain uplift *55*
  influence of biota on the environment 6–7, *7*
  paleoclimate research 132
FETKIN software package 75
Finger Mountain, Antarctica *358*
fingerprinting, mountain *173*, 176–177, *178*
fish, freshwater 9
  see also river capture hypothesis
  barrier displacement 274, *276*
  biodiversity and topography 273–274
  East African Rift System 398–399
  effect of large rivers 274–276
  montane species **275**
  Tibeto-Himalayan region **437–439**
fission-track analysis 69, 71, 76
flickering connectivity system (FCS) 172–181, *173*, *177*
floral biology, New Zealand 379–380
floral diversification, Japan. *see* Japan
flowering timing changes 189
forests, New Zealand 377–378
  see also New Zealand plants
forward genetics, environmental heterogeneity 192
fossil analysis, plant species 8, 95–96
  climate change 96–100
  CLAMP Program 100, *101*, *104*, 105, 450–451
  enthalpy 100, *102*, 103–105, *104*
  lapse rates 100, 101–103, **103**
  moist static energy 103–105
  nearest living relatives 96, 97, 98, 105
  physiognomic analysis 98–100
  San Francisco Peak *96*
  taxon-based 96, 97–98

fossil record   221
   Antarctica   362–363
   East African Rift System   398,
      401, 402, 403, 404, 405
   European Alps   418
   Japan   463–464, *465*
   mammals   202, 208–209, *209*
   rodents   217–218, 221, 222, 228
   Yunnan, China   450, 452–453, *453*
fragmentation of species
      distributions   177–178
   see also historical connectivity
   mammals   203
   Tibeto-Himalayan
      region   440–441
Fuji, Mount   *461, 469*
functional diversity, definitions   295

g

GABI. *see* Great American Biotic
      Interchange
Galápagos Islands   *20*, 25
gamma diversity,
      definitions   299–300
GBS (genotyping-by
      sequencing)   195
GCMs (global climate models)   87,
      311, 312–313
gender dimorphism, New Zealand
      plants   380
general circulation models
      (GCMs)   115, 125
genetic adaptation. *see* adaptation
genetic isolation
   drivers of   189–190
   mammals   201
genetic mapping   195
genetic variation, environmental
      heterogeneity   188, 189
genome-wide scans   195
genomic divergence   177, 190–191,
      192, *193*, 195
genotyping   195
geoconservation   145, 147, 156
geodiversity hot spots. *see also*
      mountain geodiversity
   Iberian Peninsula   156
   thermal structure of
      mantle   25
   volcanoes   141
geodiversity index (GI)   156,
      157, 168
   Vorarlberg, Austria   158–161,
      159, *160*, 168
geodiversity mapping   8, 155–157,
      167–168
   Au West study   161–167, *162*,
      163, *165*, **166**
   definitions/concepts   155–156

Vorarlberg, Austria   157–161,
      158, **159**, *160*, **161**
geodynamic processes. *see also*
      dynamic topography; geologic
      history; mountain building;
      plate tectonics
   crust thickening processes   17
   phytopaleoaltimetry   95
   rodents   218–220
   stable isotope paleoaltimetry   81
Geographic Information
      Systems (GIS)
   geodiversity mapping   156, 157,
      159–161, 167
   historical connectivity   174
   Yunnan, China   *450*
geographic-range shifts,
      mammals   211
geoheritage   138, 144, 145
geological age, river capture
      hypothesis   282, 285, *286*
geological dynamics. *see* geodynamic
      processes
geological record, dynamic
      topography   42
geologic history, in relation to
      ecology   9, 231–232, 239
   Andes case study   235, 235–239,
      236, *237, 238, 239*
   and biodiversity   231–235, 253
   climate change   234
   drivers of diversification   263
   mountain building   232
   stability   234–235
   topographic complexity   232, 233,
      236–239
geologic time   *502 see also inside
      cover*
geomorphology   137–138
   see also geodiversity
   Japan   461–463, *462, 467*
   plate tectonics and mountain
      building   8
   river capture hypothesis   277
   Yunnan, China   449–450
geomorphosites, definitions/
      concepts   157
geophysical relief,
      reconstructions   60
geosphere-biosphere-atmosphere-
      hydrosphere interactions   6–7, *7*
geothermal areas,
      Antarctica   366–367
GI. *see* geodiversity index
GIS. *see* Geographic Information
      Systems
glacial inheritance effect   62
glacial lake outburst floods
      (GLOFs)   141

glaciation/glaciation cycles
   Antarctica   361–362
   erosion/mountain uplift   51, *52*,
      55–56, 61, 62
   European Alps   418–421, 424
   historical connectivity   172,
      177–178, 179
   Isthmus of Panama   323,
      326, 327
   mountain geodiversity
      141–143, *142*
   paleoclimate research   131
   Tibeto-Himalayan region   432,
      440–441
global climate models (GCMs)   87,
      311, 312–313
   global cooling, European Alps
      418, 420
   see also climate change
global diversity, mammals   *204*
global meteoric water line
      (GMWL)   82
global scale mountain
      geodiversity   138–141, *139*
GLOFs (glacial lake outburst
      floods)   141
Gondwana landmass, breaking up
      of   358–360
graben, East African Rift
      System   393
Great American Biotic Interchange
      (GABI)   323–325, *326*, 328–331,
      330, 331
Great Plains, North America
   mammals   201
   rodents   220–228, *223, 227*
greenhouse gases
   paleoclimate research
      123–124, 126
   removal by silicate weathering   51
greenhouse-icehouse transition,
      Cenozoic   129–130
growth forms, New Zealand
      plants   *379*, 379–380
Guiana Highlands (tepuis/table
      mountains)   9, 339–342, *340*,
      341–342, 350
   animal life   349–350
   Chiribiquete and Araracuara
      plateaus   *344*, 345
   climate change   346
   geology   342–345
   hydrology   345–346
   mountain areas based on
      ruggedness   *4, 174*
   Neblina Tepuis   *344*, 345
   orography   346–347
   Pantepui altitudinal transect   *347*
   phytogeography   *347*, 347–348

prominent mountain systems on
Earth *3*
Roraima tepuis **341**, 342–345, *343*
tepuis distribution *344*
Tunuí sandstone plateaus *344, 345*

h

habitat destruction/loss 9
habitat diversity
definitions 295
Tibeto-Himalayan region 430
habitat heterogeneity. *see*
environmental heterogeneity;
heterogeneous microhabitats
Harmonized World Soil Database
(HWSD) 312
Hawai'i, crust thickening
processes *20*, 25–26
HeFTy thermochronology numeric
modeling *74, 75*
Heinrich Events 131
helium analysis. *see* uranium (U-Th/
He) analyses
Hengduan Mountains, Yunnan,
China 450
hereditability, environmental
heterogeneity 188
heterogeneous microhabitats. *see also*
environmental heterogeneity
anthropogenic climate
change 193–195
and biodiversity 233
high-relief terrain, Tibeto-Himalayan
region 441
high-resolution laser altimetry
(LiDAR), land surface
parameters 157, 159
Himalaya. *see* Tibeto-Himalayan
region
historical biogeography 7, 413, 414
see also biogeography
historical connectivity, influence on
biodiversity 8, 171–172, 181
colonization 171, *173*, 176, *177*,
178–179
flickering connectivity
system 172–180, *173, 177*, 181
fragmentation of species
distributions 177–178
hybridization 179–180
intermixing 179
mountain areas based on
ruggedness *174*
mountain fingerprinting *173*,
176–177, 178
paleogeographic reconstruction
180–181
spatial reconstructions, tropical
alpine *175*

Holocene 61
East African Rift System *403*
European Alps 421, 424
mammals 211
paleoclimate research *127*, 132
Southeast Asia 478–480, *479*
hominins, East African Rift
System 402–405, *403*
hothouse climates, Mesozoic/
Paleogene 127–129
hot spots. *see* biodiversity hot spots;
geodiversity hot spots
human influence. *see* anthropogenic
pressures
Humboldt, Alexander von
xix, *2*, 6
Hutchinson's niche concept 309
HWSD (Harmonized World Soil
Database) 312
hybridization, historical
connectivity 179–180
hydrogen isotopes 82, 85, *86*, 125
see also stable isotope
paleoaltimetry
hydrology, Guiana
Highlands 345–346
hydrosphere-geosphere-biosphere-
atmosphere interactions
6–7, *7*
hydrous silicates, stable isotope
paleoaltimetry 86

i

Iberian Peninsula 156
ice age. *see* glaciation
ice caps, mountaintops
anthropogenic climate
change 188
historical connectivity
177–178
icehouse period, Carboniferous 126,
129, 130
Iceland *20*, 25, 140, *146*
immigration pathways. *see* migration
trackways
index–based approach. *see*
geodiversity index
India 42, *104, 204*, 434, 476
Indochinese Ranges, Southeast
Asia 480
Indonesia. *see* Southeast Asia
Intergovernmental Panel on Climate
Change (IPCC) 312
intermixing, endemic species 179
International Association of
Geomorphologists 157
International Standard Atmosphere
(ISA), plant fossil
analysis 101

International Union for Conservation
of Nature (IUCN), Protected
Area Management
Categories 145, 147–148
Intertropical Convergence Zone
(ITCZ) 431, 433
island arcs, plate tectonics 18, *19*,
22, 23
isostasy 18–22, *21, 23*, 24, 27, *27*,
28, 30
dynamic topography 38, 39, *40*
erosion/mountain uplift 38, 42,
52, *53*, 62
Isthmus of Panama 9, 323–325,
331–332
Caribbean Sea 327–328
Great American Biotic
Interchange 323–325, *326*,
328–331, *330*
landscape evolution
325–326, *326*
northern hemisphere
glaciation 327
terrestrial biome
reconstructions *324*
thermohaline circulation 327

j

Japan
dynamic topography 42
mountain areas based on
ruggedness *4, 174*
prominent mountain systems *3*
Japanese floral diversification 9,
459, *461*, 470
boreal plants and conifers 468,
469
distribution, montane
vegetation 459–461, *461*
evolutionary history *464*
forest distribution, relative to
geomorphology *467*
fossil record 463–464, *465*
geomorphology 461–463, *462*,
467
Miocene 465–466
montane ecosystems 468
Mt. Fuji *461, 469*
Paleogene 464–465
Pliocene *465*, 466–468, *467*
topographic barriers
469–470
Java 484–485

k

Kansas, mammals 206
Kenya, Mount *139*
Kenya Rift, mountain building
dating *74*, 77, *78, 79*

Kinabalu, Mount 483–484, 487, *487*
Kyrgyzstan, Alai Mountains *499*

l

Lake Fúquene, Columbia 174
land-cover data, species distribution models 312
landscape evolution. *see also* exhumation-rates; surface-uplift analysis
dating techniques 69
European Alps 423–424
Isthmus of Panama 325–326, *326*
Japan 461–463, *462*
New Zealand 377, *385*
river capture hypothesis 274, *276*, 287
landscape-memory 62
land surface parameters (LSPs), geodiversity mapping 156–157, 159
landsystem approach, mountain geodiversity 143
La Plata Basin, South America 113
lapse rates
European Alps 422–423, **423**
mammals 202
mountain-climate interactions 112
plant fossils 100, 101–103, *103*
Laramide orogeny, USA 59
large river systems, fish 274–276
laser ablation inductively coupled plasma mass spectrometry (LA-ICP-MS) 71
Last Glacial Maximum (LGM). *see also* glaciation
European Alps 421
historical connectivity 172
lateral viscosity variations (LVV), mantle 41–42
latitudinal gradients
birds 249–251, *250*
European Alps 422–423, **423**
Laurasian conifers, Southeast Asia *482*
leaf physiognomic (plant form) analysis 96–100, *101*, 114
see also plant fossil analysis
leaf-wax lipids 86, 98
lee waves 113
life zones. *see* biomes
linear regression models, environmental heterogeneity studies 195
Lithocarpus diversity, Southeast Asia 487–488, *489*

lithosphere
cooling 30
crust thickening *18*, 18–23, *19*, 25, *27*
dynamic topography 37
erosion and mountain uplift 62
removal 26–28, *28*
Little Ice Age 132, *142*
local scale mountain geodiversity 141–143
local representativeness, definitions 303–304
LSPs. *see* land surface parameters
LVV (lateral viscosity variations), mantle 41–42
Lydekker's Line *330*, 331

m

macroevolution *258*
see also diversification; rodent macroevolutionary dynamics
magnesium ratios, paleoclimate research 125
Malay Peninsula *477*, 480–481, *482*
mammals 8, 201–203
see also rodent macroevolutionary dynamics
adaptation 211
biogeography 210–212
diversity across continents 203–204, *204*
Ecuador 206, *207*
Guiana Highlands 349–350
New World *203*
tectonic and climate processes influencing *202*
Tibeto-Himalayan region **437–439**
topographic diversity gradients 204–210, *205*
mantle, Earth
convection 25, 37, 38, *39*, 41–42
see also dynamic topography
geodynamic processes 18, *18*
modeled mantle flow 41–42
thermal processes in mantle 30, 38, 53
thermal structure of mantle 25–28, *26*
mapping. *see* geodiversity mapping
Mariana Trench, topography 38
MCO. *see* Miocene Climatic Optimum
mean annual air temperatures (MAATs), stable isotope paleoaltimetry 87
mean annual range of temperature (MART), plant fossil analysis 98, 102, **103**

mean annual temperature (MAT), plant fossil analysis 98, 99, 100, 102, **103**
Medieval Warm Period 132
Meili Xue Shan, Yunnan, China *500*
Meratus Mountains, Borneo 483
Merida Cordillera, Venezuela 76–77, *76*
Merriam's life zones, San Francisco Peak 96, *96*
see also biomes
microhabitats. *see* heterogeneous microhabitats
microsatellite genotyping 194
Microtis spp. (orchids), migration trackways *486*
Middle Eocene Climatic Optimum (MECO) 130
Mid-Miocene Climate Transition (MMCT) 130
mid-ocean ridges *19*, 22, 37
migration trackways. *see also* dispersal/colonisation
New Zealand plants 384–385
Primula imperialis 485, *486*
Southeast Asia 476, *486*
minimum exposure age, cosmogenic nuclide analysis 73
Miocene 42, 43, *44*, *45*, *52*, 58
dating mountain building 76, 77, 79
Japanese floral diversification 465–466
mammals 208, 209, 210
mountain geodiversity *139*, 141
Southeast Asia *482*
phytopaleoaltimetry *104*, 105
stable isotope paleoaltimetry 88, 89
Tibeto-Himalayan region 436, 440
Yunnan, China *453*
Miocene Climatic Optimum (MCO)
paleoclimate research 130
rodents 220, 226, *227*, 228
moist convection, mountain-climate interactions 112
moist static energy, plant fossil analysis 103–105
Molasse rock formations, Austria 157, 161
molecular phylogenetics 217–218
Monodelphis reigi (opossum), Guiana 349
monsoon climat phylogenetics es 433–434, 452–454
monsoon strength index (MSI), Yunnan, China 452
montane fish **275**
see also river capture hypothesis

montane vegetation. *see also*
  biomes
  Antarctica   364–367
  Japanese floral diversification
    459–461, 468
  New Zealand   377, 383
  Southeast Asia   *477*, 477–478, *478*
Mont Blanc massif, European
  Alps   60
morphology, New Zealand
  plants   *379*, 379–380
mountain-belt development   8, 51,
  52, 53
mountain building   2, 7–8, 51–53,
  52, *53*, 63
  see also below
  Andean hummingbird study   *266*
  climatic controls   61–63
  see also mountain-climate
    interactions
  and evolution of life   2–3, 8
  geologic history in relation to
    ecology   232
  Guiana Highlands   346–347
  rodents   *219*
  stable isotope paleoaltimetry   81
  surface-uplift analysis   70–71
mountain building processes   8,
  17–19, 30–31
  see also isostasy
  crustal thickening   *23*, 23–25, *24*,
    26, *27*
  crustal thinning   29–30
  destruction of upland areas
    28–30
  equilibria   38
  erosion   17, 28–29
  lithosphere cooling   30
  lithosphere removal   26–28, *28*
  mantle   18, *18*
  plate tectonics   18, *19*, 22–23
  structure of earth   *18*, 19
  thermal structure of
    mantle   25–28, *26*
  world map *20*
mountain-climate interactions   8,
  61–63, 111–114, 118
  Andes   111, 113–117, *116*
  atmospheric circulation
    patterns   112–113
  climate modeling   115
  North American Cordillera   114,
    117–118, *118*
  paleoaltimetry   114–115
  precipitation   113–114
  surface-uplift analysis   115–118,
    116, *118*
mountain fingerprinting   *173*,
  176–177, 178

mountain geobiodiversity hypothesis,
  Tibeto-Himalaya   431–434,
  441–442, *442*
mountain geodiversity   8, 137–138,
  139, *142*, 147–148
  and biodiversity   156
  Breiðamerkurjökull, Iceland   *146*
  climate change   146–147
  cultural values   143–145
  definitions/concepts
    137–138, **138**
  ecosystem services   143–144, **144**
  global scale   138–141
  plate tectonics-based
    classification   **140**
  regional/local scale   141–143
  sediment cascades   **141**
mountain gravity waves   113
mountains (general)   1–4, 8–10
  definitions/concepts   3–4, **138**
  dynamic topography   42
  geosphere, biosphere, atmosphere
    and hydrosphere interactions
    6–7, *7*
  mountain areas based on
    ruggedness   *4*, *174*
  numbers of species of vascular
    plants   xx, *5*
  physiography   5–7
  plate tectonics   7–8, **140**
  prominent mountain systems on
    Earth   *3*
multivariate environmental similarity
  surfaces (MESS), species
  distribution models   317

n

NADW (North Atlantic Deep Water)
  131, 132
Namling Oiyug plant fossil site,
  Tibet   *104*, 105
natural hazards, mountain
  geodiversity   144, 146, 147
nearest living relatives (NLRs),
  plant fossil analysis   96, 97,
  98, 105
Neblina Tepuis, Guiana Highlands
  344, 345
Neogene   9, *139*, *238*
  Antarctica   360, 363
  in China. *see* Yunnan, China
  climate   220, 397, 450–452
  East African Rift System   398
  European Alps   418, *419*,
    422, 423
  fish   274, 276, 282, 287
  historical connectivity   171
  Japanese flora   459, *465*, 466
  New Zealand   384

Southeast Asia   483, 484
  Tibeto-Himalayan region
    436, 440
neotropics   171, 252
net primary productivity (NPP), and
  climate change   234
New Guinea Highlands   *3*, *4*, *174*,
  476, *479*
New World mammals   203, *203*
New Zealand, historical
  connectivity   171, 178
New Zealand plants   9, 385
  see also Southern Alps
  alpine flora   378–380, *379*, *380*,
    383–384
  alpine plant traits/growth
    form   *379*, 379–380
  biogeography   383
  endemism   380–381, *382*, **383**
  landscape evolution   377, *385*
  montane environments
    375–377, *376*
  niche-diversification/
    distribution   *380*, *381*
  plant evolution   383–385, *385*
  species radiation   380–381
  species richness   378, *379*, *380*
  vegetation   377–378
niche conservatism, birds   252
niche-diversification, New Zealand
  plants   *380*, *381*
niches, ecological   309–310
  see also species distribution
    models
NLRs. *see* nearest living relatives
non-analog conditions, species
  distribution models   315
normalized difference vegetation
  index (NDVI)   312
North America
  biomes   xx
  crust thickening processes
    20, 25
  erosion and mountain uplift   60
  mammals   203, *203*, *205*, 206, 208,
    210
  mountain-climate
    interactions   114, 117–118, *118*
  mountain geodiversity   141
  plant fossils   *102*
  rodents   220–228, *222*, *227*
  stable isotope paleoaltimetry   88
North Atlantic Deep Water
  (NADW)   131, 132
Northern Andes. *see* Andes
northern hemisphere glaciation. *see*
  glaciation
Notothlaspi rosulatum (penwiper
  plant)   378

number of migrants per generation
(Nem) *190*
nunataks 365–366, *366*

O

Obere Alpe glacial niche,
Austria *165*
Ocaña Canyon, Peru 60
ocean circulation currents 124
oceanic anoxic events
(OAEs) 129
oceanography, Isthmus of
Panama 323, 327–328
Ochotona princeps (American
pika) 212
Oligocene. *see also* Eocene–
Oligocene transition
drivers of diversification
261, 262
European Alps 418, 422, 423,
423, 424
fossil record 221, 226, 228
historical connectivity 171
Isthmus of Panama 325, 331
Japanese floral diversification
464–465, *465*
New Zealand 377, 383, 384
paleoclimate research 130
Southeast Asia *482*
stable isotope paleoaltimetry
88, 89
Tibeto-Himalayan region 436,
437, 441
one-dimensional (1D)
thermodynamic models 83, *84*,
87–88
opossum (*Monodelphis reigi*),
Guiana 349
orbital cyclicity, paleoclimate
research 127
orchids (*Microtis* spp.), migration
trackways *486*
Oregon, USA, mammals 206
organic biomarkers, stable isotope
paleoaltimetry 86
Orinoco Basin, Guiana
Highlands 339, 345–346
orogenic critically tapered
wedges 52–53, *54*, 56
orographic barrier effect. *see*
topographic barriers
oxygen isotopes 82, 85, 125, *128*
see also stable isotope
paleoaltimetry

P

Pachycladon (Brassicaceae),
diversification *381*
Pacific Ring of Fire 140

Pakaraima Mountains, Guiana 339
paleoaltimetry 97, 105, 114–115
see also fossil analysis; stable
isotope paleoaltimetry
paleoaltitude 2
European Alps 421–423
Southeast Asia *481*
Paleocene, Japanese floral
diversification 464–465
Paleocene–Eocene thermal
maximum (PETM) 129
paleoclimate research 8,
123–124, 133
coupled ocean–atmosphere
modeling *126*, 130
drivers of climate change 127
early Earth's climates 124–127
greenhouse-icehouse transition
129–130
Holocene 132
hothouse climates 127–129
ice age cycles, Quaternary 131
oxygen isotope profiles *128*
planetary orbital/astronomical
cyclicity 127
reconstruction toolbox 125
paleoelevation. *see* elevation above
sea level
paleoenvironment-dependence
diversification model
(PDDM) 263
Paleogene. *see* Eocene; Oligocene;
Palaeocene
paleogeographic maps, East African
Rift System *400*
paleotopographic reconstruction
59–60, 180–181
palynology. *see* pollen analysis
Panama Canal 325
see also Isthmus of Panama
Pantepui, Guiana *347*, 349–350
partial annealing zone/partial
retention zone concept (PAZ/
PRZ) 72
passerines 247–251, *248*
see also birds
Pebas Mega-Wetland, Peru 43
PeCUBE numeric modeling 75,
75, 76
pedogenic proxy materials 85, 114
penwiper plant (*Notothlaspi
rosulatum*) 378
Permian 126, 138, 263, 357, *358*
Perognathus parvus (Great Basin
pocket mouse) 210
Peru *24*, 42–45, *43*, *44*, *45*
see also Andes
PETM. *see* Paleocene–Eocene
thermal maximum

phenotypic plasticity
environmental heterogeneity
192–193, *194*
leaf physiognomy 99
phenotyping, environmental
heterogeneity 194
photosynthetically active radiation
(PAR), Antarctica 364
phylogenetic approaches,
diversification 260–261, 268
phylogenetic diversity,
definitions 295
phylogeny, New Zealand
plants *381*
physiognomic (plant form)
analysis 96–100, *101*, 114
see also fossil analysis
physiography, and biodiversity
5–7
phytogeography, Guiana
Highlands *347*, 347–348
phytopaleoaltimetry 97, 105
see also fossil analysis
Pinus forests, European Alps
415–416, **416**, *419*, 421
planetary cyclicity, paleoclimate
research 127
plant form analysis 96–100,
101, 114
see also fossil analysis
plant radiation. *see* species
radiation
plateaus, crust thickening
processes 18, *18*, 19, 23, *24*, 26,
27, 29
see also Ethiopian Plateau; Guiana
Highlands; Tibetan Plateau
plate tectonics 6, 7–8
see also dynamic topography
convection processes 37
crust thickening processes 18, *19*,
22–23
dynamic topography 37, 38
East African Rift System
394–396, *396*
erosion/mountain uplift 52,
52, 53, *54*, 55, *56*, *57*, 58, 59,
61, 62
Japan 461–463
mountain geodiversity 138–139,
140, 141
Southeast Asia 476–477
Yunnan, China 449–450
Pleistocene
birds 245, 246
European Alps 424
historical connectivity
171–174, *173*, *174*, 176,
177, 178–181

rodents 228
Southeast Asia 478–480, *479*
Tibeto-Himalayan
region 440–441
Pliocene
East African Rift System 402
Japanese floral diversification
465, 466–468, *467*
rodents 226–228
Tibeto-Himalayan region 440
Yunnan, China *453*
plot-based surveys 193, 194
Podoxymys roraimae (Roraima
mouse) 349
pollen analysis
European Alps 415–416, 418,
419, 420–423
paleoclimate research 125
Southeast Asia 478–479, *479*,
481–485, 487, *487*, 490
pollen-mediated gene flow 189
population genomics, environmental
heterogeneity 195
Pratt isostasy 20, 21, *21*, 22,
27, *27*
precipitation
Andes 237–239
erosion 53, 55, 71
mammals 203, *205*, 206
mountain-climate interactions
113–114
mountain influences on 69
stable isotope paleoaltimetry
81–84, *84*, 86, 87–88
Tibeto-Himalayan region 433
preservation rates, rodents 224
primary productivity, and climate
change 234
Primula imperialis, migration
trackways 485, *486*
proportional species turnover,
definitions 300
Protected Area Management
Categories (IUCN) 145,
147–148
PyRate analytical framework, rodent
macroevolutionary
dynamics 220, **221**

q
Qinghai–Tibetan Plateau. *see* Tibetan
Plateau
QTQt numeric modeling 74, 75
quartz, cosmogenic nuclide
analysis 73–74
Quaternary. *see also* Holocene;
Pleistocene
erosion and mountain uplift
51, *52*

mountain geodiversity
141–143, *142*
paleoclimate research 131
Southeast Asia 478–480, *479*

r
radiation 259–260
see also diversification
drivers of diversification 259
historical connectivity 171, 172,
174, 178–180
plant communities, New Zealand
380–381
Tibeto-Himalayan region
440, 441
rainfall. *see* precipitation
rain shadow effect 114
range shift, species distribution
models 311
Ranunculus adoneus (alpine
buttercup) 189
Raoulia eximia, New Zealand 378
Raup's kill curve 279, *279*
Rayleigh-type distillation, stable
isotope paleoaltimetry
82, *84*, 87
RCH. *see* river capture hypothesis
realized niches 309
receiver operating characteristic
(ROC) plots 313–315
reciprocal transplant experiments,
environmental heterogeneity
studies 194, *194*
red queen scenario (RQS),
diversification drivers *258*,
258–259, 261, *262*, **267**,
268, 269
refugia
Antarctica 361–366
European Alps 424
Tibeto-Himalayan
region 440–441
regional scale mountain geodiversity
141–143
relict landscapes, reconstruction 59
representative concentration
pathways (RCPs) 312, *314*
reproductive traits, New Zealand
plants 379–380
residual topography 39, *40*,
40–41
see also dynamic topography
reverse genetics, environmental
heterogeneity 193
Rhododendron spp. 487–489, *488*
rift lakes, East African 391, 393,
393, 398–402, *400*
rift zones 393–394, *394*
see also East African Rift System

river capture hypothesis (RCH) 9,
276, 276–279, 287
erosion 277–278, *279*
experimental discussion
285–287
experimental results 282,
283–284
geographic area, age, species
richness *286*
geomorphological processes 277
global distribution of river capture
events *285*
literature references for data
used **280–281**
low-frequency, high-impact
events 278–279, *279*
river elevation profiles *278*
study methodology 279–282
and topographic relief *278*, *279*
veil line 287
river water isotopes, stable isotope
paleoaltimetry 83, *84*
Rocky Mountains, North
America *3*
crust thickening processes 26
erosion and mountain
uplift 61
mammals 201, 203,
204–206, *205*
mountain areas based on
ruggedness *4*
mountain-climate
interactions 111, 112
prominent mountain systems on
Earth *3*
rodents 223, 223–228, *227*
stable isotope paleoaltimetry 88
Rodentia (rodents) 210
rodent macroevolutionary
dynamics 8, 217–218,
226–229
birth–death models 223–225,
225
fossil record 217–218, 220–221,
223, 229
and geological
dynamics 218–220
mountains, climate, and
biodiversity *219*
mountains compared with
lowlands 226–228
North America case study
220–228, 223, 224, **225**
PyRate analytical framework **222**
speciation/extinction
estimates *222*, *224*, *227*
Roraima Formation, Guiana 345
Roraima mouse (*Podoxymys
roraimae*) 349

Roraima tepuis, Guiana **341**, 342–345, *343*
RPANDA model, drivers of diversification 263, 264
ruggedness, mountain areas based on *4, 174*

S
Sadler effect 58, 62
Salix herbacea (creeping willow) 189–193, *190*, **192**
SALLJ. *see* South American low-level jet
sampling, definitions of diversity 302–304
sampling effort 302–303
sampling representativeness 303–304
satellite imagery, species distribution models 312
scanning electron microscopy (SEM), fossil analysis 98
Schwaner Mountains, Borneo 482
SDM algorithms **310**, 311–316, *314*
    see also species distribution models
sea surface temperature (SST), mountain-climate interactions 117
sediment flux
    erosion 55, *55*, 57–59, 62
    mountain geodiversity 141, **141**
seed dispersal 189
Seiser Alm, Dolomites, Italy *498*
selection, environmental heterogeneity 191, **192**, 195
Sequoia spp. (coast redwood) 453
Shannon entropy diversity index 298–299
shear zones, stable isotope paleoaltimetry 86, *86*
shells, stable isotope paleoaltimetry 85
Shen Kuo (Chinese philosopher) 97
shield volcanoes, crust thickening processes 26
Shuttle Radar Topographic Mission (SRTM) 311, 340
Sierra Nevada, California 60–61
silicate weathering, temperature dependence 51
Simpson diversity index 299
single-nucleotide polymorphism (SNPs), environmental heterogeneity 192
Skadar Lake, Montenegro *501*
sky islands 6
    birds 245
    dispersal 8
    historical connectivity 176, 178, 179, 181

snowball Earth 124, 126
snowmelt timing 189, 190, *190*, **192**
software packages, dating techniques 74–75
soil properties, species distribution models 312
solar radiation, Antarctica 364
Soricomorpha (shrews, moles) 210
South America. *see also* Amazon Basin; Andes; Eastern Cordillera; Guiana Highlands
    mammals 203, *203*
    Peruvian case study 42–45, *43, 44, 45*
South American low-level jet (SALLJ) 113, 115–117
Southeast Asia 9, 475–477, *476*, 490–491
    dispersal/evolution 485–490
    dynamic topography 42
    Eocene–Oligocene transition 481
    Lithocarpus diversity 487–488, *489*
    migration trackways 476, *486*
    montane forests *477*, 477–478, *478*
    Malay Peninsula *477, 478*
    Mount Kinabalu endemic species 487, *487*
    Oligocene/Miocene *482*
    paleoaltitudes *481*
    Quaternary vegetation dynamics 478–480, *479*
    Rhododendron spp. 487–489, *488, 488*
    species diversity *489*
    surface-uplift analysis 480–485
    volcanoes 484–485
Southern Alps, New Zealand
    crust thickening processes *20*, 24, 29
    erosion and mountain uplift 61–62
    mountain areas based on ruggedness *4, 174*
    plant communities 375–377, *376*
    prominent mountain systems on Earth *3*
South Pacific, plate tectonics *19*
space-for-time (SFT) substitution 193
spatial heterogeneity, and biodiversity 156
spatial projections, species distribution models *315*, 316–317
spatial representativeness, definitions 304–305

speciation
    birds 250–251
    geologic history in relation to ecology 234–235
    mammals 211–212
        see also rodent macroevolutionary dynamics
    plate tectonics and mountain building 8
    river capture hypothesis 274, 276, *285*
    Tibeto-Himalayan region 430, 440–441
    topographic complexity 233
species distribution
    historical connectivity 172, 176, 177–178
    Japanese floral diversification 459–461, *461*
species distribution models (SDMs) 9
    abiotic data variables 311–313
    accuracy 313–315
    applications 313
    complexity 315
    ensemble model 315–316
    SDM algorithms **310**, 311–316, *314*
    species ecology 316
    species niches 309–310
    species occurrence data 310–311
    time/space projections using *315*, 316–317
    Vaccinium phillyreoides, Borneo *314*
species diversification. *see* diversification
species diversity 295, 296, *297, 489*
    see also biodiversity; diversity (definitions of)
species occurrence, species distribution models 310–311
species pumps 172, 430, 442
species richness. *see also* species diversity
    definitions of diversity 296, 297, *297*
    New Zealand 378, *379, 380*
species turnover, definitions 300–301, *303*
SST. *see* sea surface temperature
SRTM. *see* Shuttle Radar Topographic Mission
stability hypothesis 233, 234–235
stable isotope paleoaltimetry 8, 81–82, 89
    applications 84–85

authigenic/pedogenic proxy
    materials 85
clumped isotope thermometry 87
definitions/concepts 82–84
faults/shear zones 86, *87*
leaf-wax lipids 85, 98
mountain-climate
    interactions 114
North American example 88
one-dimensional models 83, *84*,
    87–88
oxygen and hydrogen
    isotopes 82
precipitation 81–84, *84*, *86*,
    87–88
river water isotopes as a function
    of net elevation change *83, 84*
Tien Shan examples 88–89
Turkish–Anatolian Plateau
    data *83*
tooth enamel 85
volcanic glass 85
statistical approaches, environmental
    heterogeneity studies 195
Stefan–Boltzmann law 124
stream-power 53, 54
    see also erosion
strontium ratios, paleoclimate
    research 125
STRUCTURE software 195
subalpine forest zone, New
    Zealand 377–378
subduction/subduction zones 18,
    19, 22–24, 26
    Antarctica 357, 359, 360
    birds 250, 253
    dynamic topography 37, 41, 42
    Isthmus of Panama 325
    Japan 459
    mountain building 117, 124, 130
    mountain geodiversity 139, 140
    Peruvian flat-slab case
        study 42–45, *43, 44, 45*
    Southeast Asia 476, 484
    Tibeto-Himalayan region 434
Sulawesi 476, 477, 478, 485
Sumatra *479*, 484
Sumatran Track migration
    trackways *486*
summit flats, Laramide orogeny,
    USA 59
Sunda Shelf uplands, Southeast
    Asia 480, 482–483
surface-interior interactions. *see*
    dynamic topography
surface uplift analysis 8, 69–71, 79
    Andes 88
    climatic controls 61–63
    closure temperature concept 72

cosmogenic nuclides *55,* 59–62,
    69, 72–73, 79
critically tapered wedges 52–53,
    54, 56
East African Rift System case
    study 77–79, *78*
feedback loops 52–53, *55*
geologic history in relation to
    ecology 232
Guiana Highlands 345
measurement of change
    57–59, *57*
Merida Andes case study
    76–77, *76*
models/concepts 53–56
mountain-climate
    interactions 115–118, *118*
reconstruction of relief
    changes 59–61
schematic representation *70*
Southeast Asia 476, 480–485
stable isotope
    paleoaltimetry 82–87
thermochronology 69, 71–72,
    74–75, *75*, 82
Tibeto-Himalayan region
    433, 434
time-temperature models *74, 75,*
    77, 79

t

table mountains, Guiana. *see* Guiana
    Highlands
Tadrart Acacus, Libya *139*
Tamias spp. (chipmunks) 212
taxon-based plant fossil analysis 96,
    97–98
taxonomic classification
    definitions of diversity
        301–302
    species distribution models 311
tectonic aneurism model 53, 63
    see also plate tectonics
temperature
    mammals 203
    mountain-climate interactions
        111–112
    silicate weathering 51
    stable isotope paleoaltimetry 87,
        98, 99, 100, 102, **103**
    time–temperature history
        models *74, 75, 77, 79*
temporal heterogeneity, and
    biodiversity 156
Tepuihyla spp. (tree frogs) 350
tepuis, Guiana. *see* Guiana
    Highlands
terrestrial lapse rates, plant
    fossils 101–103, **103**

Tethys Ocean, dynamic
    topography 42
thermal processes, mantle
    30, 38, 53
thermochronology
    dating techniques 69, 71–72,
        74–75, *75*
    erosion/mountain uplift
        processes *55,* 58–62
    exhumation-rates/surface-uplift
        analysis 82
thermodynamic models, stable
    isotope paleoaltimetry
    83, 87–88
thermohaline circulation, Isthmus of
    Panama 327
thermokinematic modeling, relief
    reconstructions 60
Third Pole. *see* Tibeto-Himalayan
    region
thorium analysis. *see* uranium (U-Th/
    He) analyses
Tibetan Plateau. *see also* Tibeto-
    Himalayan region
    crust thickening 20, *20*, 22, 25,
        29–30, 31
    erosion and mountain
        uplift 59
    mountain areas based on
        ruggedness *4, 174*
    mountain-climate interactions 114
    north-south-oriented
        profile *432*
    stable isotope paleoaltimetry *83,*
        88–89
Tibeto-Himalayan region (THR) *3,*
    9, 429–431, 442 *see also* Tibetan
    Plateau
    atmospheric circulation
        patterns *431*
    biodiversity 429–441, **435**
    biomes 430–431
    climate change *431,* 431–434, 441
    crust thickening processes 23,
        24, 28
    diversification **437–439**,
        440–441
    dynamic topography 42
    erosion and mountain uplift 51,
        55, 61, 63
    geodiversity 431–434
    geography 431–433
    habitat diversity 430
    mammals 204
    montane ecozone changes *430*
    mountain areas based on
        ruggedness *4, 174*
    mountain-climate
        interactions 112

Tibeto-Himalayan region (THR)
    (cont'd)
    mountain geobiodiversity *139*,
        441–442, *442*
    speciation 430
    species radiation 440
    stable isotope
        paleoaltimetry 88–89
    surface-uplift analysis
        433, 434
Tien Shan, Asia *3, 20*, 23, 28,
    88–89
time projections, species distribution
    models *315*, 316–317
timescales, geologic history in
    relation to ecology 231–232,
    233, 234
time–temperature history
    models *74, 75, 77, 79*
Timor, Southeast Asia 485
tooth enamel, stable isotope
    paleoaltimetry 85
top-down inference, environmental
    heterogeneity 193
topographic barriers
    Andes 6
    birds 251
    historical connectivity 178,
        179, 180
    Japanese floral
        diversification 469–470
    phytopaleoaltimetry 103
    stable isotope paleoaltimetry 82
    Tibeto-Himalayan region 432
topographic complexity. *see also*
    environmental heterogeneity
    Andes 232, 233, 236–239
    mammals 202–204, *204, 205*,
        206–208, 210–212
topographic deformation, Yunnan,
    China 454
topographic diversity gradients,
    mammals 204–208,
        205, 212
    in deep time 208–209, *209*
    drivers of 210
topography 37–38, *38*
    see also dynamic topography
    and biodiversity 5–7
    river capture hypothesis 273–274,
        278, 279
    stable isotope paleoaltimetry
        81, 82
Torres del Paine National Park,
    Chile *501*
trackways, migration. *see* migration
    trackways

trait heritability, environmental
    heterogeneity 195
trait variation, environmental
    heterogeneity 191
transect surveys, environmental
    heterogeneity 193, 194
transplant experiments,
    environmental heterogeneity
    studies 194, *194*
tree frog (*Tepuihyla*) 350
treeline. *see* upper forest line
tree rings, paleoclimate
    research 125
tropics, birds/biodiversity 249–252
true skill statistic (TSS) 313
Tunuí sandstone plateau, Guiana
    Highlands *344*, 345
Turkish–Anatolian Plateau, stable
    isotope paleoaltimetry 83
turnover, species 300–301, *303*

u
UNESCO International Geoscience
    and Geoparks Programme 145
upper forest line (UFL)
    Andes 174, *175*, 176
    European Alps 425
    historical connectivity 179
    New Zealand 378
uranium (U-Th/He) analyses 69, 71,
    72, 76
Ursus arctos (European brown
    bear) 4

v
Vaccinium phillyreoides, species
    distribution models *314*
valley winds 113
value systems 138, 143–145,
    144, 156
vanishing refuge model 233
van Steenis' migration tracks *486*
vegetation belts. *see* biomes
vegetation history, European
    Alps 418–421
veil line, river capture
    hypothesis 287
Venezuela, Merida Andes
    76–77, *76*
vicariance
    plate tectonics and mountain
        building 8
    Tibeto-Himalayan
        region 440–441
Vienna standard mean ocean water
    (VSMOW) 82
Vireya clades, Southeast Asia *488*

volcanic mountains/volcanism
    Antarctica 360–361, *361*,
        364–367, *368*
    crust thickening processes 26
    Japan 461, 468, 470
    East African Rift System 395, *396*
    mountain geodiversity
        140–141, 143
    plate tectonics 18
    stable isotope
        paleoaltimetry 85–86, 114
Vorarlberg, Austria 157–161
    data sets **159**
    geodiversity classes **161**
    geodiversity index *160*
    outline map showing main tectonic
        zones *158*

w
Walker circulation model
    397, 398, 401
Wallacea 329, *330*, 331
Wallace Line *330*, 331
Wannengrat, Switzerland *188*
weathering thermostat 124
Wegener, Alfred 7
West Java, Quaternary vegetation
    dynamics *479*
West Papua, Bird's Head
    region 485
Whittaker's species
    turnover 300–301
wind patterns, mountain-climate
    interactions 112–113

y
Yellowstone, USA
    crust thickening processes
        20, 25
    mountain-climate
        interactions 114
    mountain geodiversity 141
Yule model, drivers of
    diversification 261, 265, **267**
Yunnan, China 9, 454, *499*
    fossil record 450, 452–453, *453*
    Geographic Information System
        map *450*
    geomorphology/tectonics/plant
        diversity 449–450
    Meili Xue Shan *500*
    Neogene climates 450–452
    plant diversity 449–450, 452–454

z
Zagros Mountains, Iran *139*
zircon, dating techniques 71, 72

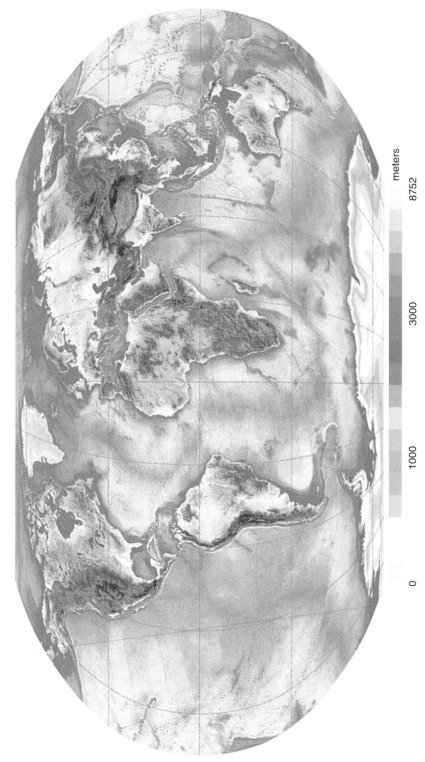

meters

0   1000   3000   8752

Topographical map of the world   See selected mountain ranges in Figure 1.2 and Plate 2. Image courtesy of Suzette Flantua.

Mountains, Climate and Biodiversity, First Edition. Edited by Carina Hoorn, Allison Perrigo and Alexandre Antonelli.
© 2018 John Wiley & Sons Ltd. Published 2018 by John Wiley & Sons Ltd.
Companion website: www.wiley.com\go\hoorn\mountains,climateandbiodiversity

Plate 1 The center section of Humboldt's classical tableau, illustrating a cross-section of the Chimborazo volcano in Ecuador, the highest mountain peak as measured from the center of the globe. This detailed drawing depicts one of the earliest studies of how the mountain biota is structured along an elevation gradient. Humboldt recognized the existence of distinct vegetation zones at different elevations with largely unique sets of species, constrained by climatic and physiological adaptations. *Source*: Humboldt & Bonpland (1807). See Chapter 1 for further information.

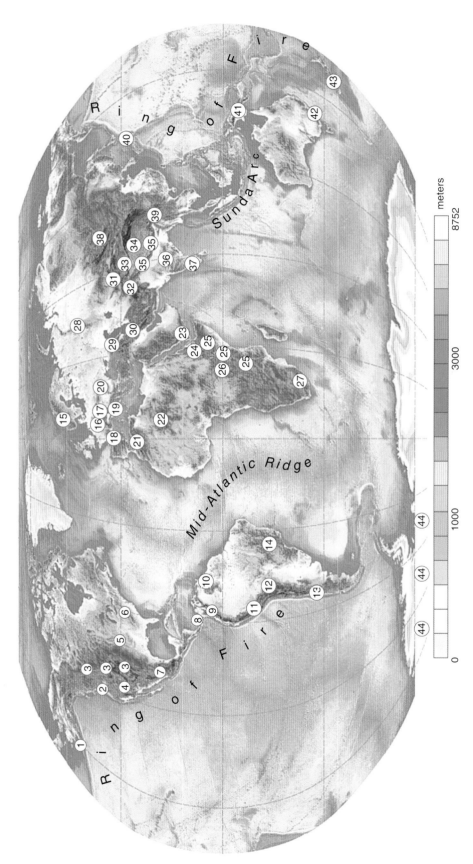

Plate 2 A selection of the most prominent mountain systems on Earth, as well as several other major geologic features and systems discussed throughout this book. Americas: 1, Aleutian Arc; 2, Cascades; 3, Rocky Mountains (Rockies); 4, Basin and Range Province; 5, Great Plains; 6, Appalachians; 7, Sierra Madre; 8, Panama Isthmus; 9, Northern Andes; 10, Guiana Highlands; 11, Central Andes; 12, Bolivian Altiplano; 13, Southern Andes; 14, Brazilian Highlands. Europe: 15, Scandinavian Mountains; 16, Jura Mountains; 17, Alps; 18, Pyrenees; 19, Apennines; 20, Carpathians. Africa and Arabia: 21, Atlas Mountains; 22, Ahaggar (Hoggar) Mountains; 23, Yemen Highlands; 24, Ethiopian Highlands; 25, East African Rift System (EARS); 26, Rwenzori Mountains; 27, Drakensberg. Asia: 28, Ural Mountains; 29, Caucasus Mountains; 30, Zagros Mountains; 31, Tien Shan; 32, Hindu Kush; 33, Kunlun Shan; 34, Tibetan Plateau; 35, Himalaya; 36, Deccan Plateau; 37, Western and Eastern Ghats; 38, Altai Mountains; 39, Hengduan Mountains; 40, Japanese Alps. Oceania: 41, New Guinea Highlands; 42, Eastern Highlands (Australia); 43, Southern Alps. Antarctica: 44, Transantarctic Mountains. Image courtesy of Suzette Flantua. See Chapter 1 for further information.

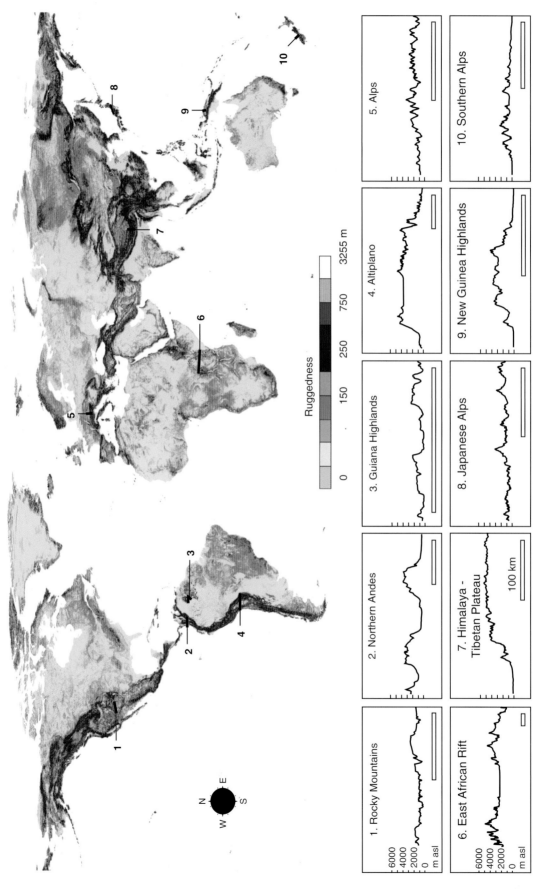

Plate 3 Mountain areas based on ruggedness, as defined by Körner et al. (2011) (maximum elevational distance between nine grid points of 30″ in 2.5′ pixel; for a 2.5′ pixel to be defined as "rugged" (i.e., mountainous), the difference between the lowest and highest of the nine points must exceed 200 m). Numbers indicate topographic profiles of selected mountain ranges around the world. The characteristic topography of a mountain directly relates to the potential impact and frequency of connectivity breaks caused by Pleistocene glacial cycles, and thus the expression of the flickering connectivity system. Bars below profiles indicate a 100 km distance proportional to the profile shown. *Source:* Adapted from Körner et al. (2011). Figure from Chapter 1. See Chapter 12. See Chapter 1 for further information.

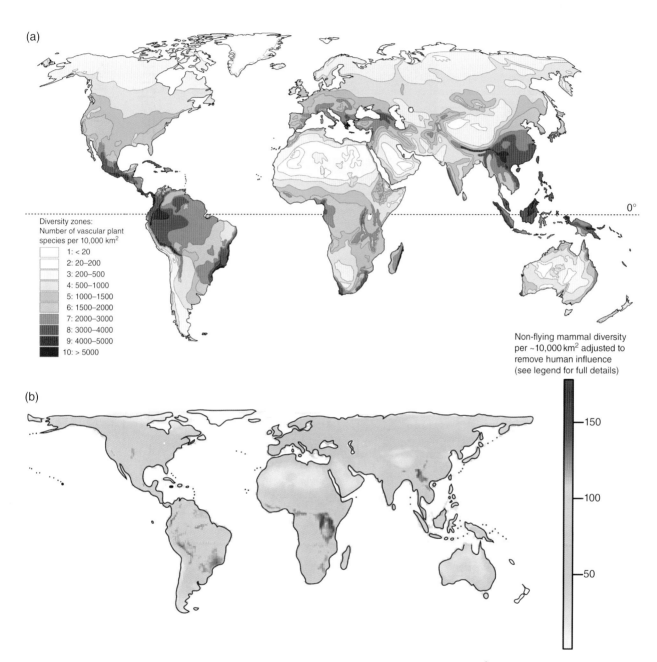

Plate 4 (a) Global plant diversity measured by estimated number of vascular plant species per 10,000 km². *Source:* Barthlott et al. (1996, 2007), with permission from Wilhelm Barthlott and Jens Mutke. (b) Estimated patterns of species diversity for terrestrial mammals. This map was constructed using estimated natural species ranges without human influences, and plotted using an equal-area Behrmann projection with colors proportional to the number of expected species. Several mountain regions have noticeably higher diversity than surrounding areas (e.g., the Rocky Mountains and Sierra Nevada in North America, the Andes in South America and, in particular, the East African Rift System). *Source:* Modified from Faurby & Svenning (2015), with permission from Søren Faurby. See Chapter 1 for further information.

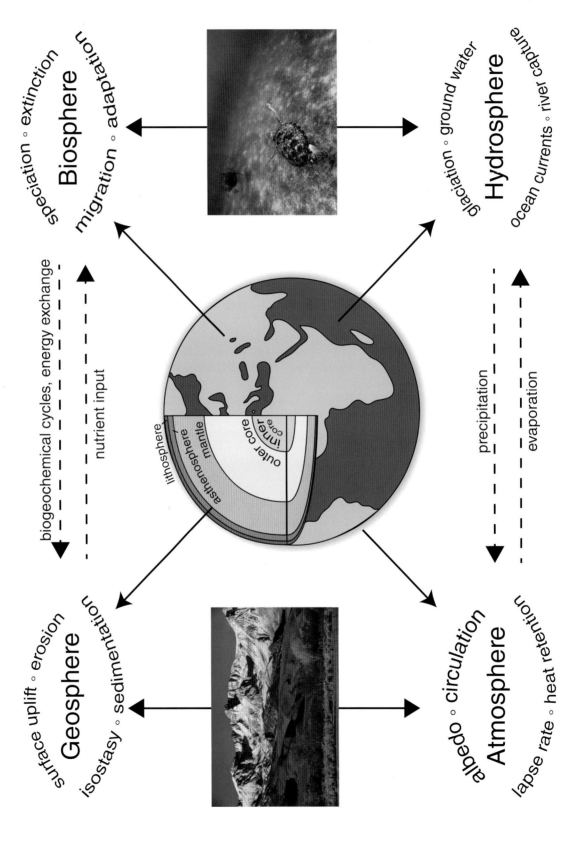

Plate 5 Schematic representation of the complex interaction between four of Earth's "spheres": the geosphere, biosphere, atmosphere and hydrosphere. Each sphere is surrounded by a selection of words indicating the terminology and processes associated with it. The Earth's layers are indicated in the cut-out, but are not drawn to scale. The lithosphere comprises the crust and the uppermost solid mantle. *Source:* Adapted from Senckenberg Magazine (2015). See Chapter 1 for further information.

Plate 6 (a) Central Andes, along the "Death Road" between La Paz and Coroico, Yungas, Bolivia. This is a highly diverse, but still poorly explored region and under increasing pressure by human activities. (b) Table Mountain, Kirstenbosch National Botanical Garden, South Africa. In the foreground is a member of the Proteaceae family, which is particularly diverse in the Cape Floristic Region where it constitutes a characteristic element. (c) Southern Andes, Torres del Paine National Park, Chile. The famous granite peaks seen in the background extend up to 2,500 metres above sea level and are surrounded by glaciers, lakes and rivers. Guanacos, a non-domesticated species of llama, are often seen grazing the open areas. (d) Isla Navarino, Cabo de Hornos, Chile. This is the southernmost populated island of South America, and despite a harsh climate the island is home to a rich diversity of bryophytes, lichens and insects, and is partly covered by *Nothofagus* forests. Photos courtesy of Alexandre Antonelli.

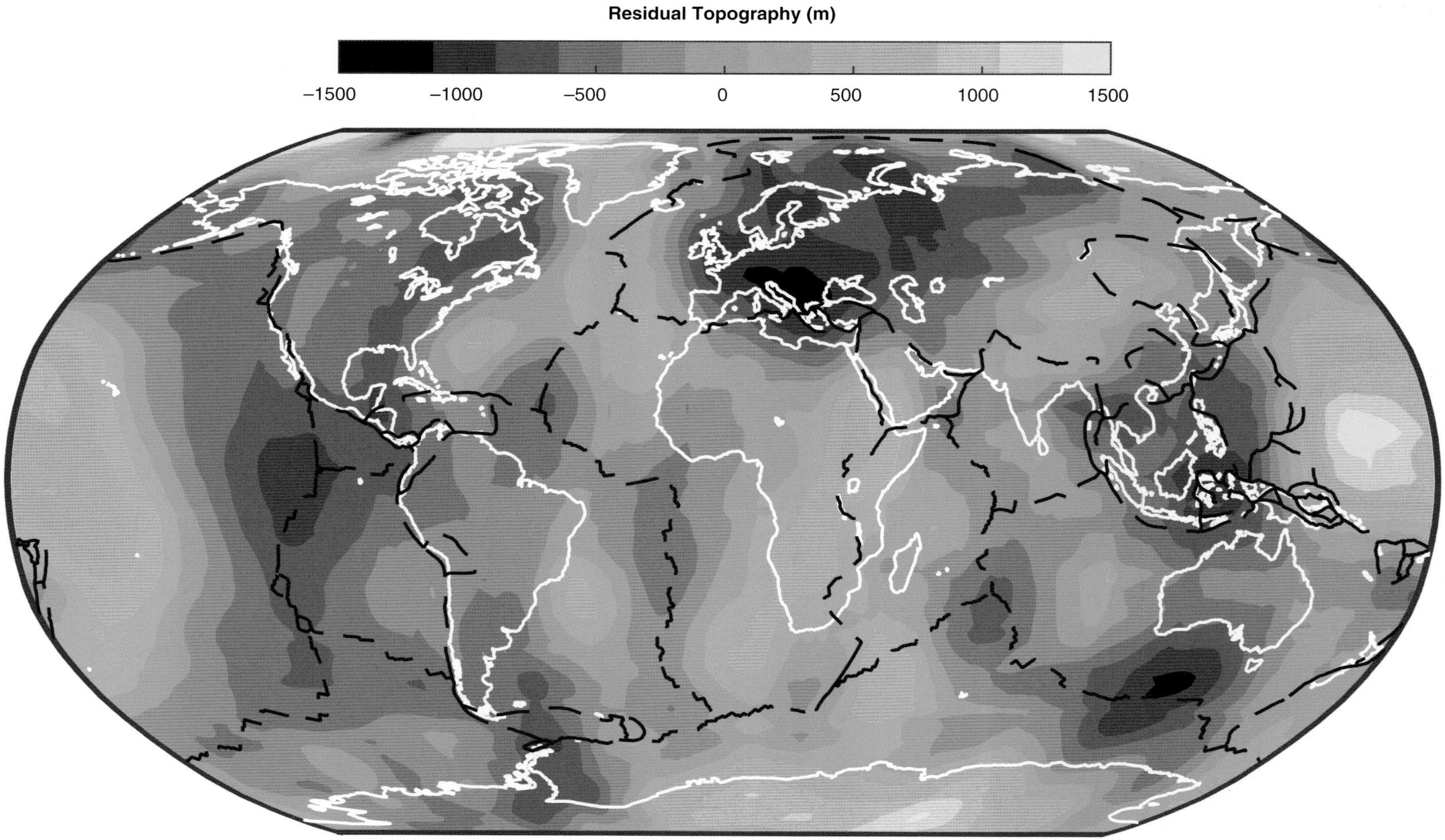

Plate 7 Map showing an example of a global residual topography calculation. The crustal and lithospheric isostatic component of topography has been subtracted from the present-day observed topography. This calculation was based on a model of lithospheric structure from Naliboff et al. (2012) and has been smoothed (averaged) over $10 \times 10°$ bins to focus on longer-wavelength features. Prominent residual highs (Light colors) are seen over the Western Pacific, the East African Rift, Antarctica, and the North Atlantic. Meanwhile, residual lows (Dark colours) are visible over Europe, the East Pacific Rise, and the Australian–Antarctic Discordance (south of Australia). Thin white lines on the map delineate land surfaces, and black dashed lines show plate boundaries. See Chapter 3 for further information.

Residual Topography (m)

−1500 −1000 −500 0 500 1000 1500

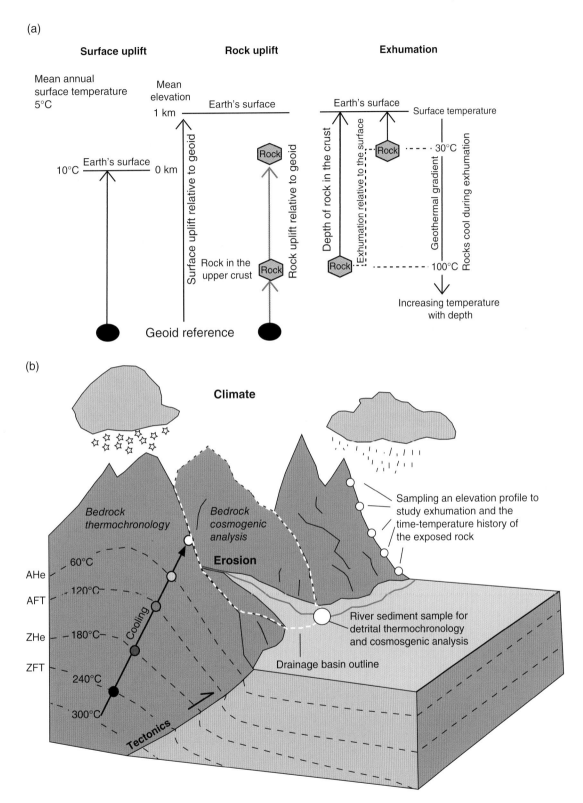

Plate 8 (a) Schematic presentation of the concept of surface uplift and rock uplift with respect to the centre of the Earth, and exhumation with respect to the Earth's surface. (b) Schematic concept showing the relationship between tectonics, surface processes, and cooling of rocks during exhumation in a convergent orogenic setting. AFT and ZFT, apatite and zircon fission-track analysis; AHe and ZHe, apatite and zircon (U-Th)/He analysis. Isotherms (indicated by black dashed lines) relate to the average closure temperatures of the different dating techniques, depending on the cooling rate (here about 15 °C/My). See Chapter 5 for further information.

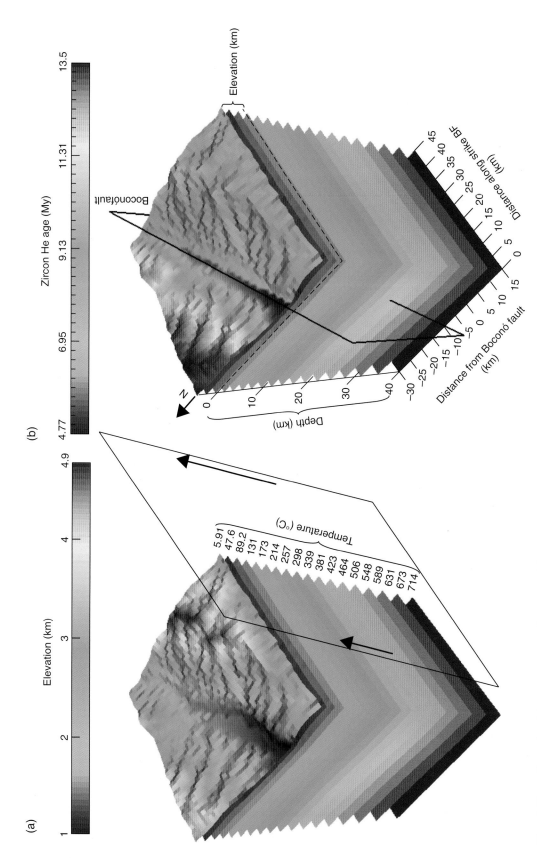

Plate 9 Three-dimensional time–temperature history modeling is useful for modeling the thermal and tectonic evolution of mountain belts. Shown here are *PeCUBE* models for (a) the topographic evolution (layers indicate isotherms, with the temperature values (°C) noted alongside) and (b) predicted zircon He ages at the surface of both sides of the Boconó fault. Inclusion of the fault in the middle of the study area permits the analysis of the exhumation of two tectonic blocks independently. See Chapter 5 for further information.

Plate 10 The northern Andes are topped by the superpáramo, the highest ecosystem, just below the snow line. It is characterized by the dominance of bedrock that provide unique azonal (dry) conditions even in humid environments. Plants struggle to survive under these conditions, often resulting in a substantial proportion of narrow endemic species. Photographs show the superpáramo of the Sierra Nevada del Cocuy National Park, in Colombia, with rocky slopes of Cretaceous (Albian-Aptian) sedimentary quartzitic sandstone bedrock and shales with occasional limestone inclusions. a) *Espeletia lopezii* Cuatrec. growing in a pit at the base of the vertical slopes of the Ritacuba Blanco peak. b) *Espeletia cleefii* Cuatrec., typically in small clusters on the bedrock of the eastern side of Cocuy. c) A zonal area with a patch of páramo vegetation dominated by grasses and rosettes (*Espeletia*), in an azonal landscape at 4,300 m; d) *Senecio adglacialis* Cuatrec. emerging from the glacial moraines at 4,650 m. e) *Lupinus alopecuroides* Desr. (Fabaceae; palmate leaves, slate blue flowers) and *Senecio niveoaureus* Cuatrec. (Asteraceae; white leaves, yellow flowers) on the superpáramo slopes of the Ritacuba Blanco (east side), at 4,638 m. Photos courtesy of Mauricio Diazgranados Cadelo.

Plate 11 Near the start of the Langtang Treck, Nepal, at an altitude of around 2500 m looking east up the Langtang Valley. The slope on the right is north facing and has more moisture/shade loving plants than on the left (south facing) slope. Photo courtesy of Robert A. Spicer.

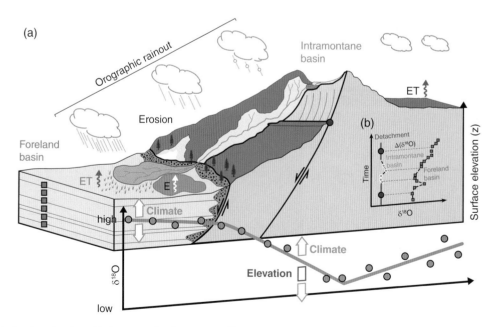

Plate 12 Practical approaches in stable isotope paleoaltimetry. (a) $\delta^{18}O$ values of precipitation decrease systematically during orographic rainout and associated cooling and condensation of water vapor as a function of elevation, z. Increasing $\delta^{18}O$ values in the lee of the mountain range result from evaporative ^{18}O enrichment in precipitation. Over geologic time, $\delta^{18}O$ values of advected water vapor may change due to changing climate conditions (arrows). (b) Reconstructing differences in the oxygen isotopic composition of precipitation, $\Delta(\delta^{18}O)$, between low-elevation sites (e.g., samples from foreland basin, squares) and sites at unknown high elevation (intramontane basins or faults/detachments, circles) eliminates some of the uncertainties associated with single-site stable isotope paleoaltimetry. ET, evapotranspiration; E, evaporation. *Source:* Adapted from Campani et al. (2012) and Mulch (2016). See Chapter 6 for further information.

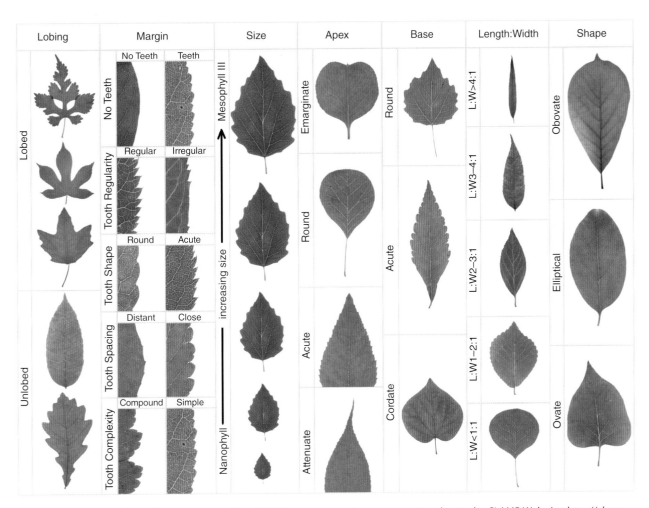

Plate 13 Summary of the leaf characters used in a CLAMP analysis. For the scoring protocols, see the CLAMP Web site: http://clamp. ibcas.ac.cn. See Chapter 7 for further information.

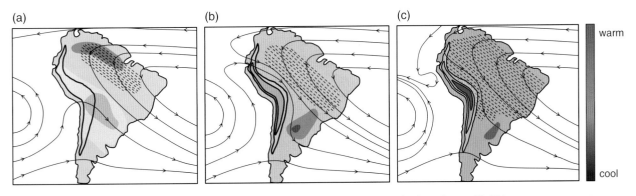

Plate 14 Schematic of South American climatology with (a) low, (b) medium and (c) high surface uplift. Thin vectors represent lower-level winds. Diagonal hatching indicates regions with mean annual precipitation greater than 150 cm. Dark shading indicates cooler temperatures and light shading indicates warmer temperatures. See Chapter 8 for further information.

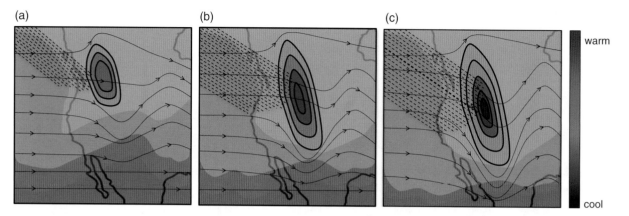

Plate 15 Schematic of North American climatology with (a) low, (b) medium and (c) high surface uplift. Thin vectors represent lower-level winds. Diagonal hatching indicates regions with mean annual precipitation greater than 150 cm. Dark shading indicates cooler temperatures and light shading indicates warmer temperatures. See Chapter 8 for further information.

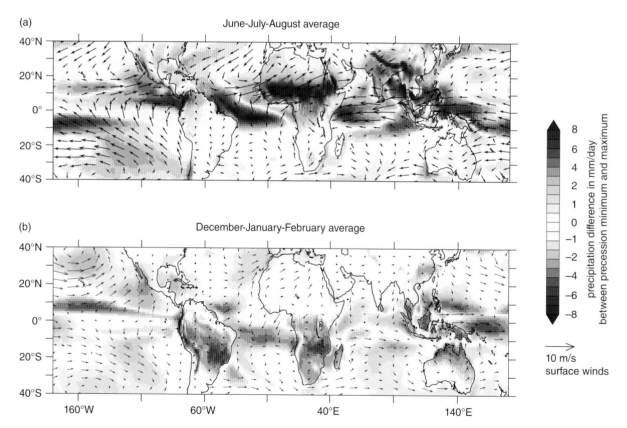

Plate 16 Model results from the coupled ocean–atmosphere weather model ECBilt (Bosmans 2014), simulating the impact of precession minima and maxima on precipitation. Shown are the differences in average precipitation (mm/day) and surface winds (m/s) for precession minima minus precession maxima for average northern hemisphere (a) summer and (b) winter periods. The red areas on the northern hemisphere continents in their summer and the hatched red areas on the southern hemisphere continents in their summer exemplify the intensified monsoon activity during precession minima and precession maxima, respectively. Note the large longitudinal variation in precipitation and winds within a single precession phase, highlighting the impossibility of interpreting astronomical forcing at Earth's surface directly from solar insolation changes at the top of the atmosphere. *Source:* Adapted from a version produced by Joyce Bosmans (Utrecht University, Utrecht, The Netherlands). See Chapter 9 for further information.

Plate 17 Geodiversity of mountain landscapes. (a) Sagarmatha National Park, Nepalese Himalaya: very high mountains of the Alpine–Himalayan belt characterized by high relief, active glacial and slope processes and shrinking valley glaciers with extensive rock debris cover. (b) The Cairngorm Mountains, Scotland: low mountains of the Caledonide belt, formed in a dissected Silurian granite intrusion with glacial landforms incised into a series of paleosurfaces. (c) The Blue Mountains, New South Wales, Australia: a dissected plateau on a passive margin uplifted during the Jurassic. (d) Allardyce Range, South Georgia: heavily glacierized mountains at sea level, comprising folded lower Cretaceous volcaniclastic sandstones and mudstones. (e) Zagros Mountains, Iran: differential erosion of folded Carboniferous–Miocene sedimentary rocks, forming a landscape of linear ridges and valleys and revealing a salt dome in the center of the image. *Source:* NASA, https://earthobservatory.nasa.gov/IOTD/view.php?id=6465. (f) Cerro Fitz Roy massif, southern Andes: granitic buttresses and towers formed in a Neogene igneous intrusion. (g) Tadrart Acacus, Libya: a monocline in Paleozoic sedimentary rocks uplifted and dissected during the Cenozoic. (h) Cotopaxi, Ecuador: an active, glacier-capped stratovolcano with flanks scarred by tracks of lahars and meltwater floods. (i) Mount Kenya: an extinct Plio–Pleistocene stratovolcano heavily dissected by glacial erosion and with small remnant glaciers. Photos courtesy of John Gordon, unless otherwise stated. See Chapter 10 for further information.

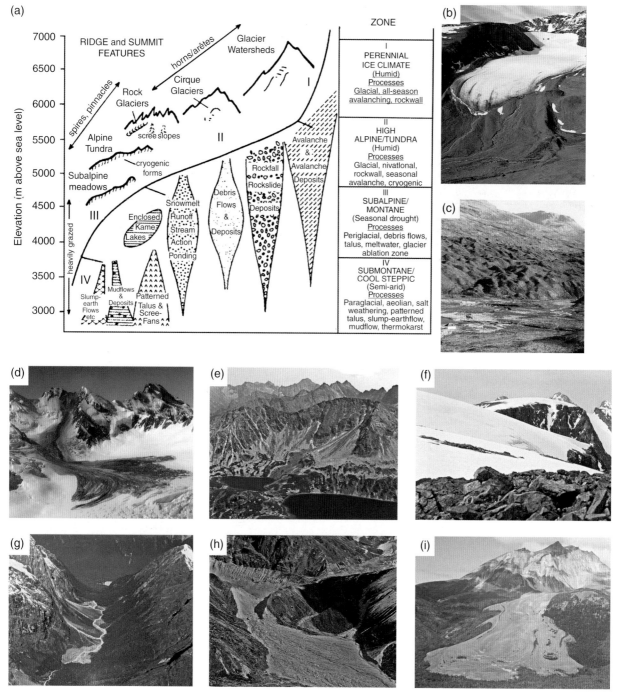

Plate 18 Geodiversity of mountain landforms and geomorphological processes. (a) Schematic representation of the altitudinal zonation of geomorphological processes and depositional environments, central Karakoram. *Source:* Hewitt (1989). Reproduced with permission of Schweizerbart Science Publishers. (b) Recession of cirque glacier from Little Ice Age moraines, Kebnekaise, Sweden. (c) Moraines formed by glaciers during the Younger Dryas, Scottish Highlands. (d) Landslide on Mount Dixon, Southern Alps, New Zealand, January 21, 2013. *Source:* Reproduced with permission of Arthur McBride, Alpine Guides (Aoraki) Ltd. (e) Rock weathering, talus formation and debris flows, Tatra Mountains, Poland. (f) Periglacial blockfield and small plateau icecaps, Lyngsalpene, northern Norway. (g) Glacial trough and alluvial fans, Southern Alps, New Zealand. (h) End moraine breach and glacial lake outburst flood (GLOF) deposits following the failure of the moraine dam at Tam Pokhari (Sabai Tsho) glacial lake, Hinku Valley, Nepal, September 3, 1998. (i) Rock glacier, Wrangell Mountains, Alaska. Photos courtesy of John Gordon, unless otherwise stated. See Chapter 10 for further information.

Plate 19 Breiðamerkurjökull, Vatnajökull National Park, Iceland. Many glaciers worldwide are popular (geo)tourism destinations where visitors can both enjoy the spectacular scenery and appreciate the effects of climate change, which are clearly demonstrated in glacier recession. Photo courtesy of John Gordon. See Chapter 10 for further information.

Plate 20 Arenal Volcano in Costa Rica. Photo courtesy of Steve Meyer.

Plate 21 (a) Yellowstone National Park, USA (b) Rocky Mountain National Park, USA (c) Eastern Sierra near Bishop, California, USA (d) Buffalo grazing in front of geothermal activity at Yellowstone National Park, USA. Photos courtesy of Paul Babb..

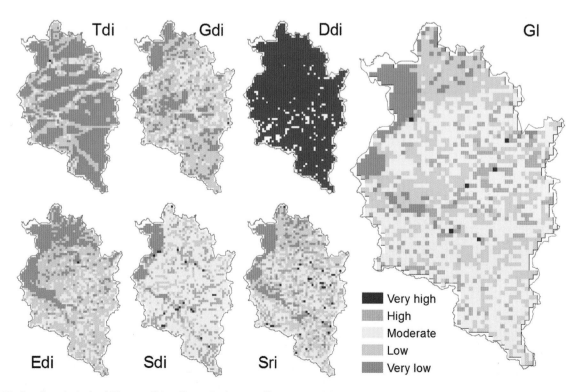

Plate 22 Geodiversity index (GI) map of Vorarlberg. Grid size in all maps is 1 × 1 km. Geodiversity for each cell follows the five-class scheme presented in the figure, ranging from very low to very high for each index, with the exception of the map for Ddi, where each cell is marked solely on presence (blue) or absence (white) of drainage. Tdi, tectonic diversity index; Gdi, geological diversity index; Ddi, drainage diversity index; Edi, elevation diversity index; Sdi, slope diversity index; Sri, solar radiation diversity index; GI, geodiversity index. See Chapter 11 for further information.

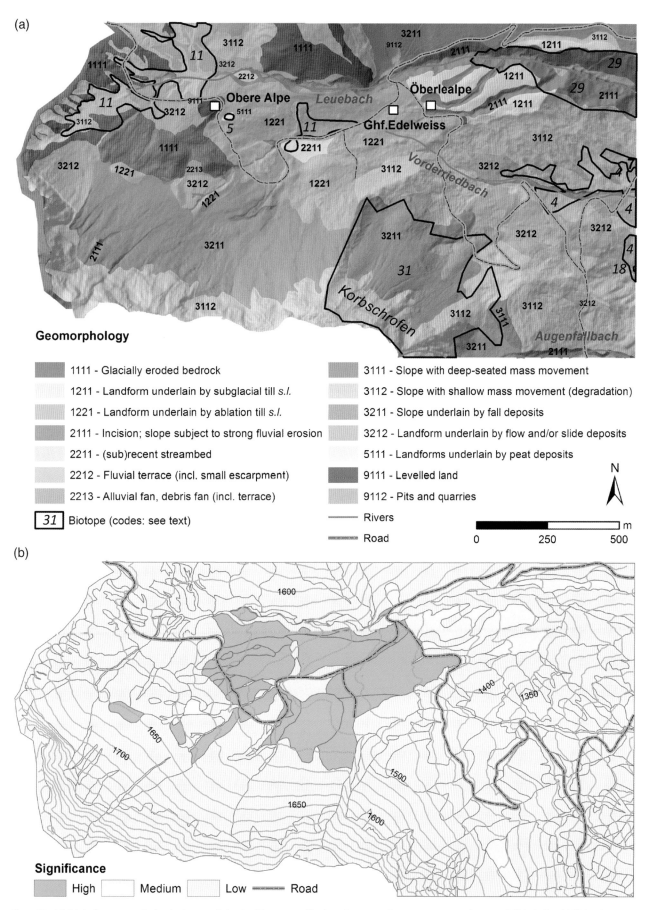

(a)

Geomorphology

	1111 - Glacially eroded bedrock		3111 - Slope with deep-seated mass movement
	1211 - Landform underlain by subglacial till *s.l.*		3112 - Slope with shallow mass movement (degradation)
	1221 - Landform underlain by ablation till *s.l.*		3211 - Slope underlain by fall deposits
	2111 - Incision; slope subject to strong fluvial erosion		3212 - Landform underlain by flow and/or slide deposits
	2211 - (sub)recent streambed		5111 - Landforms underlain by peat deposits
	2212 - Fluvial terrace (incl. small escarpment)		9111 - Levelled land
	2213 - Alluvial fan, debris fan (incl. terrace)		9112 - Pits and quarries
31	Biotope (codes: see text)	—	Rivers
		═══	Road

N

0   250   500 m

(b)

Significance

High   Medium   Low   ══ Road

Plate 23 (a) Digital geomorphological map of the Au West area, displayed as a semi-transparent overlay on a DEM-derived hillshade map. The location of the biotope units is indicated. (b) Potential geoconservation map of the Au West area, shown on a background of 25 m-contour lines. Refer to Chapter 11, Section 11.4.3 for further explanation. See Chapter 11 for further information.

(a)

(b)

Plate 24 (a) View to the south of part of the Obere Alpe glacial niche. Rockfall and debris-flow deposits cover the lower slopes of the headwall, which is part of the Gungern-Klippern mountain range. The hummocky topography around and to the left of Obere Alpe (center-right) is formed by an intricate pattern of glacially eroded bedrock and ablation tills. (b) View to the south-west of the glacial landscape in the central part of the study area. The east-sloping surface to the left of Ghf. Edelweiss is underlain by subglacial till (exposed in the flank of the Leuebach incision in the lower-central part of the photo). The steep Korbschrofen cliff (left) produces abundant scree. The central-right cliff, with an apron of talus, is the headwall of the Obere Alpe glacial niche. See Chapter 11 for further information.

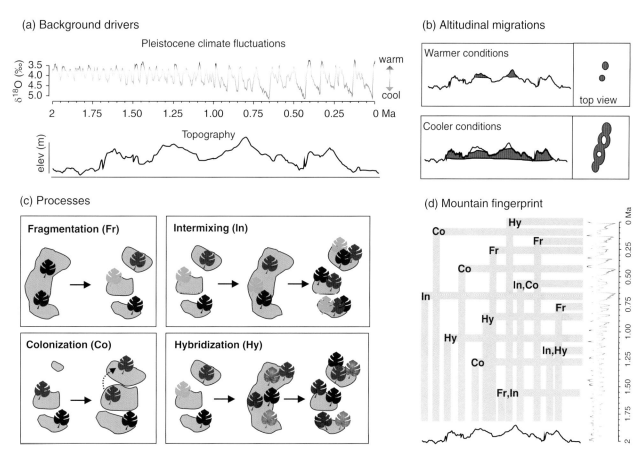

Plate 25 Conceptual framework of the flickering connectivity system (FCS). (a) The background drivers of speciation are the large Pleistocene climate fluctuations and highly complex montane topography. The $\delta^{18}O$ curve is based on composite stable oxygen isotope ratios from benthic foraminifera and is an indicator of global ice volume and temperature (Lisiecki & Raymo 2005). (b) Altitudinal migrations of hypothetical high-mountain biota, shown in a simple two-phase setting reflecting warmer and cooler conditions. (c) Schematic representation of the intrinsic processes of the FCS as a result of changes in connectivity: fragmentation (Fr), colonization (Co), intermixing (In) and hybridization (Hy). (d) The "mountain fingerprint" is defined by the interaction between climate and topography. It is a unique mountain identifier in which the processes of (a) occur in a spatially and temporally complex way, and therefore cause different timings and patterns of species diversification when comparing between mountains. See Chapter 12 for further information.

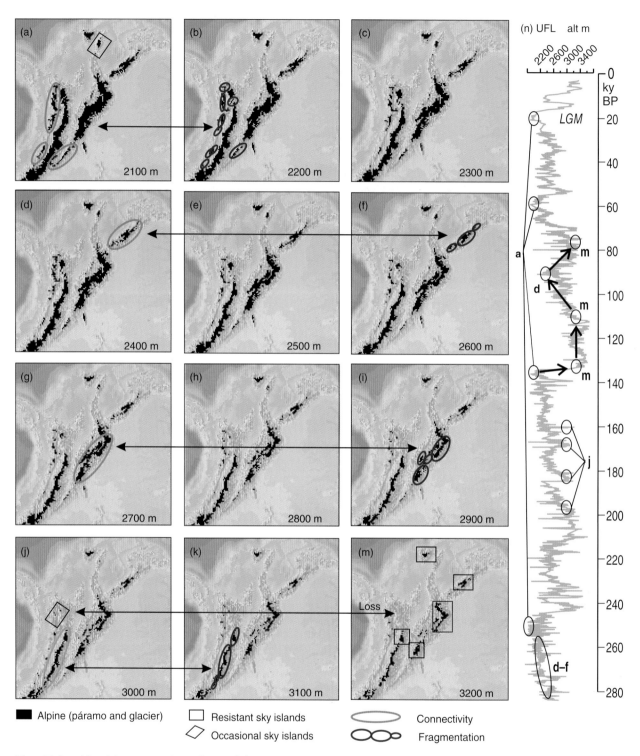

Plate 26 (a–m) Spatial reconstructions of tropical alpine systems (páramo and glaciers; black) in the northern Andes during the last 280 ky, showing the upper forest line (UFL) moving between elevations of 2100 and 3200 m. Each map represents a simplified reconstruction of the distribution of the alpine Andean ecosystem (the páramo) using a digital elevation model. (n) Estimated elevations of the UFL are inferred from the Fúquene-9C pollen record (Bogotá-Angel et al. 2011; Groot et al. 2011). Letters correspond to the maps. Low UFL reflects cooler periods, such as the Last Glacial Maximum (LGM), while a higher UFL reflects warmer periods (interglacial conditions, such as the present). Different regions experience alpine system connectivity and fragmentation at different moments in time. Some páramo areas persist continuously (resistant sky islands), while others appear and disappear (occasional sky islands). See Chapter 12 for further information.

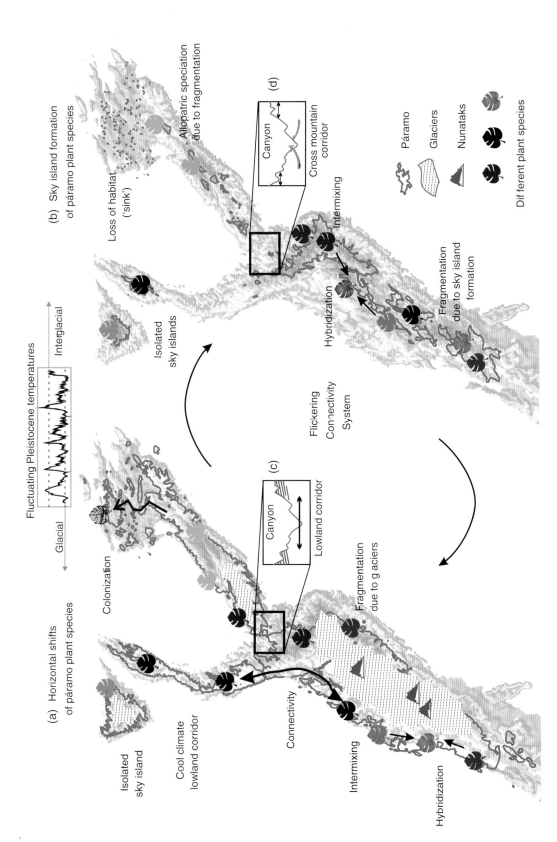

Plate 27 Spatial representation of the four intrinsic processes of the FCS in the Eastern Cordillera of the Colombian Andes. The potential distribution of páramo is shown during (a) cooler and (b) warmer conditions. The figure shows how different processes can occur at different locations throughout a mountain system, and as a result cause a spatially complex biogeographic pattern. The many possible intermediate configurations are shown in Plate 26 (see also Chapter 12, Figure 12.3). (c) Cool climate corridor of alpine species through a mid-elevation or lowland canyon. Glaciers are seen on the mountain tops. The arrow indicates the direction of connectivity. (d) Cross-mountain corridor between populations on either side of a mountain. Alpine species are restricted here to high elevation, and connectivity is reduced. See Chapter 12 for further information.

Plate 28 Mosaic of snowbeds and exposed ridges in the spring (May 2011) on Wannengrat, Switzerland. See Chapter 13 for further information.

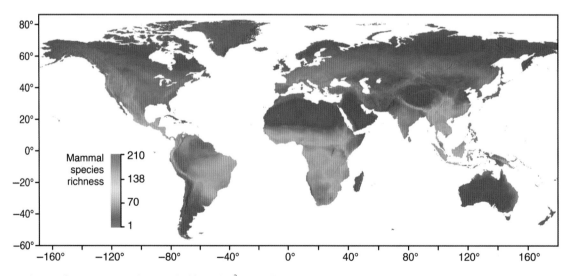

Plate 29 Richness of extant mammals, compiled for 10 km² grid cells. Overprinting the latitudinal diversity gradient is a strong topographic diversity gradient on all continents, excluding Antarctica. Mammalian species richness is elevated in the montane west of North America, along the Andes of South America, in the Alps of Europe, within the topographically complex East African Rift System, over mountainous regions in India and south-eastern Asia and along the eastern coastal ranges of Australia. *Source:* Adapted from IUCN (2015). See Chapter 14 for further information.

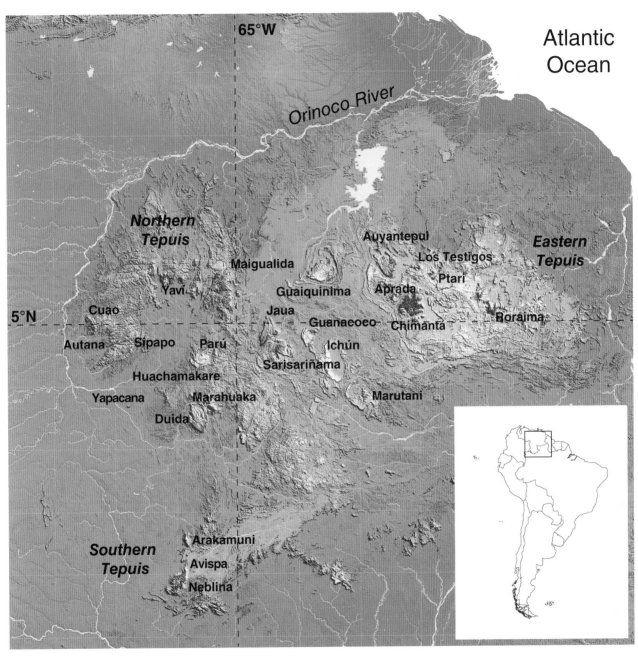

Plate 36 Map showing the distribution and locations of the main mountain systems in the Pantepui Province of the Guiana Highlands. *Source:* Courtesy of Charles Brewer-Carías. See Chapter 23 for further information.

Plate 37 (a) View of Cañón del Diablo (Devil's Valley), with Salto Angel (Angel Falls) on the left, falling from Auyantepui in Venezuela. *Source:* Courtesy of Charles Brewer-Carías. (b) Roraima sandstone geology in the Chimantá massif, Venezuela. *Source:* Gérard Vigo, https://commons.wikimedia.org/wiki/File:Roraima_Rocks7.JPG. Licensed under CC BY-SA 3.0. (c) Mt. Roraima and the Eastern Tepui chain. *Source:* Courtesy of Charles Brewer-Carías. (d) *Bonnetia crassa* Gleason, endemic treelet of the northern Pantepui province. (e) *Oreophrynella quelchii* Boulenger, endemic toad of the summit of Mt. Roraima. *Source:* Courtesy of Javier Mesa. See Chapter 23 for further information.

Plate 38 (a) North-east face of Finger Mountain, Victoria Land, Antarctica, showing three key elements of the stratigraphy: 1) the Devonian Beacon quartz sandstone (horizontal massive light-colored rocks) grading into thin late Devonian fish-bearing sandstone and siltstone; 2) a thin white feldspathic sandstone, marking an unconformity in which the Carboniferous System is missing, and which forms the base of the Permian Glossopteris-bearing coal measures; and 3) a largely sand-filled channel cut into Devonian sandstone by pre-Permian glaciers. The dark brown horizontal layer (lower middle) is a 200 m-thick sill of Ferrar Dolerite, which connects with an inclined sheet of dolerite reaching the summit from the right. *Source:* Courtesy of J. Smellie. (b) Fossilized leaves of *Glossopteris* from Permian sediments, Allan Hills, Transantarctic Mountains. The large leaf is approximately 14 cm long. *Source:* Courtesy of J. Francis. (c) Large piece of fossilized tree trunk, latest Cretaceous age, Seymour Island, Antarctic Peninsula. The log is approximately 60 cm in height. *Source:* Courtesy of J. Francis. See Chapter 24 for further information.

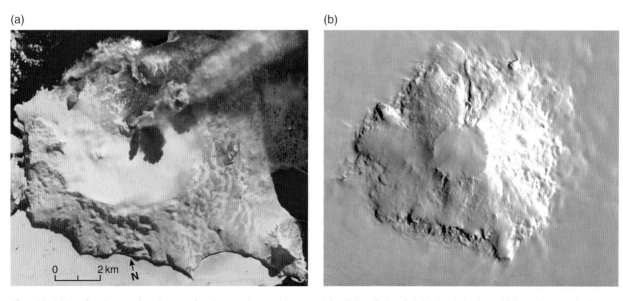

Plate 39 (a) Landsat image showing a volcanic eruption on Montagu Island, South Sandwich Islands, in 2005. Although extensively ice-covered and thought to be dead, with no historical records of activity, the volcano unexpectedly burst into life in a long-lived eruption between 2001 and 2007. It thus probably supports currently undetected subglacial geothermal areas possibly capable of sustaining life during glacials. *Source:* Patrick & Smellie (2013). Reproduced with permission of Antarctic Science Ltd. (b) GoogleEarth satellite image showing Mt. Takahe, a large stratovolcano in Marie Byrd Land, lying within the WARS. Although ice-covered and seemingly dead, this volcano is the source of numerous ash layers recovered in regional boreholes, representing eruptions extending back hundreds of thousands of years. It is thus merely quiescent, and, although no surface heat has currently been detected, probably supports subglacial heat under the extensive ice cover. See Chapter 24 for further information.

Plate 40 Lijiang, Yunnan, China. Nakhi people carrying the typical baskets of the region. Scene from a public performance in Jade Dragon Snow Mountain Open Air Theatre. Photo by CEphoto, Uwe Aranas.

Plate 41 Alai Mountains in Kyrgyzstan. Photo courtesy of Allison Perrigo.

Plate 42 Species-corrected weighted (CWE) endemism for 579 alpine species comprising 50 117 georeferenced samples. CWE statistically analyses and corrects for species range sizes to provide a quantitative estimate of endemism (Laffan et al. 2016). Increasing CWE is shown by grid cells transitioning from light yellow to dark blue. Grid cells with higher endemism than expected are noted with a plus (+), while those with lower endemism than expected are marked with a dot (·). Light blue overlay: Last Glacial Maximum (LGM) ice distribution. *Source:* Provided by David Barrell (2011), Geological and Nuclear Sciences, Dunedin, New Zealand. See Chapter 25 for further information.

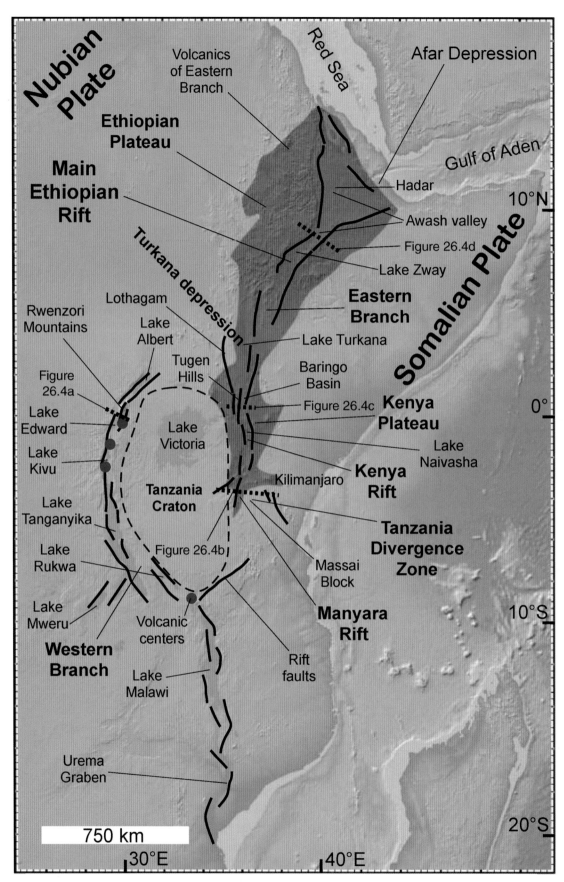

Plate 43 The East African Rift System (EARS), superimposed on the Ethiopian and Kenyan plateaus, collectively known as the East African Plateau (www.geomapapp.org). The EARS comprises series of individual graben that link up to form the Western and Eastern branches. The rift floors of the Eastern Branch, with vast volcanic rock accumulations (in red shading), have high elevations, and only the Turkana graben of this branch has a lower elevation. Localities mentioned in the text are indicated, and dotted lines for cross-sections depicted in Chapter 26, Figure 26.4 are shown. See Chapter 26 for further information.

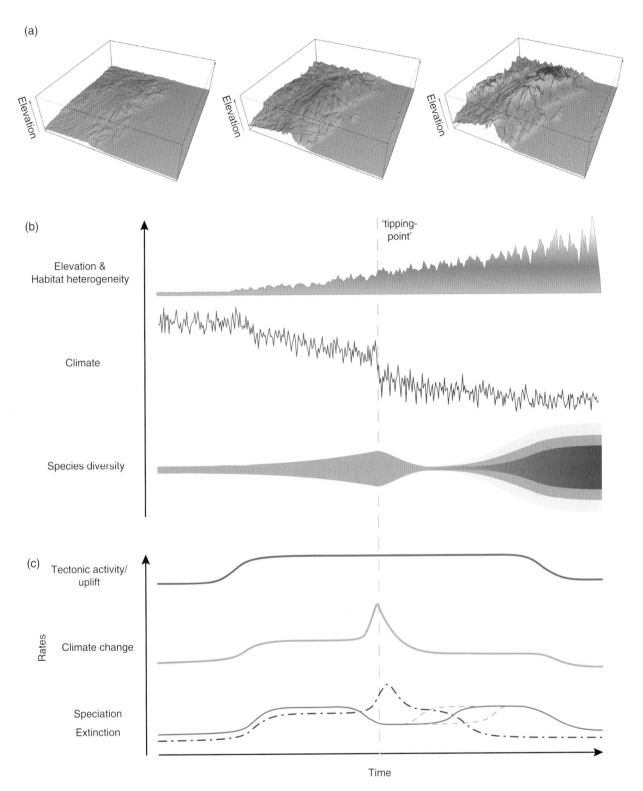

Plate 30 Relationships between mountains, climate and biodiversity. (a) Uplift of a mountain range through time. (b) Associated changes in elevation, habitat heterogeneity, climate and species diversity. (c) Rates of change corresponding to the different processes involved, namely the rate of tectonic activity/uplift, the rate of climatic change and the rates of speciation (blue line) and local extinction (dashed line). The gradual change of the relief increases topographic complexity, creates novel habitats and affects regional climatic conditions. Such changes, even if moderate, will likely affect the rates of speciation (afforded by adaptation to the novel conditions and divergence) and local extinction (if lineages fail to adapt). Continued tectonic change may result in a state shift ("tipping point"), causing profound and rapid climatic changes. A corresponding peak in species extinction is followed by an increase in immigration and in situ speciation, for example by pre-adapted lineages from other regions. As clades diversify, they fill ecological niche space, and the rate of speciation slows again. See Chapter 15 for further information.

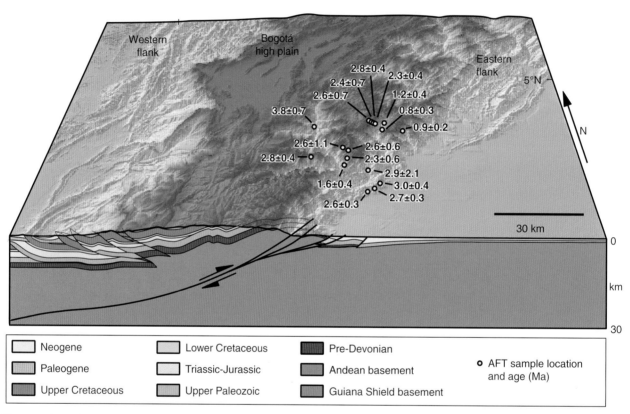

Plate 31 Three-dimensional (3D) model depicting the topography (SRTM-90) and geological structure of the Eastern Cordillera at ~4°N latitude. Numbers indicate apatite fission track ages (including 1σ error), which broadly correspond to the times when rocks cropping out at the surface were buried at a temperature of >120°C and thus illustrate the time span required for the erosion of 3–4 km of rocks. See Chapter 16, Figure 16.1. *Source:* Mora et al. (2008). Reproduced with permission of The Geological Society of America. See Chapter 16 for further information.

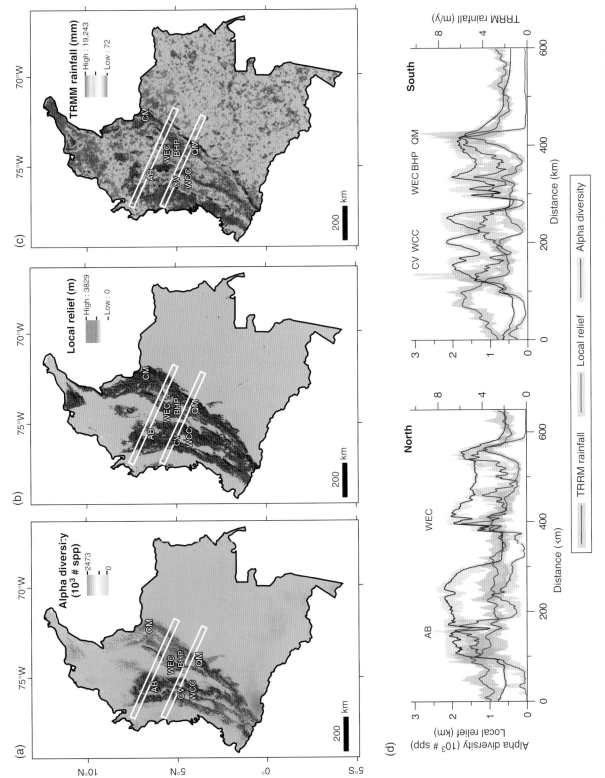

Plate 32 (a) Alpha diversity map for 5400 plant and animal species in Colombia. (b) Local relief map calculated with a 3 km moving window using SRTM-90 m topographic data. (c) Radar-based precipitation, from Tropical Rainfall Measuring Mission (TRMM) data. (d) Data extracted from 30 km-wide swath profiles along northern and central Colombia (indicated by white rectangles in (a), (b) and (c)) for alpha diversity, local relief and TRMM rainfall. Abbreviations as indicated in Chapter 16, Table 16.1. See Chapter 16 for further information. AB: Antioquia Batholith; BHP: Bogota high plain; CM: Cocuy Massif; CV: Cauca Valley; QM: Quetame Massif; WCC: Western side of Central Cordillera; WEC: Western side of Eastern Cordillera.

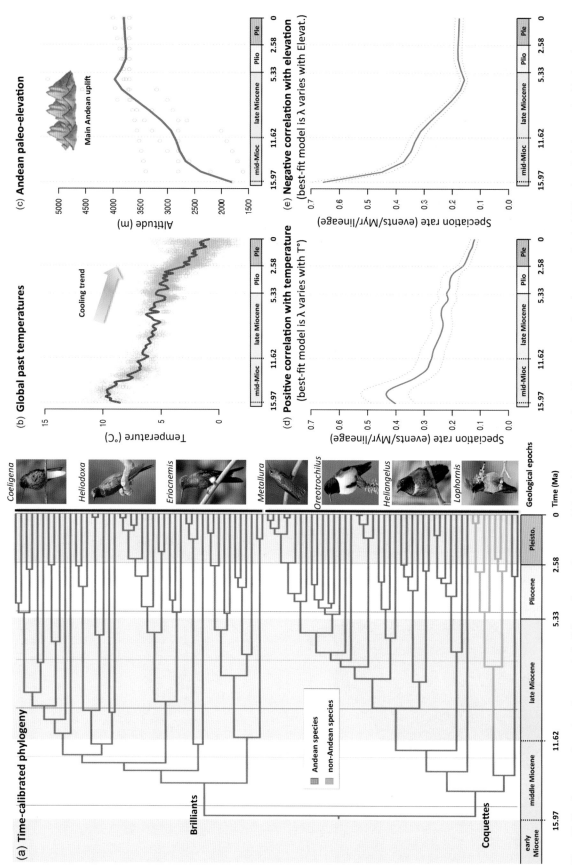

Plate 33 Teasing apart the roles of mountain building, climate change and biotic drivers in the diversification of Andean hummingbirds. (a) The dated phylogeny highlights the Andean and non-Andean species of the two hummingbird clades (Brilliants and Coquettes, with extant exemplars on the right). Two past environmental variables that may have influenced the hummingbirds' diversification are temperature (b) and elevation (c). The best-fitting diversification models varying with temperature (d) and elevation (e) are shown, as determined by the analytical framework described in the text (see Chapter 18, Table 18.1 for all tested models). *Source:* Adapted from McGuire 2014. See Chapter 18 for further information.

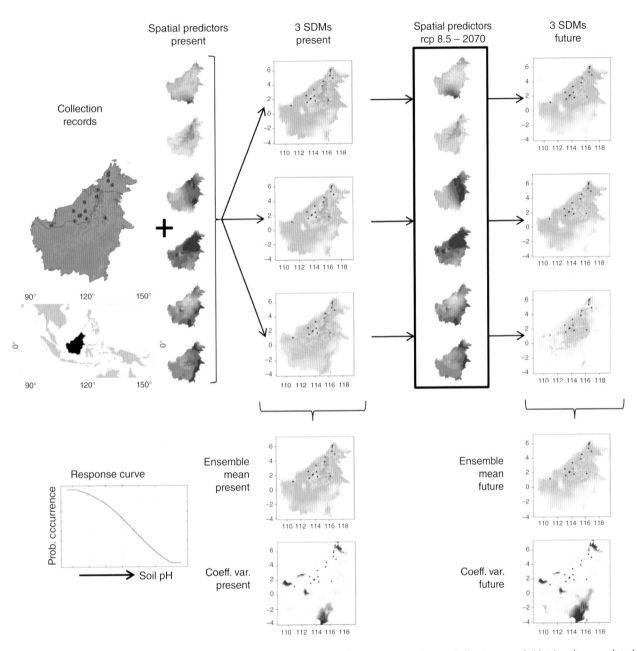

Plate 34 Species distribution model (SDM) workflow for *Vaccinium phillyreoides* occurring on Borneo. Collection records (dots) and uncorrelated spatial predictors of present conditions are used to create three different SDMs using different algorithms (Table 21.2 in Chapter 21); white indicates high probability of occurrence. An ensemble (mean) of the three SDMs shows where the models agree and mapping of the coefficient of variation identifies areas where predictions are least consistent (dark gray). The SDMs are then projected to future climatic conditions (here, scenario RCP8.5, the most pessimistic climate change scenario, where greenhouse gas emissions continue increasing after the year 2100), resulting in three individual future projections. These are assembled in an ensemble mean forecast. The lower left corner shows a response curve of probability of occurrence decreasing with increasing Soil pH. See Chapter 21 for further information.

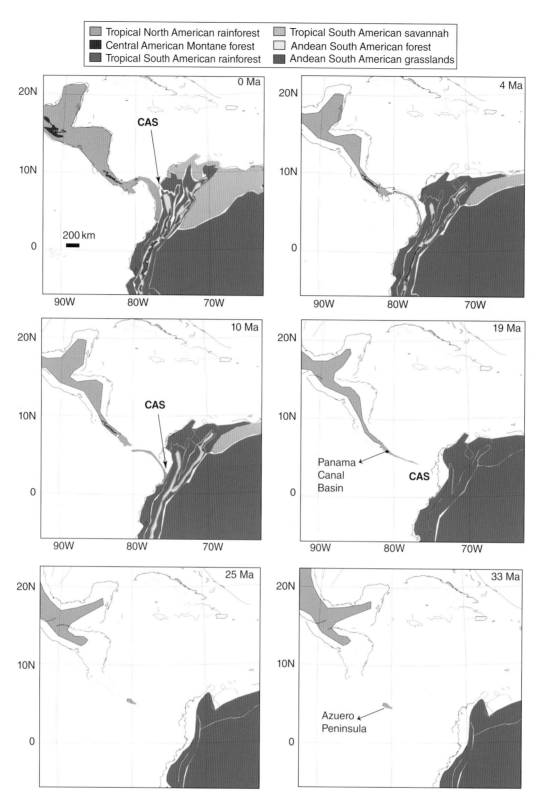

Plate 35 Terrestrial biome reconstruction for the past 36 My of the Isthmus of Panama. The reconstruction is an orthographic projection based on the plate tectonic model of GPlates 1.5.0, using the plate reconstruction of Seton et al. (2012). Terrestrial biomes include the tropical rainforest from South America, the tropical rainforest of North America (corresponding to the Central American rainforest), the Central American montane forests (forest >2000 m elevation), the Andean South American forest (forest >2000 m elevation) and the Andean South American grasslands (grasslands above the tree line in the Andes of South America). The first development of terrestrial landscape occurred during the late Eocene, as an island spanning from the present site of central Panama to the Azuero Peninsula. A second major build-up of the terrestrial landscape occurred during the early Miocene. A third occurred during the late Miocene. The Central American Seaway (CAS; defined as the ocean gap along the tectonic boundary between the South American plate and the Panama microplate) was closed by 10 Ma. From 10.0 to 3.5 Ma, there were intermittent Caribbean–Pacific connections through pathways other than the CAS. At 3.5 Ma, there was a complete closure of the Isthmus. A movie of the GPlates landscape reconstruction can be found in online Appendix 22.1. *Source:* Terrestrial biomes adapted from Jaramillo & Cardenas (2013), the exhumation evolution of the Isthmus of Panama from the Montes models (Farris et al. 2011; Montes et al. 2012a,b, 2015), the terrestrial plant fossil record of the region from Stern & Eyde (1963), Jaramillo et al. (2006, 2014), Herrera et al. (2010, 2012a,b, 2014a,b), Jaramillo & Cardenas (2013) and Rodriguez-Reyes et al. (2014), and the terrestrial vertebrate record from (Whitmore & Stewart 1965; Slaughter 1981; MacFadden & Higgins 2004; MacFadden 2006a,b, 2009, 2010; Cadena et al. 2012; Head et al. 2012; MacFadden et al. 2012; Rincon et al. 2012, 2013; Hastings et al. 2013). See Chapter 22 for further information.

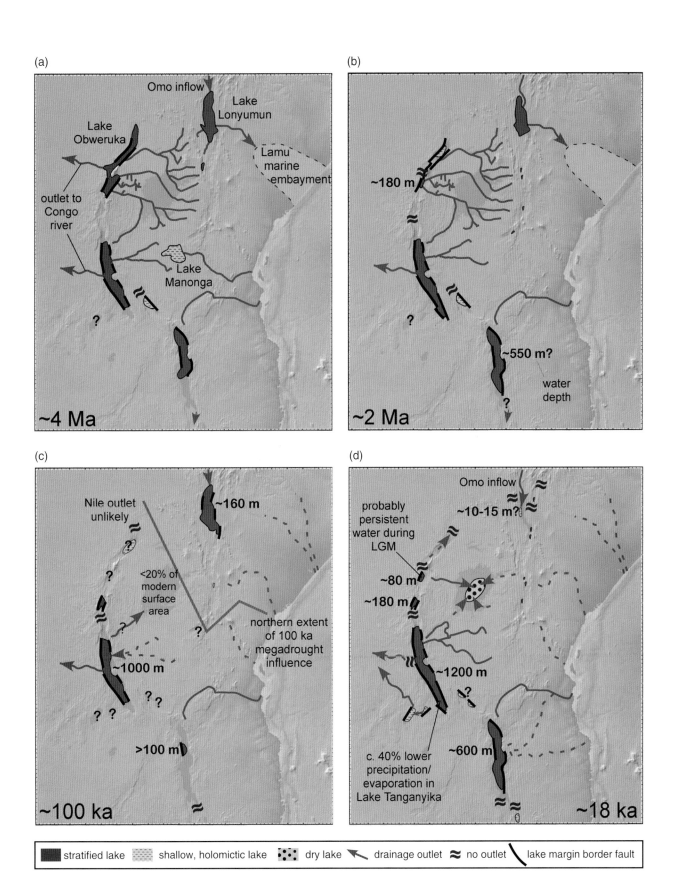

Plate 44 Paleogeographic maps of the EARS lakes, showing hydrographic configurations of the major lakes and rivers for four periods over the past 4 My: (a) ~4 Ma, (b) ~2 Ma, (c) ~100 ka and (d) ~18 ka. Solid lines indicate major perennial rivers; dashed lines show intermittent rivers. The main drainage until ~18 ka (Last Glacial Maximum, LGM) was to the west, into the Western Branch depocenters and ultimately the Atlantic Ocean. The Lamu marine embayment was controlled by a Cretaceous rift, which also controls the Turkana depression. Note the dramatic changes in hydrography at 18 ka. *Source:* Adapted from Salzburger et al. (2014). See Chapter 26 for further information.

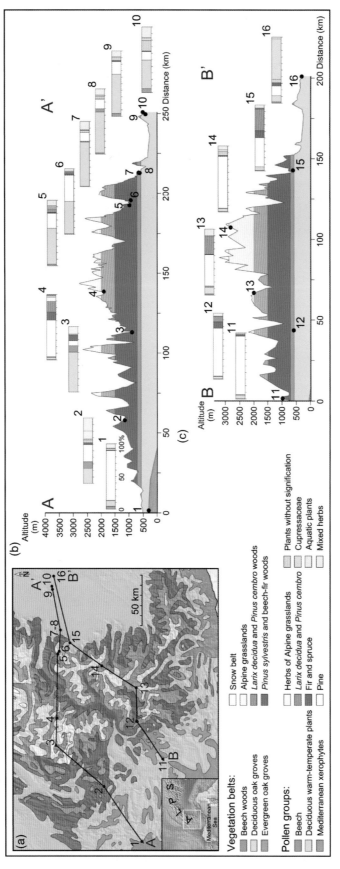

Plate 45 Relationship between vegetation belts positioned on a topographic profile, following studies by Noirfalise et al. (1987) and Ozenda (2002), and modern pollen data collected from moss polsters. (a) Simplified vegetation map of the Southwestern Alps (from Noirfalise et al. 1987), on which the transects are shown (dots correspond to the moss polsters). Pollen localities and altitudes are indicated below. (b) Transect AA′: 1, Nyons (300 m); 2, Luz la Croix haute (1100 m); 3, Le Freney (800 m); 4, Col du Lautaret (1850 m); 5,6, Pinerolese (Lago Villaretto di Roure and Balma di Roure, 980 and 890 m, respectively); 7,8, Cumiana Giaveno (612 and 605 m, respectively); 9,10, Baldissero Torinese (Val Samfrà and Superga road, 383 and 537 m, respectively). (c) Transect BB′: 11, Taulanne (995 m); 12, Sisteron (540 m); 13, Saint-Léger (2000 m); 14, Agnel (2800 m); 15, Pinerolese, Inverso Pinasca (560 m); 16, Moncucco Torinese, Borelli (315 m). *Source:* Adapted from Fauquette et al. (2015). See Chapter 27 for further information.

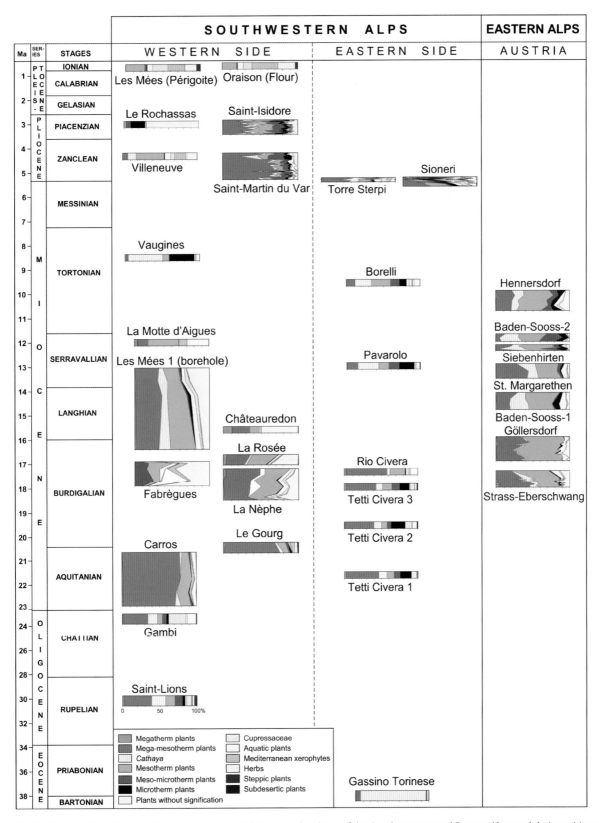

Plate 46 Synthetic pollen diagrams from late Eocene to Pleistocene localities of the Southwestern and Eastern Alps, and their positions in a chronostratigraphic frame. Note the change of time scale at the Paleogene–Neogene boundary. Taxa have been arranged into 13 different groups (*Pinus* excluded) based on ecological criteria to visualize the main changes in the paleovegetation cover. Plants without signification indicates families (e.g., Rosaceae, Ranunculaceae, Rutaceae) in which pollen identification is difficult at the genus level and includes cosmopolitan taxa and/or taxa with wide ecological requirements, plus unidentified pollen grains. Some groups are classified according to thermic requirements with respect to the Nix (1982) classification: equatorial and tropical forests are inhabited by megatherm plants (mean annual temperature, MAT >24 °C), subtropical forests by mega-mesotherm plants (MAT 20–24 °C), temperate deciduous forests by mesotherm plants (MAT 14–20 °C), boreal coniferous forests successively by meso-microtherm (MAT 12–14 °C) and microtherm (MAT <12 °C) plants and tundra by microtherm plants. *Source:* Adapted from Fauquette et al. (2015) and Jiménez-Moreno et al. (2008). See Chapter 27 for further information.

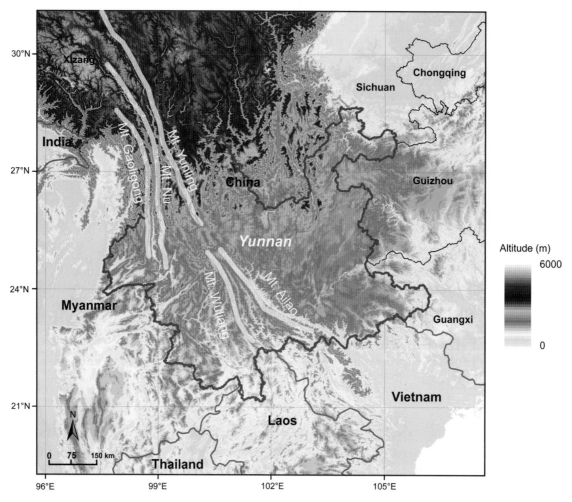

Plate 47 Geographic Information System (GIS) map showing the geographical position and topography of Yunnan, China. *Source:* Created with ArcGIS 9.0 (Environmental Systems Research Institute, Inc.). See Chapter 29 for further information.

Plate 48 Mt. Fuji (3776 m) and its lateral volcanoes (on the right side, 1424–1566 m), which erupted in AD 864–866, with basaltic lava flow covering the north-western foot. See Chapter 30 for further information.

Plate 49 Evergreen coniferous forest in Aokigahara (900–1500 m) at the north-western foot of Mt. Fuji: an edaphic climax forest that developed on a lava flow formed in AD 864–866. The mountains in the far background are the southern part of the Akaishi Mountains (2500–3000 m). See Chapter 30 for further information.

Plate 50 (a) Meili Xue Shan (Kawagebo; 6740 m) and Mingyong Glacier, Deqin, Yunnan, China. (b) Daxueshan range (4719 m), north of Shangrila, Yunnan, China. Photos courtesy of Adrien Favre.

Plate 51 Skadar Lake in Montenegro. Skadar Lake is a hotspot of freshwater biodiversity and contains a high number of cryptic species. Photo courtesy of Piotr Gadawski.

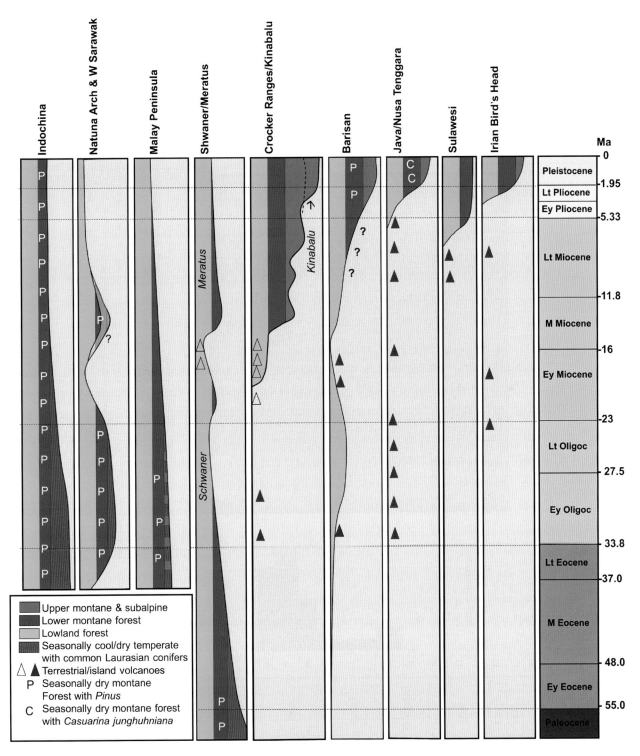

Plate 52 Timing and paleoaltitudes of Southeast Asian mountains and suggested paleovegetation. See Chapter 31 for further information.

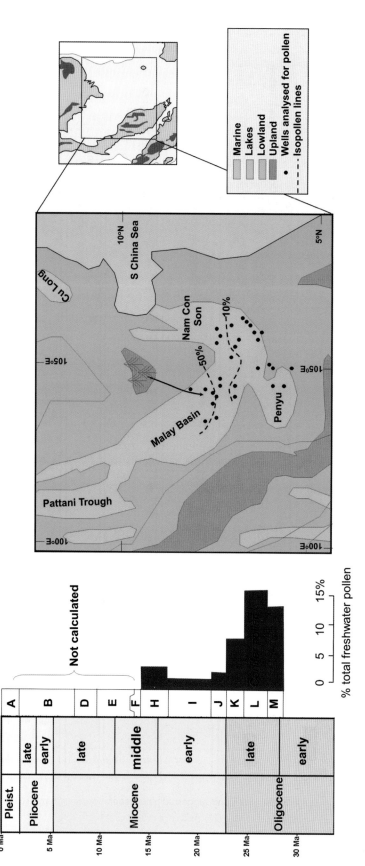

Plate 53 (a) Averaged percentage abundance of Laurasian conifers in Malay Basin and West Natuna petroleum exploration wells during the Oligocene and early Miocene. (b) Paleogeographic map for the late Oligocene of Sunda Shelf (ca. 24 Ma). Pollen of Laurasian conifers from exploration wells (black circles) is presented in terms of total freshwater pollen, calculated for each Malay Basin seismic group (stratigraphic packages applied to Malay Basin geology where traditional use of formational terminology fails) and their lateral equivalents in the West Natuna Basin (Morley 2014). Isopollen lines demonstrate that pollen is transported from upland in the region of the Con Son Rise to the north, and the black arrow shows the direction of transportation. The area of the inset differs from the area of the map, because this region has compressed since the Oligocene due to the rotation of Borneo and the extrusion of Indochina. *Source:* Morley (2014). Reproduced with permission of the Indonesian Petroleum Association. Basemap from de Bruyn et al. (2012). See Chapter 31 for further information.

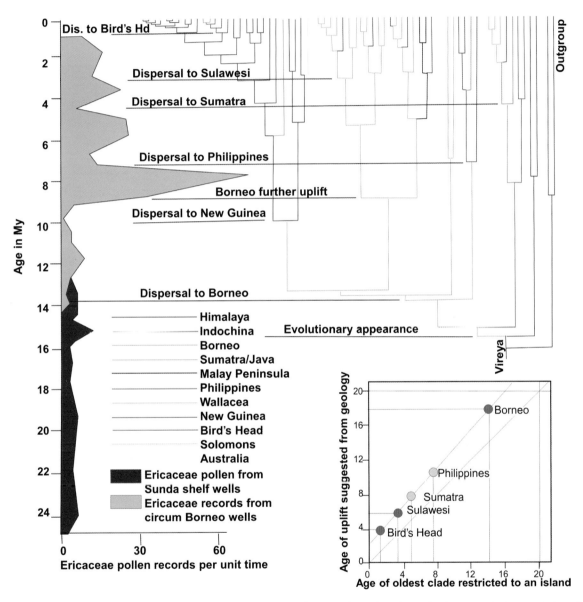

Plate 54 Timing of entry of *Vireya* clades across insular Southeast Asia. The inset shows the relationship between the known timing of the uplift and appearance of *Vireya* spp. (green circles) and the predicted timing of the uplift of uplands for which there is incomplete geological evidence based on the first appearance of *Vireya* (light blue circles). *Source:* Adapted from Webb & Ree (2012). See Chapter 31 for further information.

Plate 55 (a) Torres del Paine National Park, Chile (b) Altiplano, Bolivia (c) Guanaco, Torres del Paine National Park, Chile (d) Laguna Hedionda, Bolivia (e) Salar de Uyuni, Bolivia. Photos courtesy of Peter Wright.

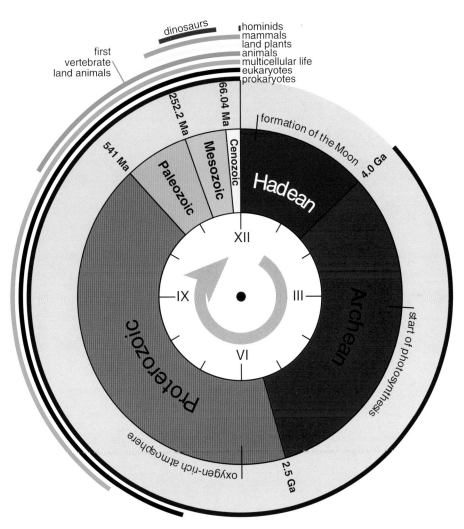

Plate 56 Geologic time demonstrated as a clock analogy. Assume that the 4.6 Gy of Earth's history is scaled to 12 hours on this clock ("the small/hour hand"). The first primitive life forms appeared more than 10 hours ago, the extinction of the dinosaurs about 10 min ago and the emergence of early humans happened ~20s ago. The colour coding of the different eons (Hadean, Archean and Proterozoic) and eras (Paleozoic, Mesozoic and Cenozoic) is according to the commission for the Geological Map of the World (CGMW), Paris, France. Courtesy of Klaudia Kuiper.